ALLE ZEIT WACH 1842

The cover illustrations portray the fundamental subdivision of viscoelastic behavior into rheodictic and arrheodictic response. Somewhat whimsically, they represent, on the left, an uncrosslinked and, on the right, a crosslinked polymer. The strings of beads simulate polymer chains. The small figures are "Maxwell's Demons": imaginary sentient, intelligent beings of atomic dimensions. In the illustration on the right the chains form a three-dimensionally crosslinked network. If a Demon finds itself on any portion of any chain, it can reach any other point on the network simply by walking along. In the illustration on the left the chains do not form a network. Here, the Demon is forced to jump or climb from one chain to the next. A crosslinked network cannot exhibit steady-state flow: therefore, the behavior is arrheodictic. An uncrosslinked network can flow: hence, the response is rheodictic. Both illustrations also contain 'entanglements'. The Demon is puzzled: are these crosslinks or not? Can it walk across them?

Cover illustrations: Christopher A. Tschoegl

Nicholas W. Tschoegl

The Phenomenological Theory of Linear Viscoelastic Behavior

An Introduction

With 227 Figures and 25 Tables

Springer-Verlag Berlin Heidelberg New York
London Paris Tokyo

Nicholas W. Tschoegl

Professor of Chemical Engineering, Emeritus
California Institute of Technology
1201 East California Blvd.
Pasadena, CA 91125/USA

ISBN-13:978-3-642-73604-9 e-ISBN-13:978-3-642-73602-5
DOI: 10.1007/978-3-642-73602-5

Library of Congress Cataloging-in-Publication Data.
Tschoegl, Nicholas W., 1918– .
The phenomenological theory of linear viscoelastic behavior :
an introduction / Nicholas W. Tschoegl.
p. cm. . Includes index.
ISBN-13:978-3-642-73604-9
1. Viscoelasticity. 2. Rheology. I. Title.
QA931.T765 1989 532'.053--dc19 88-30800

Softcover reprint of the hardcover 1st edition 1989

Typesetting: Asco Trade Typesetting Ltd., Hong Kong

2152/3020-543210 – Printed on acid-free paper

This book is dedicated in deep gratitude to
Sophie
without whom I would not have become a scientist,
and to those who made me one:
Alex, Eric, John and Thor

[Gott . . .]

Tu mir kein Wunder zulieb.
Gib deinen Gesetzen recht
die von Geschlecht zu Geschlecht
sichtbarer sind.

Das Stundenbuch
R.M. Rilke

Preface

One of the principal objects of theoretical research in any department of knowledge is to find the point of view from which the subject appears in its greatest simplicity.

J. Willard Gibbs

This book is an outgrowth of lectures I have given, on and off over some sixteen years, in graduate courses at the California Institute of Technology, and, in abbreviated form, elsewhere. It is, nevertheless, not meant to be a textbook. I have aimed at a full exposition of the phenomenological theory of linear viscoelastic behavior for the use of the practicing scientist or engineer as well as the academic teacher or student. The book is thus primarily a reference work.

In accord with the motto above, I have chosen to describe the theory of linear viscoelastic behavior through the use of the Laplace transformation. The treatment of linear time-dependent systems in terms of the Laplace transforms of the relations between the excitation and response variables has by now become commonplace in other fields. With some notable exceptions, it has not been widely used in viscoelasticity. I hope that the reader will find this approach useful.

Elementary calculus and the rudiments of complex variable theory is the basic mathematical apparatus required for a profitable use of this book. The elements of transformation calculus are summarized in an Appendix. It introduces the notation used in this book and serves as a convenient reference. It also contains a discussion of those special functions, the delta, step, slope, ramp, and gate functions, which are indispensable in the theory. The reader is advised to scan the Appendix before he sets out on the book itself. This will show him whether he feels comfortable with what he knows about transformation calculus. If he does not, he should read one or the other of the many texts on the subject.

The compass of the book is outlined on the pages entitled Scope which follow this Preface. As pointed out there, the linear theory of viscoelastic behavior is a specialized form of general linear response theory. The theory has much beauty owing to the symmetry resulting from the interchangeability of the variables which are considered excitation and response, respectively. I placed much emphasis on the development of this duality, sometimes at the risk of being repetitive. I did this because there are pitfalls in interchanging the excitation and response, and there is the tendency, all too prevalent, of arguing by analogy. This can easily lead to erroneous statements or equations. A further advantage of fully working out both sides of the theory, i.e. stating it in terms of *relaxation behavior* (response to strain) as well as in terms of *retardation (or creep) behavior* (response to stress) is that the book becomes a compendium in which most of the important relations are readily available. A quite detailed subject index should further aid this purpose. I attempted to include in it all those terms that the reader – in my idio-syncratic opinion – might be most likely to wish to look up. It is clearly impossible to do this so that it will satisfy everyone. My apologies if I slipped too often. Clearly, every occurrence of every term could not be referenced. Particularly with 'bread-and-butter' terms such as *behavior, excitation, model, response, viscoelastic, etc.*, I had to concentrate on listing primarily those occurrences which refer to matters I thought important. I hope I did not miss too many items that should have been included. The entries are fairly extensively cross-indexed to minimize irritating *'see-under . . .'*-s but I could not avoid these completely.

A list of symbols and an author index precede the subject index. The main body of the book is subdivided into eleven chapters and the Appendix. The various parts of the book are headed by one or more mottos. In addition there is also a prefatory quotation, a prologue and an epilogue. Notes and literary references to these follow the Appendix.

The chapters and the Appendix are divided into sections, and the latter are often further divided into subsections. In numbering these divisions, the chapter number is separated by a decimal point from the section number in the first, and the subsection number in the second, decimal position. The numbering of equations, figures, and tables begins anew with each section. The number is preceded by the chapter and section number and is separated from them by a dash. Thus, Eq. (8.2–5) is Equation 5 of Section 2 of Chapter 8.

When two equations are placed on the same line they are assigned the same number and are distinguished from each other in references by numerical subcripts appended to the equation number. The same device is used to distinguish portions of concatenated equations. Thus, when an equation takes the form

$$f(x)=g(x)=h(x) \tag{1}$$

Equation $(1)_1$ refers to $f(x)=g(x)$. Equation $(1)_2$ may be $g(x)=h(x)$ or $f(x)=h(x)$. The context always makes clear which is meant.

In the interest of brevity certain abbreviations were used routinely. Thus, 'step strain' is simply short for 'strain as a step function of time'. Similary, 'harmonic stress excitation' stands for 'excitation consisting of a stress in the sinusoidal steady-state'. I trust that these shortcuts will be self-explanatory everywhere.

Each chapter contains several fully worked problems. These are collected at the end of the chapter for easier cross-referencing and to avoid interrupting the flow of the exposition. Many are essential to a full understanding of the theory. Others clarify or amplify mathematical details. Still others are designed to develop and test the manipulative skill of the reader.

References to the work of others, indicated by numbers in brackets, and compiled at the end of each chapter, have been used sparingly. I would have liked greatly to follow the historical development of the subject in detail. However, the book was long in writing and would have taken even longer if this kind of scholarly research had been added. Therefore, I made reference only to the earliest work whenever this seemed appropriate. Otherwise, the literature is referred to merely when I thought it necessary for the sake of further clarification or an extension of the text. This restraint applies equally to my own papers. I apologize to all that feel left out.

I have added, in footnotes, short biographical comments to the names (capitalized in the subject index) of the more often quoted scientists, physicists, mathematicians, and engineers where they are first mentioned in the text. The way foreign names are commonly pronounced by English speakers unfamiliar with the spelling conventions (and their aberrations) of foreign tongues is all too often horrifying to others. I have therefore tried to render these names in the footnotes by their nearest American English phonemes, indicating the accent by an underscore. I hope that I have not erred here myself too often. Concerning my own name, for those who care, the letter combination 'Tsch' should be pronounced as the 'Ch' in Churchill. As for the rest of the name, the pronounciation has been clarified by Professor R. B. Bird of the University of Wisconsin, in the limerick:*

An eminent linguist called Tschoegl
at an age when he barely could gurgle
knew Turkish and Frisian
and Old Indonesian
and that the German for birds is Vögel.

During the writing of this book I was more then once reminded of the bewildered cry for help of a young warrior during Hungary's struggle against the Turks: "törököt fogtam, de nem ereszt"–'I caught a Turk but he doesn't let go of me'. It is thus with pleasure and pride that I acknowledge the contributions made by many of my students and collaborators who have taken part in working out specific details, checking problems, and reading and correcting parts of the manuscript. I would like to mention particularly (and in alphabetical order) R. Bloch, W.V. Chang, R.E. Cohen, M. Cronshaw, D.G. Fesko, R.W. Fillers, Çiğdem Gürer, L. Heymans, K. Jud, W.K. Moonan, S.C. Sharda, G. Ward, and K. Yagii.

Heartfelt thanks go also to several of my colleagues at Caltech who helped me with specific problems. These are especially professors Tom Apostol, Paco Lagerstrom, Willem Luxembourg, Charles de Prima, and John Todd.

Much of Chap. 5 was written during my two months tenure of a visiting professorship at the Technische Hogeschool, Delft, The Netherlands. Chapters 6, 7, and 9 as well as several sections in other chapters were worked out during a six month stay at the Johannes Gutenberg University in Mainz, Federal Republic of Germany, as the recipient of a U.S. Senior Scientist Award from the Alexander von Humboldt Foundation. It is a pleasure to mention that this award, also called the Humboldt Prize, was instituted by the German Federal Government in recognition of aid received from the United States after World War II.

Further work was done during my tenure of a visiting professorship at the Eidgenössische Technische Hochschule in Zurich, Switzerland, and during two months spent at the Centre de Recherche sur les Macromolécules in Strasbourg, France. Chapters 10 and 11 were drafted largely during a stay of three months at Edvard Kardelj University in Ljubljana, Yugoslavia. My sincere thanks go to all

* Not to be outdone, I offer advice on the writing of Professor Bird's name as follows:

Another great linguist called Bird
by his friends was once overheard
to mutter: in Chinese
for my name the sign is
the same as for bird. How absurd!

those who made these stays possible, particularly professors Hermann Janeschitz-Kriegl, Erhard Fischer, Joachim Meissner, Henri Benoit, and Igor Emri.

Last but not least I wish to acknowledge the dedicated work of a succession of very able secretaries, particularly (and, this time, in chronological order) the late Mrs. Eileen Walsh-Finke, Mrs. Kim Engel, Mrs. Lorraine Peterson, Mrs. Helen Seguine, Mrs. Rita Mendelson.

A great deal is being said in this book about models. I though it appropriate, therefore, to append as a Prologue Jorge Luis Borge's delightful little piece *Del rigor en la sciencia,* dealing with mankind's original model, the map.

The Epilogue at the end of the text admirably expresses my own feelings at the completion of my labor of many years.

Pasadena, January 1989 Nicholas W. Tschoegl

Scope

Il concetto vi dissi. Or,
ascoltate com'egli è svolto.

Leoncavallo: I Pagliacci

This book is concerned with the phenomenological description of the behavior of materials when these are deformed mechanically. Its subject matter is therefore a particular aspect of the science of *rheology,* that branch of mechanics which deals with the deformation and flow of matter. Material behavior is governed by *rheological equations of state* or *constitutive equations.* A constitutive equation links a dynamic quantity, the *stress,* with a kinematic quantity, the *strain,* through one or more parameters or functions which represent the characteristic response of the material per unit volume regardless of size or shape. In dealing with material behavior we may be concerned primarily with one or the other of two complementary aspects of the constitutive equation. Thus, we may be interested primarily in the stress-strain relations taking the material properties as given, or we may be concerned primarily with the material properties and not with the particular stresses and strains to which a given body of matter is subjected. The prediction of stresses and strains resulting from the imposition of prescribed tractions and/or displacements on a material body is the subject of *stress analysis* and is discussed in texts on solid and fluid mechanics. In this book we shall be concerned with the alternative way of viewing the constitutive equation, that of *material behavior.* An example may make the distinction clearer. The extension of a rod and the deflection of a cantilever beam fashioned of the same material represent different stress analysis problems. Both deformations, however, depend on the same material property. For a purely elastic material this is simply its Young's modulus. The determination of the modulus from either deformation assumes the stress-strain relations to be known. Conversely, the prediction of the extension of the rod or the imposed force requires that the modulus be known. Clearly, material behavior and stress analysis cover complementary aspects of deformation and flow.

Material behavior is termed *viscoelastic* if the material stores part of the deformational energy elastically as potential energy, and dissipates the rest simultaneously through viscous forces. We shall distinguish the *theory of viscoelastic behavior* as a discipline concerned with material behavior, from the *theory of viscoelasticity* which is concerned with stress analysis problems involving materials that are neither purely viscous nor purely elastic.

The rheological properties of a viscoelastic material are time-dependent. Although, in principle, all real materials are viscoelastic, this property becomes most prominent when the time required for the full development of a response is comparable with the time scale of the experiment performed to elicit it. The condition is notably present in polymeric materials which are thus the viscoelastic materials *par excellence*. Although we shall deal here more specifically with the theory as developed for polymeric materials, most of it is applicable, *mutatis mutandis*, to other materials such as metals and ceramics, inasmuch as they exhibit viscoelastic behavior. Furthermore, the theory, with suitable modifications in notation, is applicable also to time-dependent material behavior other than rheological. In particular, the theory of dielectric behavior is quite closely related to that of viscoelastic behavior. The two theories have, in fact, developed in close parallel and allusions to this will be made several times.

The foregoing has served to clarify provisionally the meaning of the words *viscoelastic behavior* in the title of the book. This will be enlarged upon in Chap. 2. However, some further comments appear in order.

By *phenomenological theory* I mean that I have tried to formulate a general framework which, in principle, applies to all viscoelastic materials regardless of their molecular structure. The viscoelastic behavior of polymers in relation to their structure has been described in several excellent books (see, e.g. [1, 2]). Similar books exist in the field of metals (see e.g. [3]). These have generally made use, without developing it explicitly, of the underlying phenomenological theory.

Confining the discussions in this book to *linear* behavior restricts it to behavior in deformations which are small enough so that a doubling of the excitation will elicit twice the response from the material. This restriction results in very considerable simplification and allows formulation of the subject as a unified theory applicable to all aspects of deformation and flow within the limitation to linear response. Moreover, the theory thus becomes another branch of *general linear response theory*. This permits us to use results worked out in other fields of physics concerned with linear response (e.g., electric circuit theory, the most highly developed of all linear systems theories) by applying the appropriate analogies (cf. Chap. 3).

Another important restriction ist that viscoelastic behavior is discussed in this book under the assumption that the thermodynamic variables, temperature and pressure, are constant. Thus I am dealing here exclusively with isothermal and isobaric viscoelastic behavior. I hope to present a discussion of the effects of temperature and of pressure elsewhere at another time.

Finally, the word *Introduction* in the subtitle refers to the level of presentation. In accordance with my aims stated in the Preface, I have foregone mathematical rigor without, I hope, sacrificing precision and clarity. The emphasis is thus on an understanding of the structure of the theory as it is applied in practice. The reader wishing to go on to more exacting treatments is referred to the excellent axiomatic presentations of Gurtin and Sternberg [4], and of Leitman and Fischer [5].

References

1. J.D. Ferry, *Viscoelastic Properties of Polymers*, 3rd ed., Wiley, New York, 1980.
2. R. Byron Bird, Robert C. Armstrong, and Ole Hassager, *Dynamics of Polymeric Liquids: Volume I, Fluid Dynamics;* the previous authors and C.F. Curtiss, *Volume II, Kinetic Theory*, Wiley, New York, 1977.
3. C.M. Zener, *Elasticity and Anelasticity of Metals*, University of Chicago Press, Chicago, 1948.
4. M.E. Gurtin and E. Sternerg, *On the Linear Theory of Viscoelasticity*, Arch. Rat. Mech. Anal. **11:** 291–356 (1962).
5. M.J. Leitman and G.M.C. Fischer, *The Linear Theory of Viscoelasticity*, in: S. Flügge, ed., *Encyclopedia of Physics*, Vol. VIa/3, Springer, Berlin, Heidelberg, New York 1973, pp. 1–123.

Prologue

Del rigor en la cienca

. . . En aquel Imperio, el Arte de la Cartografía logró tal Perfección que el Mapa de una sola Provincia ocupaba toda una Ciudad, y el Mapa del Imperio, toda una Provincia. Con el tiempo, estos Mapas Desmesurados no satisficieron y los Colegios de Cartógrafos levantaron un Mapa del Imperio, que tenía el Tamaño del Imperio y coincidía puntualmente con él. Menos Adictas al Estudio de la Cartografía, las Generaciones Siguientes entendieron que ese dilatado Mapa eta Inútil y no sin Impiedad lo entregaron a las Inclemencias del Sol y de los Inviernos. En los Desiertos del Oeste perduran despedazadas Ruinas del Mapa, habitadas por Animales y por Mendigos; en todo el País no hay otra reliquia de las Disciplinas Geográficas.

Of Exactitude in Science

. . . In that Empire, the craft of Cartography attained such Perfection that the Map of a Single province covered the space of an entire City, and the Map of the Empire itself an entire Province. In the course of Time, these Extensive maps were found somehow wanting, and so the College of Cartographers evolved a Map of the Empire that was of the same Scale as the Empire and that coincided with it point for point. Less attentive to the Study of Cartography, succeeding Generations came to judge a map of such Magnitude cumbersome, and, not without Irreverence, they abandoned it to the Rigours of sun and Rain. In the western Deserts, tattered Fragments of the Map are still to be found, Sheltering an occasional Beast or beggar; in the whole Nation, no other relic is left of the Discipline of Geography.

Contents

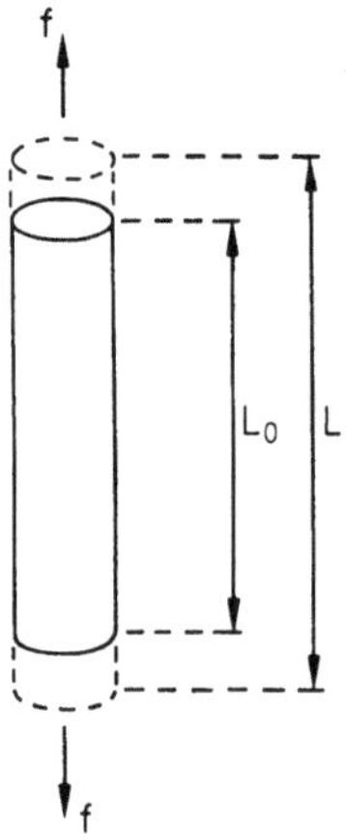

Fig. 1.1-1. Elongation of a cylindrical rod

$L - L_0$ is not too large, it will be directly proportional to the force f. We can express this in the form $f = k\,\Delta L$.

The question now arises: is k a material parameter? Evidently k depends on the nature of the material. The same force will stretch a rubber rod much more than a steel rod. However, k depends also on the dimensions of the rod. Thus k is not a material parameter. To obtain a material parameter we must make the proportionality constant independent of the rod geometry.

The two parameters which determine the size and shape of a cylindrical rod of uniform cross-section are its length and cross-sectional area. Experiments show that, for elongations of reasonable magnitude, the force required to produce a given elongation is directly proportional to the original cross-sectional area, A_0, and inversely proportional to the original length, L_0, of the rod. Moreover, this is true whatever the precise shape of the cross-section is as long as it is uniform. We can take this into account by writing

$$f = EA_0\,\Delta L/L_0 \tag{1.1-1}$$

where $E = kL_0/A_0$. Since E now no longer depends on the geometry of the rod, it represents a characteristic material property.

It is convenient to rearrange Eq. (1.1-1). We call f/A_0, the force per unit original area, the *tensile stress*, σ, and $\Delta L/L_0$, the fractional elongation, the *tensile strain*, γ. For a purely elastic material E is a material property and is known as the *modulus of elasticity*. We may thus rewrite Eq. (1.1-1) as

$$\sigma = E\gamma\ . \tag{1.1-2}$$

Even though it describes a particular type of deformation (uniaxial extension) only, Eq. (1.1-2) is commonly referred to as a constitutive equation because it is independent of the size and shape of the test piece. In a strict interpretation, however, a constitutive equation should express the relation between a general deformation and a general set of forces.

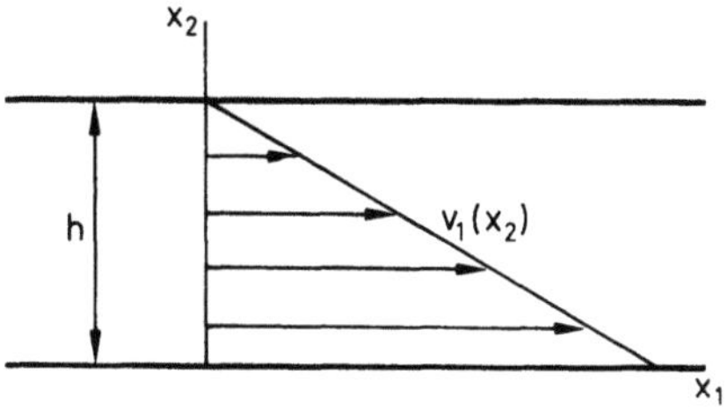

Fig. 1.1-2. Laminar shear flow of a liquid

The first constitutive equation for solids was formulated by Hooke* in his famous pronouncement [1] "*ut tensio sic vis*" (as the tension, so the strength), first published as an anagram (CEIIINOSSSTTUU). In Hooke's own later words [2]: "the power of any spring is in the same proportion with the tension thereof". Hooke understood spring to mean any elastic body. Power or strength, and tension (or extension), were used for what we now call stress and strain. The proportionality between stress and strain expressed by Eq. (1.1-2) is consequently called *Hooke's law* and a material obeying this rheological equation of state is referred to as a *Hookean solid.*

Instead of the tensile deformation of a solid we may consider the laminar shear flow of a purely viscous liquid. This is the simplest flow a liquid can evince. In another thought experiment let a liquid be contained between two large parallel plates and let the lower plate move in a fixed direction with a constant velocity. If the velocity is not too large, the velocity profile attained after the flow has reached a steady-state will be as shown in Fig. 1.1-2 where h is the separation between the two plates. The velocity at any point is given by

$$v_1 = v_1(x_2) \,, \qquad v_2 = v_3 = 0 \,. \tag{1.1-3}$$

If f is the frictional force resisting the motion of two adjacent parallel layers of the liquid past each other, and a is the area of contact, then the force f will be proportional to the area and to the velocity decrease in the distance h, i.e. to the *shear gradient* dv_1/dx_2. We may express this in form of the constitutive equation

$$f = \eta a \frac{dv_1}{dx_2} \tag{1.1-4}$$

in which the proportionality constant η is called the *coefficient of shear viscosity* or *viscosity* for short. The viscosity is a material property, independent of geometry. Equation (1.1-4) may be recast as

$$\sigma = \eta\dot{\varepsilon} \,. \tag{1.1-5}$$

Hence $\sigma = f/a$ is the *shear stress*, and $\dot{\varepsilon} = dv_1/dx_2$, having the dimension of reciprocal time, is the *rate of shear* [or more precisely (see Eq. (1.3-22), the rate of the amount of shear]. The formal analogy between Eqs. (1.1-1) and (1.1-4), and Eqs. (1.1-2) and (1.1-5) is immediately apparent.

* Robert Hooke, 1635–1703, English experimental physicist, generally regarded as the founder of the theory of elasticity.

This constitutive equation expressing the proportionality between stress and the rate of strain was introduced [3] by Newton* when he considered the properties of the hypothetical fluid which, according to the belief of his time, filled all space. Although Newton never experimented with any real liquids, Eq. (1.1-5) is called *Newton's law*, and a material which obeys this constitutive equation is termed a *Newtonian liquid.*

Equations (1.1-2) and (1.1-5) are both constitutive equations in which a particular kind of load (expressed as tensile or shear stress) is linked to a deformation (expressed as strain) or a flow (expressed as rate of strain) through a characteristic material property (an elastic modulus or a coefficient of viscosity), independent of the size and shape of the material which it describes. Extension to constitutive equations which accommodate a general set of forces (load) eliciting purely elastic or purely viscous response will be undertaken in Sects. 1.4 and 1.5 of this chapter. In Chap. 2 we will be concerned with constitutive equations for materials which are neither purely elastic nor purely viscous. In these and subsequent chapters we will restrict ourselves to linear response.

In the foregoing discussion we have tacitly assumed that our materials were homogeneous and isotropic, and that end-effects could be neglected. A material is *homogeneous* if it has identical properties at any point. It is *isotropic* if a given property is the same in all directions at a point. Certain types of *anisotropic* behavior will be discussed in Sects. 1.4 and 11.1. The behavior of non-homogeneous (or inhomogeneous) materials is outside our scope. It should be noted, however, that materials which are, in fact, non-homogeneous, can be treated as homogeneous if they may be regarded to be so within randomly selected, sufficiently small regions of the body, and all such regions have identical properties. What is "sufficiently small" depends on the property. Thus a dispersion of finely divided carbon black in rubbers may be regarded as homogeneous in an experiment designed to find the stretch modulus, but inhomogeneous with respect to the propagation of high frequency sound waves.

End effects arise because of the finite size of the body on which a material property is being determined. The boundaries of the body introduce inhomogeneities in the stress, strain, or rate of strain which persist to certain distances into the interior of the body. In the elongation of a rod, for instance, such stress and strain inhomogeneity will result from the restraints put on the ends of the test piece when it is clamped to allow the application of the force. Extension forces the material to contract along the length of the rod. It is, however, prevented from contracting at the clamps. Similarly, when we wish to determine the viscosity of a newtonian material between parallel plates, the flow field will become distorted around the edges of the (finite) plates.

End effects are determined through *stress analysis*. Throughout this book we shall assume that end effects are negligible or can be taken into account in some suitable way when stresses and strains are measured.

---

* Sir Isaac Newton, 1642–1727, English physical scientist and mathematician, one of the greatest figures in all of science. Independent co-discoverer, with Gottfried Wilhelm Leibnitz, pron. Libe-nits, 1646–1716, German philosopher and mathematician, of the infinitesimal calculus.

It is the concepts of stress, strain, and rate of strain which enable us to formulate material behavior in terms of constitutive equations. In the preceding discussion of the elongation of an elastic rod or the shear flow of a viscous liquid we have introduced these concepts in a heuristic way. In the next two sections we will consider them in a more general way. This will not be an exhaustive treatment. Rather it will restrict itself to those aspects which we will need in later parts of this book. More extensive discussions are available to the interested reader in many texts on continuum mechanics and strength of materials.

1.2 Stress

Stress, i.e. force per unit area, is a macroscopic concept which looses its meaning at the molecular level. To apply the concept of stress we must consider the material to be distributed continuously throughout its volume. It should be noted that this does not mean that the density of the material must be the same everywhere. Rather, it means that the material must be everywhere dense, i.e. it must fill the space it occupies completely. This concept, referred to as the *continuum hypothesis*, allows us to formulate the mathematical description of material behavior in terms of piecewise continuous functions of the space coordinates and time.

To introduce the concept of stress, let us consider a material continuum B occupying a region R of space, and subjected to a set of surface forces $f_j^{(r)}$ and body forces $b_j^{(s)}$ $(j = 1, 2, 3; r, s = 1, 2, \ldots)$. Consider further a closed surface S within B, comprising an arbitrary volume V. The material within V interacts with the material outside of it. Now consider a small element of surface with area ΔS comprising the point P as shown in Fig. 1.2-1. The forces and moments exerted by the material outside of V on the material inside V across ΔS can be represented by the resultant force element $\Delta f_j^{(n_i)}$ and resultant moment element $\Delta M_j^{(n_i)}$. Clearly, the resolution will depend on the orientation of ΔS. This is characterized by the unit vector n_i $(i = 1, 2, 3)$ normal to the element of surface, and this is indicated by the superscripts in parentheses.

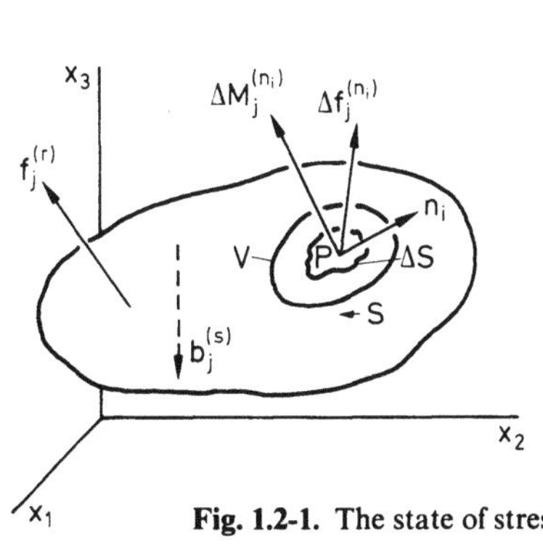

Fig. 1.2-1. The state of stress at the point P

for the directions specified. At the same time, as ΔV tends to dV, the $b_i \cdot dV$ term may be neglected against the other terms because the volume shrinks more rapidly than the surfaces, being an order higher in dimensions. Thus, after division by dS, we have

$$t_j^{(n_i)} = n_i t_j^{(e_i)} \quad (1.2\text{-}9)$$

or, using Eq. (1.2-4),

$$t_j^{(n_i)} = n_i \sigma_{ij} \quad (1.2\text{-}10)$$

which is the desired relation connecting the stress vector at the point P contained in an arbitrary plane with unit normal n_i with the stress tensor σ_{ij} at the same point.

Now, the consideration that the body being deformed should not undergo translation, requires that

$$\int_S t_j^{(n_i)} dS + \int_V b_j dV = 0 \ , \quad (1.2\text{-}11)$$

i.e. that the surface and body forces producing the deformation should be in equilibrium. But by Eq. (1.2-10)

$$\int_S t_j^{(n_i)} dS = \int_S n_i \sigma_{ij} dS \quad (1.2\text{-}12)$$

and, by the divergence theorem [5]

$$\int_S n_i \sigma_{ij} dS = \int_V \sigma_{ij,i} dV \quad (1.2\text{-}13)$$

where $\sigma_{ij,i}$ is the divergence of σ_{ij}, the comma followed by an index signifying partial differentiation with respect to the coordinate indicated by the index. Combining Eqs. (1.2-11), (1.2-12), and (1.2-13) gives

$$\int_V (\sigma_{ij,i} + b_j) dV = 0 \ . \quad (1.2\text{-}14)$$

But the volume is arbitrary, and, hence, the equilibrium of forces requires that

$$\sigma_{ij,i} + b_j = 0 \ . \quad (1.2\text{-}15)$$

Equation (1.2-15) is called the *equation of equilibrium*.

We also require that the forces producing the deformation should not produce a rotation of the body as a whole. There will be no such rotation if the moments produced by the applied forces are also in equilibrium. The moment about the x_i-axis is given by

$$m_i = x_j[t_k^{(n_p)} + b_k] - x_k[t_j^{(n_p)} + b_j] \ , \qquad i \neq j \neq k \ . \quad (1.2\text{-}16)$$

In the absence of couple stresses, therefore, the requirement that there should be no rigid rotation, leads to

$$\int_S e_{ijk} x_j t_k^{(n_p)} dS + \int_V e_{ijk} x_j b_k dV = 0 \quad (1.2\text{-}17)$$

where

$$e_{ijk} = \begin{cases} 1, & \text{when ijk} = 123, 231, \text{or } 312 \\ 0, & \text{when two or all subscripts are repeated} \\ -1, & \text{when ijk} = 132, 213, \text{or } 321 \ . \end{cases} \tag{1.2-18}$$

Using Eq. (1.2-10) we obtain

$$\int_S e_{ijk} x_j n_p \sigma_{pk} dS + \int_V e_{ijk} x_j b_k dV = 0 \ . \tag{1.2-19}$$

Introduction of the divergence theorem gives

$$\int_V e_{ijk} x_{j,p} \sigma_{pk} dV + \int_V e_{ijk} x_j \sigma_{pk,p} dV + \int_V e_{ijk} x_j b_k dV = 0 \ . \tag{1.2-20}$$

But the last two terms add to zero by Eq. (1.2-15), and the first will be zero unless $p = j$. Hence, Eq. (1.2-20), and therefore Eq. (1.2-16), reduce to

$$\int_V e_{ijk} \sigma_{jk} dV = 0 \ . \tag{1.2-21}$$

Equation (1.2-21) thus embodies the requirement that the applied forces should not produce a rotation of the body as a whole.

Again, the volume is arbitrary, and, therefore, we must have

$$e_{ijk} \sigma_{jk} = 0 \ . \tag{1.2-22}$$

But Eq. (1.2-21) can only be true if

$$\sigma_{ij} = \sigma_{ji} \ . \tag{1.2-23}$$

Hence, the stress tensor is symmetric, as asserted earlier. The stress matrix thus becomes

$$[\sigma_{ij}] = \begin{bmatrix} \sigma_{11} & \sigma_{12} & \sigma_{13} \\ \cdot & \sigma_{22} & \sigma_{23} \\ \cdot & \cdot & \sigma_{33} \end{bmatrix} \tag{1.2-24}$$

where the dots indicate symmetrical components.

1.3 Strain and Rate of Strain

We first turn our attention to the derivation of an appropriate measure of strain. Since we shall be concerned only with the theory of the mechanical behavior of materials when the deformations are very small (in principle, infinitesimal) we shall consider only infinitesimal strain in this chapter. Finite (non-infinitesimal) strains require a more elaborate formalism.

Consider two points, P and Q, in a continuum which are connected by the elemental vector dr_i as shown in Fig. 1.3-1. The position vectors of the two points are r_i and $r_i + dr_i$, respectively. Let the points P and Q become displaced to the points P′ and Q′ where the continuum is deformed through the application of a load. In general, the two *displacements* will be functions of the position vectors.

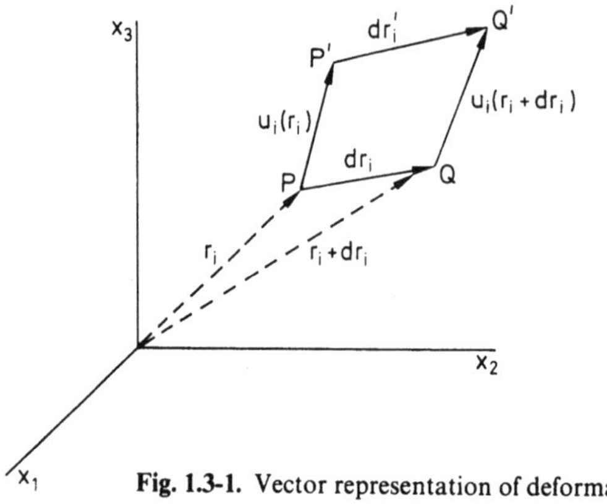

Fig. 1.3-1. Vector representation of deformation in a continuum

Hence

$$\overrightarrow{PP'} = u_i(r_i) \tag{1.3-1}$$

and

$$\overrightarrow{QQ'} = u_i(r_i + dr_i) \ . \tag{1.3-2}$$

We further have

$$u_i(r_i) + dr_i' = dr_i + u_i(r_i + dr_i) \tag{1.3-3}$$

which we may rearrange to

$$dr_i' = dr_i + u_i(r_i + dr_i) - u_i(r_i) \ . \tag{1.3-4}$$

If we restrict ourselves to small deformations, $u_i(r_i + dr_i)$ may be expanded in a Taylor series

$$u_i(r_i + dr_i) = u_i(r_i) + dr_j \frac{\partial\, u_i(r_i)}{\partial r_j} + \text{higher terms} \tag{1.3-5}$$

in which the higher terms may be neglected. Substituting the truncated form of Eq. (1.3-5) into Eq. (1.3-4) yields

$$dr_i' = dr_i + dr_j \frac{\partial\, u_i(r_i)}{\partial r_j} \ . \tag{1.3-6}$$

Equation (1.3-6) relates the elemental separation vector, dr_i', in the deformed body to that in the undeformed body, dr_i, through

$$\frac{\partial\, u_i(r_i)}{\partial r_j} = \frac{\partial u_i}{\partial x_j} \ . \tag{1.3-7}$$

The right hand side of Eq. (1.3-7) is obtained by recognizing that the coordinates x_j are the components of the position vector r_j. The nine components $\partial u_i/\partial x_j$ transform according to the rules of tensor transformation and thus form the com-

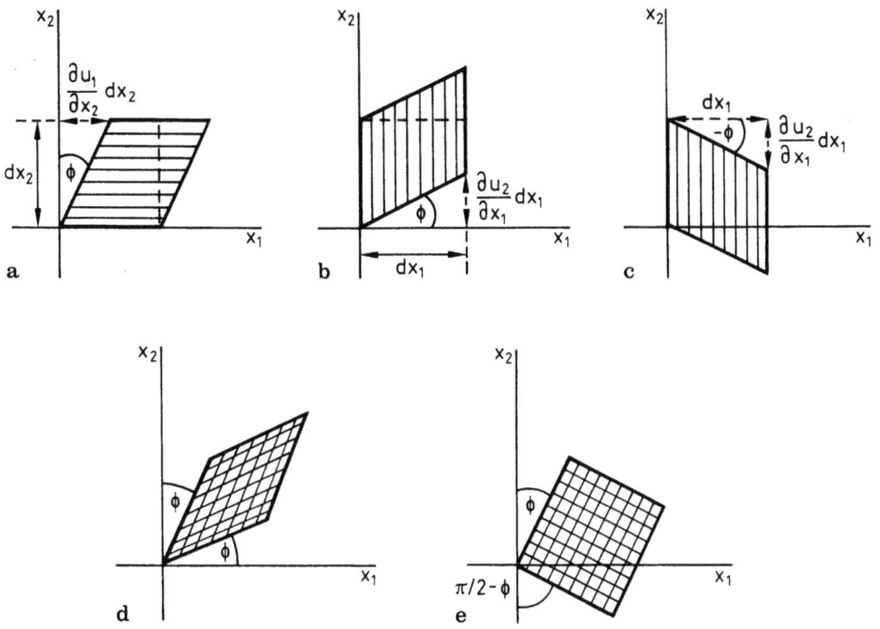

Fig. 1.3-2a-e. Simple shears and their superposition

ponents of a tensor of order two which we call the *displacement gradient tensor*. The question now arises: is this tensor a suitable measure of strain?

A strain measure which will serve our purpose must vanish for a motion of the body as a whole. The displacement gradient tensor certainly fulfills that condition for a translational motion of the body. In that case u_i is necessarily the same everywhere in the body, i.e. it is independent of r_i. Under these conditions the tensor vanishes and $dr_i' = dr_i$ as required. The situation is different with respect to a rotation of the body as a whole.

To understand the issues involved we consider the deformations shown diagrammatically in Fig. 1.3-2. Figure 1.3-2a represents a shear in the x_1, x_2-plane parallel to the x_1-axis. The amount by which any infinitesimally thin layer parallel to the x_1-axis is moved is $(\partial u_1/\partial x_2)dx_2$. The deformation can therefore be characterized by

$$\phi \simeq \tan\phi = \partial u_1/\partial x_2 \ . \tag{1.3-8}$$

Replacement of tan ϕ by the angle ϕ is justified here because the displacements are considered to be infinitesimally small. In this particular case, therefore, the deformation is correctly given by the displacement gradient tensor whose only non-zero component here is $\partial u_1/\partial x_2$. Similar considerations apply to the two shears parallel to the x_2-axis shown in Fig. 1.3-2b, and Fig. 1.3-2c for which we have

$$\phi \simeq \tan\phi = \partial u_2/\partial x_1 \tag{1.3-9}$$

and

represents this linearity between tensor components only. In the general case the linear relationship will be expressed by

$$\sigma_{ij} = C_{ijkl}\gamma_{kl} \tag{1.4-1}$$

where C_{ijkl}, the *modulus tensor*, is a tensor of the fourth order, the 81 components of which are material constants known as the *elastic coefficients*. Fortunately, the components of the modulus tensor are not all independent. Since σ_{ij} and γ_{kl} are symmetric tensors, interchange of the indices i and j, and k and l, in Eq. (1.4-1) does not change the equation. We have $C_{ijkl} = C_{jikl}$ and $C_{ijkl} = C_{ijlk}$. Thus, at most $6 \times 6 = 36$ independent constants are required.

A further reduction results from the following considerations. Let W represent the (Helmholz*) *free energy of deformation* per unit volume**. For infinitesimal strains we then have

$$W = \int \sigma_{ij} d\gamma_{ij} = \int C_{ij}\gamma_{kl} d\gamma_{ij} = \tfrac{1}{2} C_{ijkl}\gamma_{ij}\gamma_{kl} \tag{1.4-2}$$

and

$$\sigma_{ij} = \partial W/\partial\gamma_{ij} \; , \tag{1.4-3}$$

i.e. the components of the stress tensor may be derived from W by partial differentiation with respect to the components of the strain tensor. From Eq. (1.4-3) it follows that $C_{ijkl} = C_{klij}$ and thus the maximum number of independent material constants is further reduced to 21.

Since representation of a fourth order tensor in matrix form requires four dimensions (a $3 \times 3 \times 3 \times 3$ matrix in this case) we need a mathematical device that will make it possible to handle Eq. (1.4-1) in a convenient way. The device consists in renumbering*** the six independent components of the stress and strain tensors matrices [cf. Eqs. (1.2-24) and (1.3-16)] according to

$$[\sigma_{ij}] = \begin{bmatrix} \sigma_1 & \sigma_4 & \sigma_6 \\ \cdot & \sigma_2 & \sigma_5 \\ \cdot & \cdot & \sigma_3 \end{bmatrix} \quad \text{and} \quad [\gamma_{ij}] = \begin{bmatrix} \gamma_1 & \gamma_4 & \gamma_6 \\ \cdot & \gamma_2 & \gamma_5 \\ \cdot & \cdot & \gamma_3 \end{bmatrix} \tag{1.4-4}$$

assembling them into 6-component column vectors, $[\sigma_k]$ and $[\gamma_k]$. We can then represent Eq. (1.4-1) by

$$[\sigma_k] = [C_{kl}][\gamma_l] \; , \qquad k, l = 1, 2, \ldots, 6 \tag{1.4-5}$$

where $[C_{kl}] = [C_{lk}]$ is a symmetric 6×6 matrix which may be called the *modulus matrix*. Clearly, $[\sigma_k]$ and $[\gamma_k]$ are column vectors only in a mathematical, not a physical, sense. The modulus matrix for a general infinitesimal elastic deformation is then

* Hermann Ludwig Ferdinand von Helmholtz, pron. Helmholts, 1821–1894, German physicist, anatomist, and physiologist.

** This is also called the *elastic potential*, or the *strain energy density function*.

*** Some authors number the components differently.

$$[C_{kl}] = \begin{bmatrix} C_{11} & C_{12} & C_{13} & C_{14} & C_{15} & C_{16} \\ \cdot & C_{22} & C_{23} & C_{24} & C_{25} & C_{26} \\ \cdot & \cdot & C_{33} & C_{34} & C_{35} & C_{36} \\ \cdot & \cdot & \cdot & C_{44} & C_{45} & C_{46} \\ \cdot & \cdot & \cdot & \cdot & C_{55} & C_{56} \\ \cdot & \cdot & \cdot & \cdot & \cdot & C_{66} \end{bmatrix} . \quad (1.4\text{-}6)$$

The components of the modulus tensor C_{ijkl} are readily obtained from these matrix components by replacing 1 by 11, 2 by 22, 3 by 33, 4 by 12, 5 by 23, 6 by 13.

With the matrix as given by Eq. (1.4-6), Eq. (1.4-5) represents the constitutive relation for fully anisotropic materials. The behavior of an anisotropic material under a transformation of coordinates can be illustrated by a simple example. Consider a homogeneous elastic body referred to a system of axes x_i to which we apply a simple tension p along the x_1-axis. Then $\sigma_1 = p$, and $\sigma_k = 0$ otherwise.

Let us now reorient the stress so as to reflect a simple tension p along the axis x'_1 of a second coordinate system x'_i as shown in Fig. 1.4-1. Let $x'_i = l_{ij}x_j$ where the l_{ij}'s are the direction cosines (the reader is assumed to be familiar with the principles of coordinate transformations). Then, $\sigma'_1 = p$, and $\sigma'_k = 0$ otherwise, as before, and

$$\sigma'_k = C'_{kl}\gamma'_l \ . \quad (1.4\text{-}7)$$

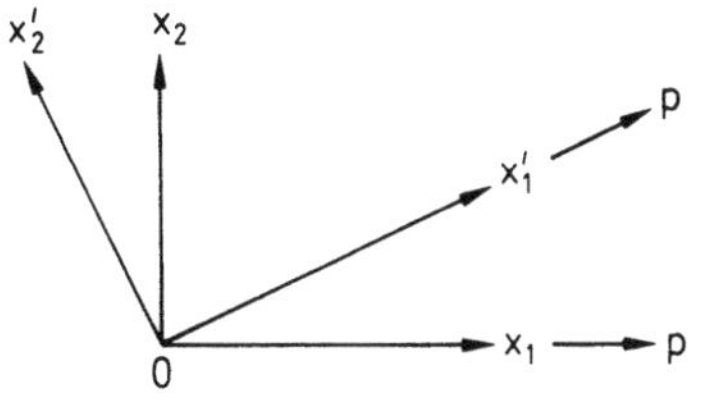

Fig. 1.4-1. Change of coordinate system

The physical situation is the same: a tensile stress along one axis. The stress components are the same also. However, $\gamma'_l \neq \gamma_l$, i.e., the strain components are not the same if the material is anisotropic. This follows because

$$C'_{ijkl} = \frac{\partial x_i \partial x_j \partial x_k \partial x_l}{\partial x'_r \partial x'_s \partial x'_t \partial x'_p} C_{rstp} = l_{ir} l_{js} l_{kt} l_{lp} C_{rstp} \neq C_{ijkl} \ , \quad (1.4\text{-}8)$$

i.e. C_{ijkl} (or $[C_{kl}]$) is not in general invariant under a transformation of coordinates. Few materials are completely anisotropic. Most possess at least some degree of symmetry. Symmetry implies invariance of C_{ijkl} under certain coordinate transformations. Crystalline materials and their symmetries are described in many textbooks [5, 6]. Here we shall consider materials with non-crystalline symmetries which are likely to play a role in the theory of linear viscoelastic behavior. We shall begin with materials having a single plane of symmetry, and shall introduce progressively more symmetry until we arrive at the *isotropic behavior* with which this book is chiefly concerned. Introduction of symmetry reduces the number of material constants required to represent the behavior. Isotropic materials possess only two elastic constants.

1.4.1 Single-Plane Symmetry

Certain composite materials, (e.g. laminates such as plywood) possess a single symmetry element, viz. a plane of symmetry. Consider an elastic material symmetric with respect to the x_1, x_3-plane as shown in Fig. 1.4-2. This symmetry is expressed by the statement that $[C_{kl}]$ is invariant under the transformation $x_1 = x'_1$, $x_2 = -x'_2$, $x_3 = x'_3$. The matrix of direction cosines l_{ij} of this transformation is

$$l_{ij} = \begin{bmatrix} 1 & 0 & 0 \\ 0 & -1 & 0 \\ 0 & 0 & 1 \end{bmatrix} . \tag{1.4-9}$$

The transformed stress components are, therefore, given by

$$\sigma'_{ij} = l_{ik} l_{jl} \sigma_{kl} \tag{1.4-10}$$

and the strain components by

$$\gamma'_{ij} = l_{ik} l_{jl} \gamma_{kl} \ . \tag{1.4-11}$$

Using Eqs. (1.4-4), matrix multiplication yields

$$\begin{bmatrix} \sigma'_1 & \sigma'_4 & \sigma'_6 \\ \cdot & \sigma'_2 & \sigma'_5 \\ \cdot & \cdot & \sigma'_3 \end{bmatrix} = \begin{bmatrix} \sigma_1 & -\sigma_4 & \sigma_6 \\ \cdot & \sigma_2 & -\sigma_5 \\ \cdot & \cdot & \sigma_3 \end{bmatrix} \tag{1.4-12}$$

and

$$\begin{bmatrix} \gamma'_1 & \gamma'_4 & \gamma'_6 \\ \cdot & \gamma'_2 & \gamma'_5 \\ \cdot & \cdot & \gamma'_3 \end{bmatrix} = \begin{bmatrix} \gamma_1 & -\gamma_4 & \gamma_6 \\ \cdot & \gamma_2 & -\gamma_5 \\ \cdot & \cdot & \gamma_3 \end{bmatrix} \tag{1.4-13}$$

so that

$$[\sigma'_k] = \begin{bmatrix} \sigma_1 \\ \sigma_2 \\ \sigma_3 \\ -\sigma_4 \\ -\sigma_5 \\ \sigma_6 \end{bmatrix} \quad \text{and} \quad [\gamma'_k] = \begin{bmatrix} \gamma_1 \\ \gamma_2 \\ \gamma_3 \\ -\gamma_4 \\ -\gamma_5 \\ \gamma_6 \end{bmatrix} . \tag{1.4-14}$$

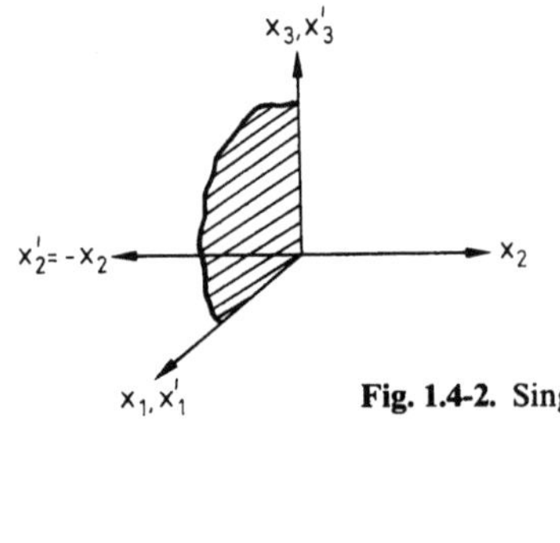

Fig. 1.4-2. Single plane of symmetry

Substitution of Eqs. (1.4-14) into Eq. (1.4-7) and then rewriting the result in form of Eq. (1.4-5) gives $[C_{kl}]$ as

$$[C_{kl}] = \begin{bmatrix} C_{11} & C_{12} & C_{13} & -C_{14} & -C_{15} & C_{16} \\ \cdot & C_{22} & C_{23} & -C_{24} & -C_{25} & C_{26} \\ \cdot & \cdot & C_{33} & -C_{34} & -C_{35} & C_{36} \\ \cdot & \cdot & \cdot & C_{44} & C_{45} & -C_{46} \\ \cdot & \cdot & \cdot & \cdot & C_{55} & -C_{56} \\ \cdot & \cdot & \cdot & \cdot & \cdot & C_{66} \end{bmatrix} . \qquad (1.4\text{-}15)$$

The invariance of the C_{kl} under the considered transformation [cf. Eq. (1.4-6)] requires that

$$C_{14} = C_{15} = C_{24} = C_{25} = C_{34} = C_{35} = C_{46} = C_{56} = 0 \ . \qquad (1.4\text{-}16)$$

Thus, the modulus matrix of a material with a single plane of symmetry becomes

$$[C_{kl}] = \begin{bmatrix} C_{11} & C_{12} & C_{13} & 0 & 0 & C_{16} \\ \cdot & C_{22} & C_{23} & 0 & 0 & C_{26} \\ \cdot & \cdot & C_{33} & 0 & 0 & C_{36} \\ \cdot & \cdot & \cdot & C_{44} & C_{45} & 0 \\ \cdot & \cdot & \cdot & \cdot & C_{55} & 0 \\ \cdot & \cdot & \cdot & \cdot & \cdot & C_{66} \end{bmatrix} . \qquad (1.4\text{-}17)$$

In the case of single-plane symmetry, therefore, the number of independent elastic constants is reduced from 21 to 13. The matrix $[C_{kl}]$ contains 20 non-zero components.

1.4.2 Three-Plane Symmetry or Orthotropy

Materials possessing three planes of symmetry as their only symmetry elements are called *orthotropic* or *orthorhombic*. There are three orthogonal planes of symmetry through each point of the material as shown in Fig. 1.4-3. The elastic properties of

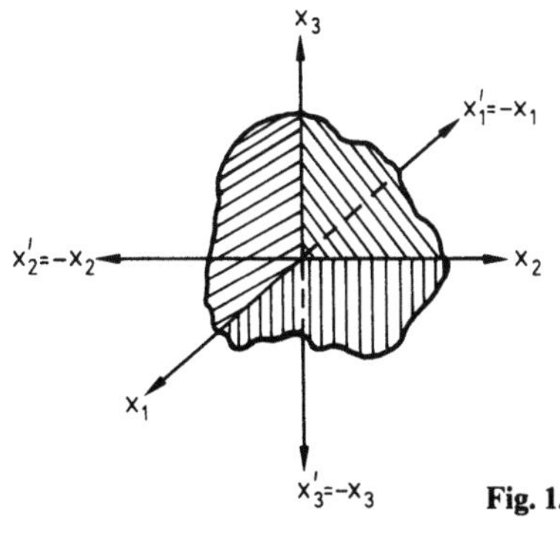

Fig. 1.4-3. Three planes of symmetry

homogeneous (or nearly homogeneous) sections of wood can often be described with good approximation as orthotropic.

Polymer films may also possess orthotropic symmetry if their orientation is unbalanced, i.e., if they were obtained by rolling or by drawing in a single direction. In such materials the elastic coefficients C_{kl} must be invariant to the transformations defined by

$$l_{ij} = \begin{bmatrix} -1 & 0 & 0 \\ 0 & 1 & 0 \\ 0 & 0 & 1 \end{bmatrix} \quad \text{and} \quad l_{ij} = \begin{bmatrix} 1 & 0 & 0 \\ 0 & 1 & 0 \\ 0 & 0 & -1 \end{bmatrix} \tag{1.4-18}$$

in addition to that defined by Eq. (1.4-9). Using Eqs. (1.4-10) and (1.4-11), Eq. $(1.4\text{-}18)_1$ leads to

$$[\sigma'_k] = \begin{bmatrix} \sigma_1 \\ \sigma_2 \\ \sigma_3 \\ -\sigma_4 \\ \sigma_5 \\ -\sigma_6 \end{bmatrix} \quad \text{and} \quad [\gamma'_k] = \begin{bmatrix} \gamma_1 \\ \gamma_2 \\ \gamma_3 \\ -\gamma_4 \\ \gamma_5 \\ -\gamma_6 \end{bmatrix}. \tag{1.4-19}$$

We calculate $[C_{kl}]$ again and use the already established relations contained in Eqs. (1.4-16) to obtain

$$[C_{kl}] = \begin{bmatrix} C_{11} & C_{12} & C_{13} & 0 & 0 & -C_{16} \\ \cdot & C_{22} & C_{23} & 0 & 0 & -C_{26} \\ \cdot & \cdot & C_{33} & 0 & 0 & -C_{36} \\ \cdot & \cdot & \cdot & C_{44} & -C_{45} & 0 \\ \cdot & \cdot & \cdot & \cdot & C_{55} & 0 \\ \cdot & \cdot & \cdot & \cdot & \cdot & C_{66} \end{bmatrix}. \tag{1.4-20}$$

Comparison with Eq. (1.4-6) shows that we must now also have

$$C_{16} = C_{26} = C_{36} = C_{45} = 0 \tag{1.4-21}$$

in addition to Eqs. (1.4-16). Equation $(1.4\text{-}18)_2$ furnishes no new relations. Indeed, the existence of two orthogonal planes of symmetry in a homogeneous material implies the existance of a third which is orthogonal to the other two. Therefore, the modulus matrix of an orthotropic material becomes

$$[C_{kl}] = \begin{bmatrix} C_{11} & C_{12} & C_{13} & 0 & 0 & 0 \\ \cdot & C_{22} & C_{23} & 0 & 0 & 0 \\ \cdot & \cdot & C_{33} & 0 & 0 & 0 \\ \cdot & \cdot & \cdot & C_{44} & 0 & 0 \\ \cdot & \cdot & \cdot & \cdot & C_{55} & 0 \\ \cdot & \cdot & \cdot & \cdot & \cdot & C_{66} \end{bmatrix} \tag{1.4-22}$$

and the number of independent elastic constants reduces to 9. The matrix $[C_{kl}]$ contains 12 non-zero components.

1.4.3 Axisymmetry, or Transverse Isotropy

Materials who possess an axis of symmetry normal to a plane of symmetry as shown in Fig. 1.4-4 are called *axisymmetric* or *transversely isotropic*. The elastic properties are unchanged by a rotation about the axis of symmetry, i.e. the elastic properties are the same in every direction in the plane which is perpendicular to this axis. Such materials are represented by drawn polymer fibers and some polymer films, if, in the latter case, the draw directions are at right angles and of equal magnitude.

Axisymmetry requires that the C_{kl}'s be invariant under the transformation of coordinates defined by the direction cosines

$$l_{ij} = \begin{bmatrix} \cos\alpha & \sin\alpha & 0 \\ -\sin\alpha & \cos\alpha & 0 \\ 0 & 0 & 1 \end{bmatrix} \tag{1.4-23}$$

in addition to the transformation defined by Eq. (1.4-9). Equation (1.4-23) takes the x_3-coordinate for the axis of symmetry. The angle of rotation is designated by α. From Eqs. (1.4-10) and (1.4-11), then,

$$[\sigma_k'] = \begin{bmatrix} \sigma_1\cos^2\alpha + \sigma_2\sin^2\alpha + 2\sigma_4\sin\alpha\cos\alpha \\ \sigma_1\sin^2\alpha + \sigma_2\cos^2\alpha - 2\sigma_4\sin\alpha\cos\alpha \\ \sigma_3 \\ (\sigma_2 - \sigma_1)\sin\alpha\cos\alpha + \sigma_4(\cos^2\alpha - \sin^2\alpha) \\ \sigma_5\cos\alpha - \sigma_6\sin\alpha \\ \sigma_5\sin\alpha + \sigma_6\cos\alpha \end{bmatrix} \tag{1.4-24}$$

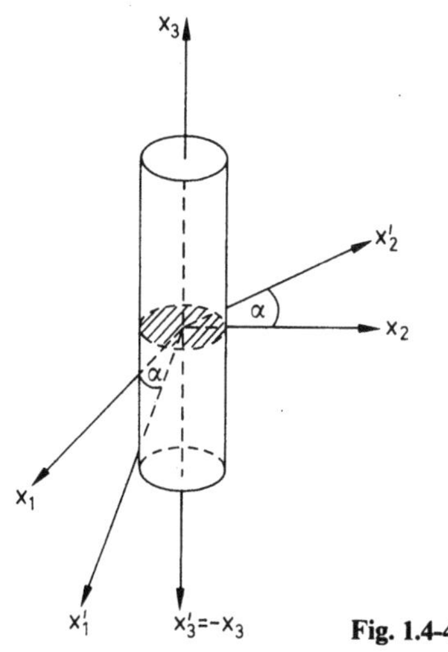

Fig. 1.4-4. Plane and axis of symmetry

and

$$[\gamma'_{kl}] = \begin{bmatrix} \gamma_1\cos^2\alpha + \gamma_2\sin^2\alpha + 2\gamma_4\sin\alpha\cos\alpha \\ \gamma_1\sin^2 + \gamma_2\cos^2\alpha - 2\gamma_4\sin\alpha\cos\alpha \\ \gamma_3 \\ (\gamma_2 - \gamma_1)\sin\alpha\cos\alpha + \gamma_4(\cos^2\alpha - \sin^2\alpha) \\ \gamma_5\cos\alpha - \gamma_6\sin\alpha \\ \gamma_5\sin\alpha + \gamma_6\cos\alpha \end{bmatrix}. \tag{1.4-25}$$

Now, Eqs. (1.4-24) and (1.4-25) must be valid for *any* angle. We can, therefore, lighten the arithmetic without loss of generality by selecting for it an arbitrary value, say, $\frac{\pi}{2}$. Equations (1.4-24) and (1.4-25) then reduce to

$$[\sigma'_k] = \begin{bmatrix} \sigma_2 \\ \sigma_1 \\ \sigma_3 \\ -\sigma_4 \\ -\sigma_6 \\ \sigma_5 \end{bmatrix} \quad \text{and} \quad [\gamma'_k] = \begin{bmatrix} \gamma_2 \\ \gamma_1 \\ \gamma_3 \\ -\gamma_4 \\ -\gamma_6 \\ \gamma_5 \end{bmatrix}. \tag{1.4-26}$$

Proceeding as before, we find

$$[C_{kl}] = \begin{bmatrix} C_{22} & C_{12} & C_{23} & 0 & 0 & 0 \\ \cdot & C_{11} & C_{13} & 0 & 0 & 0 \\ \cdot & \cdot & C_{33} & 0 & 0 & 0 \\ \cdot & \cdot & \cdot & C_{44} & 0 & 0 \\ \cdot & \cdot & \cdot & \cdot & C_{66} & 0 \\ \cdot & \cdot & \cdot & \cdot & \cdot & C_{55} \end{bmatrix} \tag{1.4-27}$$

and it follows that

$$C_{22} = C_{11} \tag{1.4-28}$$

$$C_{66} = C_{55} \tag{1.4-29}$$

and

$$C_{23} = C_{13}\,. \tag{1.4-30}$$

Hence, the modulus matrix of an axisymmetric material is

$$[C_{kl}] = \begin{bmatrix} C_{11} & C_{12} & C_{13} & 0 & 0 & 0 \\ \cdot & C_{11} & C_{13} & 0 & 0 & 0 \\ \cdot & \cdot & C_{33} & 0 & 0 & 0 \\ \cdot & \cdot & \cdot & C_{44} & 0 & 0 \\ \cdot & \cdot & \cdot & \cdot & C_{55} & 0 \\ \cdot & \cdot & \cdot & \cdot & \cdot & C_{55} \end{bmatrix}. \tag{1.4-31}$$

We see that σ_4' depends not only on σ_4 but also on $\sigma_2 - \sigma_1$. Equations (1.4-24) and (1.4-25) suggest, therefore, that the elastic coefficients are not all independent. Using Eq. (1.4-31) in Eq. (1.4-5), it is easily shown that

$$\sigma_2 - \sigma_1 = (C_{11} - C_{12})(\gamma_2 - \gamma_1) \; . \tag{1.4-32}$$

Inserting this into Eq. $(1.4\text{-}24)_4$ gives

$$\sigma_4' = (C_{11} - C_{12})(\gamma_2 - \gamma_1)\cos\alpha \sin\alpha + \sigma_4(\cos^2\alpha - \sin^2\alpha) \; . \tag{1.4-33}$$

But

$$\sigma_4' = C_{44}\gamma_4' \quad \text{and} \quad \sigma_4 = C_{44}\gamma_4 \; . \tag{1.4-34}$$

Introducing these relations and substituting for γ_4' from Eq. $(1.4\text{-}25)_4$ leads to

$$C_{44} = C_{11} - C_{12} \; . \tag{1.4-35}$$

The matrix $[C_{kl}]$ again contains 12 non-zero elements but the number of independent elastic constants of axisymmetric materials is 5. We may choose the five elastic constants in such a way that the axisymmetric relations reduce to the isotropic relations to be introduced in Sect. 1.4.4 when three of the five constants reduce to zero. To this end, we replace the five independent elastic coefficients by linear combinations of five other constants, λ, G, ρ, δ, and β. We let

$$C_{11} = \lambda + 2G \tag{1.4-36}$$

$$C_{12} = \lambda + 2\rho \tag{1.4-37}$$

$$C_{13} = \lambda + \delta \tag{1.4-38}$$

$$C_{33} = \lambda + \beta + 2G \tag{1.4-39}$$

$$C_{55} = 2G \tag{1.4-40}$$

and rewrite Eq. (1.4-31) as

$$C_{kl} = \begin{bmatrix} \lambda + 2G & \lambda + 2\rho & \lambda + \delta & 0 & 0 & 0 \\ \cdot & \lambda + 2G & \lambda + \delta & 0 & 0 & 0 \\ \cdot & \cdot & \lambda + 2G + \beta & 0 & 0 & 0 \\ \cdot & \cdot & \cdot & 2(G - \rho) & 0 & 0 \\ \cdot & \cdot & \cdot & \cdot & 2G & 0 \\ \cdot & \cdot & \cdot & \cdot & \cdot & 2G \end{bmatrix} . \tag{1.4-41}$$

Comparing Eq. (1.4-41) with Eq. (1.4-54) to be derived in the next subsection, it will be seen that a transversely isotropic material becomes isotropic as β, ρ, and δ approach zero.

1.4.4 Isotropy

Complete isotropy is obtained by requiring that $[C_{kl}]$ be invariant not only to the transformations leading to transverse isotropy but to further rotations around any second arbitrarily selected axis. Without loss of generality we may make this the x_1-axis. The matrix of direction cosines for rotations of angle β around the x_1-axis is

$$l_{ij} = \begin{bmatrix} 1 & 0 & 0 \\ 0 & \cos\beta & \sin\beta \\ 0 & -\sin\beta & \cos\beta \end{bmatrix} \tag{1.4-42}$$

and we find

$$[\sigma'_k] = \begin{bmatrix} \sigma_1 \\ \sigma_2\cos^2\beta + \sigma_3\sin^2\beta + 2\sigma_5\sin\beta\cos\beta \\ \sigma_2\sin^2\beta + \sigma_3\cos^2\beta - 2\sigma_5\cos\beta\sin\beta \\ \sigma_4\cos\beta + \sigma_6\sin\beta \\ (\sigma_3 - \sigma_2)\sin\beta\cos\beta + \sigma_5(\cos^2\beta - \sin^2\beta) \\ -\sigma_4\sin\beta + \sigma_6\cos\beta \end{bmatrix} \tag{1.4-43}$$

and

$$[\gamma'_k] = \begin{bmatrix} \gamma_1 \\ \gamma_2\cos^2\beta + \gamma_3\sin^2\beta + 2\gamma_5\sin\beta\cos\beta \\ \gamma_2\sin^2\beta + \gamma_3\sin\beta\cos^2\beta - 2\gamma_5\sin\beta\cos\beta \\ \gamma_4\cos\beta + \gamma_6\sin\beta \\ (\gamma_3 - \gamma_2)\sin\beta\cos\beta + \gamma_5(\cos^2\beta - \sin^2\beta) \\ -\gamma_4\sin\beta + \gamma_6\cos\beta \end{bmatrix}. \tag{1.4-44}$$

With $\beta = \frac{\pi}{2}$ we have

$$[\sigma'_k] = \begin{bmatrix} \sigma_1 \\ \sigma_3 \\ \sigma_2 \\ \sigma_6 \\ -\sigma_5 \\ -\sigma_4 \end{bmatrix} \quad \text{and} \quad [\gamma'_k] = \begin{bmatrix} \gamma_1 \\ \gamma_3 \\ \gamma_2 \\ \gamma_6 \\ -\gamma_5 \\ -\gamma_4 \end{bmatrix}, \tag{1.4-45}$$

and $[C_{kl}]$ becomes

$$[C_{kl}] = \begin{bmatrix} C_{11} & C_{13} & C_{12} & 0 & 0 & 0 \\ \cdot & C_{33} & C_{13} & 0 & 0 & 0 \\ \cdot & \cdot & C_{11} & 0 & 0 & 0 \\ \cdot & \cdot & \cdot & C_{55} & 0 & 0 \\ \cdot & \cdot & \cdot & \cdot & C_{55} & 0 \\ \cdot & \cdot & \cdot & \cdot & \cdot & C_{44} \end{bmatrix}. \tag{1.4-46}$$

Reconciliation with Eq. (1.4-6) yields

$$C_{13} = C_{12} \tag{1.4-47}$$

$$C_{33} = C_{11} \tag{1.4-48}$$

and

$$C_{55} = C_{44} = C_{11} - C_{12} \tag{1.4-49}$$

so that the modulus matrix of an isotropic material emerges in the form

$$[C_{kl}] = \begin{bmatrix} C_{11} & C_{12} & C_{12} & 0 & 0 & 0 \\ \cdot & C_{11} & C_{12} & 0 & 0 & 0 \\ \cdot & \cdot & C_{11} & 0 & 0 & 0 \\ \cdot & \cdot & \cdot & C_{44} & 0 & 0 \\ \cdot & \cdot & \cdot & \cdot & C_{44} & 0 \\ \cdot & \cdot & \cdot & \cdot & \cdot & C_{44} \end{bmatrix} . \tag{1.4-50}$$

For a completely isotropic material the number of material constants is thus finally reduced to the minimum number of two.

Letting

$$C_{11} = \lambda + 2G \tag{1.4-51}$$

$$C_{12} = \lambda \tag{1.4-52}$$

and

$$C_{11} - C_{12} = C_{44} = 2G \tag{1.4-53}$$

Eq. (1.4-50) becomes

$$[C_{kl}] = \begin{bmatrix} \lambda + 2G & \lambda & \lambda & 0 & 0 & 0 \\ \cdot & \lambda + 2G & \lambda & 0 & 0 & 0 \\ \cdot & \cdot & \lambda + 2G & 0 & 0 & 0 \\ \cdot & \cdot & \cdot & 2G & 0 & 0 \\ \cdot & \cdot & \cdot & \cdot & 2G & 0 \\ \cdot & \cdot & \cdot & \cdot & \cdot & 2G \end{bmatrix} \tag{1.4-54}$$

where λ and G are called the *first* and *second Lamé* constants.*

We now proceed to a more detailed examination of the constitutive equation of linear isotropic purely elastic materials.

1.4.4.1 The Generalized Hooke's Law and the Elastic Moduli

With Eq. (1.4-54), Eq. (1.4-1) becomes

$$\sigma_{ij} = \lambda \Delta \delta_{ij} + 2G\gamma_{ij} \tag{1.4-55}$$

where δ_{ij} is the unit tensor, also called Kronecker's delta, and

$$\Delta = \gamma_{11} + \gamma_{22} + \gamma_{33} \tag{1.4-56}$$

is the trace of the strain tensor matrix. But

---

* Gabriel Lamé, pron. Lummay, 1795–1870, French mathematician.

$$V/V_0 - 1 = (1 + \gamma_{11})(1 + \gamma_{22})(1 + \gamma_{33}) - 1 \tag{1.4-57}$$

where V and V_0 are the deformed and undeformed volume, respectively. Hence, for infinitesimal strains (whose products may be neglected), Δ is the *relative volume change* or *cubical strain.* Δ is also called the *dilatation* when the volume change is positive and the *contraction* when it is negative.

Equation (1.4-55) is the general constitutive equation for an isotropic elastic body subjected to infinitesimal strains. It is a generalization of Hooke's law (cf. Sect. 1.1) to a general state of stress. We therefore call it the *generalized Hooke's law.* We observe that the stress tensor σ_{ij} is not directly proportional to the strain tensor γ_{ij}. If, however, we define a tensor γ'_{ij} such that

$$\gamma'_{ij} = (\Delta/3)\delta_{ij} = \gamma_{ij} - \gamma''_{ij} \tag{1.4-58}$$

we may rewrite Eq. (1.4-55) as

$$\sigma_{ij} = 3K\gamma'_{ij} + 2G\gamma''_{ij} \tag{1.4-59}$$

where

$$K = \lambda + \tfrac{2}{3}G\ . \tag{1.4-60}$$

If we now also define another tensor σ'_{ij} such that

$$\sigma'_{ij} = (\Sigma/3)\delta_{ij} = \sigma_{ij} - \sigma''_{ij} \tag{1.4-61}$$

where

$$\Sigma = \sigma_{11} + \sigma_{22} + \sigma_{33} \tag{1.4-62}$$

is the trace of the stress tensor matrix, then we may write

$$\sigma'_{ij} = 3K\gamma'_{ij} \tag{1.4-63}$$

and

$$\sigma''_{ij} = 2G\gamma''_{ij}\ . \tag{1.4-64}$$

We have thus decomposed the stress and the strain tensors into the sums of two tensors each, which are respectively proportional. The primed tensors are known as *isotropic, dilatational, spherical,* or *mean normal tensors,* since $\Sigma/3$ is the mean normal stress, representing an isotropic (hydrostatic) tension or pressure, and $\Delta/3$ is the mean normal strain, i.e. 1/3 of the dilatation or contraction. The double-primed tensors are called *deviatoric tensors* or *deviators.* It is obvious from the definitions, Eqs. (1.4-58), and (1.4-61), that the mean normal tensors have components only on the main diagonal of the matrix, and that all three of the components are equal. We have, explicitly,

$$[\sigma'_{ij}] = \frac{1}{3}\begin{bmatrix} \Sigma & 0 & 0 \\ \cdot & \Sigma & 0 \\ \cdot & \cdot & \Sigma \end{bmatrix} \tag{1.4-65}$$

and

$$[\gamma'_{ij}] = \frac{1}{3}\begin{bmatrix} \Delta & 0 & 0 \\ \cdot & \Delta & 0 \\ \cdot & \cdot & \Delta \end{bmatrix} . \tag{1.4-66}$$

These tensors have the traces Σ, and Δ, respectively.

Since an isotropic pressure produces a change in size only, leaving the shape unaltered, K is the material constant expressing the response of the material to change in size. It is therefore called the *bulk modulus*.

The deviatoric tensors become, by Eqs. (1.4-58) and (1.4-61),

$$[\sigma''_{ij}] = \begin{bmatrix} (2\sigma_{11}-\sigma_{22}-\sigma_{33})/3 & \sigma_{12} & \sigma_{13} \\ \cdot & (2\sigma_{22}-\sigma_{11}-\sigma_{33})/3 & \sigma_{23} \\ \cdot & \cdot & (2\sigma_{33}-\sigma_{11}-\sigma_{22})/3 \end{bmatrix} \tag{1.4-67}$$

and

$$[\gamma''_{ij}] = \begin{bmatrix} (2\gamma_{11}-\gamma_{22}-\gamma_{33})/3 & \gamma_{12} & \gamma_{13} \\ \cdot & (2\gamma_{22}-\gamma_{11}-\gamma_{33})/3 & \gamma_{23} \\ \cdot & \cdot & (2\gamma_{33}-\gamma_{11}-\gamma_{22})/3 \end{bmatrix} . \tag{1.4-68}$$

Their traces vanish identically. Because the deviators are those portions of the total tensors from which the mean normal tensors (representing changes in size) have been subtracted, they are seen to represent changes in shape. The material constant, G, called the *shear modulus* (or *modulus of rigidity*, or *torsional modulus*) expresses the response of the material to a change in shape.

Using Eqs. (1.4-60) and (1.4-55), the generalized Hooke's law may be rewritten as

$$\sigma_{ij} = (K - \tfrac{2}{3}G)\Delta\delta_{ij} + 2G\gamma_{ij} . \tag{1.4-69}$$

The numerical factors in Eqs. (1.4-63) and (1.4-64) and hence, in Eqs. (1.4-55) and (1.4-69) arise because of the traditional definitions of the moduli. The bulk modulus is traditionally defined through the equation

$$p = -K\Delta \tag{1.4-70}$$

where p is the (isotropic) pressure. Letting $p = -\sigma_{11}$, and $\Delta = 3\gamma_{11}$, we regain Eq. (1.4-63). The shear modulus is traditionally defined through the equation

$$\sigma = G\varepsilon \tag{1.4-71}$$

where σ is the shear stress and ε is the amount of shear. If σ_{12} is the shear component of the stress tensor (i.e. the material is sheared in the 1,2-plane), then $\sigma = \sigma_{12}$, and $\varepsilon = 2\gamma_{12}$ by Eq. (1.3-19), γ_{12} being the shear component of the strain tensor. Hence $\sigma_{12} = 2G\gamma_{12}$ in accordance with Eq. (1.4-64).

Only for infinitesimal strains can the total deformation be separated neatly into changes in size (dilatation or contraction) and in shape (distortion), respectively. The simplification achieved in the limit of infinitesimal strains is one of the useful aspects of the linear theory of elasticity. The bulk and shear moduli are fundamental moduli in the sense of this decomposition. As stated by Eqs. (1.4-70) and (1.4-71), they are the moduli appropriate to pressure/volume, and shear stress/strain experiments. Different moduli may be useful in other experiments. These will, however, represent changes of size as well as shape, and may always be expressed in terms of the fundamental moduli, K and G.

The most often used of the "derived" moduli expressing the response of the material in experimental situations which involve simultaneous changes of shape and size are the *stretch modulus* (*tensile*, *Young's* modulus* or *modulus of elasticity*), E, and the *wave modulus* (or *longitudinal bulk modulus*) M.

The stretch modulus arises in experiments in simple (or uniaxial) tension or compression (cf. Sect. 1.1). In such experiments we apply a force in one direction only so that the stress tensor matrix becomes

$$[\sigma_{ij}] = \begin{bmatrix} \sigma_{11} & 0 & 0 \\ \cdot & 0 & 0 \\ \cdot & \cdot & 0 \end{bmatrix} \qquad (1.4\text{-}72)$$

if the pull is in the 1-direction. There is a resultant elongation (stretch) in the direction in which the force acts, and this is represented by the component γ_{11} of the strain tensor. Simultaneously, however, an isotropic material contracts equally in the two perpendicular directions, so that the strain tensor matrix becomes

$$[\gamma_{ij}] = \begin{bmatrix} \gamma_{11} & 0 & 0 \\ \cdot & \gamma_{22} & 0 \\ \cdot & \cdot & \gamma_{22} \end{bmatrix} . \qquad (1.4\text{-}73)$$

We define the stretch modulus as the ratio of the stress component in the direction of the force to the strain component in the same direction, i.e. we write

$$E = \sigma_{11}/\gamma_{11} . \qquad (1.4\text{-}74)$$

To obtain the relation between E, K and G, we must solve the two simultaneous equations

$$\lambda(\gamma_{11} + 2\gamma_{22}) + 2G\gamma_{11} = \sigma_{11} \qquad (1.4\text{-}75)$$

$$\lambda(\gamma_{11} + 2\gamma_{22}) + 2G\gamma_{22} = 0 \qquad (1.4\text{-}76)$$

obtained from Eq. (1.4-55), for γ_{11} and substitute the resulting expression into Eq. (1.4-74). The result is

$$E = \frac{9KG}{3K + G} . \qquad (1.4\text{-}77)$$

In many materials (e.g. most rubbers) the shear modulus is very much smaller (by about three to four orders of magnitude) than the bulk modulus. Such a material appears incompressible compared to the ease with which it is sheared. A material for which $K \gg G$ is (not quite correctly) called incompressible. An incompressible

---

* Thomas Young, 1775–1829, English physicist and physician.

material possesses the single modulus, G. Letting $K \to \infty$, Eq. (1.4-77) yields the relation $E = 3G$ for such a material.

Solving Eq. (1.4-76) for $-\gamma_{22}/\gamma_{11}$ and using Eq. (1.4-60) furnishes an expression for the ratio of the lateral contraction to the elongation. This ratio is known as *Poisson's* ratio*, μ. We have

$$\mu = \frac{-\gamma_{22}}{\gamma_{11}} = \frac{3K - 2G}{6K + 2G} . \tag{1.4-78}$$

For an incompressible material, i.e. for $K \to \infty$, Eq. (1.4-78) gives $\mu = \frac{1}{2}$ for Poisson's ratio.

The wave modulus arises when a force applied in one direction is accompanied by a deformation in the same direction, but any contraction or expansion in the perpendicular directions is prevented. This occurs, e.g., when the material is placed into a strong walled cylinder closed at the bottom, and is then compressed by a tight-fitting piston. It also occurs in the propagation of acoustical waves through the material. Hence the name wave modulus. Since there is no deformation in the directions perpendicular to the direction of the force, stresses are set up in these two directions. We have

$$[\sigma_{ij}] = \begin{bmatrix} \sigma_{11} & 0 & 0 \\ \cdot & \sigma_{22} & 0 \\ \cdot & \cdot & \sigma_{22} \end{bmatrix} \tag{1.4-79}$$

and

$$[\gamma_{ij}] = \begin{bmatrix} \gamma_{11} & 0 & 0 \\ \cdot & 0 & 0 \\ \cdot & \cdot & 0 \end{bmatrix} . \tag{1.4-80}$$

From Eqs. (1.4-55) and (1.4-60), then,

$$M = \sigma_{11}/\gamma_{11} = K + \tfrac{4}{3}G \tag{1.4-81}$$

and the ratio of the lateral to the imposed stress is given by

$$\frac{\sigma_{22}}{\sigma_{11}} = \frac{3K - 2G}{3K + 4G} . \tag{1.4-82}$$

Thus, for an incompressible material the stress is isotropic.

Any elastic constant can always be expressed as a function of any two others. The relations between the elastic constants, K, G, E, and μ are assembled in Table 1.4-1 for convenience. The Lamé constant λ and the wave modulus M can be expressed in terms of the other constants through Eqs. (1.4-60) and (1.4-81), respectively, using the relations tabulated in Table 1.4-1.

Further elastic constants could be defined to cover specific experimental conditions but have not been found particularly useful. Problems 1.4-1 through 1.4-3 deal with this approach by way of illustration.

---

* Simeon Denis Poisson, pronounced Pwasson, 1781–1840, French physicist.

Table 1.4-1. Relations between the elastic constants K, G, E, and μ

Elastic constant	Expressed as function of:					
	G and E	G and μ	E and μ	K and E	K and μ	K and G
K	$\frac{GE}{9G-3E}$	$\frac{2G(1+\mu)}{3(1-2\mu)}$	$\frac{E}{3(1-2\mu)}$	—	—	—
G	—	—	$\frac{E}{2(1+\mu)}$	$\frac{3KE}{9K-E}$	$\frac{3K(1-2\mu)}{2(1+\mu)}$	—
E	—	$2G(1+\mu)$	—	—	$3K(1-2\mu)$	$\frac{9KG}{3K+G}$
μ	$\frac{E}{2G}-1$	—	—	$\frac{1}{2}-\frac{E}{6K}$	—	$\frac{3K-2G}{6K+2G}$

$\lambda = K - \frac{2}{3}G$, $M = K + \frac{4}{3}G$

We close this section by considering the form of the generalized Hooke's law for an incompressible material. When $K \gg G$ we may replace $\lambda\Delta\delta_{ij}$ by $-p\delta_{ij}$ where p is an arbitrary isotropic pressure. We have

$$\sigma_{ij} = -p\delta_{ij} + 2G\gamma_{ij} , \qquad (1.4\text{-}83)$$

i.e. the stress is defined only within an arbitrary isotropic pressure.

1.4.4.2 The Generalized Hooke's Law and the Elastic Compliances

The foregoing considerations were based on Eqs. (1.4-1) or (1.4-5) with Eq. (1.4-50). The relations between the strain and stress tensor components of an elastic material may, however, be stated equally in the form

$$\gamma_{ij} = S_{ijkl}\sigma_{kl} \qquad (1.4\text{-}84)$$

where S_{ijkl} is the *compliance tensor*. We can then further define

$$[\gamma_k] = [S_{kl}][\sigma_l] , \qquad k, l = 1, 2, \ldots, 6 \qquad (1.4\text{-}85)$$

where, for an isotropic material,

$$[S_{kl}] = \begin{bmatrix} S_{11} & S_{12} & S_{12} & 0 & 0 & 0 \\ \cdot & S_{11} & S_{12} & 0 & 0 & 0 \\ \cdot & \cdot & S_{11} & 0 & 0 & 0 \\ \cdot & \cdot & \cdot & S_{44} & 0 & 0 \\ \cdot & \cdot & \cdot & \cdot & S_{44} & 0 \\ \cdot & \cdot & \cdot & \cdot & \cdot & S_{44} \end{bmatrix} . \qquad (1.4\text{-}86)$$

The matrix $[S_{kl}]$ is called the *compliance matrix*. We let (see Problems 1.4-5, 1.4-6, and 1.4-7)

$$S_{11} = \tfrac{1}{9}B + \tfrac{1}{3}J \tag{1.4-87}$$

$$S_{12} = \tfrac{1}{9}B - \tfrac{1}{6}J \tag{1.4-88}$$

and

$$S_{44} = \tfrac{1}{2}J \tag{1.4-89}$$

where B and J are defined by

$$\gamma'_{ij} = \tfrac{1}{3}B\sigma'_{ij} \tag{1.4-90}$$

and

$$\gamma''_{ij} = \tfrac{1}{2}J\sigma''_{ij} \tag{1.4-91}$$

respectively. The *bulk compliance* $B = 1/K$ relates to changes in size. The *shear compliance* $J = 1/G$ relates to changes in shape. The *dual form* of the generalized Hooke's law then becomes (cf. Problem 1.4-3)

$$\gamma_{ij} = (\tfrac{1}{9}B - \tfrac{1}{6}J)\Sigma\delta_{ij} + \tfrac{1}{2}J\sigma_{ij} \tag{1.4-92}$$

for a compressible material, and

$$\gamma_{ij} = \tfrac{1}{2}J\sigma''_{ij} \tag{1.4-93}$$

for an incompressible material, for which $B \to 0$.

We also define a *stretch* or *tensile compliance* $D = 1/E$. The relations between the three compliances and Poisson's ratio, μ, are tabulated in Table 1.4-2.

Some examples of the use of the components of the compliance matrix are given in Sect. 1.6 (Problems).

Table 1.4-2. Relations between the elastic constants B, J, D, and μ

Elastic constant	Expressed as Function of:					
	J and D	J and μ	D and μ	B and D	B and μ	B and J
B	$9D - 3J$	$\dfrac{3J(1-2\mu)}{2(1+\mu)}$	$3D(1-2\mu)$	—	—	—
J	—	—	$2D(1+\mu)$	$3D - B/3$	$\dfrac{2B(1+\mu)}{3(1-2\mu)}$	—
D	—	$\dfrac{J}{2(1+\mu)}$	—	—	$\dfrac{B}{3(1-2\mu)}$	$B/9 + J/3$
μ	$\dfrac{J}{2D} - 1$	—	—	$\dfrac{1}{2} - \dfrac{B}{6D}$	—	$\dfrac{3J-2B}{6J+2B}$

1.5 Purely Viscous Linear Response

In the preceding section we have discussed the constitutive equations of isotropic elastic materials in infinitesimal deformations. These equations link the stress and the strain through appropriate material constants, the elastic constants. In a material showing purely viscous response the stress is linked not to the strain but to the rate of strain (cf. Sect. 1.1). Although anisotropic liquids exist (e.g. liquid crystals), we shall consider here only isotropic ones.

1.5.1 The Generalized Newton's Laws

As we have done with the strain tensor, we now split the *rate of strain tensor*, $\dot{\gamma}_{ij}$, into two parts, the *isotropic rate of strain tensor*, $\dot{\gamma}_{ij}$, and the *deviatoric rate of strain tensor*, $\dot{\gamma}''_{ij}$ [cf. Eq. (1.3-20)]. We have

$$\dot{\gamma}_{ij} = \dot{\gamma}'_{ij} + \dot{\gamma}''_{ij}, \tag{1.5-1}$$

and, for a change in shape we obtain

$$\sigma''_{ij} = 2\eta\dot{\gamma}''_{ij} \tag{1.5-2}$$

in analogy with Eq. (1.4-64). Here the proportionality constant η is called the *coefficient of* (*shear*) *viscosity*, or *viscosity* for short. Its reciprocal

$$\phi = 1/\eta \tag{1.5-3}$$

is the *coefficient of* (*shear*) *fluidity*, or simply the *fluidity*.

For a change in size we have

$$\sigma'_{ij} = 3\xi\dot{\gamma}'_{ij} \tag{1.5-4}$$

in analogy to Eq. (4-63). Here ξ is the *coefficient of bulk viscosity*. Its reciprocal, the *bulk fluidity*, is rarely used. Denoting the trace of $\dot{\gamma}'_{ij}$ by $\dot{\Delta}$, we have

$$\sigma_{ij} = (\xi - \tfrac{2}{3}\eta)\dot{\Delta}\delta_{ij} + 2\eta\dot{\gamma}_{ij} \tag{1.5-5}$$

where the rate of dilatation

$$\dot{\Delta} = \dot{\gamma}_{11} + \dot{\gamma}_{22} + \dot{\gamma}_{33}\ . \tag{1.5-6}$$

Equation (1.5-5) may be compared with Eq. (1.4-69).

For a shear in the 1,2-plane we have

$$\sigma_{12} = \sigma \qquad \text{and} \qquad \dot{\gamma}_{12} = \dot{\varepsilon}/2 \tag{1.5-7}$$

by Eq. (1.3-22). This leads directly to Eq. (1.1-5). Hence, we call Eq. (1.5-5) the *generalized Newton's law*. Derivation of Eq. (1.1-4) from Eq. (1.5-5) is left to Problem 1.5-1.

The viscous analog of the stretch or Young's modulus is the *stretch*, *tensile*, or *elongational viscosity* or *Trouton's* coefficient of viscosity*, ζ. This is defined by

---

* Frederick Thomas Trouton, 1863–1922, Irish English physicist, who determined the elongational viscosity of pitches.

$$\zeta = \sigma_{11}/\dot{\gamma}_{11} \tag{1.5-8}$$

for a pull in the 1-direction. It is related to ξ and η by

$$\zeta = \frac{9\xi\eta}{3\xi + \eta} \ . \tag{1.5-9}$$

If $\xi \gg \eta$, we have $\zeta = 3\eta$ in accordance with Trouton's original observation. Equation (1.5-9) is the viscous analog of Eq. (1.4-77).

The absence of a term in γ''_{ij} in Eq. (1.5-2) is in accordance with the fact that a purely viscous material does not sustain a shear. However, a material which is purely viscous in its response to shear may still respond elastically to an isotropic pressure. If this is taken into account, Eqs. (1.5-4) and (1.5-5) become

$$\sigma'_{ij} = 3K\gamma'_{ij} + 3\xi\dot{\gamma}'_{ij} \tag{1.5-10}$$

and

$$\sigma_{ij} = [K\Delta + (\xi - \tfrac{2}{3}\eta)\dot{\Delta}]\delta_{ij} + 2\eta\dot{\gamma}_{ij} \tag{1.5-11}$$

respectively. The elongational viscosity then becomes

$$\zeta = \frac{9\xi\eta}{3\xi + \eta(1 - 3K\Delta/\sigma_{11})} \tag{1.5-12}$$

which again reduces to $\zeta = 3\eta$ for $\xi \gg \eta$.

For an incompressible body we may replace the first term in Eq. (1.5-11) by $-p\delta_{ij}$, where p is the isotropic pressure, and write

$$\sigma_{ij} = -p\delta_{ij} + 2\eta\dot{\gamma}_{ij} \ . \tag{1.5-13}$$

Thus, the stress is then known only within an arbitrary isotropic pressure, p, and we can only work with normal stress *differences*. In the uniaxial elongation of an isotropic material the stresses, strains, and rates of strain, are identical in the two transverse directions, and there is then only one normal stress difference. We find

$$\sigma_{11} - \sigma_{22} = 2\eta(\dot{\gamma}_{11} - \dot{\gamma}_{22}) \ . \tag{1.5-14}$$

But, since the volume remains constant, the rate of dilatation, $\dot{\Delta}$, now vanishes, and we have

$$\dot{\Delta} = \dot{\gamma}_{11} + 2\dot{\gamma}_{22} = 0 \ . \tag{1.5-15}$$

Then $\dot{\gamma}_{22} = -\dot{\gamma}_{11}/2$ and, using the earlier relation $\zeta = 3\eta$, we obtain

$$\sigma_{11} - \sigma_{22} = 6\eta\dot{\gamma}_{11} = 3\zeta\dot{\gamma}_{11} \tag{1.5-16}$$

and find

$$\zeta = \frac{\sigma_{11} - \sigma_{22}}{\dot{\gamma}_{11}} \tag{1.5-17}$$

for the stretch, or elongational viscosity of an *incompressible* liquid. In a solid, σ_{22} would be zero in uniaxial extension.

Equations (1.5-11) and (1.5-13) are special forms of the generalized Newton's law.

1.6 Problems

Problem 1.1-1. Three specimens are fashioned from a rubber having a stretch modulus, E, of 10^5 Nm^{-2}. The first specimen is 0.2 m long with a cross-sectional area of 0.025×10^{-3} m^2. The second and third specimens are 0.15 m long and have cross-sectional areas of 0.025×10^{-3} m^2 and 0.0125×10^{-3} m^2, respectively.

All three specimens are extended by hanging 0.05 kg weights onto them. Compare the extended lengths.

Problem 1.1-2. A piece of rubber is elongated by the application of a uniaxial stress equal to its modulus of elasticity. What is the percent elongation?

Problem 1.2-1. Find the equation for the unit vector in the direction of the stress vector on the *octahedral plane*. The latter is the plane which cuts the coordinates at the points $X_1 = X_2 = X_3 = 1$ in Fig. 1.2-3.

Problem 1.2-2. Let the state of stress at a point in a body be given by the stress tensor matrix

$$[\sigma_{ij}] = \begin{bmatrix} \sigma & a\sigma & b\sigma \\ a\sigma & \sigma & c\sigma \\ b\sigma & c\sigma & \sigma \end{bmatrix} \tag{1}$$

where σ is some stress value and a, b, and c are constants. What values must these constants assume so that the stress vector on the octahedral plane (see Problem 1.2-1) vanishes? Write the matrix for the stress tensor with these values of the constants.

Problem 1.3-1. Let a displacement field be given by $u_1 = (x_1 - x_3)^2$, $u_2 = (x_2 + x_3)^2$, and $u_3 = -x_1 x_2$. Determine the matrices of the displacement gradient tensor, the strain tensor, and the rotation tensor. Then find the matrices of the spherical and deviatoric strain tensors.

Problem 1.4-1. In analogy to Young's modulus in uniaxial tension, define a modulus, H, for uniform biaxial tension.

Express H, and μ_H, the ratio of the contraction in the free direction to the extension in either of the other two directions, in terms of K and G.

What is H and μ_H for $K \to \infty$?

Problem 1.4-2. Define a modulus, U, for a constrained biaxial deformation (pure shear) in which a stress is applied in the 1-direction but the material is constrained so that there is no deformation in the 2-direction. Define the ratio of the lateral contraction to the extension in the 1-direction as μ_U.

Give U and μ_U in terms of K and G. What is the ratio of the stress in the 2-direction, set up as a result of the constraint, to the stress applied in the

1-direction? Express the ratio in terms of K and G. What are the limiting cases for an incompressible material?

Problem 1.4-3. Derive Eq. (1.4-92)

$$\gamma_{ij} = (\tfrac{1}{9}B - \tfrac{1}{6}J)\Delta\delta_{ij} + \tfrac{1}{2}J\sigma_{ij} \ . \tag{1}$$

Problem 1.4-4. Find Young's modulus and Poisson's ratio for an isotropic material in terms of the components of the compliance matrix.

Problem 1.4-5. Find S_{11}, S_{12}, and S_{44} for an isotropic material in terms of C_{11}, C_{12} and C_{44}. Then express C_{11}, C_{12}, and C_{44} in terms of S_{11}, S_{12}, and S_{44}.

Problem 1.4-6. Show that $S_{11} = \frac{1}{9}B + \frac{1}{3}J$, $S_{12} = \frac{1}{9}B - \frac{1}{6}J$, and $S_{44} = \frac{1}{2}J$.

Problem 1.4-7. Show that, in terms of the elastic compliances, S_{ij},

$$K = \frac{1}{3(S_{11} + 2S_{12})} \tag{1}$$

and

$$G = \frac{1}{2(S_{11} - S_{12})} \ . \tag{2}$$

Problem 1.4-8. Show that

$$\gamma_{11} = \frac{1}{E}\sigma_{11} - \frac{\mu}{E}(\sigma_{22} + \sigma_{33}) \ . \tag{1}$$

Problem 1.4-9. Consider the uniaxial compression of a solid material which is confined laterally. Let σ_{11} be the applied stress. Express the modulus M appropriate for this mode of deformation in terms of the bulk modulus K and Poisson's ratio μ. Discuss the results for $\mu = 1/2$ and $\mu = 1/3$.

Problem 1.4-10. An axisymmetric elastic material is characterized by five material parameters. Define these in terms of an axial, a transverse, and a torsional compliance, and an axial and a transverse Poisson's ratio.

Problem 1.4-11. Three Poisson's ratios may be defined for an axisymmetric material. Referring to Fig. 1.4-4, obtain these ratios in terms of the constants λ, G, β, γ, and δ introduced by Eqs. (1.4-36) through (1.4-40). Satisfy yourself that all three equations reduce to Eq. (1.4-78) as $\beta, \gamma, \rho \to 0$. Then obtain the expressions for the case that $\lambda \to \infty$ (incompressibility assumption).

Problem 1.4-12. Referring to Fig. 1.4-4, obtain the moduli E_1 and E_3 for an axisymmetric material in terms of the components of the modulus matrix $[C_{kl}]$.

Express the moduli in terms of the parameters λ, G, β, ρ and δ. Show that $E_1 \to E_3 \to E$ as β, ρ, $\delta \to 0$. Find the expressions for E_1 and E_3 as $\lambda \to \infty$ (incompressible material).

Problem 1.5-1. Show that the generalized Newton's law, Eq. (1.5-5), reduces to Newton's equation, Eq. (1.1-4), for a one-dimensional laminar shear flow.

Problem 1.5-2. Derive Eq. (1.5-12)

$$\zeta = \frac{9\xi\eta}{3\xi + \eta(1 - 3K\Delta/\sigma_{11})} . \tag{1}$$

References (Chap. 1)

1. R. Hooke, *A Description of Helioscopes and Some Other Instruments*, London, 1676; R.T. Gunther, *Early Science in Oxford*, *8*:119 (1931).
2. R. Hooke, *Lectures de Potentia Restitutiva*, or *Of Explaining the Power of Springing Bodies*, London, 1678; R.T. Gunther, *Early Science in Oxford*, *8*:331 (1931).
3. I. Newton, *Philosophiae Naturalis Principia Mathematica*, London, 1687; 3rd ed.; ed. H. Pemberton, London, 1726; repr. Glasgow, 1871; trans. A. Motte, *Sir Issac Newton's Mathematical Principles of Natural Philosophy, and his System of the World*, London, 1729; revised ed., Berkeley, 1934.
4. A.L. Cauchy, *Oeuvres*, (2) *7*:60, 82 (1827); (2) *8*:195 (1828).
5. J.F. Nye, *Physical Properties of Crystals*, Clarendon Press, Oxford, 1957.
6. I.S. Sokolnikoff, *Mathematical Theory of Elasticity*, McGraw-Hill, New York, 1956.

2. Linear Viscoelastic Response

Time is the principal character in this book.

Harold deWitt Smith

God created time so that things would not happen all at once.

Graffito

Quid est tempus? Si nemo a me quaerit, scio; si quaerenti explicare velim, nescio.

St. Augustine

2.0 Introduction

When a stress or a strain is impressed upon a body, rearrangements take place inside the material by which it responds to the imposed excitation. In any real material these rearrangements necessarily require a finite time. The time required, however, may be very short or very long. When the changes take place so rapidly that the time is negligible compared with the time scale of the experiment, we regard the material as purely viscous. In a purely viscous material, all the energy required to produce the deformation is dissipated as heat. When the material rearrangements take virtually infinite time, we speak of a purely elastic material. In a purely elastic material the energy of deformation is stored and may be recovered completely upon release of the forces acting on it. Water comes close to being a purely viscous material; and steel, if deformed to no more than a percent or two, behaves in an almost completely elastic fashion. In principle, however, all real materials are *viscoelastic*. Some energy may always be stored during the deformation of a material under appropriate conditions, and energy storage is always accompanied by dissipation of some energy.

In a typically viscoelastic material the time necessary for the material rearrangements to take place is comparable with the time scale of the experiment. The relation between the two time scales can be expressed conveniently by a dimensionless number [1] introduced by Reiner*. The number, N_D, is the ratio of the time scale of the material rearrangements, τ_{mat}, to the experimental time scale, τ_{exp}. In symbols

$$N_D = \frac{\tau_{mat}}{\tau_{exp}} . \qquad (2.0\text{-}1)$$

Reiner termed his number the *Deborah number* after a passage in the victory song of the prophetess Deborah (Judges V, 5) which may be interpreted to mean

* Marcus Reiner, pron. Riner, 1886–1976, Austrian born Israeli civil engineer, who did much for the development of rheology as a separate branch of mechanics.

that "the mountains flow before the Lord". This simile vividly expresses the fact that objects which are seemingly unchangeable against the time scale of the human life span, deform when measured on a geological time scale.

Equation (2.0-1) expresses the relativity of the material and experimental time scales. When the Deborah number approaches zero, the material is purely viscous, when it approaches infinity, the material is purely elastic. For a Deborah number of unity the material is typically viscoelastic. The concept of the Deborah number is fairly easy to grasp intuitively but it is much more difficult to express mathematically because suitable definitions for both the material rearrangement times and the experimental time scale have to be found.

As a consequence of the material rearrangements taking place on a time scale comparable to that of the experiment in which they are observed, the relations between stress and strain or rate of strain cannot be expressed by *material constants* as in the case of purely elastic or purely viscous materials. The rheological behavior of viscoelastic materials is characterized by time dependent *material functions.* In the most general case the stress becomes a *functional** of the strain, i.e. it depends on the strain history. We denote this by writing

$$\sigma_{ij}(t) = \mathfrak{S}\left[\gamma_{ij}\overset{t}{\underset{-\infty}{(u)}}\right] \tag{2.0-2}$$

where t is the *present* or *current time* (or time of derivation) and u is the *past, previous,* or *historic time.* The notation emphasizes that the stress at the present time, t, depends on the strains at all past times, u. In an analogous way, the strain may be regarded as a functional of the stress, and we may express this by writing

$$\gamma_{ij}(t) = \mathfrak{G}\left[\sigma_{ij}\overset{t}{\underset{-\infty}{(u)}}\right] . \tag{2.0-3}$$

For sufficiently small deformations the functionals can be expressed by linear differential equations with constant coefficients, or, equivalently, by convolution integrals with difference kernels. The theory of the large deformation of viscoelastic materials, however, presents considerable theoretical and practical complexity.

In metals rheology, the term viscoelasticity is often replaced by *anelasticity.* Some authors distinguish between the viscoelasticity of solids and the *elastoviscosity* of liquids. The first investigations of viscoelastic behavior were made by

* We say that in $y = \mathfrak{F}[g(\overset{b}{\underset{a}{u}})]$, y is a functional of the function g(u) in the interval (a, b) when y depends on all values taken by g(u) for u varying in the interval (a, b) [2]. An alternative definition states: y is a functional of the function g(u) if a rule is given by which to every function g(u) defined within the interval (a, b) there can be made to correspond one and only one quantity y [2].

The argument of a functional is a *function.* Thus, as an example of a functional, the definite integral $y = \int_0^t g(u)du$ has the function g(u) as its argument, i.e., for the stated interval of integration, y depends on g(u) but not on u. As another example, the differential equation $dy/dt = g(t)$ also defines a functional. For a given boundary condition, say, $y = 0$ for $t = 0$, y depends only on the form of the function g(t).

A functional must not be confused with a composite function although there is a superficial resemblance. When $y = f(z)$, $z = g(x)$, we may express y in the form of the composite function $y = f[g(x)]$. This makes it appear that y is a function of g(x). We could, however, write $y = h(x)$ instead, and thus recognize that y is a function of x, i.e. the true argument of y is the *variable* x, and not the function g(x).

Weber* [3], and were continued by R. Kohlrausch and his son F. Kohlrausch**. Weber and R. Kohlrausch studied materials that could be used as suspension filaments for galvanometers, notably glass, silk, and various metals, particularly silver and copper. F. Kohlrausch, in addition to these, also made the first study of a polymeric material, natural rubber [4]. An excellent account of the early history of viscoelasticity has been given by H. Leaderman*** [5].

For a summary of the fundamental features of the linear thoery see Sect. 11.4.

2.1 Linear Time-dependent Behavior

In this section we will be concerned with time-dependent stresses or strains. Here, and in most other chapters we will disregard the tensorial character of stress and strain, and will formulate our relations as they apply to a time-dependent shear stress, say $\sigma_{12}(t)$, which we will simply call the stress and will represent by $\sigma(t)$, and to the corresponding time-dependent amount of shear, $\varepsilon_{12}(t)$ (cf. Sect. 1.3), which we will simply call the strain and denote by $\varepsilon(t)$. Generalization to the full tensorial representation will be discussed in Chap. 11.

In Sects. 1.4 and 1.5 we showed that purely elastic and purely viscous linear behavior is expressed typically by constitutive equations of the form

$$\sigma = G\varepsilon \tag{2.1-1}$$

for the elastic case, and

$$\sigma = \eta\frac{d\varepsilon}{dt} \tag{2.1-2}$$

for the viscous case. One would expect that linear viscoelastic behavior should be described by a constitutive equation which is a combination of Eqs. (2.1-1) and (2.1-2), such as, e.g., an equation of the form

$$\sigma = G\varepsilon + \eta\frac{d\varepsilon}{dt}\ . \tag{2.1-3}$$

This equation had indeed been proposed as early as 1874 by O. Meyer**** [6] as a generalization of the equation of classical elasticity theory.

However, for reasons that will become clear in Sect. 3.3, Eq. (2.1-3) does not completely describe viscoelastic behavior. It will be shown there that the simplest

* Wilhelm Edward Weber, pron. Vayber, 1804–1891, German physicist, noted for his contributions to the study of electricity and electromagnetism. The SI unit of magnetic flux is named after him.

** Rudolph Kohlrausch, pron. Kolrowsh, 1809–1858, and Friedrich Wilhelm Georg Kohlrausch, 1840–1910, German physicists. F. Kohlrausch is particularly noted for his researches on electromagnetism and electrolytic conductivity.

*** Herbert Leaderman, (1913–1965), English-born U.S. physicist, who greatly contributed to the development of the molecular theory of viscoelastic behavior [5].

**** Oskar Emil Meyer, 1834–1909, German physicist who contributed to the theory of elasticity.

constitutive equation which adequately describes the infinitesimal deformation of a viscoelastic body is

$$\sigma + a\frac{d\sigma}{dt} = b\varepsilon + c\frac{d\varepsilon}{dt} \ . \tag{2.1-4}$$

This is a linear differential equation with constant coefficients containing only first derivatives. An adequate representation of the behavior of real materials generally requires differential equations containing higher derivatives of both the stress and the strain up to any order. In a general deformation, moreover, the constitutive equation may not be linear, and it may not contain constant coefficients. However, in the limit of infinitesimal deformation, viscoelastic behavior can be described by *linear* differential equations with *constant* coefficients. Such behavior is termed *linear time-dependent* or *linear viscoelastic behavior*. In practice it is found that most materials show linear time dependent behavior even in finite deformation as long as the strain remains below a certain limit, the *linear viscoelastic limit*. This limit, i.e. the magnitude of the strain above which linear viscoelastic behavior is no longer observed, varies from material to material and is thus a material property.

If the response to an excitation is to be linear, the functionals introduced with Eqs. (2.0-2) and (2.0-3) must satisfy two conditions. First, an increase in the stimulus by an arbitrary factor α must increase the response by the same factor. Thus we must have

$$\mathfrak{S}[\alpha\,\varepsilon(\overset{t}{\underset{-\infty}{u}})] = \alpha\,\mathfrak{S}[\varepsilon(\overset{t}{\underset{-\infty}{u}})] = \alpha\,\sigma(t) \tag{2.1-5}$$

and, analogously,

$$\mathfrak{E}[\alpha\,\sigma(\overset{t}{\underset{-\infty}{u}})] = \alpha\,\mathfrak{E}[\sigma(\overset{t}{\underset{-\infty}{u}})] = \alpha\,\varepsilon(t) \ . \tag{2.1-6}$$

Linearity in accordance with Eqs. (2.1-5) and (2.1-6), is commonly called *stress-strain linearity*.

Second, an arbitrary sequence of stimuli must elicit a response which is equal to the sum of the responses which would have been obtained if all stimuli had acted independently. Thus

$$\mathfrak{S}\left[\sum_{n=1}^{\infty} \varepsilon_n(t - \overset{t}{\underset{-\infty}{u_n}})\right] = \sum_{n=1}^{\infty} \mathfrak{S}[\varepsilon_n(t - \overset{t}{\underset{-\infty}{u_n}})] \tag{2.1-7}$$

and

$$\mathfrak{E}\left[\sum_{n=1}^{\infty} \sigma_n(t - \overset{t}{\underset{-\infty}{u_n}})\right] = \sum_{n=1}^{\infty} \mathfrak{E}[\sigma_n(t - \overset{t}{\underset{-\infty}{u_n}})] \tag{2.1-8}$$

where $t - u$ is the *elapsed time*. The notation we have used emphasizes that the stimuli may be imposed at different times. The linearity expressed by Eqs. (2.1-7) and (2.1-8) may be called *time dependence linearity*. It is also referred to as *time shift invariance* because a shift of the stimulus along the time scale results in a corresponding shift of the response without changing it in any other way (see Problem 2.3-4).

Clearly, Eqs. (2.1-7) and (2.1-8) contain Eqs. (2.1-5) and (2.1-6) if the stimuli are simply interpreted appropriately. The two conditions for linear behavior in viscoelastic materials must, nevertheless, be carefully distinguished.

2.1.1 Differential Representations: The Operator Equation

Linear differential equations with constant coefficients satisfy the conditions expressed by Eqs. (2.1-5) through (2.1-8). The general equation linking the time-dependent stress and strain is

$$\sum_{n=0}^{\infty} u_n \frac{d^n\sigma(t)}{dt^n} = \sum_{m=0}^{\infty} q_m \frac{d^m\varepsilon(t)}{dt^m} \tag{2.1-9}$$

where u_n and q_m are the constant coefficients.* Equation (2.1-9) together with the appropriate initial conditions, describes the time-dependent (viscoelastic) behavior of an isotropic material subjected to an infinitesimally small shear deformation. It can be cast conveniently in the form of an *operator equation*

$$\mathscr{U}[\sigma(t)] = \mathscr{Q}[\varepsilon(t)] \tag{2.1-10}$$

where

$$\mathscr{U}[\quad] = \sum_{n=0}^{\infty} u_n \frac{d^n}{dt^n}[\quad] \quad \text{and} \quad \mathscr{Q}[\quad] = \sum_{m=0}^{\infty} q_m \frac{d^m}{dt^m}[\quad] \tag{2.1-11}$$

are linear differential operators operating on the stress, $\sigma(t)$, and strain, $\varepsilon(t)$, respectively. The summations do not have to extend to infinity. In Sect. 3.7 the coefficients u_n and q_m of the operator equation will be related to the parameters of the rheological models to be introduced in Chap. 3.

The manipulation of differential, integral, or integro-differential equations is greatly facilitated by the use of integral transform methods.** For our purposes the *one-sided Laplace*** transformation* (cf. Appendix A3) is most suitable. The (one-sided) Laplace transform of the function f(t) is given by

$$\bar{f}(s) = \int_0^{\infty} f(t)\exp(-st)dt \tag{2.1-12}$$

where s is the Laplace transform variable. Equation (2.1-12) assumes that the time domain is restricted to the interval between zero and infinity. This causes no problem as long as we may consider the state in which the material is at $t = 0$ to be a *reference state* in which it is free of the effects of any stress or strain history it might have experienced prior to $t = 0$. If the material has a preferred configuration,

* The symbol u will be used both for the coefficients in the differential equation (when it usually appears with a subscript) and for the historic time (when it usually appears without a subscript). This should cause no confusion.

** The reader is urged to consult the Appendix on Transformation Calculus.

*** Pierre Simon, marquis de Laplace, pron. Laplahs, 1749–1827, French mathematician, astronomer, and physicist, celebrated for his work on gravitation.

this reference state is the *undeformed state*. In Chap. 3 the term *arrheodictic* will be introduced for such a material. Some solids, such as e.g., rubber vulcanisates, are arrheodictic. By contrast, most common liquids, including viscoelastic ones such as, e.g., polymer melts, are *rheodictic*. Such materials do not have a preferred configuration and it makes no sense to talk about an undeformed state. However, experiments designed to elicit linear viscoelastic response from rheodictic materials can generally be arranged so that the material may be considered to have been completely at rest for $t < 0$, so that the strain at $t = 0$ can still be taken as the reference state. This is the view-point adopted in this book. In fluid mechanics it is, however, often desirable to consider a liquid as having been in motion (say, in steady laminar shear) for an indefinite period of time in the past before any observation is made. In this case the strain (or rate of strain) at the present time, t, must serve as the reference state. The way in which this can be done is illustrated in Problem 2.6-3.

Taking the reference strain to be the strain at $t = 0$ allows us to utilize the simplicity afforded by the (one-sided) Laplace transformation. Through it, a differential equation with constant coefficients in the *time domain* becomes an algebraic equation in the *complex domain*. This equation may be manipulated algebraically and the result can then be retransformed into the real time domain.

Transformation of the operator equation leads to

$$\bar{u}(s)\bar{\sigma}(s) = \bar{q}(s)\bar{\varepsilon}(s) \tag{2.1-13}$$

where $\bar{\sigma}(s)$ and $\bar{\varepsilon}(s)$ are the *stress* and *strain transforms*, respectively, and

$$\bar{u}(s) = \sum_n u_n s^n \quad \text{and} \quad \bar{q}(s) = \sum_m q_m s^m \tag{2.1-14}$$

are polynomials in the transform variable s. Equations (2.1-14) are valid for *zero initial conditions*. We shall take these conditions for granted because we can almost always define our stress or strain history in an appropriate way. There is no loss of generality in this simplification because initial conditions can be incorporated into the formulations when required. Suitable methods may be found in numerous texts on electric circuit theory. Whenever necessary, we shall draw attention to non-zero initial conditions.*

From Eqs. (2.1-13) and (2.1-14) we obtain

$$\bar{\sigma}(s) = \bar{Q}(s)\bar{\varepsilon}(s) \quad \text{and} \quad \bar{\varepsilon}(s) = \bar{U}(s)\bar{\sigma}(s) \tag{2.1-15}$$

where

$$\bar{Q}(s) = \bar{q}(s)/\bar{u}(s) \quad \text{and} \quad \bar{U}(s) = \bar{u}(s)/\bar{q}(s)\ . \tag{2.1-16}$$

Equations (2.1-15) form the basis upon which we shall develop the linear theory of viscoelastic behavior. $\bar{Q}(s)$ and $\bar{U}(s)$ will be termed the *operational* (shear) *relaxance* and *retardance* [7], respectively. The corresponding relations between stress and rate of strain are the *operational* (shear) *impedance* and *admittance*, respectively, in analogy to the familiar terms of electric circuit theory [8]. Just as impedance and admittance are given the common name *immittance*, we may refer to the relaxance

---

* This will arise mainly in connection with the response to step excitations (see Sects. 2.3, 9.1.1 and 9.2.1).

and retardance as *respondances*. The reasons for the choice of the names *relaxance* and *retardance* will become clear in Sect. 3.3.

It is seen from Eqs. (2.1-16) with Eqs. (2.1-14) that description of linear viscoelastic behavior by linear differential equations with constant coefficients leads to operational respondances which are ratios of polynomials in the transform variable s. The operational respondances are thus rational algebraic fractions. From this we may deduce some important consequences concerning the behavior of the respondances in the complex plane. Since they are rational algebraic fractions, the only singularities which they will contain are poles. They will not contain essential singularities or branch points. We will show later (Sect. 3.4.5) that physical considerations require that the poles be simple poles lying to the left of the imaginary axis of the finite complex plane, excepting a possible simple pole at the origin.

Equations (2.1-15) also form the basis for the important *correspondence* or *equivalence principle* [12]. According to this principle, if an elastic solution to a boundary value problem (stress analysis problem) is known, substitution of the appropriate Laplace transforms for the quantities employed in the elastic analysis furnishes the viscoelastic solution in the transform plane. The time-dependent viscoelastic solution is then obtained by inverting the transform. The principle can only be applied if the boundaries themselves do not change with time. We shall make use of it in Sects. 5.6.2.2 and 11.1.

2.1.2 Integral Representation: The Boltzmann Superposition Integrals

The operator equation, Eq. (2.1-10), with its transform, Eq. (2.1-13), is *one* form of the general constitutive equation for linear viscoelastic behavior. It expresses the *principle of superposition* (or *additivity*) of the response to an arbitrary train of excitations. This additivity is an essential feature of linear behavior. It is implicit in the operator equation but is perhaps more easily recognized if the basic stress-strain relations are formulated not as differential equations but as (equivalent) integral equations. It was in this integral form that the principle of superposition was first enunciated by Boltzmann,* and it is therefore known as the *Boltzmann superposition principle* [9].

In the time domain the principle can be stated in terms of the *Boltzmann superposition integrals*

$$\sigma(t) = \int_0^t Q(t-u)\varepsilon(u)du \quad \text{and} \quad \varepsilon(t) = \int_0^t U(t-u)\sigma(u)du \ . \qquad (2.1\text{-}17)$$

Equation $(2.1\text{-}17)_1$ may be read: the stress at the *present time* t under an arbitrary strain history is a linear superposition (integral or sum) of all strains applied at *previous (historic) times* u, multiplied by the values of a *weighting function* Q(t)

* Ludwig Boltzmann, 1844–1906, Austrian physicist, whose greatest achievement was the statistical mechanical explanation of the second law of thermodynamics. His superposition principle, actually stated by him essentially in the form given by Eq. $(2.3\text{-}16)_2$, was his only foray into the field of materials behavior.

corresponding to the time intervals (t − u) which have elapsed since imposition of the respective strain. Equation $(2.1\text{-}17)_2$ is then interpreted similarly by interchanging stress and strain, and the weighting functions Q(t) and U(t). The latter represent the material response and are termed *characteristic material functions*. They are also called *heredity* or *memory* functions because they link the present state of the material to its previous history.

In Eqs. (2.1-17) the strain or stress history is arbitrary. By introducing the appropriately chosen experimental stress or strain history, the corresponding response can be derived from the Boltzmann superposition integrals. Integrals of this mathematical form are known as *convolution integrals* [10a]. We may regard

$$\int_0^t Q(t-u)[\quad]du \qquad \text{and} \qquad \int_0^t U(t-u)[\quad]du \tag{2.1-18}$$

as integral transform operators (convolution operators) operating on the strain, ε(t), and the stress, σ(t), respectively. Upon Laplace transformation Eqs. (2.1-17) yield Eqs. (2.1-15). This establishes the connection between the operator equation and the superposition integrals. Accordingly, Q(t) is termed the (shear) *relaxance*, and U(t) the (shear) *retardance* of the material. The simplification achieved by the transformation from the real t-axis into the complex s-plane is immediately apparent. Convolution, a relatively complicated operation on the real t-axis, becomes simple multiplication in the complex s-plane.

The arguments of the kernels in Eqs. (2.1-17) have the simple form of a difference. The integrals thus correspond to linear differential equations with constant coefficients [11]. For linear superposition it is not required that the kernels have this particular form, or that the differential equations have constant coefficients. However, the mathematical handling of the superposition is thus greatly facilitated and simple relations can be found between the constant coefficients of the corresponding linear differential equations and the parameters of the mechanical series-parallel models used to represent linear viscoelastic behavior (cf. Chap. 3).

Equations (2.1-15) may be looked upon as the viscoelastic forms of Hooke's law for shear, Eq. (2.1-1), in the transform plane. It follows from these equations that

$$\bar{Q}(s)\bar{U}(s) = 1 \ , \tag{2.1-19}$$

i.e. that the operational respondances are reciprocal to each other. This, of course, is not true of the corresponding functions in the time domain, Q(t) and U(t), for which we obtain, using Eqs. (A3-16),

$$\int_0^t Q(t-u)U(u)du = \int_0^t Q(u)U(t-u)du = \delta(t) \ . \tag{2.1-20}$$

The two forms of Eq. (2.1-20) result from the *commutativity* of the convolution operation. Equation $(2.1\text{-}20)_2$ may be obtained from $(2.1\text{-}20)_1$ by a change of variable. Denoting the *elapsed* time, t − u, by w, and carrying out the change, we find

$$\int_0^t Q(t-u)U(u)du = \int_0^t Q(w)U(t-w)dw \ . \tag{2.1-21}$$

But w is a dummy variable for which we may substitute u. This gives Eq. $(2.1\text{-}20)_2$.

It should be noted, however, that the physical meaning of u in the two forms of Eq. (2.1-20) is different. In Eq. $(2.1\text{-}20)_1$ u denotes the *historic time*, in Eq. $(2.1\text{-}20)_2$ it signifies the *elapsed time.*

Care must be taken when using Eqs. (2.1-20) because of the peculiar nature of $\delta(t)$ which is not a function in the ordinary sense. As explained in Appendix A2.1, although the delta-function may generally be manipulated formally as if it were an ordinary function, formulations involving it are strictly correct only if they all appear as integrals over time.

We note that Eqs. (2.1-17) may similarly be written in the equivalent forms

$$\sigma(t) = \int_0^t Q(u)\varepsilon(t-u)du \qquad \text{and} \qquad \varepsilon(t) = \int_0^t U(u)\sigma(t-u)du \tag{2.1-22}$$

because of the commutativity of the convolution operation.

The symmetric form of the operator equation and of the superposition integrals (or of their transforms) shows that it is immaterial from a formal viewpoint whether we regard stress as the response to a strain excitation, or strain as a response to a stress excitation. In either cases the theory is formulated in exactly parallel fashion. This *dualism* is a distinctive feature of the theory of linear viscoelastic behavior.

2.1.3 Excitation and Response in the Transform Plane

In the preceding subsection it was pointed out that Eqs. (2.1-15) represent *Hooke's law in the complex plane.* In Eq. $(2.1\text{-}15)_1$ $\bar{\sigma}(s)$ is the *response transform* and $\bar{\varepsilon}(s)$ is the *excitation transform.* In Eq. $(2.1\text{-}15)_2$ their role is reversed. The response and excitation transforms are connected through the respondances, $\bar{Q}(s)$ and $\bar{U}(s)$, which may also be called *material transforms* because they embody the properties of the material. In the preceding section no particular form of the excitation transforms had been specified. Now, any excitation is, in some sense, a function of time, whether it is constant with time, increases linearly with it, or takes any other form. Purely elastic or purely viscous behavior, being characterized by material constants, remains the same whatever may be the history (or time dependence) of the excitation which produces the response. By contrast, viscoelastic behavior is time-dependent. Consequently, the behavior of a viscoelastic material will differ according to what form the time-dependence of the excitation chosen to elicit the response may take. If the time dependence is linear, then any of the multitude of possible material responses, if known over the entire time or frequency domain, completely determines the material behavior. Consequently, at least in principle, any linear material response can be converted into any other and this will be discussed in Chap. 8.

The criterion for the selection of a suitable excitation for the purpose of determining material behavior is experimental and theoretical simplicity. The method of choice is the imposition of a stress or a strain as a step function of time (see Appendix A2.2), and this will be further examined in Sect. 2.3. Response to a *step excitation* generally allows linear viscoelastic behavior to be determined over about 1.5 to 4.5 logarithmic decades of time. Many viscoelastic materials, however, require much larger time spans for a reasonable characterization. An extension of the time scale, again by about 1.5 to 4.5 decades, is possible by combining the response to step excitations with the response to *harmonic (sinusoidal steady-state) excitations* over a range of the radian frequency, ω. This will be discussed in Sect. 2.5. Since $t \sim 1/\omega$, harmonic response extends step response into the short-time region of the time scale. The step and harmonic excitations will be referred to as *standard excitations*. Optionally, the response to a strain or stress as a slope function of time (see Appendix A2.4) may be added to the standard excitations. The *slope excitation* will be discussed in Sect. 2.4.

The response to a stress or strain impulse, i.e. to a stress or strain as a delta function of time (see Appendix A2.1) is not used experimentally for the elicitation of linear viscoelastic response. It has, however, theoretical importance and will therefore be discussed in Sect. 2.2.

The impulse, step, and harmonic excitations are, strictly speaking, not realizable experimentally. An impulse (cf. Appendix A2.1) would have to be infinitely large for an infinitely short time. A step excitation (cf. Appendix A2.2) would have to be applied instantaneously and a steady-state sinusoidal excitation would have had to be applied an infinite time ago. Nevertheless, any of these excitations can be approximated experimentally so closely that any further improvement in the approximation could not be detected. This justifies the use of the mathematically simpler but physically unrealizable excitations instead of the mathematically more complicated ones by which they are approximated.

The response to *unit standard excitation*, whether it be the response to the impulse, step, slope or harmonic excitation, defines a characteristic material response function. The discussion in the remainder of this chapter is confined to a general discussion of the standard excitations, the characteristic material functions associated with them, and the relation of the latter to the respondances. Discussion of the *form* of the characteristic viscoelastic response will be deferred until the next chapter. Other (non-standard) excitations and the response to them will be taken up in Chap. 7.

In deriving the characteristic material responses to unit standard stimuli we shall make use of the simplification afforded by the Laplace transformation. The stimulus will be formulated first in the time domain and will then be transformed into the complex domain. The resulting excitation transform will then be substituted into Eqs. (2.1-15). The material transform resulting from this substitution for unit excitation will then be retransformed from the complex to the time domain. In this process we shall automatically relate the new functions to the (operational) respondances, $\bar{Q}(s)$ and $\bar{U}(s)$, which thus form a focal point of the theory in the presentation adopted in this book. Their role arises from the simplicity of the basic relations involving them, and not from their experimental importance.

2.2 The Impulse Response Functions

Equations (2.1-15) furnish further insight into the nature of the weighting functions $Q(t)$ and $U(t)$. Let the excitation be an impulsive strain of strength $\hat{\varepsilon}_0$, applied at

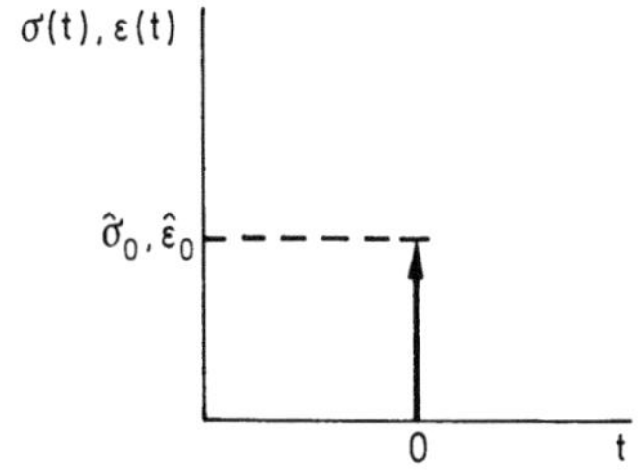

Fig. 2.2-1. Impulse excitation

$t = 0$, as shown schematically in Fig. 2.2-1, where the arrow denotes the impulse. The figure also illustrates an impulsive stress of strength $\hat{\sigma}_0$.

The strain excitation may now be written as

$$\varepsilon(t) = \hat{\varepsilon}_0 \delta(t) \qquad (2.2\text{-}1)$$

where $\delta(t)$ is the impulse or delta function (see Appendix A2.1). Laplace transformation, using LTP (1), yields

$$\bar{\varepsilon}(s) = \hat{\varepsilon}_0 \qquad (2.2\text{-}2)$$

and substituting this into Eq. $(2.1\text{-}15)_1$ we obtain

$$\bar{Q}(s) = \bar{\sigma}(s)/\hat{\varepsilon}_0 \qquad (2.2\text{-}3)$$

for the operational relaxance. For the retardance we find in a similar way

$$\bar{U}(s) = \bar{\varepsilon}(s)/\hat{\sigma}_0 \qquad (2.2\text{-}4)$$

where $\hat{\sigma}_0$ is the strength of the stress impulse. Retransformation gives

$$Q(t) = \sigma(t)/\hat{\varepsilon}_0 \qquad (2.2\text{-}5)$$

and

$$U(t) = \varepsilon(t)/\hat{\sigma}_0 \ . \qquad (2.2\text{-}6)$$

$Q(t)$ and $U(t)$ thus turn out to be the response of the material to unit impulses of strain and stress, respectively.

We may now derive the Boltzmann superposition principle by considering an arbitrary stress history as consisting of a sequence of impulses as shown in Fig. 2.2-2. The interval of interest is $[0, t]$, where t is the *present time.* We subdivide this interval into n sub-intervals each of equal width. This simplifies the derivation without loss of generality. Letting u be as before the *historic time*, Δu is the interval between two impulses and $u = k\,\Delta u$ where k is the number of impulses. The strength of an impulse at $k\,\Delta u$ is the area of the rectangle with base Δu and height $\sigma(k\,\Delta u)$. An approximation to the total stress is then obtained from the totality (or sum) of all these impulses. We have

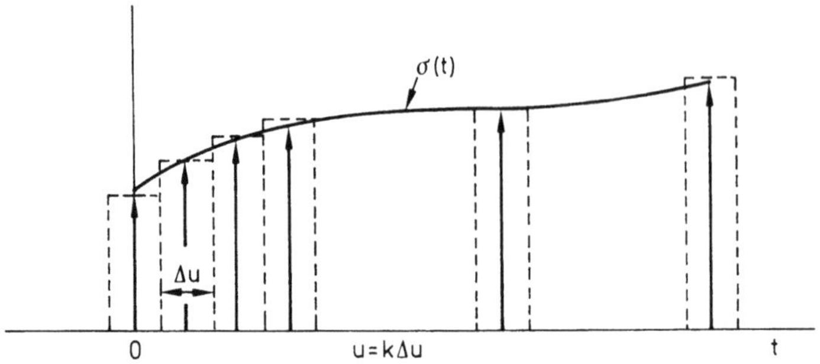

Fig. 2.2-2. Boltzmann superposition principle. Stress as a superposition of impulses

$$\begin{aligned}\sigma(t) \simeq\ & \sigma(0)\Delta u\ \delta(t) + \sigma(\Delta u)\Delta u\ \delta(t - \Delta u) + \sigma(2\ \Delta u)\Delta u\ \delta(t - 2\ \Delta u) + \cdots \\ & + \sigma(k\ \Delta u)\Delta u\ \delta(t - k\ \Delta u) + \cdots \\ & + \sigma(n\ \Delta u)\Delta u\ \delta(t - n\ \Delta u) = \sum_{k=0}^{n} \sigma(k\ \Delta u)\Delta u\ \delta(t - k\ \Delta u)\ .\end{aligned} \tag{2.2-7}$$

If we now let the response to a unit (shifted) stress impulse, $\delta(t - u)$, be $U(t - u)$, then the response, $\Delta\ \varepsilon(u)$, to an impulse of strength $\sigma(u)\Delta u$ is

$$\Delta\ \varepsilon(u) = \sigma(u)\Delta u\ U(t - u) \tag{2.2-8}$$

and the total strain $\varepsilon(t)$ is given approximately by

$$\varepsilon(t) \simeq \sum_{k=0}^{n} \Delta\ \varepsilon(k\ \Delta u) = \sum_{k=0}^{n} \sigma(k\ \Delta u)\Delta u\ U(t - k\ \Delta u)\ , \tag{2.2-9}$$

i.e., an approximation to the total response is obtained by summing the $\Delta\varepsilon(k\ \Delta u)$'s.

We may obtain the exact response by proceeding to the limit. We do this by making Δu approach 0 and n approach ∞ while keeping $u = k\ \Delta u$ constant. In other words: we let the number of intervals k become very large, simultaneously reducing Δu to very small values in such a way that their product does not change. We then have

$$\varepsilon(t) = \lim_{\Delta u \to 0,\, n \to \infty} \sum_{k=0}^{n} \sigma(k\ \Delta u)\Delta u\ U(t - k\ \Delta u) = \int_0^\infty \sigma(u)U(t - u)du\ . \tag{2.2-10}$$

But $\sigma(u) = 0$ when $u > t$ (i.e., there is no stress imposed after time t), and therefore

$$\varepsilon(t) = \int_0^t U(t - u)\sigma(u)du = \int_0^t U(u)\sigma(t - u)du\ . \tag{2.2-11}$$

Equations (2.2-11) are identical with Eqs. $(2.1\text{-}17)_2$. The two forms arise because of the commutativity of the convolution operator.

According to the principle of duality the stress as a response to a sequence of strain impulses can be derived in an entirely analogous way. We thus obtain the relations

$$\sigma(t) = \int_0^t Q(t-u)\varepsilon(u)du = \int_0^t Q(u)\varepsilon(t-u)du \tag{2.2-12}$$

which are the duals of Eqs. (2.2-11).

The relaxance, Q(t), thus represents the response of a material to a unit impulse of strain. Similarly, the retardance, U(t), represents the response to a unit impulse of stress. In practive, however, impulse excitations are rarely, if ever, used to determine viscoelastic properties.*

2.3 The Step Response Functions

One of the most widely used experimental excitations is the sudden imposition of a strain or stress at time t = 0 which is then held constant. Mathematically such an excitation can be modeled by a step function (cf. Appendix A2.2). This, like the delta function, is a mathematical abstraction. Nevertheless, it is possible to obtain experimental responses (see Sect. 7.12) which conform quite closely to a step excitation which we therefore proceed to examine. Let the excitation be a step of strain of height ε_0. It may then be expressed as

$$\varepsilon(t) = \varepsilon_0 h(t) \tag{2.3-1}$$

where h(t) is the unit step function. The excitation is shown graphically in Fig. 2.3-1 which also serves to illustrate the analogous stress excitation.

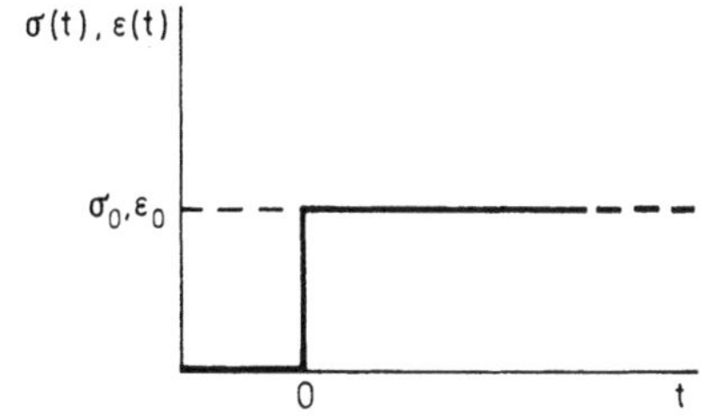

Fig. 2.3-1. Step excitation

Laplace transformation of Eq. (2.3-1), using LTP (4), gives

$$\bar{\varepsilon}(s) = \varepsilon_0/s \ . \tag{2.3-2}$$

Substituting this into the viscoelastic form of Hooke's law, Eq. $(2.1\text{-}15)_1$, we find

$$\bar{\sigma}(s) = \frac{\bar{Q}(s)}{s}\varepsilon_0 \tag{2.3-3}$$

and retransformation gives

* A true delta function impulse is a mathematical abstraction. Besides, at very short loading times inertial effects cause considerable complications.

$$\sigma(t) = \varepsilon_0 \mathscr{L}^{-1}\bar{Q}(s)/s \ . \tag{2.3-4}$$

The inverse transform of $\bar{Q}(s)/s$ has the dimensions of a modulus. $\bar{Q}(s)/s$ therefore is the transform of the modulus connecting the (time-dependent) stress with a unit step of strain and accordingly, we define the transform of the modulus as

$$\bar{G}(s) = \frac{\bar{Q}(s)}{s} \ . \tag{2.3-5}$$

Inserting into Eq. (2.3-3) and retransforming we find

$$G(t) = \sigma(t)/\varepsilon_0 \ . \tag{2.3-6}$$

We call G(t) the (shear) *relaxation modulus.* The relation of its transform to the relaxance, $\bar{Q}(s)$, is given by Eq. (2.3-5). In the time domain we have

$$G(t) = \mathscr{L}^{-1}\frac{\bar{\sigma}(s)}{\varepsilon_0} = \mathscr{L}^{-1}\frac{\bar{Q}(s)}{s} = \int_0^t Q(u)du \ . \tag{2.3-7}$$

Equation (2.3-5) gives us an alternative definition of $\bar{Q}(s)$. Since $\bar{Q}(s) = s\bar{G}(s)$, we may define it as the s-multiplied Laplace transform (or the Carson transform, see Appendix A3.4) of the relaxation modulus.

In a similar way we may define the transform of the (shear) strain response to a unit step of stress as

$$\bar{J}(s) = \frac{\bar{U}(s)}{s} \ . \tag{2.3-8}$$

In analogy to Eq. (2.3-6) we have

$$J(t) = \varepsilon(t)/\sigma_0 \tag{2.3-9}$$

We call J(t) the (shear) *creep compliance.* We also have

$$J(t) = \mathscr{L}^{-1}\frac{\bar{\varepsilon}(s)}{\sigma_0} = \mathscr{L}^{-1}\frac{\bar{U}(s)}{s} = \int_0^t U(u)du \ . \tag{2.3-10}$$

Again, Eq. (2.3-8) shows that $\bar{U}(s)$ is the Carson transform of J(t).

Retransforming $\bar{Q}(s) = s\bar{G}(s)$ and $\bar{U}(s) = s\bar{J}(s)$ gives [cf. Eq. (A3-10)]

$$Q(t) = \frac{d\,G(t)}{dt} + G(0)\delta(t) \tag{2.3-11}$$

and

$$U(t) = \frac{d\,J(t)}{dt} + J(0)\delta(t) \tag{2.3-12}$$

where the delta function terms represent the initial conditions. The values of G(t) and J(t) at $t = 0$, G(0) and J(0), are the *instantaneous modulus* and *compliance*, respectively. G(0) and J(0) are often denoted by G_0 and J_0. We will follow the practice of polymer rheology and write G_g and J_g for them. The subscript g is taken from the word *glassy* because amorphous polymers behave like glasses at very short loading times. Metal rheologists prefer the terms *unrelaxed modulus*, G_u, and *unrelaxed compliance*, J_u, for the *glassy modulus* and *glassy compliance* of polymer rheology.

It follows from Eqs. (2.3-5), (2.3-8) and (2.1-19) that

$$\bar{G}(s)\bar{J}(s) = 1/s^2 \tag{2.3-13}$$

that is, using Eqs. (A3-16),

$$\int_0^t G(t-u)J(u)du = \int_0^t G(u)J(t-u)du = t \ . \tag{2.3-14}$$

Equations (2.3-13) and (2.3-14) should be compared with Eqs. (2.1-19) and (2.1-20). Equation (2.3-13) shows that the s-multiplied Laplace transforms of G(t) and J(t) are reciprocal to each other.

Hooke's law in the complex plane, Eqs. (2.1-15) can, of course, be expressed in terms of $\bar{G}(s)$ and $\bar{J}(s)$ instead of $\bar{Q}(s)$ and $\bar{U}(s)$. Inserting $\bar{Q}(s) = s\bar{G}(s)$ into Eq. $(2.1\text{-}15)_1$ and $\bar{U}(s) = s\bar{J}(s)$ into Eq. $(2.1\text{-}15)_2$ we obtain

$$\bar{\sigma}(s) = s\bar{G}(s)\bar{\varepsilon}(s) \qquad \text{and} \qquad \bar{\varepsilon}(s) = s\bar{J}(s)\bar{\sigma}(s) \ . \tag{2.3-15}$$

Inversion of the transform yields expressions for the Boltzmann superposition integral in which the response to a unit step excitation is used instead of the unit impulse response. These expressions are obtained in four equivalent forms because the differentiation implied in the complex plane by multiplication by s in Eqs. (2.3-15) can be carried out in the time domain either on the material function or on the excitation function, and because each of the two forms obtained in this way has two variants because of the commutativity of the convolution, i.e. according to whether the historic time or the elapsed time is considered to be the variable of integration.

To invert the Laplace transform of Eq. $(2.3\text{-}15)_1$ we use Eqs. (A3-17). We let $F_1(t) = G(t)$, and $F_2(t) = \varepsilon(t)$, and make use of $\varepsilon(0) = 0$ since we can always define t so that there is no stress or strain for $t < 0$. With $G(0) = G_g$, we find

$$\sigma(t) = \begin{cases} G_g\varepsilon(t) - \int_0^t \dfrac{d\,G(t-u)}{du}\varepsilon(u)du \\[2ex] G_g\varepsilon(t) + \int_0^t \dfrac{d\,G(u)}{du}\varepsilon(t-u)du \\[2ex] \int_0^t G(t-u)\dfrac{d\,\varepsilon(u)}{du}du \\[2ex] -\int_0^t G(u)\dfrac{d\,\varepsilon(t-u)}{du}du \ . \end{cases} \tag{2.3-16}$$

From Eq. $(2.3\text{-}15)_2$ four analogous expressions can be derived for the shear strain as a superposition of shear stresses. With $J(0) = J_g$, we have

$$
\varepsilon(t) = \begin{cases} J_g\sigma(t) - \int_0^t \frac{d\,J(t-u)}{du}\sigma(u)du \\ J_g\sigma(t) + \int_0^t \frac{d\,J(u)}{du}\sigma(t-u)du \\ \int_0^t J(t-u)\frac{d\,\sigma(u)}{du}du \\ -\int_0^t J(u)\frac{d\,\sigma(t-u)}{du}du\ . \end{cases} \tag{2.3-17}
$$

We note that

$$
\int_0^t G(t-u)\frac{d\,\varepsilon(u)}{du}du = \int_0^t G(t-u)d\,\varepsilon(u)
$$

and (2.3-18)

$$
\int_0^t J(t-u)\frac{d\,\sigma(u)}{du}du = \int_0^t J(t-u)d\,\sigma(u)
$$

where the convolutions on the right are called Stieltjes* convolutions. These are often useful when discussing excitations which contain discontinuities. [13].

We may now derive the superposition integral by considering an arbitrary stress history as consisting of a sequence of steps as shown in Fig. 2.3-2.

Much of the notation is as before. The height of the step at $u = k\,\Delta u$ is given by $\sigma(k\,\Delta u)$ and an approximation to the total stress is obtained by summing over all steps. Thus

$$
\begin{aligned} \sigma(t) &\simeq \Delta\,\sigma(0)h(t) + \Delta\,\sigma(\Delta u)h(t-\Delta u) + \Delta\,\sigma(2\,\Delta u)h(t - 2\,\Delta u) + \cdots \\ &\quad + \Delta\,\sigma(k\,\Delta u)h(t-k\,\Delta u) + \cdots \\ &\quad + \Delta\,\sigma(n\,\Delta u)h(t-n\,\Delta u) = \sum_{k=0}^{n} \Delta\,\sigma(k\,\Delta u)h(t-k\,\Delta u)\ . \end{aligned} \tag{2.3-19}
$$

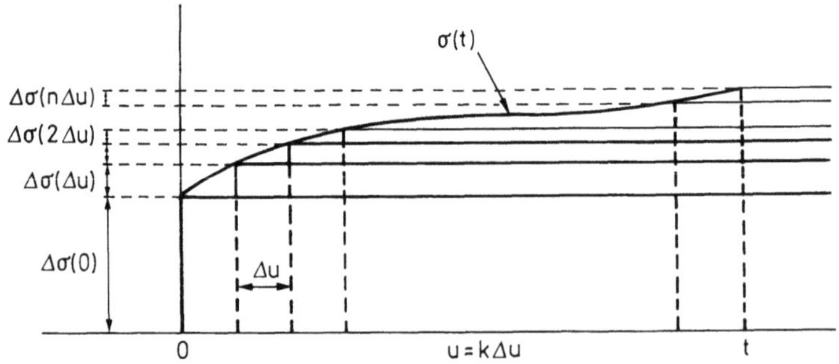

Fig. 2.3-2. Boltzmann superposition principle. Stress as a superposition of steps

* Thomas Jean Stieltjes, pron. Steelchus, 1856–1894, Dutch-born French mathematician.

In Eq. (2.3-19) $\Delta\,\sigma(k\,\Delta u)$ is the stress increment at $u = k\,\Delta u$. If we now let the response to a unit step of stress, $h(t-u)$, be $J(t-u)$, then the response, $\Delta\,\varepsilon(u)$, to a step of height $\Delta\,\sigma(u)$ is

$$\Delta\,\varepsilon(u) = \Delta\,\sigma(u)J(t-u) \tag{2.3-20}$$

and an approximation to the total strain is obtained by summing over all steps

$$\varepsilon(t) \simeq \sum_{k=0}^{n} \Delta\,\varepsilon(k\,\Delta u) = \sum_{k=0}^{n} \Delta\,\sigma(k\,\Delta u)J(t-k\,\Delta u)\ . \tag{2.3-21}$$

The exact expression may be obtained again by going to the limit as $\Delta u \to 0, n \to \infty$, while keeping $u = k\,\Delta u$ constant. We have

$$\varepsilon(t) = \lim_{\Delta u \to 0, n \to \infty} \sum_{k=0}^{n} \Delta\,\sigma(k\,\Delta u)J(t-k\,\Delta u) = \int_0^{\infty} J(t-u)d\,\sigma(u)\ . \tag{2.3-22}$$

The right hand side of Eq. (2.3-22) is a Stieltjes integral. But $\sigma(u) = 0$ for $u > t$, because the material cannot be aware of its own future. Hence,

$$\varepsilon(t) = \int_0^t J(t-u)d\,\sigma(u)du = \int_0^t J(t-u)\frac{d\,\sigma(u)}{du}du\ . \tag{2.3-23}$$

By Eq. $(2.3\text{-}18)_2$ this is identical with Eq. $(2.3\text{-}17)_3$.

Again, by the principle of duality, an analogous derivation can be made for the stress. The equation which is dual to Eq. (2.3-23) then is identical with Eq. $(2.3\text{-}16)_3$.

We shall now show that $G(t)$ and $J(t)$ may also be considered to be the responses of the material to unit impulses of the rate of strain and rate of stress, respectively. Since the transform of the rate of strain and the rate of stress, given zero initial conditions, are

$$\bar{\dot{\varepsilon}}(s) = s\bar{\varepsilon}(s) \qquad \text{and} \qquad \bar{\dot{\sigma}}(s) = s\bar{\sigma}(s)\ , \tag{2.3-24}$$

substitution into Eqs. (2.3-15) yields

$$\bar{\sigma}(s) = \bar{G}(s)\bar{\dot{\varepsilon}}(s) \qquad \text{and} \qquad \bar{\varepsilon}(s) = \bar{J}(s)\bar{\dot{\sigma}}(s)\ . \tag{2.3-25}$$

For a unit impulse of the rate of strain and the rate of stress we find $\bar{\sigma}(s) = \bar{G}(s)$ and $\bar{\varepsilon}(s) = \bar{J}(s)$, respectively.

2.4 The Slope Response Functions

A frequently discussed excitation is the *slope excitation* (see Appendix A2.4). This consists in the imposition of a strain or a stress at a constant rate and is, therefore, also referred to as *constant rate of strain* (or *stress*) *excitation*. Because in such an excitation the linear viscoelastic limit may be exceeded quite soon after imposition of the excitation, the slope excitation is not generally used to elicit linear response. We include it among the standard excitations because an understanding of the linear response to slope excitations forms the basis of an understanding of the non-linear response to a frequently used excitation. The slope excitation is shown in Fig. 2.4-1.

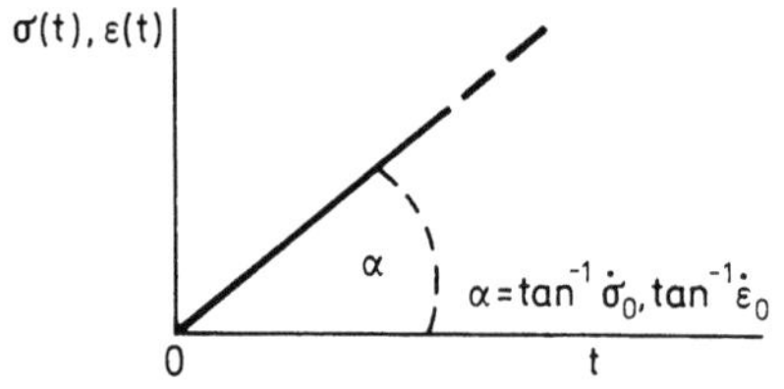

Fig. 2.4-1. Slope excitation

We denote the constant rate of stress by $\dot{\sigma}_0$ and the constant rate of strain by $\dot{\varepsilon}_0$. For a constant rate of strain imposed at $t = 0$ we have

$$\varepsilon(t) = \dot{\varepsilon}_0 p(t) = \dot{\varepsilon}_0 t \tag{2.4-1}$$

where p(t) is the slope function. Laplace transformation, using LTP (6), gives

$$\bar{\varepsilon}(s) = \dot{\varepsilon}_0/s^2 \ . \tag{2.4-2}$$

Substituting this into Eq. $(2.1\text{-}15)_1$ yields

$$\bar{\sigma}(s) = \frac{\bar{Q}(s)}{s^2}\dot{\varepsilon}_0 \ . \tag{2.4-3}$$

But

$$\frac{\bar{\sigma}(s)}{\dot{\varepsilon}_0} = \bar{\eta}(s) \tag{2.4-4}$$

where $\bar{\eta}(s)$ is the transform of the time-dependent *shear viscosity* (cf. Sect. 1.5) connecting the stress transform with a constant rate of strain transform. We thus have

$$\bar{\eta}(s) = \frac{\bar{Q}(s)}{s^2} = \frac{\bar{G}(s)}{s} = \mathscr{L}\eta(t) = \mathscr{L}\frac{\sigma(t)}{\dot{\varepsilon}_0} \tag{2.4-5}$$

the second of Eqs. (2.4-5) following from Eq. (2.3-5). Retransforming to the time domain, Eqs. (2.4-4) and (2.4-5) yield

$$\eta(t) = \frac{\sigma(t)}{\dot{\varepsilon}_0} = \int_0^t G(u)du = \int_0^t \int_0^v Q(u)du\, dv \ . \tag{2.4-6*}$$

Since we also have

$$\bar{Q}(s) = s\bar{G}(s) = s^2\bar{\eta}(s) \tag{2.4-7}$$

we find, by retransformation,

$$Q(t) = \frac{d\,G(t)}{dt} + G(0)\delta(t) = \frac{d^2\eta(t)}{dt^2} + \eta'(0)\delta(t) + \eta(0)\delta'(t) \tag{2.4-8}$$

and

---

* The last integral has the three equivalent forms $\int_0^t u\,Q(t-u)du = \int_0^t (t-u)Q(u)du = \int_0^t \int_0^v Q(u)du\,dv$ [cf. Eq. (A3-18)].

$$G(t) = \frac{d\,\eta(t)}{dt} + \eta(0)\delta(t) \ . \tag{2.4-9}$$

The terms in the delta function, $\delta(t)$, and its derivative, $\delta'(t)$, take care of the initial conditions (cf. Problem 2.3-1). We note that $G(0) = \eta'(0) = G_g$ and $\eta(0) = 0$.

Analogous expressions can be introduced for the response to a constant rate of stress (or slope stress). We have

$$\bar{\chi}(s) = \frac{\bar{\varepsilon}(s)}{\dot{\sigma}_0} = \frac{\bar{U}(s)}{s^2} = \frac{\bar{J}(s)}{s} \tag{2.4-10}$$

with

$$\chi(t) = \frac{\varepsilon(t)}{\dot{\sigma}_0} = \int_0^t J(u)du = \int_0^t u\,U(t-u)du \ . \tag{2.4-11}$$

Also,

$$\bar{U}(s) = s\bar{J}(s) = s^2\bar{\chi}(s) \ . \tag{2.4-12}$$

Hence

$$U(t) = \frac{d\,J(t)}{dt} + J(0)\delta t = \frac{d^2\chi(t)}{dt^2} + \chi'(0)\delta(t) + \chi(0)\delta'(t) \tag{2.4-13}$$

and

$$J(t) = \frac{d\,\chi(t)}{dt} + \chi(0)\delta(t) \ . \tag{2.4-14}$$

Again, $J(0) = \chi'(0) = J_g$, and $\chi(0) = 0$.

It follows from Eqs. $(2.4\text{-}5)_1$, $(2.4\text{-}10)_2$, and $(2.1\text{-}19)_2$ that

$$\bar{\eta}(s)\bar{U}(s) = \bar{\chi}(s)\bar{Q}(s) = 1/s^2 \ . \tag{2.4-15}$$

Retransformation gives

$$\int_0^t \eta(t-u)U(u)du = \int_0^t \chi(t-u)Q(u)du = t \tag{2.4-16}$$

and

$$\int_0^t \eta(u)U(t-u)du = \int_0^t \chi(u)Q(t-u)du = t \tag{2.4-17}$$

which may be compared with Eqs. (2.1-20) and (2.1-21). Dividing Eqs. (2.4-15) by s we obtain

$$\bar{\eta}(s)\bar{J}(s) = \bar{\chi}(s)\bar{G}(s) = 1/s^3 \tag{2.4-18}$$

and from this we derive

$$2\int_0^t \eta(t-u)J(u)du = 2\int_0^t \chi(t-u)G(u)du = t^2 \tag{2.4-19}$$

and

$$2\int_0^t \eta(u)J(t-u)du = 2\int_0^t \chi(u)G(t-u)du = t^2 \tag{2.4-20}$$

by Eq. (A3-16) and LTP (12).

We now turn to an alternative interpretation of the respondances, $\bar{Q}(s)$ and $\bar{U}(s)$. In Sect. 2.3 we considered the strain or stress arising as the response to unit step excitations of stress or strain. We now examine what form the *rate of strain* or *stress* takes in response to the step excitations. We introduce a function

$$\bar{\phi}(s) = \frac{\bar{\dot{\varepsilon}}(s)}{\sigma_0} \tag{2.4-21}$$

as the transform of the response of the rate of strain to a step function of stress. Substituting $s\bar{\varepsilon}(s) = \bar{\dot{\varepsilon}}(s)$ and $\bar{\sigma}(s) = \sigma_0/s$ into $\bar{\varepsilon}(s) = \bar{U}(s)\,\bar{\sigma}(s)$ we discover that

$$\bar{\phi}(s) = \bar{U}(s) \ . \tag{2.4-22}$$

Thus the retardance may be looked upon as the response of the rate of strain to a unit step of stress as well as the response of the strain to a unit impulse of stress. It will be convenient for later use to retain the separate symbol $\bar{\phi}(s)$ for the transform of the response of the rate of strain to a unit step function of stress. The inverse transform, $\phi(t)$, is a time-dependent fluidity defined as

$$\phi(t) = \dot{\varepsilon}(t)/\sigma_0 \ . \tag{2.4-23}$$

Similarly, we may define a function

$$\bar{\kappa}(s) = \frac{\bar{\dot{\sigma}}(s)}{\varepsilon_0} \tag{2.4-24}$$

as the transform of the response of the rate of stress to a unit step function of strain. Again, substitution of $s\bar{\sigma}(s) = \bar{\dot{\sigma}}(s)$ and $\bar{\varepsilon}(s) = \varepsilon_0/s$ into $\bar{\sigma}(s) = \bar{Q}(s)\bar{\varepsilon}(s)$ yields

$$\bar{\kappa}(s) = \bar{Q}(s) \tag{2.4-25}$$

where the last term follows from the definition of the relaxance. Thus, in turn, the relaxance may be looked upon as the response of the rate of stress to a unit step of strain as well as the response of the stress to a unit impulse of strain. Again, we retain the separate symbol to denote the response of the rate of stress to a unit step-function of strain, and write

$$\kappa(t) = \dot{\sigma}(t)/\varepsilon_0 \ . \tag{2.4-26}$$

We mention, for the sake of completeness, that Hooke's law in the complex domain may be written in terms of the transforms of the time-dependent viscosity, $\eta(t)$, and the response function $\chi(t)$ as

$$\bar{\sigma}(s) = s^2\bar{\eta}(s)\bar{\varepsilon}(s) \qquad \text{and} \qquad \bar{\varepsilon}(s) = s^2\bar{\chi}(s)\bar{\sigma}(s) \ . \tag{2.4-27}$$

These equations may be compared with Eqs. (2.3-15) on the one hand, and with Eqs. (2.1-15) on the other.

2.5 The Harmonic Response Functions

Another widely used form of excitation is the *harmonic* or *sinusoidal steady-state* excitation, which is often referred to as *dynamic excitation* for no very good reason. Again we first consider a harmonic *strain* excitation. Mathematically it can be represented in the two equivalent forms

$$\varepsilon(t) = \varepsilon_0 \sin \omega t \qquad \text{and} \qquad \varepsilon(t) = \varepsilon_0 \cos \omega t \tag{2.5-1}$$

where ε_0 is the strain amplitude and ω is the radian frequency, so that the period, T, is given by $2\pi/\omega$. The two forms are illustrated in Fig. 2.5-1. They are 90° out-of-phase with respect to one another but, in the steady-state, the choice of representation is arbitrary. Thus, we may consider Curve 1 to represent $\varepsilon_0 \sin \omega t$, and then Curve 2 must be interpreted as $\varepsilon_0 \sin(\omega t + \frac{\pi}{2}) = \cos \omega t$. Or, we interpret Curve 1 as $\varepsilon_0 \cos \omega t$, and in that case Curve 2 would have to be considered to be $\varepsilon_0 \cos(\omega t - \frac{\pi}{2}) = \varepsilon_0 \sin \omega t$. To enable us to handle both forms of the excitation simultaneously, we write it in complex form as

$$\varepsilon(t) = \varepsilon_0(\cos \omega t + j \sin \omega t) = \varepsilon_0 \exp(j\omega t) \tag{2.5-2}$$

and call $\varepsilon(t)$ defined in this way the *generalized harmonic strain.* A *generalized harmonic stress* excitation with stress amplitude σ_0 can be defined in an analogous way and this has been indicated in Fig. 2.5-1.

We now derive the response to $\varepsilon(t)$ in the usual way. Its Laplace transform, using LTP (13), becomes

$$\bar{\varepsilon}(s) = \varepsilon_0/(s - j\omega) \ . \tag{2.5-3}$$

Inserting this into Eq. $(2.1\text{-}15)_1$ we find

$$\bar{\sigma}(s) = \frac{\varepsilon_0 \bar{Q}(s)}{s - j\omega} \ . \tag{2.5-4}$$

Retransformation to the real time axis would yield the total stress response to the sinusoidal steady-state excitation. The imposition of such an excitation does not immediately elicit a steady-state response (cf. Problem 3.4-5). The total response

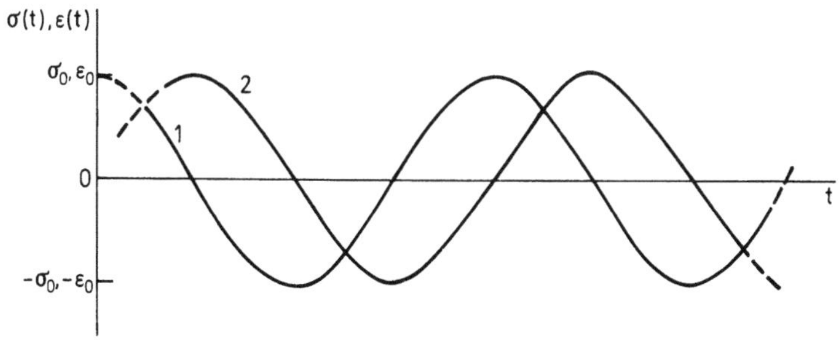

Fig. 2.5-1. Harmonic excitation

consists of two parts: a periodic function of time representing the *steady-state response*, and a non-periodic function representing the *transient response*. The stress transform will, therefore, also contain not only the steady-state, but the transient response of the material. It will become clear in Chap. 3 (as an example, see Problem 3.4-5) that the transient response consists of decaying exponentials arising from the poles of the *material transforms*, $\bar{Q}(s)$ and $\bar{U}(s)$, while the steady-state response arises from the poles of the *excitation* or *driving transforms*. To obtain the steady-state response alone [10b], we may use Eq. $(2.1\text{-}16)_1$ and rewrite Eq. (2.5-4) with its help as

$$\frac{\bar{\sigma}(s)}{\varepsilon_0} = \frac{\bar{q}(s)}{(s - j\omega)\bar{u}(s)} . \tag{2.5-5}$$

Decomposing into partial fractions we have

$$\frac{\bar{\sigma}(s)}{\varepsilon_0} = \frac{A}{s - j\omega} + \frac{B(s)}{\bar{u}(s)} \tag{2.5-6}$$

where A is not a function of s because $s - j\omega$ is linear in s, and B(s) is, in general, a polynomial in s of degree one less than the degree of $\bar{u}(s)$.

The first of the partial fractions in Eq. (2.5-6) represents the steady-state response arising from the pole of the driving transform. The second partial fraction represents the transient response arising here from the zeros of the denominator $\bar{u}(s)$, i.e. from the poles of the relaxance $\bar{Q}(s)$. We must now find A in terms of $\bar{Q}(s)$. Combining Eqs. (2.5-5) and (2.5-6) and rearranging gives

$$A\,\bar{u}(s) + (s - j\omega)B(s) = \bar{q}(s) \tag{2.5-7}$$

or

$$A = \bar{Q}(s) - (s - j\omega)B(s)/\bar{u}(s) . \tag{2.5-8}$$

Since we are interested in the steady-state response, we seek the meaning of A when $s = j\omega$. Now, $\bar{U}(s)$ has no zero for $s = j\omega$. Hence, making the substitution, we find

$$A = [\bar{Q}(s)]_{s=j\omega} = \bar{Q}(j\omega) . \tag{2.5-9}$$

Thus A is the operational relaxance in which the transform variable s is replaced by $j\omega$.

For the steady state, therefore, we find

$$\bar{\sigma}_{ss}(s)/\varepsilon_0 = \frac{\bar{Q}(j\omega)}{s - j\omega} \tag{2.5-10}$$

where we have used the subscript ss to distinguish the steady state transform from the total stress transform. Retransformation yields

$$\sigma_{ss}(t) = \bar{Q}(j\omega)\varepsilon_0 \exp(j\omega t) \tag{2.5-11}$$

where $\sigma_{ss}(t)$ denotes the generalized harmonic stress response. An alternative derivation of Eq. (2.5-11) which makes use of the Residue Theorem of Laplace transformation is given in Problem 2.5-2.

In the steady-state the variable of interest is no longer the time t because the response will vary with t in a periodic manner. The response will, however, vary with the *frequency* of the periodic excitation which therefore becomes the variable of interest in the steady-state. We may, consequently, write $\varepsilon(\omega)$ for $\varepsilon_0\exp(j\omega t)$, and $\sigma(\omega)$ for $\sigma_{ss}(t)$ whenever we are interested in the dependence of the steady-state stress on the radian frequency rather than on time. Eq. (2.5-11) then gives

$$\sigma(\omega) = \bar{Q}(j\omega)\varepsilon(\omega) = G^*(\omega)\varepsilon(\omega) \ . \tag{2.5-12}$$

The quantity $\bar{Q}(j\omega)$ has the dimensions of a modulus. In polymer mechanics, it is usually denoted by $G^*(\omega)$ (in shear) and is called the *complex (shear) modulus*. Because of its relation to the harmonic operator, $j\omega$, $G^*(\omega)$ may also be called the *harmonic relaxance*. As indicated by Eq. $(2.5\text{-}12)_2$, $G^*(\omega)$ is defined as the sinusoidal steady-state stress response to a sinusoidal steady-state strain of unit amplitude.

By Eq. (2.5-12) the relation between the harmonic and the operational relaxance is given by

$$G^*(\omega) = \bar{Q}(s)|_{s=j\omega} \tag{2.5-13}$$

[cf. Eq. (A4-19)]. We have derived this relation here by separating out the contribution of the driving transform to the total response transform by a decomposition into partial fractions. With respect to the nature of the material transform the method assumes that the transform possesses an unspecified number of simple poles as its only singularities. This follows from the fact (see the end of Sect. 2.2.1) that the operational respondances are rational algebraic fractions in s. The nature of the relaxance will be examined in detail in Chap. 3. In Chap. 4 (see Problem 4.1-1) we will briefly outline yet another, more general method for deriving Eq. (2.5-13) which is based on Cauchy's Integral Formula. The derivation of Eq. (2.5-13) from the Boltzmann superposition integral is given in Sect. 2.6.

In a completely analogous fashion one may derive the *harmonic retardance* or *complex (shear) compliance*. Letting

$$\sigma(t) = \sigma_0(\cos \omega t + j \sin \omega t) = \sigma_0\exp(j\omega t) \tag{2.5-14}$$

be the generalized harmonic stress, we have, in shear,

$$\varepsilon_{ss}(t) = \bar{U}(j\omega)\sigma_0\exp(j\omega t) \tag{2.5-15}$$

where $\varepsilon_{ss}(t)$ denotes the generalized harmonic stress response. Equation (2.5-15) may, of course, also be derived along the lines of Problem 2.5-2 using the Residue Theorem. In analogy to Eq. (2.5-12) we further have

$$\varepsilon(\omega) = \bar{U}(j\omega)\sigma(\omega) = J^*(\omega)\sigma(\omega) \ . \tag{2.5-16}$$

In Eq. $(2.5\text{-}16)_2$ we have introduced the symbol $J^*(\omega)$ for the complex shear compliance, defined as the response of a sinusoidal steady-state strain to a sinusoidal steady-state shear stress of unit amplitude. We thus have

$$J^*(\omega) = \bar{U}(s)|_{s=j\omega} \tag{2.5-17}$$

[cf. Eq. (A4-19)]. For an alternative derivation using Cauchy's Integral Formula

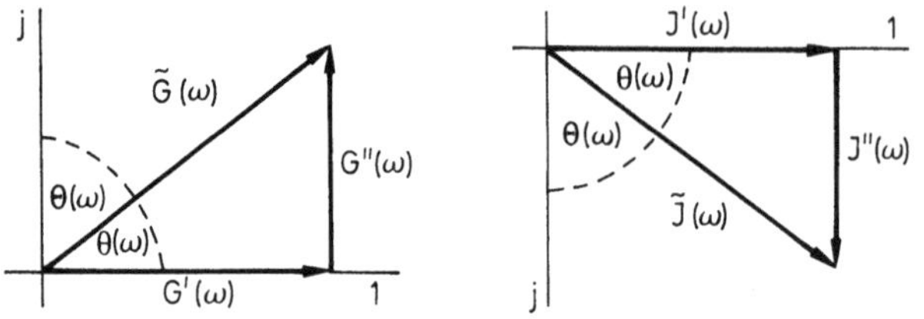

Fig. 2.5-2. Vectorial resolution of the components of the complex shear modulus and compliance

see Problem 4.1-2. The derivation from the Boltzmann superposition integral is again given in Sect. 2.6.

Equations (2.5-12) and (2.5-16) show that the response elicited by a steady-state sinusoidal excitation from a linear viscoelastic material is again sinusoidal, and has the same frequency as the excitation.

Since $G^*(\omega)$ and $J^*(\omega)$ are complex quantities, we may decompose them into cartesian and polar components as

$$G^*(\omega) = G'(\omega) + j\, G''(\omega) = \tilde{G}(\omega)\exp[j\, \theta(\omega)] \tag{2.5-18}$$

and

$$J^*(\omega) = J'(\omega) - j\, J''(\omega) = \tilde{J}(\omega)\exp[-j\, \theta(\omega)] \;. \tag{2.5-19}$$

The reason for the negative sign in Eq. $(2.5\text{-}19)_1$ and the fact that the same phase angle appears in both equations will become clear presently. The vectorial resolution of the components is displayed in Fig. 2.5-2.

As will be shown in Chap. 9, $G'(\omega)$ and $J'(\omega)$ are proportional to the average energy stored during a cycle of deformation per unit volume of the material and are therefore called *storage modulus* and *storage compliance*, respectively. $G''(\omega)$ and $J''(\omega)$ are proportional to the energy dissipated per unit volume of the material over a cycle and are consequently termed *loss modulus* and *loss compliance*, respectively. $\tilde{G}(\omega)$ and $\tilde{J}(\omega)$ are known as the *absolute modulus* and *absolute compliance*, while $\theta(\omega)$ is referred to as the *loss angle*.

In terms of the cartesian components, the absolute moduli and compliances are given by

$$\tilde{G}(\omega) = \{[G'(\omega)]^2 + [G''(\omega)]^2\}^{1/2} \quad \text{and} \quad \tilde{J}(\omega) = \{[J'(\omega)]^2 + [J''(\omega)]^2\}^{1/2} \tag{2.5-20}$$

and the *loss tangent* by

$$\tan\theta(\omega) = \frac{G''(\omega)}{G'(\omega)} = \frac{J''(\omega)}{J'(\omega)} \;. \tag{2.5-21}$$

Conversely, the cartesian components are given in terms of the polar ones by

$$G'(\omega) = \tilde{G}(\omega)\cos\theta(\omega) \quad \text{and} \quad G''(\omega) = \tilde{G}(\omega)\sin\theta(\omega) \tag{2.5-22}$$

and by

$$J'(\omega) = \tilde{J}(\omega)\cos\theta(\omega) \quad \text{and} \quad J''(\omega) = \tilde{J}(\omega)\sin\theta(\omega) \;. \tag{2.5-23}$$

Let us now inquire further into the meaning of the polar and cartesian components of the harmonic response functions. Eq. (25-11) may be rewritten as

$$\sigma(\omega) = G^*(\omega)\varepsilon_0 \exp(j\omega t) \ . \tag{2.5-24}$$

Combining this with Eq. (2.5-18) we have

$$\sigma(\omega) = \tilde{G}(\omega)\varepsilon_0 \exp\{j[\omega t + \theta(\omega)]\} \quad \text{and} \quad \sigma(\omega) = \sigma_0(\omega)\exp\{j[\omega t + \theta(\omega)]\} \tag{2.5-25}$$

where we have introduced the relation

$$\tilde{G}(\omega) = \sigma_0(\omega)/\varepsilon_0 \ . \tag{2.5-26}$$

Thus, the absolute modulus is the ratio of the peak amplitude of the stress to the peak amplitude of the strain. We can similarly obtain

$$\tilde{J}(\omega) = \varepsilon_0(\omega)/\sigma_0 \ , \tag{2.5-27}$$

i.e. the absolute modulus and compliance represent the ratio of the peak amplitude of the response to that of the excitation. We have written the peak response amplitude as $\sigma_0(\omega)$ and $\varepsilon_0(\omega)$ to indicate that they are functions of the frequency for a given peak amplitude ε_0 or σ_0 of the excitation.

The loss angle $\theta(\omega)$ is the phase angle between the steady-state stress and strain. Since $\tan\theta(\omega)$ is the ratio of two moduli [cf. Eq. (2.5-21)], the phase angle is always positive. As shown in Fig. 2.5-3, therefore, the stress always reaches its peak value earlier than the strain, i.e. the *stress leads the strain*. Conversely, the *strain lags the stress*.

We now examine the steady-state response further. By Eq. $(2.5\text{-}25)_2$ we have

$$\sigma(\omega) = \sigma_0(\omega)\{\cos[\omega t + \theta(\omega)] + j \sin[\omega t + \theta(\omega)]\} \ . \tag{2.5-28}$$

Applying the trigonometric formulas for the sum of two angles, we may write

$$\sigma(\omega) = \sigma'(\omega) + \sigma''(\omega) \tag{2.5-29}$$

where

$$\sigma'(\omega) = \begin{cases} \sigma_0(\omega)\cos\theta(\omega)[\cos\omega t + j\sin\omega t] \\ \varepsilon_0 G'(\omega)[\cos\omega t + j\sin\omega t] \end{cases} \tag{2.5-30}$$

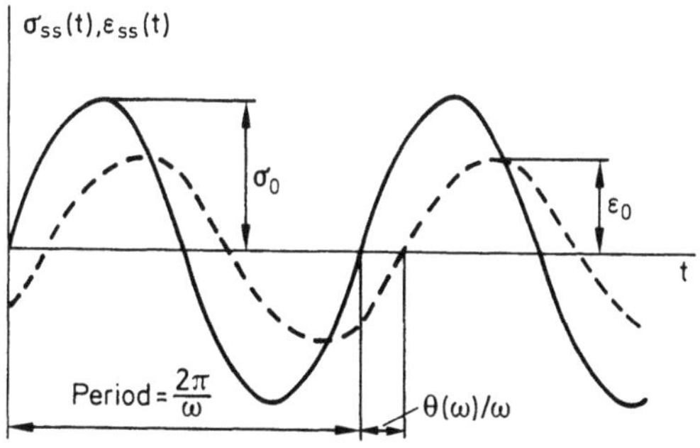

Fig. 2.5-3. Steady-state stress and strain as functions of time

and

$$\sigma''(\omega) = \begin{cases} -\sigma_0(\omega)\sin\theta(\omega)[\sin\omega t - j\cos\omega t] \\ -\varepsilon_0 G''(\omega)[\sin\omega t - j\cos\omega t] \end{cases} \tag{2.5-31}$$

because

$$\sigma_0(\omega)\cos\theta(\omega) = \varepsilon_0 G'(\omega) \quad \text{and} \quad \sigma_0(\omega)\sin\theta(\omega) = \varepsilon_0 G''(\omega) \; . \tag{2.5-32}$$

Since the strain is $\varepsilon(t) = \cos\omega t + j\sin\omega t$, $\sigma'(\omega)$ is that part of the stress response which is *in phase* with the strain excitation and $\sigma''(\omega)$ is that part which is *in quadrature* (90° out of phase) with the strain. This is shown graphically in Fig. 2.5-4. The top part of the figure shows the sinusoidal steady state strain as a function of

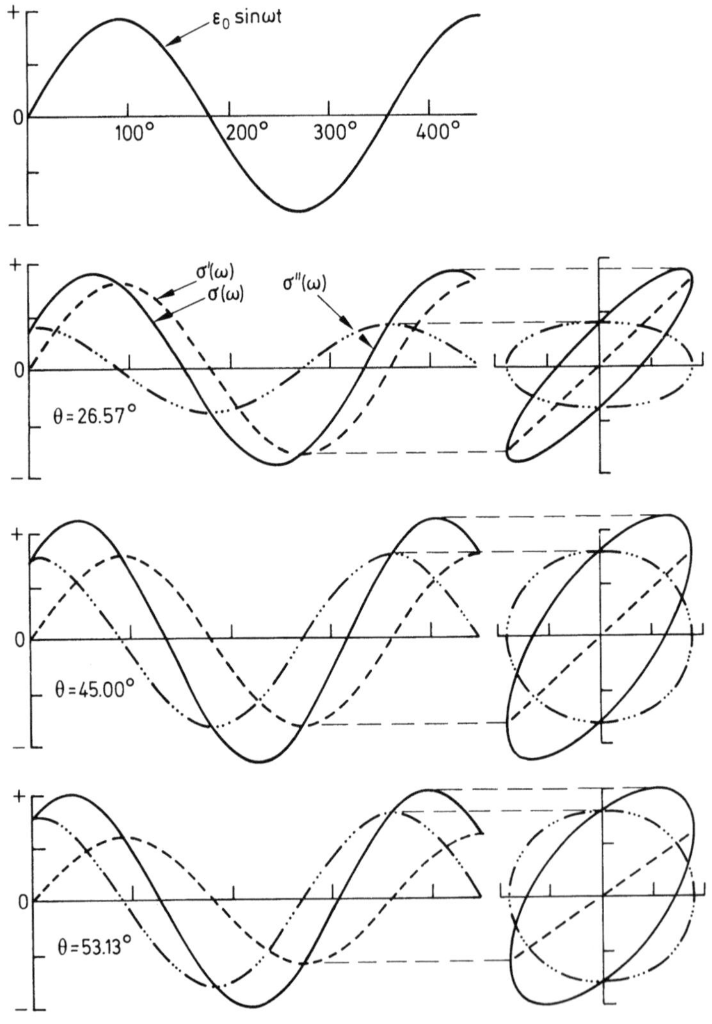

Fig. 2.5-4. In-phase and out-of-phase components

time. Below, on the left, are plotted the corresponding (imaginary) parts of $\sigma(\omega)$, $\sigma'(\omega)$, and $\sigma''(\omega)$ for three different values of $\theta(\omega)$.

The right hand side of Fig. 2.5-4 shows the Lissajous* figures obtained by plotting the stresses not against time but against strain. Use will be made of the Lissajous ellipse in Sect. 9.3.

The in-phase component of the stress, $\sigma'(\omega)$, is often called the *elastic component* because of its direct proportionality to the strain. The out-of-phase component, $\sigma''(\omega)$, which is directly proportional to the rate of strain

$$\dot{\varepsilon}(t) = d\,\varepsilon(t)/dt = -\varepsilon_0\omega[\sin\omega t - j\cos\omega t] \tag{2.5-33}$$

is called the *viscous component*.

An analogous derivation, with

$$\varepsilon(\omega) = \varepsilon_0(\omega)\exp j[\omega t - \theta(\omega)] \tag{2.5-34}$$

and

$$\varepsilon_0(\omega)\cos\theta(\omega) = \sigma_0 J'(\omega) \qquad \text{and} \qquad \varepsilon_0(\omega)\sin\theta(\omega) = \sigma_0 J''(\omega) \tag{2.5-35}$$

leads to

$$\varepsilon'(\omega) = \begin{cases} \varepsilon_0(\omega)\cos\theta(\omega)[\cos\omega t + j\sin\omega t] \\ \sigma_0 J'(\omega)[\cos\omega t + j\sin\omega t] \end{cases} \tag{2.5-36}$$

and

$$\varepsilon''(\omega) = \begin{cases} \varepsilon_0(\omega)\sin\theta(\omega)[\sin\omega t - j\cos\omega t] \\ \sigma_0 J''(\omega)[\sin\omega t - j\cos\omega t] \end{cases} \tag{2.5-37}$$

for the in-phase and out-of-phase components of the steady-state strain.

It is often useful to define a *complex viscosity* as the response of the sinusoidal steady-state stress, $\sigma(\omega)$, to the sinusoidal steady-state rate of strain, $\dot{\varepsilon}(\omega)$. Remembering that $\varepsilon(\omega) = \varepsilon_0\exp(j\omega t)$, we have

$$\dot{\varepsilon}(\omega) = \frac{d\,\varepsilon(\omega)}{dt} = j\omega\,\varepsilon(\omega) \tag{2.5-38}$$

and hence,

$$\eta^*(\omega) = \frac{\sigma(\omega)}{j\omega\,\varepsilon(\omega)} = \frac{G^*(\omega)}{j\omega}\,. \tag{2.5-39}$$

The complex viscosity is the rheological analog of the mechanical impedance, defined as the ratio of the sinusoidal steady-state force to the sinusoidal steady-state velocity.

Decomposition of the complex viscosity into cartesian and polar components yields

$$\eta^*(\omega) = \eta'(\omega) - j\,\eta''(\omega) = \tilde{\eta}(\omega)\exp[-j\,\Theta(\omega)] \tag{2.5-40}$$

---

* Jules Antoine Lissajous, (pron. Lisser<u>zhoo</u>), 1822–1880, French physicist who investigated the curves named after him in the 1850's [14].

where

$$\eta'(\omega) = G''(\omega)/\omega \qquad \text{and} \qquad \eta''(\omega) = G'(\omega)/\omega \ , \tag{2.5-41}$$

$$\tilde{\eta}(\omega) = \{[\eta'(\omega)]^2 + [\eta''(\omega)]^2\}^{1/2} \tag{2.5-42}$$

and

$$\tan \Theta(\omega) = \frac{\eta''(\omega)}{\eta'(\omega)} = \frac{G'(\omega)}{G''(\omega)} \tag{2.5-43}$$

where $\Theta(\omega)$ is the *storage angle.* The loss angle is given by

$$\tan \theta(\omega) = \frac{\eta'(\omega)}{\eta''(\omega)} = \frac{G''(\omega)}{G'(\omega)} \tag{2.5-44}$$

and the two angles are connected by the relation

$$\Theta(\omega) + \theta(\omega) = \tfrac{\pi}{2} \ . \tag{2.5-45}$$

Hence

$$\tan \Theta(\omega) = \cot \theta(\omega) = \tan[\tfrac{\pi}{2} - \theta(\omega)] \ . \tag{2.5-46}$$

These relations are shown graphically in Fig. 2.5-5. While $\theta(\omega)$ is the phase angle between the stress and strain, $\Theta(\omega)$ is the phase angle between stress and rate of strain (cf. Problem 2.5-3).

Thus the relation between the cartesian and the polar components is

$$\eta'(\omega) = \tilde{\eta}(\omega)\cos \Theta(\omega) = \tilde{\eta}(\omega)\sin \theta(\omega) \tag{2.5-47}$$

and

$$\eta''(\omega) = \tilde{\eta}(\omega)\sin \Theta(\omega) = \tilde{\eta}(\omega)\cos \theta(\omega) \tag{2.5-48}$$

while

$$\tilde{\eta}(\omega) = \frac{\sigma_0(\omega)}{\dot{\varepsilon}_0} = \frac{\sigma_0(\omega)}{\omega\, \varepsilon_0} \ . \tag{2.5-49}$$

We leave it to Problem 2.5-1 to show that

$$\sigma(\omega) = -\omega\varepsilon_0[\eta'(\omega)\sin \omega t - \eta''(\omega)\cos \omega t] + j\omega\, \varepsilon_0[\eta'(\omega)\cos \omega t + \eta''(\omega)\sin \omega t] \ . \tag{2.5-50}$$

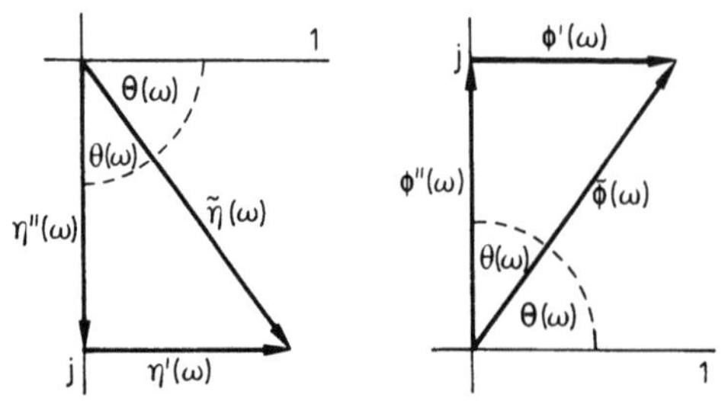

Fig. 2.5-5. Vectorial resolution of the components of the complex shear viscosity and fluidity

We may similarly define a *complex fluidity* as the reciprocal of the complex viscosity, i.e. as the response of the sinusoidal steady-state rate of strain to a sinusoidal steady-state stress of unit amplitude. We have

$$\phi^*(\omega) = \frac{1}{\eta^*(\omega)} = \frac{j\omega\,\varepsilon(\omega)}{\sigma(\omega)} = j\omega\,J^*(\omega) \tag{2.5-51}$$

and therefore

$$\phi^*(\omega) = \phi'(\omega) + j\,\phi''(\omega) = \tilde{\phi}(\omega)\exp[j\,\Theta(\omega)] \tag{2.5-52}$$

where $\Theta(\omega)$ is again the storage angle.

It should be noted that it is the imaginary parts of the complex viscosity and fluidity which are related to the energy stored during the deformation while the real parts are related to the energy dissipated. These could therefore be referred to as *storage* and *loss viscosity*, and *storage* and *loss fluidity*.

To complete the symmetrical pattern of the response functions to a sinusoidal steady-state excitation, we now define

$$\chi^*(\omega) = \frac{\varepsilon(\omega)}{\dot{\sigma}(\omega)} = \frac{\varepsilon(\omega)}{j\omega\,\sigma(\omega)} = \frac{J^*(\omega)}{j\omega} \tag{2.5-53}$$

as the response of the steady-state strain to a steady-state rate of stress of unit amplitude, and

$$\kappa^*(\omega) = \frac{\dot{\sigma}(\omega)}{\varepsilon(\omega)} = \frac{j\omega\,\sigma(\omega)}{\varepsilon(\omega)} = j\omega\,G^*(\omega) \tag{2.5-54}$$

as the response of the steady-state rate of stress to a steady-state strain of unit amplitude.

We have

$$G^*(\omega) = [s\bar{G}(s)]_{s=j\omega} \quad \text{and} \quad J^*(\omega) = [s\bar{J}(s)]_{s=j\omega} \tag{2.5-55}$$

$$\eta^*(\omega) = [s\bar{\eta}(s)]_{s=j\omega} \quad \text{and} \quad \phi^*(\omega) = [s\bar{U}(s)]_{s=j\omega} \tag{2.5-56}$$

and

$$\kappa^*(\omega) = [s\bar{Q}(s)]_{s=j\omega} \quad \text{and} \quad \chi^*(\omega) = [s\bar{\chi}(s)]_{s=j\omega} \;. \tag{2.5-57}$$

By Eq. (2.1-19) we find

$$G^*(\omega)J^*(\omega) = \eta^*(\omega)\phi^*(\omega) = \kappa^*(\omega)\chi^*(\omega) = 1 \;. \tag{2.5-58}$$

The reciprocity expressed by Eqs. (2.5-58) requires the minus sign in Eqs. (2.5-19) and (2.5-40) for a consistent definition of the components of the complex functions.

2.6 Excitation and Response in the Time Domain

In the preceding sections we have derived the *standard response functions* from Hooke's law in the complex plane, Eqs. (2.1-15), by finding the Laplace transform of the desired excitation function, inserting it into Eq. $(2.1\text{-}15)_1$ or $(2.1\text{-}15)_2$, and defining an appropriate standard response function as the response to unit excitation. Thus, for the response to the standard strain excitations we obtained

$$G(t) = \frac{\sigma(t)}{\varepsilon_0} = \mathscr{L}^{-1}\frac{\bar{Q}(s)}{s} \tag{2.6-1}$$

$$\eta(t) = \frac{\sigma(t)}{\dot{\varepsilon}_0} = \mathscr{L}^{-1}\frac{\bar{Q}(s)}{s^2} \tag{2.6-2}$$

and

$$G^*(\omega) = \frac{\sigma(\omega)}{\varepsilon(\omega)} = [\bar{Q}(s)]_{s=j\omega} \tag{2.6-3}$$

in terms of the relaxance $\bar{Q}(s)$, the transform of the response to a unit impulse of strain. The responses to the standard stress excitations were derived as

$$J(t) = \frac{\varepsilon(t)}{\sigma_0} = \mathscr{L}^{-1}\frac{\bar{U}(s)}{s} \tag{2.6-4}$$

$$\chi(t) = \frac{\varepsilon(t)}{\dot{\sigma}_0} = \mathscr{L}^{-1}\frac{\bar{U}(s)}{s^2} \tag{2.6-5}$$

and

$$J^*(\omega) = \frac{\varepsilon(\omega)}{\sigma(\omega)} = [\bar{U}(s)]_{s=j\omega} \tag{2.6-6}$$

in terms of the retardance, $\bar{U}(s)$, the transform of the response to a unit impulse of stress. Because of the simplicity of the various interrelations between these functions in the transform plane this is a convenient way of showing the structure of the interdependence of the viscoelastic response functions.

Evidently, however, the response functions can equally well be derived from the superposition integrals. For those who prefer this, the derivations will now be given. First we shall extricate the unit impulse responses themselves from under the convolution integral. Let

$$\varepsilon(t) = \hat{\varepsilon}_0\delta(t) \tag{2.6-7}$$

and substitute this into Eq. $(2.1\text{-}22)_1$, i.e.

$$\sigma(t) = \int_0^t Q(u)\varepsilon(t-u)du\ . \tag{2.6-8}$$

We have

$$\sigma(t) = \hat{\varepsilon}_0 \int_{-\infty}^{\infty} Q(u)\delta(t-u)du \tag{2.6-9}$$

and hence

$$\frac{\sigma(t)}{\hat{\varepsilon}_0} = Q(t) \tag{2.6-10}$$

as required. Equation (2.6-10) follows from Eq. (2.6-9) because the value of $\delta(t-u)$ is zero when $t \neq u$. Hence the limits can be extended from t to ∞, and from 0 to $-\infty$ without changing the value of the integral which becomes simply Q(t) [cf. Eq. (A2-15)]. A parallel derivation leads to the corresponding equation for U(t).

To derive the response to unit step of strain, let

$$\varepsilon(t) = \varepsilon_0 h(t) \tag{2.6-11}$$

and substitute into Eq. (2.6-8). Then

$$\sigma(t) = \varepsilon_0 \int_0^t Q(t-u)h(u)du = \varepsilon_0 \int_0^t Q(u)h(t-u)du \ . \tag{2.6-12}$$

But $h(t-u)$ is unity because $t > u$ and hence $t-u$ is always positive. Consequently

$$\frac{\sigma(t)}{\varepsilon_0} = \int_0^t Q(u)du = G(t) \tag{2.6-13}$$

where G(t) is introduced as the new material response function. One can, of course, also start from Eqs. (2.3-16) (cf. Problem 2.3-2) and arrive directly at

$$\frac{\sigma(t)}{\varepsilon_0} = G(t) \ . \tag{2.6-14}$$

J(t) can be derived by analogous reasoning.

For a slope strain we have

$$\varepsilon(t) = \dot{\varepsilon}_0 t \tag{2.6-15}$$

and hence

$$\frac{\sigma(t)}{\dot{\varepsilon}_0} = \int_0^t Q(u)(t-u)du = \eta(t) \tag{2.6-16}$$

where $\eta(t)$ is the new material response function.

Integration by parts yields the alternative form (cf. Problem 2.4-2)

$$\frac{\sigma(t)}{\dot{\varepsilon}_0} = \int_0^t \int_0^v Q(u)du\, dv = \eta(t) \tag{2.6-17}$$

We could also have inserted Eq. (2.6-15) into any of Eqs. (2.3-16). This would have given (cf. Problem 2.6-2)

$$\frac{\sigma(t)}{\dot{\varepsilon}_0} = \int_0^t G(u)du = \eta(t) \tag{2.6-18}$$

directly.

The response to unit steady-state sinusoidal stress is obtained by setting

$$\varepsilon(t) = \varepsilon_0 \exp(j\omega t) \tag{2.6-19}$$

and substituting into Eq. (2.6-8). This gives

$$\sigma(t) = \varepsilon_0 \int_0^t Q(u)\exp[j\omega(t-u)]du = \varepsilon(t) \int_0^t Q(u)\exp(-j\omega u)du \ . \tag{2.6-20}$$

The steady state is reached in the limit of large values of t, i.e.

$$\frac{\sigma(\omega)}{\varepsilon(\omega)} = \lim_{t\to\infty} \int_0^t Q(u)\exp(-j\omega u)du \tag{2.6-21}$$

where we have, as before, used $\sigma(\omega)$ and $\varepsilon(\omega)$ to denote the sinusoidal steady state stress and strain respectively.

Changing the variable of integration to t, we find the response to a sinusoidal steady-state strain to be given by

$$\frac{\sigma(\omega)}{\varepsilon(\omega)} = \int_0^\infty Q(t)\exp(-j\omega t)dt = G^*(\omega) = \mathscr{F}[Q(t);j\omega] \tag{2.6-22}$$

where $G^*(\omega)$ is the new material function. We note that $G^*(\omega)$ is the Fourier* transform of Q(t) (see Appendix A4).

Starting from

$$\sigma(t) = \sigma_0\exp(j\omega t) \tag{2.6-23}$$

we arrive at

$$\frac{\varepsilon(\omega)}{\sigma(\omega)} = \int_0^\infty U(t)\exp(-j\omega t)dt = J^*(\omega) = \mathscr{F}[U(t);j\omega] \tag{2.6-24}$$

showing that the complex compliance is the Fourier transform of the response to unit impulse of stress just as the complex modulus is the Fourier transform of the response to unit impulse of strain.

2.7 Problems

Problem 2.3-1. Show that

$$G(t) = \int_0^t Q(u)du \tag{1}$$

by integrating Eq. (2.3-11). What would have been the result of omitting the delta function term (i.e. the initial condition) from that equation?

Problem 2.3-2. Derive the unit step response

$$\sigma(t)/\varepsilon_0 = G(t) \tag{1}$$

from Eqs. (2.3-16).

Problem 2.3-3. The Boltzmann superposition integral expresses the stress as a function of a continuous strain history. Derive an equation for the superposition of a discrete set of step strains in terms of the relaxation modulus.

* Jean Baptiste Joseph, baron Fourier, pron. Fooryeh, 1768–1830, French mathematician and physicist, famous for his work on the representation of functions by trigonometric series.

Problem 2.3-4. Compare the stress response to a step strain imposed at $t = 0$ to that resulting from the same strain imposed at $t = t_1$. Work in the complex plane and translate only the final results into the time domain.

Problem 2.4-1. Given Eq. $(2.4\text{-}8)_1$, show that

$$\eta(t) = \int_0^t \int_0^v Q(u)du\,dv \ . \tag{1}$$

Problem 2.4-2. Starting from the Boltzmann superposition integral

$$\sigma(t) = \int_0^t Q(t-u)\varepsilon(u)du \tag{1}$$

show that

$$\eta(t) = \int_0^t \int_0^v Q(u)du\,dv \ . \tag{2}$$

Problem 2.5-1. Derive Eq. (2.5-50).

Problem 2.5-2. Derive Eq. (2.5-11) from the Residue Theorem (cf. Appendix A3.2.1). (*Hint*: in the steady state only the poles of the excitation function need to be obtained.)

Problem 2.5-3. Show that the storage angle, $\Theta(\omega)$, is the phase angle between the stress and the rate of strain.

Problem 2.6-1. Show that $\eta^*(\omega)$ is the Fourier transform of $G(t)$ by deriving it along the lines of the derivation of $G^*(\omega)$ from $Q(t)$ in Sect. 2.6.

Problem 2.6-2. Derive Eq. (2.6-18) from Eq. (2.3-16).

Problem 2.6-3 Consider the strain defined by

$$\varepsilon_t(s) = \varepsilon(t-s) - \varepsilon(t) \tag{1}$$

where $0 \le s \le \infty$ is the *elapsed time* extending backward into the indefinite past. Thus, $\varepsilon_t(s)$ is the strain at time s referred to the strain, $\varepsilon(t)$, at the present time, t.

(A) Express the stress-strain relation, Eq. $(2.3\text{-}16)_3$ in terms of $\varepsilon_t(s)$. Then use this new relation to express the response (B) to a step strain, and (C) to a slope strain.

Note: The elapsed times s and w [cf. Eq. (2.1-21)] are not identical. They range over different intervals ($0 \le s \le \infty$ and $0 \le w \le t$, respectively). Also, the view point is different. When using $w = t - u$ we look *forward* from the past time, u, to the present time, t. When using s we look *backward* from the present time, t, into the indefinite past.

References (Chap. 2)

1. M. Reiner, Physics Today, January, 1964, p. 62.
2. V. Volterra, *Theory of Functionals and of Integral and Integro-differential Equations*, Blackie, London, 1930; Dover, New York.
3. W.E. Weber, Pogg. Ann. Phys. (2) *4*:247 (1835); (2) *24*:1 (1841).
4. F. Kohlrausch, Pogg. Ann. Phys. (4) *29*:337 (1863); (5) *8*:1,207,399 (1866); (6) *8*:337 (1876).
5. H. Leaderman, *Elastic and Creep Properties of Filamentous Materials and Other High Polymers*, The Textile Foundation, Washington, DC, 1943. (Available as xerograph from University Microfilms, Ann Arbor, Michigan).
6. O. Meyer, J. rein. u. angew. Math. *78*:130 (1874).
7. N.W. Tschoegl, Kolloid-Z., *174*:113 (1961).
8. H.W. Bode, *Network Analysis and Feedback Amplifier Design*, Van Nostrand, New York, 1945.
9. L. Boltzmann, Sitzungsber. Kgl. Akad. Wiss. Wien, Math.-Naturw. Classe, (2A) *70*:275 (1874); Pogg. Ann. Phys. (Ergänzungsband) 7:624 (1876).
10. M.F. Gardner and J.L. Barnes, *Transients in Linear Systems*, Vol. 1, Wiley, New York, 1942; (a) pp. 228 and 364; (b) p. 176.
11. J.N. Franklin, *Matrix Theory*, Prentice-Hall, Englewood Cliffs, N.J., 1968, p. 62.
12. E.H. Lee, Quart. Appl. Math., *13*:183 (1955).
13. T.M. Apostol, *Mathematical Analysis*, 2nd ed., Addison-Wesley, Reading, MA, 1975, p. 140.
14. J.A. Lissajous, Ann. chim. phys. ser. 3, *51*:147 (1857).

3. Representation of Linear Viscoelastic Behavior by Series-Parallel Models

There are many good books on linear viscoelastic behavior. You find them listed under electric circuit theory.

Attributed to H. Leaderman

3.0 Introduction

In the preceding chapter we have treated the formal aspects of the impulse, step, slope, and harmonic responses but have said nothing about the actual form of the response. In this chapter and in Chap. 5 we will discuss the representation of viscoelastic behavior by mechanical, or, more precisely, rheological models.

The so-called *series-parallel models* furnish a convenient and didactically useful way of modelling the response of linear viscoelastic materials to an arbitrary stimulus. They form the topic of this chapter. The so-called *ladder models* will be discussed in Chap. 5. The necessary background in the theory of mechanical models is developed in the first section of this chapter.

3.1 The Theory of Model Representation

The theory of mechanical models is the mechanical analog of electric circuit theory. In the latter the underlying physical system is replaced by idealized counterparts of its actual constitutive elements. The interconnection of these elements is shown in graphical form in a manner which greatly facilitates the analysis of the system. It is less well known that mechanical systems may be analyzed in an entirely analogous fashion. The purpose of this section is the development of the rudiments of the theory of mechanical model diagrams. For more detail the interested reader is referred to, e.g., the excellent textbook by Gardner and Barnes [1]. Electric circuit theory has had a venerable history and is discussed in a much larger number of textbooks. Most students of the physical sciences have some familarity with it. We therefore preface our discussion of mechanical model diagrams with a brief review of the elements of electric circuit theory.

3.1.1 The Elements of Electric Circuit Analysis

In all physical systems we distinguish active and passive elements. Active elements are sources of energy. In passive elements energy is either stored or dissipated. In an electrical system the energy sources are sources of either voltage or current. The dualism between excitation and response in a linear system allows us interchangeably to regard either voltage as the excitation and current as the response or current as the excitation and voltage as the response. For this reason it makes no difference in principle whether voltage or current is selected as the active element. Sometimes the nature of the problem favors current as the source or excitation. Generally, however, voltage is chosen for that purpose.

The passive elements of a linear electric system are *inductors*, *capacitors*, and *resistors*. Energy is stored kinetically in the first, potentially in the second, and is dissipated in the third. The amount of energy stored in an inductor or a capacitor is measured by its *inductance* and *capacitance*, respectively. Similarly, *resistance* is a measure of the ability of a resistor to dissipate energy.

In the model representation of an electrical system the various elements are treated as idealized elements. Thus, the internal inductance, capacitance, or resistance of the source is either neglected, or represented separately by the appropriate passive elements. The passive model elements are pure elements. While e.g., an inductor always possesses some capacitance as well as resistance apart from inductance, these again are either neglected or represented by separate (pure) elements. It is the same, *mutatis mutandis*, with other passive elements.

The simplest electrical system which contains one of each kind of element is the electric harmonic oscillator shown in Fig. 3.1-1. This schematic diagram represents an electric system in which a sinusoidal steady-state voltage $V(\omega)$ of radian frequency ω produces a sinusoidal steady-state current $I(\omega)$ of the same frequency. The current is pictured as a flow of electrons through the passive elements having harmonic inductance $j\omega L$, resistance R, and capacitance $j\omega C$. The quantities $j\omega L$, R, and $1/j\omega C$ are termed *impedances*. The quantities $1/j\omega L$, $1/R$, and $j\omega C$ are called *admittances*. The passive elements are represented by the well known conventionalized symbols shown in Fig. 3.1-1. These circuit elements are all *two-terminal* elements: the current enters at one terminal and leaves at the other. The active and passive elements are connected by conducting wires represented schematically by the solid lines in Fig. 3.1-1. The wires are considered to possess no inductance, resistance, or capacitance themselves. If necessary, separate elements can be added to account for such properties. The current flows in closed circuits through the

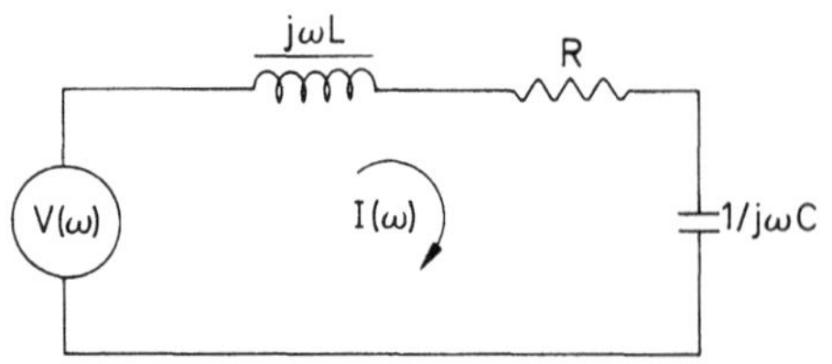

Fig. 3.1-1. The electric harmonic oscillator: a single-mesh electric circuit

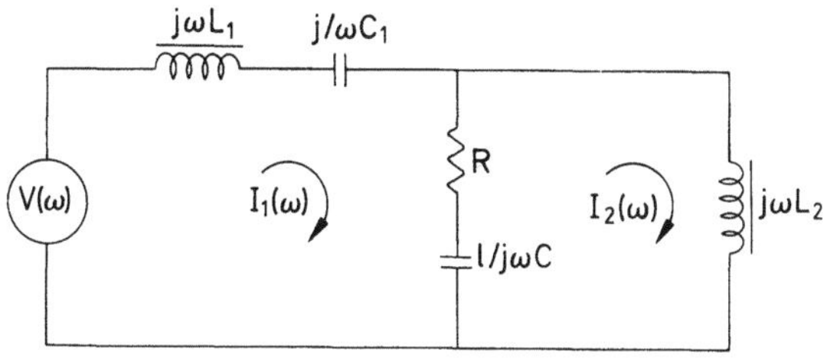

Fig. 3.1-2. A two-mesh electric circuit

elements. Figure 3.1-1 represents a single-mesh electric circuit. A two-mesh electric circuit is shown in Fig. 3.1-2.

The so-called *mesh equations*, representing the equations of motion of the alternating current, are found simply by applying *Kirchhoff's* voltage law* [2] which, for our purposes, we may state as follows: around any closed circuit (mesh) the instantaneous impressed voltage equals the product of the instantaneous mesh current and mesh impedance. In determining the mesh impedance, we observe the *combination rules* that *impedances add in series, admittances add in parallel.* These rules result from the fact that current flows *through* the passive elements but voltage (more precisely voltage drop) is measured *across* them.

Thus, from Fig. 3.1-1 we obtain

$$V(\omega) = (j\omega L + R + 1/j\omega C)I(\omega) \tag{3.1-1}$$

or

$$V(\omega) = Z(\omega)I(\omega) \tag{3.1-2}$$

where

$$Z(\omega) = R + j(\omega L - 1/\omega C) \tag{3.1-3}$$

is the impedance of the circuit. Its real part is the *resistance* R, while its imaginary part is the *reactance* $\omega L - 1/\omega C$.

From Fig. 3.1-2 we obtain

$$V(\omega) = (j\omega L + 1/j\omega C_1 + R + 1/j\omega C)I_1(\omega) - (R + 1/j\omega C)I_2(\omega) \tag{3.1-4}$$

and

$$0 = (R + 1/j\omega C + j\omega L_2)I_2(\omega) - (R + 1/j\omega C)I_1(\omega) \tag{3.1-5}$$

where Eqs. (3.1-4) and (3.1-5) are the mesh equations of the first and second mesh, respectively. The currents $I_1(\omega)$ and $I_2(\omega)$ are pictured as flowing clockwise through the two meshes. In the *common branch* containing the impedance

$$Z_c(\omega) = R - j/\omega C \tag{3.1-6}$$

the two currents thus oppose each other. Consequently, the voltages represented by $Z_c(\omega)I_1(\omega)$ and $Z_c(\omega)I_2(\omega)$ are subtracted in the two mesh equations. The currents $I_1(\omega)$ and $I_2(\omega)$ are found by solving these two simultaneous equations.

---

* Gustaf Robert Kirchhoff, pronounced Keerkhof, 1824–1887, German physicist.

Table 3.1-1. Electric circuit analysis

Harmonic quantities	Operational quantities
$V(\omega)$	$\bar{V}(s)$
$I(\omega)$	$\bar{I}(s)$
$j\omega L$	Ls
R	R
$1/j\omega C$	$1/Cs$

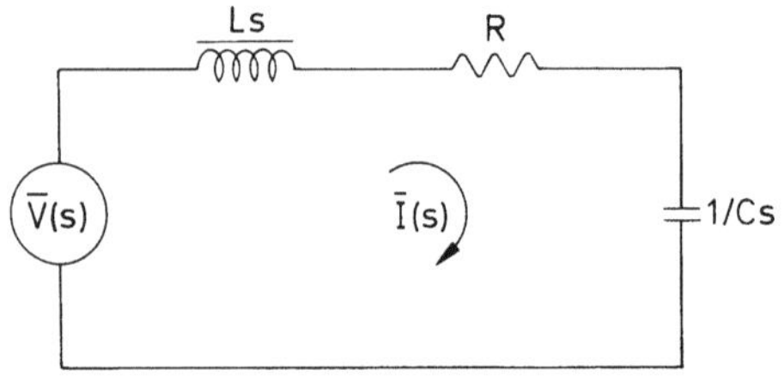

Fig. 3.1-3. Single-mesh electric circuit in operational form

We have dealt here with the sinusoidal steady-state. The circuit analysis is, however, easily extended to the general case through the Laplace transformation using the methods of the *operational calculus* summarized in the Appendix. We simply replace the harmonic quantities by the corresponding operational ones as listed in Table 3.1-1. Thus, Fig. 3.1-1 becomes Fig. 3.1-3 and Eq. (3.1-1) changes to

$$\bar{V}(s) = \bar{Z}(s)\bar{I}(s) = (Ls + R + 1/Cs)\bar{I}(s) . \tag{3.1-7}$$

The two mesh equations for the similarly modified circuit of Fig. 3.1-2 are

$$\bar{V}(s) = (Ls + 1/C_1 s + R + 1/Cs)\bar{I}_1(s) - (R + 1/Cs)\bar{I}_2(s) \tag{3.1-8}$$

and

$$0 = (R + 1/Cs + Ls)\bar{I}_2(s) - (R + 1/Cs)\bar{I}_1(s) . \tag{3.1-9}$$

The essence of electric circuit analysis consists in the fact that the transform equations (and, hence, the differential equations, or the sinusoidal steady-state relations) linking voltage and current through the impedances or admittances in any mesh of the circuit can be found immediately from a set of simple rules once the electric circuit representing the system has been drawn [1, 3]. In the following section we develop an analogous theory for mechanical systems.

3.1.2 The Elements of Mechanical Model Analysis

Familiar examples of passive mechanical elements are the *spring*, in which energy is stored potentially, and the *dashpot*, or *damper*, in which energy is dissipated as heat. We shall use the symbol of a spring, shown in Fig. 3.1-4a, to represent storage of potential energy. Similarily, we shall use the symbol of a dashpot, or damper, shown in Fig. 3.1-4b, to represent the dissipation of mechanical energy. We shall

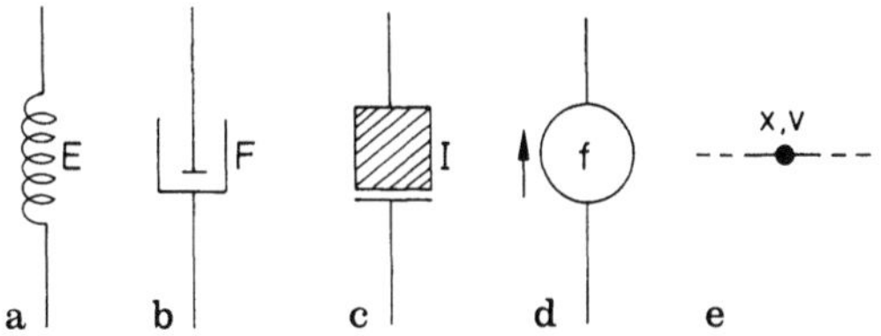

Fig. 3.1-4. Conventionalized symbols for mechanical model diagrams: (**a**) spring, (**b**) dashpot, (**c**) mass, (**d**) force, (**e**) displacement or velocity

further assume that the spring is a Hookean (i.e. linear) spring (cf. Sect. 1.1), and that the dashpot is a Newtonian (i.e. linear) dashpot (cf. Sect. 1.1). We also assume that the response of the spring is purely elastic (i.e. no energy is dissipated in it and its mass is negligible), and that the response of the dashpot is purely viscous (i.e. it is rigid and its mass is again negligible). Our spring and dashpot thus are idealized elements as are their electric counterparts.

Under the action of a force, f, applied to the two ends of a spring the spring extends, resulting in a displacement, x, of its ends. The proportionality constant between the force and the displacement will be called *elastance* and will be denoted by E. Under the action of a tensile force applied to the two ends of the dashpot, the two ends move apart with constant velocity. The proportionality constant between force and the velocity will be called *frictance* and will be denoted by F. The displacement or the velocity are thus measured *across* these two passive elements. Both elements are *two-terminal* elements just as their electric counterparts. The elastance, E, is a measure of the energy stored potentially, and the frictance, F, is a measure of the energy dissipated in the mechanical system.

Energy may be stored not only potentially, but also kinetically. We shall use as our passive element representing kinetic energy storage an idealized *mass*, i.e. a mass whose response to an applied force is purely inertial (i.e. free from elastic or viscous effects). We also require that the response of our idealized mass be linear, i.e. that it respond to an applied force with a constant acceleration. Thus the mass must be rigid and move without friction. We shall call the proportionality constant between the force and the acceleration the *inertance* and denote it by I. The mass element again must be symbolized as a two-terminal element because its motion is observed with reference to an inertial frame (usually ground). We denote our inertial element by the symbol shown in Fig. 3.1-4c where the line connected to the rectangle representing the mass symbolizes the point of application of the force, while the line stopping short of the rectangle symbolizes the connection of the element to the reference point. It is a peculiarity of mechanical systems that the motion of masses must always be referred to ground as a reference point. This results in severe restrictions on the possible combinations of mechanical elements (see Problem 3.2-1).

We shall find it convenient to discuss the energetics of mechanical systems in terms of force and displacement rather than velocity. Just as in the electrical case, because of the duality between excitation and response in a linear system, either force or displacement may be taken as the active source. In most cases force comes naturally to be regarded as the active element. We shall represent a force also by a two-terminal element as shown in Fig. 3.1-4d. The two-terminal nature of force is in accord with Newton's law of action and reaction. We must take into account not only the point of application of the force, i.e. its point of action, but also the point of reaction. The arrow indicating the direction of the force will henceforth usually be omitted.

Since we wish to represent the relations between force and displacement in a linear mechanical system in the greatest generality, we shall formulate them in the transform plane. We denote the force transform by $\bar{f}(s)$, and the displacement transform by $\bar{x}(s)$. Thus we can describe potential energy storage by

$$\bar{f}(s) = E\bar{x}(s) , \tag{3.1-10}$$

energy dissipation by

$$\bar{f}(s) = Fs\bar{x}(s) , \tag{3.1-11}$$

and kinetic energy storage by

$$f(s) = Is^2\bar{x}(s) , \tag{3.1-12}$$

provided that the system is initially at rest, i.e.,

$$x(0) = \dot{x}(0) = \ddot{x}(0) \equiv 0 \tag{3.1-13}$$

where the dots indicate differentiation with respect to time. Thus, we assume *quiescent* or *zero initial conditions* unless stated otherwise. Equations (3.1-10), (3.1-11), and (3.1-12) are examples of the application of *d'Alembert's principle** which is the mechanical analog of Kirchhoff's voltage law. Equations analogous to Eqs. (3.1-10), (3.1-11), and (3.1-13) apply, of course, to linear electric systems.

We call E, Fs, and Is^2 the *(operational) mechanical relaxances* and their reciprocals the *(operational) mechanical retardances*. The term *respondance* will refer to relaxance *or* retardance when it is not necessary or not desired to make the distinction between the two forms. The term respondance will thus be used in the same sense in which *immittance* is used in electric circuit analysis to comprise both impedance and admittance.

We are now ready to apply these concepts to the analysis of mechanical systems. As a simple example consider a slab of mass (inertance) I supported on a spring with elastance E and a dashpot with frictance F as shown in Fig. 3.1-5.

We are interested in the displacement x of the slab to which the force f is applied. To find the equation of motion we combine the information contained in the representation of the actual system shown in Fig. 3.1-5 into a diagram connecting the corresponding symbols discussed below. The resulting diagram, which we shall call a *mechanical model diagram*, is shown in Fig. 3.1-6.

To construct the diagram, we note that the point of reaction of the mass is the laboratory ground symbolized by the bottom line. The slab to which the force is applied acts as a rigid connection for the transmission of the force to the spring and the dashpot which rest on the ground. In the idealized representation of the model

---

* Jean le Rond d'Alembert, pron. Dullambare, 1717–1783, French mathematician and philosopher. Ascription of this principle to d'Alembert does not seem to be fully justified historically [4].

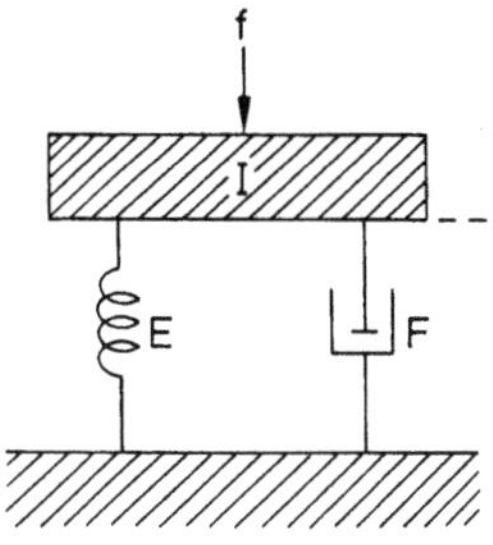

Fig. 3.1-5. Mechanical system consisting of spring, dashpot, and mass

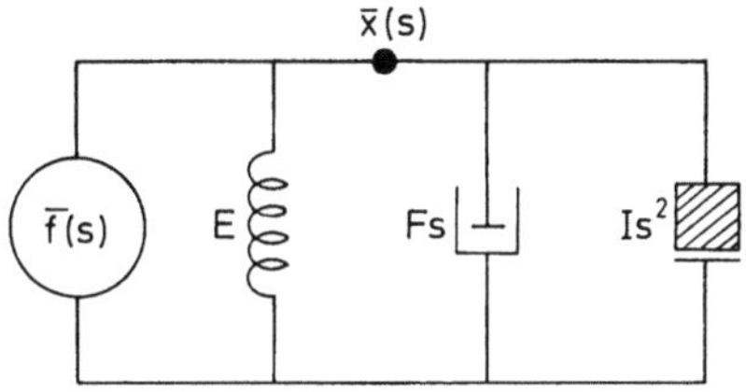

Fig. 3.1-6. Mechanical model diagram of assembly shown in Fig. 3.1-5

diagram the function of the slab as a connecting element and its property of possessing the mass I are separated. The mass is represented by the shaded square attached to the upper solid line in Fig. 3.1-6. The latter symbolizes the rigid connection and may be pictured as a rigid but weightless bar to which the mass of inertance, I, as well as the spring of elastance, E, and the dashpot of frictance, F, are attached. The horizontal line under the shaded square is connected to the bottom line in accordance with the fact that the motion of the mass is observed with reference to the laboratory ground. The displacement of the rigid bar is idealized as occuring at the *node* symbolized by the small full circle (cf. Fig. 3.1-4e) which may be placed anywhere along the bar. Figure 3.1-5 shows clearly that the same force acts *through* all three passive elements, and their displacement is measured *across* them. Thus the relaxances are additive and we obtain the *node equation* as

$$\bar{f}(s) = (Is^2 + Fs + E)\bar{x}(s) = \bar{Q}_t(s)\bar{x}(s) \tag{3.1-14}$$

where $\bar{Q}_t(s)$ is the (translational) mechanical relaxance of the model. Equation (3.1-14) is the transform of the equation of motion. Retransforming $(3.1\text{-}14)_1$ we have

$$f(t) = I\frac{d^2x(t)}{dt^2} + F\frac{d\,x(t)}{dt} + E\,x(t)\ . \tag{3.1-14'}$$

The mechanical model diagram drawn by inspection of the mechanical assembly which it represents, thus allows us to derive the differential equations governing the behavior of the assembly by *node analysis*, the counter part of *mesh analysis* applied to an electric circuit diagram. We may, therefore, formulate d'Alembert's principle for our purposes as follows: at any node the impressed instantaneous force equals the product of the instantaneous node displacement and node relaxance. The analogy with Kirchhoff's voltage law is complete. In determining the node relaxance we observe the *combination rules* that *relaxances add in parallel, retardances add in series.* Relaxances shared by two nodes are analogous to impedances shared by two meshes. The forces represented by the two products of displacement and relaxance are subtracted in the two node equations (cf. Problem 3.1-1).

The combination rules are a consequence of the additivity of forces in a parallel combination of passive elements, and of the additivity of the displacements (or velocities) in a series combination. In a parallel combination of passive elements all elements experience the same displacement (or velocity) and thus the forces in the elements are additive. Conversely, the same force acts through all elements in a series combination. In such a combination, therefore, it is the displacements (or velocities) which are additive.

If we had considered velocity instead of displacement, the operational form of the equation of motion would have become

$$\bar{f}(s) = (Is + F + E/s)\bar{v}(s) \ . \tag{3.1-15}$$

Equation (3.1-15) is easily obtained by inserting the transform relation

$$\bar{v}(s) = s\bar{x}(s) \tag{3.1-16}$$

(valid for quiescent initial conditions), into Eq. $(3.1\text{-}14)_1$. It may be noted that the form of Eq. (3.1-15) is identical with that of Eq. $(3.1\text{-}7)_2$. However, Figs. 3.1-3 and 3.1-6 differ in the interconnection of the elements. In Fig. 3.1-3 these are in series, in Fig. 3.1-6 they are in parallel. This is a consequence of having chosen voltage, which is measured *across* the passive electrical elements, as the excitation in one case, and force, which acts *through* the mechanical elements, in the other case. The choice of voltage in the electric case, and of force in the mechanical one, as the excitation, preserves the interconnection of the elements so that both diagrams can be drawn directly from an inspection of the actual physical interconnection of the elements in the two systems.

Occasionally it is convenient to consider not force but displacement (or velocity) as the excitation. Mechanical model diagrams may be drawn with displacement or velocity as the active element. In such diagrams the interconnection of the elements is reversed and the analysis is of the mesh type [1]. However, in most cases the diagram can be drawn as explained above with the force producing the displacement or velocity formally constituting the active element. (See Problems 3.1-4 and 3.1-7.)

Figure 3.1-6 represents a single-node mechanical model diagram. An example of a two-node diagram is discussed in Problem 3.1-1. In both cases the force and displacement are translational. The extension to rotational cases is straightforward (cf. Problem 3.1-7). The transforms of force and displacement are replaced by those of the torque, $\bar{M}(s)$, and the angular displacement $\bar{\theta}(s)$. The translational mechanical relaxances and retardances become rotational mechanical relaxances and retardances. Thus, while inertance is the mass in translational motion, it is the moment of inertia in the rotational motion of an ideal rigid body. Translational frictance is the mechanical (viscous or frictional) resistance of an ideal dashpot. Rotational frictance is the moment of mechanical resistance. Elastance is the stiffness of an ideal spring. Rotational elastance, again, is the moment of stiffness (or torsional stiffness). The reciprocal of the elastance, E, is the *mechanical compliance*, S. The name *glidance* (and the symbol Φ) will be used for the reciprocal of the frictance, F. No name or symbol will be needed for the reciprocal inertance. Mechanical model diagrams are useful for the derivation of the equations of motion of rheological apparatus. This is illustrated in Sect. 3.8 (see Problems 3.1-3 to 3.1-7).

In the mechanical model diagram of rheological apparatus the respondance of the specimen must be shown as a *mechanical* respondance. This can usually be done through the use of a *shape* or *form factor*, H, which converts force/displacement or torque/angular displacement into stress/strain. We have

$$\sigma/\varepsilon = H_t(f/x) = H_r(M/\theta) \ . \tag{3.1-17}$$

H_t has the dimension of a reciprocal length, H_r that of a reciprocal volume. The subscript will be omitted when the meaning is clear. The reciprocal of the shape factor is variously called the *specimen*, or *apparatus*, or *cell constant* or *coefficient*. It will be denoted by b. With the use of H or b the operational and harmonic mechanical relaxances may be represented by b $\bar{Q}(s)$ or b $G^*(\omega)$, respectively. The mechanical compliances become H $\bar{U}(s)$ and H $J^*(\omega)$. In the model diagram such respondances will be shown as exemplified below.

$$\boxed{b\, G^*(\omega)}$$

In the analysis of mechanical models, just as in the analysis of electrical networks, a distinction must be made between systems whose behavior can be formulated in terms of ordinary differential equations and systems that require partial differential equations for their description. The former are customarily referred to as systems with *lumped parameters*, the latter as systems with *distributed parameters*. The analysis of systems with distributed parameters will be discussed in Sect. 5.6 using the concept of a *material transmission line*.

3.2 Electromechanical Analogies

In the preceding section we had occasion several times to point to analogies between electrical networks and mechanical models. In fact, as comparison of Eqs. (3.1-7) and (3.1-15) shows, although Figs. 3.1-3 and 3.1-6 describe two different physical systems, one an electrical, the other a mechanical system, the differential (or the transform) equations have the same form. By equating voltage with force, current with velocity, inductance with inertance, resistance with frictance, and capacitance with compliance (reciprocal elastance), one system can be described in terms of the physical variables of the other. Because of the ease with which electrical networks can be built and analyzed, and because of the advanced state of electric circuit analysis, other physical systems such as mechanical, acoustic, hydraulic or other systems, are often represented and analyzed in terms of their electrical analogs [1, 5, 6, 11]. In particular, electromechanical (or more precisely, electrorheological) analogies have been used to analyze [7, 8, 9], and even to simulate [10] viscoelastic behavior. In this section, therefore, we briefly discuss electromechanical analogies. These are easily extended to electrorheological analogies.

We deal with these analogies mainly because of the insight they provide into the general nature of linear physical systems. Mechanical model analysis makes it unnecessary to treat mechanical or rheological problems in terms of electric analogs which, in most instances, complicate rather than help the analysis. This not

Table 3.2-1. Electromechanical analogies

Electrical concept	Mechanical concept		Electrical concept
Electrostatic analogy	Translational motion	Rotational motion	Electromagnetic analogy
Flux $\bar{\varphi}(s)$	Momentum $\bar{m}(s)$	Angular momentum $\bar{\mu}(s)$	Charge q(s)
Voltage $\bar{V}(s)$	Force $\bar{f}(s)$	Torque $\bar{M}(s)$	Current $\bar{I}(s)$
Current $\bar{I}(s)$	Velocity $\bar{v}(s)$	Angular velocity $\bar{\omega}(s)$	Voltage V(s)
Charge $\bar{q}(s)$	Displacement $\bar{x}(s)$	Angular displacement $\bar{\theta}(s)$	Flux $\varphi(s)$
Impedance $\bar{Z}(s)$	Tr. impedance $\bar{Z}_t(s)$	Rot. impedance $\bar{Z}_r(s)$	Admittance Y(s)
Inductance Ls	Tr. inertance $I_t s$	Rot. inertance $I_r s$	Capacitance Cs
Resistance R	Tr. frictance F_t	Rot. frictance F_r	Conductance G
Recipr. capacitance 1/Cs	Tr. elastance E_t/s	Rot. elastance E_r/s	Recipr. inductance 1/Ls
Defective *ES* analogy	Translational motion	Rotational motion	Defective *EM* analogy
Flux $\bar{\varphi}(s)$	Momentum $\bar{m}(s)$	Angular momentum $\bar{\mu}(s)$	Charge $\bar{q}(s)$
Voltage $\bar{V}(s)$	Force $\bar{f}(s)$	Torque $\bar{M}(s)$	Current $\bar{I}(s)$
Current $\bar{I}(s)$	Displacement $\bar{x}(s)$	Angular displacement $\bar{\theta}(s)$	Voltage $\bar{V}(s)$
Charge $\bar{q}(s)$	—	—	Flux $\bar{\varphi}(s)$
Impedance $\bar{Z}(s)$	Tr. relaxance $\bar{Q}_t(s)$	Rot. relaxance $\bar{Q}_r(s)$	Admittance $\bar{Y}(s)$
—	Tr. inertance $I_t s^2$	Rot. inertance $I_r s^2$	—
Inductance Ls	Tr. frictance $F_t s$	Rot. frictance $F_r s$	Capacitance Cs
Resistance R	Tr. elastance E_t	Rot. elastance E_r	Conductance G
Recipr. capacitance 1/Cs	—	—	Recipr. inductance 1/Ls

withstanding, familiarity with the methods of electric circuit analysis and a proper understanding of electromechanical analogies may suggest useful methods of solution in certain cases. Thus, we will use such an analogy to good advantage in Sect. 5.6, and in Problems 5.6-1 and 5.6-2.

The use of electromechanical analogies is encumbered by the fact that at least four different analogies exist and have been used by various authors. The correspondence between physical variables according to these analogies are assembled in Table 3.2-1.

The electrical quantities do not require comment. Operational mechanical (translational or rotational) impedance and relaxance are defined as the ratios of the force or torque transform to the (angular) velocity or displacement transform, respectively.

Two *complete* electromechanical analogies are possible, according to whether one equates force with voltage or with current. The former may be termed electrostatic analogy and the latter electromagnetic analogy [11], because force is proportional to voltage in electrostatic coupling whilst it is proportional to current in electromagnetic coupling. Two more analogies may be set up equating displacement instead of velocity with either current or voltage. These analogies may be called *defective* because, as can be seen from Table 3.2-1, they cannot handle inertances. In fact the defective electromagnetic analogy was proposed originally [7] for rheological models neglecting inertia. There has been some discussion in the earlier literature as to which of these analogies is the correct one. It must be emphasized that they are all correct. Depending on circumstances, or preferences, one or the other may be more convenient to use.

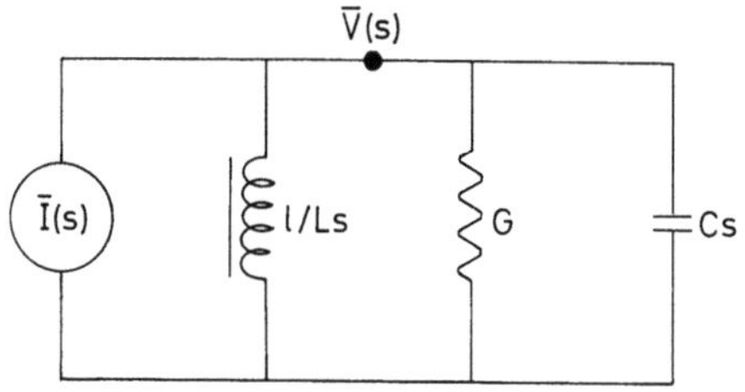

Fig. 3.2-1. Equivalent circuit of mechanical model shown in Fig. 3.1-6 (electromagnetic analogy)

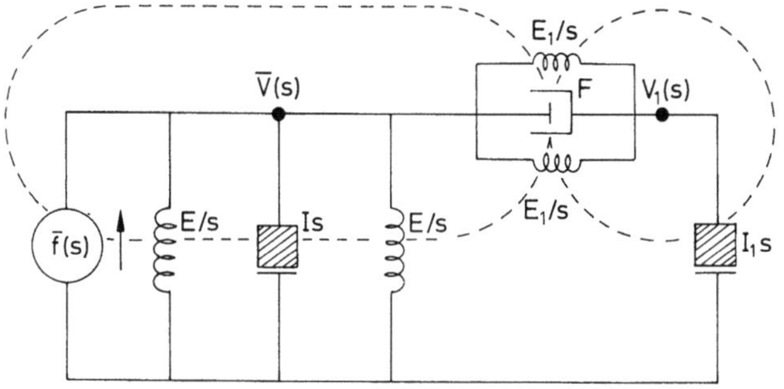

Fig. 3.2-2. Mechanical model diagram of damped vibration absorber

The electromagnetic analogies preserve the interconnection between the active and passive elements. In the electrostatic analogies the interconnection is reversed, i.e. parallel coupling in the mechanical model becomes series coupling in the equivalent electric circuit, and *vice versa*. This change in coupling is immediately apparent upon comparing Figs. 3.1-3 and 3.1-6. By contrast, Fig. 3.2-1 presents the circuit equivalent to the mechanical model shown in Fig. 3.1-6 on the basis of the electromagnetic analogy.

The equivalent circuit of Fig. 3.2-1, in which current is the active source, is analyzed most simply by *node analysis* [1, 3]. This yields at once

$$I(s) = (Cs - G + 1/Ls)\bar{V}(s) = \bar{Y}(s)\bar{V}(s) \tag{3.2-1}$$

where $\bar{Y}(s)$ is the operational admittance of the circuit. Equation (3.2-1) is, by the electromagnetic analogy, equivalent to Eq. (3.1-15).

For more complex models or networks graphical methods may be used to rewrite the coupling as required by the electrostatic analogy. Thus, Fig. 3.2-2 shows the mechanical model diagram of the damped vibration absorber discussed in Problem 3.1-1.

Figure 3.2-2 represents a two-node diagram. To derive the equivalent electric circuit on the basis of the electrostatic analogy [12] we draw circles, as shown by the dotted lines in Fig. 3.2-2, around each node shown, connecting all circuit elements attached to that node. Each circle represents a mesh of the equivalent electric circuit. Elements through which two circles pass are common to two meshes. The equivalent electric circuit can now be redrawn as shown in Fig. 3.2-3 making the appropriate substitutions according to Table 3.2-1.

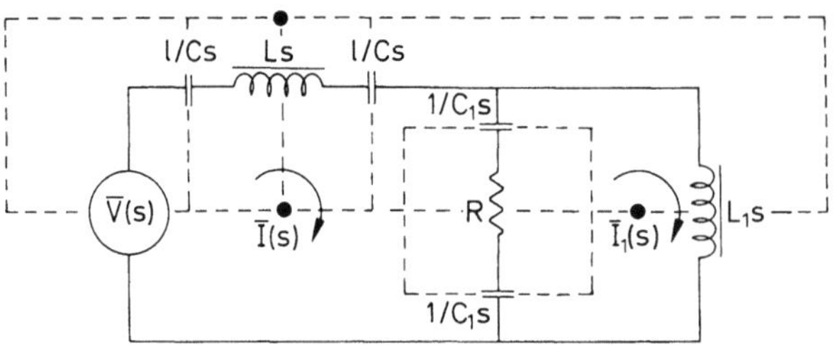

Fig. 3.2-3. Equivalent circuit of mechanical model shown in Fig. 3.2-2 (electrostatic analogy)

In the reverse procedure [1], finding the mechanical model diagram equivalent to a given electric circuit, a dot is placed inside each mesh, another is placed outside the circuit, and the dots are connected by lines passing through the circuit elements as shown in Fig. 3.3-3. The dots represent the nodes of the dual model diagram, the outside dot being the reference or ground node. The diagram is then redrawn with the appropriate substitutions for the active and passive elements.

3.3 The Elementary Rheological Models

We now begin to discuss the application of mechanical models to the representation of viscoelastic behavior. Poynting and Thomson [13] in 1902 first used a mechanical model in an attempt to describe the *creep*, then referred to as *elastic after-effect*, of metals. As we shall see the models are a convenient graphical form of the representation of the differential equations which govern viscoelastic behavior. They fulfill a useful didactic role by interpreting viscoelastic behavior in easily grasped mechanical terms. It must be emphasized at the outset, however, that the representation of viscoelastic behavior by springs and dashpots must not be taken to imply that these mechanical elements in any manner reflect molecular or supramolecular mechanisms in the material whose behavior they model. In fact, we shall show that the observed behavior can generally be represented by a multiplicity of models — all equivalent. If this point is understood, considerable insight into the phenomenology of viscoelastic behavior can be gained through the visualization which the models afford.

The application of mechanical models to the representation of viscoelastic behavior requires certain changes [14] in the formalism introduced in Sect. 3.1. Thus, although we keep the graphical representation unchanged, in the rheological models we replace the translational mechanical concepts of force, displacement (or velocity), elastance and frictance, or their rotational analogs by the rheological concepts stress, strain (or rate of strain), modulus, and viscosity. Formally, one may define *inertivity* as the proportionality constant between stress and acceleration of strain just as viscosity is the proportionality constant between stress and rate of strain, and as modulus is that between stress and strain. While, however, viscosity

Table 3.3-1. Dimensions of some rheological and mechanical quantities

Rheological quantity	Dimension	Translational mechanical quantity	Dimension	Rotational mechanical quantity	Dimension
Stress	$ml^{-1}t^{-2}$	Force	mlt^{-2}	Torque	ml^2t^{-2}
Strain	—	Displacement	l	Ang. Displacement	—
Rate of strain	t^{-1}	Velocity	lt^{-1}	Ang. velocity	t^{-1}
Modulus	$ml^{-1}t^{-2}$	Tr. elastance	mt^{-2}	Rot. elastance	ml^2t^{-2}
Viscosity	$ml^{-1}t^{-1}$	Tr. frictance	mt^{-1}	Rot. frictance	ml^2t^{-1}
(Inertivity	ml^{-1})	Tr. inertance	m	Rot. inertance	ml^2
Energy*	$ml^{-1}t^{-2}$	Energy	ml^2t^{-2}	Energy	ml^2t^{-2}
Power*	$ml^{-1}t^{-3}$	Power	ml^2t^{-3}	Power	ml^2t^{-3}

* per unit volume

and modulus are material parameters (i.e. they are independent of the size and shape of a piece of material,) an inertivity as just defined would not, and, therefore, would not qualify as a rheological concept. To see this, consider the elongation of a rod of length L_0 and cross-sectional area A_0 as shown in Fig. 1.1-1. In translational motion the inertance is the mass of the rod and the shape factor, H_t, is L_0/A_0 [cf. Eq. (1.1-1)]. Thus, the inertivity would be $H_t m = mL_0/A_0 = \rho L_0^2$ where ρ is the density. Since the latter is a material parameter, the former is not because it depends on the squared length of the rod.

As its name implies, the theory of linear viscoelastic behavior is not concerned with inertial effects. It is useful because such effects are demonstrably negligible in the response of a material to a large class of excitations. Under circumstances when the inertia cannot be neglected, i.e. at short loading times (as in impact testing) or high frequencies (as in wave propagation), the theory requires an extension in which the density is introduced as the material parameter related to the storage of kinetic energy. This theory will be called the theory of the *material transmission line* and will be discussed in Sect. 5.6. Almost everywhere else in this book we shall tacitly assume that the inertia of the material is negligible and in this chapter we therefore consider exclusively models in which springs and dashpots are the only passive elements.

For convenience, and to further clarify the difference between mechanical and rheological concepts, Table 3.3-1 lists the dimensions of the most important rheological and mechanical quantities. Mass is denoted by m, length by l, and time by t.

3.3.1 The Spring and Dashpot Elements

Purely elastic behavior can be represented as shown in Fig. 3.3-1a. Purely viscous behavior is represented by the model shown in Fig. 3.3-1b. These two models contain simple elements: one, a spring, the other a dashpot. By the rules expounded in Sect. 3.1 we have

$$\bar{\sigma}(s) = G\,\bar{\varepsilon}(s) \tag{3.3-1}$$

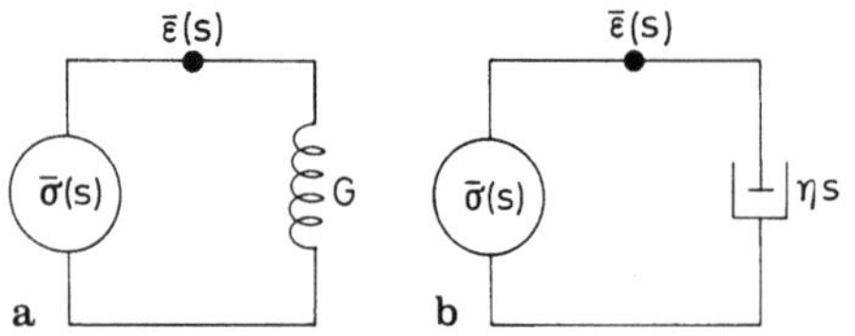

Fig. 3.3-1a,b. Model representation of purely elastic and purely viscous behavior

Table 3.3-2. Elementary respondances

Model parameter	Operational respondance	Harmonic respondance
G	G	G
J	J	J
η	ηs	$j\omega\eta$
ϕ	ϕ/s	$\phi/j\omega$

for the first, and

$$\bar{\sigma}(s) = \eta\, s\bar{\varepsilon}(s) = \eta\, \bar{\dot{\varepsilon}}(s) \tag{3.3-2}$$

for the second model. The two models are graphical representations of Eqs. (2.1-1) and (2.1-2). As before, we shall always assume initially quiescent systems unless stated otherwise. Comparison with Eq. $(2.1\text{-}15)_1$ shows that in the first case the relaxance, $\bar{Q}(s)$, is simply G, the (shear) modulus. Similarly, in the second case $\bar{Q}(s)$ is ηs. G and ηs are the *operational relaxances* for purely elastic and purely viscous behavior, respectively. Analogously, J and ϕ/s are termed the *operational retardances*. We call G and η the parameters of the models. These quantities are listed in Table 3.3-2 which also shows the *harmonic respondances* appropriate for the sinusoidal steady-state. These are obtained from the operational respondances simply by replacing s by $j\omega$.

3.3.2 The Maxwell and Voigt Units

We now turn to a more complex model. The model which represents the behavior described by Eq. (2.1-3) is shown in Fig. 3.3-2.

The mechanical model diagram of Fig. 3.3-2 represents a parallel combination of a spring and a dashpot. In a parallel combination all elements see the same strain and the stresses in the elements are additive. In a series combination the strains are

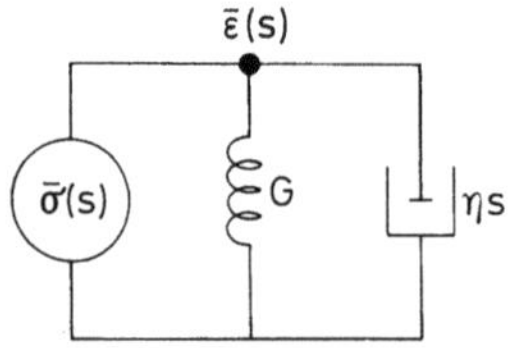

Fig. 3.3-2. The Voigt unit

additive and the same stress acts through all of the elements. Thus, although we now consider stresses and strains instead of forces (or torques) and displacements (or angular displacements), the *combination rules* that relaxances add in parallel and retardances add in series, remain valid. Thus we obtain

$$\bar{Q}(s) = G + \eta s \tag{3.3-3}$$

and, hence,

$$\bar{\sigma}(s) = (G + \eta s)\bar{\varepsilon}(s) \tag{3.3-4}$$

which is the Laplace transform of Eq. (2.1-3). Although that equation was originally proposed [15] by O. Meyer in 1874 (cf. Sect. 2.1), it was reintroduced [16] in 1892 by W. Voigt*. The parallel combination of a spring and dashpot is therefore commonly referred to as a Voigt model. We shall show that this model does not represent viscoelastic behavior adequately. It is, however, an important building block in other, more complex models. We shall therefore refer to it as the *Voigt unit*. It usually occurs in series with other elements or Voigt units. Since retardances add in series, we note that the retardance of the Voigt unit is

$$\bar{U}_V(s) = \frac{J}{1 + \tau_V s} \tag{3.3-5}$$

where τ_V is called the *retardation time*. It is given by

$$\tau_V = J/\phi \tag{3.3-6}$$

where $\phi = 1/\eta$. The reason for the choice of the name retardation time will become clear later.

The next simplest model is the series combination of a spring and a dashpot shown in Fig. 3.3-3. From the combination rules we obtain the retardance of this model as

$$\bar{U}(s) = J + \phi/s \tag{3.3-7}$$

where we have again used J for 1/G and ϕ for $1/\eta$. Hence,

$$\bar{\varepsilon}(s) = (J + \phi/s)\bar{\sigma}(s) \tag{3.3-8}$$

which leads to the differential equation

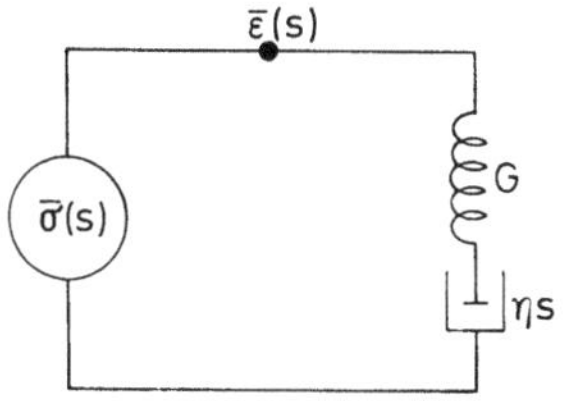

Fig. 3.3-3. The Maxwell unit

* Woldemar Voigt, pron. Fokt, 1850–1919, German physicist and crystallographer.

$$\frac{d\varepsilon}{dt} = \frac{1}{\eta}\sigma + \frac{1}{G}\frac{d\sigma}{dt} . \qquad (3.3\text{-}9)$$

This equation was proposed [17] for the representation of viscoelastic behavior by Maxwell* in 1876 as a result of his theoretical study of the behavior of gases. Like the Voigt unit, it does not, by itself, satisfactorily represent viscoelastic behavior, but is again an important building block in more complex models. We shall refer to it as the *Maxwell unit*. It occurs mostly in parallel combination. Introducing the *relaxation time*

$$\tau_M = \eta/G \qquad (3.3\text{-}10)$$

the relaxance of the Maxwell unit may be written as

$$\bar{Q}_M(s) = \frac{G\tau_M s}{1 + \tau_M s} . \qquad (3.3\text{-}11)$$

Again the choice of the term relaxation time will be made clear below. Making use of the relaxation and retardation time, we obtain the operational retardance of the Maxwell unit as

$$\bar{U}_M(s) = J + \phi/s = J(1 + 1/\tau_M s) \qquad (3.3\text{-}12)$$

and the operational relaxance of the Voigt unit as

$$\bar{Q}_V(s) = G + \eta s = G(1 + \tau_V s) . \qquad (3.3\text{-}13)$$

It will be seen later, however, that when one deals with more complex models it is almost always the relaxance of the Maxwell unit and the retardance of the Voigt unit which is actually needed.

The two combinations: series and parallel (or Maxwell and Voigt) give the basic building blocks out of which more complex models can be constructed. In themselves, neither the Maxwell nor the Voigt unit can represent viscoelastic behavior because they do not respond in the required manner to *both* stress and strain excitations. A viscoelastic material shows both *retardation of strain* (i.e. *creep*), and *relaxation of stress* depending on the excitation. The two phenomena are simply dual expressions of the fact that the molecular rearrangements are time dependent. The first is exhibited when the stress is regarded as the excitation. The second is shown when the strain is taken to be the excitation.

The Voigt unit is capable of expressing strain retardation but not stress relaxation. The reverse is true for the Maxwell unit. To see this more clearly let us examine the response of these two basic building blocks to an excitation either in the form of a stress or a strain. We first turn to the Maxwell unit. Since, by Eq. (2.3-3), the response to a step of strain is given by

$$\bar{\sigma}(s) = \bar{Q}(s)\varepsilon_0/s , \qquad (3.3\text{-}14)$$

we obtain, by substituting Eq. (3.3-11) into Eq. (3.3-14) and inverting the transform,

---

* James Clark Maxwell, British physicist, 1831–1879, particularly noted for his great contributions to electromagnetic theory.

$$\sigma(t) = \varepsilon_0 G \exp(-t/\tau_M) \tag{3.3-15}$$

for the stress in a Maxwell unit upon imposition of a constant strain of magnitude ε_0. But $G\varepsilon_0 = \sigma_0$, the *initial* value of the stress at $t = 0$. Therefore,

$$\sigma(t) = \sigma_0 \exp(-t/\tau_M) \ . \tag{3.3-16}$$

This shows that the stress induced in a Maxwell unit upon application of a constant strain relaxes with time until it reaches zero. The choice of the name relaxation time and relaxance is now clear. In a purely elastic material (for which a spring would be the model) the stress would stay constant. We would, therefore, simply have $\sigma(t) = \sigma_0$. This is equivalent to saying that the relaxation time is infinite, i.e. the stress would take an infinite time to relax. Conversely, in a purely viscous material (for which a dashpot would be the model) the stress would relax instantaneously. We would, therefore, simply have $\sigma(t) = 0$. This is equivalent to saying that the relaxation time is zero (see also Problem 9.1-2).

Equation (3.3-16) also serves to give a more precise physical interpretation to the relaxation time. For $t = \tau_M$

$$\sigma(\tau_M) = \sigma_0/e = 0.369\sigma_0 \ . \tag{3.3-17}$$

The relaxation time, therefore, is that time for which the stress has fallen to $1/e$ of its initial value. This is shown graphically in Fig. 3.3-4. The strain across each of the elements of the Maxwell unit in response to a step, slope, or harmonic excitation depends on the relaxation time. This is discussed in Problems 9.1-2, 9.1-5, and 9.1-10.

Turning now to the Voigt unit, the strain transform in response to a step of stress is

$$\bar{\varepsilon}(s) = \bar{U}(s)\sigma_0/s \tag{3.3-18}$$

and retransformation after substitution of Eq. (3.5-5) gives

$$\varepsilon(t) = \sigma_0 J[1 - \exp(-t/\tau_V)] \tag{3.3-19}$$

for the strain as a function of time when a constant stress of magnitude σ_0 is imposed. But $J\sigma_0 = \varepsilon_\infty$, the *final* value of the strain at $t = \infty$, and hence

$$\varepsilon(t) = \varepsilon_\infty[1 - \exp(-t/\tau_V)] \ . \tag{3.3-20}$$

This shows that the strain developed in a Voigt unit upon application of a constant stress reaches its final value only at infinite time. The strain is thus retarded

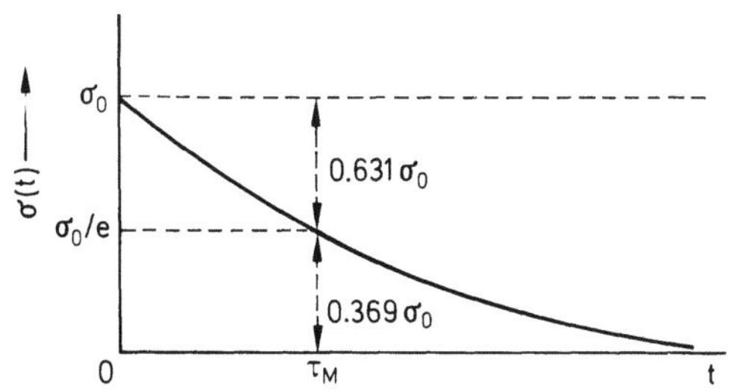

Fig. 3.3-4. Relaxation time of Maxwell unit. Stress response to a step strain

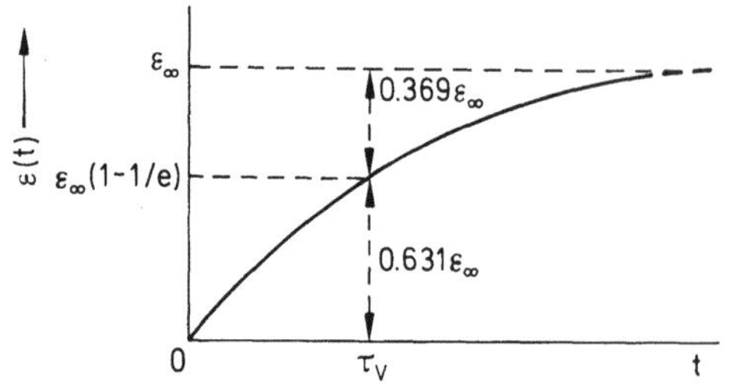

Fig. 3.3-5. Retardation time of Voigt unit. Strain response to a step stress

and the choice of the term retardation time and retardance becomes clear. In a purely elastic material, we would, of course, have $\varepsilon(t) = \varepsilon_\infty$, i.e., the strain would reach its final value immediately. This is equivalent to saying that the retardation time in an elastic material is zero. Conversely, in a purely viscous material no strain could develop because the stress would be dissipated instantaneously. Formally, one can say that the retardation time in a purely viscoelastic material is infinite (see also Problem 9.2-1).

Looking now at the definition of the retardation time we see that for $t = \tau_V$

$$\varepsilon(\tau_V) = \varepsilon_\infty(1 - 1/e) = 0.631\varepsilon_\infty \qquad (3.3\text{-}21)$$

and recognize the retardation time as that time for which the strain has risen to a point where it is within 1/e of its final value. This is shown in Fig. 3.3-5.

The dependence on the retardation time of the stress acting through each of the elements of a Voigt unit is discussed in Problems 9.2-1, 9.2-2, and 9.2-6.

The two *respondance times*, the relaxation time and the retardation time, differ in their physical meaning. Relaxation times arise when the stress is a response to a strain excitation. Conversely, retardation times arise when the strain is a response to a stress excitation. In the theory of linear viscoelasticity it is not customary to use different symbols for relaxation and retardation time and we too will henceforth drop the subscripts which have hitherto been used to distinguish them. When necessary, we shall write τ_M for the relaxation, and τ_V for the retardation time.

We have seen the response of the Voigt unit to a step input of stress and that of the Maxwell unit to a step input of strain. If these two units were "dual" in their responses, the Voigt unit would respond to a step input of strain as the Maxwell unit does, and conversely, the Maxwell unit would respond to a step input of strain in the manner of the Voigt unit. In fact, they do not. That this is so can be seen at once by considering the mechanical behavior of the spring and dashpot combinations represented by the units. One of the reasons for the usefulness of rheological models is the physical intuition into viscoelastic processes that one can derive from consideration of their physical behavior.

If a series combination of a spring and a dashpot is suddenly extended to a fixed length, the stress set up in the spring will gradually relax and fade away as the piston of the dashpot overcomes the resistance of the damping fluid. This was shown in Fig. 3.3-4. If, instead, a constant force is applied there will be an immediate extension of the spring, but then the piston will keep moving at a constant rate

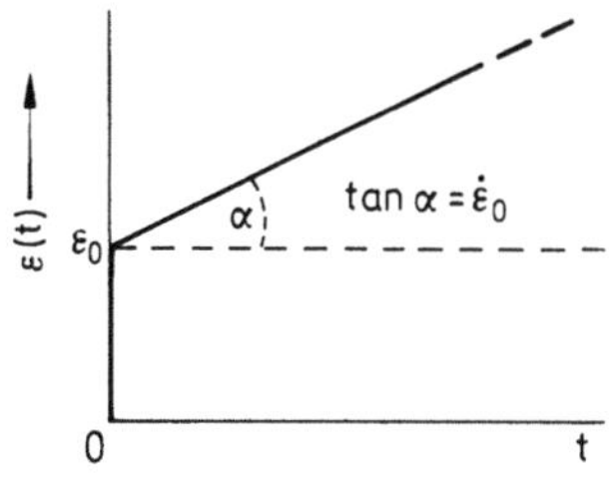

Fig. 3.3-6. Response of a Maxwell unit to a step stress

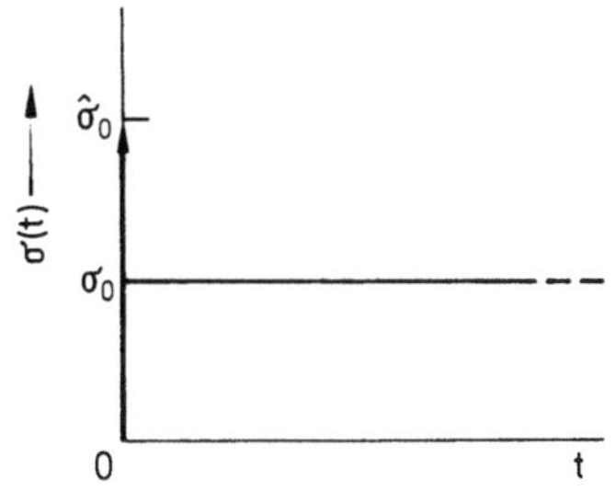

Fig. 3.3-7. Response of a Voigt unit to a step strain

(if the damping fluid is newtonian) and there can be no retardation of strain. This behavior is shown in Fig. 3.3-6 and is seen to be quite different from that shown in Fig. 3.3-5.

Analogously, if a constant force is applied to a parallel combination of a spring and a dashpot, the dashpot prevents the spring from extending immediately to the length it would take up under the influence of the force in the absence of the dashpot. The strain is thus retarded as was shown in Fig. 3.3-5. Extension to a fixed length at $t = 0$ is, however, simply not possible because the resistance of the dashpot cannot be overcome in an instant as would be required for the response to a step function of time. This is shown in Fig. 3.3-7 which is clearly quite different from Fig. 3.3-4. The arrow at $t = 0$ represents the impulse of strength $\hat{\sigma}_0$ which would be needed to bring the parallel spring-dashpot combination to the required elongation instantaneously.

To understand these relations more clearly, let us examine them mathematically. The retardance of the Maxwell model is given by Eq. (3.3-12). Substituting this into Eq. $(2.1\text{-}15)_2$ with $\bar{\sigma}_0(s) = \sigma_0/s$ gives

$$\bar{\varepsilon}(s) = \sigma_0(J/s + \phi/s^2) \tag{3.3-22}$$

or, inverting the transforms,

$$\varepsilon(t) = \sigma_0(J + t/\eta) = \varepsilon_0 + \dot{\varepsilon}_0 t \ . \tag{3.3-23}$$

At $t = 0$ there is an instantaneous strain $\varepsilon_0 = J\sigma_0$, indicating the sudden elongation of the spring. From then on the dashpot gives way linearly at the (constant) rate of strain $\dot{\varepsilon}_0 = \sigma_0/\eta$. Equation (3.3-23) may be compared with Eq. (3.3-20) with which it would have to be formally identical if the responses of the two units were, in fact, dual. The behavior is shown in Fig. 3.3-6.

The relaxance of the Voigt unit is given by Eq. (3.3-13). Inserting Eq. $(3.3\text{-}13)_1$ into Eq. $(2.1\text{-}15)_1$ with $\bar{\varepsilon}(s) = \varepsilon_0/s$ gives

$$\bar{\sigma}(s) = \varepsilon_0(\eta + G/s) \tag{3.3-24}$$

and inversion of the transforms yields

$$\sigma(t) = \varepsilon_0[\eta\ \delta(t) + G] = \hat{\sigma}_0\delta(t) + \sigma_0 \ . \tag{3.3-25}$$

The first term indicates that a stress impulse of strength $\hat{\sigma}_0 = \eta\varepsilon_0$ is required to produce a strain ε_0 in the combination at $t = 0$, in order to overcome the resistance of the dashpot which cannot elongate instantaneously as a spring can. When this strain is reached, the stress remains constant at $\sigma_0 = G\varepsilon_0$. Equation (3.3-25) would have to be formally identical with Eq. (3.3-16) if the responses were dual. Equations $(3.3\text{-}23)_2$ and $(3.3\text{-}25)_2$ clearly show that the Maxwell unit does not represent strain retardation and that the Voigt unit does not represent stress relaxation. Thus these units model viscoelastic behavior only in response to some stimuli, but not to the excitations which are the duals of those stimuli (cf. Problems 3.3-1).

3.4 Models with the Minimum Number of Elements

It has become clear in the foregoing discussion that the Voigt and Maxwell units cannot represent viscoelastic behavior adequately. It is for this reason that we call them units rather than models. In this section we shall show that an adequate model of *solid-like* linear viscoelastic behavior requires a minimum of three elements (two springs and one dashpot) while *liquid-like* behavior must be modeled by no less than four elements (two springs and two dashpots). Addition of a spring, a dashpot, or another Voigt unit in *parallel* to a Voigt unit, or a spring, a dashpot, or another Maxwell unit in *series* to a Maxwell unit, does not change the nature of the added-to unit because two elements of the same kind in parallel or in series with each other can always be replaced by a single element of the same kind but different numerical value. Hence, we always add to a Voigt unit in series and to a Maxwell unit in parallel. Since retardances add in series and relaxances add in parallel, this is the reason why we have chosen to represent the Voigt unit by retardances and the Maxwell unit by relaxances.

In this section we examine those rheological models which adequately reproduce the essential features of rheological behavior but do this with the minimum number of elements required by theory.

3.4.1 The Standard Three- and Four-Parameter Series-Parallel Models

The 3-parameter models obtained by adding a spring in series with a Voigt unit or in parallel with a Maxwell unit will be called *standard series-parallel models.* They are designated as standard models because, as will be shown later, they respond to an excitation as a step function of time in a physically realistic way. The *non-standard* 3-parameter models obtained by adding a dashpot in series with a Voigt unit or in parallel with a Maxwell unit, do not.

Three-parameter models are always series-parallel models. Two springs and two dashpots, however, can be combined into series-parallel as well as other models. In addition, here too, we must distinguish between standard and non-standard models.

Whenever a Maxwell unit is in parallel with others or with an isolated spring or dashpot, the relaxance of the model can be written down immediately because relaxances add in parallel. Models constructed in this way will be called *Maxwell models*. In such models the elements will be represented as relaxances. When a Voigt unit is in series with others or with an isolated spring or dashpot, the models will be called *Voigt models*. In such models it is the retardance which can be obtained at once. In Voigt models the elements will therefore be represented as retardances. This convention is purely a matter of convenience. It helps avoid confusion and generally simplifies the writing of the equations. Models containing a large number of either Maxwell or Voigt units will be discussed in Sect. 3.5. In this section we shall consider the subset consisting of models which contain the minimum number of elements required to produce viscoelastic response whether the excitation is a strain *or* a stress.

3.4.1.1 *The Models of the Standard Linear Solid*

Consider the model shown in Fig. 3.4-1. This is obtained by adding a spring in series with a Voigt unit and will be called the *standard 3-parameter Voigt model*. Using the combination rules in conjunction with Eq. (3.3-5), its retardance is readily obtained as

$$\bar{U}_{3V}(s) = J_g + \frac{J}{1 + \tau s} \tag{3.4-1}$$

where $\tau = J/\phi$ is the retardation time of the model. For a step input of stress, then,

$$\bar{\varepsilon}(s) = \sigma_0\left(\frac{J_g}{s} + \frac{J}{s(1 + \tau s)}\right) \tag{3.4-2}$$

or, upon inversion,

$$\varepsilon(t) = \sigma_0\{J_g + J[1 - \exp(-t/\tau)]\} = \varepsilon_0 + \varepsilon[1 - \exp(-t/\tau)] \ . \tag{3.4-3}$$

The response of this model to a constant stress is thus a sudden strain of magnitude $\varepsilon_0 = J_g\sigma_0$ followed by a retarded or delayed strain which reaches an equilibrium strain $\varepsilon_e = \varepsilon_0 + \varepsilon = J_e\sigma_0$ at $t = \infty$ as shown in Fig. 3.4-2. Such behavior is commonly referred to as creep. Hence, the name *creep compliance* for $\varepsilon(t)/\sigma_0 = J(t)$. J_g is the instantaneous compliance because it is the compliance at $t = 0$. J_e is the *equilibrium compliance* given by

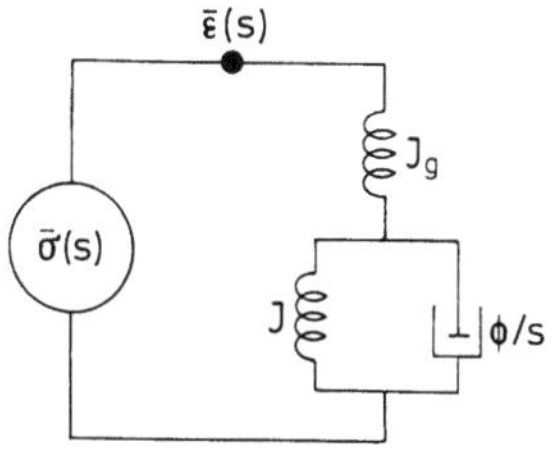

Fig. 3.4-1. Standard 3-parameter Voigt model

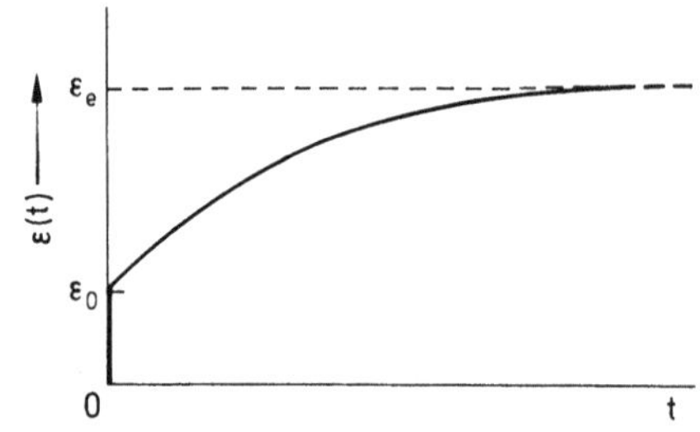

Fig. 3.4-2. Response of the standard 3-parameter models to a step stress

$$J_e = J_g + J \ . \tag{3.4-4}$$

J_e is called the equilibrium compliance because it is the compliance when equilibrium is reached, i.e. at $t = \infty$. J is called the *retarded* or *delayed compliance.*

We now investigate the response of the model to a step input of strain. The relaxance of the model, by the relation $\bar{Q}(s)\bar{U}(s) = 1$, and using Eq. (3.4-4), becomes

$$\bar{Q}_{3V}(s) = \frac{1 + \tau s}{J_e + J_g \tau s} \ . \tag{3.4-5}$$

Hence, for a constant strain, the stress transform will be

$$\bar{\sigma}(s) = \varepsilon_0 \frac{1 + \tau s}{s(J_e + J_g \tau s)} \ . \tag{3.4-6}$$

By partial fraction decomposition

$$\bar{\sigma}(s) = \varepsilon_0 \left(\frac{1}{J_e s} + \frac{J_e - J_g}{J_e J_g} \frac{\tau'}{1 + \tau' s} \right) \tag{3.4-7}$$

where

$$\tau' = J_g \tau / J_e \ . \tag{3.4-8}$$

Inversion of the transform yields

$$\sigma(t) = \varepsilon_0 \left(\frac{1}{J_e} + \frac{J_e - J_g}{J_e J_g} \exp(-t/\tau') \right) = \sigma_e + \sigma \exp(-t/\tau') \ . \tag{3.4-9}$$

The model therefore responds to a unit step of strain with an initial stress

$$\sigma(0) = \sigma_0 = \sigma_e + \sigma = \varepsilon_0 \left(\frac{1}{J_e} + \frac{J_e - J_g}{J_e J_g} \right) = \varepsilon_0 / J_g \tag{3.4-10}$$

at $t = 0$. It then relaxes to an equilibrium value $\sigma_e = \varepsilon_0 / J_e$. The behavior is shown in Fig. 3.4-3.

Let us now turn to the model shown in Fig. 3.4-4. It is obtained by adding a spring in parallel to a Maxwell unit and will be called the *standard* 3-*parameter Maxwell model.* We obtain its relaxance as

$$\bar{Q}_{3M}(s) = G_e + \frac{G \tau s}{1 + \tau s} \tag{3.4-11}$$

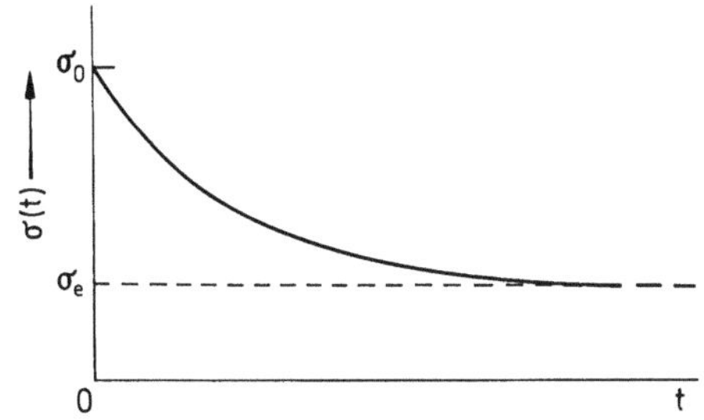

Fig. 3.4-3. Response of the standard 3-parameter models to a step strain

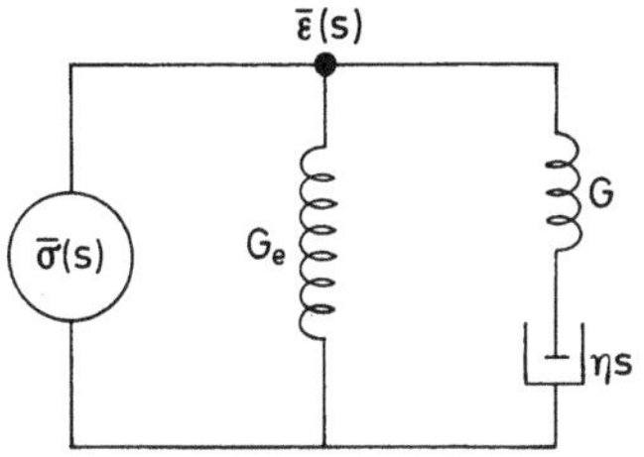

Fig. 3.4-4. Standard 3-parameter Maxwell model

by the rule that relaxances add in parallel [see Eq. (3.3-11)]. For a step strain of height ε_0 we then find

$$\bar{\sigma}(s) = \varepsilon_0\left(\frac{G_e}{s} + \frac{G\tau}{1+\tau s}\right) \tag{3.4-12}$$

or, inverting

$$\sigma(t) = \varepsilon_0[G_e + G\exp(-t/\tau)] = \sigma_e + \sigma\exp(-t/\tau)\ . \tag{3.4-13}$$

Thus at $t = 0$ the stress is $\sigma_0 = \sigma_e + \sigma = G_g\varepsilon_0$, where

$$G_g = G_e + G \tag{3.4-14}$$

is the instantaneous modulus, G_e is the *equilibrium modulus*, and G may be called the *relaxing modulus*. At $t = \infty$ we have $\sigma_e = G_e\varepsilon_0$ which shows that it is indeed the isolated spring (to which we had assigned the modulus G_e) which represents the equilibrium modulus. The behavior is thus identical with that shown in Fig. 3.4-3 for the standard 3-parameter Voigt model. It represents relaxation of stress and accounts for the name *relaxation modulus* for $\sigma(t)/\varepsilon_0 = G(t)$. The equivalence of the response of the two models to a step input of strain finds expression, of course, in the formal identity of Eqs. (3.4-7) and (3.4-12), or (3.4-9) and (3.4-13). We note that, with the appropriate choice of the parameters, the behavior of the two models is not only qualitatively, but even quantitatively identical (see Sect. 3.7).

We now show that the standard 3-parameter Maxwell model responds to a step of stress in the same way as the standard 3-parameter Voigt model. The retardance of the model shown in Fig. 3.4-4 is

$$\bar{U}_{3M}(s) = \frac{1+\tau s}{G_e + G_g\tau s} \tag{3.4-15}$$

where we have used Eq. (3.4-14). For a step stress of height σ_0 then,

$$\bar{\varepsilon}(s) = \sigma_0\frac{1+\tau s}{s(G_e + G_g\tau s)} \tag{3.4-16}$$

which can be rearranged to

$$\bar{\varepsilon}(s) = \sigma_0\left(\frac{1}{G_g s} + \frac{G_g - G_e}{G_g G_e}\frac{1}{s(1 + \tau' s)}\right) \tag{3.4-17}$$

where

$$\tau' = G_g \tau / G_e \ . \tag{3.4-18}$$

Inversion of the transform then yields

$$\varepsilon(t) = \sigma_0\left(\frac{1}{G_g} + \frac{G_g - G_e}{G_g G_e}[1 - \exp(-t/\tau')]\right) = \varepsilon_0 + \varepsilon[1 - \exp(-t/\tau')] \ . \tag{3.4-19}$$

The standard 3-parameter Maxwell model therefore responds to a step input of stress with an initial strain $\varepsilon_0 = \sigma_0/G_g$ at $t = 0$. The strain then gradually increases to an equilibrium value

$$\varepsilon(\infty) = \varepsilon_e = \sigma_0\left(\frac{1}{G_g} + \frac{G_g - G_e}{G_g G_e}\right) = \sigma_0/G_e \tag{3.4-20}$$

at $t = \infty$. The behavior is thus identical with that shown in Fig. 3.4-2. With the proper choice of parameters both models again represent quantitatively identical behavior. This is clear, of course, from the formal identity of Eqs. (3.4-2) and (3.4-17), or (3.4-3) and (3.4-19).

Pairs of Maxwell and Voigt models which, by the proper choice of the numerical values of their parameters, respond identically to the same excitation, are called *conjugate models*. Which of two conjugate models is chosen is a matter of convenience. Because Voigt models are "retardance models", they are generally easier to use when stress is the excitation. By the same token, Maxwell models, being "relaxance models", are usually more convenient to use when the applied stimulus is a strain.

There are general rules for the construction of the conjugate model from a given primitive. These rules are known as *Alfrey's rules* [18] and may be stated as follows:

1. The number of elements of each kind (spring and dashpot) must be the same in the conjugate model.
2. A parallel combination of two elements of different kind is replaced by a series combination and vice versa.
3. The absence (presence) of an isolated element of one kind requires the presence (absence) of an isolated element of the other kind in the conjugate model.

Alfrey's rules are valid only for the series-parallel models. Indeed, the term *conjugate models* has meaning only within the set of these models.

Starting from the 3-parameter Voigt model shown in Fig. 3.4-1 we see that the conjugate model must consist of two springs and one dashpot. Of these, one spring and the dashpot must be in series, and the conjugate model must contain an isolated spring because of the absence of an isolated dashpot in the primitive model. These requirements fully determine the conjugate 3-parameter Maxwell model shown in Fig. 3.4-4. For the 3-parameter models this exercise appears to be trivial. Its usefulness will become apparent in conjunction with the 4-parameter models.

As shown by Figs. 3.4-2 and 3.4-3, the strain in the models represented by Figs. 3.4-1 and 3.4-4 reaches an equilibrium value under a constant stress, and the stress relaxes to an equilibrium value under a constant strain. These models therefore represent the behavior of materials which reach a fixed equilibrium value when sufficient time has elapsed after the imposition of an excitation. Such materials are exemplified by crosslinked polymers. Inasmuch as the absence of steady-state flow is typical of the behavior of a solid, the standard 3-parameter models are sometimes referred to as models of the *standard linear solid.* They will also be referred to as the *standard arrheodictic models* for reasons that will become clear in the next section.

3.4.1.2 The Models of the Standard Linear Liquid

In response to a step input of strain the stress in an uncrosslinked polymer, and in most metals, rocks, etc., relaxes to zero. The strain under a constant stress eventually increases linearly with time (as long as the behavior is linearly viscoelastic). Such materials are said to exhibit steady-state flow. We shall call such materials *rheodictic.** By contrast, a material which does not evince steady-state flow, will be called *arrheodictic.* To make the arrheodictic 3-parameter Voigt model display rheodictic behavior, it must be modified by the series addition of an isolated dashpot to which we assign the fluidity ϕ_f. The resulting model is represented by Fig. 3.4-5. Its conjugate Maxwell model, as derived by Alfrey's rules, is represented by Fig. 3.4-6. It differs from the model shown in Fig. 3.4-4 by having a dashpot in series with the isolated spring. The modulus of the former isolated spring, G_e, now becomes G_0, the added dashpot has viscosity η_0, and the parameters of the original Maxwell unit are distinguished by the subscript 1.

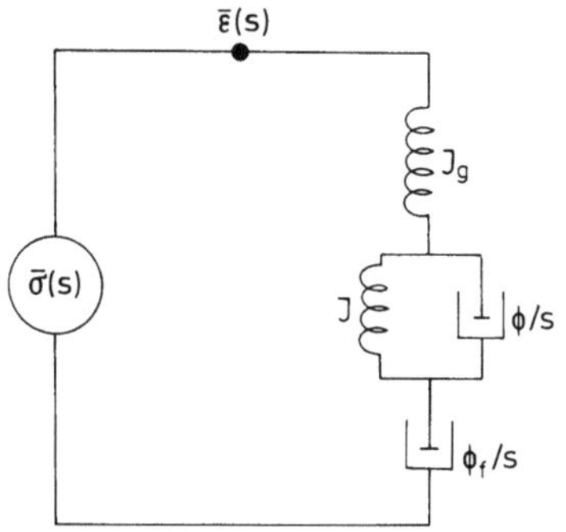

Fig. 3.4-5.
Standard 4-parameter Voigt model

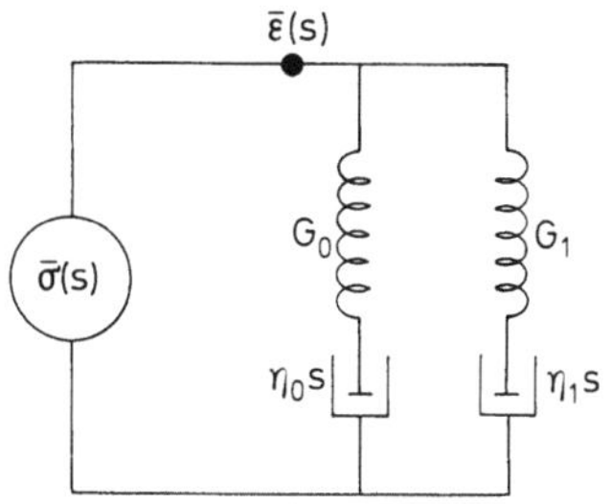

Fig. 3.4-6.
Standard 4-parameter Maxwell model

The retardance of the *standard 4-parameter Voigt* model (Fig. 3.4-5) is

$$\bar{U}_{4V}(s) = J_g + \frac{J}{1 + \tau s} + \frac{\phi_f}{s} \tag{3.4-21}$$

* From the Greek *rhéos*, a flow, and *deiktikós*, to be able to show.

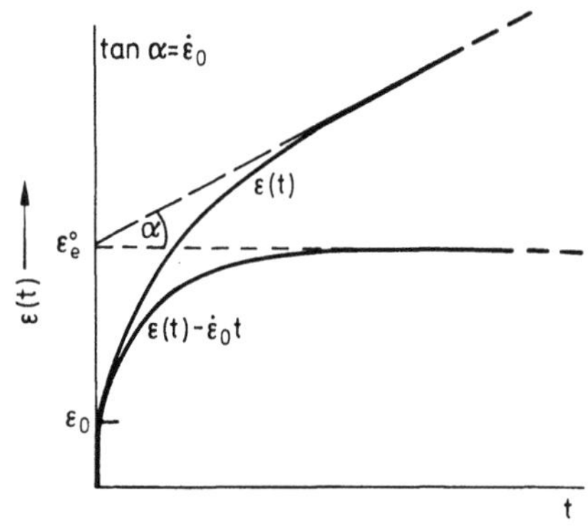

Fig. 3.4-7. Response of the standard 4-parameter models to a step stress

and we obtain, for a step stress of height σ_0, the strain as

$$\varepsilon(t) = \sigma_0\{J_g + J[1 - \exp(-t/\tau)] + \phi_f t\} \tag{3.4-22}$$

or

$$\varepsilon(t) = \varepsilon_0 + \varepsilon[1 - \exp(-t/\tau)] + \dot{\varepsilon}_0 t \ . \tag{3.4-23}$$

The presence of the isolated dashpot now allows the model to be strained in steady-state flow once the springs are fully extended. The behavior is shown in Fig. 3.4-7.

The response curve now consists of three distinct portions. As with the standard 3-parameter model, application of a step input of stress results in the sudden strain $\varepsilon_0 = J_g\sigma_0$ of the isolated spring of compliance J_g (the instantaneous compliance) at $t = 0$. This initial strain is followed by the delayed strain, $\varepsilon = J\sigma_0$, associated with the Voigt unit. In the standard 4-parameter model, however, the strain does not reach a finite value at long times but attains a steady state in which it becomes proportional to the time. This is the third portion of the response curve. The proportionality constant, i.e. the slope of the asymptote represented by the broken line in Fig. 3.4-7, is the (constant) rate of strain, $\dot{\varepsilon}_0 = \phi_f\sigma_0$. The parameter ϕ_f is the *steady-state* or *steady-flow fluidity*, and its reciprocal is the *steady-state* or *steady-flow viscosity*, η_f. The subscript f distinguishes these parameters as responsible for the steady-state flow.

A rheodictic material evidently cannot possess an equilibrium strain. It is clear, however, from Eq. (3.4-23) that at long times $\varepsilon(t) - \dot{\varepsilon}_0 t$ approaches a limiting value

$$\varepsilon_e^\circ = \lim_{t\to\infty} [\varepsilon(t) - \dot{\varepsilon}_0 t] = \varepsilon_0 + \varepsilon \tag{24}$$

which may be called the *pseudo-equilibrium*, or, preferably, the *steady-state strain*. It represents the strain corresponding to the maximum extension of the springs. We distinguish the steady-state strain by the superscript ° from the equilibrium strain ε_e. The course of $\varepsilon_0 + \varepsilon[1 - \exp(-t/\tau)]$ is shown by the lower curve in Fig. 3.4-7. The model thus possesses an instantaneous compliance $J_g = \varepsilon_0/\sigma_0$, and a *steady-state compliance*, $J_e^\circ = \varepsilon_e^\circ/\sigma_0$. It is sometimes useful to use the *steady-state modulus*, G_e°, defined as the reciprocal of the steady-state compliance.

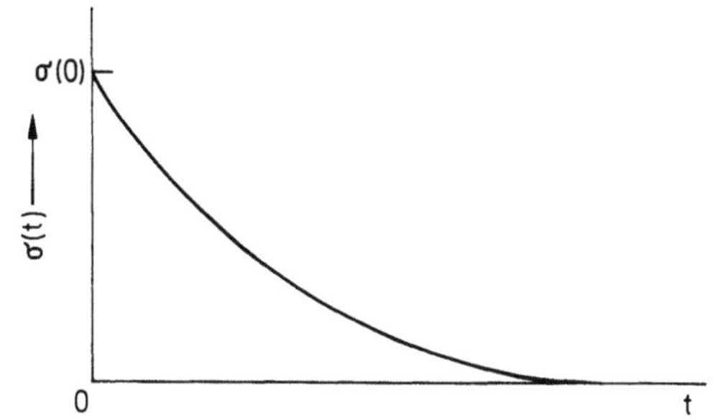

Fig. 3.4-8. Response of the standard 4-parameter models to a step strain

The separation of the response into a purely elastic, viscoelastic, and (if it is present) a purely viscous portion is characteristic of the *retardance*. The relaxance does not show the separation into additive terms.

The relaxance of the *standard 4-parameter Maxwell* model shown in Fig. 3.4-6 is

$$\bar{Q}_{4M}(s) = \frac{G_0\tau_0 s}{1+\tau_0 s} + \frac{G_1\tau_1 s}{1+\tau_1 s} \tag{3.4-25}$$

and we have, for a step strain of height ε_0,

$$\sigma(t) = \varepsilon_0[G_0\exp(-t/\tau_0) + G_1\exp(-t/\tau_1)] \tag{3.4-26}$$

or

$$\sigma(t) = \sigma_0\exp(-t/\tau_0) + \sigma_1\exp(-t/\tau_1) \tag{3.4-27}$$

where $\sigma_0 = \varepsilon_0 G_0$ and $\sigma_1 = \varepsilon_0 G_1$. The behavior is shown in Fig. 3.4-8. At $t=0$, $\sigma(0) = \sigma_0 + \sigma_1$. The stress relaxes from the initial value, $\sigma(0)$, at $t=0$, to zero at $t=\infty$.

The four-parameter Voigt model shows the same response to a strain excitation as its conjugate Maxwell model. To demonstrate this, we merely have to show that the relaxance of the four-parameter Voigt model has the same form as the relaxance of the four-parameter Maxwell model. The relaxance, $\bar{Q}(s)$, is the reciprocal of the retardance, $\bar{U}(s)$, given by Eq. (3.4-21). We have

$$\bar{Q}_{4V}(s) = \frac{(1+\tau s)s}{\phi_f + (J_g + J + \phi_f\tau)s + J_g\tau s^2} \tag{3.4-28}$$

which we may rewrite as

$$\bar{Q}_{4V}(s) = \frac{(1+\tau s)s}{\phi_f(1+\tau_0 s)(1+\tau_1 s)} \tag{3.4-29}$$

where

$$\tau_0 + \tau_1 = J_g/\phi_f + J/\phi_f + J/\phi \tag{3.4-30}$$

and

$$\tau_0\tau_1 = J_g J/\phi_f\phi\ . \tag{3.4-31}$$

By decomposing into partial fractions, we obtain

$$\bar{Q}_{4V}(s) = \frac{s}{\phi_f(\tau_0 - \tau_1)}\left(\frac{\tau_0 - \tau}{1+\tau_0 s} + \frac{\tau - \tau_1}{1+\tau_1 s}\right)\ . \tag{3.4-32}$$

Equation (3.4-32) is of the same form as Eq. (3.4-25).

In a similar fashion we may show that the retardance of the four-parameter Maxwell model has the same form as that of the conjugate Voigt model. Taking the reciprocal of Eq. (3.4-25) yields

$$\bar{U}_{4M}(s) = \frac{(1 + \tau_0 s)(1 + \tau_1 s)}{s[\eta_0 + \eta_1 + (\eta_1 \tau_0 + \eta_0 \tau_1)s]} \tag{3.4-33}$$

where we have made use of the relation $G\tau = \eta$ in the denominator. To bring Eq. (3.4-33) into the required form we let

$$\tau = \frac{\eta_0 \tau_1 + \eta_1 \tau_0}{\eta_0 + \eta_1} \tag{3.4-34}$$

and rewrite Eq. (3.4-33) in the form

$$\bar{U}_{4M}(s) = \frac{1}{\eta_0 + \eta_1} \frac{1 + (\tau_0 + \tau_1)s + \tau_0 \tau_1 s^2}{s(1 + \tau s)} . \tag{3.4-35}$$

By the method of partial fractions we find

$$\bar{U}_{4M}(s) = \frac{1}{\eta_0 + \eta_1}\left(\frac{\tau_0 \tau_1}{\tau} + \frac{\tau_0 + \tau_1 - \tau - \tau_0 \tau_1/\tau}{1 + \tau s} + \frac{1}{s}\right) \tag{3.4-36}$$

which is identical in form with Eq. (3.4-21).

With the proper choice of parameters, these two conjugate models again describe quantitatively identical behavior. The relation between the parameters of conjugate models will be discussed in Sect. 3.7.

Because the exhibition of steady-state flow is typical of the behavior of liquids, the standard 4-parameter models are sometimes referred to as models of the *standard linear liquid*. We shall also refer to them as the *standard rheodictic models*.

3.4.2 The Non-Standard Three- and Four-Parameter Series-Parallel Models

Without the guidance provided by Alfrey's rules one might have inferred erroneously that the model conjugate to the standard 4-parameter Voigt model is not the model shown in Fig. 3.4-6 but that shown in Fig. 3.4-9.

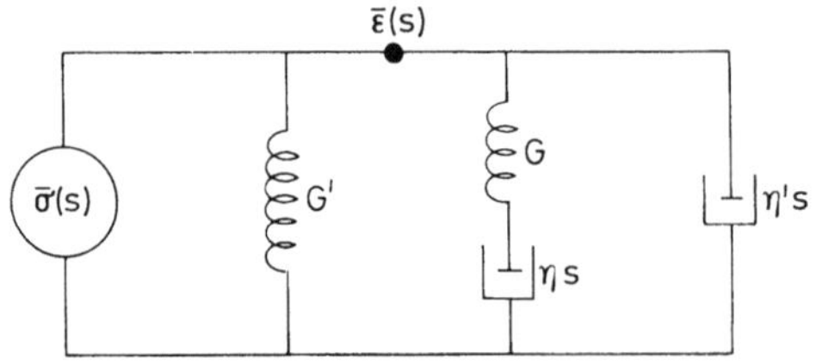

Fig. 3.4-9. Non-standard 4-parameter Maxwell model

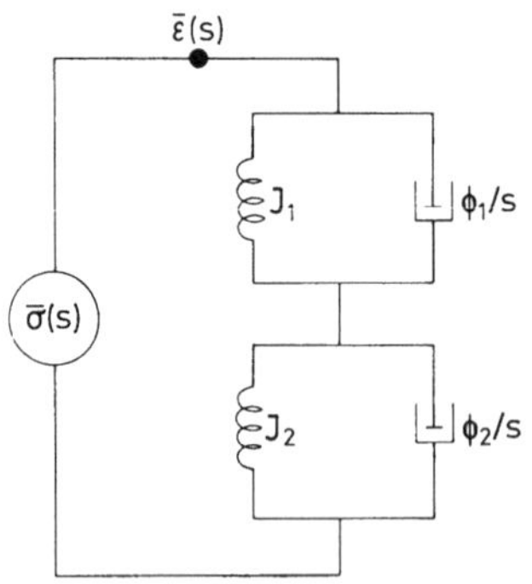

Fig. 3.4-10. Non-standard 4-parameter Voigt model

This model, however, as well as its conjugate, the non-standard 4-parameter Voigt model shown in Fig. 3.4-10, do not represent the same behavior as the previous two, whose equivalence we have just demonstrated.

3.4.2.1 The Non-Standard Three-Parameter Models

To understand more clearly the issues involved we first examine the addition of an isolated dashpot to a Voigt or Maxwell unit. This leads to the two three-parameter models shown in Figs. 3.4-11 and 3.4-13. These models can be derived from each other by Alfrey's rules and are therefore conjugate.

The retardance of a Voigt unit with added dashpot, shown in Fig. 3.4-11, is simply

$$\bar{U}_{3V}(s) = \frac{J}{1+\tau s} + \frac{\phi_f}{s} \,. \tag{3.4-37}$$

where $\tau = J/\phi_f$. The response to a step function of stress, therefore, is obtained from

$$\bar{\varepsilon}(s) = \sigma_0\left(\frac{J}{s(1+\tau s)} + \frac{\phi_f}{s^2}\right) \tag{3.4-38}$$

as

$$\varepsilon(t) = J\sigma_0[1 - \exp(-t/\tau)] + \sigma_0\phi_f t = \varepsilon_0[1 - \exp(-t/\tau)] + \dot{\varepsilon}_0 t \tag{3.4-39}$$

and is shown in Fig. 3.4-12.

At long times the curve approaches asymptotically to the broken line whose slope is $\dot{\varepsilon}_0 = \phi_f\sigma_0$. The parameter ϕ_f is, of course, again the steady-flow or steady-state fluidity and its reciprocal is the steady-flow or steady-state viscosity. Because of the absence of an isolated spring, the steady-state strain ε_e° is simply equal to the delayed strain, ε, which is also approached asymptotically. The model thus possesses a steady-state compliance $J_e^\circ = \varepsilon/\sigma_0$ but has no instantaneous compliance.

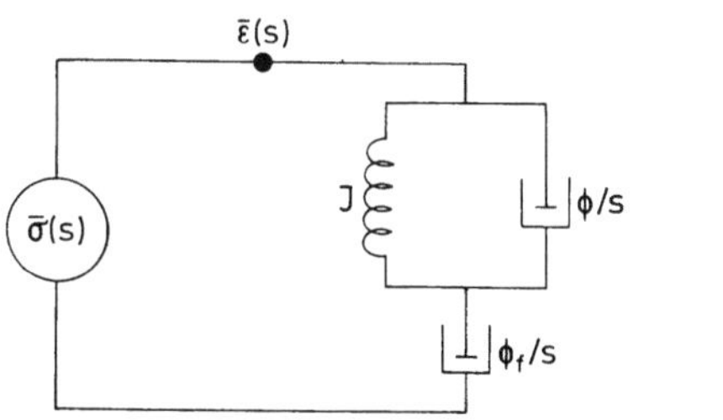

Fig. 3.4-11. Non-standard 3-parameter Voigt model

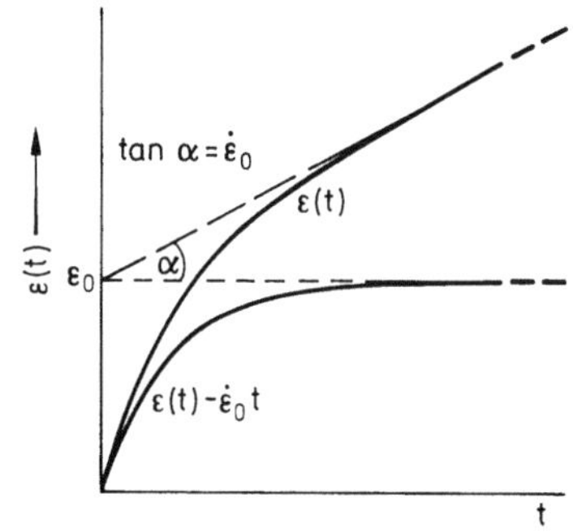

Fig. 3.4-12. Response of the non-standard 3-parameter models to a step stress

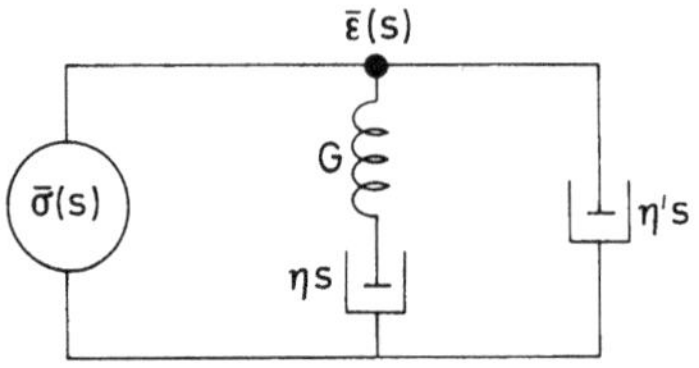

Fig. 3.4-13. Non-standard 3-parameter Maxwell model

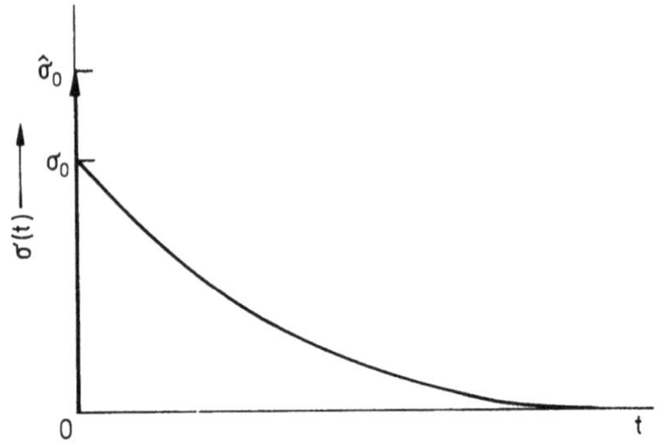

Fig. 3.4-14. Response of the non-standard 3-parameter models to a step strain

For the Maxwell unit with a dashpot in parallel, shown in Fig. 3.4-13, we have

$$\bar{Q}_{3M}(s) = \eta' s + \frac{G\tau s}{1 + \tau s} \tag{3.4-40}$$

where $\tau = \eta/G$. The behavior in response to a step of strain is obtained from

$$\bar{\sigma}(s) = \varepsilon_0\left(\eta' + \frac{G\tau}{1 + \tau s}\right) \tag{3.4-41}$$

as

$$\sigma(t) = \varepsilon_0[\eta'\delta(t) + G\exp(-t/\tau)] = \hat{\sigma}_0\delta(t) + \sigma_0\exp(-t/\tau) \tag{3.4-42}$$

and is shown in Fig. 3.4-14. The arrow at $t = 0$ indicates the stress impulse, $\hat{\sigma}_0 = \eta'\varepsilon_0$, which is needed at $t = 0$ to overcome the resistance of the dashpot. The stress then decays from $\sigma_0 = G\varepsilon_0$ to zero. Since there is no isolated spring, there is also no equilibrium modulus.

The stress response of a Voigt unit with added dashpot (Fig. 3.4-11) to a step of strain can be obtained from

$$\sigma(t) = \varepsilon_0\mathscr{L}^{-1}\frac{\bar{Q}(s)}{s} = \varepsilon_0\mathscr{L}^{-1}1/s\bar{U}(s) \tag{3.4-43}$$

which, after some algebra, yields

$$\sigma(t) = \varepsilon_0\left[\frac{\delta(t)}{\phi + \phi_f} + \frac{1}{J}\left(\frac{\phi}{\phi + \phi_f}\right)^2\exp(-t/\tau')\right] \tag{3.4-44}$$

where

$$\tau' = J/\phi + J/\phi_f\ . \tag{3.4-45}$$

The behavior is thus identical with that of a Maxwell unit with added dashpot shown in Fig. 3.4-14, with the appropriate change of the parameters of Eq. (3.4-44) to those of Eq. (3.4-42).

Proceeding in an analogous way, the strain response of a Maxwell unit with added dashpot (Fig. 3.4-13) to a step function of stress is found to be

$$\varepsilon(t) = \frac{\sigma_0}{G}\left(\frac{\eta}{\eta' + \eta}\right)^2 [1 - \exp(-t/\tau')] + \frac{\sigma_0 t}{\eta' + \eta} \tag{3.4-46}$$

where

$$\tau' = (G/\eta + G/\eta')^{-1} \; . \tag{3.4-47}$$

It is clear that for $t = 0$, $\varepsilon(0) = 0$, i.e., the instantaneous compliance of the model is zero. With the appropriate change in the parameters of Eq. (3.4-39) to those of Eq. (3.4-46), the behavior of a Maxwell unit with added dashpot is thus seen to be identical with that of a Voigt unit with added dashpot.

3.4.2.2 The Non-Standard Four-Parameter Models

We now return to the models shown in Figs. 3.4-9 and 3.4-10. From Fig. 3.4-10 we obtain

$$\bar{U}_{4V}(s) = \frac{J_1}{1 + \tau_1 s} + \frac{J_2}{1 + \tau_2 s} \; . \tag{3.4-48}$$

This model again has no instantaneous compliance. The retardance can be brought into the form

$$\bar{U}_{4V}(s) = J_1 + J_2 - \frac{J_1\tau_1 s}{1 + \tau_1 s} - \frac{J_2\tau_2 s}{1 + \tau_2 s} \tag{3.4-49}$$

from which the response to a step input of stress results as

$$\begin{aligned}\varepsilon(t) &= \sigma_0[J_1 + J_2 - J_1\exp(-t/\tau_1) - J_2\exp(-t/\tau_2)]\\ &= \varepsilon_e - \varepsilon_1\exp(-t/\tau_1) - \varepsilon_2\exp(-t/\tau_2) \; .\end{aligned} \tag{3.4-50}$$

The retardance of the model shown in Fig. 3.4-9 is

$$\bar{U}_{4M}(s) = \frac{1 + \tau s}{G'(1 + \tau_1' s)(1 + \tau_2' s)} \tag{3.4-51}$$

where $\tau = \eta/G$, $\tau_1'\tau_2' = \tau\tau'$, $\tau_1' + \tau_2' = \tau + \tau' + G\tau/G'$, and $\tau' = \eta'/G'$. Partial fraction decomposition and rearranging yields

$$\bar{U}_{4M}(s) = \frac{1}{G'}\left(1 - \frac{(\tau_1 - \tau)\tau_1 s}{(\tau_1 - \tau_2)(1 + \tau_1 s)} - \frac{(\tau - \tau_2)\tau_2 s}{(\tau_1 - \tau_2)(1 + \tau_2 s)}\right) . \tag{3.4-52}$$

Hence, a step excitation of stress will cause the strain

$$\varepsilon(t) = \frac{\sigma_0}{G'}\left(1 - \frac{\tau_1 - \tau}{\tau_1 - \tau_2}\exp(-t/\tau_1) - \frac{\tau - \tau_2}{\tau_1 - \tau_2}\exp(-t/\tau_2)\right) . \tag{3.4-53}$$

Comparing Eqs. (3.4-50) and (3.4-53) we note that the strain is zero at $t = 0$ as required, and reaches the equilibrium values $\sigma_0(J_1 + J_2)$ and σ_0/G', respectively,

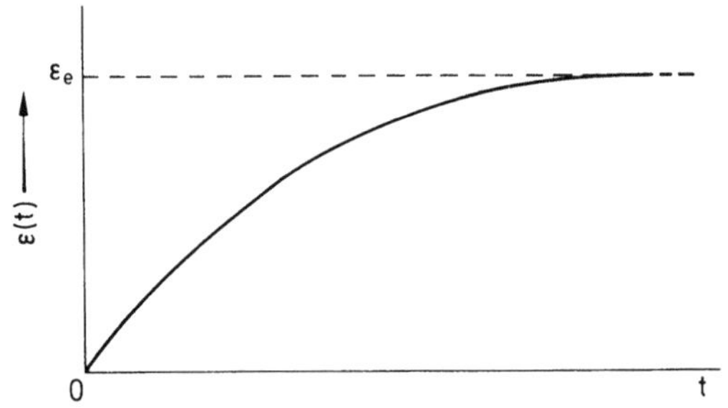

Fig. 3.4-15. Response of the non-standard 4-parameter series-parallel models to a step stress

as t approaches infinity. The response of the non-standard 4-parameter Maxwell and Voigt models to a step of stress is shown in Fig. 3.4-15.

The reciprocal of $\bar{U}_{4V}(s)$ is

$$\bar{Q}_{4V}(s) = \frac{1 + (\tau_1 + \tau_2)s + \tau_1\tau_2 s^2}{J_1 + J_2 + (J_1\tau_2 + J_2\tau_1)s} . \tag{3.4-54}$$

Letting

$$\tau'' = \frac{J_1\tau_2 + J_2\tau_1}{J_1 + J_2} = \frac{1/\phi_1 + 1/\phi_2}{1/J_1 + 1/J_2} , \tag{3.4-55}$$

we obtain

$$\bar{Q}_{4V}(s) = \frac{1}{J_1 + J_2}\left(\frac{1 + (\tau_1 + \tau_2)s + \tau_1\tau_2 s^2}{1 + \tau'' s}\right) . \tag{3.4-56}$$

Since the numerator is of higher degree than the denominator, we divide the former by the latter and rearrange the relaxance into the form

$$\bar{Q}_{4V}(s) = \frac{1}{J_1 + J_2}\left(\frac{\tau_1\tau_2 s}{\tau''} + 1 + \frac{(\tau_1 + \tau_2 - \tau_1\tau_2/\tau'' - \tau'')s}{1 + \tau'' s}\right) . \tag{3.4-57}$$

For a step input of strain, therefore, the stress becomes

$$\sigma(t) = \varepsilon_0\left(\frac{\delta(t)}{\phi_1 + \phi_2} + \frac{1}{J_1 + J_2} + \frac{\tau_1 + \tau_2 - \tau'' - \tau_1\tau_2/\tau''}{(J_1 + J_2)\tau''}\exp(-t/\tau'')\right) . \tag{3.4-58}$$

From Fig. 3.4-9 we find

$$\bar{Q}_{4M}(s) = G' + \eta' s + \frac{G\tau s}{1 + \tau s} . \tag{3.4-59}$$

For a step input of strain we have

$$\sigma(t) = \varepsilon_0[\eta'\delta(t) + G' + G\exp(-t/\tau)] = \hat{\sigma}_0\delta(t) + \sigma_e + \sigma\exp(-t/\tau) . \tag{3.4-60}$$

Equations (3.4-58) and (3.4-60) may be compared with each other. Both again require an impulsive stress at $t = 0$ to extend the combination of springs and dashpots to the fixed extension, ε_0. As t approaches infinity, the stress decays to the equilibrium values, $\varepsilon_0/(J_1 + J_2)$ and $\varepsilon_0 G'$, respectively. The behavior of the two

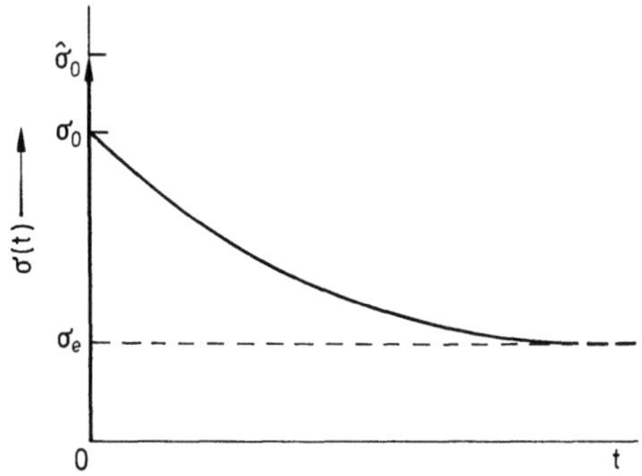

Fig. 3.4-16. Response of the non-standard 4-parameter series-parallel models to a step strain

non-standard 4-parameter series-parallel models in response to a step excitation of strain is shown in Fig. 3.4-16.

It should be noted that, among the non-standard series-parallel models, it is the 3-parameter models which exhibit rheodictic behavior while the 4-parameter ones do not. In the standard series-parallel models the reverse is true.

The non-standard three and four-parameter models are conjugate and, with the appropriate choice of the parameters, describe identical behavior. They differ from the four standard series-parallel models in that their instantaneous compliance is zero. These models therefore require an impulsive stress to assume a constant strain instantaneously at $t = 0$. Models with the same defect also arise among the ladder models to be discussed in Chap. 5. One may ask whether this defect should at all be a matter of concern in view of the fact that a step function is a mathematical concept which can only be approximated experimentally but can never be realized strictly. The experimental approximation to a step excitation is the ramp excitation discussed in more detail in Sect. 7.1.2. Models not possessing an instantaneous compliance do respond in a realistic way to such an excitation (cf. Problem 7.1-7) and are sometimes used to represent the linear viscoelastic behavior of liquids and solutions when one is not concerned with the response to a step stimulus.

It will be shown in Sect. 3.4.4 [cf. Eq. (3.4-123)] that the instantaneous compliance is the reciprocal of the instantaneous modulus, i.e. that $J_g = 1/G_g$. Thus, the absence of an instantaneous compliance implies an infinite instantaneous modulus. Although the experimental determination of the instantaneous compliance and modulus is a difficult matter (they are operationally ill-defined), a zero instantaneous compliance and an infinite instantaneous modulus appear to be physically unrealistic. For this reason only the standard series-parallel models are generally used to represent linear viscoelastic behavior.

3.4.3 Other Four-Parameter Models

The four models shown in Figs. 3.4-1, 4, 11, and 13 are the only 3-parameter models that can be constructed from two elements of one kind, and one element of the other kind. By contrast, the four models shown in Figs. 3.4-5, 6, 9, and 10 are not the only

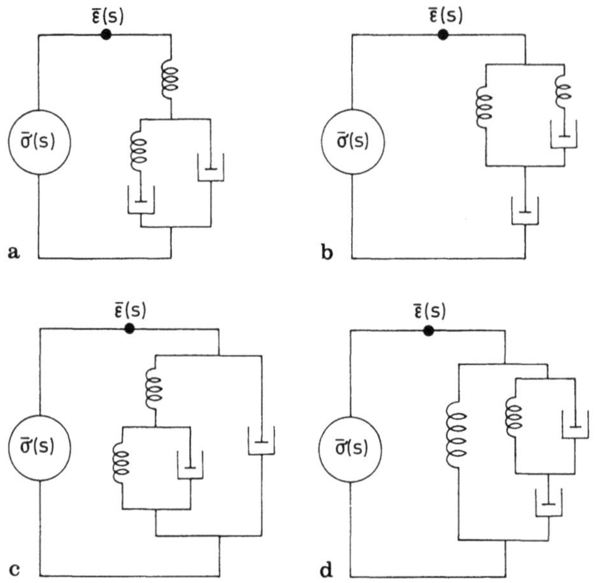

Fig. 3.4-17a-d. Other 4-parameter models with two springs and two dashpots

4-parameter models one may construct from two elements of each kind. There exist four more 4-parameter models constructed from two springs and two dashpots. They are shown in Fig. 3.4-17.

Following the procedure employed in the preceding section, one can ascertain (cf. Problems 3.4-7 and 3.4-8) that models (a) and (b) in Fig. 3.4-17 show, with the appropriate choice of the model parameters, identical behavior with that exhibited by the models shown in Figs. 3.4-5 and 3.4-6. Models (c) and (d) in Fig. 3.4-17, on the other hand, are identical in behavior with those shown in Figs. 3.4-9 and 10, i.e. they possess no instantaneous compliance and need an impulse at time $t = 0$ to extend the springs in response to a step of strain (cf. Problem 3.4-9).

If there are several models which show identical behavior, which should one select? The answer is: we choose those models which are simplest to analyze. A relatively simple analysis will only be possible if the models show certain regularities which facilitate their analysis. The *generalized series-parallel models* to be introduced in Sect. 3.5 constitute one such group. Another are the *regular ladder models* which will be discussed in Chap. 5. In constructing series-parallel models we follow the rule always to add further elements (springs or dashpots) or further units (i.e. Maxwell or Voigt units) in parallel to the first Maxwell unit or in series with the first Voigt unit. This rule is based on the combination rules that relaxances add in parallel, and retardances add in series. The simplicity in analysis which makes the series-parallel models particularly attractive, stems from these rules.

The models shown in Fig. 3.4-17 clearly do not qualify as series-parallel models within the meaning of these rules. In Chap. 5 it will be shown that they may be regarded as ladder models. The 3-parameter models may be regarded simultaneously either as series-parallel, or as ladder models. Of the 4-parameter models four are series-parallel and four are ladder models. If the number of elements in a model is larger than four, there will always be models which cannot be classed either as series-parallel or as ladder models. We note that, of the models shown in Fig. 3.4-17, no two are conjugates in the sense of Alfrey's rules which apply only to series-parallel models.

The rule for the construction of series-parallel models which we have stated above, also removes the ambiguity which one may otherwise feel concerning the models shown in Figs. 3.4-5 and 3.4-9. Does Fig. 3.4-5 represent a Voigt unit with added spring and dashpot or is it a Maxwell unit in series with a Voigt unit? Our rule that we add to a Voigt unit in series and to a Maxwell unit in parallel, makes it clear that the model in Fig. 3.4-5 is derived from a Voigt unit. We have emphasized this in the way in which the model is drawn. The added spring and dashpot in Fig. 3.4-5 are separated by the Voigt unit to show that we are not considering them to be elements of a Maxwell unit. Analogous considerations apply to Fig. 3.4-9. The way in which the models are drawn does not, however, affect the respondances. Thus retardances add in series and the particular sequence is immaterial. In a series combination (Voigt model) the *strains* are additive and the same stress acts through all elements or units in series. In a parallel combination (Maxwell model) the *stresses* are additive and all elements or units see the same strain.

3.4.4 Behavior of the Standard Models in the Complex Plane. Initial and Final Values

We now turn to an examination of the properties of the respondances of the standard series-parallel models in the complex plane. Using subscripts 3 and 4 to distinguish the 3- and 4-parameter models, and subscripts V and M to distinguish the Voigt and Maxwell models, the retardances of the first two are

$$\bar{U}_{3V}(s) = J_g + \frac{J}{1 + \tau s} \tag{3.4-61}$$

and

$$\bar{U}_{4V}(s) = J_g + \frac{J}{1 + \tau s} + \frac{\phi_f}{s} \tag{3.4-62}$$

while the relaxances of the last two are given by

$$\bar{Q}_{3M}(s) = G_e + \frac{G\tau s}{1 + \tau s} \tag{3.4-63}$$

and

$$\bar{Q}_{4M}(s) = \frac{G_0 \tau_0 s}{1 + \tau_0 s} + \frac{G_1 \tau_1 s}{1 + \tau_1 s} \; . \tag{3.4-64}$$

As discussed in Chap. 2, the transform variable s is a variable in the complex transform plane. It is of interest, therefore, to examine the singularities of the respondances of the four standard models. To this end, we first rearrange the respondances of the standard models to

$$\bar{U}_{3V}(s) = \frac{(J_g + J)\phi + J_gJs}{\phi + Js} \tag{3.4-65}$$

$$\bar{Q}_{3M}(s) = \frac{G_eG + (G_e + G)\eta s}{G + \eta s} \tag{3.4-66}$$

$$\bar{U}_{4V}(s) = \frac{\phi\phi_f + (J_g\phi + \phi + J\phi_f)s + J_gJs^2}{s(\phi + Js)} \tag{3.4-67}$$

and

$$\bar{Q}_{4M}(s) = \frac{[(\eta_0 + \eta_1)G_0G_1 + \eta_0\eta_1(G_0 + G_1)s]s}{(G_0 + \eta_0 s)(G_1 + \eta_1 s)} . \tag{3.4-68}$$

This demonstrates what we have already stated at the end of Sect. 2.2.1, viz. that the respondances are rational algebraic fractions, i.e. ratios of polynomials, of the complex transform variable s. Their only singularities, therefore, are poles. We have [cf. Eqs. (2.1-17)]

$$\bar{U}_V(s) = \frac{\bar{u}_v(s)}{\bar{q}_V(s)} \tag{3.4-69}$$

and

$$\bar{Q}_M(s) = \frac{\bar{q}_M(s)}{\bar{u}_M(s)} . \tag{3.4-70}$$

The poles are the roots of the polynominals in the denominators, $\bar{q}_V(s)$, and $\bar{u}_M(s)$. The roots of the numerators, $\bar{u}_V(s)$, and $\bar{q}_M(s)$ are the zeros. It follows from $\bar{U}(s)\bar{Q}(s) = 1$ [cf. Eq. (2.1-19)], that the zeros of the retardances are the poles of the relaxances, and *vice versa*. This is a necessary condition for the two three-parameter and the two 4-parameter models to be conjugate. The Voigt and Maxwell units have no zeros and, therefore cannot be each others' conjugates. It is clear from Eqs. (3.4-65) through (3.4-68) that the poles and zeros of the respondances of the standard models are simple poles and zeros.

Let us now examine the poles and zeros of the standard models more closely. We shall use subscripts ∞ and 0 on s to denote poles and zeros, respectively, and shall use superscripts 3V, 4V, 3M and 4M to distinguish the models.

From $\bar{q}_{3V}(s) = 0$ we find the pole

$$s_\infty^{3V} = -\phi/J \tag{3.4-71}$$

and from $\bar{u}_{3V}(s) = 0$ we have the zero

$$s_0^{3V} = -J_e\phi/J_gJ \tag{3.4-72}$$

where $J_e = J_g + J$ is the equilibrium modulus.

Now, from $\bar{q}_{4V}(s) = 0$ we derive the two poles

$$s_{\infty 0}^{4V} = 0 \tag{3.4-73}$$

and

$$s_{\infty 1}^{4V} = -\phi/J \ . \tag{3.4-74}$$

From $\bar{u}_{4V}(s) = 0$ we obtain the two zeros

$$s_{01,0}^{4V} = -\varepsilon(1 \pm \sqrt{1-\delta}) \tag{3.4-75}$$

where

$$2\varepsilon = \phi_f/J_g + \phi/J + \phi/J_g \tag{3.4-76}$$

and

$$\delta = \frac{4J_g J\phi\phi_f}{(J_e^\circ \phi + J\phi_f)^2} \tag{3.4-77}$$

where $J_e^\circ = J_g + J$ is now the steady-state compliance. When $\delta \ll 1$, then $\sqrt{1-\delta} \simeq 1 - \delta/2$, and

$$s_{01}^{4V} \simeq -\varepsilon(1 - \delta/4) \simeq -2\varepsilon \tag{3.4-78}$$

while

$$s_{00}^{4V} \simeq -\delta\varepsilon/2 \ . \tag{3.4-79}$$

Turning now to the Maxwell models, $\bar{u}_{3M}(s) = 0$ gives the pole

$$s_{\infty}^{3M} = -G/\eta \tag{3.4-80}$$

and from $\bar{q}_{3M}(s) = 0$ we obtain the zero

$$s_0^{3M} = -G_e G/G_g \eta \ . \tag{3.4-81}$$

Now $\bar{u}_{4M}(s) = 0$ yields the two poles

$$s_{\infty 0}^{4M} = -G_0/\eta_0 \tag{3.4-82}$$

and

$$s_{\infty 1}^{4M} = -G_1/\eta_1 \ . \tag{3.4-83}$$

From $\bar{q}_{4M}(s) = 0$ we have the two zeros

$$s_{00}^{4M} = 0 \tag{3.4-84}$$

and

$$s_{01}^{4M} = -G_0 G_1 \eta_f / G_g \eta_0 \eta_1 \tag{3.4-85}$$

where $\eta_f = \eta_0 + \eta_1$ is the steady-flow viscosity.

As an example, we select a set of realistic values which will be used in Sect. 3.4.5 as well as elsewhere in this book. These are

$$\begin{aligned} J_g &= 10^{-9}\ \mathrm{m^2/N} \\ J &= 10^{-6}\ \mathrm{m^2/N} \\ \phi &= 10^{-5}\ \mathrm{m^2/Ns} \end{aligned} \tag{3.4-86}$$

and

$$\phi_f = 10^{-7}\ m^2/Ns\ .$$

We derive $\tau = J/\phi = 10^{-1}$ s and $J_e = J_e^\circ = J_g + J = 1.001 \times 10^{-6}\ m^2/N$. From these values, using Eqs. (3.4-71) through (3.4-79), we find

$$s_\infty^{3V} = -10\ s^{-1} \tag{3.4-87}$$

$$s_0^{3V} = -1.001 \times 10^4\ s^{-1} \tag{3.4-88}$$

for the three-parameter model, and

$$s_{\infty 0}^{4V} = 0 \tag{3.4-89}$$

$$s_{\infty 1}^{4V} = -10\ s^{-1} \tag{3.4-90}$$

$$s_{00}^{4V} = -0.9891 \times 10^{-1}\ s^{-1} \tag{3.4-91}$$

and

$$s_{01}^{4V} = -1.011 \times 10^4\ s^{-1} \tag{3.4-92}$$

for the 4-parameter model. The pole-zero patterns in the s-plane are shown in Fig. 3.4-18 where the poles are indicated by crosses and the zeros by circles.

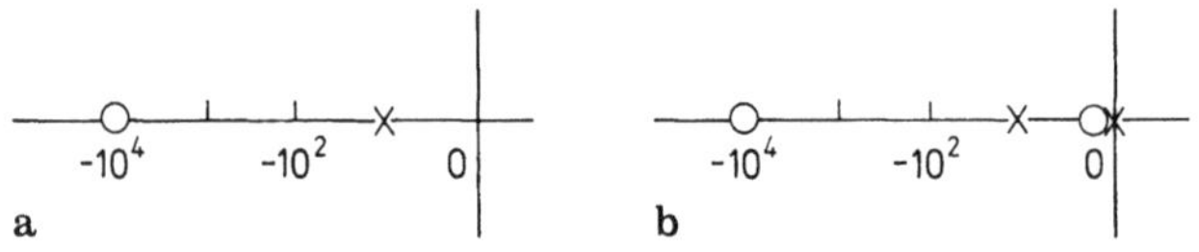

Fig. 3.4-18. Pole-zero patterns for the standard Voigt models: (**a**) 3-parameter, (**b**) 4-parameter model

The parameters of the conjugate Maxwell models may be obtained from those of the Voigt models. By equations to be derived in Sect. 3.7, we have

$$G_e = 1/J_e = 10^6/1.001\ N/m^2$$

$$G = J/J_gJ_e = 10^9/1.001\ N/m^2 \tag{3.4-93}$$

and

$$\eta = J^2/J_e^2\phi = 10^5/1.001^2\ Ns/m^2$$

for the three-parameter Maxwell model. We find $G_g = G_e + G = 10^9\ N/m^2$ and $\tau = 10^{-4}/1.001$ s. The latter is the *relaxation* time of that Maxwell model which exhibits the same behavior as the conjugate Voigt model which has a *retardance* time of 10^{-1} s.

From Eqs. (3.4-80) and (3.4-81) we find

$$s_\infty^{3M} = -1.001 \times 10^4\ s^{-1} \tag{3.4-94}$$

and

$$s_0^{3M} = -10\ s^{-1}\ . \tag{3.4-95}$$

The parameters of the 4-parameter Maxwell model may be obtained from Eqs. (3.7-32) through (3.7-38). The calculations yield

$$G_0 = 0.9794 \times 10^6\ \mathrm{N/m^2}$$

$$G_1 = 0.9991 \times 10^9\ \mathrm{N/m^2}$$

$$\eta_0 = 0.9901 \times 10^7\ \mathrm{Ns/m^2} \tag{3.4-96}$$

and

$$\eta_1 = 0.9881 \times 10^5\ \mathrm{Ns/m^2}\ .$$

We have $G_g = G_0 + G_1 = 10^9\ \mathrm{N/m^2}$, $\eta_f = \eta_0 + \eta_1 = 10^7\ \mathrm{Ns/m^2}$, $\tau_0 = 1.011 \times 10^1$ s, and $\tau_1 = 0.9891 \times 10^{-4}$ s. Equations (3.4-91) through (3.4-94) then give

$$s_{\infty 0}^{4M} = -0.9891 \times 10^{-1}\ \mathrm{s^{-1}} \tag{3.4-97}$$

$$s_{\infty 1}^{4M} = -1.011 \times 10^{4}\ \mathrm{s^{-1}} \tag{3.4-98}$$

$$s_{00}^{4M} = 0 \tag{3.4-99}$$

and

$$s_{01}^{4M} = -10\ \mathrm{s^{-1}}\ . \tag{3.4-100}$$

The pole-zero patterns are shown in Fig. 3.4-19 and may be compared with those displayed in Fig. 3.4-18.

We note that, for a model to describe rheodictic behavior, its retardance must have a pole at the origin, while, conversely, its relaxance must possess a zero at that point.

Once the poles and zeros of a respondance are known, it is completely determined except for a scale factor. The four respondances of the standard series-parallel models can all be derived from

$$\bar{R}(s) = A\frac{(s - s_{00})(s - s_{01})}{(s - s_{\infty 0})(s - s_{\infty 1})} \tag{3.4-101}$$

where A is a scale factor which is known as soon as any value of the function (other than at a zero or pole) has been determined. The zeros and poles completely determine the analytic properties of the respondances. Thus, e.g., $\bar{R}(s) = \bar{U}_{3v}(s)$ when $s_{00} = s_{\infty 0} \equiv 0$, s_{01} and $s_{\infty 1}$ are given by Eqs. (3.4-72) and (3.4-71), and $A = J_g$.

Decomposition of Eq. (3.4-101) into partial fractions gives

$$\bar{R}(s) = A + \frac{b_0}{s - s_{\infty 0}} + \frac{b_1}{s - s_{\infty 1}} \tag{3.4-102}$$

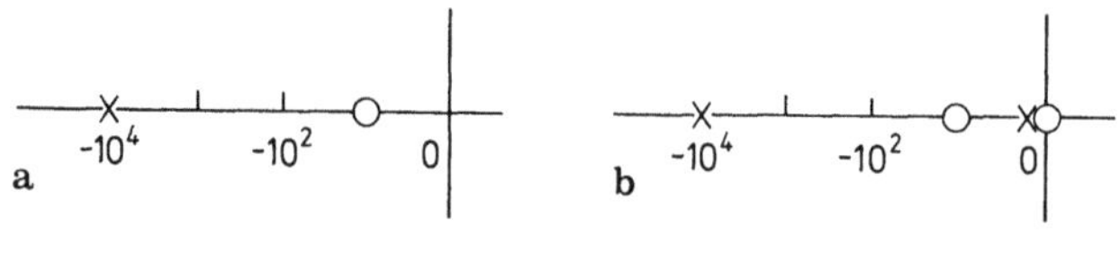

Fig. 3.4-19. Pole-zero patterns for the standard Maxwell models: (**a**) 3-parameter, (**b**) 4-parameter model

where

$$b_0 = A\frac{(s_{\infty 0} - s_{01})(s_{\infty 0} - s_{01})}{s_{\infty 0} - s_{\infty 1}} \tag{3.4-103}$$

and

$$b_1 = A\frac{(s_{\infty 1} - s_{00})(s_{\infty 1} - s_{01})}{s_{\infty 1} - s_{\infty 0}} \ . \tag{3.4-104}$$

Thus the respondances of the standard models are resolved into a constant plus two terms representing the contributions from the poles. Since b_0 and b_1 are also constants, the time-dependent behavior is determined entirely by the poles.

Equation (3.4-102) shows that the only singularities which the respondances of the standard models possess are simple poles on the negative real axis of the complex plane. We had stated already in Chap. 2 that the respondances representing the behavior of any real material, being rational algebraic fractions, do not contain essential singularities or branch points. Earlier in this section we had asserted that physically meaningful behavior requires a finite instantaneous compliance. We now demonstrate that this requirement is tantamount to the demand that the retardance of the model should not vanish, or alternatively, that its relaxance should not diverge, as s approaches infinity.

To show this, we turn to the *initial value theorem* (see Appendix A3.1.7). As applied to the creep compliance, this theorem states that

$$\lim_{t\to 0} J(t) = \lim_{s\to\infty} s\bar{J}(s) \ , \tag{3.4-105}$$

i.e. the behavior of the model at $t = 0$ approached from $t > 0$, can be obtained from $s\bar{J}(s)$ by letting s approach infinity. Now

$$\lim_{t\to 0} J(t) = J(0) = J_g \tag{3.4-106}$$

and, since $s\bar{J}(s) = \bar{U}(s)$, we have

$$J_g = \lim_{s\to\infty} \bar{U}(s) \ . \tag{3.4-107}$$

Thus our requirement may be stated by demanding that the retardance of the model should not vanish as s approaches infinity. Now, because $\bar{U}(s) = 1/\bar{Q}(s)$,

$$\lim_{s\to\infty} \bar{U}(s) = 0 \tag{3.4-108}$$

would entail

$$\lim_{s\to\infty} \bar{Q}(s) = \infty \ . \tag{3.4-109}$$

Now, by the initial value theorem applied to the relaxation modulus,

$$\lim_{t\to 0} G(t) = \lim_{s\to\infty} s\bar{G}(s) \ . \tag{3.4-110}$$

But, $s\bar{G}(s) = \bar{Q}(s)$, and, therefore

$$\lim_{t \to 0} G(t) = G(0) = G_g \ . \tag{3.4-111}$$

Hence

$$G_g = \lim_{s \to \infty} \bar{Q}(s) \ . \tag{3.4-112}$$

Thus, stated in other terms, the requirement demands that the relaxance of the model should not possess a pole at infinity, i.e., that its instantaneous modulus be finite.

Equations (3.4-107) and (3.4-112) are quite general. The initial value theorem thus furnishes a convenient way to ascertain whether a given model is physically meaningful with respect to its behavior at $t = 0$. In a similar way, the *final value theorem* (see Appendix A3.1.7) allows us to investigate the behavior of a model at $t = \infty$. It should thus give information on whether the model describes rheodictic behavior or not.

By the final value theorem as applied to the relaxation modulus we have

$$\lim_{t \to \infty} G(t) = \lim_{s \to 0} s\bar{G}(s) \ . \tag{3.4-113}$$

But

$$\lim_{t \to \infty} G(t) = G(\infty) = G_e \tag{3.4-114}$$

and therefore

$$G_e = \lim_{s \to 0} \bar{Q}(s) \ . \tag{3.4-115}$$

Vanishing of the limit therefore shows that the model does not have an equilibrium modulus and, thus, is rheodictic, i.e. exhibits steady-state flow.

From

$$\lim_{t \to \infty} J(t) = \lim_{s \to 0} s\bar{J}(s) \tag{3.4-116}$$

with

$$\lim_{t \to \infty} J(t) = J(\infty) = J_e \tag{3.4-117}$$

we find

$$J_e = \lim_{s \to 0} \bar{U}(s) \ . \tag{3.4-118}$$

Because $\bar{U}(s) = 1/\bar{Q}(s)$, the absence of an equilibrium compliance (i.e. the existence of steady-state flow) is indicated by divergence of the limit.

The steady-flow viscosity is defined by

$$\eta_f = \lim_{t \to \infty} \eta(t) \tag{3.4-119}$$

where

$$\eta(t) = \mathscr{L}^{-1}\bar{\eta}(s) = \mathscr{L}^{-1}\bar{Q}(s)/s^2 \ . \tag{3.4-120}$$

By the final value theorem, therefore,

$$\eta_f = \lim_{s\to 0} \bar{Q}(s)/s = \lim_{s\to 0} \bar{G}(s) \tag{3.4-121}$$

But $s\bar{G}(s) = 1/s\bar{J}(s)$ and therefore the steady-state fluidity is obtained from

$$\phi_f = \lim_{s\to 0} s^2\,\bar{J}(s) = \lim_{s\to 0} s\bar{U}(s)\ . \tag{3.4-122}$$

Equations (3.4-107), (3.4-112), (3.4-115), (3.4-118), (3.4-121), and (3.4-122) are useful in examining the limiting behavior of a given model at $t = 0$, and $t = \infty$. We shall find further application for them in Chap. 5, where we will discuss ladder models.

Since the product of two limits is the limit of the products, we have

$$G_gJ_g = 1\ , \qquad G_eJ_e = 1\ , \quad \text{and} \quad \eta_f\phi_f = 1\ , \tag{3.4-123}$$

i.e. the instantaneous and the equilibrium moduli and compliances, and the steady-flow viscosity and fluidity, are reciprocals of each other [see also Eq. (3.4-159) and Sect. 4.6].

3.4.5 Response of the Standard Models to the Standard Excitations

Figures 3.4-2, 3, 7, and 8 presented a qualitative picture of the stress and strain, respectively, as displayed by the standard series-parallel models in response to step inputs of strain, or stress. In this section we examine the quantitative response of these models to the standard experimental excitation functions, the step functions, and the harmonic functions. The response to the slope excitation will be left to Sect. 3.8 (Problems).

To be realistic, we assign to the elements of the models the values given by Eqs. (3.4-86), (3.4-93), and (3.4-96). These represent compliance and modulus values which are not too different from those commonly found in many synthetic polymers. Because the ranges of the compliance and modulus values extend over several decades, they are customarily plotted logarithmically. The time or frequency is also plotted logarithmically. Figure 3.4-20 shows the relaxation modulus, $G_{3M}(t)$, the storage and loss moduli, $G'_{3M}(\omega)$ and $G''_{3M}(\omega)$, and the loss tangent, $\tan\theta_{3M}(\omega)$, of the 3-parameter Maxwell model plotted in this way. The frequency is plotted as $-\log\omega$ for easier comparison with the time-dependent behavior. The frequency dependent curves are, therefore, mirror images of those shown in Fig. 3.4-20 and the rest of the figures in this section. The plots must be pictured as normalized with respect to unit modulus, loss tangent, time, or frequency. Thus $\log G_{3M}(t)$ and $\log t$ should be taken to mean

$$\log G_{3M}(t) = \log G_{3M}(t) - \log[G] \tag{3.4-124}$$

and

$$\log t = \log t - \log[t] \tag{3.4-125}$$

where [G] and [t] signify unit modulus, and unit time, respectively.

The equations represented by the curves in Fig. 3.4-20 are the standard response functions obtained, in this case, from the relaxance of the standard 3-parameter

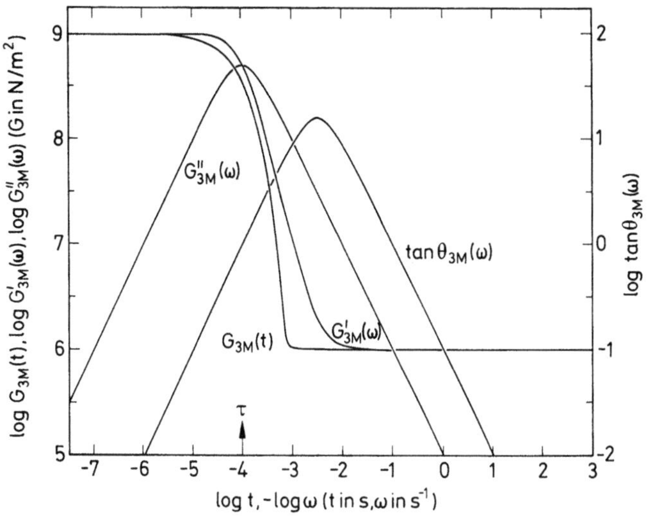

Fig. 3.4-20. Response of the 3-parameter Maxwell model to the standard stimuli

Maxwell model, Eq. (3.4-11),

$$\bar{Q}_{3M}(s) = G_e + \frac{G\tau s}{1 + \tau s} \; . \tag{3.4-126}$$

The relations by which the standard response functions are obtained from the respondances as discussed in Chap. 2 are listed in Table 3.4-1 for convenience.

Table 3.4-1. Response to the standard stimuli

From $\bar{U}(s)$	From $\bar{Q}(s)$
$J(t) = \mathscr{L}^{-1}\bar{U}(s)/s$	$G(t) = \mathscr{L}^{-1}\bar{Q}(s)/s$
$J^*(\omega) = [\bar{U}(s)]_{s=j\omega} = \bar{U}(j\omega)$	$G^*(\omega) = [\bar{Q}(s)]_{s=j\omega} = \bar{Q}(j\omega)$
$J'(\omega) = \mathscr{R}e\,\bar{U}(j\omega)$	$G'(\omega) = \mathscr{R}e\,\bar{Q}(j\omega)$
$J''(\omega) = -\mathscr{I}m\,\bar{U}(j\omega)$	$G''(\omega) = \mathscr{I}m\,\bar{Q}(j\omega)$
$\tan\theta(\omega) = J''(\omega)/J'(\omega)$	$\tan\theta(\omega) = G''(\omega)/G(\omega)$

From Eq. (3.4-126) we find

$$G_{3M}(t) = G_e + G\exp(-t/\tau) \tag{3.4-127}$$

$$G'_{3M}(\omega) = G_e + \frac{G\omega^2\tau^2}{1 + \omega^2\tau^2} \tag{3.4-128}$$

$$G''_{3M}(\omega) = \frac{G\omega\tau}{1 + \omega^2\tau^2} \tag{3.4-129}$$

and

$$\tan \theta_{3M}(\omega) = \frac{G\omega\tau}{G_e + G_g\omega^2\tau^2} \ . \tag{3.4-130}$$

At short times, i.e. high frequencies, $\log G_{3M}(t)$ and $\log G'_{3M}(\omega)$ exhibit the plateau corresponding to $\log G_g$. They then drop through transition zones whose inflection points are located in the vicinity of $\log t = \log \tau = -\log \omega$, to the plateau corresponding to $\log G_e$ at long times, i.e., low frequencies. By contrast, $\log G''_{3M}(\omega)$ and $\log \tan \theta_{3M}(\omega)$ exhibit maxima. We now examine the positions of the inflection points and maxima. To find these we note that

$$\frac{d \log f(x)}{d \log x} = \frac{d \ln f(x)}{d \ln x} = \frac{x}{f(x)} \frac{d f(x)}{dx} \tag{3.4-131}$$

and, therefore, the condition for an extremum in logarithmic coordinates is

$$f'(x) = 0 \tag{3.4-132}$$

where the prime denotes differentiation with respect to the argument. Further,

$$\frac{d^2 \log f(x)}{d \log^2 x} = \mu x \frac{d}{dx}\left(\frac{x f'(x)}{f(x)}\right)$$
$$= \frac{\mu x f'(x)}{f(x)}\left(1 - \frac{x f'(x)}{f(x)} + \frac{x f''(x)}{f'(x)}\right) \tag{3.4-133}$$

where $\mu = \ln 10 = 2.303$. Therefore, the condition for an inflection point in logarithmic coordinates is

$$1 - \frac{x f'(x)}{f(x)} + \frac{x f''(x)}{f'(x)} = 0 \ . \tag{3.4-134}$$

We find that t_{i3M}, the value of t at which the inflection on the $\log G(t)$ vs. $\log t$ plot occurs for the 3-parameter Maxwell model, is given by

$$t_{i3M}/\tau = 1 + (G/G_e)\exp(-t_{i3M}/\tau) \ . \tag{3.4-135}$$

This is a transcendental equation. Solution with the values given by Eqs. (3.4-93) gives $\log t_{i3M} = -3.266$. It follows from Eq. (3.4-135) that t_{i3M} is always greater than τ, i.e. the inflection point lies to the right of the arrow indicating the position of τ on the log(t)-axis in Fig. 3.4-20. The amount by which $\log t_{i3M}$ is shifted to the right of $\log \tau$ depends on the ratio G/G_e. It should be noted that the occurrence of an inflection point is a consequence of the use of logarithmic coordinates. It is easily ascertained that in arithmetic coordinates the condition for an inflection point, $f''(x) = 0$, would be satisfied only for $t_{i3M} = \infty$, i.e. there is no inflection.

Proceeding identically for the storage modulus gives

$$(\omega_{i3M}\tau)^4 = G_e/G_g \tag{3.4-136}$$

where ω_{i3M} is the radian frequency at which the inflection occurs in a plot of $\log G'(\omega)$ vs. $\log \omega$ in the 3-parameter Maxwell model. In Fig. 3.4-20 the inflection point is thus located at $-\log \omega = \log \tau + 0.25 \log(G_g/G_e)$. With the values appropriate for Fig. 3.4-20 we find $\log \omega_{i3M} = 3.25$.

The point of inflection, ω_{i3M}, must be less than $1/\tau$, i.e. it must lie to the right of τ in Fig. 3.4-20. As before, the value of ω_i depends upon the values of the instantaneous and equilibrium moduli.

The location of the maximum in the $\log G''(\omega)$ vs. $\log \omega$ curve is found by seeking that value of ω for which the first derivative is zero and the second derivative is negative. We find

$$\omega_{m3M}\tau = 1 \tag{3.4-137}$$

where ω_{m3M} is the frequency at which the maximum occurs. Thus, in Fig. 3.4-20, $\log \omega_{m3M} = 4.0$. The value of ω_{m3M} depends only on τ. Furthermore, it is the same in logarithmic and arithmetic coordinates. The latter is also true for the maximum in the loss tangent, $\theta_{3M}(\omega)$. The maximum in $\log \tan \theta_{3M}(\omega)$ or $\tan \theta_{3M}(\omega)$ is shifted to the right of that in $\log G''_{3M}(\omega)$ or $G''_{3M}(\omega)$ because of the division by $G'_{3M}(\omega)$. From Eq. (3.4-130), proceeding as before, we find the condition for the occurrence of the maximum in the $\log \theta_{3M}(\omega)$ vs. $-\log \omega$ or in the $\tan \theta_{3M}(\omega)$ vs. ω curve to be fulfilled for

$$(\omega_{\theta 3M}\tau)^2 = G_e/G_g \; . \tag{3.4-138}$$

Thus in logarithmic coordinates the maximum lies at $-\log \omega_{\theta 3M} = \log \tau + 0.5 \log(G_g/G_e)$, i.e. at $\log \omega = 2.5$ in Fig. 3.4-20.

The standard response functions for the 3-parameter Voigt model are obtained from

$$\bar{U}_{3V}(s) = J_g + \frac{J}{1 + \tau s} \; . \tag{3.4-139}$$

We have

$$J_{3V}(t) = J_g + J[1 - \exp(-t/\tau)] \tag{3.4-140}$$

$$J'_{3V}(\omega) = J_g + \frac{J}{1 + \omega^2\tau^2} \tag{3.4-141}$$

$$J''_{3V}(\omega) = \frac{J\omega\tau}{1 + \omega^2\tau^2} \tag{3.4-142}$$

and

$$\tan \theta_{3V} = \frac{J\omega\tau}{J_e + J_g\omega^2\tau^2} \; . \tag{3.4-143}$$

The curves, calculated with the values given in Eqs. (3.4-86), are shown in Fig. 3.4-21. They are seen to be virtually mirror images, with respect to the logarithmic time or frequency axis, of the curves in Fig. 3.4-20. The inflections occur for

$$t_{i3V}/\tau = 1 - (J/J_e)\exp(-t_{i3V}/\tau) \tag{3.4-144}$$

and

$$(\omega_{i3V}\tau)^4 = J_e/J_g \; , \tag{3.4-145}$$

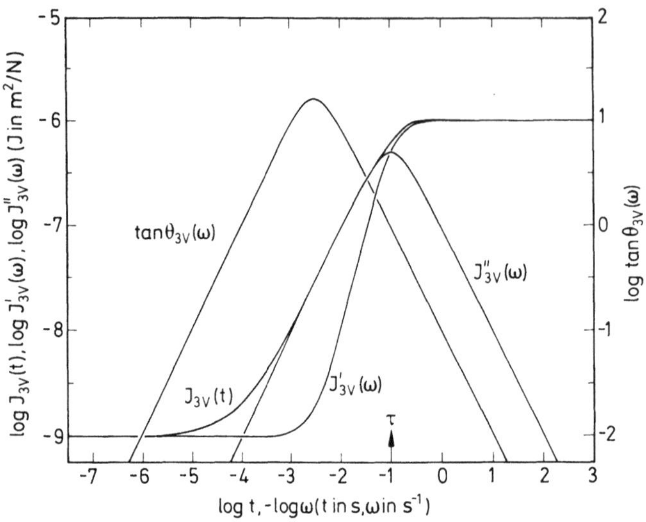

Fig. 3.4-21. Response of the 3-parameter Voigt model to the standard stimuli

i.e. at $\log t_{i3V} = -2.356$, and at $-\log \omega_{i3V} = \log \tau + 0.25 \log(J_g/J_e)$, i.e. at $\log \omega_{i3V} = 1.75$. The maximum in the loss compliance occurs at

$$\omega_{m3V}\tau = 1 \tag{3.4-146}$$

which gives $\log \omega_{m3V} = 1.0$. For the loss tangent we find

$$(\omega_{\theta 3V}\tau)^2 = J_e/J_g \ , \tag{3.4-147}$$

i.e. $\log \omega_{\theta 3V} = -\log \tau + 0.5 \log(J_e/J_g) = 2.5$. Thus

$$\omega_{\theta 3M} = \omega_{\theta 3V} = \omega_{\theta 3} \tag{3.4-148}$$

so that

$$\omega_{m3V} < \omega_{\theta 3} < \omega_{m3M} \ . \tag{3.4-149}$$

This is, of course, a consequence of Eqs. (3.4-123) and the fact to be derived in Sect. 3.6 and already utilized in obtaining Eqs. (3.4-93) from Eqs. (3.4-86), that $\tau_{3V} = (G_g/G_e)\tau_{3M}$.

Since two conjugate models represent the same behavior it is quite generally true that

$$J''_V(\omega)/J'_V(\omega) = G''_M(\omega)/G'_M(\omega) = \tan \theta(\omega) \tag{3.4-150}$$

in accordance with Eq. (2.5-21). Hence, quite generally,

$$\omega''_{mJ} < \omega_\theta < \omega''_{mG} \ , \tag{3.4-151}$$

i.e. the maximum in the loss compliance lies to the left, and that in the loss modulus lies to the right of the maximum in the loss tangent when these are plotted as functions of log ω.

The retardance of the 4-parameter Voigt model is

$$\bar{U}_{4V}(s) = J_g + \frac{J}{1 + \tau s} + \phi_f/s \tag{3.4-152}$$

and we obtain

$$J_{4V}(t) = J_g + J[1 - \exp(-t/\tau)] + \phi_f t \tag{3.4-153}$$

$$J'_{4V}(\omega) = J_g + \frac{J}{1 + \omega^2\tau^2} \tag{3.4-154}$$

$$J''_{4V}(\omega) = \frac{J\omega\tau}{1 + \omega^2\tau^2} + \phi_f/\omega \tag{3.4-155}$$

and

$$\tan\theta_{4V}(\omega) = \frac{J\omega^2\tau + \phi_f(1 + \omega^2\tau^2)}{\omega(J_e + J_g\omega^2\tau^2)} \ . \tag{3.4-156}$$

The curves calculated with the values given by Eqs. (3.4-86) are shown in Fig. 3.4-22.

The course of $\log J'_{4V}(\omega)$ is identical with that of $\log J'_{3V}(\omega)$ and has the same inflection point. This is, of course, a consequence of the identity of Eqs. (3.4-141) and (3.4-154). At short times and at high frequencies the behavior of the other three curves is virtually identical with those of the 3-parameter model. At long times and at low frequencies, however, the added dashpot changes the behavior. There are now two inflection points in $J_{4V}(t)$, one, slightly to the right from that in $\log J_{3V}(t)$, at $\log t = -2.164$, the other at $\log t = -0.208$. In the limit as t approaches infinity, $\log J_{4V}(t)$ diverges because of the presence of the steady-flow term in Eq. (3.4-153). By Eq. (3.4-117), $J(\infty) = \infty$. However, the model possesses a *steady-state compliance* [cf. Eq. (3.4-24)] given by

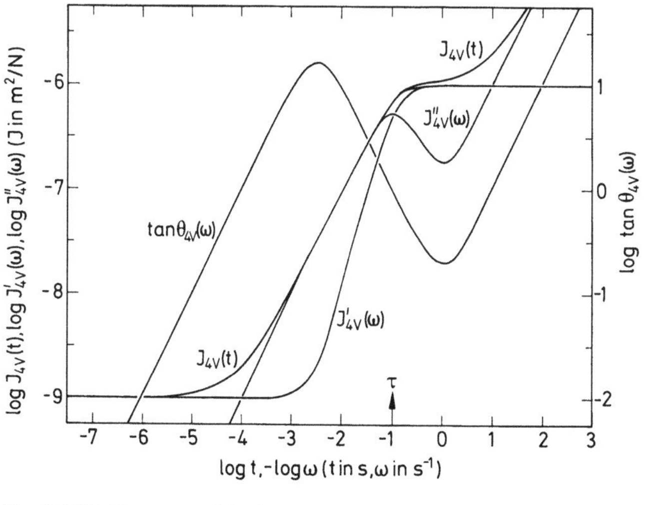

Fig. 3.4-22. Response of the 4-parameter Voigt model to the standard stimuli

$$J_e^\circ = \lim_{t\to\infty} [J_{4V}(t) - \phi_f t] = \lim_{t\to\infty} \{J_g + J[1 - \exp(-t/\tau)]\} = J_g + J \ . \quad (3.4\text{-}157)$$

In terms of the retardance we have

$$J_e^\circ = \lim_{s\to 0} [\bar{U}(s) - \phi_f/s] \ . \quad (3.4\text{-}158)$$

The reciprocal of J_e° is the *steady state modulus*, G_e°, and we have

$$J_e^\circ G_e^\circ = 1 \quad (3.4\text{-}159)$$

[cf. Eq. (3.4-123), and Sect. 4.6].

The loss functions, $J''_{4V}(\omega)$, and $\tan\theta_{4V}(\omega)$, now have minima at $\log\omega = 0.006$ and $\log\omega = 0.0022$, and inflection points at $\log\omega = 0.499$, and $\log\omega = 1.249$, respectively. The maximum in $\log\tan\theta_{4V}(\omega)$ appears at $\log\omega = 2.500$, as in $\log\tan\theta_{3V}(\omega)$, but the maximum in $\log J''_{4V}(\omega)$ is shifted to $\log\omega = 0.991$ from $\log\omega \doteq 1$ in $\log J''_{3V}(\omega)$.

Turning now to the 4-parameter Maxwell model, we note that its relaxance is

$$\bar{Q}_{4M}(s) = \frac{G_0\tau_0 s}{1 + \tau_0 s} + \frac{G_1\tau_1 s}{1 + \tau_1 s} \quad (3.4\text{-}160)$$

from which we have

$$G_{4M}(t) = G_0\exp(-t/\tau_0) + G_1\exp(-t/\tau_1) \quad (3.4\text{-}161)$$

$$G'_{4M}(\omega) = \frac{G_0\omega^2\tau_0^2}{1 + \omega^2\tau_0^2} + \frac{G_1\omega^2\tau_1^2}{1 + \omega^2\tau_1^2} \quad (3.4\text{-}162)$$

$$G''_{4M}(\omega) = \frac{G_0\omega\tau_0}{1 + \omega^2\tau_0^2} + \frac{G_1\omega\tau_1}{1 + \omega^2\tau_1^2} \quad (3.4\text{-}163)$$

and

$$\tan\theta_{4M}(\omega) = \frac{\eta_0 + \eta_1 + (\eta_0\tau_1^2 + \eta_1\tau_0^2)\omega^2}{(\eta_0\tau_0 + \eta_1\tau_1)\omega + (\eta_0\tau_1 + \eta_1\tau_0)\tau_0\tau_1\omega^3} \ . \quad (3.4\text{-}164)$$

The curves calculated with the values given by Eqs. (3.4-96) are shown in Fig. 3.4-23.

At short times and high frequencies the behavior is virtually identical with part of the 3-parameter Maxwell model. The inflections in $\log G_{4M}(t)$ and $\log G'_{4M}(\omega)$ occur at $\log t = -3.269$ and $\log\omega = 3.253$, and the maxima in $\log G''_{4M}(\omega)$ and $\log\theta_{4M}(\omega)$ are found at $\log\omega = 4.005$ and 2.500, respectively. At long times and low frequencies the behavior of the 3- and 4-parameter Maxwell models differs. The curves for $\log G_{4M}(t)$ and $\log G'_{4M}(\omega)$ do not remain level but begin to drop at around $\log t = -\log\omega = 0$. The curve for $\log G''_{4M}(\omega)$ now shows a second maximum at $\log\omega = -0.996$ with a minimum at $\log\omega = 0.011$, and two inflection points at $\log\omega = 2.000$ and -0.503 respectively. The logarithm of the loss tangent, $\log\theta_{4M}(\omega)$, now has an inflection point at $\log\omega = 1.249$, followed by a minimum at $\log\omega = 0.995$. The curve then rises to approach ∞ instead of $-\infty$ for very low values of the frequency. We note that the maxima in the $\log G'_{4M}(\omega)$ curve occur

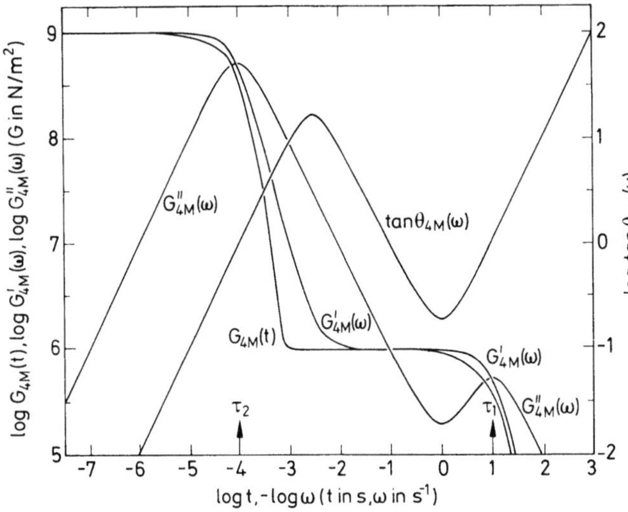

Fig. 3.4-23. Response of the 4-parameter Maxwell model to the standard stimuli

very nearly at $-\log \omega = \log \tau_2$ and $\log \tau_1$, respectively. We will show in Sect. 3.5.1.2 that this behavior can be expected only when the relaxation times are spaced sufficiently far from each other. The effects of closely spaced relaxation times superpose and interact.

We close this section by a brief examination of the response of the standard Maxwell models to a slope excitation. For the 3-parameter Maxwell model we have

$$\eta_{3M}(t) = G_e t + G\tau[1 - \exp(-t/\tau)] \tag{3.4-165}$$

$$\eta'_{3M}(\omega) = \frac{G\tau}{1 + \omega^2\tau^2} \tag{3.4-166}$$

and

$$\eta''_{3M}(\omega) = \frac{G_e}{\omega} + \frac{G\omega\tau^2}{1 + \omega^2\tau^2} \,. \tag{3.4-167}$$

The behavior is shown in Fig. 3.4-24.

For the 4-parameter Maxwell model we obtain

$$\eta_{4M}(t) = G_0\tau_0[1 - \exp(-t/\tau_0)] + G_1\tau_1[1 - \exp(-t/\tau_1)] \tag{3.4-168}$$

$$\eta'_{4M}(\omega) = \frac{G_0\tau_0}{1 + \omega^2\tau_0^2} + \frac{G_1\tau_1}{1 + \omega^2\tau_1^2} \tag{3.4-169}$$

and

$$\eta''_{4M}(\omega) = \frac{G_0\omega\tau_0^2}{1 + \omega^2\tau_0^2} + \frac{G_1\omega\tau_1^2}{1 + \omega^2\tau_1^2} \,. \tag{3.4-170}$$

The behavior is shown in Fig. 3.4-25.

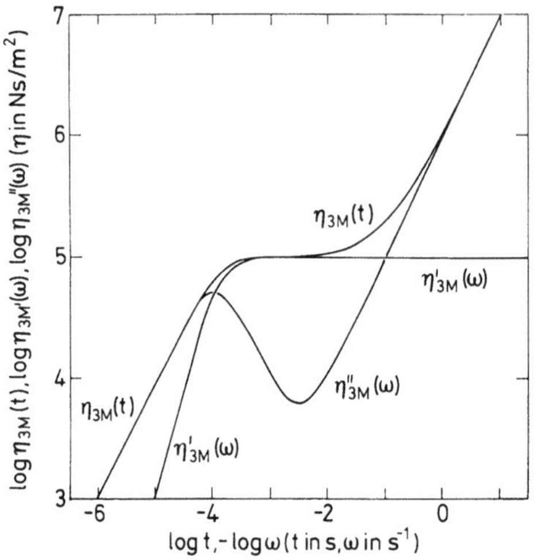

Fig. 3.4-24. Response of the 3-parameter Maxwell model to a slope strain

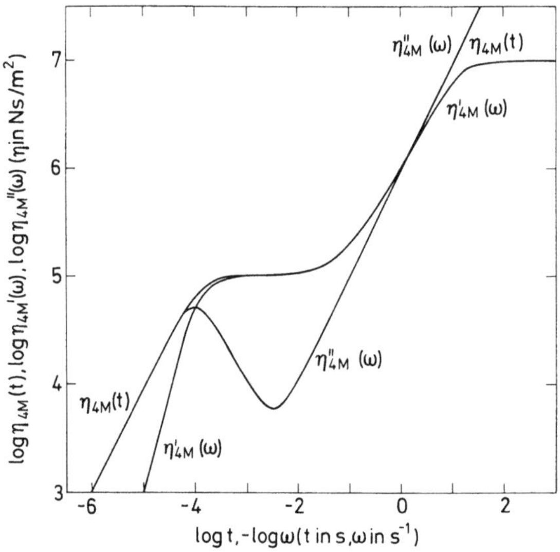

Fig. 3.4-25. Response of the 4-parameter Maxwell model to a slope strain

The response of the two models is seen to be qualitatively very similar at short times and at high frequencies.

3.5 Models with Large Numbers of Elements

The standard 3- and 4-parameter series-parallel models allow us to gain insight into the physical meaning of such typical viscoelastic phenomena as stress relaxation, strain retardation or creep, steady flow viscosity and fluidity, instantaneous and equilibrium modulus and compliance, and concepts such as relaxation and retardation time. The models are useful also in modelling the response to more complex stimuli (see Chap. 7) because the equations describing their behavior are simple, requiring at most the solution of cubic equations.

Occasionally, the standard 3- or 4-parameter models may describe observed behavior with reasonably good approximation. Generally, however, the representation of the behavior of most viscoelastic materials requires models with a large or even infinite number of elements. Such models can easily be derived from an extension, or generalization, of the 3- and 4-parameter series-parallel models. The generalized series-parallel models (just like the 3- and 4-parameter models which obviously form a subset of the generalized ones) have two forms: the generalized Maxwell models which are useful when the excitation is a strain, and the generalized Voigt models which are to be used when the excitation is a stress. The two forms are interrelated by Alfrey's rules (cf. Sect. 3.4.1.1). The generalized series-parallel models will be discussed in Sect. 3.5.1 below. Clearly, these models in their turn, form only a subset of all possible linear models. Some comments pertaining to non-series-parallel models with a large number of parameters will be made in Sect. 3.5.2.

3.5.1 The Generalized Series-Parallel Models

Only *standard* series-parallel models will be considered here. As discussed in Sect. 3.4, standard models possess a finite instantaneous modulus or, equivalently, a non-vanishing instantaneous compliance. The non-standard forms with infinite instantaneous modulus or zero instantaneous compliance will be discussed briefly in Sect. 3.5.2.

3.5.1.1 The Wiechert and Kelvin Models

By adding further Maxwell units to the standard 4-parameter model we obtain the *generalized Maxwell model* shown in Fig. 3.5-1.

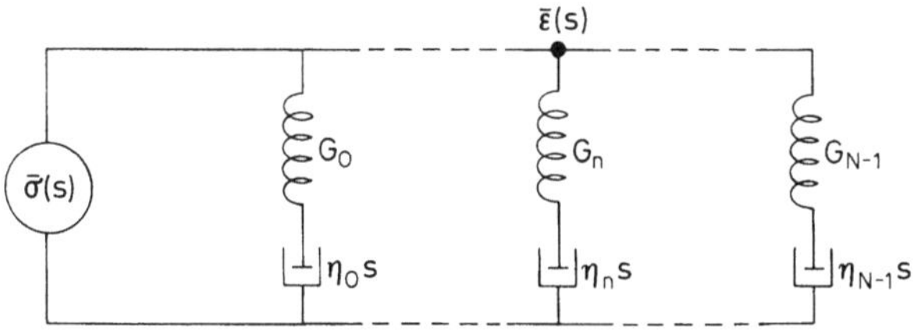

Fig. 3.5-1. Generalized Maxwell or Wiechert model describing rheodictic behavior

We write the relaxance of the model as

$$\bar{Q}(s) = \sum_n \frac{G_n \tau_n s}{1 + \tau_n s} \tag{3.5-1}$$

where the summation index n runs from 0 to N − 1. The notation includes, of course, the 4-parameter model as the special case with N = 2. The relaxation modulus corresponding to the relaxance given by Eq. (3.5-1) is

$$G(t) = \sum_n G_n \exp(-t/\tau_n) \ . \tag{3.5-2}$$

G(t) thus represents a superposition of exponentials. Mathematically such a series is known as a Prony* or Dirichlet** series. The view that viscoelastic relaxation may be regarded as a superposition of elementary processes in which the stress relaxes exponentially, had been proposed around the turn of the century by [19] J.J. Thomson*** and, particularly [20], Wiechert.**** The generalized Maxwell model may, therefore, appropriately be called *Wiechert model.*

This model, just like the standard 4-parameter Maxwell model, represents rheodictic behavior and, hence, has no equilibrium modulus. When a model representing arrheodictic behavior is required, an extra isolated spring must be added in parallel as shown in Fig. 3.5-2.

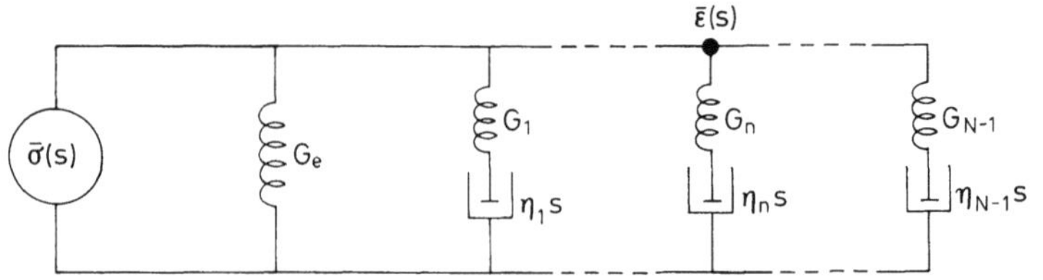

Fig. 3.5-2. Generalized Maxwell or Wiechert model describing arrheodictic behavior

This new model is an extension or generalization of the standard 3-parameter Maxwell model. Its relaxance is

$$\bar{Q}(s) = G_e + \sum_n \frac{G_n \tau_n s}{1 + \tau_n s} \ . \tag{3.5-3}$$

The summation index now runs from 1 to N − 1, and the 3-parameter model is obtained by letting N = 2.

By the choice of the way in which the elements are numbered, N denotes the number of springs in all standard generalized Maxwell models with a finite number of parameters. In the models describing rheodictic behavior the number of dashpots

* Gaspart Clair François Marie Riche, baron de Prony, pron. Pronee, 1755–1839, French engineer and mathematician.

** Gustaf Lejeune Dirichlet, pron. Deereeklay, 1805–1859, German mathematician.

*** Sir Joseph John Thomson, 1856–1940, British physicist who obtained the Nobel prize for his work on the conduction of electricity through gases.

**** Johann Emil Wiechert, pron. Veekhert, 1861–1928, German geophysicist.

is equal to the number of springs. In the models representing arrheodictic behavior, the number of springs always exceeds the number of dashpots by one. An arrheodictic Maxwell model is thus transformed into the corresponding rheodictic one by redesignating the spring labelled G_e as G_0 and inserting a dashpot labelled η_0 in series with it.

It will be convenient to introduce notation which will allow us to treat rheodictic and arrheodictic behavior simultaneously. For this purpose we use braces around the parameters of the element which must be either present or absent according to whether or not the model evinces steady-state flow. With this notation the relaxance of the Wiechert model becomes

$$\bar{Q}(s) = \{G_e\} + \sum_n \frac{G_n \tau_n s}{1 + \tau_n s} \; . \tag{3.5-4}$$

It will also be convenient to introduce another form of the relaxance. By the initial value theorem [cf. Eq. (3.4-112)] we have

$$G_g = \lim_{s\to\infty} \bar{Q}(s) = \{G_e\} + \sum_n G_n \tag{3.5-5}$$

which allows us to write

$$\bar{Q}_\bullet(s) = G_g - \sum_n \frac{G_n}{1 + \tau_n s} \; . \tag{3.5-6}$$

This *dot-form* of the relaxance is sometimes useful. We note that $\bar{Q}(s) \equiv \bar{Q}_\bullet(s)$ and shall use the dot subscript to distinguish the two forms merely for bookkeeping.

From the relationships tabulated in Table 3.4-1 we obtain the relaxation modulus of the Wiechert model as

$$G(t) = \{G_e\} + \sum_n G_n \exp(-t/\tau_n) \quad \text{or} \quad G_\bullet(t) = G_g - \sum_n G_n[1 - \exp(-t/\tau_n)] \; . \tag{3.5-7}$$

The complex modulus becomes

$$G^*(\omega) = \{G_e\} + \sum_n \frac{G_n j\omega\tau_n}{1 + j\omega\tau_n} \quad \text{or} \quad G_\bullet^*(\omega) = G_g - \sum_n \frac{G_n}{1 + j\omega\tau_n} \tag{3.5-8}$$

and its two components are

$$G'(\omega) = \{G_e\} + \sum_n \frac{G_n \omega^2 \tau_n^2}{1 + \omega^2 \tau_n^2} \quad \text{or} \quad G'_\bullet(\omega) = G_g - \sum_n \frac{G_n}{1 + \omega^2 \tau_n^2} \tag{3.5-9}$$

and

$$G''(\omega) = \sum_n \frac{G_n \omega \tau_n}{1 + \omega^2 \tau_n^2} \; . \tag{3.5-10}$$

The loss modulus has only one form. Integration of either G(t) or $G_\bullet(t)$ [cf. Eq. $(2.4\text{-}6)_2$] leads to

$$\eta(t) = \{G_e\}t + \sum_n G_n \tau_n [1 - \exp(-t/\tau_n)] = \{G_e\}t + \eta_{\{f\}} - \sum_n \eta_n \exp(-t/\tau_n) \tag{3.5-11}$$

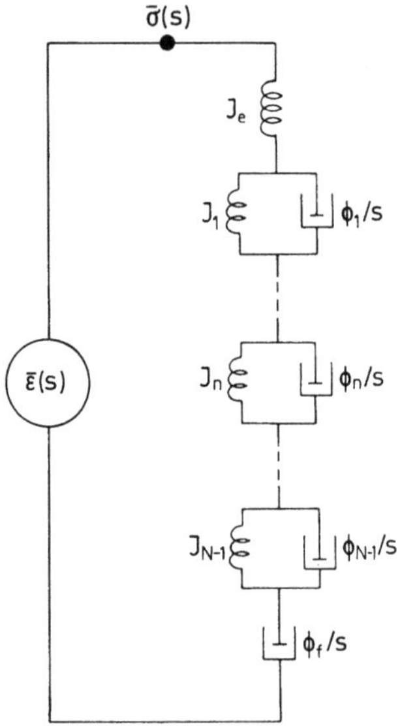

Fig. 3.5-3. Generalized Voigt or Kelvin model describing rheodictic behavior

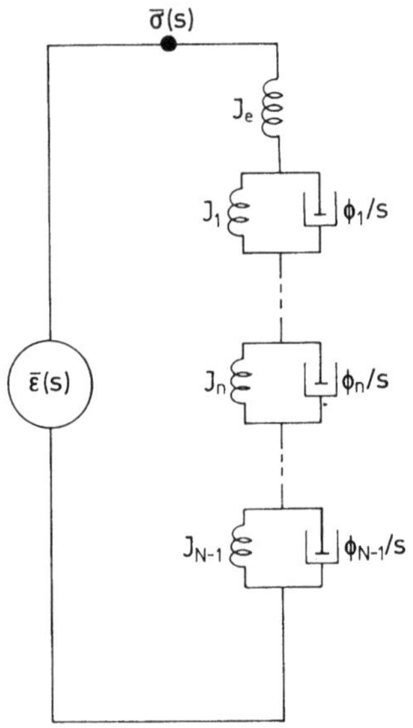

Fig. 3.5-4. Generalized Voigt or Kelvin model describing arrheodictic behavior

where

$$\eta_{\{f\}} = \sum_n \eta_n = \sum_n G_n \tau_n \ . \tag{3.5-12}$$

Thus, $\eta_{\{f\}}$ is the sum of the viscosities of all dashpots. When the behavior is rheodictic, the braces are omitted because the sum of the viscosities is then the steady-flow viscosity, η_f. When the behavior is arrheodictic, the sum does not represent the steady-flow viscosity (which would be infinite) and is then written simply as η.

By application of Alfrey's rules to the Wiechert models we obtain their conjugates, shown in Figs. 3.5-3 and 3.5-4.

These models are extensions or generalizations of the 3- and 4-parameter Voigt models (cf. Figs. 3.4-1 and 3.4-5) obtained by the addition of further Voigt units. They will be called *generalized Voigt* or *Kelvin models*, after Lord Kelvin* who contributed significantly to an early understanding of viscoelastic phenomena [21]. The retardance of these models is

$$\bar{U}(s) = J_g + \sum_n \frac{J_n}{1 + \tau_n s} + \{\phi_f/s\} \tag{3.5-13}$$

where the summation index runs from 1 to N − 1 and the braces around the last term indicate that this term is absent in the arrheodictic form of the model which does not show steady-state flow.

---

* Lord Kelvin, formerly (Sir) William Thomson, 1824–1907, British physicist.

Again we can derive an alternate form. By the final value theorem [cf. Eqs. (3.4-118) and (3.4-158)]

$$J_e^{\{o\}} = \lim_{s \to 0} [\bar{U}(s) - \{\phi_f/s\}] = J_g + \sum_n J_n \tag{3.5-14}$$

where $J_e^{\{o\}}$ is the equilibrium compliance, J_e, in the absence of steady-state flow, and the steady-state compliance, J_e°, in the presence of it. With Eq. (3.5-13) we have the dot-form

$$\bar{U}_\bullet(s) = J_e^{\{o\}} - \sum_n \frac{J_n \tau_n s}{1 + \tau_n s} + \{\phi_f/s\} \ . \tag{3.5-15}$$

Again, $\bar{U}(s) \equiv \bar{U}_\bullet(s)$ and the dot will be used merely for bookkeeping. We note the similarities of the $\bar{Q}(s)$ and $\bar{U}_\bullet(s)$ forms on the one hand, and the $\bar{Q}_\bullet(s)$ and $\bar{U}(s)$ forms on the other hand.

By the relations tabulated in Table 3.4-1 we obtain the creep compliance of the Kelvin model as

$$J(t)_\bullet = J_g + \sum_n J_n[1 - \exp(-t/\tau_n)] + \{\phi_f t\}$$

or

$$J_\bullet(t) = J_e^{\{o\}} - \sum_n J_n \exp(-t/\tau_n) + \{\phi_f t\} \ . \tag{3.5-16}$$

The complex compliance becomes

$$J^*(\omega) = J_g + \sum_n \frac{J_n}{1 + j\omega\tau_n} + \{\phi_f/j\omega\}$$

or

$$J_\bullet^*(\omega) = J_e^{\{o\}} - \sum_n \frac{J_n j\omega\tau_n}{1 + j\omega\tau_n} + \{\phi_f/j\omega\} \tag{3.5-17}$$

and the real and imaginary components are

$$J'(\omega) = J_g + \sum_n \frac{J_n}{1 + j\omega^2\tau_n^2}$$

or

$$J'_\bullet(\omega) = J_e^{\{o\}} - \sum_n \frac{J_n \omega^2\tau_n^2}{1 + \omega^2\tau_n^2} \tag{3.5-18}$$

and

$$J''(\omega) = \sum_n \frac{J_n \omega\tau_n}{1 + \omega^2\tau_n^2} + \{\phi_f/\omega\} \tag{3.5-19}$$

respectively. Again we note certain formal corespondences between the alternative forms of the time-dependent and frequency-dependent moduli and compliances.

Since the equations introduced in this section contain the standard 3- and 4-parameter series-parallel models as the special cases with $N = 2$, these models also possess the alternate forms derivable from $\bar{Q}_\bullet(s)$ and $\bar{U}_\bullet(s)$.

The pole-zero patterns of the Wiechert and Kelvin models shown in Figs. 3.5-1 to 3.5-4 are essentially the same as those of the standard 3- and 4-parameter series-parallel models (cf. Figs. 3.4-18 and 3.4-19) but the number of poles and zeros is larger. The number of poles of a rheodictic model is evidently equal to the number of dashpots, N. In an arrheodictic model the number is N − 1. The number of zeros equals that of the poles. A rheodictic retardance necessarily has a pole at the origin. This is associated with the state-state fluidity. Equivalently, a rheodictic relaxance possesses a zero at the origin. Equation (3.4-101) can be generalized to

$$\bar{R}(s) = A \prod_n \frac{s - s_{0n}}{s - s_{\infty n}} \tag{3.5-20}$$

where, again, $\bar{R}(s)$ is the respondance, A is a numerical scale factor, and s_{0n} and $s_{\infty n}$ are the nth zero, and n^{th} pole, respectively. Any linear combination of linear springs and dashpots always leads to Eq. (3.5-20). The behavior predicted by Eq. (3.5-20) is in excellent qualitative agreement with the observed behavior of viscoelastic materials in infinitesimal deformations. This forms, ultimately, the justification for representing linear viscoelastic behavior by rheological models.

It is clear from Eq. (3.5-20) that the zeros of the relaxance (i.e. of the Maxwell model) are the poles of the retardance (i.e. of the conjugate Voigt model) and *vice versa*. The two models are simply the most convenient way to represent relaxance and retardance, respectively. Evidently, the zeros and poles alternate. This is displayed graphically in Figs. 3.5-5 through 3.5-8. Because the only singularities of the respondances are poles on the negative real axis of the transform plane, we may plot them as functions of the transform variable, s. Figure 3.5-5 shows a schematic plot of the relaxance, $\bar{Q}^{\circ}(s)$, of a rheodictic model.* A similar plot of a rheodictic retardance, $\bar{U}^{\circ}(s)$, is shown in Fig. 3.5-6.

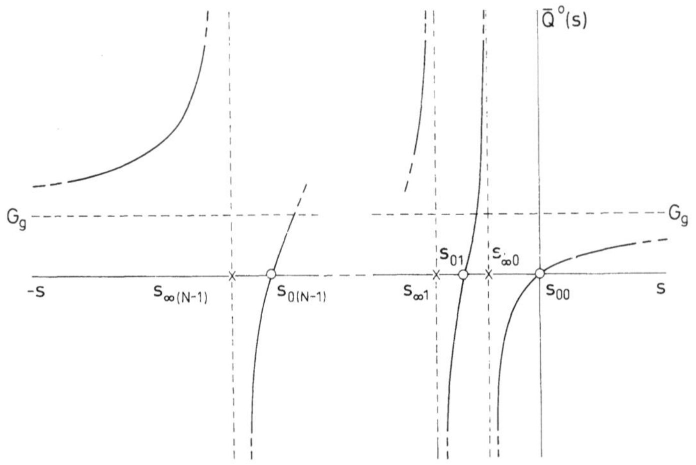

Fig. 3.5-5. $\bar{Q}^{\circ}(s)$ as a function of s

* The superscripts ° and × will be used to designate rheodictic and arrheodictic respondances whenever desirable.

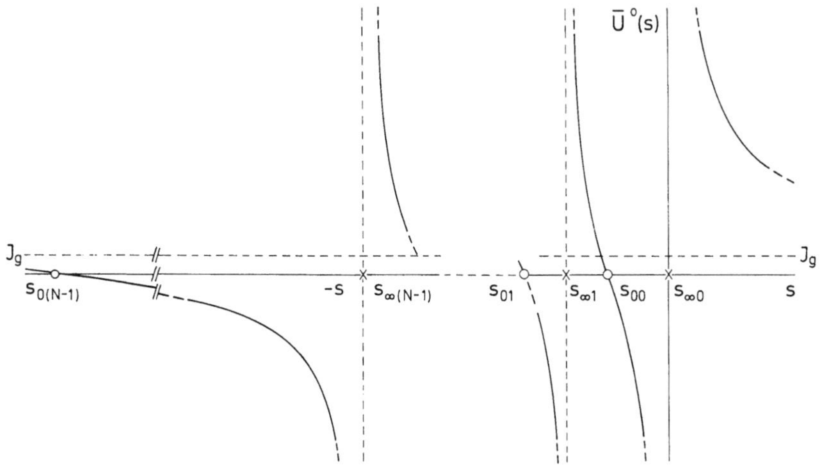

Fig. 3.5-6. $\bar{U}^{\circ}(s)$ as a function of s

Both figures may be compared with Figs. 3.4-19b and 3.4-18b. The relaxance of the rheodictic Wiechert model has essentially the same pole-zero pattern as the 4-parameter Maxwell model while the retardance of the rheodictic Kelvin model is comparable to that of the 4-parameter Voigt model. Ranking the poles and zeros according to their magnitude (i.e. their location along the negative real axis of the plane) we otain

$$s_{00} > s_{\infty 0} > s_{01} > s_{\infty 1} > \cdots > s_{0n} > s_{\infty n} > \cdots > s_{0(N-1)} > s_{\infty(N-1)} \tag{3.5-21}$$

for the relaxance. For the retardance we have

$$s_{\infty 0} > s_{00} > s_{\infty 1} > s_{01} > \cdots > s_{\infty n} > s_{0n} > \cdots > s_{\infty(N-1)} > s_{0(N-1)} \ . \tag{3.5-22}$$

Figures 3.5-7 and 3.5-8 show similar plots for an arrheodictic relaxance, $\bar{Q}^{\times}(s)$, and an arrheodictic retardance, $\bar{U}^{\times}(s)$. The pole-zero patterns of these figures may be compared with those of Figs. 3.4-19a and 3.4-18b for the 3-parameter Maxwell and Voigt models.

We now have

$$s_{01} > s_{\infty 1} > \cdots > s_{0n} > s_{\infty n} > \cdots > s_{0(N-1)} \tag{3.5-23}$$

for the relaxance and

$$s_{\infty 1} > s_{01} > \cdots > s_{\infty n} > s_{0n} > \cdots > s_{\infty(N-1)} > s_{0(N-1)} \tag{3.5-24}$$

for the retardance. It is thus clear that poles and zeros alternate in the respondances. Now, for the relaxances (the Maxwell models) the poles are the negative reciprocals of the relaxation times, i.e. $s_{\infty n} = -1/\tau_{Mn}$. The zeros are the negative reciprocals of the retardation times of the conjugate Voigt models, i.e. $s_{0n} = -1/\tau_{Vn}$, where the subscripts M and V distinguish between the two respondance times. For the retardances these relations are evidently reversed, i.e. $s_{\infty n} = -1/\tau_{Vn}$, and $s_{0n} = -1/\tau_{Mn}$.

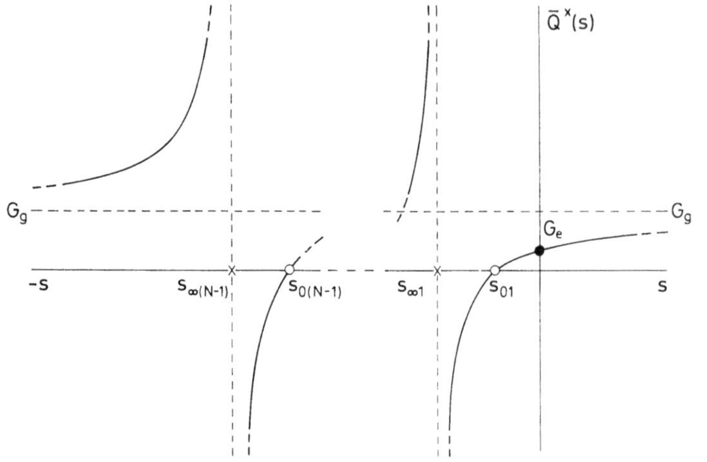

Fig. 3.5-7. $\bar{Q}^{\times}(s)$ as a function of s

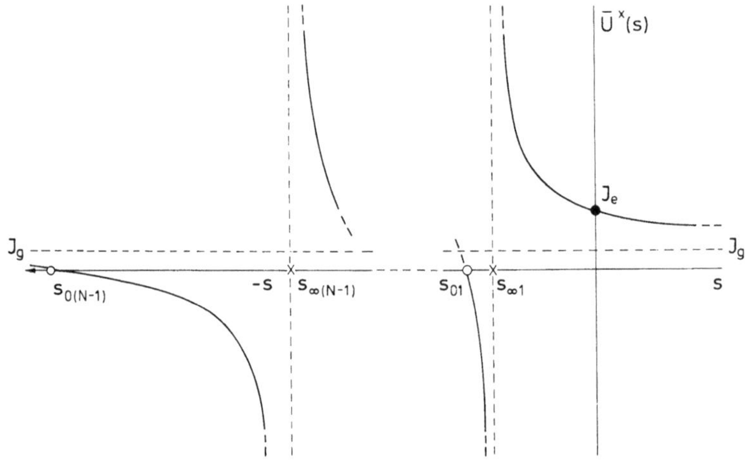

Fig. 3.5-8. $\bar{U}^{\times}(s)$ as a function of s

We also recognize that the retardation time associated with the steady-flow term, i.e. with $s_{\infty 0}$ of $\bar{U}^{\circ}(s)$, is infinite. Leaving this out, the respondance times become ranked as

$$\{\tau_{M0}\} > \tau_{V1} > \tau_{M1} > \cdots > \tau_{Vn} > \tau_{Mn} > \cdots > \tau_{V(N-1)} > \tau_{M(N-1)} \qquad (3.5\text{-}25)$$

where τ_{M0} is the *terminal* (or *longest*, or *zeroth*) *relaxation time* of a rheodictic model. For an arrheodictic model the largest respondance time is always the first retardation time. The smallest respondance time is always a relaxation time. Retardation and relaxation times alternate.

3.5.1.2 Spectral Strength and Time Dependence in Wiechert and Kelvin Models

We are now ready to examine the effect of particular sets, i.e. *distributions*, of model elements on the viscoelastic response of the series-parallel models. We first introduce the concept of *spectral strength*. The summations in Eqs. (3.5-1) through (3.5-4) imply that each relaxation time, τ_n, is associated with a definite modulus, G_n. We may thus talk of a *spectrum* of moduli. This idea will, indeed, be developed further in Chap. 4. We call G_n the *spectral strength* at $t = \tau_n$.

Instead of considering the spectral strength at $t = \tau_n$ to arise from a single modulus, G_n, we may alternatively consider it as resulting from a total of G_n distinct relaxation times, all of *unit* spectral strength, degenerating to superpose at $t = \tau_n$ on the time scale. For convenience, and without loss of generality, we assume here that G_n is an integer. When applied to a 3-parameter Maxwell model these two interpretations correspond to the two alternatives shown in Fig. 3.5-9.

Figure 3.5-9a represents the standard 3-parameter model consisting of an isolated spring for the equilibrium modulus, G_e, and a Maxwell unit with spring G and dashpot η. The relaxance is thus given by

$$\bar{Q}_a(s) = G_e + \frac{G\tau s}{1 + \tau s} \, . \tag{3.5-27}$$

In Fig. 3.5-9b, the equilibrium modulus is represented by G_e springs, all of unit modulus, superposing at $t = 0$, and the single Maxwell unit with spring G and dashpot η is replaced by G units each having springs of unit modulus and dashpots of viscosity η. The relaxance is thus

$$\bar{Q}_b(s) = \sum_1^{G_e} 1 + \sum_1^{G} \frac{\tau s}{1 + \tau s} = G_e + \frac{G\tau s}{1 + \tau s} \, . \tag{3.5-28}$$

It is often useful to think of *relaxation strength* in the sense of Eq. $(3.5\text{-}28)_1$.

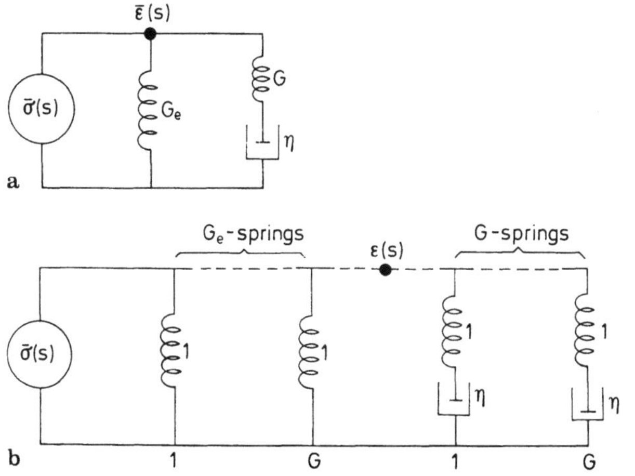

Fig. 3.5-9a,b. Two equivalent Maxwell models illustrating the concept of spectral strength

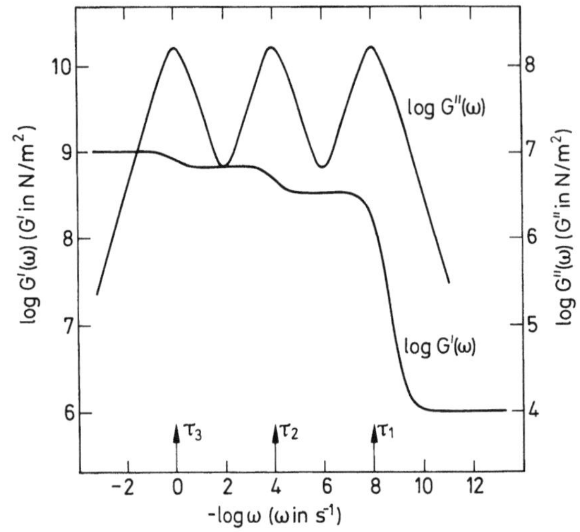

Fig. 3.5-10. Plots of G′(ω) and G″(ω) against frequency in logarithmic coordinates for a spectrum containing three equally weighted relaxation times with a spacing of four decades between neighboring relaxation times

In relation to a Kelvin model we may consider the *retardation strength* to be given either by a compliance J_n associated with the retardation time τ_n, or by the sum of J_n distinct retardation times, all associated with unit compliance, superposing at $t = \tau_n$. Only Wiechert models will be considered further on. Application of analogous considerations to Kelvin models is straightforward.

Let us now consider the effect of changing the spectral strength distribution in a Wiechert model. Figures 3.5-10 and 3.5-11 show plots of G′(ω) and G″(ω) in logarithmic coordinates for a discrete spectrum of three relaxation times separated from each other by four decades on the frequency scale. In Fig. 3.5-10, $G_e = 10^6$ N/m^2, and the three G_n were all chosen equal to 0.333×10^9 N/m^2 while in Fig. 3.5-11 the values of the G_n were 0.9×10^9, 0.9×10^8, and 0.9×10^7 N/m^2, respectively, for the corresponding τ_n in order of increasing relaxation time. Comparison of these two figures reveals that the numerical value of G_n can have a significant effect on the shape of the response curves. Using the concept of degenerate relaxation times superposing at certain points along the time scale the spectrum giving rise to Fig. 3.5-11 can be considered as containing 0.9×10^9 relaxation times of unit spectral strength at $t = \tau_3$, 0.9×10^8 relaxation times at $t = \tau_2$, and 0.9×10^7 relaxation times at $t = \tau_1$. We might say that the spectrum is weighted a hundred times more at τ_3, and ten times more at τ_2, then it is at τ_1. This is symbolized by the unequal lengths of the arrows.

In Figs. 3.5-10 and 3.5-11 the relaxation times τ_1, τ_2, and τ_3 were spaced sufficiently far from each other on the time scale to allow the associated spectral strengths to exert their influence effectively independently. We now proceed to consider the effect of a closer spacing of the relaxation times.

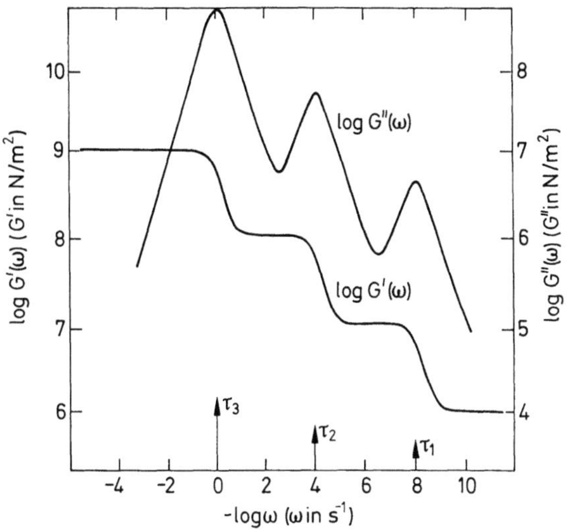

Fig. 3.5-11. Plots of G′(ω) and G″(ω) for a spectrum containing unequally weighted relaxation times with a spacing of four decades; shorter relaxation times are weighted more heavily to give equal drops in the G′(ω) curve

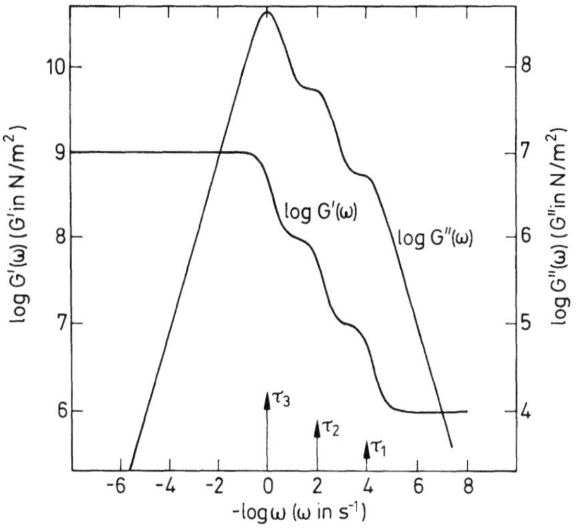

Fig. 3.5-12. Plots of G′(ω) and G″(ω) for a spectrum as in Fig. 3.5-11 with a spacing of two decades

The curves in Figs. 3.5-12 and 3.5-13 were generated using a spectrum weighted as in Fig. 3.5-11, but decreasing the spacing between relaxation times. The separation was four decades in Fig. 3.5-11, two decades in Fig. 3.5-12, and one decade in Fig. 3.5-13. It is apparent from Fig. 3.5-11 that a spacing of four decades is sufficient for the effect of the relaxation times to be completely independent of each other. The storage moduli drop from instantaneous to equilibrium values in equal steps

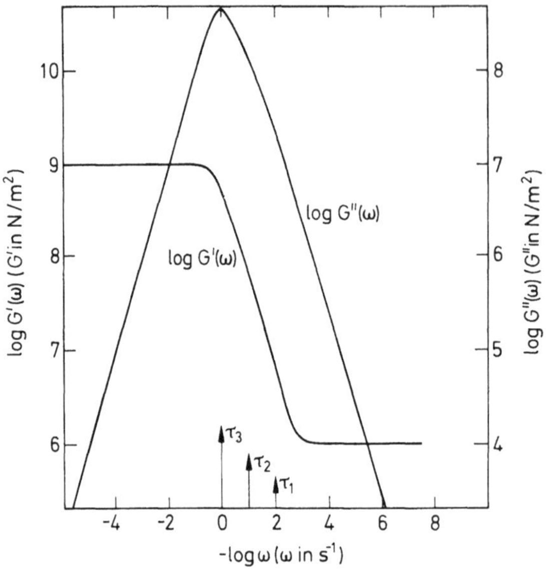

Fig. 3.5-13. Plots of $G'(\omega)$ and $G''(\omega)$ for a spectrum as in Fig. 3.5-11 with a spacing of one decade

with a distinct plateau between each step. The loss modulus has three distinct peaks with the highest peak occurring at $t = \tau_3$ and the two smaller peaks occurring at $t = \tau_2$ and $t = \tau_1$, respectively.

Moving the relaxation times to a smaller spacing of two decades results in the beginning of interaction between neighboring relaxation times. The plateaus between drops in the storage modulus are no longer well defined. The loss modulus now shows shoulders at each of the two highest relaxation times with a distinct peak remaining at the lowest relaxation time, τ_3. Moving the relaxation times to a spacing of one decade results in behavior qualitatively similar to that observed for a single relaxation time. The loss modulus has a single peak at the lowest relaxation time, and the storage modulus passes through its transition in a single step. The shapes of the curves are different from those obtained from a single relaxation time, the multiple transition-time curves being extended to the right.

We may now consider the effect of adding relaxation times to the spectrum. Figure 3.5-14 shows plots of the storage and loss moduli for three cases. Curves A were calculated for a single relaxation time, τ_3, with $G_3 = 1 \times 10^7$ N/m^2. Curves B had a second relaxation time, τ_2, added to the spectrum at a distance of one decade upscale, with $G_3 = 0.969 \times 10^9$ and $G = 0.309 \times 10^8$ N/m^2. Curves C were generated using three values of τ_i spaced one decade from each other with $G_3 = 0.9 \times 10^9$, $G_2 = 0.9 \times 10^8$, and $G_1 = 0.9 \times 10^7$ N/m^2. All of the curves are nearly identical below $t = \tau_3$, i.e. below $-\log \omega = 0$. However, above τ_3 the curves are shifted to the right by adding to the spectrum upscale. The peak in $\log G''(\omega)$ always appears at $t = \tau_3$, but the curves are broadened to the right by relaxation times added upscale. Again, we may consider that we are not dealing with three single relaxation times. Due to the manner in which the G_n were chosen, the time corresponding to

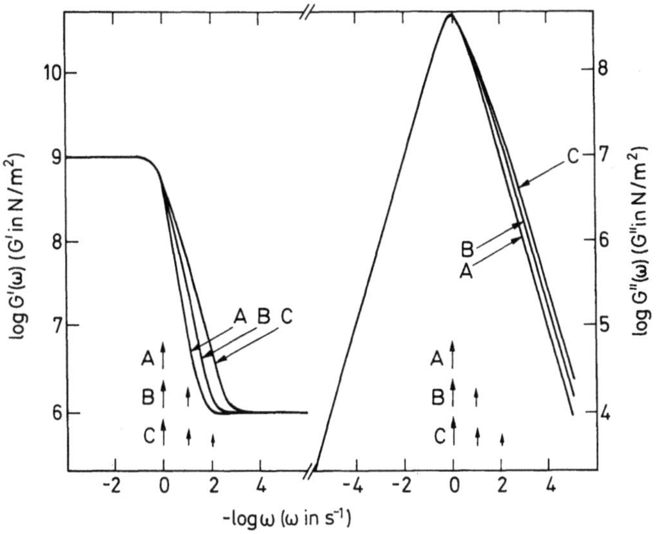

Fig. 3.5-14. Plots of G′(ω) and G″(ω) for three spectra showing the effect on the spectrum of adding longer relaxation times upscale

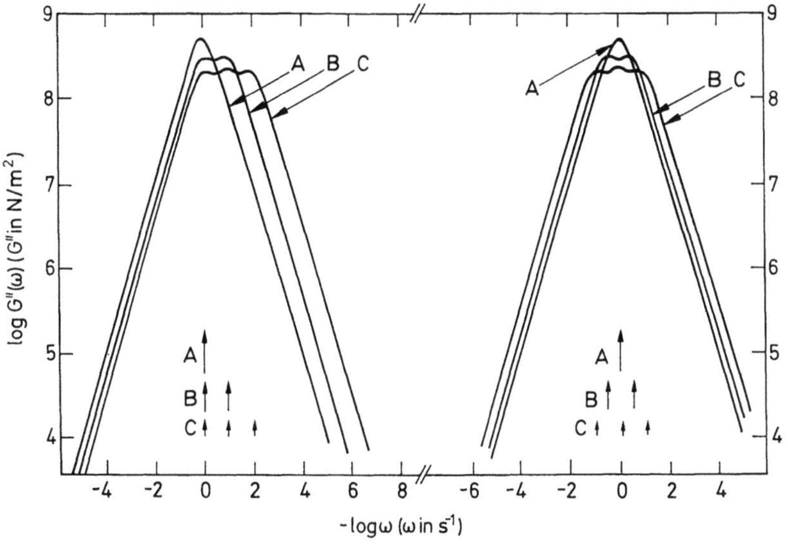

Fig. 3.5-15. Plots of G″(ω) for three spectra showing the effect on the spectrum of adding equally weighted relaxation times upscale (left), and symmetrically (right)

τ_3 may be considered to contain many more relaxation times of unit spectral strength associated with it than the others do.

Equally weighted relaxation times, i.e. relaxation times associates with the same spectral strengths, were used to generate the curves shown in Figs. 3.5-15 and 3.5-16.

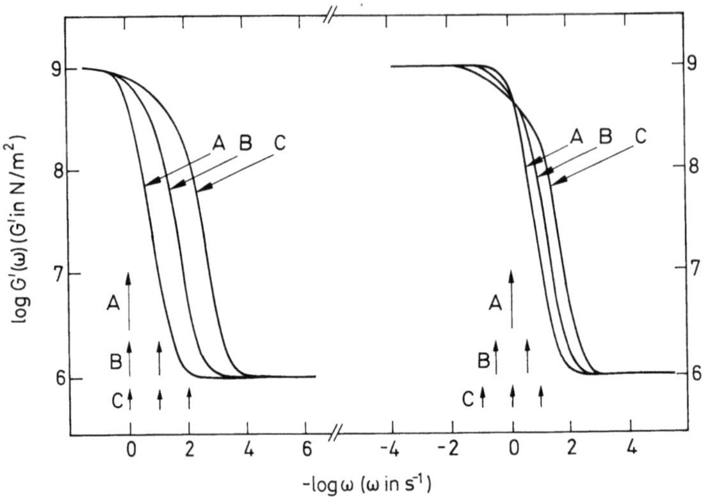

Fig. 3.5-16. Plots of G'(ω) for three spectra showing the effect on the spectrum of adding equally weighted relaxation times upscale (left), and symmetrically (right)

Curves A result from a single modulus, $G_3 = 10^9$ N/m². Curves B had $G_2 = G_3 = 0.05 \times 10^9$ N/m² and Curves C had $G_1 = G_2 = G_3 = 0.333 \times 10^9$ N/m². As in all the other figures, the curves represent arrheodictic behavior with $G_e = 10^6$ N/m². The relaxation times had values as indicated by the arrows.

From Fig. 3.5-15 it is apparent that adding relaxation times of equal strength *upscale* in the spectrum changes the position of the loss modulus peak as well as the shape of the curve. However, adding relaxation times of equal strength *sym-*

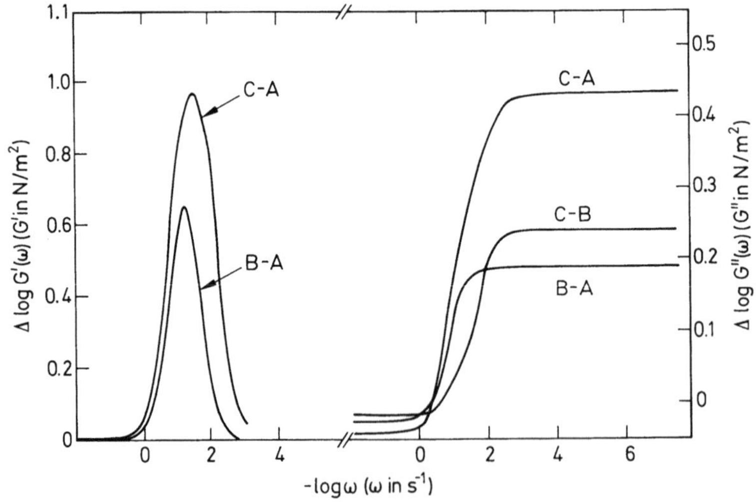

Fig. 3.5-17. Differences between the G'(ω) and between the G''(ω) curves of Fig. 3.5-14 (additional relaxation times placed upscale)

metrically about a position on the frequency axis (Fig. 3.5-15b) results in a stationary peak with symmetric broadening of the curve. Figure 3.5-16 shows the effect of the same spectral changes on the storage modulus. The curves are displaced upscale by any addition of relaxation times, symmetric or not.

From Figs. 3.5-14, 3.5-15, and 3.5-16, it is possible to obtain the differences between the curves calculated for the various spectra. Plotting this difference shows the region in which an additional relaxation time makes significant changes in the shape of the curves. The differences C-A and B-A between the curves in Fig. 3.5-14 are shown in Fig. 3.5-17.

The spectra for these curves had been weighted so that most of the relaxation times were positioned at the lowest relaxation time, τ_3. For the storage modulus (Fig. 3.5-17a), there is little effect below τ_3, i.e. below $-\log \omega = 0$. However, as shown by the B-A curve of Fig. 3.5-17, the influence of the relaxation time added upscale begins to increase rapidly at $t = \tau_3$. This influence passes through a maximum and is small for times greater than one decade above τ_2, i.e. above $-\log \omega = 2$. Similarly, the effect of two relaxation times added upscale is shown by the C-A curve. The loss moduli of Fig. 3.5-14 show (below τ_3) a slight decrease in value upon addition of relaxation times upscale (Fig. 3.5-17b). Above τ_3 the difference between the curves for one and two relaxation times (B-A) rises steeply and then levels off to a constant value around one decade above τ_2. Similar behavior is observed for the difference between curve C and curve B.

Figure 3.5-18 shows the results of a similar analysis performed on the curves of Figs. 3.5-15 and 16 for the case in which the equally weighted relaxation times were

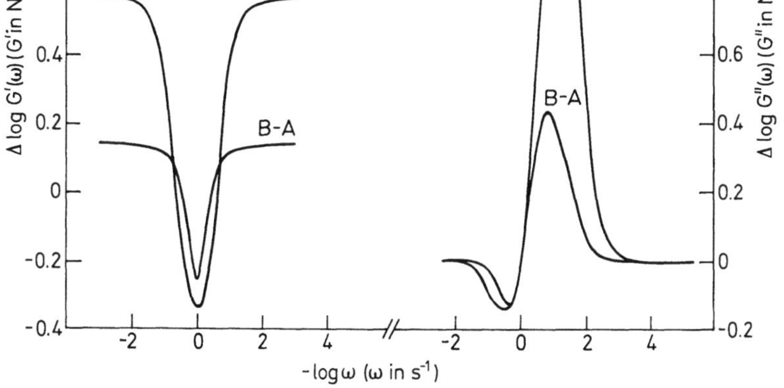

Fig. 3.5-18. Differences between the G′(ω) and between the G″(ω) curves of Figs. 3.5-15 and 3.5-16 (relaxation times placed symmetrically)

placed symmetrically about the point $-\log \omega = 0$. The effect on $G'(\omega)$ is negative below zero and then passes through a maximum. The behavior of $G''(\omega)$ is different from that observed in Fig. 3.5-17. Here the difference between curves passes through a minimum at $-\log \omega = 0$ and then rises steeply to reach a constant value for times above and below this point.

From both Figs. 3.5-17 and 3.5-18, it can be seen that the addition of one relaxation time shows a strong effect on the shape of both $G'(\omega)$ and $G''(\omega)$ over a relatively narrow interval (about two logarithmic decades) on the time scale. Outside of this range, the effect is not necessarily zero, but levels off to some constant value which is maintained over the rest of the time scale. Thus, it may be concluded that the effective range of influence of a single relaxation time is around ± 1 decade from its own position.

Considerations analogous to these may be based on retardation instead of relaxation times.

3.5.1.3 The Canonical Models

The Wiechert and Kelvin models we have just discussed contain a large but finite number of elements. It is possible and, indeed, generally useful, to let the number of elements or (more precisely, the number of Maxwell or Voigt units) approach infinity. The formulations in this section which contain summations from 0 or 1 to $N - 1$ remain valid as written but the summation index is now made to run from 0 to ∞. When this is done, it is necessary to restrict the distribution of moduli or compliances in such a manner that $G_g - \{G_e\}$ and $J_e^{\{o\}} - J_g$ remain bounded, i.e. that the sums $\Sigma_n G_n$ and $\Sigma_n J_n$ in Eqs. (3.5-5) and (3.5-14) be finite. Thus, we must have

$$\sum_{n=1}^{n=\infty} G_n = G_g - \{G_e\} \tag{3.5-29}$$

and

$$\sum_{n=1}^{n=\infty} J_n = J_e^{\{o\}} - J_g \; . \tag{3.5-30}$$

If this restriction is not imposed, Wiechert or Kelvin models with an infinite number of Maxwell or Voigt units, respectively, would not be standard models, i.e. they would have an infinite instantaneous modulus and a vanishing instantaneous compliance.

Generalized Maxwell (Wiechert) and Voigt (Kelvin) models with an infinite number of parameters, subject to restrictions Eqs. (3.5-29) and (3.5-30), respectively, will be called *canonical models*.

3.5.2 Non-Standard Series-Parallel Models

Standard generalized Maxwell of Voigt models may become non-standard by increasing the number of parameters (or elements) to infinity without restricting the sets of moduli or compliances according to Eqs. (3.5-29) or (3.5-30).

Generalization of non-standard 3- and 4-parameter series-parallel models leads to a generalized Maxwell model with relaxance

$$\bar{Q}(s) = \{G'\} + \sum_n \frac{G_n \tau_n s}{1 + \tau_n s} + \eta' s \tag{3.5-31}$$

and to a generalized Voigt model with retardance

$$\bar{U}(s) = \sum_n \frac{J_n}{1 + \tau_n s} + \left\{\frac{\phi_f}{s}\right\} . \tag{3.5-32}$$

These models are non-standard because

$$G_g = \lim_{s \to \infty} \bar{Q}(s) = \infty \tag{3.5-33}$$

and

$$J_g = \lim_{s \to \infty} \bar{U}(s) = 0 . \tag{3.5-34}$$

For rheodictic behavior $G' = 0$, and $\phi_f \neq 0$. For arrheodictic behavior G' is the equilibrium modulus, and $\Sigma_n J_n$ is the equilibrium compliance. The steady-state viscosity is given by

$$\eta_f = \lim_{s \to 0} \bar{Q}(s)/s = \sum_n \eta_n + \eta' \tag{3.5-35}$$

where $\eta_n = G_n \tau_n$.

3.5.3 Non-Series-Parallel Models with Large Numbers of Elements

Series-parallel and ladder models (see Chap. 5) form special subsets of all possible linear combinations of springs and dashpots. When the number of elements exceeds four, there always exists a number of models which are neither series-parallel nor ladder models. Their number increases rapidly with the number of elements. Section 4.5 outlines a procedure based on the calculus of residues by which any given model can be reduced to an *equivalent* series-parallel model. It will become clear in that section that the maximum number of poles that a model with a finite number of elements may possess is the number of poles possessed by the equivalent series-parallel model. If follows that, if a model contains a number of elements in excess of the number required to represent the equivalent series-parallel model, then the excess number of elements is redundant. Reduction to the series-parallel form (see Sect. 4.5) automatically eliminates the redundant elements. Simple examples of reduction to series-parallel form using the method of matching coefficients in the respondances instead of the calculus of residues are given in Problems 3.4-8, 3.4-9, 3.5-1, and 3.5-2.

The most important characteristics of any given model are:

(1) whether it is standard or non-standard; and
(2) whether it is rheodictic or arrheodictic.

This can be decided by application of the limit value theorems. These, however, require knowledge of the form of the respondances. When the number of parameters exceeds four and the model is neither a series-parallel, nor a *regular* ladder model, calculation of the respondances may be tedious. It is therefore worth knowing that the nature of any model may be ascertained simply by inspection. Two simple rules apply.

(1) If the model can be traversed from source to ground node through rigid connections and *dashpots* only, the model is *non-standard.*
(2) If the model can be traversed from source to ground node through rigid connections and *springs* only, the model is *arrheodictic.*

Examples may be found in Sect. 3.8 (Problems).

3.6 Model Fitting

In Sects. 3.5.1.1 and 3.5.1.2 we discussed the qualitative aspects of the Wiechert and Kelvin models. These models were introduced because the models of the standard solid and the standard liquid are generally inadequate for modelling the linear viscoelastic behavior of real materials. The question arises: given a set of experimental data, can one derive from these numerical values for the parameters of the Wiechert-Kelvin models, or, in other words, can one fit the models to describe observed behavior? In this section we discuss three procedures which were developed for this purpose using Wiechert-Kelvin models with a limited number of elements (about 4 to 10 or so). The procedures will be referred to as Procedure X, the Collocation Method, and the Multidata Method, respectively. A fourth method is briefly mentioned in Sect. 3.6.4.

3.6.1 Procedure X

This procedure was introduced by Tobolsky and Murakami [22, 23] and named by them Procedure X for no obvious reason. It is not applicable to the harmonic response functions. Taking the relaxation modulus first, we write Eq. (3.5-7) in logarithmic form and separate out the term with the longest relaxation time. This gives

$$\log[G(t) - \{G_e\}] = \log\left[\sum_n G_n \exp(-t/\tau_n) + G_p \exp(-t/\tau_p)\right] \tag{3.6-1}$$

where $p = 0$ for rheodictic, and 1 for arrheodictic behavior and n now runs from $p + 1$ to $N - 1$. If the largest relaxation time, τ_p, is spaced sufficiently far from the others, a plot of $\log[G(t) - \{G_e\}]$ vs. t will produce a curve whose asymptote is

$$\log[G(t) - \{G_e\}] = \log G_p - t/2.303\tau_p \,. \tag{3.6-2}$$

G_p and τ_p may be obtained from the intercept and slope, respectively, of this asymptote. One may then set

$$\log[G(t) - \{G_e\} - G_p\exp(-t/\tau_p)]$$
$$= \log\left[\sum_n G_n\exp(-t/\tau_n) + G_{p+1}\exp(-t/\tau_{p+1})\right] . \qquad (3.6\text{-}3)$$

Thus a plot of the left hand side of Eq. (3.6-3) vs. t will have an asymptote from which G_{p+1} and τ_{p+1} may be obtained. In principle, this procedure may be repeated. To yield a good fit the plots must be straight lines over a sufficiently long region to allow a satisfactory extrapolation and, hence, an adequate determination of the slope. The relaxation times thus obtained should be neither too widely nor too closely spaced (not closer than what one may reasonably expect to be resolvable). The validity of Procedure X must be checked by the reconstruction of the original data.

The procedure cannot be used with response functions which do not contain exponential kernels. It can, however, be applied to the response to a slope excitation (see Problem 3.6-1), and to creep data. The analog of Eq. (3.6-1) in the latter case is

$$\log[J_e^{\{o\}} - J_\bullet(t) + \{\phi_f t\}] = \log\left[\sum_n J_n\exp(-t/\tau_n) + J_1\exp(-t/\tau_1)\right] . \qquad (3.6\text{-}4)$$

Thus, if the data allow determination of $J_e^{\{o\}}$ first, the same procedure can be followed as for the modulus.

From an experimental point of view, Procedure X suffers from the fact that it starts from those data in a given set which are usually least reliably determined. Also, generally the relaxation or retardation times are not spaced sufficiently widely to allow more than a few of them to be determined at most. Furthermore, because of computational uncertainties, the method is not unique.

3.6.2 Collocation Method

In contrast to Procedure X, the collocation method is general and applicable to all response functions although it, too, is not unique. Again, we choose G(t) to demonstrate it.

Suppose that a discrete set of moduli, G_j, have been obtained at the times t_j. On the basis of Eq. (3.5-7), we may then write

$$G_j = G(t_j) = \{G_e\} + \sum_k G_k\exp(-t_j/\tau_k) , \qquad j = 1, 2, \ldots, K . \qquad (3.6\text{-}5)$$

Equations (3.6-5) *collocates* the observed values G_j with the values $G(t_j)$ of the theoretical expression which represents linear viscoelastic behavior. The problem is to derive an appropriate set of G_k's and τ_k's from Eqs. (3.6-5). K can always be made even. It is then, in principle, possible to solve Eqs. (3.6-5) to furnish K/2 values of G_k and τ_k. This is the basis of the collocation method [24, 25].

The set of K non-linear equations is difficult to solve. Schapery [24] has suggested a method of solution based on relating the relaxation times, τ_k, to the times, t_k, through the equation

$$\tau_k = at_k \tag{3.6-6}$$

where a is a proportionality constant which must be determined in a suitable way. Schapery's relation eliminates the τ_k's as unknowns and there remain K G_k's. To find them we rewrite Eqs. (3.6-5) in matrix form, making use, at the same time, of Eq. (3.6-6). This yields

$$A_j = B_{jk}G_k \tag{3.6-7}$$

where

$$A_j = G_j - \{G_e\} \tag{3.6-8}$$

and

$$B_{jk} = \exp(-t_j/at_k) \ . \tag{3.6-9}$$

Inversion of the matrix gives

$$G_k = (B^{-1})_{kj}A_j \tag{3.6-10}$$

as the sought-for relation for the G_k's in terms of the G_j's and t_j's.

Experimental observations on linear viscoelastic materials are usually made at time intervals which are spaced equally on a logarithmic scale. Letting this spacing be p, we have

$$t_j = 10^{m+jp} \ , \qquad j = 1, 2, \ldots, K \tag{3.6-11}$$

where m marks the beginning of the time scale, the first observation being taken at $t_1 = 10^{m+p}$.

As with Procedure X, a successful set of G_k's and τ_k's can only be derived by collocation if the collocation points are neither too widely nor too closely spaced. Generally, therefore, only a subset of the complete set of observations is collocated. Thus, if the observations are actually spaced by 0.2 logarithmic units, p may be chosen to be unity. Such a decadic spacing is usually satisfactory [26] and discussion of the collocation method will be restricted here to this spacing. The equations are easily modified for different spacings. Combining Eqs. (3.6-6) and (3.6-11) gives

$$\tau_k = a\,10^{m+k} \ . \tag{3.6-12}$$

It is seen that $\tau_k < \tau_{k+1}$ which is not in agreement with Eq. (3.5-25). This, however, is of no consequence since the τ_k's, as well as the G_k's, are curve fitting parameters which have nothing to do with the "true" relaxation times or moduli. Their ranking according to Eq. (3.6-12) has been chosen purely as a matter of convenience and could, indeed, be changed without affecting the validity of the method.

Substitution of Eq. (3.6-11) (with p = 1) into Eq. (3.6-9) yields

$$B_{jk} = \exp[-(10^{j-k}/a)] \tag{3.6-13}$$

as the generating equation for the elements of the matrix. The choice of the constant 'a' is somewhat arbitrary. Different sets of values for G_k and τ_k will be obtained for different choices of 'a'. However, whatever value is chosen, the curves reconstructed from

$$G_r(t) = \{G_e\} + \sum_k G_k \exp(-t/\tau_k) \tag{3.6-14}$$

where the subscript r refers to the curve obtained with the rth set of G_k and τ_k values, will always pass through the collocation points. Between collocation points the value of $G_r(t)$ depends only slightly on the value of 'a' if 'a' has been chosen so that the matrix does not become seriously ill-conditioned. It is usually satisfactory to choose a so that the terms on the principal diagonal of the matrix become 0.5. In the present case, Eq. (3.6-13) furnishes $a = 1/\ln 2$ when $B_{jj} = 0.5$.

In addition to a, the collocated G_k's and τ_k's also depend on the choice of m and of p for a given set of observations. The G_k do not necessarily satisfy the condition

$$\sum_k G_k = G_g - \{G_e\} \tag{3.6-14}$$

[cf. Eq. (3.5-29)]. To ensure that this condition be fulfilled, we augment the column vector A_j to contain an element

$$A_0 = G_g - \{G_e\} \tag{3.6-15}$$

and border the matrix B_{jk} with a zeroth row and a zeroth column, the elements of which are

$$\begin{aligned} B_{0k} &= 1 \ , & k &\geq 0 \\ B_{j0} &= \exp(-10^j/a) \ , & j &> 0 \ . \end{aligned} \tag{3.6-16}$$

Thus, when calculating the elements of the matrix, k is made to run from 0 to K and the elements in the first row all become unity. Equation $(3.6\text{-}16)_1$ follows from setting $t_j = t_0 = 0$ in Eq. (3.6-9). Equation $(3.6\text{-}16)_2$ is obtained by letting $k = 0$ in Eq. (3.6-13). By Eq. (3.6-12), $\tau_0 = a \cdot 10^m$ gives the relaxation time associated with G_0. Clearly, m must be chosen so that $G(t_0) \simeq G_g$.

Equation (3.6-17) below shows the matrix B_{jk} for $K = 3$ and $a = 1/\ln 2$. We have

$$B_{jk} = \begin{bmatrix} 1.000 & 1.000 & 1.000 & 1.000 \\ 0.001 & 0.500 & 0.933 & 0.993 \\ 0 & 0.001 & 0.500 & 0.933 \\ 0 & 0 & 0.001 & 0.500 \end{bmatrix} . \tag{3.6-17}$$

If zeros are substituted for the values below the principal diagonal which are finite but very small, the matrix becomes triangular. This facilitates solution of Eq. (3.6-10).

The method is easily extended to the harmonic responses. We again assume that $p = 1$ and that both j and k run from 1 to K. For the storage modulus we then have

$$G'_j = G'(\omega_j) = \{G_e\} + \sum_k \frac{G_k \omega_j^2 \tau_k^2}{1 + \omega_j^2 \tau_k^2} . \tag{3.6-18}$$

Letting

$$\omega_j = 10^{m+j} \tag{3.6-19}$$

and

$$\tau_k = 1/a\omega_k = 1/a\ 10^{m+k} \tag{3.6-20}$$

we have

$$G_k = (B^{-1})_{k_j}A_j \tag{3.6-21}$$

where

$$A_j = G'_j - \{G_e\} \tag{3.6-22}$$

and

$$B_{jk} = 1/(1 + a^2\omega_k^2/\omega_j^2) = 1/(1 + a^2\ 100^{k-j})\ . \tag{3.6-23}$$

To satisfy the condition $\Sigma_k G_k = G_g - \{G_e\}$, A_j is now augmented by the term

$$A_{K+1} = G_g - \{G_e\} \tag{3.6-24}$$

and the matrix is bordered by a $(K + 1)^{st}$ row and $(K + 1)^{st}$ column containing the elements

$$B_{j(K+1)} = 1/(1 + a^2\ 100^{K+1-j}) \tag{3.6-25}$$

and

$$B_{(K+1)k} = 1 \tag{3.6-26}$$

since we must now set $\omega_j = \omega_{K+1} = \infty$ in Eq. $(3.6\text{-}23)_1$. Again, m must be chosen so that $G'(\omega_{K+1}) \simeq G_g$. Thus, when calculating B_{jk}, k now runs from 1 to K + 1 and the elements in the last, i.e. the $(K + 1)^{st}$ row, are unity.

The condition $B_{jj} = 0.5$ is now satisfied with a = 1. For K = 3 the matrix becomes

$$B_{jk} = \begin{bmatrix} 0.500 & 0.010 & 0 & 0 \\ 0.990 & 0.500 & 0.010 & 0 \\ 1.000 & 0.990 & 0.500 & 0.010 \\ 1.000 & 1.000 & 1.000 & 1.000 \end{bmatrix} . \tag{3.6-27}$$

For the loss modulus we write

$$G''_j = G''(\omega_j) = \sum_k \frac{G_k \omega_j \tau_k}{1 + \omega_j^2 \tau_k^2}\ . \tag{3.6-28}$$

Using Eqs. (3.6-19) and (3.6-20) we find

$$G_k = (B^{-1})_{kj}A_j \tag{3.6-29}$$

where

$$A_j = G''_j \tag{3.6-30}$$

and

$$B_{jk} = 1/(a\omega_k/\omega_j + \omega_j/a\omega_k) = 1/(a\ 10^{k-j} + 10^{j-k}/a)\ . \tag{3.6-31}$$

The collocation should be made by choosing the maximum in the experimental data (which may have to be estimated) as one of the collocation points. Choosing K = 3 and a = 1, the matrix now becomes

$$B_{jk} = \begin{bmatrix} 0.500 & 0.099 & 0.010 \\ 0.099 & 0.500 & 0.099 \\ 0.010 & 0.099 & 0.500 \end{bmatrix} . \tag{3.6-32}$$

It should be noted that the condition $\Sigma_k G_k = G_g - \{G_e\}$ will, in general, not be satisfied.

We now turn to the compliances. For the creep compliance we write

$$J_j = J(t_j) = J_g + \sum_k [1 - \exp(-t_j/\tau_k)] + \{\phi_f t_j\} . \tag{3.6-33}$$

The τ_k's are still given by Eq. (3.6-12) but

$$J_k = (B^{-1})_{kj} A_j , \tag{3.6-34}$$

where

$$A_j = J_j - J_g - \{\phi_f t_j\} , \qquad j = 1, 2, \ldots, K \tag{3.6-35}$$

$$A_{K+1} = J_e^{\{o\}} - J_g \tag{3.6-36}$$

$$B_{jk} = 1 - \exp(-t_j/at_k) = 1 - \exp(-10^{j-k}/a) , \qquad k = 1, 2, \ldots, K, K+1 \tag{3.6-37}$$

and

$$B_{(K+1)k} = 1 . \tag{3.6-38}$$

The last equation results from setting $t_j = t_{K+1} = \infty$ in Eq. $(3.6\text{-}37)_1$. The condition $B_{jj} = 0.5$ gives $a = 1/\ln 2$ as before.

The matrix obtained for K = 3 is

$$B_{jk} = \begin{bmatrix} 0.500 & 0.067 & 0.007 & 0 \\ 0.999 & 0.500 & 0.067 & 0.007 \\ 1.000 & 0.999 & 0.500 & 0.067 \\ 1.000 & 1.000 & 1.000 & 1.000 \end{bmatrix} . \tag{3.6-39}$$

For the storage compliance we let

$$J'_j = J'(\omega) = J_g + \sum_k \frac{J_k}{1 + \omega_j^2 \tau_k^2} \tag{3.6-40}$$

where the ω_j's and the τ_k's are given by Eqs. (3.6-19) and (3.6-20), respectively, and

$$J_k = (B^{-1})_{kj} A_j . \tag{3.6-41}$$

Thus now

$$A_0 = J_e^{\{o\}} - J_g \tag{3.6-42}$$

$$A_j = J'_j - J_g , \qquad j = 1, 2, \ldots, K \tag{3.6-43}$$

$$B_{0k} = 1 , \qquad k = 0, 1, \ldots, K \tag{3.6-44}$$

and

$$B_{jk} = 1/(1 + \omega_j^2/a^2\omega_k^2) = 1/(1 + 100^{j-k}/a^2) \ . \tag{3.6-45}$$

Equation (3.6-44) is obtained from Eq. $(3.6\text{-}45)_1$ by setting $\omega_j = \omega_0 = 0$. For $K = 3$ and $a = 1$ the matrix becomes

$$B_{jk} = \begin{bmatrix} 1.000 & 1.000 & 1.000 & 1.000 \\ 0.010 & 0.500 & 0.990 & 1.000 \\ 0 & 0.010 & 0.500 & 0.990 \\ 0 & 0 & 0.010 & 0.500 \end{bmatrix} . \tag{3.6-46}$$

In case of the loss compliance we have

$$J_j'' = J''(\omega_j) = \sum_k \frac{J_k \omega_j \tau_k}{1 + \omega_j^2 \tau_k^2} + \{\phi_f/\omega_j\} \tag{3.6-47}$$

and

$$J_k = (B^{-1})_{kj} A_j \tag{3.6-48}$$

where

$$A_j = J_j'' - \{\phi_f/\omega_j\} \tag{3.6-49}$$

and B_{jk} is given by Eq. (3.6-31) and, for $K = 3$ and $a = 1$, by Eq. (3.6-32). Again, in general, the condition $\Sigma_k J_k = J_e^{\{o\}} - J_g$ will not be satisfied.

A collocation may be performed not only to experimental data but to, say, calculated values of the respondances (see e.g., Sect. 8.1.1). For the relaxance then,

$$\bar{Q}_j = \bar{Q}(s_j) = \{G_e\} + \sum_k \frac{G_k \tau_k s_j}{1 + \tau_k s_j} \tag{3.6-50}$$

where

$$\begin{aligned} s_j &= 10^{m+j} \\ \tau_k &= 1/as_k = 1/a\ 10^{m+k} \end{aligned} \tag{3.6-51}$$

and

$$G_k = (B^{-1})_{kj} A_j \tag{3.6-52}$$

so that

$$A_j = \bar{Q}_j - \{G_e\} \ , \qquad j = 1, 2, \ldots, K \tag{3.6-53}$$

$$A_{K+1} = G_g - \{G_e\} \ ,$$

$$B_{jk} = 1/(1 + as_k/s_j) = 1/(1 + a\ 10^{k-j}) \qquad k = 1, 2, \ldots, K, K+1 \tag{5.6-54}$$

and

$$B_{(K+1)k} = 1 \tag{3.6-55}$$

because $s_j = s_{K+1} = \infty$. For $K = 3$ and $a = 1$ the matrix becomes

$$B_{jk} = \begin{bmatrix} 0.500 & 0.091 & 0.010 & 0 \\ 0.909 & 0.500 & 0.091 & 0.010 \\ 0.990 & 0.909 & 0.500 & 0.091 \\ 1.000 & 1.000 & 1.000 & 1.000 \end{bmatrix} . \tag{3.6-56}$$

For the retardance we have

$$\bar{U}_j = \bar{U}(s_j) = J_g + \sum_k \frac{J_k}{1 + \tau_k s_j} + \{\phi_f/s_j\} \tag{3.6-57}$$

where

$$J_k = (B^{-1})_{kj} A_j \tag{3.6-58}$$

$$A_0 = J_e^{\{o\}} - J_g \tag{3.6-59}$$

$$A_j = \bar{U}_j - J_g - \{\phi_f/s_j\} \ , \qquad j = 1, 2, \ldots, K \tag{3.6-60}$$

$$B_{jk} = 1/(1 + s_j/as_k) = 1/(1 + 10^{j-k}/a) \ , \qquad k = 0, 1, \ldots, K \tag{3.6-61}$$

and

$$B_{0k} = 1 \tag{3.6-62}$$

since we must set $s_j = s_0 = 0$ in Eq. (3.6-61). For $K = 3$ and $a = 1$ the matrix is

$$B_{jk} = \begin{bmatrix} 1.000 & 1.000 & 1.000 & 1.000 \\ 0.091 & 0.500 & 0.909 & 0.990 \\ 0.010 & 0.091 & 0.500 & 0.909 \\ 0 & 0.010 & 0.091 & 0.500 \end{bmatrix} . \tag{3.6-63}$$

It has been pointed out that collocation is not unique, i.e. different sets of moduli or compliances and relaxation or retardation times may be obtained from a given set of experimental data according to the values of m, p, or a that have been selected. It is always necessary to reconstruct the original curve with the collocated parameters to ascertain whether the reconstruction produces a curve which is adequate for the purpose at hand. It may happen that some of the calculated values become negative. This results from the fact that the matrix inversion typified by Eq. (3.6-10) is mathematically an ill-posed problem. Small variations on one side of the matrix equation may result in large changes on the other side. Nevertheless, when used with caution, and always bearing in mind that the moduli, compliances, and respondance times obtained by collocation are mere curve-fitting parameters, the method does have its uses. The use of collocation in converting one viscoelastic response function into another will be discussed in Chap. 8. Examples of collocations are given in Problems 3.6-2 and 8.1-1.

3.6.3 Multidata Method

Out of a set of M observations the collocation method as described in the preceding section utilizes only a subset consisting of $K < M$ data points to determine the moduli, G_k, or compliances, J_k, corresponding to the selected values of the respondance times, τ_k. We now discuss a method, referred to as the *multidata method* [26] which can utilize all of the observations. This results in the reduction of error in the determination of the G_k's or J_k's.

To illustrate the method, let us again consider, as we have done with the collocation method, the determination of a Dirichlet-Prony series [cf. Eq. (3.5-2)] to represent the relaxation modulus, G(t), of which a given set of observations, $\{G(t_m)\}$ (where m = 1, 2, ..., M), constitutes a sample. We write the series as

$$G_K(t) = \{G_e\} + \sum_k G_k \exp(-t/\tau_k) \tag{3.6-64}$$

where k = 1, 2, ..., K. The mean-square error, E^2, arising from any mismatch between G(t), as represented by $G(t_m)$, and $G_K(t)$, is

$$E^2 = \sum_m [G(t_m) - G_K(t_m)]^2 \ . \tag{3.6-65}$$

Since the τ_k are predetermined, the error needs to be minimized with respect to the G_k's only. Since $d\,G_K(t_m)/dG_k = \exp(-t_m/\tau_k)$, the condition $dE^2/dG_k = 0$ yields

$$2\sum_m \{[G(t_m) - G_K(t_m)]\exp(-t_m/\tau_k)\} = 0 \tag{3.6-66}$$

which constitutes K equations for the K unknown G_k's. We now introduce, in analogy to Eq. (3.6-8), the column vector

$$A_m = G(t_m) - \{G_e\} \ , \tag{3.6-67}$$

and the matrices

$$B_{mk} = \exp(-t_m/\tau_k) \tag{3.6-68}$$

and

$$B^T_{lm} = \exp(-t_m/\tau_l) \ . \tag{3.6-69}$$

B_{mk} is the exponential in Eq. (3.6-64). B^T_{lm} is its transpose and is the exponential in Eq. (3.6-66). We have changed the running index in the latter from k to l = 1, 2, ..., K for proper bookkeeping. B_{mk} is an M × K, and B^T_{lm} is a K × M matrix where K < M. With the help of these singular matrices and the column vectors A_m and G_k the K equations resulting from the minimization of the error can be written in matrix notation as

$$B^T_{lm}A_m = B^T_{lm}B_{mk}G_k \tag{3.6-70}$$

were repeated indices again imply summation. Equation (3.6-70) is the basic equation of the multidata method as applied to relaxation data. We note that, when M = K, both matrices become K × K square matrices. Premultiplication by the inverse of B^T_{lm}, B^{-T}_{jl}, then yields $A_j = B_{jk}G_k$, i.e. we recover Eq. (3.6-10) of the collocation method.

All the elements of A_m, B^T_{lm}, and B_{mk} are known. Introducing the column vector

$$F_l = B^T_{lm}A_m \tag{3.6-71}$$

and the square matrix

$$C_{lk} = B^T_{lm} B_{mk} \tag{3.6-72}$$

we find

$$G_k = (C^{-1})_{kl} F_l \tag{3.6-73}$$

as the solution for the unknown G_k's.

3.6.4 Algorithm of Emri and Tschoegl

Unfortunately too late to be properly included here, an iterative computer algorithm [27] has recently become available for calculating an underlying spectral distribution from given experimental data. The main advantage of the algorithm lies in its iterative nature. It offers several advantages over the collocation method. Thus, negative spectral lines, which may result from fitting by collocation, can be completely avoided. The distribution is obtained with only a small loss of information even if the data are truncated at each end. The relaxation moduli of the standard solid and liquid models lead, respectively, to singlet and doublet spectra with the correct relaxation times and their associated moduli. Experimental error in the source data is, of course, reflected in the calculated distribution but can be substantially reduced if the data are smoothed first. Experimental response curves are faithfully reproduced from the spectral distribution generated by the algorithm.

3.7 Series-Parallel Models and the Operator Equation

In Sect. 2.1.1 we introduced the *operator equation* [28] whose transform was

$$\bar{u}(s)\bar{\sigma}(s) = \bar{q}(s)\bar{\varepsilon}(s) \tag{3.7-1}$$

where

$$\bar{u}(s) = \sum_n u_n s^n \qquad \text{and} \qquad \bar{q}(s) = \sum_m q_m s^m \; . \tag{3.7-2}$$

We shall now show that the coefficients u_n and q_n of the polynomials $\bar{u}(s)$ and $\bar{q}(s)$ can be related in a straight-forward way to the parameters of the series-parallel models. We have

$$\bar{Q}(s) = \frac{\bar{q}(s)}{\bar{u}(s)} = \frac{\sum_m q_m s^m}{\sum_n u_n s^n} \qquad \text{and} \qquad \bar{U}(s) = \frac{\bar{u}(s)}{\bar{q}(s)} = \frac{\sum_n u_n s^n}{\sum_m q_m s^m} \; . \tag{3.7-3}$$

Thus, determination of the respondances of the model and equating the coefficients furnishes the q_m's and u_n's in terms of the parameters of the model. An example will make this clear.

Let us write the relaxance of the simple 3-parameter Maxwell model given by Eq. (3.4-11) as

$$\bar{Q}_{3M}(s) = \frac{G_e + (G_e + G)\tau s}{1 + \tau s} = \frac{G_e + G_g \tau s}{1 + \tau s} \ . \tag{3.7-4}$$

Comparing Eqs. $(3.7\text{-}4)_2$ and $(3.7\text{-}3)_1$ we find $q_0 = G_e$, $q_1 = (G_e + G)\tau = G_g\tau$, $q_2 = \cdots = q_m = 0$, $u_0 = 1$, $u_1 = \tau$, and $u_2 = \cdots = u_n = 0$. The differential equation thus becomes

$$G_g \tau \frac{d\,\varepsilon(t)}{dt} + G_e \varepsilon(t) = \tau \frac{d\,\sigma(t)}{dt} + \sigma(t) \tag{3.7-5}$$

and the operators are

$$\mathscr{U}[\quad] = \left(\tau\frac{d}{dt} + 1\right)[\quad] \qquad \text{and} \qquad \mathscr{Q}[\quad] = \left(G_g\tau\frac{d}{dt} + G_e\right)[\quad] \tag{3.7-6}$$

respectively. It is thus clear that the (constant) coefficients of the linear differential equations which describe linear viscoelastic behavior are related in a simple way to the parameters of the series-parallel models. The rheological models may therefore be regarded as graphical representations of the differential equations.

The operator equation can also be used to relate the coefficients of two conjugate models. Because the models are conjugate, we have

$$\bar{Q}_M(s)\bar{U}_V(s) = 1 \tag{3.7-7}$$

and hence,

$$\bar{q}_M(s)\bar{u}_V(s) = \bar{u}_M(s)\bar{q}_V(s) \ . \tag{3.7-8}$$

Equating coefficients of identical powers of s on each side of Eq. (3.7-8) yields n equations for the required n relations.

As an example we consider the models of the standard linear solid. We have

$$\bar{q}_M(s) = G_e + G_g\tau_M s \qquad \text{and} \qquad \bar{u}_M(s) = 1 + \tau_M s \tag{3.7-9}$$

as well as

$$\bar{u}_V(s) = J_e + J_g\tau_V s \qquad \text{and} \qquad \bar{q}_V(s) = 1 + \tau_V s \ . \tag{3.7-10}$$

This gives

$$(G_e + G_g\tau_M s)(J_e + J_g\tau_V s) = (1 + \tau_M s)(1 + \tau_V s) \tag{3.7-11}$$

which leads to the three relations

$$G_e J_e = 1 \tag{3.7-12}$$

$$G_e J_g \tau_V + J_e G_g \tau_M = \tau_M + \tau_V \tag{3.7-13}$$

$$G_g J_g = 1 \ . \tag{3.7-14}$$

These may be solved for either G_e, G, and η, or for J_g, J, and ϕ. We note that

$$\tau_V = G_g\tau_M/G_e \qquad \text{and} \qquad \tau_M = J_g\tau_V J_e \tag{3.7-15}$$

and find

$$G_e = 1/J_e \quad (3.7\text{-}16)$$

$$G = J/J_g J_e \quad (3.7\text{-}17)$$

$$\eta = J^2/J_e^2 \phi \quad (3.7\text{-}18)$$

for the parameters of the standard 3-parameter Maxwell model in terms of those of the conjugate Voigt model, and

$$J_g = 1/G_g \quad (3.7\text{-}19)$$

$$J = G/G_e G_g \quad (3.7\text{-}20)$$

$$\phi = G^2/G_g^2 \eta \quad (3.7\text{-}21)$$

for the parameters of the standard 3-parameter Voigt model in terms of those of the conjugate Maxwell model.

The equations furnishing the parameters of any of the models of the standard linear liquid in terms of those of its conjugate can be found in a similar manner. Inserting Eqs. (3.4-67) and (3.4-68) into Eq. (3.7-7), multiplying both sides of the equation by the denominator, and equating coefficients yields the following four equations:

$$J_g = 1/(G_0 + G_1) \quad (3.7\text{-}22)$$

$$\phi_f = 1/(\eta_0 + \eta_1) \quad (3.7\text{-}23)$$

$$\eta_0 \eta_1 \phi_f / J_g + G_0 G_1 (J_g + J)/\phi_f = G_0 \eta_1 + G_1 \eta_0 \quad (3.7\text{-}24)$$

and

$$\eta_0 \eta_1 (\phi + \phi_f)/J_g + G_0 G_1 J_g/\phi_f = G_0 \eta_1 + G_1 \eta_0 \ . \quad (3.7\text{-}25)$$

Equations (3.7-22) and (3.7-23) were used to simplify Eqs. (3.7-24) and (3.7-25). Subtracting these last equations from each other gives

$$\eta_0 \eta_1 / G_0 G_1 = J_g J / \phi \phi_f \ . \quad (3.7\text{-}26)$$

Dividing Eq. (3.7-24) by $G_0 G_1$ and using Eq. (3.7-26) leads to

$$\eta_0/G_0 + \eta_1/G_1 = J_g/\phi_f + J/\phi + J/\phi_f \ . \quad (3.7\text{-}27)$$

The last two equations had already been established in Sect. 3.4.2 [cf. Eqs. (3.4-30) and (3.4-31) where $\tau_0 = \eta_0/G_0$ and $\tau_1 = \eta_1/G_1$]. Solving Eqs. (3.7-26) and (3.7-27) for J and ϕ gives

$$J = \frac{(G_0 \eta_1 - G_1 \eta_0)^2}{G_0 G_1 (G_0 + G_1)(\eta_0 + \eta_1)^2} \quad (3.7\text{-}28)$$

and

$$\phi = \frac{(G_0 \eta_1 - G_1 \eta_0)^2}{\eta_0 \eta_1 (\eta_0 + \eta_1)(G_0 + G_1)^2} \ . \quad (3.7\text{-}29)$$

Thus the steady-flow viscosity and the instantaneous compliance of the model of the standard linear liquid is the sum of the viscosities of the two dashpots, and the sum of the moduli of the two springs, respectively. The delayed compliance and fluidity are more involved functions of the four relaxances. However, the retardation time can be expressed as

$$\tau = \frac{\eta_1\tau_0 + \eta_0\tau_1}{\eta_0 + \eta_1} = \frac{1/G_0 + 1/G_1}{1/\eta_0 + 1/\eta_1}\ , \tag{3.7-30}$$

i.e. the retardation time of the 4-parameter Voigt model is the ratio of the sum of the compliances of the conjugate Maxwell model to the sum of the fluidities. Solving Eq. (3.7-24) for $J_e^\circ = J_g + J$ we obtain

$$J_e^\circ = \frac{\eta_0\tau_0 + \eta_1\tau_1}{(\eta_0 + \eta_1)^2}\ . \tag{3.7-31}$$

To solve for G_0, G_1, η_0, and η_1, we let

$$J_g J/\phi\phi_f = \gamma^2 = \tau_0\tau_1 \tag{3.7-32}$$

and

$$J_g/\phi_f + J/\phi + J/\phi_f = 2\alpha = \tau_0 + \tau_1 \tag{3.7-33}$$

and first solve for the two relaxation times. Letting $\tau_0 > \tau_1$, and

$$\alpha^2 - \gamma^2 = \beta^2 \tag{3.7-34}$$

we have

$$\tau_0 = \alpha + \beta \quad \text{and} \quad \tau_1 = \alpha - \beta\ . \tag{3.7-35}$$

We can then use Eq. (3.7-23) together with Eq. $(3.7\text{-}30)_1$ in the form

$$\eta_1\tau_0 + \eta_0\tau_1 = \tau/\phi_f \tag{3.7-36}$$

to solve for η_0 and η_1. This leads to

$$\eta_0 = \frac{\tau_0 - \tau}{\phi_f(\tau_0 - \tau_1)} \quad \text{and} \quad \eta_1 = \frac{\tau - \tau_1}{\phi_f(\tau_0 - \tau_1)}\ . \tag{3.7-37}$$

Finally, the moduli are obtained as

$$G_0 = \eta_0/\tau_0 = \frac{\tau_0 - \tau}{\phi_f\tau_0(\tau_0 - \tau_1)} \quad \text{and} \quad G_1 = \eta_1/\tau_1 = \frac{\tau - \tau_1}{\phi_f\tau_1(\tau_0 - \tau_1)}\ . \tag{3.7-38}$$

In principle, the conjugate forms of models with a finite number of elements can be interrelated using the same procedure. However, any but the standard models lead to polynomials in s which are of third or higher degree. Thus, in practice, only numerical solutions could be attempted. Another method, based on the calculus of residues, is outlined in Sect. 4.5.

3.8 Problems

Problem 3.1-1. In the mechanical system shown schematically in Fig. P3.1-1a, the *damped vibration absorber* consisting of the small mass I_1, the springs E_1, and the dashpot F_1, if properly tuned, damps the vibrations of the large mass I riding on the springs E.

(A) By inspection, simply tracing the actual connection of the elements and bearing in mind that force and mass are represented by 2-terminal symbols and that one terminal of the latter must always be referred to ground, find the mechanical model diagram.

(B) Determine the node equations. Consider the displacements x and x_1 to be determined with respect to their equilibrium positions in the absence of the applied force, f.

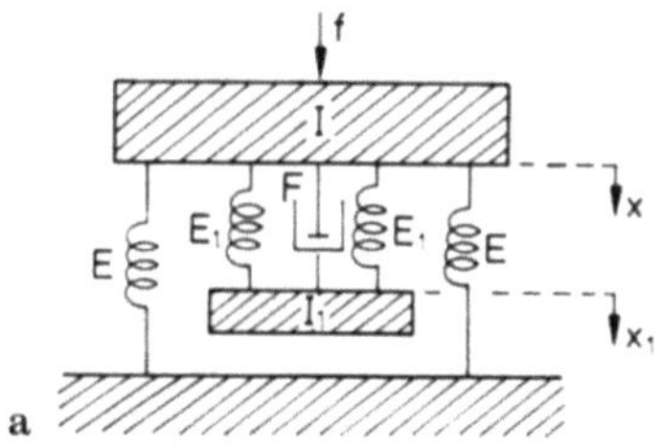

Fig. P3.1-1a. Damped vibration absorber

Problem 3.1-2. Consider the combinations of mechanical elements displayed in Figs. P3.1-2a and 2b. The mass of the rigid bar to which the force f is applied is deemed to be negligible. Assume zero initial conditions.

(A) Draw the mechanical model diagrams for the combinations (a) and (b), and find the node equations and mechanical relaxances, $\bar{Q}_{ma}(s)$, and $\bar{Q}_{mb}(s)$.

(B) Determine the conditions, if any, for which $\bar{Q}_{ma}(s) = \bar{Q}_{mb}(s)$.

(C) What conclusion can you draw from your findings?

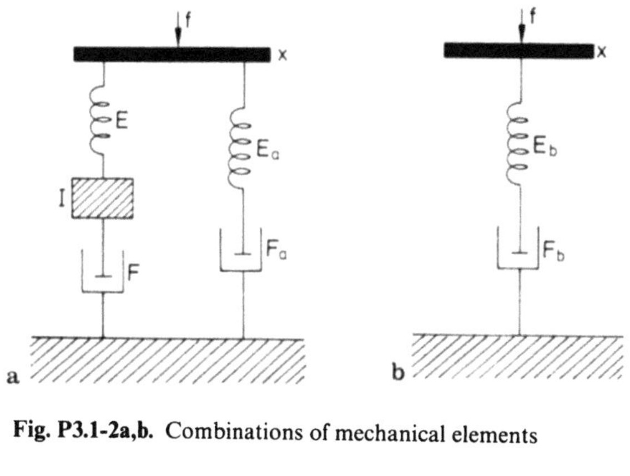

Fig. P3.1-2a,b. Combinations of mechanical elements

Problem 3.1-3. Shown in Fig. P3.1-3a is a schematic representation of an apparatus for measuring the shear properties of two identical specimens as a function of frequency in the steady state.

a 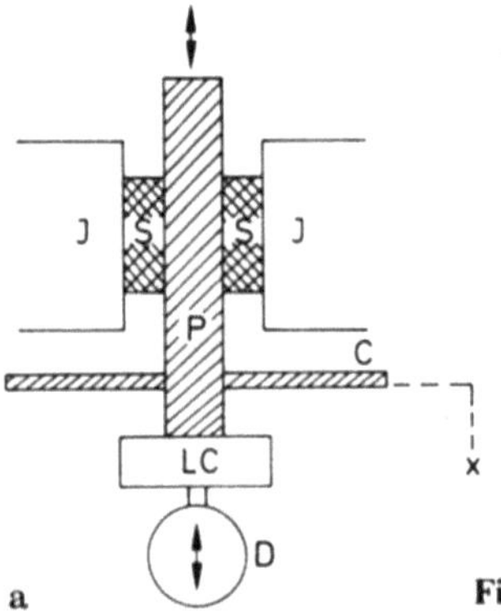

Fig. P3.1-3a. Schematic of dynamic shear rheometer

The specimens (S) are held between two jaws (J) and a shearing member consisting of a shearing plate (P) and a crossbeam (C) connected to a travelling potentiometer which measures the displacement, x, of the plate. The shearing member performs harmonic oscillations. Its inertance, I, is assumed to be known. It is driven (D) through a load cell (LC) which measures the force, f, required to drive the shearing member and shear the specimens. Consider the force to be given by

$$f(\omega) = f_0 \exp(j\omega t) \tag{1}$$

where f_0 is the peak amplitude. Assume, as not relevant in the present context, that the factors converting the recorded electrical signals into force and displacement, respectively, have been determined. Assume, further, that the inertia of the specimens is negligible at the frequency of measurement, and that the shearing plate is rigid and frictional effects can be neglected.

Using the method of mechanical model diagrams, find the equations through which the storage and loss shear moduli, $G'(\omega)$ and $G''(\omega)$, are related to the amplitude ratio, $A_r(\omega) = f_0/x_0(\omega)$, where $x_0(\omega)$ is the peak amplitude of the displacement, and the phase angle, $\delta(\omega)$, between the force, $f(\omega)$, and displacement, $x(\omega)$, respectively. Also derive the expression for the loss tangent of the material.

In drawing the mechanical model diagram, the storage and loss moduli may be represented as elastances by multiplying them by the *specimen coefficient*, b. If both specimens are of height h and cross-sectional area A_0, the specimen coefficient will be $b = 2A_0/h$.

Problem 3.1-4. Consider an experimental arrangement in which a driving unit, consisting e.g. of a loudspeaker coil, produces a harmonic force $f_1(\omega)$ which is transmitted through a specimen with complex tensile modulus $E^*(\omega)$ to a force pick-up consisting of a load cell with internal elastance E_T. The internal frictance of the load cell is negligible. The specimen is held between two clamps connected by rods to the driver, and the pick-up, respectively. The ratio, $A_r(\omega)$, of the peak amplitudes, and the phase angle, $\delta(\omega)$, between the displacement, $x_1(\omega)$, generated at the driver, and the force picked up by the load cell, $f_2(\omega)$, are measured. The arrangement is shown schematically below:

Since the displacement, $x_1(\omega)$, is measured, the force, $f_1(\omega)$, which generates it, does not need to be known. The apparatus may be considered to be *displacement driven*, i.e. the excitation is a displacement. The mechanical model diagram may, nevertheless, be drawn as if the force, $f_1(\omega)$, were the excitation.

Develop the equations for the real and imaginary parts of $E^*(\omega)$, neglecting the mass of the moving parts. The frequency is low enough so that the specimen inertia may also be neglected.

[*Hint*: when drawing the mechanical model diagram, consider that the force exerted by the driver is transmitted through the harmonic specimen elastance $bE^*(\omega)$ to the internal elastance E_T, then use the relation $f_2(\omega) = E_T x_2(\omega)$.]

Problem 3.1-5. (A) Prove that, in the dynamic tensile rheometer discussed in Problem 3.1-4, the displacement $x_2(\omega)$ is negligible with respect to the displacement $x_1(\omega)$ if the internal elastance, E_T, of the pick-up is large with respect to the elastance presented by the absolute tensile modulus, $\tilde{E}(\omega)$, of the material.

(B) Show that, in the latter case,

$$E'(\omega) = HA_r(\omega)\cos\delta(\omega) \tag{1}$$

and

$$E''(\omega) = HA_r(\omega)\sin\delta(\omega)\ . \tag{2}$$

Problem 3.1-6. In Problems 3.1-4 and 3.1-5 the analysis was carried out under the assumption that a possible effect of the inertia of the moving parts could be neglected. Repeat the analysis of Problem 3.1-5 without this assumption. State the condition under which the means of the moving parts can, in fact, be neglected. Suggest means for determining the inertance experimentally.

Problem 3.1-7. Consider the torsional rheometer represented schematically in Fig. P3.1-7a. A cylindrical specimen (Sp) having an operational mechanical relaxance $b\bar{Q}(s)$ can be twisted through the known angle θ_1. A rotational elastance (E) of value E_A is inserted in series with the specimen. The torque, M_2, required to twist the combination is measured through a torsional load cell (LC) having the internal elastance E_L. The purpose of the added elastance is to keep the torque seen by the load cell within desired limits. The *specimen coefficient*, b, is given by $b = \pi D^4/32L$ where D is the diameter, and L is the length of the specimen. Inertial effects are deemed negligible. All connections between the various elements are perfectly rigid.

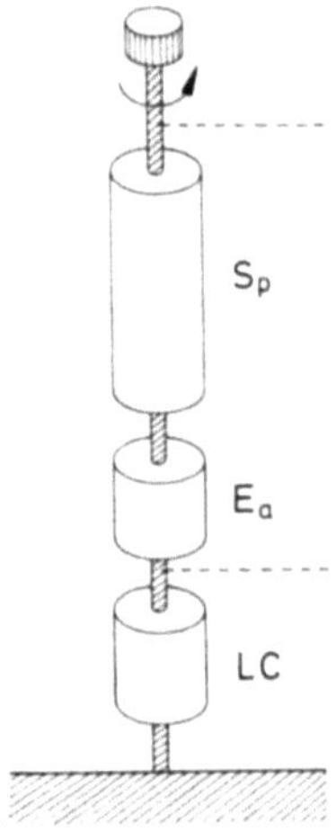

a Fig. P3.1-7a. Schematic of torsional rheometer

(A) Determine the relation by which the shear relaxation modulus, G(t), may be obtained from the time-dependent torque, $M_2(t)$, if the twist is applied as a step function of time.

(B) Determine the relation by which the complex shear modulus, $G^*(\omega)$, may be obtained from the amplitude ratio and phase angle between the angular displacement, $\theta_1(\omega)$, and the torque, $M_2(\omega)$, if the specimen is twisted harmonically.

Problem 3.2-1. Demonstrate that the electric circuit shown in Fig. P3.2-1a does not have a physically realizable mechanical equivalent. Use both (A) the electrostatic and (B) the electromechanical analogy. (C) Can the defective analogies be used?

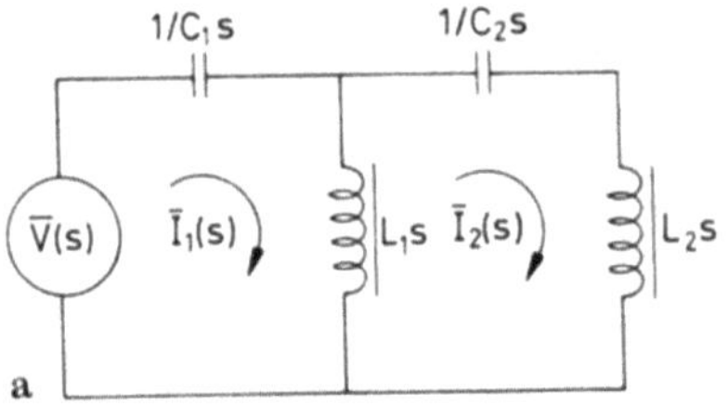

a Fig. P3.2-1a. Two-mesh electric circuit

Problem 3.3-1. Using the relation $\bar{Q}(s)\bar{U}(s) = 1$ show that the Maxwell and Voigt units cannot be equivalent representations of material behavior.

(*Hint*: the relation must be valid for all values of s.)

Problem 3.4-1. (A) Find Q(t) for the standard linear solid from the equation for G(t) and then obtain $\eta(t)$ from Q(t).

(B) Next obtain G(t) both from Q(t) and from $\eta(t)$.

(C) Finally, obtain Q(t) from $\eta(t)$. These derivations should be made without using the Laplace transformation.

Problem 3.4-2. Derive the expression for the time-dependent stress in response to a step strain for the standard linear solid using Eq. $(2.3\text{-}16)_3$

$$\sigma(t) = \int_0^t G(t-u)\frac{d\,\varepsilon(u)}{du}du \ . \tag{1}$$

Problem 3.4-3. (A) Working entirely in the time domain, find U(t) for the standard linear liquid from the equation for J(t).

(B) Now, obtain U(t) from J(t).

(C) Next, obtain $\chi(t)$ from J(t).

(D) Finally, check the solution arrived at under (C) by working in the complex plane.

Problem 3.4-4. Find the stress as a function of the general strain history, $\varepsilon(t)$, for the standard arrheodictic model:

(A) by integrating the operator equation, and

(B) from the Boltzmann superposition integral. Compare the two approaches. [*Hint*: use the integrating factor, $\exp(t/\tau)$, in the integration in (A); use the step response in (B).]

Problem 3.4-5. A material whose behavior corresponds to that of the standard solid model is subjected to a sinusoidal excitation $\varepsilon(t) = \varepsilon_0 \sin \omega t$. Determine

(A) the total stress, $\sigma(t)$, including transients, if any;

(B) the steady-state stress $\sigma(\omega)$, and its in-phase and out-of-phase components.

Problem 3.4-6. Find the equations for the asymptotes, $\varepsilon_{as}(t)$, shown as the broken lines in Figs. 3.4-7 and 3.4-12, and compare the two.

Problem 3.4-7. Consider the model shown in Fig. 3.4-17a. Designate the isolated spring by G′, the isolated dashpot by η', and the elements of the Maxwell unit by G and η, respectively.

(A) Obtain the relaxance of the model;

(B) then ascertain whether it represents rheodictic or arrheodictic behavior; and

(C) whether its instantaneous modulus is finite;

(D) Use the "inspection" method of Sect. 3.5.3.

[*Hint*: use the limit value theorem in (B) and (C).]

Problem 3.4-8. In Problem 3.4-7 we found that the model shown in Fig. 3.4-17a had a finite modulus and that it represented rheodictic behavior. Since it contains the same numbers of springs and dashpots as the model of the standard linear liquid, it must represent identical behavior.

Show that the relaxance of the model, $\bar{Q}_{4a}(s)$, can be brought into the same form as $\bar{Q}_{4M}(s)$ and express the parameters of the springs and dashpots of $\bar{Q}_{4M}(s)$ in terms of those of $\bar{Q}_{4a}(s)$.

Problem 3.4-9. Consider the model shown in Fig. 3.4-17c. Denote the spring and dashpot of the Voigt unit by G_1 and η_1, the isolated spring by G_2 and the isolated dashpot by η_2. Obtain the relaxance of the model. Is its instantaneous modulus finite or infinite? Does it represent rheodictic behavior? What are the parameters of the equivalent series-parallel model in terms of G_1, G_2, η_1, and η_2?

Problem 3.4-10. Using Laplace transformation show that, if

$$G(t) = G_e + G \exp(-t/\tau_M) \tag{1}$$

then

$$J(t) = J_g + J[1 - \exp(-t/\tau_V)] \tag{2}$$

where $J_g = 1/G_g$, $J = 1/G_e - 1/G_g$, and $\tau_V = G_g\tau_M/G_e$.

Problem 3.5-1. Find the series-parallel model with the minimum number of elements which is equivalent to the model shown in Fig. P3.5-1.

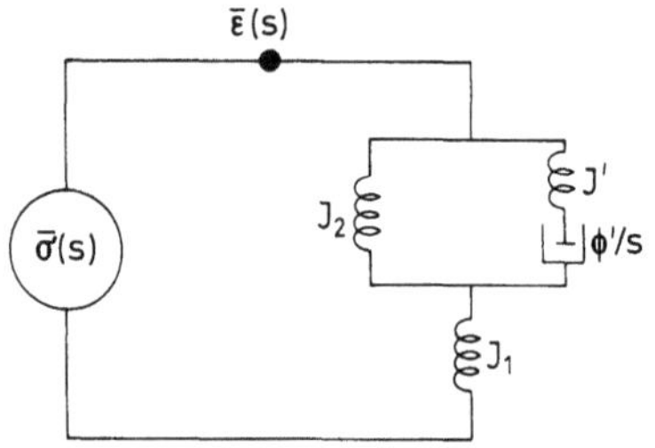

Fig. P3.5-1. Model consisting of three springs and a dashpot

Problem 3.5-2. The model shown in Fig. P3.5-2 is redundant because the number of springs exceeds that of the dashpots by more than one. Using the limit value theorems and the fact that the poles of the two models must be identical if they are to represent the same behavior, prove that the model shown above is equivalent to the standard solid model with the following correspondances:

$$G_e = G_1 + \frac{G_2 G'}{G_2 + G'} \tag{1}$$

$$G = \frac{G_2^2}{G_2 + G'} \tag{2}$$

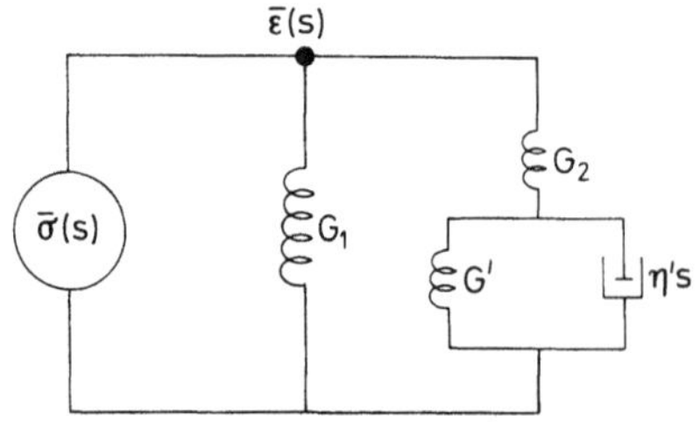

Fig. P3.5-2. Model consisting of three springs and a dashpot

and

$$\eta = \left(\frac{G_2}{G_2 + G'}\right)^2 \eta' \,. \tag{3}$$

Problem 3.5-3. (A) Prove that the model shown in Fig. P3.5-3 possesses an equilibrium modulus but that its instantaneous modulus is infinite.

(B) How could the model be modified to insure that its instantaneous modulus be finite?

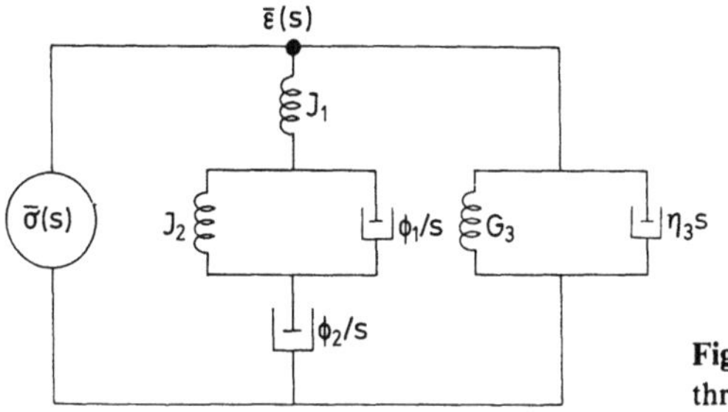

Fig. P3.5-3. Model consisting of three springs and three dashpots

Problem 3.6-1. Procedure X is to be used to estimate the terminal relaxation time, τ_0, of a rheodictic material from measurements of the time-dependent viscosity, $\eta(t)$. Develop the equation from which τ_0 can be found.

Problem 3.6-2. The following values were read from a plot of log G(t) vs. log t:

(j)	$\log t_j$ (s)	$\log G(t_j)$ (N/m^2)
1	−7	8.26
2	−6	8.11
3	−5	7.70
4	−4	7.04
5	−3	6.34
6	−2	6.11

The instantaneous and equilibrium moduli were estimated to be 196×10^6 and 1.1×10^6 N/m^2.

Using the collocation method with $a = 1/\ln 2$, obtain the set of modulus values, G_k, and relaxation times, τ_k. Then calculate and plot the curve of log G(t) vs. log t using these values. Show the experimental values tabulated above as full circles on the same graph to assess the goodness of fit of the calculated curve.

The matrix, $\mathbf{B}_{jk}$, is nearly triangular. Let any term ≥ 0.999 equal unity, and any term ≤ 0.001 equal zero. The matrix will then be triangular and the $\mathbf{B}_k$ can be calculated without inversion. Compare the G_k you have calculated in this way with those obtained by computer inversion of the complete matrix which yielded 0.92, 11.80, 100.16, 64.05, 16.15, 1.44, 0.38 for $G_k \times 10^{-6}$, $k = 0, 1, \ldots, 6$.

Problem 3.7-1. Derive the operator equation for the 4-parameter Voigt model.

References (Chap. 3)

1. M.F. Gardner and J.L. Barnes, *Transients in Linear Systems*, Vol. I, John Wiley and Sons, New York, 1953.
2. G. Kirchhoff, Ann. Phys. Chem. (2) *72*:497–574 (1847). English Transl. Trans. I.R.E. Vol. CT-5 (March): 4 (1958).
3. S. Goldman, *Transformation Calculus and Electrical Transients*, Constable and Co., London, 1949.
4. C. Truesdell and R.A. Toupin, *The Classical Field Theories*, in: S. Flügge, ed., Encyclopedia of Physics, Vol. III/1, p. 532, Springer, Berlin, 1960.
5. F.A. Firestone, *American Institute of Physics Handbook*, McGraw-Hill, New York, 1957.
6. H.F. Olsen, *Dynamical Analogies*, Van Nostrand, New York, 1945.
7. J.M. Burgers in: *First Report on Viscosity and Plasticity*, 2nd ed., p. 5, Noord-Hollandsche, Amsterdam, 1939.
8. T. Alfrey, *Mechanical Behavior of High Polymers*, Interscience, New York, 1948, p. 193.
9. B. Gross and R.M. Fuoss, J. Polymer Sci. *19*:39 (1956).
10. R. Stambaugh, Ind. Eng. Chem. *44*:1590 (1952).
11. J. Miles, J. Acoust. Soc. Amer. *14*:183 (1943).
12. N.W. Tschoegl, Australian J. Phys., *14*:307 (1961).
13. J.H. Poynting and J.J. Thomson, *Properties of Matter*, C. Griffin and Co., London, 1902.
14. N.W. Tschoegl, Kolloid-Z. *174*:113 (1961).
15. O. Meyer, J. reine u. angew. Math. *78*, 130 (1874).
16. W. Voigt, Ann. d. Phys. *47*:671 (1892).
17. J.C. Maxwell, Phy. Trans. Roy. Soc. London A157:52 (1867): Papers *2*:26–78, Dover, 1952.
18. T. Alfrey and P. Doty, J. Appl. Phys. *16*, 700 (1945).
19. J.J. Thomson, *Applications of Dynamics to Physics and Chemistry*, Chapter viii, MacMillan and Co., London, 1888.
20. E. Wiechert, Ann. d. Phys. u. Chem. *50*:335 (1893).
21. W. Thomson (Lord Kelvin), "*Elasticity*" in Encyclopedia Britannica, 9th ed., Vol. VII, p. 803d, Charles Scribner's, New York, 1878.
22. A.V. Tobolsky and K. Murakami, J. Polymer Sci., *40*:443 (1959).
23. A.V. Tobolsky, *Properties and Structure of Polymers*, Wiley and Sons, New York, 1960, pp. 186–195.
24. R.A. Schapery, Proc. Fourth U.S. Nat. Congress Appl. Mech., *2*:1075 (1962).
25. M.H. Gradowczyk and F. Moavenzadeh, Trans. Soc. Rheol. *13*:173 (1969).
26. T.L. Cost and E.B. Becker, Intern. J. Num. Methods in Engg. *2*:207 (1970).
27. I. Emri and N.W. Tschoegl, to be submitted to Rheol. Acta.
28. T. Alfrey, Quart. Appl. Math. *2*:113 (1944).

4. Representation of Linear Viscoelastic Behavior by Spectral Response Functions

Near enough is good enough

Popular Australian saying

4.0 Introduction

The fundamental equations of the linear theory of viscoelastic behavior, $\bar{\sigma}(s) = \bar{Q}(s)\bar{\varepsilon}(s)$, and $\bar{\varepsilon}(s) = \bar{U}(s)\bar{\sigma}(s)$, can be condensed into one by writing

$$\bar{\mathscr{R}}(s) = \bar{\mathscr{M}}(s)\bar{\mathscr{E}}(s) \tag{4.0-1}$$

where $\bar{\mathscr{R}}(s)$ is called the *response transform*, $\bar{\mathscr{M}}(s)$ the *material* or *system transform*, and $\bar{\mathscr{E}}(s)$ the *excitation, stimulus*, or *driving transform*. The *material function* corresponding to the material transform is determined by applying an excitation and observing the response. The applied excitation is necessarily a function of time: a step function, slope function, sinusoidal steady-state function (Chap. 2), or any other function of time (Chap. 7). Thus the observed response and the material function derived from it always carry the signature of the time regime chosen to elicit the response. In principle, therefore, an infinity of material functions exist: one for every conceivable driving function. If known over the entire time (or frequency) scale from zero to infinity all material functions contain the same information although this is weighted differently in different regions of the time scale according to the manner in which the driving function depends on time, i.e. on its *time regime*. In Chap. 8 we shall be concerned with the interconversion of the standard excitation functions.

Clearly, it would be of interest to possess material functions which would be independent of the time regime of the driving function. Such material functions can indeed be defined and it is the purpose of the present chapter to investigate them. It is evident that they cannot be obtained from experiment directly because there is no driving function which would not be associated with a particular time regime in the very general sense in which this concept must be understood in our context. The material response functions which we will introduce in the next section are called *spectral distribution functions*, or simply *spectra*. We distinguish between *relaxation* and *retardation spectra* according to whether they are derived from the response to a strain or a stress stimulus. We further distinguish between *continuous* and *line spectra*, and between *time* and *frequency spectra*. Although we can, in general, determine their actual form from experimental data only approximately through the application of mathematical techniques to be discussed in later sections of this chapter, the concept of spectral distribution functions is extremely useful in the theory of linear viscoelastic behavior. Because of their independence of any particular time regime, they occupy a central position in the theory.

The concept of a distribution of relaxation times was first proposed by Wiechert [1] in 1893 in his classical studies of the "elastic after-effect". The idea was adopted by later workers in the theory of dielectrics. With the revival of interest in linear viscoelastic behavior brought about by the development of synthetic polymers, it was introduced [2] into polymer science by W. Kuhn* in 1939. While Wiechert had used the lognormal distribution [cf. Eq. (6.3-1)] to model the distribution of relaxation times in glasses, Kuhn left it as an unspecified function and it has remained essentially that ever since. Representation of the spectra by mathematical models will be treated in Chap. 6. Molecular theories of the behavior of synthetic macromolecules (cf. Sect. 6.3.1) predict discrete distributions of relaxation times. Such distributions will be discussed in Sect. 4.5.

4.1 The Continuous Spectral Response Functions

The concept of continuous spectral response functions to characterize linear viscoelastic behavior is based on the fact, pointed out in the introduction to Sect. 3.5, that attempts at a satisfactory description of the observed behavior by series-parallel models generally require a very large number of elements. In such a set of elements the relaxation or retardation times become so closely spaced that the sum of the discrete contributions of the individual terms in the equations of Sect. 3.5 may be replaced by the integral over appropriate continuous functions.

4.1.1 Continuous Time Spectra

We investigate the time spectra first and begin with relaxation behavior. From Eq. (3.5-4) we obtain the relaxance $\bar{Q}(s)$ as

$$\bar{Q}(s) = \{G_e\} + \int_0^\infty \hat{Q}(\tau)\frac{\tau s}{1 + \tau s}\, d\tau \qquad (4.1\text{-}1)$$

where $\hat{Q}(\tau)$ is a continuous function of the relaxation time, τ. Its integral with respect to τ replaces the discrete sum of the G_n's. The function $\hat{Q}(\tau)$ is often referred to as a *distribution* of *relaxation times*. In fact, it is a *modulus density on relaxation time*, having the same dimensions as Q(t), i.e. modulus/time. To obtain a function which has the dimensions of modulus we define a new function.

* Werner Kuhn, (pron. Coon), 1899–1963, Swiss physical chemist, famous for his work on the behavior of long chain molecules. Not to be confused with the younger (and unrelated) Hans Kuhn, born 1919, a German physical chemist with whom he also collaborated.

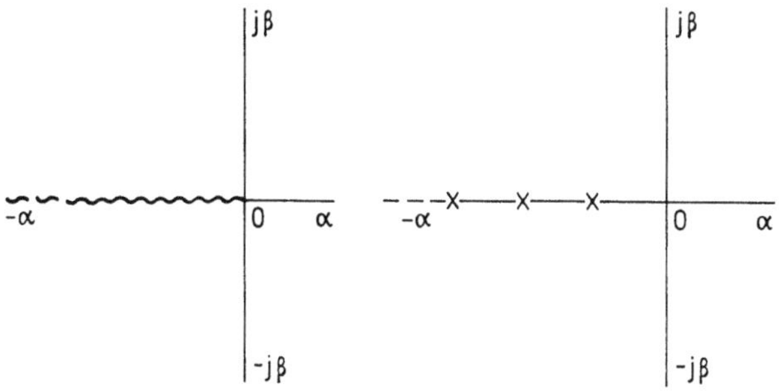

Fig. 4.1-1. Singularities of $\bar{Q}(s)$

$$\hat{G}(\tau) = H(\tau) = \tau\hat{Q}(\tau) \tag{4.1-2}$$

which we call the *relaxation spectrum.* $\hat{G}(\tau)$ is also commonly called a distribution of relaxation times but is in fact a distribution of moduli. It is referred to as a time spectrum because it is a function of the relaxation *times.* Instead of $\hat{G}(\tau)$ the symbol $H(\tau)$ has been widely adopted in polymer rheology and will be used exclusively from now on. Note that $\hat{G}(\tau)$ and $G(t)$ are different functions of their respective argument although both have the dimensions of modulus.

Introducing $H(\tau)$ into Eq. (4.1-1) we obtain

$$\bar{Q}(s) = \{G_e\} + \int_{-\infty}^{\infty} H(\tau)\frac{\tau s}{1 + \tau s}\, d\ln\tau \ . \tag{4.1-3}$$

We note that, when $\bar{Q}(s)$ is expressed by Eqs. (4.1-1) or (4.1-3) it possesses a cut along the entire negative real axis of the complex plane between the branch points $s = 0$ and $s = -\infty$. This cut replaces the discrete set of poles on that axis which represent the singularities of $\bar{Q}(s)$ when it is given by Eq. (3.5-4). The two patterns are illustrated in Fig. 4.1-1, the branch cut being indicated by a wavy line.

We remark that the simple method of partial fraction decomposition which was used in Chap. 2 to obtain the harmonic response functions from the respondances is not applicable when the distribution of respondance times is continuous. Problems 4.1-1 and 4.1-2 discuss a more general method valid for continuous as well as for discrete respondance time distributions.

In expressions where the symbol $\ln\tau$ appears, it should be interpreted as $\ln(\tau/[\tau])$ where $[\tau]$ symbolizes unit relaxation (or retardation) time (cf. Sect. 3.4.5). The differential, $d\ln\tau$, may be considered to be a shorthand symbol for $d\tau/\tau$. It will sometimes be useful to replace it by $d\tau/\tau$, changing the lower limit of integration from $-\infty$ to 0. We note, further, that $H(\tau)$ represents a function of τ, not of $\ln\tau$, even though the latter may be considered to be the variable of integration. When we have occasion to represent the spectrum as a function of $\ln\tau = z$, we write $\mathscr{H}(z)$, using the script symbol because $\mathscr{H}(z)$ is not the same function of z as $H(\tau)$ is of τ. Finally, we remark that $H(\tau)d\ln\tau$ may be considered to be the modulus contribution associated with relaxation times whose logarithms lie between $\ln\tau$ and $\ln\tau + d\ln\tau$.

Response functions written in terms of the spectra will be said to be expressed in the *canonical representation* if they are the continuous analogs of the discrete forms associated with the canonical models (cf. Sect. 3.5.1.3).

From Eq. (4.1-3), the transform of the relaxation modulus in the canonical representation becomes

$$\bar{G}(s) = \frac{\{G_e\}}{s} + \int_{-\infty}^{\infty} H(\tau)\frac{\tau}{1 + \tau s}\, d\ln\tau \,. \tag{4.1-4}$$

Equation (4.1-4) in its turn leads to

$$G(t) = \{G_e\} + \int_{-\infty}^{\infty} H(\tau)\exp(-t/\tau)\, d\ln\tau \tag{4.1-5}$$

for the step response, and to

$$G^*(\omega) = \{G_e\} + \int_{-\infty}^{\infty} H(\tau)\frac{j\omega\tau}{1 + j\omega\tau}\, d\ln\tau \tag{4.1-6}$$

for the harmonic response functions, respectively. The real and imaginary components of the complex modulus become

$$G'(\omega) = \{G_e\} + \int_{-\infty}^{\infty} H(\tau)\frac{\omega^2\tau^2}{1 + \omega^2\tau^2}\, d\ln\tau \tag{4.1-7}$$

and

$$G''(\omega) = \int_{-\infty}^{\infty} H(\tau)\frac{\omega\tau}{1 + \omega^2\tau^2}\, d\ln\tau \,. \tag{4.1-8}$$

For reference and later use we list the equivalent representations based on the $\bar{Q}_{\bullet}(s)$ form of the relaxance. We recall (cf. Sect. 3.5) that the dot-forms allow us to write equations for the responses to strain excitations and to stress excitations in such a manner that their viscoelastic portions are expressed in formally analogous ways. [cf., e.g., Eqs. (4.1-5) and (4.1-28)]. In particular, the behavior in the complex plane of the integrals in $\bar{Q}_{\bullet}(s)$ and $\bar{U}(s)$, and in $\bar{Q}(s)$ and $\bar{U}_{\bullet}(s)$ are identical [cf. Eqs. (4.1-10) and (4.1-18), and Eqs. (4.1-3) and (4.1-26)]. Derivations which depend on this behavior then become analogous. This simplifies the presentation and emphasizes the dual nature of the theory of linear viscoelastic behavior (cf. Sect. 2.1.2). Such use of the dot-forms will be made in Sect. 4.1.2 and elsewhere in this book.

To obtain the dot forms of the moduli in the canonical representations, we start from

$$\bar{Q}_{\bullet}(s) = G_g - \int_{0}^{\infty} \hat{Q}(\tau)\frac{1}{1 + \tau s}\, d\tau \tag{4.1-9}$$

to obtain first

$$\bar{Q}_{\bullet}(s) = G_g - \int_{-\infty}^{\infty} H(\tau)\frac{1}{1 + \tau s}\, d\ln\tau \tag{4.1-10}$$

and

$$\bar{G}_{\bullet}(s) = \frac{G_g}{s} - \int_{-\infty}^{\infty} H(\tau)\frac{1}{s(1+\tau s)}\, d\ln\tau \ , \tag{4.1-11}$$

and then

$$G_{\bullet}(t) = G_g - \int_{-\infty}^{\infty} H(\tau)[1 - \exp(-t/\tau)]d\ln\tau \tag{4.1-12}$$

and

$$G^*_{\bullet}(\omega) = G_g - \int_{-\infty}^{\infty} H(\tau)\frac{1}{1+j\omega\tau}\, d\ln\tau \tag{4.1-13}$$

whose real part is

$$G'_{\bullet}(\omega) = G_g - \int_{-\infty}^{\infty} H(\tau)\frac{1}{1+\omega^2\tau^2}\, d\ln\tau \ . \tag{4.1-14}$$

The imaginary part is, of course, identical with Eq. (4.1-8).

For the transform of the time-dependent viscosity we get

$$\bar{\eta}(s) = \frac{\bar{G}(s)}{s} = \left\{\frac{G_e}{s^2}\right\} + \int_{-\infty}^{\infty} H(\tau)\frac{\tau}{s(1+\tau s)}\, d\ln\tau \tag{4.1-15}$$

and for the viscosity itself we have

$$\eta(t) = \{G_e t\} + \int_{-\infty}^{\infty} \tau H(\tau)[1 - \exp(-t/\tau)]d\ln\tau \ . \tag{4.1-16}$$

For $\eta^*(\omega)$, $\eta'(\omega)$, and $\eta''(\omega)$ see Problem 4.1-4.

Turning now to the canonical representation of creep behavior, we obtain the retardance $\bar{U}(s)$ from Eq. (3.5-13) as

$$\bar{U}(s) = J_g + \int_{0}^{\infty} \hat{U}(\tau)\frac{1}{1+\tau s}\, d\tau + \{\phi_f/s\} \ . \tag{4.1-17}$$

This becomes

$$\bar{U}(s) = J_g + \int_{-\infty}^{\infty} L(\tau)\frac{1}{1+\tau s}\, d\ln\tau + \{\phi_f/s\} \tag{4.1-18}$$

with the new function

$$\hat{J}(\tau) = \tau\hat{U}(\tau) = L(\tau) \tag{4.1-19}$$

which has the dimension of a compliance and is known as the *retardation time spectrum* or simply the *retardation spectrum*. The symbol $L(\tau)$ has been adopted in polymer rheology and will be used from now on. $\hat{U}(\tau)$ is again commonly referred to as *distribution of retardation times* but is in fact a *compliance density on retardation time*. The remarks made earlier in regard to $H(\tau)$ apply equally to $L(\tau)$.

The transform of the creep compliance is now

$$\bar{J}(s) = \frac{J_g}{s} + \int_{-\infty}^{\infty} L(\tau)\frac{1}{s(1+\tau s)}\, d\ln\tau + \{\phi_f/s^2\} \tag{4.1-20}$$

from which we have

$$J(t) = J_g + \int_{-\infty}^{\infty} L(\tau)[1 - \exp(-t/\tau)]d \ln \tau + \{\phi_f t\} \tag{4.1-21}$$

and

$$J^*(\omega) = J_g + \int_{-\infty}^{\infty} L(\tau)\frac{1}{1 + j\omega\tau} d \ln \tau + \{\phi_f/j\omega\} , \tag{4.1-22}$$

the real and imaginary parts of which are

$$J'(\omega) = J_g + \int_{-\infty}^{\infty} L(\tau)\frac{1}{1 + \omega^2\tau^2} d \ln \tau \tag{4.1-23}$$

and

$$J''(\omega) = \int_{-\infty}^{\infty} L(\tau)\frac{\omega\tau}{1 + \omega^2\tau^2} d \ln \tau + \{\phi_f/\omega\} . \tag{4.1-24}$$

The equivalent dot-forms derived from $\bar{U}_\bullet(s)$ first give

$$\bar{U}_\bullet(s) = J_e^{\{o\}} - \int_0^{\infty} \hat{U}(\tau)\frac{\tau s}{1 + \tau s} d\tau + \{\phi_f/s\} \tag{4.1-25}$$

and

$$\bar{U}_\bullet(s) = J_e^{\{o\}} - \int_{-\infty}^{\infty} L(\tau)\frac{\tau s}{1 + \tau s} d \ln \tau + \{\phi_f/s\} \tag{4.1-26}$$

which then lead to

$$\bar{J}_\bullet(s) = \frac{J_e^{\{o\}}}{s} - \int_{-\infty}^{\infty} L(\tau)\frac{\tau}{1 + \tau s} d \ln \tau + \{\phi_f/s^2\} . \tag{4.1-27}$$

From Eq. (4.1-27) we obtain

$$J_\bullet(t) = J_e^{\{o\}} - \int_{-\infty}^{\infty} L(\tau)\exp(-t/\tau)d \ln \tau + \{\phi_f t\} \tag{4.1-28}$$

and from Eq. (4.1.-26) we have

$$J_\bullet^*(\omega) = J_e^{\{o\}} - \int_{-\infty}^{\infty} L(\tau)\frac{j\omega\tau}{1 + j\omega\tau} d \ln \tau + \{\phi_f/j\omega\} \tag{4.1-29}$$

whose real part is

$$J_\bullet'(\omega) = J_e^{\{o\}} - \int_{\infty}^{\infty} L(\tau)\frac{\omega^2\tau^2}{1 + \omega^2\tau^2} d \ln \tau \tag{4.1-30}$$

while the imaginary part is identical with Eq. (4.1-24).

The response to a constant rate of stress (the dual of the time-dependent viscosity) has not been widely discussed, no doubt because it is experimentally relatively difficult to produce. The canonical representations of the response to a constant rate of stress are, however, readily derived. The equation for $\chi(t)$ which is the dual of Eq. (4.1-16) forms the subject of Problem 4.1-6.

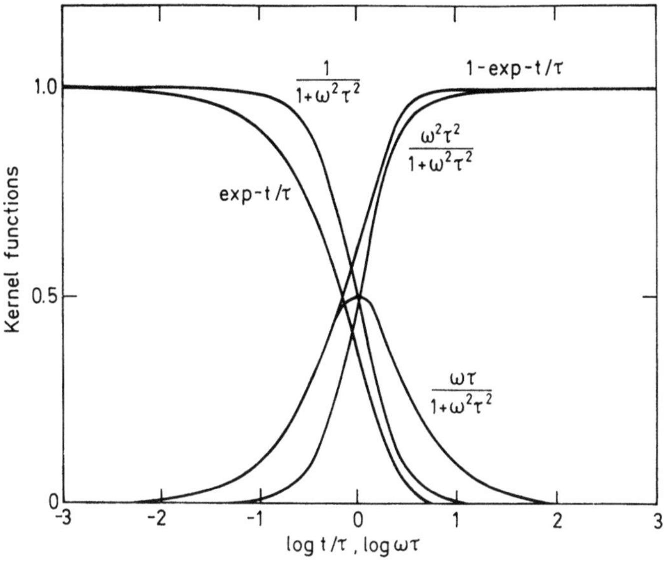

Fig. 4.1-2. Kernel functions

After subtraction of the terms containing the viscoelastic constants $\{G_e\}$, G_g, J_g, $J_e^{\{o\}}$, and ϕ_f, the various experimental functions may thus be considered to be integrals over the continuous spectral distribution functions multiplied by certain kernel functions characteristic of the time regime of the stimulus which elicited the response. The kernel functions are plotted semilogarithmically in Fig. 4.1-2.

In concluding our presentation of the canonical representations of the *experimental response functions* we point out that they are all *functionals* (see Introduction to Chap. 2). This means that every given value of a response function depends on *all* values of the spectral function over its entire domain of definition from $\tau = 0$ to $\tau = \infty$. It is also to be noted that the spectral functions determine only the time-(or frequency-)dependent part of the response function. This point will be taken up in Sect. 4.6.

4.1.2 Continuous Frequency Spectra

The relaxation and retardation spectra may also be expressed as functions of the *relaxation* or *retardation frequencies*, ζ, which are the reciprocals of the relaxation and retardation times. As the relaxation and retardation times, the frequencies are not distinguished by separate symbols when no confusion is likely to arise. The difference in the physical significance of the frequencies and times must not be overlooked, however. We define

$$M(\zeta) = H(\zeta^{-1}) = H(\tau)|_{\tau=\zeta^{-1}} \tag{4.1-31}$$

and

$$N(\zeta) = L(\zeta^{-1}) = L(\tau)|_{\tau=\zeta^{-1}} \tag{4.1-32}$$

where $M(\zeta)$ and $N(\zeta)$ are the relaxation and retardation *frequency* spectra, having the dimensions of modulus and compliance, respectively. These spectra, when displayed as functions of $\log \zeta$, are mirror images, reflected in the $M(\zeta)$ and $N(\zeta)$ axis, of $H(\tau)$ and $L(\tau)$, respectively, because $\log \zeta = -\log \tau$.

We further define two more spectral frequency distribution functions which have the dimensions of modulus $\times$ time and compliance $\times$ time, respectively.

These are

$$\Xi(\zeta) = \zeta^{-1}H(\zeta^{-1}) = \tau H(\tau)|_{\tau=\zeta^{-1}} \tag{4.1-33}$$

and

$$\Lambda(\zeta) = \zeta^{-1}L(\zeta^{-1}) = \tau L(\tau)|_{\tau=\zeta^{-1}} \; . \tag{4.1-34}$$

The two sets of frequency spectra are therefore related through $M(\zeta) = \zeta\Xi(\zeta)$ and $N(\zeta) = \zeta\Lambda(\zeta)$, respectively.

We now introduce the frequency spectra into the expressions for the respondances. Carrying out the change of variable $\tau = \zeta^{-1}$ from relaxation time to relaxation frequency in Eq. (4.1-10) we obtain

$$\bar{Q}_\bullet(s) = G_g - \int_0^\infty \frac{M(\zeta)}{\zeta + s}\, d\zeta \; . \tag{4.1-35}$$

But the integral is the Stieltjes transform (see Appendix A5.1) of $M(\zeta)$. Thus

$$G_g - \bar{Q}_\bullet(s) = \mathscr{S}[M(\zeta); s] = \tilde{M}(s) \tag{4.1-36}$$

and $G_g - \bar{Q}_\bullet(s)$ is recognized as the Stieltjes transform of the relaxation frequency spectrum. In a similar manner, Eq. (4.1-18) yields

$$\bar{U}(s) = J_g + \int_0^\infty \frac{N(\zeta)}{\zeta + s}\, d\zeta + \{\phi_f/s\} \tag{4.1-37}$$

and

$$\bar{U}(s) - J_g - \{\phi_f/s\} = \mathscr{S}[N(\zeta); s] = \tilde{N}(s) \; , \tag{4.1-38}$$

i.e. $\bar{U}(s) - J_g - \{\phi_f/s\}$ is the Stieltjes transform of the retardation frequency spectrum. Conversely, $M(\zeta)$ and $N(\zeta)$ are recognized as the inverse Stieltjes transforms of $G_g - \bar{Q}_\bullet(s)$ and $U(s) - J_g - \{\phi_f/s\}$, respectively. We express this as

$$\mathscr{S}^{-1}\{[G_g - \bar{Q}_\bullet(s)]; \zeta\} = M(\zeta) = H(\tau)|_{\tau=\zeta^{-1}} \tag{4.1-39}$$

and

$$\mathscr{S}^{-1}\{[\bar{U}(s) - J_g - \{\phi_f/s\}]; \zeta\} = N(\zeta) = L(\tau)|_{\tau=\zeta^{-1}} \; . \tag{4.1-40}$$

From Eq. (4.1-3) we obtain

$$\bar{Q}(s) = \{G_e\} + s\int_0^\infty \frac{\Xi(\zeta)}{\zeta + s}\, d\zeta \tag{4.1-41}$$

and thus

$$\bar{Q}(s) - \{G_e\} = s\,\mathscr{S}[\Xi(\zeta); s] = s\tilde{\Xi}(s) \; . \tag{4.1-42}$$

Accordingly, $\bar{Q}(s) - \{G_e\}$ is the s-multiplied Stieltjes transform of $\Xi(\zeta)$. Inversion gives

$$\mathscr{S}^{-1}\{[\bar{G}(s) - \{G_e\}/s]; \zeta\} = \Xi(\zeta) = \tau H(\tau)|_{\tau=\zeta^{-1}} \; , \tag{4.1-43}$$

i.e. $\Xi(\zeta)$ is the inverse Stieltjes transform of $\bar{G}(s) - \{G_e/s\}$.

Equation (4.1-26) analogously yields

$$\bar{U}_\bullet(s) = J_e^{\{o\}} - s\int_0^\infty \frac{\Lambda(\zeta)}{\zeta + s}\,d\zeta + \{\phi_f/s\} \tag{4.1-44}$$

and

$$J_e^{\{o\}} - \bar{U}_\bullet(s) + \{\phi_f/s\} = s\,\mathscr{S}[\Lambda(\zeta); s] = s\tilde{\Lambda}(s) \; . \tag{4.1-45}$$

Thus, $J_e^{\{o\}} - \bar{U}_\bullet(s) + \{\phi_f/s\}$ is the s-multiplied Stieltjes transform of $\Lambda(\zeta)$, and, by

$$\mathscr{S}^{-1}\{[J_e^{\{o\}}/s - \bar{J}_\bullet(s) + \{\phi_f/s^2\}]; \zeta\} = \Lambda(\zeta) = \tau L(\tau)|_{\tau=\zeta^{-1}} \tag{4.1-46}$$

$\Lambda(\zeta)$ is recognized as the inverse Stieltjes transform of $J_e^{\{o\}}/s - \bar{J}_\bullet(s) + \{\phi_f/s^2\}$. We note that, because $\bar{Q}(s) \equiv \bar{Q}_\bullet(s)$,

$$s\tilde{\Xi}(s) + \tilde{M}(s) = G_g - \{G_e\} \tag{4.1-47}$$

and, because $\bar{U}(s) \equiv \bar{U}_\bullet(s)$

$$s\tilde{\Lambda}(s) + \tilde{N}(s) = J_e^{\{o\}} - J_g \; . \tag{4.1-48}$$

The experimental response functions can all be expressed in terms of $M(\zeta)$, $\Xi(\zeta)$, $N(\zeta)$, and $\Lambda(\zeta)$ instead of $H(\tau)$ and $L(\tau)$. Thus, from Eq. (4.1-41)

$$\bar{G}(s) = \frac{\{G_e\}}{s} + \int_0^\infty \frac{\Xi(\zeta)}{\zeta + s}\,d\zeta \tag{4.1-49}$$

and, therefore,

$$G(t) = \{G_e\} + \int_0^\infty \Xi(\zeta)\exp(-\zeta t)\,d\zeta \; . \tag{4.1-50}$$

The relaxation modulus, after subtraction of $\{G_e\}$, may then be recognized as the Laplace transform of $\Xi(\zeta)$ with the time, t, as the Laplace transform variable. The variable is here restricted to values on the positive real axis of the complex plane including the origin. We write

$$G(t) - \{G_e\} = \mathscr{L}[\Xi(\zeta); t] = \bar{\Xi}(t) \tag{4.1-51}$$

from which

$$\mathscr{L}\{[G(t) - \{G_e\}]; s\} = \mathscr{L}\{\mathscr{L}[\Xi(\zeta); t]; s\} = \mathscr{S}[\Xi(\zeta); s] \tag{4.1-52}$$

in accordance with the nature of the Stieltjes transformation as a repeated Laplace transformation (cf. Appendix A5.1). Equation (4.1-52) is, of course, identical with Eq. (4.1-49). Inversion of Eq. (4.1-51) yields

$$\mathscr{L}^{-1}\{[G(t) - \{G_e\}];\zeta\} = \Xi(\zeta) = \tau H(\tau)|_{\tau=\zeta^{-1}} \ . \tag{4.1-53}$$

In parallel fashion we may obtain

$$\bar{J}_\bullet(s) = \frac{J_e^{\{o\}}}{s} - \int_0^\infty \frac{\Lambda(\zeta)}{\zeta + s} d\zeta + \{\phi_f/s^2\} \tag{4.1-54}$$

from Eq. (4.1-44) and write

$$J_\bullet(t) = J_e^{\{o\}} - \int_0^\infty \Lambda(\zeta)\exp(-\zeta t) d\zeta + \{\phi_f t\} \tag{4.1-55}$$

which gives

$$J_e^{\{o\}} - J_\bullet(t) + \{\phi_f t\} = \mathscr{L}[\Lambda(\zeta);t] = \bar{\Lambda}(t) \tag{4.1-56}$$

and

$$\mathscr{L}\{[J_e^{\{o\}} - J_\bullet(t) + \{\phi_f t\}];s\} = \mathscr{L}\{\mathscr{L}[\Lambda(\zeta);t];s\} = \mathscr{S}[\Lambda(\zeta);s] \ . \tag{4.1-57}$$

Transform inversion then yields

$$\mathscr{L}^{-1}\{[J_e^{\{o\}} - J_\bullet(t) + \{\phi_f t\}];\zeta\} = \Lambda(\zeta) = \tau L(\tau)|_{\tau=\zeta^{-1}} \ . \tag{4.1-58}$$

The same results can be obtained starting from Eqs. (4.1-35) and (4.1-37), respectively, instead of Eqs. (4.1-41) and (4.1-44) (cf. Problem 4.1-7). The step response functions are thus seen to be essentially Laplace transforms of the τ-multiplied relaxation and retardation time spectra.

We now turn to the harmonic response functions. From Eq. (4.1-35), replacing s by $j\omega$, we obtain

$$G_\bullet^*(\omega) = G_g - \int_0^\infty \frac{M(\zeta)}{\zeta + j\omega} d\zeta \ . \tag{4.1-59}$$

But the integral is the Fourier-Laplace transform (see Appendix A5.3) of $M(\zeta)$. Thus

$$G_g - G_\bullet^*(\omega) = \mathscr{I}[M(\zeta);j\omega] = \tilde{M}^*(\omega) \tag{4.1-60}$$

where $\tilde{M}^*(\omega) = \tilde{M}'(\omega) - j\,\tilde{M}''(\omega)$, with

$$\tilde{M}'(\omega) = \int_0^\infty \frac{\zeta M(\zeta)}{\zeta^2 + \omega^2} d\zeta \quad \text{and} \quad \tilde{M}''(\omega) = \int_0^\infty \frac{\omega\, M(\zeta)}{\zeta^2 + \omega^2} d\zeta \ . \tag{4.1-61}$$

From Eq. (4.1-41) we obtain, proceeding in a similar fashion,

$$G^*(\omega) = \{G_e\} + j\omega \int_0^\infty \frac{\Xi(\zeta)}{\zeta + j\omega} d\zeta \tag{4.1-62}$$

and, therefore,

$$G^*(\omega) - \{G_e\} = j\omega\, \mathscr{I}[\Xi(\zeta);j\omega] = j\omega\, \tilde{\Xi}^*(\omega) \tag{4.1-63}$$

where $\tilde{\Xi}^*(\omega) = \tilde{\Xi}'(\omega) - j\,\tilde{\Xi}''(\omega)$, with

$$\tilde{\Xi}'(\omega) = \int_0^\infty \frac{\zeta\Xi(\zeta)}{\zeta^2 + \omega^2} d\zeta \quad \text{and} \quad \tilde{\Xi}''(\omega) = \int_0^\infty \frac{\omega\, \Xi(\zeta)}{\zeta^2 + \omega^2} d\zeta \ . \tag{4.1-64}$$

For the complex compliance, we have, starting from Eq. (4.1-37)

$$J^*(\omega) = J_g + \int_0^\infty \frac{N(\zeta)}{\zeta + j\omega} d\zeta + \{\phi_f/j\omega\} , \tag{4.1-65}$$

i.e.

$$J^*(\omega) - J_g - \{\phi_f/j\omega\} = \mathscr{I}[N(\zeta); j\omega] = \tilde{N}^*(\omega) \tag{4.1-66}$$

where $N^*(\omega) = \tilde{N}'(\omega) - j\,\tilde{N}''(\omega)$, with

$$\tilde{N}'(\omega) = \int_0^\infty \frac{\zeta N(\zeta)}{\zeta^2 + \omega^2} d\zeta \quad \text{and} \quad \tilde{N}''(\omega) = \int_0^\infty \frac{\omega\, N(\zeta)}{\zeta^2 + \omega^2} d\zeta . \tag{4.1-67}$$

Finally, from Eq. (4.1-44) we obtain

$$J^*_\bullet(\omega) = J_e^{\{o\}} - j\omega \int_0^\infty \frac{\Lambda(\zeta)}{\zeta + j\omega} d\zeta + \{\phi_f/j\omega\} \tag{4.1-68}$$

or

$$J_e^{\{o\}} - J^*_\bullet(\omega) + \{\phi_f/j\omega\} = j\omega\, \mathscr{I}[\Lambda(\zeta); j\omega] = j\omega\, \tilde{\Lambda}^*(\omega) \tag{4.1-69}$$

where $\tilde{\Lambda}^*(\omega) = \tilde{\Lambda}'(\omega) - j\,\tilde{\Lambda}''(\omega)$, with

$$\tilde{\Lambda}'(\omega) = \int_0^\infty \frac{\zeta \Lambda(\zeta)}{\zeta^2 + \omega^2} d\zeta \quad \text{and} \quad \tilde{\Lambda}''(\omega) = \int_0^\infty \frac{\omega\, \Lambda(\zeta)}{\zeta^2 + \omega^2} d\zeta . \tag{4.1-70}$$

Thus, the complex harmonic response functions are Fourier-Laplace transforms of the frequency spectra. Their inversion relations are assembled below:

$$\mathscr{I}^{-1}\{[G_g - G^*_\bullet(\omega)]; \zeta\} = M(\zeta) = H(\tau)|_{\tau=\zeta^{-1}} \tag{4.1-71}$$

$$\mathscr{I}^{-1}\{[J^*(\omega) - J_g - \{\phi_f/j\omega\}]; \zeta\} = N(\zeta) = L(\tau)|_{\tau=\zeta^{-1}} \tag{4.1-72}$$

$$\mathscr{I}^{-1}\{(j\omega)^{-1}[G^*(\omega) - \{G_e\}]; \zeta\} = \Xi(\zeta) = \tau H(\tau)|_{\tau=\zeta^{-1}} \tag{4.1-73}$$

$$\mathscr{I}^{-1}\{(j\omega)^{-1}[J_e^{\{o\}} - J^*_\bullet(\omega) + \{\phi_f/j\omega\}]; \zeta\} = \Lambda(\zeta) = \tau L(\tau)|_{\tau=\zeta^{-1}} . \tag{4.1-74}$$

The real and imaginary components of the complex response functions can easily be related to the appropriate components of the transforms. Thus

$$G'(\omega) = \{G_e\} + \omega\tilde{\Xi}''(\omega) \quad \text{or} \quad G'_\bullet(\omega) = G_g - \tilde{M}'(\omega) \tag{4.1-75}$$

$$G''(\omega) = \omega\tilde{\Xi}'(\omega) = \tilde{M}''(\omega) \tag{4.1-76}$$

$$J'_\bullet(\omega) = J_e^{\{o\}} - \omega\tilde{\Lambda}''(\omega) \quad \text{or} \quad J'(\omega) = J_g + \tilde{N}'(\omega) \tag{4.1-77}$$

and

$$J''(\omega) = \omega\tilde{\Lambda}'(\omega) + \{\phi_f/\omega\} = \tilde{N}''(\omega) + \{\phi_f/\omega\} . \tag{4.1-78}$$

The components $\tilde{M}'(\omega)$, etc, can, with suitable transformations of the variables, be brought into the form of Stieltjes transforms. This may be done by two different choices of variables. We first let $\tau^2 = \zeta^{-2} = \xi^{-1}$ and $\omega^2 = v$.

Using Eqs. $(4.1\text{-}75)_2$ and $(4.1\text{-}61)_1$ we then obtain

$$G_g - G'_\bullet(\omega) = \tilde{M}'(\omega) = \frac{1}{2} \int_0^\infty \frac{M(\xi^{1/2})}{\xi + v} d\xi \bigg|_{v=\omega^2} . \tag{4.1-79}$$

Equations (4.1-76) with either $(4.1\text{-}64)_1$ or $(4.1\text{-}61)_2$ give

$$G''(\omega) = \omega\tilde{\Xi}'(\omega) = \tilde{M}''(\omega) = \frac{\omega}{2}\int_0^\infty \frac{\xi^{-1/2}M(\xi^{1/2})}{\xi + v}\,d\xi\bigg|_{v=\omega^2} \tag{4.1-80}$$

and Eq. $(4.1\text{-}75)_1$ with $(4.1\text{-}64)_2$ yields

$$G'(\omega) - \{G_e\} = \omega\tilde{\Xi}''(\omega) = \frac{\omega^2}{2}\int_0^\infty \frac{\xi^{-1}M(\xi^{1/2})}{\xi + v}\,d\xi\bigg|_{v=\omega^2}. \tag{4.1-81}$$

Analogously, we find

$$J'(\omega) - J_g = \tilde{N}'(\omega) = \frac{1}{2}\int_0^\infty \frac{N(\xi^{1/2})}{\xi + v}\,d\xi\bigg|_{v=\omega^2} \tag{4.1-82}$$

from Eq. $(4.1\text{-}77)_2$ with $(4.1\text{-}67)_1$,

$$J''(\omega) - \{\phi_f/\omega\} = \omega\tilde{\Lambda}'(\omega) = \tilde{N}''(\omega) = \frac{\omega}{2}\int_0^\infty \frac{\xi^{-1/2}N(\xi^{1/2})}{\xi + v}\,d\xi\bigg|_{v=\omega^2} \tag{4.1-83}$$

from Eqs. (4.1-78) with either Eq. $(4.1\text{-}70)_1$ or $(4.1\text{-}67)_2$, and

$$J_e^{\{o\}} - J'_\bullet(\omega) = \omega\tilde{\Lambda}''(\omega) = \frac{\omega^2}{2}\int_0^\infty \frac{\xi^{-1}N(\xi^{1/2})}{\xi + v}\,d\xi\bigg|_{v=\omega^2} \tag{4.1-84}$$

from Eq. $(4.1\text{-}77)_1$ with $(4.1\text{-}70)_2$.

We can condense these relations into the single equation

$$\tilde{\Phi}(v) = \int_0^\infty \frac{\Phi(\xi)}{\xi + v}\,d\xi = \mathscr{S}[\Phi(\xi); v] \tag{4.1-85}$$

where

$$\tilde{\Phi}(v)|_{v=\omega^2} = \begin{cases} 2[G_g - G'_\bullet(\omega)] \\ 2G''(\omega)/\omega \\ 2[G'(\omega) - \{G_e\}]/\omega^2 \\ 2[J'(\omega) - J_g] \\ 2[J''(\omega) - \{\Phi_f/\omega\}]/\omega \\ 2[J_e^{\{o\}} - J'_\bullet(\omega)]/\omega^2 \end{cases} \quad \text{and} \quad \Phi(\xi)|_{\xi=\zeta^2} = \begin{cases} M(\zeta) = \zeta\Xi(\zeta) \\ \zeta^{-1}M(\zeta) = \Xi(\zeta) \\ \zeta^{-2}M(\zeta) = \zeta^{-1}\,\Xi(\zeta) \\ N(\zeta) = \zeta\Lambda(\zeta) \\ \zeta^{-1}N(\zeta) = \Lambda(\zeta) \\ \zeta^{-2}N(\zeta) = \zeta^{-1}\Lambda(\zeta) \end{cases} \tag{4.1-86}$$

or, in terms of the time spectra,

$$\Phi(\xi)|_{\xi=\tau^{-2}} = \begin{cases} H(\tau) \\ \tau H(\tau) \\ \tau^2 H(\tau) \\ L(\tau) \\ \tau L(\tau) \\ \tau^2 L(\tau). \end{cases} \tag{4.1-86'}$$

Similar results can be obtained by a different choice of variables. Let $\tau^2 = \zeta^{-2} = \lambda$ and $\omega^2 = \mu^{-1}$. We then find

$$\tilde{\Phi}(\mu) = \int_0^\infty \frac{\Phi(\lambda)}{\lambda + \mu}\, d\lambda = \mathscr{S}[\Phi(\lambda); \mu] \tag{4.1-87}$$

where

$$\tilde{\Phi}(\mu)|_{\mu=\omega^{-2}} = \begin{cases} 2\omega^2[G_g - G'_\bullet(\omega)] \\ 2\omega\, G''(\omega) \\ 2[G'(\omega) - \{G_e\}] \\ 2\omega^2[J'(\omega) - J_g] \\ 2\omega[J''(\omega) - \{\phi_f/\omega\}] \\ 2[J_e^{\{o\}} - J'_\bullet(\omega)] \end{cases} \quad \text{and} \quad \Phi(\lambda)|_{\lambda=\zeta^{-2}} = \begin{cases} \zeta^2 M(\zeta) = \zeta^3 \Xi(\zeta) \\ \zeta M(\zeta) = \zeta^2 \Xi(\zeta) \\ M(\zeta) = \zeta \Xi(\zeta) \\ \zeta^2 N(\zeta) = \zeta^3 \Lambda(\zeta) \\ \zeta N(\zeta) = \zeta^2 \Lambda(\zeta) \\ N(\zeta) = \zeta \Lambda(\zeta) \end{cases} \tag{4.1-88}$$

or, again in terms of the time spectra,

$$\Phi(\lambda)|_{\lambda=\tau^2} = \begin{cases} \tau^{-2}H(\tau) \\ \tau^{-1}H(\tau) \\ H(\tau) \\ \tau^{-2}L(\tau) \\ \tau^{-1}L(\tau) \\ L(\tau)\ . \end{cases} \tag{4.1-88'}$$

The importance of the frequency spectra lies in the fact that they are related, in relatively simple ways, to transforms with well-known inversion properties. These relations thus constitute a starting point for the computation of the time spectra from the experimental response functions.

4.1.3 Determination of the Continuous Spectra

It has been pointed out in the introduction to this chapter that the spectral response functions cannot be obtained directly by any experiment. They must be extricated by mathematical means from under the integrals appearing in the canonical representations of the experimental response functions. The integrals involving the continuous time spectra which we have discussed in Sect. 4.1.1 are Fredholm* integral equations of the first kind which can, in principle, be solved explicitly for $H(\tau)$ or $L(\tau)$. However, the solution of integral equations of the type with which

* Erik Ivar Fredholm (1866–1927), Swedish mathematician, who laid the foundations of the modern theory of integral equations.

we are concerned here is mathematically an "ill-posed" problem because small variations in, say, the function G(t), can cause large oscillations in H(τ) when an attempt is made to recover it [3]. Methods for obtaining approximations to the spectra by solution of the integral equations have been proposed but have not found wide-spread acceptance. An adequate treatment of these methods would have to be lengthy. For these reasons they will not be discussed here. Instead, we shall examine in detail approximations to the spectra obtained by processes of differentiation of the experimental data. These approximations are relatively simple and are being used widely. Differentiation does not, of course, turn an ill-posed into a well-posed problem. The reason why these methods work in practice will be discussed briefly in Sect. 4.4.

In Sect. 4.2 we consider approximations to the continuous spectra from the step response functions. Section 4.3 will deal with approximations from the harmonic response functions. The approximations to the continuous spectral functions will be compared in Sect. 4.4. In Sect. 4.5 we discuss the discrete spectral functions. The viscoelastic constants, G_e, G_g, η_f, J_g, $J_e^{\{o\}}$ and ϕ_f, which are intimately related to the spectral functions, will form the topic of Sect. 4.6.

4.2 Methods for Deriving the Continuous Spectra from the Step Responses

The concept of obtaining approximations to the spectral functions by differentiation of the step response functions is linked to the inversion properties of the Laplace transform and is introduced on this basis in Sect. 4.2.1. Approximations derived from a more general method are investigated in Sect. 4.2.2. A procedure based on assuming a particular form for the spectrum is discussed in Sect. 4.2.3. Finally, in Sect. 4.2.4, we replace differentiation by finite differencing.

4.2.1 The Transform Inversion Method: The Approximations of Schwarzl and Staverman

It was shown in Sect. 4.1 [cf. Eqs. (4.1-51) and (4.1-56)] that the unit step response functions, G(t) and J(t), are essentially Laplace transforms of the frequency spectra $\Xi(\zeta)$ and $\Lambda(\zeta)$. We can thus use the inversion properties of the Laplace transform (see Appendix A3.2.2) to obtain approximations to the spectra from the step response functions. To allow simultaneous handling of the approximations to be obtained from the relaxation modulus and the creep compliance, we introduce the notation [cf. Eqs. (4.1-51), (4.1-56), $(4.1\text{-}33)_2$ and $(4.1\text{-}34)_2$]

$$\overline{\psi}(t) = \begin{cases} \bar{\Xi}(t) = G(t) - \{G_e\} \\ \bar{\Lambda}(t) = J_e^{\{o\}} - [J_\bullet(t) - \{\phi_f t\}] \end{cases} \tag{4.2-1}$$

and

$$\psi(\zeta) = \begin{cases} \Xi(\zeta) = \tau H(\tau)|_{\tau=\zeta^{-1}} \\ \Lambda(\zeta) = \tau L(\tau)|_{\tau=\zeta^{-1}} \end{cases} \tag{4.2-2}$$

where

$$\bar{\psi}(t) = \int_0^\infty \psi(\zeta)\exp(-\zeta t)d\zeta = \mathscr{L}[\psi(\zeta); t] \ . \tag{4.2-3}$$

Equation (4.2-3) is a condensation of Eqs. (4.1-50) and (4.1-55). Because the transform variable, t, is restricted to positive real values, we cannot use Eq. (A3-28) for the transform inversion since this requires a contour integration in the complex plane. Instead, we use the formula for the inversion of the transform on the real time axis, given in the Appendix as Eq. (A3-45). We have

$$\psi(\zeta) = \lim_{k\to\infty} \frac{(-1)^k t}{k!} D_t^{(k)} \bar{\psi}(t)|_{t=k\zeta^{-1}} \tag{4.2-4}$$

where

$$D_t^{(k)} = D_t(D_t - 1)(D_t - 2)\ldots(D_t - k + 1) \tag{4.2-5}$$

is the *factorial function* introduced in Appendix A5.2.2, and

$$D_t^k = \frac{d^k}{d\ln^k t} \ . \tag{4.2-6}$$

According to this formula $\bar{\psi}(t)$ may be inverted by a process of logarithmic differentiation of infinite order. This, of course, is not practical. We may, however, obtain an *approximation of order* k, which we denote by $\psi_k(\zeta)$, by stopping short of the limit. We may thus write

$$\zeta\psi_k(\zeta) = \frac{(-1)^k}{(k-1)!} D_t^{(k)} \bar{\psi}(t)|_{t=k\zeta^{-1}} \ . \tag{4.2-7}$$

The approximations to the time spectra are then obtained from

$$F_k(\tau) = \frac{(-1)^k}{(k-1)!} D_t^{(k)} f(t)|_{t=k\tau} \tag{4.2-8}$$

where

$$F_k(\tau) = \begin{cases} H_k(\tau) \\ L_k(\tau) \end{cases} \tag{4.2-9}$$

and

$$f(t) = \begin{cases} G(t) \\ -[J(t) - \{\phi_f t\}] \ . \end{cases} \tag{4.2-10}$$

It should be noted that, in Eq. $(4.2\text{-}10)_2$, the flow term $\{\phi_f t\}$ must be subtracted from J(t). The contribution of the constants, $\{G_e\}$, and J_g, vanishes in the differentiations. The first order approximation to the relaxation spectrum from the relaxation modulus results as

$$H_1(\tau) = -\left.\frac{d\,G(t)}{d\ln t}\right|_{t=\tau} \tag{4.2-11}$$

and is therefore obtained by taking the negative slope of G(t) as a function of ln t. We note that the first order approximation involves the first logarithmic derivative of the source function G(t).

Approximations of higher order may be obtained by letting k assume increasingly higher values. Thus

$$H_2(\tau) = -\frac{d\,G(t)}{d\ln t} + \frac{d^2G(t)}{d\ln^2 t}\bigg|_{t=2\tau} \tag{4.2-12}$$

and

$$H_3(\tau) = -\frac{d\,G(t)}{d\ln t} + \frac{3}{2}\frac{d^2G(t)}{d\ln^2 t} - \frac{1}{2}\frac{d^3G(t)}{d\ln^3 t}\bigg|_{t=3\tau} . \tag{4.2-13}$$

The order of the logarithmic derivative thus equals the order of the approximation.

Still higher approximations can be obtained in the same way but experimental data are generally not precise enough to allow taking more than the third derivative at most.

For the retardation spectrum the transform inversion method gives

$$L_1(\tau) = \frac{d\,J(t)}{d\ln t} - \{\phi_f t\}\bigg|_{t=\tau} \tag{4.2-14}$$

for the first order,

$$L_2(\tau) = \frac{d\,J(t)}{d\ln t} - \frac{d^2J(t)}{d\ln^2 t}\bigg|_{t=2\tau} \tag{4.2-15}$$

for the second order, and

$$L_3(\tau) = \frac{d\,J(t)}{d\ln t} - \frac{3}{2}\frac{d^2J(t)}{d\ln^2 t} + \frac{1}{2}\frac{d^3J(t)}{d\ln^3 t}\bigg|_{t=3\tau} \tag{4.2-16}$$

for the third order approximation. The contributions from the flow terms cancel in the last two approximations because

$$\frac{d^k t}{d\ln^k t} = t . \tag{4.2-17}$$

When creep experiments on a rheodictic material have not been carried far enough to determine ϕ_f with confidence, Eq. (4.2-14) may not be used.

We note that the approximations of the same order, derived from the relaxation modulus and the creep compliance, respectively, are identical in form, except for a change in sign and the possible appearance of the flow term in $L_k(\tau)$.

Let us now examine the extent to which the "accessible" approximations $F_k(\tau)$ differ from the "inaccessible" true spectra $F(\tau)$. By mathematical induction we infer that, for any order k,

$$F_k(\tau) = \mathscr{P}_k[f(t)]|_{t=\beta_k\tau} \tag{4.2-18}$$

where $\mathscr{P}_k$ is a logarithmic differential operator given by

$$\mathscr{P}_k = \sum_{n=1}^{n=k} a_n \frac{d^n}{d \ln^n t} \tag{4.2-19}$$

in which the a_n are numerical constants. We could have inferred also that $\beta_k = k$ but we will see later that this is true only for the particular choice of the a_n arising from the use of the inversion formula, Eq. (4.2-4).

Applying the operator $\mathscr{P}_k$ to both sides of the canonical representation of the relaxation modulus,

$$G(t) = \{G_e\} + \int_{-\infty}^{\infty} H(\tau)\exp(-t/\tau)d \ln \tau \tag{4.2-20}$$

gives

$$\mathscr{P}_k[G(t)] = \int_{-\infty}^{\infty} H(\tau)\mathscr{P}_k[\exp(-t/\tau)]d \ln \tau \ . \tag{4.2-21}$$

We now introduce the function

$$I_k(t/\tau) = \mathscr{P}_k[\exp(-t/\tau)] = -\mathscr{P}_k[1 - \exp(-t/\tau)] \tag{4.2-22}$$

which results from applying the operator $\mathscr{P}_k$ to the kernel functions of the step responses. Then, from Eqs. (4.2-18) and $(4.2\text{-}22)_1$ we obtain

$$H_k(\tau) = \int_{-\infty}^{\infty} H(\tau)I_k(t/\tau)d \ln \tau|_{t=\beta_k\tau} \ . \tag{4.2-23}$$

Similarly, application of the operator $\mathscr{P}_k$ to both sides of

$$J(t) - \{\phi_f t\} = J_g + \int_{-\infty}^{\infty} L(\tau)[1 - \exp(-t/\tau)]d \ln \tau \tag{4.2-24}$$

yields

$$\mathscr{P}_k[J(t) - \{\phi_f t\}] = \int_{-\infty}^{\infty} L(\tau)\mathscr{P}_k[1 - \exp(-t/\tau)]d \ln \tau \ . \tag{4.2-25}$$

Using Eqs. (4.2-18) and $(4.2\text{-}22)_2$ we find

$$L_k(\tau) = \int_{-\infty}^{\infty} L(\tau)I_k(t/\tau)d \ln \tau|_{t=\beta_t\tau} \ . \tag{4.2-26}$$

Equations (4.2-23) and (4.2-26) may be condensed into

$$F_k(\tau) = P_k[f(t)]|_{t=\beta_k\tau} = \int_{-\infty}^{\infty} F(\tau)I_k(t/\tau)d\ln \tau|_{t=\beta_k\tau} \ . \tag{4.2-27}$$

Thus, to obtain the approximations, $F_k(\tau)$, for a given τ, we must integrate on the right of Eq. (4.2-27) over all τ's from 0 to ∞, and then change t to $\beta_k\tau$. The role of the kernel, $I_k(t/\tau)$, is to pick out from the true spectrum, $F(\tau)$, those contributions which will make up the approximations at the point where $\tau = t/\beta_k$. For this reason, $I_k(t/\tau)$ is termed the *intensity function.* If $I_k(t/\tau)$ were the delta function, and β_k were unity, the intensity function would pick only a single contribution at exactly $\tau = t$ and no approximation would be involved. This is seen more clearly upon transforming variables. Let z = ln τ, and n = ln t. Then Eq. (4.2-27) becomes

$$\mathscr{F}_k(z) = \int_{-\infty}^{\infty} \mathscr{F}(z)\mathscr{I}_k(n - z)dz|_{n=z+\ln\beta_k} \tag{4.2-28}$$

where

$$\mathscr{F}_{(k)}(z) = F_{(k)}(\tau)|_{\tau=\exp(z)} \tag{4.2-29}$$

whether we are dealing with the exact spectrum or the kth approximation. We also have

$$\mathscr{I}_k(n - z) = I_k(t/\tau)|_{t/\tau=\exp(n-z)} \quad . \tag{4.2-30}$$

We note that $\mathscr{F}_{(k)}(z)$ is the same function of z as F(τ) is of ln τ, and $\mathscr{I}_k(n - z)$ is the same function of n − z as $I_k(t/\tau)$ is of ln t/τ.

Now consider the result of letting the intensity function become the delta function, i.e. replacing $\mathscr{I}_k(n - z)$ by δ(n − z). We then [cf. Eq. (A2-8)] have

$$\mathscr{F}_k(z) = \mathscr{F}(n)|_{n=z+\ln\beta_k} \tag{4.2-31}$$

and $\mathscr{F}_k(n)$ becomes identical with $\mathscr{F}(z)$ for $\beta_k = 1$. In general $I_k(t/\tau)$ will not be a delta function although it can be shown that it approaches the delta function as $k \to \infty$ in Eq. (4.2-22). The approximation $F_k(\tau)$ will therefore be "muddied" by contributions selected to the left and right of $\tau = t/\beta_k$ instead of *at* $\tau = t/\beta_k$ only. For any approximation to be useful, $I_k(t/\tau)$ as a function of ln t/τ must be a bell-shaped curve (cf., e.g., Fig. 4.2-1) differing from zero only over a small range of values to the left and right of its maximum. It is clear that the approximation will be the better the more closely the intensity function resembles the delta function. It is also obvious that the intensity function must be normalized (i.e. the area under the curve must equal unity) since otherwise the resulting approximation would be multiplied by an arbitrary factor equal to the area.

Finally, β_k may now be interpreted as that value of t/τ at which $I_k(t/\tau)$ would be located along the (t/τ)-axis if it were a delta function. If $\beta_k \neq 1$, the delta function would pick contributions from F(τ) not at t/τ = 1 but at $t/\tau = \beta_k$. Consequently, β_k, called the *spectral shift factor*, is that factor by which the time scale of the contribution must be corrected for this fact. This can be indicated by the two equivalent expressions

$$F_k(\tau) = \mathscr{P}_k[f(t)]|_{t=\beta_k\tau} \quad \text{and} \quad F_k(\tau/\beta_k) = \mathscr{P}_k[f(t)]|_{t=\tau} \quad . \tag{4.2-32}$$

If $I_k(t/\tau)$ is not a delta function, it is not clear how an appropriate value for β_k might be selected. Formally, the intensity function is a distribution function. One may, therefore, select, e.g., the mean, the median, or the mode of $I_k(t/\tau)$ as a suitable measure of β_k. The median is that value of t/τ at which the ordinate divides the area under the distribution curve into two equal parts. The mode is the value at which the distribution has a maximum. We shall adopt this value as our measure of β_k. For a symmetrical intensity function the three definitions coincide.

To assess the goodness of the approximations of the first three orders we now examine the associated intensity functions. To this end we rewrite Eq. (4.2-22) in a more convenient form. We recognize that

Table 4.2-1. Stirling numbers, S_x^y, of the second kind

x \ y	1	2	3	4
1	1			
2	1	1		
3	1	3	1	
4	1	7	6	1

$$\frac{d^n}{d\ln^n x} = \sum_{m=1}^{m=n} S_n^m x^m \frac{d^m}{dx^m} \tag{4.2-33}$$

where the S_n^m are the Stirling* numbers of the second kind [4], the first few of which are tabulated in Table 4.2-1.

Using Eq. (4.2-33) and noting that

$$\frac{d^m \exp(-x)}{dx^m} = (-1)^m \exp(-x) \tag{4.2-34}$$

we write Eq. (4.2-22) in the form

$$I_k(t/\tau) = \exp(-t/\tau) \sum_{n=1}^{n=k} a_n \sum_{m=1}^{m=n} S_n^m (-t/\tau)^m \; . \tag{4.2-35}$$

By reversing the order in which the terms are obtained, Eq. (4.2-35) may be rearranged to

$$I_k(t/\tau) = \exp(-t/\tau) \sum_{n=1}^{n=k} (-t/\tau)^n \sum_{m=n}^{m=k} S_m^n a_m \; . \tag{4.2-36}$$

The intensity functions corresponding to the approximation of the first order can be obtained from Eq. (4.2-36) by letting $a_1 = -1$. We will show in Sect. 4.2.2 that this is a necessary requirement for all intensity functions derivable from the step responses. To obtain the intensity functions associated with the approximations of order $k = 2$ we let $a_2 = 1$, and for $k = 3$ let $a_2 = 3/2$ and $a_3 = -1/2$. The reasons for these choices will become clear in the next section. Accepting them for the moment, we find

$$I_1(t/\tau) = (t/\tau)\exp(-t/\tau) \tag{4.2-37}$$

$$I_2(t/\tau) = (t/\tau)^2\exp(-t/\tau) \tag{4.2-38}$$

and

$$I_3(t/\tau) = 0.5(t/\tau)^3\exp(-t/\tau) \tag{4.2-39}$$

respectively. The three functions are compared in Fig. 4.2-1.

The comparison shows that, as expected, the curves become narrower and more peaked as k increases. The maxima are located at $t/\tau = 1, 2,$ and 3, i.e. at $\ln t/\tau = 0$,

* James Stirling, 1692–1770, Scottish mathematician

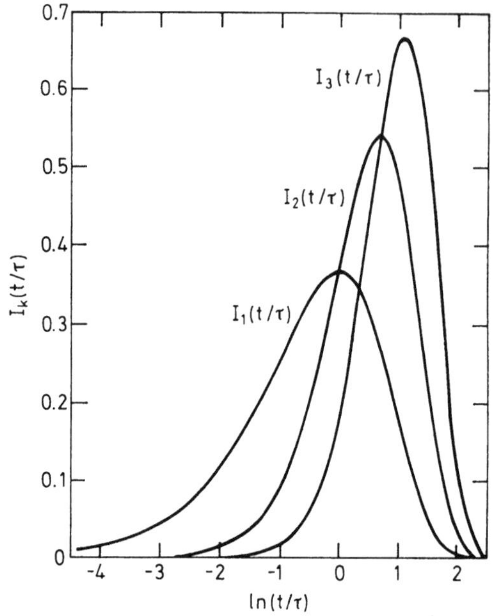

Fig. 4.2-1. Intensity functions $I_1(t/\tau)$, $I_2(t/\tau)$, and $I_3(t/\tau)$ as functions of ln t/τ

0.693, and 1.099, respectively. The approximations associated with these intensity functions were introduced [5] by Schwarzl and Staverman* and are often referred to under their names. The Schwarzl and Staverman approximations of higher order have the intensity functions

$$I_k(t/\tau) = \frac{(t/\tau)^k}{(k-1)!} \exp(-t/\tau) , \qquad k = 1, 2, \ldots \tag{4.2-40}$$

and the maxima are found from $d\, I_k(t/\tau)/d(t/\tau) = 0$. This gives $\beta_k = k$ for all of these approximations.

4.2.2 The Differential Operator Method

The question now arises whether the approximations obtained by transform inversion are the only ones applicable to the step response functions. The intensity functions associated with the approximations of Schwarzl and Staverman were obtained through a particular choice of the coefficients, a_n, of the differential operator, Eq. (4.2-19). Do other choices lead to other approximations and, if so, how do these compare with the former? We proceed to answer this question and begin by establishing the result that $a_1 = -1$ for any intensity function given by Eq. (4.2-22). This is so because a_1 is fixed by the requirement that the intensity

* Pron. Shvartsl and Staffermun

function be normalized, i.e. that the area under the curve be unity when $I_k(t/\tau)$ is plotted as a function of $\ln(t/\tau)$. The normalization requirement, therefore, may be stated as

$$\int_{-\infty}^{\infty} I_k(t/\tau) d \ln t/\tau = 1 \ . \tag{4.2-41}$$

Substitution of Eq. (4.2-22), into Eq. (4.2-41) and separating out the first term of the operator, we obtain

$$a_1 \int_{-\infty}^{\infty} \frac{d}{d \ln t}[\exp(-t/\tau)] d \ln t/\tau + \sum_{n=2}^{n=k} a_n \frac{d^{n-1}}{d \ln^{n-1} t} \int_{-\infty}^{\infty} \frac{d}{d \ln t}[\exp(-t/\tau)] d \ln t/\tau = 1 \ . \tag{4.2-42}$$

But

$$\int_{-\infty}^{\infty} \frac{d}{d \ln t}[\exp(-t/\tau)] d \ln t/\tau = -1 \ . \tag{4.2-43}$$

Thus, the higher terms contribute nothing, and $a_1 = -1$ regardless of the value of k. Hence, there is only one approximation of first order.

When k exceeds unity, the coefficients of the operator, $\mathscr{P}_k$, may be selected in more than one way. To ensure that these selections result in bell-shaped intensity functions, we require that they be determined in such a manner that $I_k(t/\tau)$ have a single maximum only. This will be the case if

$$\frac{d\, I_k(t/\tau)}{d \ln t/\tau} = 0 \tag{4.2-44}$$

has only one positive real root. The general form of the polynomial resulting from Eq. (4.2-44) is

$$\sum_{n=0}^{n=k} c_n (t/\tau)^n = 0 \ , \tag{4.2-45}$$

the coefficients c_n being given by

$$c_n = a_n + \sum_{m=n}^{m=k-1} S_{m+2}^{n+1} a_{m+1} \ , \qquad a_0 = 0 \tag{4.2-46}$$

where the S_m^n are the Stirling numbers (see Table 4.2-1). The sum vanishes when the lower index exceeds the upper one. Now, Eq. (4.2-44) will have a single positive real root if Eq. (4.2-45) is reduced to $(t/\tau)^m$ = constant, where $m = 1, 2, \ldots, k$. This can clearly be done by letting $k - 1$ out of the k coefficients $c_0, \ldots, c_{k-1}$ equal zero. This furnishes the necessary $k - 1$ relations for determining the $k - 1$ coefficients, $a_2, \ldots, a_k$, a_1 being fixed by the normalization requirement. Evidently, c_k cannot be made to vanish because this would lower the order of the approximation. Letting the vanishing coefficients be $c_0, \ldots, c_{k-2}$ leads to $(t/\tau) = c_k/c_{k-1}$ and reproduces the intensity functions associated with the Schwarzl and Staverman approximations obtainable by transform inversion.

Other choices of the coefficients to be equated to zero, which include c_{k-1}, lead to $(t/\tau)^m$ = constant where $2 \leq m \leq k$. There are $k - 1$ intensity functions of order k which can be obtained in this way. Thus it appears that the *differential operator method* just outlined is capable of supplying more approximations of order $k > 1$ than the transform inversion method. Are these approximations useful? It turns out that the intensity functions generated by $\mathscr{P}_k$ from the kernels of the step responses always consist of more than one term and thus are linear combinations of the Schwarzl and Staverman functions up to order k. They are therefore necessarily broader than the latter because they contain smaller contributions from the highest derivative. We consider the case $k = 2$ as an example. For this we have $c_0 = a_1 + a_2$ and $c_1 = a_1 + 3a_2$. But $a_1 = -1$. Letting $c_0 = 0$ gives $a_2 = 1$ and leads to Eq. (4.2-38). The choice $c_1 = 0$ gives $a_2 = 1/3$ and, therefore, furnishes

$$\begin{aligned} I_{2L}(t/\tau) &= \tfrac{1}{3}[2(t/\tau) + (t/\tau)^2]\exp(-t/\tau) \\ &= \tfrac{2}{3}I_1(t/\tau) + \tfrac{1}{3}I_2(t/\tau) \end{aligned} \qquad (4.2\text{-}47)$$

which has a maximum for $t/\tau = \sqrt{2}$ so that $\beta_{2L} = \sqrt{2}$. The subscript L indicates that the maximum occurs on the left of the maximum in $I_2(t/\tau)$ when $I_{2L}(t/\tau)$ is plotted as a function of ln t/τ.

We note that Eq. (4.2-47) is not the only possible two-term intensity function of order 2. In fact, if the intensity function consists of more than one term an infinite number of functions is possible because the set of coefficients is underdetermined if we merely require that the functions be normalized and have a single peak (cf. Problem 4.2-3). Only single-term intensity functions are unique because the only coefficient, a_k, is fixed through the normalization requirement.

Clearly, the requirement that the polynomial which is obtained by equating the first derivative of the intensity function to zero possess a single positive real root is not strong enough to ensure the generation of intensity functions with optimum properties by the differential operator method. The requirement must be reformulated to demand that the polynomial reduce to a single term of order k. This is equivalent to requiring that Eq. (4.2-45) be reduced to (t/τ) = constant. In the case of the approximations derivable by the operator method from the step responses this requirement produces only one approximation of order k, namely the Schwarzl-Staverman approximation of the same order. In the case of the storage functions the same requirement will yield k approximations of order k. In the case of the loss functions, $k + 1$ approximations of order k will be obtained.

Once the stronger requirement that the intensity function consist of a single term of order k has been adopted, it is not necessary to obtain the $a_2, \ldots, a_k$ from $c_2, \ldots, c_{k-1} = 0$. The intensity functions of order $k > 1$ can be obtained directly from Eq. (4.2-36) by demanding that all but the highest term vanish. This furnishes the $k - 1$ relations between $k - 1$ coefficients $a_2, \ldots, a_k$ needed to obtain the approximation $F_k(\tau)$ from Eqs. (4.2-32). We have, in fact, used this procedure in Sect. 4.2.1 to derive the coefficients a_2 and a_3 for the approximations of the second and third order (see Problem 4.2-4).

We conclude that the Schwarzl-Staverman approximations, $F_k(\tau)$, may be obtained either by the transform inversion or by the differential operator method.

The latter consists in using the operator $\mathscr{P}_k$ as specified by Eq. (4.2-32). The coefficients appearing in $\mathscr{P}_k$ [cf. Eq. (4.2-19)] are found by demanding that the intensity functions, defined by Eq. (4.2-22), be normalized and consist of a single term of order k.

Before closing this section we derive a result which will be needed in Sect. 4.2.4. Let us set $t/\tau = x$ for simplicity. We then claim that $I_k(x)$ can be expressed in terms of its derivatives evaluated at $x = 0$, i.e. in terms of

$$I_k^{(r)}(0) = \left.\frac{d^r I_k(x)}{dx^r}\right|_{x=0} = (-1)^r \sum_{n=1}^{n=k} a_n r^n \ . \tag{4.2-48}$$

The formula, which can be derived by mathematical induction, is given by

$$I_k(x) = \exp(-x) \sum_{p=1}^{p=k} \frac{1}{(k-p)!} \sum_{r=k-p+1}^{r=k} \frac{x^r}{(r-k+p)!} I_k^{(r-k+p)}(0) \tag{4.2-49}$$

and represents a sum of truncated Taylor expansions around $x = 0$. The approximations of Schwarzl and Staverman are obtained from Eq. (4.2-49) by requiring that the $I_k^{(r)}(0)$ up to order $k - 1$ vanish, i.e. that

$$I_k^{(r)}(0) = 0 \ , \qquad r = 1, 2, \ldots, k-1 \ . \tag{4.2-50}$$

Equation (4.2-50) is the result to be used in Sect. 4.2.4. We see that, using Eq. (4.2-50), Eq. (4.2-49) reduces to

$$I_k(x) = \frac{x^k}{k!} I_k^{(k)}(0) \exp(-x) \tag{4.2-51}$$

where $I_k^{(k)}(0)$ is given by Eq. (4.2-48). Equation (4.2-50) and the normalization condition, Eq. (4.2-41), provide k relations from which the $a_1, \ldots, a_k$ and, hence, $I_k^{(k)}(0)$, can be determined.

An example will make this clear. Let $k = 3$. Then Eq. (4.2-49) gives

$$\begin{aligned} I_3(x) = [&\tfrac{1}{2}x^3 I_3^{(1)}(0) + x^2 I_3^{(2)}(0) + \tfrac{1}{2}x^3 I_3^{(3)}(0) \\ &+ x\, I_3^{(1)}(0) + \tfrac{1}{2}x^2 I_3^{(2)}(0) + \tfrac{1}{6}x^3 I_3^{(3)}(0)]\exp(-x) \end{aligned} \tag{4.2-52}$$

which, by Eq. (4.2-50), reduces to

$$I_3(x) = \frac{x^3}{6} I_3^{(3)}(0) \exp(-x) \ . \tag{4.2-53}$$

But, by Eq. $(4.2\text{-}48)_2$,

$$I_3^{(3)}(0) = -3(a_1 + 3a_2 + 9a_3) \tag{4.2-54}$$

and, by Eq. (4.2-50), using Eq. $(4.2\text{-}48)_2$, we obtain

$$I_3^{(2)}(0) = 2(a_1 + 2a_2 + 4a_3) = 0 \tag{4.2-55}$$

and

$$I_3^{(1)}(0) = -(a_1 + a_2 + a_3) = 0 \ . \tag{4.2-56}$$

In addition, $a_1 = -1$ from Eq. (4.2-41). Solving Eqs. $(4.2\text{-}55)_2$ and $(4.2\text{-}56)_2$, we find $a_2 = 3/2$, and $a_3 = -1/2$ as before. Hence, $I_3^{(3)}(0) = 3$, and Eq. (4.2-51) reduces to Eq. (4.2-39).

It may be seen (cf. Problem 4.2-5) that equating $k - 1$ of the $I_k^{(r)}(0)$, *including* that of highest order, $I_k^{(k)}(0)$, to zero, leads to another selection of the coefficients a_n. However, the intensity functions derived in this way are again linear combinations of the intensity functions associated with the approximations of Schwarzl and Staverman, and again offer no advantage over them.

4.2.3 The Power Law Method: The Approximations of Williams and Ferry

In Chap. 6 we will discuss the representation of the spectral functions by mathematical models. One such model is that of the *wedge distribution* (cf. Sect. 6.3) which assumes that the relaxation and retardation spectra can be represented over the range of interest of the respondance times by a *power law* expression which takes the form

$$H(\tau) = H_0(\tau/\tau_0)^{-m} \tag{4.2-57}$$

for the relaxation spectrum, and

$$L(\tau) = L_0(\tau/\tau_0)^{m} \tag{4.2-58}$$

for the retardation spectrum. H_0, L_0, and m are constants.

Williams and Ferry [6] derived approximations to the spectra which are based on these power law expressions. We consider the relaxation spectrum first. The exponent m is usually positive since $H(\tau)$ is generally a decreasing function of τ in the range of interest. We insert Eq. (4.2-57) into Eq. (4.1-5) and carry out a change of variable $\tau = t/x$. This leads to

$$G(t) = \{G_e\} + H_0(t/\tau_0)^{-m} \int_0^\infty x^{m-1}\exp(-x)dx = \{G_e\} + H_0(t/\tau_0)^{-m}\Gamma(m) \tag{4.2-59}$$

where $\Gamma(m)$ is the gamma function.

The first approximation is given by Eq. (4.2-11). Inserting Eq. (4.2-59) we find

$$H_1(\tau) = \Gamma(m+1)H_0(\tau/\tau_0)^{-m} \ . \tag{4.2-60}$$

The first approximation therefore differs by the factor $\Gamma(m + 1)$ from the original assumption for the spectrum given by Eq. (4.2-57). Assuming that very nearly the same error will occur when the approximation is obtained from experimental data rather than from Eq. (4.2-59), we obtain the Williams-Ferry approximation for the relaxation spectrum from the relaxation modulus as

$$H_{WF}(\tau) = M(m)H_1(\tau) \tag{4.2-61}$$

where

$$M(m) = 1/\Gamma(m+1) \ . \tag{4.2-62}$$

$H_{WF}(\tau)$ is a better approximation than $H_1(\tau)$.

The procedure is confined to positive values of m. One first obtains $H_1(\tau)$ from the first approximation, Eq. (4.2-60). One then plots $\log H_1(\tau)$ against $\log \tau$ and obtains m from

$$m = -\frac{d \log H_1(\tau)}{d \log \tau} . \tag{4.2-63}$$

Finally the first approximation values $H_1(\tau)$ are multiplied by M(m) as indicated by Eq. (4.2-61).

The same procedure is followed for $L(\tau)$ when calculating the retardation spectrum from the creep compliance, except that now m is given by

$$m = \frac{d \log L_1(\tau)}{d \log \tau} \tag{4.2-64}$$

and the approximation is obtained as

$$L_{WF}(\tau) = M(-m)L_1(\tau) . \tag{4.2-65}$$

Values of M(m) and M(−m) have been tabulated [6, 6a].

The labor involved in obtaining these approximations is the same as that needed for the second approximations of Schwarzl and Staverman. They have been discussed here because they have been widely used in the literature. They also serve as an example for the use of mathematical models in deriving approximations to the spectra.

4.2.4 The Finite Difference Operator Method: The Approximations of Yasuda and Ninomiya, and of Tschoegl

The approximations discussed in the preceding sections require graphical or numerical differentiation of the source data. We now turn to a discussion of approximations based on finite differencing rather than continuous differentiation. These approximations are particularly useful for computer calculations.

In the calculus of finite differences one defines [4] a *translation* or *shift operator* E by

$$E f(x) = f(x + h) \tag{4.2-66}$$

where h is some given number, most often a positive integer. The effect of applying the operator E to the function f(x) is to produce a new function which is identical with the original function in which the independent variable is shifted by the amount h. Successive application of E leads to

$$E^n f(x) = f(x + nh) . \tag{4.2-67}$$

One further defines a *difference operator* Δ by

$$\Delta f(x) = (E - 1)f(x) = f(x + h) - f(x) \tag{4.2-68}$$

and a *difference quotient*

$$\frac{\Delta f(x)}{h} = \frac{E-1}{h} f(x) = \frac{f(x+h)-f(x)}{h} \; . \tag{4.2-69}$$

Repeated application yields

$$\frac{\Delta^n f(x)}{h^n} = \frac{(E-1)^n}{h^n} f(x) = h^{-n} \sum_{m=0}^{m=n} (-1)^m \binom{n}{m} E^{n-m} f(x) \tag{4.2-70}$$

where the last expression on the right is obtained by the binomial expansion, and

$$\binom{n}{m} = \frac{n!}{(n-m)!\,m!} \tag{4.2-71}$$

are the Euler symbols representing the coefficients of the expansion. The difference quotient is the analog, in the calculus of finite differences, of the continuous derivative of infinitesimal calculus, and reduces to it in the limit as $h \to 0$. Thus

$$\lim_{h \to 0} \frac{\Delta^n f(x)}{h^n} = \frac{d^n f(x)}{dx^n} \; . \tag{4.2-72}$$

We now introduce a logarithmic difference quotient in analogy to logarithmic differentiation. We first define a *multiplication operator* or *logarithmic shift* or *translation operator* L by

$$L\, f(x) = f(xh) \; . \tag{4.2-73}$$

The operator L has the property of translating, or shifting the values of f(x) by the amount ln h along ln x, i.e. on the logarithmic scale of x. The *logarithmic difference operator* Λ is defined by

$$\Lambda\, f(x) = (L-1) f(x) = f(xh) - f(x) \tag{4.2-74}$$

and the *logarithmic difference quotient* by

$$\frac{\Lambda\, f(x)}{\ln h} = \frac{L-1}{\ln h} f(x) = \frac{f(xh)-f(x)}{\ln h} \; . \tag{4.2-75}$$

Repeated application leads to

$$L^n f(x) = f(xh^n) \tag{4.2-76}$$

$$\Lambda^n f(x) = (L-1)^n f(x) \tag{4.2-77}$$

and

$$\frac{\Lambda^n f(x)}{\ln^n h} = \frac{(L-1)^n}{\ln^n h} f(x) = \ln^{-n} h \sum_{m=0}^{m=n} (-1)^m \binom{n}{m} L^{n-m} f(x) \; . \tag{4.2-78}$$

The logarithmic difference quotient reduces to the logarithmic derivative as $h \to 1$. Thus

$$\lim_{h \to 1} \frac{\Lambda^n f(x)}{\ln^n h} = \frac{d^n f(x)}{d \ln^n x} \; . \tag{4.2-79}$$

We note that

$$\sum_{m=0}^{m=n} (-1)^m \binom{n}{m} L^{n-m} = \sum_{m=0}^{m=n} (-1)^{n-m} \binom{n}{n-m} L^m \tag{4.2-80}$$

because replacement of m by n − m in the summation merely reverses the order in which the terms appear in the expansion. With the help of Eqs. (4.2-78), (4.2-79) and (4.2-80) we obtain the finite difference analog of the continuous logarithmic operator $\mathscr{P}_k$, [cf. Eq. (4.2-19)], as

$$\mathscr{P}_{kh} = \sum_{n=1}^{n=k} \frac{a_n}{\ln^n h} \sum_{m=0}^{m=n} (-1)^{n-m} \binom{n}{m} L^m \tag{4.2-81}$$

where we have also used the identity

$$\binom{n}{n-m} = \binom{n}{m} . \tag{4.2-82}$$

The upper limit of summation, n, in the second sum may be replaced by k if we stipulate that

$$\binom{n}{m} = 0 , \qquad n < m . \tag{4.2-83}$$

This allows us to interchange the order of summation and we may write

$$\mathscr{P}_{kh} = \sum_{m=0}^{m=k} b_m L^m \tag{4.2-84}$$

where

$$b_m = \sum_{n=1}^{n=k} (-1)^{n-m} \binom{n}{m} \frac{a_n}{\ln^n h} \tag{4.2-85}$$

and the restriction (4.2-83) applies. Thus we have

$$\begin{aligned} b_0 &= -\frac{a_1}{\ln h} \\ b_1 &= \frac{a_1}{\ln h} \end{aligned} \tag{4.2-86}$$

for the approximations of first order,

$$\begin{aligned} b_0 &= -\frac{a_1}{\ln h} + \frac{a_2}{\ln^2 h} \\ b_1 &= \frac{a_1}{\ln h} - \frac{2a_2}{\ln^2 h} \\ b_2 &= \frac{a_2}{\ln^2 h} \end{aligned} \tag{4.2-87}$$

for the approximations of second order, and

$$
\begin{aligned}
b_0 &= -\frac{a_1}{\ln h} + \frac{a_2}{\ln^2 h} - \frac{a_3}{\ln^3 h} \\
b_1 &= \frac{a_1}{\ln h} - \frac{2a_2}{\ln^2 h} + \frac{3a_3}{\ln^3 h} \\
b_2 &= \frac{a_2}{\ln^2 h} - \frac{3a_3}{\ln^3 h} \\
b_3 &= \frac{a_3}{\ln^3 h}
\end{aligned}
\tag{4.2-88}
$$

for those of third order. The coefficients for higher order approximations are easily derived if needed. Equations (4.2-86), (4.2-87), and (4.2-88) will be utilized in this section as well as in Sect. 4.3.5. We note (see Problem 4.2-6) that

$$\sum_{m=0}^{m=k} b_m = 0 \tag{4.2-89}$$

for approximations of any order from the step response functions. This relation provides a useful check on the b_m.

The intensity functions of the approximations from the step response functions are obtained from

$$
\begin{aligned}
I_{kh}(t/\tau) &= \mathscr{P}_{kh}[\exp(-t/\tau)] = -\mathscr{P}_{kh}[1 - \exp(-t/\tau)] \\
&= \sum_{m=0}^{m=k} b_m \exp(-h^m t/\tau) \ .
\end{aligned}
\tag{4.2-90}
$$

Equation (4.2-90) is the analog of Eq. (4.2-22). We now turn to the determination of the coefficients a_n in terms of which the b_m are defined by Eq. (4.2-85). We obtain these coefficients essentially on the basis of the method summarized at the end of Sect. 4.2.2. The normalization requirement, Eq. (4.2-41), again yields $a_1 = -1$, regardless of the order k, or of the magnitude of h. To show this, we write the intensity function of order k in the form

$$
\begin{aligned}
I_{kh}(t/\tau) &= \frac{a_1}{\ln h}[\exp(-ht/\tau) - \exp(-t/\tau)] \\
&\quad + \sum_{n=2}^{n=k} \frac{a_n}{\ln^n h} \sum_{m=0}^{m=n} (-1)^{n-m} \binom{n}{m} \exp(-h^m t/\tau) \ .
\end{aligned}
\tag{4.2-91}
$$

Equation (4.2-91) is obtained by applying the operator, Eq. (4.2-81), to the kernel $\exp(-t/\tau)$ and separating out the first term. Now, by a known definite integral [7], we have

$$\int_{-\infty}^{\infty} [\exp(-h^{r+1}x) - \exp(-h^r x)] d \ln x = \ln h \ . \tag{4.2-92}$$

Hence,

$$\int_{-\infty}^{\infty} \sum_{m=0}^{m=n} (-1)^{n-m} \binom{n}{m} \exp(-h^m t/\tau) d \ln t/\tau = 0 \tag{4.2-93}$$

for any $n > 1$ because, when $n > 1$, the sum in Eq. (4.2-93) can always be developed into a series of integrals like that on the left hand side of Eq. (4.2-92). Since the number of such integrals is necessarily even, and they alternate in sign, the series is self-canceling. Consequently, Eq. (4.2-91) reduces to

$$\int_{-\infty}^{\infty} I_{kh}(t/\tau)\,d\ln t/\tau = -a_1 \tag{4.2-94}$$

and, by Eq. (4.2-41), $a_1 = -1$.

Equating the derivative of $I_{kh}(t/\tau)$ to zero and reducing the resulting polynomial to a single term does not lead to useful relations for determining the b_m. We therefore resort to the result derived at the end of Sect. 4.2.2. According to this result, the requirement that the intensity functions of order k defined by Eq. (4.2-90) have a single peak only, will be satisfied [cf. Eq. (4.2-50)] if their derivatives up to order $k-1$ vanish for $t/\tau = 0$, i.e. if

$$\left.\frac{d^r I_{kh}(t/\tau)}{d(t/\tau)^r}\right|_{t/\tau=0} = 0\ , \qquad r = 1, 2, \ldots, k-1\ . \tag{4.2-95}$$

This requirement was shown to be valid for the approximations of Schwarzl and Staverman and is equally valid for their finite difference analogs. It establishes $k-1$ relations

$$\sum_{m=0}^{m=k} b_m h^{mr} = 0\ , \qquad r = 1, 2, \ldots, k-1 \tag{4.2-96}$$

which, together with $a_1 = -1$, suffice to determine the k coefficients a_n and, hence, the coefficients b_n. We note that Eq. (4.2-96) includes Eq. (4.2-89) for the case of $r = 0$.

For the approximations of order 1 we only need a_1. Thus Eqs. (4.2-86) immediately give

$$\begin{aligned} b_0 &= \frac{1}{\ln h} \\ b_1 &= -\frac{1}{\ln h} \end{aligned} \tag{4.2-97}$$

and the intensity function becomes

$$I^*_{1h}(t/\tau) = \frac{1}{\ln h}[\exp(-t/\tau) - \exp(-ht/\tau)]\ . \tag{4.2-98}$$

The asterisk distinguishes the intensity functions obtained in this way from others to be introduced later.

For the second order approximations Eq. (4.2-96) furnishes the relation

$$b_0 + hb_1 + h^2 b_2 = 0 \tag{4.2-99}$$

and Eqs. (4.2-87) then yield

$$a_2 = \frac{\ln h}{h-1} \tag{4.2-100}$$

so that

$$b_0 = \frac{h}{(h-1)\ln h}$$

$$b_1 = -\frac{h+1}{(h-1)\ln h} \tag{4.2-101}$$

$$b_2 = \frac{1}{(h-1)\ln h} \ .$$

The intensity function results as

$$I_{2h}^*(t/\tau) = \frac{1}{(h-1)\ln h}[h \exp(-t/\tau) - (h+1)\exp(-ht/\tau) + \exp(-h^2 t/\tau)] \ . \tag{4.2-102}$$

With respect to the third order approximations Eq. (4.2-96) furnishes

$$b_0 + hb_1 + h^2 b_2 + h^3 b_3 = 0$$

and (4.2-103)

$$b_0 + h^2 b_1 + h^4 b_2 + h^6 b_3 = 0$$

which allow us to solve Eqs. (4.2-88) to yield

$$a_2 = \frac{(h+2)\ln h}{h^2 - 1}$$

and (4.2-104)

$$a_3 = -\frac{\ln^2 h}{(h^2-1)(h-1)} \ .$$

We find

$$b_0 = \frac{h^3}{(h^2-1)(h-1)\ln h}$$

$$b_1 = -\frac{h(h^2+h+1)}{(h^2-1)(h-1)\ln h} \tag{4.2-105}$$

$$b_2 = \frac{h^2+h+1}{(h^2-1)(h-1)\ln h}$$

and

$$b_3 = -\frac{1}{(h^2-1)(h-1)\ln h} \ .$$

Thus, the intensity function of third order becomes

$$I_{3h}^*(t/\tau) = \frac{1}{(h^2-1)(h-1)\ln h}[h^3 \exp(-t/\tau) - h(h^2+h+1)\exp(-ht/\tau) + (h^2+h+1)\exp(-h^2 t/\tau) - \exp(-h^3 t/\tau)] \ . \tag{4.2-106}$$

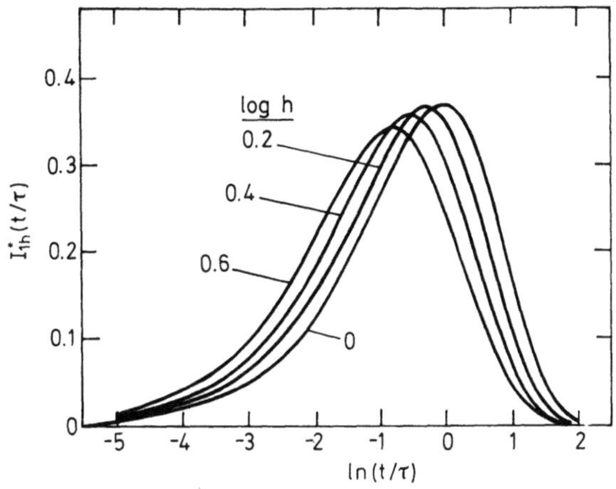

Fig. 4.2-2. Intensity functions $I^*_{1h}(t/\tau)$ for several values of log h as functins of ln t/τ

The shape of the intensity functions, as well as their location along the ln t/τ axis, depends on h. On the logarithmic time scale on which the source data are usually obtained, ht corresponds to log t + log h, h^2t to log t + 2 log h, etc. It is therefore convenient to consider the differencing interval in terms of log h. As log h → 0, i.e., h → 1, Eqs. (4.2-98), (4.2-102), and (4.2-106) become identical with Eqs. (4.2-37), (4.2-38) and (4.2-39). This is easily verified by application of L'Hospital's theorem. Figure 4.2-2 shows $I^*_{1h}(t/\tau)$ for log h = 0, 0.2, 0.4, and 0.6. The intensity function becomes steeper as log h decreases. It is, therefore, desirable to keep log h as close to zero as possible. With log h < 0.1, however, experimental errors in the source data render the results uncertain. With log h > 0.5, on the other hand, the approximation becomes less useful because the intensity function is then too broad. In practice a spacing of log h = 0.2 is generally most convenient.

Figure 4.2-3 shows the intensity functions $I^*_{1h}(t/\tau)$, $I^*_{2h}(t/\tau)$, and $I^*_{3h}(t/\tau)$ for log h = 0.2 as functions of ln t/τ.

As expected, the intensity functions become steeper as the order, k, increases. The location of the peaks, i.e. the spectral shift factor, β_{kh}, is found from

$$\frac{d\, I^*_{kh}(t/\tau)}{d\, \ln(t/\tau)} = 0 \; . \tag{4.2-107}$$

For $I^*_{1h}(t/\tau)$ we find

$$\beta^*_{1h} = \frac{\ln h}{h-1} \; . \tag{4.2-108}$$

For the approximations of higher order, the transcendental equations furnished by Eq. (4.2-107) must be solved on a computer. Table 4.2-2 assembles the spectral shift factor for log h = 0.1 to log h = 0.5 at intervals of 0.1 for convenience.

The approximations to the spectra from the step response functions are obtained from

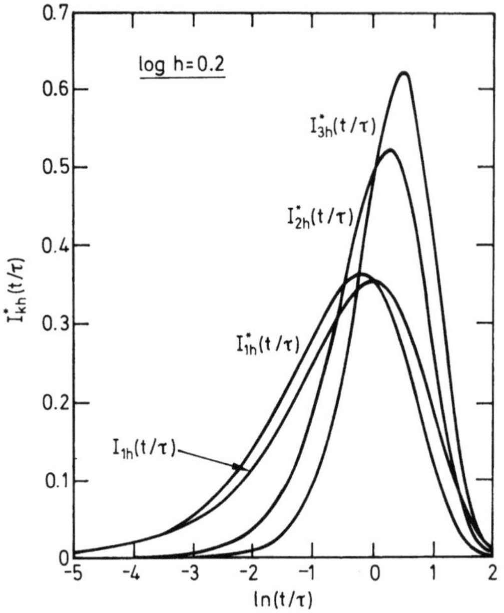

Fig. 4.2-3. Intensity functions $I^*_{1h}(t/\tau)$, $I^*_{2h}(t/\tau)$, and $I^*_{3h}(t/\tau)$ with log h = 0.2, as functions of ln t/τ. $I_{1h}(t/\tau)$ is included for comparison

Table 4.2-2. Spectral shift factors for finite difference approximations from the step function responses

log h	0.1	0.2	0.3	0.4	0.5
β^*_{1h}	0.889	0.787	0.694	0.609	0.532
β^*_{2h}	1.589	1.263	1.007	0.809	0.656
β^*_{3h}	2.139	1.550	1.150	0.878	0.689
β_{1h}	0.991	0.966	0.925	0.872	0.809
β_{2h}	2.001	2.002	2.010	2.031	2.073
β_{3h}	3.002	3.027	3.117	3.296	3.565

$$F_{kh}(\tau) = \mathscr{P}_{kh}[f(t)]|_{t=\beta_{kh}\tau} = \sum_{m=0}^{m=k} b_m f(h^m t)|_{t=\beta_{kh}\tau} \tag{4.2-109}$$

[cf. Eqs. (4.2-18), (4.2-84), and (4.2-76)] as

$$H_{kh}(\tau) = \mathscr{P}_{kh}[G(t)]|_{t=\beta_{kh}\tau} = \sum_{m=0}^{m=k} b_m G(h^m t)|_{t=\beta_{kh}\tau} \tag{4.2-110}$$

and

$$\begin{aligned} L_{kh}(\tau) &= -\mathscr{P}_{kh}[J(t) - \{\phi_f t\}]|_{t=\beta_{kh}\tau} \\ &= -\sum_{m=0}^{m=k} b_m[J(h^m t) - \{\phi_f h^m t\}]|_{t=\beta_{kh}\tau} \,. \end{aligned} \tag{4.2-111}$$

The first approximation to the relaxation spectrum from the relaxation modulus based on finite differencing is obtained from Eq. (4.2-110) as

$$H_{1h}(\tau/\beta_{1h}^*) = \left. \frac{G(t) - G(ht)}{\ln h} \right|_{t=\tau} . \tag{4.2-112}$$

Equation (4.2-112) states that an approximation to the value of the relaxation spectrum at τ/β_{1h}^* may be obtained by subtracting the value of the relaxation modulus at log t from that found at log t + log h, dividing by ln h and taking the negative inverse. The two approximations of the next higher orders are

$$H_{2h}(\tau/\beta_{2h}^*) = \left. \frac{h\,G(t) - (h + 1)G(ht) + G(h^2t)}{(h - 1)\ln h} \right|_{t=\tau} \tag{4.2-113}$$

and

$$H_{3h}(\tau/\beta_{3h}^*) = \left. \frac{h^3G(t) - h(h^2 + h + 1)G(ht) + (h^2 + h + 1)G(h^2t) - G(h^3t)}{(h^2 - 1)(h - 1)\ln h} \right|_{t=\tau} . \tag{4.2-114}$$

The appropriate values of β_{kh}^* may be selected from Table 4.2-2.

The approximations to the retardation spectrum from the creep compliance are obtained in similar manner from Eq. (4.2-111). We have

$$L_{1h}(\tau/\beta_{1h}^*) = \left. \frac{J(ht) - J(t) - \{(h - 1)\phi_f t\}}{\ln h} \right|_{t=\tau} \tag{4.2-115}$$

for the first order,

$$L_{2h}(\tau/\beta_{2h}^*) = - \left. \frac{J(h^2t) - (h + 1)J(ht) + h\,J(t)}{(h - 1)\ln h} \right|_{t=\tau} \tag{4.2-116}$$

for the second order, and

$$L_{3h}(\tau/\beta_{3h}^*) = \left. \frac{J(h^3t) - (h^2 + h + 1)J(h^2t) + h(h^2 + h + 1)J(t) - h^3J(t)}{(h^2 - 1)(h - 1)\ln h} \right|_{t=\tau} \tag{4.2-117}$$

for the third order approximation. The contributions from the flow term cancel from Eq. (4.2-111) for all approximations of higher order than the first because of Eq. (4.2-96).

The approximations given by Eqs. (4.2-112) through (4.2-117) were originally introduced in this form by Yasuda and Ninomiya [8]. We now introduce new forms of these approximations. Application of the operator L^{-p} to an intensity function does not change its shape but merely shifts it along the log(t/τ)-axis. Thus, the approximations to the spectra are unaffected, and only β_{kh} must be recalculated from Eq. (4.2-107) for the shifted intensity function. Application of the operator $L^{-k/2}$ produces approximations in which the values of $G(h^mt)$ or $J(h^mt)$ are arranged symmetrically around $t = \tau$. Thus, the second order approximation from the relaxation modulus may be shifted by L^{-1} to yield

$$H_{2h}(\tau/\beta_{2h}) = \left. \frac{h\,G(h^{-1}t) - (h + 1)G(t) + G(ht)}{(h - 1)\ln h} \right|_{t=\tau} \tag{4.2-118}$$

with the intensity function

$$I_{2h}(t/\tau) = \frac{h\exp(-h^{-1}t/\tau) - (h+1)\exp(-t/\tau) + \exp(-ht/\tau)}{(h-1)\ln h} \qquad (4.2\text{-}119)$$

from which β_{2h} must be determined. For approximations of odd orders $L^{-k/2}$ produces approximations requiring half-integral powers of h. The corresponding values of the source functions may have to be obtained by interpolation. To avoid this, one may average the approximations obtained by shifting upscale by $L^{1/2}$ and downscale by $L^{-1/2}$ after application of $L^{-k/2}$. The operator to be applied to Eqs. (4.2-112) and (4.2-114) or (4.2-115) and (4.2-117) then becomes

$$\tfrac{1}{2}(L^{1/2} + L^{-1/2})L^{-k/2} = \tfrac{1}{2}[L^{-(k-1)/2} + L^{-(k+1)/2}] \; . \qquad (4.2\text{-}120)$$

The intensity function of first order (see Fig. 4.2-3) becomes

$$I_{1h}(t/\tau) = \frac{\exp(-h^{-1}t/\tau) - \exp(-ht/\tau)}{2\ln h} \qquad (4.2\text{-}121)$$

and the equation for the spectral shift factor results as

$$\beta_{1h} = \frac{2\ln h}{h - h^{-1}} \; . \qquad (4.2\text{-}122)$$

The shift factors for log h = 0.1 to 0.5 are included in Table 4.2-2. We obtain

$$H_{1h}(\tau/\beta_{1h}) = \left.\frac{G(h^{-1}t) - G(ht)}{2\ln h}\right|_{t=\tau} \qquad (4.2\text{-}123)$$

for the relaxation spectrum. The shifted third order approximation has the intensity function

$$I_{3h}(t/\tau) = \frac{h^3\exp(-h^{-2}t/\tau) - h(h+1)\exp(-h^{-1}t/\tau) - (h^3-1)\exp(-t/\tau)}{2(h^2-1)(h-1)\ln h} + \frac{h(h+1)\exp(-ht/\tau) - \exp(-h^2t/\tau)}{2(h^2-1)(h-1)\ln h} \qquad (4.2\text{-}124)$$

yielding the spectrum

$$H_{3h}(\tau/\beta_3) = \left.\frac{h^3G(h^{-2}t) - h(h+1)G(h^{-1}t) - (h^3-1)G(t) + h(h+1)G(ht) - G(h^2t)}{2(h^2-1)(h-1)\ln h}\right|_{t=\tau} \; . \qquad (4.2\text{-}125)$$

It must be noted that, while the application of $L^{-k/2}$ does not change the shape of the intensity function, the averaging process demanded by the operator given by Eq. (4.2-120) slightly broadens it. The intensity function $I_{1h}(t/\tau)$ is plotted in Fig. 4.2-3 for comparison. The effect is not marked. A quantitative comparison is presented in Sect. 4.4.

The first order approximation to the creep compliance becomes

$$L_{1h}(\tau/\beta_{1h}) = \frac{J(ht) - J(h^{-1}t) - \{(h - h^{-1})\phi_f t\}}{2 \ln h} . \tag{4.2-126}$$

Those of the second and third order follow directly from Eqs. (4.2-118) and (4.2-124) by substituting J(t) for G(t) and changing the sign.

The paper by Yasuda and Ninomiya has not been translated into English and the approximations, which may be referred to under their names, are not widely known. The shifted approximations are due to the author and have not been reported previously.

4.3 Methods for Deriving the Continuous Spectra from the Harmonic Responses

It was demonstrated in Sect. 4.1 that the step response functions are essentially Laplace transforms of the spectral response functions. In Sect. 4.2 we showed that approximations to the spectra could be obtained by approximate inversion of the Laplace transform on the real time axis and we further showed that these approximations were the only useful ones. Now, as was also proved in Sect. 4.1, the harmonic response functions are essentially Stieltjes transforms of the spectral response functions. It may be expected, therefore, that approximations similar to those from the step responses could be obtained from the harmonic ones by approximate inversion of the Stieltjes transform on the real time axis. This is indeed true. However, in the harmonic case transform inversion does not necessarily lead to approximations whose associated intensity functions are normalized, and there is no built-in provision for correcting the time scale of the approximation when the intensity function does not have its maximum for $\omega\tau = 1$. Also, Stieltjes transform inversion cannot furnish all of the approximations which can be derived through the analog of the differential operator method developed in Sect. 4.2.2 for the step response functions. We, therefore, first discuss the operator method in Sect. 4.3.1. In Sect. 4.3.2 we show how most of the approximations can be obtained from Stieltjes transform inversion. In Sect. 4.3.3 we then briefly discuss the power law approximations which are analogous to those introduced in Sect. 4.2.3. Finally in Sect. 4.3.4 we deal with the approximations obtained by the finite difference operator method.

4.3.1 Differential Operator Method: The Approximations of Schwarzl and Staverman, and of Tschoegl

In analogy to Eqs. (4.2-9) and (4.2-10) we define

$$F_k'(\tau) = \begin{cases} H_k'(\tau) \\ L_k'(\tau) \end{cases} \quad \text{and} \quad F_k''(\tau) = \begin{cases} H_k''(\tau) \\ L_k''(\tau) \end{cases} \tag{4.3-1}$$

and

$$f'(\omega) = \begin{cases} G'(\omega) \\ -J'(\omega) \end{cases} \quad \text{and} \quad f''(\omega) = \begin{cases} G''(\omega) \\ J''(\omega) - \{\phi_f/\omega\} \end{cases} \tag{4.3-2}$$

to enable us to treat the approximations derived from the moduli and from the compliances simultaneously. We note that the primes and double primes with which the spectra are distinguished in Eqs. (4.3-1) do not signify the real and imaginary parts of a complex spectrum but are used to denote approximations derived from the real and imaginary parts of the complex harmonic response functions. The primeless symbols are, of course, reserved for approximations from the step responses.

The procedure to be followed to derive the approximations is that outlined in Sect. 4.2.2. We shall simply demand that the intensity functions be normalized and consist of a single term.

4.3.1.1 *Approximations Derived from the Storage Functions*

These approximations are obtained from the analog of Eq. (4.2-27) which takes the form

$$F'_k(\tau) = \mathscr{P}'_k[f'(\omega)]|_{\omega=\beta'_k/\tau} = \int_{-\infty}^{\infty} F(\tau)I'_k(\omega\tau)\mathrm{d}\ln\tau|_{\omega=\beta'_k/\tau} \tag{4.3-3}$$

where $F(\tau)$ stands for either $H(\tau)$ or $L(\tau)$ as before,

$$\mathscr{P}'_k = \sum_{n=1}^{n=k} a'_n \frac{d^n}{d\ln^n\omega} \tag{4.3-4}$$

is the (logarithmic) differential operator, and

$$I'_k(\omega\tau) = \mathscr{P}'_k\left[\frac{\omega^2\tau^2}{1+\omega^2\tau^2}\right] = -\mathscr{P}'_k\left[\frac{1}{1+\omega^2\tau^2}\right] \tag{4.3-5}$$

is the intensity function which, as discussed in Sect. 4.2.1, selects those contributions from the true spectrum, $F(\tau)$, which enter the approximation $F'_k(\tau)$ for each value of τ. The time scale of the spectrum must be adjusted by the factor β_k if the intensity function does not have its maximum at $\omega\tau = 1$. We now have $F'_k(\tau\beta'_k) = \mathscr{P}'_k[f'(\omega)]|_{\omega=1/\tau}$ [cf. Eq. $(4.2\text{-}32)_2$].

The two storage functions, just as the two step responses, have the same intensity functions but the approximations derived from them, although formally identical, have the opposite sign.

The coefficients of the operator $\mathscr{P}'_k$ are again determined from the normalization condition

$$\int_{-\infty}^{\infty} I'_k(\omega\tau)\mathrm{d}\ln\omega\tau = 1 \tag{4.3-6}$$

and the requirement that the intensity function consist of a single term of order k.

Normalization of $I'_k(\omega\tau)$ leads to $a'_1 = 1$ for any order k. To show this, we introduce Eq. (4.3-5) into (Eq. (4.3-6) to obtain

$$-\int_{-\infty}^{\infty} \mathscr{P}'_k\left[\frac{1}{1+\omega^2\tau^2}\right] d\ln\tau = \int_{-\infty}^{\infty} \mathscr{P}'_k\left[\frac{\omega^2\tau^2}{1+\omega^2\tau^2}\right] d\ln\tau = 1 \ . \tag{4.3-7}$$

Separating out the first term of the operator leads to

$$a'_1 \int_{-\infty}^{\infty} \frac{d}{d\ln\omega}\left[\frac{1}{1+\omega^2\tau^2}\right] d\ln\tau + \sum_{n=2}^{n=k} a'_n \frac{d^{n-1}}{d\ln^{n-1}\omega} \int_{-\infty}^{\infty} \frac{d}{d\ln\omega}\left[\frac{1}{1+\omega^2\tau^2}\right] d\ln\tau = -1 \tag{4.3-8}$$

and

$$a'_1 \int_{-\infty}^{\infty} \frac{d}{d\ln\omega}\left[\frac{\omega^2\tau^2}{1+\omega^2\tau^2}\right] d\ln\tau + \sum_{n=2}^{n=k} a'_n \frac{d^{n-1}}{d\ln^{n-1}\omega} \int_{-\infty}^{\infty} \frac{d}{d\ln\omega}\left[\frac{\omega^2\tau^2}{1+\omega^2\tau^2}\right] d\ln\tau = 1 \ . \tag{4.3-9}$$

But

$$\int_{-\infty}^{\infty} \frac{d}{d\ln\omega}\left[\frac{1}{1+\omega^2\tau^2}\right] d\ln\tau = -\int_{-\infty}^{\infty} \frac{d}{d\ln\omega}\left[\frac{\omega^2\tau^2}{1+\omega^2\tau^2}\right] d\ln\tau = -1 \ . \tag{4.3-10}$$

Hence $a'_1 = 1$ regardless of the magnitude of k.

For the approximation of first order a'_1 is all we need. We obtain the intensity function as

$$I'_1(\omega\tau) = \frac{2\omega^2\tau^2}{(1+\omega^2\tau^2)^2} \tag{4.3-11}$$

which yields $\beta'_1 = 1$. Thus, the first order approximations become

$$H'_1(\tau) = \left.\frac{d\,G'(\omega)}{d\ln\omega}\right|_{\omega=1/\tau} \tag{4.3-12}$$

and

$$L'_1(\tau) = -\left.\frac{d\,J'(\omega)}{d\ln\omega}\right|_{\omega=1/\tau} \tag{4.3-13}$$

respectively. For k = 2, Eq. (4.3-5) yields

$$I'_2(\omega\tau) = \frac{2(1-2a'_2)\omega^4\tau^4 + 2(1+2a'_2)\omega^2\tau^2}{(1+\omega^2\tau^2)^3} \tag{4.3-14}$$

which can be reduced to a single-term function in two ways: either by letting $a'_2 = 1/2$ or by letting $a'_2 = -1/2$. With these choices for a'_2 the second order intensity functions become

$$I'_{2L}(\omega\tau) = \frac{4\omega^2\tau^2}{(1+\omega^2\tau^2)^3} \tag{4.3-15}$$

and

$$I'_{2R}(\omega\tau) = \frac{4\omega^4\tau^4}{(1+\omega^2\tau^2)^3}\;. \tag{4.3-16}$$

We find $\beta'_{2L} = 1/\sqrt{2}$ and $\beta'_{2R} = \sqrt{2}$ and the second order approximations become

$$H'_2(\tau/\sqrt{2}) = \frac{d\,G'(\omega)}{d\ln\omega} + \frac{1}{2}\frac{d^2G'(\omega)}{d\ln^2\omega}\bigg|_{\omega=1/\tau} \tag{4.3-17}$$

and

$$H'_2(\sqrt{2}\tau) = \frac{d\,G'(\omega)}{d\ln\omega} - \frac{1}{2}\frac{d^2G'(\omega)}{d\ln^2\omega}\bigg|_{\omega=1/\tau} \tag{4.3-18}$$

for the relaxation spectrum, and

$$L'_2(\tau/\sqrt{2}) = -\frac{d\,J'(\omega)}{d\ln\omega} - \frac{1}{2}\frac{d^2J'(\omega)}{d\ln^2\omega}\bigg|_{\omega=1/\tau} \tag{4.3-19}$$

and

$$L'_2(\sqrt{2}\tau) = -\frac{d\,J'(\omega)}{d\ln\omega} + \frac{1}{2}\frac{d^2J'(\omega)}{d\ln^2\omega}\bigg|_{\omega=1/\tau} \tag{4.3-20}$$

for the retardation spectrum.

$I'_1(\omega\tau)$ and $I'_{2R}(\omega\tau)$ are shown in Fig. 4.3-1. $I'_1(\omega\tau)$ is a symmetric function of $\ln\omega\tau$ but the two second order intensity functions are not. $I'_{2L}(\omega\tau)$ and $I'_{2R}(\omega\tau)$ are mirror images of each other. The subscripts L and R refer to the location of the peaks to the left and right of $\ln\omega\tau = 0$.

We note that the same approach which has led to 1 approximation of order 2 in the case of the step responses, has now led to 2 approximations of order 2. The order of the approximation (i.e. the order of the highest derivative in the operator $\mathscr{P}'_k$ which produced it) is reflected in the exponent, $k+1$, of the denominator of the associated intensity function. The question arises: which of the two approximations of order 2 should one use in a given application? We defer the answer to this question until after we have examined the approximations of order 3.

For these approximations Eq. (4.3-5) yields

$$I'_3(\omega\tau) = \frac{2(1-2a'_2+4a'_3)\omega^6\tau^6 + 4(1-8a'_3)\omega^4\tau^4 + 2(1+2a'_2+4a'_3)\omega^2\tau^2}{(1+\omega^2\tau^2)^4} \tag{4.3-21}$$

which can be reduced to a single-term equation of order 3 in exactly 3 ways by the appropriate choice of the coefficients a'_2 and a'_3.

The choice $a'_2 = 0$, $a'_3 = -1/4$, leads to the intensity function

$$I'_3(\omega\tau) = \frac{12\omega^4\tau^4}{(1+\omega^2\tau^2)^4}\;. \tag{4.3-22}$$

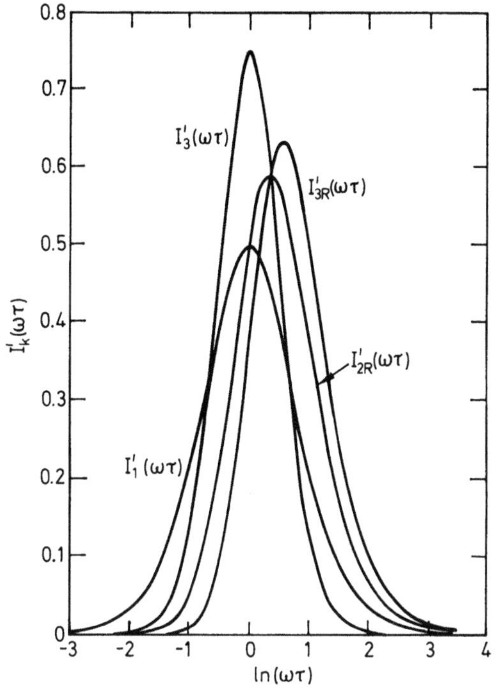

Fig. 4.3-1. Intensity functions $I'_1(\omega\tau)$, $I'_{2R}(\omega\tau)$, $I'_3(\omega\tau)$, and $I'_{3R}(\omega\tau)$ as functions of $\ln \omega\tau$

Two other choices, $a'_2 = 3/4$, $a'_3 = 1/8$, and $a'_2 = -3/4$, $a'_3 = 1/8$, give

$$I'_{3L}(\omega\tau) = \frac{6\omega^2\tau^2}{(1 + \omega^2\tau^2)^4} \tag{4.3-23}$$

and

$$I'_{3R}(\omega\tau) = \frac{6\omega^6\tau^6}{(1 + \omega^2\tau^2)^4} \tag{4.3-24}$$

respectively.

The intensity functions $I'_3(\omega\tau)$ and $I'_{3R}(\omega\tau)$ are also shown in Fig. 4.3-1. Again, $I'_{3L}(\omega\tau)$ is the mirror image of $I'_{3R}(\omega\tau)$. For $I'_3(\omega\tau)$, $\beta'_3 = 1$. For $I'_{3L}(\omega\tau)$ and $I'_{3R}(\omega\tau)$ we find $\beta'_{3L} = 1/\sqrt{3}$ and $\beta'_{3R} = \sqrt{3}$, respectively. The third order approximations to the relaxation spectrum thus become

$$H'_3(\tau/\sqrt{3}) = \frac{d\,G'(\omega)}{d\ln\omega} + \frac{3}{4}\frac{d^2G'(\omega)}{d\ln^2\omega} + \frac{1}{8}\frac{d^3G'(\omega)}{d\ln^3\omega}\bigg|_{\omega=1/\tau} \tag{4.3-25}$$

$$H'_3(\tau) = \frac{d\,G'(\omega)}{d\ln\omega} - \frac{1}{4}\frac{d^3G'(\omega)}{d\ln^3\omega}\bigg|_{\omega=1/\tau} \tag{4.3-26}$$

and

$$H'_3(\sqrt{3}\tau) = \frac{d\,G'(\omega)}{d\ln\omega} - \frac{3}{4}\frac{d^2G'(\omega)}{d\ln^2\omega} + \frac{1}{8}\frac{d^3G'(\omega)}{d\ln^3\omega}\bigg|_{\omega=1'\tau} \; . \tag{4.3-27}$$

For the retardation spectrum we have

$$L'_3(\tau/\sqrt{3}) = -\frac{d\,J'(\omega)}{d\ln\omega} - \frac{3}{4}\frac{d^2J'(\omega)}{d\ln^2\omega} - \frac{1}{8}\frac{d^3J'(\omega)}{d\ln^3\omega}\bigg|_{\omega=1/\tau} \tag{4.3-28}$$

$$L'_3(\tau) = -\frac{d\,J'(\omega)}{d\ln\omega} + \frac{1}{4}\frac{d^3J'(\omega)}{d\ln^3\omega}\bigg|_{\omega=1/\tau} \tag{4.3-29}$$

and

$$L'_3(\sqrt{3}\tau) = -\frac{d\,J'(\omega)}{d\ln\omega} + \frac{3}{4}\frac{d^2J'(\omega)}{d\ln^2\omega} - \frac{1}{8}\frac{d^3J'(\omega)}{d\ln^3\omega}\bigg|_{\omega=1/\tau}. \tag{4.3-30}$$

The approximations of odd order which have symmetric intensity functions were introduced by Schwarzl and Staverman [9] and are generally referred to under their names. The other approximations discussed here were introduced by the author [10].

The question concerning the proper choice from among the alternative forms of the approximations of order 2 and 3 is best decided by replacing the canonical representation of the storage modulus by a suitably chosen mathematical model from among those which will be discussed in Chap. 6. If the storage modulus is expressed in terms of such a model, an "exact" spectrum may be obtained through the special property of Stieltjes transform inversion discussed at the end of Sect. A5.2.1 of the Appendix. That an "exact" spectrum may be obtained in this way is not in conflict with our earlier assertion that an exact inversion of the canonical representations is not possible because they are Fredholm integral equations of the first kind whose solution is in general an ill-posed problem. The resolution of this seeming paradox is simple. An exact spectrum can be obtained through Stieltjes transform inversion only when a suitable model replaces the canonical representations of the storage or loss functions. For reasons that will become clear in Chap. 6, however, a mathematical model is necessarily only an *approximation* to the canonical representation. Hence, the spectrum derived from it is *exact* with respect to the function but is still an approximation to the *true* spectrum.

Nevertheless, for our present purposes the use of an analytic test function is ideal. Apart from allowing us to compare the approximations calculated from the test function with the "exact" spectrum derived from it, it also frees us from the vagaries of experimental errors and, furthermore, allows us to obtain the approximations by direct, rather then numerical or graphical differentiation.

The particular test function we shall be using is Eq. (6.1-66) of Chap. 6 and has the form

$$G'(\omega) = G_e + \frac{G_g - G_e}{1 + (\omega_0/\omega)^m} \tag{4.3-31}$$

where $\log G_e = 6.39$, $\log G_g = 10.30$, $\log \tau_0 = 9.180$, and $m = 0.595$. These values are based on realistic data [11]. The "exact" spectrum, $H'(\tau)$, was obtained from

$$H'(\tau) = \frac{2(G_g - G_e)\sin m\pi/2}{\pi[(\tau_0/\tau)^m + 2\cos \pi m/2 + (\tau/\tau_0)^m]} \tag{4.3-32}$$

where $\tau_0 = 1/\omega_0$ [cf. Eq. (6.3-129]. Figure 4.3-2 compares the exact spectrum

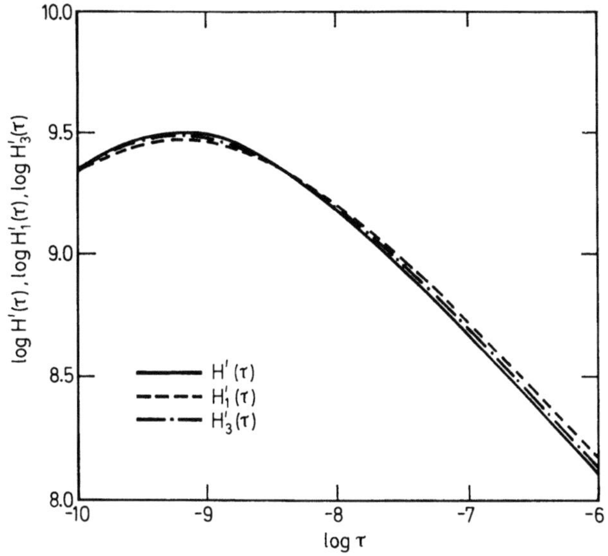

Fig. 4.3-2. log $H'(\tau)$, log $H'_1(\tau)$, and log $H'_3(\tau)$ as functions of log τ

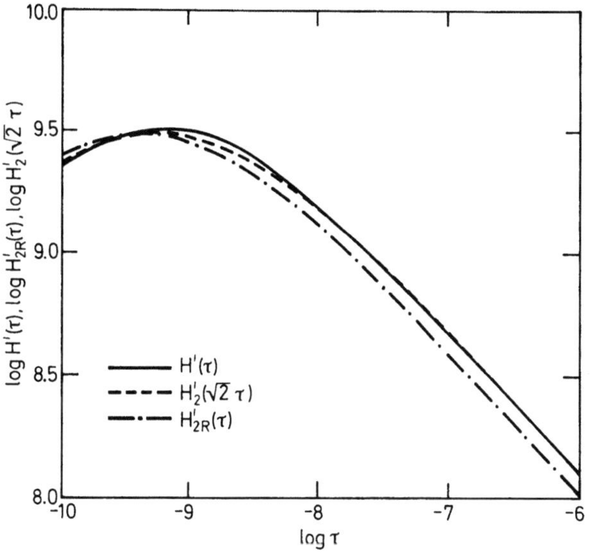

Fig. 4.3-3. log $H'(\tau)$, log $H'_{2R}(\tau)$, and log $H'_2(\sqrt{2}\tau)$ as functions of log τ

with the first and third order approximations whose intensity functions are centered on $\ln \omega\tau = 0$ and which, therefore, do not require adjustment of the time scale of the spectrum.

Figures 4.3-3 and 4.3-4 compare the exact spectrum with the approximations calculated for the intensity functions $I'_{2R}(\omega\tau)$ and $I'_{3R}(\omega\tau)$, with and without adjust-

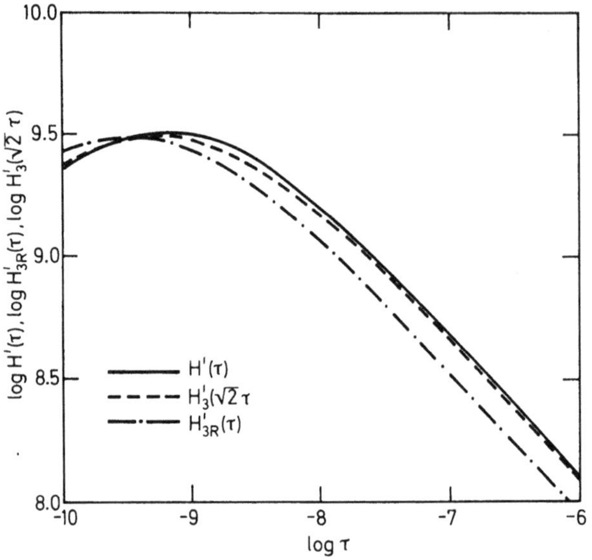

Fig. 4.3-4. log $H'(\tau)$, log $H'_{3R}(\tau)$, and log $H'_3(\sqrt{3}\tau)$ as functions of log τ

ments of the time scale. Clearly, the approximations are unsatisfactory without the shift but become excellent when the appropriate shift is introduced. In fact, the second order approximation is as good as the third order one, except around the maximum in $H'(\tau)$. This is at first sight surprising in view of the fact that theoretically the true spectrum is obtained only when $k \to \infty$. The explanation is, of course, to be found in the mildness of the test function. If the source function is not a smooth, slowly varying function of the frequency, approximations with much steeper and narrower intensity functions are needed to derive a satisfactory approximation to the spectrum.

It is an interesting aspect of the comparison between Figs. 4.3-2 and 4.3-4 that $H'_3(\sqrt{3}\tau)$ appears to be a better approximation, at least to the right of the maximum in $H'(\tau)$, than $H'_3(\tau)$. To understand this we consider Fig. 4.3-5. This shows $H'(\tau)$ along with a qualitative representation of the intensity functions $I'_3(\omega\tau)$ and $I'_{3R}(\omega\tau)$ so placed that their peaks are located at $\log \tau = \log \tau_0$ [the skewness of $I'_{3R}(\omega\tau)$ is exaggerated]. $I'_{3R}(\omega\tau)$ leads to a better approximation than $I'_3(\omega\tau)$ when log $H'(\tau)$ is a rapidly decreasing function of log τ because contributions to the approximation taken from the true spectrum to the left of τ_0 (i.e. on the high side) are weighted less and contributions taken from the spectrum to the right of τ_0 (i.e. on the low side) are weighted more by $I'_{3R}(\omega\tau)$. Thus a more accurate approximation to the true spectrum is achieved. Similar considerations apply to $I'_{3L}(\omega\tau)$ when the slope of log $H'(\tau)$ vs. log τ is positive. That this is so, is shown in Fig. 4.3-6 in which log $H'_2(\sqrt{2}\tau)$ and log $H'_2(\tau/\sqrt{2})$ are plotted with the exact spectrum, log $H'(\tau)$.

We see that $H'_2(\sqrt{2}\tau)$ works well where the slope of log $H(\tau)$ is negative and $H'_2(\tau/\sqrt{2})$ is most appropriate for a positive slope. The accuracy of the approximations in certain areas are partly coincidental, however, since a steeper or shallower

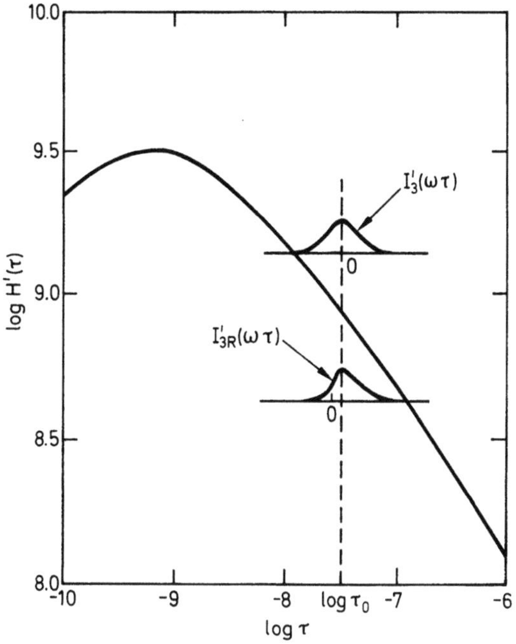

Fig. 4.3-5. Dependence of the approximations on the slope of the spectrum

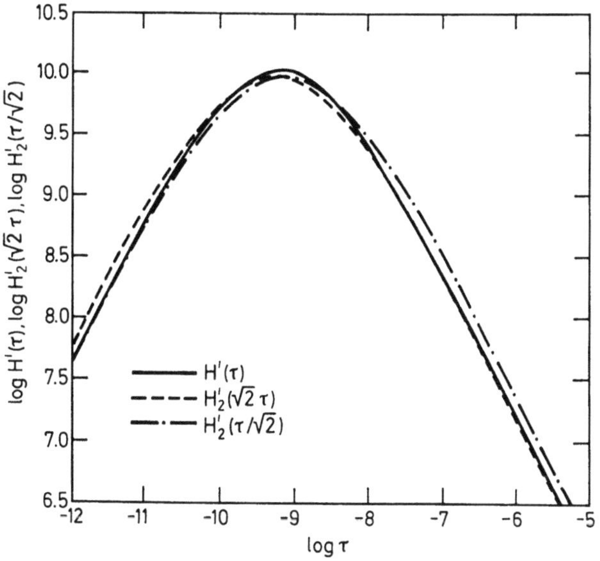

Fig. 4.3-6. log $H'(\tau)$, log $H_2'(\sqrt{2}\tau)$, and log $H_2'(\tau/\sqrt{2})$ as functions of log τ

slope would shift the approximation above or below the exact values respectively. This becomes apparent near the peak of the spectrum. Of course, at the peak, any approximation of practical usefulness will give a low value because of the low order of differentiation.

As a general rule, therefore, approximations whose intensity functions have a tail at higher values of τ should be used when the slope of $\log H_1'(\tau)$ or $\log L_1'(\tau)$ vs. $\log \tau$ is negative and vice versa. It should be noted that the approximations with symmetric intensity functions provide better overall approximations for both positive and negative slopes of the spectrum, and are preferable when the spectrum is flat.

4.3.1.2 Approximations Derived from the Loss Functions

Turning now to the loss functions, we have

$$F_k''(\tau) = \mathscr{P}_k''[f''(\omega)]|_{\omega=\beta_k''/\tau} = \int_{-\infty}^{\infty} F(\tau) I_k''(\omega\tau) d\ln\tau|_{\omega=\beta_k''/\tau} \tag{4.3-33}$$

where $F_k''(\tau)$ and $f''(\omega)$ are given by Eqs. $(4.3\text{-}1)_2$ and $(4.3\text{-}2)_2$,

$$\mathscr{P}_k'' = a_0'' + \sum_{n=1}^{n=k} a_n'' \frac{d^n}{d\ln^n\omega} , \tag{4.3-34}$$

and the associated intensity function is

$$I_k''(\omega\tau) = \mathscr{P}_k''\left[\frac{\omega\tau}{1+\omega^2\tau^2}\right] . \tag{4.3-35}$$

The coefficient a_0' takes care of the fact that the loss functions possess an approximation of order zero, because the kernel of the loss functions is itself a crude approximation to the delta function (cf. Fig. 4.1-2).

Since the two loss functions have the same kernel, they also have the same intensity functions and the approximations derived from them for the relaxation and retardation spectrum are identical in form, except possibly for a contribution from the flow term, $\{\phi_f/\omega\}$. The coefficients of $\mathscr{P}_k''$ are again found from the normalization condition

$$\int_{-\infty}^{\infty} I_k''(\omega\tau) d\ln\omega\tau = 1 \tag{4.3-36}$$

and the requirement that the intensity function consist of a single term.

Normalization now yields $a_0'' = 2/\pi$ for any order k. To show this, we substitute Eq. (4.3-35) into Eq. (4.3-36) and obtain

$$\int_{-\infty}^{\infty} \mathscr{P}_k''\left[\frac{\omega\tau}{1+\omega^2\tau^2}\right] d\ln\tau = 1 . \tag{4.3-37}$$

Separating out the zeroth term of the operator gives

$$a_0'' \int_{-\infty}^{\infty}\left[\frac{\omega\tau}{1+\omega^2\tau^2}\right] d\ln\tau + \sum_{n=1}^{n=k} a_n'' \frac{d^{n-1}}{d\ln^{n-1}\omega} \int_{-\infty}^{\infty} \frac{d}{d\ln\omega}\left[\frac{\omega\tau}{1+\omega^2\tau^2}\right] d\ln\tau = 1 . \tag{4.3-38}$$

But

$$\int_{-\infty}^{\infty} \frac{\omega\tau}{1+\omega^2\tau^2} d\ln\tau = \frac{\pi}{2} \tag{4.3-39}$$

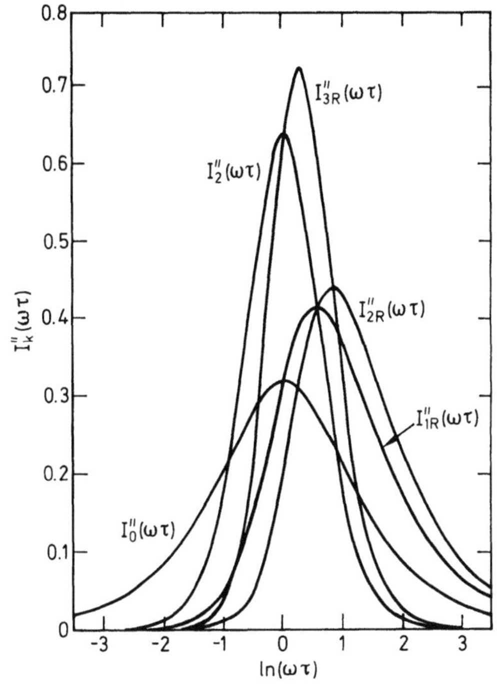

Fig. 4.3-7. The intensity functions $I''_0(\omega\tau)$, $I''_{1R}(\omega\tau)$, $I''_2(\omega\tau)$, $I''_{2R}(\omega\tau)$, and $I''_{3R}(\omega\tau)$ as functions of $\ln \omega\tau$

and

$$\int_{-\infty}^{\infty} \frac{d}{d \ln \omega}\left[\frac{\omega\tau}{1+\omega^2\tau^2}\right] d \ln \tau = 0 \ . \tag{4.3-40}$$

Hence $a''_0 = 2/\pi$ regardless of the magnitude of k.

It is seen immediately that the intensity function associated with the approximation of order zero is simply

$$I''_0(\omega\tau) = \frac{2\omega\tau}{\pi(1+\omega^2\tau^2)} \ . \tag{4.3-41}$$

The function has a maximum at $\ln \omega\tau = 0$ and is symmetric with respect to the $\ln(\omega\tau)$-axis as shown in Fig. 4.3-7.

The spectra are

$$H''_0(\tau) = \frac{2}{\pi} G''(\omega)|_{\omega=1/\tau} \tag{4.3-42}$$

and

$$L''_0(\tau) = \frac{2}{\pi}[J''(\omega) - \{\phi_f/\omega\}]|_{\omega=1/\tau} \tag{4.3-43}$$

respectively.

For k = 1, Eq. (4.3-35) yields

$$I_1''(\omega\tau) = \frac{(a_0'' - a_1'')\omega^3\tau^3 + (a_0'' + a_1'')\omega\tau}{(1 + \omega^2\tau^2)^2} \tag{4.3-44}$$

which can be reduced to a single term either by letting $a_1'' = a_0''$ or by letting $a_1'' = -a_0''$.

The two intensity functions thus become

$$I_{1L}''(\omega\tau) = \frac{4\omega\tau}{\pi(1 + \omega^2\tau^2)^2} \tag{4.3-45}$$

and

$$I_{1R}''(\omega\tau) = \frac{4\omega^3\tau^3}{\pi(1 + \omega^2\tau^2)^2} \tag{4.3-46}$$

where the subscripts again refer to the location of the maxima at $\omega\tau = 1/\sqrt{3}$ and $\sqrt{3}$ respectively. The intensity functions are not symmetric and are again mirror images of each other. $I_{1R}''(\omega\tau)$ is shown in Fig. 4.3-7 together with $I_0''(\omega\tau)$.

With the appropriate shift in the spectra to take into account the fact that the intensity functions do not have their maxima at $\omega\tau = 1$, the first order approximations then become

$$H_1''(\tau/\sqrt{3}) = \frac{2}{\pi}\left[G''(\omega) + \frac{d\,G''(\omega)}{d\ln\omega}\right]\Bigg|_{\omega=1/\tau} \tag{4.3-47}$$

$$H_1''(\sqrt{3}\tau) = \frac{2}{\pi}\left[G''(\omega) - \frac{d\,G''(\omega)}{d\ln\omega}\right]\Bigg|_{\omega=1/\tau} \tag{4.3-48}$$

for the relaxation spectrum and

$$L_1''(\tau/\sqrt{3}) = \frac{2}{\pi}\left[J''(\omega) + \frac{d\,J''(\omega)}{d\ln\omega}\right]\Bigg|_{\omega=1/\tau} \tag{4.3-49}$$

and

$$L_1''(\sqrt{3}\tau) = \frac{2}{\pi}\left[J''(\omega) - \frac{d\,J''(\omega)}{d\ln\omega} - \{2\phi_f/\omega\}\right]\Bigg|_{\omega=1/\tau} \tag{4.3-50}$$

for the retardation spectrum. There are, thus, only two first order approximations to the spectra from the loss modulus or loss compliance.

The situation is different for the approximations of second order. Equation (4.3-35) gives the intensity function as

$$I_2''(\omega\tau) = \frac{(a_0'' - a_1'' + a_2'')\omega^5\tau^5 + 2(a_0'' - 3a_2'')\omega^3\tau^3 + (a_0'' + a_1'' + a_2'')\omega\tau}{(1 + \omega^2\tau^2)^3}. \tag{4.3-51}$$

Equation (4.3-51) may be reduced to a single-term equation of order 2 through the three choices of setting either $a_1'' = 4a_0''/3$ and $a_2'' = a_0''/3$, or $a_1'' = 0$ and $a_2'' = -a_0''$, or $a_1'' = -4a_0''/3$ and $a_2'' = a_0''/3$. These lead to

$$I''_{2L}(\omega\tau) = \frac{16\omega\tau}{3\pi(1+\omega^2\tau^2)^3} \tag{4.3-52}$$

$$I''_2(\omega\tau) = \frac{16\omega^3\tau^3}{\pi(1+\omega^2\tau^2)^3} \tag{4.3-53}$$

and

$$I''_{2R}(\omega\tau) = \frac{16\omega^5\tau^5}{3\pi(1+\omega^2\tau^2)^3} \tag{4.3-54}$$

with maxima at $\omega\tau = 1/\sqrt{5}$, 1, and $\sqrt{5}$, respectively. $I''_2(\omega\tau)$ and $I''_{2R}(\omega\tau)$ are plotted in Fig. 4.3-7. $I''_{2L}(\omega\tau)$ is the mirror image of $I''_{2R}(\omega\tau)$. The approximations to the relaxation spectrum become

$$H''_2(\tau/\sqrt{5}) = \frac{2}{\pi}\left[G''(\omega) + \frac{4}{3}\frac{d\,G''(\omega)}{d\ln\omega} + \frac{1}{3}\frac{d^2G''(\omega)}{d\ln^2\omega}\right]\Bigg|_{\omega=1/\tau} \tag{4.3-55}$$

$$H''_2(\tau) = \frac{2}{\pi}\left[G''(\omega) - \frac{d^2G''(\omega)}{d\ln^2\omega}\right]\Bigg|_{\omega=1/\tau} \tag{4.3-56}$$

and

$$H''_2(\sqrt{5}\tau) = \frac{2}{\pi}\left[G''(\omega) - \frac{4}{3}\frac{d\,G''(\omega)}{d\ln\omega} + \frac{1}{3}\frac{d^2G''(\omega)}{d\ln^2\omega}\right]\Bigg|_{\omega=1/\tau} . \tag{4.3-57}$$

For the retardation spectrum we have

$$L''_2(\tau/\sqrt{5}) = \frac{2}{\pi}\left[J''(\omega) + \frac{4}{3}\frac{d\,J''(\omega)}{d\ln\omega} + \frac{1}{3}\frac{d^2J''(\omega)}{d\ln^2\omega}\right]\Bigg|_{\omega=1/\tau} \tag{4.3-58}$$

$$L''_2(\tau) = \frac{2}{\pi}\left[J''(\omega) - \frac{d^2J''(\omega)}{d\ln^2\omega}\right]\Bigg|_{\omega=1/\tau} \tag{4.3-59}$$

$$L''_2(\sqrt{5}\tau) = \frac{2}{\pi}\left[J''(\omega) - \frac{4}{3}\frac{d\,J''(\omega)}{d\ln\omega} + \frac{1}{3}\frac{d^2J''(\omega)}{d\ln^2\omega} - \{8\phi_f/3\omega\}\right]\Bigg|_{\omega=1/\tau} \tag{4.3-60}$$

since the steady-flow term cancels in Eqs. (4.3-58) and (4.3-59) but not in Eq. (4.3-60).

Following the same procedure for $k = 3$, we find four single-term intensity functions. Of these, only the two whose maxima lie closer to $\ln\omega\tau = 0$ need to be considered. The other two are broader and more skewed (cf. Sect. 4.4) and offer no advantage. The intensity functions associated with the two useful approximations of order 3 are

$$I''_{3L}(\omega\tau) = \frac{32\omega^3\tau^3}{\pi(1+\omega^2\tau^2)^4} \tag{4.3-61}$$

$$I''_{3R}(\omega\tau) = \frac{32\omega^5\tau^5}{\pi(1+\omega^2\tau^2)^4} . \tag{4.3-62}$$

The maxima occur at $\omega\tau = 1/\sqrt{5/3}$ and $\sqrt{5/3}$. There is thus no third order intensity function which is symmetric with respect to the ln $\omega\tau$ axis and whose maximum occurs at $\omega\tau = 1$. $I''_{3R}(\omega\tau)$ is plotted in Fig. 4.3-7 along with the others.

We note that the order of the intensity functions associated with the approximations from the loss functions is again revealed by the exponent, k + 1, in the denominator.

The approximations of order 3 to the relaxation spectrum then become

$$H''_3(\tau/\sqrt{5/3}) = \frac{2}{\pi}\left[G''(\omega) + \frac{1}{3}\frac{d\,G''(\omega)}{d\ln\omega} - \frac{d^2G''(\omega)}{d\ln^2\omega} - \frac{1}{3}\frac{d^3G''(\omega)}{d\ln^3\omega}\right]\Bigg|_{\omega=1/\tau} \tag{4.3-63}$$

$$H''_3(\sqrt{5/3}\tau) = \frac{2}{\pi}\left[G''(\omega) - \frac{1}{3}\frac{d\,G''(\omega)}{d\ln\omega} - \frac{d^2G''(\omega)}{d\ln^2\omega} + \frac{1}{3}\frac{d^3G''(\omega)}{d\ln^3\omega}\right]\Bigg|_{\omega=1/\tau} . \tag{4.3-64}$$

For approximations to the retardation spectrum we simply replace $H''_k(\tau/\beta''_k)$ by $-L''_k(\tau/\beta''_k)$, and $G''(\omega)$ by $J''(\omega)$. The contributions from the flow term cancel, except in the equation corresponding to Eq. (4.3-64) where this term is $-\{4\phi_f/\pi\omega\}$.

The decision as to which of the alternative approximations of the same order should be used in a given case is again governed by the considerations developed at the close of Sect. 4.3.2.1.

Only approximations of even order from the loss functions possess intensity functions which are symmetric on the log(ωt)-axis. These approximations were introduced by Schwarzl and Stavermann [9] and are therefore referred to under their name. The asymmetric approximations were introduced by the author [10].

4.3.2 The Transform Inversion Method

In this section we show that approximations to the spectra from the harmonic response functions can also be derived through inversion of the Stieltjes transform according to Eqs. (A5-48) and (A5-49) of the Appendix. This justifies the differential operator method as being an extension of the transform inversion method. Applying the inversion formulae to Eq. (4.1-85) in the form

$$\Phi(\xi) = \mathscr{S}^{-1}\bar{\Phi}(\nu)|_{\nu=\xi} \tag{4.3-65}$$

we have, for approximations of *even* order, i.e. for k = 2n,

$$\Phi_{2n}(\xi) = C(D_\nu + n)^{(2n)}\bar{\Phi}(\nu)|_{\nu=\xi} , \qquad n = 1, 2, \ldots \tag{4.3-66}$$

and for approximations of *odd* order, i.e. for k = 2n + 1, we find

$$\Phi_{2n+1}(\xi) = C(D_\nu + n + 1)^{(2n+1)}\bar{\Phi}(\nu)|_{\nu=\xi} , \qquad n = 1, 2, \ldots . \tag{4.3-67}$$

In Eqs. (4.3-66) and (4.3-67) the factorial function is defined by

$$(D + \alpha)^{(r)} = (D + \alpha)(D + \alpha - 1)\ldots(D + \alpha - r + 1) \tag{4.3-68}$$

and we also have

$$D_\nu^k = \frac{d^k}{d\ln^k\nu} . \tag{4.3-69}$$

The constant C is, in principle, a function of k as given by Eqs. (A5-37) and (A5-45) of the Appendix. However, it turns out that approximations derived through the Stieltjes transform inversion do not necessarily have normalized intensity functions. The normalization constant must be determined separately for each approximation from the requirement that the area under the intensity function be unity. Hence, the value of the constant at this stage is irrelevant and, in fact, during the derivation we will conveniently subsume other constants under C.

To obtain useful approximations, we must now change from $\Phi(\xi)$ to $F'(\tau)$ and $F''(\tau)$, and from $\tilde{\Phi}(\nu)$ to $f'(\omega)$ and $f''(\omega)$. These functions are defined by Eqs. (4.3-1) and (4.3-2). We note that Eqs. (4.1-86′) may be condensed into the forms

$$\Phi(\xi) = \tau^m F(\tau)|_{\tau=\xi^{-1/2}} , \qquad m = 0, 1, 2, \ldots \tag{4.3-70}$$

and

$$\tilde{\Phi}(\nu) = c_m \omega^{-m} f(\omega)|_{\omega=\nu^{1/2}} , \qquad m = 0, 1, 2, \ldots \tag{4.3-71}$$

where $F(\tau)$ and $f(\omega)$ stand for $F'(\omega)$ and $f'(\omega)$ when $m = 0$ or 2, for $F''(\tau)$ and $f''(\omega)$ when $m = 1$. The constant c_m is -2 when $m = 0$, and 2 when $m = 1$ or 2. From Eq. (4.3-69) we have

$$D_\nu = D_\omega/2 \tag{4.3-72}$$

because $\nu = \omega^2$. Equations (4.3-70), (4.3-71), and (4.3-72) must now be substituted into Eqs. (4.3-66) and (4.3-67). Before we do so, however, we consider the equation

$$\Phi(\lambda) = \mathscr{S}^{-1}\tilde{\Phi}(\mu)|_{\mu=\lambda} \tag{4.3-73}$$

[cf. Eq. (4.1-87)]. Equation (4.3-73) yields

$$\Phi_{2n}(\lambda) = C(D_\mu + n)^{(2n)}\tilde{\Phi}(\mu)|_{\mu=\lambda} , \qquad n = 1, 2, \ldots \tag{4.3-74}$$

for approximations of *even* order, and

$$\Phi_{2n+1}(\lambda) = C(D_\mu + n + 1)^{(2n+1)}\tilde{\Phi}(\mu)|_{\mu=\lambda} , \qquad n = 1, 2, \ldots \tag{4.3-75}$$

for those of *odd* order. The factorial function is again defined by Eq. (4.3-68). We have

$$D_\mu^k = \frac{d^k}{d\ln^k\mu} \tag{4.3-76}$$

and the same remarks as before apply to the constant C.

Equation (4.1-88′) yields

$$\Phi(\lambda) = \tau^{-m} F(\tau)|_{\tau=\lambda^{1/2}} , \qquad m = 0, 1, 2, \ldots \tag{4.3-77}$$

and

$$\tilde{\Phi}(\mu) = c_m \omega^m f(\omega)|_{\omega=\mu^{-1/2}} , \qquad m = 0, 1, 2, \ldots \tag{4.3-78}$$

and we also have

$$D_\mu = -D_\omega/2 \tag{4.3-79}$$

because $\mu = \omega^{-2}$.

We are now ready to substitute Eqs. (4.3-70), (4.3-71), and (4.3-72) on the one hand, and Eqs. (4.3-77), (4.3-78), and (4.3-79) on the other hand, into Eqs. (4.3-66) and (4.3-67). To do so, we make use of the relation $\tau = \omega^{-1}$ together with the identity

$$(\pm D_\omega/2 + \alpha)^{(r)}\omega^{\pm m}f(\omega) = \omega^{\pm m}(\pm D_\omega/2 + \alpha - m/2)^{(r)}f(\omega) \tag{4.3-80}$$

and obtain

$$F_{2n}(\tau/\beta) = C(\pm D_\omega/2 + n - m/2)^{(2n)}f(\omega)|_{\omega=\tau^{-1}}\ , \qquad n = 1, 2, \ldots \tag{4.3-81}$$

and

$$F_{2n+1}(\tau/\beta) = C(\pm D_\omega/2 + n + 1 - m/2)^{(2n+1)}f(\omega)|_{\omega=\tau^{-1}}\ , \qquad n = 1, 2, \ldots . \tag{4.3-82}$$

We may also write

$$F_{2n}(\tau/\beta) = C(\pm D_\omega + 2n - m)^{(2n)}f(\omega)|_{\omega=\tau^{-1}}\ , \qquad n = 1, 2, \ldots \tag{4.3-83}$$

and

$$F_{2n+1}(\tau/\beta) = C(\pm D_\omega + 2n + 2 - m)^{(2n+1)}f(\omega)|_{\omega=\tau^{-1}}\ , \qquad n = 1, 2, \ldots \tag{4.3-84}$$

where the factorial function is now defined by

$$(D + \alpha)^{(r)} = (D + \alpha)(D + \alpha - 2)(D + \alpha - 4)\ldots(D + \alpha - 2r + 2)\ . \tag{4.3-85}$$

In Eqs. (4.3-81) through (4.3-84) c_m has been subsumed into C and we have introduced β to take account of any adjustment of the time scale which might be necessary when the associated intensity functions do not peak at $\ln \omega\tau = 0$.

The approximations to the spectra from the harmonic response functions obtained through the Stieltjes transform inversion may be generated either from Eqs. (4.3-81) and (4.3-82) or, equivalently, from Eqs. (4.3-83) and (4.3-84). The two sets of equations are the analogs of Eq. (4.2-8). The latter led to approximations whose intensity functions were normalized. It also provided for the adjustment of the time scale which is required because the intensity functions do not peak at $\ln(t/\tau) = 0$. By contrast, the Stieltjes transform inversion leads to approximations whose intensity functions are not necessarily normalized and it does not furnish the value of β necessary to make the appropriate shift in the time scale when required. Thus, C must be determined through normalization according to Eqs. (4.3-6) and (4.3-36), and β must be obtained in the usual way by finding the location of the maximum.

In addition, strictly speaking, approximations of zeroth and of first order cannot be generated by Stieltjes transform inversion because the lowest value of n in Eqs. (4.3-66), (4.3-67), (4.3-74), and (4.3-75) is 1 which generates second order approximations from Eqs. (4.3-81) or (4.3-83), and third order approximations from Eqs. (4.3-82) or (4.3-84). The introduction of the factorial function into the inversion formulae of Hirschman, and of Hirschman and Widder (cf. Appendix A5.2.2), allows us to extend the range of n down to 0 if we, following accepted practice, let $D^{(0)} = 1$. Equation (4.3-83) will then correctly predict the zeroth order approximation from the loss functions for $m = 1$ but the cases $m = 0$ and $m = 2$ must be excluded.

Table 4.3-1. Approximation up to order 3 obtainable through Stieltjes transform inversion

Spectrum	Eqs.[a]	$\pm D_\infty$	n	m	C
$F_1'(\tau)$	12, 13	$\pm$	0	2	± 1
$F_2'(\tau/\sqrt{2})$	17, 19	+	1	0, 2	1/2, −1/2
$F_2'(\sqrt{2}\tau)$	18, 20	−	1	0, 2	−1/2, 1/2
$F_3'(\tau/\sqrt{3})$	25, 28	+	1	0	1/8
$F_3'(\tau)$	26, 29	$\pm$	1	2	$\mp 1/4$
$F_3'(\sqrt{3}\tau)$	27, 30	−	1	0	−1/8
$F_0''(\tau)$	42, 43	$\pm$	0	1	$2/\pi$
$F_1''(\tau/\sqrt{3})$	47, 49	+	0	1	$2/\pi$
$F_1''(\sqrt{3}\tau)$	48, 50	−	0	1	$2/\pi$
$F_2''(\tau/\sqrt{5})$	55, 58		not available		
$F_2''(\tau)$	56, 59	$\pm$	1	1	$-2/\pi$
$F_0''(\sqrt{5}\tau)$	57, 60		not available		
$F_3''(\tau\sqrt{5/3})$	63	+	1	1	$-2/3\pi$
$F_3''(\sqrt{5/3}\tau)$	64	−	1	1	$-2/3\pi$

[a] Equation numbers refer to Sect. 4.3 and should be read as Eq. (4.3–)

Similarly, Eq. (4.3-84) will correctly give the first order approximations if $m = 0$ is excluded. We also note that Eq. (4.3-83) cannot yield $F_2''(\tau/\sqrt{5})$ and $F_2''(\sqrt{5}/\tau)$. The operator method discussed in Sect. 4.3.2 is thus the more general one. This method is based on the assumption that approximations to the spectral response functions can be obtained as the algebraic sums of terms containing the logarithmic derivatives of the source functions. The Stieltjes transform inversion shows that this assumption is justified.

Table 4.3-1 lists the approximations up to order 3 which may be obtained from the harmonic response functions through Stieltjes transform inversion with $n = 0$ and $n = 1$.

The first two columns of the table contain the symbol for the approximated spectrum and the equation number under which it appears in Sect. 4.3.1. The next three columns list the parameters with which the spectrum is generated from Eqs. (4.3-81) through (4.3-84), respectively, as the case may be. The last column lists the value of the constant C which must be inserted into Eqs. (4.3-81) through (4.3-84) to yield the spectrum with the appropriately normalized intensity function.

4.3.3 The Power Law Method: The Approximations of Williams and Ferry

To obtain the Williams-Ferry approximation (see Sect. 4.2.3) for the storage modulus, we insert the assumed spectral representation

$$H(\tau) = H_0(\tau/\tau_0)^{-m} \tag{4.3-86}$$

into Eq. (4.1-7) and carry out a change of variable by letting $\omega\tau = x^{1/2}$. Using $-m/2 = n - 1$ yields

$$G'(\omega) = \{G_e\} + \frac{1}{2} H_0(\omega\tau_0)^m \int_0^\infty \frac{x^{n-1}}{1+x} dx \; . \tag{4.3-87}$$

But

$$\int_0^\infty \frac{x^{n-1}}{1+x} dx = \frac{\pi}{\sin n\pi} \; , \qquad 0 < n < 1 \tag{4.3-88}$$

is a standard definite integral. We thus obtain

$$G'(\omega) = \{G_e\} + H_0(\omega\tau_0)^m \frac{\pi/2}{\sin m\pi/2} \; . \tag{4.3-89}$$

The first order approximation is

$$H'_1(\tau) = \left. \frac{d\, G'(\omega)}{d \ln \omega} \right|_{\omega = 1/\tau} \; . \tag{4.3-90}$$

Substituting Eq. (4.3-89) into Eq. (4.3-90) yields

$$H'_1(\tau) = H_0(\tau/\tau_0)^{-m} \frac{m\pi/2}{\sin m\pi/2} \tag{4.3-91}$$

which differs from the assumed spectral representation, Eq. (4.3-86), by the factor

$$A(m) = \frac{\sin m\pi/2}{m\pi/2} \; . \tag{4.3-92}$$

Thus we obtain

$$H'_{WF}(\tau) = A(m) H'_1(\tau) \tag{4.3-93}$$

for the Williams-Ferry approximation. It is calculated by first determining $H'_1(\tau)$ from Eq. (4.3-90), and then m from

$$m = -\frac{d \log H'_1(\tau)}{d \log \tau} \; . \tag{4.3-94}$$

For the storage compliance we obtain by an analogous derivation

$$L'_{WF}(\tau) = -A(m) L'_1(\tau) \tag{4.3-95}$$

where $L'_1(\tau)$ is given by Eq. (4.3-13), and m is obtained from

$$m = \frac{d \log L'_1(\tau)}{d \log \tau} \; . \tag{4.3-96}$$

Because of the restriction on n in Eq. (4.3-87) the procedure is valid only for values of m between 0 and 2, a condition which is normally satisfied.

The Williams-Ferry approximations for the loss modulus and loss compliance are derived in an analogous way. Inserting Eq. (4.3-86) into Eq. (4.1-8) leads, with the change of variable $\omega\tau = x^{1/2}$, and setting $-(m + 1)/2 = n - 1$, to

$$G''(\omega) = H_0(\omega\tau_0)^m \frac{\pi/2}{\cos m\pi/2} . \tag{4.3-97}$$

The first order approximation is

$$H_1''(\tau) = \frac{2}{\pi}\left[G''(\omega) \pm \frac{d\,G''(\omega)}{d\ln\omega}\right]\bigg|_{\omega=\beta/\tau} \tag{4.3-98}$$

where $\beta = \beta_1''$ is $\sqrt{3}$ for the approximation with the negative, and $1/\sqrt{3}$ for the approximation with the positive sign. Upon insertion of Eq. (4.3-97) we find

$$H_1''(\tau) = \frac{(1 \pm m)\beta^m}{\cos m\pi/2} H_0(\tau/\tau_0)^{-m} \tag{4.3-99}$$

and the approximation becomes

$$H''_{WF}(\tau) = B(m)H_1''(\tau) \tag{4.3-100}$$

where

$$B(m) = \frac{\cos m\pi/2}{(1 \pm m)\beta^m} . \tag{4.3-101}$$

This also takes into account the shift in the time scale of the spectrum which is inherent in Eq. (4.3-98).

The derivation given here differs from that presented in the original reference [6] because Eq. (4.3-99) was not available to Williams and Ferry. In consequence, the correct values of B(m) differ from those in the earlier references [6, 6a] by the factor $0.5\pi/3^{m/2}$. As an example, for a slope of 1/2 the value in the original reference is too small by 0.838.

We have derived Eq. (4.3-101) on the assumption that m is positive when the slope $d\log H_1''(\tau)/d\log\tau$ is negative. It was suggested in Sect. 4.3.2 that of the two choices available for the first order approximation, the one with the negative sign be used when the slope is negative, i.e. when m is positive, and vice versa. Hence we have

$$B(m) = \frac{\cos m\pi/2}{3^{|m|/2}(1 - |m|)} \tag{4.3-102}$$

where

$$m = \frac{d\log H_1''(\tau)}{d\log\tau} . \tag{4.3-103}$$

For the retardation spectrum we have, quite analogously

$$L''_{WF}(\tau) = B(m)L_1''(\tau) \tag{4.3-104}$$

where $L_1''(\tau)$ is given by Eqs. (4.3-49) and (4.3-50). The restriction imposed by Eq. (4.3-88) implies that the procedure is valid only for values of m between -1 and 1.

4.3.4 The Finite Difference Operator Method: The Approximations of Ninomiya and Ferry, and of Tschoegl

Approximations to the spectral functions from the harmonic response functions may also be obtained by the method of finite differences developed in Sect. 4.2.4 for approximations from the step function responses.

4.3.4.1 Approximations Derived from the Storage Functions

We first consider approximations derived from the storage functions, $G'(\omega)$ and $J'(\omega)$, which we will again lump under the symbol $f'(\omega)$ [cf. Eq. $(4.3\text{-}2)_1$]. The approximations are derived by replacing the continuous logarithmic differential operator $\mathscr{P}'_k$ defined by Eq. (4.3-4) with its finite difference analog

$$\mathscr{P}'_{kh} = \sum_{n=1}^{n=k} \frac{a'_n}{\ln^n h} \sum_{m=0}^{m=n} (-1)^{n-m} \binom{n}{m} L^m \tag{4.3-105}$$

[cf. Eq. (4.2-81)]. We have, in analogy to Eqs. (4.2-84) and (4.2-85),

$$\mathscr{P}'_{kh} = \sum_{m=0}^{m=k} b'_m L^m \tag{4.3-106}$$

where

$$b'_m = \sum_{n=1}^{n=k} (-1)^{n-m} \binom{n}{m} \frac{a'_n}{\ln^n h} \tag{4.3-107}$$

and the restriction expressed by Eq. (4.2-83) again applies. The operator $\mathscr{P}'_{kh}$ is formally analogous to $\mathscr{P}'_k$ (see Eq. (4.2-84)) except that the constants a and b now appear as a′ and b′. Hence, Eqs. (4.2-85), (4.2-86), (4.2-87), and (4.2-88) apply in the primed forms. In analogy to Eqs. (4.2-90) and (4.3-5) we have

$$I'_{kh}(\omega\tau) = \mathscr{P}'_{kh}\left[\frac{\omega^2\tau^2}{1+\omega^2\tau^2}\right] = \sum_{m=0}^{m=k} \frac{b'_m h^{2m}\omega^2\tau^2}{1+h^{2m}\omega^2\tau^2} \tag{4.3-108}$$

or

$$I'_{kh}(\omega\tau) = \mathscr{P}'_{kh}\left[\frac{1}{1+\omega^2\tau^2}\right] = -\sum_{m=0}^{m=k} \frac{b'_m}{1+h^{2m}\omega^2\tau^2} \tag{4.3-109}$$

(see Problem 4.3-4). The spectra are given, in analogy to Eq. (4.2-109), by

$$F'_{kh}(\tau) = \mathscr{P}'_{kh}[f'(\omega)]\big|_{\omega=\beta'_{kh}/\tau} = \sum_{m=0}^{m=k} b'_m f'(\omega h^m)\big|_{\omega=\beta'_{kh}/\tau} \tag{4.3-110}$$

where $F'(\tau)$ is defined by Eq. $(4.3\text{-}1)_1$.

We now show that normalization of $I'_{kh}(\omega\tau)$ leads to $a_1 = 1$ regardless of the order k of the approximation. We apply the operator $\mathscr{P}'_{kh}$, Eq. (4.3-105), to the kernel, $\omega^2\tau^2/(1+\omega^2\tau^2)$, separating out the first term. This gives

$$I'_{kh}(\omega\tau) = \frac{a'_1}{\ln h}\left(-\frac{\omega^2\tau^2}{1+\omega^2\tau^2} + \frac{\omega^2\tau^2}{h^{-2}+\omega^2\tau^2}\right)$$
$$+ \sum_{n=2}^{n=k} \frac{a'_n}{\ln^n h} \sum_{m=0}^{m=n} (-1)^{n-m}\binom{n}{m}\frac{\omega^2\tau^2}{h^{-2m}+\omega^2\tau^2} \;. \tag{4.3-111}$$

But

$$\int_{-\infty}^{\infty} \left[\frac{\omega^2\tau^2}{h^{-2r}+\omega^2\tau^2} - \frac{\omega^2\tau^2}{h^{-2r+2}+\omega^2\tau^2}\right] d\ln\omega\tau = \ln h \tag{4.3-112}$$

and therefore

$$\int_{-\infty}^{\infty} \sum_{m=0}^{m=n} (-1)^{n-m}\binom{n}{m}\frac{\omega^2\tau^2}{h^{-2m}+\omega^2\tau^2}\, d\ln\omega\tau = 0 \tag{4.3-113}$$

for any $n > 1$, because, when $n > 1$, the sum in Eq. (4.3-113) can always be developed into a series of integrals like that on the left hand side of Eq. (4.3-112). Since the number of such integrals is necessarily even, and they alternative in sign, the series is self-canceling. Hence, it follows that

$$\int_{-\infty}^{\infty} I'_{kh}(\omega\tau)d\ln\omega\tau = a'_1 \tag{4.3-114}$$

and, therefore, $a'_1 = 1$ as asserted.

The intensity function of the approximation of the first order is obtained from either Eq. (4.3-108) or (4.3-109), taking b'_0 and b'_1 from Eqs. (4.2-86), and substituting $a'_1 = 1$. This leads to

$$I^*_{1h}(\omega\tau) = \frac{(h^2-1)\omega^2\tau^2}{(1+\omega^2\tau^2)(1+h^2\omega^2\tau^2)\ln h} \;. \tag{4.3-115}$$

The intensity function is symmetric on the $\ln(\omega t)$-axis but its peak does not lie at $\ln\omega\tau = 0$. A centered (but slightly broadened) intensity function is obtained by applying the shift operator $(1 + L^{-1})/2$ as explained in Sect. 4.2.4. This produces

$$I'_{1h}(\omega\tau) = \frac{(h^2-h^{-2})\omega^2\tau^2}{2(h^2+\omega^2\tau^2)(h^{-2}+\omega^2\tau^2)\ln h} \tag{4.3-116}$$

and the corresponding spectra are obtained from Eq. (4.3-110) as

$$F'_{1h}(\tau) = -\left.\frac{f'(\omega h^{-1}) - f'(\omega h)}{2\ln h}\right|_{\omega=1/\tau} \tag{4.3-117}$$

since $\beta'_{1h} = 1$ as is easily ascertained from $d\,I'_{1h}(\omega\tau)/d\omega\tau = 0$. $I'_{1h}(\omega\tau)$ is plotted in Fig. 4.3-8 as a function of $\ln\omega\tau$ for a differencing interval of $\log h = 0.2$.

The intensity function of the approximation of the second order is obtained from Eq. (4.3-109), making use of Eq. (4.2-89), as

$$I'_{2h}(\omega\tau) = \frac{c'_2 h^2\omega^2\tau^2 + c'_4 h^4\omega^4\tau^4}{(1+\omega^2\tau^2)(1+h^2\omega^2\tau^2)(1+h^4\omega^2\tau^2)} \tag{4.3-118}$$

where

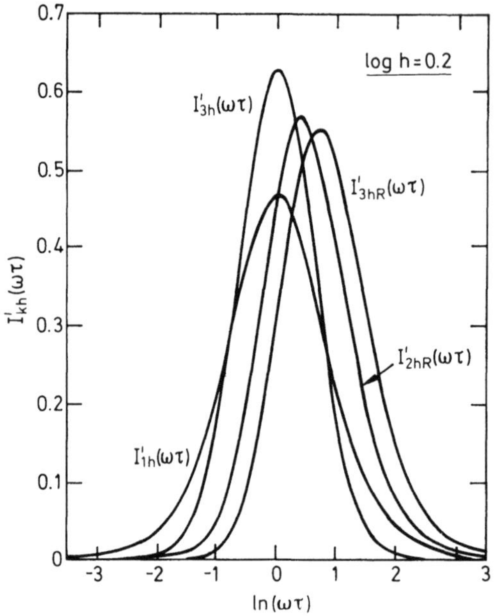

Fig. 4.3-8. The intensity functions $I'_{1h}(\omega\tau)$, $I'_{2hR}(\omega\tau)$, $I'_{3h}(\omega\tau)$, and $I'_{3hR}(\omega\tau)$ as functions of ln ωτ

$$c'_2 = b'_0 h^{-2} + b'_1 + b'_2 h^2 \tag{4.3-119}$$

and

$$c'_4 = b'_0(1 + h^{-2}) + b'_1(h^2 + h^{-2}) + b'_2(h^2 + 1) \ . \tag{4.3-120}$$

Equation (4.3-118) can be reduced to a single-term equation by setting either c'_2 or c'_4 equal to zero. Taking b'_0, b'_1, and b'_2 from Eqs. (4.2-87), $c'_2 = 0$ yields an equation from which a'_2 is obtained as

$$a'_2 = \frac{h^{-1} \ln h}{h - h^{-1}} \ . \tag{4.3-121}$$

With the use of Eq. (4.3-121), Eq. (4.3-120) gives

$$c'_4 = \frac{h^2 - h^{-2}}{\ln h} \tag{4.3-122}$$

and the intensity function follows as

$$I'^{*}_{2hR}(\omega\tau) = \frac{(h^2 - h^{-2}) h^4 \omega^4 \tau^4}{(1 + \omega^2\tau^2)(1 + h^2\omega^2\tau^2)(1 + h^4\omega^2\tau^2) \ln h} \ . \tag{4.3-123}$$

The subscript R is justified because

$$\lim_{h \to 1} I'^{*}_{2hR}(\omega\tau) = \frac{4\omega^4\tau^4}{(1 + \omega^2\tau^2)^3} = I'_{2R}(\omega\tau) \tag{4.3-124}$$

[cf. Eq. (4.3-16)].

Table 4.3-2. Spectral shift factors for finite difference approximations from the storage functions

log h	0.1	0.2	0.3	0.4	0.5
β'_{2hR}	1.431	1.482	1.568	1.691	1.849
β'_{3hR}	1.801	2.006	2.348	2.835	3.485

Shifting by L^{-1} to obtain functions in which the values of $f'(\omega h^m)$ are arranged symmetrically around $\ln \omega\tau = 0$, produces the intensity function

$$I'_{2hR}(\omega\tau) = \frac{(h^2 - h^{-2})\omega^4\tau^4}{(h^2 + \omega^2\tau^2)(1 + \omega^2\tau^2)(h^{-2} + \omega^2\tau^2)\ln h} . \tag{4.3-125}$$

To derive the corresponding spectra we use Eq. (4.3-110), obtaining b'_0, b'_1, and b'_2 from Eq. (4.2-87) with the appropriate values of a'_1 and a'_2. This yields

$$F'_{2hR}(\tau\beta'_{2hR}) = -\left.\frac{hf'(\omega h^{-1}) - (h + h^{-1})f'(\omega) + h^{-1}f'(\omega h)}{(h - h^{-1})\ln h}\right|_{\omega=1/\tau} . \tag{4.3-126}$$

The value of β'_{2hR} is found from $dI'_{2hR}(\omega\tau)/d\omega\tau = 0$ to be given by

$$(\beta'_{2hR})^6 - (h^2 + 1 + h^{-2})(\beta'_{2hR})^2 - 2 = 0 . \tag{4.3-127}$$

It is tabulated for various values of h in Table 4.3-2.

If, instead of c'_2 we equate c'_4 to zero, then, proceeding as before, we now find

$$a'_2 = \frac{h \ln h}{h - h^{-1}} \tag{4.3-128}$$

and

$$c'_2 = \frac{h^2 - h^{-2}}{\ln h} . \tag{4.3-129}$$

Thus, the intensity function becomes

$$I'^{*}_{2hL}(\omega\tau) = \frac{(h^2 - h^{-2})h^2\omega^2\tau^2}{(1 + \omega^2\tau^2)(1 + h^2\omega^2\tau^2)(1 + h^4\omega^2\tau^2)\ln h} \tag{4.3-130}$$

which is written with the subscript L because

$$\lim_{h\to 1} I'_{2hL}(\omega\tau) = \frac{4\omega^2\tau^2}{(1 + \omega^2\tau^2)^3} = I'_{2L}(\omega\tau) . \tag{4.3-131}$$

Shifting by L^{-1} gives

$$I'_{2hL}(\omega\tau) = \frac{(h^2 - h^{-2})\omega^2\tau^2}{(h^2 + \omega^2\tau^2)(1 + \omega^2\tau^2)(h^{-2} + \omega^2\tau^2)\ln h} . \tag{4.3-132}$$

$I'_{2hR}(\omega\tau)$ and $I'_{2hL}(\omega\tau)$ are not symmetric but are mirror images of each other, as are $I'_{2R}(\omega\tau)$ and $I'_{2L}(\omega\tau)$. $I'_{2hR}(\omega\tau)$ is plotted for log h = 0.2 in Fig. 4.3-8 for comparison with $I'_{1R}(\omega\tau)$.

The spectra take the form

$$F'_{2hL}(\tau\beta'_{2hL}) = -\frac{h^{-1}f'(\omega h^{-1}) - (h + h^{-1})f'(\omega) + hf'(\omega h)}{(h - h^{-1})\ln h} \tag{4.3-133}$$

where $\beta'_{2hL} = 1/\beta'_{2hR}$.

Proceeding in an analogous manner, the intensity function of the approximations of order 3 is found to be

$$I'_{3hL}(\omega\tau) = \frac{c'_2 h^3\omega^2\tau^2 + c'_4 h^6\omega^4\tau^4 + c'_6 h^9\omega^6\tau^6}{(1 + \omega^2\tau^2)(1 + h^2\omega^2\tau^2)(1 + h^4\omega^2\tau^2)(1 + h^6\omega^2\tau^2)} \tag{4.3-134}$$

where

$$c'_2 = b'_0 h^{-3} + b'_1 h^{-1} + b'_2 h + b'_3 h^3 \tag{4.3-135}$$

$$c'_4 = b'_0 h^{-2}(h^2 + 1 + h^{-2}) + b'_1 h^{-1}(h^3 + h + h^{-3}) + b'_2 h(h^3 + h^{-1} + h^{-3}) + b'_3 h^2(h^2 + 1 + h^{-2}) \tag{4.3.136}$$

$$c'_6 = b'_0 h^{-1}(h^2 + 1 + h^{-2}) + b'_1(h^3 + h^{-1} + h^{-3}) + b'_2(h^3 + h + h^{-3}) + b'_3 h(h^2 + 1 + h^{-2})\ . \tag{4.3-137}$$

To obtain intensity functions consisting of a single term we must set two out of the three coefficients c'_2, c'_4, and c'_6 simultaneously equal to zero. Since the general procedure is now well established we omit most of the details of the derivation. Letting $c'_2 = c'_6 = 0$, and shifting the resulting intensity function by $(L^{-1} + L^{-2})/2$, we find

$$I'_{3h}(\omega\tau) = \frac{(h^4 - h^{-4})(h^2 + 1 + h^{-2})\omega^4\tau^4}{2(h^4 + \omega^2\tau^2)(h^2 + \omega^2\tau^2)(h^{-2} + \omega^2\tau^2)(h^{-4} + \omega^2\tau^2)\ln h} \tag{4.3-138}$$

and

$$F'_{3h}(\tau) = \frac{f'(\omega h^{-2}) - (h^2 + h^{-2})f'(\omega h^{-1}) + (h^2 + h^{-2})f'(\omega h) - f'(\omega h^2)}{2(h - h^{-1})^2\ln h} \tag{4.3-139}$$

because $\beta'_{3h} = 1$. Hence, $I'_{3h}(\omega\tau)$ is centered on $\ln \omega\tau = 0$ and is symmetric on the $\ln(\omega\tau)$-axis. It is again plotted for $\log h = 0.2$ in Fig. 4.3-8.

Setting $c'_4 = c'_6 = 0$, and $c'_2 = c'_4 = 0$, respectively, and using the same shift operator, we further obtain

$$I'_{3hL}(\omega\tau) = \frac{(h^3 - h^{-3})[h + h^{-1} + (h^3 + h^{-3})\omega^2\tau^2]\omega^2\tau^2}{2(h^4 + \omega^2\tau^2)(h^2 + \omega^2\tau^2)(1 + \omega^2\tau^2)(h^{-2} + \omega^2\tau^2)(h^{-4} + \omega^2\tau^2)\ln h} \tag{4.3-140}$$

and

$$I'_{3hR}(\omega\tau) = \frac{(h^3 - h^{-3})[h^3 + h^{-3} + (h + h^{-1})\omega^2\tau^2]\omega^6\tau^6}{2(h^4 + \omega^2\tau^2)(h^2 + \omega^2\tau^2)(1 + \omega^2\tau^2)(h^{-2} + \omega^2\tau^2)(h^{-4} + \omega^2\tau^2)\ln h}\ . \tag{4.3-141}$$

$I'_{3hL}(\omega\tau)$ is the mirror image of $I'_{3hR}(\omega\tau)$, and $\beta'_{3hL} = 1/\beta'_{3hR}$. $I'_{3h}(\omega\tau)$ is plotted in Fig. 4.3-8.

The spectra follow as

$$F'_{3hL}(\tau\beta'_{3hL}) = A\,f'(\omega h^{-2}) + B\,f'(\omega h^{-1}) + C\,f'(\omega) + B'f'(\omega h) + A'f'(\omega h^2) \tag{4.3-142}$$

and

$$F'_{3hR}(\tau\beta'_{3hR}) = -A'\,f'(\omega h^{-2}) + B\,f'(\omega h^{-1}) - C\,f'(\omega) + B'\,f'(\omega h) - A\,f'(\omega h^2) \tag{4.3-143}$$

where

$$\begin{aligned} A &= -h^{-3}/2(h^2 - h^{-2})(h - h^{-1})\ln h \\ B &= 1/2(h^2 - 2 + h^{-2})\ln h \\ C &= (h^2 + 1 + h^{-2})/2(h^2 - h^{-2})\ln h \end{aligned} \tag{4.3-144}$$

and $A' = -h^6 A$ while $B' = -B$. Values of β'_{3hR} for different values of h are listed in Table 4.3-2.

The symmetric approximations $F'_{1h}(\tau)$ and $F'_{3h}(\tau)$ were introduced by Ninomiya and Ferry [12]. The others are new. The choice of a particular second or third order approximation is governed by the same considerations which have been discussed in Sect. 4.3.2 relative to the approximations obtained by continuous differentiation.

4.3.4.2 *Approximations Derived from the Loss Functions*

We now turn to finite difference approximations derived from the loss functions, $G''(\omega)$ and $J''(\omega)$, lumped together as $f''(\omega)$ [cf. Eq. $(4.3\text{-}2)_2$]. The logarithmic finite difference operator is

$$\mathscr{P}''_{kh} = \sum_{n=0}^{n=k} \frac{a''_n}{\ln^n h} \sum_{m=0}^{m=n} (-1)^{n-m} \binom{n}{m} L^m \tag{4.3-145}$$

which we use in the form

$$\mathscr{P}''_{kh} = \sum_{m=0}^{m=k} b''_m L^m \tag{4.3-146}$$

where

$$b''_m = \sum_{n=0}^{n=k} (-1)^{n-m} \binom{n}{m} \frac{a''_n}{\ln^n h} \tag{4.3-147}$$

subject to the restriction expressed by Eq. (4.2-83). We note that the lower limit of the summation index in $\mathscr{P}''_{kh}$ is zero, in contrast to that of $\mathscr{P}'_{kh}$ and $\mathscr{P}_{kh}$ in which the summation begins with n = 1. As a result the sum of the b''_m is

$$\sum_{m=0}^{m=k} b''_m = 2/\pi \tag{4.3-148}$$

while that of the b'_m is 0 (cf. Problem 4.3-5). Thus we now have

$$I''_{kh}(\omega\tau) = \mathscr{P}''_{kh}\left[\frac{\omega\tau}{1+\omega^2\tau^2}\right] = \sum_{m=0}^{m=k} \frac{b''_m h^m \omega\tau}{1+h^{2m}\omega^2\tau^2} \tag{4.3-149}$$

and

$$F''_{kh}(\tau) = \mathscr{P}''_{kh}[f''(\omega)]|_{\omega=\beta''_{kh}/\tau} = \sum_{m=0}^{m=k} b_m f''(\omega h^m)|_{\omega=\beta''_{kh}/\tau} \tag{4.3-150}$$

for the intensity functions and the spectra, respectively. $F''(\tau)$ stands for $H''(\tau)$ and $L''(\tau)$ as the case may be [cf. Eq. $(4.3\text{-}1)_2$].

We now show that normalization of $I''_{kh}(\omega\tau)$ leads to $a''_0 = 2/\pi$ regardless of the order k of the approximation. We apply the operator $\mathscr{P}''_{kh}$ given by Eq. (4.3-145) to the kernel $\omega\tau/(1+\omega^2\tau^2)$, separating out the first term. This gives

$$I''_{kh}(\omega\tau) = \frac{a''_0\omega\tau}{1+\omega^2\tau^2} + \sum_{n=1}^{n=k} \frac{a''_n}{\ln^n h} \sum_{m=0}^{m=n} (-1)^{n-m}\binom{n}{m}\frac{h^{-m}\omega\tau}{h^{-2m}+\omega^2\tau^2} \,. \tag{4.3-151}$$

But

$$\int_{-\infty}^{\infty} \frac{h^{-m}\omega\tau}{h^{-2m}+\omega^2\tau^2}\, d\ln\omega\tau = \frac{\pi}{2} \tag{4.3-152}$$

for any m. Hence

$$\int_{-\infty}^{\infty} I''_{kh}(\omega\tau)\, d\ln\omega\tau = \int_{-\infty}^{\infty} \frac{a''_0\omega\tau}{1+\omega^2\tau^2}\, d\ln\omega\tau = \frac{\pi}{2} \tag{4.3-153}$$

and, by the normalization requirement, $a''_0 = 2/\pi$ follows.

The zeroth approximation turns out to be independent of h and is thus identical with that obtained in Sect. 4.3.1.2 [Eq. (4.3-41)].

For the approximations of first order we have

$$b''_0 = \frac{2}{\pi} - \frac{a''_1}{\ln h} \quad \text{and} \quad b''_1 = \frac{a''_1}{\ln h} \tag{4.3-154}$$

and Eq. (4.3-149) yields

$$I''_{1h}(\omega\tau) = \frac{c''_1\omega\tau + hc''_3\omega^3\tau^3}{(1+\omega^2\tau^2)(1+h^2\omega^2\tau^2)} \tag{4.3-155}$$

where

$$c''_1 = b''_0 + b''_1 h \quad \text{and} \quad c''_3 = b''_0 h + b''_1 \,. \tag{4.3-156}$$

Letting $c''_1 = 0$ gives

$$a''_1 = -\frac{2\ln h}{\pi(h-1)} \tag{4.3-157}$$

from which

$$b''_0 = \frac{2h}{\pi(h-1)} \quad \text{and} \quad b''_1 = -\frac{2}{\pi(h-1)} \,. \tag{4.3-158}$$

Hence

$$c_3'' = 2(h + 1)/\pi \tag{4.3-159}$$

and the intensity function becomes

$$I_{1hR}''^{*}(\omega\tau) = \frac{2h(h + 1)\omega^3\tau^3}{\pi(1 + \omega^2\tau^2)(1 + h^2\omega^2\tau^2)} . \tag{4.3-160}$$

As this would lead to approximations with half-integral powers in h, we apply the shift operator $(1 + L^{-1})/2$ to obtain

$$I_{1hR}''(\omega\tau) = \frac{(h + 2 + h^{-1})[h - 1 + h^{-1} + \omega^2\tau^2]\omega^3\tau^3}{\pi(h^2 + \omega^2\tau^2)(1 + \omega^2\tau^2)(h^{-2} + \omega^2\tau^2)} . \tag{4.3-161}$$

The subscript R is justified because

$$\lim_{h \to 1} I_{1hR}''(\omega\tau) = \frac{4\omega^3\tau^3}{\pi(1 + \omega^2\tau^2)^2} = I_{1R}''(\omega\tau) . \tag{4.3-162}$$

The spectra follow as

$$F_{1hR}''(\tau\beta_{1hR}'') = \left. \frac{h\, f''(\omega h^{-1}) + (h - 1)f''(\omega) - f''(\omega h)}{\pi(h - 1)} \right|_{\omega = 1/\tau} . \tag{4.3-163}$$

Values of β_{1hR}'' for different values of log h are listed in Table 4.3-3. $I_{1hR}''(\omega\tau)$ is plotted in Fig. 4.3-9 for log h = 0.2.

Table 4.3-3. Spectral shift factors for finite difference approximations from the loss functions

log h	0.1	0.2	0.3	0.4	0.5
β_{1hR}''	1.755	1.826	1.947	2.129	2.390
β_{2hR}''	2.289	2.450	2.726	3.131	3.685
β_{3hR}''	1.410	1.520	1.659	1.750	1.703

If we set $c_3'' = 0$, then

$$a_1'' = \frac{2h \ln h}{\pi(h - 1)} \tag{4.3-164}$$

and

$$b_0'' = -\frac{2}{\pi(h - 1)} \quad \text{and} \quad b_1'' = \frac{2h}{\pi(h - 1)} \tag{4.3-165}$$

so that

$$c_1'' = 2(h + 1)/\pi \tag{4.3-166}$$

again. The intensity function becomes

$$I_{1hL}''^{*}(\omega\tau) = \frac{2(h + 1)\omega\tau}{\pi(1 + \omega^2\tau^2)(1 + h^2\omega^2\tau^2)} \tag{4.3-167}$$

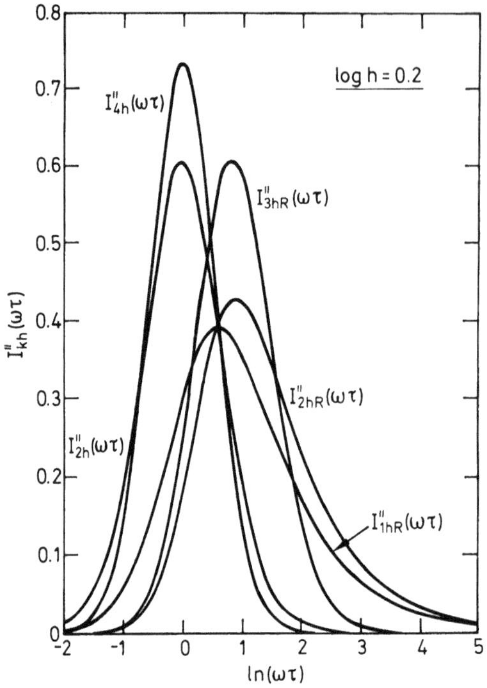

Fig. 4.3-9. The intensity functions $I''_{1hR}(\omega\tau)$, $I''_{2R}(\omega\tau)$, $I''_{2hR}(\omega\tau)$, $I''_{3hR}(\omega\tau)$, and $I''_{4h}(\omega\tau)$ as functions of ln ωτ

and, upon shifting by $(1 + L^{-1})/2$, we obtain

$$I''_{1hL}(\omega\tau) = \left.\frac{(h + 2 + h^{-1})[1 + (h - 1 + h^{-1})\omega^2\tau^2]\omega\tau}{\pi(h^2 + \omega^2\tau^2)(1 + \omega^2\tau^2)(h^{-2} + \omega^2\tau^2)}\right|_{\omega=1/\tau} \tag{4.3-168}$$

and

$$F''_{1hL}(\tau\beta''_{1hL}) = -\left.\frac{f''(\omega h^{-1}) - (h - 1)f''(\omega) - hf''(\omega h)}{\pi(h - 1)}\right|_{\omega=1/\tau} . \tag{4.3-169}$$

Again, $I''_{1hL}(\tau)$ is the mirror image of $I''_{1hR}(\tau)$, and $\beta''_{1hL} = 1/\beta''_{1hR}$.

Turning now to the approximations of second order, we find

$$I''_{2h}(\omega\tau) = \frac{c''_1 h\omega\tau + c''_3 h^3\omega^3\tau^3 + c''_5 h^5\omega^5\tau^5}{(1 + \omega^2\tau^2)(1 + h^2\omega^2\tau^2)(1 + h^4\omega^2\tau^2)} \tag{4.3-170}$$

where

$$c''_1 = b''_0 h^{-1} + b''_1 + b''_2 h \tag{4.3-171}$$

$$c''_3 = b''_0(h + h^{-1}) + b''_1(h^2 + h^{-2}) + b''_2(h + h^{-1}) \tag{4.3-172}$$

$$c''_5 = b''_0 h + b''_1 + b''_2 h^{-1} . \tag{4.3-173}$$

Letting $c''_1 = c''_5 = 0$ leads, after shifting by L^{-1}, to

$$I''_{2h}(\omega\tau) = \frac{2(h + h^{-1})(h + 2 + h^{-1})\omega^3\tau^3}{\pi(h^2 + \omega^2\tau^2)(1 + \omega^2\tau^2)(h^{-2} + \omega^2\tau^2)} \tag{4.3-174}$$

and the spectrum results as

$$F''_{2h}(\tau) = -\left.\frac{h\,f''(\omega h^{-1}) - (h^2+1)f''(\omega) + h\,f''(\omega h)}{\pi(h-1)^2/2}\right|_{\omega=1/\tau}. \tag{4.3-175}$$

The approximation is "centered", i.e. its peak lies at $\ln \omega\tau = 0$. The corresponding "left" and "right" approximations of second order are obtained by letting $c''_3 = c''_5 = 0$, and $c''_1 = c''_3 = 0$, respectively. Again shifting by L^{-1}, we derive

$$I''_{2hL}(\omega\tau) = \frac{2(h^2+2h+2+2h^{-1}+h^{-2})\omega\tau}{\pi(h+1+h^{-1})(h^2+\omega^2\tau^2)(1+\omega^2\tau^2)(h^{-2}+\omega^2\tau^2)} \tag{4.3-176}$$

and

$$I''_{2hR}(\omega\tau) = \frac{2(h^2+2h+2+2h^{-1}+h^{-2})\omega^5\tau^5}{\pi(h+1+h^{-1})(h^2+\omega^2\tau^2)(1+\omega^2\tau^2)(h^{-2}+\omega^2\tau^2)} \tag{4.3-177}$$

and the spectra become

$$F''_{2hL}(\tau\beta''_{2hL}) = \left.\frac{h^{-2}f''(\omega h^{-1}) - (h+h^{-1})f''(\omega) + h^2 f''(\omega h)}{\pi(h-h^{-2})(h-1)/2}\right|_{\omega=1/\tau} \tag{4.3-178}$$

and

$$F''_{2hR}(\tau\beta''_{2hR}) = \left.\frac{h^{2}f''(\omega h^{-1}) - (h+h^{-1})f''(\omega) + h^{-2} f''(\omega h)}{\pi(h-h^{-2})(h-1)/2}\right|_{\omega=1/\tau} \tag{4.3-179}$$

respectively. The shift factor, β''_{2hR}, is listed in Table 4.3-3. The functions $I''_{2h}(\omega\tau)$ and $I''_{2hR}(\omega\tau)$ are plotted in Fig. 4.3-9. The function $I''_{2hL}(\omega\tau)$ is the mirror image of $I''_{2hR}(\omega\tau)$.

There is no "central" approximation of third order. Of the four possible approximations we shall give here only the "innermost" ones (cf. Sect. 4.4). These were shifted by $(L^{-1} + L^{-2})/2$. The intensity functions are

$$I''_{3hL}(\omega\tau) = \frac{(h^2+h+h^{-1}+h^{-2})(h+1)(h+h^{-1})[h^2+h^{-1}+(h^3+h^{-2})\omega^2\tau^2]\omega^3\tau^3}{\pi\,h(h^4+\omega^2\tau^2)(h^2+\omega^2\tau^2)(1+\omega^2\tau^2)(h^{-2}+\omega^2\tau^2)(h^{-4}+\omega^2\tau^2)} \tag{4.3-180}$$

and

$$I''_{3hR}(\omega\tau) = \frac{(h^2+h+h^{-1}+h^{-2})(h+1)(h+h^{-1})[h^3+h^{-2}+(h^2+h^{-1})\omega^2\tau^2]\omega^5\tau^5}{\pi\,h(h^4+\omega^2\tau^2)(h^2+\omega^2\tau^2)(1+\omega^2\tau^2)(h^{-2}+\omega^2\tau^2)(h^{-4}+\omega^2\tau^2)} \tag{4.3-181}$$

and the spectra become

$$F''_{3hL}(\tau\beta''_{3hL}) = A\,f''(\omega h^{-2}) + B\,f''(\omega h^{-1}) + C\,f''(\omega) + D\,f''(\omega h) + E\,f''(\omega h^2)|_{\omega=1/\tau} \tag{4.3-182}$$

and

$$F''_{3hR}(\tau\beta''_{3hR}) = E\,f''(\omega h^{-2}) + D\,f''(\omega h^{-1}) + C\,f''(\omega) + B\,f''(\omega h) + A\,f''(\omega h^2)|_{\omega=1/\tau} \quad (4.3\text{-}183)$$

where

$$\begin{aligned} A &= 1/\pi(h-1)^3(h+1+h^{-1}) \\ B &= -h(h^2+1-h^{-1}+h^{-2})/\pi(h-1)^3(h+1+h^{-1}) \\ C &= h(h^2+1+h^{-2})/\pi(h-1)^2(h+1+h^{-1}) \\ D &= h^2(h^2-h+1+h^{-2})/\pi(h-1)^3(h+1+h^{-1}) \end{aligned} \quad (4.3\text{-}184)$$

and

$$E = -1/\pi(1-h^{-1})^3(h+1+h^{-1}) = -h^3A\ .$$

$I''_{3hR}(\omega\tau)$ is plotted in Fig. 4.3-9 along with the others. It is the mirror image of $I''_{3hL}(\omega\tau)$. The shift factor, β''_{3hR}, may be found in Table 4.3-3.

Finally, again without detailed derivation, we list the intensity function and spectra for the "centered" approximation of fourth order [12],

$$I''_{4h}(\omega\tau) = \frac{2(h^2+h^{-2})(h+2+h^{-1})^2(h+h^{-1})^2(h-1+h^{-1})\omega^5\tau^5}{\pi(h+1+h^{-1})(h^4+\omega^2\tau^2)(h^2+\omega^2\tau^2)(1+\omega^2\tau^2)(h^{-2}+\omega^2\tau^2)(h^{-4}+\omega^2\tau^2)}\ . \quad (4.3\text{-}185)$$

The corresponding spectra are

$$F''_{4h}(\tau) = A\,f''(\omega h^{-2}) + B\,f''(\omega h^{-1}) + C\,f''(\omega) + B\,f''(\omega h) + A\,f''(\omega h^2)|_{\omega=1/\tau} \quad (4.3\text{-}186)$$

where

$$\begin{aligned} A &= 2h^4/\pi(h-1)^4(h^2+h+1)^2 \\ B &= -2h(h^2+1)(h^4+1)/\pi(h-1)^4(h^2+h+1)^2 \end{aligned} \quad (4.3\text{-}187)$$

and

$$C = 2(h^4+1)(h^4+h^2+1)/\pi(h-1)^4(h^2+h+1)^2\ .$$

The approximations $F''_{2R}(\tau)$ and $F''_{4h}(\tau)$ were introduced by Ninomiya and Ferry [12]. The others are due to the author.

4.4 Comparison of the Approximations to the Continuous Spectra

In the two preceding sections we discussed the estimation of the spectral from the experimental response functions by various methods of approximation, based essentially on differentiation of the source function. The question naturally arises: how good is any given approximation, and how does it compare with any other?

Some general observations have already been made in Sects. 4.2 and 4.3. Problems 4.2-2 and 4.3-3 are also instructive. Thus, the higher the order of any approximation of a given type (e.g., continuous differentiation of the relaxation modulus, or finite differencing of the loss compliance, etc.) the better is the approximation because it will approach a delta function more closely (cf. Figs. 4.2-1, 4.3-1, and 4.3-7). It has also been noted that the necessity of using low order approximations is likely to result in an underestimation of the height of the peak of a spectrum but has a relatively minor effect on the slope (recall the discussion at the beginning of Sect. 4.1.3 and at the end of Sect. 4.3.1.1). The reason that the differentiation methods for the estimation of the spectral functions, which is mathematically an ill-posed problem, work at all, ultimately resides in the fact that the spectra are generally slowly changing functions of the respondence time. Rapidly oscillating spectra could not be estimated by these methods.

Another general observation concerns the comparison of a given continuous differentiation method with its finite difference analog. It has been pointed out that the latter methods reduce to the former as $h \to 1$. Since h is always larger than 1, a finite difference approximation necessarily produces a broader intensity function than does the continuous differentiation approximation of the same order (cf. Fig. 4.2-2).

It is evidently desirable to make these comparisons quantitative. Formally, the intensity functions are normalized distribution functions. Distribution functions may be characterized by the moments of the distribution. The pth moment about the origin of a continuous distribution, F(x), is given by

$$\mu_p = \int_0^\infty x^p F(x) dx \ . \tag{4.4-1}$$

The first moment about the origin is called the *mean* and is denoted by $\bar{x}$. Thus, we have

$$\bar{x} = \int_0^\infty x F(x) dx \tag{4.4-2}$$

as the equation of the mean. It determines the *location* of the distribution along the axis of x, and differs from the *mode*, which is that value of x at which the distribution peaks, or from the *median* which divides the area under the curve into two equal halves.

Higher moments are particularly useful when computed about the mean rather than the origin. The second moment around the mean is called the *variance* and is denoted by σ_x^2. The square root, σ_x, is the *standard deviation*. This is a measure of the *spread*, or *breadth*, of the distribution. The equation for the variance is

$$\sigma_x^2 = \int_0^\infty x^2 F(x) dx - \bar{x}^2 \ . \tag{4.4-3}$$

The third moment about the mean, when normalized through division by σ^3, is called the *skewness*, and is a measure of the *asymmetry* of the distribution. The skewness, γ_x, is given by

$$\gamma_x = \sigma_x^{-3} \int_0^\infty x^3 F(x) dx - 3\bar{x}/\sigma_x - (\bar{x}/\sigma_x)^3 \ . \tag{4.4-4}$$

For a symmetric distribution, $\gamma_x = 0$. If the distribution has a tail on the right of the mean, γ_x is positive. Negative skewness implies a tail on the left of $\bar{x}$. Moments higher than the third can be obtained but location, spread, and asymmetry are sufficient for a characterization of our intensity functions.

To compare the various intensity functions, we cast them in the forms $\bar{I}_k(z)$, $\bar{I}'_k(z)$, and $\bar{I}''_k(z)$ where $z = \ln t/\tau$ when the original function is unprimed, and $z = \ln \omega\tau$, when it is primed or double-primed. For the approximations of lowest order obtained by continuous differention we have

$$\bar{I}_1(z) = \exp[z - \exp(z)] \tag{4.4-5}$$

$$\bar{I}'_1(z) = (1/2)\mathrm{sech}^2 z \tag{4.4-6}$$

$$\bar{I}''_0(z) = (1/\pi)\mathrm{sech}\, z \tag{4.4-7}$$

and

$$\bar{I}''_{1L}(z) = (1/\pi)\exp(-z)\mathrm{sech}^2 z\ . \tag{4.4-8}$$

These functions are plotted as functions of z in Fig. 4.4-1. $\bar{I}''_{1L}(z)$ has been shifted by $\ln \beta''_{1L} = -\ln\sqrt{3}$ so that all functions would peak at $z = 0$. It is seen that the intensity functions of order 1 are roughly comparable as to their spread. $\bar{I}'_1(z)$ is the only symmetric one among them. $\bar{I}''_0(z)$ which is also symmetric, is considerably broader than $\bar{I}'_1(z)$.

For qualitative comparison the moments of the functions were calculated from

$$\mu_p = \int_{-\infty}^{z_L} z^p\, \bar{I}_k(z)dz + \int_{z_L}^{z_U} z^p\, \bar{I}_k(z)dz + \int_{z_U}^{\infty} z^p\, \bar{I}_k(z)dz \tag{4.4-9}$$

and its primed and double-primed analogs. The center integral was evaluated numerically using Simpson's rule, keeping the lower and upper limits of integration as wide apart as was reasonable. The other two integrals represent corrections terms

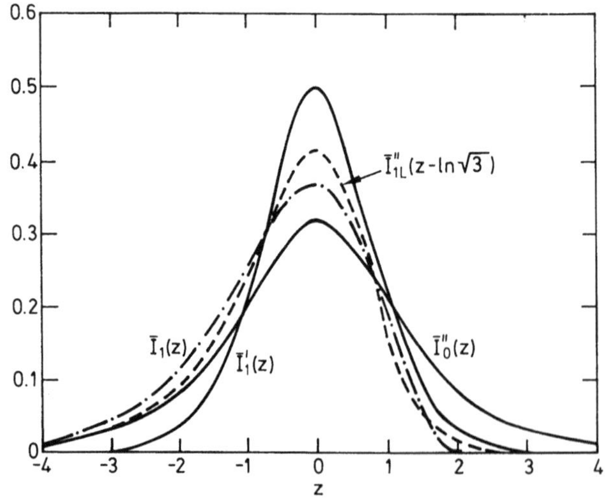

Fig. 4.4-1. Comparison of the intensity functions of first and zeroth order

which were evaluated analytically using suitable approximations for the integrand. The final results are accurate in at least the third decimal place.

Tables 4.4-1, 4.4-2, and 4.4-3 contain the results for the intensity functions of order 1, 2, and 3, respectively. Only the values for log h = 0.2 have been tabulated for the functions obtained by finite differencing. The results for the zeroth-order

Table 4.4-1. Characteristic parameters of the approximations of first order (log h = 0.2)

Type	$\bar{z}$	σ_z	γ_z	B	z_{max}
$I_1(z)$	−0.577	1.283	−1.140	1.114	0
$I^*_{1h}(z)$	−0.808	1.289	−1.120	1.120	−0.240
$I_{1h}(z)$	−0.577	1.310	−1.069	1.138	−0.035
$I'_1(z)$	0	0.907	0	0.788	0
$I'_{1h}(z)$	0	0.945	0	0.821	0
$I''_{1L}(z)$	−1.000	1.211	−1.125	1.052	−0.549
$I''_{1hL}(z)$	−1.018	1.240	−1.048	1.077	−0.562

Table 4.4-2. Characteristic parameters of the approximations of second order (log h = 0.2)

Type	$\bar{z}$	σ_z	γ_z	B	z_{max}
$I_2(z)$	0.423	0.803	−0.780	0.700	0.693
$I^*_{2h}(z)$	−0.020	0.825	−0.722	0.716	0.233
$I_{2h}(z)$	0.440	0.825	−0.721	0.716	0.694
$I'_{2L}(z)$	−0.500	0.757	−0.577	0.657	−0.347
$I'_{2hL}(z)$	−0.535	0.779	−0.525	0.677	−0.393
$I''_2(z)$	0	0.684	0	0.594	0
$I''_{2h}(z)$	0	0.709	0	0.616	0
$I''_{2L}(z)$	−1.333	1.165	−1.313	1.012	−0.805
$I''_{2hL}(z)$	−1.402	1.179	−1.265	1.024	−0.896

Table 4.4-3. Characteristic parameters of the approximations of third order (log h = 0.2)

Type	$\bar{z}$	σ_z	γ_z	B	z_{max}
$I_3(z)$	0.923	0.628	−0.621	0.546	1.099
$I^*_{3h}(z)$	0.285	0.668	−0.425	0.580	0.438
$I_{3h}(z)$	0.975	0.707	−0.441	0.614	1.108
$I'_3(z)$	0	0.568	0	0.493	0
$I'_{3h}(z)$	0	0.654	0	0.568	0
$I'_{3L}(z)$	−0.750	0.714	−0.773	0.620	−0.549
$I'_{3hL}(z)$	−0.852	0.783	−0.583	0.680	−0.709
$I''_{3L}(z)$	−0.333	0.570	−0.348	0.519	−0.255
$I''_{3hL}(z)$	−0.385	0.679	−0.234	0.589	−0.344
$I''_{3LL}(z)$	−1.533	1.148	−1.387	0.997	−0.973

approximation, $\bar{I}''_0(z)$, are $\bar{z} = \gamma_x = 0$, $\sigma_z = 1.571$, and $B = 1.364$. The parameter B is given by

$$B = \log \exp(2\sigma_x) \tag{4.4-10}$$

and thus represents twice the standard deviation on the (decadic) logarithmic scale on which both the experimental response functions and the approximation to the spectral functions would normally be plotted. B is correlated with the minimum distance, $\Delta \log \tau$, required to resolve the separation of two peaks in the approximation of the kth order when the underlying true spectrum is a doublet line spectrum (see Sect. 4.5).

The last columns contain z_{max}. This is the location of the *mode* when the intensity function has not been shifted.

Figure 4.4-1 as well as the three tables show unambiguously that, for approximations of the same order, those obtained from the harmonic responses are always better than those from the step responses; and, of the former, approximations from the storage functions are always better than those from the loss functions.

There are no approximations from the step responses with symmetric intensity functions. The intensity functions $\bar{I}_k(z)$ all show negative skewness. By contrast, the approximations from the harmonic responses, when they are asymmetric, always have two forms which are each others mirror images and possess negative or positive skewness, respectively.

Concerning the starred and unstarred forms of the approximations derived from the step responses, it is clear that application of the operator L^{-1} to $\bar{I}_2^*(z)$ to obtain $\bar{I}_2(z)$ causes no change in either broadness or skewness. Application of the operator $(1/2)[L^{-(k-1)/2} + L^{-(k+1)/2}]$ to the intensity functions of odd order, $\bar{I}_1^*(z)$, and $\bar{I}_3^*(z)$, increases broadness and skewness but only slightly.

Table 4.4-3 lists parameters for the intensity function $\bar{I}''_{3LL}(z)$, i.e. for

$$I''_{3LL}(\omega\tau) = \frac{32\omega\tau}{5\pi(1 + \omega^2\tau^2)^4} \tag{4.4-11}$$

whose maximum lies at $\omega\tau = 1/\sqrt{7}$, i.e. to the left of the intensity function $I''_{3L}(\omega\tau)$. It is clear that the former is both much broader and more skewed than the latter. In fact, a comparison of $\bar{I}''_{3LL}(z)$ with $\bar{I}''_{2L}(z)$ shows that $\bar{I}''_{3LL}(z)$ is only marginally better than $\bar{I}''_{2L}(z)$. It is for this reason that $I''_{3LL}(\omega\tau)$ and its mirror image, $I''_{3RR}(\omega\tau)$ were omitted from the discussion of third order approximations in Sects. 4.3.1.2 and 4.1.4.2.

Finally, it must be recognized that the range of a spectral function obtained by finite differencing is always narrower than the range of the source function because the calculation of the value of the approximation at the point of interest also entails contributions taken to the left and right of the latter.

4.5 The Discrete Spectral Response Functions

The introduction of continuous spectral distribution functions in Sect. 4.1 was justified on the grounds that the description of experimentally observed viscoelastic behavior by series-parallel models generally requires a very large (practically infinite) number of Maxwell or Voigt units whose respondance times are so closely spaced that resolution becomes impossible. However, viscoelastic behavior can often be described with satisfactory approximation by models containing a finite, generally small, number of elements. We will show in this section that such models must be associated with *discrete spectral distribution functions* or *line spectra.* Such line spectra have been indicated by arrows in Figs. 3.5-10 through 3.5-16. Molecular theories of the viscoelastic behavior of polymers (cf. Sect. 6.3.1) also lead to discrete distribution functions. We may, therefore, ask ourselves how such discrete distribution functions are to be defined and how they are related to the respondances, $\bar{Q}(s)$ and $\bar{U}(s)$.

In Sect. 4.1 we showed that the frequency spectra are essentially inverse Stieltjes transforms of the respondances, i.e. that [cf. Eqs. $(4.1\text{-}36)_1$ and $(4.1\text{-}38)_1$]

$$M(\zeta) = \mathscr{S}^{-1}\{[G_g - \bar{Q}_\bullet(s)]; \zeta\} \tag{4.5-1}$$

and

$$N(\zeta) = \mathscr{S}^{-1}\{[\bar{U}(s) - J_g - \{\phi_f/s\}]; \zeta\} \ . \tag{4.5-2}$$

These equations are valid whether $\bar{Q}(s)$ or $\bar{U}(s)$ are continuous or discrete functions of $\tau = 1/\zeta$. Equation (4.5-1) with

$$\bar{Q}_\bullet(s) = G_g - \int_0^\infty \frac{\hat{Q}(\tau)}{1 + \tau s}\, d\tau \tag{4.5-3}$$

leads to $M(\zeta)$ as a continuous function of ζ. With

$$\bar{Q}_\bullet(s) = G_g - \sum_n \frac{G_n}{1 + \tau_n s} \tag{4.5-4}$$

we obtain a line spectrum as we shall show in this section. The line spectrum is a consequence of the fact that in the relaxance defined by Eq. (4.5-4) the relaxation time takes on the discrete values τ_n. Thus $\bar{Q}(s)$ possesses a discrete set of poles located at $s = -1/\tau_n$ along the negative real axis of the complex plane (cf. Figs. 3.5-5 and 3.5-7). By contrast, the relaxance defined by Eq. (4.5-3) has no poles but a continuous cut along the entire length of the negative real axis because τ is now a continuous variable (cf. Fig. 4.1-1).

In Sect. 4.3.2 we used Stieltjes transform inversion for calculating continuous spectra. Here we use it to obtain discrete spectra. Letting $s = -\zeta \pm j\beta$, and taking $M(\zeta)$ first, Eq. (A5-14) gives

$$M(\zeta) = \frac{1}{2\pi j} \lim_{\beta \to 0} [\tilde{M}(-\zeta - j\beta) - \tilde{M}(-\zeta + j\beta)] \ . \tag{4.5-5}$$

But, by Eq. $(4.1\text{-}36)_2$, the singularities of $\tilde{M}(s)$ are the same as those of $\bar{Q}_\bullet(s)$ and, if $\bar{Q}(s)$ is given by Eq. (4.5-4), they will be a series of simple poles along the negative real axis of the transform plane. Now, it is shown in complex variable theory (see, e.g., Refs. [13] or [14]) that $\tilde{M}(s)$ may then be expanded in the Laurent* series

* Pierre Alphonse Laurent, pron. Lor<u>an</u>, 1813–1854, French mathematician.

$$\tilde{M}(s) = \frac{A_{-1}}{s - z} + A_0 + A_1(s - z) + A_2(s - z)^2 + \cdots \tag{4.5-6}$$

around each point z at which $\tilde{M}(s)$ has a pole. The coefficients are given by

$$A_m = \frac{1}{2\pi j} \oint \frac{\tilde{M}(s)}{(s - z)^{m+1}} ds \ , \tag{4.5-7}$$

the path of integration being a closed contour around z, along which $\tilde{M}(s)/(s - z)^{m+1}$ is analytic, and which includes no other singularity. One can show further [13, 14] that then $A_m = 0$ when $m = 0, 1, 2, \ldots$ but that

$$A_{-1} = \frac{1}{2\pi j} \oint \tilde{M}(s)ds = \mathrm{Res}(\tilde{M}) \tag{4.5-8}$$

where $\mathrm{Res}(\tilde{M})$ is the *residue* associated with the pole at z. Now, $\tilde{M}(s)$ possesses a series of such poles and their contributions are additive. Hence, we may rewrite Eq. (4.5-6) in the form

$$\tilde{M}(s) = \sum_n \frac{\mathrm{Res}_n(\tilde{M})}{s + \zeta_n} \tag{4.5-9}$$

where $\mathrm{Res}_n(\tilde{M}) = A_{-1}$ is the residue associated with the pole at $-\zeta_n$. This may be obtained from (see Appendix A3.2.1 and Refs. [13, 14] to Chap. 4)

$$\mathrm{Res}_n(\tilde{M}) = \lim_{s \to -\zeta_n} (s + \zeta_n)\tilde{M}(s) = \left.\frac{\tilde{m}(s)}{\tilde{v}'(s)}\right|_{s=-\zeta_n} \tag{4.5-10}$$

where $\tilde{m}(s)$ and $\tilde{v}(s)$ are the numerator and denominator of $\tilde{M}(s)$, and the prime denotes differentiation with respect to the argument. Substituting Eq. (4.5-9) into Eq. (4.5-5) we have

$$M(\zeta) = \frac{1}{2\pi j} \sum_n \mathrm{Res}_n(\tilde{M}) \lim_{\beta \to 0} \left(\frac{1}{\zeta_n - \zeta - j\beta} - \frac{1}{\zeta_n - \zeta + j\beta} \right) \tag{4.5-11}$$

which reduces to

$$M(\zeta) = \sum_n \mathrm{Res}_n(\tilde{M}) \frac{1}{\pi} \lim_{\beta \to 0} \left(\frac{\beta}{(\zeta - \zeta_n)^2 + \beta^2} \right) . \tag{4.5-12}$$

But, by Eq. (A2-10),

$$\lim_{\beta \to 0} \left(\frac{\beta}{(\zeta - \zeta_n)^2 + \beta^2} \right) = \pi\, \delta(\zeta - \zeta_n) \tag{4.5-13}$$

and, therefore,

$$M(\zeta) = \sum_n \mathrm{Res}_n(\tilde{M}) \delta(\zeta - \zeta_n) . \tag{4.5-14}$$

Thus, when the relaxance is the sum of discrete terms with the relaxation times τ_n, the frequency spectrum, $M(\zeta)$, is a *line spectrum* consisting of a series of delta functions of strength $\mathrm{Res}_n(\tilde{M})$.

To find $H(\tau)$, we return to Eq. (4.5-5) and change the variable from ζ to τ. With $H(\tau)$ for $M(\zeta)$ and $G_g - \bar{Q}_\bullet(-1/\tau \pm j\beta)$ for $M(-\zeta \pm j\beta)$ we have

$$H(\tau) = \frac{1}{2\pi j} \lim_{\beta \to 0} [\bar{Q}_\bullet(-1/\tau + j\beta) - \bar{Q}_\bullet(-1/\tau - j\beta)] \ . \tag{4.5-15}$$

Now, by Eqs. (4.5-1) and (4.5-9),

$$\bar{Q}_\bullet(s) = G_g + \sum_n \frac{\mathrm{Res}_n(\bar{Q}_\bullet)}{s + 1/\tau_n} \tag{4.5-16}$$

where

$$\mathrm{Res}_n(\bar{Q}_\bullet) = \lim_{s \to -1/\tau_n} (s + 1/\tau_n)\bar{Q}_\bullet(s) = \left.\frac{\bar{q}(s)}{\bar{u}'(s)}\right|_{s=-1/\tau_n} \ . \tag{4.5-17}$$

We note that

$$\mathrm{Res}_n(\bar{Q}_\bullet) = -\mathrm{Res}_n(\tilde{M})|_{\zeta_n = 1/\tau_n} = -G_n/\tau_n \ . \tag{4.5-18}$$

Introducing Eq. (4.5-16) into Eq. (4.5-15) leads to

$$H(\tau) = \frac{1}{2\pi j} \sum_n \mathrm{Res}_n(\bar{Q}_\bullet) \lim_{\beta \to 0} \left(\frac{1}{1/\tau_n - 1/\tau + j\beta} - \frac{1}{1/\tau_n - 1/\tau - j\beta} \right) \tag{4.5-19}$$

which reduces to

$$H(\tau) = -\sum_n \mathrm{Res}_n(\bar{Q}_\bullet) \frac{\tau\tau_n}{\pi} \lim_{\beta \to 0} \left(\frac{\gamma}{(\tau - \tau_n)^2 + \gamma^2} \right) \tag{4.5-20}$$

where $\gamma = \tau\tau_n\beta$. As β approaches zero, so does γ. Hence

$$H(\tau) = -\sum_n \mathrm{Res}_n(\bar{Q}_\bullet)\tau_n^2 \delta(\tau - \tau_n) = \sum_n G_n \tau_n \delta(\tau - \tau_n) \ . \tag{4.5-21}$$

Turning to $N(\zeta)$, we have

$$N(\zeta) = \frac{1}{2\pi j} \lim_{\beta \to 0, \beta > 0} [\tilde{N}(-\zeta - j\beta) - \tilde{N}(-\zeta + j\beta)] \tag{4.5-22}$$

and obtain

$$\tilde{N}(s) = \sum_n \frac{\mathrm{Res}_n(\tilde{N})}{s + \zeta_n} \tag{4.5-23}$$

where

$$\mathrm{Res}_n(\tilde{N}) = \lim_{s \to -\zeta_n} (s + \zeta_n)\tilde{N}(s) = \left.\frac{\tilde{n}(s)}{\tilde{\mu}'(s)}\right|_{s=-\zeta_n} \tag{4.5-24}$$

and $\tilde{n}(s)$ and $\tilde{\mu}(s)$ are the numerator and denominator of $\tilde{N}(s)$. Substituting Eq. (4.5-23) into Eq. (4.5-22) and carrying out the limiting process leads to the *line spectrum*

$$N(\zeta) = \sum_n \mathrm{Res}_n(\tilde{N})\delta(\zeta - \zeta_n) \ . \tag{4.5-25}$$

Now replacing $N(\zeta)$ by $L(\tau)$ and $\tilde{N}(s)$ by $\bar{U}(s) - J_g - \{\phi_f/s\}$ we find

$$L(\tau) = \frac{1}{2\pi j} \lim_{\beta\to 0, \beta>0} [\bar{U}(-1/\tau - j\beta) - \bar{U}(-1/\tau + j\beta)] \tag{4.5-26}$$

because the contribution from the flow term, $\{\phi_f/s\}$, cancels in the limiting process. Into Eq. (4.5-26) we must introduce

$$\bar{U}(s) = J_g + \sum_n \frac{\mathrm{Res}_n(\bar{U})}{s + 1/\tau_n} + \mathrm{Res}_0(\bar{U})/s \tag{4.5-27}$$

where

$$\mathrm{Res}_n(\bar{U}) = \mathrm{Res}_n(\tilde{N})|_{\zeta_n=1/\tau_n} = J_n/\tau_n\ , \qquad n \neq 0 \tag{4.5-28}$$

and

$$\mathrm{Res}_0(\bar{U}) = \{\phi_f\}\ . \tag{4.5-29}$$

The limiting process leads to

$$L(\tau) = \sum_n \mathrm{Res}_n(\bar{U})\tau_n^2\delta(\tau - \tau_n) = \sum_n J_n\tau_n\delta(\tau - \tau_n) \tag{4.5-30}$$

where

$$\mathrm{Res}_n(\bar{U}) = \lim_{s\to -1/\tau_n} (s + 1/\tau_n)\bar{U}(s) = \left.\frac{\bar{u}(s)}{\bar{q}'(s)}\right|_{s=-1/\tau_n}\ , \qquad s \neq 0\ . \tag{4.5-31}$$

Equations (4.5-21) and (4.5-30) are the *line spectra* to which the respondances of the series-parallel models give rise. The series of delta functions of which the spectra consist may be represented graphically by a series of arrows located at $\tau = \tau_n$ along the axis of respondance times, and having the lengths G_n, or J_n, respectively. For series parallel models with a finite number of elements $n = \{0\}, 1, 2, \ldots, N-1$ where n may take the value zero only for rheodictic models. For the retardance, $\bar{U}(s)$, $n = 0$ is then a pole at the origin representing the contribution from the isolated dashpot. For the relaxance, $\bar{Q}(s)$, which does not have an isolated dashpot, it is not. For canonical models with an infinite number of elements n increases without bounds and the spectra consist of an infinite number of lines.

Making use of the relations [cf. Eqs. (4.1-42) and (4.1-45)]

$$\Xi(\zeta) = \mathscr{S}^{-1}[\bar{G}(s) - \{G_e\}/s]; \zeta\} \tag{4.5-32}$$

and

$$\Lambda(\zeta) = \mathscr{S}^{-1}[J_e^{\{o\}}/s - \bar{U}(s) + \{\phi_f/s^2\}]; \zeta\} \tag{4.5-33}$$

it is possible to show, by entirely analogous developments, that

$$\Xi(\zeta) = \sum_n \mathrm{Res}_n(\tilde{\Xi})\delta(\zeta - \zeta_n) \qquad \text{and} \qquad \Lambda(\zeta) = \sum_n \mathrm{Res}_n(\tilde{\Lambda})\delta(\zeta - \zeta_n) \tag{4.5-34}$$

and that

$$H(\tau) = \sum_n \mathrm{Res}_n(\bar{G})\tau_n\delta(\tau - \tau_n) \qquad \text{and} \qquad L(\tau) = -\sum_n \mathrm{Res}_n(\bar{J}_\bullet)\tau_n\delta(\tau - \tau_n) \tag{4.5-35}$$

where

$$\mathrm{Res}_n(\bar{\Xi}) = \mathrm{Res}_n(\bar{G})|_{\tau=\zeta^{-1}} \quad \text{and} \quad \mathrm{Res}_n(\bar{\Lambda}) = -\mathrm{Res}_n(\bar{J}_\bullet)|_{\tau=\zeta^{-1}} \ . \tag{4.5-36}$$

We have just shown that series-parallel models with discrete distributions of respondence times give rise to line spectra. We now claim that all linear lumped parameter models (see Sect. 3.1.2) entail line spectra. To see this we merely need to realize that an equivalent series-parallel model exists for any non-series-parallel one. Since the behavior of any of our models is determined solely by its poles and their associated residues in addition to the viscoelastic constants, G_g, and $\{G_e\}$, or J_g, $J_e^{\{o\}}$, and $\{\phi_f\}$, we need only to determine these quantities for non-series-parallel models to represent the same behavior by an equivalent series-parallel model. The respondance times will be given by

$$\tau_n = -1/s_n \tag{4.5-37}$$

and the moduli and compliances by

$$G_n = -\tau_n \mathrm{Res}_n(\bar{Q}) \qquad \text{and} \qquad J_n = \tau_n \mathrm{Res}_n(\bar{U}) \ , \tag{4.5-38}$$

respectively. The poles are the solutions of the equations

$$\bar{u}(s_n) = 0 \qquad \text{and} \qquad \bar{q}(s_n) = 0 \tag{4.5-39}$$

where $\bar{u}(s)$ and $\bar{q}(s)$ are the denominators of $\bar{Q}(s)$ and $\bar{U}(s)$, respectively. The residues are then calculated from Eqs. (4.5-17) or (4.5-31), as the case may be. Finally, the viscoelastic constants are obtained from the *limit value theorems* (see Appendix A3.1.7), which have been introduced in Sect. 3.4.4 and are grouped together for convenience in Sect. 4.6. We remark that, since we have

$$\mathrm{Res}_0(\bar{U}) = \lim_{s\to 0} s\bar{U}(s) = \phi_f \ , \tag{9.5-40}$$

by Eq. (4.5-31) with $\tau_0 = \infty$, determination of ϕ_f as the residue associated with the pole at the origin or from the final value theorem amounts to the same thing.

In Chap. 5 we shall use the *method of residues* which we have just discussed, to determine the parameters of the series-parallel models which are equivalent to the *regular ladder models*. Some other examples may be found in Sect. 4.7 (Problems) of this chapter.

4.6 The Viscoelastic Constants

In the limit of very short times or very high frequencies a linear viscoelastic material shows purely elastic response. In the limit of very long times or very low frequencies a rheodictic linear material shows purely viscous response while an arrheodictic one will again show purely elastic response. The material constants G_g, G_e, J_g, $J_e^{\{o\}}$, and $\{\phi_f\}$ which characterize this behavior may collectively be called viscoelastic constants. In the canonical representations of the experimental response functions introduced in Sect. 4.1.1 they accompany the spectral response functions. The latter determine the time-varying part of linear viscoelastic behavior to the exclusion of the purely elastic or purely viscous behavior governed by the time-independent viscoelastic constants.

Other quantities exist which also have the nature of a constant in that they do not depend on time and which contain useful information on linear viscoelastic behavior. These quantities do not appear explicitly in the canonical representations but may be related to the spectral response functions. Such quantities are the momenta of the respondance time distributions, the sum of the fluidities in the canonical models of linear viscoelastic behavior, the sum of the viscosities in the canonical models of arrheodictic behavior, and others more. It is, therefore, appropriate that we conclude this chapter on the representation of linear viscoelastic behavior by spectral response functions with an examination of the relationships between these functions and the viscoelastic constants. We begin with a brief review of the relations between the constants and the manner in which they can be obtained from the experimental response functions.

It was demonstrated in Chap. 3, while discussing the standard series-parallel models,* that

$$G_e^{\{o\}} J_e^{\{o\}} = 1 \tag{4.6-1}$$

$$G_g J_g = 1 \tag{4.6-2}$$

and

$$\eta_f \phi_f = 1 \tag{4.6-3}$$

[cf. Eqs. (3.4-123)]. We emphasize here that these relations are quite general and must hold whatever representation of linear viscoelastic behavior is chosen. For completeness we also list the relations derived from the *limit value theorems* [cf. Eqs. (3.4-105) through (3.4-122)], adding, however, the appropriate harmonic forms. Thus, we have

$$G_e = \lim_{s\to 0} \bar{Q}(s) = \lim_{t\to\infty} G(t) = \lim_{\omega\to 0} G'(\omega) \tag{4.6-4}$$

$$G_g = \lim_{s\to\infty} \bar{Q}(s) = \lim_{t\to 0} G(t) = \lim_{\omega\to\infty} G'(\omega) \tag{4.6-5}$$

$$J_e^{\{o\}} = \lim_{s\to 0} [\bar{U}(s) - \{\phi_f/s\}] = \lim_{t\to\infty} [J(t) - \{\phi_f t\}] = \lim_{\omega\to 0} J'(\omega) \tag{4.6-6}$$

$$J_g = \lim_{s\to\infty} \bar{U}(s) = \lim_{t\to 0} J(t) = \lim_{\omega\to\infty} J'(\omega) \tag{4.6-7}$$

$$\phi_f = \lim_{s\to 0} s\bar{\phi}(s) = \lim_{t\to\infty} \phi(t) = \lim_{\omega\to\infty} \phi'(\omega) \tag{4.6-8}$$

and

$$\eta_f = \lim_{s\to 0} s\bar{\eta}(s) = \lim_{t\to\infty} \eta(t) = \lim_{\omega\to 0} \eta'(\omega) \tag{4.6-9}$$

where we have used $\bar{\phi}(s) = \bar{U}(s)$ [cf. Eq. (2.4-22)] to point up the symmetry in these

* For G_e^o see Sect. 3.4.1.2.

relations. Equations (4.6-8) and (4.6-9) may, of course, also be recognized in the forms

$$\phi_f = \lim_{s\to 0} \bar{J}(s) = \lim_{t\to\infty} \frac{d\,J(t)}{dt} = \lim_{\omega\to 0} \omega J''(\omega) \tag{4.6-10}$$

and

$$\eta_f = \lim_{s\to 0} \bar{G}(s) = \lim_{t\to\infty} \int_0^t G(u)du = \lim_{\omega\to 0} G''(\omega)/\omega \ . \tag{4.6-11}$$

Turning now to the relations between the constants and the spectral functions, we recall that the functions $\hat{Q}(\tau)$ and $\hat{U}(\tau)$ introduced with Eqs. (4.1-1) and (4.1-17) are *distribution functions.* As we shall demonstrate presently, use of the moments of these distributions allows us to define averages of the respondance times which can also be considered as viscoelastic constants and can be related to those mentioned earlier. To derive these averages we proceed as follows.

We first recognize that $\hat{Q}(\tau)$ and $\hat{U}(\tau)$ are not normalized. Rewriting Eqs. (4.1-5) and (4.1-28) as

$$G(t) = \{G_e\} + \int_0^\infty \hat{Q}(\tau)\exp(-t/\tau)d\tau \tag{4.6-12}$$

and

$$J_\bullet(t) = J_e^{\{o\}} - \int_0^\infty \hat{U}(\tau)\exp(-t/\tau)d\tau + \{\phi_f t\} \tag{4.6-13}$$

and letting $t \to 0$, we find

$$\int_0^\infty \hat{Q}(\tau)d\tau = G_g - \{G_e\} \tag{4.6-14}$$

and

$$\int_0^\infty \hat{U}(\tau)d\tau = J_e^{\{o\}} - J_g \tag{4.6-15}$$

because $G(0) = G_g$ and $J_\bullet(0) = J_g$ by Eqs. $(4.6\text{-}5)_2$ and $(4.6\text{-}7)_2$. Thus the differences between the instantaneous and equilibrium moduli and compliances may, respectively, be used as normalizing factors. The moments of the normalized distributions around the origin then are

$$\langle\tau_M^\alpha\rangle = [G_g - \{G_e\}]^{-1} \int_0^\infty \tau^\alpha \hat{Q}(\tau)d\tau \tag{4.6-16}$$

and

$$\langle\tau_V^\alpha\rangle = [J_e^{\{o\}} - J_g]^{-1} \int_0^\infty \tau^\alpha \hat{U}(\tau)d\tau \tag{4.6-17}$$

where α takes on integral values only and where we have used the subscripts M and V to distinguish between average relaxation times and average retardation times, respectively. We recognize the normalizing factors $G_g - \{G_e\}$ and $J_e^{\{o\}} - J_g$ as the zeroth moments of the distributions, and note that, formally, $\langle\tau_M^0\rangle = \langle\tau_V^0\rangle = 1$.

The moments of $\hat{Q}(\tau)$ and $\hat{U}(\tau)$ are related to those of the other spectral functions introduced in preceding sections of this chapter by

$$\int_0^\infty \tau^{\alpha+1}\hat{Q}(\tau)d\tau = \begin{cases} \int_0^\infty \tau^{\alpha}H(\tau)d\tau \\ \int_0^\infty \zeta^{-\alpha-2}M(\zeta)d\zeta \\ \int_0^\infty \zeta^{-\alpha-1}\Xi(\zeta)d\zeta \end{cases} \tag{4.6-18}$$

and

$$\int_0^\infty \tau^{\alpha+1}\hat{U}(\tau)d\tau = \begin{cases} \int_0^\infty \tau^{\alpha}L(\tau)d\tau \\ \int_0^\infty \zeta^{-\alpha-2}N(\zeta)d\zeta \\ \int_0^\infty \zeta^{-\alpha-1}\Lambda(\zeta)d\zeta\ . \end{cases} \tag{4.6-19}$$

Letting $\alpha = -1$, Eqs. $(4.6\text{-}18)_1$ and (4.6-14) lead to

$$\int_{-\infty}^{\infty} H(\tau)d\ln\tau = \sum_n G_n = G_g - \{G_e\} \tag{4.6-20}$$

and Eqs. $(4.6\text{-}19)_1$ and (4.6-15) give

$$\int_{-\infty}^{\infty} L(\tau)d\ln\tau = \sum_n J_n = J_e^{\{o\}} - J_g \tag{4.6-21}$$

where we have included the interpretation of $G_g - \{G_e\}$ and $J_e^{\{o\}} - J_g$ in terms of the parameters of the series-parallel models for completeness. These equations may be generalized to

$$\int_{-\infty}^{\infty} \tau^{\alpha}H(\tau)d\ln\tau = \sum_n G_n\tau_n^{\alpha} = \langle\tau_M^{\alpha}\rangle[G_g - \{G_e\}] \tag{4.6-22}$$

and

$$\int_{-\infty}^{\infty} \tau^{\alpha}L(\tau)d\ln\tau = \sum_n J_n\tau_n^{\alpha} = \langle\tau_V^{\alpha}\rangle[J_e^{\{o\}} - J_g] \tag{4.6-23}$$

where $\alpha = -1, 0, 1, \ldots$. Equations (4.6-22) and (4.6-23) relate the respondence time averages to the relaxation and retardation spectra, respectively. We recognize, in particular, that

$$\int_{-\infty}^{\infty} \tau H(\tau)d\ln\tau = \sum_n G_n\tau_n = \sum_n \eta_n = \eta_{\{f\}} \tag{4.6-24}$$

where the subscript $\{f\}$ indicates that η_f is the steady-flow viscosity for a rheodictic material, while, for an arrheodictic material, η simply denotes the sum (integral) over all viscous contributions. The sum of the fluidities, $\sum_n \phi_n = \sum_n J_n/\tau_n$, or the corresponding integral, Eq. $(4.6\text{-}23)_1$, with $\alpha = -1$, will receive no special symbol. However, a short-hand notation is sometimes useful for the sum or integral represented by Eq. (4.6-23) with $\alpha = 1$, and we accordingly write

$$\int_{-\infty}^{\infty} \tau L(\tau) d \ln \tau = \sum_n J_n \tau_n = \chi \ . \tag{4.6-25}$$

The *average relaxation time* $\langle \tau_M \rangle$ and the *average retardation time* $\langle \tau_V \rangle$ are the *means* of the distribution and determine the location of the time spectra along the axis of relaxation or retardation times. In the same manner, $\langle \tau_M^{-1} \rangle = \langle \zeta_M \rangle$ and $\langle \tau_V^{-1} \rangle = \langle \zeta_V \rangle$ locate the frequency spectra along the axis of relaxation or retardation frequencies. The higher moments may be used to characterize the spread and asymmetry of the distributions. Thus, the *standard deviation*, σ, of the time spectra, a measure of their spread, is the square root of the variance, σ^2, obtained from

$$\sigma^2 = \langle \tau^2 \rangle - \langle \tau \rangle^2 \ . \tag{4.6-26}$$

The measure of asymmetry, the *skewness*, γ, is given by

$$\gamma = \langle \tau^3 \rangle / \sigma^3 - 3\langle \tau \rangle / \sigma - \langle \tau \rangle^3 / \sigma^3 \ . \tag{4.6-27}$$

A brief discussion of these statistics is contained in Sect. 4.4 (see also Sect. 6.1).

It remains to relate the moments, $\langle \tau^\alpha \rangle$, of the normalized distributions to the experimental response functions. For the step response functions we have

$$\Gamma^{-1}(\alpha) \int_{-\infty}^{\infty} t^\alpha [G(t) - \{G_e\}] d \ln t = \langle \tau_M^\alpha \rangle [G_g - \{G_e\}] \tag{4.6-28}$$

and

$$\Gamma^{-1}(\alpha) \int_{-\infty}^{\infty} t^\alpha [J(t) - J_g - \{\phi_f t\}] d \ln t = \langle \tau_V^\alpha \rangle [J_e^{\{o\}} - J_g] \tag{4.6-29}$$

where $\Gamma^{-1}(\alpha)$ is the reciprocal of the gamma function of argument α, and $\alpha = 1, 2, \ldots$. These relations may be verified by substituting Eqs. (4.1-5) and (4.1-21) for G(t) and J(t), and integrating after an interchange of the order of integration. This leads to Eqs. (4.6-22) and (4.6-23).

For the harmonic response functions we find

$$\frac{2}{\pi} \int_{-\infty}^{\infty} \omega [G_g - G'_\bullet(\omega)] d \ln \omega = \langle \tau_M^{-1} \rangle [G_g - \{G_e\}] \tag{4.6-30}$$

$$\frac{2}{\pi} \int_{-\infty}^{\infty} G''(\omega) d \ln \omega = G_g - \{G_e\} \tag{4.6-31}$$

$$\frac{2}{\pi} \int_{-\infty}^{\infty} \omega^{-1} [G'(\omega) - \{G_e\}] d \ln \omega = \langle \tau_M \rangle [G_g - \{G_e\}] \tag{4.6-32}$$

and

$$\frac{2}{\pi} \int_{-\infty}^{\infty} \omega^{-2} [G''(\omega) - d\, G''(\omega)/d \ln \omega] d \ln \omega = \langle \tau_M^2 \rangle [G_g - \{G_e\}] \tag{4.6-33}$$

from the storage and loss moduli, and

$$\frac{2}{\pi}\int_{-\infty}^{\infty} \omega[J'(\omega) - J_g]d\ln\omega = \langle\tau_V^{-1}\rangle[J_e^{\{o\}} - J_g] \tag{4.6-34}$$

$$\frac{2}{\pi}\int_{-\infty}^{\infty} [J''(\omega) - \{\phi_f/\omega\}]d\ln\omega = J_e^{\{o\}} - J_g \tag{4.6-35}$$

$$\frac{2}{\pi}\int_{-\infty}^{\infty} \omega^{-1}[J_e^{\{o\}} - J'_{\bullet}(\omega)] = \langle\tau_V\rangle[J_e^{\{o\}} - J_g] \tag{4.6-36}$$

and

$$\frac{2}{\pi}\int_{-\infty}^{\infty} \omega^{-2}[J''(\omega) - d\,J''(\omega)/d\ln\omega]d\ln\omega = \langle\tau_V^2\rangle[J_e^{\{o\}} - J_g] \tag{4.6-37}$$

from the storage and loss compliances. Table 4.6-1 summarizes the available expressions. Higher moments may be derived if desired.

Table 4.6-1. Moments of the spectral functions obtainable from the standard response functions

	τ^{-1}	τ^0	τ^1	τ^2
G(t), J(t)	–	–	×	×
G′(ω), J′(ω)	×	–	×	–
G″(ω), J″(ω)	–	×	–	×

We close this section by introducing the relations

$$\eta_{\{f\}} = [G_g - \{G_e\}]\langle\tau_M\rangle \tag{4.6-38}$$

and

$$J_e^\circ = \eta_f^{-2} G_g \langle\tau_M^2\rangle\ . \tag{4.6-39}$$

The first of these is obtained from Eqs. (4.6-24) and (4.6-22). The second can be derived from Eq. $(4.6\text{-}6)_3$ by substituting $G'(\omega)/\tilde{G}^2(\omega)$ for $J'(\omega)$ [see Eq. $(8.1\text{-}6)_1$], and again using Eqs. (4.6-24) and (4.6-22) (see Problem 4.6-6). It follows from these relations using Eqs. (4.6-22), (4.6-28), and (4.6-32) or (4.6-33), that

$$\eta_{\{f\}} = \begin{cases} \int_{-\infty}^{\infty} \tau H(\tau)d\ln\tau \\ \int_{-\infty}^{\infty} t[G(t) - \{G_e\}]d\ln t \\ (2/\pi)\int_{-\infty}^{\infty} \omega^{-1}[G'(\omega) - \{G_e\}]d\ln\omega \end{cases} \tag{4.6-40}$$

and

$$
J_e^o = \begin{cases} \eta_f^{-2} \int_{-\infty}^{\infty} \tau^2 H(\tau) d \ln \tau \\ \eta_f^{-2} \int_{-\infty}^{\infty} t^2 G(t) d \ln t \\ \eta_f^{-2}(2/\pi) \int_{-\infty}^{\infty} \omega^{-2}[G''(\omega) - d\, G''(\omega)/d \ln \omega] d \ln \omega \ . \end{cases} \tag{4.6-41}
$$

For a rheodictic material, $\eta_f = 1/\phi_f$ and J_e^o are readily obtained if the responses to the standard stress excitations are given [cf. Eqs. (4.1-28), (4.1-30), and (4.1-24)]. By Eqs. (4.6-40) and (4.6-41) these parameters can also be obtained from the responses to the standard strain excitations. For a discrete distribution of moduli we have

$$
J_e^o = \eta_f^{-2} \sum_n G_n \tau_n^2 = \eta_f^{-2} \sum_n \eta_n \tau_n \tag{4.6-42}
$$

where η_f is given by Eq. (4.6-24).

Some of the viscoelastic constants can be related to the maxima in the response functions which contain them. Let max[f(x)] denote the value of the function f(x) which it assumes at its maximum. By the usual method of finding maxima we obtain

$$
G_g - \{G_e\} = \begin{cases} 2 \max[G''(\omega)] \\ 2 \max[d\, G'(\omega)/d \ln \omega] \\ 2.718 \max[-d\, G(t)/d \ln t] \end{cases} \tag{4.6-43}
$$

using Eq. (4.6-20), and

$$
J_e^{\{o\}} - J_g = \begin{cases} 2 \max[J''(\omega) - \{\phi_f/\omega\}] \\ 2 \max[-d\, J'(\omega)/d \ln \omega] \\ 2.718 \max[d\, J(t)/d \ln t - \{\phi_f t\}] \end{cases} \tag{4.6-44}
$$

making use of Eq. (4.6-21).

Maxima may also be used to obtain $\eta_{\{f\}}$. Proceeding in a manner similar to that used above, we find

$$
\eta_{\{f\}} = \begin{cases} -2 \max[d\, \eta'(\omega)/d \ln \omega] = 2 \max[G''(\omega)/\omega - d\, G''(\omega)/d\omega] \\ 2 \max[\eta''(\omega) - \{G_e/\omega\}] = 2 \max[(G'(\omega) - \{G_e\})/\omega] \\ 2.718 \max[d\, \eta(t)/d \ln t - \{G_e t\}] = 2.718 \max[t\, G(t) - \{G_e t\}] \ . \end{cases} \tag{4.6-45}
$$

Unfortunately, it is usually not easy to obtain good estimates of the values of the maxima.

We notice that Eqs. (4.6-38), (4.6-40), and (4.6-45) remain valid whether $\eta_{\{f\}}$ represents the steady-flow viscosity, η_f, of a rheodictic material, or the sum, η, of the viscosities in arrheodictic response. Equations $(4.6\text{-}9)_3$, and $(4.6\text{-}11)_3$, also remain valid. Expressions $(4.6\text{-}9)_1$, $(4.6\text{-}9)_2$, $(4.6\text{-}11)_1$, $(4.6\text{-}11)_2$ are readily extended to include η. We have, for all three of Eqs. (4.6-9),

$$\eta_{\{f\}} = \begin{cases} \lim\limits_{s\to 0} [s\bar{\eta}(s) - \{G_e/s\}] \\ \lim\limits_{t\to\infty} [\eta(t) - \{G_e t\}] \\ \lim\limits_{\omega\to 0} \eta'(\omega) \end{cases} \tag{4.6-46}$$

and note that $s\bar{\eta}(s) = \bar{G}(s)$, $\eta(t) = \int_0^t G(u)du$, and $\eta'(\omega) = G''(\omega)/\omega$. Substitution of these relations yields the extended Eqs. (4.6-11).

Equations (4.6-45) and (4.6-46) allow the determination, at least in theory, of $\eta_{\{f\}}$. Equation (4.6-38) can then be used to obtain the average relaxation time, $\langle\tau_M\rangle$, for both rheodictic and arrheodictic behavior. By Eq. (4.6-38) this is

$$\langle\tau_M\rangle = \eta_{\{f\}}/[G_g - \{G_e\}] \ . \tag{4.6-47}$$

We recognize that $\langle\tau_M\rangle$ is *not* the arithmetic mean of the relaxation times. For a discrete distribution it is given by

$$\langle\tau_M\rangle = \sum_n G_n\tau_n/\sum_n G_n \tag{4.6-48}$$

and is thus an average of the relaxation times weighted by the moduli of the Maxwell units. In case of a continuous distribution the weighting is done with the help of the spectral function.

The question now arises: can the average retardation time, $\langle\tau_V\rangle$, be found in a similar manner? We proceed to show that it can. From Eqs. (4.6-25) and (4.6-23) we have

$$\langle\tau_V\rangle = \chi/[J_e^{\{o\}} - J_g] \ . \tag{4.6-49}$$

For a discrete distribution of retardation times this is equivalent to

$$\langle\tau_V\rangle = \sum_n J_n\tau_n\Big/\sum_n J_n \ . \tag{4.6-50}$$

For a continuous distribution we derive

$$\chi = \begin{cases} -2\max[d\,\chi'(\omega)/d\ln\omega + \{2\phi_f/\omega^2\}] = 2\max[J''(\omega)/\omega - d\,J''(\omega)/d\omega - \{2\phi_f/\omega^2\}] \\ 2\max[J_e^{\{o\}}/\omega - \chi''(\omega)] = 2\max[(J_e^{\{o\}} - J_\bullet(\omega)] \\ 2.718\max[J_e^{\{o\}}t - d\,\chi(t)/d\ln t + \{\phi_f t^2\}] = 2.718\max[J_e^{\{o\}}t - t\,J_\bullet(t) + \{\phi_f t^2\}] \ . \end{cases} \tag{4.6-51}$$

We also have, in analogy to Eqs. (4.6-46),

$$\chi = \begin{cases} \lim\limits_{s\to 0} [J_e^{\{o\}}/s - s\bar{\chi}(s) + \{\phi_f/s^2\}] \\ \lim\limits_{t\to\infty} [J_e^{\{o\}}t - \chi(t) - \{\phi_f t^2/2\}] \\ \lim\limits_{\omega\to 0} [\chi'(\omega) - \{\phi_f/\omega^2\}] \ . \end{cases} \tag{4.6-52}$$

Again, we note that $s\bar{\chi}(s) = \bar{J}(s)$, $\chi(t) = d\,J_\bullet(t)/dt$, and $\chi'(\omega) = J''(\omega)/\omega$. Substitution of the value of χ found in any of these ways into Eq. (4.6-49) yields $\langle\tau_V\rangle$.

We remark that χ is not the sum of the fluidities. That sum is related to the average reciprocal retardation time. By definition, the sum of the fluidities is given by

$$\Phi = \sum_n \phi_n + \{\phi_f\} \ . \tag{4.6-53}$$

Then, by Eq. (4.6-23)

$$\Phi = \langle \tau_V^{-1} \rangle [J_e^{\{o\}} - J_g] + \{\phi_f\} \ . \tag{4.6-54}$$

4.7 Problems

Problem 4.1-1. In Sect. 2.5 we derived the relation

$$G^*(\omega) = \bar{Q}(s)|_{s=j\omega} \tag{1}$$

by separating the contribution to $\bar{Q}(s)/(s - j\omega)$ [see Eq. (2.5-4)] arising from the pole of the driving transform $\varepsilon_0/(s - j\omega)$ from the contributions associated with the poles of the material function $\bar{Q}(s)$. When $\bar{Q}(s)$ is considered to possess a finite or infinite number of simple poles as discussed in Chap. 3 (i.e., when the distribution of relaxation times is discrete; see Sect. 4.5), a partial fraction decomposition leads to the desired separation. This way is not available, however, when $\bar{Q}(s)$ is expressed in terms of a continuous distribution of relaxation times (see Sect. 4.1). In this case $\bar{Q}(s)$ possesses a branch cut along the entire negative real axis of the s-plane.

Show that Eq. (1) can be obtained in this case by a more general derivation which, using Cauchy's Integral Formula and Residue Theorem [13, 14] proceeds through the demonstration that

$$\bar{Q}(j\omega) = \frac{1}{2\pi j} \oint \frac{\bar{Q}(s)}{s - j\omega} \, ds \ , \tag{2}$$

the integral being taken along a contour which includes the pole at $s = j\omega$ but excludes the branch cut along the negative real axis.

Problem 4.1-2. Derive the relation

$$J^*(\omega) = \bar{U}(s)|_{s=j\omega} \tag{1}$$

by showing that

$$\bar{U}(j\omega) = \frac{1}{2\pi j} \oint \frac{\bar{U}(s)}{s - j\omega} \, ds \tag{2}$$

where the contour of integration is the same as that used to obtain $\bar{Q}(j\omega)$ in Problem 4.1-1.

Problem 4.1-3. Using canonical representations show that

$$\bar{Q}(s) = G_g + \bar{\dot{G}}(s) \tag{1}$$

where $\bar{\dot{G}}(s)$ is the Laplace transform of the derivative of G(t).

Problem 4.1-4. Derive the equation for $\eta^*(\omega)$ and its real and imaginary parts in terms of $H(\tau)$.

Problem 4.1-5. Show that $\tau H(\tau)$ can be considered to be a viscosity density on relaxation time. Can $\tau L(\tau)$ then be looked upon as a fluidity density on retardation time? (*Hint*: use $\eta = G\tau$.)

Problem 4.1-6. The canonical representation of the stress response to a constant rate of strain is given by Eq. (4.1-16). Find the representation of the strain response to a constant rate of stress.

Problem 4.1-7. Starting from Eq. (4.1-35),

$$\bar{Q}_{\bullet}(s) = G_g - \int_0^\infty \frac{M(\zeta)}{\zeta + s} d\zeta \ , \tag{1}$$

and using the relation

$$\int_{-\infty}^{\infty} M(\zeta) d \ln \zeta = G_g - \{G_e\} \tag{2}$$

(cf. Problem 4.6-3) show that

$$G(t) = \{G_e\} + \int_{-\infty}^{\infty} M(\zeta) \exp(-\zeta t) d \ln \zeta \tag{3}$$

which is identical with Eq. (4.1-50) since $\zeta^{-1} M(\zeta) = \Xi(\zeta)$.

Problem 4.1-8. In experiments at constant rate of strain it is sometimes convenient to use a *constant strain rate modulus*, $R(t)$, defined as the ratio of the stress, $\sigma(t)$, to the strain, $\varepsilon(t) = \dot{\varepsilon}_0 t$ (cf. Problems 11.3-3 through 11.3-6). Thus

$$R(t) = \frac{\sigma(t)}{\varepsilon(t)} = \frac{\sigma(t)}{\dot{\varepsilon}_0 t} \ . \tag{1}$$

(A) what is $G(t)$ if $R(t)$ is given?

(B) Express the Laplace transform of $R(t)$ in terms of $\bar{Q}(s)$.

(C) Find $\bar{R}(s)$ in terms of $H(\tau)$ and transform the resulting expression into the time domain.

(D) Suggest a simple way of obtaining the last result.

[*Hint*: $\mathscr{L}\, f(t)/t = \int_s^\infty f(x) dx$.]

Problem 4.2-1. In Sect. 4.2 the first order approximation

$$H_1(\tau) = -\left.\frac{d\,G(t)}{d \ln t}\right|_{t=\tau} \tag{1}$$

was derived through Laplace transform inversion. Show that it can also be obtained by approximating the kernel of Eq. (4.1-5), $\exp(-t/\tau)$, by the unit step function, $h(\tau - t)$.

Problem 4.2-2. The kth approximation from the relaxation modulus of the four-parameter Maxwell model is

$$H_{k,4M}(\tau/k) = \mathscr{P}_k[G_{4M}(t)]|_{t=\tau} \quad (1)$$

by Eq. $(4.2\text{-}32)_2$. By Eq. (3.4-161), then,

$$H_{k,4M}(\tau/k) = G_0\mathscr{P}_k[\exp(-t/\tau_0)] + G_1\mathscr{P}_k[\exp(-t/\tau_1)]|_{t=\tau} \quad (2)$$

and, by Eq. (4.2-22),

$$H_{k,4M}(\tau/k) = G_0 I_k(t/\tau_0) + G_1 I_k(t/\tau_1)|_{t=\tau} \quad (3)$$

where $I_k(t/\tau)$ is given by Eq. (4.2-40).

Using the values for the moduli and the relaxation times specified by Eqs. (3.4-96) plot $\log H_{k,4M}(\tau/k)$ as a function of $\log \tau$ for $k = 1, 2, 3$. Also show the true *line* spectrum (cf. Sect. 4.5)

$$H(\tau) = \sum_n G_n\tau_n\delta(\tau - \tau_n) \quad (4)$$

by arrows of height $\log G_n$ located at $\log \tau = \log \tau_n$. What conclusions can you draw from these plots?

Problem 4.2-3. (A) Prove that multi-term intensity functions of the type

$$I_k(x) = \exp(-x)\sum_{n=1}^{n=k} b_n x^n \quad (1)$$

where

$$b_n = \sum_{m=n}^{m=k} (-1)^n S_m^n a_m \quad (2)$$

[cf. Eq. (4.2-36)] are not unique.

(B) Obtain the relation which connects the coefficients b_n through the normalization requirement. Then, illustrate the lack of uniqueness taking $I_2(x)$ as an example.

[*Hint*: the requirement that $I_k(x)$ have a single maximum, implies that $b_n > 0$, $b_1 \ldots b_{n-1} \geq 0$. The only other restriction on the coefficients is that $I_k(x)$ be normalized.]

Problem 4.2-4. Using Eq. (4.2-36) determine the intensity functions for the Schwarzl and Staverman approximations of second and third order.

Problem 4.2-5. What intensity functions of second and third order, other than those corresponding to the Schwarzl and Staverman approximations, can be obtained from Eq. (4.2-49)? How do they compare with the intensity functions given by Eqs. (4.2-38) and (4.2-39)?

Problem 4.2-6. Show that

$$\sum_{m=0}^{m=k} b_m = 0 \quad (1)$$

[cf. Eq. (4.2-89)].

Problem 4.2-7. Prove that

$$\lim_{h\to 1} H_{2h}(\tau/\beta_{2h}) = H_2(\tau/2) \; . \tag{1}$$

Problem 4.3-1. Show that the first order approximation from the storage modulus,

$$H_1'(\omega) = \left.\frac{d\, G'(\omega)}{d \ln \omega}\right|_{\omega=1/\tau} \tag{1}$$

can be obtained by approximating the kernel of Eq. (4.1-7), $\omega^2\tau^2/(1+\omega^2\tau^2)$, by the unit step function, $h(\tau - 1/\omega)$.

Problem 4.3-2. Show that

$$\mathscr{P}_k'\left[\frac{\omega^2\tau^2}{1+\omega^2\tau^2}\right] = -\mathscr{P}_k'\left[\frac{1}{1+\omega^2\tau^2}\right] \tag{1}$$

[cf. Eq. (4.3-5)].

Problem 4.3-3. Obtain the approximations of the lowest three orders to the relaxation spectrum from the storage and loss moduli of the four-parameter Maxwell model using the values given by Eqs. (3.4-96). Compare these approximations with each other and with the approximations from the relaxation modulus (Problem 4.2-2). Calculate only the symmetric approximation where others also exist. Indicate the true line spectra by arrows [cf. Eq. (4) of Problem 4.2-2].

Problem 4.3-4. By Eqs. (4.3-108) and (4.3-109),

$$\sum_{m=0}^{m=k} \frac{b_m' h^{2m}\omega^2\tau^2}{1+h^{2m}\omega^2\tau^2} = -\sum_{m=0}^{m=k} \frac{b_m'}{1+h^{2m}\omega^2\tau^2} \; . \tag{1}$$

Show that this is true.

Problem 4.3-5. Show that

$$\sum_{m=0}^{m=k} b_m'' = 2/\pi \tag{1}$$

[cf. Eq. (4.3-148)].

Problem 4.5-1. Let

$$\bar{Q}_\bullet(s) = G_g - \frac{G_1}{1+\tau_1 s} \; . \tag{1}$$

(A) Find the residues of $\bar{Q}_\bullet(s)$. Obtain $H(\tau)$ and then recover $\bar{Q}_\bullet(s)$.
(B) Find the residues of $\tilde{M}(s)$, obtain $M(\zeta)$ and again recover $\bar{Q}_\bullet(s)$.

Problem 4.5-2. Let

$$\bar{Q}(s) = G_e + \frac{G_1\tau_1 s}{1+\tau_1 s} \; . \tag{1}$$

(A) Find the residues of $\bar{Q}(s)$. Obtain $H(\tau)$ and then recover $\bar{Q}(s)$.
(B) Find the residues of $\tilde{\bar{\Xi}}(s)$, obtain $\Xi(\zeta)$ and again recover $\bar{Q}(s)$.

Problem 4.5-3. In Sect. 3.7 the parameters of the 4-parameter Maxwell and Voigt models were interrelated through the *operator equation*. The same task can be accomplished by the *method of residues* supplemented by the use of the *limit value theorems*.

As an example, given the relaxance of the 4-parameter Maxwell model [cf. Eq. (3.4-25)], find the parameters of the conjugate Voigt model.

Problem 4.5-4. Problems 3.4-7 and 3.4-8 considered the parameters of the series-parallel model which is the equivalent of the non-series parallel model represented by Fig. 3.4-17a. Find the same parameters by the method of residues.

Problem 4.5-5. Given the model represented by Fig. P4.5-5, find the parameters of the equivalent 4-parameter series-parallel model by the method of residues.

Note that this model is identical with that shown in Fig. 3.4-17b but is here redrawn as a ladder model (cf. Chap. 5).

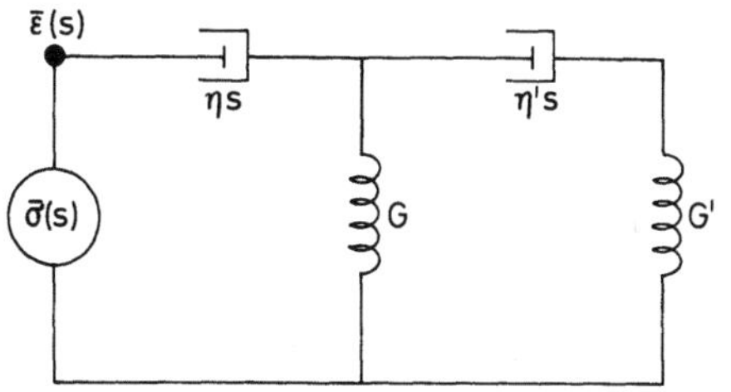

Fig. P4.5-5. 4-parameter ladder model

Problem 4.6-1. Show that

$$\tilde{M}(s) + s\tilde{\bar{\Xi}}(s) = \int_{-\infty}^{\infty} M(\zeta) d \ln \zeta \tag{1}$$

and

$$\tilde{N}(s) + s\tilde{\Lambda}(s) = \int_{-\infty}^{\infty} N(\zeta) d \ln \zeta \; . \tag{2}$$

Problem 4.6-2. Show that

$$\tilde{M}(s) + s\tilde{\bar{\Xi}}(s) = \int_{0}^{\infty} \Xi(\zeta) d\zeta \tag{1}$$

and

$$\tilde{N}(s) + s\tilde{\Lambda}(s) = \int_{0}^{\infty} \Lambda(\zeta) d\zeta \tag{2}$$

by using the formula for the multiplication of the Stieltjes transform by the transform variable, STP (1).

Problem 4.6-3. Show that

$$\int_{-\infty}^{\infty} M(\zeta)d\ln\zeta = \int_{-\infty}^{\infty} \zeta\Xi(\zeta)d\ln\zeta = G_g - \{G_e\} \tag{1}$$

and that

$$\int_{-\infty}^{\infty} N(\zeta)d\ln\zeta = \int_{-\infty}^{\infty} \zeta\Lambda(\zeta)d\ln\zeta = J_e^{\{o\}} - J_g\ . \tag{2}$$

Problem 4.6-4. Derive the following four relations between the real and imaginary components of the Fourier-Laplace transforms of the frequency spectra:

$$\tilde{M}'(\omega) + \omega\tilde{\Xi}''(\omega) = G_g - \{G_e\} \tag{1}$$

$$\tilde{M}''(\omega) - \omega\tilde{\Xi}'(\omega) = 0 \tag{2}$$

$$\tilde{N}'(\omega) + \omega\tilde{\Lambda}''(\omega) = J_e^{\{o\}} - J_g \tag{3}$$

$$\tilde{N}''(\omega) - \omega\tilde{\Lambda}'(\omega) = 0\ . \tag{4}$$

Problem 4.6-5. Verify Eqs. (4.6-28) and (4.6-29) by showing that

$$\int_{-\infty}^{\infty} t^{\alpha}[G(t) - \{G_e\}]d\ln t = \Gamma(\alpha) \int_{-\infty}^{\infty} \tau^{\alpha}H(\tau)d\ln\tau \tag{1}$$

where $\alpha = 0, 1, 2, \ldots$.

(*Hint*: make use of the change of variable $t = k\tau$.)

Problem 4.6-6. Use the relations

$$J'(\omega) = G'(\omega)/\tilde{G}^2(\omega) \tag{1}$$

[cf. Eq. (8.1-6)$_1$] and

$$J_e^{\circ} = \lim_{\omega\to 0} J'(\omega) \tag{2}$$

[cf. Eq. (4.6-6)] to obtain Eq. (4.6-39).

Problem 4.6-7. Express J_e° in terms of the frequency spectra, $M(\zeta)$, and $\Xi(\zeta)$, respectively.

Problem 4.6-8. Derive the steady-flow viscosity, η_f, of the standard linear liquid from Eq. (4.6-40)$_3$.

Problem 4.6-9. Derive the steady-state compliance, J_e°, of the standard linear liquid from Eq. (4.6-41)$_3$.

Problem 4.6-10. Show that the standard deviation of the distribution of relaxation times of a rheodictic material is

$$\sigma_{\tau} = \eta_f\sqrt{(J_e^{\circ} - J_g)J_g}\ . \tag{1}$$

References (Chap. 4)

1. E. Wiechert, Ann. Phys. *50*:335, 546 (1893).
2. W. Kuhn, Z. Angew. Chem. *52*:289 (1939).
3. J.F. Clauser and W.G. Knauss, Trans. Soc. Rheol. *12*:143 (1968).
4. M.R. Spiegel, *Finite Differences* and *Difference Equations*, McGraw-Hill, New York, 1971.
5. F. Schwarzl and A.J. Staverman, Physica, *18*:791 (1952).
6. J.D. Ferry and M.L. Williams, J. Colloid Sci. *7*:347 (1952); (a) J.D. Ferry, *Viscoelastic Properties of Polymers*, 3rd ed., Wiley, New York, 1980.
7. G.A. Korn and T.M. Korn, *Mathematical Handbook for Scientists and Engineers*, 2nd ed., McGraw-Hill, New York, 1968, p. 977.
8. G. Yasuda and K. Ninomiya, Nihon Gomu Kyôkaishi (Proc. Japan Rubber Assoc.) *39*:81 (1966).
9. F. Schwarzl and A.J. Staverman, Appl. Sci. Res. *A4*:127 (1953).
10. N.W. Tschoegl, Rheol, Acta, *10*:582 (1971); *12*:82 (1973).
11. E.R. Fitzgerald, C.E. Grandine, and J.D. Ferry, J. Appl. Phys. *24*:650 (1953).
12. K. Ninomiya and J.D. Ferry, J. Colloid Sci. *14*:36 (1959).
13. S. Goldman, *Transformation Calculus* and *Electrical Transients*, Constable, London, 1949.
14. T.M. Apostol, *Mathematical Analysis*, 2nd ed., Addison-Wesley, Reading, Mass., 1974.

5. Representation of Linear Viscoelastic Behavior by Ladder Models

And he dreamed, and behold a ladder set up on the earth, and the top of it reached to heaven.

Genesis 88:12

5.0 Introduction

In Sect. 3.5 of we discussed the representation of viscoelastic behavior by *series-parallel models*. Because of the orderly arrangement of the elements in these models their respondences are easily determined even if the number of elements is large. Another orderly arrangement is available in the so-called *ladder models.**

5.1 General Ladder Models

A ladder model with 2N + 1 elements is shown in Fig. 5.1-1. To distinguish it from the *converse ladder model* to be introduced later, we shall call it the *obverse ladder model.*

As drawn in Fig. 5.1-1, the model permits one to visualize the motion at the nodes most easily. Thus, merely from inspection, one may deduce that

$$J_g = J_0 \tag{5.1-1}$$

and

$$\eta_f = \sum_{n=1}^{n=N} \eta_n \,. \tag{5.1-2}$$

Since it is not possible to traverse the model through dashpots only, it is a *standard model*, i.e. possesses a finite, non-zero glass compliance and glass modulus (cf. Sect. 3.5.3). It will be convenient as well as instructive to redraw it in the equivalent form shown in Fig. 5.1-2. This form displays the ladder nature of the model.

In the Wiechert model (cf. Figs. 3.5-1 and 3.5-2) in which Maxwell units are assembled in parallel, the elements were denoted as relaxances because relaxances

---

* The theory of ladder models has analogies in the theory of electric filters.

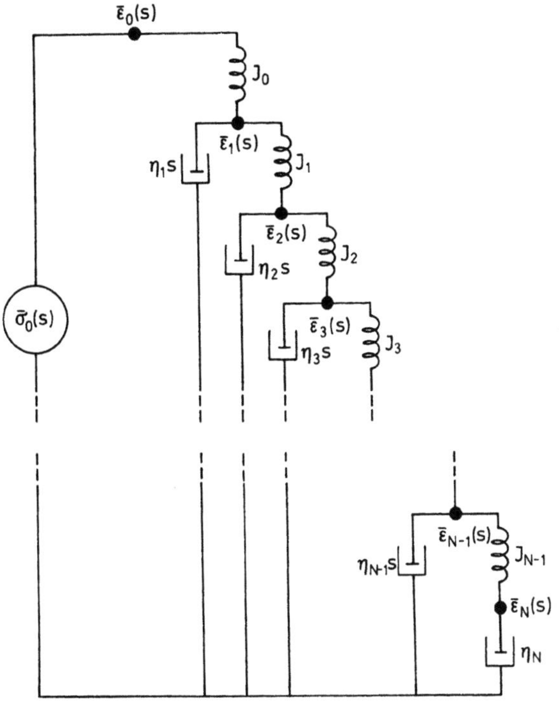

Fig. 5.1-1. Obverse standard ladder model

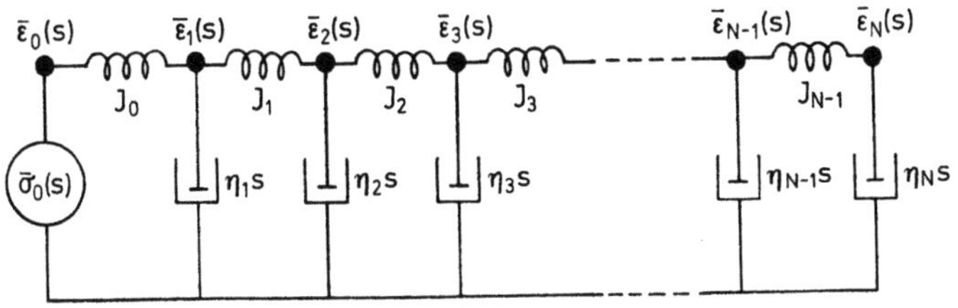

Fig. 5.1-2. Obverse standard rheodictic ladder model

add in parallel. Similarly, in the Kelvin model (cf. Figs. 3.5-3 and 3.5-4) in which Voigt units are combined in series, the elements were denoted as retardances because retardances add in series. In the ladder models, for similar reasons, the elements on the rungs of the ladder will be displayed as relaxances and the elements on the struts as retardances.

The obverse standard ladder model representing arrheodictic behavior is drawn in Fig. 5.1-3. In this model the Nth dashpot has been replaced by a rigid connection. As a result of this rigid connection, the Nth node now coincides with the ground node. In other words, there is no motion at that node and, therefore, $\phi_f = 0$. We note that the model may be traversed through springs only. This reveals its arrheodictic nature.

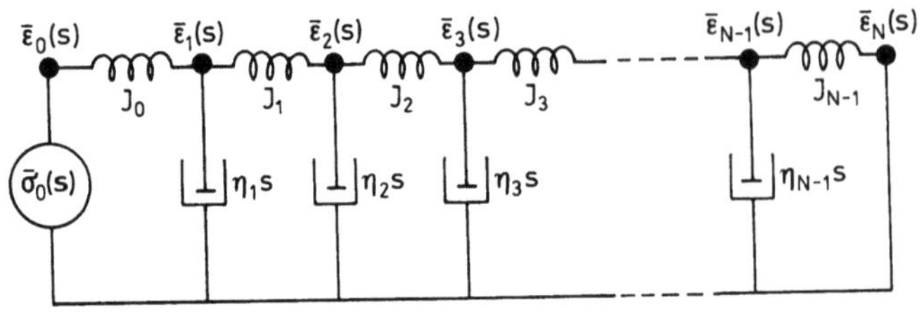

Fig. 5.1-3. Obverse standard arrheodictic ladder model

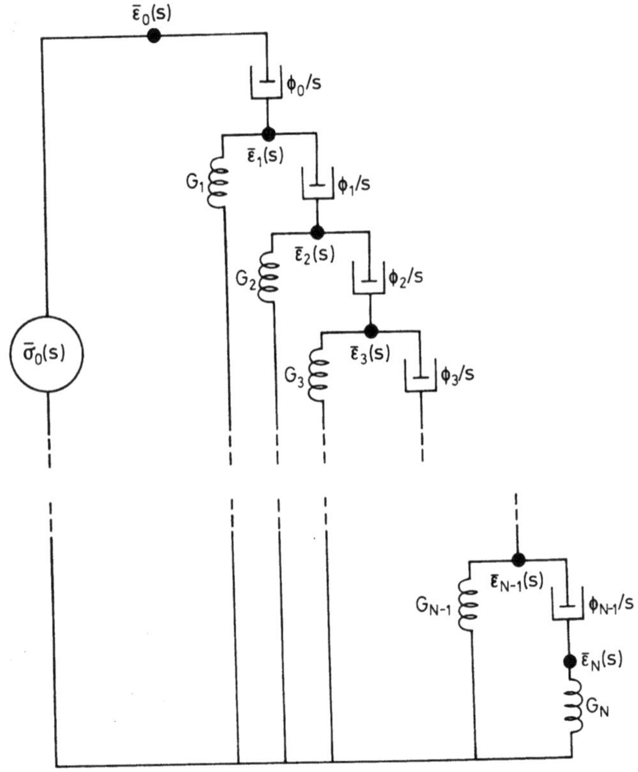

Fig. 5.1-4. Converse standard ladder model

An alternative general ladder model is shown in Fig. 5.1-4. We shall call this model, which is derived from the model of Fig. 5.1-1 by interchanging springs and dashpots, the *converse ladder model*. The converse ladder model shown in Fig. 5.1-4 is also a *standard model* since it cannot be traversed through dashpots only. Just as the obverse ladder model of Fig. 5.1-1, the converse model also allows for rheodictic behavior. Redrawn, it appears in the form of Fig. 5.1-5.

The arrheodictic form is displayed in Fig. 5.1-6. To obtain this form, the zeroth dashpot has now been replaced by a rigid connection. In this form the zeroth and first nodes coincide.

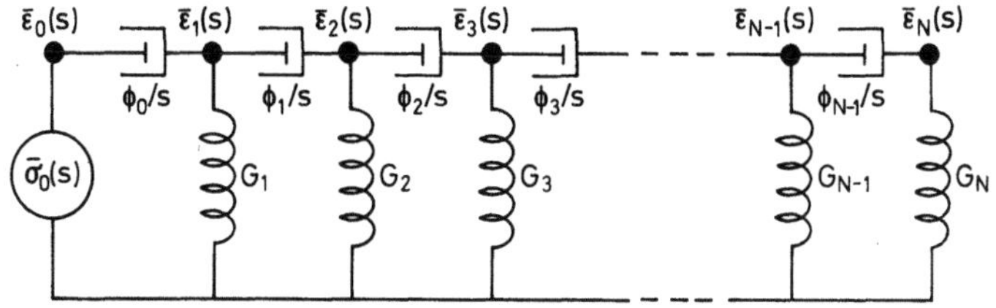

Fig. 5.1-5. Converse standard rheodictic ladder model

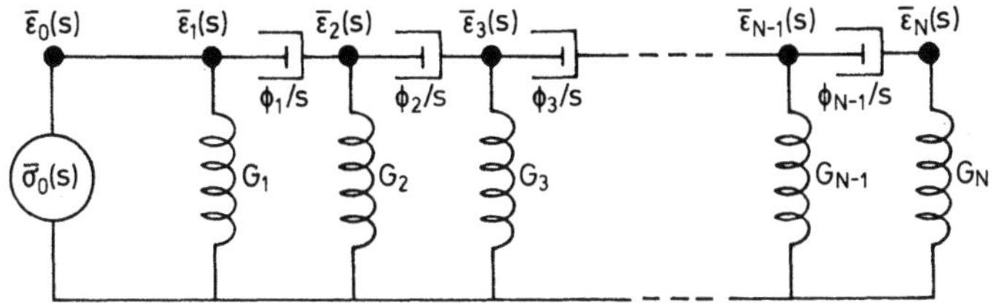

Fig. 5.1-6. Converse standard arrheodictic ladder model

Inspection of Figs. 5.1-4 or 5.1-5 shows that

$$\phi_f = \phi_0 \tag{5.1-3}$$

and

$$G_g = \sum_{n=1}^{n=N} G_n \ . \tag{5.1-4}$$

These relations may be compared with Eqs. (5.1-1) and (5.1-2).

To complete the pattern, four more models could be drawn in which the springs J_0 in Figs. 5.1-2 and 5.1-3 would be replaced by a rigid connection or an additional dashpot would be added at the ends of the models in Figs. 5.1-5 and 5.1-6. These models would be *non-standard models*, i.e. they would have an infinite instantaneous modulus and a zero instantaneous compliance and would, therefore, not represent physically realistic behavior.

By redrawing the four 3-parameter series-parallel models shown in Figs. 3.4-1, 3.4-4, 3.4-11 and 3.4-13, in the equivalent forms displayed in Fig. 5.1-7, one recognizes that these simple models may be regarded alternatively as series-parallel or as ladder models. Models (a) and (b) are the two standard 3-parameter models while (c) and (d) represent the two non-standard ones.

The four series-parallel 4-parameter models (Figs. 3.4-5, 3.4-6, 3.4-9, and 3.4-10) cannot be redrawn as ladder models. However, the 4-parameter models displayed in Fig. 3.4-17 can be redrawn as shown in Fig. 5.1-8 in which the letters correspond to those in Fig. 3.4-17. It turns out that of the eight 4-parameter models one half are series-parallel and the other half are ladder models. With more elements than four there will be series-parallel, ladder, and other models.

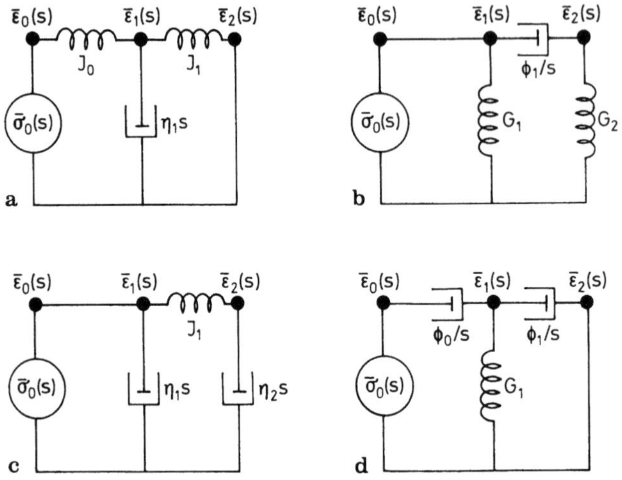

Fig. 5.1-7a-d. The four series-parallel 3-parameter models redrawn as ladder models

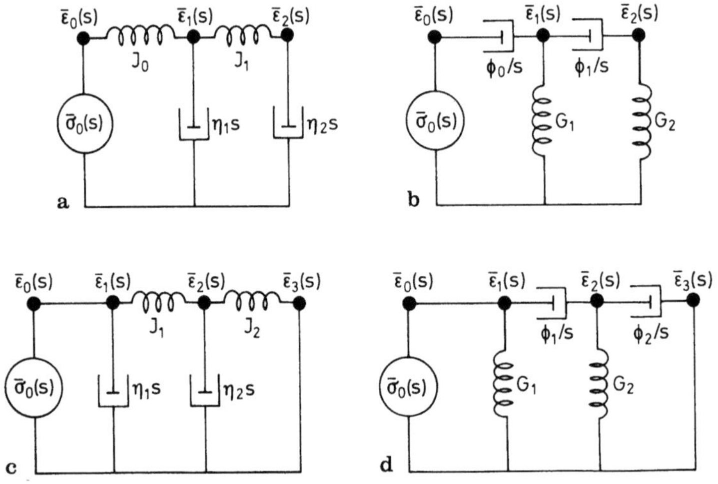

Fig. 5.1-8a-d. The four non-series-parallel 4-parameter models redrawn as general ladder models

In principle, ladder models can be analyzed in the same way as all models by writing down the node equations for each node, and solving the resulting set of simultaneous equations (see Problem 5.1-1). However, with a large number of nodes the analysis becomes excessively complicated as it does with all except the series-parallel models. The advantage that the orderly arrangement of elements in the ladder models have, comes to the fore only if all springs and all dashpots are identical. If, in the obverse models shown in Figs. 5.1-2 and 5.1-3 we let

$$J_0 = J_1 = J_2 = \cdots = J_{N-1} = J \tag{5.1-5}$$

and

$$\eta_1 = \eta_2 = \eta_3 = \cdots = \eta_{N-1}\{= \eta_N\} = \eta \tag{5.1-6}$$

then, as we shall see in the next section, solution of the node equations is greatly simplified. Similarly, we may set

$$G_1 = G_2 = \cdots = G_N = G \tag{5.1-7}$$

and

$$\{\phi_0 =\}\phi_1 = \phi_2 = \cdots = \phi_{N-1} = \phi \tag{5.1-8}$$

in the converse models shown in Figs. 5.1-5 and 5.1-6. Ladder models in which all springs have the same compliances or moduli, and all dashpots have the same viscosities or fluidities, will be called *regular ladder models.* The two kinds of standard regular ladder models composed of a finite number of elements will be discussed in Sects. 5.2 and 5.3. It will be shown there that these models are equivalent to *regular series-parallel models,* i.e. to series-parallel models in which a definite (regular) relationship exists between the values of the compliances, moduli, viscosities, or fluidities of the springs and dashpots, respectively.

Because they embody *distributions* of respondance times, the regular ladder models represent viscoelastic behavior more realistically than do the standard linear solid and the standard linear liquid models. In addition, ladder models with an infinite number of elements form a convenient basis for the theory of the *viscoelastic transmission line.* This important concept will be discussed in Sect. 5.6. Problem 5.5-6 discusses instances of the application of ladder model theory to the analysis of the *bead-and-spring model* of viscoelastic behavior.

5.2 Regular Ladder Models with a Finite Number of Elements: The Gross-Marvin Models

The regular ladder model corresponding to that shown in Fig. 5.1-3, representing arrheodictic behavior, was introduced in a slightly different form (see Problem 5.2-1) by Gross [1]. The rheodictic form was proposed by Marvin [2]. Accordingly, we shall call the obverse standard regular ladder models which have springs along the strut, and dashpots on the rungs of the ladder, the *Gross-Marvin models.* Where necessary, we shall use the subscript GM to distinguish these models from others.

The analysis of any regular ladder model takes advantage of the fact that such a model can be subdivided into an *initial section,* a *propagation section,* and a *terminal* or *final section.* The propagation section consists of a number of identical repeating sections. It is this feature of the regular ladder models which facilitates their analysis. The repeating section can be drawn in either of two equivalent ways. If the section is chosen in such a manner that its elements are arranged in the form of a "T", the section is called a T-section. In the alternative representation the elements are arranged in the form of a "Π", and the section is then called a Π-section. The choice of the repeating section as either a T- or Π-section determines the form of the initial and final sections also. Because it leads to simpler forms for these, the

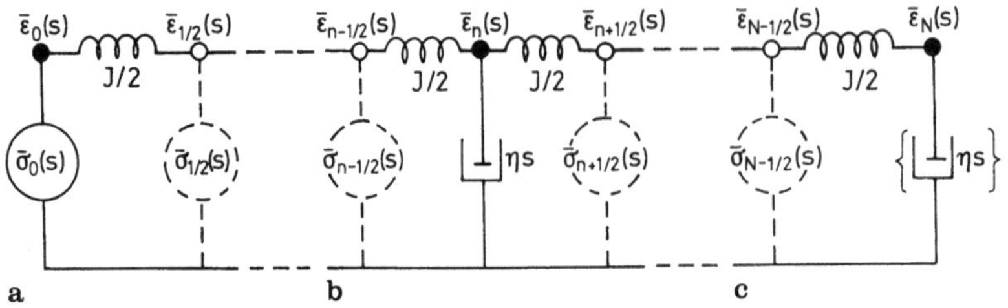

Fig. 5.2-1a-c. Representation of the Gross-Marvin model by T-sections

analysis of the Gross-Marvin model is slightly easier by T-section representation of the repeating section. The Π-section analysis of the model is left as an exercise (Problem 5.2-2).

The three sections of the Gross-Marvin models are shown in Fig. 5.2-1. Only one of the $N-1$ identical T-sections is displayed. The stress output of the initial section becomes the stress input to the first repeating unit of the propagation section, i.e. the first T-section. The stress output of the latter becomes the input to the next one and so forth, until eventually, the output of the last T-section becomes the input to the final section. Thus, letting $N = 2$, we have $n = N - 1 = 1$, and we recover the models shown in Figs. 5.1-7a and 5.1-8a, depending on whether the dashpot surrounded by braces is replaced by a rigid connection or not.

The nodes denoted by open circles are fictitious nodes introduced merely for the purpose of analysis. They are needed because in the T-section analysis the elements along the strut of the ladder (i.e. the springs in the case discussed) are halved. It is for this reason that we assign half-integral values to these nodes. For the input node of the first T-section $n = 1$ and the node thus coincides with the output node of the initial section. Similarly, since there are $N - 1$ T-sections, for the last node of the propagation section $n = N - 1$ and the output node of the last T-section coincides with the input node of the final section.

5.2.1 The Respondances of the Gross-Marvin Models

To obtain the relaxance or retardance of the models, we deal separately with the three sections. The relaxance of the final section is $Q_F(s) = \bar{\sigma}_{N-(1/2)}(s)/\bar{\varepsilon}_{N-(1/2)}(s) = 1/(J/2 + \{1/\eta s\})$ and this becomes

$$Q_F^{\times}(s) = 2/J \tag{5.2-1}$$

for the arrheodictic, and

$$Q_F^{\circ}(s) = \frac{\eta s}{1 + \theta s/2} \tag{5.2-2}$$

for the rheodictic form of the model. In Eq. (5.2-2) we have introduced the symbol θ for the product $J\eta$ of the two material parameters of the regular obverse ladder model. We call θ the *characteristic time* of the model because, as we shall see in Sect. 5.4, a change in θ shifts the response of the models along the logarithmic time-scale.

For the final section the relaxance is all we shall need. The other two sections must be handled in a different manner. For these sections we write down the node equations as discussed in Sect. 3.1.2, solve these for the input stress and the input strain, respectively, and express the relation between these and the output stress and strain in matrix form.

For the initial section we have

$$\bar{\sigma}_0(s) = (2/J)\bar{\varepsilon}_0(s) - (2/J)\bar{\varepsilon}_{1/2}(s) \tag{5.2-3}$$

$$0 = -(2/J)\bar{\varepsilon}_0(s) + (2/J)\bar{\varepsilon}_{1/2}(s) + \bar{\sigma}_{1/2}(s) \ . \tag{5.2-4}$$

At the (fictitious) output node there is no input stress. The output stress appears on the right hand side of Eq. (5.2-4). Solving for $\bar{\sigma}_0(s)$ and $\bar{\varepsilon}_0(s)$ in terms of $\bar{\sigma}_{1/2}(s)$ and $\bar{\varepsilon}_{1/2}(s)$ and writing the result in matrix form yields

$$\begin{bmatrix} \bar{\sigma}_0(s) \\ \bar{\varepsilon}_0(s) \end{bmatrix} = \begin{bmatrix} 1 & 0 \\ J/2 & 1 \end{bmatrix} \begin{bmatrix} \bar{\sigma}_{1/2}(s) \\ \bar{\varepsilon}_{1/2}(s) \end{bmatrix} \ . \tag{5.2-5}$$

The node equations for the repeat section the (T-section) become

$$\bar{\sigma}_{n-(1/2)}(s) = (2/J)\bar{\varepsilon}_{n-(1/2)}(s) - (2/J)\bar{\varepsilon}_n(s) \tag{5.2-6}$$

$$0 = -(2/J)\bar{\varepsilon}_{n-(1/2)}(s) + (4/J + \eta s)\bar{\varepsilon}_n(s) - (2/J)\bar{\varepsilon}_{n+(1/2)}(s) \tag{5.2-7}$$

$$0 = -(2/J)\bar{\varepsilon}_n(s) + (2/J)\bar{\varepsilon}_{n+(1/2)}(s) + \bar{\sigma}_{n+(1/2)}(s) \ . \tag{5.2-8}$$

We use Eq. (5.2-7) to eliminate $\bar{\varepsilon}_n(s)$ from Eqs. (5.2-6) and (5.2-8) and then solve for $\bar{\sigma}_{n-(1/2)}(s)$ and $\bar{\varepsilon}_{n-(1/2)}(s)$ in terms of $\bar{\sigma}_{n+(1/2)}(s)$ and $\bar{\varepsilon}_{n+(1/2)}(s)$. Writing the result in matrix form again we obtain

$$\begin{bmatrix} \bar{\sigma}_{n-(1/2)}(s) \\ \bar{\varepsilon}_{n-(1/2)}(s) \end{bmatrix} = \begin{bmatrix} 1 + \theta s/2 & \eta s \\ J(1 + \theta s/4) & 1 + \theta s/2 \end{bmatrix} \begin{bmatrix} \bar{\sigma}_{n+(1/2)}(s) \\ \bar{\varepsilon}_{n+(1/2)}(s) \end{bmatrix} \tag{5.2-9}$$

where we made use again of the characteristic time of the model. Since the propagation section consists of $N - 1$ T-sections, the matrix equation for the complete propagation section becomes

$$\begin{bmatrix} \bar{\sigma}_{1/2}(s) \\ \bar{\varepsilon}_{1/2}(s) \end{bmatrix} = \begin{bmatrix} 1 + \theta s/2 & \eta s \\ J(1 + \theta s/4) & 1 + \theta s/2 \end{bmatrix}^{N-1} \begin{bmatrix} \bar{\sigma}_{N-(1/2)}(s) \\ \bar{\varepsilon}_{N-(1/2)}(s) \end{bmatrix} \ . \tag{5.2-10}$$

For the final section we write

$$\begin{bmatrix} \bar{\sigma}_{N-(1/2)}(s) \\ \bar{\varepsilon}_{N-(1/2)}(s) \end{bmatrix} = \begin{bmatrix} Q_F(s) \\ 1 \end{bmatrix} \bar{\varepsilon}_{N-(1/2)}(s) \tag{5.2-11}$$

and, combining all three sections, we have

$$\begin{bmatrix} \bar{\sigma}_0(s) \\ \bar{\varepsilon}_0(s) \end{bmatrix} = \begin{bmatrix} 1 & 0 \\ J/2 & 1 \end{bmatrix} \begin{bmatrix} 1 + \theta s/2 & \eta s \\ J(1 + \theta s/4) & 1 + \theta s/2 \end{bmatrix}^{N-1} \begin{bmatrix} Q_F(s) \\ 1 \end{bmatrix} \bar{\varepsilon}_{N-(1/2)}(s) \ . \tag{5.2-12}$$

We must now raise the matrix of the T-section to the $(N - 1)^{st}$ power. This is accomplished by diagonalizing the matrix and is done in detail in Problems 5.2-12 and 5.2-13. Here it suffices to point out that the eigenvalues of the matrix are exp[a(s)] and exp[−a(s)] where

$$a(s) = \ln(1 + \theta s/2 + \sqrt{\theta s(4 + \theta s)}/2) \; . \tag{5.2-13}$$

The function a(s) will be called the *propagation function*. Since exp[a(s)] = sinh a(s) + cosh a(s), we have

$$1 + \theta s/2 = \cosh a(s) \tag{5.2-14}$$

and

$$\sqrt{\theta s(4 + \theta s)}/2 = \sinh a(s) \; . \tag{5.2-15}$$

We further introduce

$$Q_T(s) = 2\eta s/\sqrt{\theta s(4 + \theta s)} \tag{5.2-16}$$

as the *characteristic relaxance*, and its reciprocal, $U_T(s)$, as the *characteristic retardance* of the T-section. These respondances arise naturally when the matrix is raised to a given power (see Problem 5.2-13). It is shown in Problem 5.2-8 that, if the propagation matrix is terminated in its characteristic respondance, the input respondance of the matrix is the characteristic respondance itself. If the analysis is carried out in terms of Π-sections (see Problem 5.2-2), the same is true of $Q_\Pi(s)$, the characteristic respondance of the Π-section.

With the help of Eqs. (5.2-14), (5.2-15), and (5.2-16) we find

$$\eta s = Q_T(s)\sinh a(s) \tag{5.2-17}$$

and

$$J(1 + \theta s/4) = U_T(s)\sinh a(s) \; . \tag{5.2-18}$$

These equations allow us to express the matrix of the propagation section in terms of the propagation function and the characteristic respondances. We obtain

$$\begin{bmatrix} 1 + \theta s/2 & \eta s \\ J(1 + \theta s/4) & 1 + \theta s/2 \end{bmatrix} = \begin{bmatrix} \cosh a(s) & Q_T(s)\sinh a(s) \\ U_T(s)\sinh a(s) & \cosh a(s) \end{bmatrix} \tag{5.2-19}$$

and it can be seen that a(s) governs the way in which the stress and strain transforms are transmitted from one node of the ladder to the next. It is now easily shown (by mathematical induction, see Problem 5.2-3) that

$$\begin{bmatrix} 1 + \theta s/2 & \eta s \\ J(1 + \theta s/4) & 1 + \theta s/2 \end{bmatrix}^{N-1} = \begin{bmatrix} \cosh[(N-1)a(s)] & Q_T(s)\sinh[(N-1)a(s)] \\ U_T(s)\sinh[(N-1)a(s)] & \cosh[(N-1)a(s)] \end{bmatrix} . \tag{5.2-20}$$

By virtue of this last equation, Eq. (5.2-12) becomes

$$\begin{bmatrix} \bar{\sigma}_0(s) \\ \bar{\varepsilon}_0(s) \end{bmatrix} = \begin{bmatrix} 1 & 0 \\ J/2 & 1 \end{bmatrix} \begin{bmatrix} \cosh[(N-1)a(s)] & Q_T(s)\sinh[(N-1)a(s)] \\ U_T(s)\sinh[(N-1)a(s)] & \cosh[(N-1)a(s)] \end{bmatrix} \times \begin{bmatrix} Q_F(s) \\ 1 \end{bmatrix} \bar{\varepsilon}_{N-(1/2)}(s) \; . \tag{5.2-21}$$

Matrix multiplication and the use of $\bar{Q}_{GM}(s) = \bar{\sigma}_0(s)/\bar{\varepsilon}_0(s)$ finally leads to

$$\bar{Q}_{GM}(s) = \frac{Q_F(s)\cosh[(N-1)a(s)] + Q_T(s)[\sinh(N-1)a(s)]}{[JQ_F(s)/2+1]\cosh[(N-1)a(s)] + [Q_F(s)U_T(s) + JQ_T(s)/2]\sinh[(N-1)a(s)]} . \tag{5.2-22}$$

We must now specialize Eq. (5.2-22) for the rheodictic and for the arrheodictic case. To enable us to do this we note that

$$\cosh a(s)/2 = \sqrt{4+\theta s}/2 \quad \text{and} \quad \sinh a(s)/2 = \sqrt{\theta s}/2 . \tag{5.2-23}$$

We then also have

$$Q_T(s)\cosh a(s)/2 = \eta s/\sqrt{\theta s} = \sqrt{\theta s}/J \tag{5.2-24}$$

and

$$J\,Q_T(s)/2 = \tanh a(s)/2 . \tag{5.2-25}$$

Now, for rheodictic behavior

$$Q_F^{\circ}(s) = Q_T(s)\tanh a(s) \tag{5.2-26}$$

by Eqs. (5.2-2), (5.2-17), and (5.2-14). Substituting this into Eq. (5.2-22) and using Eqs. (5.2-25) and (5.2-24) then leads to

$$\bar{Q}_{GM}^{\circ}(s) = \frac{\sqrt{\theta s}\,\sinh N\,a(s)}{J\cosh[(N+\frac{1}{2})a(s)]} \tag{5.2-27}$$

or

$$\bar{Q}_{GM}^{\circ}(s) = (\eta s/2)\{\sqrt{1+4/\theta s}\,\tanh[(N+\tfrac{1}{2})a(s)] - 1\} \tag{5.2-28}$$

and

$$\bar{U}_{GM}^{\circ}(s) = \frac{J\cosh[(N+\frac{1}{2})a(s)]}{\sqrt{\theta s}\,\sinh N\,a(s)} \tag{5.2-29}$$

or

$$\bar{U}_{GM}^{\circ}(s) = (J/2)[\sqrt{1+4/\theta s}\,\coth N\,a(s) + 1] \tag{5.2-30}$$

where we have used the superscript $^{\circ}$ to indicate rheodictic behavior as this is not obvious from the form of the equations. Equations (5.2-28) and (5.2-30) are obtained with the help of Eqs. (5.2-23). We note that the same result would have been obtained if we had made the final section a rigid connection (which would have an infinite relaxance) and had increased the number of nodes of the propagation section by one (cf. Problem 5.2-1).

For arrheodictic behavior

$$Q_F^{\times}(s) = Q_T(s)\coth a(s)/2 \tag{5.2-31}$$

by Eqs. (5.2-1) and (5.2-25). Substitution into Eq. (5.2-22) eventually gives

$$\bar{Q}^{\times}_{GM}(s) = \frac{\sqrt{\theta s}\cosh[(N - \frac{1}{2})a(s)]}{J \sinh N\, a(s)} \tag{5.2-32}$$

or

$$\bar{Q}^{\times}_{GM}(s) = (\eta s/2)(\sqrt{1 + 4/\theta s}\coth N\, a(s) - 1) \tag{5.2-33}$$

and

$$\bar{U}^{\times}_{GM}(s) = \frac{J \sinh N\, a(s)}{\sqrt{\theta s}\cosh[(N - \frac{1}{2})a(s)]} \tag{5.2-34}$$

or

$$\bar{U}^{\times}_{GM}(s) = (J/2)\{\sqrt{1 + 4/\theta s}\tanh[(N - \tfrac{1}{2})a(s)] + 1\} \tag{5.2-35}$$

where the superscript $^{\times}$ indicates arrheodictic behavior.

For large N, $(N \pm \frac{1}{2}) \rightarrow N$ is a permissible approximation in Eqs. (5.2-28) and (5.2-35) but not in Eqs. (5.2-27), (5.2-29), (5.2-32), and (5.2-34). In the latter, this substitution produces equations which evince non-standard behavior as demonstrated in Problem 5.2-4.

Equations (5.2-27) through (5.2-30) and (5.2-32) through (5.2-35) give the respondances of the Gross-Marvin models. In principle the step and slope responses could now be found from these equations through division by s and s^2 followed by inversion of the transforms (see Problem 5.2-5), and the harmonic responses by replacement of s by $j\omega$. However, the inverse transforms are not available and separation of the real and imaginary parts of $\bar{Q}_{GM}(j\omega)$ turns out to be forbiddingly complicated. We therefore turn to the *method of residues* discussed in Sect. 4.5. There it was shown that any linear combination of springs and dashpots can always be replaced by an equivalent series-parallel model which shows the same responses as the original. Once this is available, the experimental response functions can be found easily in the already established manner. We therefore proceed now to find the series-parallel models which are equivalent to the Gross-Marvin ladder models.

5.2.2 The Equivalent Series-Parallel Models

The procedure consists in determining the poles and associated residues of the ladder models, together with the necessary limit values which will become the viscoelastic constants of the equivalent series-parallel models. We begin with the latter.

5.2.2.1 Limit Values

The respondances of the equivalent series-parallel models require the equilibrium modulus for the relaxance, and the instantaneous compliance as well as the steady-state fluidity for the retardance.

We first turn our attention to the relaxances. The equilibrium modulus is obtained from Eq. (3.4-115). Using the relation $\eta s = \theta s/J$, we rewrite Eq. (5.2-33) in the form

$$\bar{Q}^{\times}_{GM}(s) = \frac{\sqrt{\theta s(1 + \theta s/4)}}{J \tanh N\, a(s)} - \frac{\eta s}{2} \tag{5.2-36}$$

and obtain

$$G_e = \lim_{s \to 0} \bar{Q}^{\times}_{GM}(s) = \frac{1}{J} \lim_{s \to 0} \frac{\sqrt{\theta s}}{\tanh N\, a(s)} \ . \tag{5.2-37}$$

The limit is indeterminate. Writing $\sqrt{\theta s} = x$ and using Eq. $(5.2\text{-}23)_2$ we find

$$\lim_{x \to 0} \frac{x}{\tanh[2N \sinh^{-1}(x/2)]} = \frac{1}{N} \tag{5.2-38}$$

by applying L'Hospital's theorem (see Problem 5.2-6).

It follows that

$$G_e = 1/JN \tag{5.2-39}$$

for the arrheodictic model. For the rheodictic model G_e must, of course, vanish. This follows at once by applying the final value theorem to Eq. (5.2-27) because $a(0) = 0$.

Turning now to the retardances, we note that we have already deduced, from inspection, that $J_g = J_0$ for the obverse ladder model in both its rheodictic and arrheodictic form. By Eq. (5.1-5) the instantaneous compliance of the Gross-Marvin models becomes

$$J_g = J \ . \tag{5.2-40}$$

Equations (5.1-2) and (5.1-6) immediately furnish the relation

$$\phi_f = 1/\eta N \tag{5.2-41}$$

for the steady-flow fluidity. These equations are easily verified. Applying Eq. (3.4-107) to Eqs. (5.2-30) and (5.2-35), both equations yield

$$J_g = \lim_{s \to \infty} \bar{U}_{GM}(s) = (J/2) \lim_{s \to \infty} (\sqrt{1 + 4/\theta s} + 1) = J \tag{5.2-42}$$

because $a(s) \to \infty$ as $s \to \infty$ and $\tanh(\infty) = \coth(\infty) = 1$. The steady-state fluidity is obtained from Eq. (3.4-122). From Eq. (5.2-30)

$$s\bar{U}^{\circ}_{GM}(s) = (1/2\eta)[\sqrt{\theta s(4 + \theta s)} \coth N\, a(s) + 1] \tag{5.2-43}$$

since $Js = \theta s/\eta$. We thus have

$$\phi_f = \lim_{s \to 0} s\bar{U}_{GM}(s) = (1/\eta) \lim_{s \to 0} \frac{\sqrt{\theta s}}{\tanh N\, a(s)} \tag{5.2-44}$$

and Eq. (5.2-41) follows by Eqs. (5.2-37) and (5.2-38).

5.2.2.2 *Poles and Residues*

The next step consists in determining the poles and associated residues. The only singularities which any model consisting of linear combinations of springs and dashpots can possess are simple poles along the negative real axis of the transform plane. However, the occurrence of the factors $\sqrt{\theta s}$ and $\sqrt{1 + 4/\theta s}$ in the equations for the respondances of the Gross-Marvin models suggest the possible existence of branch points at $s = 0$, $s = -4/\theta$, and $s = \infty$. The demonstration that these points are *not* branch points is left as an exercise (see Problem 5.2-7).

The residues associated with the poles, s_n, of the respondances are found from the formulae [cf. Eqs. (4.5-17) and (4.5-31)]

$$\mathrm{Res}_n(\bar{Q}) = \left.\frac{\bar{q}_{GM}(s)}{\bar{u}'_{GM}(s)}\right|_{s=s_n} \quad \text{and} \quad \mathrm{Res}_n(\bar{U}) = \left.\frac{\bar{u}_{GM}(s)}{\bar{q}'_{GM}(s)}\right|_{s=s_n} \tag{5.2-45}$$

where $\bar{u}_{GM}(s)$ and $\bar{q}_{GM}(s)$ are the numerator and denominator of $\bar{U}_{GM}(s)$, and the prime denotes differentiation with respect to s.

We begin by finding the poles of $\bar{U}^{\times}_{GM}(s)$. Since the behavior is arrheodictic $s = 0$ is a regular point. Equation (5.2-34) shows that, therefore, $\bar{U}^{\times}_{GM}(s)$ will have poles for those values of s for which the equation $\cosh[(N - \frac{1}{2})a(s)] = 0$ is satisfied, i.e. for

$$a(s_n) = j\pi(2n - 1)/(2N - 1)\ , \qquad n = 0, 1, 2, \ldots\ . \tag{5.2-46}$$

Substitution of Eq. (5.2-46) into Eq. $(5.2\text{-}23)_2$ leads to

$$s_n = -\frac{4}{\theta}\sin^2\frac{\pi(2n-1)}{2(2N-1)}\ , \qquad n = 1, 2, \ldots, N - 1 \tag{5.2-47}$$

as the equation for the poles. The index ranges from 1 to $N - 1$ because (see Problem 5.2-9) the retardance is regular at $s_N = -4/\theta$, $s_0 = s_1$, and $n > N$ does not generate new poles. Thus, the poles are simple poles along the negative real axis of the complex plane in the interval $-4\theta < s_n < 0$.

To find the residues associated with these poles we rewrite Eq. (5.2-35) as

$$\bar{U}^{\times}_{GM}(s) = \frac{J\sqrt{\theta s}\cosh[(N - \frac{1}{2})a(s)] + J\sqrt{4 + \theta s}\sinh[(N - \frac{1}{2})a(s)]}{2\sqrt{\theta s}\cosh[(N - \frac{1}{2})a(s)]}\ . \tag{5.2-48}$$

We then find

$$\bar{q}^{\times\prime}_{GM}(s_n) = \left.\frac{\theta\cosh[(N - \frac{1}{2})a(s)]}{\sqrt{\theta s}} + \frac{(2N-1)\theta\sinh[(N - \frac{1}{2})a(s)]}{\sqrt{4 + \theta s}}\right|_{s=s_n} \tag{5.2-49}$$

where we have used the derivative

$$\frac{d\,a(s)}{ds} = \frac{\theta}{\sqrt{\theta s(4 + \theta s)}}\ . \tag{5.2-50}$$

But the residues must be evaluated for $s = s_n$ and the s_n satisfy $\cosh[(N - \frac{1}{2})a(s)] = 0$. By Eq. $(5.2\text{-}45)_2$ then

$$\text{Res}_n[\bar{U}^{\times}_{GM}(s)] = \frac{J(4+\theta s_n)}{(2N-1)\theta}, \qquad n = 1, 2, \ldots, N-1 . \tag{5.2-51}$$

Using Eq. (5.2-47) we finally obtain the equation for the residues as

$$\text{Res}_n(\bar{U}^{\times}_{GM}) = \frac{4J}{(2N-1)\theta}\cos^2\frac{\pi(2n-1)}{2(2N-1)}, \qquad n = 1, 2, \ldots, N-1 . \tag{5.2-52}$$

The relaxance is handled similarly to the retardance. Equation (5.2-32) shows that $\bar{Q}^{\times}_{GM}(s)$ will have poles for those values of s for which sinh Na(s) = 0. This will be satisfied for

$$a(s_n) = j\pi n/N, \qquad n = 0, 1, \ldots . \tag{5.2-53}$$

Equation $(5.2\text{-}20)_2$ then gives

$$s_n = -\frac{4}{\theta}\sin^2\frac{\pi n}{2N}, \qquad n = 1, 2, \ldots, N-1 \tag{5.2-54}$$

as the equation for the poles. In Eq. (5.2-54) n ranges from 1 to $N-1$ because $n > N$ does not furnish new poles and the relaxance is regular at the origin and at $s = -4/\theta$. The first is obvious from Eq. (5.2-37) and the second is also readily shown along the lines of Problem 5.2-9. Thus the poles again lie in the interval $-4/\theta < s_n < 0$.

To find the residues we rewrite Eq. (5.2-33) as

$$\bar{Q}^{\times}_{GM}(s) = \frac{\eta s\sqrt{4+\theta s}\cosh N\,a(s) - \eta s\sqrt{\theta s}\sinh N\,a(s)}{2\sqrt{\theta s}\sinh N\,a(s)} . \tag{5.2-55}$$

Hence

$$\bar{u}^{\times\prime}_{GM}(s_n) = \left.\frac{\theta\sinh N\,a(s)}{\sqrt{\theta s}} + \frac{2N\theta\cosh N\,a(s)}{\sqrt{4+\theta s}}\right|_{s=s_n} . \tag{5.2-56}$$

Since the s_n satisfy sinh N a(s) = 0, the residues become

$$\text{Res}_n(\bar{Q}^{\times}_{GM}) = \frac{\eta s(4+\theta s_n)}{2N\theta}, \qquad n = 1, 2, \ldots, N-1 . \tag{5.2-57}$$

From Eq. (5.2-54) we obtain the equation for the residues as

$$\text{Res}_n(\bar{Q}^{\times}_{GM}) = -\frac{2}{JN\theta}\sin^2\frac{\pi n}{N}, \qquad n = 1, 2, \ldots, N-1 . \tag{5.2-58}$$

We now turn to the model allowing for steady-state flow. A comparison of Eqs. (5.2-29) and (5.2-32) shows that the poles of $\bar{U}^{\circ}_{GM}(s)$ are identical with those of $\bar{Q}^{\times}_{GM}(s)$, except that there is now also a pole at $s = 0$ because $\bar{U}^{\circ}_{GM}(s)$ is not regular at the origin. This follows directly from Eq. (5.2-29) and the fact that $a(0) = 0$. We then have

$$s_n = -\frac{4}{\theta}\sin^2\frac{\pi n}{2N}, \qquad n = 1, 2, \ldots, N-1 \tag{5.2-59}$$

as the equation for the poles which are seen to lie in the interval $-4/\theta < s_n \leq 0$.

To obtain the residues, we rewrite Eq. (5.2-60) as

$$\bar{U}^{\circ}_{GM}(s) = \frac{J\sqrt{\theta s}\sinh N\,a(s) + J\sqrt{4+\theta s}\cosh N\,a(s)}{2\sqrt{\theta s}\sinh N\,a(s)} \;. \tag{5.2-60}$$

and find

$$\bar{q}^{\circ\prime}_{GM}(s_n) = \left.\frac{\theta\sinh N\,a(s)}{\sqrt{\theta s}} + \frac{2N\theta\cosh N\,a(s)}{\sqrt{4+\theta s}}\right|_{s=s_n} . \tag{5.2-61}$$

But $s_0 = 0$, and, hence, $\bar{q}^{\circ\prime}_{GM}(s_0)$ is indeterminate because of the first term on the right of Eq. (5.2-61). Applying L'Hospital's theorem we obtain

$$\bar{q}^{\circ\prime}_{GM}(s_0) = \frac{4N\theta\cosh N\,a(s_0)}{\sqrt{4+\theta s_0}} \tag{5.2-62}$$

and thus

$$\mathrm{Res}_0(\bar{U}^{\circ}_{GM}) = \frac{J(4+\theta s_0)}{4N\theta} = \frac{J}{N\theta} = \frac{1}{\eta N} \;. \tag{5.2-63}$$

This residue is associated with the steady-flow fluidity.

For the other residues we find

$$\mathrm{Res}_n(\bar{U}^{\circ}_{GM}) = \frac{J(4+\theta s_n)}{2N\theta} \;, \qquad n = 1, 2, \ldots, N-1 \tag{5.2-64}$$

because now the first term in Eq. (5.2-61) is zero. This gives

$$\mathrm{Res}_n(\bar{U}^{\circ}_{GM}) = \frac{2J}{N\theta}\cos^2\frac{\pi n}{2N} \;, \qquad n = 1, 2, \ldots, N-1 \;. \tag{5.2-65}$$

Equations (5.2-63) and (5.2-65) together are the equations for the residues.

Let us now consider $\bar{Q}^{\circ}_{GM}(s)$. From Eq. (5.2-27) we see that the poles result from $\cosh[(N+\frac{1}{2})a(s)] = 0$ which is satisfied for

$$a(s_n) = j\pi(2n+1)/(2N+1) \;, \qquad n = 0, 1, \ldots \;. \tag{5.2-66}$$

Using Eq. $(5.2\text{-}23)_2$ leads to

$$s_n = -\frac{4}{\theta}\sin^2\frac{\pi(2n+1)}{2(2N+1)} \;, \qquad n = 0, 1, \ldots, N-1 \tag{5.2-67}$$

as the equation for the poles because the relaxance is regular at the origin and at $s = -4/\theta$.

To obtain the residues, we express Eq. (5.2-28) in the form

$$\bar{Q}^{\circ}_{GM}(s) = \frac{\eta s\sqrt{4+\theta s}\sinh[(N+\frac{1}{2})a(s)] - \eta s\sqrt{\theta s}\cosh[(N+\frac{1}{2})a(s)]}{2\sqrt{\theta s}\cosh[(N+\frac{1}{2})a(s)]} \;. \tag{5.2-68}$$

Therefore

$$\bar{u}^{\circ\prime}_{GM}(s_n) = \left.\frac{\theta\cosh[(N+\frac{1}{2})a(s)]}{\sqrt{\theta s}} + \frac{(2N+1)\theta\sinh[(N+\frac{1}{2})a(s)]}{\sqrt{4+\theta s}}\right|_{s=s_n} . \tag{5.2-69}$$

Since $s_0 \neq 0$, the first term in Eq. (5.2-69) vanishes and the residues are obtained from

$$\mathrm{Res}_n(\bar{Q}^{\circ}_{GM}) = \frac{\eta s_n(4 + \theta s_n)}{(2N+1)\theta} , \qquad n = 0, 1, \ldots, N-1 \tag{5.2-70}$$

which leads to

$$\mathrm{Res}_n(\bar{Q}^{\circ}_{GM}) = -\frac{4}{J(2N+1)\theta} \sin^2 \frac{\pi(2n+1)}{2N+1} , \quad n = 0, 1, \ldots, N-1 . \tag{5.2-71}$$

We are now in possession of the limit values as well as the poles and residues of all four of the respondances of the Gross-Marvin models and are ready, therefore, to establish the respondances of the corresponding series–parallel models.

5.2.2.3 *Respondances*

The equivalent series-parallel models which will show the same response as the Gross-Marvin ladder models, can be given either in the Maxwell or in the Voigt representation, i.e. either as Wiechert or as Kelvin models, according to whether the relaxance or the retardance is required.

Let us consider the Kelvin models first. By Eq. (3.5-13) the retardance of these models will be

$$\bar{U}_{GM}(s) = J_g + \sum_n \frac{J_n}{1 + \tau_n s} + \{\phi_f/s\} . \tag{5.2-72}$$

Our task is to express the parameters of the series-parallel models in terms of the parameters J and η (or θ) of the ladder models. Now, by Eq. (5.2-40) the instantaneous compliance, J_g, is J. The retardation times are obtained as $\tau_n = -1/s_n$ according to Eq. (4.5-37), and the compliances as $J_n = \tau_n \mathrm{Res}_n(\bar{U}_{GM})$ according to Eq. $(4.5\text{-}38)_2$. For the arrheodictic model we find

$$\tau_n = \frac{\theta}{4} \csc^2 \frac{\pi(2n-1)}{2(2N-1)} , \qquad n = 1, 2, \ldots, N-1 \tag{5.2-73}$$

by Eq. (5.2-47), and

$$J_n = \frac{J}{2N-1} \cot^2 \frac{\pi(2n-1)}{2(2N-1)} , \qquad n = 1, 2, \ldots, N-1 \tag{5.2-74}$$

by Eq. (5.2-52), so that

$$\bar{U}_{GM}(s) = J + \sum_{n=1}^{N-1} \frac{J_n}{1 + \tau_n s} . \tag{5.2-75}$$

We have omitted the superscript on $\bar{U}_{GM}(s)$ in Eq. (5.2-75) because it is clear from the absence of the flow term that $\bar{U}_{GM}(s)$ is the retardance of a model displaying arrheodictic behavior.

Let us now determine J_e. Applying the final value theorem, Eq. (3.4-118), to Eq. (5.2-75) gives

$$J_e = \lim_{s\to 0} \bar{U}_{GM}(s) = J + \sum_n J_n = J + \frac{J}{2N-1}\sum_{n=1}^{N-1} \cot^2\frac{\pi(2n-1)}{2(2N-1)} \ . \tag{5.2-76}$$

But the sum equals $(N-1)(2N-1)$ and, therefore,

$$J_e = JN \ . \tag{5.2-77}$$

Thus, in the absence of steady-state flow, $N = J_e/J_g$. It therefore represents the ratio of the equilibrium to the instantaneous compliance.

We have now established the result that the series-parallel model which has the same retardance as the arrheodictic Gross-Marvin model is a Kelvin model with N springs and $N-1$ dashpots in which the isolated spring has the compliance $J_g = J$, the others have the compliances given by Eq. (5.2-74), and the dashpots have fluidities given by

$$\phi_n = \frac{4}{(2N-1)\eta}\cos^2\frac{\pi(2n-1)}{2(2N-1)}\ , \qquad n = 1, 2, \ldots, N-1 \ . \tag{5.2-78}$$

The Gross-Marvin model also contains N springs and $N-1$ dashpots but in the ladder model all springs have the same modulus and all dashpots have the same viscosity. We recall that the expression for $\bar{U}^{\times}_{GM}(s)$, Eq. (5.2-34), is equivalent to a set of N simultaneous node equations (cf. Fig. 5.1-3). These become "uncoupled" by the transformation from $\bar{U}^{\times}_{GM}(s)$ to $\bar{U}_{GM}(s)$. This transformation is thus seen to be equivalent to a transformation to normal coordinates [1].

For the rheodictic model we find

$$\tau_n = \frac{\theta}{4}\csc^2\frac{\pi n}{2N}\ , \qquad n = 1, 2, \ldots, N-1 \tag{5.2-79}$$

for the retardation times,

$$J_n = \frac{J}{2N}\cot^2\frac{\pi n}{2N}\ , \qquad n = 1, 2, \ldots, N-1 \tag{5.2-80}$$

for the compliances, and, by Eq. (5.2-41),

$$\phi_f = 1/\eta N \ . \tag{5.2-81}$$

for the steady-state fluidity. Thus, Eq. (5.2-72) now becomes

$$\bar{U}_{GM}(s) = J + \sum_{n=1}^{N-1}\frac{J_n}{1+\tau_n s} + \frac{1}{\eta N s}\ . \tag{5.2-82}$$

We recognize $\bar{U}_{GM}(s)$ as the retardance of a Kelvin model with 2N elements in which the isolated spring has the compliance $J_g = J$, the other $N-1$ springs have the compliances given by Eq. (5.2-80), the isolated dashpot has the fluidity given by Eq. (5.2-81) and the other dashpots have fluidities

$$\phi_n = \frac{2}{\eta N}\cos^2\frac{\pi n}{2N}\ . \tag{5.2-83}$$

It is instructive to inquire into the nature of the steady-state compliance, J_e°. By the final value theorem, [see Eq. (3.4-118)], we have

$$J_e^o = \lim_{s\to 0}\,[\bar{U}_{GM}(s) - \phi_f/s] = J + \sum_n J_n = J + \frac{J}{2N}\sum_{n=1}^{N-1} \cot^2\frac{\pi n}{2N} \ . \tag{5.2-84}$$

Now, the sum is $(2N - 1)(N - 1)/3$. Hence,

$$J_e^o = J[2N^2 + 3N + 1]/6N \ . \tag{5.2-85}$$

Let us now turn to the representation of the Gross-Marvin models by Wiechert models. By Eq. (3.5-4) the relaxance of these models will be

$$\bar{Q}_{GM}(s) = \{G_e\} + \sum_n \frac{G_n \tau_n s}{1 + \tau_n s} \ . \tag{5.2-86}$$

For the arrheodictic model $G_e = 1/JN$ by Eq. (5.2-39). We now have $G_n = -\tau_n\,\mathrm{Res}_n(\bar{Q}_{GM})$ by Eq. $(4.5\text{-}38)_1$. By Eq. (5.2-54) the relaxation times become

$$\tau_n = \frac{\theta}{4}\csc^2\frac{\pi n}{2N}\ , \qquad n = 1, 2, \ldots, N - 1 \tag{5.2-87}$$

and the moduli result as

$$G_n = \frac{2}{JN}\cos^2\frac{\pi n}{2N}\ , \qquad n = 1, 2, \ldots, N - 1 \tag{5.2-88}$$

from Eqs. (5.2-57) and (5.2-54). Thus,

$$\bar{Q}_{GM}(s) = \frac{1}{JN} + \sum_{n=1}^{N-1} \frac{G_n \tau_n s}{1 + \tau_n s} \ . \tag{5.2-89}$$

To obtain the meaning of N again, we now determine G_g. Application of the initial value theorem, Eq. (3.4-112), to Eq. (5.2-86) yields

$$G_g = \lim_{s\to\infty} \bar{Q}_{GM}(s) = \frac{1}{JN} + \sum_n G_n = \frac{1}{JN} + \frac{2}{JN}\sum_{n=1}^{N-1}\cos^2\frac{\pi n}{2N} = \frac{1}{J} \tag{5.2-90}$$

since the sum equals $(N - 1)/2$. Thus, as indeed expected from internal consistency and the reciprocity between G_g and J_g, and between G_e and J_e, N is again a scaling factor such that $N = G_g/G_e$.

Equation (5.2-89) represents the relaxance of a Wiechert model with N springs and $N - 1$ dashpots in which the isolated spring has the modulus $G_e = 1/JN$, the others have the moduli given by Eq. (5.2-88), and the dashpots have viscosities which are given by

$$\eta_n = G_n \tau_n = \frac{\eta}{2N}\cot^2\frac{\pi n}{2N}\ , \qquad n = 1, 2, \ldots, N - 1 \ . \tag{5.2-91}$$

For the rheodictic model $G_e = 0$ and we find

$$\tau_n = \frac{\theta}{4}\csc^2\frac{\pi(2n + 1)}{2(2N + 1)}\ , \qquad n = 0, 1, \ldots, N - 1 \tag{5.2-92}$$

for the relaxation times, and

$$G_n = \frac{4}{J(2N+1)} \cos^2 \frac{\pi(2n+1)}{2(2N+1)}, \qquad n = 0, 1, \ldots, N-1 \tag{5.2-93}$$

for the moduli. Hence,

$$\bar{Q}_{GM}(s) = \sum_{n=0}^{N-1} \frac{G_n \tau_n s}{1 + \tau_n s} . \tag{5.2-94}$$

Note that the lower limit of the sum is zero here while it was unity for all the other respondances. Equation (5.2-94) is the relaxance of a Wiechert model with 2N elements in which the springs have the moduli given by Eq. (5.2-93) and the dashpots have viscosities given by

$$\eta_n = \frac{\eta}{2N+1} \cot^2 \frac{\pi(2n+1)}{2(2N+1)}, \qquad n = 0, 1, \ldots, N-1 . \tag{5.2-95}$$

The instantaneous modulus is obtained as

$$G_g = \sum_n G_n = \frac{4}{J(2N+1)} \sum_{n=0}^{N-1} \cos^2 \frac{\pi(2n+1)}{2(2N+1)} = 1/J \tag{5.2-96}$$

because the sum is $(2N + 1)/4$.

By Eq. (5.2-41), we expect the steady flow viscosity to be given by $\eta_f = \eta N$. As a check, we apply Eq. (3.4-121). This gives

$$\eta_f = \lim_{s \to 0} \bar{U}^{\circ}_{GM}(s)/s = \sum_n G_n \tau_n = \frac{\eta}{2N+1} \sum_{n=0}^{N-1} \cot^2 \frac{\pi(2n+1)}{2(2N+1)} = \eta N \tag{5.2-97}$$

since the sum equals $N(2N + 1)$.

5.3 Regular Ladder Models with a Finite Number of Elements: The Regular Converse Ladder Models

The (standard) regular converse ladder models are obtained from Figs. 5.1-5 and 5.1-6 by letting all dashpots and all springs assume, respectively, the same fluidities and moduli, as specified by Eqs. (5.1-7) and (5.1-8). It may be noted that the rheodictic obverse and converse model can be obtained from each other by an interchange of springs and dashpots. This interchange would, however, turn the standard arrheodictic obverse model into a non-standard rheodictic converse model, since that model could be traversed through dashpots only, and could not be traversed through springs alone.

We shall obtain the respondances of both forms of the regular converse model by Π-section analysis. Only one of the N − 1 identical Π-sections is shown in Fig. 5.3-1. As with the regular obverse ladder models (i.e. the Gross-Marvin models), we represent the elements on the strut of the ladder as retardances and those on the

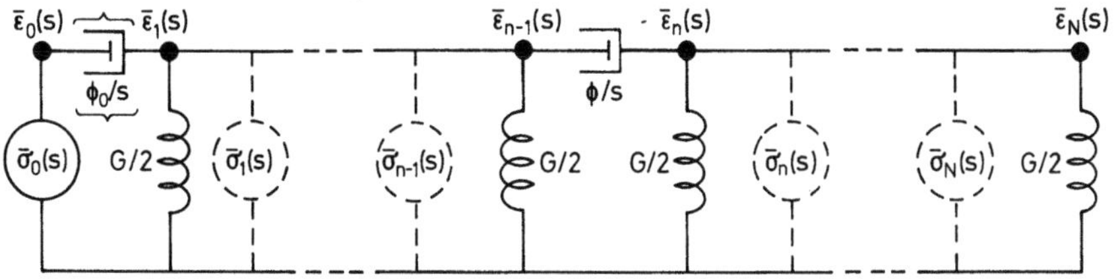

Fig. 5.3-1. Representation of the regular converse ladder model by Π-sections

rungs as relaxances. To distinguish them from other models, the subscript RC will be used for the regular converse ladder models.

5.3.1 The Respondances of the Regular Converse Ladder Models

In these models the final section is the same, whether the rheodictic or arrheodictic form is used. The relaxance of this section is

$$\hat{Q}_F(s) = \bar{\sigma}_N(s)/\bar{\varepsilon}_N(s) = G/2 \ . \tag{5.3-1}$$

The difference in the two forms of the model now appears in the initial section. The node equations for the rheodictic form are

$$\bar{\sigma}_0(s) = (s/\phi)\bar{\varepsilon}_0(s) - (s/\phi)\bar{\varepsilon}_1(s) \tag{5.3-2}$$

$$0 = -(s/\phi)\bar{\varepsilon}_0(s) + (G/2 + s/\phi)\bar{\varepsilon}_1(s) + \bar{\sigma}_1(s) \tag{5.3-3}$$

and the matrix equation becomes

$$\begin{bmatrix} \bar{\sigma}_0(s) \\ \bar{\varepsilon}_0(s) \end{bmatrix} = \begin{bmatrix} 1 & G/2 \\ \phi/s & 1 + 1/2\hat{\theta}s \end{bmatrix} \begin{bmatrix} \bar{\sigma}_1(s) \\ \bar{\varepsilon}_1(s) \end{bmatrix} \tag{5.3-4}$$

where $\hat{\theta} = 1/G\phi$ is the *characteristic time* of the regular converse model. For the arrheodictic form we have

$$\bar{\varepsilon}_0(s) = \bar{\varepsilon}_1(s) \tag{5.3-5}$$

$$\bar{\sigma}_0(s) = (G/2)\bar{\varepsilon}_1(s) + \bar{\sigma}_1(s) \tag{5.3-6}$$

and

$$\begin{bmatrix} \bar{\sigma}_0(s) \\ \bar{\varepsilon}_0(s) \end{bmatrix} = \begin{bmatrix} 1 & G/2 \\ 0 & 1 \end{bmatrix} \begin{bmatrix} \bar{\sigma}_1(s) \\ \bar{\varepsilon}_1(s) \end{bmatrix} \ . \tag{5.3-7}$$

The node equations for the Π-section are obtained as

$$\bar{\sigma}_{n-1}(s) = (G/2 + s/\phi)\bar{\varepsilon}_{n-1}(s) - (s/\phi)\bar{\varepsilon}_n(s) \tag{5.3-8}$$

$$0 = -(s/\phi)\bar{\varepsilon}_{n-1}(s) + (G/2 + s/\phi)\bar{\varepsilon}_n(s) + \bar{\sigma}_n(s) \tag{5.3-9}$$

and the matrix equation of the propagation section follows as

$$\begin{bmatrix} \bar{\sigma}_{n-1}(s) \\ \bar{\varepsilon}_{n-1}(s) \end{bmatrix} = \begin{bmatrix} 1 + 1/2\hat{\theta}s & G(1 + 1/4\hat{\theta}s) \\ \phi/s & 1 + 1/2\hat{\theta}s \end{bmatrix} \begin{bmatrix} \bar{\sigma}_N(s) \\ \bar{\varepsilon}_N(s) \end{bmatrix} . \tag{5.3-10}$$

The propagation section contains N − 1 identical Π-sections. Combination with the initial section and using Eq. (5.3-1) then gives

$$\begin{bmatrix} \bar{\sigma}_0(s) \\ \bar{\varepsilon}_0(s) \end{bmatrix} = \begin{bmatrix} 1 & G/2 \\ \phi/s & 1 + 1/2\hat{\theta}s \end{bmatrix} \begin{bmatrix} 1 + 1/2\hat{\theta}s & G(1 + 1/4\hat{\theta}s) \\ \phi/s & 1 + 1/2\hat{\theta}s \end{bmatrix}^{N-1} \begin{bmatrix} G/2 \\ 1 \end{bmatrix} \bar{\varepsilon}_N(s) . \tag{5.3-11}$$

We now introduce the *propagation function*

$$\hat{a}(s) = \ln(1 + 1/2\hat{\theta}s + \sqrt{1 + 4\hat{\theta}s}/2\hat{\theta}s) \tag{5.3-12}$$

and the *characteristic relaxance* of the Π-section.

$$\hat{Q}_\Pi(s) = G\sqrt{1 + 4\hat{\theta}s}/2 , \tag{5.3-13}$$

with its reciprocal, the *characteristic retardance*, $\hat{U}_\Pi(s)$. We then have

$$1 + 1/2\hat{\theta}s = \cosh \hat{a}(s) \tag{5.3-14}$$

and

$$\sqrt{1 + 4\hat{\theta}s}/2\hat{\theta}s = \sinh \hat{a}(s) . \tag{5.3-15}$$

With the help of these relations we find

$$\phi/s = \hat{U}_\Pi(s) \sinh \hat{a}(s) \tag{5.3-16}$$

and

$$G(1 + 1/4\hat{\theta}s) = \hat{Q}_\Pi(s) \sinh \hat{a}(s) \tag{5.3-17}$$

and the propagation matrix becomes

$$\begin{bmatrix} 1 + 1/2\hat{\theta}s & G(1 + 1/4\hat{\theta}s) \\ \phi/s & 1 + 1/2\hat{\theta}s \end{bmatrix}^{N-1} = \begin{bmatrix} \cosh[(N-1)\hat{a}(s)] & \hat{Q}_\Pi(s) \sinh[(N-1)\hat{a}(s)] \\ \hat{U}_\Pi(s) \sinh[(N-1)\hat{a}(s)] & \cosh[(N-1)\hat{a}(s)] \end{bmatrix} . \tag{5.3-18}$$

The complete matrix equation for the arrheodictic model therefore is

$$\begin{bmatrix} \bar{\sigma}_0(s) \\ \bar{\varepsilon}_0(s) \end{bmatrix} = \begin{bmatrix} 1 & G/2 \\ 0 & 1 \end{bmatrix} \begin{bmatrix} \cosh[(N-1)\hat{a}(s)] & \hat{Q}_\Pi(s) \sinh[(N-1)\hat{a}(s)] \\ \hat{U}_\Pi(s) \sinh[(N-1)\hat{a}(s)] & \cosh[(N-1)\hat{a}(s)] \end{bmatrix} \begin{bmatrix} G/2 \\ 1 \end{bmatrix} \bar{\varepsilon}_N(s) . \tag{5.3-19}$$

We note that

$$\cosh \hat{a}(s)/2 = \sqrt{4 + 1/\hat{\theta}s}/2 \quad \text{and} \quad \sinh \hat{a}(s)/2 = 1/2\sqrt{\theta s} , \tag{5.3-20}$$

and that

$$\hat{Q}_\Pi(s)/\cosh \hat{a}(s)/2 = G\sqrt{\hat{\theta}s} = s/\phi\sqrt{\hat{\theta}s} \tag{5.3-21}$$

$$G/2 = \hat{Q}_\Pi(s) \tanh \hat{a}(s)/2 \tag{5.3-22}$$

and

$$G\,\hat{U}_{\Pi}(s)/2 = \tanh \hat{a}(s)/2 \ . \tag{5.3-23}$$

With these equations we obtain

$$\bar{Q}^{\times}_{RC}(s) = \frac{G\sqrt{\hat{\theta}s}\,\sinh N\,\hat{a}(s)}{\cosh[(N-\frac{1}{2})\hat{a}(s)]} \tag{5.3-24}$$

or

$$\bar{Q}^{\times}_{RC}(s) = (G/2)\{\sqrt{1+4\hat{\theta}s}\,\tanh[(N-\tfrac{1}{2})\hat{a}(s)] + 1\} \tag{5.3-25}$$

for the relaxance, and

$$\bar{U}^{\times}_{RC}(s) = \frac{\cosh[(N-\frac{1}{2})\hat{a}(s)]}{G\sqrt{\hat{\theta}s}\,\sinh N\,\hat{a}(s)} \tag{5.3-26}$$

or

$$\bar{U}^{\times}_{RC}(s) = (\phi/2s)[\sqrt{1+4\hat{\theta}s}\,\coth N\,\hat{a}(s) - 1] \tag{5.3-27}$$

for the retardance of the arrheodictic regular converse ladder model.

The complete matrix for the rheodictic form becomes

$$\begin{bmatrix}\bar{\sigma}_0(s)\\ \bar{\varepsilon}_0(s)\end{bmatrix} = \begin{bmatrix}1 & G/2\\ \phi/s & \cosh\hat{a}(s)\end{bmatrix} \times \begin{bmatrix}\cosh[(N-1)\hat{a}(s)] & \hat{Q}_{\Pi}(s)\sinh[(N-1)\hat{a}(s)]\\ \hat{U}_{\Pi}(s)\sinh[(N-1)\hat{a}(s)] & \cosh[(N-1)\hat{a}(s)]\end{bmatrix}\begin{bmatrix}G/2\\ 1\end{bmatrix}\bar{\varepsilon}_N(s) \tag{5.3-28}$$

and we have

$$\bar{Q}^{\circ}_{RC}(s) = \frac{G\sqrt{\hat{\theta}s}\,\sinh N\,\hat{a}(s)}{\cosh[(N+1/2)\hat{a}(s)]} \tag{5.3-29}$$

or

$$\bar{Q}^{\circ}_{RC}(s) = (G/2)\{\sqrt{1+4\hat{\theta}s}\,\tanh[(N+\tfrac{1}{2})\hat{a}(s)] - 1\} \tag{5.3-30}$$

and the retardance follows as

$$\bar{U}^{\circ}_{RC}(s) = \frac{\cosh[(N-\frac{1}{2})\hat{a}(s)]}{G\sqrt{\hat{\theta}s}\,\sinh N\,\hat{a}(s)} \tag{5.3-31}$$

or

$$\bar{U}^{\circ}_{RC}(s) = (\phi/2s)[\sqrt{1+4\hat{\theta}s}\,\coth N\,\hat{a}(s) + 1] \ . \tag{5.3-32}$$

Equations (5.3-24) to (5.3-27), and (5.3-29) to (5.3-32) represent, respectively, the respondances of the arrheodictic and the rheodictic forms of the regular converse ladder model.

5.3.2 The Equivalent Series-Parallel Models

The series-parallel models which will evince the same behavior as the standard regular converse ladder models can be derived in almost complete analogy to the procedure employed in Sect. 5.2.2. Therefore, we content ourselves here with simply listing the poles, residues and limit values of these models. From these the equivalent series-parallel forms are readily derived. An example is given in Problem 5.3-1.

The only procedural difference in the derivation of the poles and residues concerns the sequence in which they are ordered. As it results from the method of Sect. 5.2.2.2 this is $s_1 < s_2 < \cdots$. To conform with the inequality (3.5-23) we substitute N -- n for n. This merely reverses the order of the poles without changing their numerical values.

The poles of $\bar{Q}^{\times}_{RC}(s)$ then are

$$s_n = -\frac{1}{4\theta}\sec^2\frac{\pi n}{2N-1}\,, \qquad n = 1, 2, \ldots, N-1 \tag{5.3-33}$$

and the residues become

$$\mathrm{Res}_n(\bar{Q}^{\times}_{RC}) = -\frac{G\tan^2\pi n/(2N-1)}{4(2N-1)\hat{\theta}\cos^2\pi n/(2N-1)}\,, \qquad n = 1, 2, \ldots, N-1\,. \tag{5.3-34}$$

For $\bar{U}^{\times}_{RC}(s)$ we find

$$s_n = -\frac{1}{4\hat{\theta}}\sec^2\frac{\pi n}{2N}\,, \qquad n = 1, 2, \ldots, N-1 \tag{5.3-35}$$

and

$$\mathrm{Res}_n(\bar{U}^{\times}_{RC}) = \frac{1}{2GN\hat{\theta}}\tan^2\frac{\pi n}{2N}\,, \qquad n = 1, 2, \ldots, N-1\,. \tag{5.3-36}$$

The relaxance of the rheodictic model, $\bar{Q}^{\circ}_{RC}(s)$, has the poles

$$s_n = -\frac{1}{4\hat{\theta}}\sec^2\frac{\pi(n+1)}{2N+1}\,, \qquad n = 0, 1, \ldots, N-1 \tag{5.3-37}$$

where we have replaced n by N − 1 − n to reverse the order. The residues become

$$\mathrm{Res}_n(\bar{Q}^{\circ}_{RC}) = -\frac{G\tan^2\pi(n+1)/(2N+1)}{4(2N+1)\hat{\theta}\cos^2\pi(N+1)/(2N+1)}\,, \qquad n = 0, 1, \ldots, N-1\,. \tag{5.3-38}$$

The retardance $\bar{U}^{\circ}_{RC}(s)$ has a pole at the origin associated with the residue

$$\mathrm{Res}_0(\bar{U}^{\circ}_{RC}) = \phi\,. \tag{5.3-39}$$

Its other poles are identical with those of $\bar{U}^{\times}_{RC}(s)$ and are given by Eq. (5.3-35). The residues associated with those poles, however, are

$$\mathrm{Res}_n(\bar{U}^{\circ}_{RC}) = \frac{1}{2GN\hat{\theta}}\tan^2\frac{\pi n}{2N}\,, \qquad n = 1, 2, \ldots, N-1\,. \tag{5.3-40}$$

The relaxation times follow directly from $\tau_n = -1/s_n$, the moduli from $G_n = -\tau_n \mathrm{Res}_n(\bar{Q}_{RC})$, and the viscosities from $\eta_n = G_n\tau_n$. The equilibrium modulus of the arrheodictic model is G. The instantaneous modulus is GN for either model, the sum required for this evaluation being $(2N-1)(N-1)$ for the arrheodictic, and $N(2N+1)$ for the rheodictic model. The steady-flow viscosity of the latter is $1/\phi$, the sum required being $(2N+1)/4$.

The retardation times again follow from $\tau_n = -1/s_n$, the compliances from $J_n = \tau_n \mathrm{Res}_n(\bar{U}_{RC})$, and the fluidities from $\phi_n = J_n/\tau_n$. The instantaneous compliance of both models is 1/GN. The equilibrium and the steady-state compliances are both 1/G and are found using the sum $(N-1)/2$ in either case. The steady-flow fluidity of the rheodictic model is ϕ.

5.4 Comparison of the Obverse and Converse Regular Ladder Models. Model Fitting

Ladder models, particularly the regular standard ladder models, can be useful in modelling viscoelastic behavior. In Sect. 3.4 we learned that the simplest models, the standard linear solid, and the standard linear liquid model, embody all the essential features of viscoelastic behavior. While this is certainly true, these models possess only a single respondance time.* Therefore they cannot represent the behavior of materials which are properly characterized by a distribution of respondance times. The mathematical models to be discussed in Sect. 6.1 are able to emulate the effect of a distribution of respondance times with only a minimal number of parameters, generally two to three, in addition to the viscoelastic constants. However, the stress-strain relations resulting from these equations do not conform with the requirement discussed in Chap. 2 that linear viscoelastic behavior be described by constant-coefficient linear differential equations or, equivalently, by convolution integrals. Consequently, knowledge of the parameters with which any of these mathematical models describes the response to a given excitation does not generally permit one to predict the response to another excitation.

The canonical models require specifications of the distribution of moduli or compliances before they can be used for modelling. This is precisely what the *regular* ladder models provide. We have shown in the previous two sections that these models are equivalent to series-parallel models with specific discrete distributions of moduli or compliances. In terms of the spectral functions these distributions are expressed [cf. Eqs. $(4.5\text{-}21)_2$ and $(4.5\text{-}30)_2$] by

$$H(\tau) = \sum_n G_n\tau_n\delta(\tau - \tau_n) \tag{5.4-1}$$

* The 4-parameter Maxwell model contains two relaxation times. However, these may be considered to represent two independent relaxation mechanisms which, in a more elaborate model, would each be described by a single distribution of relaxation times.

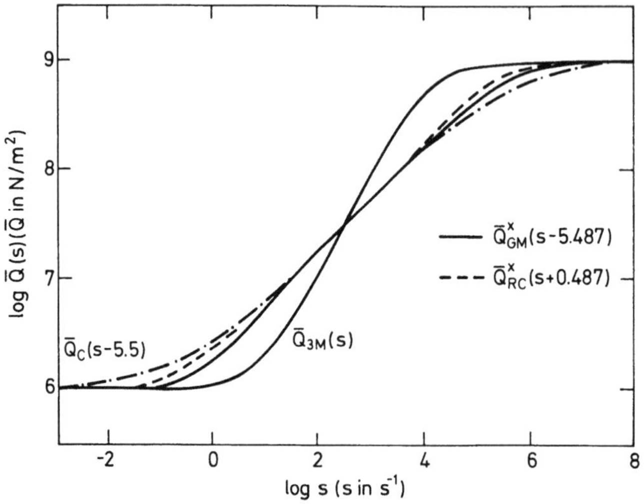

Fig. 5.4-1. $\bar{Q}_{3M}(s)$, $\bar{Q}_C(s)$, $\bar{Q}^{\times}_{GM}(s)$, and $\bar{Q}^{\times}_{RC}(s)$ as functions of s

and

$$L(\tau) = \sum_n J_n \tau_n \delta(\tau - \tau_n) \ . \tag{5.4-2}$$

The moduli, compliances, and respondance times are given in terms of only three parameters J, θ, and N, or G, $\hat{\theta}$, and N. Once either set of parameters is selected, the response to any excitation can be predicted in a manner which is more realistic than that generated from the standard solid or liquid models, and is without the shortcomings of the mathematical models.

As an example, $\bar{Q}^{\times}_{GM}(s)$ and $\bar{Q}^{\times}_{RC}(s)$ are compared in Fig. 5.4-1 with each other, with the linear solid model,

$$\bar{Q}_{GM}(s) = G_g - \frac{G_g - G_e}{1 + \tau s} \tag{5.4-3}$$

and with Eq. (6.1-81), i.e. with

$$\bar{Q}_C(s) = G_g - \frac{G_g - G_e}{1 + \sqrt{s/s_0}} \ , \tag{5.4-4}$$

as an example of one of the mathematical models referred to above. For G_g, G_e, and τ the values given by Eqs. (3.4-93) were used. The value of s_0 was selected so that Eqs. (5.4-3) and (5.4-4) coincide at $\bar{Q}(s) = \sqrt{G_g G_e}$, and is given by $\log s_0 = -5.5$. For the relaxances of the ladder models we have $N = G_g/G_e$, $J = 1/G_g$, and $G = G_e$. The parameter N is 1001. The parameters θ and $\hat{\theta}$ determine the location of the response curves along the log (s)-axis. They were found by plotting $\log \bar{Q}^{\times}_{GM}(s)$ and $\log \bar{Q}^{\times}_{RC}(s)$ against log s with $\theta = \hat{\theta} = 1$ and determining the shift along log s which is required to bring the curves into superposition with $\log \bar{Q}(s)$ at $\bar{Q}(s) = \sqrt{G_g G_e}$. These shift distances are $\log \theta = -5.487$ and $\log \hat{\theta} = 0.487$, respectively.

We observe that the relaxances of the arrheodictic models, although rather similar, are not identical as they would be if the models were conjugate. With both models the transition from the instantaneous to the equilibrium modulus is much less abrupt as it is with $\bar{Q}_{GM}(s)$. This is a reflection of the fact that the relaxances of the ladder models contain a distribution of relaxation times or, rather, moduli. The response of the mathematical model, $\bar{Q}_C(s)$, coincides with the response of the ladder models over most of the transition but is somewhat broader still near the ends. As $s \to 0$ or $s \to \infty$ all four curves coincide because G_e and G_g, respectively, are the same for all four.

In the arrheodictic models we treated θ and $\hat{\theta}$ (or, equivalently, η and ϕ) as adjustable (model fitting) parameters. By contrast, the rheodictic models are fully determined if $J_g = 1/G_g$, and J_e° are given. We then have $N = J_e^\circ/J_g$, $J = J_g$, $\eta = \eta_f/N$, and $G = 1/J_gN$, $\phi = 1/\eta_f$. Consequently, $\theta = J_g\eta_f/N$ and $\hat{\theta} = J_g\eta_f N$. We have $\log\theta = -5.000$ and $\log\hat{\theta} = 1.000$. Figure 5.4-2 compares $\bar{Q}^\circ_{GM}(s)$ and $\bar{Q}^\circ_{RC}(s)$ with

$$\bar{Q}_{4M}(s) = \frac{G_0\tau_0 s}{1+\tau_0 s} + \frac{G_1\tau_1 s}{1+\tau_1 s}\,, \tag{5.4-5}$$

the relaxance of the standard liquid model. The various parameters were given by, or derived from, Eqs. (3.4-96), J_e° being $1.001 \times 10^{-6}\ \mathrm{m^2/N}$.

Again, the responses of the two ladder models is closely similar but not identical. $\bar{Q}_{4M}(s)$ clearly shows the effect of two well separated relaxation times. The ladder models match its behavior at high values of s (because all three relaxances have the same instantaneous modulus, G_g) and at low values of s (because all three models have the same steady-flow viscosity, η_f). The match at intermediate values of s is greatly improved if the two relaxation times of the standard liquid

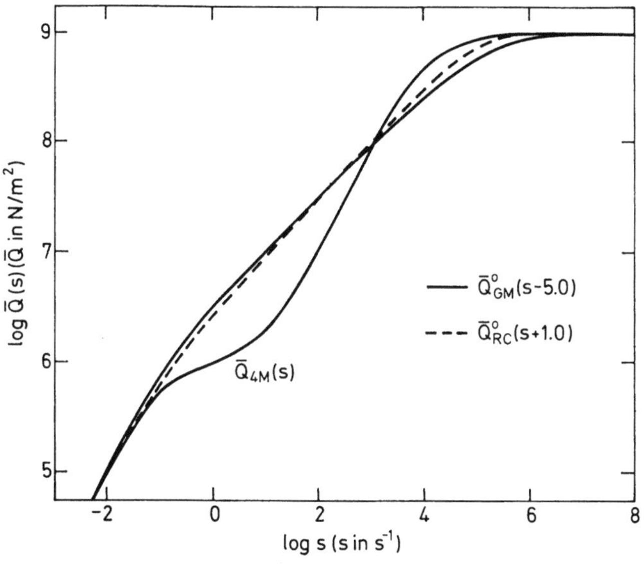

Fig. 5.4-2. $\bar{Q}_{4M}(s)$, $\bar{Q}^\circ_{GM}(s)$, and $\bar{Q}^\circ_{RC}(s)$ as functions of s

model are replaced by two independent distributions. This will be discussed in Chap. 10.

5.5 Regular Ladder Models with an Infinite Number of Elements

In the preceding section we have considered regular ladder models with a finite number of elements. We showed that these are equivalent to the generalized series-parallel models with the same numbers of springs and of dashpots, except that, because of the *regular* nature of the equivalent ladder models, the respondances represented by the elements of the series-parallel models form finite series.

In Sect. 3.5.1.3 we extended the standard generalized series-parallel models to the *canonical models* containing infinite numbers of elements. In this section we shall discuss ladder models with an infinite number of elements (or, equivalently, nodes). For brevity's sake we shall refer to such ladder models as *extended* models. In Sect. 5.5.1 we examine the consequences of increasing the number of nodes of the Gross-Marvin ladder without limit. One consequence of this is that the extended Gross-Marvin ladder is a non-standard model, i.e. its instantaneous compliance vanishes. In Sect. 5.5.2 we therefore introduce the 3-parameter Marvin-Oser ladder, a simple modification of the Gross-Marvin ladder, whose extended form possesses a finite instantaneous compliance. In Sect. 5.5.3 we demonstrate that the extended regular converse ladder is not suitable for the representation of linear viscoelastic behavior. Finally, in Sect. 5.5.4 we show that the extended forms of the seven other possible 3-parameter ladders are all non-standard.

5.5.1 The Extended Gross-Marvin Models

Extension of the generalized series-parallel models to the canonical ones merely required a change of the range of the summation in the expressions for the respondances. To extend the Gross-Marvin models from a finite to an infinite number of nodes we redefine J and η as the *compliances* and *viscosities per node* and then let the number of nodes approach infinity.

Let us first consider the rheodictic Gross-Marvin model. In this model the number of nodes (not counting, of course, the zeroth and the ground node) is N and the number of springs and dashpots is the same. Letting $J_T^\circ = JN$ and $\eta_T^\circ = \eta N$ denote the sum of all compliances and of all viscosities in the model, we see that $J = J_T^\circ/N$ is the *compliance per node* and $\eta = \eta_T^\circ/N$ is the *viscosity per node* in the rheodictic model. We now inquire into the fate of N a(s) as $N \to \infty$. Equation (5.2-13), in which $\theta = J\eta$, can now be written as

$$N\,a(s) = N \ln(1 + Z/2) \tag{5.5-1}$$

where

$$Z = J_T^\circ \eta_T^\circ s/N^2 + \sqrt{4J_T^\circ \eta_T^\circ s + (J_T^\circ \eta_T^\circ s/N)^2}/N\ . \tag{5.5-2}$$

Expanding the logarithm [3] we find

$$N\ln(1+Z/2)=\frac{NZ}{2+Z}+\frac{N}{2}\left(\frac{Z}{2+Z}\right)^2+\frac{N}{3}\left(\frac{Z}{2+Z}\right)^3+\cdots . \tag{5.5-3}$$

But

$$\lim_{N\to\infty} Z=0\ ,\qquad \lim_{N\to\infty} NZ=2\sqrt{J_T^\circ\eta_T^\circ s} \tag{5.5-4}$$

and

$$\lim_{N\to\infty} NZ^p=0\ ,\qquad p=2,3,\ldots . \tag{5.5-5}$$

The limit as $N\to\infty$ remains the same if $(N\pm\frac{1}{2})a(s)$ is substituted for $N\,a(s)$. Thus, we have

$$\lim_{N\to\infty} N\,a(s)=\lim_{N\to\infty}\,[(N\pm\tfrac{1}{2})a(s)]=\sqrt{\theta_\infty s}=\Gamma(s)\ . \tag{5.5-6}$$

We shall call $\Gamma(s)$ the *transmittance* of the ladder. It governs the way in which the stress and strain transforms are transmitted from one end of an infinite ladder to the other (cf. also Problem 5.5-9). $\Gamma(s)$ depends on the characteristic time, θ_∞. For the rheodictic Gross-Marvin model we have $\Gamma^\circ(s)=\sqrt{\theta_\infty^\circ s}$ and $\theta_\infty^\circ=J_T^\circ\eta_T^\circ$.

In the arrheodictic model the number of springs is still N but the number of nodes is $N-1$ and so is the number of dashpots. We have $J=J_T^\times/N$ and $\eta=\eta_T^\times/(N-1)$. The value of $N\,a(s)$ as $N\to\infty$ is again given by Eq. (5.5-6) except that we now have $\Gamma^\times(s)=\sqrt{\theta_\infty^\times s}$ for the propagation function and $\theta_\infty^\times=J_T^\times\eta_T^\times$ for the characteristic time.

We now express the total compliances, J_T, in terms of $J_e^{\{o\}}$ and the total viscosities, η_T, in terms of $\eta_{\{f\}}$. By Eq. (5.2-97) $\eta_T^\circ=\eta_f$ and by Eq. (5.2-77) $J_T^\times=J_e$. Further, $J_T^\circ=JN$ where $J=6NJ_e^\circ/(2N+1)(N+1)$ by Eq. (5.2-85). Hence,

$$J_T^\circ=6N^2J_e^\circ/(2N+1)(N+1)\simeq 3J_e^\circ\ , \tag{5.5-7}$$

the last equation being valid for large N. Now, $\eta_T^\times=\eta(N-1)$. But, by Eq. (5.2-91), $\eta=2N\eta_\times/\Sigma_n\cot^2(\pi n/2N)$, where $\eta_\times=\Sigma_n\eta_n$. Earlier [cf. Eqs. (3.5-12) and (4.6-24)] we had introduced the symbol η for the sum of the viscosities of the dashpots of an arrheodictic Wiechert model. We add the subscript $^\times$ here to avoid confusion with the viscosity η assigned to the dashpots of the ladder model. Since the sum over the squared cotangents [cf. Eq. (5.2-84)] is $(2N-1)(N-1)/3$, we find

$$\eta_T^\times=6N\eta_\times/(2N-1)\simeq 3\eta_\times\ , \tag{5.5-8}$$

the last equation again being valid for large N. We note that $\theta_\infty^\circ=J_T^\circ\eta_T^\circ=3J_e^\circ\eta_f$, and $\theta_\infty^\times=J_T^\times\eta_T^\times=3J_e\eta_\times$.

It is now a simple matter to obtain the respondances. We rewrite Eqs. (5.2-28), (5.2-30), (5.2-33), and (5.2-35) as

$$\bar{Q}_{GM}^\circ(s)=\sqrt{(\eta_T^\circ s/2N)^2+\eta_T^\circ s/J_T^\circ}\{\tanh[(N+\tfrac{1}{2})a(s)]-\eta_T^\circ s/2N\} \tag{5.5-9}$$

$$\bar{U}_{GM}^\circ(s)=\sqrt{(J_T^\circ/2N)^2+J_T^\circ/\eta_T^\circ s}\{\coth N\,a(s)+J_T^\circ/2N\} \tag{5.5-10}$$

$$\bar{Q}^{\times}_{GM}(s) = \sqrt{(\eta^{\times}_{T} s/2N)^2 + \eta^{\times}_{T} s/J^{\times}_{T}}\,\{\coth N\,a(s) - \eta^{\times}_{T} s/2N\} \tag{5.5-11}$$

and

$$\bar{U}^{\times}_{GM}(s) = \sqrt{(J^{\times}_{T}/2N)^2 + J^{\times}_{T}/\eta^{\times}_{T} s}\,\{\tanh[(N - \tfrac{1}{2})a(s)] + J^{\times}_{T}/2N\}\ . \tag{5.5-12}$$

Taking the limit as $N \to \infty$ we find, for the rheodictic model,

$$\bar{Q}^{\circ}_{GM\infty}(s) = Q^{\circ}_{C}(s)\tanh \Gamma^{\circ}(s) \quad \text{and} \quad \bar{U}^{\circ}_{GM\infty}(s) = U^{\circ}_{C}(s)\coth \Gamma^{\circ}(s) \tag{5.5-13}$$

and, for the arrheodictic one,

$$\bar{Q}^{\times}_{GM\infty}(s) = Q^{\times}_{C}(s)\coth \Gamma^{\times}(s) \quad \text{and} \quad \bar{U}^{\times}_{GM\infty}(s) = U^{\times}_{C}(s)\tanh \Gamma^{\times}(s) \tag{5.5-14}$$

where the characteristic respondances are

$$Q^{\circ}_{C}(s) = \sqrt{\eta_f s/3J^{\circ}_{e}} \quad \text{and} \quad U^{\circ}_{C}(s) = \sqrt{3J^{\circ}_{e}/\eta_f s} \tag{5.5-15}$$

$$Q^{\times}_{C}(s) = \sqrt{3\eta_{\times} s/J_e} \quad \text{and} \quad U^{\times}_{C}(s) = \sqrt{J_e/3\eta_{\times} s} \tag{5.5-16}$$

and the transmittances are given by

$$\Gamma^{\circ}(s) = \sqrt{3J^{\circ}_{e}\eta_f s} \quad \text{and} \quad \Gamma^{\times}(s) = \sqrt{3J_e\eta_{\times} s}\ . \tag{5.5-17}$$

Equations (5.5-14) are new. Equations (5.5-13) were introduced by Marvin and Oser [2].

It is easy to show by expansion of the hyperbolic functions that Eqs. (5.5-13) and (5.5-14) do not contain branch points. We can now apply the same method of analysis that we applied to the models with a finite number of parameters in Sect. 5.2. We rewrite Eq. $(5.5\text{-}14)_2$ as

$$\bar{U}^{\times}_{GM\infty}(s) = \frac{J_e \sinh \Gamma^{\times}(s)}{\Gamma^{\times}(s)\cosh \Gamma^{\times}(s)} \tag{5.5-18}$$

and note that

$$\lim_{s\to 0} \bar{U}^{\times}_{GM\infty}(s) = J_e \lim_{s\to 0} \frac{\sinh\sqrt{\theta^{\times}_{\infty} s}}{\sqrt{\theta^{\times}_{\infty} s}} = J_e \tag{5.5-19}$$

as required. Thus the poles arise from the relation $\cosh \Gamma^{\times}(s) = 0$ and this is satisfied when

$$s_n = -\frac{(2n-1)^2\pi^2}{4\theta^{\times}_{\infty}}\,, \qquad n = 1, 2, \ldots\ . \tag{5.5-20}$$

We note that the number of poles is infinite but there is no pole at infinity. The poles are simple poles lying in the interval $-\infty < s_n \le -\pi^2/4\theta^{\times}_{\infty}$ on the negative real axis of the complex plane.

The retardation times follow immediately as

$$\tau_n = \frac{4\theta^{\times}_{\infty}}{(2n-1)^2\pi^2}\,, \qquad n = 1, 2, \ldots\,, \tag{5.5-21}$$

Calculation of the residues is simplified by making use of the expansion [3]

$$\tanh(\sqrt{\theta s}) = 2\sqrt{\theta s} \sum_{n=1}^{\infty} \frac{1}{(n - 1/2)^2\pi^2 + \theta s} \ . \tag{5.5-22}$$

This gives

$$\bar{U}^{\times}_{GM\infty}(s) = 2J_e \sum_{n=1}^{\infty} \frac{1}{(n - 1/2)^2\pi^2 + \theta^{\times}_{\infty} s} \ . \tag{5.5-23}$$

Each term in the sum gives rise to a residue. By Eq. $(5.2\text{-}45)_2$ we have then $\mathrm{Res}_n(\bar{U}^{\times}_{GM\infty}) = 2J_e/\theta^{\times}_{\infty}$. The compliances become

$$J_n = \frac{8J_e}{(2n - 1)^2\pi^2} \ , \qquad n = 1, 2, \ldots \tag{5.5-24}$$

and the retardance follows as

$$\bar{U}_{GM\infty}(s) = \sum_{n=1}^{\infty} \frac{J_n}{1 + \tau_n s} \ . \tag{5.5-25}$$

Equation (5.5-25) shows that $J_g = 0$. Thus, increasing the number of nodes without limit has turned the original Gross-Marvin model from a standard into a non-standard one. This is so, of course, because the distribution given by Eq. (5.5-23) is not subject to the restriction expressed by Eq. (3.5-30). Lack of conformance with Eqs. (3.5-29) and (3.5-30) is a common feature of all the distributions of respondance times discussed in this section. $\bar{U}^{\times}_{GM\infty}(s)$ is seen to be equivalent to the retardance of an infinite number of Voigt units in series, whose springs have compliances given by Eq. (5.5-24) while the dashpots all have the same fluidity, given by $\phi_n = 2J_e/\theta^{\times}_{\infty}$.

To deal with the relaxance of the arrheodictic model, we rewrite Eq. (5.5-15) as

$$\bar{Q}^{\times}_{GM\infty}(s) = \frac{\Gamma^{\times}(s)\cosh \Gamma^{\times}(s)}{J_e \sinh \Gamma^{\times}(s)} \ . \tag{5.5-26}$$

The relaxance will have poles for those values of s which satisfy the equation $\sinh \Gamma^{\times}(s) = 0$, i.e. for

$$s_n = -\frac{n^2\pi^2}{\theta^{\times}_{\infty}} \ , \qquad n = 1, 2, \ldots \ . \tag{5.5-27}$$

The poles are again simple poles and lie in the interval $-\infty \le s_n \le -\pi^2/\theta^{\times}_{\infty}$. We note that there is now a pole also at infinity. The relaxation times follow as

$$\tau_n = \frac{\theta^{\times}_{\infty}}{n^2\pi^2} \ , \qquad n = 1, 2, \ldots \ . \tag{5.5-28}$$

Again, the calculation of the residues is simple. We use the expansion [3]

$$\coth\sqrt{\theta s} = \frac{1}{\sqrt{\theta s}} + 2\sqrt{\theta s} \sum_{n=1}^{\infty} \frac{1}{n^2\pi^2 + \theta s} \tag{5.5-29}$$

which immediately gives

$$\bar{Q}^{\times}_{GM\infty}(s) = \frac{1}{J_e} + \frac{2}{J_e} \sum_{n=1}^{\infty} \frac{\theta^{\times}_{\infty} s}{n^2\pi^2 + \theta^{\times}_{\infty} s} \ . \tag{5.5-30}$$

Thus, the residues become $\operatorname{Res}_n(\bar{Q}^{\times}_{GM\infty}) = 2s_n/J_e$, and the moduli follow as $G_n = 2/J_e$. Then,

$$\bar{Q}_{GM\infty}(s) = \frac{1}{J_e} + \frac{2}{J_e} \sum_{n=1}^{\infty} \frac{\tau_n s}{1 + \tau_n s} . \tag{5.5-31}$$

Clearly, $G_e = 1/J_e$ as it should, and $G_g = \infty$, as expected from the existence of a pole at infinity. Thus $\bar{Q}^{\times}_{GM\infty}(s)$ is equivalent to the relaxance of an isolated spring and an infinite number of Maxwell units in parallel. The isolated spring has the modulus $1/J_e$. The springs of the Maxwell units have the moduli already given and the dashpots have the viscosities

$$\eta_n = \frac{2\theta^{\times}_{\infty}}{n^2\pi^2 J_e}, \qquad n = 1, 2, \ldots . \tag{5.5-32}$$

Turning now to the rheodictic model, we see from Eq. $(5.5\text{-}13)_2$ that $\bar{U}^{\circ}_{GM\infty}(s)$ will have a pole at the origin which is associated with the steady-flow term in the corresponding series-parallel model. The other poles arise from $\sinh \Gamma^{\circ}(s) = 0$. Hence we have

$$s_n = -\frac{n^2\pi^2}{\theta^{\circ}_{\infty}}, \qquad n = 0, 1, 2, \ldots . \tag{5.5-33}$$

These simple poles lie in the interval $-\infty < s_n \leq 0$ along the negative real axis including the origin but there is no pole at infinity.

Using Eq. (5.5-29) in Eq. $(5.5\text{-}14)_2$ gives

$$\bar{U}^{\times}_{GM\infty}(s) = 6J^{\circ}_e \sum_{n=1}^{\infty} \frac{1}{n^2\pi^2 + \theta^{\circ}_{\infty} s} + \frac{1}{\eta_f s} . \tag{5.5-34}$$

The retardation times are

$$\tau_n = \frac{\theta^{\circ}_{\infty}}{n^2\pi^2}, \qquad n = 1, 2, \ldots \tag{5.5-35}$$

and the residues not associated with the flow term become $\operatorname{Res}_n(\bar{U}^{\circ}_{GM\infty}) = 6J^{\circ}_e/\theta^{\circ}_{\infty}$. Hence, the compliances result as

$$J_n = \frac{6J^{\circ}_e}{n^2\pi^2}, \qquad n = 1, 2, \ldots \tag{5.5-36}$$

and we have

$$\bar{U}_{GM\infty}(s) = \sum_{n=1}^{\infty} \frac{J_n}{1 + \tau_n s} + \frac{1}{\eta_f s} . \tag{5.5-37}$$

We note that $J_g = 0$ again and that, as it should,

$$\lim_{s\to 0} [\bar{U}_{GM\infty}(s) - 1/\eta_f s] = \sum_{n=1}^{\infty} J_n = \frac{6J^{\circ}_e}{\pi^2} \sum_{n=1}^{\infty} \frac{1}{n^2} = J^{\circ}_e \tag{5.5-38}$$

because [4]

$$\sum_{n=1}^{\infty} \frac{1}{n^2} = \frac{\pi^2}{6} \,. \tag{5.5-39}$$

Thus $\bar{U}^{\circ}_{GM\infty}(s)$ is equivalent to the retardance of an isolated dashpot with fluidity $\phi_f = 1/\eta_f$ in series with an infinite number of Voigt units the springs of which have the compliances given by Eq. (5.5-36), the dashpots having the fluidities $\phi_n = 2/\eta_f$.

The poles of $\bar{Q}^{\circ}_{GM\infty}(s)$ result from $\cosh \Gamma^{\circ}(s) = 0$ [cf. Eq. (5.5-20)]. We write

$$s_n = -\frac{(2n+1)^2\pi^2}{4\theta^{\circ}_{\infty}} \,, \qquad n = 0, 1, \ldots \tag{5.5-40}$$

starting the series from zero as required by Eq. (5.2-94). The poles lie in the interval $-\infty \leq s_n \leq 0$ along the negative real axis of the complex plane including the origin. Again, there is a pole at infinity resulting in an infinite instantaneous modulus and, hence, non-standard behavior of the model. The poles give rise to the relaxation times

$$\tau_n = \frac{4\theta^{\circ}_{\infty}}{(2n+1)^2\pi^2} \,, \qquad n = 0, 1, \ldots \,. \tag{5.5-41}$$

Using Eq. (5.5-22) gives

$$\bar{Q}^{\circ}_{GM\infty}(s) = 2n_f s \sum_{n=0}^{\infty} \frac{1}{(n+1/2)^2\pi^2 + \theta^{\circ}_{\infty} s} \tag{5.5-42}$$

and the residues follow as $\mathrm{Res}_n(\bar{Q}^{\circ}_{GM\infty}) = 2s_n/3J^{\circ}_e$. Hence, the moduli become $G_n = 2/3J^{\circ}_e$ and we have

$$\bar{Q}^{\circ}_{GM\infty}(s) = \frac{2}{3J^{\circ}_e} \sum_{n=0}^{\infty} \frac{\tau_n s}{1 + \tau_n s} \,. \tag{5.5-43}$$

Evidently, $G_g = \infty$, and $G_e = 0$. As required, we also have

$$\lim_{s \to 0} \frac{\bar{Q}^{\circ}_{GM\infty}(s)}{s} = \frac{8n_f}{\pi^2} \sum_{n=0}^{\infty} \frac{1}{(2n+1)^2} = \eta_f \tag{5.5-44}$$

because [4]

$$\sum_{n=0}^{\infty} \frac{1}{(2n+1)^2} = \frac{\pi^2}{8} \,. \tag{5.5-45}$$

$\bar{Q}^{\circ}_{GM\infty}(s)$ is thus equivalent to the relaxance of an infinite number of Maxwell units in parallel in which all springs have the moduli already given and the dashpots have the viscosities

$$\eta_n = \frac{8\eta_f}{(2n+1)^2\pi^2} \,, \qquad n = 0, 1, \ldots \,. \tag{5.5-46}$$

In the foregoing we have discussed the consequences of increasing the number of nodes of the Gross-Marvin models without limit. Had we done this in the case of the Gross models we would have obtained the same results (see Problem 5.5-1). In the transition from a finite to an infinite number of nodes the initial section is lost. Letting J = 0 in Eq. (5.2-22) and using Eq. (5.5-6) (the limit remains the same when N − 1 is substituted for N) we obtain the relaxance of the extended models, i.e. the models with an infinite number of nodes, as

$$\bar{Q}_\infty(s) = Q_C(s)\frac{Q_F(s)\cosh\Gamma(s) + Q_C(s)\sinh\Gamma(s)}{Q_C(s)\cosh\Gamma(s) + Q_F(s)\sinh\Gamma(s)} , \tag{5.5-47}$$

where $\Gamma(s)$ is the transmittance and $Q_C(s)$ is the characteristic relaxance. Equation (5.5-47) predicts rheodictic behavior for $Q_F(s) = 0$, and arrheodictic behavior for $Q_F(s) = \infty$. It is seen that the model consists of an infinite number of propagation sections which are terminated either into an infinite or into a zero respondance. The transmittance and the characteristic respondance are found from the propagation section. It is easily ascertained that it makes no difference whether this is taken as a T- or a Π-section (see Problem 5.5-2).

For the Gross-Marvin models the transmittance and the characteristic respondances have already been discussed. It should be noted, however, that Eq. (5.5-47) is valid regardless of the actual form of the strut retardance and the rung relaxance. Denoting by $U_s(s)$ the strut retardance, and by $Q_r(s)$ the rung relaxance, in which the material parameters have been replaced by their T-subscripted forms, we may write

$$\Gamma(s) = \lim_{N\to\infty} N\,a(s) = \sqrt{Q_r(s)U_s(s)} \tag{5.5-48}$$

and

$$Q_C(s) = \lim_{N\to\infty} Q_T(s) = \lim_{N\to\infty} Q_\Pi(s) = \sqrt{Q_r(s)/U_s(s)} = \Gamma(s)/U_s(s) . \tag{5.5-49}$$

For the Gross-Marvin models we have $Q_r(s) = \eta_T s$ and $U_s(s) = J_T$ where η_T and J_T are the total viscosity and the total compliance, respectively.

The retardance, $\bar{U}_\infty(s)$, is, of course, the reciprocal of $\bar{Q}_\infty(s)$. It is given explicitly in Problem 5.5-10.

We note that the harmonic respondances can be obtained in closed form because the substitution $s = j\omega$ is now a relatively easy matter (see Problem 5.5-4).

5.5.2 The Extended Marvin-Oser Model

The models with an infinite number of nodes discussed in Sect. 5.5.1 are unrealistic in the sense that their relaxances have poles at infinity. Consequently, they are non-standard models, i.e. their instantaneous modulus is infinite and their instantaneous compliance is zero. To remedy this deficiency Marvin and Oser [2] introduced the model shown in Fig. 5.5-1 which differs from the Gross-Marvin model described in Sect. 5.2 by the insertion of springs of modulus G in series with the dashpots. This model contains three independent material parameters. When it is treated as a model with an infinite number of nodes, the inclusion of the springs with modulus G removes the pole at infinity. The model is easy to analyze. We use the principle outlined at the end of Sect. 5.5.1, i.e. we use Eq. (5.5-47) with Eqs. (5.5-48) and (5.5-49), simply replacing, in these equations, the rung relaxance of the Gross-Marvin model, η, by the new rung relaxance $\eta s/(1 + \delta s)$ where $\delta = \eta/G$.

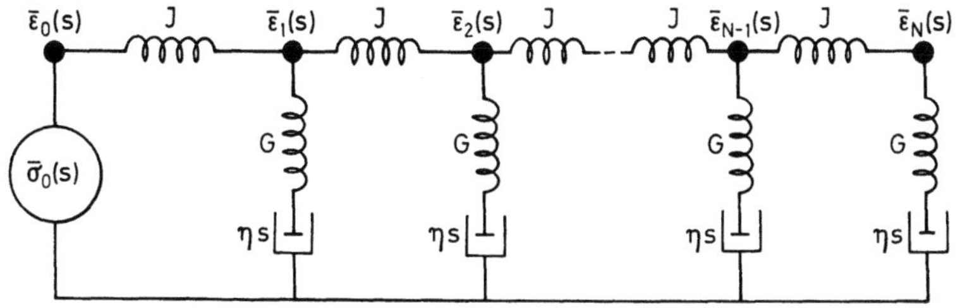

Fig. 5.5-1. The Marvin-Oser model

Again as before, we replace η by η_T/N and, in addition, G by G_T/N where G_T is the sum of the rung moduli. As $N \to \infty$ we then have $\delta = \eta_T/G_T$. One problem immediately arises, however. The model is clearly redundant (see Sect. 3.5.3) in that the number of springs exceeds the number required to represent its behavior by the equivalent series-parallel models. While it is reasonable to assume (and this will indeed be justified later) that $\eta_T^\times = 3\eta_\times$ and $\eta_T^\circ = \eta_f$ as with the Gross-Marvin models, it is not clear whether $J_T^\times$ and J_T° will still be equal to J_e and $3J_e^\circ$, respectively, and it is not clear at this point what meaning is to be assigned to G_T. For this reason we shall carry out the analysis in terms of J_T, η_T, and G_T and ascertain the meaning of all three later when we are in a better position to do so. We shall write, therefore,

$$\Gamma(s) = \sqrt{\frac{J_T\eta_T s}{1 + \eta_T s/G_T}} = \sqrt{\frac{\theta_\infty s}{1 + \delta s}} \tag{5.5-50}$$

for the transmittance, and

$$Q_C(s) = \frac{1}{J_T}\sqrt{\frac{J_T\eta_T s}{1 + \eta_T s/G_T}} = \frac{1}{J_T}\sqrt{\frac{\theta_\infty s}{1 + \delta s}} \tag{5.5-51}$$

for the characteristic relaxance.

We first turn to the arrheodictic form of the model. Equation $(5.5\text{-}14)_2$ immediately furnishes

$$\bar{U}^\times_{MO\infty}(s) = J_T^\times \sqrt{\frac{1 + \delta^\times s}{\theta^\times_\infty s}} \tanh \sqrt{\frac{\theta^\times_\infty s}{1 + \delta^\times s}} \tag{5.5-52}$$

which has the poles

$$s_n = -\frac{(2n-1)^2\pi^2}{4\theta^\times_\infty + (2n-1)^2\pi^2\delta^\times}\,, \qquad n = 1, 2, \ldots\,. \tag{5.5-53}$$

The retardance does not vanish at infinity. In fact, since $\delta^\times = \eta_T^\times/G_T^\times$ and $\theta^\times_\infty = J_T^\times\eta_T^\times$, the instantaneous compliance is found to be

$$J_g = \sqrt{J_T^\times/G_T^\times}\,\tanh\sqrt{J_T^\times G_T^\times}\,. \tag{5.5-54}$$

Using Eq. (5.5-22) we now obtain the retardance in the form

$$\bar{U}^{\times}_{MO\infty}(s) = 2J^{\times}_{T} \sum_{n=1}^{\infty} \frac{1}{(n-\frac{1}{2})^2\pi^2 + \theta^{\times}_{\infty}/(1+\delta^{\times}s)} \tag{5.5-55}$$

from which the residues follow at once as

$$\mathrm{Res}_n(\bar{U}^{\times}_{MO\infty}) = \frac{8J^{\times}_{T}(1+\delta^{\times}s_n)}{4\theta^{\times}_{\infty} + (2n-1)^2\pi^2\delta^{\times}}, \qquad n = 1, 2, \ldots . \tag{5.5-56}$$

The retardation times are

$$\tau_n = \delta^{\times} + \frac{4\theta^{\times}_{\infty}}{(2n-1)^2\pi^2}, \qquad n = 1, 2, \ldots \tag{5.5-57}$$

and the associated compliances become

$$J_n = 8\,J^{\times}_{T}\left(\frac{1}{(2n-1)^2\pi^2} - \frac{1}{4J^{\times}_{T}G^{\times}_{T} + (2n-1)^2\pi^2}\right), \qquad n = 1, 2, \ldots . \tag{5.5-58}$$

Thus, $\bar{U}^{\times}_{MO\infty}(s)$ is equivalent to a Kelvin model with an isolated spring and an infinite number of Voigt units in series. The compliances of the springs are given by Eq. (5.5-58) and the dashpots have fluidities given by

$$\phi_n = \frac{2}{\eta^{\times}_{T}}\left(\frac{4J^{\times}_{T}G^{\times}_{T}}{4J^{\times}_{T}G^{\times}_{T} + (2n-1)^2\pi^2}\right)^2, \qquad n = 1, 2, \ldots . \tag{5.5-59}$$

Turning now to the relaxance, Eq. $(5.5\text{-}14)_1$ gives

$$\bar{Q}^{\times}_{MO\infty}(s) = \frac{1}{J^{\times}_{T}}\sqrt{\frac{\theta^{\times}_{\infty}s}{1+\delta^{\times}s}}\coth\sqrt{\frac{\theta^{\times}_{\infty}s}{1+\delta^{\times}s}}, \tag{5.5-60}$$

the poles of which are

$$s_n = -\frac{n^2\pi^2}{\theta^{\times}_{\infty} + n^2\pi^2\delta^{\times}}, \qquad n = 1, 2, \ldots . \tag{5.5-61}$$

There is no pole at infinity. In fact, it is easily ascertained that

$$\lim_{s\to\infty} \bar{Q}^{\times}_{MO\infty}(s) = \sqrt{G^{\times}_{T}/J^{\times}_{T}}\,\coth\sqrt{G^{\times}_{T}J^{\times}_{T}} = G_g\,. \tag{5.5-62}$$

With Eq. (5.5-29) we obtain

$$\bar{Q}^{\times}_{MO\infty}(s) = \frac{1}{J^{\times}_{T}} + \frac{2\theta^{\times}_{\infty}s}{J^{\times}_{T}(1+\delta^{\times}s)} \sum_{n=1}^{\infty} \frac{1}{n^2\pi^2 + \theta^{\times}_{\infty}s/(1+\delta^{\times}s)} \tag{5.5-63}$$

and the residues become

$$\mathrm{Res}_n(\bar{Q}^{\times}_{MO\infty}(s) = \frac{2G^{\times}_{T}s_n}{J^{\times}_{T}G^{\times}_{T} + n^2\pi^2}, \qquad n = 1, 2, \ldots . \tag{5.5-64}$$

From these, the moduli follow at once as

$$G_n = \frac{2G^{\times}_{T}}{J^{\times}_{T}G^{\times}_{T} + n^2\pi^2}, \qquad n = 1, 2, \ldots . \tag{5.5-65}$$

The relaxation times, of course, are

$$\tau_n = \delta^\times + \frac{\theta_\infty^\times}{n^2\pi^2}\,, \qquad n = 1, 2, \ldots\,. \tag{5.5-66}$$

We see that $\bar{Q}_{MO\infty}^\times(s)$ is equivalent to the relaxance of a spring with modulus $G_e = 1/J_T$ in parallel with an infinite number of Maxwell units in which the springs have the moduli given by Eq. (5.5-65), and the dashpots have the viscosities

$$\eta_n = \frac{2\eta_T^\times}{n^2\pi^2}\,, \qquad n = 1, 2, \ldots\,. \tag{5.5-67}$$

We are now ready to assign meaning to the parameters with subscript T. It is clear from Eq. (5.5-63) that $J_T^\times = J_e$. Equation (5.5-67), recalling Eq. (5.5-39), shows that $\eta_T^\times = \Sigma_n \eta_n = 3\eta_\times$. Finally, from Eq. (5.5-62), we have

$$G_g = \sqrt{G_T^\times G_e}\,\coth\sqrt{G_T^\times/G_e} \simeq \sqrt{G_T^\times G_e}\,. \tag{5.5-68}$$

The simplified expression is an excellent approximation because it is clear that $G_T^\times$ must be larger than G_g and therefore $\coth\sqrt{G_T^\times J_T^\times} \simeq 1$. Thus $G_T^\times \simeq G_g^2/G_e$ will be very satisfactory in most applications. An exact value can be found by solving Eq. $(5.5\text{-}68)_1$ with known values of G_g and G_e.

The retardance of the arrheodictic form of the extended Marvin-Oser model is thus equivalent to

$$\bar{U}_{MO\infty}(s) = J_g + \sum_{n=1}^{\infty} \frac{J_n}{1 + \tau_n s} \tag{5.5-69}$$

where

$$\tau_n = \theta_\infty^\times \left[\left(\frac{J_g}{J_e}\right)^2 + \frac{4}{(2n-1)^2\pi^2}\right] \tag{5.5-70}$$

and

$$J_n = 8\,J_e\left(\frac{1}{(2n-1)^2\pi^2} - \frac{1}{4(J_e/J_g)^2 + (2n-1)^2\pi^2}\right) \tag{5.5-71}$$

while its relaxance is equivalent to

$$\bar{Q}_{MO\infty}(s) = G_e + \sum_{n=1}^{\infty} \frac{G_n \tau_n s}{1 + \tau_n s} \tag{5.5-72}$$

where

$$\tau_n = \theta_\infty^\times \left[\left(\frac{G_e}{G_g}\right)^2 + \frac{1}{n^2\pi^2}\right] \tag{5.5-73}$$

and

$$G_n = G_e \frac{2(G_g/G_e)^2}{(G_g/G_e)^2 + n^2\pi^2}\,. \tag{5.5-74}$$

We now examine the rheodictic form. From Eq. (5.5-13)$_2$ the retardance is

$$\bar{U}^{\circ}_{MO\infty}(s) = J^{\circ}_T \sqrt{\frac{1+\delta^{\circ}s}{\theta^{\circ}_{\infty}s}} \coth \sqrt{\frac{\theta^{\circ}_{\infty}s}{1+\delta^{\circ}s}} \tag{5.5-75}$$

where $\delta^{\circ} = \eta^{\circ}_T/G^{\circ}_T$ and $\theta^{\circ}_{\infty} = J^{\circ}_T\eta^{\circ}_T$. The poles are

$$s_n = -\frac{n^2\pi^2}{\theta^{\circ}_{\infty} + n^2\pi^2\delta^{\circ}}, \qquad n = 0, 1, \ldots \tag{5.5-76}$$

and Eq. (5.5-29) gives

$$\bar{U}^{\circ}_{MO\infty}(s) = \frac{1}{G^{\circ}_T} + 2J^{\circ}_T \sum_{n=1}^{\infty} \frac{1}{n^2\pi^2 + \theta^{\circ}_{\infty}s/(1+\delta^{\circ}s)} + \frac{1}{\eta^{\circ}_T s}. \tag{5.5-77}$$

Using the initial value theorem and then applying Eq. (5.5-29) gives

$$J_g = \frac{1}{G^{\circ}_T} + 2J^{\circ}_T \sum_{n=1}^{\infty} \frac{1}{n^2\pi^2 + J^{\circ}_T G^{\circ}_T} = \sqrt{J^{\circ}_T/G^{\circ}_T} \coth \sqrt{J^{\circ}_T G^{\circ}_T} \tag{5.5-78}$$

for the instantaneous compliance. The residues become

$$\begin{aligned} &\mathrm{Res}_0(\bar{U}^{\circ}_{MO\infty}) = 1/\eta^{\circ}_T \\ &\mathrm{Res}_n(\bar{U}^{\circ}_{MO\infty}) = \frac{2J^{\circ}_T(1+\delta^{\circ}s_n)}{\theta^{\circ}_{\infty} + n^2\pi^2\delta^{\circ}}, \qquad n = 1, 2, \ldots \end{aligned} \tag{5.5-79}$$

and it is immediately clear that $\eta^{\circ}_T = \eta_f$ as expected. The retardation times are

$$\tau_n = \delta^{\circ} + \frac{\theta^{\circ}_{\infty}}{n^2\pi^2}, \qquad n = 1, 2, \ldots \tag{5.5-80}$$

and the compliances associated with the poles not at the origin become

$$J_n = 2\,J^{\circ}_T\left(\frac{1}{n^2\pi^2} - \frac{1}{J^{\circ}_e G^{\circ}_T + n^2\pi^2}\right), \qquad n = 1, 2, \ldots . \tag{5.5-81}$$

We see that $\bar{U}^{\circ}_{MO\infty}(s)$ is equivalent to the retardance of a Kelvin model with an isolated spring of compliance J_g in series with an infinite number of Voigt elements and an isolated dashpot with fluidity $\phi_f = 1/\eta_f$. The compliance of the Voigt units are given by Eq. (5.5-81). The fluidities are found to be

$$\phi_n = \frac{2}{\eta^{\circ}_T}\left(\frac{J^{\circ}_T G^{\circ}_T}{J^{\circ}_T G^{\circ}_T + n^2\pi^2}\right)^2, \qquad n = 1, 2, \ldots . \tag{5.5-82}$$

Let us now consider the relaxance. By Eq. (5.5-13)$_1$

$$\bar{Q}^{\circ}_{MO\infty}(s) = \frac{1}{J^{\circ}_T}\sqrt{\frac{\theta^{\circ}_{\infty}s}{1+\delta^{\circ}s}} \tanh \sqrt{\frac{\theta^{\circ}_{\infty}s}{1+\delta^{\circ}s}} \tag{5.5-83}$$

which has poles at

$$s_n = -\frac{(2n+1)^2\pi^2}{4\theta^{\circ}_{\infty} + (2n+1)^2\pi^2\delta^{\circ}}, \qquad n = 0, 1, \ldots . \tag{5.5-84}$$

Again there is no pole at infinity since, by Eq. (5.5-78),

$$\lim_{s\to\infty} \bar{Q}^{\circ}_{MO\infty}(s) = \sqrt{G^{\circ}_T/J^{\circ}_T}\,\tanh\sqrt{G^{\circ}_T J^{\circ}_T} = G_g\ . \tag{5.5-85}$$

Using Eq. (5.5-22) gives

$$\bar{Q}^{\circ}_{MO\infty}(s) = \frac{8\theta^{\circ}_{\infty}s}{J^{\circ}_T(1+\delta^{\circ}s)}\sum_{n=1}^{\infty}\frac{1}{(2n+1)^2\pi^2 + 4\theta^{\circ}_{\infty}s/(1+\delta^{\circ}s)} \tag{5.5-86}$$

and the residues follow as

$$\mathrm{Res}_n(\bar{Q}^{\circ}_{MO\infty}) = \frac{8\eta^{\circ}_T s_n}{4\theta^{\circ}_{\infty} + (2n+1)^2\pi^2\delta^{\circ}}\ , \qquad n = 0, 1, \ldots \tag{5.5-87}$$

from which the moduli are obtained as

$$G_n = \frac{8G^{\circ}_T}{4J^{\circ}_T G^{\circ}_T + (2n+1)^2\pi^2}\ , \qquad n = 0, 1, \ldots\ . \tag{5.5-88}$$

The relaxation times are now

$$\tau_n = \delta^{\circ} + \frac{4\theta^{\circ}_{\infty}}{(2n+1)^2\pi^2}\ , \qquad n = 0, 1, \ldots\ . \tag{5.5-89}$$

Evidently, $\bar{Q}^{\circ}_{MO\infty}(s)$ is equivalent to the relaxance of a Wiechert model consisting of an infinite number of Maxwell units in parallel. The units have the moduli given by Eq. (5.5-88) and the viscosities

$$\eta_n = \frac{8\eta^{\circ}_T}{(2n+1)^2\pi^2}\ , \qquad n = 0, 1, \ldots\ . \tag{5.5-90}$$

We now turn to the meaning of the parameters J°_T, and G°_T. We have already seen that $\eta^{\circ}_T = \eta_f$. To find the meaning of J°_T, we obtain J°_e. We use

$$J^{\circ}_e = \lim_{s\to 0}\,[\bar{U}^{\circ}_{MO\infty}(s) - 1/\eta_f s] \tag{5.5-91}$$

which, with Eq. (5.5-77), gives

$$J^{\circ}_e = \frac{1}{G^{\circ}_T} + 2J^{\circ}_T\sum_{n=1}^{\infty}\frac{1}{n^2\pi^2} = J^{\circ}_T/3 + 1/G^{\circ}_T\ . \tag{5.5-92}$$

Hence,

$$J^{\circ}_T = 3(J^{\circ}_e - 1/G^{\circ}_T)\ . \tag{5.5-93}$$

Equations (5.5-93) and (5.5-85) may be solved for J°_T and G°_T. If we accept the approximation $\tanh\sqrt{G^{\circ}_T J^{\circ}_T} = 1$, which will be excellent for most applications, then

$$G^{\circ}_T \simeq G_g^2 J^{\circ}_T \tag{5.5-94}$$

and substituting Eq. (5.5-94) into (5.5-93) gives

$$J^{\circ}_T \simeq 1.5J^{\circ}_e[1 + \sqrt{1 - (4J_g/3J^{\circ}_e)^2}]\ . \tag{5.5-95}$$

The expression in brackets will usually be extremely close to 2 (it is 1.96 for

$J_e^\circ/J_g = 10$). With this further simplification we finally obtain $J_T^\circ \simeq 3J_e^\circ$ and $G_T^\circ \simeq 3G_g^2 J_e^\circ$.

With the use of these relations we may represent the retardance of the rheodictic form of the extended Marvin-Oser model as

$$\bar{U}_{MO\infty}(s) = J_g + \sum_{n=1}^{\infty} \frac{J_n}{1 + \tau_n s} + \frac{\phi_f}{s} \tag{5.5-96}$$

where

$$\tau_n = \theta_\infty^\circ \left[\left(\frac{J_g}{3J_e^\circ} \right)^2 + \frac{1}{n^2\pi^2} \right], \qquad n = 1, 2, \ldots \tag{5.5-97}$$

and

$$J_n = 6\, J_e^\circ \left(\frac{1}{n^2\pi^2} - \frac{1}{(3J_e^\circ/J_g)^2 + n^2\pi^2} \right), \qquad n = 1, 2, \ldots . \tag{5.5-98}$$

The relaxance then becomes

$$\bar{Q}_{MO\infty}(s) = \sum_{n=1}^{\infty} \frac{G_n \tau_n s}{1 + \tau_n s} \tag{5.5-99}$$

where

$$\tau_n = \theta_\infty^\circ \left[\left(\frac{J_g}{3J_e^\circ} \right)^2 + \frac{4}{(2n+1)^2\pi^2} \right], \qquad n = 0, 1, \ldots \tag{5.5-100}$$

and

$$G_n = G_e^\circ \frac{24(G_g/G_e^\circ)^2}{(6G_g/G_e)^2 + (2n+1)^2\pi^2}, \qquad n = 0, 1, \ldots \tag{5.5-101}$$

where $G_e^\circ = 1/J_e^\circ$ is the steady-state modulus.

Again, the harmonic respondances may be obtained in closed form although the expressions are somewhat cumbersome (see Problem 5.5-7).

Comparing the expression for the respondance times, moduli, compliances, etc., of the Gross-Marvin and Marvin-Oser models with an infinite number of nodes it is clear that the latter, as they should, reduce to the former for $G_g = \infty$ (or, equivalently, $\delta = 0$). In particular, we note that addition of the springs with modulus G in series with the dashpots results in the simple addition of δ to the respondance times of the Gross-Marvin models.

Both models have been helpful in theoretical investigations of linear viscoelastic behavior. The extended Marvin-Oser model is particularly useful. We will discuss it again in Chap. 10.

5.5.3 The Extended Regular Converse Ladder Model

Let us now examine the result of increasing the number of nodes without limit in the regular converse ladder model. Letting $G = G_T/N$ and $\phi = \phi_T/N$, and $\hat{\theta}_\infty =$

$1/G_T\phi_T$, Eq. (5.5-47) takes the form

$$\bar{Q}_{RC\infty}(s) = \hat{Q}_C(s)\frac{Q_F(s)\cosh\hat{\Gamma}(s) + Q_C(s)\sinh\hat{\Gamma}(s)}{Q_C(s)\cosh\hat{\Gamma}(s) + Q_F(s)\sinh\hat{\Gamma}(s)} \tag{5.5-102}$$

where $\hat{\Gamma}(s) = 1/\sqrt{\hat{\theta}_\infty s}$ by Eq. (5.5-48) and $Q_C(s) = G_T\sqrt{\hat{\theta}_\infty s}$ by Eq. (5.5-49). We have

$$\bar{Q}^{\times}_{RC\infty}(s) = G_T\sqrt{\hat{\theta}_\infty(s)}\coth(1/\sqrt{\hat{\theta}_\infty s}) \tag{5.5-103}$$

and

$$\bar{Q}^{\circ}_{RC\infty}(s) = G_T\sqrt{\hat{\theta}_\infty s}\tanh(1/\sqrt{\hat{\theta}_\infty s}) \ . \tag{5.5-104}$$

However, for the arrheodictic form,

$$G_e = \lim_{s\to 0}\bar{Q}^{\times}_{RC\infty}(s) = G_T\lim_{s\to 0}\sqrt{\hat{\theta}_\infty s} = 0 \ , \tag{5.5-105}$$

i.e. the equilibrium modulus vanishes. For the rheodictic form

$$\eta_f = \lim_{s\to 0}\frac{\bar{Q}_{RC\infty}(s)}{s} = \frac{1}{\phi_T}\lim_{s\to 0}\frac{\tanh(1/\sqrt{\hat{\theta}_\infty s})}{\sqrt{\hat{\theta}_\infty s}} = \infty \ , \tag{5.5-106}$$

i.e. the steady-flow viscosity becomes infinite. The extended regular converse ladder model is thus unsuitable for the description of linear viscoelastic behavior.

5.5.4 Other Extended Regular Ladder Models

The addition of a spring in series with the dashpot in the Gross-Marvin model leads to the Marvin-Oser model which retains a finite, non-zero instantaneous compliance as the number of nodes, N, is increased without limit. The question then arises: are there any other regular extended ladder models with three distinct elements which have a similar compliance and are useful for the modelling of linear viscoelastic behavior?

Figure 5.5-2 shows the propagation sections of the eight models which have either a Maxwell or a Voigt unit on either the strut or the rungs of the ladder. The propagation section for the Marvin-Oser model is included for completeness.

By the initial value theorem the instantaneous modulus is obtained from

$$J_g = \lim_{s\to\infty}\bar{U}(s) \ . \tag{5.5-107}$$

But (cf. Problem 5.5–10), for a rheodictic model with an infinite number of nodes

$$\bar{U}^{\circ}_{\infty}(s) = U_s(s)\frac{\coth\Gamma(s)}{\Gamma(s)} \tag{5.5-108}$$

and for an arrheodictic one

$$\bar{U}^{\times}_{\infty}(s) = U_s(s)\frac{\tanh\Gamma(s)}{\Gamma(s)} \ . \tag{5.5-109}$$

$U_s(s)$, $Q_r(s)$, and $\Gamma(s) = \sqrt{Q_r(s)U_s(s)}$ for the eight models are assembled in Table

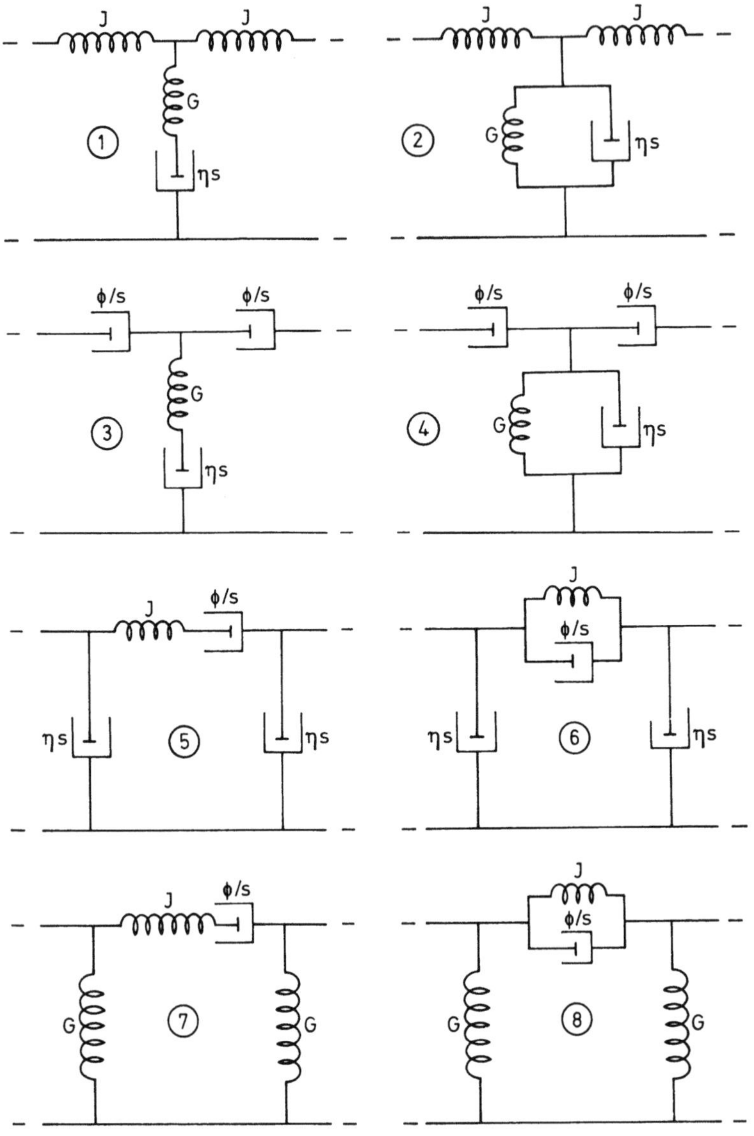

Fig. 5.5-2. Repeat sections containing either Maxwell or Voigt units

5.5-1. The subscripts T on J, η, G, and φ, and the subscript ∞ on θ and $\hat{\theta}$ have been omitted for simplicity.

The last column can be filled in almost by inspection except for models No. 3 and 8 for which the limit becomes indeterminate. In these cases we use

$$J_g = \left(\lim_{s\to\infty} U_s(s)\right)\left(\lim_{s\to\infty} \frac{\tanh \Gamma(s)}{\Gamma(s)}\right) \quad \text{or} \quad G_g = \left(\lim_{s\to\infty} \frac{1}{U_s(s)}\right)\left(\lim_{s\to\infty} \frac{\tanh \Gamma(s)}{\Gamma(s)}\right) \tag{5.5-110}$$

Table 5.5-1. Characteristic parameters of extended regular ladder models containing either a Maxwell or a Voigt unit

No.	$Q_r(s)$	$U_s(s)$	$\Gamma(s)$	$\Gamma(\infty)$	J_g
1	$\frac{G\eta s}{G+\eta s}$	J	$\frac{1}{\sqrt{1/JG+1/\theta s}}$	$\sqrt{JG}$	>0
2	$G+\eta s$	J	$\sqrt{JG+\theta s}$	∞	0
3	$\frac{G\eta s}{G+\eta s}$	ϕ/s	$\frac{1}{\sqrt{1/\eta\phi+\hat{\theta}s}}$	0	0
4	$G+\eta s$	ϕ/s	$\sqrt{\eta\phi+1/\hat{\theta}s}$	$\sqrt{\eta\phi}$	0
5	ηs	$J+\phi/s$	$\sqrt{\eta\phi+\theta s}$	∞	0
6	ηs	$\frac{J\phi}{\phi+Js}$	$\frac{1}{\sqrt{1/\eta\phi+1/\theta s}}$	$\sqrt{\eta\phi}$	0
7	G	$J+\phi/s$	$\sqrt{JG+1/\hat{\theta}s}$	$\sqrt{JG}$	>0
8.	G	$\frac{J\phi}{\phi+Js}$	$\frac{1}{\sqrt{1/JG+\hat{\theta}s}}$	0	0

where

$$\lim_{s\to\infty}\frac{\tanh\Gamma(s)}{\Gamma(s)}=\lim_{s\to\infty}\operatorname{sech}^2\Gamma(s)=\operatorname{sech}^2(0)=1\ . \tag{5.5-111}$$

Apart from the Marvin-Oser model, only model No. 7 possesses a non-zero instantaneous compliance. The instantaneous modulus of the arrheodictic form becomes

$$G_g=\lim_{s\to\infty}\frac{\Gamma(s)}{U_s(s)\tanh\Gamma(s)}=\sqrt{G/J}\coth\sqrt{GJ} \tag{5.5-112}$$

and that of the rheodictic form is

$$G_g=\lim_{s\to\infty}\frac{\Gamma(s)}{U_s(s)\coth\Gamma(s)}=\sqrt{G/J}\tanh\sqrt{GJ}\ . \tag{5.5-113}$$

As $s\to 0$, $\Gamma(s)\to\infty$, and $\coth\Gamma(s)$ and $\tanh\Gamma(s)\to 1$, the equilibrium modulus of both forms of the model therefore turns out to be

$$G_e=\lim_{s\to 0}\frac{\Gamma(s)}{U_s(s)}=0\ . \tag{5.5-114}$$

This is right for the rheodictic form but incorrect for the arrheodictic one. The steady-flow viscosity of both models is obtained as

$$\eta_f=\lim_{s\to 0}\frac{\Gamma(s)}{sU_s(s)}=\infty \tag{5.5-115}$$

which is now the right value for the arrheodictic form but the wrong one for the rheodictic form of the model. Hence the model is not suitable for describing linear viscoelastic behavior and we conclude that the extended Marvin-Oser model is the only one which possesses the proper characteristics.

5.6 The Continuous Ladder or Material Transmission Line

In Chap. 4 we extended the representation of linear viscoelastic behavior by discrete distributions of moduli or compliances to its representation by continuous distributions. This extension led to the concept of the spectral response functions. Here, we shall be concerned with a somewhat similar transition from a discrete to a continuous case. We shall extend the concept of a *lumped* or *discrete ladder* in which viscoelastic elements are associated with (or lumped at) a discrete number of nodes to that of a *distributed* or *continuous ladder* in which the viscoelastic parameters are distributed continuously over the ladder. This extension leads to another significant concept, that of the *material transmission line.* The theory of the material transmission line allows us to deal with cases where the simple theory of linear viscoelastic behavior cannot be applied because inertial effects cannot be neglected, i.e. at excitations of high frequency or of short loading times. In particular, the theory of the *lossy line* (cf. Sect. 5.6.2.2) furnishes the relations required to obtain the complex modulus or compliance from measurements of the propagation of stress or displacement waves in a linear viscoelastic material.

We begin discussion of the theory by considering the concept of a continuous ladder. The material whose relaxance we wish to model is assumed to be a body of uniform cross-section* and length L as shown in Fig. 5.6-1. An arbitrary point along the line is denoted by l or by d = L − l depending on whether it is though to be approached from the beginning (or entrance) or the end (or termination) of the line.

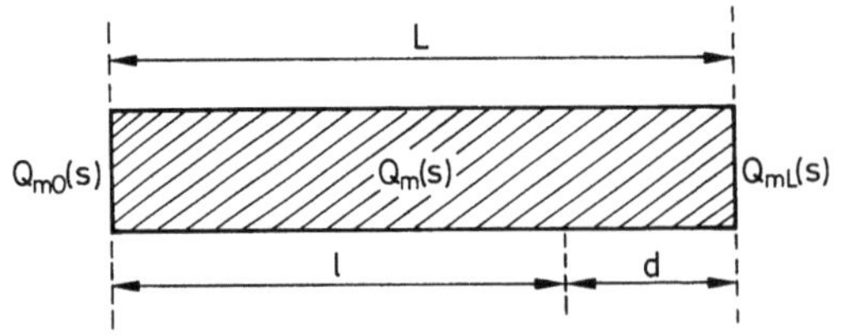

Fig. 5.6-1. Schematic representation of the material transmission line

For reasons that will become clear in Sect. 5.6.2 we discuss the continuous ladder in terms of (generalized) force and displacement rather than stress and strain. Accordingly, the material properties are represented as (operational) mechanical relaxances or retardances. We recall from Sect. 3.1.2 that the mechanical relaxance, $Q_m(s)$, equals bQ(s) and the mechanical retardance, $U_m(s)$, is equal to HU(s) where

* The treatment of non-uniform cross-sections requires no new principle as far as the material behavior is concerned.

$H = 1/b$ is the *shape factor* (see Sect. 3.1.2). The mechanical relaxance along the line will be designated as $Q_m(s, l)$, and that at the two ends of the line by $Q_{mo}(s) = Q_m(s, 0)$ and $Q_{mL}(s) = Q_m(s, L)$ as shown in Fig. 5.6-1.

Let us now denote the length of an arbitrarily small but finite section of the line by Δl. If we then imagine the body to consist of a large number of such sections, each having the same material properties, we can represent its behavior by a ladder model in which each node is replaced by one of the sections.

5.6.1 The Continuous Gross-Marvin Ladder or Inertialess Material Transmission Line

We now apply these concepts to the Gross-Marvin ladder. Choosing representation by Π-sections for the analysis*, the propagation section becomes the mechanical analog of that depicted in Fig. P5.2-2. It takes the form shown in Fig. 5.6-2. The transform of the force at the input end of the section at the point l is denoted by $\bar{f}(s, l)$ and the corresponding transform at the output end by $\bar{f}(s, l + \Delta l)$. The displacement transforms at the two ends then become $\bar{x}(s, l)$ and $\bar{x}(s, l + \Delta l)$. S_l and F_l stand for the *mechanical compliance*, and the *frictance, per unit length*, respectively. The node equations are obtained in the usual way and become

$$\bar{f}(s, l) = (F_l s\, \Delta l/2 + 1/S_l \Delta l)\bar{x}(s, l) - (1/S_l \Delta l)\bar{x}(s, l + \Delta l) \tag{5.6-1}$$

and

$$0 = -(1/S_l \Delta l)\bar{x}(s, l) + (F_l s\, \Delta l/2 + 1/S_l \Delta l)\bar{x}(s, l + \Delta l) + \bar{f}(s, l + \Delta l). \tag{5.6-2}$$

Adding Eqs. (5.6-1) and (5.6-2) yields

$$\bar{f}(s, l) = (F_l s\, \Delta l/2)[\bar{x}(s, l) + \bar{x}(s, l + \Delta l)] + \bar{f}(s, l + \Delta l) \tag{5.6-3}$$

and subtracting Eq. (5.6-2) from (5.6-1) gives

$$\bar{f}(s, l) = (F_l s\, \Delta l/2 + 2/S_l \Delta l)[\bar{x}(s, l) - \bar{x}(s, l + \Delta l)] - \bar{f}(s, l + \Delta l). \tag{5.6-4}$$

But Δl is small and, therefore,

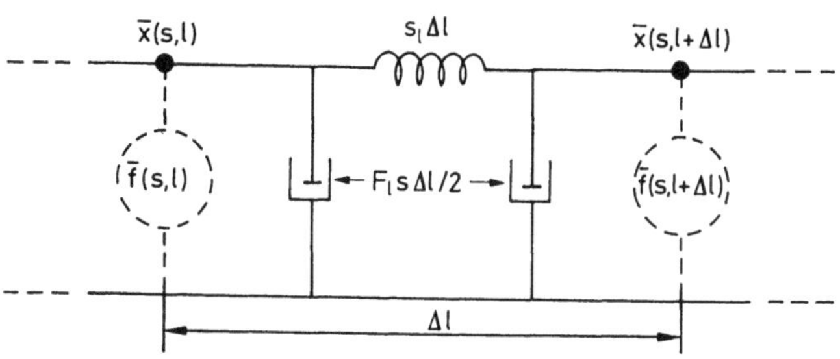

Fig. 5.6-2. Mechanical model diagram of an elemental section of the distributed Gross-Marvin model

---

* T-section analysis leads, of course, to the same result.

$$\bar{f}(s,l+\Delta l) = \bar{f}(s,l) + \frac{\partial\,\bar{f}(s,l)}{\partial l}\Delta l \tag{5.6-5}$$

and

$$\bar{x}(s,l+\Delta l) = \bar{x}(s,l) + \frac{\partial\,\bar{x}(s,l)}{\partial l}\Delta l\ . \tag{5.6-6}$$

Introducing Eqs. (5.6-5) and (5.6-6) into the preceding two leads to

$$-\frac{\partial\,\bar{f}(s,l)}{\partial l} = F_l s\,\bar{x}(s,l) + (F_l s/2)\frac{\partial\,\bar{x}(s,l)}{\partial l}\Delta l \tag{5.6-7}$$

and

$$-\frac{\partial\,\bar{x}(s,l)}{\partial l} = \frac{S_l\bar{f}(s,l) + \dfrac{\partial\,\bar{f}(s,l)}{\partial l}S_l\Delta l/2}{1 + S_l F_l s(\Delta l/2)^2}\ . \tag{5.6-8}$$

Up to this point we have considered the body to consist of a finite number N of discrete sections of length Δl so that $N\,\Delta l = L$. A section in our present treatment is equivalent to a propagation section in our earlier treatment of the Gross-Marvin model in Sect. 5.2. We now execute the transition from a finite to an infinite number of sections. This is essentially the same step as that involved in going from a finite to an infinite number of nodes (see Sect. 5.5.1) but now is also equivalent to a transition from a discrete to a continuous handling of the material properties, i.e. it is equivalent to the transition from the (extended) *lumped*, or *discrete*, to the *distributed* or *continuous* Gross-Marvin model. We let N approach infinity while simultaneously allowing Δl to shrink to zero in such a manner that L remains constant. In the limit as $\Delta l \to 0$, Eqs. (5.6-7) and (5.6-8) reduce to

$$-\frac{\partial\,\bar{f}(s,l)}{\partial l} = F_l s\,\bar{x}(s,l) \tag{5.6-9}$$

and

$$-\frac{\partial\,\bar{x}(s,l)}{\partial l} = S_l\bar{f}(s,l)\ . \tag{5.6-10}$$

Eliminating $\bar{x}(s,l)$ or $\bar{f}(s,l)$ between these two equations yields

$$\frac{\partial^2\bar{f}(s,l)}{\partial l^2} = \Gamma_l^2(s)\bar{f}(s,l) \tag{5.6-11}$$

and

$$\frac{\partial^2\bar{x}(s,l)}{\partial l^2} = \Gamma_l^2(s)\bar{x}(s,l) \tag{5.6-12}$$

where

$$\Gamma_l(s) = \sqrt{S_l F_l s}\ , \tag{5.6-13}$$

which has the physical dimensions of a reciprocal length, is the *propagation function* of the continuous or distributed ladder. It governs the way in which the force and displacement transforms are transmitted along the length of the piece of material from one section of unit length to the next. Its role is therefore analogous to that of the dimensionless function a(s) in the theory of the discrete or lumped ladder (cf. Sect. 5.2.1) which governs the way in which the stress and strain transforms are transmitted from one node to the next. We note that Eqs. (5.6-11) and (5.6-12) remain valid if $\bar{f}(s, l)$ and $\bar{x}(s, l)$ are changed to $\bar{\sigma}(s, l)$ and $\bar{\varepsilon}(s, l)$, and the appropriate product of compliance and viscosity is substituted for $S_l F_l$ in $\Gamma_l(s)$, as we shall do further below.

Equations (5.6-11) and (5.6-12) are partial differential equations. Their solutions become

$$\bar{f}(s, l) = f_1(s)\exp[-l\,\Gamma_l(s)] + f_2(s)\exp[l\,\Gamma_l(s)] \tag{5.6-14}$$

and

$$\bar{x}(s, l) = x_1(s)\exp[-l\,\Gamma_l(s)] + x_2(s)\exp[l\,\Gamma_l(s)] \tag{5.6-15}$$

where $f_1(s)$, $f_2(s)$, $x_1(s)$, and $x_2(s)$ are functions of s (or of t in the time domain) but not functions of l. They are not independent of each other but are interrelated through Eqs. (5.6-9) and (5.6-10). Substituting Eqs. (5.6-14) and (5.6-15) into Eq. (5.6-9) yields

$$\begin{aligned} &f_1(s)\Gamma_l(s)\exp[-l\,\Gamma_l(s)] - f_2(s)\Gamma_l(s)\exp[l\,\Gamma_l(s)] \\ &\quad = F_l s\{x_1(s)\exp[-l\,\Gamma_l(s)] + x_2(s)\exp[l\,\Gamma_l(s)]\}\ . \end{aligned} \tag{5.6-16}$$

But this is an identity valid for all values of l. Hence, the coefficients of the same exponentials on both sides of the equation must be equal and we must have

$$f_1(s) = Q_{mC}(s)x_1(s) \qquad \text{and} \qquad f_2(s) = -Q_{mC}(s)x_2(s) \tag{5.6-17}$$

where

$$Q_{mC}(s) = \sqrt{F_l s/S_l} \qquad \text{and} \qquad U_{mC}(s) = \sqrt{S_l/F_l s} \tag{5.6-18}$$

are, respectively, the *characteristic mechanical relaxance* and *characteristic mechanical retardance* of the distributed ladder. The same results would have been obtained had we substituted Eqs. (5.6-14) and (5.6-15) into Eq. (5.6-10). $\Gamma_l(s)$ and $Q_{mC}(s)$ are the *characteristic functions* of the continuous ladder. Equations (5.6-17) relate $x_1(s)$ and $x_2(s)$ to $f_1(s)$ and $f_2(s)$. It remains to determine the latter from the boundary conditions. Clearly, at the beginning of the ladder, i.e. at $l = 0$,

$$\bar{f}(s, 0) = f_1(s) + f_2(s) \tag{5.6-19}$$

and

$$\bar{x}(s, 0) = [f_1(s) - f_2(s)]U_{mC}(s) \tag{5.6-20}$$

from Eqs. (5.6-14), (5.6-15), and (5.6-17). Solving for $f_1(s)$ and $f_2(s)$ leads to

$$f_1(s) = \tfrac{1}{2}\bar{f}(s, 0) + \tfrac{1}{2}Q_{mC}(s)\bar{x}(s, 0) \tag{5.6-21}$$

and

$$f_2(s) = \tfrac{1}{2}\bar{f}(s,0) - \tfrac{1}{2}Q_{mC}(s)\bar{x}(s,0) \ . \tag{5.6-22}$$

Substitution of these relations into Eqs. (5.6-14) and (5.6-15) gives the transforms of the force and of the displacement at the point l as

$$\bar{f}(s,l) = \bar{f}(s,0)\cosh l\,\Gamma_l(s) - Q_{mC}(s)\bar{x}(s,0)\sinh l\,\Gamma_l(s) \tag{5.6-23}$$

and

$$\bar{x}(s,l) = \bar{x}(s,0)\cosh l\,\Gamma_l(s) - U_{mC}(s)\bar{f}(s,0)\sinh l\,\Gamma_l(s) \ . \tag{5.6-24}$$

These equations may be called the *canonical equations* of the transmission line. Letting $l = L$, we have, at the end of the line,

$$\bar{f}(s,L) = \bar{f}(s,0)\cosh L\,\Gamma_l(s) - Q_{mC}(s)\bar{x}(s,0)\sinh L\,\Gamma_l(s) \tag{5.6-25}$$

and

$$\bar{x}(s,L) = \bar{x}(s,0)\cosh L\,\Gamma_l(s) - U_{mC}(s)\bar{f}(s,0)\sinh L\,\Gamma_l(s) \ . \tag{5.6-26}$$

Solving for $\bar{f}(s,0)$ and $\bar{x}(s,0)$ with the aid of the relation $\cosh^2 x - \sinh^2 x = 1$ we find the force and displacement transforms at the beginning of the line to be

$$\bar{f}(s,0) = \bar{f}(s,L)\cosh L\,\Gamma_l(s) + Q_{mC}(s)\bar{x}(s,L)\sinh L\,\Gamma_l(s) \tag{5.6-27}$$

and

$$\bar{x}(s,0) = \bar{x}(s,L)\cosh L\,\Gamma_l(s) + U_{mC}(s)\bar{f}(s,L)\sinh L\,\Gamma_l(s) \ . \tag{5.6-28}$$

Let us now introduce the relations

$$\bar{f}(s,0) = Q_{m0}(s)\bar{x}(s,0) \qquad \text{and} \qquad \bar{x}(s,0) = U_{m0}(s)\bar{f}(s,0) \tag{5.6-29}$$

where $Q_{m0}(s)$ and $U_{m0}(s)$ are the *initial* or *input mechanical respondances*, and

$$\bar{f}(s,L) = Q_{mL}(s)\bar{x}(s,L) \qquad \text{and} \qquad \bar{x}(s,L) = U_{mL}(s)\bar{f}(s,L) \tag{5.6-30}$$

where $Q_{mL}(s)$ and $U_{mL}(s)$ are the *terminal* or *output mechanical respondances*. With the help of these we find

$$Q_{m0}(s) = Q_{mC}(s)\frac{Q_{mL}(s)\cosh L\,\Gamma_l(s) + Q_{mC}(s)\sinh L\,\Gamma_l(s)}{Q_{mC}(s)\cosh L\,\Gamma_l(s) + Q_{mL}(s)\sinh L\,\Gamma_l(s)} \tag{5.6-31}$$

and

$$U_{m0}(s) = U_{mC}(s)\frac{U_{mL}(s)\cosh L\,\Gamma_l(s) + U_{mC}\sinh L\,\Gamma_l(s)}{U_{mC}(s)\cosh L\,\Gamma_l(s) + U_{mL}\sinh L\,\Gamma_l(s)} \ . \tag{5.6-32}$$

We note that, if the line is infinitely long, $Q_{m0}(s) = Q_{mC}(s)$, i.e. such a line appears to be terminated in its characteristic relaxance. This sheds light on the meaning of $Q_{mC}(s)$.

We are now ready to make the transition from mechanical to rheological concepts. As mentioned earlier, we may substitute stress for force and strain for displacement in the preceding equations simply by introducing the appropriate material parameters for the mechanical ones. We first note that $S_l = S/L$ and $F_l = F/L$ where S is the total mechanical compliance, and F is the total frictance of the body. But $bS = J$ and $HF = \eta$ where $H = 1/b$ is the shape factor and J and η are the compliance and the viscosity of the material. To connect the distributed with the discrete extended Gross-Marvin model, we identify J and η with J_T and η_T, the total compliance and viscosity defined in Sect. 5.5.1. It then follows that $S_l F_l = J_T \eta_T / L^2$ and $F_l/S_l = \eta_T/J_T$. Hence, L $\Gamma_l(s)$ becomes $\Gamma(s)$ and, since $Q_m(s) = bQ(s)$, $Q_{mC}(s)$ becomes $bQ_C(s)$, $Q_{mL}(s)$ becomes $bQ_F(s)$, and $Q_{m0}(s)$ becomes $b\bar{Q}(s)$. Thus, Eq. (5.6-31) reduces to Eq. (5.5-47).

Clearly, the distributed Gross-Marvin ladder is the continuous analog of the discrete extended Gross-Marvin model. The relation between the two is thus, in a certain sense, analogous to that between the (continuous) *canonical representation* and the (discrete) *canonical models*. Just as the introduction of the canonical representation led to the new concept of the spectral distribution function, a new concept enters with the introduction of the continuous ladder. The regular lumped ladders discussed in the preceding sections of this chapter simply served as spring-and-dashpot models with specific distributions of respondance times in contrast to the series-parallel models in which the distribution is not specified. As we have just shown, the distributed ladder can be used for the same purpose. However, there is more to it than that. The relations we have derived for the continuous ladder allow us to model the propagation or *transmission* of an excitation (and the response to it) along the length of the ladder. The canonical equations (5.6-23) and (5.6-24) are the equations of a *material transmission line* analogous to the equations of an electric transmission line along which electrical signals are transmitted. Indeed, the continuous Gross-Marvin ladder is the rheological analog of the non-inductive, leakage-free cable well-known from electric transmission line theory (cf. Problem 5.6-1), and may thus also be called the *inertialess material transmission line*.

In electrical transmission line theory one is concerned with very long, relatively thin, conductors or *lines* (cables). In the mechanical counterpart of the theory the term *line* is not really appropriate but we shall adopt it for convenience to refer simply to a suitably shaped piece of material such as a thin rod of uniform circular cross-section (a cylinder) along which a mechanical disturbance in the form of a shear deformation is propagated. What shear deformation can be propagated along a thin cylinder will be discussed in Sect. 11.3.6. There, propagation of disturbances in some other geometries and propagation in deformations other than shear will also be considered.

To develop the theory of the material transmission line further, let us divide the canonical equations, Eqs. (5.6-23) and (5.6-24) by each other. Using Eq. $(5.6\text{-}29)_1$ we find

$$Q_m(s, l) = Q_{mC}(s)\frac{Q_{m0}(s) - Q_{mC}(s)\tanh l\,\Gamma_l(s)}{Q_{mC}(s) - Q_{m0}(s)\tanh l\,\Gamma_l(s)}\,. \tag{5.6-33}$$

Elimination of $Q_{m0}(s)$ between Eqs. (5.6-33) and (5.6-31) leads to

$$Q_m(s, l) = Q_{mC}(s)\frac{Q_{mL}(s) + Q_{mC}(s)\tanh[(L - l)\Gamma_l(s)]}{Q_{mC}(s) + Q_{mL}(s)\tanh[(L - l)\Gamma_l(s)]}\,. \tag{5.6-34}$$

$Q_m(s, l)$ is the mechanical relaxance looking from the point l in the line toward L. Thus, it is really the *input relaxance at the point l*, i.e. the relaxance one would observe if the portion of the line between 0 and l were taken away. Similarly, $Q_m(s,d)$ is the input relaxance at the point $d = L - l$ looking towards the beginning of the line. It is thus the relaxance at d which one would obtain if the portion of the line between d and L were removed. In this case the initial and terminal relaxances must be interchanged and l is replaced by d. We then have

$$Q_m(s,d) = Q_{mC}(s)\frac{Q_{mL}(s) - Q_{mC}(s)\tanh d\,\Gamma_l(s)}{Q_{mC}(s) - Q_{mL}(s)\tanh d\,\Gamma_l(s)} \tag{5.6-35}$$

and

$$Q_m(s,d) = Q_{mC}(s)\frac{Q_{m0}(s) + Q_{mC}(s)\tanh[(L-d)\Gamma_l(s)]}{Q_{mC}(s) + Q_{m0}(s)\tanh[(L-d)\Gamma_l(s)]} . \tag{5.6-35'}$$

We note that $Q_m(s,d=0) = Q_m(s,l=L)$ and $Q_m(s,d=L) = Q_m(s,l=0)$.

It is often convenient to express the initial and terminal respondances in terms of the characteristic ones using the transformations

$$Q_{m0}(s) = Q_{mC}(s)\coth\Gamma_0 \qquad \text{or} \qquad U_{m0}(s) = U_{mC}(s)\tanh\Gamma_0 \tag{5.6-36}$$

and

$$Q_{mL}(s) = Q_{mC}(s)\coth\Gamma_L \qquad \text{or} \qquad U_{mL}(s) = U_{mC}(s)\tanh\Gamma_L . \tag{5.6-37}$$

Γ_0 and Γ_L express the boundary conditions at the two ends of the line. If the end is clamped, we have $Q_{mL}(s) = \infty$ and $U_{mL}(s) = 0$. Both conditions are satisfied when $\Gamma_L = 0$. If the line is open, $Q_{mL}(s) = 0$ and $U_{mL}(s) = \infty$. These conditions are satisfied when $\Gamma_L = j\pi/2$. Using Eq. $(5.6\text{-}37)_1$, Eq. (5.6-34) simplifies to

$$Q_m(s,l) = Q_{mC}(s)\coth[\Gamma_L + (L-l)\Gamma_l(s)] . \tag{5.6-38}$$

The mechanical retardances, $U_m(s,l)$ and $U_m(s,d)$ are, of course, simply the reciprocals of $Q_m(s,l)$ and $Q_m(s,d)$. Sometimes the *mechanical impedance* or the *mechanical admittance* is desired. These are obtained from

$$Z_m(s,l) = s^{-1}Q_m(s,l) \qquad \text{and} \qquad Y_m(s,l) = sU_m(s,l) . \tag{5.6-39}$$

As an instance of the use of the theory of the inertialess mechanical transmission line we mention its application [5] in recasting Blizard's theory of the relaxance of networks formed from bead-and-spring assemblies (cf. Problem 5.5-6). The theory was originally formulated in terms of the electromagnetic analogy (cf. Problem 5.6-2). The direct application of mechanical transmission line theory makes this detour unnecessary [6].

Although the theory of the inertialess material transmission line is useful in modelling respondances, a little reflection shows that it is not suitable for modelling the propagation of an excitation (or the response to it) through a viscoelastic material. There are two main reasons for this. First, in such a propagation the inertia presented by each successive section of the material cannot be neglected. Second, the propagation section cannot be modelled as in Fig. 5.6-2 because the viscoelastic behavior of a section is not properly represented by the combination of a single spring and a single dashpot (shown as two halves in Fig. 5.6-2 purely for convenience). Each section, in fact, possesses the relaxance $\bar{Q}(s)$, which the entire line was intended to model. In the next section we turn to the proper representation of viscoelastic behavior on the basis of the material transmission line with inertia. The effort undertaken in Sect. 5.6.1 is by no means wasted, however, because the formalism remains much the same.

5.6.2 The Material Transmission Line with Inertia

The theory of the inertial line follows closely on that of the inertialess line. We shall develop it first for the so-called *lossless line*, i.e. for the line which contains no dissipative elements. The conditions for the lossless line are approximated in the limit of infinitely high frequencies or infinitely short loading times. Extension to the more general *lossy line* is straight-forward.

The inertial material transmission line is the mechanical analog of the *inductive electrical transmission line*. Because of its importance in power transmission as well as communication, the theory of the electric transmission line is richly developed. Many of its results can be transferred to the material line by analogy.* A comparison of the basic features of electrical and material transmission line theory brings out the similarities as well as the essential differences, and facilitates the handling of the analogies. We therefore undertake this comparison first.

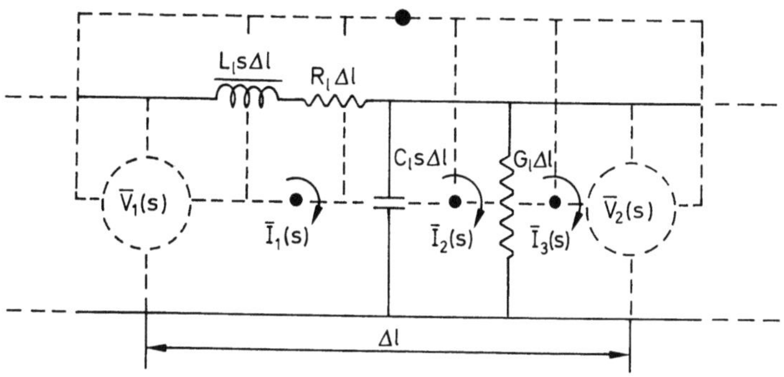

Fig. 5.6-3. Circuit diagram representing an elemental section of the electrical transmission line

The circuit diagram of an infinitesimal section of the electrical transmission line may be represented as shown in Fig. 5.6-3. $\bar{V}(s)$ and $\bar{I}(s)$ are the voltage and the current transforms, and L_l, R_l, C_l, and G_l are, respectively, the inductance, series resistance, capacitance, and shunt (or leakage) conductance of the line per unit length. It is assumed that neither of these immittances are functions of time or, equivalently, frequency. The model diagram representing the mechanical analog of the electric line, using the electrostatic analogy is easily derived from Fig. 5.6-3 by the method discussed in Sect. 3.2 (see Fig. 3.2-3). We obtain the diagram Fig 5.6-4. In accordance with the principles discussed in Sect. 3.2, series connections become parallel ones, and *vice versa*. The passive mechanical elements are represented as

* The reader may recall the motto at the head of Chap. 3.

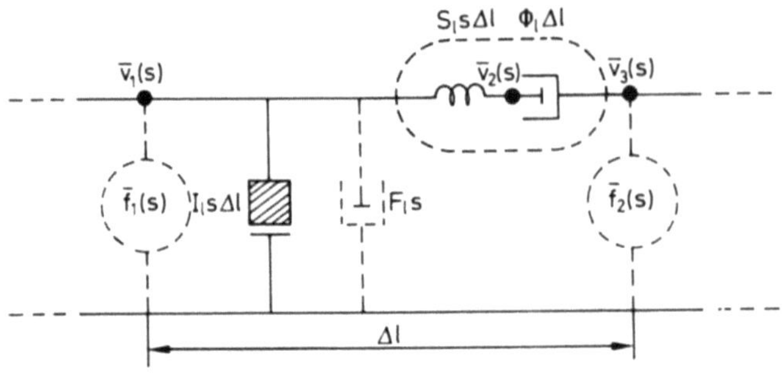

Fig. 5.6-4.
Mechanical model diagram representing the mechanical dual of the electric circuit shown in Fig. 5.6-3

mechanical immittances. Hence, $\bar{v}(s) = s\bar{x}(s)$ is the velocity transform, $\bar{f}(s)$, F_l, and S_l have their previous significance, Φ_l, the glidance (reciprocal frictance) per unit length, is the analog of the leakage conductance, G_l, and I_l, the inertance per unit length, is the analog of the inductance, L_l, of the electric circuit. We can now state why we elected to formulate the theory of the material transmission line in terms of mechanical rather than rheological concepts. The material parameter related to the inertance is the density and this cannot be represented as a proportionality between the stress and the second derivative of the strain without introducing a squared length, i.e. a geometric factor, in addition. We note also that the representation of the inertial line as shown in Fig. 5.6-4 is unique in the sense that the inertance must be featured as a rung relaxance because the inertance must always be referred to ground.

The analog of the series resistance per unit length, R_l, a frictance per unit length, F_l, is absent in the material line and the dashpot representing it has therefore been drawn with broken lines. This is the first of the essential differences between the two transmission lines.

The second essential difference stems from the fact that the nature of the mechanical strut admittance is quite different from that of the electrical shunt admittance. While the latter may usually be represented satisfactorily by a parallel combination of a capacitance and resistance (an electrical Voigt unit), the former cannot be represented by a Maxwell unit for the reasons discussed in Chap. 3. The strut admittance represents the properties of an infinitesimal section of a viscoelastic material and must therefore be shown as $\bar{Y}_{ml}(s) = b\, s\bar{U}_l(s)$, where b is the reciprocal of the shape factor H and $\bar{U}_l(s)$ is the retardance per unit length. This, in general, would be modelled by a Kelvin model. Attention has been drawn to this fact in Fig. 5.6-4 by surrounding the Maxwell unit by a broken-line oval (cf. Sect. 3.1.2).

The presence of the series resistance and the fact that the shunt admittance can be modelled satisfactorily by the parallel combination of a capacitance and a conductance, impart a symmetry to the theory of the electric line which is absent in the material analog. It is a consequence of this lack of symmetry that there is no material counterpart of the so-called distortionless line of the Heaviside type for which $R_l/L_l = G_l/C_l$.

A third (but not essential) difference arises from the fact that we prefer to cast our relations in terms of force and displacement instead of force and velocity.

Once these differences are understood, it becomes a relatively easy matter to switch between the electric and the material transmission line theories. It also becomes clear now why the inertialess material line is not suitable for modelling the propagation of an excitation and the response to it through a viscoelastic medium: it constitutes an attempt to model $\bar{U}(s)$, the strut retardance of the material line, rather than the line itself.

The theory of the material transmission line with inertia is essentially a means for handling inertial effects in the infinitesimal deformation of a viscoelastic material. It is clearly the more comprehensive concept, and subsumes linear viscoelastic theory in which inertial effects are disregarded. Criteria for deciding when this is possible will be discussed further below. Problem 5.6-3 considers the reduction of a problem formulated on the basis of transmission line theory to the simpler situation when inertial effects are negligible. Problem 5.6-4 discusses a simple "lumped mass" approximation which may be useful when the inertial effects are not large. For now we turn to a discussion of the material line with inertia but no loss.

5.6.2.1 *The Lossless Line*

To model the lossless material line, we consider, as in Sect. 5.6.1, a piece of material of length L and of uniform cross-sectional area A, but we now stipulate that the piece consist of a homogeneous, isotropic, purely elastic material having shear modulus G (or, equivalently, shear compliance J). Thus its relaxance, $\bar{Q}(s) = G$, is known. This is in accord with our aim to use transmission line theory not to model a respondance but to model the transmission of an excitation or the response to it through the material. The mechanical model diagram for the elemental section of the line is shown in Fig. 5.6-5.

The diagram is identical with that in Fig. 5.6-4 except that $\bar{x}(s)$ now replaces $\bar{v}(s)$, and that the passive elements are therefore represented as mechanical respondances rather than immittances. In addition, Φ_l, and, of course, F_l, are absent. We note that the lossless mechanical line has an exact counterpart in the lossless

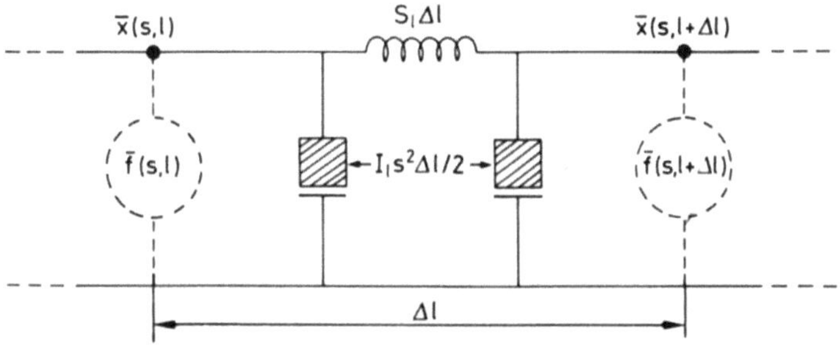

Fig. 5.6-5. Mechanical model diagram of the elemental section of the lossless inertial line

electrical line in which $R_l = G_l = 0$. The representation of the rung relaxance, $I_l s^2$, as two halves is purely a matter of convenience for the purpose of analysis.

Comparing Fig. 5.6-5 with Fig. 5.6-2 we see that the *inertances per unit length*, $I_l s^2$, now appear on the rungs of the ladder. Thus, the derivation of the line equations parallels that in Sect. 5.6.1 with $I_l s^2$ everywhere replacing $F_l s$. As a result, the propagation function is given by $\Gamma_l(s) = s\sqrt{I_l S_l}$, and the characteristic mechanical relaxance by $Q_{mC}(s) = s\sqrt{I_l/S_l}$. I_l and S_l depend on the cross-sectional area, A, of the line. Now, $I_l L$ is the total inertance of the line, and AL is its volume. Hence, I_l/A is the density, ρ. Similarly, $S_l L$ is the total mechanical compliance, S, of the line. But $bS = J$, the shear compliance, and $b = A/L$. We thus have $I_l = A\rho$ and $S_l = J/A$. With these substitutions the propagation function becomes $s\sqrt{\rho J}$. We introduce a separate symbol for the propagation function of the inertial line, and write

$$\gamma(s) = s\sqrt{\rho J}\ . \tag{5.6-40}$$

Of course, $\gamma(s)$, just as $\Gamma_l(s)$, has the dimension of a reciprocal length. The characteristic mechanical relaxance becomes $sA\sqrt{\rho G}$. It is convenient to introduce the *characteristic mechanical relaxance per unit cross-sectional area* which is given by

$$\mathscr{Q}_C(s) = s\sqrt{\rho G}\ . \tag{5.6-41}$$

The script symbol makes it unnecessary to attach the subscript m. We note that

$$\gamma(s)\mathscr{Q}_C(s) = s^2\rho\ . \tag{5.6-42}$$

Since $\bar{\sigma}(s) = \bar{f}(s)/A$, we obtain the analog of Eq. (5.6-31) as

$$\mathscr{Q}_0(s) = \mathscr{Q}_C(s)\frac{\mathscr{Q}_L(s)\cosh L\,\gamma(s) + \mathscr{Q}_C(s)\sinh L\,\gamma(s)}{\mathscr{Q}_C(s)\cosh L\,\gamma(s) + \mathscr{Q}_L(s)\sinh L\,\gamma(s)} \tag{5.6-43}$$

where $\mathscr{Q}_0(s)$ and $\mathscr{Q}_L(s)$ are the *input* and *output mechanical relaxances per unit cross-sectional area*, defined by

$$\bar{\sigma}(s,0) = \mathscr{Q}_0(s)\bar{x}(s,0) \qquad \text{and} \qquad \bar{\sigma}(s,L) = \mathscr{Q}_L(s)\bar{x}(s,L)\ . \tag{5.6-44}$$

Equations analogous to Eqs. (5.6-33) through (5.6-38) are obtained simply by replacing $Q_m(s)$ by $A\mathscr{Q}(s)$ and writing $\gamma(s)$ for $\Gamma_l(s)$.

Let us now consider how a displacement applied at the input end of the lossless line is propagated along it. Equation (5.6-15) becomes

$$\bar{x}(s,l) = x_1(s)\exp[-l\,\gamma(s)] + x_2(s)\exp[l\,\gamma(s)]\ . \tag{5.6-45}$$

For simplicity we assume the line to be infinitely long. Since the displacement applied at $l = 0$ is necessarily finite, it cannot become infinite at $l = L$ and we must have $x_2(s) = 0$. It then follows that $x_1(s) = \bar{x}(s,0)$ and, using Eq. (5.6-40), we obtain

$$\bar{x}(s,l) = \bar{x}(s,0)\exp(-ls\sqrt{\rho J})\ . \tag{5.6-46}$$

Retransformation into the time domain with the aid of the Translation Theorem, Eq. (A3-6), gives

$$x(t,l) = x(t - l\sqrt{\rho J}, 0)\ . \tag{5.6-47}$$

Therefore, the displacement applied at the beginning of the lossless line at time t appears at the point l along its length at time $t + l\sqrt{\rho J}$. Thus, it is *delayed* (shifted) by the amount $l\sqrt{\rho J}$ but is not changed in any other way. In particular, it is neither *attenuated* nor *distorted*. Equation (5.6-47) applies to *any* displacement as long as its Laplace transform exists. Thus, a truly lossless infinitely long line is a *distortionless line*.

In the special case when the displacement is harmonic, i.e. when $x(t, 0) = x_0 \exp(j\omega t)$, we have

$$x(t, l) = x_0 \exp[j\omega(t - l\sqrt{\rho J})] \ . \tag{5.6-48}$$

It is shown in any physics text that this is the equation of a harmonic displacement wave travelling from $l = 0$ toward $l = L$ with velocity* $v = \sqrt{G/\rho}$.

Let us now consider a line of finite length. Clearly, in the frequency domain Eq. (5.6-45) represents the superposition of two waves, one travelling in the forward direction towards $l = L$, the other travelling in the opposite direction, towards $l = 0$. Since the source of excitation initiates a harmonic wave speeding towards L, the backward wave must come into existence through a process of reflection at the far end of the line** and its amplitude and phase must depend on the terminal relaxance, $\mathcal{Q}_L(s)$. We may define an (*operational*) *reflection coefficient*, $r_{xL}(s)$, as the ratio of the transform of the reflected (backward) wave, $\bar{x}_r(s, l) = x_2(s)\exp[l\,\gamma(s)]$, to the incident (forward) wave, $\bar{x}_i(s, l) = x_1(s)\exp[-l\,\gamma(s)]$, at $l = L$. To understand the meaning of $r_{xL}(s)$ we must, however, inquire first into the meaning of $x_1(s)$ and $x_2(s)$.

To do so, we rewrite Eq. (5.6-24) to make it applicable to the lossless line. This leads to

$$\bar{x}(s, l) = \bar{x}(s, 0)\cosh l\,\gamma(s) - \mathcal{U}_C(s)\bar{\sigma}(s, 0)\sinh l\,\gamma(s) \tag{5.6-49}$$

where $\mathcal{U}_C(s) = 1/\mathcal{Q}_C(s)$. Now, by Eqs. $(5.6\text{-}44)_1$ and (5.6-43),

$$\mathcal{U}_C(s)\bar{\sigma}(s, 0) = \frac{\mathcal{Q}_L(s)\cosh L\,\gamma(s) + \mathcal{Q}_C(s)\sinh L\,\gamma(s)}{\mathcal{Q}_C(s)\cosh L\,\gamma(s) + \mathcal{Q}_L(s)\sinh L\,\gamma(s)}\bar{x}(s, 0) \tag{5.6-50}$$

and substitution of this into the preceding equation gives

$$\bar{x}(s, l) = \bar{x}(s, 0)\frac{\mathcal{Q}_C(s)\cosh[(L - l)\gamma(s)] + \mathcal{Q}_L(s)\sinh[(L - l)\gamma(s)]}{\mathcal{Q}_C(s)\cosh L\,\gamma(s) + \mathcal{Q}_L(s)\sinh L\,\gamma(s)} \ . \tag{5.6-51}$$

Separating the hyperbolic functions in the numerator into the exponentials, collecting those with positive and with negative arguments respectively, and comparing the result with Eq. (5.6-45) shows that

$$x_1(s) = X(s)[\mathcal{Q}_C(s) + \mathcal{Q}_L(s)]\exp[L\,\gamma(s)] \tag{5.6-52}$$

---

* It is convenient to recall that, for a travelling wave, $\omega = 2\pi/T$, $\beta = 2\pi/\lambda$, and $\omega/\beta = v = \lambda/T$ where ω is the radian frequency, λ is the wave length, T is the period, v is the wave, or phase, velocity, and β is called the phase constant. Equation (5.6-48) can therefore be written in the equivalent forms $x(t, l) = x_0 \exp[j(\omega t - \beta l)] = x_0 \exp[2\pi j(t/T - l/\lambda)]$.

** Application of an excitation at the input end of the line at $t = 0$ results in backward and forward reflections until a steady-state is reached (cf. Problem 5.6-5). Consideration of these multiple reflections does not change the form of Eq. (5.6-57) below but provides an alternative interpretation for $x_1(s)$.

and

$$x_2(s) = X(s)[\mathcal{Q}_C(s) - \mathcal{Q}_L(s)]\exp[-L\,\gamma(s)] \tag{5.6-53}$$

where

$$X(s) = \frac{\bar{x}(s,0)}{2\,\mathcal{Q}_C(s)\cosh L\,\gamma(s) + 2\,\mathcal{Q}_L(s)\sinh L\,\gamma(s)} = \frac{\bar{x}(s,L)}{2\,\mathcal{Q}_C(s)} . \tag{5.6-54}$$

The last result follows because, by Eqs. (5.6-28) and $(5.6\text{-}44)_2$,

$$\bar{x}(s,0) = \bar{x}(s,L)\mathcal{U}_C(s)[\mathcal{Q}_C(s)\cosh L\,\gamma(s) + \mathcal{Q}_L(s)\sinh L\,\gamma(s)] . \tag{5.6-55}$$

We can now give meaning to the reflection coefficient. It becomes

$$r_{xL}(s) = \frac{\bar{x}_r(s,L)}{\bar{x}_i(s,L)} = \frac{\mathcal{Q}_C(s) - \mathcal{Q}_L(s)}{\mathcal{Q}_C(s) + \mathcal{Q}_L(s)} \tag{5.6-56}$$

and we may recast Eq. (5.6-45) in the form

$$\bar{x}(s,l) = x_1(s)\{\exp[-l\,\gamma(s)] + \exp[-2L\,\gamma(s)]r_{xL}(s)\exp[l\,\gamma(s)]\} . \tag{5.6-57}$$

The reflection coefficient depends both on the characteristic relaxance, a property of the material of the line, and on the nature of the terminal relaxance. This may, e.g., be the internal relaxance of a transducer placed at the end of the line. Particularly simple relations result if the line is either *clamped* or *open.* The latter corresponds to a short-circuited electric line, while the former is the analog of an open-circuit line. For a clamped line, $\mathcal{Q}_L(s) = \infty$, and for an open line $\mathcal{Q}_L(s) = 0$. We shall now consider wave propagation in the frequency domain on a clamped lossless line. Problem 5.6-6 deals with the open line.

When $\mathcal{Q}_L(s) = \infty$, we have $x_1(s) = \bar{x}(s,0)\exp L\,\gamma(s)/2\sinh L\,\gamma(s)$ by Eq. (5.6-52) and $r_{xL}(s) = -1$ by Eq. (5.6-56). Equation (5.6-57) thus becomes

$$\bar{x}(s,l) = \bar{x}(s,0)\frac{\sinh[(L-l)\gamma(s)]}{\sinh L\,\gamma(s)} . \tag{5.6-58}$$

If the excitation is harmonic, $\bar{x}(s,l)$ becomes $x(t,l) = x_0\exp(j\omega t)$, and $\gamma(s)$ becomes $j\beta$ where $\beta = \omega/v$. In the steady state, then,

$$x(t,l) = x_0\exp(j\omega t)\frac{\sin\beta(L-l)}{\sin\beta L} \tag{5.6-59}$$

which, taking the real part of $\exp(j\omega t)$, becomes

$$x(t,l) = x_0'\{\sin[\omega t + \beta(L-l)] - \sin[\omega t - \beta(L-l)]\} \tag{5.6-60}$$

where $x_0' = x_0/2\sin\beta L$. The forward and backward waves have equal amplitudes but opposite signs. Their phase relation is such that $x(t,L) = 0$ at all times, as required by the boundary condition imposed by clamping.

Whenever two waves of equal frequency proceed in opposing directions in the same medium, the phenomenon of *interference* occurs, resulting in *standing waves.* Let us examine the situation along a clamped, lossless line. Equation (5.6-60) tells us that, at any instant of time, t', the displacement at the point $d = L - l$, is given by

$$x(t', d) = 2x'_0 \cos \omega t' \sin \beta d \ . \tag{5.6-61}$$

There will thus be points along the line at which the displacement is zero at all times. These points, called *nodes*, occur when $d_n = \pi n/\beta$ $(n = 0, 1, \ldots)$. The end of a clamped line is necessarily a node. At the points $d_n = \pi(n + 1/2)/\beta$, called *antinodes*, the displacement assumes its largest absolute value. This is twice that of either the forward or the backward wave. At the antinodes the two waves combine with *constructive interference*. The interference at the nodes is *destructive interference*. Equation (5.6-61) is the equation of a *standing wave* because the wave pattern along the line is fixed (the location of the nodes and antinodes is independent of time). It is only the magnitude of the amplitude which depends on the time chosen to inspect the pattern. (See Problem 5.6-6 for the location of nodes and antinodes on an open line.) The analysis of standing wave patterns is of considerable importance, particularly in low-loss lines (see next section), but the matter cannot be pursued here further.

Another matter of practical as well as theoretical importance on which we can only touch here is the phenomenon of *resonance*. Let us first investigate this phenomenon in the clamped line again. By Eq. (5.6-43) the input relaxance of such a line is

$$\mathcal{Q}_0(s) = \mathcal{Q}_C(s) \coth L\,\gamma(s) \ . \tag{5.6-62}$$

In the harmonic steady state $\gamma(s)$ becomes $j\beta = 2\pi j/\lambda$ [cf. Eq. (5.6-59)] and $\mathcal{Q}_C(s)$ becomes $j\omega\sqrt{\rho G} = j\rho\omega v$. Hence

$$\mathcal{Q}_0(\omega) = \rho\omega v \cot(2\pi L/\lambda) \tag{5.6-63}$$

and the input relaxance will increase without limit whenever $L/\lambda = n/2$ $(n = 1, 2, \ldots)$, i.e. when the length of the line is a half-integral multiple of the wave length. Thus $L/\lambda = n/2$ is the condition of resonance. For an open line we find

$$\mathcal{Q}_0(\omega) = \rho\omega v \tan(2\pi L/\lambda) \tag{5.6-64}$$

and the condition of resonance is $L/\lambda = (n + 1/2)/2$.

We have discussed the lossless line primarily in terms of displacement. Analogous considerations apply, *mutatis mutandis*, to stress. Recalling Eqs. (5.6-17) we recognize that

$$\sigma_1(s) = \mathcal{Q}_C(s) x_1(s) \qquad \text{and} \qquad \sigma_2(s) = -\mathcal{Q}_C(s) x_2(s) \tag{5.6-65}$$

and therefore,

$$\bar{\sigma}(s, l) = \mathcal{Q}_C(s)\{x_1(s)\exp[-l\,\gamma(s)] - x_2(s)\exp[l\,\gamma(s)]\} \ . \tag{5.6-66}$$

This equation should be compared with Eq. (5.6-45). The difference in sign in the relation between the transforms of the excitation and the response to it is fundamental in any transmission line. As one consequence of this difference we note that the transforms of the reflection coefficients for stress waves and for displacement waves are connected by the relation $r_{\sigma L}(s) + r_{xL}(s) = 0$.

Before proceeding to a discussion of the lossy line, we must address a question of fundamental importance in the theory of linear viscoelastic behavior: under what conditions can inertial effects in a viscoelastic material be neglected? Problem 3.1-4 describes an experimental arrangement in which a harmonic force is applied at one end of the specimen and the force transmitted through it is picked up at the other end. The experiment is analyzed under the assumption that the frequency is low enough for inertial effects to be negligible. What is the criterion, and how does it arise? To answer this question we return to Eq. (5.6-59). Suppose we demand that the arguments of the sine-functions be small enough so that they may replace the functions themselves. In that case we have

$$x(t,l) = \frac{L-l}{L} x_0 \exp(j\omega t) \tag{5.6-67}$$

and it follows that the strain,

$$\varepsilon(t) = \frac{\partial\, x(t,l)}{\partial l} = -\frac{x_0}{L} \exp(j\omega t) \tag{5.6-68}$$

does not depend on l and is therefore uniform over the length (or thickness), L. But then so is the stress because the stress-strain relation is assumed to be linear. This uniformity justifies the simple treatment used in solving Problem 3.1-4 (cf. also Problems 5.6-3 and 5.6-4). If the arguments are not small enough, the treatment is more complicated. That, however, is a matter of stress analysis and does not concern us here.

To develop the desired criterion further, we note that $|L - l| \leq L$, and that $\sin x \simeq x$ when $x \leq 0.1$ so that we may formulate our criterion by letting the critical length be given by $L_c\beta = 0.1$. Thus the length of the line should be equal to or less than

$$L_c = 0.1/\beta = 0.1\, v/\omega = 0.1\, \lambda/2\pi = 0.1\sqrt{G/\rho}/\omega\ . \tag{5.6-69}$$

L_c is generally considered to be too restrictive and the critical length is usually given as $L'_c = \sqrt{G/\rho}/f$ where f is the frequency in Hz.

The criterion just discussed has been developed for a harmonic excitation. Any non-harmonic periodic excitation can be decomposed by Fourier analysis into a sum of an infinite number of frequency components and the criterion must then be based on the highest frequency that must be retained to assure a reasonable representation of the form of the excitation. If the latter is non-periodic, the Fourier integral takes the place of the sum.

5.6.2.2 *The Lossy Line*

Although, in the limit of very high (theoretically infinite) or very low (theoretically zero) frequency, a material may behave almost as if it were purely elastic, any real material dissipates at least a few percent of the deformational energy in the form of heat at any finite, non-zero frequency. Although useful as a didactic tool, the lossless line therefore serves at best as an approximation to the theory of the behavior of a real material at high frequencies or short loading times. We are accordingly com-

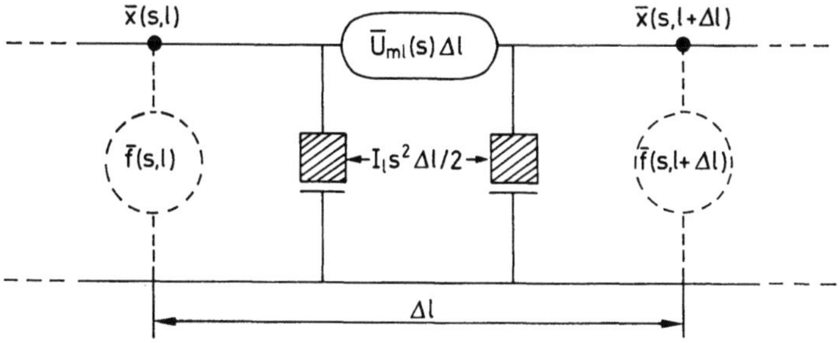

Fig. 5.6-6. Mechanical model diagram of an elemental section of the lossy inertial line

pelled to consider the theory of the so-called *lossy line*. Extension of the theory of the lossless to that of the lossy line is simple but results in a number of important differences.

To effect the change we simply invoke the *correspondence principle* (cf. Sect. 2.1.2) and replace the mechanical compliance per unit length, S_l, by its viscoelastic equivalent in the complex domain, the mechanical retardance per unit length, $\bar{U}_{ml}(s)$. The elemental section of the lossy line is shown in Fig. 5.6-6.

By the arguments used with respect to S_l in case of the lossless line, we have $\bar{U}_{ml}(s) = \bar{U}(s)A$, and the characteristic line functions, the propagation function per unit length, $\gamma(s)$, and the characteristic mechanical relaxance per unit area, $\mathcal{Q}_C(s)$, of the lossy line become

$$\gamma(s) = s\sqrt{\rho\,\bar{U}(s)} \tag{5.6-70}$$

and

$$\mathcal{Q}_C(s) = s\sqrt{\rho\,\bar{Q}(s)}\ . \tag{5.6-71}$$

In the harmonic steady state these equations take the form

$$\gamma^*(\omega) = j\omega\sqrt{\rho\,J^*(\omega)} \tag{5.6-72}$$

and

$$\mathcal{Q}_C^*(\omega) = j\omega\sqrt{\rho\,G^*(\omega)} \tag{5.6-73}$$

where $J^*(\omega)$ and $G^*(\omega)$ are, respectively, the complex compliance and complex modulus of the material. These take the place of the constants J and G of the lossless line. Hence, in the harmonic steady state the displacement along a lossy line of infinite length [cf. Eq. (5.6-48)] becomes

$$x(t,l) = x_0\exp(j\omega t)\exp[-j\omega l\sqrt{\rho\,J^*(\omega)}]\ . \tag{5.6-74}$$

We let

$$\gamma^*(\omega) = \alpha(\omega) + j\,\beta(\omega) \tag{5.6-75}$$

where

$$\alpha(\omega) = \omega\sqrt{\rho\,\tilde{J}(\omega)}\sin[\theta(\omega)/2] \tag{5.6-76}$$

is the *attenuation function*, and

$$\beta(\omega) = \omega\sqrt{\rho\, \mathfrak{J}(\omega)}\cos[\theta(\omega)/2] \tag{5.6-77}$$

is the *phase function*, in which $\mathfrak{J}(\omega)$ and $\theta(\omega)$ are the absolute compliance and the loss angle of the material [cf. Eq. (2.5-19)]. The attenuation function has no counterpart [$\alpha(\omega) = 0$] in the theory of the lossless line. The phase function, $\beta(\omega)$, takes the place of the phase constant, β. Thus

$$x(t, l) = x_0 \exp[-l\,\alpha(\omega)]\exp\{j[\omega t - l\,\beta(\omega)]\} \tag{5.6-78}$$

and we see that a harmonic displacement applied at $l = 0$ to a lossy line is not only *delayed* [by the amount $l\,\beta(\omega)$], but also *attenuated* or *damped* (by the amount $l\,\alpha(\omega)$) and both effects are functions of the frequency. By contrast, in the lossless line there is no attenuation and the delay does not depend on the frequency. Numerically, attenuation is expressed in *nepers* per length. A displacement or stress wave is said to experience an attenuation of N nepers when its magnitude changes by a factor of $\exp(-N)$ as the wave travels between two points on the line. The word neper is the Latin form of Napier*. Another measure of attenuation, more commonly used in electric transmission line theory and electronics, is the *decibel* (db). A signal experiences an attenuation of D decibels when its magnitude changes by a factor of $10^{-D/20}$ between two points. One decibel is a tenth of a *bel*, named after Alexander Graham Bell**.

As a consequence of the phenomenon of damping, resulting from the viscous losses in the material, a lossy line is effectively a line of infinite length if the attenuation is large enough so that the forward (incident) wave is effectively damped out before it can generate a backward (reflected) wave at the end of the line. Under these circumstances the storage and loss moduli or compliances of the material can be obtained from either characteristic line function. If the damping is such that the amplitude remains measurable over a distance equivalent to a few wave lengths, one may use the propagation function for this purpose. Combining Eqs. (5.6-72) and (5.6-75) and separating real and imaginary parts gives

$$J'(\omega) = \frac{\beta^2(\omega) - \alpha^2(\omega)}{\omega^2\rho} \tag{5.6-79}$$

and

$$J''(\omega) = \frac{2\,\alpha(\omega)\beta(\omega)}{\omega^2\rho} \tag{5.6-80}$$

where $\alpha(\omega) = 1/x_0(\omega)$, $x_0(\omega)$ being the distance within which the amplitude falls off by a factor of $1/e$, and $\beta(\omega) = \omega v$ where v is the phase velocity. Since $G^*(\omega) = 1/J^*(\omega)$, the moduli become

$$G'(\omega) = \frac{\omega^2\rho[\beta^2(\omega) - \alpha^2(\omega)]}{[\beta^2(\omega) + \alpha^2(\omega)]^2} \tag{5.6-81}$$

---

* John Napier, Laird of Merchiston, 1550–1617, Scottish inventor of logarithms.

** Alexander Graham Bell, 1847–1922, Scottish born USA audiologist, inventor of the telephone.

and

$$G''(\omega) = \frac{2\omega^2\rho\,\alpha(\omega)\beta(\omega)}{[\beta^2(\omega) + \alpha^2(\omega)]^2} \; . \tag{5.6-82}$$

The substitutions $\beta(\omega) = 2\pi/\lambda$, or $v = \omega/\beta(\omega)$, or $r = \alpha(\omega)/\beta(\omega)$ furnish alternative expressions in terms of the experimentally observable quantities, i.e. the damping and the wave length or the velocity.

If the damping is quite rapid (as in many liquids, cf. Problem 5.6-8) one may take advantage of the characteristic mechanical relaxance per unit area, Eq. (5.6-73). By Eq. (5.6-62) the input relaxance of a line of infinite length is equal to its characteristic relaxance. Hence, in this case,

$$\mathscr{Q}_0^*(\omega) = j\omega\sqrt{\rho\, G^*(\omega)} = \mathscr{Q}_C^*(\omega) \; . \tag{5.6-83}$$

Using Eq. (5.6-39) we derive

$$\mathscr{Z}_0^*(\omega) = \sqrt{\rho\, G^*(\omega)} = \mathscr{Z}_C^*(\omega) \tag{5.6-84}$$

where $\mathscr{Z}_C^*(\omega)$ is the *characteristic mechanical impedance per unit cross sectional area* of the line. In the case of rapid damping it can simply be measured as the ratio of the harmonic force (per unit area) to the harmonic velocity at the input end of the line. In terms of its real and imaginary parts we have

$$[\mathscr{R}_C(\omega) + j\,\mathscr{X}_C(\omega)]^2 = \rho[G'(\omega) + j\,G''(\omega)] \tag{5.6-85}$$

where* $\mathscr{R}_C(\omega) = \mathscr{Z}_C'(\omega)$ and $\mathscr{X}_C(\omega) = \mathscr{Z}_C''(\omega)$. Solving Eq. (5.6-85) for the moduli gives

$$G'(\omega) = \frac{\mathscr{R}_C^2(\omega) - \mathscr{X}_C^2(\omega)}{\rho} \tag{5.6-86}$$

and

$$G''(\omega) = \frac{2\mathscr{R}_C(\omega)\mathscr{X}_C(\omega)}{\rho} \; . \tag{5.6-87}$$

It should be noted that whether the propagation function or the characteristic impedance form the basis, the determination of the moduli and compliances does not require knowledge of a shape factor. This is, of course, a consequence of the fact that the medium is considered to be effectively of infinite extent.

If the damping is small, so that the incident wave will be reflected at the end of the line, standing waves or resonances can be utilized to determine the complex moduli or compliances. A discussion of the experimental techniques is, however, outside of the scope of this book.

In either case, the critical length now becomes

$$L_c = 1/\beta(\omega) = 0.1\sqrt{\tilde{G}(\omega)/\rho} \tag{5.6-88}$$

which is usually relaxed to $L_c' = \sqrt{G'(\omega)/\rho}/f$.

---

* In the literature the symbols $\mathscr{R}$ and $\mathscr{X}$ usually replace the more systematic $\mathscr{Z}_C'$ and $\mathscr{Z}_C''$.

We have seen that a harmonic excitation propagated along a lossy line is both *delayed* and *attenuated* but it is not *distorted*, i.e. its shape is not changed apart from the decrease in amplitude. However, *any other* excitation (and, of course, the response to it) travelling along a lossy line also becomes distorted, i.e. its shape is not preserved. A lossless line, by contrast, is free of distortion for any excitation [cf. Eq. (5.6-47)]. To examine the phenomenon of distortion, let us consider a *gate pulse train* (i.e. a series of repeated gate functions) as shown in the top portion of Fig. 5.6-7. The unit gate function is discussed in Appendix A2.3. Its Laplace transform, by LTP (8), is $[1 - \exp(-t's)]/s$. As shown in Sect. 7.3 [cf. Eq. (7.3-2)], the transform of a displacement gate pulse train is then

$$\bar{x}(s) = \frac{x_0[1 - \exp(-t's)]}{s[1 - \exp(-Ts)]} \tag{5.6-89}$$

where T is the period. Now, for a well-damped (effectively infinite) line we have [cf. Eq. (5.6-46)]

$$\bar{x}(s, l) = \bar{x}(s, 0)\exp[-l\,\gamma(s)] \ . \tag{5.6-90}$$

With Eq. (5.6-89) for the transform of the excitation applied at $l = 0$, we obtain

$$\bar{x}(s, l) = x_0 \frac{[1 - \exp(-t's)]\exp[-l\,\gamma(s)]}{s[1 - \exp(-Ts)]} \ . \tag{5.6-91}$$

We now invert the transform with the help of the Residue Theorem (see Appendix A3.2.1). We have

$$x(t, l) = \sum_k \mathrm{Res}_k[\bar{x}(s, l)\exp(ts)] \ , \tag{5.6-92}$$

the residues associated with the poles $s_k = 2\pi kj/T$ $(k = 0, \pm 1, \pm 2, \ldots)$, being obtained from

$$\mathrm{Res}_k = x_0 \lim_{s \to s_k} (s - s_k) \frac{[1 - \exp(-t's)]\exp[-l\,\gamma(s)]}{s[1 - \exp(-Ts)]} \exp(ts) \ . \tag{5.6-93}$$

Taking the residue at $s = 0$ first, we find

$$\mathrm{Res}_0 = x_0 \lim_{s \to 0} \frac{1 - \exp(-t's)}{1 - \exp(-Ts)} = \frac{x_0 t'}{T} \ . \tag{5.6-94}$$

The remaining residues then become

$$\mathrm{Res}_k = x_0 \frac{\{\exp(j\omega_k t) - \exp[j\omega_k(t - t')]\}\exp[-l\,\gamma^*(\omega_k)]}{2\pi kj} \tag{5.6-95}$$

where we have used

$$\lim_{s \to s_k} \frac{s - s_k}{1 - \exp(-Ts)} = \lim_{s \to s_k} \frac{1}{T\exp(-Ts_k)} = \frac{1}{T} \tag{5.6-96}$$

and $\omega_k = 2\pi k/T$ $(k = \pm 1, \pm 2, \ldots)$. From Eq. (5.6-75) we have

$$\exp[-l\,\gamma^*(\omega_k)] = \exp[-l\,\alpha(\omega_k)]\exp[-jl\,\beta(\omega_k)] \tag{5.6-97}$$

and, therefore, find

$$\mathrm{Res}_k = x_0 \exp[-l\,\alpha(\omega_k)] \frac{\exp\{j[\omega_k t - l\,\beta(\omega_k)]\} - \exp\{j[\omega_k(t-t') - l\,\beta(\omega_k)]\}}{2\pi k j} . \tag{5.6-98}$$

Equation (5.6-92) becomes

$$x(t,l) = \mathrm{Res}_0 + \sum_{k=1}^{\infty} (\mathrm{Res}_k + \mathrm{Res}_{-k}) , \qquad k = 1, 2, \dots . \tag{5.6-99}$$

But $\alpha(\omega_k)$ is an even and $\beta(\omega_k)$ is an odd function of ω_k. Hence

$$x(t,l) = x_0 \Big[t'/T + \sum_{k=1}^{\infty} (\pi k)^{-1} \exp[-l\,\alpha(\omega_k)] \{\sin[\omega_k t - l\,\beta(\omega_k)] - \sin[\omega_k(t-t') - l\,\beta(\omega_k)]\} \Big] . \tag{5.6-100}$$

The series converges fairly rapidly because the contribution of the higher frequency terms to the sum are suppressed by the $(\pi k)^{-1}$ factor. The excitation, $x(t,l)$, is plotted as a function of t for several values of l in Fig. 5.6-7 with $t' = 1$ second and $T = 3t'$. $J^*(\omega)$ was taken as the complex compliance of the standard linear solid with the parameters given by Eqs. (3.4-86). Thirty to forty terms in the sums appears to be sufficient.

Because the system is in the steady state (the excitation is a pulse *train*), there is no attenuation with time. Also, t_0 is chosen arbitrarily. The total phase shift is $\Delta t = mT + \Delta' t$ and of this only the shift $\Delta' t$ relative to mT shows up in the plot. Thus, $x(t, 100)$ appears to be shifted less than $x(t, 50)$ but this is not a valid inference because the number of periods, m, involved in the two cases is not the same. The

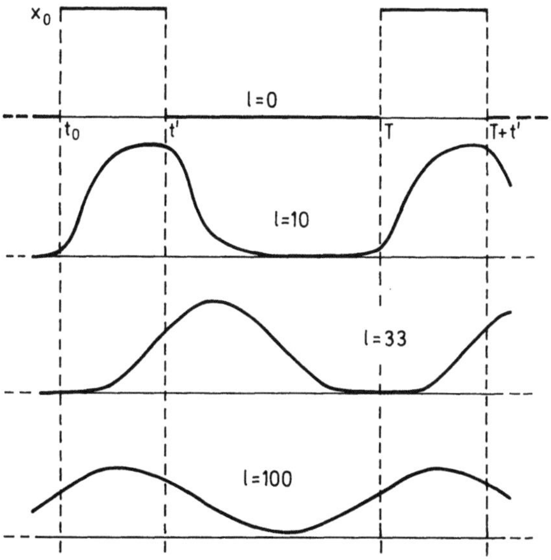

Fig. 5.6-7. Propagation of a displacement gate pulse train as a function of time for several values of l

amount of delay as function of l can be seen unambiguously only in the case of a single pulse. Nevertheless, the plot does show the phenomenon of delay, and it also shows rather clearly the phenomena of attentuation and distortion. It is also obvious that both increase with the distance, l, along the line, and that individual pulses may merge into each other as l increases.

The explanation of these phenomena is straight-forward. As any periodic function, the gate pulse train can be decomposed by Fourier analysis into a spectrum of harmonic waves which superpose in such a manner that the original function is reproduced through just the right constructive and destructive interference. As shown in Problem 5.6-10 the Fourier expansion of the unit gate pulse train is

$$h_{t'}(t;T) = t'/T + \sum_{k=1}^{\infty} (\pi k)^{-1}[\sin \omega_k t - \sin \omega_k(t - t')] \ . \tag{5.6-101}$$

Each harmonic wave of frequency ω_k travels through the medium with its own velocity, $v_k = \omega_k/\beta(\omega_k)$, and attenuation, $\alpha(\omega_k)$, because the medium is *dispersive*, i.e. its properties depend on frequency. As a consequence of the superposition of the totality of waves the original shape of the pulse is not preserved. Moreover, the change of shape, or *distortion*, depends on the distance, l, over which the pulse has travelled.

Pulse methods offer many experimental advantages and are widely used to measure material properties at high frequencies. Pulse techniques often involve reflection (echoing) of the pulse from suitably chosen terminations. In measurements based on pulse techniques the experimenter usually observes the attenuation and the velocity of propagation of the pulse or a quantity related to the latter. This may be, e.g., the *transit time*, i.e. the time the pulse requires to traverse a known distance. The attenuation per unit length is obtained from

$$a = \frac{\ln(h_1/h_2)}{l_2 - l_1} \tag{5.6-102}$$

where h_1 and h_2 are the pulse heights observed at l_1 and l_2, respectively. The velocity is

$$v = \frac{l_2 - l_1}{\Delta t} \tag{5.6-103}$$

where Δt is the transit time. These quantities express the true attenuation and velocity of the pulse and can furnish valuable information on the behavior of the material. However, the velocity with which the pulse travels is not a *phase* but a *group velocity*. It represents the velocity of propagation of the energy contained in the "packet" of harmonic waves, each travelling with its own phase velocity and frequency, which together make up the pulse. Because of the lack of a unique frequency the components of the complex modulus, $G^*(\omega)$, or the complex compliance, $J^*(\omega)$, which require specification of the radian frequency, ω, cannot be obtained from such measurements in general. However, if the excitation consists of a train of gated high frequency oscillations, the measured properties closely reflect the behavior of the material at the *carrier frequency* provided the pulse duration is long enough so that an approximate steady state is reached, and small enough to prevent the development of standing waves. The harmonic respondances can then be obtained from Eqs. (5.6-79) to (5.6-82) or (5.6-86) and (5.6-87).

5.7 Problems

Problem 5.1-1. Obtain the relaxance, $\bar{Q}(s) = \bar{\sigma}_0(s)/\bar{\varepsilon}_0(s)$, of the ladder model shown in Fig. P5.1-1 by the method of node equations (see Sect. 3.1.2).

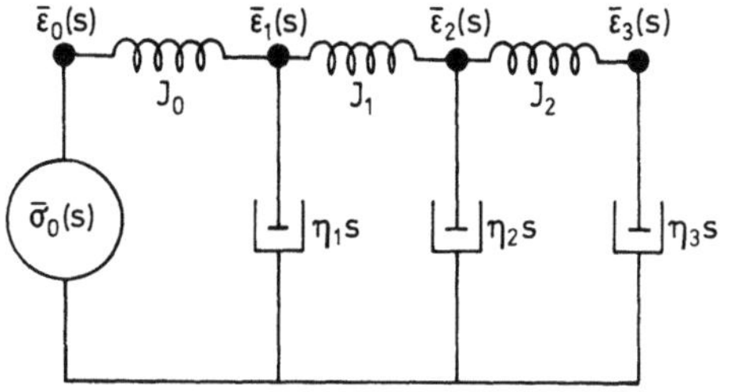

Fig. P5.1-1. General obverse ladder model with 6 elements

Problem 5.1-2. Find the relaxance of the model displayed in Fig. P5.1-1 by applying the *combination rules* (cf. Sect. 3.1.2), instead of the method of node equations used in Problem 5.1-1.

[*Hint:* use continued fractions.]

Problem 5.2-1. Gross [1] analyzed the regular obverse ladder model in the form shown in Fig. P5.2-1 in which the first and last springs have only half the compliance of the others.

(A) Find the respondances of this model as well as the parameters of the equivalent Wiechert and Kelvin models.

(B) The model is arrheodictic. What are the respondances of the rheodictic form?

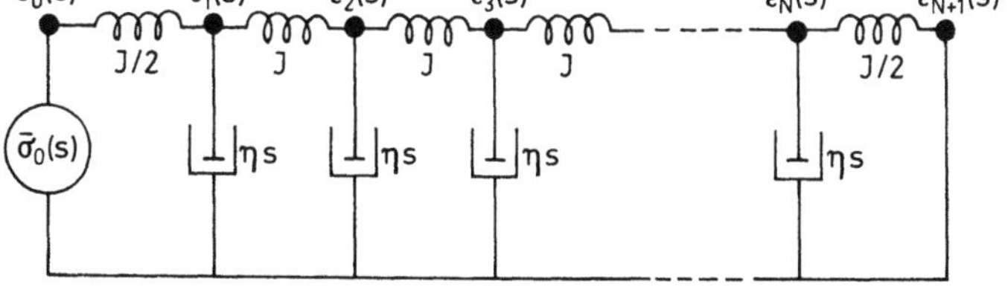

Fig. P5.2-1. Ladder model of Gross

Problem 5.2-2. Investigate the properties of the Gross-Marvin models by Π-section analysis.

Problem 5.2-3. Show that

$$\begin{bmatrix} \cosh a & U_T \sinh a \\ Q_T \sinh a & \cosh a \end{bmatrix}^N = \begin{bmatrix} \cosh Na & U_T \sinh Na \\ Q_T \sinh Na & \cosh Na \end{bmatrix} \tag{1}$$

provided that $Q_T U_T = 1$.

Problem 5.2-4. According to Eq. (5.2-32)

$$\bar{Q}^{\times}_{GM}(s) = \frac{\sqrt{\theta s}\cosh[(N - \frac{1}{2})a(s)]}{J \sinh N\, a(s)} . \tag{1}$$

But, by Eqs. (5.2-39) and (5.2-90), $N = G_g/G_e$. Thus, realistically, N is of the order of 1000. One is then tempted to write

$$\bar{Q}^{\times}_{GM}(s) = \frac{\sqrt{\theta s}}{J} \coth N\, a(s) \tag{2}$$

What are the consequences of this? Compare the behavior of $\bar{Q}^{\times}_{GM}(s)$ defined by Eqs. (1) and (2) as $s \to \infty$.

Problem 5.2-5. Find $G^{\circ}_{GM}(t)$ from

$$\bar{Q}^{\circ}_{GM}(s) = \frac{\sqrt{\theta s}\sinh N\, a(s)}{J \cosh[(N + \frac{1}{2})a(s)]} \tag{1}$$

[cf. Eq. (5.2-27)] by inversion of $\bar{G}^{\circ}_{GM}(s)$.

Problem 5.2-6. Show that

$$\lim_{x \to 0} \frac{x}{\tanh[2N \sinh^{-1}(x/2)]} = \frac{1}{N} \tag{1}$$

Problem 5.2-7. Prove that the respondances of the Gross-Marvin models, Eqs. (5.2-27) to (5.2-30), and (5.2-32) to (5.2-35), do *not* have branch points at $s = 0$, $s = -4/\theta$ and $s = \infty$.

{*Hint*: it is sufficient to show that

$$f_1(s) = \sqrt{(4 + \theta s)/\theta s}\, \coth N\, a(s) \tag{1}$$

[cf. Eq. (5.2-30)] and

$$f_2(s) = \sqrt{\theta s(4 + \theta s)}\, \coth N\, a(s) \tag{2}$$

do not contain branch points because, if this is true, $\bar{U}^{\circ}_{GM}(s)$ and $\bar{Q}^{\times}_{GM}(s)$ will have no branch points and neither will their reciprocals, $\bar{Q}^{\circ}_{GM}(s)$ and $\bar{U}^{\times}_{GM}(s)$ [cf. Eq. (5.2-33) and note that $\eta s\sqrt{1 + 4/\theta s} = (1/J)\sqrt{\theta s(4 - \theta s}]$. To prove the point, show that a complete circuit along a (suitably chosen) closed contour around the suspected branch points does not lead to a change in the values of either $f_1(s)$ or $f_2(s)$.}

Problem 5.2-8. Show that the input relaxance of the T-section of the Gross-Marvin models is the characteristic relaxance, $Q_T(s)$, itself, if the T-section is terminated in $Q_T(s)$.

Problem 5.2-9. Show that $\bar{U}^{\times}_{GM}(s)$ is regular at $s_n = -4/\theta$.

Problem 5.2-10. Find the instantaneous modulus of the Gross-Marvin model in terms of the compliance parameter J by applying the limit value theorem to Eqs. (5.2-28) and (5.2-33).

Problem 5.2-11. Obtain the steady-flow viscosity of the rheodictic Gross-Marvin model by applying the limit value theorem to Eq. (5.2-28).

Problem 5.2-12. Find the eigenvalues of the matrix

$$\begin{bmatrix} 1 + \theta s/2 & \eta s \\ J(1 + \theta s/4) & 1 + \theta s/2 \end{bmatrix} .$$

Problem 5.2-13. Raise the matrix

$$M' = \begin{bmatrix} 1 + \theta s/2 & \eta s \\ J(1 + \theta s)/4 & 1 + \theta s/2 \end{bmatrix} \quad (1)$$

to the Zth power.

Problem 5.3-1. Find the series-parallel model which has the same retardance as $\bar{U}^{\circ}_{RC}(s)$.

Problem 5.4-1. Express $G^{\times}_{GM}(t)$ and $G^{\times}_{RC}(t)$ in terms of G_e, and of θ or $\hat{\theta}$ respectively.

Problem 5.5-1. Find the retardance of the arrheodictic form of the Gross model as the number of nodes is increased without limit. Compare the result with Eq. $(5.5\text{-}14)_2$.

Problem 5.5-2. Show that both characteristic relaxances, $Q_T(s)$ and $Q_\Pi(s)$, reduce to the same expression as the number of nodes is increased without limit.

Problem 5.5-3. Obtain the relaxation moduli and creep compliances of both the rheodictic and arrheodictic forms of the extended Gross-Marvin models.

Problem 5.5-4. (A) Obtain the storage and loss modulus of the arrheodictic extended Gross-Marvin model in closed form.

(B) Find the equilibrium modulus from the storage modulus.

Problem 5.5-5. (A) Show that

$$\theta^{\circ}_{\infty} = 3\langle\tau^2_M\rangle/\langle\tau_M\rangle \ . \quad (1)$$

(B) Express the characteristic time, θ°_{∞}, in terms of the relaxation modulus, $G(t)$, and in terms of the components of the complex modulus, $G^*(\omega)$.

Problem 5.5-6. Synthetic polymers are materials composed of flexible long-chain molecules. The long-range cooperative motions required for long-chain molecules to accommodate to an imposed stress or strain result in relatively long respondance times. Thus, the *dispersion zone*, i.e. the transition from short time to long time behavior is normally within the time scale of ordinary experience. Synthetic polymers therefore are viscoelastic materials *par excellence.*

In molecular theories of the behavior of synthetic polymers extensive use has been made of the so-called *bead-and-spring model.* The molecular chain is replaced by a linear assembly of massless beads linked by Newtonian springs. Energy dissipation during deformation or during recovery from deformation is modelled by the frictional forces the beads experience as they move through their surroundings. Energy storage, on the other hand, is modelled through the deformation of the springs. A schematic two-dimensional representation of the bead-and-spring model is shown in Fig. P5.5-6a.

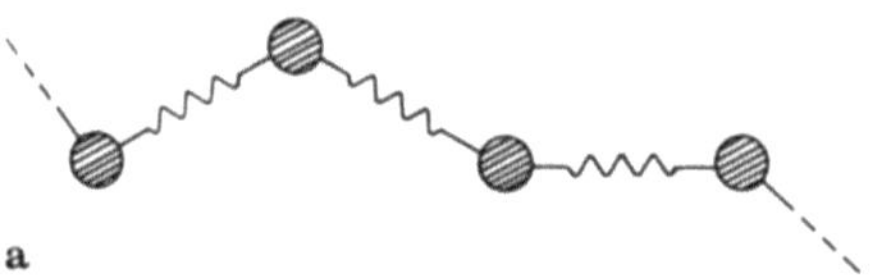

a

Fig. P5.5-6a. Bead-and-spring model

In the seminal theory of Rouse [7] the relaxance of the model in the customary limit of an infinite number of beads, is obtained as

$$\bar{Q}_R(s) = G \sum_{n=1}^{\infty} \frac{\tau_n s}{1 + \tau_n s} \tag{1}$$

where $\tau_n = \theta_R / n^2 \pi^2$, $G = 2/5J_e^\circ$ and $\theta_R = 15J_e^\circ \eta_f$.

(A) Using the method described in Sect. 3.1.2, draw the mechanical model diagram corresponding to the bead-and-spring assembly and from it obtain the expression for the relaxance in closed form.

(B) Express the relaxance as given by the Rouse theory in closed form also and compare the two results.

Problem 5.5-7. The real and imaginary components of the complex modulus and the complex compliance of the extended Marvin-Oser model can be obtained without too much trouble. Upon substitution of $j\omega$ for s in Eqs (5.5-13) to (5.5-17) it is found that they contain either the form $(\alpha + j\beta)\tanh(\alpha + j\beta)$ or the form $(\alpha + j\beta)\coth(\alpha + j\beta)$.

(A) Find the real and imaginary parts of these forms.

(B) Determine α and β for the extended Marvin-Oser model and indicate, as an example, how $G'_{MO\infty}(\omega)$ would be obtained.

(C) Find α and β for the extended Gross-Marvin model. How do these compare with those used in Problem 5.5-4?

Problem 5.5-8. With Eq. (5.5-92) we showed that the steady-state compliance of the extended Marvin-Oser model is

$$J_e^\circ = J_T^\circ/3 + 1/G_T^\circ \tag{1}$$

where J_T° is the sum of the compliances along the strut of the ladder and G_T° is the sum of the moduli on its rungs. Equation (1) was obtained from

$$J_e^\circ = \lim_{s\to 0} [\bar{U}_{MO\infty}(s) - 1/\eta_f s] \tag{2}$$

making use of Eq. (5.5-77). There are, of course, several other ways of obtaining J_e°. As a comparative exercise, find J_e° (A) from Eq. (2) but without using Eq. (5.5-77), (B) from the equivalent Kelvin model using the relation

$$J_e^\circ = J_g + \sum_n J_n\ , \tag{3}$$

and (C) from the equivalent Wiechert model using the relation

$$J_e^\circ \eta_f^2 = \sum_n \tau_n \eta_n \tag{4}$$

[cf. Eq. (4.6-42)].

Problem 5.5-9. Write the matrix equation for a ladder with an infinite number of nodes, using the characteristic respondances and the transmittance in the matrix for the propagation section. Then derive Eq. (5.5-47).

Problem 5.5-10. Find the retardance, $\bar{U}_\infty(s)$, of the extended ladder models having an infinite number of nodes.

Problem 5.6-1. Modify the mechanical model diagram of the elemental section of the inertialess viscoelastic transmission line shown in Fig. 5.6-2 so that the nodes represent velocity instead of displacement. Then derive the analog electric circuit using the electrostatic analogy (see Sect. 3.2).

Problem 5.6-2. Modify the diagram in Fig. 5.6-2 as instructed in Problem 5.6-1. Then derive the analog electric circuit using the electromagnetic analogy.

Problem 5.6-3. You are required to analyze the experimental arrangement described in Problem 3.1-4 *without* the assumption that the specimen inertia can be neglected.

(A) Draw the mechanical model diagram showing the initial, propagation, and final section.

(B) Derive the equation furnishing the relation between the measured quantities, $f(\omega, L)$, and $x(\omega, 0)$.

(C) Reduce both the model diagram and the equation to the case when the inertia *can* be neglected.

Problem 5.6-4. When inertial effects can be neglected, $G^*(\omega) = Q_0^*(\omega)$ where $Q_0^*(\omega)$ is the harmonic relaxance measured at the input end of the line. Equation (5.6-20) presents a criterion for the critical length of the line below which it becomes possible to neglect the effect of specimen inertia by substituting their arguments for the sine functions themselves in Eq. (5.6-59). Retaining the first *two* terms in the expansion of the sine functions instead of the first only, $G^*(\omega)$ can be related to $Q_0^*(\omega)$ through a *lumped mass* approximation. Derive this relation and explain the term "lumped mass".

[*Hint*: use $\sigma(\omega) = G^*(\omega)\varepsilon(\omega)$ and note that $\sigma(\omega)$ is the stress exerted by the source of excitation on the sample, while the stress (force) used in deriving the canonical equations of the line is the stress presented by it to the source.]

Problem 5.6-5. The harmonic steady-state form of Eq. (5.6-57) shows that the displacement along the line is the result of the superposition of two waves: one travelling out from the source of excitation (the driving point), the other a reflected wave originating at the termination of the line. Now, it is clear that the reflected wave must again be reflected at the driving point and this process must be repeated until a steady state is reached.

Derive the displacement transform in terms of these multiple reflections and show that the phenomenon of multiple reflection does not affect the form of Eq. (5.6-45) but provides an alternative interpretation of the factor $x_1(s)$ in that equation.

[*Hint*: introduce the mechanical relaxance at the driving point, $Q_{mD}(s)$, as shown in Fig. P5.6-5, where $\bar{f}_D(s)$ is the driving transform, $\bar{x}_1(s, 0)$ is the value at $l = 0$ of the transform of the initial outgoing wave and $Q_{m0}(s)$ is the mechanical relaxance of the line at the same point. Then consider the superposition of successive waves reflected from both ends of the line.]

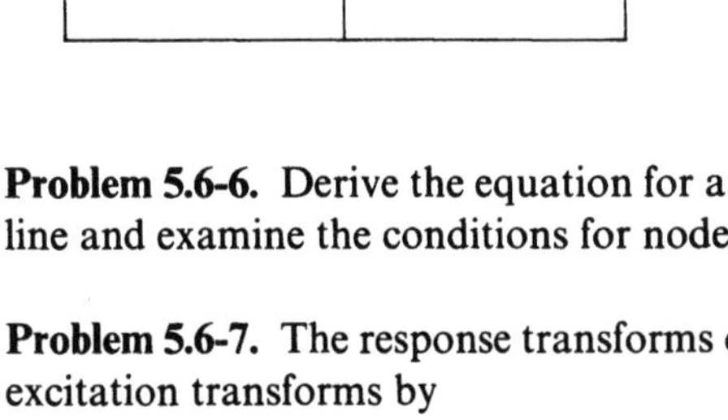

Fig. P5.6-5. Mechanical model diagram representing a source of excitation

Problem 5.6-6. Derive the equation for a harmonic displacement wave on an open line and examine the conditions for nodes and antinodes along the line.

Problem 5.6-7. The response transforms of the inertial line are given in terms of the excitation transforms by

$$\bar{\sigma}(s, l) = \mathcal{Q}_C(s)\frac{\mathcal{Q}_L(s)\cosh d\,\gamma(s) + \mathcal{Q}_C(s)\sinh d\,\gamma(s)}{\mathcal{Q}_C(s)\cosh L\,\sigma(s) + \mathcal{Q}_L(s)\sinh L\,\sigma(s)}\bar{x}(s, 0) \tag{1}$$

and by

$$\bar{x}(s, l) = \mathcal{U}_C(s)\frac{\mathcal{U}_L(s)\cosh d\,\sigma(s) + \mathcal{U}_C(s)\sinh d\,\sigma(s)}{\mathcal{U}_C(s)\cosh L\,\gamma(s) + \mathcal{U}_L(s)\sinh L\,\gamma(s)}\bar{\sigma}(s, 0) \tag{2}$$

where $d = L - l$. Derive these alternative forms of the *canonical equations* of the line.

Problem 5.6-8. Show that the viscosity of a purely viscous liquid obtained from wave propagation measurements is

$$\eta = \omega\rho/2\alpha^2(\omega) \tag{1}$$

where $\alpha(\omega)$ is the attenuation function at the frequency of measurement.

Problem 5.6-9. Express the attenuation function, $\alpha(\omega)$, and the phase function, $\beta(\omega)$, in terms of the real and imaginary parts of $\mathscr{L}_C^*(\omega)$, and *vice versa*.

Problem 5.5-10. Obtain the Fourier series expansion of the train of unit gate pulses shown in the top portion of Fig. 5.6-7, letting $x_0 = 1$.

References (Chap. 5)

1. B. Gross, J. Polymer Sci., *20*: 123 (1956).
2. R.S. Marvin and H. Oser, J. Res. Nat. Bureau Standards, *66B*: 171 (1962).
3. I.S. Gradshteyn and I.M. Ryzhik, *Tables of Integrals, Series and Products*, Academic Press, New York, (1965).
4. L.B.W. Jolley, *Summation of Series*, Dover Publications, New York (1961).
5. A.B. Blizard, J. Appl. Phys. *22*: 730 (1951).
6. V.W. Chang and N.W. Tschoegl, unpublished work.
7. P.E. Rouse, J. Chem. Phys. *21*: 1272 (1953).

6. Representation of Linear Viscoelastic Behavior by Mathematical Models

He had been eight years upon a project for extracting sunbeams out of cucumbers, which were to be put into vials hermetically sealed, and let out to warm the air in raw, inclement summers.

Jonathan Swift: A Voyage to Laputa

This is not the real McCoy

Popular saying

6.0 Introduction

The representations discussed in the preceding three chapters are all related to the Boltzmann superposition principle which forms the analytic basis of the theory of linear viscoelastic behavior. In this chapter we discuss the representation of such behavior by empirical mathematical models which bear no relationship to the superposition principle. These models are useful when it is desired to describe linear viscoelastic behavior qualitatively by a small, easily manageable, number of parameters. The standard linear solid and liquid models satisfy the latter requirement but they do not model the *distribution* of respondance times so characteristic of viscoelastic behavior.

If the number of Maxwell or Voigt units is increased to the minimum number required for a series-parallel model to represent such a distribution at all adequately, the simplicity of the standard models is lost and, in addition, arbitrary decisions must be made in assigning suitable values to the model elements. The regular ladder models portray a distribution of respondance times quite satisfactorily but are not flexible enough for many modelling applications.

The notion of mathematical models which emulate the effect of a distribution of respondance times without containing such a distribution explicitly, has been introduced briefly in Sect. 4.5. Being essentially empirical curve-fitting equations containing adjustable parameters, these mathematical models are more flexible than the ladder models. However, they suffer from the disadvantage, stemming from their empirical nature, that, if one of these mathematical models has been found to represent the response of a linear viscoelastic material to a given excitation, it is, as a rule, not possible to find, by standard procedures, the form of the model which will represent the response to another excitation*. By contrast, representation of linear viscoelastic behavior by models based on the superposition integrals are, at least in principle, *interconvertible* into one another. This subject will be discussed

* But see Sect. 6.3.2.

extensively in Chap. 8. In applications in which interconvertibility is not an issue, the simplicity of the mathematical models outweighs their lack of a proper foundation in the theory of linear viscoelastic behavior.

In this chapter we deal with these empirical mathematical models in the framework of a unified theory. One way of deriving such models is by what will be called the *method of matching functions*. This will be discussed in Sect. 6.1. Another set of simple models with a somewhat limited range of properties can be derived through *fractional derivation*. This is the subject of Sect. 6.2. Modelling of the spectral response functions requires special considerations. It will be treated in Sect. 6.3.

6.1 Modelling by the Use of Matching Functions

Each viscoelastic response function possesses certain general features which any model function chosen to represent it must also possess as a minimum requirement. These features are properties of, and may be derived from, the kernel functions of the canonical representations whose behavior as function of their arguments is illustrated in Fig. 4.1-2. In this section we discuss first the theory of model functions possessing these minimum requirements. We then deal with the way in which these model functions can be used to model the experimental response functions, $G(t)$, $\eta(t)$, $G'(\omega)$, $G''(\omega)$, $J(t)$, $J'(\omega)$, and $J''(\omega)$; and the respondances, $\bar{Q}(s)$, and $\bar{U}(s)$.

To facilitate the discussion we introduce *dimensionless response functions* which have the same general features as the corresponding kernel functions. The dimensionless functions are obtained with the aid of the *averaging operator*

$$\langle k(x,\tau)\rangle_F = \frac{\int_{-\infty}^{\infty} F(\tau)k(x,\tau)d\ln\tau}{\int_{-\infty}^{\infty} F(\tau)d\ln\tau} \tag{6.1-1}$$

where x stands for t or ω, while $k(x,\tau)$ is the kernel function. The subscript indicates the distribution function which serves to form the average. Since we are using the spectral functions for this purpose, $F(\tau)$ stands for either $H(\tau)$ or $L(\tau)$. In Sect. 4.6 the operator was used to form averages of the respondance times, their squares, and their reciprocals. These averages were constants. Making $\langle\cdots\rangle_F$ operate on the kernel functions results in averages for each and every value of the argument $0 \le x \le \infty$. The operator, therefore, generates functions of the argument.

For the step response functions we find

$$Z_H(t) = \langle\exp(-t/\tau)\rangle_H = \frac{G(t) - \{G_e\}}{G_g - \{G_e\}} \tag{6.1-2}$$

$$Z_L(t) = \langle\exp(-t/\tau)\rangle_L = \frac{J_e^{\{o\}} - J_\bullet(t) + \{\phi_f t\}}{J_e^{\{o\}} - J_g} \tag{6.1-3}$$

and

$$S_H(t) = \langle 1 - \exp(-t/\tau)\rangle_H = \frac{G_g - G_\bullet(t)}{G_g - \{G_e\}} \tag{6.1-4}$$

$$S_L(t) = \langle 1 - \exp(-t/\tau)\rangle_L = \frac{J(t) - J_g - \{\phi_f t\}}{J_e^{\{o\}} - J_g} \ . \tag{6.1-5}$$

For the storage functions we obtain

$$Z_H(\omega) = \left\langle \frac{1}{1+\omega^2\tau^2} \right\rangle_H = \frac{G_g - G'_\bullet(\omega)}{G_g - \{G_e\}} \tag{6.1-6}$$

$$Z_L(\omega) = \left\langle \frac{1}{1+\omega^2\tau^2} \right\rangle_L = \frac{J'(\omega) - J_g}{J_e^{\{o\}} - J_g} \tag{6.1-7}$$

and

$$S_H(\omega) = \left\langle \frac{\omega^2\tau^2}{1+\omega^2\tau^2} \right\rangle_H = \frac{G'(\omega) - \{G_e\}}{G_g - \{G_e\}} \tag{6.1-8}$$

$$S_L(\omega) = \left\langle \frac{\omega^2\tau^2}{1+\omega^2\tau^2} \right\rangle_L = \frac{J_e^{\{o\}} - J'_\bullet(\omega)}{J_e^{\{o\}} - J_g} \ . \tag{6.1-9}$$

Finally, for the loss functions, we have

$$\Lambda_H(\omega) = \left\langle \frac{\omega\tau}{1+\omega^2\tau^2} \right\rangle_H = \frac{G''(\omega)}{G_g - \{G_e\}} \tag{6.1-10}$$

$$\Lambda_L(\omega) = \left\langle \frac{\omega\tau}{1+\omega^2\tau^2} \right\rangle_L = \frac{J''(\omega) - \{\phi_f/\omega\}}{J_e^{\{o\}} - J_g} \ . \tag{6.1-11}$$

Just as the respective kernel functions, the functions $Z(x)$ are characterized by the relation $1 \geq Z(x) \geq 0$, i.e. by $Z(x_1) \geq Z(x_2)$ for $x_1 < x_2$. Thus, these functions are monotone non-increasing functions of their arguments. The symbol Z has been chosen for them because their shape resembles that of a Z which has been stretched out by pulling its two ends in opposing directions. At the limits of the domain of definition of x,

$$\lim_{x\to 0} Z(x) = 1 \quad \text{and} \quad \lim_{x\to\infty} Z(x) = 0 \ . \tag{6.1-12}$$

Analogous considerations apply to the functions $S(x)$. For these, $0 \leq S(x) \leq 1$, i.e. $S(x_1) \leq S(x_2)$ for $x_1 < x_2$. Thus, these functions are monotone non-decreasing functions of their arguments ressembling a stretched-out S. We have

$$\lim_{x\to 0} S(x) = 0 \quad \text{and} \quad \lim_{x\to\infty} S(x) = 1 \tag{6.1-13}$$

at the limits of the time, respectively the frequency, scale.

The functions $\Lambda(x)$ are neither monotone non-increasing, nor monotone non-decreasing. They vanish at both ends of the x-scale, i.e.

$$\lim_{x\to 0,\infty} \Lambda(x) = 0 \tag{6.1-14}$$

but have maxima at a characteristic value x_{max}, i.e.

$$\Lambda(x \neq x_{max}) < \Lambda(x_{max}) \ . \tag{6.1-15}$$

The symbol Λ reflects these properties. The maxima are obtained from

$$\left.\frac{d\,\Lambda(x)}{d\ln x}\right|_{x=x_{max}} = 0 \ . \tag{6.1-16}$$

We use them to introduce new functions $\bar{\Lambda}(\omega)$ with the property $\bar{\Lambda}(\omega_{max}) = 1$. These functions are

$$\bar{\Lambda}_H(\omega) = \frac{\Lambda_H(\omega)}{G''_{max}}[G_g - \{G_e\}] = \frac{G''(\omega)}{G''_{max}}$$

and (6.1-17)

$$\bar{\Lambda}_L(\omega) = \frac{\Lambda_L(\omega)}{J''_{max}}[J_e^{\{o\}} - J_g] = \frac{J''(\omega) - \{\phi_f/\omega\}}{J''_{max}}$$

where $G''_{max} = G''(\omega_{max})$ and $J''_{max} = J''(\omega_{max})$ are the maxima in the loss functions (cf. Figs. 3.4-20 through 3.4-23).

Our task now consists in finding mathematical models subject to the appropriate set of conditions which will *match* the dimensionless response functions as closely as possible with the minimum number of parameters. Each model must contain at least two adjustable parameters: one for locating it along the time or frequency axis, and another to regulate its spread.* Clearly, a very large number of such functions can be constructed. Thus, every unimodal probability density distribution satisfies Eqs. (6.1-14) and (6.1-15) and may serve to model $\bar{\Lambda}(\omega)$ when adjusted to possess a peak value of unity. Similarly, every cumulative unimodal probability density distribution may model Z(x). To distinguish the matching functions from the dimensionless response functions, we shall denote the former by $\hat{Z}(x)$, $\hat{S}(x)$, and $\hat{\Lambda}(x)$, respectively, using x as a general variable. Later, we shall let x equal t, ω, s, or τ, as needed.

6.1.1 Matching Functions of the Z- and the S-Type

We begin with some general comments concerning the matching functions $\hat{Z}(x)$ and $\hat{S}(x)$. Given $\hat{Z}(x)$, we also have

$$\hat{S}(x) = 1 - \hat{Z}(x) \qquad \text{and} \qquad \hat{Z}(x) = 1 - \hat{S}(x) \ . \tag{6.1-18}$$

For every monotone non-increasing function $\hat{Z}(x)$, a monotone non-decreasing function $\hat{S}(x)$ with the required characteristics can be found by a reflection in the horizontal axis followed by a translation along the vertical axis. Being symmetrical, the transformation (6.1-18) of course also produces a monotone non-increasing function $\hat{Z}(x)$ from a monotone non-decreasing function $\hat{S}(x)$.

* See also Sects. 4.4 and 4.6.

We note that $S_F(x) = 1 - Z_F(x)$. Hence, the matches $S_F(x) = \hat{S}(x)$ and $Z_F(x) = \hat{Z}(x)$ necessarily furnish the same mathematical model. Therefore, only one of the matchings needs to be specified. We let this be $\hat{Z}(x)$.

Equations (6.1-18) are not the only possible transformations for generating a function of type S from one of type Z and *vice versa*. The transformations

$$\hat{s}(x) = \hat{Z}(1/x) \quad \text{and} \quad \hat{z}(x) = \hat{S}(1/x) \tag{6.1-19}$$

produce *associated functions* $\hat{s}(x)$ and $\hat{z}(x)$ with the same general characteristics as $\hat{S}(x)$ and $\hat{Z}(x)$. These new functions represent a reflection in the vertical axis followed by a translation along the logarithmic horizontal axis. Successive application of the transformations (6.1-18) and (6.1-19) shows that $\hat{z}(x) = 1 - \hat{Z}(1/x)$ and $\hat{s}(x) = 1 - \hat{S}(1/x)$ and we note that $\hat{s}(x) = 1 - \hat{z}(x)$ and $\hat{z}(s) = 1 - \hat{s}(x)$. Hence, one of the associated functions suffices. The choice of $\hat{s}(x)$ is consistent with our choice of $\hat{Z}(x)$ as the primary function. In general, $\hat{z}(x)$ and $\hat{Z}(x)$, and $\hat{s}(x)$ and $\hat{S}(x)$, are different functions of their arguments. If $\hat{z}(x) = \hat{Z}(x)$, or $\hat{s}(x) = \hat{S}(x)$, the upper and lower halves of the curves when plotted against log x, may be brought into superposition with each other by reflection and translation. A mathematical model possessing this property will be called *self-congruent*.

It was pointed out in the previous section that every cumulative unimodal probability density distribution possesses the properties required of a matching function of the Z-type. The cumulative error distribution has, in fact, been used to model linear viscoelastic behavior [1]. Here, we shall discuss another, more widely used, class of matching functions of the S- and the Z-type. These can be derived from the kernel functions of the canonical representations of the response functions by a simple modification. This consists in introducing into the kernel an exponent which has the effect of broadening the spread of the kernel function along the logarithmic time scale (or, equivalently, of the logarithmic frequency scale, or, for the respondances, the scale of the transform variable, s). Thus, behavior similar to that resulting from a distribution of respondance times is achieved by a rather simple expedient.

Matching equations of the modified kernel type will be discussed first. We then introduce, as an example, a matching equation which has been selected simply because it has an appropriate form. These equations contain two adjustable parameters each.* We end this section by discussing a mathematical model possessing three such parameters.

The functions will be distinguished from one another by capital letter subscripts which represent the initial or, if that has been pre-empted, the second consonant of the name of the function. The parameter(s) governing the spread of the function become the lower case letters corresponding to the subscript. When the distinction is not necessary, the generic symbol m will generally denote the parameter used to modify the kernel function. This parameter usually lies between zero and unity. The parameter which determines the location of the function along the log x axis will be denoted by the same symbol, x_0, in all cases. We now turn to a consideration of

* The viscoelastic constants, G_e, J_g, ϕ_f, etc., will not be considered adjustable parameters.

various functions $\hat{Z}(x)$ and $\hat{s}(x)$. Most have been used extensively in the literature on linear viscoelastic or dielectric behavior.

As early as 1863, F. Kohlrausch [2] used essentially the function

$$\hat{Z}_K(x) = \exp[-(x/x_0)^k] \tag{6.1-20}$$

to describe the elastic after-effect in the torsion of glass fibers. Even earlier, R. Kohlrausch had used a similar expression for the analogous dielectric problem. We will call it the *Kohlrausch function*. Some authors, apparently unaware of the earlier contribution of the Kohlrausches, call it the *Williams-Watts function.*

By Eq. (6.1-19) the *associated* matching function takes the form

$$\hat{s}_K(x) = \exp[-(x_0/x)^k] \ . \tag{6.1-21}$$

$\hat{Z}_K(x)$ is seen to be a modification of the kernel function appearing in the canonical representation of G(t) [cf. Eq. (4.1-5)], obtained by replacing t/τ by $(x/x_0)^k$. In Eq. (6.1-20) x_0 is the value of x at which $\hat{Z}_K(x)$ assumes (1/e)th of its initial value. Thus, when $x = t$, t_0 is a characteristic time of the nature of a respondance time which locates $\hat{Z}_K(t)$ along the logarithmic time axis. The exponent k governs the spread of $\hat{Z}_K(x)$. As shown in Fig. 6.1-1, the spread is the broader the smaller k is. The curve with $k = 1$ represents the unmodified kernel function. $\hat{Z}_K(x)$ and $\hat{s}_K(x)$ are, of course, mirror images of each other, reflected in the ordinate at $\log(x/x_0) = 0$.

$\hat{Z}_K(x)$ is not self-congruent. It decreases relatively gradually for $x/x_0 < 1$ but plunges more steeply for $x/x_0 > 1$. This behavior is similar to that of the kernel function $\exp(-t/\tau)$ in Fig. 4.1-2. We note that the associated function, $\hat{s}_K(x)$, has no counterpart among the kernel functions. Modification of the kernel of J(t) gives $1 - \exp[-(t/t_0)^k]$, which is $\hat{S}_K(t)$.

Another matching function can be derived from the kernel function of the canonical representation of $J'(\omega)$ [cf. Eq. (4.1-23)]. Substitution of $(x/x_0)^c$ for $(\omega\tau)^2$ yields

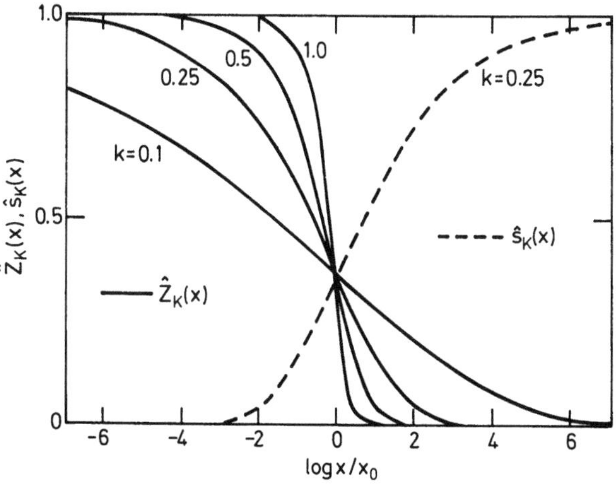

Fig. 6.1-1. The Kohlrausch functions for several values of k

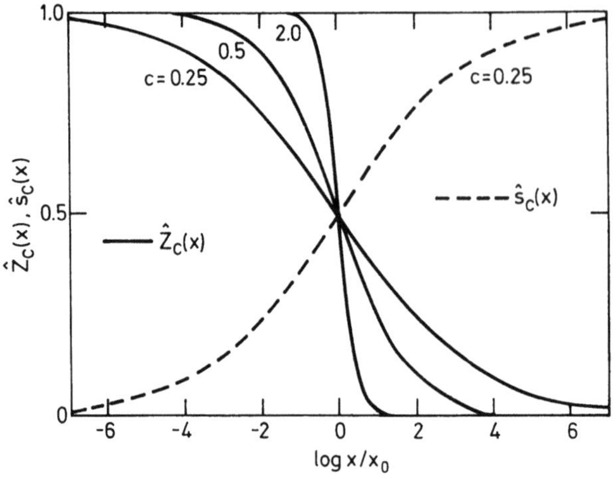

Fig. 6.1-2. The Cole-Cole functions for several values of c

$$\hat{Z}_C(x) = \frac{1}{1 + (x/x_0)^c} = \hat{z}_C(x) \; . \tag{6.1-22}$$

This function is self-congruent. Its associated function is

$$\hat{s}_C(x) = \frac{1}{1 + (x_0/x)^c} = \hat{S}_C(x) \; . \tag{6.1-23}$$

$\hat{S}_C(x)$ may be derived from the kernel of $G'(\omega)$ by the same substitution for $(\omega\tau)^2$. We shall call Eq. (6.1-22) the *Cole-Cole function.* Its complex form [cf. Eq. (6.1-87)] was proposed in 1941 by Cole and Cole [3] to represent the complex dielectric constant.

The function assumes one-half of its original value when $x = x_0$. Again, the function broadens as c decreases (Fig. 6.1-2).

Yet another matching function may be obtained from the kernel of $\bar{U}(s)$ [see Eq. (4.1-18)] by raising the denominator to an exponent which determines the spread. Letting $\tau s = x/x_0$, we have

$$\hat{Z}_B(x) = \frac{1}{(1 + x/x_0)^b} \tag{6.1-24}$$

with the associated function

$$\hat{s}_B(x) = \frac{1}{(1 + x_0/x)^b} \; . \tag{6.1-25}$$

$\hat{S}_B(x)$, which may be derived from the kernel of $\bar{Q}(s)$ by raising it to the exponent b, was used by Kobeko* [4] in 1937 to represent creep data. We will call $\hat{Z}_B(x)$ the *Kobeko function* (Fig. 6.1-3). Its complex form [cf. Eq. (6.1-92)] is known in dielectrics as the *Cole-Davidson function.* $\hat{Z}_B(x)$ is self-congruent only when $b = 1$.

* A.L. Kobeko (pron. Cubbyaka)

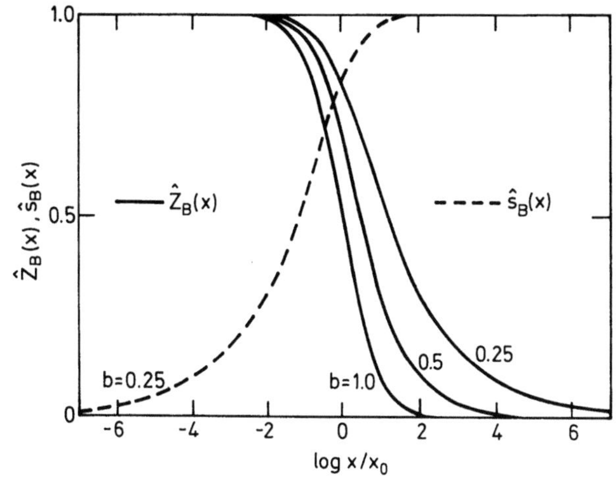

Fig. 6.1-3. The Kobeko functions for several values of b

As an example of a matching function which cannot be related to one of the kernel functions, we introduce

$$\hat{Z}_P(x) = \tanh(x_0/x)^p \tag{6.1-26}$$

with its associated function

$$\hat{s}_P(x) = \tanh(x/x_0)^p \ . \tag{6.1-27}$$

Letting $p = 1/2$, $x = s$, and $x_0 = 1/\theta$, we have $\hat{s}_P(s) = \tanh\sqrt{\theta s}$ and perceive a kinship to the ladder models [cf. Eq. $(5.5\text{-}13)_1$]. The equations are not self-congruent. They all assume the same value, 0.762, when $x = x_0$. Equation (6.1-26) will be called the *hyperbolic tangent function* (Fig. 6.1-4).

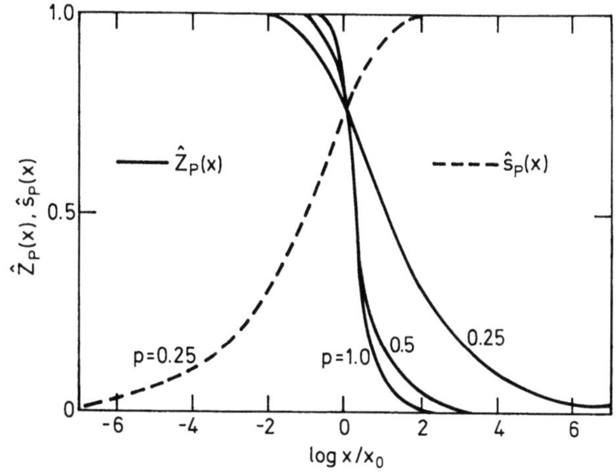

Fig. 6.1-4. The hyperbolic tangent functions for several values of p

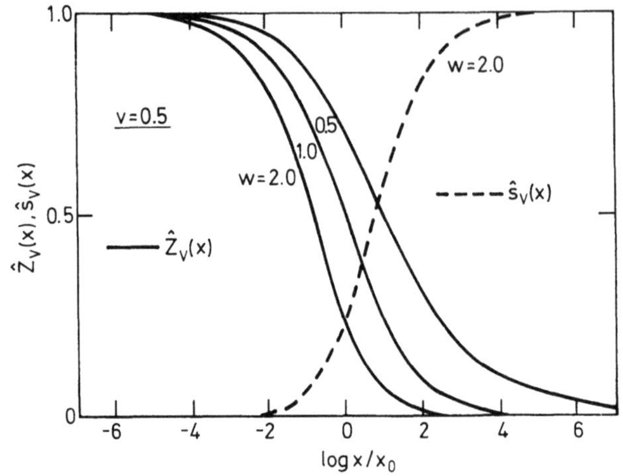

Fig. 6.1-5. The Havriliak-Negami functions for various values of w when v = 0.5

As an example of a matching function with three instead of two adjustable parameters we list the equation of Havriliak and Negami [5],

$$\hat{Z}_V(x) = \frac{1}{[1 + (x/x_0)^v]^w}, \tag{6.1-28}$$

with its associated function

$$\hat{s}_V(x) = \frac{1}{[1 + (x_0/x)^v]^w}. \tag{6.1-29}$$

These functions may be considered to be combinations of the Cole-Cole and the Kobeko functions. As such, they are self-congruent when w = 1. Figure 6.1-5 shows their shape for v = 0.5 and several values of w.

6.1.2 Matching Functions of the Λ-Type

To allow it to be matched with the functions $\bar{\Lambda}(\omega)$, a matching function of the Λ-type must satisfy the condition $\hat{\Lambda}(\omega_{max}) = 1$ in addition to Eqs. (6.1-14) and (6.1-15). Such a function may be derived from a modification of the kernel of the loss functions. Before we consider this function, however, we briefly discuss functions of the Λ-type which are derived from the functions of the Z- or the S-type. The derivation is based on the fact that differentiation of any monotone non-increasing or monotone non-decreasing function with a single inflection point produces a function with a single extremum. Thus, a matching function of the Λ-type can be derived from any of the functions of the Z- or the S-type* by logarithmic differentiation with respect

* We note that the various intensity functions discussed in Chap. 4, having been derived by differentiation of the kernel functions of the Z- and S-type kernels, all provide matching functions of the Λ-type.

to x. The derivative will contain x_0. By equating the second derivative to zero, we find a relation between x_0 and x_{max}, the location of the extremum. After substituting x_{max} for x_0 we determine the peak value by letting x equal x_{max} and then use this value to normalize the function so that it satisfies the condition $\hat{\Lambda}(x_{max}) = 1$. By virtue of Eqs. (6.1-18), derivation from $\hat{Z}(x)$ and $\hat{S}(x)$ necessarily leads to the same result. Derivation from the associated functions does so only when the primary function is self-congruent. When it is not, the derivation produces a matching function which is the mirror image, on the logarithmic frequency scale, of that obtained from the primary function.

We demonstrate the procedure using the Cole-Cole function. From Eq. (6.1-22) we obtain

$$\frac{d\,\hat{Z}_C(x)}{d\ln x} = -\frac{c}{(x/x_0)^c + 2 + (x_0/x)^c}\,. \tag{6.1-30}$$

Setting the second derivative equal to zero we find that the first derivative assumes its peak value when $x = x_0$. Thus, the location of the peak, x_{max}, is, in this case, equal to x_0. Substitution of x_{max} for x_0 consequently will locate the peak at x_{max}. Letting now x equal x_{max}, we find that the peak value is $-c/4$. Division by this value finally yields

$$\hat{\Lambda}_C(x) = \frac{4}{(x/x_{max})^c + 2 + (x_{max}/x)^c}\,. \tag{6.1-31}$$

Since the Cole-Cole function is self-congruent, $\hat{\Lambda}_C(x)$ will be symmetric on the log(x)-axis around the ordinate at $x = x_{max}$. Derivation from a non-self-congruent matching function yields an equation of the Λ-type which is asymmetric. Thus, from Kobeko's function, Eq. (6.1-24), we obtain

$$\hat{\Lambda}_B(x) = (x/x_{max})\left(\frac{1+b}{x/x_{max}+b}\right)^{1+b} \tag{6.1-32}$$

which is symmetric only when $b = 1$, in which case it reduces to $\hat{\Lambda}_C(x)$ with $c = 1$. Derivation from the associated function, $\hat{s}_B(x)$, leads to

$$\hat{\lambda}_B(x) = (x_{max}/x)\left(\frac{1+b}{x_{max}/x+b}\right)^{1+b} \tag{6.1-33}$$

which, on the log(x)-axis, is the mirror image of $\hat{\Lambda}_B(x)$, reflected in the ordinate at $x = x_{max}$.

Omission of the addend 2 in the denominator of Eq. (6.1-31) clearly leaves its relevant properties unaffected. After renormalization we thus have the function

$$\hat{\Lambda}_R(x) = \frac{2}{(x_{max}/x)^r + (x/x_{max})^r}\,. \tag{6.1-34}$$

Equation (6.1-34) will be called the Lorentzian after H.A. Lorentz* who, in his

* Hendrick Antoon Lorentz, 1853–1928, Dutch mathematical physicist, famous for the relativistic *Lorentz transformation*.

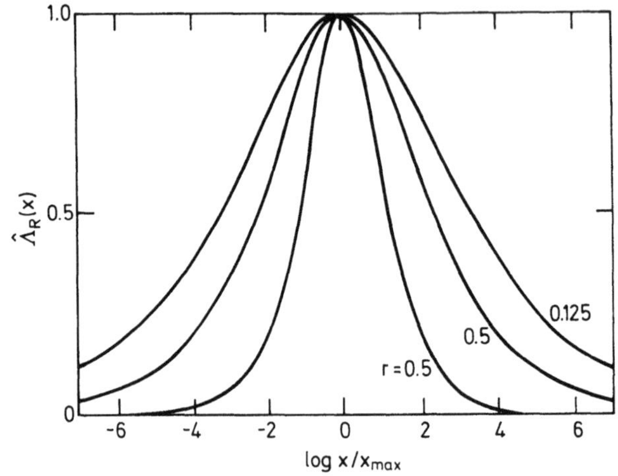

Fig. 6.1-6. The Lorentzian for various values of r

classical work [6] of 1906, showed that (when $r = 1$) it describes the shape of the spectral lines of absorption and emission spectra. Equation (6.1-34) may be written in the form

$$\hat{\Lambda}_R(x) = \frac{2(x/x_{max})^r}{1 + (x/x_{max})^{2r}} \quad . \tag{6.1-34'}$$

Letting $r = 1$, and $x/x_{max} = \omega\tau$, we recognize the kernel of the loss functions [cf. Eqs. (4.1-8) and (4.1-24)]. The equation was proposed in 1941 by Fuoss and Kirkwood [7] for the dielectric loss function [cf. Eq. (6.1-67)].

The Lorentzian is symmetric. It is plotted in Fig. 6.1-6 for several values of r. Again, the function broadens as r decreases.

An asymmetric form of the Lorentzian can be developed from Eq. (6.1-34). We first rewrite it in the form

$$\hat{\Lambda}(x) = \frac{2}{(x_0/x)^n + (x/x_0)^{n'}} \tag{6.1-35}$$

where $n' \neq n$. Equating the derivative to zero shows that

$$x_0 = x_{max}(n'/n)^{1/(n+n')} \quad . \tag{6.1-36}$$

Introducing this into Eq. (6.1-35) yields

$$\hat{\Lambda}(x) = \frac{2}{r'(x_{max}/x)^n + r(x/x_{max})^{n'}} \tag{6.1-37}$$

where

$$r = (n/n')^{1/(1+n/n')} \qquad \text{and} \qquad r' = (n'/n)^{1/(1+n'/n)} \quad . \tag{6.1-38}$$

Division by $\hat{\Lambda}(x_{max})$ then yields

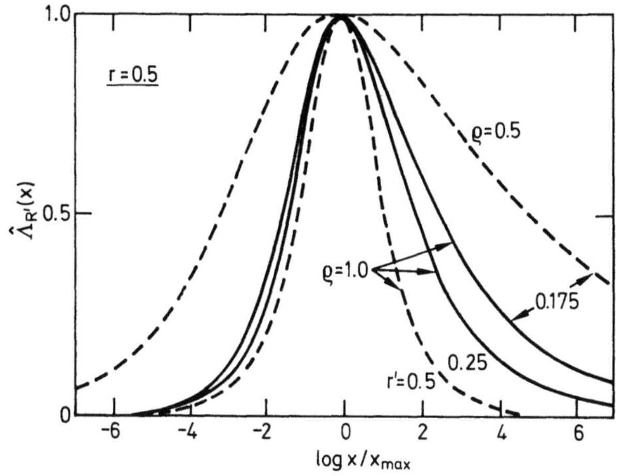

Fig. 6.1-7. The asymmetric Lorentzian for $r = 0.5$ and various values of r' with $\rho = 1$, and for $r' = 0.175$ with $\rho = 0.5$

$$\hat{\Lambda}_{R'}(x) = \frac{r + r'}{r'(x_{max}/x)^{n} + r(x/x_{max})^{n'}} \, . \tag{6.1-39}$$

This asymmetric form of the Lorentzian contains two adjustable parameters (x_{max} is not considered to be adjustable) since r and r′ are functions of n and n′, as shown by Eqs. (6.1-38). However, again setting the derivative of $\hat{\Lambda}_{R'}(x)$ to zero reveals that the extremum condition merely requires that $nr' = n'r$. This condition is satisfied if we put $n = \rho r$ and $n' = \rho r'$ where ρ is an arbitrary real constant. Thus, we may consider n and n′ to be functions of r and r′ and write

$$\hat{\Lambda}_{R'}(x) = \frac{r + r'}{r'(x_{max}/x)^{\rho r} + r(x/x_{max})^{\rho r'}} \, . \tag{6.1-40}$$

This form of the Lorentzian contains three adjustable parameters. An equation of this type was apparently first used as a matching function by Tuijnman* et al. [8] in the form

$$\hat{\Lambda}_{N}(z) = \frac{r + r'}{r \exp(r'z) + r' \exp(-rz)} \, . \tag{6.1-41}$$

Tuijnman's function reduces to Eq. (6.1-40) with $\rho = 1$ upon substitution of $\ln(x/x_{max})$ for z (Fig. 6.1-7). $\hat{\Lambda}_{R'}(x)$ is determined mainly by r for small values, and by r′ for large values of x. The steepness on either side of the maximum increases as r or r′ increase. The parameter ρ governs the overall spread of the function, independently of r and r′. Equation (6.1-40) thus possesses considerable flexibility. Formally, it may be regarded as a combination of the Lorentzian with Tuijnman's

* pron. Toynmun

equation. It reduces to the latter when $\rho = 1$, and to the former when $r = r'$ in addition.

6.1.3 Modelling of the Experimental Response Functions

The representation of the experimental response functions by mathematical models consists quite simply in equating the appropriate dimensionless response function, $Z_F(t)$, $S_F(t)$, $Z_F(\omega)$, $S_F(\omega)$, or $\hat{\Lambda}(\omega)$, with a suitably chosen matching function of the same type. The purpose of this is twofold. In the first place, one may merely wish to simulate typical linear viscoelastic behavior. In this case, selection of the matching function is largely a matter of taste or preference. The exponent m is then commonly taken as 1/2, and x_0 or x_{max} are arbitrary so that the responses are conveniently displayed as functions of $\log x/x_0$ or $\log x/x_{max}$ as in Figs. 6.1-1 through 6.1-7. In the second place, matching may serve to obtain a mathematical description of a set of experimental data. In this case the choice of the matching function may be influenced by the nature of the data, and m and x_0 or x_{max} become curve-fitting parameters to be determined from the data.

The common choice of $m = 1/2$ in theoretical simulations has several reasons. First, when derived from experimental data, m is usually fairly close to 1/2. Second, theoretical distributions of relaxation times in polymers, such as the Rouse distribution (see Sect. 6.3), predict this value for m. Last, but not least, many matching functions can be Laplace transformed conveniently only when $m = 1/2$.

6.1.3.1 The Step Responses

For the *step responses* the match $Z_F(t) = \hat{Z}_M(t)$ yields

$$G_M(t) = \{G_e\} + [G_g - \{G_e\}]\hat{Z}_M(t) \tag{6.1-42}$$

and

$$J_M(t) = J_e^{\{o\}} - [J_e^{\{o\}} - J_g]\hat{Z}_M(t) + \{\phi_f t\} \tag{6.1-43}$$

where M stands for K, C, B, etc. The match $S_F(t) = \hat{s}_M(t)$ furnishes

$$G_m(t) = G_g - [G_g - \{G_e\}]\hat{s}_M(t) \tag{6.1-44}$$

and

$$J_m(t) = J_g + [J_e^{\{o\}} - J_g]\hat{s}_M(t) + \{\phi_f t\} \tag{6.1-45}$$

where the lower case subscript distinguishes these relations as derived from the *associated* functions $\hat{s}(x)$. When the matching function is self-congruent, $G_M(t)$ and $G_m(t)$ are of course, identical, and so are $J_M(t)$ and $J_m(t)$.

We make the procedure explicit by an example. Choosing the Kohlrausch function, which is not self-congruent, as the matching function, we have

$$G_K(t) = \{G_e\} + [G_g - \{G_e\}]\exp[-(t/t_0)^k] \tag{6.1-46}$$

$$J_K(t) = J_e^{\{o\}} - [J_e^{\{o\}} - J_g]\exp[-(t/t_0)^k] + \{\phi_f t\} \tag{6.1-47}$$

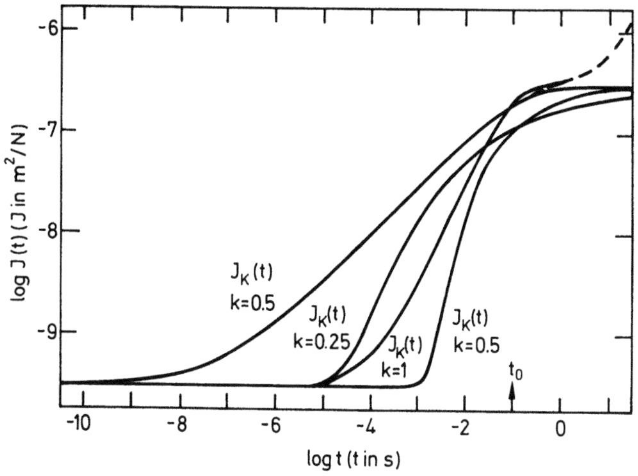

Fig. 6.1-8. Creep compliance simulated by the Kohlrausch function

and

$$G_k(t) = G_g - [G_g - \{G_e\}]\exp[-(t_0/t)^k] \tag{6.1-48}$$

$$J_k(t) = J_g + [J_e^{\{o\}} - J_g]\exp[-(t_0/t)^k] + \{\phi_f t\} \ . \tag{6.1-49}$$

In Fig. 6.1-8, Eqs. (6.1-47) and (6.1-49) are used to simulate the creep compliance of a viscoelastic material. The dashed portions represent the departure of rheodictic from arrheodictic behavior. The curves were calculated with the values for the viscoelastic constants, J_g, $J_e^{\{o\}}$, and ϕ_f, given by Eqs. (3.4-86), letting $t_0 = \tau = 0.1$ second. The curve identified as $k = 1$ reproduces the equations for the standard linear models, Eqs. (3.4-140) and (3.4-153), plotted in Figs. 3.4-21 and 3.4-22. In Fig. 6.1-8 all the qualitative features of the standard models are correctly rendered but $J_K(t)$ with $k = 0.5$ is a better representation of the creep compliance of a linear viscoelastic material than are $J_{3V}(t)$ and $J_{4V}(t)$. With $k = 0.5$, $J_k(t)$ is too steep, and rather small values of k are needed to achieve a more gradual ascent.

Figure 6.1-9 shows a similar plot for the relaxation modulus. The curve with $k = 1$ reproduces the arrheodictic behavior of $G_{3M}(t)$ shown in Fig. 3.4-20. The exponent k must be decreased considerably to obtain a less precipitous descent. Now it is $G_k(t)$ which, with $k = 0.5$, gives a reasonable representation of the relaxation modulus of an arrheodictic material. With $k < 0.5$, the descent steepens. The qualitative features of the standard linear *solid* model are again reproduced correctly. This is not true for the standard linear *liquid* model as a comparison with Fig. 3.4-23 reveals at once. The reason is readily seen. $G_{3M}(t)$, $J_{3V}(t)$, and $J_{4V}(t)$ are the step responses of models containing a *single* Maxwell or Voigt unit. Those among Eqs. (6.1-42) to (6.1-45) which model corresponding behavior do this by modifying the response of these single units. By contrast, $G_{4M}(t)$ is the response of a model containing *two* Maxwell units while $G_M(t) = G_g\hat{Z}_M(t)$ or $G_m(t) = G_g\hat{Z}_M(t)$ represent responses obtained from modifications of a single Maxwell unit only. One

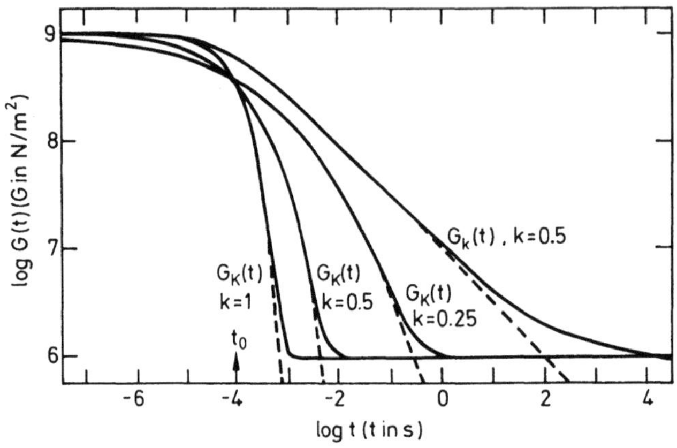

Fig. 6.1-9. Relaxation modulus simulated by the Kohlrausch function

must be aware of this deficiency when using Eqs. (6.1-42) and (6.1-44) to model rheodictic behavior. The equations can be amended, and this will be done in Chap. 10, in which we discuss the modelling of multimodal distributions of respondance times.

Figures 6.1-8 and 6.1-9 contain examples of the *simulation* of linear viscoelastic behavior by the use of mathematical models. We now give an example of the use of such models for representing experimental data. The data were obtained [9] on an uncrosslinked polymer, polyisobutylene, in stress relaxation in simple tension. To conform with the practice adopted in this book they were changed to shear relaxation using the approximate relation $E(t) = 3G(t)$ [Eq. (11.4-6)], and were expressed in units of N/m^2 instead of $dyne/cm^2$. The unit of time was also changed, using seconds instead of hours. Since the behavior is rheodictic, the *caveat* voiced in the preceding paragraph applies. Therefore, only that portion of the data has been used which resembles arrheodictic behavior. The full set is tabulated in Table 6.1-1 and will be matched in Chap. 10 (see Sect. 10.1.4 and Problem 10.1-6).

Since these are stress relaxation data, one is inclined to choose the Kohlrausch equation as the matching function. However, the data (see Fig. 6.1-10) are nearly self-congruent. Hence, Eq. (6.1-22) was selected as the matching function. Equation (6.1-42) then gives

$$G_C(t) = G_e + \frac{G_g - G_e}{1 + (t/t_0)^c} \, . \tag{6.1-50}$$

The logarithms of the instantaneous and the (pseudo-)equilibrium modulus were taken as 9.00 and 5.40, respectively. The parameters c and t_0 were determined from the slope and intercept of a plot, shown in Fig. 6.1-10, of

$$\log r(t) = \log \frac{G_g - G_C(t)}{G_C(t) - G_e} = c \log t - c \log t_0 \tag{6.1-51}$$

obtained by rearranging Eq. (6.1-50). This yields $c = 0.680$ and $\log t_0 = -8.413$.

Table 6.1-1. Relaxation modulus and creep compliance of NBS polyisobutylene at 25°C

log t (sec.)	log G(t) (N/m^2)	log J(1.26t) (m^2/N)	log t (sec.)	log G(t) (N/m^2)	log J(1.26t) (m^2/N)
−10.84	9.00	−9.00	−2.24	5.50	−5.53
−10.64	9.00	−8.99	−2.04	5.48	−5.49
−10.44	8.98	−8.98	−1.84	5.46	−5.47
−10.24	8.97	−8.97	−1.64	5.44	−5.45
−10.04	8.96	−8.96	−1.44	5.43	−5.43
−9.84	8.95	−8.94	−1.24	5.42	−5.42
−9.64	8.93	−8.92	−1.04	5.42	−5.41
−9.44	8.91	−8.91	−0.84	5.41	−5.41
−9.24	8.89	−8.88	−0.64	5.40	−5.40
−9.04	8.86	−8.85	−0.44	5.40	−5.40
−8.84	8.82	−8.81	−0.24	5.39	−5.39
−8.64	8.77	−8.77	−0.04	5.38	−5.38
−8.44	8.72	−8.73	0.16	5.37	−5.37
−8.24	8.67	−8.67	0.36	5.36	−5.35
−8.04	8.59	−8.65	0.56	5.34	−5.34
−7.84	8.50	−8.54	0.76	5.33	−5.32
−7.64	8.40	−8.46	0.96	5.31	−5.31
−7.44	8.29	−8.37	1.16	5.30	−5.29
−7.24	8.17	−8.27	1.36	5.27	−5.27
−7.04	8.04	−8.17	1.56	5.25	−5.25
−6.84	7.91	−8.06	1.76	5.23	−5.22
−6.64	7.78	−7.95	1.96	5.20	−5.20
−6.44	7.64	−7.84	2.16	5.17	−5.17
−6.24	7.51	−7.72	2.36	5.14	−5.14
−6.04	7.38	−7.68	2.56	5.11	−5.11
−5.84	7.25	−7.48	2.76	5.07	−5.07
−5.64	7.12	−7.35	2.96	5.02	−5.02
−5.44	6.99	−7.23	3.16	5.02	−5.02
−5.24	6.85	−7.10	3.36	4.97	−4.98
−5.04	6.71	−6.97	3.56	4.91	−4.92
−4.84	6.57	−6.84	3.76	4.85	−4.87
−4.64	6.44	−6.71	3.96	4.78	−4.80
−4.44	6.32	−6.58	4.16	4.69	−4.73
−4.24	6.21	−6.46	4.36	4.60	−4.65
−4.04	6.10	−6.34	4.56	4.49	−4.57
−3.84	6.00	−6.22	4.76	4.37	−4.47
−3.64	5.90	−6.10	4.96	4.22	−4.36
−3.44	5.81	−5.99	5.16	4.06	−4.25
−3.24	5.73	−5.88	5.36	3.86	−4.12
−3.04	5.66	−5.79	5.56	3.70	−3.99
−2.84	5.60	−5.70	5.76	3.42	−3.84
−2.64	5.56	−5.63	5.96	3.02	—
−2.44	5.52	−5.57	6.16	2.52	—

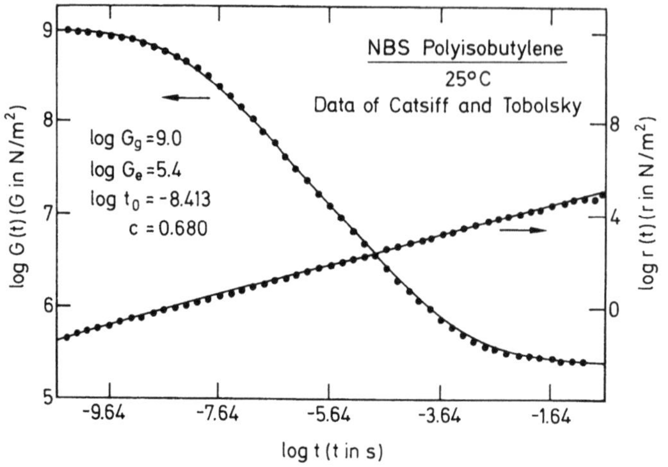

Fig. 6.1-10. Matching of experimental data to Eq. (6.1-50)

The curve in Fig. 6.1-10 represents Eq. (6.1-50) with these values. The full circles represent the data points. The fit is quite satisfactory.

The original stress relaxation data were then converted to tensile creep compliance data [10] by the method discussed in Sect. 8.1.3.2. These data, again converted to shear and adjusted as for G(t), could be fitted by the equation

$$J_C(t) = J_e - \frac{J_e - J_g}{1 + (t/t_0)^c} \tag{6.1-52}$$

with $c = 0.676$ and $\log t_0 = -2.837$ obtained from the relation

$$\log \frac{J_C(t) - J_g}{J_e - J_C(t)} = c \log t - c \log t_0 \ . \tag{6.1-53}$$

It is important to realize that $G_M(t)$ and $J_M(t)$, or $G_m(t)$ and $J_m(t)$, are not related to each other through their Carson transforms while the relaxation modulus, G(t), and the creep compliance, J(t), are so related [cf. Eq. (2.3-13)]. This is so because Eqs. (6.1-18) are not relations between Carson transforms but simply generating functions which generate from each other mathematical models possessing the appropriate minimum requirements. For the same reasons $G_M(t)$ and $J_M(t)$, or $G_m(t)$ and $J_m(t)$, will not generally represent data on the same material with the same value of m, as we have just seen. We note that the relations $G_gJ_g = 1$, $G_eJ_e = 1$, and $\eta_f\phi_f = 1$ remain, of course, valid since the kernel modification affects the time-dependence, not the viscoelastic constants.

6.1.3.2 The Slope Responses

Let us now turn to the modelling of the *slope responses*. We shall consider only the response to a constant rate of strain. The case of a constant rate of stress can be treated analogously. On the basis of Eq. (2.4-6) one might expect that $\eta(t)$ would be modelled adequately by the relation

$$\eta_M(t) = \{G_e t\} + [G_g - \{G_e\}] \int_0^t \hat{z}_M(u) du \; . \tag{6.1-54}$$

This may be successful (see Problem 6.1-1). However, the theoretical restriction

$$\eta_{\{f\}} = \lim_{t\to\infty} [\eta(t) - \{G_e t\}] < \infty \tag{6.1-55}$$

[cf. Eq. $(4.6\text{-}9)_2$] is generally violated (see Problem 6.1-2). This is a direct consequence of the lack of a proper foundation of the empirical mathematical models in the theory of linear viscoelastic behavior. The converse is also true: differentiation of $\eta_M(t)$ does not generally yield acceptable forms for $G_M(t)$ (see Problem 6.1-3).

To derive models by direct matching we proceed via the operator

$$\langle k(x,\tau)\rangle_{\tau F} = \frac{\int_{-\infty}^{\infty} \tau F(\tau) k(x,\tau) d \ln \tau}{\int_{-\infty}^{\infty} \tau F(\tau) d \ln \tau} \tag{6.1-56}$$

[cf. Eq. (6.1-1)]. For a slope strain this gives the dimensionless function

$$S_{\tau H}(t) = \langle 1 - \exp(-t/\tau)\rangle_{\tau H} = \frac{\eta(t) - \{G_e t\}}{\eta_{\{f\}}} \tag{6.1-57}$$

and, therefore, leads to

$$\eta_M(t) = \{G_e t\} + \eta_{\{f\}} \hat{S}_M(t) \; , \tag{6.1-58}$$

or, alternatively, to

$$\eta_m(t) = \{G_e t\} + \eta_{\{f\}} \hat{s}_M(t) \; , \tag{6.1-59}$$

where, by Eqs. $(6.1\text{-}18)_1$ and $(6.1\text{-}19)_1$, $\hat{S}_M(t) = 1 - \hat{Z}_M(t)$ and $\hat{s}_M(t) = \hat{Z}_M(1/t)$. From Eq. (6.1-22), for example, we obtain

$$\eta_C(t) = \frac{\eta_f}{1 + (t_0/t)^c} \tag{6.1-60}$$

for rheodictic behavior.

6.1.3.3 *The Harmonic Responses*

Matching of the complex functions will be discussed in Sect. 6.1.4.1. For the *storage functions*, $Z_F(\omega) = \hat{Z}_M(\omega)$ furnishes

$$G'_M(\omega) = G_g - [G_g - \{G_e\}] \hat{Z}_M(\omega) \tag{6.1-61}$$

and

$$J'_M(\omega) = J_g + [J_e^{\{o\}} - J_g] \hat{Z}_M(\omega) \tag{6.1-62}$$

while $S_F(\omega) = \hat{s}_M(\omega)$ leads to

$$G'_m(\omega) = \{G_e\} + [G_g - \{G_e\}]\hat{s}_M(\omega) \tag{6.1-63}$$

and

$$J'_m(\omega) = J_e^{\{o\}} - [J_e^{\{o\}} - J_g]\hat{s}_M(\omega) \ . \tag{6.1-64}$$

For the *loss functions*, $\bar{\Lambda}_F(\omega) = \hat{\Lambda}_M(\omega)$ gives

$$G''_M(\omega) = G''_{max}\hat{\Lambda}_M(\omega) \qquad \text{and} \qquad J''_M(\omega) = J''_{max}\hat{\Lambda}_M(\omega) + \{\phi_f/\omega\} \ . \tag{6.1-65}$$

As an example of a model for $G'(\omega)$, Eq. (6.1-22) gives

$$G'_C(\omega) = \{G_e\} + \frac{G_g - \{G_e\}}{1 + (\omega_0/\omega)^c} \ . \tag{6.1-66}$$

The arrheodictic form of this equation was used in Sect. 4.3.1.1 where reference is made to its fitting to experimental data. As an example of a model for $G''(\omega)$, Eq. (6.1-34) produces

$$G''_R(\omega) = \frac{2G''_{max}}{(\omega_{max}/\omega)^m + (\omega/\omega_{max})^m} = G''_{max}\,\mathrm{sech}(m \ln \omega/\omega_{max}) \ . \tag{6.1-67}$$

Similar forms may be obtained for the components of the complex viscosity, $\eta'(\omega)$ and $\eta''(\omega)$. From Eq. (6.1-56) we obtain the dimensionless functions

$$Z_{\tau H}(\omega) = \langle 1/(1 + \omega^2\tau^2)\rangle_{\tau H} = \frac{\eta'(\omega)}{\eta_{\{f\}}} \tag{6.1-68}$$

and

$$\Lambda_{\tau H}(\omega) = \langle \omega\tau/(1 + \omega^2\tau^2)\rangle_{\tau H} = \frac{\eta''(\omega) - \{G_e/\omega\}}{\eta_{\{f\}}} \ . \tag{6.1-69}$$

The match $Z_{\tau H}(\omega) = \hat{Z}_C(\omega)$ then gives

$$\eta'_C(\omega) = \frac{\eta_{\{f\}}}{1 + (\omega/\omega_0)^c} \tag{6.1-70}$$

and $\Lambda_{\tau H}(\omega) = \hat{\Lambda}_R(\omega)$ supplies

$$\eta''_R(\omega) = \{G_e/\omega\} + \frac{2\eta_{\{f\}}}{(\omega_{max}/\omega)^m + (\omega/\omega_{max})^m} \ . \tag{6.1-71}$$

We note that as indeed indicated by the different subscripts, $G'_C(\omega)$ and $G''_R(\omega)$, and $\eta'_C(\omega)$ and $\eta''_R(\omega)$, are not self-consistent, i.e. they are not the real and imaginary parts, respectively, of the same complex function. Self-consistent functions will be derived in the next section. These are obtained from models for the complex functions, $G^*(\omega)$ or $J^*(\omega)$, rather than their components.

Rheodictic models for $G'(\omega)$ and $G''(\omega)$ have the same deficiencies as those for $G(t)$, and for the same reason. However, rheodictic models for $G'(\omega)$ and $G''(\omega)$ with the correct properties can be obtained through inversion of models for $J^*(\omega)$. An example of this will be given in Sect. 6.1.4.

6.1.4 Modelling of the Respondances

Mathematical models may be used to model the respondances, $\bar{Q}(s)$ and $\bar{U}(s)$. If the transform can be inverted, $\mathscr{L}^{-1}\ \bar{Q}(s)/s$ and $\mathscr{L}^{-1}\ \bar{U}(s)/s$ then yield the step responses, G(t) and J(t), $\mathscr{L}^{-1}\ \bar{Q}(s)/s^2$ and $\mathscr{L}^{-1}\ \bar{U}(s)/s^2$ yield the slope responses, $\eta(t)$ and $\chi(t)$, and replacement of s by jω gives the complex modulus, $G^*(\omega)$, and the complex compliance, $J^*(\omega)$. Thus, the experimental responses can, in this case, be modelled in a consistent way, in accordance with the linear theory of viscoelastic behavior, even though the models chosen for $\bar{Q}(s)$ or $\bar{U}(s)$ do not conform to it. Essentially two ways are open for doing this: $\bar{Q}(s)$ or $\bar{U}(s)$ may be obtained from an already postulated model for one of the experimental responses, or such a model may be proposed directly. Discussion of the first case properly belongs in Chap. 8 on interconversion (but see Problem 6.1-4). We shall concern ourselves here only with the direct modelling of the respondances.

We use the operator, Eq. (6.1-1), to derive the dimensionless functions

$$Z_H(s) = \langle 1/(1 + \tau s)\rangle_H = \frac{G_g - \bar{Q}_\bullet(s)}{G_g - \{G_e\}} \tag{6.1-72}$$

$$S_H(s) = \langle \tau s/(1 + \tau s)\rangle_H = \frac{\bar{Q}(s) - \{G_e\}}{G_g - \{G_e\}} \tag{6.1-73}$$

$$Z_L(s) = \langle 1/(1 + \tau s)\rangle_L = \frac{\bar{U}(s) - J_g - \{\phi_f/s\}}{J_e^{\{o\}} - J_g} \tag{6.1-74}$$

and

$$S_L(s) = \langle \tau s/(1 + \tau s)\rangle_L = \frac{J_e^{\{o\}} - \bar{U}_\bullet(s) + \{\phi_f/s\}}{J_e^{\{o\}} - J_g}\ . \tag{6.1-75}$$

Remembering Eqs. (6.1-18) and (6.1-19), matching yields

$$\bar{Q}_M(s) = G_g - [G_g - \{G_e\}]\hat{Z}_M(s) \tag{6.1-76}$$

$$\bar{U}_M(s) = J_g + [J_e^{\{o\}} - J_g]\hat{Z}_M(s) + \{\phi_f/s\} \tag{6.1-77}$$

and

$$\bar{Q}_m(s) = \{G_e\} + [G_g - \{G_e\}]\hat{s}_M(s) \tag{6.1-78}$$

$$\bar{U}_m(s) = J_e^{\{o\}} - [J_e^{\{o\}} - J_g]\hat{s}_M(s) + \{\phi_f/s\}\ . \tag{6.1-79}$$

As an example we list

$$\bar{Q}_C(s) = G_g - \frac{G_g - \{G_e\}}{1 + (s/s_0)^c}\ . \tag{6.1-80}$$

The arrheodictic form of this equation, with c = 1/2, was used in Chap. 5 [cf. Eq. (5.4-4)]. For the modelling of $\bar{\eta}(s)$ see Problem 6.1-5.

6.1.4.1 *Harmonic Response Models from Respondance Models*

Models for the harmonic responses are easily derived from Eqs. (6.1-76) to (6.1-79) by substituting jω for s. Thus,

$$G^*_M(\omega) = G_g - [G_g - \{G_e\}]\hat{Z}^*_M(\omega) \tag{6.1-81}$$

$$J^*_M(\omega) = J_g + [J_e^{\{o\}} - J_g]\hat{Z}^*_M(\omega) + \{\phi_f/j\omega\} \tag{6.1-82}$$

and

$$G^*_m(\omega) = \{G_e\} + [G_g - \{G_e\}]\hat{s}^*_M(\omega) \tag{6.1-83}$$

$$J^*_m(\omega) = J_e^{\{o\}} - [J_e^{\{o\}} - J_g]\hat{s}^*_M(\omega) + \{\phi_f/j\omega\} \tag{6.1-84}$$

where

$$\hat{Z}^*_M(\omega) = \hat{Z}_M(j\omega) = \hat{Z}'_M(\omega) - j\,\hat{Z}''_M(\omega) \tag{6.1-85}$$

and

$$\hat{s}^*_M(\omega) = \hat{s}_M(j\omega) = \hat{s}'_M(\omega) + j\,\hat{s}''_M(\omega)\ . \tag{6.1-86}$$

These equations yield self-consistent models for the storage and loss functions within the response to stress or strain excitations, respectively. As an example,

$$\hat{Z}^*_C(\omega) = \frac{1}{1 + (j\omega/\omega_0)^c} \tag{6.1-87}$$

has the real part

$$\hat{Z}'_C(\omega) = \frac{(\omega_0/\omega)^c + \cos \pi c/2}{(\omega_0/\omega)^c + 2\cos \pi c/2 + (\omega/\omega_0)^c} \tag{6.1-88}$$

and the imaginary part

$$Z''_C(\omega) = \frac{\sin \pi c/2}{(\omega_0/\omega)^c + 2\cos \pi c/2 + (\omega/\omega_0)^c}\ . \tag{6.1-89}$$

Hence, we obtain

$$G'_C(\omega) = G_g - [G_g - \{G_e\}]\frac{(\omega_0/\omega)^c + \cos \pi c/2}{(\omega_0/\omega)^c + 2\cos \pi c/2 + (\omega/\omega_0)^c} \tag{6.1-90}$$

and

$$G''_C(\omega) = [G_g - \{G_e\}]\frac{\sin \pi c/2}{(\omega_0/\omega)^c + 2\cos \pi c/2 + (\omega/\omega_0)^c}\ . \tag{6.1-91}$$

Problem 6.1-7 introduces G''_{max} into Eq. (6.1-91). For a "symmetric" form of Eq. (6.1-90) see Problem 6.1-10.

For the real and imaginary parts of

$$\hat{Z}^*_B(\omega) = \frac{1}{(1 + j\omega/\omega_0)^b} \tag{6.1-92}$$

see Problem 6.1-8. For $\hat{Z}^*_V(\omega)$ see Sect. 8.3.1.3. Problem 6.1-6 discusses the modelling of $\eta^*(\omega)$.

$G'_M(\omega)$ and $G''_M(\omega)$, etc., consisting of the real and imaginary parts of a complex function, are connected with one another by the Hilbert transformation, just as $G'(\omega)$ and $G''(\omega)$, or $J'(\omega)$ and $J''(\omega)$ are. This will also be discussed in Sect. 8.3.1.3.

Making use of the simple interconversion between the complex modulus and the complex compliance, $G^*(\omega) = 1/J^*(\omega)$ [cf. Eq. (2.5-58)$_1$], rheodictic models can be obtained for $G'(\omega)$ and $G''(\omega)$ which do not suffer from the shortcomings discussed in the preceding section. We illustrate this by inverting $J_C^*(\omega)$ to obtain $G'_{C'}(\omega)$ and $G''_{C'}(\omega)$ where the prime on the subscripts indicates that the model has been derived through interconversion. By Eqs. (6.1-82) and (6.1-87)

$$J_C^*(\omega) = J_g + \frac{J_e^{\{o\}} - J_g}{1 + (j\omega/\omega_0)^c} + \{\phi_f/j\omega\} \tag{6.1-93}$$

which has the components

$$J'_C(\omega) = J_g + [J_e^{\{o\}} - J_g]\frac{(\omega_0/\omega)^c + \cos \pi c/2}{(\omega_0/\omega)^c + 2\cos \pi c/2 + (\omega/\omega_0)^c} \tag{6.1-94}$$

and

$$J''_C(\omega) = [J_e^{\{o\}} - J_g]\frac{\sin \pi c/2}{(\omega_0/\omega)^c + 2\cos \pi c/2 + (\omega/\omega_0)^c} + \{\phi_f/\omega\} \tag{6.1-95}$$

[cf. Eqs. (6.1-85), (6.1-88), and (6.1-89)]. Letting $G_{C'}^*(\omega) = 1/J_C^*(\omega)$, we find, after working through the complex algebra,

$$G'_{C'}(\omega) = \frac{(\omega\eta_f)A(\omega)}{(\omega\eta_f)^{-1}B(\omega) + 2\,C(\omega) + (\omega\eta_f)D(\omega)} \tag{6.1-96}$$

and

$$G''_{C'}(\omega) = \frac{B(\omega) + (\omega\eta_f)C(\omega)}{(\omega\eta_f)^{-1}B(\omega) + 2\,C(\omega) + (\omega\eta_f)D(\omega)} \tag{6.1-97}$$

where

$$\begin{aligned}
A(\omega) &= G_e^{\{o\}}(\omega/\omega_0)^c + [G_g + G_e^{\{o\}}]\cos \pi c/2 + G_g(\omega/\omega)^c \\
B(\omega) &= G_g G_e^{\{o\}}[(\omega/\omega_0)^c + 2\cos \pi c/2 + (\omega_0/\omega)^c] \\
C(\omega) &= [G_g - G_e^{\{o\}}]\sin \pi c/2 \\
D(\omega) &= [G_e^{\{o\}}/G_g](\omega/\omega_0)^c + 2\cos \pi c/2 + [G_g/G_e^{\{o\}}](\omega_0/\omega)^c
\end{aligned} \tag{6.1-98}$$

and $G_e^{\{o\}} = 1/J_e^{\{o\}}$, $G_g = 1/J_g$, $\eta_f = 1/\phi_f$. For $c = 1$, Eqs. (6.1-96) and (6.1-97) become identical with the equations for the standard linear solid, Eqs. (3.4-128) and (3.4-129), when $\eta_f = \infty$, and with those for the standard linear liquid, Eqs. (3.4-162) and (3.4-163), when η_f is finite. This is easily checked numerically. In the first, i.e. the *arrheodictic* case, Eqs. (6.1-96) and (6.1-97) reduce, as they should, to those for $G'_C(\omega)$ and $G''_C(\omega)$ given by Eqs. (6.1-90) and (6.1-91) upon multiplication of ω_0 by $(G_g/G_e)^c$ to shift $G'_{C'}(\omega)$ and $G''_{C'}(\omega)$ along the log(ω)-axis by the appropriate amount (see Problem 6.1-9).

The situation is different, however, for the *rheodictic* case, i.e. when η_f is finite. Equations (6.1-96) and (6.1-97) then do not reduce to Eqs. (6.1-90) and (6.1-91). They

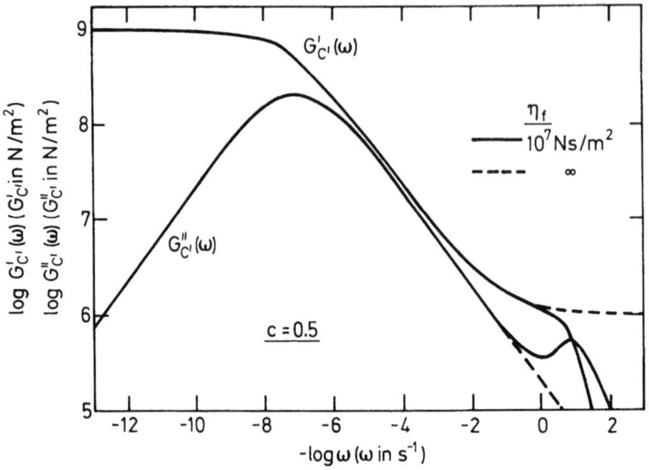

Fig. 6.1-11. $G'_{C'}(\omega)$ and $G''_{C'}(\omega)$ as functions of ω plotted in logarithmic coordinates

are plotted for $\eta_f = \infty$ and $\eta_f = 10^7$ Ns/m², respectively, in Fig. 6.1-11 with $c = 1/2$ and the values specified by Eqs. (3.4-86) for $G_g = 1/J_g$, $G_e^\circ = 1/J_e^\circ$, $\phi_f = 1/\eta_f$, and $\tau = 1/\omega_0$.

When η_f is finite, $G'_{C'}(\omega)$ shows a shoulder, and $G''_{C'}(\omega)$ has a maximum at low frequencies. The moduli are plotted as functions of the *negative* logarithm of the frequency to facilitate comparison with Figs. 3.4-20 and 3.4-23 which display the behavior of the standard solid and liquid models. The comparison shows that the transition region is broadened in Fig. 6.1-11 because $c < 1$. The high frequency peak in $G''_{C'}(\omega)$ is shifted from $\log\omega = 4$ to $\log\omega = 7$. We also note that $G'_{C'}(\omega) > G''_{C'}(\omega)$ while $G'_{3M}(\omega) < G''_{3M}(\omega)$. The location of the minimum and of the low frequency maximum in the rheodictic form of $G''_{C'}(\omega)$ is unchanged although the minimum is higher and the maximum is lower. A decrease in c, i.e. a broadening of the dispersion, tends to reduce the difference in the values of the minimum and maximum. An increase in the steady-flow viscosity shifts the low frequency maximum to lower frequencies without changing its magnitude. This is brought out in Fig. 6.1-12 which compares the cases with $\eta_f = 10^7$ and $\eta_f = 10^9$ Ns/m² on an expanded scale. This effect of the steady-flow viscosity on the low frequency (long time) behavior of a rheodictic model will be discussed further in Sect. 10.1.4 and in Problem 10.1-6.

It is instructive to compare the loss angles $\phi_C(\omega)$ and $\phi_{C'}(\omega)$ for the rheodictic models. From Eqs. (6.1-96) and (6.1-97)

$$\tan\phi_{C'}(\omega) = \frac{C(\omega)}{A(\omega)} + \frac{B(\omega)}{(\omega\eta_f)A(\omega)} . \tag{6.1-99}$$

As $\omega \to 0$, the first term vanishes and the second term diverges. Thus, the loss tangent becomes infinite as it should. By contrast, since $G_e = 0$, Eqs. (6.1-90) and (6.1-91) give

$$\tan\phi_C(\omega) = \frac{\sin\pi c/2}{\cos\pi c/2 + (\omega/\omega_0)^c} \tag{6.1-100}$$

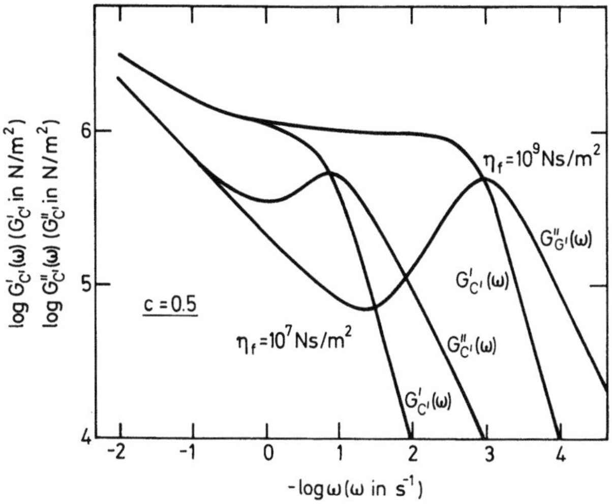

Fig. 6.1-12.
Comparison of $G'_{C'}(\omega)$ and $G''_{C'}(\omega)$ as functions of ω in logarithmic coordinates for two values of η_f

which takes on the finite value $\tan(\pi c/2)$. This incorrect result is, of course, merely another consequence of the inadequacy of Eqs. (6.1-90) and (6.1-91) for the representation of rheodictic behavior.

6.1.4.2 Step Response Models from Respondance Models

Models for the step responses can be derived from Eqs. (6.1-76) to (6.1-79) using the relations

$$G_M(t) = G_g - [G_g - \{G_e\}]\mathscr{L}^{-1}s^{-1}\,\hat{Z}_M(s)$$

$$J_M(t) = J_g + [J_e^{\{o\}} - J_g]\mathscr{L}^{-1}s^{-1}\,\hat{Z}_M(s) + \{\phi_f t\} \qquad (6.1\text{-}102)$$

and

$$G_m(t) = \{G_e\} + [G_g - \{G_e\}]\mathscr{L}^{-1}s^{-1}\,\hat{s}_M(s) \qquad (6.1\text{-}103)$$

$$J_m(t) = J_e^{\{o\}} - [J_e^{\{o\}} - J_g]\mathscr{L}^{-1}s^{-1}\,\hat{s}_M(s) + \{\phi_f t\}\ . \qquad (6.1\text{-}104)$$

Unfortunately, the transform inversion is generally difficult, and even then the resulting equations are of a fairly complicated form which is not overly useful for modelling. Of the five matching functions listed in Sect. 6.1.1, only the inverse transforms of $\hat{Z}_K(s)/s$, $\bar{Z}_C(s)/s$, and $\hat{Z}_B(s)/s$ can be obtained in closed form, and these only when $m = \frac{1}{2}$. From LTP (27) we find

$$\mathscr{L}^{-1}\,\hat{Z}_B(s)/s = \mathrm{Erf}(\sqrt{t/t_0}) \qquad (6.1\text{-}105)$$

where

$$\mathrm{Erf}(x) = \frac{2}{\sqrt{\pi}} \int_0^x \exp(-u^2)du \tag{6.1-106}$$

is the *Gaussian error function** or *normal error integral* [11]. This function is extensively tabulated and is also relatively easily calculated because its series expansion

$$\mathrm{Erf}(x) = \frac{2}{\sqrt{\pi}} \sum_{n=1}^{\infty} (-1)^{n-1} \frac{x^{2n-1}}{(n-1)!(2n-1)} \tag{6.1-107}$$

converges quite rapidly. Sometimes the expansion [18]

$$\mathrm{Erf}(x) = 1 - \frac{1}{\pi} \exp(-x^2)\left(\frac{1}{x} - \frac{1}{2x^2} + \frac{1 \cdot 3}{2^2 x^5} - \frac{1 \cdot 3 \cdot 5}{2^3 x^7} \pm \cdots\right) \tag{6.1-107'}$$

may be useful.

The inverse transforms of $\hat{Z}_K(s)/s$ and $\hat{Z}_C(s)/s$ can be obtained from LTPs (25) and (26), again only when $m = 1/2$. These transforms contain the *complementary error function*

$$\mathrm{Erfc}(x) = 1 - \mathrm{Erf}(x) = \frac{2}{\sqrt{\pi}} \int_x^{\infty} \exp(-u^2)du \ . \tag{6.1-108}$$

The inverse transforms of $\hat{s}_K(s)/s$ and $\hat{s}_B(s)/s$ cannot be stated in a practically useful form even when $m = 1/2$. Since $\hat{s}_C(s)$ is self-congruent, the inverse transform of $\hat{s}_C(s)/s$ with $c = 1/2$ can also be obtained from LTP (26) but this does not, of course, lead to a new matching. We have

$$G_c(t) = \{G_e\} + [G_g - \{G_e\}]\exp(t/t_0)\mathrm{Erfc}(\sqrt{t/t_0}) \tag{6.1-109}$$

when $c = 1/2$. As $t \to \infty$, $\exp(t/t_0)\mathrm{Erf}(\sqrt{t/t_0})$ approaches $(\pi t/t_0)^{-1/2}$.

We remark that an approximate inversion of the Laplace transforms is possible through the collocation (or through the multidata) method discussed in Sect. 3.6. Thus, collocation to $\bar{Q}_C(s)$, say, furnishes a set of moduli, G_k, and relaxation times, τ_k, from which $G_C(t)$ can be reconstructed by Eq. $(3.5\text{-}7)_1$.

6.2 Models Based on Fractional Differentiation (Power Laws)

Another group of mathematical models can be derived in a manner which relates them to the concept of fractional differentiation. Just as the extended Gross-Marvin ladder models discussed in Sect. 5.5.1, these models are non-standard. Like the former they can, nevertheless, be useful if attention is paid to their shortcomings.

The models to be introduced arise from a generalization of the behavior of a purely elastic and a purely viscous material. For the former we have $\bar{\sigma}(s) = G\,\bar{\varepsilon}(s)$ and, therefore, $\bar{Q}(s) = G$. For the latter, $\bar{\sigma}(s) = \eta\,\bar{\dot{\varepsilon}}(s) = G\tau\, s\bar{\varepsilon}(s)$ and, consequently, $\bar{Q}(s) = G\tau s$. Let this be generalized to

---

* Named after its discoverer, Johann Carl Friedrich Gauss, pron. Gouse (rhyming with house), 1777–1855, German mathematician and scientist, often called the founder of modern mathematics.

$$\bar{\sigma}(s) = \bar{Q}_D(s)\bar{\varepsilon}(s) \tag{6.2-1}$$

where

$$\bar{Q}_D(s) = G \cdot (\tau s)^m \ . \tag{6.2-2}$$

G and $(\tau s)^m$ have been separated by a period to indicate that no functional dependence of G on $(\tau s)^m$ is implied. This period notation will be used in this section and elsewhere whenever it appears useful. With $m = 0$ we recover the purely elastic, and with $m = 1$ the purely viscous case.

It might then be expected that, with $0 < m < 1$, these equations would express various degrees of viscoelasticity. Clearly, we are not dealing here with a matching function in the sense of Sect. 6.1. We chose the subscript D to identify the new models as obtained through fractional differentiation. Their relation to the latter is seen by realizing that

$$\mathscr{L}^{-1} s^m \, \bar{\varepsilon}(s) = \frac{d^m \varepsilon(t)}{dt^m} \tag{6.2-3}$$

as long as m lies between zero and unity and $\varepsilon(0) = 0$. We call $d^m f(t)/dt^m$, where m is non-integral, the *derivative of fractional order* m. Further on we shall express it in terms of more common mathematical operations. These are not needed, however, to derive the equations for the experimental responses from Eq. (6.2-2). Let us first introduce

$$\bar{U}_D(s) = J \cdot (\tau s)^{-m} \tag{6.2-4}$$

where $J = 1/G$. We have

$$\bar{G}_D(s) = G \cdot (\tau s)^m / s \ , \qquad \bar{J}_D(s) = J \cdot (\tau s)^{-m} / s \tag{6.2-5}$$

and

$$G_D^*(\omega) = G \cdot (j\omega\tau)^m \ , \qquad J_D^*(\omega) = J \cdot (j\omega\tau)^{-m} \ . \tag{6.2-6}$$

The step responses are obtained from Eqs. (6.2-5) with the help of LTP (12) as

$$G_D(t) = G \, \Gamma^{-1}(1-m)(t/t_0)^{-m} \ , \qquad J_D(t) = J \, \Gamma^{-1}(1+m)(t/t_0)^m \tag{6.2-7}$$

where $t_0 = \tau$. The transform pair is applicable as long as m is restricted to $0 < m < 1$. The harmonic responses follow from Eqs. (6.2-6) as

$$G_D'(\omega) = G \cos(\pi m/2)(\omega/\omega_0)^m \ , \qquad J_D'(\omega) = J \cos(\pi m/2)(\omega/\omega_0)^{-m} \tag{6.2-8}$$

and

$$G_D''(\omega) = G \sin(\pi m/2)(\omega/\omega_0)^m \ , \qquad J_D''(\omega) = J \sin(\pi m/2)(\omega/\omega_0)^{-m} \tag{6.2-9}$$

where $\omega_0 = 1/\tau$. Since G and J are adjustable parameters, the gamma-functions as well as the sine and cosine terms may be subsumed into them. Therefore, we may write

$$G_D(t) = G_0 \cdot (t/t_0)^{-m} \ , \qquad J_D(t) = J_0 \cdot (t/t_0)^m \tag{6.2-10}$$

$$G_D'(\omega) = G_0' \cdot (\omega/\omega_0)^m \ , \qquad J_D'(\omega) = J_0' \cdot (\omega/\omega_0)^{-m} \tag{6.2-11}$$

and

$$G''_D(\omega) = G''_0 \cdot (\omega/\omega_0)^m , \qquad J''_D(\omega) = J''_0 \cdot (\omega/\omega_0)^{-m} . \tag{6.2-12}$$

$G_0, G'_0, G''_0, J_0, J'_0, J''_0, t_0, \omega_0$, and m are constants which are to be chosen in such a way that Eqs. (6.2-10) to (6.2-12) adequately model linear viscoelastic behavior. It turns out that these models are rather limited in their ability to do so. $G_D(0) = G'_D(\infty) = G''_D(\infty)$ and $J_D(\infty) = J'_D(0) = J''_D(0)$ diverge, while $G_D(\infty) = G'_D(0) = G''_D(0)$ and $J_D(0) = J'_D(\infty) = J''_D(\infty)$ vanish. Neither $G''_D(\omega)$ nor $J''_D(\omega)$ possess a maximum. The first is monotone non-decreasing while the second is monotone non-increasing. Also, the storage and loss functions differ from one another by the constant factor $\cos(\pi m/2)$.

Their limitations make Eqs. (6.2-10) to (6.2-12) useful only when it is desired to model merely those regions of the response functions in which they might be represented by a straight line on a plot in logarithmic coordinates. The parameter m is then the absolute value of the slope of such a plot. In Fig. 6.1-10 for instance, G(t) is adequately described by Eq. $(6.2\text{-}10)_1$ with $m = 0.68$ over a region of four logarithmic decades of time.

Equations (6.2-10) to (6.2-12) do not contain any of the viscoelastic constants. These could, of course, simply be added. The equations which would result from this will, however, be derived more rigorously from the wedge distribution of respondance times in Sect. 6.3.1.2 [see Eqs. (6.3-56) to (6.3-58) and (6.3-84) to (6.3-86)]. This derivation also furnishes expressions in terms of m and either $G_g - \{G_e\}$ or $J_e^{\{o\}} - J_g$ for the modulus or compliance constants in Eqs. (6.2-7) to (6.2-9) and (6.2-10) to (6.2-12). We emphasize that the concept of fractional differentiation is thus *not* required to obtain these equations.

Let us now return to Eq. (6.2-3) in order to assign meaning to the fractional derivative. By LTP (12) and Eq. $(A3\text{-}16)_2$ we have

$$\mathscr{L}^{-1} s^{m-1} \bar{f}(s) = \Gamma^{-1}(1-m) \int_0^t u^{-m} f(t-u) du . \tag{6.2-13}$$

But multiplication by s in the complex domain is equivalent to differentiation in the time domain. Hence

$$\mathscr{L}^{-1} s^{m} \bar{f}(s) = \Gamma^{-1}(1-m) \frac{d}{dt} \int_0^t u^{-m} f(t-u) du . \tag{6.2-14}$$

The differentiation is accomplished with the help of the relation [12]

$$\frac{d}{dt} \int_0^t \phi(t,u) du = \phi(t,t) + \int_0^t \frac{d\,\phi(t,u)}{dt} du . \tag{6.2-15}$$

Since $\phi(t,u) = u^{-m} f(t-u)$, Eq. (6.2-14) becomes

$$\mathscr{L}^{-1} s^{m} \bar{f}(s) = \Gamma^{-1}(1-m) \left(f(0) t^{-m} + \int_0^t u^{-m} \frac{d\,f(t-u)}{dt} du \right) \tag{6.2-16}$$

and, if $f(0) = 0$, we have

$$\frac{d^m f(t)}{dt^m} = \Gamma^{-1}(1-m) \int_0^t u^{-m} \frac{d\,f(t-u)}{dt} du . \tag{6.7-17}$$

Equation (6.2-17), subject to the restrictions that f(0) be 0 and that m lie between 0 and 1, expresses the fractional derivative of order m in terms of a convolution integral. Problem 6.2-1 is concerned with obtaining $G_D(t)$ from Eq. (6.2-17).

Equations of the type considered here are often referred to as *power laws*. A power law was apparently first proposed for the description of material properties by Nutting [13]. His equation takes the form $x = at^n f^m$ where x is the displacement, f is the force, and a, n, and m are constants. Nutting also derived his equation by generalizing the behavior of a purely elastic and a purely viscous material but did not recognize the connection with fractional derivation.

The respondances modelled by the Cole-Cole function, Eq. (6.2-22), are readily related to fractional differentiation. Letting $s_0 = 1/\tau$, and $G_e > 0$, Eq. (6.1-80) may be recast as

$$\bar{Q}_C(s) = \frac{G_e + G_g(\tau s)^c}{1 + (\tau s)^c} . \tag{6.2-18}$$

Now recall the operator equation discussed in Sects. 2.1 and 3.7. In the transform plane the equation is

$$\bar{q}(s)\bar{\varepsilon}(s) = \bar{u}(s)\bar{\sigma}(s) . \tag{6.2-19}$$

But $\bar{Q}_C(s) = \bar{q}_C(s)/\bar{u}_C(s)$ [cf. Eq. $(2.1\text{-}16)_1$] and we have

$$G_e\bar{\varepsilon}(s) + G_g(\tau s)^c\,\bar{\varepsilon}(s) = \bar{\sigma}(s) + (\tau s)^c\,\bar{\sigma}(s) , \tag{6.2-20}$$

or, in the time domain,

$$G_e\varepsilon(t) + G_g\tau^c\frac{d^c\varepsilon(t)}{dt^c} = \tau^c\frac{d^c\sigma(t)}{dt^c} + \sigma(t) . \tag{6.2-21}$$

This should be compared with Eq. (3.7-5). For $c = 1$, Eq. (6.2-21) reduces to the operator equation of the standard linear solid, i.e. the three-parameter Maxwell model.

From Eq. (6.2-21) we may now derive expressions for the step responses which are self-consistent with those for the harmonic ones [cf. Eqs. (6.1-90) and (6.1-91)]. Unfortunately, they contain convolution integrals. Using Eq. (6.2-17), Eq. (6.2-21) becomes

$$\sigma(t) + \frac{\tau^c}{\Gamma(1-c)}\int_0^t \frac{d\,\sigma(t-u)}{dt}\frac{du}{u^c} = G_e\varepsilon(t) + \frac{G_g\tau^c}{\Gamma(1-c)}\int_0^t \frac{d\,\varepsilon(t-u)}{dt}\frac{du}{u^c} . \tag{6.2-22}$$

For $\varepsilon(t) = \varepsilon_0 h(t)$, since $\sigma(t) = \varepsilon_0 G(t)$, we find

$$\Gamma(1-c)G(t) + \int_0^t \frac{d\,G(t-u)}{dt}(u/\tau)^{-c}du = \Gamma(1-c)G_e + G_g(t/\tau)^{-c} . \tag{6.2-23}$$

For the retardance,

$$\bar{U}_C(s) = \frac{J_eJ_g[1 + (\tau s)^c]}{J_g + J_e(\tau s)^c} , \tag{6.2-24}$$

we obtain

$$\Gamma(1-c)J(t) + (J_e/J_g)\int_0^t \frac{d\,J(t-u)}{dt}(u/\tau)^c du = \Gamma(1-c)J_e + J_e(t/\tau)^{-c} \ . \qquad (6.2\text{-}25)$$

In concluding this section we remark that the power laws discussed here will find repeated application in Chap. 8.

6.3 Modelling of the Spectral Response Functions

It was pointed out in Chap. 4 that the spectral response functions occupy a central place in the theory of linear viscoelastic behavior since all the experimental response functions can be obtained from them. The respondances play a similar role. However, while always feasible theoretically, derivation of the experimental responses from the respondances can be quite difficult and, as we have seen, is not always possible in closed form. By contrast, from the spectral responses the experimental responses are obtained by integration. When this cannot be done analytically, it can always be accomplished numerically.

We shall first discuss some propositions for modelling the spectra directly. We shall then consider models for the spectra which are derived from one or the other of the models proposed for the experimental response functions.

6.3.1 Direct Modelling

It would obviously be desirable to develop expressions for the spectral response functions from molecular or statistical mechanical considerations. As early as 1893, Wiechert proposed [14] the spectrum

$$H(\tau) = H_{max} \exp\left[-\frac{1}{2}\left(\frac{\ln \tau/\tau_{max}}{\sigma}\right)^2\right] , \qquad (6.3\text{-}1)$$

derived from the logarithmic normal distribution (σ is the standard deviation) of the relaxation times of glasses by considering that all states of relaxation were equally likely. Other considerations led to different formulations [15] but none so far proved lasting with the exception of the so-called Rouse distribution [16] which is derived from a statistical mechanical analysis of the behavior of polymer chains in infinitely dilute solution. The Rouse distribution is a (non-standard) discrete distribution. It has the form

$$H(\tau) = G \sum_{n=1}^{N} \tau_n \, \delta(\tau - \tau_n) \qquad (6.3\text{-}2)$$

where, when N is large, $\tau_n/\tau_1 = 1/n^2$, and G and τ_1 find interpretation in molecular terms [17]. The Rouse distribution and various extensions and modifications of it have been used to model the behavior of undiluted bulk polymers [17]. We note that the arrheodictic extended Gross-Marvin model possesses the same distribution of relaxation times but a different G and τ_1. The regular ladder models all possess known, fixed, discrete distributions of respondance times. The spectra are obtained from Eqs. (5.4-1) and (5.4-2) by inserting the appropriate distribution. Here, however, we are concerned with mathematical models which are more flexible and represent the general shape of observed relaxation and retardation spectra by relatively simple continuous functions.

To derive such models we consider that the behavior of the spectral response functions is qualitatively the same as that of the loss functions. Indeed, the zeroth approximation to the spectra from the loss functions is the function itself, multiplied by $2/\pi$ [cf. Eqs. (4.3-42) and (4.3-43)]. Thus, all matching functions of the Λ-type can serve to model the spectral functions through the relation $F(\tau) = F_{max}\,\hat{\Lambda}(\tau)$. The expression derived from the asymmetric Lorentzian, Eq. (6.1-40), which takes the form

$$F(\tau) = \frac{F_{max}(r + r')}{r'(\tau_{max}/\tau)^{\rho r} + r(\tau/\tau_{max})^{\rho r'}}\,, \tag{6.3-3}$$

is probably the most flexible (see Problem 6.3-1 for the normalized form of this spectrum). It has not been widely used because it cannot, in general, be integrated analytically.

Two other functions, which can be so integrated, have become known as the *box* and the *wedge distribution*. The box distribution is a rather crude approximation to the spectrum. It cannot be related to any of the models already treated. The wedge distribution is of the form of the power laws discussed in Sect. 6.2. It can handle only one side of a typical spectrum. The two distributions are sometimes used in combination (see Sect. 10.2.2).

6.3.1.1 *The Box Distribution*

We discuss the box distribution first. It assumes that the spectrum may be represented by a rectangular distribution, i.e. that

$$F(\tau) = \begin{cases} F_0, & \tau_{min} < \tau < \tau_{max} \\ 0, & \tau < \tau_{min},\ \tau > \tau_{max} \end{cases} \tag{6.3-4}$$

or, equivalently,

$$F(\tau) = F_0[h(\tau - \tau_{min}) - h(\tau - \tau_{max})] \tag{6.3-4'}$$

(cf. Appendix A2.3). The distribution is shown in Fig. 6.3-1.

The spectrum thus has the constant value F_0 over the closed interval $[\tau_{min}, \tau_{max}]$, and the value of zero outside the interval. The value F_0 can be expressed in terms of τ_{min}, τ_{max}, and the viscoelastic constants through the normalization conditions.

For the relaxation spectrum, the normalization condition, Eq. (4.6-20), requires that

$$\int_{-\infty}^{\infty} H(\tau)\,d\ln\tau = H_0 \int_{\tau_{min}}^{\tau_{max}} d\ln\tau = G_g - \{G_e\} \tag{6.3-5}$$

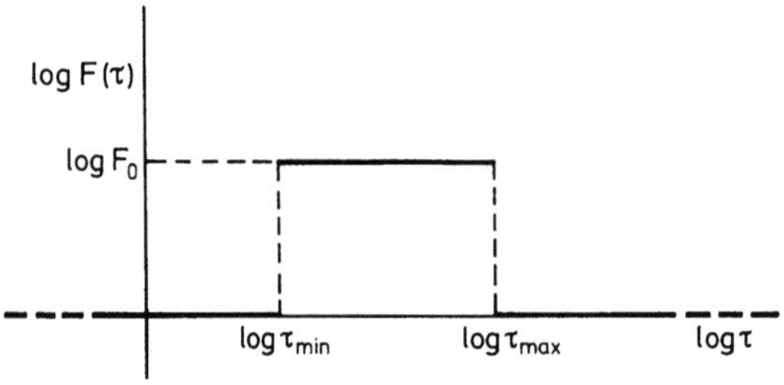

Fig. 6.3-1. Box distribution of respondance times

so that

$$H_0 = \frac{G_g - \{G_e\}}{\ln(\tau_{max}/\tau_{min})} \tag{6.3-6}$$

and, hence*,

$$H_b(\tau) = \begin{cases} \dfrac{G_g - \{G_e\}}{\ln(\tau_{max}/\tau_{min})}, & \tau_{min} < \tau < \tau_{max} \\ 0, & \tau < \tau_{min}, \tau > \tau_{max} \end{cases} . \tag{6.3-7}$$

To find the relaxance for the box distribution, $\bar{Q}_b(s)$, we substitute Eq. (6.3-7) into Eq. (4.1-3). This gives

$$\bar{Q}_b(s) = \{G_e\} + \frac{G_g - \{G_e\}}{\ln(\tau_{max}/\tau_{min})} \int_{\tau_{min}}^{\tau_{max}} \frac{\tau s}{1 + \tau s} d \ln \tau \tag{6.3-8}$$

and integration leads to

$$\bar{Q}_b(s) = \{G_e\} + \frac{G_g - \{G_e\}}{\ln(\tau_{max}/\tau_{min})} \ln \frac{1 + \tau_{max} s}{1 + \tau_{min} s} . \tag{6.3-9}$$

From $\bar{G}(s) = \bar{Q}(s)/s$ we have

$$\bar{G}_b(s) = \frac{\{G_e\}}{s} + \frac{G_g - \{G_e\}}{\ln(\tau_{max}/\tau_{min})} \frac{1}{s} \ln \frac{1 + \tau_{max} s}{1 + \tau_{min} s} . \tag{6.3-10}$$

But, by LTP (28)

$$\mathscr{L}^{-1}[(1/s)\ln(1 + as)] = \mathrm{Ei}(t/a) \tag{6.3-11}$$

where

$$\mathrm{Ei}(x) = \int_x^\infty u^{-1} \exp(-u) du \tag{6.3-12}$$

is the tabulated *exponential integral* function** [11] which possesses the series expansion

---

* The subscript b will be used for the box distribution. No spectral models will be derived from the associated Kobeko function, Eq. (6.1-25), for which the subscript b has been used earlier.

** Different authors use different notations for the integral in Eq. (6.3-12).

$$\mathrm{Ei}(x) = -\gamma - \ln(x) - \sum_{n=1}^{\infty} (-1)^n x^n / n\, n! \tag{6.3-13}$$

where $\gamma = 0.5772157 \ldots$ is the Euler-Mascheroni constant. Excellent numerical approximations are available [11]. From Eq. (6.3-10), then, we obtain the relaxation modulus for the box distribution as

$$G_b(t) = \{G_e\} + \frac{G_g - \{G_e\}}{\ln(\tau_{max}/\tau_{min})} [\mathrm{Ei}(t/\tau_{max}) - \mathrm{Ei}(t/\tau_{min})] . \tag{6.3-14}$$

For the complex modulus, $G^*(\omega) = \bar{Q}(s)|_{s=j\omega}$ yields

$$G_b^*(\omega) = \{G_e\} + \frac{G_g - \{G_e\}}{\ln(\tau_{max}/\tau_{min})} \ln \frac{1 + j\omega\tau_{max}}{1 + j\omega\tau_{min}} . \tag{6.3-15}$$

Separating the real and imaginary parts (see Problem 6.3-2) furnishes the storage and loss modulus from the box distribution as

$$G_b'(\omega) = \{G_e\} + \frac{G_g - \{G_e\}}{2 \ln(\tau_{max}/\tau_{min})} \ln \frac{1 + \omega^2\tau_{max}^2}{1 + \omega^2\tau_{min}^2} \tag{6.3-16}$$

and

$$G_b''(\omega) = \frac{G_g - \{G_e\}}{\ln(\tau_{max}/\tau_{min})} (\tan^{-1} \omega\tau_{max} - \tan^{-1} \omega\tau_{min}) . \tag{6.3-17}$$

The steady-flow viscosity is given by the box distribution as

$$\eta_f = H_0 \int_{\tau_{min}}^{\tau_{max}} d\tau = \frac{G_g(\tau_{max} - \tau_{min})}{\ln(\tau_{max}/\tau_{min})} . \tag{6.3-18}$$

Equations (6.3-14), (6.3-16), and (6.3-17) are plotted in Fig. 6.3-2 with $G_g = 10^9$ N/m^2, $G_e = 10^6/1.001$ N/m^2, $\tau_{min} = 10^{-4}$, and $\tau_{max} = 10^2$ seconds.

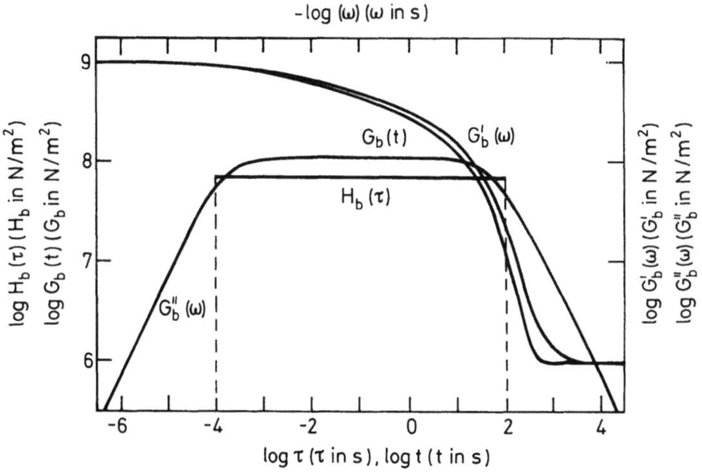

Fig. 6.3-2. Response functions calculated from box distribution of relaxation times

Turning now to the retardation spectrum, the normalization yields

$$L_b(\tau) = \begin{cases} \dfrac{J_e^{\{o\}} - J_g}{\ln(\tau_{max}/\tau_{min})}, & \tau_{min} < \tau < \tau_{max} \\ 0, & \tau < \tau_{min}, \tau > \tau_{max} \end{cases} \quad (6.3\text{-}19)$$

Hence, the retardance for the box distribution, $\bar{U}_b(s)$, becomes

$$\bar{U}_b(s) = J_e^{\{o\}} - \frac{J_e^{\{o\}} - J_g}{\ln(\tau_{max}/\tau_{min})} \int_{\tau_{min}}^{\tau_{max}} \frac{\tau s}{1 + \tau s} \, d \ln \tau + \{\phi_f/s\} \quad (6.3\text{-}20)$$

by substituting Eq. (6.3-19) into Eq. (4.1-26). Integration yields

$$\bar{U}_b(s) = J_e^{\{o\}} - \frac{J_e^{\{o\}} - J_g}{\ln(\tau_{max}/\tau_{min})} \ln \frac{1 + \tau_{max} s}{1 + \tau_{min} s} + \{\phi_f/s\} \quad (6.3\text{-}21)$$

from which we obtain

$$J_b(t) = J_e^{\{o\}} - \frac{J_e^{\{o\}} - J_g}{\ln(\tau_{max}/\tau_{min})} [\mathrm{Ei}(t/\tau_{max}) - \mathrm{Ei}(t/\tau_{min})] + \{\phi_f t\} \quad (6.3\text{-}22)$$

for the creep compliance,

$$J_b'(\omega) = J_e^{\{o\}} - \frac{J_e^{\{o\}} - J_g}{2 \ln(\tau_{max}/\tau_{min})} \ln \frac{1 + \omega^2 \tau_{max}^2}{1 + \omega^2 \tau_{min}^2} \quad (6.3\text{-}23)$$

for the storage compliance, and

$$J_b''(\omega) = \frac{J_e^{\{o\}} - J_g}{\ln(\tau_{max}/\tau_{min})} (\tan^{-1} \omega\tau_{max} - \tan^{-1} \omega\tau_{min}) + \{\phi_f/\omega\} \quad (6.3\text{-}24)$$

for the loss compliance, respectively. Plots of $J_b(t)$, $J_b'(\omega)$, and $J_b''(\omega)$ are not shown since they are analogous, *mutatis mutandis*, with $G_b(t)$, $G_b'(\omega)$, and $G_b''(\omega)$.

6.3.1.2 The Wedge and Associated Power Law Distribution

The box distribution is clearly not very realistic. It is often used in conjunction with the *wedge distribution* which is based on the experimental observation that, in the region of prime interest, a plot of log $F(\tau)$ vs. log τ often yields an approximately straight line with a negative slope for the relaxation spectrum, and a positive slope for the retardation spectrum. For the wedge distribution

$$F(\tau) = \begin{cases} F_0 \cdot (\tau/\tau_0)^{\pm m}, & \tau_{min} < \tau < \tau_{max} \\ 0, & \tau < \tau_{min}, \tau > \tau_{max} \end{cases} \quad (6.3\text{-}25)$$

or, equivalently,

$$F(\tau) = F_0 \cdot (\tau/\tau_0)^{\pm m} [h(\tau - \tau_{min}) - h(\tau - \tau_{max})] \quad (6.3\text{-}25')$$

where F_0, τ_0, and m are constants. The distribution is shown schematically in Fig. 6.3-3.

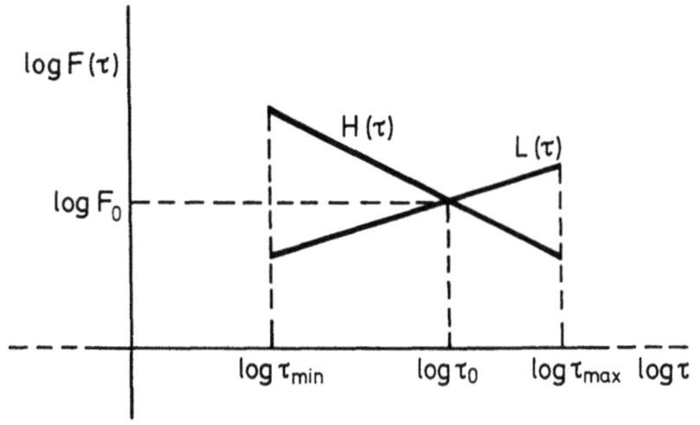

Fig. 6.3-3. Wedge distribution of respondance times

The parameter m is the slope of log F(τ) in the region in which it is a nearly linear function of log τ. The slope of log H(τ) is negative, and that of log L(τ) is positive. Consequently, we define

$$m = -\frac{d \log H(\tau)}{d \log \tau} \quad \text{and} \quad m = \frac{d \log L(\tau)}{d \log \tau} \tag{6.3-26}$$

so that m is always positive (m > 0) [cf. Eqs. (4.2-63) and (4.2-64)].

For the relaxation spectrum we use m with the minus sign in Eq. (6.3-25). The normalization condition

$$\int_{-\infty}^{\infty} H(\tau) d \ln \tau = H_0 \tau_0^m \int_{\tau_{min}}^{\tau_{max}} \tau^{-m} d \ln \tau = G_g - \{G_e\} \tag{6.3-27}$$

yields

$$H_0 = \frac{[G_g - \{G_e\}]m}{1 - (\tau_{min}/\tau_{max})^m} (\tau_0/\tau_{min})^{-m} \tag{6.3-28}$$

and hence

$$H_w(\tau) = \begin{cases} \dfrac{[G_g - \{G_e\}]m}{1 - (\tau_{min}/\tau_{max})^m} (\tau/\tau_{min})^{-m}, & \tau_{min} \le \tau \le \tau_{max} \\ 0, & \tau < \tau_{min}, \tau > \tau_{max} . \end{cases} \tag{6.3-29}$$

The relaxance for the wedge distribution becomes

$$\bar{Q}_w(s) = \{G_e\} + \frac{[G_g - \{G_e\}]m}{1 - (\tau_{min}/\tau_{max})^m} \int_{\tau_{min}}^{\tau_{max}} (\tau/\tau_{min})^{-m} \frac{\tau s}{1 + \tau s} d \ln \tau \tag{6.3-30}$$

through substitution of Eq. (6.3-29) into Eq. (4.1-3). With the change of variable $\tau s = 1/x$ we obtain

$$\bar{Q}_w(s) = \{G_e\} + \frac{[G_g - \{G_e\}]m}{1 - (\tau_{min}/\tau_{max})^m} (\tau_{min} s)^m \times [B(m, 1 - m; 1/\tau_{min} s) - B(m, 1 - m; 1/\tau_{max} s)] . \tag{6.3-31}$$

where

$$B(p,q;x) = \int_0^x \frac{u^{p-1}}{(1+u)^{p+q}} du \tag{6.3-32}$$

is the *incomplete beta-function* [11].

Now, by LTP (29) we have

$$\mathscr{L}^{-1}[s^{m-1}B(m,1-m;1/\tau s) = \gamma(m;t/\tau)t^{-m} \tag{6.3-33}$$

where

$$\gamma(m;x) = \int_0^x u^{m-1}\exp(-u)du \tag{6.3-34}$$

is the *incomplete gamma-function* [11] which has the series expansion

$$\gamma(m;x) = -x^{m-1}\sum_{n=1}^{\infty}\frac{(-x)^n}{(n+m-1)(n-1)!} . \tag{6.3-35}$$

Hence, the wedge distribution gives the relaxation modulus as

$$G_w(t) = \{G_e\} + \frac{[G_g - \{G_e\}]m}{1-(\tau_{min}/\tau_{max})^m}[\gamma(m;t/\tau_{min}) - \gamma(m;t/\tau_{max})](t/\tau_{min})^{-m} . \tag{6.3-36}$$

For the complex modulus we have

$$G_w^*(\omega) = \{G_e\} + \frac{[G_g - \{G_e\}]m}{1-(\tau_{min}/\tau_{max})^m}(j\omega\tau_{min})^m$$

$$\times [B(m,1-m;1/j\omega\tau_{min}) - B(m,1-m;1/j\omega\tau_{max})] . \tag{6.3-37}$$

But (see Problem 6.3-3)

$$j^m B(m,1-m;1/j\omega\tau) = \tfrac{1}{2}B(m/2,1-m/2;1/\omega^2\tau^2)$$

$$+\frac{j}{2}B(\tfrac{1}{2}+m/2,\tfrac{1}{2}-m/2;1/\omega^2\tau^2) \tag{6.3-38}$$

and, hence, separation of the real and imaginary parts yields

$$G_w'(\omega) = \{G_e\} + \frac{[G_g - \{G_e\}](m/2)(\omega\tau_{min})^m}{1-(\tau_{min}/\tau_{max})^m}$$

$$\times [B(m/2,1-m/2;1/\omega^2\tau_{min}^2) - B(m/2,1-m/2;1/\omega^2\tau_{max}^2)] \tag{6.3-39}$$

for the storage modulus, and

$$G_w''(\omega) = \frac{[G_g - \{G_e\}](m/2)(\omega\tau_{min})^m}{1-(\tau_{min}/\tau_{max})^m}$$

$$\times [B(\tfrac{1}{2}+m/2,\tfrac{1}{2}-m/2;1/\omega^2\tau_{min}^2) - B(\tfrac{1}{2}+m/2,\tfrac{1}{2}-m/2;1/\omega^2\tau_{max}^2)] \tag{6.3-40}$$

for the loss modulus, derived from the wedge distribution.

The steady-flow viscosity follows as

$$\eta_f = H_0 \int_{\tau_{min}}^{\tau_{max}} d\tau = \frac{G_g m}{1-m}\frac{\tau_{max}^{1-m} - \tau_{min}^{1-m}}{\tau_{min}^{-m} - \tau_{max}^{-m}} . \tag{6.3-41}$$

For m = 1/2, the response functions can be calculated with relative ease. Equation $(6.3\text{-}35)_1$ becomes

$$\gamma(\tfrac{1}{2};x) = -\frac{2}{\sqrt{x}}\sum_{n=1}^{\infty}\frac{(-x)^n}{(2n-1)(n-1)!} \tag{6.3-42}$$

which converges rapidly for $x < 10$ and is close to $\Gamma(\tfrac{1}{2}) = \sqrt{\pi}$ for $x \geq 10$.

Letting τ stand for either τ_{min} or τ_{max}, the beta-functions required in Eq. (6.3-39) become

$$B(\tfrac{1}{4},\tfrac{3}{4};1/\omega^2\tau^2) = 4\int_0^{1/\sqrt{\omega\tau}}\frac{dx}{1+x^4} \tag{6.3-43}$$

where we have used Eq. (6.3-32) with the change of variable $u = x^4$. But [12]

$$\int\frac{dx}{1+x^4} = \frac{1}{2\sqrt{2}}\left(\frac{1}{2}\ln\frac{x^2+x\sqrt{2}+1}{x^2-x\sqrt{2}+1} + \tan^{-1}\frac{x\sqrt{2}}{1-x^2}\right) \tag{6.3-44}$$

and, therefore, for m = 1/2

$$G'_w(\omega) = \{G_e\} + \frac{[G_g-\{G_e\}]\sqrt{\omega\tau_{min}/2}}{2(1-\sqrt{\tau_{min}/\tau_{max}})}\left|\frac{1}{2}\ln\frac{\omega\tau+\sqrt{2\omega\tau}+1}{\omega\tau-\sqrt{2\omega\tau}+1} + \tan^{-1}\frac{\sqrt{2\omega\tau}}{\omega\tau-1}\right|_{\tau_{max}}^{\tau_{min}} \tag{6.3-45}$$

where

$$|f(\tau)|_{\tau_{max}}^{\tau_{min}} = f(\tau_{min}) - f(\tau_{max}) \ . \tag{6.3-46}$$

The beta-functions needed in Eq. (6.3-40) become

$$B(\tfrac{3}{4},\tfrac{1}{4};1/\omega^2\tau^2) = 4\int_0^{1/\sqrt{\omega\tau}}\frac{x^2dx}{1+x^4} \tag{6.3-47}$$

where [12]

$$\int\frac{x^2dx}{1+x^4} = \frac{1}{2\sqrt{2}}\left(\frac{1}{2}\ln\frac{x^2-x\sqrt{2}+1}{x^2+x\sqrt{2}+1} + \tan^{-1}\frac{x\sqrt{2}}{1-x^2}\right) . \tag{6.3-48}$$

Hence, for m = 1/2

$$G''_w(\omega) = \frac{[G_g-\{G_e\}]\sqrt{\omega\tau_{min}/2}}{2(1-\sqrt{\tau_{min}/\tau_{max}})}\left|\frac{1}{2}\ln\frac{\omega\tau-\sqrt{2\omega\tau}+1}{\omega\tau_m+\sqrt{2\omega\tau}+1} + \tan^{-1}\frac{\sqrt{2\omega\tau}}{\omega\tau-1}\right|_{\tau_{max}}^{\tau_{min}} . \tag{6.3-49}$$

When $\omega\tau < 1$, the angles in Eqs. (6.3-45) and (6.3-48) lie in the second quadrant and π must be added to the inverse tangent.

The steady-flow viscosity derived from the wedge distribution takes the form $\eta_f = G_g\sqrt{\tau_{max}\tau_{min}}$ when m = 1/2.

$G_w(t)$, $G'_w(\omega)$, and $G''_w(\omega)$ for m = 1/2 are shown by the solid lines in Fig. 6.3-4 below. They were calculated with the same parameters that were used in constructing Fig. 6.3-2. $G_w(t)$ lies below $G'_w(\omega)$ throughout.

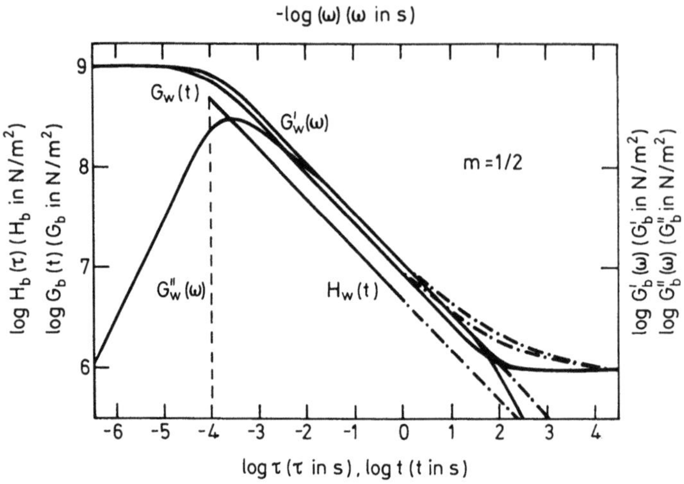

Fig. 6.3-4.
Response functions calculated from the indicated wedge distribution of relaxation times for m = 1/2

Equations (6.3-36), (6.3-39) and (6.3-40) can be simplified. Letting $\tau_{max} = \infty$ furnishes

$$H(\tau) = \begin{cases} H_0 \cdot (\tau/\tau_0)^{-m}, & \tau \geq \tau_{min} \\ 0, & \tau < \tau_{min} \end{cases} \tag{6.3-50}$$

or, equivalently,

$$H(\tau) = H_0 \cdot (\tau/\tau_0)^{-m} h(\tau - \tau_{min}) \ . \tag{6.3-50'}$$

$H(\tau)$ so defined is sometimes referred to as the *cut-off power law distribution* for the relaxation spectrum. It is displayed schematically in Fig. 6.3-5.

Equation (6.3-31) now becomes

$$\bar{Q}_w(s) = \{G_e\} + [G_g - \{G_e\}] \, m(\tau_{min}s)^m B(m, 1-m; 1/\tau_{min}s) \tag{6.3-51}$$

and we have

$$G_w(t) = \{G_e\} + [G_g - \{G_e\}] m \, \gamma(m; t/\tau_{min})(t/\tau_{min})^{-m} \tag{6.3-52}$$

$$G_w'(\omega) = \{G_e\} + [G_g - \{G_e\}](m/2)(\omega\tau_{min})^m B(m/2, 1-m/2; 1/\omega^2\tau_{min}^2) \tag{6.3-53}$$

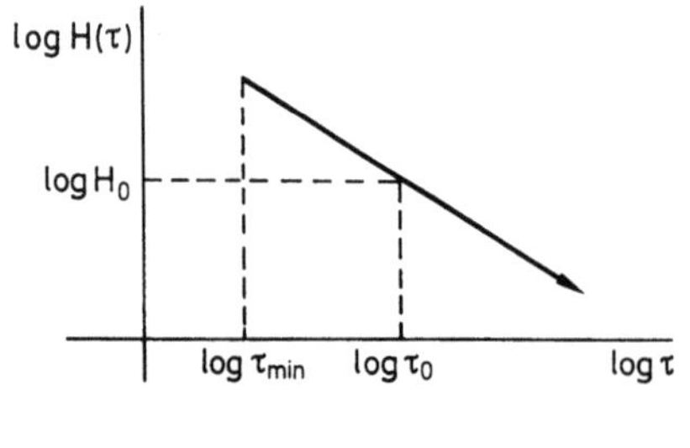

Fig. 6.3-5. Wedge distribution of relaxation times cut-off at the short time end of the spectrum

and

$$G''_w(\omega) = [G_g - \{G_e\}](m/2)(\omega\tau_{min})^m B(\tfrac{1}{2} + m/2, \tfrac{1}{2} - m/2; 1/\omega^2\tau_{min}^2) \; . \qquad (6.3\text{-}54)$$

The cut-off wedge functions are also displayed in Fig. 6.3-4. They deviate (dashed-dotted lines) from the wedge functions (solid lines) for $\log \tau > 0$.

The models represented by Eqs. (6.3-31), (6.3-36), (6.3-39), and their simplified versions, Eqs. (6.3-51), (6.3-52), and (6.3-53) are standard, i.e. they possess a finite, non-zero instantaneous modulus (see Problem 6.3-4).

If τ_{min} is small, we may further simplify Eqs. (6.3-51) to (6.3-53) by approximating $\gamma(m; t/\tau_{min})$ by $\gamma(m; \infty) = \Gamma(m)$ and $B(p, q; 1/\omega^2\tau_{min}^2)$ by $B(p, q)$ where

$$B(p,q) = \int_0^\infty \frac{u^{p-1}}{(1+u)^{p+q}} du = \frac{\Gamma(p)\Gamma(q)}{\Gamma(p+q)} \qquad (6.3\text{-}55)$$

is the *(complete) beta-function* [11]. Now τ_0 may take any value within the closed interval $[\tau_{min}, \tau_{max}]$. We may, therefore, let $\tau_{min} = \tau_0$. Replacing τ_0 in turn by t_0 or $1/\omega_0$ as needed, Eqs. (6.3-51) to (6.3-53) become

$$G_p(t) = \{G_e\} + [G_g - \{G_e\}]\Gamma(1+m)(t/t_0)^{-m} \qquad (6.3\text{-}56)$$

$$G'_p(\omega) = \{G_e\} + [G_g - \{G_e\}](\pi m/2)(\omega/\omega_0)^m \csc \pi m/2 \qquad (6.3\text{-}57)$$

and

$$G''_p(\omega) = [G_g - \{G_e\}](\pi m/2)(\omega/\omega_0)^m \sec \pi m/2 \; . \qquad (6.3\text{-}58)$$

These should be compared with the equations obtained through fractional differentiation in Sect. 6.2, i.e. with Eqs. $(6.2\text{-}7)_1$ to $(6.2\text{-}12)_1$. With

$$G = [G_g - \{G_e\}]\pi m/\sin \pi m = H_0 \pi \csc \pi m \qquad (6.3\text{-}59)$$

$$G_0 = [G_g - \{G_e\}]\Gamma(1+m) = H_0\Gamma(m) \qquad (6.3\text{-}60)$$

$$G'_0 = [G_g - \{G_e\}](\pi m/2)/\sin \pi m/2 = H_0 \tfrac{\pi}{2} \csc \pi m/2 \qquad (6.3\text{-}61)$$

and

$$G''_0 = [G_g - \{G_e\}](\pi m/2)/\cos \pi m/2 = H_0 \tfrac{\pi}{2} \sec \pi m/2 \qquad (6.3\text{-}62)$$

the latter become identical with Eqs. (6.3-56) to (6.3-58) except for the viscoelastic constants. The last two equations follow because $\Gamma(m)\Gamma(1-m) = \pi/\sin \pi m$ and $\Gamma(\tfrac{1}{2}+m)\Gamma(\tfrac{1}{2}-m) = \pi/\cos \pi m$. We note that $G'_0 = G_0\Gamma(1-m)\cos m\pi/2$, $G''_0 = G_0\Gamma(1-m)\sin m\pi/2$, and $G = G_0\Gamma(1-m)$.

H_0 in the equations above is $[G_g - \{G_e\}]m$. This is obtained by letting $\tau_{min} = \tau_0$ and $\tau_{max} = \infty$ in Eq. (28). With these two substitutions the wedge distribution becomes the *power law distribution* [hence the subscript p in Eqs. (6.3-56) to (6.3-58)] with

$$H_p(\tau) = H_0 \cdot (\tau/\tau_0)^{-m} \qquad (6.3\text{-}63)$$

as the *power law relaxation spectrum*. This will be used in Chap. 8 because its product with the kernels of the response functions is easily integrated. Equations (6.3-56) to

(6.3-58) have, in fact, been obtained in Sects. 4.2.3 and 4.3.3 by introducing the power law spectrum into the response functions.

The constant τ_0 is often subsumed into H_0 and the power law spectrum is written in the form $H(\tau) = k\tau^{-m}$. This, however, makes the dimensions of k depend on m and is not recommended.

For the retardation spectrum we take m with the plus sign in Eq. (6.3-25). The normalization condition

$$\int_{-\infty}^{\infty} L(\tau)d\ln\tau = L_0\tau_0^{-m}\int_{\tau_{min}}^{\tau_{max}} \tau^m d\ln\tau = J_e^{\{o\}} - J_g \tag{6.3-64}$$

yields

$$L_0 = \frac{[J_e^{\{o\}} - J_g]m}{1-(\tau_{min}/\tau_{max})^m}(\tau_0/\tau_{max})^m \tag{6.3-65}$$

and we have

$$L_w(\tau) = \begin{cases} \dfrac{[J_e^{\{o\}} - J_g]m}{1-(\tau_{min}/\tau_{max})^m}(\tau/\tau_{max})^m, & \tau_{min} \le \tau \le \tau_{max} \\ 0, & \tau > \tau_{min}, \tau > \tau_{max}\ . \end{cases} \tag{6.3-66}$$

To obtain the retardance, $\bar{U}_w(s)$, for the wedge distribution, we substitute Eq. (6.3-66) into Eq. (4.1-18). The change of variable $\tau s = 1/x$ then yields

$$\bar{U}_w(s) = J_g + \frac{[J_e^{\{o\}} - J_g]m}{1-(\tau_{min}/\tau_{max})^m}(\tau_{max}s)^{-m} \times [B(1-m, m; 1/\tau_{min}s) - B(1-m, m; 1/\tau_{max}s)] + \{\phi_f/s\} \tag{6.3-67}$$

and the complex compliance becomes

$$J_w^*(\omega) = J_g + \frac{[J_e^{\{o\}} - J_g]m}{1-(\tau_{min}/\tau_{max})^m}(j\omega\tau_{max})^{-m} \times [B(1-m, m; 1/j\omega\tau_{min}) - B(1-m, m; 1/j\omega\tau_{max})] + \{\phi_f/j\omega\}\ . \tag{6.3-68}$$

Separation of the real and imaginary parts with the help of the relation

$$j^{-m}B(1-m, m; 1/j\omega\tau) = \tfrac{1}{2}B(1-m/2, m/2; 1/\omega^2\tau^2) + \frac{j}{2}B(\tfrac{1}{2}-m/2, \tfrac{1}{2}+m/2; 1/\omega^2\tau^2) \tag{6.3-69}$$

leads to

$$J_w'(\omega) = J_g + \frac{[J_e^{\{o\}} - J_g](m/2)(\omega\tau_{max})^{-m}}{1-(\tau_{min}/\tau_{max})^m} \times [B(1-m/2, m/2; 1/\omega^2\tau_{min}^2) - B(1-m/2, m/2; 1/\omega^2\tau_{max}^2)] \tag{6.3-70}$$

for the storage compliance, and to

$$J''_w(\omega) = \frac{[J_e^{\{o\}} - J_g](m/2)(\omega\tau_{max})^{-m}}{1-(\tau_{min}/\tau_{max})^m}$$
$$\times [B(\tfrac{1}{2}-m/2, \tfrac{1}{2}+m/2; 1/\omega^2\tau_{min}^2) - B(\tfrac{1}{2}-m/2, \tfrac{1}{2}+m/2; 1/\omega^2\tau_{max}^2)]$$
$$+ \{\phi_f/\omega\} \tag{6.3-71}$$

for the loss compliance.

For $m = 1/2$, the beta-functions required in $J'_w(\omega)$ become identical with those needed in $G''_w(\omega)$, given by Eq. (6.3-47), and those in $J''_w(\omega)$ are the same as those needed in $G'_w(\omega)$, given by Eq. (6.3-43). For $m = 1/2$, using Eq. (6.3-46),

$$J'_w(\omega) = J_g + \frac{[J_e^{\{o\}} - J_g]/\sqrt{2\omega\tau_{max}}}{2(1-\sqrt{\tau_{min}/\tau_{max}})} \left| \frac{1}{2}\ln\frac{\omega\tau - \sqrt{2\omega\tau}+1}{\omega\tau+\sqrt{2\omega\tau}+1} + \tan^{-1}\frac{\sqrt{2\omega\tau}}{\omega\tau - 1}\right|_{\tau_{max}}^{\tau_{min}} \tag{6.3-72}$$

and

$$J''_w(\omega) = \frac{[J_e^{\{o\}} - J_g]/\sqrt{2\omega\tau_{max}}}{2(1-\sqrt{\tau_{min}/\tau_{max}})} \left| \frac{1}{2}\ln\frac{\omega\tau + \sqrt{2\omega\tau}+1}{\omega\tau-\sqrt{2\omega\tau}+1} + \tan^{-1}\frac{\sqrt{2\omega\tau}}{\omega\tau - 1}\right|_{\tau_{max}}^{\tau_{min}} + \{\phi_f/\omega\} . \tag{6.3-73}$$

To obtain the creep compliance derived from the wedge distribution, we substitute Eq. (6.3-66) directly into Eq. (4.1-28) (for an alternative derivation, see Problem 6.3-5). This yields

$$J_w(t) = J_e^{\{o\}} - \frac{[J_e^{\{o\}} - J_g]m}{1-(\tau_{min}/\tau_{max})^m}[\gamma(-m; t/\tau_{min}) - \gamma(-m; t/\tau_{max})](t/\tau_{max})^m + \{\phi_f t\} . \tag{6.3-74}$$

The series expansion of the incomplete gamma function, Eq. (6.3-35), is valid as long as m is restricted by the relation $0 < |m| < 1$. For large values of t, the middle term in Eq. (6.3-74) contains the product of a small and a large number. Computational difficulties can be avoided by introducing the function

$$\theta(m; x) = 1/m + \gamma(-m; x)x^m = \sum_{n=1}^{\infty} \frac{(-x)^n}{(n-m)n!} . \tag{6.3-75}$$

This leads to

$$J_w(t) = J_g + \frac{[J_e^{\{o\}} - J_g]m}{1-(\tau_{min}/\tau_{max})^m}$$
$$\times [\theta(m; t/\tau_{max}) - (\tau_{min}/\tau_{max})^m \theta(m; t/\tau_{min})] + \{\phi_f t\} \tag{6.3-76}$$

from which $J_w(t)$ is readily calculated. The reader may wish to show that in the limit as $t \to \infty$, $J_w(t)$ becomes $J_e^{\{o\}}$ as it should (see Problem 6.3-6).

For $m = 1/2$, we have

$$J_w(t) = J_g + \frac{[J_e^{\{o\}} - J_g]}{2(1-\sqrt{\tau_{min}/\tau_{max}})}[\theta(\tfrac{1}{2}; t/\tau_{max}) - \sqrt{\tau_{min}/\tau_{max}}\,\theta(\tfrac{1}{2}; t/\tau_{min})] + \{\phi_f t\} \tag{6.3-77}$$

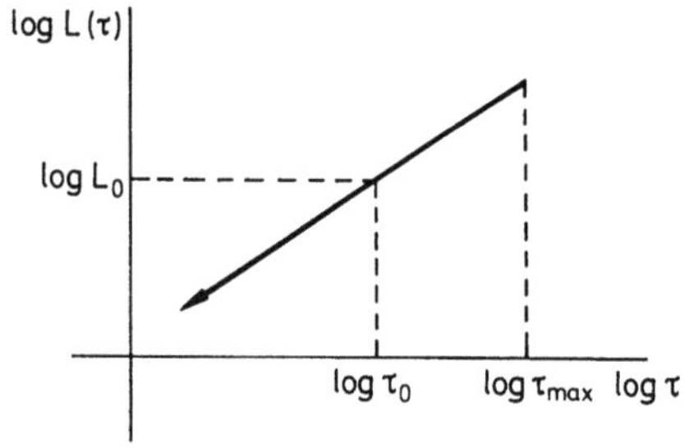

Fig. 6.3-6. Wedge distribution of retardation times cut-off at the long time end of the spectrum

in which

$$\theta(\tfrac{1}{2}; x) = 2 \sum_{n=1}^{\infty} \frac{(-x)^n}{(2n-1)n!} . \qquad (6.3\text{-}78)$$

The series converges rapidly for $x < 10$, and closely approaches $2(1 - \sqrt{\pi x})$ for $x \geq 10$. The latter expression is obtained from Eq. $(6.3\text{-}75)_1$ noting that, for large values of x, $\gamma(-m; x)$ approaches $\Gamma(-1/2) = -2\sqrt{\pi}$.

Again, Eqs. (6.3-74) or (6.3-76), and (6.3-72) and (6.3-73) can be simplified. We now let $\tau_{min} = 0$ and obtain the spectrum

$$L(\tau) = \begin{cases} L_0 \cdot (\tau/\tau_0)^m, & \tau \leq \tau_{max} \\ 0, & \tau > \tau_{max} \end{cases} \qquad (6.3\text{-}79)$$

or, equivalently,

$$L(\tau) = L_0 \cdot (\tau/\tau_0)^m h(\tau_{max} - \tau) . \qquad (6.3\text{-}79')$$

This is the *cut-off power law distribution* for the retardation spectrum show below (Fig. 6.3-6).

Equation (6.3-74) now takes the form

$$J_w(t) = J_e^{\{o\}} - [J_e^{\{o\}} - J_g] m [\Gamma(-m) - \gamma(-m; t/\tau_{max})] (t/\tau_{max})^m + \{\phi_f t\} \qquad (6.3\text{-}80)$$

and, using Eq. (6.3-75) in Eq. (6.3-80), Eq. (6.3-76) becomes

$$J_w(t) = J_g + [J_e^{\{o\}} - J_g] m [\theta(m; t/\tau_{max}) - \Gamma(-m)(t/\tau_{max})^m] + \{\phi_f t\} . \qquad (6.3\text{-}81)$$

For the storage and loss compliances we obtain

$$J'_w(\omega) = J_g + [J_e^{\{o\}} - J_g](m/2) \times [\pi \csc \pi m/2 - B(1 - m/2, m/2; 1/\omega^2\tau_{max}^2)](\omega\tau_{max})^{-m} \qquad (6.3\text{-}82)$$

and

$$J''_w(\omega) = [(J_e^{\{o\}} - J_g)(m/2) \times [\pi \sec \pi m/2 - B(\tfrac{1}{2} - m/2, \tfrac{1}{2} + m/2; 1/\omega^2\tau_{max}^2)](\omega\tau_{max})^{-m} + \{\phi_f/\omega\} . \qquad (6.3\text{-}83)$$

where we have used the relations $B(1 - m/2, m/2) = \Gamma(1 - m/2)\Gamma(m/2) = \pi/\sin \pi m$, and $B(\tfrac{1}{2} - m/2, \tfrac{1}{2} + m/2) = \Gamma(\tfrac{1}{2} - m/2)\Gamma(\tfrac{1}{2} + m/2) = \pi/\cos \pi m$. The compliance models discussed so far are all standard models just as the corresponding modulus models. Plots of the compliance models are not shown since they are, *mutatis mutandis*, analogous to the modulus models.

Equations (6.3-81) to (6.3-83) can again be simplified further by considering τ_{max} to be large enough for the theta-function in Eq. (6.3-81) and the beta-functions in Eqs. (6.3-82) and (6.3-83) to vanish. Letting τ_{max} then equal t_0 or $1/\omega_0$ as required, we find

$$J_p(t) = J_g + [J_e^{\{o\}} - J_g]\Gamma(1 - m)(t/t_0)^m + \{\phi_f t\} \tag{6.3-84}$$

$$J'_p(\omega) = J_g + [J_e^{\{o\}} - J_g](\pi m/2)(\omega/\omega_0)^{-m} \csc \pi m/2 \tag{6.3-85}$$

and

$$J''_p(\omega) = [J_e^{\{o\}} - J_g](\pi m/2)(\omega/\omega_0)^{-m} \sec \pi m/2 + \{\phi_f/\omega\} \ . \tag{6.3-86}$$

These again become identical with Eqs. $(6.2\text{-}7)_2$ to $(6.2\text{-}12)_2$ (except for the viscoelastic constants) if, in the latter, we let

$$J = [J_e^{\{o\}} - J_g]\pi m/\sin \pi m = L_0 \pi \csc \pi m \ . \tag{6.3-87}$$

$$J_0 = [J_e^{\{o\}} - J_g]\Gamma(1 - m) = -L_0\Gamma(-m) \tag{6.3-88}$$

$$J'_0 = [J_e^{\{o\}} - J_g](\pi m/2)/\sin \pi m/2 = L_0(\pi/2)\csc \pi m/2 \tag{6.3-89}$$

and

$$J''_0 = [J_e^{\{o\}} - J_g](\pi m/2)/\cos \pi m/2 = L_0(\pi/2)\sec \pi m/2 \ . \tag{6.3-90}$$

Here, $L_0 = [J_e^{\{o\}} - J_g]m$, and $J'_0 = J_0\Gamma(1 + m)\cos m\pi/2$, $J''_0 = J_0\Gamma(1 + m)\sin m\pi/2$, and $J = J_0\Gamma(1 + m)$.

The *power law retardation spectrum* becomes

$$L_p(\tau) = L_0 \cdot (\tau/\tau_0)^m \ . \tag{6.3-91}$$

It is often found in the form $L(\tau) = k\tau^m$ analogous to $H(\tau) = k\tau^{-m}$.

6.3.2 Spectra Derived from Models for the Experimental Responses

There are general rules by which the spectral functions associated with a given mathematical model for one of the experimental response functions can be found. As will be shown in Sect. 6.3.2.2, this is always possible from models for the harmonic response functions although the spectrum may have some undesirable features (see, e.g., Problem 6.3-12). From models of the step responses it is not always possible to obtain a convenient closed-form expression for the spectrum. Their case will be discussed in Sect. 6.3.2.1.

The expressions to be derived in the following two sections can serve to model the spectral functions just as the models discussed in the preceding section. In addition, however, the new models can be useful when the response to a given excitation has been modelled and it is desired to predict the response to another excitation. One can then proceed via the canonical representations (see Problem 6.3-13).

6.3.2.1 *Spectra Derived from Models for the Step Responses*

We have seen in Sect. 4.1 that the step response functions are essentially Laplace transforms of the spectral functions. Consequently, it should be possible to derive the spectra corresponding to the mathematical models for the step response functions through the inverse Laplace transformation. While this is true, it is convenient only in those, unfortunately few, cases where the inverse transform can be given in closed form. When this is not possible, an approximation to the spectrum can usually be obtained by the methods of Sect. 4.2 (see Problem 6.3-8).

Let us now look at the procedure for finding the spectra from step response models by the method of transform inversion. By Eq. (4.1-53)

$$H(\tau) = \zeta \mathscr{L}^{-1}\{[G(t) - \{G_e\}]; \zeta\}|_{\zeta=1/\tau} \; . \tag{6.3-92}$$

Equation (6.1-2) then allows us to write

$$H_M(\tau) = [G_g - \{G_e\}]\zeta \mathscr{L}^{-1}[\hat{Z}_M(t); \zeta]_{\zeta=1/\tau} \; . \tag{6.3-93}$$

By Eq. (4.1-58) we have

$$L(\tau) = \zeta \mathscr{L}^{-1}\{[J_e^{\{o\}} - J_\bullet(t) + \{\phi_f t\}]; \zeta\}_{\zeta=1/\tau} \tag{6.3-94}$$

and Eq. (6.1-3) leads to

$$L_M(\tau) = [J_e^{\{o\}} - J_g]\zeta \mathscr{L}^{-1}[\hat{Z}_M(t); \zeta]_{\zeta=1/\tau} \; . \tag{6.3-95}$$

If the equation of Kohlrausch, i.e. $\hat{Z}_K(t)$, or the Cole-Cole equation, i.e. $\hat{Z}_C(t)$, is chosen, the inverse transformation is possible in closed form only when $m = 1/2$ (use LTP's (31) and (32), respectively). We now turn to Kobeko's equation, i.e. $\hat{Z}_B(t)$, for which the inverse transformation can be given for the general case when $0 < b < 1$. Letting $t_0 = a = 1/\zeta_0$, the use of LTP (30) (where, here, t is s and ζ is t) leads to

$$\mathscr{L}^{-1}[\hat{Z}_B(t); \zeta] = \mathscr{L}^{-1}[1/(1 + t/t_0)^b; \zeta] = \zeta^{-1}\Gamma^{-1}(b)(\zeta/\zeta_0)^b \exp(-\zeta/\zeta_0) \tag{6.3-96}$$

and Eq. (6.3-93) then gives

$$H_B(\tau) = H_0 \cdot (\tau_0/\tau)^b \exp(-\tau_0/\tau) \tag{6.3-97}$$

where $H_0 = [G_g - \{G_e\}]/\Gamma(b)$. $H_B(\tau)$ is shown schematically in Fig. 6.3-7. It has a

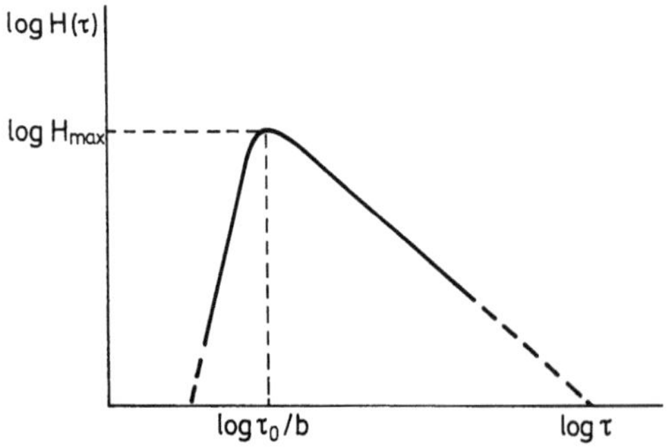

Fig. 6.3-7. Modified Wedge distribution of relaxation times

maximum at $\tau_{max} = \tau_0/b$ where its value is $H_0 b^b \exp(-b)$. It rises sharply on the left of the maximum and drops more slowly to the right of it where the slope of the asymptote is $-b$. A comparison of Figs. 6.3-7 and 6.3-5, and of Eqs. (6.3-97) and (6.3-50′) shows that Eq. (6.3-97) may be regarded as a modification of Eq. (6.3-50′) in which the exponential factor $\exp(-\tau_0/\tau)$ avoids the artificial cut-off at τ_{min}. For this reason Eq. (6.3-97) is sometimes referred to as the *modified wedge* or *modified power law* distribution. In Problem 6.3-7 we introduce an alternative form of Eq. (6.3-97) and show that it satisfies the normalization condition, Eq. (6.3-27).

The relaxance, $\bar{Q}_B(s)$, of the modified wedge distribution is obtained by introducing Eq. (6.3-97) into Eq. (4.1-3). The change of variable $\tau = 1/\zeta$ reveals the integral to be a Stieltjes transform. Using STP (1) (letting ζ be t and a be $1/\zeta_0$) gives

$$\bar{Q}_B(s) = \{G_e\} + [G_g - \{G_e\}](s/s_0)^b \Gamma(1 - b; s/s_0)\exp(s/s_0) \tag{6.3-98}$$

where $s_0 = 1/\zeta_0$ and

$$\Gamma(m; x) = \Gamma(m) - \gamma(m; x) = \int_x^{\infty} u^{m-1} \exp(-u) du \tag{6.3-99}$$

is the *complementary incomplete gamma-function* [11]. The latter can be calculated using Eq. (6.3-35). Through division by s and inversion of the transform with the aid of LTP (24), and letting $s_0 = a = 1/t_0$, we obtain

$$G_B(t) = \{G_e\} + [G_g - \{G_e\}](1 + t/t_0)^{-b} \tag{6.3-100}$$

which, as it should be, is none other than Eq. (6.1-42) with $\hat{Z}_B(t)$. By contrast, the harmonic responses cannot be obtained in a simple closed form from $G_B^*(\omega) = \bar{Q}_B(j\omega)$.

One might expect that an analogous derivation based on the *associated* Kobeko function, $\hat{s}_B(t)$, would lead to a modified power law expression for $L(\tau)$. The attempt fails, however, because the Laplace transformation cannot be carried out in closed form.

6.3.2.2 Spectra Derived from Models for the Harmonic Responses

In the preceding section we have seen that the spectrum corresponding to a given model for one of the step responses can be found only in the favorable case when the inverse Laplace transform exists and can be obtained in closed form. The situation is different with regard to the harmonic response functions. It was shown in Sect. 4.1.2 that the spectra are essentially Stieltjes transforms of these functions. The Stieltjes transform is inverted by the limiting operation discussed in Appendix A5.2.1. In Sect. 4.5 this procedure was used to invert rational functions and we arrived thus at the line spectra characteristic of series-parallel models (or their equivalents, such as the ladder models). The mathematical models we are concerned with here are not rational functions. In their case, therefore, the limiting operation can be "anticipated" (see Appendix A5.2.1) and the inversion is even easier. The result is that, given a model for any of the harmonic response functions, the corresponding spectrum can always be found.

We begin by considering the complex harmonic responses because several models were originally proposed for the complex responses. The complex responses are Fourier-Laplace transforms (see Appendix A5.3) but, these being simply special cases of the Stieltjes transform, their inversion properties are the same. We merely need to remember that the transform variable is now $j\omega$.

Let us look at the complex modulus. By Eq. (4.1-71), the relaxation spectrum is

$$H(\tau) = \mathscr{I}^{-1}\{[G_g - G_\bullet^*(\omega)]; \zeta\}\,|_{\zeta=1/\tau} \; . \tag{6.3-101}$$

The inversion of the Fourier-Laplace transform is accomplished with the use of Eq. (A5-19), replacing the transform variable, $j\omega$, by $\zeta \exp(\pm j\pi)$. We obtain*

$$H^*(\tau) = \pm\pi^{-1}\mathscr{Im}\, G_\bullet^*[\tau^{-1}\exp(\pm j\pi)] \; . \tag{6.3-102}$$

The instantaneous modulus, G_g, does not appear in the equation because only the imaginary part resulting from the substitution for $j\omega$ is needed.

By a completely analogous derivation from Eq. (4.1-72) we find

$$L^*(\tau) = \mp\pi^{-1}\mathscr{Im}\, J^*[\tau^{-1}\exp(\pm j\pi)] \tag{6.3-103}$$

since the flow terms cancels in the limiting operation (cf. Sect. 4.5). The same relations would have been obtained had we started from Eqs. (4.1-73) and (4.1-74).

Introducing Eqs. (6.1-81) to (6.1-84) we obtain

$$H_M^*(\tau) = \mp\pi^{-1}[G_g - \{G_e\}]\mathscr{Im}\, \hat{Z}_M^*[\tau^{-1}\exp(\pm j\pi)] \tag{6.3-104}$$

$$H_m^*(\tau) = \pm\pi^{-1}[G_g - \{G_e\}]\mathscr{Im}\, \hat{s}_M^*[\tau^{-1}\exp(\pm j\pi)] \tag{6.3-105}$$

$$L_M^*(\tau) = \mp\pi^{-1}[J_e^{\{o\}} - J_g]\mathscr{Im}\, \hat{Z}_M^*[\tau^{-1}\exp(\pm j\pi)] \tag{6.3-106}$$

and

$$L_m^*(\tau) = \pm\pi^{-1}[J_e^{\{o\}} - J_g]\mathscr{Im}\, \hat{s}_M^*[\tau^{-1}\exp(\pm j\pi)] \; . \tag{6.3-107}$$

It must be remembered that $\tau^{-1}\exp(\pm j\pi)$ replaces $j\omega$, not ω alone.

As an example let us find $H^*(\tau)$ corresponding to $\hat{Z}_C^*(\omega)$. Substituting Eq. (6.1-87) into Eq. (6.3-104) gives

$$H_C^*(\tau) = \mp\pi^{-1}[G_g - \{G_e\}]\mathscr{Im}\frac{1}{1 + (\tau_0/\tau)^c\exp(\pm j\pi c)} \tag{6.3-108}$$

where $\tau_0 = 1/\omega_0$. Separating into real and imaginary parts and retaining the latter one only, yields

$$H_C^*(\tau) = \pi^{-1}[G_g - \{G_e\}]\frac{\sin \pi c}{(\tau_0/\tau)^c + 2\cos\pi c + (\tau/\tau_0)^c} \; . \tag{6.3-109}$$

We now consider the model spectra corresponding to the storage and loss functions. We turn to Eq. (4.1-85) in the form

---

* We shall use the superscripts *, ′, and ″, to indicate the harmonic response function from whose model the spectrum was derived. Thus, when appended to H or to L, these symbols do not designate a complex function or its real or imaginary part.

$$\Phi(\xi) = \mathscr{S}^{-1}[\tilde{\Phi}(\nu); \xi] \ , \tag{6.3-110}$$

which, using Eq. (A5-19) again, becomes

$$\Phi(\xi) = \mp\pi^{-1} \mathscr{I}\!m\, \tilde{\Phi}[\xi \exp(\pm j\pi)] \ . \tag{6.3-111}$$

Now, by the pair of Eqs. $(4.1\text{-}86)_1$,

$$\Phi(\xi) = H(\tau)\,|_{\tau=1/\sqrt{\xi}} \tag{6.3-112}$$

and

$$\tilde{\Phi}[\xi \exp(\pm j\pi)] = 2[G_g - G'_\bullet(\omega)]\,|_{\omega=\sqrt{\xi}\exp(\pm j\pi/2)} \ . \tag{6.3-113}$$

Substituting these into Eq. (6.3-111) leads to

$$H'(\tau) = \pm(2/\pi)\mathscr{I}\!m\, G'[\tau^{-1}\exp(\pm j\pi/2)] \ . \tag{6.3-114}$$

The use of the pair of Eqs. $(4.1\text{-}86)_3$ would, of course, have yielded the same result.

Using Eqs. (6.1-61) and (6.1-63) we find

$$H'_M(\tau) = \mp(2/\pi)[G_g - \{G_e\}]\mathscr{I}\!m\, \hat{Z}_M[\tau^{-1}\exp(\pm j\pi/2)] \tag{6.3-115}$$

and

$$H'_m(\tau) = \pm(2/\pi)[G_g - \{G_e\}]\mathscr{I}\!m\, \hat{s}_M[\tau^{-1}\exp(\pm j\pi/2)] \ . \tag{6.3-116}$$

By the pair of Eqs. $(4.1\text{-}86)_2$,

$$\Phi(\xi) = \tau H(\tau)\,|_{\tau=\sqrt{\xi}} \tag{6.3-117}$$

and

$$\tilde{\Phi}[\xi \exp(\pm j\pi)] = 2\, G''(\omega)/\omega\,|_{\omega=\sqrt{\xi}\exp(\pm j\pi/2)} \ . \tag{6.3-118}$$

Hence,

$$\tau H(\tau) = \mp(2/\pi)\mathscr{I}\!m[\omega^{-1}G''(\omega)]\,|_{\omega=\tau^{-1}\exp(\pm j\pi/2)} \ . \tag{6.3-119}$$

But $\omega^{-1} = \tau\exp(\mp j\pi/2) = \mp j\tau$ and $\mathscr{I}\!m\, j(\alpha + j\beta) = \mathscr{R}\!e(\alpha + j\beta)$. Therefore,

$$H''(\tau) = (2/\pi)\mathscr{R}\!e\, G''[\tau^{-1}\exp(\pm j\pi/2)] \tag{6.3-120}$$

and, introducing Eq. $(6.1\text{-}65)_1$,

$$H''_M(\tau) = (2/\pi)G''_{max}\mathscr{R}\!e\, \hat{\Lambda}_M[\tau^{-1}\exp(\pm j\pi/2)] \ . \tag{6.3-121}$$

In these equations, $\tau^{-1}\exp(\pm j\pi/2)$ replaces ω.

Analogous derivation yields

$$L'(\tau) = \mp(2/\pi)\mathscr{I}\!m\, J'[\tau^{-1}\exp(\pm j\pi/2)] \tag{6.3-122}$$

or

$$L'_M(\tau) = \mp(2/\pi)[J_e^{\{o\}} - J_g]\mathscr{I}\!m\, \hat{Z}_M[\tau^{-1}\exp(\pm j\pi/2)] \tag{6.3-123}$$

and

$$L'_m(\tau) = \pm(2/\pi)[J_e^{\{o\}} - J_g]\mathscr{I}\!m\, \hat{s}_M[\tau^{-1}\exp(\pm j\pi/2)] \tag{6.3-124}$$

and

$$L''(\tau) = (2/\pi)\mathcal{R}e\, J''(\tau^{-1}\exp(\pm j\pi/2)] \tag{6.3-125}$$

or

$$L''_M(\tau) = (2/\pi)J''_{max}\mathcal{R}e\, \hat{\Lambda}_M[\tau^{-1}\exp(\pm j\pi/2)] \ . \tag{6.3-126}$$

The same relations could have been derived also starting from Eq. (4.1-87).

As examples, let us find $H'_C(\tau)$ corresponding to $\hat{S}_C(\omega)$, and $H''_R(\tau)$ corresponding to $\hat{\Lambda}_R(\omega)$. We have [cf. Eq. (6.1-23)]

$$\hat{S}_C(\omega) = \frac{1}{1 + (\omega_0/\omega)^c} = \hat{s}_C(\omega) \tag{6.3-127}$$

and Eq. (6.3-116) gives

$$H'_C(\tau) = \pm(2/\pi)[G_g - \{G_e\}]\mathcal{I}m \frac{1}{1 + (\tau/\tau_0)^c \exp(\mp j\pi c/2)} \tag{6.3-128}$$

which becomes

$$H'_C(\tau) = \frac{(2/\pi)[G_g - \{G_e\}]\sin \pi c/2}{(\tau_0/\tau)^c + 2\cos \pi c/2 + (\tau/\tau_0)^c} \ . \tag{6.3-129}$$

This equation was used in Chap. 4 as Eq. (4.3-32).

In the second case we start from

$$\hat{\Lambda}_R(\omega) = \frac{2}{(\omega/\omega_{max})^r + (\omega_{max}/\omega)^r} \tag{6.3-130}$$

[cf. Eq. (6.1-35)] and obtain

$$H''_R(\tau) = (4/\pi)G''_{max}\mathcal{R}e \frac{1}{(\tau/\tau_{max})^r \exp(\mp j\pi r/2) + (\tau_{max}/\tau)^r \exp(\pm j\pi r/2)} \tag{6.3-131}$$

which, after a little algebra, gives

$$H''_R(\tau) = (4/\pi)G''_{max}\cos \pi r/2 \frac{(\tau/\tau_{max})^r + (\tau_{max}/\tau)^r}{(\tau/\tau_{max})^{2r} + 2\cos \pi r + (\tau_{max}/\tau)^{2r}} \ . \tag{6.3-132}$$

Because $\hat{S}_C(\omega)$ is self-congruent, Eq. (6.3-129) could equally have been obtained from $\hat{Z}_C(\omega)$ using Eq. (6.1-115). We note that Eq. (6.3-129) is not the same as Eq. (6.3-109). This is so because $\hat{Z}_C(\omega)$ is not the real part of $\hat{Z}^*_C(\omega)$. If we let $\hat{Z}_M(\omega)$ be $\hat{Z}'_C(\omega)$, given by Eq. (6.1-88), Eq. (6.3-115) yields the same expression for the spectrum as does $\hat{Z}^*_C(\omega)$. This is shown in Problem 6.3-9 where the spectrum corresponding to $\hat{Z}''_C(\omega)$ is also obtained. For this we replace Eq. (6.3-121) by

$$H''_M(\tau) = (2/\pi)[G_g - \{G_e\}]\mathcal{R}e\, \hat{Z}''_M[\tau^{-1}\exp(\pm j\pi/2)] \tag{6.3-133}$$

obtained by the use of Eq. $(6.1\text{-}17)_1$ and letting $\hat{\Lambda}_M(\omega) = \hat{Z}''_M(\omega)$. For the retardation spectrum we have

$$L''_M(\tau) = (2/\pi)[J_e^{\{o\}} - J_g]\mathcal{R}e\, \hat{Z}''_M[\tau^{-1}\exp(\pm j\pi/2)] \tag{6.3-134}$$

when it is obtained from the imaginary part of $\hat{Z}^*_M(\omega)$.

Analogous relations for the spectra corresponding to models for the complex viscosity and its real and imaginary parts can also be obtained. These form the subjects of Problems 6.3-10 and 6.3-11.

6.4 Problems

Problem 6.1-1. Integrate $G_k(t)$ to obtain a mathematical model for $\eta(t)$. Examine its behavior as $t \to \infty$. Then specialize the integral for the rheodictic case with $k = 1/2$.

Problem 6.1-2. Integrate $G_C(t)$ with $c = 1/2$ to obtain a mathematical model for $\eta(t)$. Show that the integration is not successful because Eq. (6.1-55) is violated.

Problem 6.1-3. Show that differentiation of $\eta_k(t)$ and $\eta_C(t)$ leads to non-standard models for $G_k(t)$ and $G_C(t)$.

Problem 6.1-4. If a matching function, $\hat{Z}_M(t)$ or $\hat{s}_M(t)$, has been found to represent data on the step responses, G(t) or J(t), then the corresponding respondances are given by

$$\bar{Q}_M(s) = \{G_e\} + [G_g - \{G_e\}]s\,\mathscr{L}\,\hat{Z}_M(t) \tag{1}$$

$$\bar{U}_M(s) = J_e^{\{o\}} - [J_e^{\{o\}} - J_g]s\,\mathscr{L}\,\hat{Z}_M(t) + \{\phi_f/\tau\}\ . \tag{2}$$

and

$$\bar{Q}_m(s) = G_g - [G_g - \{G_e\}]s\,\mathscr{L}\,\hat{s}_M(t) \tag{3}$$

$$\bar{U}_m(s) = J_g + [J_e^{\{o\}} - J_g]s\,\mathscr{L}\,\hat{s}_M(t) + \{\phi_f/s\}\ . \tag{4}$$

Demonstrate the correctness of these equations.

[*Hint*: begin by showing that $Z_F(s) = s\mathscr{L}S_F(t)$.]

Problem 6.1-5. Show that, for a rheodictic material, $\bar{\eta}(s)$ can be modelled by

$$\bar{\eta}(s) = \eta_f/s - J_e^o\eta_f^2\,\hat{Z}_M(s)\ . \tag{1}$$

Problem 6.1-6. Model the complex viscosity, $\eta^*(\omega)$, using the matching functions $\hat{Z}_M^*(\omega)$, and then show that

$$\eta_C^*(\omega) = \frac{\eta_f}{1 + (j\omega/\omega_0)^c} \tag{1}$$

for a rheodictic material.

Problem 6.1-7. Rewrite Eq. (6.1-91) in terms of G''_{max}.

Problem 6.1-8. Obtain the real and imaginary parts of $J_B^*(\omega)$.

Problem 6.1-9. Show that $G'_{C'}(\omega)$ and $G''_{C'}(\omega)$ given by Eqs. (6.1-96) and (6.1-97) reduce, in the arrheodictic case, to $G'_C(\omega)$ and $G''_C(\omega)$ given by Eqs. (6.1-90) and (6.1-91) upon multiplication of ω_0 by the factor $(G_g/G_e)^c$.

Problem 6.1-10. Show that Eqs. (6.1-90) and (6.1-91) have the alternative forms

$$G'_C(\omega) = \frac{G_g + \{G_e\}}{2} + \frac{G_g - \{G_e\}}{2}\left(\frac{\sinh cz}{\cosh cz + \cos \pi c/2}\right)\Bigg|_{\ln z=\omega/\omega_0} \tag{1}$$

and

$$G''_C(\omega) = \frac{G_g - \{G_e\}}{2}\left(\frac{\sin \pi c/2}{\cosh cz + \cos \pi c/2}\right)\Bigg|_{\ln z=\omega/\omega_0} \tag{2}$$

while Eqs. (6.1-94) and (6.1-95) have the alternative forms

$$J'_C(\omega) = \frac{J_e^{\{o\}} + J_g}{2} + \frac{J_e^{\{o\}} - J_g}{2}\left(\frac{\sinh cz}{\cosh cz + \cos \pi c/2}\right)\Bigg|_{\ln z=\omega/\omega_0} \tag{3}$$

and

$$J''_C(\omega) = \frac{J_g - J_e^{\{o\}}}{2}\left(\frac{\sin \pi c/2}{\cosh cz + \cos \pi c/2}\right)\Bigg|_{\ln z=\omega/\omega_0} + \{\phi_f/\omega\} \ . \tag{4}$$

Problem 6.1-11. Show that

$$\eta''_{C,\max} = \tfrac{1}{2}\eta_f \tan \pi c/2 \ . \tag{1}$$

Then express $\eta''_C(\omega)$ in terms of $\eta''_{C,\max}$, and show that

$$\eta'_C(\omega_{\max}) = \eta''_{C,\max} \cot \pi c/4 \ . \tag{2}$$

[*Hint*: $\eta^*_C(\omega)$ is given by Eq. (1) of Problem 6.1-6.]

Problem 6.2-1. Obtain $G_D(t)$, Eq. $(6.2\text{-}7)_1$, from $\bar{\sigma}(s) = \bar{Q}_D(s)\bar{\varepsilon}(s)$, Eq. (6.2-1), using Eq. (6.2-17).

Problem 6.3-1. The normalized form of the spectrum, $H(\tau)$, given by Eq. (6.3-3), is

$$H(\tau) = \frac{(\rho/\pi)(r + r')[G_g - \{G_e\}]\sin[r\pi/(r + r')]}{(\tau_{\max}/\alpha\tau)^{\rho r} + (\alpha\tau/\tau_{\max})^{\rho r'}} \tag{1}$$

where

$$\alpha = (r/r')^{1/\rho(r+r')} \ . \tag{2}$$

Derive this form. Then give the analogous expression for $L(\tau)$.

Problem 6.3-2. Separate the real and imaginary parts of $G^*_b(\omega)$ given by Eq. (6.3-15).

Problem 6.3-3. Obtain Eq. (6.3-38).
[*Hint*: use the substitution $ju = x^{1/2}$ in Eq. (6.3-32).]

Problem 6.3-4. Show that Eq. (6.3-36) is the equation of a standard model.

Problem 6.3-5. Obtain the creep compliance from the retardance, $\bar{U}_w(s)$, of the wedge distribution.

Problem 6.3-6. Demonstrate that $J_w(t)$ becomes $J_e^{\{o\}}$ as $t \to \infty$.

Problem 6.3-7. Derive

$$H_B(\tau) = \Gamma^{-1}(b)[G_g - \{G_e\}](b\tau_{max}/\tau)^b \exp(-b\tau_{max}/\tau) \tag{1}$$

as an alternate form of Eq. (6.3-97) and show that it satisfies the normalization condition, Eq. (6.3-27).

Problem 6.3-8. What is the first order approximation to the relaxation spectrum derived from Eq. (6.1-50)?

Problem 6.3-9. Find $H'_C(\tau)$ and $H''_C(\tau)$ from the real and imaginary parts of $\hat{Z}^*_C(\omega)$, given by Eqs. (6.1-88) and (6.1-89), respectively and show that $H^*_C(\tau) = H'_C(\tau) = H''_C(\tau)$.

Problem 6.3-10. Obtain the relaxation spectrum in terms of the complex viscosity, $\eta^*(\omega)$, and its real and imaginary components, $\eta'(\omega)$ and $\eta''(\omega)$.

Problem 6.3-11. Find the relaxation spectrum if the complex viscosity, $\eta^*(\omega)$, is modelled by Eq. (1) of Problem 6.1-6.

Problem 6.3-12. Find the retardation spectrum associated with the complex Kobeko equation, $\hat{Z}^*_B(\omega)$.

Problem 6.3-13. Suppose it is found that the storage compliance of a given material is well represented by the real part of the compliance model

$$J^*_C(\omega) = J_g + \frac{J_e + J_g}{1 + (j\omega/\omega_0)^c} \tag{1}$$

which is based on the complex Cole-Cole function [cf. Eqs. (6.1-82), (6.1-87), and (6.1-88)]. Can one predict the corresponding creep compliance, $J(t)$?

[*Hint*: work through the retardation spectrum, $L^*_C(\tau)$.]

References (Chap. 6)

1. J. Bischoff, E. Catsiff, and A. V. Tobolsky, J. Am. Chem. Soc., *74*:3378 (1952).
2. F. Kohlrausch, Pogg. Ann. Physik (4) *29*:337 (1863).
3. K.S. Cole and R.H. Cole, J. Chem. Phys. *9*:341 (1941).
4. P. Kobeko, E. Kuvshinskij and G. Gurevitch, Techn. Phys. USSR *4*:622 (1937).
5. S. Havriliak and S. Negami, J. Polymer Sci., Part C, No. 14, pp. 99–117 (1966).

6. H.A. Lorentz, Proc. Acad. Sci. Amsterdam, *8*:591 (1906).
7. R.M. Fuoss and J.G. Kirkwood, J. Am. Chem. Soc. *63*:385 (1941).
8. C.A.F. Tuijnman, S. Kruyer, and Th.N. Zwietering, Proc. 5th Intern. Congress Rheol., Vol. I., pp. 361–379, University Park Press, Baltimore, Maryland, (1969).
9. E. Catsiff and A.V. Tobolsky, J. Colloid Sci., *10*:375 (1955).
10. I.L. Hopkins and R.W. Hamming, J. Appl. Phys. *28*:906 (1957).
11. M. Abramowitz and I. Stegun, eds., *Handbook of Mathematical Functions*, U.S. Department of Commerce, National Bureau of Standards, AMS 55, Washington, D.C., (1966).
12. T.M. Apostol, *Mathematical Analysis*, Addison-Wesley, Reading, MA, (1957), p. 220.
13. T.G. Nutting, J. Franklin Inst., *191*:679 (1921).
14. E. Wiechert, Ann. d. Phys. u. Chem. *50*:546 (1893).
15. B. Gross, *Mathematical Structure of the Theories of Viscoelasticity*, Herman et C[ie], Paris, (1953), p. 66.
16. P. Rouse, Jr., J. Chem. Phys. *21*:1272 (1953).
17. J.D. Ferry, Viscoelastic Properties of Polymers, 3rd ed., Wiley, New York, (1980).
18. E. Kreyszig, Advanced Engineering Mathematics, 4th ed., Wiley, New York, (1979), p. 831.

7. Response to Non-Standard Excitations

"The time has come," the walrus said,
"to speak of many things . . . ."

Lewis Carrol: Through the Looking Glass

7.0 Introduction

Chapter 2 dealt with the response to step, slope, and harmonic excitations. These are the stimuli used most often to elicit responses for the purpose of characterizing linear viscoelastic behavior. They were, therefore, called *standard excitations*. They share the common feature of being simple stimuli, applied once only, at $t = 0$. Viscoelastic response may be elicited, however, by a variety of other stimuli which we shall discuss, as a matter of convenience, under the heading of *non-standard excitations*. They fall naturally into two groups:

(1) the response to the *removal* or the *reversal of direction* of a stimulus, and
(2) the response to *repeated* stimuli.

Excitations in the first group are described in Sect. 7.1. Excitations in the second group are produced by the repeated imposition and removal, or the reversal of the direction, of a stimulus. Such excitations would not normally be applied in the characterization of linear viscoelastic behavior. They do appear, however, in a number of important applications such as, e.g., fatigue testing and other types of periodic excitations. In these, the response is generally not linearly viscoelastic because the excitation exceeds the linear viscoelastic limit. It is, nevertheless, desirable to understand the theory of the response to repeated stimuli in the linear viscoelastic range because this forms the basis of the non-linear response. A general method of dealing with the response to repeated stimuli will be described in Sect. 7.2.

The excitations discussed in this chapter share the common feature of being *sectionally continuous excitations*. A function $f(t)$ is sectionally, or piecewise, continuous on a finite or a countably infinite number of subintervals if it is possible to subdivide each interval into a finite number of subintervals in each of which $f(t)$ is continuous and has finite limits as t approaches either end point of the subinterval from the interior. Sectionally continuous functions may contain discontinuities. These are of the type known as ordinary points of discontinuity where the value of the function makes a finite jump. The response to repeated stimuli typically consist of a series of sectionally smooth functions, each defined in an appropriate interval. If the intervals are of equal duration, and the same function is repeated in each, the excitation is called *periodic*. We shall call a periodic excitation in which the stimulus returns to zero at the end of the interval, a *cyclic* excitation.

The mathematical treatment of sectionally continuous functions is greatly simplified by the Laplace transformation since the transform will generally be of closed form.* In addition, the transform of a periodic function can be obtained immediately if the transform of the repeat function is known.

7.1 Response to the Removal or the Reversal of a Stimulus

The removal of a stimulus elicits a response just as the imposition of a stimulus does. Two important non-standard excitations are of this type. In the first, to be discussed in Sect. 7.1.1, a constant stress is suddenly applied for a finite interval of time after which it is just as suddenly removed. In the second, to be discussed in Sect. 7.1.2, a constant rate of strain is applied and the deformation is stopped when the strain has reached a predetermined value. There are two important variants of this excitation. In one, commonly applied to a rheodictic material, the material has reached steady-state flow when the excitation is removed. In the other, the rate of strain is applied rapidly, so that the resulting deformation approximates a step of strain.

In Sect. 7.1.3 we discuss an excitation in which the direction of a constant rate of strain is suddenly reversed. This excitation forms the basis of a common method of determining hysteresis, to be examined in detail in Chap. 9.

7.1.1 Creep Recovery

The response of a viscoelastic material to a stress excitation applied in the form of a step of unit magnitude is the creep compliance introduced in Sect. 2.3. Creep is a time dependent deformation which may reach a steady-state when the material is rheodictic, or an equilibrium value when it is arrheodictic. If the stress is removed after the creep has been in progress for some time, the deformation is completely recovered in an arrheodictic, and partially recovered in a rheodictic material. The response to the removal of the stress may be expressed as the application, at time $t = t'$, of another step, equal in magnitude with, but of opposite sign to, the step imposed at $t = 0$.

The excitation is represented in Fig. 7.1-1 and may be expressed with the help of the *unit gate function* (see Appendix A2.3)

$$\sigma(t) = \sigma_0 h_{t'}(t) \tag{7.1-1}$$

---

* For an excellent collection of the Laplace transforms of sectionally continuous functions (represented pictorially) see, e.g., Ref. [1].

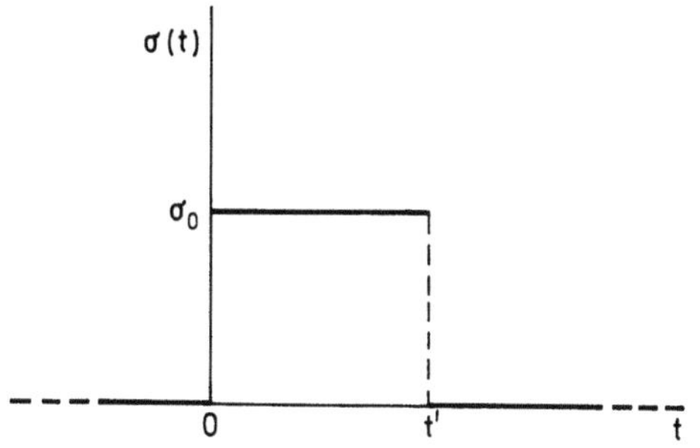

Fig. 7.1-1. Stress as a gate function of time

or, in terms of unit step functions, as

$$\sigma(t) = \sigma_0 h(t) - \sigma_0 h(t - t') \ . \tag{7.1-2}$$

The first term on the right of Eq. (7.1-2) represents the imposition of a step of stress of magnitude σ_0 at $t = 0$, while the second term represents the removal of the stress, expressed as the imposition of a step of equal height but opposite direction, delayed by the amount t'.

To obtain the response of the material to a gate stress we proceed as in Chap. 2, i.e., we substitute the stress transform into $\bar{\varepsilon}(s) = \bar{U}(s)\bar{\sigma}(s)$, Eq. (2.3-15). The transform [see LTP (8)] is

$$\bar{\sigma}(s) = \sigma_0 \bar{h}_{t'}(s) = \sigma_0[1 - \exp(-t's)]/s \tag{7.1-3}$$

and the substitution yields [see Eq. (2.3-8)]

$$\bar{\varepsilon}(s) = \sigma_0 \bar{U}(s)[1 - \exp(-t's)]/s = \sigma_0 \bar{J}(s)[1 - \exp(-t's)] \ . \tag{7.1-4}$$

Using the *translation theorem*, Eq. (A3-6), retransformation gives

$$\varepsilon(t) = \sigma_0[J(t)h(t) - J(t - t')h(t - t')] \tag{7.1-5}$$

which may be expressed equivalently in the form

$$\varepsilon(t) = \begin{cases} \sigma_0 J(t), & 0 < t < t' \\ \sigma_0[J(t) - J(t - t')], & t > t' \ . \end{cases} \tag{7.1-6}$$

Thus, the material response at $t > t'$ may be considered to consist of the response, $\sigma_0 J(t)$, to a step of stress from which the same response, shifted by t' along the time axis, is subtracted as shown in Fig. 7.1-2. Clearly, this represents an illustration of the *principle of superposition* or *time shift invariance*.

By Eq. (4.1-21),

$$J(t) = J_g + \int_{-\infty}^{\infty} L(\tau)[1 - \exp(-t/\tau)]d \ln \tau + \{\phi_f\}t \ . \tag{7.1-7}$$

Substitution into Eq. (7.1-5) and dividing by σ_0 leads to

$$\begin{aligned} J(t) = J_g[h(t) - h(t - t')] + \int_{-\infty}^{\infty} L(\tau)\Big(& [1 - \exp(-t/\tau)]h(t) \\ & - \{1 - \exp[-(t - t')]/\tau\}h(t - t')\Big) d \ln \tau \\ & + \{\phi_f[th(t) - (t - t')h(t - t')]\} \ . \end{aligned} \tag{7.1-8}$$

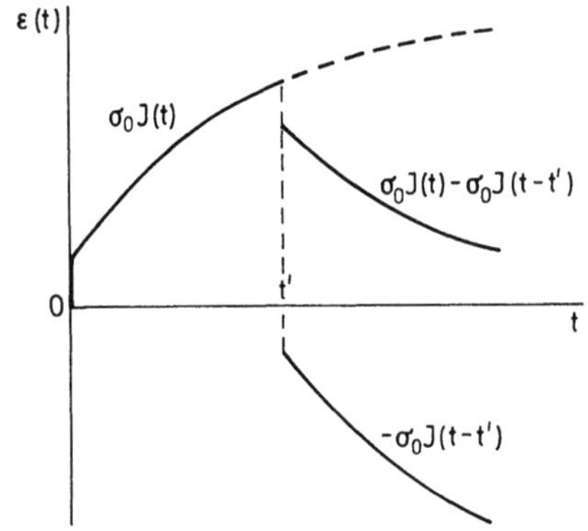

Fig. 7.1-2. Response to a stress as a gate function of time

Introducing the *shifted time*

$$\theta = t - t' \tag{7.1-9}$$

we obtain

$$J_{t'}(\theta) = \int_{-\infty}^{\infty} L(\tau)[1 - \exp(-t'/\tau)]\exp(-\theta/\tau)d \ln \tau + \{\phi_f\}t' \; . \tag{7.1-10}$$

where $J_{t'}(\theta) = J(\theta + t') - J(\theta)$ is the response of the material, as a function of θ, to the removal of a constant stress at $\theta = 0$. This response is called *creep recovery.* When t' is large enough so that $J(t') - \{\phi_f\}t'$ is sensibly equal to $J_e^{\{o\}}$, then

$$J_{t'}(\theta) - \{\phi_f\}t' \simeq \int_{-\infty}^{\infty} L(\tau)\exp(-\theta/\tau)d \ln \tau \; . \tag{7.1-11}$$

It follows that $\sigma_0[J_{t'}(\theta) - \{\phi_f\}t']$ represents a *relaxation of strain.* Substituting θ for t in Eq. (7.1-7) and eliminating the integral between the resulting expression and Eq. (7.1-11) leads to

$$J(\theta) - \{\phi_f\}\theta = J_e^{\{o\}} - J_{t'}(\theta) + \{\phi_f\}t' \tag{7.1-12}$$

as the relation between the creep compliance, $J(\theta)$, and the response in creep recovery, $J_{t'}(\theta)$, when t' is large. For an arrheodictic material Eq. (7.1-12) reduces to $J(\theta) = J_e - J_{t'}(\theta)$. Clearly, if t' is large enough, the creep and creep recovery curves can be brought into superposition by reflecting either one around the t-axis and then shifting them along both axes. Thus, creep and creep recovery contain the same information on the time-dependent behavior of a linear viscoelastic material. We now examine these relations in more detail.

Figure 7.1-3 qualitatively illustrates the response of an arrheodictic material in creep and creep recovery. Two recovery curves are shown. For the first, $J(t_1') < J_e$, for the second, $J(t_2') \simeq J_e$. In either case the original undeformed state is completely recovered [i.e. $J(\theta) \to 0$] as $\theta \to \infty$. This is in accordance with the fact that $\phi_f = 0$ for an arrheodictic material. The instant at which the stress is removed is a point

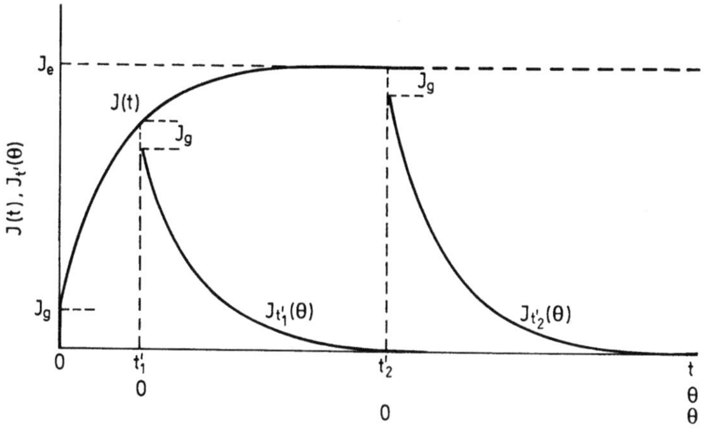

Fig. 7.1-3. Creep and creep recovery in an arrheodictic material

of discontinuity at which J(t) drops by the amount representing the magnitude of the glassy compliance. To understand the difference between $J_{t_1'}(\theta)$ and $J_{t_2'}(\theta)$, it is instructive to compare the *creep rates*

$$\dot{J}(t) = \int_{-\infty}^{\infty} \tau^{-1}L(\tau)\exp(-t/\tau)d\ln\tau + \{\phi_f\} \tag{7.1-13}$$

and

$$\dot{J}_{t'}(\theta) = -\int_{-\infty}^{\infty} \tau^{-1}L(\tau)[1 - \exp(-t'/\tau)]\exp(-\theta/\tau)d\ln\tau \tag{7.1-14}$$

for any linearly viscoelastic material. For an arrheodictic material $\phi_f = 0$. Hence, when $J(t_2') \simeq J_e$, $\dot{J}(t)$ and $\dot{J}_{t'}(\theta)$ have opposite signs but are equal in magnitude for the same values of t and θ. Accordingly, J(t) and $J_{t_2'}(\theta)$ can be brought into coincidence by flipping $J_{t_2'}(\theta)$ around the abscissa and shifting it up by J_e and to the left by t_2'. By contrast, when $J(t_1') < J_e$, the slopes of J(t) and $J_{t_1'}(\theta)$ will attain equal magnitude only as $\theta \to \infty$. The two curves can then not be made to coincide.

The behavior of a rheodictic material is illustrated qualitatively in Fig. 7.1-4. Here again, the point $t = t'$ is a point of discontinuity. However, for a rheodictic material, $\phi_f \neq 0$. Accordingly, as $\theta \to \infty$, $J_{t'}(\theta) \to \phi_f t'$ and the undeformed state is only partially recovered. The material thus suffers a *permanent set* of magnitude $\sigma_0\phi_f t'$ representing the deformation produced by irreversible flow. Because of the permanent set, the slope of $J_{t'}(\theta)$ must now be compared with that of $J(t) - \phi_f t$. When $[J(t_2') - \phi_f t_2'] \simeq J_e^\circ$, the creep recovery curve can be brought into coincidence with $J(t) - \phi_f t$ by translating the reflected curve upwards by the amount $J_e^\circ - \phi_f t_2'$, and to the left by t_2'. Thus, the two curves are equivalent. When $[J(t_1') - \phi_f t_1'] < J_e^\circ$, the curves cannot be shifted into coincidence.

Creep recovery experiments are a convenient way to ascertain whether a material is rheodictic or not. Determination of ϕ_f from the permanent set provides a check of its determination from the slope of the asymptote

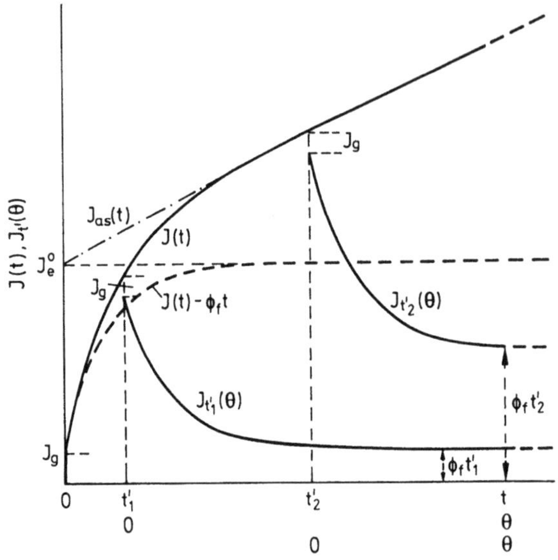

Fig. 7.1-4. Creep and creep recovery in a rheodictic material

$$J_{as}(t) = J_e^\circ + \phi_f t \qquad (7.1\text{-}15)$$

of the creep compliance [cf. Eq. (3.4-24)].

7.1.2 Response to the Removal of a Constant Rate of Strain

In another common non-standard mode of excitation a constant rate of strain is removed at an instant $t' > 0$ and the strain reached at that instant is then held constant. This process may be described as the application of a strain as a *ramp function* of time. Thus, the excitation may be represented as

$$\varepsilon(t) = \varepsilon_0 r_{t'}(t) \qquad (7.1\text{-}16)$$

where $r_{t'}(t)$ is the *ramp function* discussed in Appendix A2.5, and

$$\varepsilon_0 = \dot{\varepsilon}_0 t' \qquad (7.1\text{-}17)$$

is the strain reached at $t = t'$, $\dot{\varepsilon}_0$ being the (constant) rate of strain. The strain history defined by Eq. (7.1-16) is illustrated in Fig. 7.1-5.

By Eq. (A2-40) we may rewrite Eq. (7.1-16) in the form

$$\varepsilon(t) = \dot{\varepsilon}_0[th(t) - (t - t')h(t - t')] \ . \qquad (7.1\text{-}18)$$

Thus, a strain $\dot{\varepsilon}_0 t$ is imposed at $t = 0$ and removed again at $t = t'$. Laplace transformation [see LTP (9)] yields

$$\bar{\varepsilon}(s) = \varepsilon_0 \bar{r}_{t'}(s) = \dot{\varepsilon}_0[1 - \exp(-t's)]/s^2 \ . \qquad (7.1\text{-}19)$$

Inserting Eq. (7.1-19) into $\bar{\sigma}(s) = \bar{Q}(s)\bar{\varepsilon}(s)$ gives the stress transform as

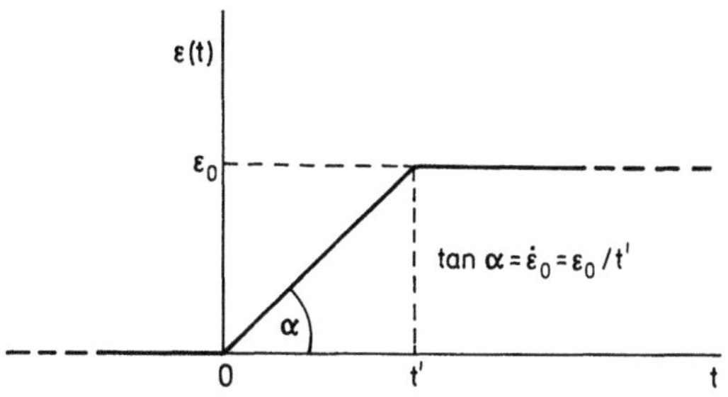

Fig. 7.1-5.
Strain as a ramp function of time

$$\bar{\sigma}(s) = \dot{\varepsilon}_0\bar{Q}(s)[1 - \exp(-t's)]/s^2 = \dot{\varepsilon}_0\bar{\eta}(s)[1 - \exp(-t's)] \ . \quad (7.1\text{-}20)$$

Retransformation leads to

$$\sigma(t) = \dot{\varepsilon}_0[\eta(t)h(t) - \eta(t - t')h(t - t')] \quad (7.1\text{-}21)$$

which may be written equivalently as

$$\sigma(t) = \begin{cases} \dot{\varepsilon}_0\eta(t), & 0 < t < t' \\ \dot{\varepsilon}_0[\eta(t) - \eta(t - t')], & t > t' \ . \end{cases} \quad (7.1\text{-}22)$$

We may therefore consider the material response at $t > t'$ to consist of the response to a constant rate of strain, $\dot{\varepsilon}_0\eta(t)$, from which the same response, shifted by t' along the time scale, is subtracted as illustrated in Fig. 7.1-6. This situation is analogous to that shown in Fig. 7.1-2.

Now, according to Eqs. (4.1-16) and (4.6-24), $\eta(t)$ is given by

$$\eta(t) = \{G_e\}t + \eta_{\{f\}} - \int_{-\infty}^{\infty} \tau H(\tau)\exp(-t/\tau)d\ln\tau \ . \quad (7.1\text{-}23)$$

Inserting this into Eq. (7.1-21) and using Eq. (7.1-17) we find

$$\eta(t) = [\{G_e\}t + \eta_{\{f\}}][h(t) - h(t - t')] + \{G_e\}t'h(t - t')$$
$$- \int_{-\infty}^{\infty} \tau H(\tau)\exp(-t/\tau)\{h(t) - [\exp(t'/\tau)]h(t - t')\}d\ln\tau \ . \quad (7.1\text{-}24)$$

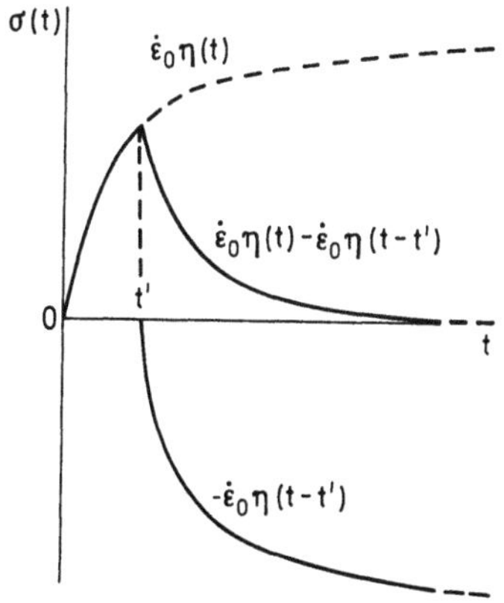

Fig. 7.1-6.
Response to a strain as a ramp function of time

Introducing the *shifted time*, $\theta = t - t'$, gives

$$\eta_{t'}(\theta) = \{G_e\}t' + \int_{-\infty}^{\infty} \tau H(\tau)[1 - \exp(-t'/\tau)]\exp(-\theta/\tau)d \ln \tau \tag{7.1-25}$$

where $\eta_{t'}(\theta) = \eta(\theta + t') - \eta(\theta)$ is the response of the material, as a function of θ, to the removal of a constant rate of strain at $\theta = 0$. This response is called appropriately, albeit in a somewhat cumbersome manner, *stress relaxation after deformation at constant rate of strain.* It must be noted that the stress, $\sigma_{t'}(\theta) = \dot{\varepsilon}_0\eta_{t'}(\theta)$, does not relax in the same manner in which the stress, $\sigma(\theta) = \varepsilon_0 G(\theta)$, does in response to the imposition of a strain as a step function of time. This is so partly because of the presence of the factor $1 - \exp(-t'/\tau)$ in the integrand, although this approaches unity when t' is sufficiently large. Stress relaxation in response to the removal of a constant rate of strain differs from stress relaxation in response to the imposition of a constant strain primarily because of the appearance, under the integral sign in Eq. (7.1-25), of $\tau H(\tau) = \Xi(\tau)$ instead of $H(\tau)$. Compared to relaxation in step response, the stress is thus weighted in favor of the longer relaxation times.

When t' is large enough so that $\eta(t') - \{G_e\}t'$ sensibly equals $\eta_{\{f\}}$, then

$$\eta_{t'}(\theta) - \{G_e\}t' \simeq \int_{-\infty}^{\infty} \tau H(\tau)\exp(-\theta/\tau)d \ln \tau \ . \tag{7.1-26}$$

For a rheodictic material the $G_e t'$-term is absent. The behavior is then commonly called *stress relaxation after cessation of steady-state flow*. It is illustrated in Fig. 7.1-7. Formally, this behavior is analogous to that shown in Fig. 7.1-3 for $t > t_2'$ (but note the absence of discontinuities at $t = 0$ and $t = t'$).

Substituting θ for t in Eq. (7.1-23) and eliminating the integrals between the resulting expression and Eq. (7.1-26) gives

$$\eta(\theta) - \{G_e\}\theta = \eta_{\{f\}} - \eta_{t'}(\theta) + \{G_e\}t' \tag{7.1-27}$$

which may be compared with Eq. (7.1-12). For a rheodictic material Eq. (7.1-27) reduces to $\eta(\theta) = \eta_f - \eta_{t'}(\theta)$. Again, the two responses contain the same information on the time-dependent behavior. Later on we shall consider a method of reconstructing $\eta(t)$ from measured values of $\eta_{t'}(\theta)$.

It is generally not possible or not convenient to remove the excitation at large values of t'. We therefore now examine the consequences of keeping t' small. This is

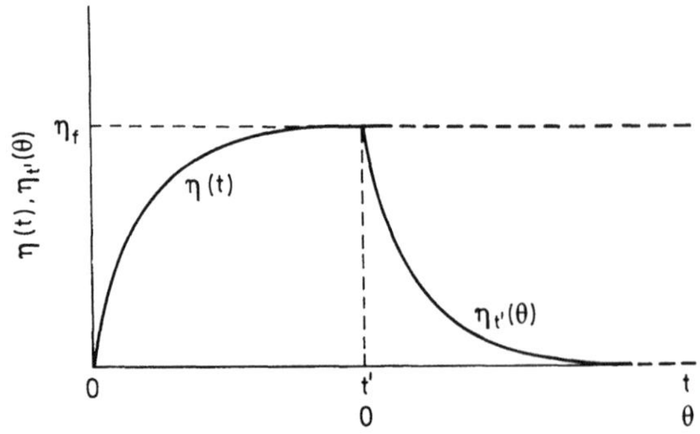

Fig. 7.1-7. Stress relaxation after cessation of steady-state flow

of importance in two respects. First, the application of a strain as a ramp function of time having a relatively short rise time t′ is a convenient way to obtain the time-dependent viscosity, $\eta(t)$, on *solidlike* (*highly viscous*) *rheodictic* and on arrheodictic materials alike. In such materials it is difficult to obtain $\eta(t)$ directly because the deformation must be kept below the linear viscoelastic limit at all times. Second, it is experimentally impossible to apply a strain as a true step function of time. In practice a ramp strain with short rise time is invariably imposed. Hence, we first examine the difference between the stress, $\sigma_{t'}(\theta) = \sigma(\theta + t') - \sigma(\theta)$, resulting from the imposition of a strain as a ramp function of time, and the stress, $\sigma(\theta)$, obtained in response to the imposition of a strain as a step function of time at $\theta = 0$. We shall then derive a method by which $\eta(t)$ can be obtained from $\eta_{t'}(\theta)$, and, finally, discuss the derivation of G(t) from $\eta(t)$.

Multiplication of Eq. (7.1-25) by $\dot{\varepsilon}_0$ and remembering Eq. (7.1-17) leads to

$$\sigma_{t'}(\theta) = \varepsilon_0\{G_e\} + \varepsilon_0 \int_{-\infty}^{\infty} H(\tau)(\tau/t')[1 - \exp(-t'/\tau)]\exp(-\theta/\tau)\mathrm{d}\ln\tau \ . \quad (7.1\text{-}28)$$

Now, the ramp function approaches the step function as $t' \to 0$ [cf. Eq. (A2-44) and Problem 7.1-6] and we have

$$\lim_{t'\to 0} \sigma_{t'}(\theta) = \sigma(\theta) = \varepsilon_0\{G_e\} + \varepsilon_0 \int_{-\infty}^{\infty} H(\tau)\exp(-\theta/\tau)\mathrm{d}\ln\tau \ . \quad (7.1\text{-}29)$$

The difference between the two stresses is

$$\sigma(\theta) - \sigma_{t'}(\theta) = \dot{\varepsilon}_0 \int_{-\infty}^{\infty} H(\tau)\{t' - \tau[1 - \exp(-t'/\tau)]\}\exp(-\theta/\tau)\mathrm{d}\ln\tau \ . \quad (7.1\text{-}30)$$

But $\tau[1 - \exp(-t'/\tau)]$ is a monotonic non-decreasing function of τ for any positive t′ and

$$\lim_{\tau\to 0} \tau[1 - \exp(-t'/\tau)] = 0 \quad \text{while} \quad \lim_{\tau\to\infty} \tau[1 - \exp(-t'/\tau)] = t' \ . \quad (7.1\text{-}31)$$

Hence, the term in braces in Eq. (7.1-30) is positive semi-definite. Therefore, $\sigma(\theta) \geq \sigma_{t'}(\theta)$ for $0 \leq \theta < \infty$. The equality applies only in the limit that $t' \to 0$ or $\theta \to \infty$. Thus, except as $\theta \to \infty$, the stress produced in response to the removal of a constant rate of strain is always smaller than the stress resulting from the imposition of a strain as a step function of time when the strains have the same magnitude, i.e. when $\varepsilon_0 = \dot{\varepsilon}_0 t'$ at $\theta = 0$. The two responses are compared schematically in Fig. 7.1-8.

We have already pointed out that $\sigma(\theta) - \sigma_{t'}(\theta)$ decreases as θ increases. When t′ is small enough, the difference may be negligible within experimental error, and the ramp response then becomes an acceptable approximation to the true, experimentally unattainable, step response. The relaxation modulus can then be obtained from

$$G(\theta) = \sigma_{t'}(\theta)/\dot{\varepsilon}_0 t' \ , \qquad \theta > mt' \ . \quad (7.1\text{-}32)$$

The restriction $\theta > mt'$ has been added because the initial portion of the $\sigma_{t'}(\theta)$-curve, from $\sigma_{t'}(0)$ to $\sigma_{t'}(mt')$, where m is an empirical multiplier, is often rejected to ensure that $\sigma_{t'}(\theta)$ has approached $\sigma(\theta)$ closely enough for Eq. (7.1-32), to be valid. A better approximation, however, is

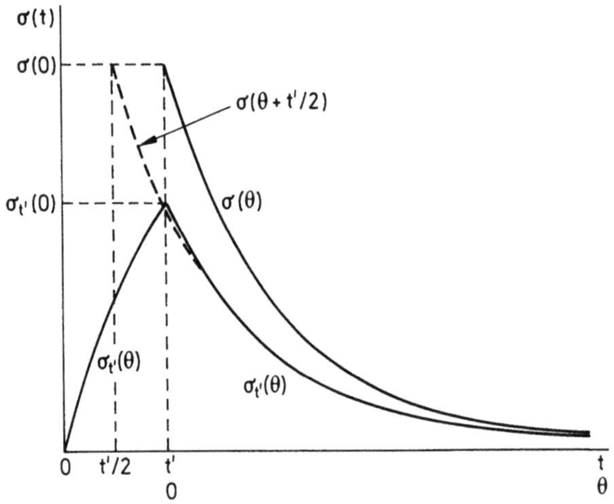

Fig. 7.1-8. Response to a strain as a ramp and as a step function of time

$$G(\theta + t'/2) = \sigma_{t'}(\theta)/\dot{\varepsilon}_0 t' \tag{7.1-33}$$

which may be obtained directly from the Boltzmann superposition integral (see Problem 7.1-1). The dashed curve in Fig. 7.1-8 represents $\sigma(\theta + t'/2) = \varepsilon_0\, G(\theta + t'/2)$ calculated with the same parameters that were used to construct $\sigma(\theta)$.

By Eq. (2.4-9), the step response can be derived from the slope response by differentiation. We, therefore, now turn our attention to the reconstruction of the slope response, $\eta(t)$, from measured values of $\eta_{t'}(\theta)$. Let $\bar{\eta}_r(s)$ be the Laplace transform of the (total) response to a ramp excitation. We seek an operator, $\bar{T}(s)$, which will transform $\bar{\eta}_r(s)$ into $\bar{\eta}(s)$ according to

$$\bar{\eta}(s) = \bar{T}(s)\bar{\eta}_r(s) \ . \tag{7.1-34}$$

But $\bar{\eta}(s) = \bar{\sigma}(s)/\dot{\varepsilon}_0$ and

$$\bar{\eta}_r(s) = \bar{\eta}(s)\bar{h}_{t'}(s) = \bar{\sigma}(s)[1 - \exp(-t's)]/\dot{\varepsilon}_0 \ . \tag{7.1-35}$$

Hence,

$$\bar{T}(s) = \frac{1}{1 - \exp(-t's)} = \sum_{n=0}^{n=\infty} \exp(-nt's) \tag{7.1-36}$$

and, therefore,

$$\bar{\eta}(s) = \bar{\eta}_r(s) \sum_{n=0}^{n=\infty} \exp(-nt's) \ . \tag{7.1-37}$$

This, upon retransformation using Eq. (A3-6), yields

$$\eta(t) = \sum_{n=0}^{n=\infty} \eta_r(t - nt')h(t - nt') \ . \tag{7.1-38}$$

Now let the data be given as a set of values $\eta_r(it') = \sigma_r(it')/\dot{\varepsilon}_0$ where $i = 1, 2, \ldots$. Thus, the data are assumed to be equally spaced along the time axis by integral

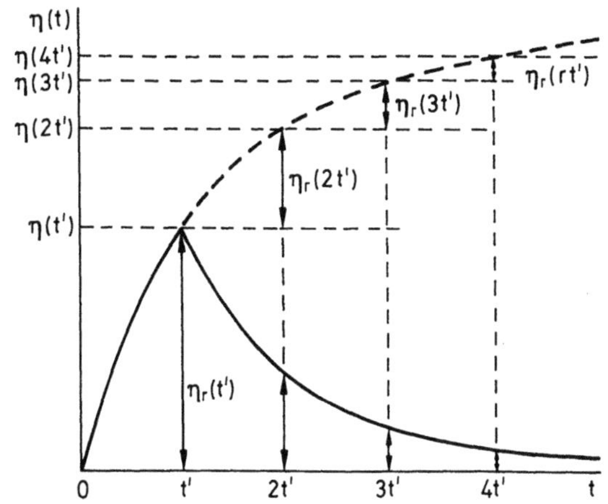

Fig. 7.1-9.
Reconstruction of $\eta(t)$ from $\eta_r(t)$

multiples of t'. We may then write

$$\eta(it') = \sum_{n=0}^{n=\infty} \eta_r[(i-n)t']h[(i-n)t'] \ . \tag{7.1-39}$$

But $\eta_r(0) = 0$ and $h[(i-n)t'] = 0$ when $n > i$. Therefore,

$$\eta(it') = \sum_{n=0}^{n=i-1} \eta_r[(i-n)t'] = \sum_{n=1}^{n=i} \eta_r(nt') \ . \tag{7.1-40}$$

Hence, from the set of data $\eta_r(it')$ a set $\eta(it')$, representing the time-dependent viscosity, $\eta(t)$, may be constructed. We note that Eq. (7.1-40) is equivalent to the recurrence relation

$$\eta(it') = \eta[(i-1)t'] + \eta_r(it') \ , \tag{7.1-41}$$

the starting value being given by $\eta(0) = 0$. The procedure is illustrated in Fig. 7.1-9. It should be remembered that $\eta_r(t)$ is the *total* ramp response, so that $\eta_r(t') = \eta_{t'}(0)$.

Once $\eta(t)$ has been constructed, G(t) may be obtained from it by graphical differentiation. Alternatively, one may take advantage of the fact that $\eta(t)$ is available as a set of equally spaced values $\eta(it')$, and obtain G(t) through the rules of the calculus of finite differences. We have

$$G[(i-1/2)t'] = \{\eta(it') - \eta[(i-1)t']\}/t' \tag{7.1-42}$$

and using Eq. (7.1-47) leads to

$$G[(i-1/2)t'] = \eta_r(it')/t' \ . \tag{7.1-43}$$

Therefore, the relaxation modulus at $t = (i-1/2)t'$ is obtained through division of $\eta_r(it')$ by t'. Alternatively, we may let

$$G[(j+1/2)t'] = \eta_{t'}(jt')/t' = \sigma_{t'}(jt')/\dot{\varepsilon}_0 t' \tag{7.1-44}$$

where $j = 0, 1, \ldots$, and jt' is the shifted time, θ. Equation (7.1-33) is at once recognized as the continuous form of Eq. (7.1-44).

7.1.3 Response to the Reversal of Direction of a Constant Rate of Strain

In the preceding two sections we examined the response to excitations in which a stimulus was first applied and then removed. We now discuss an excitation in which a stimulus is imposed, then its direction is reversed, and finally it is removed. This excitation is the *triangular pulse*. Use will be made of it in the next section on repeated stimuli and in Chap. 9, more particularly in Sect. 9.3 on hysteresis.

In the general triangular pulse excitation shown in Fig. 7.1-10a, the strain increases at a constant rate $\dot{\varepsilon}_0' = \varepsilon_0/t'$ for $0 < t < t'$ and thereupon decreases at the rate $\dot{\varepsilon}_0'' = \varepsilon_0/t''$ for $t' < t < (t' + t'')$. It is zero for $t > (t' + t'')$. In such a pulse, therefore, the direction of the stimulus is reversed at $t = t'$. Here ε_0 is termed the pulse height.

In a *regular triangular strain pulse* $\dot{\varepsilon}_0' = \dot{\varepsilon}_0'' = \dot{\varepsilon}_0$ and $t'' = t'$. Such a pulse is shown in Fig. 7.1-10b. In terms of slope functions, the strain may be expressed as

$$\varepsilon(t) = \dot{\varepsilon}_0[p(t) - 2p(t - t') + p(t - 2t')] \ . \qquad (7.1\text{-}45)$$

In terms of step functions we have

$$\varepsilon(t) = \dot{\varepsilon}_0[th(t) - 2(t - t')h(t - t') + (t - 2t')h(t - 2t')] \qquad (7.1\text{-}46)$$

which transforms to

$$\begin{aligned}\bar{\varepsilon}(s) &= (\dot{\varepsilon}_0/s^2)[1 - 2\exp(-t's) + \exp(-2t's)] \\ &= (\dot{\varepsilon}_0/s^2)[1 - \exp(-t's)]^2 \ .\end{aligned} \qquad (7.1\text{-}47)$$

Half of the second term in the center portion of Eqs. (7.1-45) and (7.1-46) annihilates the effect of the first term. The third term counteracts the effect of the second half of the second term for $t > 2t'$.

From $\bar{\sigma}(s) = s^2\bar{\eta}(s)\bar{\varepsilon}(s)$ [see Eq. (2.4-27)$_1$] we obtain

$$\bar{\sigma}(s) = \dot{\varepsilon}_0\bar{\eta}(s)[1 - 2\exp(-t's) + \exp(-2t's)] \qquad (7.1\text{-}48)$$

which retransforms to

$$\sigma(t) = \dot{\varepsilon}_0[\eta(t)h(t) - 2\eta(t - t')h(t - t') + \eta(t - 2t')h(t - 2t')] \qquad (7.1\text{-}49)$$

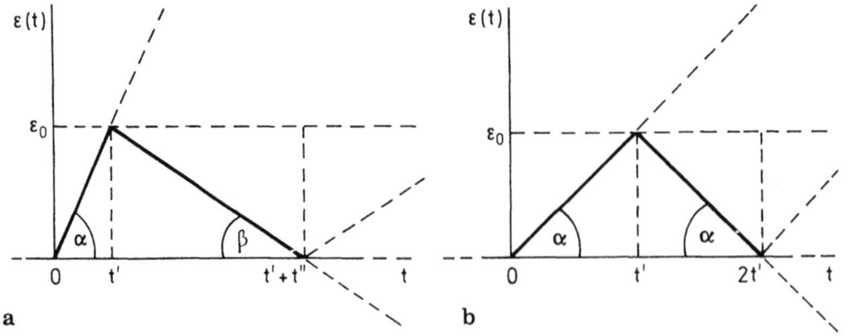

Fig. 7.1-10a,b. Triangular strain pulse

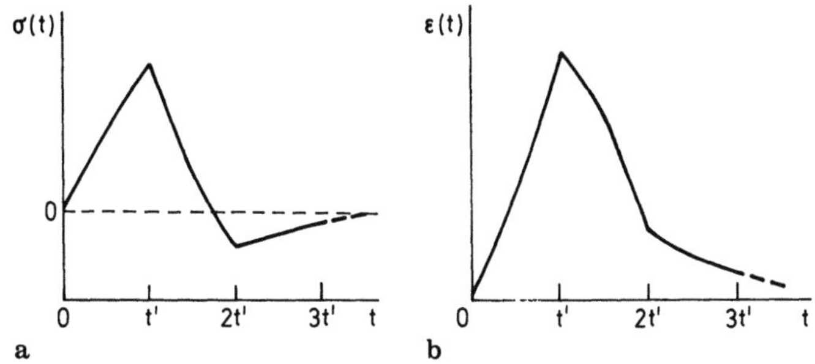

Fig. 7.1-11. Typical response of an arrheodictic material to a regular triangular strain pulse (**a**), and stress pulse (**b**)

by Eqs. (A3-5) and (A3-6). Thus the stress is given by

$$\sigma(t) = \begin{cases} \dot{\varepsilon}\,\eta(t), & 0 \le t \le t' \\ \dot{\varepsilon}_0[\eta(t) - 2\eta(t - t')], & t' \le t \le 2t' \\ \dot{\varepsilon}_0[\eta(t) - 2\eta(t - t') + \eta(t - 2t')], & t \ge 2t' \;. \end{cases} \tag{7.1-50}$$

Equations (7.1-49) and (7.1-50) may be compared with Eqs. (7.1-21) and (7.1-22).

An analogous treatment of the regular triangular *stress* pulse leads to

$$\varepsilon(t) = \dot{\sigma}_0[\phi(t)h(t) - 2\phi(t - t')h(t - t') + \phi(t - 2t')h(t - 2t')] \tag{7.1-51}$$

where $\dot{\sigma}_0$ is the rate of stress.

The typical response of an arrheodictic material [2] to a regular triangular strain and stress pulse is shown in Fig. 7.1-11. The direction of the stress is reversed during the latter part of the strain excitation (see Fig. 7.1-11a). The reversed stress eventually creeps to zero. At t = 2t′, when the stress excitation is completely removed (see Fig. 7.1-11b), the strain is still finite but gradually relaxes to zero.

7.2 Response to Repeated Non-Cyclic Excitations

We now turn our attention to a class of excitations in which a stimulus, or a sequence of stimuli, is reapplied at regular intervals. If the nature of the stimuli is cumulative, i.e. the stimulus imposed in each successive interval adds algebraically to that applied in the preceding interval, we speak of a *periodically repeated*, but *non-cyclic, excitation*. More particularly, such a superposition of repeated stimuli may be called an *ascending*, or a *descending staircase excitation* according to whether the stimuli increase or decrease from interval to interval. The Laplace transform of such an excitation is obtained by the use of the *repetition theorem* (see Appendix A3.1.9). Denoting the *repeat function* by $\varepsilon_r(t)$ when the excitation is a *strain* staircase excitation, and by $\sigma_r(t)$ when it is a *stress* staircase, we have

$$\bar{\varepsilon}(s) = \frac{\bar{\varepsilon}_r(s)}{1 - \exp(-t's)} = \bar{\varepsilon}_r(s) \sum_{n=0}^{n=\infty} \exp(-nt's) \tag{7.2-1}$$

and

$$\bar{\sigma}(s) = \frac{\bar{\sigma}_r(s)}{1 - \exp(-t's)} = \bar{\sigma}_r(s) \sum_{n=0}^{n=\infty} \exp(-nt's) \tag{7.2-2}$$

where t' is the duration of the interval after which the stimuli are reapplied. If the number, n, of the intervals is finite, as it will have to be in practice, the series in Eqs. (7.2-1) and (7.2-2) are terminated with the appropriate interval.

7.2.1 Staircase Excitations

We shall here discuss only the step staircase. As another example of a staircase excitation, the ramp staircase is the subject of Problem 7.2-1. Although we are concerned primarily with *regular* (periodic) excitations in this chapter, it will be instructive to begin by considering a *general* staircase *strain* excitation.

The general ascending step staircase excitation consists of a superposition of N step excitations, each of different magnitude, ε_n, and duration, t_n. The strain is therefore given by

$$\varepsilon(t) = \sum_{n=0}^{N} \varepsilon_n h(t - t_n) \ . \tag{7.2-3}$$

where $t_0 = 0$. The sudden changes in strain at $t = t_n$ produce a strain function with N discontinuities in the time domain. Laplace transformation [LTP (5)] yields the strain transform

$$\bar{\varepsilon}(s) = \sum_{n=0}^{N} \varepsilon_n[\exp(-t_n s)]/s \ . \tag{7.2-4}$$

In the *regular* staircase excitation illustrated in the top portion of the left half of Fig. 7.2-1, the step heights and intervals are all equal. Thus $\varepsilon_0 = \varepsilon_1 = \cdots = \varepsilon_N$, $t_1 = t_2 = \cdots = t_N = t'$, and Eqs. (7.2-3) and (7.2-4) become

$$\varepsilon(t) = \varepsilon_0 \sum_{n=0}^{N} h(t - nt') \tag{7.2-5}$$

and

$$\bar{\varepsilon}(s) = \varepsilon_0 \sum_{n=0}^{N} [\exp(-nt's)]/s \ . \tag{7.2-6}$$

Since the transform of a step strain of height ε_0 is ε_0/s, Eq. (7.2-6) could have been obtained by substituting $\bar{\varepsilon}_r(s) = \varepsilon_0/s$ into Eq. $(7.2\text{-}1)_2$. Use of the repetition theorem thus eliminates the need for formulating the excitation in terms of translated step functions.

The response, i.e. the stress transform, is obtained from $\bar{\sigma}(s) = s\bar{G}(s)\bar{\varepsilon}(s)$ [see Eq. $(2.3\text{-}15)_1$] as

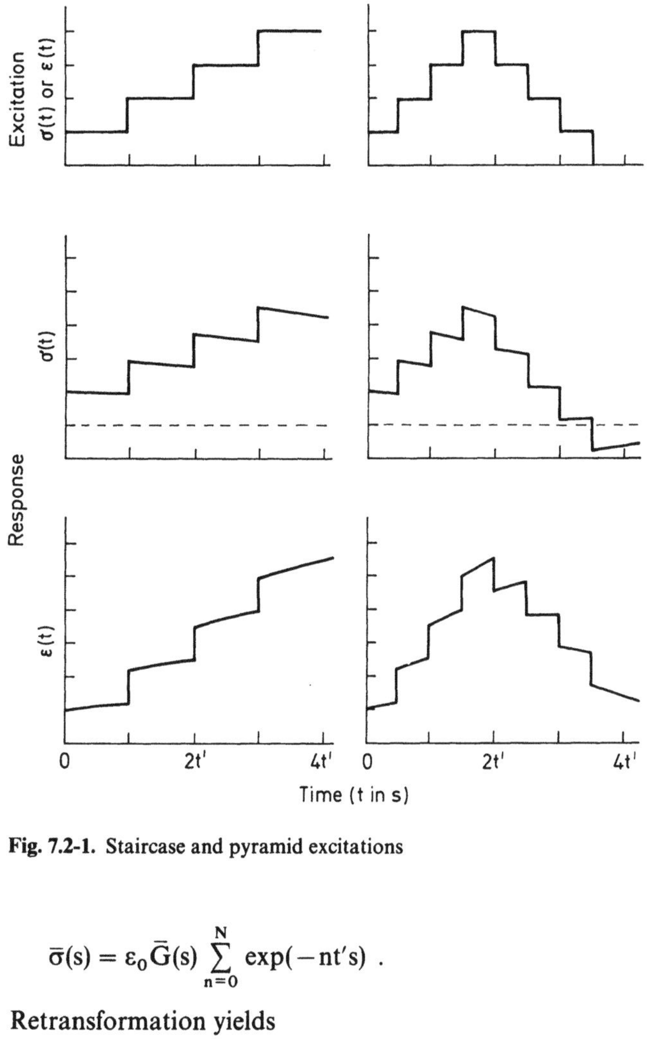

Fig. 7.2-1. Staircase and pyramid excitations

$$\bar{\sigma}(s) = \varepsilon_0 \bar{G}(s) \sum_{n=0}^{N} \exp(-nt's) \; . \tag{7.2-7}$$

Retransformation yields

$$\sigma(t) = \varepsilon_0 \sum_{n=0}^{N} G(t - nt')h(t - nt') \; . \tag{7.2-8}$$

Consideration of a regular stair case *stress* excitation leads to

$$\varepsilon(t) = \sigma_0 \sum_{n=0}^{N} J(t - nt')h(t - nt') \; . \tag{7.2-9}$$

Equations (7.2-8) and (7.2-9) were derived here through the Laplace transformation to illustrate the general method. In these simple cases it is just as easy — or even easier — to derive the equations by substitution of Eq. (7.2-5) or its stress analog into the Boltzmann superposition integral (see Problem 7.2-2).

Typical responses to ascending [2] step staircase excitations are shown in the left half of Fig. 7.2-1. It should be noted that the rate of creep and the rate of stress relaxation increase as the stress or the strain increase in each successive interval t'.

7.2.2 Pyramid Excitations

The regular pyramid excitation shown in the top portion of the right half of Fig. 7.2-1 consists of the combination of an ascending and a descending regular step staircase excitation. For the regular step pyramid strain excitation, therefore,

$$\varepsilon(t) = \varepsilon_0\left(\sum_{n=0}^{M} h(t - nt') - \sum_{n=M+1}^{N} h(t - nt')\right) \tag{7.2-10}$$

which transforms to

$$\bar{\varepsilon}(s) = \varepsilon_0 \sum_{n=0}^{M} [\exp(-nt's)]/s - \varepsilon_0 \sum_{n=M+1}^{N} [\exp(-nt's)]/s \ . \tag{7.2-11}$$

An analogous relation can be written for the regular step pyramid stress excitation. The resulting response functions become

$$\sigma(t) = \varepsilon_0\left(\sum_{n=0}^{M} G(t - nt')h(t - nt') - \sum_{n=M+1}^{N} G(t - nt')h(t - nt')\right) \tag{7.2-12}$$

and

$$\varepsilon(t) = \sigma_0\left(\sum_{n=0}^{M} J(t - nt')h(t - nt') - \sum_{n=M+1}^{N} J(t - nt')h(t - nt')\right) . \tag{7.2-13}$$

Typical responses are shown in the right half of Fig. 7.2-1.

It should be noted that the strain in response to the stress excitation still increases after the first decrease in the stress at $t = 4t'$, albeit at a reduced level, and is almost level after the following decrease. Only after the third decrease does the creep curve turn into a typical recovery curve. Although the stress is removed completely at $t = nt'$, the creep recovery continues for several intervals before the strain is finally reduced to zero. Similar observations can be made on the response to the pyramidal strain excitation. Following the last decrease in the strain the stress actually reverses direction before it relaxes to zero several intervals later. The behavior is qualitatively similar for any linear viscoelastic material and is a clear demonstration of the viscoelastic "persistence of memory."

The responses shown here are idealized linear viscoelastic responses. In practice it is difficult to remain within the linear response region when the excitation is increased by several increments. Thus, in this type of excitation a steady-state cannot develop.

7.3 Response to Cyclic Excitations

In the staircase excitations discussed in Sect. 7.2 the effect of the periodically reapplied stimuli was cumulative. We now examine the case where this is not so because each stimulus is returned to zero before the next one is applied. This is achieved by restricting the *repeat function* to the finite interval $0 \leq t \leq T$. In that case f(t) is *periodic* with *period* T and satisfies the relation

$$f(t) = f(t + nT) \ . \tag{7.3-1}$$

An excitation which is restricted in this manner will be called a *cyclic excitation* to distinguish it from the non-cyclic excitations discussed in Sect. 7.2. The standard cyclic excitation is, of course, the harmonic excitation introduced in Sect. 2.5.

The Laplace transform of a cyclic excitation is obtained by the use of the *periodicity theorem* (see Appendix A3.1.9). This yields

$$\bar{\varepsilon}(s) = \frac{\bar{\varepsilon}_T(s)}{1 - \exp(-Ts)} = \bar{\varepsilon}_T(s) \sum_{n=0}^{n=\infty} \exp(-nTs) \tag{7.3-2}$$

for a *periodic strain excitation*, and

$$\bar{\sigma}(s) = \frac{\bar{\sigma}_T(s)}{1 - \exp(-Ts)} = \bar{\sigma}_T(s) \sum_{n=0}^{n=\infty} \exp(-nTs) \tag{7.3-3}$$

for a *periodic stress excitation*. In Eqs. (7.3-2) and (7.3-3) $\bar{\varepsilon}_T(s)$ and $\bar{\sigma}_T(s)$, if they exist, are transforms of the *repeat functions*. We note that it is not necessary that the repeat functions be cut off at $t = T$. The stimulus may be returned to zero at $t = t' < T$, thus allowing a "rest period" before the next stimulus is applied. The repeat functions then become $\varepsilon_{t'}(t)$ and $\sigma_{t'}(t)$ instead of $\varepsilon_T(t)$ and $\sigma_T(t)$, respectively, but the exponentials in Eqs. (7.3-2) and (7.3-3) remain $\exp(-Ts)$ and $\exp(-nTs)$. Mathematically, the difference between the staircase and periodic (or non-cyclic and cyclic) excitation is essentially that in the latter case the repeat functions are *finite Laplace transforms* (see Appendix A3.1.8) with either T or t' as the upper limit in the integral, while in the former case they are not. The response to a cyclic excitation does not increase or decrease with every period and thus an effective steady-state can be reached after an appropriate number of cycles. The steady-state response generally takes a simpler form than the full, non-steady-state response.

The harmonic excitation is a *continuous* cyclic excitation. As a representative *piecewise continuous* cyclic excitation consider the *sawtooth* strain excitation illustrated in Fig. 7.3-1a. Here, the stimulus is completely removed at $t = t'$. In Fig. 7.3-1a $t' = T$, in Fig. 7.3-1b $t' < T$. In the first case, we have

$$\varepsilon_{t'}(t) = \dot{\varepsilon}_0 t h_{t'}(t) \tag{7.3-4}$$

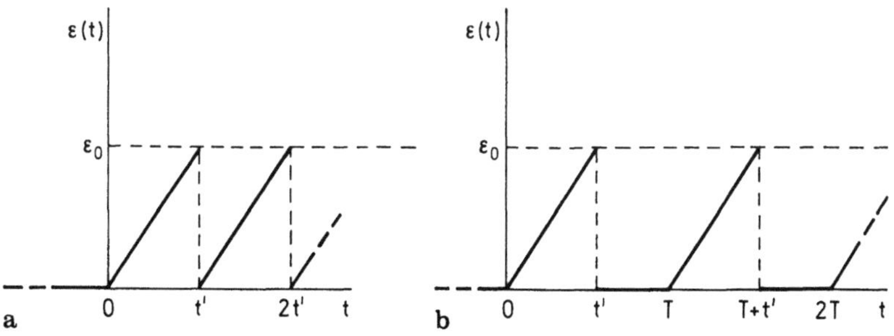

Fig. 7.3-1a,b. Sawtooth strain excitations

where $\dot{\varepsilon}_0 = \varepsilon_0/t'$. Hence, by Eq. (A3-23),

$$\bar{\varepsilon}_{t'}(s) = \dot{\varepsilon}_0 \int_0^{t'} t\,\exp(-st)dt = \frac{\dot{\varepsilon}_0}{s^2}[1 - (1 + t's)\exp(-t's)] \tag{7.3-5}$$

and substituting into Eq. $(7.3\text{-}2)_1$, using $T = t'$, yields

$$\bar{\varepsilon}(s) = \frac{\dot{\varepsilon}_0}{s^2} - \frac{\varepsilon_0 \exp(-t's)}{s[1 - \exp(-t's)]} \ . \tag{7.3-6}$$

The transform of the sawtooth excitation with rest period $T - t'$ shown in Fig. 7.3-1b results as

$$\bar{\varepsilon}(s) = \frac{\dot{\varepsilon}_0[1 - \exp(-t's)]}{s^2[1 - \exp(-Ts)]} - \frac{\varepsilon_0 \exp(-t's)}{s[1 - \exp(-Ts)]} \ . \tag{7.3-7}$$

In another form of the sawtooth excitation the direction of the stimulus is reversed, i.e. it returns to zero gradually. If the rate of return equals that at which the slope function p(t) is imposed, then $\varepsilon_{t'}(t)$ is the regular triangular pulse discussed in Sect. 7.1.3 (see Fig. 7.1-10b). We shall treat this excitation as a typical example in this section. Other examples may be found in Sect. 7.5 (Problems).

7.3.1 Non-Steady-State Response

Figure 7.3-2 shows a strain excitation of a form which we shall call the *regular triangular strain pulse train* because it consists of a sequence of regular triangular strain pulses. We note that the excitations shown in Fig. 7.3-1 are properly called sawtooth strain pulse trains. In the steady state a pulse train will be called a *wave.*

By Eq. (7.1-47) the Laplace transform of the repeat function for the regular triangular strain pulse train (*or* wave), $\varepsilon_{2t'}(t)$, becomes

$$\bar{\varepsilon}_{2t'}(s) = (\dot{\varepsilon}_0/s^2)[1 - \exp(-t's)]^2 \ . \tag{7.3-8}$$

Since the period of the excitation is $T = 2t'$, Eq. $(7.3\text{-}2)_1$ gives

$$\bar{\varepsilon}(s) = \frac{\dot{\varepsilon}_0[1 - \exp(-t's)]^2}{s^2[1 - \exp(-2t's)]} = (\dot{\varepsilon}_0/s^2)\left(1 + 2\sum_{n=1}^{\infty}(-1)^n\exp(-nt's)\right) \ . \tag{7.3-9}$$

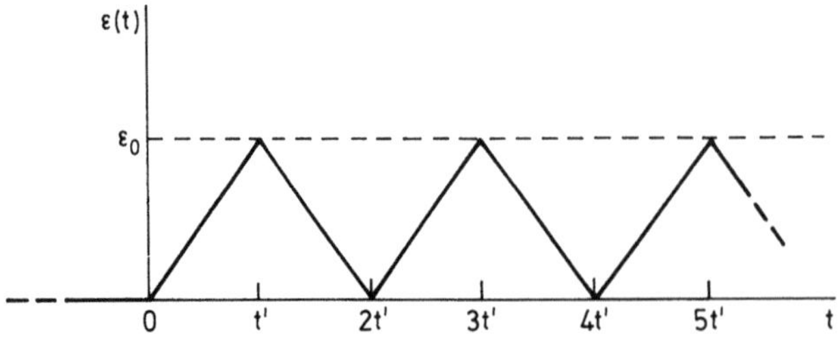

Fig. 7.3-2. Regular triangular strain pulse train

The right hand side of Eq. (7.3-9) is obtained from the center expression upon dividing and collecting terms.

The response resulting from insertion of Eq. (7.3-9) into $\bar{\sigma}(s) = s^2\bar{\eta}(s)\bar{\varepsilon}(s)$ and retransformation becomes

$$\sigma(t) = \dot{\varepsilon}_0\left(\eta(t)h(t) + 2\sum_{n=1}^{\infty}(-1)^n\eta(t-nt')h(t-nt')\right) . \tag{7.3-10}$$

Now, the stress during the Nth half-cycle, i.e. for $(N-1)t' \leq t \leq Nt'$, where $N \geq 1$, becomes

$$\sigma_N(t) = \dot{\varepsilon}_0\left(\eta(t) + 2\sum_{n=1}^{N-1}(-1)^n\eta(t-nt')\right) . \tag{7.3-11}$$

To make our equations valid even when $N = 1$ (i.e. during the first half-cycle) we adopt the convention that a term containing the summation sign vanishes when the upper limit of the summation is zero. Using

$$\eta(t) = \{G_e\}t + \eta_{\{f\}} - \int_{-\infty}^{\infty}\tau H(\tau)\exp(-t/\tau)d\ln\tau \tag{7.3-12}$$

yields the material response during the Nth half-cycle as

$$\eta_N(t) = \left(1+2\sum_{n=1}^{N-1}(-1)^n\right)\{G_e\}t - 2\sum_{n=1}^{N-1}(-1)^n n\{G_e\}t' + \left(1+2\sum_{n=1}^{N-1}(-1)^n\right)\eta_{\{f\}}$$
$$- \int_{-\infty}^{\infty}\tau H(\tau)\exp(-t/\tau)\left(1+2\sum_{n=1}^{N-1}(-1)^n\exp(nt'/\tau)\right)d\ln\tau . \tag{7.3-13}$$

Now, the material is deformed during odd half-cycles and recovers during even ones. But, for odd values of N,

$$1 + 2\sum_{n=1}^{N-1}(-1)^n = 1 \tag{7.3-14}$$

$$2\sum_{n=1}^{N-1}(-1)^n n = N - 1 \tag{7.3-15}$$

and

$$1 + 2\sum_{n=1}^{N-1}(-1)^n\exp(nt'/\tau) = \frac{2\exp[(N-1)t'/\tau] - 1 + \exp(-t'/\tau)}{1+\exp(-t'/\tau)} \tag{7.3-16}$$

[see Problem 7.3-1 for Eq. (7.3-16)]. Making use of these relations, the response during deformation becomes

$$\eta_N(t) = \{G_e\}[t-(N-1)t'] + \eta_{\{f\}}$$
$$- \int_{-\infty}^{\infty}\tau H(\tau)\frac{2\exp\{-[t+(N-1)t']/\tau\} - \exp(-t/\tau) + \exp[-(t+t')/\tau]}{1+\exp(-t'/\tau)}$$
$$\times d\ln\tau . \tag{7.3-17}$$

To obtain the response during recovery, N is replaced by N + 1 in Eqs. (7.3-11) and (7.3-13). N still being odd, we then have

$$1 + 2 \sum_{n=1}^{N} (-1)^n = -1 \tag{7.3-18}$$

$$2 \sum_{n=1}^{N} (-1)^n n = -(N + 1) \tag{7.3-19}$$

and

$$1 + 2 \sum_{n=1}^{N} (-1)^n \exp(nt'/\tau) = -\frac{2 \exp(Nt'/\tau) + 1 - \exp(-t'/\tau)}{1 + \exp(-t'/\tau)} . \tag{7.3-20}$$

Consequently,

$$\eta_{N+1}(t) = -\{G_e\}[t-(N+1)t'] - \eta_{\{f\}} + \int_{-\infty}^{\infty} \tau H(\tau) \frac{2 \exp[-(t-Nt')/\tau] + \exp(-t/\tau) - \exp[-(t+t')/\tau]}{1 + \exp(-t'/\tau)} d \ln \tau . \tag{7.3-21}$$

Equations (7.3-17) and (7.3-21) give the response during the Nth deformation half-cycle and during the subsequent recovery half-cycle, respectively. We note that Eq. (7.3-17) correctly reduces to Eq. (7.3-12) for N = 1.

A regular triangular *stress* pulse train can be treated analogously. Instead of Eq. $(7.3\text{-}9)_2$ we now have

$$\bar{\sigma}(s) = (\dot{\sigma}_0/s^2)\left(1 + 2 \sum_{n=1}^{\infty} (-1)^n \exp(-nt's)\right) . \tag{7.3-22}$$

Substitution into $\bar{\varepsilon}(s) = s^2 \bar{\chi}(s) \bar{\sigma}(s)$, where $\bar{\chi}(s)$ is the transform of the strain response to a constant rate of stress (see Eq. 2.4-10) leads to

$$\varepsilon_N(t) = \dot{\sigma}_0 \left(\chi(t) + 2 \sum_{n=1}^{N-1} (-1)^n \chi(t - nt') \right) . \tag{7.3-23}$$

as the equation of the strain during the Nth half-cycle. Integrating Eq. (4.1-28) and using Eq. (4.6-25) gives

$$\chi(t) = J_e^{\{o\}} t - \chi + \int_{-\infty}^{\infty} \tau \, L(\tau) \exp(-t/\tau) d \ln \tau + \{\phi_f\} t^2/2 . \tag{7.3-24}$$

Substituting this into Eq. (7.3-23) eventually leads to

$$\chi_N(t) = J_e^{\{o\}}[t-(N-1)t'] - \chi + \int_{-\infty}^{\infty} \tau L(\tau) \frac{2 \exp\{-[t-(N-1)t']/\tau\} - \exp(-t/\tau) + \exp[-(t+t')/\tau]}{1 + \exp(-t'/\tau)} + \{\phi_f\}(t-Nt')[t-(N-1)t']/2 + \{\phi_f\} tt'/2 \tag{7.3-25}$$

and

$$\chi_{N+1}(t) = -J_e^{\{o\}}[t-(N+1)t'] + \chi$$
$$-\int_{-\infty}^{\infty} \tau L(\tau) \frac{2\exp[-(t-Nt')/\tau] + \exp(-t/\tau) - \exp[-(t+t')/\tau]}{1+\exp(-t'/\tau)} d\ln\tau$$
$$-\{\phi_f\}(t-Nt')[t-(N+1)t']/2 + \{\phi_f\}tt'/2 \qquad (7.3\text{-}26)$$

where the relations, valid for odd values of N,

$$2\sum_{n=1}^{N-1}(-1)^n n^2 = N(N-1) \qquad (7.3\text{-}27)$$

and

$$2\sum_{n=1}^{N}(-1)^n n^2 = -N(N+1) \qquad (7.3\text{-}28)$$

were used in the calculation of the flow terms.

7.3.2 Steady-State Response

In the preceding section we demonstrated the procedure for deriving the *total* linear viscoelastic response to a pulse train. In this section we examine the *steady-state* response. This will be done both in terms of the spectral and the harmonic response functions. The first follows readily from Eqs. (7.3-17), (7.3-21), (7.3-25), and (7.3-26) by considering that the exponential terms which do not contain N represent transients which vanish rapidly as $t = Nt'$ increases. The second follows from the fact that the response to a cyclic non-sinusoidal excitation in the steady-state can always be expressed in terms of the steady-state response to a sinusoidal excitation. Basically, this is a consequence of the fact that the steady-state response to a periodic excitation is also periodic. Periodic functions can always be expanded into Fourier series (see, e.g., Ref. [3]). As we shall show, the components of the Fourier expansion contain the harmonic response functions. Thus, the steady-state response to cyclic excitations can be predicted from a knowledge of either the spectral or the harmonic responses.

In this section, as in the previous one, the procedures are demonstrated on hand of the regular triangular pulse train. Other examples may be found in Sect. 7.5 (Problems).

7.3.2.1 In Terms of the Spectral Response Functions

In the steady-state the variable of interest is the period rather than the time because the response repeats identically in each period. We, therefore, replace the time, t, by the *steady-state time*

$$\theta = t - (P-1)T, \qquad (P-1)T \le \theta \le PT \qquad (7.3\text{-}29)$$

where T is the periodic time (i.e. the duration of the period) and P is an integer equal to the number of periods. For a regular triangular pulse train as shown in Fig. 7.3-2, $T = 2t'$, and $P = (N + 1)/2$ where N is an odd number designating any arbitrary deformation half-cycle in the steady-state. Hence

$$\theta = t - (N - 1)t' , \qquad (N - 1)t' \leq \theta \leq (N + 1)t' . \tag{7.3-30}$$

Using Eq. (7.3-30) in Eqs. (7.3-17) and (7.3-21), and dropping the vanishing exponentials (i.e. those which do not contain N), we find

$$\eta_{def}(\theta) = \{G_e\}\theta + \eta_{\{f\}} - 2 \int_{-\infty}^{\infty} \tau H(\tau) \frac{\exp(-\theta/\tau)}{1 + \exp(-t'/\tau)} d \ln \tau \tag{7.3-31}$$

for the deformation half-cycle, and

$$\eta_{rec}(\theta) = -\{G_e\}(\theta - 2t') - \eta_{\{f\}} + 2 \int_{-\infty}^{\infty} \tau H(\tau) \frac{\exp[-(\theta - t')/\tau]}{1 + \exp(-t'/\tau)} d \ln \tau \tag{7.3-32}$$

for the recovery half-cycle.

For a regular triangular *stress* pulse train, substitution of Eq. (7.3-30) into Eqs. (7.3-25) and (7.3-26) yields

$$\chi_{def}(\theta) = J_e^{\{o\}}\theta - \chi + 2 \int_{-\infty}^{\infty} \tau L(\tau) \frac{\exp(-\theta/\tau)}{1 + \exp(-t'/\tau)} d \ln \tau$$
$$+ \{\phi_f\}\theta(\theta - t')/2 + \{\phi_f\}tt'/2 \tag{7.3-33}$$

and

$$\chi_{rec}(\theta) = -J_e^{\{o\}}(\theta - 2t') + \chi - 2 \int_{-\infty}^{\infty} \tau L(\tau) \frac{\exp[-(\theta - t')/\tau]}{1 + \exp(-t'/\tau)} d \ln \tau$$
$$+ \{\phi_f\}(\theta - t')(\theta - 2t')/2 + \{\phi_f\}tt'/2 . \tag{7.3-34}$$

Figure 7.3-3 exhibits the typical steady-state response of an arrheodictic material to regular triangular wave excitations [2]. Thus, in the steady-state the response to stress never reaches zero and the response to strain is a compressive stress during part of the cycle.

For a rheodictic material, Eqs. (7.3-31) and (7.3-32) apply in a strain excitation, with $G_e = 0$. However, in a stress excitation which is analogous to that shown in

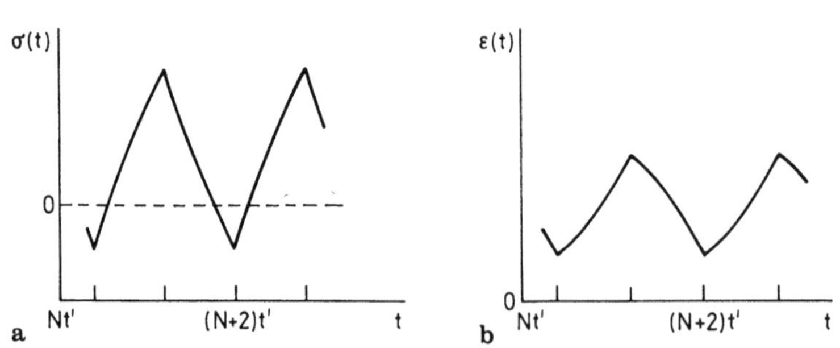

Fig. 7.3-3. Typical response of a arrheodictic material to a regular triangular strain pulse train (**a**), and stress pulse train (**b**), in the steady state

Fig. 7.3-2, a steady-state cannot develop. As shown by the last terms in Eqs. (7.3-33) and (7.3-34), the strain continuously increases with time. This is a consequence of the choice of the repeat stimulus, $\sigma_{2t'}(t)$, as shown in Fig. 7.1-10b for $\varepsilon_{2t'}(t)$. The stimulus is equivalent to a triangular wave superposed on a constant stress. A rheodictic material will flow under the effect of the latter. A different choice of the stress stimulus may allow the attainment of a steady-state. An example of this is discussed in Problem 7.3-4.

7.3.2.2 In Terms of the Harmonic Response Functions

We now consider the expansion of the steady-state response to a regular triangular pulse train into a Fourier series whose coefficients contain the harmonic response functions. The procedure is simple. We first consider the response to a strain excitation. From Chap. 2 we know that $\bar{Q}(s)$ may be expressed as the ratio of two polynomials in s, the degree of the numerator being at least one less than that of the denominator. We assume the same of the strain excitation and write the general stress-strain relation in the transform plane as

$$\bar{\sigma}(s) = \bar{Q}(s)\bar{\varepsilon}(s) = \frac{\bar{q}(s)\bar{m}_\varepsilon(s)}{\bar{u}(s)\bar{n}_\varepsilon(s)} \ . \tag{7.3-35}$$

Retransformation of $\bar{\sigma}(s)$ into the time domain may be accomplished through the use of the Residue Theorem (see Appendix A3.2.1). The equation takes the form

$$\sigma(t) = \sum_k \mathrm{Res}_k[\bar{\sigma}(s)\exp(st)] \ . \tag{7.3-36}$$

If the residues are those associated with the poles of $\bar{Q}(s)\bar{\varepsilon}(s)$, i.e. the zeros of $\bar{u}(s)\bar{n}_\varepsilon(s)$, Eq. (7.3-36) will yield the total response composed of the transient as well as the steady-state part. Now, it is shown in the theory of linear systems (see, e.g., Ref. [4]) that the steady-state response is associated with the poles of the excitation transform, i.e. the zeros of $\bar{n}_\varepsilon(s)$. Consequently, the steady-state response may be obtained from

$$\sigma_{ss}(t) = \sum_k \mathrm{Res}_k[\bar{\sigma}_{ss}(s)\exp(st)] \tag{7.3-37}$$

where

$$\bar{\sigma}_{ss}(s) = \frac{\bar{Q}(s)\bar{m}_\varepsilon(s)}{\bar{n}_\varepsilon(s)} \tag{7.3-38}$$

and the residues are calculated from the roots of $\bar{n}_\varepsilon(s) = 0$. For the steady-state response to a stress excitation we then have

$$\varepsilon_{ss}(t) = \sum_k \mathrm{Res}_k[\bar{\varepsilon}_{ss}(s)\exp(st)] \tag{7.3-39}$$

where

$$\bar{\varepsilon}_{ss}(s) = \frac{\bar{U}(s)\bar{m}_\sigma(s)}{\bar{n}_\sigma(s)} \tag{7.3-40}$$

and the residues are calculated from the roots of $\bar{n}_\sigma(s) = 0$. We note that the zeros of $\bar{n}(s)$ must not also be zeros of the derivative $\bar{n}'(s)$, a condition which is fulfilled in most practical cases.

We demonstrate the procedure here using the regular triangular wave excitation. Other examples may be found in Sect. 7.5 (Problems). From Eq. $(7.3\text{-}9)_1$ we obtain

$$\bar{\varepsilon}(s) = \frac{\dot{\varepsilon}_0[1 - \exp(-t's)]}{s^2[1 + \exp(-t's)]} = \frac{\dot{\varepsilon}_0 \sinh(t's/2)}{s^2 \cosh(t's/2)} \,. \tag{7.3-41}$$

Thus, $\bar{n}_\varepsilon(s) = s^2 \cosh(t's/2)$, and it appears that $\bar{\varepsilon}(s)$ has a pole of order two at the origin. Series expansion of the hyperbolic functions shows, however, that the pole is a simple one. By Eq. (A3-32), then, the residue associated with this pole becomes

$$\mathrm{Res}_0 = \lim_{s \to 0} \frac{\dot{\varepsilon}_0 \bar{Q}(s) \sinh(t's/2) \exp(st)}{s \cosh(t's/2)} = (\varepsilon_0/2)\bar{Q}(0) \tag{7.3-42}$$

where we have used L'Hospital's theorem together with $\dot{\varepsilon}_0 = \varepsilon_0 t'$ to obtain Eq. $(7.3\text{-}42)_2$.

The remaining poles of $\bar{\varepsilon}(s)$ are obtained from $\cosh(t's/2) = 0$. Since the excitation is periodic with period $T = 2t'$, the relation $T = 2\pi/\omega$ furnishes $t' = \pi/\omega$ where ω is the radian frequency associated with T. Inserting this, we have $\cosh(\pi s_k/2\omega) = 0$ as the relation for the poles. This is satisfied for

$$s_k = j\omega k = j\pi k/t' \,, \qquad k = \pm 1, \pm 3, \ldots \,. \tag{7.3-43}$$

The residues associated with these poles become

$$\mathrm{Res}_k = \lim_{s \to s_k} \frac{(s - s_k)\dot{\varepsilon}_0 \bar{Q}(s) \sinh(t's/2) \exp(st)}{s^2 \cosh(t's/2)} \tag{7.3-44}$$

i.e.

$$\mathrm{Res}_k = \frac{\varepsilon_0 \bar{Q}(s_k) \sinh(t's_k/2) \exp(s_k t)}{t's_k^2} \lim_{s \to s_k} \frac{1}{(t'/2)\sinh(t's/2)} \tag{7.3-45}$$

again using L'Hospital's theorem and $\dot{\varepsilon}_0 = \varepsilon_0/t'$. Thus, taking the limit, cancelling, and introducing Eqs. (7.3-43),

$$\mathrm{Res}_k = -\frac{2\varepsilon_0 \bar{Q}(j\omega k) \exp(j\omega k t)}{\pi^2 k^2} \,, \qquad k = \pm 1, \pm 3, \ldots \,. \tag{7.3-46}$$

By Eq. (7.3-37) the steady-state stress becomes

$$\sigma_{ss}(t) = (\varepsilon_0/2)\bar{Q}(0) - (2\varepsilon_0/\pi^2) \sum_k k^{-2} \bar{Q}(j\omega k) \exp(j\omega k t) \,. \tag{7.3-47}$$

Now, $\bar{Q}(0) = \{G_e\}$, and we may write

$$\sigma_{ss}(t) = (\varepsilon_0/2)\{G_e\} - (2\varepsilon_0/\pi^2) \sum_k k^{-2} [\bar{Q}(j\omega k) \exp(j\omega k t) + \bar{Q}(-j\omega k) \exp(-j\omega k t)] \tag{7.3-48}$$

the summation now extending over positive odd values of k only, i.e. $k = 1, 3, \ldots$.

But, $\bar{Q}(\pm j\omega k) = G'(\omega k) \pm j\, G''(\omega k)$ and $\exp(\pm j\omega kt) = \cos \omega kt \pm j \sin \omega kt$. Upon taking the products, the imaginary terms cancel and we finally obtain

$$\sigma_{ss}(t) = (\varepsilon_0/2)\{G_e\} - (4\varepsilon_0/\pi^2) \sum_k k^{-2}[G'(\omega k)\cos \omega kt - G''(\omega k)\sin \omega kt]_{\omega=\pi/t'} \,,$$
$$k = 1, 3, \ldots \,. \tag{7.3-49}$$

We immediately recognize this as a Fourier expansion given by

$$\mathscr{R}_{ss}(t) = \frac{a_0}{2} + \sum_{k=1,3,\ldots} (a_k \cos \omega kt + b_k \sin \omega kt) \tag{7.3-50}$$

where $\mathscr{R}_{ss}(t) = \sigma_{ss}(t)$ is the steady-state response,

$$a_0 = \varepsilon_0\{G_e\}$$
$$a_k = -(4\varepsilon_0/\pi^2 k^2)G'(\omega k)$$

and

$$b_k = (4\varepsilon_0/\pi^2 k^2)G''(\omega k) \tag{7.3-51}$$

are the coefficients of the expansion.

Treating the response to a regular triangular *stress* wave analogous to the strain excitation shown in Fig. 7.3-2 in a similar fashion, we have

$$\varepsilon_{ss}(t) = (\sigma_0/2)\bar{U}(0) - (4\sigma_0/\pi^2) \sum_k k^{-2}[J'(\omega k)\cos \omega kt + J''(\omega k)\sin \omega kt]_{\omega=\pi/t'} \,,$$
$$k = 1, 3, \ldots \,. \tag{7.3-52}$$

But now

$$\bar{U}(0) = J_e^{\{o\}} + \left\{\lim_{s\to 0} (\phi/s)\right\} \,. \tag{7.3-53}$$

Thus, for an arrheodictic material, the Fourier expansion, Eq. (7.3-50), is given by $\mathscr{R}_{ss}(t) = \varepsilon_{ss}(t)$ with

$$a_0 = \sigma_0 J_e$$
$$a_k = -(4\sigma_0/\pi^2 k^2)J'(\omega k) \tag{7.3-54}$$

and

$$b_k = -(4\sigma_0/\pi^2 k^2)J''(\omega k) \,.$$

For a rheodictic material, however, no steady-state response can develop with this stress excitation as has already been pointed out in the preceding section (but see Problem 7.3-5).

Figure 7.3-4 shows the response of the standard solid model to a regular triangular strain wave. The response is plotted as a function of the steady-state time, θ, and was calculated using the model parameters $G_e = 0.2 \times 10^9\ \mathrm{N/m^2}$, $G = 10^9\ \mathrm{N/m^2}$, and $\tau = 1$ s. The rate of strain was taken as $\dot{\varepsilon}_0 = 1\ \mathrm{s^{-1}}$, and the half-period as $t' = 5$ s.

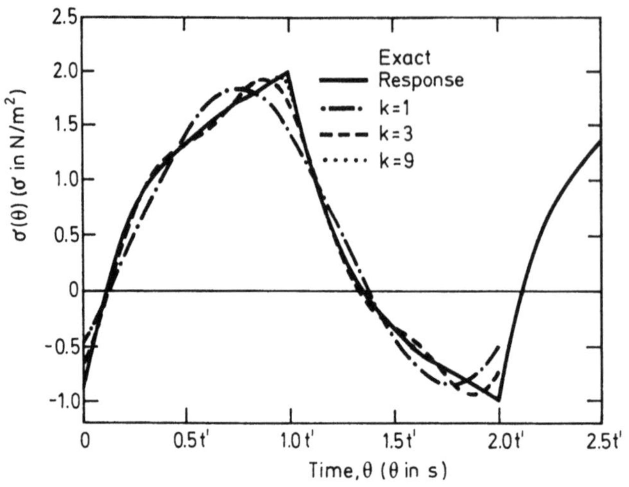

Fig. 7.3-4. Response to a regular triangular strain wave

The exact response, shown by the solid line, was calculated from the equations

$$\sigma_{def}(\theta) = \dot{\varepsilon}_0\left(G_e\theta + G\tau - \frac{2G\tau\exp(-\theta/\tau)}{1+\exp(-t'/\tau)}\right) \tag{7.3-55}$$

and

$$\sigma_{rec}(\theta) = -\dot{\varepsilon}_0\left(G_e(\theta - 2t') + G\tau - \frac{2G\tau\exp[-(\theta - t')/\tau]}{1+\exp(-t'/\tau)}\right) \tag{7.3-56}$$

[cf. Eqs. (7.3-31) and (7.3-32)]. Successive approximations with $k = 1, 3$, and 9 were calculated from

$$\sigma(\theta) = \varepsilon_0 G_e/2 - \varepsilon_0(2/\pi)^2 \sum_k k^{-2}[G'(\omega k)\cos\omega k\theta - G''(\omega k)\sin\omega k\theta] \tag{7.3-57}$$

[cf. Eq. (7.3-49)] with $\varepsilon_0 = \dot{\varepsilon}_0 t'$ and

$$G'(\omega k) = G_e + \frac{G(\omega k\tau)^2}{1+(\omega k\tau)^2} \tag{7.3-58}$$

and

$$G''(\omega k) = \frac{G\omega k\tau}{1+(\omega k\tau)^2}\,. \tag{7.3-58}$$

Analogous calculations with the same parameters ($Je = 1/G_e$, and $J = 1/G$) for the response to a regular triangular stress wave are shown in Fig. 7.3-5.

In both cases the deviation of the approximations with $k = 9$ from the exact curves is hardly noticeable except in the neighborhood of the endpoints of the half-periods. It is interesting to note that satisfactory agreement with the exact curve is reached for lower values of k in Fig. 7.3-5 than in Fig. 7.3-4. The reason is that

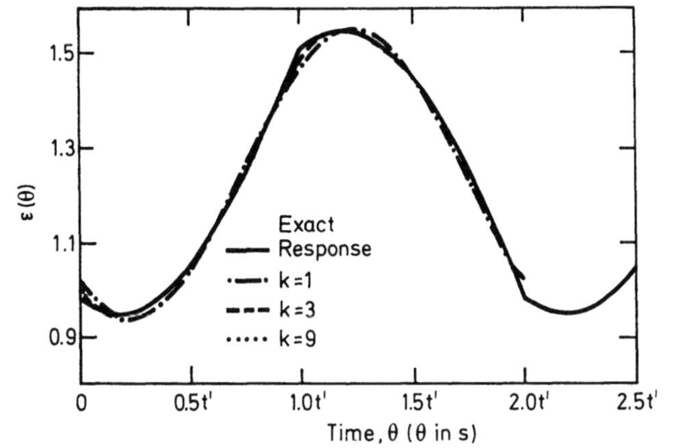

Fig. 7.3-5. Response to a regular triangular stress wave

$\varepsilon(\theta)$ resembles a sinusoid more closely than $\sigma(\theta)$ and thus for $\varepsilon(\theta)$ the Fourier series approaches the limiting form faster.

7.4 Approximations to the Spectra from Responses to Non-Standard Excitations

Chapter 4 dealt with methods for deriving approximations to the spectral functions from the responses to the standard excitations. Such approximations can also be obtained from the response to non-standard excitations. In practice, this is done only on creep recovery, and on ramp response.

By Eq. (7.1-10), creep recovery is given by

$$J(\theta) = \int_{-\infty}^{\infty} L(t)[1 - \exp(-t'/\tau)]\exp(-\theta/\tau)d\ln\tau + \{\phi_f\}t' \,. \tag{7.4-1}$$

Application of the operators $\mathscr{P}_k$ or $\mathscr{P}_{kh}$ as explained in Sects. 4.2.2 and 4.2.4, respectively, will yield an approximation to $L(\tau)[1 - \exp(-t'/\tau)]$, not to $L(\tau)$. We may, however, write

$$\hat{L}_{k(h)}(\tau) = -\left.\frac{\mathscr{P}_{k(h)}[J(\theta)]}{1 - \exp(-t'/\tau)}\right|_{\theta=\beta\tau} \tag{7.4-2}$$

and obtain an approximation which will be close to $L_{k(h)}(\tau)$.* The two become indistinguishable in practice when t' is large enough. As an example, the first approximation to the retardation spectrum is obtained from the creep recovery curve by continuous differentiation using the equation

* The subscript $_{k(h)}$ stands for either $_k$ or $_{kh}$.

$$\hat{L}_1(\tau) = - \left. \frac{d\,J(\theta)}{[1 - \exp(-t'/\tau)]d\ln\theta} \right|_{\theta=\tau} . \tag{7.4-3}$$

The negative sign in Eqs. (7.4-2) and (7.4-3) takes care of the fact that creep recovery is formally a *relaxation of strain* as pointed out in Sect. 7.1.1.

The reason that $\hat{L}_{k(h)}(\tau)$ is not identical with $L_{k(h)}(\tau)$ stems from the fact that the contributions which the intensity function $I_{k(h)}(\theta/\tau)$ selects from $L(\tau)[1 - \exp(-t'/\tau)]$ to make up the approximation, are not identical from those it would pick from $L(\tau)$ alone. Division by $1 - \exp(-t'/\tau)$ corrects for this difference only in the degree that the intensity function approaches the delta function.

An analogous situation exists when an approximation to $H(\tau)$ is to be derived from the response to a ramp excitation. Dividing Eq. (7.1-28) by $\dot{\varepsilon}_0 = \varepsilon_0/t'$ yields

$$\eta_R(\theta) = \{G_e\}t' + \int_{-\infty}^{\infty} \tau H(\tau)[1 - \exp(-t'/\tau)]\exp(-\theta/\tau)d\ln\tau . \tag{7.4-4}$$

Here, even when t' is large enough to allow cancellation of the factors $1 - \exp(-t'/\tau)$, the spectrum obtained by application of $\mathscr{P}_{k(h)}$ to $\eta_R(\theta)$ is $\tau H(\tau)$, not $H(\tau)$. We may write

$$\hat{H}_{k(h)}(\tau) = \left. \frac{\mathscr{P}_{k(h)}[\eta_R(\theta)]}{\tau[1 - \exp(-t'/\tau)]} \right|_{\theta=\beta\tau} \tag{7.4-5}$$

but must recognize that $\hat{H}_{k(h)}(\tau)$ is not identical with $H_{k(h)}(\tau)$ in principle because the two approximations will be weighted differently with respect to τ. For large t', the difference becomes smaller as the order of the approximation increases. The approximation $\hat{H}_1(\tau)$ is given by

$$\hat{H}_1(\tau) = \left. \frac{d\,\eta_R(\theta)}{\tau[1 - \exp(-t'/\tau)]d\ln\theta} \right|_{\theta=\tau} . \tag{7.4-6}$$

7.5 Problems

Problem 7.1-1. Derive Eq. (7.1-33) from the superposition integral.

Problem 7.1-2. Find the Laplace transform of the general triangular strain excitation shown in Fig. 7.1-10a. Then verify that this reduces to Eq. (7.1-47) when $t'' = t'$.

Problem 7.1-3. (A) Derive the Laplace transform of the *trapezoidal* strain excitation illustrated in Fig. P7.1-3 for which

$$\varepsilon(t) = \begin{cases} 0, & t \leq 0 \\ \varepsilon_0 t/t', & 0 \leq t \leq t' \\ \varepsilon_0, & t' \leq t \leq (t' + t'') \\ -\varepsilon_0(t - t' - t'' - t''')/t''', & (t' + t'') \leq t \leq (t' + t'' + t''') \\ 0, & t \geq (t' + t'' + t''') . \end{cases} \tag{1}$$

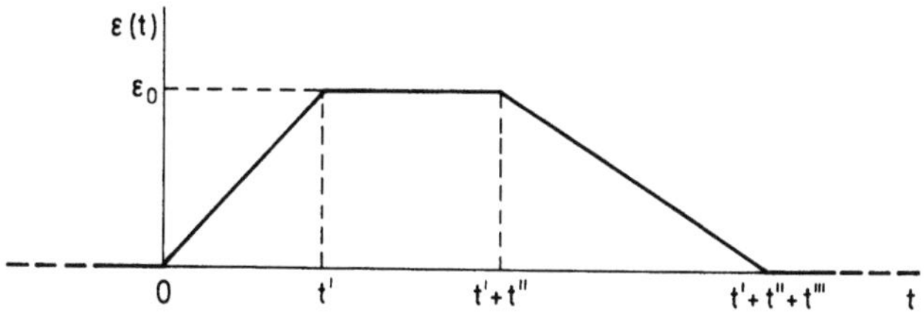

Fig. P7.1-3. Trapezoidal strain excitation

(B) Then find the Laplace transform of the *regular trapezoidal* strain excitation for which $t''' = t'' = t'$.

Problem 7.1-4. The time-dependent viscosity after cessation of steady-state flow is given by Eq. (7.1-26). Show that the area under a plot of $\eta_{t'}(\theta)$ against θ equals $J_e^\circ \eta_f^2$.

Problem 7.1-5. Show that the relaxation of the stress after cessation of steady-state flow can be expressed in terms of the stress relaxation modulus by the integral

$$\sigma(\theta) = \dot{\varepsilon}_0 \int_\theta^\infty G(t)dt \tag{1}$$

where θ is the time which has elapsed after removal of the excitation.

Problem 7.1-6. Prove that the unit ramp function reduces to the unit step function in the limit that the rise time approaches zero.

Problem 7.1-7. (A) What is the response of the non-standard 3-parameter Maxwell model to a ramp strain? After obtaining the total response, reduce it to the response at times larger than the rise time t'.

(B) Show that the total response reduces to the step response as the rise time approaches zero.

Problem 7.1-8. Suppose a viscoelastic material is subjected to a step strain at $t = 0$. It is allowed to relax up to time $t = t'$ when the stress is suddenly removed. The material now recovers its original undeformed state if it is arrheodictic. A rheodictic material will suffer a *permanent set*, i.e. it will not recover its original state although it will tend toward it.

(A) Obtain the transform relation describing the *strain recovery after partial stress relaxation* and retransform it into the time domain.

(B) Find the relation for the *permanent set* (or *residual strain*) and for the *recovered strain*.

(C) Find the strain recovery, $\varepsilon(\theta)$, using the standard linear solid models.

Problem 7.2-1. Consider the *ramp staircase* excitation shown in Fig. P7.2-1, in which the interval after which each successive ramp is imposed, is a multiple, m, of t'. Find the Laplace transform of the excitation.

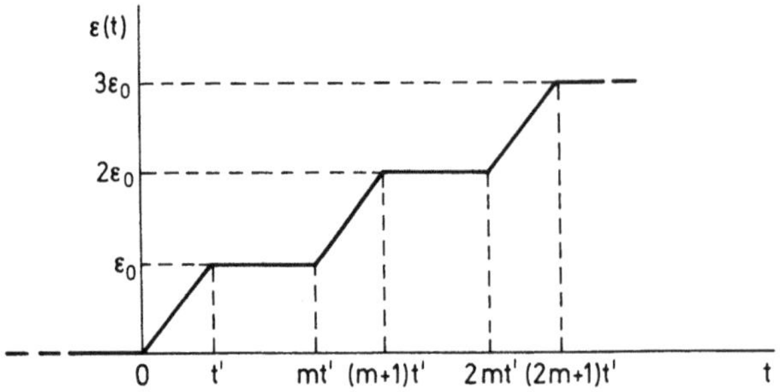

Fig. P7.2-1. Ramp staircase strain excitation

Problem 7.2-2. Obtain the equation for the stress in response to a regular staircase strain excitation working entirely in the time domain.

Problem 7.3-1. Show that, for N odd,

$$1 + 2 \sum_{n=1}^{N-1} (-1)^n E^n = \frac{2E^{N-1} + E^{-1} - 1}{E^{-1} + 1} \quad . \tag{1}$$

Problem 7.3-2. Consider the *half-sine strain pulse train* shown in Fig. P7.3-2. Find:

(A) the Laplace transform of the repeat function;
(B) the transform of the complete periodic excitation;
(C) the same when T = 2t′ (half rectified sine wave); and
(D) the same when T = t′ (full rectified sine wave).

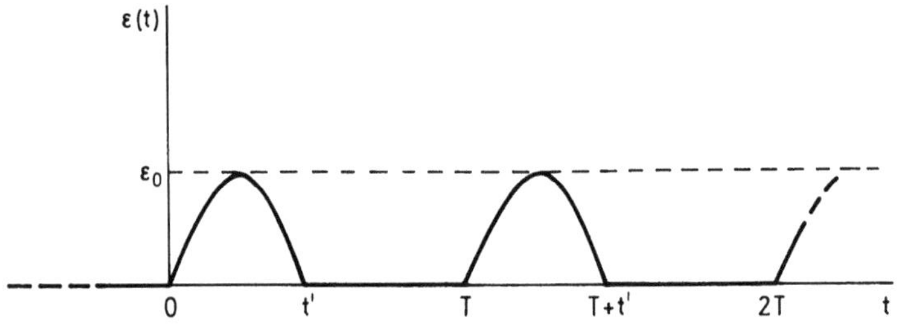

Fig. P7.3-2. Half-sine strain pulse train

Problem 7.3-3. Consider the *gate strain pulse train* shown in Fig. P7.3-3. In the steady-state this excitation is also known as a *square wave.*

(A) Find the Laplace transform of the strain for the special case when T = 2t′. Then,

(B) determine the components in the Fourier expansion of the steady-state response to the excitation in terms of the storage and loss moduli.

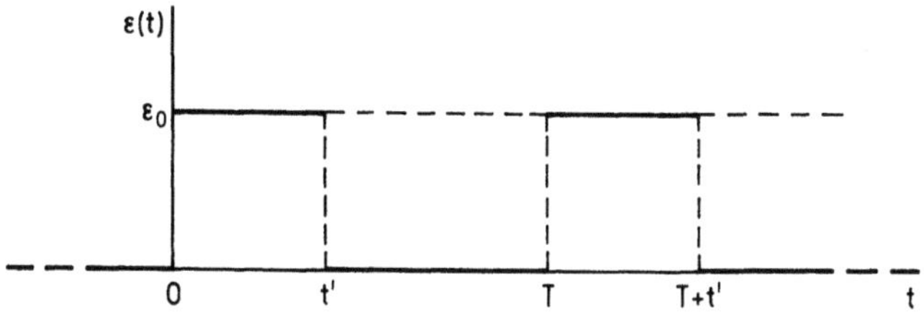

Fig. P7.3-3. Gate strain pulse train

Problem 7.3-4. In Sect. 7.3.2.1 it was shown that no steady-state can develop in the response to the triangular stress pulse train which is analogous to that shown in Fig. 7.3-2. Consider instead the stress pulse train represented in Fig. P7.3-4a and find the equations analogous to Eqs. (7.3-33) and (7.3-34). Interpret the results.

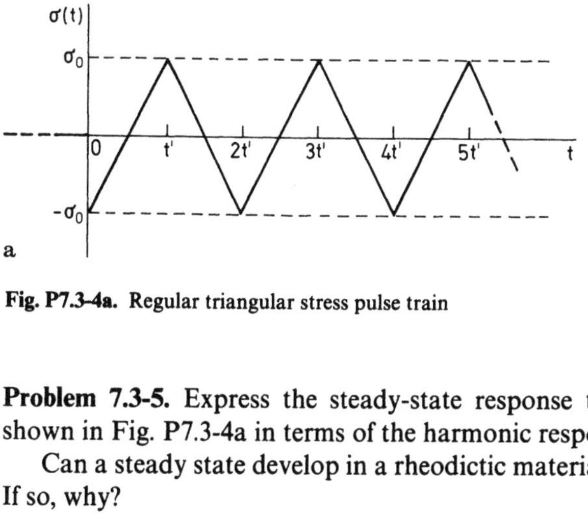

a

Fig. P7.3-4a. Regular triangular stress pulse train

Problem 7.3-5. Express the steady-state response to the triangular stress wave shown in Fig. P7.3-4a in terms of the harmonic response functions.

Can a steady state develop in a rheodictic material subjected to this excitation? If so, why?

[*Hint*: obtaining the excitation transform for the triangular stress wave forms part of Problem 7.3-4.]

References (Chap. 7)

1. G. Doetsch, *Guide to the Applications of the Laplace and Z-Transforms*, Van Nostrand Reinhold, London, 1971.
2. K. Yagii and N.W. Tschoegl, Trans. Soc. Rheol. *14*:1 (1970).
3. W. Kaplan, *Advanced Calculus*, Addison-Wesley, Reading, MA, 1952, p. 387ff.
4. M.F. Gardner and J.L. Barnes, *Transients in Linear Systems*, Vol. I, John Wiley and Sons, Inc., New York, 1942, pp. 175–176.

8. Interconversion of the Linear Viscoelastic Functions

Plus ça change, plus c'est
la même chose

Alphonse Karr

8.0 Introduction

As pointed out in the introduction to Chap. 4, all linear viscoelastic functions contain essentially the same information on the time dependent behavior of the material which they describe if they are known over the entire domain of their definition, i.e. over the complete range from zero to infinity of the time, t, frequency, ω, respondance time, τ, or transform variable, s. Thus, all linear viscoelastic response functions are, in principle, equivalent. The interrelatedness of all these responses is recognized upon the realization that they are all related to one or the other of the respondances, $\bar{Q}(s)$ or $\bar{U}(s)$, which in turn are reciprocal to each other. Thus

$$\bar{Q}(s)\bar{U}(s) = 1 \tag{8.0-1}$$

[cf. Eq. (2.1-19)] while, for the step responses,

$$G(t) = \mathscr{L}^{-1}\,\bar{Q}(s)/s \qquad \text{and} \qquad J(t) = \mathscr{L}^{-1}\,\bar{U}(s)/s\;, \tag{8.0-2}$$

[cf. Eqs. (2.3-5) and (2.3-8)] and, for the harmonic responses,

$$G^*(\omega) = [\bar{Q}(s)]|_{s=j\omega} \qquad \text{and} \qquad J^*(\omega) = [\bar{U}(s)]|_{s=j\omega} \tag{8.0-3}$$

[cf. Eqs. (2.5-13) and (2.5-17)]. Similar relations apply to $\eta(t)$, $\eta^*(\omega)$, $\chi(t)$, etc. An alternative way of recognizing the same fact arises through the realization that all experimental response functions are related to one or the other of the spectral functions, $H(\tau)$ or $L(\tau)$ (see the canonical representations of the experimental response functions in Sect. 4.1). As we shall show in Sect. 8.1.4, the spectra are, in their turn, related through Eq. (8.0-1). It follows that any linear viscoelastic function can, at least in theory, be converted into any other. These relations and interconversions will be examined in this chapter.

How does the need for interconversion arise? There are three chief reasons. First, viscoelastic responses naturally fall into two groups according to whether they arise from a strain or a stress excitation. In the former case we speak of *relaxation behavior*, in the latter of *retardation*, or more commonly, *creep behavior*. Generally,

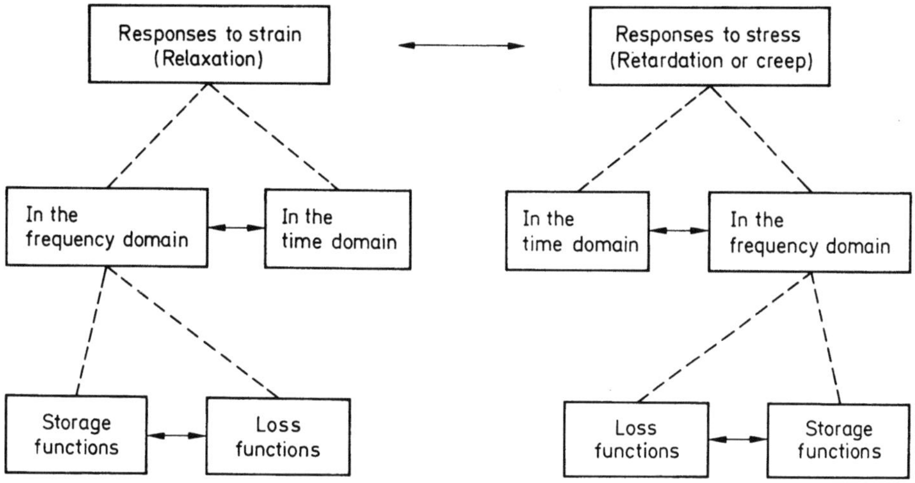

Fig. 8.0-1. Hierarchy of interconversion relations

relaxation behavior emphasizes short time processes while creep behavior weights those occuring at long times. Interconversion between these two groups of responses are therefore frequently desired. Second, for experimental reasons, a given stimulus cannot, in practice, elicit a response over the complete range of its definition. The range can, however, be extended by combining responses to different excitations. This normally requires interconversion between responses in the time and in the frequency domain. Finally, it may be desirable on occasion to interconvert between the real and imaginary parts of a complex response function within the frequency domain.

The hierarchical structure of these groups of interconversions is displayed in Fig. 8.0-1. Obviously, interconversion between and across the tiers of this hierarchy are also possible and are frequently needed. Table 8.0-1 assembles the theoretical relations between the most important of the response functions. These are entered as the complex frequency-dependent functions and the Laplace transforms of the time-dependent ones. The relations are based on Eqs. (8.0-1) to (8.0-3) as the central interconversion relations. The short hand notations

$$\bar{F}(\omega) = [\bar{F}(s)]_{s=j\omega} \qquad \text{and} \qquad F^*(s) = [F^*(j\omega)]_{j\omega=s} \tag{8.0-4}$$

are used in the table. In other words, when a Laplace transform (indicated by a bar) appears with ω instead of s as the argument, this implies that the transform variable s must be replaced by $j\omega$. Conversely, when a frequency-dependent function (indicated by an asterisk) appears with s instead of ω, $j\omega$ must be replaced by s [cf. also Eqs. (8.2-18) and (8.2-21)]

The upper left quadrant of the table contains the interrelations between the responses to strain excitations whereas the lower right quadrant lists the interrelations between responses to stress excitations. The cross-relations between these two groups are assembled in the upper right and lower left quadrants.

Table 8.0-1. Relations between the linear viscoelastic functions

Wanted \ Given	$\eta^*(\omega)$	$G^*(\omega)$	$\kappa^*(\omega)$	$\bar{\eta}(s)$	$\bar{G}(s)$	$\bar{Q}(s)$ $\bar{\kappa}(s)$	$\bar{U}(s)$ $\bar{\phi}(s)$	$\bar{J}(s)$	$\bar{\chi}(s)$	$\chi^*(\omega)$	$J^*(\omega)$	$\phi^*(\omega)$
$\eta^*(\omega)$	$\eta^*(\omega)$	$\frac{G^*(\omega)}{j\omega}$	$-\frac{\kappa^*(\omega)}{\omega^2}$	$j\omega\bar{\eta}(\omega)$	$\bar{G}(\omega)$	$\frac{\bar{Q}(\omega)}{j\omega}$	$\frac{1}{j\omega\bar{U}(\omega)}$	$\frac{1}{\omega^2\bar{J}(\omega)}$	$-\frac{1}{j\omega^3\bar{\chi}(\omega)}$	$-\frac{1}{\omega^2\chi^*(\omega)}$	$\frac{1}{j\omega J^*(\omega)}$	$\frac{1}{\phi^*(\omega)}$
$G^*(\omega)$	$j\omega\eta^*(\omega)$	$G^*(\omega)$	$\frac{\kappa^*(\omega)}{j\omega}$	$-\omega^2\bar{\eta}(\omega)$	$j\omega\bar{G}(\omega)$	$\bar{Q}(\omega)$	$\frac{1}{\bar{U}(\omega)}$	$\frac{1}{j\omega\bar{J}(\omega)}$	$-\frac{1}{\omega^2\bar{\chi}(\omega)}$	$\frac{1}{j\omega\chi^*(\omega)}$	$\frac{1}{J^*(\omega)}$	$\frac{j\omega}{\phi^*(\omega)}$
$\kappa^*(\omega)$	$-\omega^2\eta^*(\omega)$	$j\omega G^*(\omega)$	$\kappa^*(\omega)$	$-j\omega^3\bar{\eta}(\omega)$	$-\omega^2\bar{G}(\omega)$	$j\omega\bar{Q}(\omega)$	$\frac{j\omega}{\bar{U}(\omega)}$	$\frac{1}{\bar{J}(\omega)}$	$\frac{1}{j\omega\bar{\chi}(\omega)}$	$\frac{1}{\chi^*(\omega)}$	$\frac{j\omega}{J^*(\omega)}$	$-\frac{\omega^2}{\phi^*(\omega)}$
$\bar{\eta}(s)$	$\frac{\eta^*(s)}{s}$	$\frac{G^*(s)}{s^2}$	$\frac{\kappa^*(s)}{s^3}$	$\bar{\eta}(s)$	$\frac{\bar{G}(s)}{s}$	$\frac{\bar{Q}(s)}{s^2}$	$\frac{1}{s^2\bar{U}(s)}$	$\frac{1}{s^3\bar{J}(s)}$	$\frac{1}{s^4\bar{\chi}(s)}$	$\frac{1}{s^3\chi^*(s)}$	$\frac{1}{s^2J^*(s)}$	$\frac{1}{s\phi^*(s)}$
$\bar{G}(s)$	$\eta^*(s)$	$\frac{G^*(s)}{s}$	$\frac{\kappa^*(s)}{s^2}$	$s\bar{\eta}(s)$	$\bar{G}(s)$	$\frac{\bar{Q}(s)}{s}$	$\frac{1}{s\bar{U}(s)}$	$\frac{1}{s^2\bar{J}(s)}$	$\frac{1}{s^3\bar{\chi}(s)}$	$\frac{1}{s^2\chi^*(s)}$	$\frac{1}{sJ^*(s)}$	$\frac{1}{\phi^*(s)}$
$\bar{Q}(s)$ $\bar{\kappa}(s)$	$s\eta^*(s)$	$G^*(s)$	$\frac{\kappa^*(s)}{s}$	$s^2\bar{\eta}(s)$	$s\bar{G}(s)$	$\bar{Q}(s)$	$\frac{1}{\bar{U}(s)}$	$\frac{1}{s\bar{J}(s)}$	$\frac{1}{s^2\bar{\chi}(s)}$	$\frac{1}{s\chi^*(s)}$	$\frac{1}{J^*(s)}$	$\frac{s}{\phi^*(s)}$
$\bar{U}(s)$ $\phi(s)$	$\frac{1}{s\eta^*(s)}$	$\frac{1}{G^*(s)}$	$\frac{s}{\kappa^*(s)}$	$\frac{1}{s^2\bar{\eta}(s)}$	$\frac{1}{s\bar{G}(s)}$	$\frac{1}{\bar{Q}(s)}$	$\bar{U}(s)$	$s\bar{J}(s)$	$s^2\bar{\chi}(s)$	$s\chi^*(s)$	$J^*(s)$	$\frac{\phi^*(s)}{s}$
$\bar{J}(s)$	$\frac{1}{s^2\eta^*(s)}$	$\frac{1}{sG^*(s)}$	$\frac{1}{\kappa^*(s)}$	$\frac{1}{s^3\bar{\eta}(s)}$	$\frac{1}{s^2\bar{G}(s)}$	$\frac{1}{s\bar{Q}(s)}$	$\frac{\bar{U}(s)}{s}$	$\bar{J}(s)$	$s\bar{\chi}(s)$	$\chi^*(s)$	$\frac{J^*(s)}{s}$	$\frac{\phi^*(s)}{s^2}$
$\bar{\chi}(s)$	$\frac{1}{s^3\eta^*(s)}$	$\frac{1}{s^2G^*(s)}$	$\frac{1}{s\kappa^*(s)}$	$\frac{1}{s^4\bar{\eta}(s)}$	$\frac{1}{s^3\bar{G}(s)}$	$\frac{1}{s^2\bar{Q}(s)}$	$\frac{\bar{U}(s)}{s^2}$	$\frac{\bar{J}(s)}{s}$	$\bar{\chi}(s)$	$\frac{\chi^*(s)}{s}$	$\frac{J^*(s)}{s^2}$	$\frac{\phi^*(s)}{s^2}$
$\chi^*(\omega)$	$-\frac{1}{\omega^2\eta^*(\omega)}$	$\frac{1}{j\omega G^*(\omega)}$	$\frac{1}{\kappa^*(\omega)}$	$-\frac{1}{j\omega^3\bar{\eta}(\omega)}$	$-\frac{1}{\omega^2\bar{G}(\omega)}$	$\frac{1}{j\omega\bar{Q}(\omega)}$	$\frac{\bar{U}(\omega)}{j\omega}$	$\bar{J}(\omega)$	$j\omega\bar{\chi}(\omega)$	$\chi^*(\omega)$	$\frac{J^*(\omega)}{j\omega}$	$-\frac{\phi^*(\omega)}{\omega^2}$
$J^*(\omega)$	$\frac{1}{j\omega\eta^*(\omega)}$	$\frac{1}{G^*(\omega)}$	$\frac{j\omega}{\kappa^*(\omega)}$	$-\frac{1}{\omega^2\bar{\eta}(\omega)}$	$\frac{1}{j\omega\bar{G}(\omega)}$	$\frac{1}{\bar{Q}(\omega)}$	$\bar{U}(\omega)$	$j\omega\bar{J}(\omega)$	$-\omega^2\bar{\chi}(\omega)$	$j\omega\chi^*(\omega)$	$J^*(\omega)$	$\frac{\phi^*(\omega)}{j\omega}$
$\phi^*(\omega)$	$\frac{1}{\eta^*(\omega)}$	$\frac{j\omega}{G^*(\omega)}$	$-\frac{\omega^2}{\kappa^*(\omega)}$	$\frac{1}{j\omega\bar{\eta}(\omega)}$	$\frac{1}{\bar{G}(\omega)}$	$\frac{j\omega}{\bar{Q}(\omega)}$	$j\omega\bar{U}(\omega)$	$-\omega^2\bar{J}(\omega)$	$-j\omega^3\bar{\chi}(\omega)$	$-\omega^2\chi^*(\omega)$	$j\omega J^*(\omega)$	$\phi^*(\omega)$

Equations (8.2-18) to (8.2-21), to be introduced in Sect. 8.2.1, make it possible to derive the interrelations between the real and imaginary components of the complex functions and the time-dependent ones from the relations tabulated above. The procedure is made explicit in Problems 8.0-1 to 8.0-2.

For practical interconversions the theoretical relations contained in Table 8.0-1 are not useful generally. Such practical interconversion will be discussed in Sects. 8.1, 8.2, and 8.3. The first section deals with interconversions between relaxation and creep response functions including the spectral functions. Section 8.2 treats interconversion between time- and frequency-dependent functions within the two broad groups of relaxation and creep responses. Finally, Sect. 8.3 discusses interconversions within the frequency domain.

It is a common feature of all interconversion relations that they are *functional* relations (see Sect. 2.0). Consequently, a given value of the *target function* (the function *to* which one converts) will generally depend on *all* values of the *source function* (the function *from* which one converts) over the entire range of definition of its argument from 0 to ∞. Since the source function cannot be known over its entire range it must be extrapolated (perhaps graphically) to its limits. Alternatively, the integration can be carried out between two finite limits. This procedure introduces *truncation errors* which, under certain circumstances, may be quite severe.

8.1 Interconversion Between Relaxation and Creep Response Functions

The fundamental relation for these interconversions is Eq. (8.0-1). We shall consider, in turn, interconversion between the respondances, the harmonic responses, the step responses and, finally, the spectral functions.

8.1.1 Interconversion Between the Respondances

When interconversion between respondances is sought, the source function is usually in the form of a Wiechert or a Kelvin model with a finite set of moduli (or compliances) and relaxation (or retardation) times. Suppose one wishes to find $\bar{U}(s)$ when

$$\bar{Q}_{\bullet}(s) = G_g - \sum_k \frac{G_k}{1 + \tau_k s} \tag{8.1-1}$$

where the G_k and τ_k are known. They may, e.g., have been obtained from a collocation to an experimentally obtained data set (cf. Problem 3.6-2). $\bar{U}(s)$ is readily plotted out as a function of s according to Eq. (8.0-1) but this does not allow calculation of, say, J(t) or $J'(\omega)$ and $J''(\omega)$. If, therefore, it is desired to express $\bar{U}(s)$ in terms of a set of compliances, J_k, and retardation times, τ_k, it is necessary to solve the equation

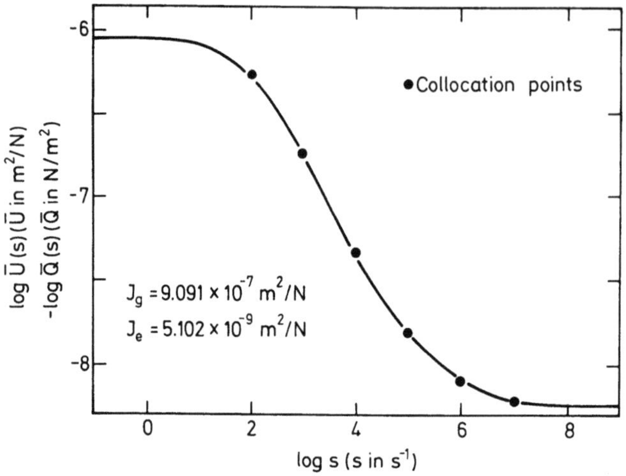

Fig. 8.1-1. $\bar{U}(s)$ obtained by collocation from $\bar{Q}(s)$

$$\bar{U}(s) = \frac{1}{G_g - \sum_k G_k/(1 + \tau_k s)} \ . \tag{8.1-2}$$

In all but the simplest cases (the standard linear solid and liquid models) this is not a practical procedure because it requires solution of an equation of the kth degree in s.

The interconversion can be accomplished, however, by fitting

$$\bar{U}(s) = J_g + \sum_k \frac{J_k}{1 + \tau_k s} \tag{8.1-3}$$

to $\bar{Q}(s) - \{s_k/\phi_f\}$ by collocation. An example of this is worked out in Problem 8.1-1. The *source function*, $\bar{Q}(s)$, as well as the *target function*, $\bar{U}(s)$, in that example are displayed in Fig. 8.1-1.

If $\bar{U}(s)$ is given and $\bar{Q}(s)$ is sought, one can proceed in an analogous manner by collocating to $\bar{U}(s)$. When the respondances are not given in the form of Eqs. (8.1-1) or (8.1-3) but by one of the mathematical models discussed in Sect. 6.1.4, one first collocates to the model and then carries out a second collocation.

Evidently, once $\bar{Q}(s)$ is known in the form of Eq. (8.1-1), the experimental response functions describing relaxation behavior are readily obtained by referring to Table 8.0-1. The same applies to $\bar{U}(s)$ when it is known in the form of Eq. (8.1-3). Thus, e.g., $\bar{\eta}(s) = \bar{Q}(s)/s^2$, $\bar{J}(s) = \bar{U}(s)/s$, etc.

8.1.2 Interconversion Between the Harmonic Responses

Interconversion between the harmonic responses is simple. From Table 8.0-1 the relation connecting the complex modulus and the complex compliance is

$$G^*(\omega)J^*(\omega) = 1 \ . \tag{8.1-4}$$

Since $G^*(\omega) = G'(\omega) + j\,G''(\omega)$ and $J^*(\omega) = J'(\omega) - j\,J''(\omega)$, we readily obtain

$$G'(\omega) = J'(\omega)/\tilde{J}^2(\omega) \qquad \text{and} \qquad G''(\omega) = J''(\omega)/\tilde{J}^2(\omega) \tag{8.1-5}$$

and, similarly,

$$J'(\omega) = G'(\omega)/\tilde{G}^2(\omega) \qquad \text{and} \qquad J''(\omega) = G''(\omega)/\tilde{G}^2(\omega) \tag{8.1-6}$$

where $\tilde{J}(\omega)$ and $\tilde{G}(\omega)$ are the absolute compliance and absolute modulus, respectively [cf. Eqs. (2.5-20)].

Other complex responses are easily interrelated using Table 8.0-1. Their real and imaginary parts can then be separated.

8.1.3 Interconversion Between the Step Responses

The interrelation connecting the stress relaxation modulus, G(t), and the creep compliance, J(t), is not as simple as that connecting $G^*(\omega)$ and $J^*(\omega)$. G(t) and J(t) are not each others reciprocal. In fact, it can be shown (see Problem 8.1-2) that

$$G(t)J(t) \leqq 1 \ , \tag{8.1-7}$$

the equality holding only in the limit as $t \to 0$ or ∞.

8.1.3.1 Theoretical Interrelations

We derive the theoretical interrelations between the time-dependent functions by contrasting the behavior in the transform plane with that in the time domain. In the former the behavior is simple. From Table 8.0-1 we have

$$\bar{G}(s)\bar{J}(s) = 1/s^2 \ , \tag{8.1-8}$$

i.e. the s-multiplied Laplace transforms of the step responses are reciprocal to each other. Multiplication by s on both sides of the equation yields

$$s\bar{G}(s)\bar{J}(s) = \bar{Q}(s)\bar{J}(s) = \bar{G}(s)\bar{U}(s) = 1/s \tag{8.1-9}$$

and a second multiplication leads to

$$s^2\bar{G}(s)\bar{J}(s) = \bar{Q}(s)\bar{U}(s) = 1 \ . \tag{8.1-10}$$

Equations (8.1-9) and (8.1-10) are, of course, included in Table 8.0-1.

How is this simple development in the transform plane paralleled in the time domain? Retransformation of Eq. (8.1-8) to the time domain using LTP (6) gives

$$\int_0^t G(t-u)J(u)du = \int_0^t G(u)J(t-u)du = p(t) \tag{8.1-11}$$

where p(t) is the slope function (see Appendix A2.4). Equations (8.1-11) represent the theoretical interrelations between the step responses.

Multiplication of a function $\bar{f}(s)$ by s in the transform plane corresponds, in the time domain, to differentiation of f(t) with respect to t, plus an additional delta function term taking care of the initial condition [cf. Eq. (A3-8)]. Hence, Eqs. (8.1-11) become

$$\frac{d}{dt}\int_0^t G(t-u)J(u)du = \frac{d}{dt}\int_0^t G(u)J(t-u)du = h(t) \tag{8.1-12}$$

since h(t) is the derivative of p(t) and the delta function terms vanish. With the help of Eq. (6.2-15), Eq. $(8.1\text{-}12)_1$ becomes

$$G_gJ(t) + \int_0^t J(u)\frac{d\,G(t-u)}{dt}du = G_gJ(t) + \int_0^t \frac{d\,G(u)}{du}J(t-u)du = h(t)\ . \tag{8.1-13}$$

These, then, are the two forms of the inverse transform of Eq. $(8.1\text{-}9)_1$. They are interrelated by a change of variable [cf. Eqs. (2.1-21)]. Retransformation into the transform plane yields

$$G_g\bar{J}(s) + [s\bar{G}(s) - G_g]\bar{J}(s) = 1/s \tag{8.1-14}$$

which, after cancellation of the $G_g\bar{J}(s)$ terms, reduces to Eq. $(8.1\text{-}9)_1$ as it should.

Alternatively, substituting

$$\frac{d\,G(t)}{dt} = Q(t) - G_g\delta(t) \tag{8.1-15}$$

[cf. Eq. (2.3-11)] into Eqs. (8.1-13) we obtain the equations

$$\int_0^t Q(t-u)J(u)du = \int_0^t Q(u)J(t-u)du = h(t) \tag{8.1-16}$$

which, upon retransformation, yield Eq. $(8.1\text{-}9)_2$.

Proceeding in a similar manner from Eq. $(8.1\text{-}12)_2$ instead of Eq. $(8.1\text{-}12)_1$ we obtain

$$J_gG(t) + \int_0^t G(u)\frac{d\,J(t-u)}{dt}du = J_gG(t) + \int_0^t G(t-u)\frac{d\,J(u)}{du}du = h(t)\ . \tag{8.1-17}$$

Retransformation again gives Eq. $(8.1\text{-}9)_1$. With

$$\frac{d\,J(t)}{dt} = U(t) - J_g\delta(t) \tag{8.1-18}$$

[cf. Eq. (2.3-12)] we obtain the equations

$$\int_0^t G(u)U(t-u)du = \int_0^t G(t-u)U(u)du = h(t) \tag{8.1-19}$$

which, retransformed, return Eq. $(8.1\text{-}9)_3$.

We now consider the equivalent, in the time domain, of a second multiplication by s in the transform plane. Differentiation of Eqs. $(8.1\text{-}13)_2$ and $(8.1\text{-}17)_2$ yields

$$\frac{d}{dt}\left(G_g J(t) + \int_0^t \frac{d\,G(u)}{du} J(t-u)du\right) + G_g J_g \delta(t)$$

$$= \frac{d}{dt}\left(J_g G(t) + \int_0^t G(t-u)\frac{d\,J(u)}{du}du\right) + J_g G_g \delta(t) = \delta(t) \tag{8.1-20}$$

since the delta function, $\delta(t)$, may be regarded as the derivative of the unit step function, $h(t)$, and $G(0)J(0) = G_g J_g$. Carrying out the indicated differentiation with the help of Eq. (6.2-15), using the identity $G_g J_g = 1$, and rearranging, gives

$$\int_0^t \frac{d\,G(u)}{du}\frac{d\,J(t-u)}{d(t-u)}du + G_g\frac{d\,J(t)}{dt} + J_g\frac{d\,G(t)}{dt}$$

$$= \int_0^t \frac{d\,G(t-u)}{d(t-u)}\frac{d\,J(u)}{du}du + G_g\frac{d\,J(t)}{dt} + J_g\frac{d\,G(t)}{dt} = 0 \tag{8.1-21}$$

as the inverse transforms of Eq. $(8.1\text{-}10)_1$. Retransformation of either Eq. $(8.1\text{-}21)_1$ or Eq. $(8.1\text{-}21)_2$ leads to

$$[s\bar{G}(s) - G_g][s\bar{J}(s) - J_g] + G_g[s\bar{J}(s) - J_g] + J_g[s\bar{G}(s) - G_g] = 0 \tag{8.1-22}$$

which, after cancellation, yields Eq. $(8.1\text{-}10)_1$.

Inserting Eqs. (8.1-15) and (8.1-18) into Eq. (8.1-21) we recover Eqs. (2.1-20),

$$\int_0^t Q(t-u)U(u)du = \int_0^t Q(u)U(t-u)du = \delta(t) \tag{8.1-23}$$

which, upon retransformation, yield Eq. $(8.1\text{-}10)_2$. Clearly, the behavior is much simpler in the complex plane than in the time domain.

Equations (8.1-11), (8.1-16), (8.1-19), (8.1-23) constitute the basic theoretical interrelations between the step and impulse responses in the time domain. The slope responses, $\eta(t)$ and $\chi(t)$, are easily related to these (cf. Problem 8.1-3).

8.1.3.2 Numerical Evaluation of the Convolution Integrals

The theoretical interconversions require that the source function be available in a Laplace transformable analytic form. How, then, do we proceed when the source function is given as a set of experimental data? In this case we may either collocate to express the data as a Prony-Dirichlet series (cf. Sect. 3.6.2), or we may evaluate the convolution integrals, notably Eqs. (8.1-11), numerically.

In the first case we make use of either the simple collocation method or the multidata method (cf. Sects. 3.6.2 and 3.6.3) to derive a set of parameters G_k and τ_k or J_k and τ_k from the source data, and then recollocate to obtain J_k and τ_k or G_k and τ_k. From these parameters we may reconstruct $J(t)$ or $G(t)$, or, indeed, any other viscoelastic function representing creep or relaxation behavior, as the case may be. In Problem 8.1-1 we have, in fact, used G_k and τ_k values obtained through collocation to $G(t)$ to calculate $\bar{Q}(s)$.

The convolution integrals may be solved numerically by the method of Hopkins and Hamming [1a, b]. In this method the interval of integration is divided into subintervals which are small enough so that a mean value of the target function over the subinterval can be taken outside of the integral. In this way a recursion relation can be set up from which the target function eventually emerges as a discrete set of values.

We shall consider first the calculation of J(t) if G(t) is given in terms of a set of n values $G(t_i)$. We use Eq. $(8.1\text{-}11)_1$, dividing the interval of integration into n equally spaced subintervals so that we obtain

$$p(t_n) = t_n = \sum_{i=1}^{i=n} \int_{t_{i-1}}^{t_i} J(u)G(t_n - u)du \tag{8.1-24}$$

where $t_0 = 0$ by definition. If the interval $t_i - t_{i-1}$ is chosen so that J(u) does not vary significantly over its range, J(u) may be taken out from under the integral sign with a suitably determined mean value. In the original version of the method, this mean value is $J(t_{i-1/2})$ where

$$t_{i-1/2} = (t_i + t_{i-1})/2 \ , \tag{8.1-25}$$

i.e. the mean value is chosen as the value of J(t) at the midpoint of the time interval. We thus obtain

$$t_n = \sum_{i=1}^{i=n} J(t_{i-1/2}) \int_{t_{i-1}}^{t_i} G(t_n - u)du \tag{8.1-26}$$

which can be integrated at once with the help of the relation

$$d\,\eta(t_n - u) = -G(t_n - u)du \tag{8.1-27}$$

[cf. Eq. (2.4-9)]. The integration yields

$$t_n = -\sum_{i=1}^{i=n} J(t_{i-1/2})[\eta(t_n - t_i) - \eta(t_n - t_{i-1})] \ . \tag{8.1-28}$$

Separating the nth term from the sum leads to

$$t_n = J(t_{n-1/2})\eta(t_n - t_{n-1}) - \sum_{i=1}^{i=n-1} J(t_{i-1/2})[\eta(t_n - t_i) - \eta(t_n - t_{i-1})] \tag{8.1-29}$$

because $\eta(0) = 0$. Solving for $J(t_{n-1/2})$ then yields the recursion relation

$$J(t_{n-1/2}) = \frac{t_n - \sum_{i=1}^{i=n-1} J(t_{i-1/2})[\eta(t_n - t_{i-1}) - \eta(t_n - t_i)]}{\eta(t_n - t_{n-1})} \tag{8.1-30}$$

for n > 1, with

$$J(t_{1/2}) = \frac{t_i}{\eta(t_i)} \tag{8.1-31}$$

as the starting value for n = 1. To obtain $\eta(t)$, we apply the trapezoidal rule of integration to

$$\eta(t_n) = \sum_{i=1}^{i=n} \int_{t_{i-1}}^{t_i} G(u)du \ . \tag{8.1-32}$$

This yields

$$\eta(t_n) = \tfrac{1}{2}\sum_{i=1}^{i=n} [G(t_i) + G(t_{i-1})](t_i - t_{i-1}) \ . \tag{8.1-33}$$

From this we obtain the recursion relation

$$\eta(t_n) = \eta(t_{n-1}) + \tfrac{1}{2}[G(t_n) + G(t_{n-1})](t_n - t_{n-1}) \tag{8.1-34}$$

with $\eta(t_0) = 0$.

Equations (8.1-30), (8.1-31), and (8.1-34) together are the equations originally proposed by Hopkins and Hamming. They are readily applied when the $G(t_i)$ are equally spaced along the time axis. Usually, however, $G(t)$ is equally spaced on the log t axis. In this case, the values of $\eta(t)$ needed in Eq. (8.1-30) must be found by interpolation using Eq. (8.1-33). The separate calculation of $\eta(t)$ can be avoided by applying the trapezoidal rule directly to Eq. (8.1-26). This yields

$$t_n = \tfrac{1}{2}\sum_{i=1}^{i=n} J(t_{i-1/2})[G(t_n - t_i) + G(t_n - t_{i-1})](t_i - t_{i-1}) \tag{8.1-35}$$

and leads to the recursion relation

$$J(t_{n-1/2}) = \frac{2t_n - \sum_{i=1}^{i=n-1} J(t_{i-1/2})[G(t_n - t_i) + G(t_n - t_{i-1})](t_i - t_{i-1})}{[G_g + G(t_n - t_{n-1})](t_n - t_{n-1})} \tag{8.1-36}$$

with

$$J(t_{1/2}) = \frac{2}{G_g + G(t_1)} \tag{8.1-37}$$

as the starting value. In these equations $G_g = G(t_0)$. Interpolation is needed to obtain the values $G(t_x)$ where $t_x = t_n - t_j$ and $j = i$ or $i - 1$. When the $G(t_j)$ are equally spaced by log h on the log(t)-axis,

$$t_x = t_n - t_j = (h^{n-j} - 1)t_j = (1 - h^{j-n})t_n \ . \tag{8.1-38}$$

Hence, t_x will lie between t_r and ht_r. The interpolation formula then is

$$G(t_x) = \frac{(t_{r+1} - t_x)G(t_r) + (t_x - t_r)G(t_{r+1})}{t_{r+1} - t_r} \ . \tag{8.1-39}$$

Since log h will lie between 0.1 and 0.5, simple rules can be given for finding t_r. For instance, with log h = 0.2, log t_r takes on the values given in the first column below when log t_n − log t_j takes the values listed in the second column

$\log t_r$	$\log t_n - \log t_j$
$\log t_j - 0.4$	0.2
$\log t_j$	0.4
$\log t_n - 0.2$	≥ 0.6

Table 6.1-1 contains values of log J(t) calculated by Eqs. (8.1-30) to (8.1-32) from the values of log G(t) in the same table. Calculations using Eqs. (8.1-36) and (8.1-37) gave identical results.

Instead of the midrange value, $J(t_{i-1/2})$, one may take out from under the integral the average value, $\frac{1}{2}[J(t_i) + J(t_{i-1})]$. When this is done and the trapezoidal rule, Eq. (8.1-33), is used for the integration, the result is the recursion relation

$$J(t_n) = -J(t_{n-1}) + \frac{4t_n - \sum_{i=1}^{i=n-1} [J(t_i) + J(t_{i-1})][G(t_n - t_i) + G(t_n - t_{i-1})](t_i - t_{i-1})}{[G_g + G(t_n - t_{n-1})](t_n - t_{n-1})} \tag{8.1-40}$$

with

$$J(t_1) = \frac{3 - G(t_1)/G_g}{G_g + G(t_1)} \tag{8.1-41}$$

as the starting value. Equation (8.1-40) is equivalent to Eqs. (8.1-30) and (8.1-36) except that it yields $J(t_n)$ instead of $J(t_{n-1/2})$.

Because the convolution integral, Eq. $(8.1\text{-}11)_1$, is symmetrical with respect to G(t) and J(t), the equivalent relations for G(t) are found simply by interchanging the two functions. As the same program can be used to obtain G(t) from J(t) and J(t) from G(t), the calculation can be checked [1b] by carrying out both computations.

8.1.3.3 *Empirical Interconversion Equations*

Before closing this section we mention three approximate relations for interrelating J(t) and G(t). The first is based on a simple shift in the time scale. The equations

$$J(t) = 1/G(t/\alpha) \qquad \text{and} \qquad G(t) = 1/J(\alpha t) \tag{8.1-42}$$

generally give satisfactory agreement. Thus, the data assembled in Table 6.1-1 gave a good fit with $\alpha = 1.828$. The difficulty lies in determining the value of α. Becker [2] proposed the relation

$$\alpha = \frac{G_g \log(1 + \sqrt{G_e/G_g})}{G_e \log(1 + \sqrt{G_g/G_e})} = \frac{J_e \log(1 + \sqrt{J_g/J_e})}{J_g \log(1 + \sqrt{J_e/J_g})} \tag{8.1-43}$$

which is derived from requiring that both sides of either of Eqs. (8.1-42) equal $\sqrt{G_g G_e} = 1/\sqrt{J_g J_e}$ for the standard linear solid. Unfortunately this does not appear to provide a satisfactory value for α when the dispersion is broader than that of the standard linear model. For the data in Table 6.1-1, Eq. (8.1-43) gives $\alpha = 15.05$ which is much too large to produce an adequate fit. Mathematical models such as the Cole-Cole functions (see Sect. 6.1.1) cannot be used to estimate α even if one makes the reasonable assumption that the *broadness parameters* of both the source and the target step responses should be the same (i.e. that $m = m'$) because there does not seem to be any reasonable way to estimate the *location parameter*, t_0'.

By the second approximate interrelation

$$G(t)J(t) = \frac{\sin \pi m}{\pi m} \ . \tag{8.1-44}$$

This aproximation is based on Eq. $(6.2\text{-}10)_1$ the Laplace transform of which is

$$\bar{G}(s) = G_0\Gamma(1 - m)(t_0 s)^{-m}/s \ . \tag{8.1-45}$$

Using the relation $\bar{J}(s) = 1/s^2\bar{G}(s)$ and retransforming leads to

$$J(t) = \frac{(t/t_0)^m}{G_0\Gamma(1 - m)\Gamma(1 + m)} = \frac{\sin \pi m}{G_0 \pi m}(t/t_0)^m \tag{8.1-46}$$

from which Eq. (8.1-44) follows [cf. Eq. $(6.2\text{-}10)_2$]. This approximation is clearly valid only in the regions in which G(t) and J(t) can be well represented by a straight line of slope $|m|$ in logarithmic coordinates.

Finally, Denby [3] found that

$$1/J(t) \simeq G(t)[1 + (\pi^2/6)m^2(t)] \tag{8.1-47}$$

where m(t) is the slope of log G(t) vs. log t.

8.1.4 Interconversion Between the Spectral Functions

The exact interrelations between the spectral response functions will be derived in Sect. 8.1.4.1. They have theoretical rather than practical significance. However, approximate interrelations may be useful at times and these are discussed in Sects. 8.1.4.2 and 8.1.4.3.

8.1.4.1 Theoretical Interrelations

The basis of the interconversion of the relaxation and retardation spectra is still the interrelation between the respondances, $\bar{Q}(s)\bar{U}(s) = 1$. By Eqs. (4.1-36) and (4.1-38)

$$\bar{Q}_\bullet(s) = G_g - \tilde{M}(s) \qquad \text{and} \qquad \bar{U}(s) = J_g + \tilde{N}(s) + \{\phi_f/s\} \ . \tag{8.1-48}$$

where $\tilde{M}(s)$ and $\tilde{N}(s)$ are the Stieltjes transforms of the frequency spectra $M(\zeta)$ and $N(\zeta)$. Multiplying these equations together, we obtain

$$[\tilde{N}(s) + \{\phi_f/s\}]G_g = [J_g + \tilde{N}(s) + \{\phi_f/s\}]\tilde{M}(s) \tag{8.1-49}$$

as the basic relation between the Stieltjes transforms of the spectral distribution functions in the complex plane. Solving for $\tilde{M}(s)$ and for $\tilde{N}(s)$ in turn, we obtain

$$\tilde{M}(s) = \frac{[\tilde{N}(s) + \{\phi_f/s\}]G_g}{J_g + \tilde{N}(s) + \{\phi_f/s\}} \tag{8.1-50}$$

for the transform of the relaxation spectrum, and

$$\tilde{N}(s) = \frac{[J_g + \{\phi_f/s\}]\tilde{M}(s) - \{\phi_f/s\}G_g}{G_g - \tilde{M}(s)} \tag{8.1-51}$$

for that of the retardation spectrum, each in terms of the other. Inversion of the transforms yields relations between the frequency spectra from which the time spectra can be obtained in turn.

To invert Eqs. (8.1-50) and (8.1-51) we use the inversion formula for the Stieltjes transform, given by Eq. (A5-18). For the relaxation spectrum this becomes

$$M(\zeta) = \frac{\tilde{M}[\zeta \exp(-j\pi)] - \tilde{M}[\zeta \exp(j\pi)]}{2\pi k} . \qquad (8.1\text{-}52)$$

Using Eq. (8.1-50) then leads to

$$M(\zeta) = \frac{G_g}{2\pi j}\left(\frac{\tilde{N}[\zeta \exp(-j\pi)] + \{\phi_f/\zeta\}}{J_g + \tilde{N}[\zeta \exp(-j\pi)] + \{\phi_f/\zeta\}} - \frac{\tilde{N}[\zeta \exp(j\pi)] + \{\phi_f/\zeta\}}{J_g + \tilde{N}[\zeta \exp(j\pi)] + \{\phi_f/\zeta\}}\right) . \qquad (8.1\text{-}53)$$

In the flow terms, $\{\phi_f/s\}$, we were able to use the relations $\zeta \exp(\pm j\pi) = \zeta$. In $\tilde{N}(s)$, because we do not know what function of s we are dealing with, we must use Eqs. (A5-11) and (A5-12). Here, these become

$$\tilde{N}[\zeta \exp(\mp j\pi)] = \fint_0^\infty \frac{N(\xi)}{\xi - \zeta} d\xi \pm j\pi N(\zeta) . \qquad (8.1\text{-}54)$$

Making use of them in Eq. (8.1-53),

$$M(\zeta) = \frac{N(\zeta)}{\left(J_g + \fint_0^\infty \frac{N(\xi)}{\xi - \zeta} d\xi + \{\phi_f/\zeta\}\right)^2 + \pi^2 N^2(\zeta)} . \qquad (8.1\text{-}55)$$

In an analogous way we derive

$$N(\zeta) = \frac{M(\zeta)}{\left(G_g - \fint_0^\infty \frac{M(\xi)}{\xi - \zeta} d\xi\right)^2 + \pi^2 M^2(\zeta)} . \qquad (8.1\text{-}56)$$

Changing back to the relaxation and retardation times from the frequencies using Eqs. (4.1-31) and (4.1-32) with $\xi = u^{-1}$ and $\zeta = \tau^{-1}$ finally gives the relations between the time spectra as

$$H(\tau) = \frac{L(\tau)}{\left(J_g + \fint_{-\infty}^\infty \frac{\tau L(u)}{\tau - u} d\ln u + \{\phi_f \tau\}\right)^2 + \pi^2 L^2(\tau)} \qquad (8.1\text{-}57)$$

and

$$L(\tau) = \frac{H(\tau)}{\left(G_g - \fint_{-\infty}^\infty \frac{\tau H(u)}{\tau - u} d\ln u\right)^2 + \pi^2 H^2(\tau)} . \qquad (8.1\text{-}58)$$

Using the relations

$$J_e^{\{o\}} - J_g = \int_0^\infty u^{-1} L(u) du \quad \text{and} \quad G_g - \{G_e\} = \int_0^\infty u^{-1} H(u) du \qquad (8.1\text{-}59)$$

[cf. Eqs. (4.6-21) and (4.6-22)] we obtain the alternative forms

$$H(\tau) = \frac{L(\tau)}{\left(J_e^{\{o\}} - \displaystyle\fint_0^\infty \frac{L(u)}{u-\tau}du + \{\phi_f\tau\}\right)^2 + \pi^2 L^2(\tau)} \tag{8.1-60}$$

and

$$L(\tau) = \frac{H(\tau)}{\left(\{G_e\} + \displaystyle\fint_0^\infty \frac{H(u)}{u-\tau}du\right)^2 + \pi^2 H^2(\tau)} \; . \tag{8.1-61}$$

All these relations are *functionals* (cf. Sect. 2.0) containing Cauchy principal values of improper integrals. To calculate one from the other, it is necessary for each point desired to evaluate the integral graphically or numerically over the entire range of the source spectrum. Unless the spectrum is given by a mathematical model, it would be known only approximately for the reasons detailed in Chap. 4. Such calculations have nevertheless, been made [4]. Generally, however, one would interconvert the experimental responses and obtain the desired spectrum from the target response by the approximations discussed in Chap. 4.

8.1.4.2 Approximate Interconversion of the Spectra

The stumbling blocks in the interconversion of the spectra are the integrals in Eqs. (8.1-60) and (8.1-61). Approximate interconversion relations can be obtained through the use of the mathematical models for the spectra (see Sect. 6.3) to evaluate the integrals. The box distribution and the wedge distribution (with m = 1/2), are particularly easy to use.

For a *box distribution* of relaxation times [cf. Eq. (6.3-7)], the integral in Eq. (8.1-61) becomes

$$H_0 \fint_{\tau_{min}}^{\tau_{max}} \frac{du}{u-\tau} = H_0 \ln\left|\frac{\tau_{max}-\tau}{\tau_{min}-\tau}\right| = H(\tau)\ln\left|\frac{\tau_{max}-\tau}{\tau_{min}-\tau}\right| \; . \tag{8.1-62}$$

The argument of the logarithm will be negative because $\tau_{min} < \tau$. However, the principal value of $\ln(-x) = \ln(x) + j(2n+1)\pi$ is $\ln(x)$. Thus, rejecting the imaginary part, we may either use the absolute value of the argument as indicated in Eq. (8.1-62) or, simpler still, change the sign of the denominator. Thus,

$$L(\tau) = \begin{cases} \dfrac{H(\tau)}{[\{G_e\} + H(\tau)\,b(\tau)]^2 + \pi^2 H^2(\tau)} \, , & \tau_{min} \le \tau \le \tau_{max} \\ 0 \, , & \tau < \tau_{min}, \tau > \tau_{max} \end{cases} \tag{8.1-63}$$

where

$$b(\tau) = \ln\frac{\tau_{max}-\tau}{\tau-\tau_{min}} \; . \tag{8.1-64}$$

A box distribution of retardation times, using Eq. (6.3-19), yields

$$H(\tau) = \begin{cases} \dfrac{L(\tau)}{[J_e^{\{o\}} - L(\tau)\, b(\tau) + \{\phi_f \tau\}]^2 + \pi^2 H^2(\tau)} \,, & \tau_{min} \leq \tau \leq \tau_{max} \\ 0 \,, & \tau < \tau_{min}, \tau > \tau_{max} \end{cases} \tag{8.1-65}$$

The *wedge distribution* with m = 1/2 [cf. Eq. (6.3-29)], again using the principal value of the logarithm, furnishes the same expressions except that b(τ) is replaced by

$$w(\tau) = \ln \frac{(\sqrt{\tau_{max}} - \sqrt{\tau})(\sqrt{\tau} + \sqrt{\tau_{min}})}{(\sqrt{\tau_{max}} + \sqrt{\tau})(\sqrt{\tau} - \sqrt{\tau_{min}})} \,. \tag{8.1-66}$$

An approximate interconversion based on the *power law spectrum* is discussed in Problem 8.1-4.

8.1.4.3 Approximate Calculation of the Spectra from the Step Responses

We examine now the problem of obtaining the retardation spectrum from relaxation data, and of obtaining the relaxation spectrum from creep data. As we have seen, it is not an easy matter to first calculate one spectrum, and then the other one from it. If the data are available in form of the storage and loss functions, one can rather easily first convert one set of data into the other (cf. Sect. 8.1.2) and then calculate the spectrum. If, however, the data are at hand in form of the relaxation modulus, or the creep compliance, one would first have to interconvert G(t) and J(t). This is not easy. In this case, therefore, one might wish to resort to a method due to Smith [5], in which one first obtains H(τ) from G(t), or L(τ) from J(t), and then uses both H(τ) and G(t), or L(τ), and J(t), respectively. We present Smith's development here in a slightly more general form.

Taking the relation for L(τ) first, we divide the range of integration in Eq. (8.1-61) into three parts and write

$$\int\!\!\!\!-_0^\infty \frac{H(u)}{u-\tau} du = -\int_0^{\tau_1} \frac{H(u)}{\tau} du + \int\!\!\!\!-_{\tau_1}^{\tau_2} \frac{H(u)}{u-\tau} du + \int_{\tau_2}^\infty \frac{H(u)}{u} du \,. \tag{8.1-67}$$

The limits τ_1 and τ_2 are to be chosen so that the assumptions used in Eq. (8.1-67), viz. that $u - \tau \simeq -\tau$ when $u < \tau_1$, and $u - \tau \simeq u$ when $u > \tau_2$, are reasonably satisfied. Next, changing the dummy variable τ in the canonical representations of G′(ω) and G″(ω) to u, we recast them in the form

$$G'(\omega)|_{\omega=1/\tau} = G'(1/\tau) = \{G_e\} + \int_0^\infty H(u) \frac{u}{u^2+\tau^2} du \tag{8.1-68}$$

and

$$G''(\omega)|_{\omega=1/\tau} = G''(1/\tau) = \int_0^\infty H(u) \frac{\tau}{u^2+\tau^2} du \,. \tag{8.1-69}$$

Now, we split the range of integration in both equations as in Eq. (8.1-67). This yields

$$G'(1/\tau) = \{G_e\} + \int_{\tau_1}^{\tau_2} \frac{uH(u)}{u^2+\tau^2} du + \int_{\tau_2}^{\infty} \frac{H(u)}{u} du \tag{8.1-70}$$

and

$$G''(1/\tau) = \int_{0}^{\tau_1} \frac{H(u)}{\tau} du + \int_{\tau_1}^{\tau_2} \frac{\tau H(u)}{u^2+\tau^2} du \tag{8.1-71}$$

where the integral from 0 to τ_1 in Eq. (8.1-70), and that from τ_2 to ∞ in Eq. (8.1-71) have been omitted because they contribute negligibly to $G'(1/\tau)$ and $G''(1/\tau)$, respectively. In addition, the approximations $u^2 + \tau^2 \simeq u^2$ when $u > \tau_2$, and $u^2 + \tau^2 = \tau^2$ when $u < \tau_1$, have also been used.

With the help of Eqs. (8.1-70) and (8.1-71), Eq. (8.1-67) becomes

$$\begin{aligned} \fint_{0}^{\infty} \frac{H(u)}{u-\tau} du &\simeq G'(1/\tau) - \{G_e\} - G''(1/\tau) + \fint_{\tau_1}^{\tau_2} \frac{H(u)}{u-\tau} du \\ &+ \int_{\tau_1}^{\tau_2} \frac{\tau H(u)}{u^2+\tau^2} du - \int_{\tau_1}^{\tau_2} \frac{u H(u)}{u^2+\tau^2} du \ . \end{aligned} \tag{8.1-72}$$

Now let the limits be equally spaced around $\tau = 1$ on the log(τ)-axis. Then $\tau_2 = \alpha\tau$ and $\tau_1 = \tau/\alpha$. Using the box distribution again to evaluate the first and third integrals, these cancel each other out whatever the value of α because both* become ln α. The value of the remaining second integral depends on the choice of α and on the model selected for evaluating the integral. Smith used $\alpha = 10$ and again used the *box distribution*. In that case the integral takes the value $I_b = 1.3715$. We use the same value for α but consider that the *wedge distribution* is a better approximation to the spectrum than the box distribution over the two-decade interval from 0.1τ to 10τ. Substitution of $H(\tau) = H_0(\tau/\tau_0)^{-m}$ into the integral allows us to take $H(\tau)$ outside upon the change of variable $u = \tau/\sqrt{x}$. We find

$$\int_{\tau_1}^{\tau_2} \frac{\tau H(u)}{u^2+\tau^2} du = H(\tau) \int_{\alpha^{-2}}^{\alpha^2} \frac{x^{(m-1)/2}}{2(x+1)} dx = I_w(m)\, H(\tau) \ . \tag{8.1-73}$$

Numerical evaluation of the integral [6], $I_w(m)$, gave the results in Table 8.1-1 from which the nearest value of $I_w(m)$ may be selected. We now have

$$\fint_{0}^{\infty} \frac{H(u)}{u-\tau} du \simeq G'(1/\tau) - \{G_e\} - G''(1/\tau) + I_w(m)\, H(\tau) \tag{8.1-74}$$

and substitution into Eq. (8.1-61) leads to

$$L(\tau) \simeq \frac{H(\tau)}{[G'(1/\tau) - G''(1/\tau) + I_w(m)\, H(\tau)]^2 + \pi^2 H^2(\tau)} \ . \tag{8.1-75}$$

In the original reference [5] I_b is used instead of $I_w(m)$.

Equation (8.1-75) is not particularly useful by itself since, if $G'(\omega)$ and $G''(\omega)$ are known, it would be simpler and more accurate to obtain $J'(\omega)$ and/or $J''(\omega)$ from

---

* In the first integral the principal value of $\ln(-\alpha)$ must be taken.

Table 8.1-1. The functions $\Psi_G(m)$, $\Psi_J(m)$, and their components

m	$I_w(m)$	$\Gamma(m)$	$\Psi_G(m)$	$\Gamma(-m)$	$\Psi_J(m)$
0.05	3.682	19.468	2.659	−20.631	−2.717
0.10	3.396	9.513	2.334	−10.687	−2.452
0.15	3.147	6.220	2.040	−7.417	−2.219
0.20	2.931	4.591	1.772	−5.821	−2.017
0.25	2.745	3.625	1.524	−4.902	−1.842
0.30	2.587	2.991	1.292	−4.327	−1.691
0.35	2.453	2.546	1.071	−3.956	−1.561
0.40	2.343	2.218	0.855	−3.723	−1.451
0.45	2.253	1.968	0.638	−3.591	−1.360
0.50	2.183	1.772	0.411	−3.545	−1.285
0.55	2.131	1.616	0.162	−3.579	−1.225
0.60	2.095	1.489	−0.124	−3.697	−1.178
0.65	2.076	1.385	−0.473	−3.917	−1.145
0.70	2.072	1.298	−0.923	−4.274	−1.122
0.75	2.082	1.225	−1.548	−4.835	−1.112
0.80	2.107	1.164	−2.489	−5.739	−1.111
0.85	2.145	1.112	−4.081	−7.318	−1.119
0.90	2.197	1.069	−7.323	−10.570	−1.135
0.95	2.262	1.035	−17.214	−20.494	−1.160

them by Eqs. (8.1-5) and calculate $L(\tau)$ from the compliances by the approximation methods discussed in Chap. 4. A more useful relation is obtained through replacement of $G'(\omega)$ and $G''(\omega)$ by $G(t)$. This can be achieved through the use of the *power law relaxation spectrum*, Eq. (6.3-63). Subtracting Eq. (6.3-56) from Eq. (6.3-57) gives

$$G'(1/\tau) = G(t)|_{t=\tau} + (G_g - \{G_e\}]m(\tau/\tau_0)^{-m}[\tfrac{\pi}{2}\csc \tau m/2 - \Gamma(m)] \quad . \qquad (8.1\text{-}76)$$

Introducing the power law spectrum

$$H(\tau) = [G_g - \{G_e\}]m(\tau/\tau_0)^{-m} \qquad (8.1\text{-}77)$$

into Eqs. (8.1-76) and (6.3-58), and substituting the results into Eqs. (8.1-75) leads to

$$L(\tau) \simeq \frac{H(\tau)}{[G(t)|_{t=\tau} + \Psi_G(m)\,(H)(\tau)]^2 + \pi^2 H^2(\tau)} \qquad (8.1\text{-}78)$$

where

$$\Psi_G(m) = \tfrac{\pi}{2}(\csc \pi m/2 - \sec \pi m/2) - \Gamma(m) + I_w(m) \quad . \qquad (8.1\text{-}79)$$

The function $\Psi_G(m)$ is given in Table 8.1-1 for convenience. Thus, in order to obtain $L(\tau)$ from $G(t)$ data, one may first calculate $H(\tau)$ by one of the approximation methods discussed in Chap. 4 and then use Eq. (8.1-78).

A similar relation may be derived for the calculation of $H(\tau)$ from $J(t)$ and $L(\tau)$. Starting from Eq. (8.1-60), using Eqs. (4.1-30) and (4.1-24), and making the same approximations as before in the integrals, we find

$$H(\tau) \simeq \frac{L(\tau)}{[J'_\bullet(1/\tau) + J''(1/\tau) - I_w(m)\,L(\tau)]^2 + \pi^2 L^2(\tau)} \quad . \qquad (8.1\text{-}80)$$

We now subtract Eq. (6.3-84) from Eq. (6.3-85) to obtain

$$J'(1/\tau) = J(t)|_{t=\tau} + [J_e^{\circ} - J_g]m(\tau/\tau_0)^m[\tfrac{\pi}{2}\csc\pi m/2 + \Gamma(-m)] - \{\phi_f\tau\} \ . \tag{8.1-81}$$

Introducing the *power law retardation spectrum* [cf. Eq. (6.3-91]

$$L(\tau) = [J_e^{\{o\}} - J_g]m(\tau/\tau_0)^m \tag{8.1-82}$$

into Eqs. (8.1-81) and (6.3-86) and substituting the results into Eq. (8.1-80) then yields

$$H(\tau) = \frac{L(\tau)}{[J(t)|_{t=\tau} + \Psi_J(m)\, L(\tau)]^2 + \pi^2 L^2(\tau)} \tag{8.1-83}$$

where

$$\Psi_J(m) = \tfrac{\pi}{2}(\csc \pi m/2 + \sec \pi m/2) + \Gamma(-m) - I_w(m) \ . \tag{8.1-84}$$

The function $\Psi_J(m)$ is also tabulated in Table 8.1-1.

Despite the various approximations used in their derivation Eqs. (8.1-78) and (8.1-83) give quite reasonable results [5].

8.2 Interconversion Between Time- and Frequency-Dependent Response Functions

The next lower tier in the hierarchy displayed in Fig. 8.0-1 refers to the interconversion of time dependent response functions into frequency dependent ones and *vice versa*. These interconversions are carried out *within* the group of responses to stress or to strain, respectively.

8.2.1 Theoretical Interrelations

The theoretical relations which furnish the harmonic responses as functions of the step responses are obtained by eliminating the respondances between Eqs. (8.0-2) and (8.0-3). Recasting Eq. $(8.0\text{-}2)_1$ in the form $\bar{Q}(s) = s\,\mathscr{L}[G(t); s]$ and inserting this into Eq. $(8.0\text{-}3)_1$ gives

$$G^*(\omega) = j\omega\,\mathscr{F}[G(t); j\omega] \tag{8.2-1}$$

because, by Eq. (A4-19), replacing the Laplace transform variable, s, by the complex variable, jω, furnishes the (generalized) Fourier transform. An analogous derivation from Eqs. $(8.0\text{-}2)_2$ and $(8.0\text{-}3)_2$ yields

$$J^*(\omega) = j\omega\,\mathscr{F}[J(t); j\omega] \ . \tag{8.2-2}$$

By Eqs. (8.2-1) and (8.2-2) the complex harmonic responses are the jω-multiplied (generalized) Fourier transforms of the step responses. Equations (8.2-1) and (8.2-2) may be condensed into the single relation

$$\Phi^*(\omega) = j\omega\, \mathscr{F}[\Phi(t); j\omega] = j\omega \int_0^\infty \Phi(t)\exp(-j\omega t)dt\ . \tag{8.2-3}$$

Separation of the real and imaginary parts according to Eq. (A4-2) leads to the (generalized) Fourier sine and cosine transforms

$$\Phi'(\omega) = \omega\, \mathscr{F}_s[\Phi(t); \omega] = \omega \int_0^\infty \Phi(t)\sin \omega t\, dt \tag{8.2-4}$$

and

$$\Phi''(\omega) = \omega\, \mathscr{F}_c[\Phi(t); \omega] = \omega \int_0^\infty \Phi(t)\cos \omega t\, dt\ . \tag{8.2-5}$$

The inverse transforms are obtained as

$$\Phi(t) = \mathscr{F}^{-1}[\Phi^*(\omega)/j\omega; t] = \frac{1}{2\pi} \int_{-\infty}^{\infty} \frac{\Phi^*(\omega)}{j\omega} \exp(j\omega t)d\omega \tag{8.2-6}$$

and as

$$\Phi(t) = \mathscr{F}_s^{-1}[\Phi'(\omega)/\omega; t] = \frac{2}{\pi} \int_0^\infty \frac{\Phi'(\omega)}{\omega} \sin \omega t\, d\omega \tag{8.2-7}$$

and

$$\Phi(t) = \mathscr{F}_c^{-1}[\Phi''(\omega)/\omega; t] = \frac{2}{\pi} \int_0^\infty \frac{\Phi''(\omega)}{\omega} \cos \omega t\, d\omega \tag{8.2-8}$$

respectively. In these equations

$$\Phi(t) = \begin{cases} G(t) \\ J(t) \end{cases} \quad \text{and} \quad \Phi^*(\omega) = \begin{cases} G^*(\omega) \\ J^*(\omega) \end{cases}$$

with (8.2-9)

$$\Phi'(\omega) = \begin{cases} G'(\omega) \\ J'(\omega) \end{cases} \quad \text{and} \quad \Phi''(\omega) = \begin{cases} G''(\omega) \\ -J''(\omega) \end{cases}$$

must be selected as corresponding pairs.

Equations (8.2-6) to (8.2-8) are not valid generally when used with Eqs. (8.2-9). Let us, for instance, obtain the creep compliance of the standard linear liquid model from its storage and its loss compliance. We find

$$J(t) = \mathscr{F}_s^{-1}\left[\frac{J_g}{\omega} + \frac{J}{\omega(1+\omega^2\tau^2)}\right] = J_g + J[1 - \exp(-t/\tau)] \tag{8.2-10}$$

and

$$J(t) = \mathscr{F}_c^{-1}\left[\frac{J\tau}{1+\omega^2\tau^2} + \frac{\phi_f}{\omega^2}\right] = J \exp(-t/\tau) - \phi_f t\ . \tag{8.2-11}$$

To effect the transform inversion, we first applied partial fraction decomposition to the second term in brackets in Eq. (8.2-10), and then used FTPs (2) and (5). In Eq. (8.2-11) FTPs (5) and (3) were used in the transform inversion. Comparison with Eq. (3.4-153) shows that in Eq. (8.2-10) the flow term is missing while Eq. (8.2-11) has the wrong sign and, in addition, lacks the steady-state compliance, $J_e^\circ = J_g + J$. The difficulty arises because $J'(\omega)$ contains no information on ϕ_f while $J''(\omega)$ contains none on J_g. Similarly, $G''(\omega)$ knows nothing about G_e.

To remedy these deficiencies we resort to an alternative derivation of Eqs. (8.2-3) and (8.2-6). This derivation is based on the elimination of the spectral functions between the canonical representations of the complex harmonic and the step responses. By Eq. $(4.1\text{-}63)_1$

$$G^*(\omega) - \{G_e\} = j\omega\, \mathscr{J}[\Xi(\zeta); j\omega)] = j\omega\, \mathscr{F}\{\mathscr{L}[\Xi(\zeta); t]; j\omega\} \;, \tag{8.2-12}$$

the last relation following because the $\mathscr{J}$-transform is a Laplace transform followed by a Fourier transform (cf. Appendix A5.3).

But

$$\mathscr{L}[\Xi(\zeta); t] = G(t) - \{G_e\} \tag{8.2-13}$$

by Eq. $(4.1\text{-}51)_1$. Hence, we obtain

$$G^*(\omega) = \{G_e\} + j\omega\, \mathscr{F}[G(t) - \{G_e\}] \tag{8.2-14}$$

which we use in lieu of Eq. (8.2-1). An entirely analogous derivation, starting from Eq. $(4.1\text{-}69)_1$ and using Eq. (4.1-56), leads to

$$J_\bullet^*(\omega) = J_e^{\{\circ\}} - j\omega\, \mathscr{F}[J_e^{\{\circ\}} - J_\bullet(t) + \{\phi t\}] + \{\phi_f/j\omega\} \tag{8.2-15}$$

as a replacement for Eq. (8.2-2). Both equations, their inverses, and the real and imaginary parts can be subsumed under Eqs. (8.2-3) to (8.2-8) by amending Eqs. (8.2-9) to read

$$\Phi(t) = \begin{cases} G(t) - \{G_e\} \\ J_e^{\{\circ\}} - J_\bullet(t) + \{\phi_f t\} \end{cases} \quad \text{and} \quad \Phi^*(\omega) = \begin{cases} G^*(\omega) - \{G_e\} \\ J_e^{\{\circ\}} - J_\bullet^*(\omega) + \{\phi_f/j\omega\} \end{cases}$$

with (8.2-16)

$$\Phi'(\omega) = \begin{cases} G'(\omega) - \{G_e\} \\ J_e^{\{\circ\}} - J'_\bullet(\omega) \end{cases} \quad \text{and} \quad \Phi''(\omega) = \begin{cases} G''(\omega) \\ J''(\omega) - \{\phi_f/\omega\} \end{cases} .$$

If we now derive the creep compliance of the standard linear liquid model from its storage and its loss compliance again, we obtain the correct relations. With Eqs. (8.2-16), Eqs. (8.2-3) to (8.2-8) are quite generally valid. Equations (8.2-16) *must* be used when the argument of the inverse Fourier transform is $J''(\omega)$, or $J'(\omega)$ in the rheodictic, or $G''(\omega)$ in the arrheodictic case. Otherwise Eqs. (8.2-9) *may* be applied.

There are no equations corresponding to Eqs. (8.2-14) and (8.2-15) from which $G_\bullet^*(\omega)$ or $J^*(\omega)$ could be obtained in terms of $G(t)$ or $J(t)$, respectively. Naturally, the inverse relations cannot be derived either. However, the relations

$$G_\bullet(t) = G_g - \frac{2}{\pi} \int_0^\infty \frac{G_g - G'_\bullet(\omega)}{\omega} \sin \omega t \, d\omega \tag{8.2-17a}$$

and

$$J(t) = J_g + \frac{2}{\pi}\int_0^\infty \frac{J'(\omega) - J_g}{\omega}\sin\omega t\, d\omega + \{\phi_f t\} \tag{8.2-17b}$$

as well as their inverses are readily obtained (see Problem 8.2-4). In addition we also have the relations

$$G_\bullet(t) = G_g - \frac{2}{\pi}\int_0^\infty \frac{G''(\omega)}{\omega}(1 - \cos\omega t)\, d\omega \tag{8.2-17c}$$

and

$$J(t) = J_g + \frac{2}{\pi}\int_0^\infty \frac{J''(\omega) - \{\phi_f/\omega\}}{\omega}(1 - \cos\omega t)\, d\omega + \{\phi_f t\} \tag{8.2-17d}$$

(see Problem 8.2-5).

It would consume too much space to list here all the possible interrelations that can be derived in the indicated way between the time-dependent and the frequency-dependent response functions. The procedure is demonstrated in Problems 8.2-1 and 8.2-3 which display the most important of these relations.

Extension of the interconversion procedures to interrelations involving response functions other than those just discussed is readily accomplished with the aid of Table 8.0-1 (see, e.g., Problem 8.2-6). We now recognize that the short hand notation introduced with Eqs. (8.0-4) is equivalent to

$$F^*(\omega) = \bar{F}(\omega) = \mathscr{F}\, F(t) \tag{8.2-18}$$

$$F'(\omega) = \mathscr{Re}\,\bar{F}(\omega) = \mathscr{F}_c\, F(t) \tag{8.2-19}$$

$$F''(\omega) = \mathscr{Im}\,\bar{F}(\omega) = -\mathscr{F}_s\, F(t) \tag{8.2-20}$$

and

$$F(t) = \mathscr{L}^{-1}F^*(s) = \begin{cases} \mathscr{F}^{-1}F^*(\omega) \\ \mathscr{F}_c^{-1}F'(\omega) \\ \mathscr{F}_s^{-1}F''(\omega)\ . \end{cases} \tag{8.2-21}$$

It turns out that $\bar{F}(\omega)$ is the (generalized) Fourier transform of $F(t)$ while $F^*(s)$ is the Laplace transform of the (generalized) inverse Fourier transform of $F^*(\omega)$. These equations extend the use of Table 8.0-1 to the real and imaginary parts of the complex response functions. Equations (8.2-19) and (8.2-20) follow from Eq. (8.2-18) by Eq. (A4-2), and Eqs. $(8.2\text{-}21)_2$ and $(8.2\text{-}21)_3$ from Eqs. $(8.2\text{-}21)_1$ by Eq. (A4-19). As an example of the use of Eq. (8.2-18) we find

$$G^*(\omega) = \bar{Q}(\omega) = \mathscr{F}\, Q(t) = \int_0^\infty Q(t)\exp(-j\omega t)dt\ . \tag{8.2-22}$$

This is Eq. (2.6-22). As anticipated in Sect. 2.6, the complex modulus is the (generalized) Fourier transform of the response to a unit impulse of strain. For the complex compliance we find

$$J^*(\omega) = \bar{U}(\omega) = \mathscr{F}\, U(t) = \int_0^\infty U(t)\exp(-j\omega t)dt \tag{8.2-23}$$

which is Eq. (2.6-24). We note that, by Eqs. (2.3-11) and (2.3-12) we have the alternate equations

$$G^*(\omega) = G_g + \int_0^{\infty} \dot{G}(t)\exp(-j\omega t)dt \tag{8.2-24}$$

and

$$J^*(\omega) = J_g + \int_0^{\infty} \dot{J}(t)\exp(-j\omega t)dt \tag{8.2-25}$$

as well as their real and imaginary parts (see also Problem 8.2-7).

When the source function is in the form of a Wiechert or a Kelvin model (see Sect. 3.5.2), interconversion is trivial. Table A4-1 shows that

$$\sum_n \frac{\omega^2\tau_n^2}{1+\omega^2\tau_n^2} = \omega\,\mathscr{F}_s \sum_n \exp(-t/\tau_n) \quad \text{and} \quad \sum_n \frac{\omega\tau_n}{1+\omega^2\tau_n^2} = \omega\,\mathscr{F}_c \sum_n \exp(-t/\tau_n) \tag{8.2-26}$$

and this is all that is needed for the interconversion between $G(t)$ and $G'(\omega)$ or $G''(\omega)$, and between $J_\bullet(t)$ and $J'_\bullet(\omega)$ or $J''(\omega)$, in either direction. Thus, one may simply apply the collocation (or multidata) method to the source data to obtain the parameters G_n, τ_n, and then use Eqs. (3.5-7), (3.5-9) and (3.5-10), or obtain J_n, τ_n and use Eqs. (3.5-16), (3.5-18) and (3.5-19).

8.2.2 Interconversions Requiring Numerical Integration

In principle, direct numerical integration of a discrete set of equally spaced source data would also furnish the desired target function. Unfortunately, the integration is fraught with difficulties. The problem has been thoroughly investigated by Schwarzl and Struik* [7] who derived *finite Fourier transforms* which give satisfactory convergence.

Alternatively, the source data may be fitted to one of the mathematical models discussed in Chap. 6 and the target function computed through numerical integration.

Marvin [8] introduced interconversion relations which are based on the fact that $G(t)$ and $G'(1/t)$, or $J(t)$ and $J'(1/t)$ are similar in shape and differ significantly only over a relatively small interval of their arguments. Subtracting $G(t)$ and $G'(\omega)$, as well as $J(t)$ and $J'(\omega)$ from each other, we obtain

$$G'(\omega) = G(t) + \int_{-\infty}^{\infty} H(\tau)K_\Delta(t/\tau)d\ln\tau\big|_{t=1/\omega} \tag{8.2-27}$$

$$G(t) = G'(\omega) - \int_{-\infty}^{\infty} H(\tau)K_\Delta(\omega\tau)d\ln\tau\big|_{\omega=1/t} \tag{8.2-28}$$

---

* pron. Shvartsl and Stroyk

$$J'(\omega) = J(t) - \int_{-\infty}^{\infty} L(\tau)K_\Delta(t/\tau)d\ln\tau - \{\phi_f t\}|_{t=1/\omega} \tag{8.2-29}$$

and

$$J(t) = J'(\omega) + \int_{-\infty}^{\infty} L(\tau)K_\Delta(\omega\tau)d\ln\tau + \{\phi_f/\omega\}|_{\omega=1/t} \tag{8.2-30}$$

where

$$K_\Delta(t/\tau) = 1/(1 + t^2/\tau^2) - \exp(-t/\tau) \tag{8.2-31}$$

and

$$K_\Delta(\omega\tau) = \omega^2\tau^2/(1 + \omega^2\tau^2) - \exp(-1/\omega\tau) \tag{8.2-32}$$

are the differences (hence the subscript Δ) of the storage and step response kernels, and will be referred to as the *difference kernels*.

$K_\Delta(t/\tau)$ and $K_\Delta(\omega\tau)$ are mirror images of each other when plotted as functions of the logarithm of their arguments (cf. Figs. 8.2-1 and 8.2-2). They have the nature of the *intensity functions* discussed at length in Chap. 4. Their maximum value is 0.1941 and this is attained when $-\log t/\tau = \log \omega\tau = 0.3375$.

From Eqs. (8.2-31) and (8.2-32) we may immediately deduce the useful inequalities $G(t) \leq G'(1/t)$, $G'(\omega) \geq G(1/\omega)$, and $J(t) \geq J'(1/t)$, $J'(\omega) \leq J(1/\omega)$. When $t = 0$, or $\omega = \infty$, both terms in Eqs. (8.2-31) and (8.2-32) are unity. As t increases, or ω decreases, the exponential decreases more rapidly than the fraction. Hence, the inequalities follow.

Since the integrals represent only relatively minor corrections, first order approximations are usually sufficient for the determination of the spectra. Nevertheless, numerical integrations are still required. Over the linear portions of the responses in logarithmic coordinates, the *power law spectra* will give reasonable approximations [9]. The resulting *Catsiff–Tobolsky relations* have, in fact, already been used in Sect. 8.1.4.3 as Eqs. (8.1-76) and (8.1-81).

8.2.3 Interconversion by Kernel Matching

We now seek to avoid the integration by replacing the integrals with judiciously selected combinations of the response functions. The difference kernel, $K_\Delta(\omega\tau)$, is plotted in Fig. 8.2-1 as function of $\log \omega\tau$ and is compared with the kernel, $K''(\omega\tau) = \omega\tau/(1 + \omega^2\tau^2)$, of the loss functions. Clearly, $K''(\omega\tau)$ is a crude approximation to the difference kernel. Replacing $K_\Delta(\omega\tau)$ by $K''(\omega\tau)$ in Eqs. (8.2-28) and (8.2-30) gives

$$G_0(t) = G'(\omega) - G''(\omega)|_{\omega=1/t} \quad \text{and} \quad J_0(t) = J'(\omega) + J''(\omega)|_{\omega=1/t}\,. \tag{8.2-33}$$

However, Fig. 8.2-1 shows that these interrelations will be satisfactory only for $\log \omega\tau > 1$ where $K''(\omega\tau)$ and $K_\Delta(\omega\tau)$ coincide. Our task is to improve the coincidence at lower values of $\omega\tau$. To this end we introduce the operator

$$\mathscr{P}''_{\Delta k} = \sum_{n=1}^{n=k} a''_n L_{b''_n} \tag{8.2-34}$$

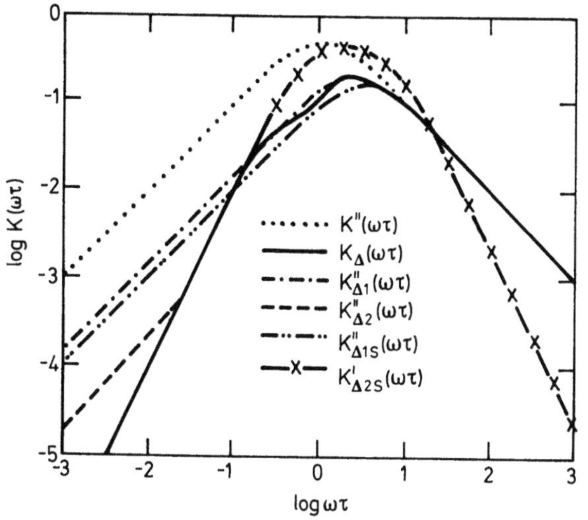

Fig. 8.2-1. Comparison of $K_\Delta(\omega\tau)$ with $K''(\omega\tau)$ and other approximation kernels

where $L_m f(x) = f(mx)$ is the multiplication operator introduced by Eq. (4.2-73), and the constants a_n'' and b_n'' are chosen so that

$$K_{\Delta k}''(\omega\tau) = \mathscr{P}_{\Delta k}''\left[\frac{\omega\tau}{1+\omega^2\tau^2}\right] = \sum_{n=1}^{n=k} a_n'' \frac{b_n''\omega\tau}{1+(b_n''\omega\tau)^2} \tag{8.2-35}$$

will form a better approximation to $K_\Delta(\omega\tau)$ than $K''(\omega\tau)$. The constant b_n'' shifts the kernel along the $\log(\omega\tau)$-axis. The constants a_n'', which are chosen to be smaller than unity, attenuate it. Substitution of $K_{\Delta k}''(\omega\tau)$ into Eq. (8.2-28) gives

$$G_k(t) = G'(\omega) - \int_{-\infty}^{\infty} H(\tau)K_{\Delta k}''(\omega\tau)d\ln\tau|_{\omega=1/\tau} = G'(\omega) - \mathscr{P}_{\Delta k}''[G''(\omega)]|_{\omega=1/t}$$

where (8.2-36)

$$\mathscr{P}_{\Delta k}''[G''(\omega)] = \sum_{n=1}^{n=k} a_n'' G''(b_n''\omega) \; . \tag{8.2-37}$$

We recognize Eq. $(8.2\text{-}33)_1$ as a sort of zeroth order approximation in which $a_0 = b_0 = 1$. We call the procedure leading to Eq. (8.2-36) *kernel matching* because it consists in matching the difference kernel in Eq. (8.2-28) with the kernel $K_{\Delta k}''(\omega\tau)$ resulting from operating on the kernel of the loss functions, $K''(\omega\tau)$, with the operator $\mathscr{P}_{\Delta k}''$. The procedure resembles that used in Sect. 4.3.4.2 [cf. Eq. (4.3-149)]. There, however, the operator had the task to turn the loss kernel into an approximation to the delta-function. Here, the role of the operator is to shape it so that it conforms as closely as possible to the difference kernel.

The method of kernel matching was introduced by Ninomiya and Ferry [10] to obtain approximations to the step responses from the harmonic ones. It was then extended to the problem of obtaining approximations to the storage functions and to the loss functions from the step responses by Yagii and Maekawa* [11] who

* pron. Yugg-ee-ee and Mah-eh-kawa.

introduced the operator

$$\mathscr{P}_{\Delta k} = \sum_{n=1}^{n=k} a_n(L_{b_n} - L_{c_n}) \tag{8.2-38}$$

where $b_n < c_n$, to this end. When this operator is applied to $\exp(-t/\tau)$, two exponentials, shifted by different amounts along the $\log(t/\tau)$-axis, are subtracted from one another. This operation turns the kernel $K(t/\tau)$, which is of the Z-type, into one of the Λ-type (cf. Sect. 6.1). Thus, the kernel

$$K_{\Delta k}(t/\tau) = \mathscr{P}_{\Delta k}[\exp(-t/\tau)] = \sum_{n=1}^{n=k} a_n[\exp(-b_n t/\tau) - \exp(-c_n t/\tau)] \tag{8.2-39}$$

will have the required bell shape. We note that $\mathscr{P}_{\Delta k}[1 - \exp(-t/\tau)] = -\mathscr{P}_{\Delta k}[\exp(-t/\tau)]$.

Further contributions were made by Schwarzl and Struik [7] and by Schwarzl [12b] who also introduced the operator

$$\mathscr{P}'_{\Delta k} = \sum_{n=1}^{n=k} a'_n(L_{b'_n} - L_{c'_n}) \tag{8.2-40}$$

in which $b_n > c'_n$. This operator furnishes an approximation

$$K'_{\Delta k}(\omega\tau) = \mathscr{P}'_{\Delta k}\left[\frac{\omega^2\tau^2}{1 + \omega^2\tau^2}\right] = \sum_{n=1}^{n=k} a'_n\left(\frac{(b'_n\omega\tau)^2}{1 + (b'_n\omega\tau)^2} - \frac{(c'_n\omega\tau)^2}{1 + (c'_n\omega\tau)^2}\right) \tag{8.2-41}$$

to $K_\Delta(\omega\tau)$ from the storage kernel, $K'(\omega\tau)$. Here, the subtraction of two storage kernels shifted with respect to one another on the $\log(\omega t)$-axis turns the storage kernel, which is of the S-type, into one of the Λ-type. We note that $\mathscr{P}'_{\Delta k}[1/(1 + \omega^2\tau^2)] = -\mathscr{P}'_{\Delta k}[\omega^2\tau^2/(1 + \omega^2\tau^2)]$. $\mathscr{P}'_{\Delta k}$ may be used in lieu of, or in conjunction with, $\mathscr{P}''_{\Delta k}$.

There does not seem to be a unique optimal way to determine the constants in the operators $\mathscr{P}_{\Delta k}$, $\mathscr{P}'_{\Delta k}$, and $\mathscr{P}''_{\Delta k}$ in general. They must therefore be obtained largely by trial and error and their choice then depends on the skill of the manipulator. The test of any one of the derived kernels $K_{\Delta k}(t/\tau)$, $K'_{\Delta k}(\omega\tau)$, or $K''_{\Delta k}(\omega\tau)$ lies in the comparison with the difference kernel as in Fig. 8.2-1. Clearly, any valid approximation must reduce as closely as possible to $K_\Delta(\omega\tau)$ for $\log \omega\tau < 1$.

To derive their first order *approximations to the step responses from the storage and loss functions*, Ninomiya and Ferry [10] let $a''_1 = b''_1$ and determined b''_1 as that value which minimizes the absolute value of the difference between $K_\Delta(\omega\tau)$ and $K''_{\Delta 1}(\omega\tau)$. From the condition

$$\frac{\partial}{\partial b''_1}\left\{\int_{-\infty}^{\infty} [K_\Delta(\omega\tau) - K''_{\Delta 1}(\omega\tau)]^2 d\ln\tau\right\} = 0 \tag{8.2-42}$$

the authors obtained $b''_1 = 0.381$. They then replaced this value by 0.40 which is close enough and has the advantage that its logarithm is -0.40. This is convenient when, as is often the case, the data are spaced by $\log h = 0.2$ on the logarithmic scale. The kernel approximation,

$$K''_{\Delta 1}(\omega\tau) = 0.40\frac{0.40\,\omega\tau}{1 + (0.40\,\omega\tau)^2}, \tag{8.2-43}$$

is compared with $K_\Delta(\omega\tau)$ in Fig. 8.2-1. It is clearly a better approximation to $K_\Delta(\omega\tau)$ than is $K''(\omega\tau)$. The approximation is quite good for $\log \omega\tau > -1$. To reduce the remaining difference, Ninomiya and Ferry added a second term to obtain the kernel approximation

$$K''_{\Delta 2}(\omega\tau) = 0.40 \frac{0.40\,\omega\tau}{1 + (0.40\,\omega\tau)^2} - 0.014 \frac{10\,\omega\tau}{1 + (10\,\omega\tau)^2} \tag{8.2-44}$$

which is also plotted for comparison in Fig. 8.2-1. Substitution of $K''_{\Delta 2}(\omega\tau)$ for $K_\Delta(\omega\tau)$ in Eq. (8.2-28) then furnishes the (second order) approximation of Ninomiya and Ferry as

$$G_2(t) = G'(\omega) - 0.40\,G''(0.40\,\omega) + 0.014\,G''(10\,\omega)|_{\omega=1/t}\ . \tag{8.2-45}$$

In analogy to Eq. (8.2-36), the compliance approximations are obtain from

$$J_k(t) = J'(\omega) + \mathscr{P}''_{\Delta k}[J''(\omega) - \{\phi_f/\omega\}] + \{\phi_f/\omega\}|_{\omega=1/t} \tag{8.2-46}$$

which is equivalent to

$$J_k(t) = J'(\omega) + \mathscr{P}''_{\Delta k}[J''(\omega)] + \{\phi_f/\omega\}\left(1 - \sum_n a_n/b_n\right)\ . \tag{8.2-47}$$

For the Ninomiya-Ferry approximation $a_1/b_1 = 1$, and $a_2/b_2 = 0.0014$. Hence, the last term in Eq. (8.2-47) can be omitted [the flow term is, of course, implicitly included in $J''(\omega)$] and we have

$$J_2(t) = J'(\omega) + 0.40\,J''(0.40\,\omega) - 0.014\,J''(10\,\omega)|_{\omega=1/t}\ . \tag{8.2-48}$$

Schwarzl and Struik [7] obtained a number of approximations of similar form. One of these [cf. their Eq. (IV.2)] is

$$G_{1S}(t) = G'(\omega) - 0.337\,G''(0.323\,\omega)|_{\omega=1/t}\ . \tag{8.2-49}$$

Its kernel approximation, $K'_{\Delta 1S}(\omega\tau)$, is included in Fig. 8.2-1. Equation (8.2-49) is only a minor improvement on the first order approximation of Ninomiya and Ferry.

As an example of an approximation to G(t) derived solely from $G'(\omega)$ with the aid of the operator $\mathscr{P}'_{\Delta k}$ [cf. Eq. (8.2-47)], we list one given by Schwarzl and Struik [7] in the form

$$G_{2S}(1.44\,t) = G'(\omega) - 0.40[G'(1.59\,\omega) - G'(0.193)]|_{\omega=1/t}\ . \tag{8.2-50}$$

The approximation to the difference kernel, $K'_{\Delta 2S}(\omega\tau)$ (see Problem 8.2-8) is also plotted in Fig. 8.2-1. In contrast to the difference kernel approximations based on the loss functions, those derived from the storage functions show excellent agreement with $K_\Delta(\omega\tau)$ at *low* frequencies but are less satisfactory at *high* frequencies.

Let us now turn to the other side of these relations, viz. the *approximation to the storage functions from the step responses.* These are derived from the exact Marvin relations, Eqs. (8.2-27) and (8.2-29), by replacing $K_\Delta(t/\tau)$ by $K_{\Delta k}(t/\tau)$. Executing this replacement in Eq. (8.2-27) and using Eq. (8.2-39) leads to

$$G'_k(\omega) = G(t) + \int_{-\infty}^{\infty} H(\tau)K_{\Delta k}(t/\tau)\,d\ln\tau|_{t=1/\omega}\ . \tag{8.2-51}$$

But the integral equals

$$\mathscr{P}_{\Delta k}[G(t) - \{G_e\}] = \mathscr{P}_{\Delta k}[G(t)] \tag{8.2-52}$$

because $\mathscr{P}_{\Delta k}[\{G_e\}] = 0$ since the second term of the operator cancels the first. Hence the approximation of kth order becomes

$$G'_k(\omega) = G(t) + \mathscr{P}_{\Delta k}[G(t)]|_{t=1/\omega} \,. \tag{8.2-53}$$

Starting from Eq. (8.2-29) we have

$$J'_k(\omega) = J(t) + \int_{-\infty}^{\infty} L(\tau)\mathscr{P}_{\Delta k}[1 - \exp(-t/\tau)]d\ln\tau - \{\phi_f t\}|_{t=1/\omega} \,. \tag{8.2-54}$$

Now the integral equals

$$\mathscr{P}_{\Delta k}[J(t) - J_g - \{\phi_f t\}] = \mathscr{P}_{\Delta k}[J(t)] - \{\phi_f t\} \sum_n a_n(b_n - c_n) \tag{8.2-55}$$

and the kth approximation becomes

$$J'_k(\omega) = J(t) + \mathscr{P}_{\Delta k}[J(t)] - \{\phi_f t\}\left[1 + \sum_n a_n(b_n - c_n)\right]\Bigg|_{t=1/\omega} \,. \tag{8.2-56}$$

Yagii and Maekawa [11] proposed first and second order approximations with the constants b_n and c_n determined in such a way that $\log b_1 = 0.20$, $\log c_1 = 0.40$, $\log b_2 = -0.60$, and $\log c_2 = -0.40$. Their second order approximations are

$$G'_2(\omega) = G(t) + 1.080[G(1.585t) - G(2.512t)] + 0.159[G(0.251t) - G(0.398t)]|_{t=1/\omega} \tag{8.2-57}$$

and

$$J'_2(\omega) = J(t) + 1.080[J(1.585t) - J(2.512t)] + 0.159[J(0.251t) - J(0.398t)] + 0.025\{\phi_f t\}|_{t=1/\omega} \,. \tag{8.2-58}$$

The kernels $K_{\Delta 1}(t/\tau)$ and $K_{\Delta 2}(t/\tau)$ are compared with $K_\Delta(t/\tau)$ in Fig. 8.2-2.

Approximation to the loss functions from the step responses can also be obtained with the help of the operator $\mathscr{P}_{\Delta k}$ with, of course, different choices for the constants. These must be selected so that $K_{\Delta k}(t/\tau)$ matches $K''(\omega\tau)|_{\omega=1/t} = K''(t/\tau)$, not $K_\Delta(t/\tau)$, as closely as possible. From

$$G''_k(\omega) = \int_{-\infty}^{\infty} H(\tau)K_{\Delta k}(t/\tau)d\ln\tau|_{t=1/\omega} \tag{8.2-59}$$

and

$$J''_k(\omega) = \int_{-\infty}^{\infty} L(\tau)K_{\Delta k}(t/\tau)d\ln\tau + \{\phi_f/\omega\}|_{t=1/\omega} \tag{8.2-60}$$

we derive

$$G''_k(\omega) = \mathscr{P}_{\Delta k}[G(t)]|_{t=1/\omega} \tag{8.2-61}$$

and

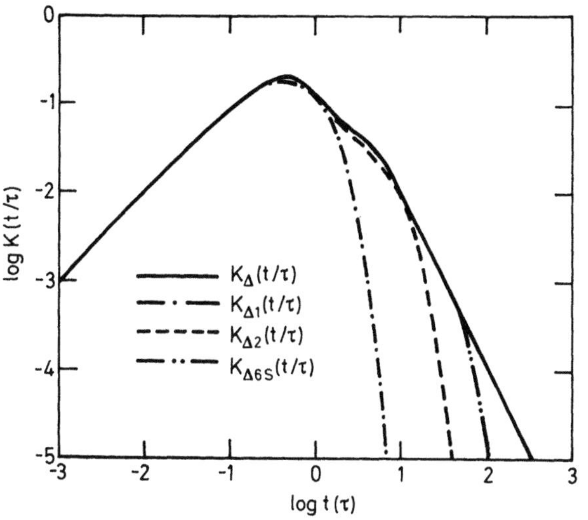

Fig. 8.2-2. Comparison of $K_\Delta(t/\tau)$ with several approximation kernels

$$J_k''(\omega) = \mathscr{P}_{\Delta k}[J(t) - \{\phi_f t\}]|_{t=1/\omega} + \{\phi_f/\omega\} \; . \tag{8.2-62}$$

The latter is equivalent to

$$J_k''(\omega) = \mathscr{P}_{\Delta k}[J(t)] + \{\phi_f t\}\left[1 - \sum_n a_n(b_n - c_n)\right]\Bigg|_{t=1/\omega} \; . \tag{8.2-63}$$

Yagii and Maekawa [11] gave first and second order approximations with the constants b_n and c_n determined so that $\log b_1 = -0.20$, $\log c_1 = 0$, $\log b_2 = -1.00$, and $\log c_2 = -0.80$. Their second order approximation for $G''(\omega)$ takes the form

$$G_2''(\omega) = 2.70[G(0.631t) - G(t)] + 0.794[G(0.100t) - G(0.159t)]|_{t=1/\omega} \; . \tag{8.2-64}$$

In the corresponding equation for $J''(\omega)$, the sign of the a_n's must be changed since $\mathscr{P}_{\Delta k}[1 - \exp(-t/\tau)] = -\mathscr{P}_{\Delta k}[\exp(-t/\tau)]$. Thus

$$J_2''(\omega) = 2.70[J(t) - J(0.631t)] + 0.794[J(0.159t) - J(0.100t)] - 0.043\{\phi_f t\}|_{t=1/\omega} \; . \tag{8.2-65}$$

The approximation kernels $K_{\Delta 1}(t/\tau)$ and $K_{\Delta 2}(t/\tau)$ are compared with $K''(\omega\tau)|_{\omega=1/t}$ in Fig. 8.2-3.

Both here and in Fig. 8.2-2 the agreement is excellent for $\log t/\tau < 0$ but becomes progressively worse as $\log t/\tau$ increases. The agreement can be improved through approximations of higher order. Unfortunately, the higher the order of the approximation, the narrower the range of the target function compared with that of the source function for the reasons pointed out in connections with the approximations to the spectral functions at the end of Sect. 4.4. There, this was not a matter of great concern. Here, the loss of range can be quite serious.

Schwarzl proposed a number of approximations of various orders for [12a] $G(t)$ from $G'(\omega)$ and $G''(\omega)$, for [12b] $J(t)$ from $J'(\omega)$ and $J''(\omega)$, for [12c] $G'(\omega)$ and $G''(\omega)$ from $G(t)$, and for [12d] $J'(\omega)$ and $J''(\omega)$ from $J(t)$. In all of these, the source

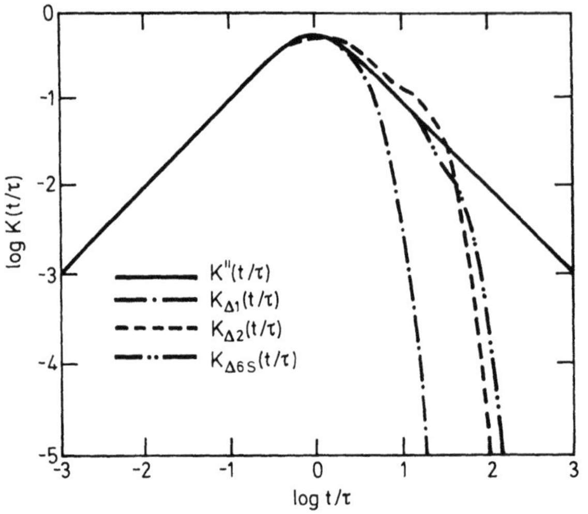

Fig. 8.2-3. Comparison of $K''(\omega\tau)|_{\omega=1/t}$ with several approximations

data are assumed to be spaced by log h = 0.301 on the logarithmic scale of their arguments. As an example, the difference kernel approximation $K_{\Delta 6S}(t/\tau)$, corresponding to the interconversion

$$G'_{6S}(\omega) = 0.142\,G(8t) - 0.860\,G(4t) + 0.674\,G(2t) + 0.942\,G(t) + 0.001\,G(t/2) + 0.101\,G(t/4) - 0.00855\,G(t/8) + 0.00855\,G(t/16) \tag{8.2-66}$$

is plotted in Fig. 8.2-2. The agreement at large values of t/τ is improved but at the expense of a loss in range. For each value of the approximation at log t contributions spanning the range from log t = −1.204 to log t = 0.903 must be selected from the source function.

Figure 8.2-3 also contains the kernel $K_{\Delta 6S}(t/\tau)$, corresponding to the approximation

$$G''_{6S}(\omega) = 0.470\,G(4t) - 2.144\,G(2t) + 1.476\,G(t) - 0.422\,G(t/2) + 0.608\,G(t/4) - 0.160\,G(t/8) + 0.172\,G(t/16)\ . \tag{8.2-67}$$

This approximation is a truncation of an infinite series proposed by Schwarzl [cf. Eq. (25) of Ref. [12c]]. The series was broken off with the G(t/16) term to give an order of approximation comparable to that in Eq. (8.2-66). The approximation requires contributions from the source function over the span from log t = −1.204 to log t = 0.603. It avoids the "overshoot" in Yagii and Maekawa's second order approximation in the interval for log $t/\tau \simeq 0.25$ to log $t/\tau \simeq 1.5$ but the agreement at large values of log t/τ is not much improved. The difference kernel approximations are both denoted by $K_{\Delta 6S}(t/\tau)$ because they both represent the operator $\mathscr{P}_{\Delta 6}[\exp(-t/\tau)]$ although the coefficients of the operator are, of course, quite different in the two cases.

Schwarzl also presented interconversion relations (spaced by log h = 0.301) for [12c] G″(ω) from G(t) and one value of G″(ω) at a high frequency, for [12b] J(t) from $J'(\omega)$ and one value of $J''(\omega)$, for $G(t)$ from $G'(\omega)$ only, and for [12a] $G(t)$ from several values of both $G'(\omega)$ and $G''(\omega)$.

In principle, any interconversion relation between moduli can be converted into one between compliances by substituting $-J(t)$ for $G(t)$, $-J'(\omega)$ for $G'(\omega)$, $J''(\omega)$ for $G''(\omega)$, and adding the flow term [cf. e.g., Eqs. (8.2-64) and (8.2-65)]. The error bounds of equations related in this way are not necessarily the same. Schwarzl and Struik [7], and Schwarzl [12], have carefully examined the error bounds of the interconversions proposed by them. The subject is, however, too lengthy to be discussed here and the reader is therefore referred to the original papers.

8.2.4 Empirical Interconversion Equations

We list here, without discussion, two compliance interrelations which may occasionally be useful. Koppelmann [13] found that

$$J(t) \simeq \mathrm{J}(\omega)|_{\omega=1/t} \tag{8.2-68}$$

where $\mathrm{J}(\omega)$ is the *absolute compliance* [cf. Eq. $(2.5\text{-}20)_2$]. The same observation was made several years later by Riande and Markovitz [14].

Plazek et. al. [15] found the relations

$$J'(\omega) \simeq [1 - m(2t)]^{0.8}[J(t) - \{\phi_f t\}]|_{t=1/\omega} \tag{8.2-69}$$

and

$$J''(\omega) - \{\phi_f/\omega\} \simeq [m(2/3t)]^{0.8}[J(t) - \{\phi_f t\}]|_{t=1/\omega} \tag{8.2-70}$$

to be useful. In these relations, $m(t)$ is the slope of $\log[J(t) - \{\phi_f t\}]$ vs. $\log t$.

The step responses and the storage functions, when plotted in logarithmic coordinates, are approximately mirror images of each other, reflected in the log modulus or log compliance axis. According to Christensen [16], the agreement can be improved by shifting the time scale. He advocates the approximate interrelations

$$J'(\omega) = J(t)|_{t=2/\pi\omega} \quad \text{and} \quad J(t) = J'(\omega)|_{\omega=2/\pi t} \tag{8.2-71}$$

and

$$G'(\omega) = G(t)|_{t=2/\pi\omega} \quad \text{and} \quad G(t) = G'(\omega)|_{\omega=2/\pi t} \tag{8.2-72}$$

(cf. Sect. 8.1.3.3).

8.3 Interconversion Within the Frequency Domain

We now turn to the third tier in Fig. 8.0-1, interconversion in the frequency domain. In Sect. 8.3.1 we discuss the relations between the real and imaginary parts of the harmonic response functions. The relations between the absolute modulus or compliance and the loss angle will be considered in Sect. 8.3.2.

8.3.1 Relations Between the Real and Imaginary Parts of the Harmonic Response Functions

The theoretical interrelations will be considered in Sect. 8.3.1.1. Approximate interrelations are introduced in Sect. 8.3.1.2. Section 8.3.1.3 discusses the interrelations resulting from mathematical models for the complex harmonic response functions.

8.3.1.1 The Kronig-Kramers Relations

The theoretical relations between the real and imaginary parts of the response to a harmonic excitation are generally referred to as the *Kronig-Kramers relations* because the first such relations were given by Kronig [17] and by Kramers [18] in studies concerned with the theory of electromagnetic radiation. As will be shown below, the real and imaginary parts of the response to a harmonic excitation are essentially *Hilbert* transforms* (see Appendix A6) of each other.

We begin by considering the relation between the storage and loss compliances. Our starting point is *Cauchy's Integral Formula* (see Problem 4.1-2) which we now write in the form

$$J^*(\omega) - J_g - \{\phi_f/j\omega\} = \frac{1}{2\pi j}\oint \frac{\bar{U}(s) - J_g - \{\phi_f/s\}}{s - j\omega}\,ds \; . \qquad (8.3\text{-}1)$$

The relations between $J'(\omega)$ and $J''(\omega)$ are obtained by equating real and imaginary parts on both sides of Eq. (8.3-1) after integration along a contour which includes the pole at $s = j\omega$ as the only singularity. Subtraction of the flow term, $\{\phi_f/s\}$, from $\bar{U}(s)$ removes the pole at the origin. The contour is shown in Fig. 8.3-1 (cf. Problem 4.1-1). The integral is obtained by shrinking the small half-circle at $s = j\omega$ to a point by letting $r \to 0$ and by expanding the large half-circle over the entire right half-plane by letting $R \to \infty$. Along segments AB and CD, $s = j\beta$. Along segment BC, $s = j\omega + r\exp(j\theta)$, and along DA, $s = R\exp(j\theta)$. Denoting the portion

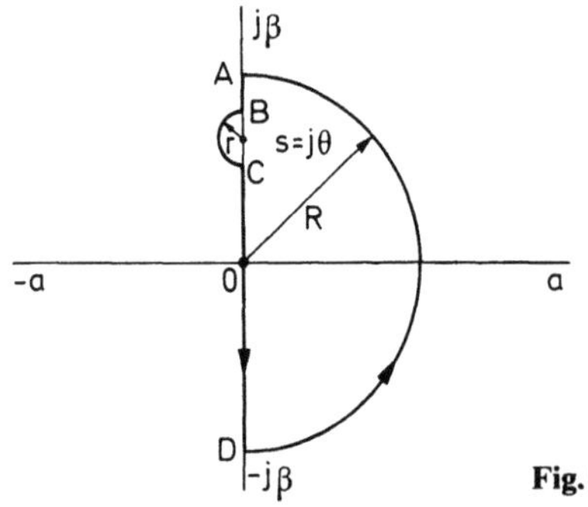

Fig. 8.3-1. Integration contour for the Kronig-Kramers relations

* David Hilbert, 1862–1943, German mathematician, famous for his development of the theory of infinitely many variables including spectral theory and the concept of Hilbert space.

of the integral along the segment AB by I_{AB}, etc., we find

$$I_{AB} = \lim_{r\to 0, R\to\infty} \frac{1}{2\pi j} \oint_{jR}^{j(\omega+r)} \frac{\bar{U}(s) - J_g - \{\phi_f/s\}}{s - j\omega} ds \Big|_{s=j\beta} \tag{8.3-2}$$

$$I_{BC} = \lim_{r\to 0} \frac{1}{2\pi j} \oint_{j(\omega+r)}^{j(\omega-r)} \frac{\bar{U}(s) - J_g - \{\phi_f/s\}}{s - j\omega} ds \Big|_{s=j\omega+r\exp(j\theta)} \tag{8.3-3}$$

$$I_{CD} = \lim_{r\to 0, R=\infty} \frac{1}{2\pi j} \oint_{j(\omega-r)}^{-jR} \frac{\bar{U}(s) - J_g - \{\phi_f/s\}}{s - j\omega} ds \Big|_{s=j\beta} \tag{8.3-4}$$

and

$$I_{DA} = \lim_{R\to\infty} \frac{1}{2\pi j} \oint_{-jR}^{jR} \frac{\bar{U}(s) - J_g - \{\phi_f/s\}}{s - j\omega} ds \Big|_{s=R\exp(j\theta)} \ . \tag{8.3-5}$$

But $\bar{U}(j\beta) = J^*(\beta)$. Thus, after taking the limits, the total contribution of the straight segments becomes

$$I_{AB,CD} = -\frac{j}{2\pi} \oint_{-\infty}^{\infty} \frac{(\omega + \beta)[J^*(\beta) - J_g + j\{\phi_f/\beta\}]}{\omega^2 - \beta^2} d\beta \tag{8.3-6}$$

where we have multiplied by $\omega + \beta$ for convenience later on.

Along segment BC $ds = jr\exp(j\theta)d\theta$. The lower limit of the integration is computed from $s_1 = j\omega + r\exp(j\theta_1) = j(\omega + r)$, i.e. $\exp(j\theta_1) = j$, as $\pi/2$. The upper limit becomes $3\pi/2$. Thus,

$$I_{BC} = \frac{1}{2\pi} \oint_{\pi/2}^{3\pi/2} [\bar{U}(j\omega) - J_g - \{\phi_f/j\omega\}]d\theta = [J^*(\omega) - J_g - \{\phi_f/j\omega\}]/2 \ . \tag{8.3-7}$$

The last contribution we must consider is that along DA. Now $ds = jR\exp(j\theta)d\theta$ and the limits of integration become $-\frac{\pi}{2}$ and $\frac{\pi}{2}$, respectively. We have

$$I_{DA} = \lim_{R\to\infty} \frac{1}{2\pi} \oint_{-\pi/2}^{\pi/2} \frac{\bar{U}[R\exp(j\theta)] - J_g - \{\phi_f\}/R\exp(j\theta)}{R\exp(j\theta) - j\omega} R\exp(j\theta)d\theta = 0 \tag{8.3-8}$$

because $\bar{U}(\infty) = J_g$.

Combining Eqs. (8.3-1), (8.3-6), and (8.3-7) leads to

$$J^*(\omega) = J_g - \frac{j}{\pi} \oint_{-\infty}^{\infty} \frac{(\omega + \beta)[J^*(\beta) - J_g + j\{\phi_f/\beta\}]}{\omega^2 - \beta^2} d\beta + \{\phi_f/j\omega\} \ . \tag{8.3-9}$$

But $J'(\beta)$ is an even, and $J''(\beta)$ an odd function of β [cf. Eqs. (4.1-23) and (4.1-24)]. Making use of this fact and of the behavior of even and odd functions when integrated over symmetric limits, separation of the real and imaginary parts of Eq. (8.3-9) yields

$$J'(\omega) = J_g + \frac{2}{\pi} \oint_{0}^{\infty} \frac{\beta[J''(\beta) - \{\phi_f/\beta\}]}{\beta^2 - \omega^2} d\beta \tag{8.3-10}$$

for the storage compliance as a function of the loss compliance, and

$$J''(\omega) = \frac{2}{\pi} \fint_0^\infty \frac{\omega[J'(\beta) - J_g]}{\omega^2 - \beta^2} d\beta + \{\phi_f/\omega\} \tag{8.3-11}$$

for the loss compliance as a function of the storage compliance.

The 'dot-form' of Eq. (8.3-10) is obtained as follows. By Eq. $(4.6\text{-}4)_3$, $J'(\omega)$ becomes $J_e^{\{o\}}$ in the limit as $\omega \to 0$. Hence, Eq. (8.3-10) yields

$$J_e^{\{o\}} - J_g = (2/\pi) \int_{-\infty}^{\infty} [J''(\beta) - \{\phi_f/\beta\}] d \ln \beta \ , \tag{8.3-12}$$

a result that had been anticipated in Sect. 4.6 [cf. Eq. (4.6-35)]. Elimination of J_g between Eqs. (8.3-10) and (8.3-12) furnishes

$$J'_\bullet(\omega) = J_e^{\{o\}} - \frac{2}{\pi} \fint_0^\infty \frac{\omega^2[J''(\beta) - \{\phi_f/\beta\}]}{\beta(\omega^2 - \beta^2)} d\beta \ . \tag{8.3-13}$$

To derive the analogous equations for the moduli we make use of *Cauchy's Integral Formula* (cf. Problem 4.1-1) in the form

$$G^*_\bullet(\omega) - G_g = \frac{1}{2\pi j} \oint \frac{\bar{Q}_\bullet(s) - G_g}{s - j\omega} ds \ . \tag{8.3-14}$$

The contour along which the integral must be evaluated is again that shown in Fig. 8.3-1. The contribution by the straight segments is

$$I_{AB,CD} = -\frac{j}{2\pi} \fint_0^\infty \frac{(\omega + \beta)[G^*_\bullet(\beta) - G_g]}{\omega^2 - \beta^2} d\beta \ . \tag{8.3-15}$$

In addition, $I_{BC} = [G^*_\bullet(\omega) - G_g]/2$ and I_{DA} again vanishes. Thus

$$G'_\bullet(\omega) = G_g - \frac{2}{\pi} \fint_0^\infty \frac{\beta G''(\beta)}{\beta^2 - \omega^2} d\beta \tag{8.3-16}$$

and

$$G''(\omega) = \frac{2}{\pi} \fint_0^\infty \frac{\omega[G_g - G'_\bullet(\beta)]}{\omega^2 - \beta^2} d\beta \ . \tag{8.3-17}$$

By Eq. $(4.6\text{-}4)_3$, $G'(\omega)$ becomes $\{G_e\}$ as $\omega \to 0$. Hence, from Eq. (8.3-16),

$$G_g - \{G_e\} = (2/\pi) \int_{-\infty}^{\infty} G''(\beta) d \ln \beta \tag{8.3-18}$$

which was anticipated in Sect. 4.6 as Eq. (4.6-31). Eliminating G_g between this and Eq. (8.3-16) gives

$$G'(\omega) = \{G_e\} + \frac{2}{\pi} \fint_0^\infty \frac{\omega^2 G''(\beta)}{\beta(\omega^2 - \beta^2)} d\beta \ . \tag{8.3-19}$$

We note that alternative forms of Eqs. (8.3-11) and (8.3-17) also exist (see Problem 8.3-1). They are

$$J''(\omega) = \frac{2}{\pi} \fint_0^\infty \frac{\omega[J_e^{\{o\}} - J'_\bullet(\beta)]}{\beta^2 - \omega^2} d\beta + \{\phi_f/\omega\} \tag{8.3-20}$$

and

$$G''(\omega) = \frac{2}{\pi} \fint_0^\infty \frac{\omega[G'(\beta) - \{G_e\}]}{\beta^2 - \omega^2} d\beta \ . \tag{8.3-21}$$

An alternative derivation of Eqs. (8.3-19) and (8.3-10) is given in Problem 8.3-2.

Equations (8.3-10), (8.3-11), (8.3-13), (8.3-16), (8.3-17), (8.3-19), (8.3-20), and (8.3-21) are the various forms of the Kronig-Kramers relations. We note that, because

$$\fint_0^\infty \frac{d\beta}{\omega^2 - \beta^2} = 0 \ , \tag{8.3-22}$$

the viscoelastic constants in the integrals of Eqs. (8.3-10), (8.3-11), (8.3-17), (8.3-20), and (8.3-21), but not Eq. (8.3-13), can be omitted, and the relations are often seen in these simplified forms.

By the same token, Eq. (8.3-22) can be used to develop other, alternative, forms of the Kronig-Kramers relations because we can add any expression under the integral sign as long as it is a constant with respect to the variable of integration. Thus Eqs. (8.3-10) and (8.3-13) can equally well be recast as

$$J'(\omega) = J_g + \frac{2}{\pi} \fint_0^\infty \frac{\beta J''(\beta) - \omega J''(\omega)}{\beta^2 - \omega^2} d\beta \tag{8.3-23}$$

and

$$J'_\bullet(\omega) = J_e^{\{o\}} - \frac{2\omega^2}{\pi} \fint_0^\infty \frac{[J''(\beta) - \{\phi_f/\beta\}]/\beta - [J''(\omega) - \{\phi_f/\omega\}]/\omega}{\omega^2 - \beta^2} d\beta \tag{8.3-24}$$

and both Eqs. (8.3-11) and (8.3-20) may be written

$$J''(\omega) = \frac{2\omega}{\pi} \fint_0^\infty \frac{J'(\beta) - J'(\omega)}{\omega^2 - \beta^2} d\beta + \{\phi_f/\omega\} \ . \tag{8.3-25}$$

In this last equation the 'dot-form' is optional.

For the relations between the moduli, we can replace Eqs. (8.3-19) and (8.3-16) by

$$G'(\omega) = \{G_e\} + \frac{2\omega^2}{\pi} \fint_0^\infty \frac{G''(\beta)/\beta - G''(\omega)/\omega}{\omega^2 - \beta^2} d\beta \tag{8.3-26}$$

and

$$G'_\bullet(\omega) = G_g - \frac{2}{\pi} \fint_0^\infty \frac{\beta G''(\beta) - \omega G''(\omega)}{\beta^2 - \omega^2} d\beta \tag{8.3-27}$$

and both Eqs. (8.3-17) and (8.3-21) by

$$G''(\omega) = \frac{2\omega}{\pi} \fint_0^\infty \frac{G'(\beta) - G'(\omega)}{\beta^2 - \omega^2} d\beta \ . \tag{8.3-28}$$

Again, it does not matter whether we use $G'(\omega)$ or $G'_\bullet(\omega)$ in Eq. (8.3-28). With the exception of Eq. (8.3-24), in all these equations the viscoelastic constants cancel under the integral sign.

We close this section by demonstrating that the storage and loss functions are Hilbert transforms of each other. With the substitutions $\omega^2 = \nu$, $\beta^2 = \mu$, Eqs. (8.3-10), (8.3-11), (8.3-13), (8.3-16), (8.3-17), (8.3-19), (8.3-20), and (8.3-21) can be condensed into the single relation

$$\hat{\Phi}(\nu) = \frac{1}{\pi} \fint_0^\infty \frac{\Phi(\mu)}{\mu - \nu} d\mu \tag{8.3-29}$$

where

$$\hat{\Phi}(\nu)|_{\nu=\omega^2} = \begin{cases} 2[G_g - G'_\bullet(\omega)] \\ 2G''(\omega)/\omega \\ 2[G'(\omega) - \{G_e\}]/\omega^2 \\ 2[J'(\omega) - J_g] \\ 2[J''(\omega) - \{\phi_f/\omega\}]/\omega \\ 2[J_e^{\{o\}} - J'_\bullet(\omega)]/\omega^2 \end{cases} \tag{8.3-30}$$

and

$$\Phi(\mu)|_{\mu=\beta^2} = \begin{cases} G''(\beta) \\ -[G_g - G'_\bullet(\omega)]/\beta, \quad \text{or} \quad [G'(\beta) - \{G_e\}]/\beta \\ -G''(\beta)/\beta^2 \\ J''(\beta) - \{\phi_f/\beta\} \\ -[J'(\beta) - J_g]/\beta, \quad \text{or} \quad [J_e^{\{o\}} - J'_\bullet(\beta)]/\beta \\ -[J''(\beta) - \{\phi_f/\beta\}]/\beta^2 \end{cases} \tag{8.3-31}$$

In these relations we have extended the range of integration from 0, ∞ to $-\infty$, ∞ (and divided by 2). This is permissible because the integrands are all even functions of β. The expressions for $\hat{\Phi}(\nu)$ and for $\Phi(\mu)$ must be selected as corresponding pairs. By Eq. (A6-1),

$$\hat{\Phi}(\nu) = \mathscr{H}[\Phi(\mu); \nu] \tag{8.3-32}$$

is the Hilbert transform of $\Phi(\mu)$.

8.3.1.2 Approximations to the Kronig-Kramers Relations

The exact Kronig-Kramers relations can generally be handled satisfactorily only if an explicit relation for the source function is assumed (cf. Problem 8.3-3). Experimental data are not readily interconverted because the integrations involve improper integrals and, furthermore, require knowledge of the data over the entire frequency range from zero to infinity. It is therefore of interest of examine the possibility of determining the imaginary part of any of the frequency-dependent response functions from the real part (or vice versa) by suitable approximate procedures.

Of particular importance are approximations that can be used to predict the loss function from the concomitant storage function. Such relations can be obtained by equating the zeroth-order approximations to the spectra from the loss functions with the first-order approximations from the storage functions. This yields

$$G''(\omega) = \frac{\pi}{2}\frac{d\,G'(\omega)}{d\ln\omega} \tag{8.3-33}$$

for the modulus relations via Eqs. (4.3-42) and (4.3-12), and

$$J''(\omega) = -\frac{\pi}{2}\frac{d\,J'(\omega)}{d\ln\omega} + \{\phi_f/\omega\} \tag{8.3-34}$$

for the compliance relations via Eqs. (4.3-43) and (4.3-13).

Equation (8.3-33) is easily changed to

$$\frac{d\log G'(\omega)}{d\log\omega} = \frac{2}{\pi}\tan\theta(\omega) \tag{8.3-35}$$

where $\tan\theta(\omega) = G''(\omega)/G'(\omega)$. Integration of Eqs. (8.3-33) and (8.3-35) yields

$$G'(\omega_2) - G'(\omega_1) = \frac{2}{\pi}\int_{\ln\omega_1}^{\ln\omega_2} G''(\omega)d\ln\omega \tag{8.3-36}$$

and

$$\ln[G'(\omega_2)/G'(\omega_1)] = \frac{2}{\pi}\int_{\ln\omega_1}^{\ln\omega_2} \tan\theta(\omega)d\ln\omega\ . \tag{8.3-37}$$

Thus, if one value of $G'(\omega)$ and either $G''(\omega)$ or the loss tangent, $\tan\theta(\omega)$, are known, another value of $G'(\omega)$ can be estimated from Eqs. (8.3-36) and (8.3-37).

Analogous relations can be obtained from Eq. (8.3-34) and its companion

$$-\frac{d\log J'(\omega)}{d\log\omega} = \frac{2}{\pi}\tan\theta(\omega) - \frac{\pi}{2}\{\phi_f/\omega\,J'(\omega)\} \tag{8.3-38}$$

but for a rheodictic material these relations are awkward to use.

Booij and Thoone* have derived the approximation [19]

$$G'(\omega) = \{G_e\} + \frac{\pi}{2}\left(G''(\omega) - \frac{d\,G''(\omega)}{d\ln\omega}\right)\ . \tag{8.3-39}$$

Comparison with Eq. (4.3-48) shows that this is equivalent to

$$G'(\omega) = \{G_e\} + \tfrac{\pi}{2}H_1''(\tau)|_{\tau=1/\sqrt{3}\omega}\ . \tag{8.3-40}$$

Consequently, the approximation must fail at higher frequencies where the spectrum decreases while the storage function is supposed to tend to G_g. Nevertheless, the authors have shown that Eq. (8.3-39) is a useful approximation over a large span of the frequency range. Its compliance analogue is

* pron. Boy and Tona

$$J'(\omega) = J_e^{\{o\}} - \frac{\pi}{2}\left(J''(\omega) - \frac{d\,J''(\omega)}{d\ln\omega} - \{2\phi_f/\omega\}\right) \tag{8.3-41}$$

which is equivalent to

$$J'(\omega) = J_e^{\{o\}} - \tfrac{\pi}{2} L_1''(\tau)|_{\tau=1/\sqrt{3}\omega} \;. \tag{8.3-42}$$

This approximation may also be expected to fail at higher frequencies but does not seem to have been tested so far.

The problem of approximate relationships between the storage and loss functions has been examined in some detail by Schwarzl and Struik [7]. As mentioned above, it is generally the loss function that one would wish to construct from a knowledge of the corresponding storage function. The authors found that conversion of $G'(\omega)$ data to $G''(\omega)$ data by their method is generally unsatisfactory. However, they present a reasonably successful relation for $J''(\omega)$ which is given by

$$J''(\omega) = -\frac{\pi}{2}\left(\frac{d\,J''(\omega)}{d\ln\omega} + 0.8225\frac{d^3\,J'(\omega)}{d\ln\omega^3}\right) \;. \tag{8.3-43}$$

They also give a relation for the derivative of $J'(\omega)$. This takes the form

$$\frac{d\,J'(\omega)}{d\ln\omega} = -\frac{\pi}{2}\left(J''(\omega) - 0.8225\frac{d^2\,J''(\omega)}{d\ln\omega^2}\right) \;. \tag{8.3-44}$$

These relations contain continuous derivatives and were derived essentially by the methods referred to in Sect. 8.2.2.

Brather* [20] offers approximations based on discrete differences for the real and imaginary parts of the complex dielectric permittivity. Expressed in terms of the complex compliance they are

$$\begin{aligned} J'(\omega) - J'(2\omega) = {} & 0.6151[J''(2\omega) + J''(\omega)] - 0.3887[J''(4\omega) + J''(\omega/4)] \\ & - 0.0086[J''(8\omega) + J''(\omega/4)] \end{aligned} \tag{8.3-45}$$

and

$$\begin{aligned} J'(\omega) - J'(2\omega) = {} & 2.718\,J''(\omega) - 1.013[J''(2\omega) + J''(\omega/2)] + 0.247[J''(4\omega) \\ & + J''(\omega/4)] - 0.152[J''(8\omega) + J''(\omega/8)] \end{aligned} \tag{8.3-46}$$

for the storage compliance in terms of the loss compliance, and

$$\begin{aligned} J''(\omega) - 0.4[J''(2\omega) + J''(\omega/2) = {} & 0.44165[J'(\omega/2) - J'(2\omega)] \\ & - 0.10701[J'(\omega/4) - J'(4\omega)] \end{aligned} \tag{8.3-47}$$

and

$$\begin{aligned} J''(\omega) - 0.2353[J''(4\omega) + J''(\omega/4)] = {} & 0.69177[J'(\omega/2) - J'(2\omega)] \\ & + 0.00902[J'(\omega/4) - J'(4\omega)] \\ & - 0.03664[J'(\omega/8) - J'(8\omega)] \end{aligned} \tag{8.3-48}$$

---

* pron. Brahter

for the loss compliance in terms of the storage compliance. Error estimates are also given in the original reference. The derivation of both the Schwarzl and Struik, and the Brather approximations are too involved to be reproduced here.

Brather [21] has also given a number of other relations between the storage and loss components of the complex modulus and of the complex compliance.

8.3.1.3 Using Mathematical Models

Replacing s by $j\omega$ in mathematical expressions designed to model the respondances, $\bar{Q}(s)$ and $\bar{U}(s)$, and separating the real and imaginary parts, leads to models for the storage and loss functions which are related to each other as Hilbert transforms, i.e. through the Kronig-Kramers relations. These models formed the subject of Sect. 6.1.4.1. Plotting loss against storage data is often a convenient way to find the parameters of the mathematical expressions with which one may wish to model the observed behavior. In this section we therefore examine the geometries resulting from plotting the imaginary against the real part of the more widely used complex models. We examine the Havriliak-Negami model [cf. Eq. (6.1-28)].

$$\hat{Z}_v^*(\omega) = \frac{1}{[1 + (j\omega/\omega_0)^v]^w} \tag{8.3-49}$$

because it contains the complex forms of both the Cole-Cole and the Kobeko models as the special cases when $w = 1$, $v = c$, and $w = b$, $v = 1$, respectively. Separating the real and imaginary parts leads to

$$\hat{Z}_v'(\omega) = \frac{(\omega_0/\omega)^{vw/2} \cos w\,\phi(\omega)}{[(\omega_0/\omega)^v + 2\cos v\pi/2 + (\omega/\omega_0)^v]^{w/2}} \tag{8.3-50}$$

and

$$\hat{Z}_v''(\omega) = \frac{(\omega_0/\omega)^{vw/2} \sin w\,\phi(\omega)}{[(\omega_0/\omega)^v + 2\cos v\pi/2 + (\omega/\omega_0)^v]^{w/2}} \tag{8.3-51}$$

with

$$\tan\phi(\omega) = \frac{\sin v\pi/2}{(\omega_0/\omega)^v + \cos v\pi/2} \,. \tag{8.3-52}$$

The reduction to the Kobeko forms [cf. Eqs. (8.3-61) and (8.3-62) below] is quite simple. To effect the reduction to the Cole-Cole forms [see Eqs. (6.1-88) and (6.1-89)] one makes use of the relations

$$\cos\phi(\omega) = \frac{(\omega_0/\omega)^{v/2} + (\omega/\omega_0)^{v/2} \cos v\pi/2}{[(\omega_0/\omega)^v + 2\cos v\pi/2 + (\omega/\omega_0)^v]^{1/2}} \tag{8.3-53}$$

and

$$\sin\phi(\omega) = \frac{(\omega/\omega_0)^{v/2} \sin v\pi/2}{[(\omega_0/\omega)^v + 2\cos v\pi/2 + (\omega/\omega_0)^v]^{1/2}} \tag{8.3-54}$$

which may be derived from Eq. (8.3-52) by standard trigonometric operations.

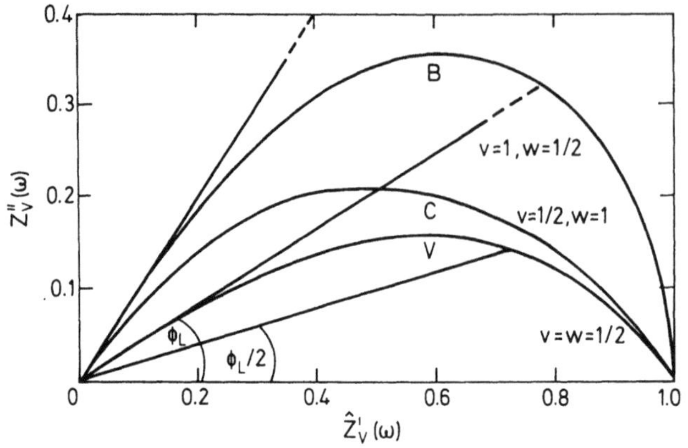

Fig. 8.3-2. Plots of the imaginary part of the Havriliak-Negami function against its real part for various values of v and w

Equations (8.3-50) and (8.3-51) are the parametric equations of a skewed semicircular arc. Several such arcs are shown in Fig. 8.3-2. Curve V represents the Havriliak-Negami equation with $w = v = 1/2$. Curve B is generated by the Kobeko equation with $v = 1$, $w = b = 1/2$, and curve C by the Cole-Cole equation with $v = c = 1/2$, $w = 1$.

We now explore the geometry of the arc to discover those features that will allow us to determine the parameters v, w, and ω_0. The first such feature is the angle, ϕ_L, between the real axis and the tangent drawn to the arc at its high frequency (i.e. *left*) end. We have

$$\tan \phi_L = \lim_{\omega \to \infty} \hat{Z}''_V(\omega)/\hat{Z}'_V(\omega) = \tan w\, \phi(\infty) \ . \tag{8.3-55}$$

From Eq. (8.3-52), $\phi(\infty) = v\pi/2$. Hence,

$$\phi_L = vw\pi/2 \ . \tag{8.3-56}$$

Next we examine the angle of the line drawn from the origin to the point at which $\omega = \omega_0$. This is given by

$$\frac{\hat{Z}''_V(\omega_0)}{\hat{Z}'_V(\omega_0)} = \tan \omega\, \phi(\omega_0) \ . \tag{8.3-57}$$

By Eq. (8.3-52) again, $\tan \phi(\omega_0) = \tan v\pi/4$ and, using Eq. (8.3-56), we see that the angle of the line results as

$$\phi_L/2 = w\, \phi(\omega_0) \ . \tag{8.3-58}$$

Hence, ω_0 is obtained as the radian frequency of the coordinates, $\hat{Z}'_V(\omega_0)$ and $\hat{Z}''_V(\omega_0)$, of the point at which the bisector $\phi_L/2$ intersects the arc.

Letting $\tilde{Z}_V(\omega_0)$ be the absolute modulus (i.e. the square root of the sum of the squares of the real and imaginary parts) of $\hat{Z}^*_V(j\omega_0)$, we find that

$$\tilde{Z}_V(\omega_0) = (2 + 2\cos v\pi/2)^{-w/2} \ . \tag{8.3-59}$$

Taking logarithms and using Eq. (8.3-56) again, gives

$$\phi_L^{-1} \log \check{Z}_V(\omega_0) = -(\pi v)^{-1} \log(2 + 2 \cos v\pi/2) \tag{8.3-60}$$

from which v can be found either through interpolation using suitably constructed tables, or by solving the transcendental equation with a root finder algorithm. When v is known, w can then be obtained from either Eq. (8.3-56) or Eq. (8.3-59).

The Kobeko equation also furnishes the parametric equations of a skewed semicircular arc. The equations are

$$\hat{Z}'_B(\omega) = \cos^b \phi(\omega) \cos b\, \phi(\omega) \tag{8.3-61}$$

and

$$\hat{Z}''_B(\omega) = \cos^b \phi(\omega) \sin b\, \phi(\omega) \tag{8.3-62}$$

since $\tan \phi(\omega) = \omega/\omega_0$ (cf. Problem 6.1-8). The arc for $b = 1/2$ is shown in Fig. 8.3-2 as curve B ($v = 1, w = 1/2$). The intersection of the arc with the line of angle $\phi_L/2$ again furnishes ω_0. Now, however, since $v = 1$, and $w = b$, we obtain b directly from Eq. (8.3-56) as $b = 2\phi_L/\pi$. (The coincidence of $\phi_L/2$ for curve B and ϕ_L for curve V is, of course, merely a consequence of our choice of parameters for the two curves.)

The situation is different for the complex form of the Cole-Cole function [cf. Eqs. (6.1-88) and (6.1-89)]. Here, it is possible to eliminate ω/ω_0 between the real and imaginary parts. This gives

$$[\hat{Z}''_C(\omega) + \tfrac{1}{2} \cot \pi c/2]^2 + [\hat{Z}_{C'}(\omega) - \tfrac{1}{2}]^2 = (\tfrac{1}{2} \csc \pi c/2)^2 \tag{8.3-63}$$

which is the equation of a circle with radius $\frac{1}{2} \csc \pi c/2$, the coordinates of the center being $\frac{1}{2}$ and $-\frac{1}{2} \cot \pi c/2$, respectively. Equation (8.3-63) can be solved explicitly for the imaginary part in terms of the real one. We find

$$\hat{Z}''_C(\omega) = \tfrac{1}{2} \cot \pi c/2 \left[\sqrt{1 + 4\, \hat{Z}'_C(\omega)[1 - \hat{Z}'_C(\omega)]\tan^2 \pi c/2} - 1\right] . \tag{8.3-64}$$

The arc for $c = 1/2$ is shown in Fig. 8.3-2 as curve C ($v = 1/2, w = 1$). Since now $\omega_0 = \omega_{max}$, the former can be found immediately from the location of the maximum in the locus of the experimental data. The coordinates of the maximum are

$$\hat{Z}'_C(\omega_{max}) = \tfrac{1}{2} \qquad \text{and} \qquad \hat{Z}''_C(\omega_{max}) = \tfrac{1}{2} \tan c\pi/4 . \tag{8.3-65}$$

Thus, the parameter c is also obtained directly from the maximum.

8.3.2 Relations Between the Absolute Modulus or Compliance and the Loss Angle

In the preceding section we discussed the interconversion between the cartesian components of the harmonic response functions, i.e. the storage and loss functions. Interconversion relations can also be established between the parameters of the polar forms, i.e. between the absolute modulus or compliance and the loss angle. These relations are useful if only one of them is known. If then the other is found by the relations to be discussed here, the absolute quantities and the loss angle can be converted into the real and imaginary parts of the harmonic response.

Any general complex quantity, $F^*(\omega)$, can be written either in cartesian or in polar form. Thus we have

$$F^*(\omega) = F'(\omega) \pm j\, F''(\omega) = \tilde{F}(\omega)\exp[\pm j\, \theta(\omega)] \tag{8.3-66}$$

where, in the general case, $F'(\omega)$ and $F''(\omega)$ are the real and imaginary parts of $F^*(\omega)$, while $\tilde{F}(\omega)$ is its amplitude and $\theta(\omega)$ its phase angle. The Kronig-Kramers relations connect the real and imaginary parts through the equations

$$F'(\omega) - F_e = \pm\frac{2\omega^2}{\pi} \int_0^\infty \frac{F''(\beta)/\beta - F''(\omega)/\omega}{\beta^2 - \omega^2} d\beta \tag{8.3-67}$$

or

$$F'(\omega) - F_g = \mp\frac{2}{\pi} \int_0^\infty \frac{\beta F''(\beta) - \omega\, F''(\omega)}{\beta^2 - \omega^2} d\beta \tag{8.3-68}$$

and

$$F''(\omega) = \pm\frac{2\omega}{\pi} \int_0^\infty \frac{F'(\beta) - F'(\omega)}{\beta^2 - \omega^2} d\beta\ . \tag{8.3-69}$$

Comparison shows that these equations are generalizations of Eqs. (8.3-23) to (8.3-28). When F stands for G, the equations must be taken with the upper signs, and F_e and F_g become $\{G_e\}$ and G_g, respectively. When F stands for J, the lower signs are the appropriate ones and F_e and F_g become $J_e^{\{o\}}$ and J_g. Also: in Eqs. (8.3-67) to (8.3-69) $F''(\omega)$ then must become $J''(\omega) - \{\phi_f/\omega\}$ to remove the singularity at the origin.

For the harmonic relations arising in linear viscoelastic theory, Eqs. (8.3-67) to (8.3-69) are quite generally valid as long as they are written in a manner which avoids complications arising from the possible existence of a singularity at the origin of the real frequency axis. To derive the interconversion relations between the amplitudes and phase angles, we take the logarithm on both sides of Eq. $(8.3\text{-}66)_2$. This yields

$$\ln F^*(\omega) = \ln \tilde{F}(\omega) \pm j\, \theta(\omega)\ . \tag{8.3-70}$$

Just as $F^*(\omega)$, $\ln F^*(\omega)$ is a complex function whose real and imaginary parts are connected by Hilbert transforms. We now have

$$\ln G^*(\omega) = \ln \tilde{G}(\omega) + j\, \theta(\omega) \tag{8.3-71}$$

for the moduli, and

$$\ln J^*(\omega) = \ln \tilde{J}(\omega) - j\, \theta(\omega) \tag{8.3-72}$$

for the compliances. For an arrheodictic material neither $\ln \tilde{G}(\omega)$ nor $\ln \tilde{J}(\omega)$ contains any singularity at the origin. Accordingly, we may write, following Booij and Thoone [19],

$$\ln \tilde{G}(\omega) = \ln G_e + \frac{2\omega^2}{\pi} \int_0^\infty \frac{\beta^{-1}\theta(\beta) - \omega^{-1}\theta(\omega)}{\omega^2 - \beta^2} d\beta \tag{8.3-73}$$

or

$$\ln \tilde{G}(\omega) = \ln G_g - \frac{2}{\pi} \fint_0^\infty \frac{\beta\theta(\beta) - \omega\theta(\omega)}{\beta^2 - \omega^2} d\beta \tag{8.3-74}$$

and

$$\theta(\omega) = \frac{2\omega}{\pi} \fint_0^\infty \frac{\ln \tilde{G}(\beta) - \ln \tilde{G}(\omega)}{\beta^2 - \omega^2} d\beta \tag{8.3-75}$$

for the relations between the absolute modulus and the loss angle. To these we add the relations

$$\ln \tilde{J}(\omega) = \ln J_e - \frac{2\omega^2}{\pi} \infty \fint_0^\infty \frac{\beta^{-1}\theta(\beta) - \omega^{-1}\theta(\omega)}{\omega^2 - \beta^2} d\beta \tag{8.3-76}$$

or

$$\ln \tilde{J}(\omega) = \ln J_g + \frac{2}{\pi} \fint_0^\infty \frac{\beta\theta(\beta) - \omega\theta(\omega)}{\beta^2 - \omega^2} d\beta \tag{8.3-77}$$

and

$$\theta(\omega) = \frac{2\omega}{\pi} \fint_0^\infty \frac{\ln \tilde{J}(\beta) - \ln \tilde{J}(\omega)}{\omega^2 - \beta^2} d\beta \tag{8.3-78}$$

between the absolute compliance and the loss angle of an arrheodictic material.

Booij and Thoone [19] also examined possible approximate relations between the real and imaginary parts of Eqs. (8.3-71) and (8.3-72). They found

$$\theta(\omega) = \frac{\pi}{2} \frac{d \ln \tilde{G}(\omega)}{d \ln \omega} \tag{8.3-79}$$

to be the only useful relation of this type involving the absolute modulus. The analogous relation

$$\theta(\omega) = -\frac{\pi}{2} \frac{d \ln \tilde{J}(\omega)}{d \ln \omega} \tag{8.3-80}$$

has not been tested. For small values of $G''(\omega)$ or $J''(\omega)$, Eqs. (8.3-79) and (8.3-80) reduce to Eqs. (8.3-35) and (8.3-38).

Equations (8.3-71) and (8.3-72) are not useful for rheodictic materials because then

$$\lim_{\omega \to 0} G^*(\omega) = 0 \quad \text{and} \quad \lim_{\omega \to 0} J^*(\omega) = \infty \tag{8.3-81}$$

and $\ln \tilde{G}(\omega)$ and $\ln \tilde{J}(\omega)$ thus possess singularities at the origin. These can be avoided, however, by using $\eta^*(\omega) = G^*(\omega)/j\omega$ and $\phi^*(\omega) = j\omega J^*(\omega)$ [cf. Eqs. (2.5-39) and (2.5-40) for the former, and Eqs. (2.5-51) and (2.5-52) for the latter]. We then have

$$\ln \eta^*(\omega) = \ln \tilde{\eta}(\omega) - j\,\Theta(\omega) \tag{8.3-82}$$

and

$$\ln \phi^*(\omega) = \ln \tilde{\phi}(\omega) + j\,\Theta(\omega) \tag{8.3-83}$$

where $\Theta(\omega)$ is the storage angle [see Eqs. (2.5-45) and (2.5-46)] defined by

$$\tan\Theta(\omega) = \cot\theta(\omega) = \frac{\eta''(\omega)}{\eta'(\omega)} = \frac{G'(\omega)}{G''(\omega)} = \frac{\phi''(\omega)}{\phi'(\omega)} = \frac{J'(\omega)}{J''(\omega)} \ . \tag{8.3-84}$$

We now have

$$\lim_{\omega\to 0} \ln\tilde{\eta}(\omega) = \ln\eta_f \tag{8.3-85}$$

and

$$\lim_{\omega\to 0} \ln\tilde{\phi}(\omega) = \ln\phi_f \ . \tag{8.3-86}$$

Let us take the complex viscosity, $\eta^*(\omega)$, first. From Eqs. (8.3-67) to (8.3-69),

$$\ln\tilde{\eta}(\omega) - \ln\tilde{\eta}(0) = \frac{2\omega^2}{\pi}\int_0^\infty \frac{\Theta(\beta)/\beta - \Theta(\omega)/\omega}{\beta^2-\omega^2}\,d\beta \tag{8.3-87}$$

or

$$\ln\tilde{\eta}(\omega) - \ln\tilde{\eta}(\infty) = \frac{2}{\pi}⨍_0^\infty \frac{\beta\Theta(\beta) - \omega\Theta(\omega)}{\beta^2-\omega^2}\,d\beta \ , \tag{8.3-88}$$

and

$$\Theta(\omega) = \frac{2\omega}{\pi}⨍_0^\infty \frac{\ln\tilde{\eta}(\beta) - \ln\tilde{\eta}(\omega)}{\omega^2-\beta^2}\,d\beta \ . \tag{8.3-89}$$

Since $\tilde{\eta}(\omega) = \tilde{G}(\omega)/\omega$, $\eta(0) = \eta_f$, and $\tilde{\eta}(\infty) = 0$, we may rewrite these relations as

$$\ln\tilde{G}(\omega) = \ln\omega\eta_f - \frac{2\omega^2}{\pi}⨍_0^\infty \frac{[\theta(\beta) - \frac{\pi}{2}]/\beta - [\theta(\omega) - \frac{\pi}{2}]/\omega}{\beta^2-\omega^2}\,d\beta \tag{8.3-90}$$

or

$$\ln\tilde{G}(\omega) = -\frac{2\omega^2}{\pi}⨍_0^\infty \frac{[\theta(\beta) - \frac{\pi}{2}]/\beta - [\theta(\omega) - \frac{\pi}{2}]/\omega}{\beta^2-\omega^2}\,d\beta \ , \tag{8.3-91}$$

and

$$\theta(\omega) = \frac{\pi}{2} - \frac{2\omega}{\pi}⨍_0^\infty \frac{\ln[\tilde{G}(\beta)/\beta] - \ln[\tilde{G}(\omega)/\omega]}{\omega^2-\beta^2}\,d\beta \ . \tag{8.3-92}$$

Turning now to the complex fluidity we find

$$\ln\tilde{\phi}(\omega) - \ln\tilde{\phi}(0) = \frac{2\omega^2}{\pi}⨍_0^\infty \frac{\Theta(\beta)/\beta - \Theta(\omega)/\omega}{\omega^2-\beta^2}\,d\beta \ , \tag{8.3-93}$$

and

$$\Theta(\omega) = \frac{2\omega}{\pi}⨍_0^\infty \frac{\ln\tilde{\phi}(\beta) - \ln\tilde{\phi}(\omega)}{\beta^2-\omega^2}\,d\beta \ . \tag{8.3-94}$$

The equation analogous to Eq. (8.3-88) is unavailable because $\tilde{\phi}(\infty) = \infty$. Using $\tilde{\phi}(\omega) = \omega\tilde{J}(\omega)$ and $\tilde{\phi}(0) = \phi_f$, however, we obtain

$$\ln \mathfrak{J}(\omega) = \ln(\phi_f/\omega) + \frac{2\omega^2}{\pi} \fint_0^\infty \frac{[\theta(\beta) - \frac{\pi}{2}]/\beta - [\theta(\omega) - \frac{\pi}{2}]/\omega}{\beta^2 - \omega^2} d\beta \tag{8.3-95}$$

and

$$\theta(\omega) = \frac{\pi}{2} - \frac{2\omega}{\pi} \fint_0^\infty \frac{\ln[\beta\mathfrak{J}(\beta)] - \ln[\omega\mathfrak{J}(\omega)]}{\beta^2 - \omega^2} d\beta \; . \tag{8.3-96}$$

The paper by Booij and Thoone [19] also contains other forms of these various equations.

8.4 Problems

Problem 8.0-1. Given

$$G(t) = G_e + G \exp(-t/r) \tag{1}$$

find $J'(\omega)$ using Table 8.0-1.

Problem 8.0-2. Derive the time-dependent viscosity, $\eta(t)$, of the standard linear solid model from the expression for its storage modulus, $G'(\omega)$.

Problem 8.1-1. Obtain

$$\bar{U}(s) = J_g + \sum_k \frac{J_k}{1 + \tau'_k s} \tag{1}$$

from

$$\bar{Q}_\bullet(s) = G_g - \sum_k \frac{G_k}{1 + \tau_k s} \; . \tag{2}$$

where the G_k and τ_k are the set of parameters developed in Problem 3.6-2, the G_k being those generated by computer inversion of the matrix.

Problem 8.1-2. Prove the proposition that $G(t)J(t) \leq 1$.

Problem 8.1-3. Find the relation between $\eta(t)$ and $J(t)$.

Problem 8.1-4. When interconversion is required only in the region in which the spectrum is a nearly linear function of the respondance time in doubly logarithmic coordinates, the viscoelastic constants may be disregarded and the *power law spectra* (see Sect. 6.3.1.2) may be used to evaluate the integrals. Making use of this possibility, show that

$$H(\tau)L(\tau) = \frac{\sin^2 \pi m}{\pi^2} \, , \tag{1}$$

is a relation between the spectral functions which is formally analogous to Eq. (8.1-44).

Problem 8.2-1. Derive the (inverse) Fourier transform relations for the storage modulus, $G'(\omega)$, and for the loss modulus, $G''(\omega)$, in terms of the relaxation modulus, $G(t)$, as well as the transform relations for the relaxation modulus in terms of the storage modulus, and the loss modulus, respectively.

Problem 8.2-2. In Problem 8.2-2 we derived the relation

$$G'(\omega) = \{G_e\} + \omega \int_0^\infty [G(t) - \{G_e\}]\sin \omega t \, dt \ . \tag{1}$$

In the derivation we used Eqs. (8.2-16). A similar derivation based on Eqs. (8.2-9) leads to the relation

$$G'(\omega) = \omega \int_0^\infty G(t)\sin \omega t \, dt \ . \tag{2}$$

Use the standard linear solid model to demonstrate that the two equations are equivalent.

Problem 8.2-3. Derive the (inverse) Fourier transform relations for the storage compliance, $J'(\omega)$, and for the loss compliance, $J''(\omega)$, in terms of the creep compliance, $J(t)$, as well as the transform relations for the creep compliance in terms of the storage compliance, and the loss compliance, respectively.

[*Hint*: the generalized Fourier cosine transform may not exist!]

Problem 8.2-4. (A) Derive Eqs. (8.2-17a) and (8.2-17b) as well as their inverses. Then (B) find $G_\bullet(t)$ and $J(t)$ for the standard linear solid models from the first two equations.

Problem 8.2-5. (A) Derive Eqs. (8.2-17c) and (8.2-17d). Then obtain $G_\bullet(t)$ and $J(t)$ for the standard linear solid models from them and compare the results with those obtained in Part (B) of Problem 8.2-4.

Problem 8.2-6. Show that the relations between the relaxation modulus, $G(t)$, and the complex viscosity and its real and imaginary parts are given by

$$G(t) = \begin{cases} \{G_e\} + \mathscr{F}^{-1}[\eta^*(\omega) - \{G_e\}/j\omega] \\ \{G_e\} + \mathscr{F}_c^{-1}[\eta'(\omega)] \\ \{G_e\} + \mathscr{F}_s^{-1}[\eta''(\omega) - \{G_e/\omega\}] \end{cases} \tag{1}$$

and

$$\eta^*(\omega) = \{G_e/j\omega\} + \mathscr{F}[G(t) - \{G_e\}] \ . \tag{2}$$

Problem 8.2-7. The transform relations

$$\dot{G}(t) = \begin{cases} \mathscr{F}^{-1}\{[G_\bullet^*(\omega) - G_g]; t\} \\ \mathscr{F}_c^{-1}\{[G'_\bullet(\omega) - G_g]; t\} \\ -\mathscr{F}_s^{-1}[G''(\omega); t] \end{cases} \tag{1}$$

and

$$\dot{J}(t) = \begin{cases} \mathscr{F}^{-1}\{[J^*(\omega) - J_g - \{\phi_f/j\omega\}]; t\} + \{\phi_f\} \\ \mathscr{F}_c^{-1}\{[J'(\omega) - J_g]; t\} + \{\phi_f\} \\ \mathscr{F}_s^{-1}\{J''(\omega) - \{\phi_f/\omega\}]; t\} + \{\phi_f\} \end{cases} \tag{2}$$

are the inverses of Eqs. (8.2-24) and (8.2-25). Verify, using the standard liquid model, that Eq. $(2)_1$ yields the correct result.

Problem 8.2-8. Find the difference kernel approximation, $K'_{\Delta 2}(\omega\tau)$, of the approximate interconversion

$$G(1.44t) = G'(\omega) - 0.40[G'(1.59\omega) - G'(1.93\omega)]_{\omega=1/t} . \tag{1}$$

Is the subscript justified, i.e. is this indeed a second order approximation?

Problem 8.3-1. Show that Eqs. (8.3-20) and (8.3-21) are alternative forms of Eqs. (8.3-11) and (8.3-17).

Problem 8.3-2. Derive Eqs. (8.3-19) and (8.3-16), i.e. the Kronig-Kramers relations for $G'(\omega)$, from the fact that the latter is the (ω-multiplied) Fourier sine transform of $G(t)$.

Compare this method of derivation with that employed in Sect. 8.3.1.1.

Problem 8.3-3. Given that

$$G'_{\bullet}(\omega) = G_g - \frac{G}{1 + \omega^2\tau^2} \tag{1}$$

find $G''(\omega)$.

Problem 8.3-4. Using the standard linear solid model, verify Eq. (8.3-77).

[*Hint*: you will need the definite integral

$$\frac{2}{\pi} \fint_{-\infty}^{+\infty} \frac{x \arctan \dfrac{ax}{b^2 + x^2}}{x^2 - y^2} dx = \ln \frac{\sqrt{(b^2 + y^2)^2 + a^2y^2}}{1 + y^2} \tag{1}$$

to solve this problem.]

References (Chap. 8)

1. I.L. Hopkins and R.W. Hamming, (a) J. Appl. Phys. *28*:906 (1957); (b) J. Appl. Phys. *29*:742 (1958).
2. G.W. Becker, Kolloid-Z., *166*:4 (1959).
3. E.F. Denby, Rheol. Acta, *14*:591 (1975).
4. J.D. Ferry, *Viscoelastic Properties of Polymers*, 3rd ed., Wiley, New York, 1980, p. 63.
5. T.L. Smith, Trans. Soc. Rheol., *2*:131 (1958).

6. I.S. Gradshteyn and I.M. Ryzhik, *Table of Integrals, Series*, and *Products*, Academic Press, New York, 1980, p. 292.
7. F.R. Schwarzl and L.C.E. Struik, Adv. Mol. Relaxation Processes *1*:201 (1967).
8. R.S. Marvin, Phys. Rev. *86*:644 (1952).
9. E. Catsiff and A.V. Tobolsky, J. Colloid. Sci., *10*:375 (1955).
10. K. Ninomiya and J.D. Ferry, J. Colloid Sci., *14*:36 (1959).
11. K. Yagii and E. Maekawa, Nippon Gomu Kyokaishi, *40*:46 (1967).
12. F. Schwarzl, Rheol. Acta, (a) *14*:581 (1975); (b) *9*:382 (1970); (c) *10*:166 (1971); (d) *8*:6 (1969).
13. J. Koppelmann, Rheol. Acta, 1:20 (1958).
14 E. Riande and H. Markovitz, J. Polymer Sci., Polymer Phys. Ed., *13*:947 (1975).
15. D.J. Plazek, N. Ragupathi, and S.J. Orbon, J. Rheol., *23*:477 (1979).
16. R.M. Christensen, Theory of Viscoelasticity, 2nd ed., Academic Press, New York, 1982, Section 4.6.
17. R. de L. Kronig, J. Opt. Soc. Amer., *12*:547 (1926).
18. H.A. Kramers, Atti Cong. Intern. dei Fisici, Como, 1927, p. 545.
19. H.C. Booij and G.P.J.M. Thoone, Rheol. Acta, *21*:15 (1982).
20. A. Brather, Colloid & Polymer Sci., *257*:467 (1978).
21. A. Brather, Rheol. Acta, *17*:325 (1978).

9. Energy Storage and Dissipation in a Linear Viscoelastic Material

There is no free lunch.

Life is tough.

President J. Carter's formulations of the first two laws of thermodynamics

9.0 Introduction

During the deformation of a viscoelastic body, part of the total work of deformation is dissipated as heat through viscous losses but the remainder of the deformational energy is stored elastically. It is frequently of interest to determine, for a given piece of material in a given mode of deformation, the total work of deformation as well as the amount of energy stored and the amount dissipated. Similarly, one may wish to know the rate at which the energy of deformation is absorbed by the material or the rate at which it is stored or dissipated.

The rate at which energy is absorbed per unit volume of a viscoelastic material during deformation is equal to the *stress power*, i.e. the rate at which work is performed. The stress power at time t is

$$\dot{W}(t) = \sigma(t)\dot{\varepsilon}(t) \ , \qquad (9.0\text{-}1)$$

i.e. it is the product of the instantaneous stress and rate of strain. The electrical analog of Eq. (9.0-1) is the well-known relation which states that the electrical power equals the product of the instantaneous voltage and current.

The total work of deformation or, in other words, the mechanical energy absorbed per unit volume of material in the deformation up to time t, results as

$$W(t) = \int_0^t \dot{W}(u)du = \int_0^t \sigma(u)\dot{\varepsilon}(u)du \ . \qquad (9.0\text{-}2)$$

The energy stored, $W_s(t)$, and the energy dissipated, $W_d(t)$, combine to make up the total deformational energy. Thus

$$W(t) = W_s(t) + W_d(t) \ . \qquad (9.0\text{-}3)$$

All energy or work terms and their derivatives will henceforth refer to unit volume of the material even when this is not explicitly stated.

Elastically stored energy is potential energy. Energy may also be stored inertially as kinetic energy (cf. Sect. 3.1.2). Such energy storage may be encountered in fast loading experiments, e.g. in response to impulsive excitation, or in wave propagation at high frequency. In the linear theory of viscoelastic behavior, however, inertial energy storage plays no role (cf. Sect. 3.3). Equation (9.0-3) may be looked upon as the definition of a viscoelastic material. According to this definition a viscoelastic material is one in which the total absorbed energy is partly stored elastically and partly dissipated as heat. Since energy can only be stored or dissipated, Eq. (9.0-3) is quite generally true. What makes it a viscoelastic relation is the interpretation of the stored energy as purely potential energy. How much of the total energy is stored and how much is dissipated, i.e. the precise form of Eq. (9.0-3), depends, of course, on the nature of the material on the one hand, and on the type of deformation on the other.

Differentiation of Eq. (9.0-3) yields

$$\dot{W}(t) = \dot{W}_s(t) + \dot{W}_d(t) \tag{9.0-4}$$

i.e. the rate at which energy is absorbed by the body during the deformation at time t equals the sum of the rates at which energy is stored, and dissipated, respectively. All terms in Eqs. (9.0-3) and (9.0-4) are positive. We are concerned here only with their magnitude. In thermodynamics, the proper signs have to be chosen in accordance with the convention adopted.

Equations (9.0-1) through (9.0-4) are not restricted to linear behavior. Our further development, however, does rest on the assumption that the behavior is linear. This further development is conveniently based on the representation of linear viscoelastic behavior by series-parallel models (cf. Chap. 3). It must be emphasized, however, that the treatment adopted here does *not* depend on the assumption that the behavior can be modeled by springs and dashpots. If one prefers, one may simply think of energy storing and energy dissipating mechanisms without identifying them with mechanical models, and modify the arguments as needed.

We begin by considering energy dissipation in an appropriate rheological model. By definition, energy is dissipated in the dashpots and the dashpots alone. Also, *all* of the energy required to deform the dashpots is dissipated. Denoting the rate at which energy is dissipated at time t by the nth dashpot as $\dot{w}_{dn}(t)$, we have, from Eq. (9.0-1),

$$\dot{w}_{dn}(t) = \sigma_{dn}(t)\dot{\varepsilon}_{dn}(t) \tag{9.0-5}$$

where $\sigma_{dn}(t)$ and $\dot{\varepsilon}_{dn}(t)$ are the stress, and the rate of strain, respectively, in the nth dashpot at time t. By the defining relation of a dashpot as a mechanical element [cf. Eq. (3.3-2)] we have

$$\sigma_{dn}(t) = \eta_n\dot{\varepsilon}_{dn}(t) \qquad \text{and} \qquad \dot{\varepsilon}_{dn}(t) = \phi_n\sigma_{dn}(t) \tag{9.0-6}$$

where η_n is the coefficient of viscosity of the nth dashpot and ϕ_n is its coefficient of fluidity. Thus

$$\dot{w}_{dn}(t) = \eta_n\dot{\varepsilon}_{dn}^2(t) = \phi_n\sigma_{dn}^2(t) \ . \tag{9.0-7}$$

Summing over all dashpots then yields

$$\dot{W}_d(t) = \sum_n \eta_n\dot{\varepsilon}_{dn}^2(t) = \sum_n \phi_n\sigma_{dn}^2(t) \ . \tag{9.0-8}$$

Turning now to energy storage, we consider that energy is stored in the springs, and in the springs alone. Furthermore, *all* the energy required to deform the springs is stored. By the defining equation for a spring as a mechanical element [Eq. (3.3-1)] we have

$$\sigma_{sn}(t) = G_n \varepsilon_{sn}(t) \qquad \text{and} \qquad \varepsilon_{sn}(t) = J_n \sigma_{sn}(t) \tag{9.0-9}$$

where G_n is the modulus of the nth spring, J_n is its compliance, and $\sigma_{sn}(t)$ and $\varepsilon_{sn}(t)$ are the stress, and the strain, respectively, in the nth spring at time t. Denoting the energy stored in the nth spring up to time t by $w_{sn}(t)$, we have, from Eqs. $(9.0\text{-}9)_1$ and $(9.0\text{-}2)_2$

$$w_{sn}(t) = G_n \int_0^t \dot{\varepsilon}_{sn}(u)\varepsilon_{sn}(u)du = G_n \int_{\varepsilon_{sn}(0)}^{\varepsilon_{sn}(t)} \varepsilon_{sn}(u)d\varepsilon_{sn}(u) \tag{9.0-10}$$

where $\dot{\varepsilon}_{sn}(t)$ is the rate of strain in the nth spring at time t. Since $\varepsilon_{sn}(0) = 0$ by definition, the integration yields

$$w_{sn}(t) = \tfrac{1}{2}G_n\varepsilon_{sn}^2(t) = \tfrac{1}{2}J_n\sigma_{sn}^2(t) \tag{9.0-11}$$

and summing over all springs gives

$$W_s(t) = \tfrac{1}{2}\sum_n G_n\varepsilon_{sn}^2(t) = \tfrac{1}{2}\sum_n J_n\sigma_{sn}^2(t) \; . \tag{9.0-12}$$

Equations $(9.0\text{-}11)_2$ and $(9.0\text{-}12)_2$ follow from $(9.0\text{-}11)_1$ and $(9.0\text{-}12)_1$ through substitution of the derivative of Eq. $(9.0\text{-}9)_2$ into (9.0-2) by an argument similar to that used to obtain Eqs. $(9.0\text{-}11)_1$ and $(9.0\text{-}12)_1$.

Equations (9.0-12) and (9.0-8) are the basic relations for determining energy storage and dissipation, respectively, during a particular deformation. They are given meaning by finding the stresses, strains, or rates of strain in the springs and dashpots of the series-parallel models in the given mode of deformation. The nature of the material is reflected in the distribution of the parameters G_n and η_n, or J_n and ϕ_n, with respect to the relaxation or retardation time, respectively.

According to the general scheme we have followed in this book so far, we should now find the energies stored or dissipated in response to a given excitation by determining the appropriate stress or strain transforms, introduce these into the Laplace transforms of Eqs. (9.0-8) and (9.0-12), and then retransform to obtain $W_s(t)$ and $W_d(t)$. This is perfectly feasible. However, Eqs. (9.0-8) and (9.0-12) contain the squares of time-dependent functions which become convolutions in the complex plane upon Laplace transformation. Thus, in the case at hand, transformation at this point does not simplify the further development (cf. Problem 9.1-1 as an example). We shall, therefore, find the appropriate stress and strain transforms by our usual procedure but shall then retransform them into the time domain for substitution into the equations for $W_s(t)$ and $\dot{W}_d(t)$. The following sections will deal with the more important excitations in this manner.

9.1 Strain Excitation (Stress Relaxation Behavior)

We begin by considering energy storage and dissipation in a viscoelastic body in response to a strain excitation, i.e. in stress relaxation behavior. The model which

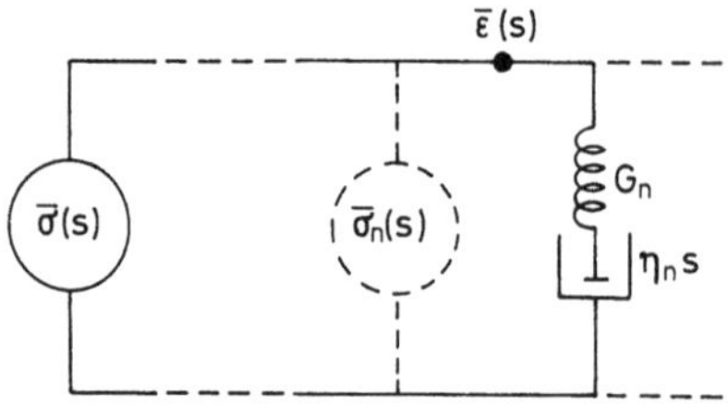

Fig. 9.1-1.
The nth Maxwell unit of a Wiechert model

most conveniently represents this behavior is the Wiechert model (cf. Figs. 3.5-1 and 3.5-2). This model consists essentially of Maxwell units in parallel. Different stresses act through the different units but the stress through the spring and dashpot of each unit is the same because they are coupled in series. Thus, the strains across the two elements of each unit will be different although the same strain is impressed across all units and the isolated spring, if present. Hence it suffices to consider a single Maxwell unit to establish the general relation, valid for each unit, which apportions the strain across the nth spring, and the nth dashpot. Once $\bar{\varepsilon}_{sn}(s)$ and $\bar{\varepsilon}_{dn}(s)$, the transforms of the strains across the nth spring and across the nth dashpot, respectively, have been found in terms of the overall strain transform, $\bar{\varepsilon}(s)$, the resulting relations may be specialized to any particular stimulus, of which we shall consider here only the step, slope and harmonic strains. The ramp strain forms the subject of Problem 9.1-9.

The nth Maxwell unit of a Wiechert model is shown in Fig. 9.1-1. By the rules of model analysis expounded in Sects. 3.1.2 and 3.3 we have

$$\bar{\sigma}_n(s) = \bar{Q}_n(s)\bar{\varepsilon}(s) \tag{9.1-1}$$

where $\bar{\sigma}_n(s)$ is the transform of the stress acting through the nth unit, and

$$\bar{Q}_n(s) = \frac{G_n \tau_n s}{1 + \tau_n s} \tag{9.1-2}$$

is the relaxance of the nth unit. In Eq. (9.1-2) $\tau_n = \eta_n/G_n$ is the nth relaxation time. The same stress acts through both elements of the unit. Hence, the transform of the strain in the spring is obtained directly as

$$\bar{\varepsilon}_{sn}(s) = \bar{\sigma}_n(s)/G_n \tag{9.1-3}$$

and that in the dashpot as

$$\bar{\varepsilon}_{dn}(s) = \bar{\sigma}_n(s)/\eta_n s \ . \tag{9.1-4}$$

Expressing $\bar{\sigma}_n(s)$ in terms of $\bar{\varepsilon}(s)$ from Eqs. (9.1-1) and (9.1-2), we obtain

$$\bar{\varepsilon}_{sn}(s) = \frac{\tau_n s}{1 + \tau_n s}\bar{\varepsilon}(s) \tag{9.1-5}$$

and

$$\bar{\varepsilon}_{dn}(s) = \frac{1}{1 + \tau_n s}\bar{\varepsilon}(s) \ . \tag{9.1-6}$$

The two strains (as well as their transforms) are additive, and we have

$$\varepsilon_{sn}(t) + \varepsilon_{dn}(t) = \varepsilon(t) \tag{9.1-7}$$

whatever the type of strain excitation. Equations (9.1-5) and (9.1-6) are the sought-for expressions for the transforms of the strain in the nth spring and the nth dashpot, respectively. For an isolated spring, if present, we have simply $\bar{\varepsilon}_{se}(s) = \bar{\varepsilon}(s)$. This follows from Eq. (9.1-5) by letting $\tau_n = \tau_e \to \infty$. We have distinguished the strain in the isolated spring by subscript e because the strain is associated with the equilibrium modulus, G_e.

From Eq. $(9.0\text{-}12)_1$ the energy stored in the response to a strain excitation up to time t becomes

$$W_s(t) = \tfrac{1}{2}\{G_e\}\varepsilon^2(t) + \tfrac{1}{2}\sum_n G_n\varepsilon_{sn}^2(t) \tag{9.1-8}$$

where

$$\varepsilon_{sn}(t) = \mathscr{L}^{-1}\frac{\tau_n s\bar{\varepsilon}(s)}{1 + \tau_n s} \tag{9.1-9}$$

and τ_n is the relaxation time of the nth Maxwell unit or the nth relaxation mechanism. For a continuous distribution of relaxation times (more precisely, for a continuous distribution of moduli) we have

$$W_s(t) = \tfrac{1}{2}\{G_e\}\varepsilon^2(t) + \tfrac{1}{2}\int_{-\infty}^{\infty} H(\tau)\varepsilon_s^2(t;\tau)\,d\ln\tau \tag{9.1-10}$$

where

$$\varepsilon_s(t;\tau) = \mathscr{L}^{-1}\frac{\tau s\bar{\varepsilon}(s)}{1 + \tau s}\ . \tag{9.1-11}$$

The rate at which energy is dissipated in the response to a strain excitation is obtained, for a discrete distribution of relaxation times, as

$$\dot{W}_d(t) = \sum_n \eta_n\dot{\varepsilon}_{dn}^2(t) \tag{9.1-12}$$

from Eq. $(9.0\text{-}8)_1$, where

$$\dot{\varepsilon}_{dn}(t) = \mathscr{L}^{-1}\frac{s\bar{\varepsilon}(s)}{1 + \tau_n s}\ . \tag{9.1-13}$$

For a continuous distribution we have

$$\dot{W}_d(t) = \int_{-\infty}^{\infty} \tau H(\tau)\dot{\varepsilon}_d^2(t;\tau)\,d\ln\tau \tag{9.1-14}$$

where

$$\dot{\varepsilon}_d(t;\tau) = \mathscr{L}^{-1}\frac{s\bar{\varepsilon}(s)}{1 + \tau s}\ . \tag{9.1-15}$$

$W_s(t)$ and $\dot{W}_d(t)$ are the *primary* energy functions. $\dot{W}_s(t)$ and $W_d(t)$ are most conveniently derived by differentiation of $W_s(t)$ or integration of $\dot{W}_d(t)$, respectively. $\dot{W}(t)$ and $W(t)$ can then be obtained by simple addition [cf. Eqs. (9.0-3) and (9.0-4)], although they can also be obtained directly from Eqs. (9.0-2) and (9.0-1). This is useful as a check on the calculation.

We are now ready to consider particular modes of strain excitation.

9.1.1 Step Response

For a step strain $\varepsilon(t) = \varepsilon_0$, and, hence, $\bar{\varepsilon}(s) = \varepsilon_0/s$ [cf. Eq. (2.3-2)]. The response of the elements of a Maxwell unit to a step strain are discussed in Problem 9.1-2. For a continuous distribution of relaxation times we have

$$\varepsilon_s(t;\tau) = \mathscr{L}^{-1}\frac{\varepsilon_0\tau}{1+\tau s} = \varepsilon_0 \exp(-t/\tau) \tag{9.1-16}$$

and substitution into Eq. (9.1-10) gives

$$W_s(t) = (\varepsilon_0^2/2)\left[\{G_e\} + \int_{-\infty}^{\infty} H(\tau)\exp(-2t/\tau)\,d\ln\tau\right]. \tag{9.1-17}$$

Thus the energy stored upon imposition of a strain as a step function of time is initially $(\varepsilon_0^2/2)G_g$ since, for $t = 0$, the integral reduces to $G_g - \{G_e\}$ by Eq. (4.6-20).

The rate at which the energy is stored is obtained by differentiating Eq. (9.1-17). When this is done following the ordinary rules of differential calculus, the result is identical with Eq. (9.1-19) below, but without the delta function term. This would be inconsistent with the result obtained by applying the rules of transformation calculus (cf. Problem 9.1-3). The reason is that in the latter calculus the initial conditions become part of the equations while they do not do so in the former. We can incorporate the initial condition, $W_s(0) = (\varepsilon_0^2/2)G_g$, in the equations derived by the ordinary rules of differentiation by taking into account the special nature of Eq. (9.1-17) which implies that the energy is stored instantaneously at $t = 0$ as the step of strain is imposed. Consequently, $W_s(t)$ is zero for $t < 0$, and is, therefore, discontinuous at the origin. Thus, it is properly written as $W_s(t)h(t)$ where $h(t)$ is the unit step function (see Appendix A2.2). By the rule for the differentiation of a product of functions we have, then,

$$\dot{W}_s(t) = \frac{d\,W_s(t)}{dt} + W_s(0)\delta(t) \tag{9.1-18}$$

where we have omitted the symbol $h(t)$ because it is understood that $dW_s(t)/dt$ is zero for $t < 0$, and where we have written $W_s(0)$ for $W_s(t)$ in the second term on the right because $W_s(0)$ is the only value of $W_s(t)$ which survives integration over the product with the delta function [see Eq. (A2-8)]. Now, since $W_s(0) = (\varepsilon_0^2/2)G_g$, we obtain

$$\dot{W}_s(t) = \varepsilon_0^2\left[G_g\delta(2t) - \int_{\infty}^{\infty} \tau^{-1}H(\tau)\exp(-2t/\tau)\,d\ln\tau\right] \tag{9.1-19}$$

where we have used the identity $\delta(t)/2 = \delta(2t)$ [see Eq. (A2-16)]. The first term in

Eq. (9.1-19) represents the impulsive (instantaneous) storage of a "packet" of energy, of magnitude $(\varepsilon_0^2/2)G_g$, at $t = 0$. This energy is transferred from the driving system to the body at an infinitely fast rate over an infinitely small time interval. Thereupon the rate at which the energy remains stored will decrease and eventually reach zero.

The rate of strain in the dashpot associated with the relaxation time τ is obtained by introducing $\bar{\varepsilon}(s) = \varepsilon_0/s$ into Eq. (9.1-13). Thus

$$\dot{\varepsilon}_d(t;\tau) = \mathscr{L}^{-1}\frac{\varepsilon_0}{1+\tau s} = \varepsilon_0\tau^{-1}\exp(-t/\tau) \ . \tag{9.1-20}$$

By Eq. (9.1-14), then,

$$\dot{W}_d(t) = \varepsilon_0^2 \int_{-\infty}^{\infty} \tau^{-1}H(\tau)\exp(-2t/\tau)d\ln\tau \ . \tag{9.1-21}$$

$\dot{W}_d(t)$ decays to zero from an initial value represented by

$$\dot{W}_d(0) = \varepsilon_0^2 \int_{-\infty}^{\infty} \tau^{-1}H(\tau)d\ln\tau \ . \tag{9.1-22}$$

Integration of Eq. (9.1-21) yields the energy dissipated up to time t as

$$W_d(t) = (\varepsilon_0^2/2) \int_{-\infty}^{\infty} H(\tau)[1-\exp(-2t/\tau)]d\ln\tau \ . \tag{9.1-23}$$

Thus the energy dissipated increases in the same manner in which the energy stored decreases. By Eq. (9.0-4) the rate at which energy is absorbed is the algebraic sum of the rates at which energy is stored and dissipated. Thus

$$\dot{W}(t) = \varepsilon_0^2 G_g\delta(2t) = (\varepsilon_0^2/2)G_g\delta(t) \ , \tag{9.1-24}$$

i.e. only the impulsive storage rate at time $t = 0$ survives the summation. At time $t > 0$ the rate at which the stored energy decreases, and the rate at which energy is dissipated, exactly balance and the net rate of energy absorption is zero.

From Eq. (9.0-3) the total work of deformation, resulting from the imposition of a step of strain at $t = 0$, is

$$W(t) = (\varepsilon_0^2/2)G_g \ . \tag{9.1-25}$$

It is thus constant with time and represents only the energy stored at $t = 0$. No energy is dissipated in deforming the body, the dissipation occuring only after the body has been deformed. In fact, Eq. (9.1-17) shows that

$$W_s(0) = (\varepsilon_0^2/2)G_g \qquad \text{and} \qquad W_s(\infty) = (\varepsilon_0^2/2)\{G_e\} \tag{9.1-26}$$

while Eq. (9.1-23) reveals that

$$W_d(0) = 0 \qquad \text{and} \qquad W_d(\infty) = (\varepsilon_0^2/2)[G_g - \{G_e\}] \ . \tag{9.1-27}$$

$W_s(\infty)$, of course, represents the energy which remains stored after the stress in the body has completely relaxed, while $W_d(\infty)$ represents the total amount of energy which is eventually dissipated.

It is often convenient to express the energies and their rates in terms of the response functions instead of the spectral functions. We have

$$W_s(t) = (\varepsilon_0^2/2)G(2t) \tag{9.1-28}$$

$$W_d(t) = (\varepsilon_0^2/2)[G_g - G(2t)] \tag{9.1-29}$$

$$\dot{W}_s(t) = \varepsilon_0^2[G_g\delta(2t) + \dot{G}(2t)] = \varepsilon_0^2 Q(2t) \tag{9.1-30}$$

and

$$\dot{W}_d(t) = -\varepsilon_0^2\dot{G}(2t) = \varepsilon_0^2[G_g\delta(2t) - Q(2t)]\ . \tag{9.1-31}$$

where $\dot{G}(t) = d\,G(t)/dt$.

In equations (9.1-24) and (9.1-25) can be derived also from Eqs. (9.0-1) and (9.0-2) but this requires special care (see Problem 9.1-4).

9.1.2 Slope Response

In slope response the strain is $\varepsilon(t) = \dot{\varepsilon}_0 t$, where $\dot{\varepsilon}_0$ is the (constant) rate of strain. Thus $\bar{\varepsilon}(s) = \dot{\varepsilon}_0/s^2$, and insertion into Eqs. (9.1-11) and (9.1-15) gives

$$\varepsilon_s(t;\tau) = \mathscr{L}^{-1}\frac{\dot{\varepsilon}_0\tau}{s(1+\tau s)} = \dot{\varepsilon}_0\tau[1 - \exp(-t/\tau)] \tag{9.1-32}$$

and

$$\dot{\varepsilon}_d(t;\tau) = \mathscr{L}^{-1}\frac{\dot{\varepsilon}_0}{s(1+\tau s)} = \dot{\varepsilon}_0[1 - \exp(-t/\tau)] \tag{9.1-33}$$

(cf. Problem 9.1-5). Consequently, the energy stored up to time t becomes

$$W_s(t) = (\dot{\varepsilon}_0^2/2)\left[\{G_e\}t^2 + \int_{-\infty}^{\infty} \tau^2 H(\tau)[1 - \exp(-t/\tau)]^2 d\ln\tau\right] \tag{9.1-34}$$

for a continuous distribution while the rate at which energy is dissipated at time t results as

$$\dot{W}_d(t) = \dot{\varepsilon}_0^2 \int_{-\infty}^{\infty} \tau H(\tau)[1 - \exp(-t/\tau)]^2 d\ln\tau\ . \tag{9.1-35}$$

The rate at which energy is stored at time t follows as

$$\dot{W}_s(t) = \dot{\varepsilon}_0^2\left[\{G_e\}t + \int_{-\infty}^{\infty} \tau H(\tau)[1 - \exp(-t/\tau)]\exp(-t/\tau) d\ln\tau\right] \tag{9.1-36}$$

and the amount of energy dissipated up to time t becomes

$$W_d(t) = (\dot{\varepsilon}_0^2/2)\int_{-\infty}^{\infty} \tau H(\tau)\left(2t + \tau\{1 - [2 - \exp(-t/\tau)]^2\}\right) d\ln\tau\ . \tag{9.1-37}$$

Finally, the energy absorbed in the deformation up to time t is

$$W(t) = (\dot{\varepsilon}_0^2/2)\left[\{G_e\}t^2 + 2\int_{-\infty}^{\infty} \tau H(\tau)\{t - \tau[1 - \exp(-t/\tau)]\} d\ln\tau\right] \tag{9.1-38}$$

and the rate at which the energy is absorbed at time t becomes

$$\dot{W}(t) = \dot{\varepsilon}_0^2\left[\{G_e\}t + \int_{-\infty}^{\infty} \tau H(\tau)[1 - \exp(-t/\tau)]d\ln\tau\right]. \tag{9.1-39}$$

As expected from this mode of deformation, no energy is absorbed, stored, or dissipated, and all rates vanish at $t = 0$. A body consisting of an arrheodictic material cannot, of course, be deformed indefinitely. For a rheodictic material (indicated by a superscript circle) we have, however, $\dot{W}_s^\circ(\infty) = 0$, $W^\circ(\infty) = W_d^\circ(\infty) = \infty$, while

$$\dot{W}^\circ(\infty) = \dot{W}_d^\circ(\infty) = \dot{\varepsilon}_0^2\eta_f \tag{9.1-40}$$

and

$$W_s^\circ(\infty) = (\dot{\varepsilon}_0^2/2)\int_{-\infty}^{\infty} \tau^2 H(\tau)d\ln\tau = (\dot{\varepsilon}_0^2/2)J_e^\circ\eta_f^2 \tag{9.1-41}$$

[cf. Eqs. (4.6-40), and $(4.6\text{-}41)_1$]. Thus, in a rheodictic material, at very long times, the energy stored remains constant at the value given by Eq. (9.1-41) which represents the energy stored in the fully extended springs; the rate at which energy is absorbed equals the rate at which it is dissipated, these rates being proportional to the steady-flow viscosity on the one hand, and the square of the rate of strain on the other; finally, the amounts of energy absorbed and dissipated increase without limit.

Equation (9.0-1) leads directly to

$$\dot{W}(t) = \dot{\varepsilon}_0^2\eta(t) \tag{9.1-42}$$

which is identical with Eq. (9.1-39) by Eq. (4.1-16). Integration of Eq. (9.1-42), or using Eq. (9.0-2) with $\varepsilon(t) = \dot{\varepsilon}_0 t$ and $\sigma(t) = \dot{\varepsilon}_0\eta(t)$, gives

$$W(t) = \dot{\varepsilon}_0^2\int_0^t \eta(u)du = \dot{\varepsilon}_0^2\psi(t) \tag{9.1-43}$$

which is, of course, identical with Eq. (9.1-38). The function $\psi(t)$ may be obtained experimentally by numerical or graphical integration of $\eta(t)$. Equation (9.1-43) also implies (see Problem 9.1-6) that the energy absorbed by the material is represented by the area under the stress-strain curve. Hence, $\psi(t)$ can be obtained also from that area. In terms of $\psi(t)$, $W_s(t)$ and $W_d(t)$ can be expressed as

$$W_s(t) = (\dot{\varepsilon}_0^2/2)[\psi(2t) - 2\psi(t)] \tag{9.1-44}$$

and

$$W_d(t) = (\dot{\varepsilon}_0^2/2)[4\,\psi(t) - \psi(2t)] \tag{9.1-45}$$

and can thus be determined directly from experiment without the need to know the relaxation spectrum. We also have

$$\dot{W}_s(t) = \dot{\varepsilon}_0^2[\eta(2t) - \eta(t)] \tag{9.1-46}$$

and

$$\dot{W}_d(t) = \dot{\varepsilon}_0^2[2\,\eta(t) - \eta(2t)] \tag{9.1-47}$$

for the rates at which energy is stored and dissipated.

9.1.3 Harmonic Response

Let the excitation be given by $\varepsilon(t) = \varepsilon_0 \sin \omega t$. Then $\bar{\varepsilon}(s) = \varepsilon_0\omega/(s^2 + \omega^2)$ by LTP (14) and we have

$$\bar{\varepsilon}_s(s;\tau) = \frac{\varepsilon_0 \omega \tau s}{(s^2 + \omega^2)(1 + \tau s)} \tag{9.1-48}$$

by Eq. (9.1-11). Substitution of Eq. (9.1-48) into Eq. (9.1-10) would yield the energy stored in the transient as well as in the steady-state (harmonic) part of the response. We are only interested in the latter. Decomposition into partial fractions yields

$$\bar{\varepsilon}_s(s;\tau) = \frac{\varepsilon_0 \omega(\tau s + \omega^2\tau^2)}{(s^2 + \omega^2)(1 + \omega^2\tau^2)} - \frac{\varepsilon_0 \omega \tau^2}{(1 + \tau s)(1 + \omega^2\tau^2)} \ . \tag{9.1-49}$$

The poles of the first term on the right ($s = \pm j\omega$) arise from the excitation transform. This term, therefore, is associated with the steady-state response (cf. Sects. 2.5 and 7.3.2.2). The poles of the second term ($s = -1/\tau$) arise from the material transform and are associated with the transient response. Rejecting the latter, and retransforming yields

$$\varepsilon_s^{ss}(t;\tau) = \frac{\varepsilon_0 \omega \tau}{1 + \omega^2\tau^2}(\cos \omega t + \omega\tau \sin \omega t) \tag{9.1-50}$$

where the superscript ss symbolizes the steady-state. Substituting Eq. (9.1-50) into Eq. (9.1-10) leads to

$$\begin{aligned} W_s(t) = (\varepsilon_0^2/4)\Bigg\{\{G_e\} &+ \int_{-\infty}^{\infty} H(\tau)\frac{\omega^2\tau^2}{1 + \omega^2\tau^2} d\ln\tau \\ &- \left[\{G_e\} - \int_{-\infty}^{\infty} H(\tau)\frac{\omega^2\tau^2(1 - \omega^2\tau^2)}{(1 + \omega^2\tau^2)^2} d\ln\tau\right]\cos 2\omega t \\ &+ \int_{-\infty}^{\infty} H(\tau)\frac{2\omega^3\tau^3}{(1 + \omega^2\tau^2)^2} d\ln\tau \cdot \sin 2\omega t\Bigg\} \ . \end{aligned} \tag{9.1-51}$$

It is easily ascertained that

$$G'(\omega) - \frac{d\,G'(\omega)}{d\ln\omega} = \{G_e\} - \int_{-\infty}^{\infty} H(\tau)\frac{\omega^2\tau^2(1 - \omega^2\tau^2)}{(1 + \omega^2\tau^2)^2} d\ln\tau \tag{9.1-52}$$

and

$$G''(\omega) - \frac{d\,G''(\omega)}{d\ln\omega} = \int_{-\infty}^{\infty} H(\tau)\frac{2\omega^3\tau^3}{(1 + \omega^2\tau^2)^2} d\ln\tau \ . \tag{9.1-53}$$

Equation (9.1-51) thus reduces to

$$\begin{aligned} W_s(t) = (\varepsilon_0^2/4)\Bigg[G'(\omega) &- \left(G'(\omega) - \frac{d\,G'(\omega)}{d\ln\omega}\right)\cos 2\omega t \\ &+ \left(G''(\omega) - \frac{d\,G''(\omega)}{d\ln\omega}\right)\sin 2\omega t\Bigg] \end{aligned} \tag{9.1-54}$$

and it is interesting to note that the energy stored in a harmonic deformation depends not only on the storage modulus, $G'(\omega)$, but also on the loss modulus, $G''(\omega)$, as well as on their derivatives. This is a consequence of the lack of phase coherence among the energy storage mechanisms and will be discussed further later on. We also note that

$$W_s(0) = (\varepsilon_0^2/4)\frac{d\,G'(\omega)}{d\ln\omega}\ , \tag{9.1-55}$$

i.e. the energy stored does not vanish at $t = 0$. This is a consequence of the arbitrariness involved in selecting a reference point in the steady state which extends indefinitely back into time. Later we shall make Eq. (9.1-55) consistent with the definition of $t = 0$ as any time at which the total energy absorbed vanishes, i.e. we shall ensure that $W(0) = 0$.

Differentiation of Eq. (9.1-54) yields the rate at which energy is stored at time t as

$$\dot{W}_s(t) = (\varepsilon_0^2/2)_\omega\left[\left(G'(\omega) - \frac{d\,G'(\omega)}{d\ln\omega}\right)\sin 2\omega t + \left(G''(\omega) - \frac{d\,G''(\omega)}{d\ln\omega}\right)\cos 2\omega t\right]\ . \tag{9.1-56}$$

To see the meaning of Eqs. (9.1-54) and (9.1-56) more clearly we transform both by setting [cf. Eqs. (9.1-52) and (9.1-53)]

$$G'(\omega) - \frac{d\,G'(\omega)}{d\ln\omega} = R_s'(\omega) = \tilde{R}_s(\omega)\cos 2\mu_s(\omega) \tag{9.1-57}$$

and

$$G''(\omega) - \frac{d\,G''(\omega)}{d\ln\omega} = R_s''(\omega) = \tilde{R}_s(\omega)\sin 2\mu_s(\omega) \tag{9.1-58}$$

where

$$\tilde{R}_s(\omega) = \sqrt{[R_s'(\omega)]^2 + [R_s''(\omega)]^2} \tag{9.1-59}$$

and

$$\tan 2\mu_s(\omega) = R_s''(\omega)/R_s'(\omega)\ . \tag{9.1-60}$$

With these substitutions, Eq. (9.1-54) becomes

$$W_s(t) = (\varepsilon_0^2/4)\{G'(\omega) - \tilde{R}_s(\omega)\cos 2[\omega t + \mu_s(\omega)]\}\ . \tag{9.1-61}$$

and Eq. (9.1-56) takes the form

$$\dot{W}_s(t) = (\varepsilon_0^2/2)\omega\tilde{R}_s(\omega)\sin 2[\omega t + \mu_s(\omega)]\ . \tag{9.1-62}$$

The stored energy is seen to oscillate with twice the frequency of the excitation and lags the latter by the angle $\mu_s(\omega)$. The lag arises from the fact that the various energy storing mechanisms do not store energy coherently. Storage in the nth mechanism is out of phase with the excitation by the angle $\mu_n(\omega) = \arctan(1/\omega\tau_n)$ where μ_n is the relaxation time associated with the nth mechanism, or n^{th} Maxwell unit. The angle $\mu_s(\omega)$ represents the combined effect of all mechanisms.

Integration of Eq. (9.1-62) shows that there is no net energy storage over any complete cycle of either the excitation or of the stored energy. The statement has appeared in the literature many times (see, e.g., Refs. [1, 2]) that the maximum energy stored per unit volume of material in a cycle is given by

$$W_s(\omega)_{max} = \int_0^{\varepsilon_0} G'(\omega)\varepsilon(t)d\varepsilon(t) = (\varepsilon_0^2/2)G'(\omega) = [\varepsilon_0\sigma_0(\omega)/2]\cos\theta(\omega) \ . \qquad (9.1\text{-}63)$$

This, however, is incorrect. By a change of variable from $\varepsilon(t) = \varepsilon_0 \sin \omega t$ to t we obtain

$$W_s(\omega)_{max} = \int_0^{\pi/2\omega} (\varepsilon_0^2/2)G'(\omega)\sin 2\omega t \, dt \ . \qquad (9.1\text{-}64)$$

This is equivalent to assuming that the rate of energy storage is given by the integrand in Eq. (9.1-64) (cf. below) and then integrating over the first quarter of a cycle of the excitation. However, the correct rate of energy storage is given by Eqs. (9.1-56) or (9.1-62). We leave it to Problem 9.1-7 to show that $W_s(\omega)_{max}$ as given by Eq. $(9.1\text{-}63)_2$ is the maximum *coherently* storable energy, i.e. the amount of energy that could be stored if all storage mechanisms were in phase with each other and with the excitation. This, however, would be true only for a purely elastic material.

To obtain the maximum energy which is *actually* stored in a cycle we observe that $W_s(t)$, as given by Eq. (9.1-61) has minima when

$$\omega t = \pi n - \mu_s(\omega) \ , \qquad n = 0, 1, \ldots \qquad (9.1\text{-}65)$$

and maxima when

$$\omega t = \pi(2n+1)/2 - \mu_s(\omega) \ , \qquad n = 0, 1, \ldots \ . \qquad (9.1\text{-}66)$$

Introducing Eq. (9.1-66) into Eq. (9.1-61) yields the maximum energy which is actually stored per unit volume of the material in the steady state as

$$W_s(\omega)_{act} = (\varepsilon_0^2/4)[G'(\omega) + \tilde{R}_s(\omega)] \ . \qquad (9.1\text{-}67)$$

Substitution into Eq. (9.1-61) of Eq. (9.1-65), on the other hand, shows that the stored energy is never zero but that there is a *residual* energy, given by

$$W_s(\omega)_{res} = (\varepsilon_0^2/4)[G'(\omega) - \tilde{R}_s(\omega)] \ . \qquad (9.1\text{-}68)$$

which, as a consequence of the incoherence of the storage mechanisms, always remains stored in the steady state, even though the energy stored in each individual mechanism disappears at the appropriate time. The difference between $W_s(\omega)_{act}$ and $W_s(\omega)_{res}$ is the maximum energy which is cycled between the material and the driving system. This quantity,

$$W_s(\omega)_{cyc} = (\varepsilon_0^2/2)\tilde{R}_s(\omega) \ , \qquad (9.1\text{-}69)$$

may, of course, also be obtained by integrating $\dot{W}_s(\omega)$ between the limits $t_1 = \pi n/\omega - \mu_s(\omega)/\omega$ and $t_2 = \pi(2n+1)/2\omega - \mu_s(\omega)/\omega$.

When the losses in the material are small, $\tilde{R}_s(\omega) \to G'(\omega)$, and then

$$W_s(\omega)_{act} = W_s(\omega)_{cyc} = W_s(\omega)_{max} \ . \qquad (9.1\text{-}70)$$

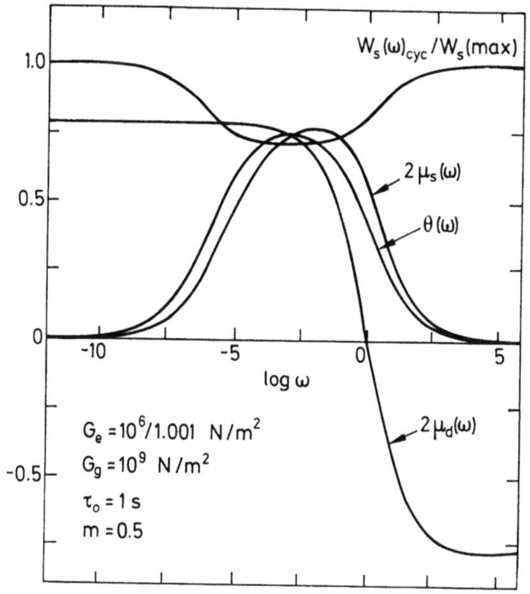

Fig. 9.1-2. $W_s(\omega)_{act}/W_s(\omega)_{max}$, $\theta(\omega)$, $2\mu_s(\omega)$, and $2\mu_d(\omega)$ as functions of log ω for typical arrheodictic behavior

This occurs at low frequencies as $\tilde{R}_s(\omega) \to \{G_e\}$, and at high frequencies as $\tilde{R}_s(\omega) \to G_g$. Thus

$$\lim_{\omega \to 0} W_s(\omega)_{act} = (\varepsilon_0^2/2)\{G_e\} \quad \text{and} \quad \lim_{\omega \to \infty} W_s(\omega)_{act} = (\varepsilon_0^2/2)G_g \,. \tag{9.1-71}$$

Clearly, the key role in determining the various forms of the stored energy in the steady state is played by $\tilde{R}_s(\omega)$. The ratio

$$\frac{W_s(\omega)_{cyc}}{W_s(\omega)_{max}} = \frac{\tilde{R}_s(\omega)}{G'(\omega)} \tag{9.1-72}$$

is plotted in Fig. 9.1-2 as function of log ω. $G'(\omega)$ and $G''(\omega)$ were calculated for a hypothetical arrheodictic material modelled by Eqs. (6.1-96) and (6.1-97) with $\omega_0 = 1\ s^{-1}$ and $c = 0.5$ as parameters. For G_e and G_g the usual values of $10^6/1.001$ and 10^9 N/m^2 were used. The loss angle, $\theta(\omega)$, and the angles, $2\mu_s(\omega)$ and $2\mu_d(\omega)$, are also plotted for comparison. The angle $\mu_d(\omega)$ will be discussed presently.

The ratio $W_s(\omega)_{cyc}/W_s(\omega)_{max}$ reaches its minimum value when the loss is highest. At this frequency only about 70% of the maximum coherently storable energy is pumped in and out of our hypothetical material. It further follows from

$$\frac{W_s(\omega)_{act}}{W_s(\omega)_{max}} = \frac{1}{2}\left(1 + \frac{\tilde{R}_s(\omega)}{G'(\omega)}\right) \tag{9.1-73}$$

that about 85% of the maximum coherently storable energy is actually stored. By

$$\frac{W_s(\omega)_{res}}{W_s(\omega)_{max}} = \frac{1}{2}\left(1 - \frac{\tilde{R}_s(\omega)}{G'(\omega)}\right) \tag{9.1-74}$$

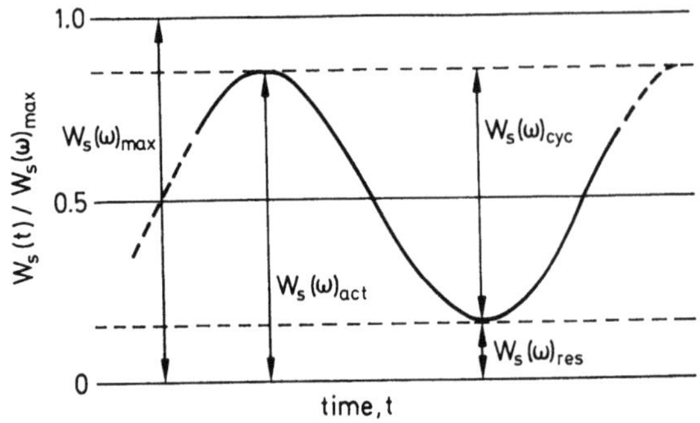

Fig. 9.1-3. Stored energy in the steady-state

about 15% of $W_s(\omega)_{max}$ remains stored in the material in the steady-state at the frequency of maximum loss. This potential energy is accumulated in the body before the steady-state is reached, and is released only when the excitation is removed. The relative magnitudes for the quantities $W_s(\omega)_{max}$, $W_s(\omega)_{act}$, $W_s(\omega)_{cyc}$, and $W_s(\omega)_{res}$, normalized to $W_s(\omega)_{max} = 1$, are indicated in Fig. 9.1-3 for our hypothetical material at the frequency when the losses are largest. Thus, in any linear viscoelastic material, the energy cycled between the driving system and the material is always less or equal to the maximum coherently storable energy, i.e.

$$W_s(\omega)_{cyc} \leq W_s(\omega)_{max} \, . \tag{9.1-75}$$

The equality applies strictly only when $\omega = 0$ or ∞. In practice, it is applicable with good approximation at sufficiently low or high frequencies.

We now turn our attention to the energy dissipated. Substitution of $\bar{\varepsilon}(s) = \varepsilon_0\omega/(s^2 + \omega^2)$ into Eq. (9.1-15) gives

$$\bar{\dot{\varepsilon}}_d(s;\tau) = \frac{\varepsilon_0 \omega s}{(s^2 + \omega^2)(1 + \tau s)} \tag{9.1-76}$$

and decomposition into partial fractions leads to

$$\bar{\dot{\varepsilon}}_d(s;\tau) = \frac{\varepsilon_0\omega(s + \omega^2\tau)}{(s^2 + \omega^2)(1 + \omega^2\tau^2)} - \frac{\varepsilon_0\omega\tau}{(1 + \tau s)(1 + \omega^2\tau^2)} \, . \tag{9.1-77}$$

Rejecting the second part and retransforming furnishes

$$\dot{\varepsilon}_d^{ss}(t;\tau) = \frac{\varepsilon_0\omega}{1 + \omega^2\tau^2}(\cos\omega t + \omega\tau \sin\omega t) \tag{9.1-78}$$

and Eq. (9.1-14) yields

$$\dot{W}_d(t) = (\varepsilon_0^2/2)\omega\left[\int_{-\infty}^{\infty} H(\tau)\frac{\omega\tau}{1 + \omega^2\tau^2}\, d\ln\tau + \int_{-\infty}^{\infty} H(\tau)\frac{2\omega^2\tau^2}{(1 + \omega^2\tau^2)^2}\, d\ln\tau \cdot \sin 2\omega t \right.$$
$$\left. + \int_{-\infty}^{\infty} H(\tau)\frac{\omega\tau(1 - \omega^2\tau^2)}{(1 + \omega^2\tau^2)^2}\, d\ln\tau \cdot \cos 2\omega t\right] \, . \tag{9.1-79}$$

But

$$\frac{d\,G'(\omega)}{d\ln\omega} = \int_{-\infty}^{\infty} H(\tau)\frac{2\omega^2\tau^2}{(1+\omega^2\tau^2)^2}\,d\ln\tau \tag{9.1-80}$$

and

$$\frac{d\,G''(\omega)}{d\ln\omega} = \int_{-\infty}^{\infty} H(\tau)\frac{\omega\tau(1-\omega^2\tau^2)}{(1+\omega^2\tau^2)^2}\,d\ln\tau\ . \tag{9.1-81}$$

Thus, we obtain

$$\dot{W}_d(t) = (\varepsilon_0^2/2)\omega\left[G''(\omega) + \frac{d\,G''(\omega)}{d\ln\omega}\cos 2\omega t + \frac{d\,G'(\omega)}{d\ln\omega}\sin 2\omega t\right] \tag{9.1-82}$$

for the rate at which energy is dissipated at time t in the steady state.

The energy dissipated up to time t must be obtained by integrating Eq. (9.1-82) subject to a suitably chosen initial condition. We select this so that the total energy absorbed be zero at $t = 0$. Thus, by Eq. (9.0-3), we have $W(0) = W_s(0) + W_d(0)$ and, by Eq. (9.1-55), obtain the initial condition as

$$W_d(0) = -W_s(0) = -(\varepsilon_0^2/4)\frac{d\,G'(\omega)}{d\ln\omega}\ . \tag{9.1-83}$$

Integration then yields

$$W_d(t) = (\varepsilon_0^2/4)\left[2\omega t\,G''(\omega) + \frac{d\,G''(\omega)}{d\ln\omega}\sin 2\omega t - \frac{d\,G'(\omega)}{d\ln\omega}\cos 2\omega t\right]\ . \tag{9.1-84}$$

Equations (9.1-82) and (9.1-84) may be brought into another form with the aid of the substitutions

$$\frac{d\,G'(\omega)}{d\ln\omega} = R_d'(\omega) = \tilde{R}_d(\omega)\cos 2\,\mu_d(\omega) \tag{9.1-85}$$

and

$$\frac{d\,G''(\omega)}{d\ln\omega} = R_d''(\omega) = \tilde{R}_d(\omega)\sin 2\,\mu_d(\omega) \tag{9.1-86}$$

where

$$\tilde{R}_d(\omega) = \sqrt{[R_d'(\omega)]^2 + [R_d''(\omega)]^2} \tag{9.1-87}$$

and

$$\tan 2\,\mu_d(\omega) = R_d''(\omega)/R_d'(\omega)\ . \tag{9.1-88}$$

The angle $2\,\mu_d(\omega)$ is plotted for comparison with $2\,\mu_s(\omega)$ in Fig. 9.1-2 for the specified material. With $\tilde{R}_d(\omega)$ and $\mu_d(\omega)$ we obtain

$$W_d(t) = (\varepsilon_0^2/4)\{2\omega t G''(\omega) - \tilde{R}_d(\omega)\cos 2[\omega t + \mu_d(\omega)]\} \tag{9.1-89}$$

and

$$\dot{W}_d(t) = (\varepsilon_0^2/2)\omega\{G''(\omega) + \tilde{R}_d(\omega)\sin 2[\omega t + \mu_d(\omega)]\}\ . \tag{9.1-90}$$

Thus the rate at which energy is dissipated oscillates between the limits $(\varepsilon_0^2/2)\omega[G''(\omega) \pm \tilde{R}_d(\omega)]$ with twice the frequency of excitation. The maximum rate of energy dissipation which would result if all dashpots would dissipate energy coherently, is

$$\dot{W}_d(\omega)_{max} = \varepsilon_0^2\omega\, G''(\omega) = \varepsilon_0\sigma_0(\omega)\omega \sin\theta(\omega) \tag{9.1-91}$$

(cf. Problem 9.1-8). The energy dissipated is seen to increase continuously with time. Oscillations of twice the frequency of excitation are superposed on a straight line of constant slope. When ωt assumes the values $\pi(2n+1)/4 - \mu_d(\omega)$ (where $n = 0, 1, \ldots$), $W_d(t)$ takes on the values $(\varepsilon_0^2/2)\omega t\, G''(\omega)$. By Eq. (9.1-89) these points lie on the straight line. The same points satisfy $\ddot{W}_d(t) = 0$ and are, therefore, inflection points of $W_d(t)$.

Again, the statement is frequently found in the literature (see, e.g., Refs. [1,2]) that the energy dissipated over a quarter cycle of the excitation is

$$W_d(\omega)_{qtr} = \int_0^{\varepsilon_0} \omega^{-1}G''(\omega)\dot{\varepsilon}(t)d\,\varepsilon(t) = (\varepsilon_0^2/4)\pi\, G''(\omega) = [\varepsilon_0\sigma_0(\omega)/4]\pi \sin\theta(\omega)\ . \tag{9.1-92}$$

This is erroneous. By the change of variable from $\varepsilon(t) = \varepsilon_0 \sin\omega t$ to t, Eq. (9.1-92) can be brought into the form

$$W_d(\omega)_{qtr} = \int_0^{\pi/2\omega} (\varepsilon_0^2/2)\omega\, G''(\omega)(1 + \cos 2\omega t)dt\ . \tag{9.1-93}$$

Equation (9.1-92) is thus equivalent to assuming that the rate of energy dissipation is given by the integrand in Eq. (9.1-93) and then integrating over the first quarter of a cycle of the excitation. However, the integrand represents the rate of *coherent* energy dissipation, i.e., the rate at which energy would be dissipated if all dissipating mechanisms were in phase. The correct rate, by contrast, is given by Eqs. (9.1-82) or (9.1-90).

The total energy absorbed up to time t, $W(t)$, and the rate at which it is absorbed at time t, $\dot{W}(t)$, can be obtained either from Eqs. (9.0-3) and (9.0-4), or from Eqs. (9.0-2) and (9.0-1). The first choice immediately yields

$$W(t) = (\varepsilon_0^2/4)[G''(\omega)(2\omega t + \sin 2\omega t) + G'(\omega)(1 - \cos 2\omega t)] \tag{9.1-94}$$

from Eqs. (9.1-54) and (9.1-84), and

$$\dot{W}(t) = (\varepsilon_0^2/2)\omega[G''(\omega)(1 + \cos 2\omega t) + G'(\omega)\sin 2\omega t] \tag{9.1-95}$$

from Eqs. (9.1-56) and (9.1-82). The derivatives of the moduli thus cancel and both equations are composed of two terms which are associated with the storage modulus, $G'(\omega)$, and the loss modulus, $G''(\omega)$, respectively. Alternatively, $W(t)$ and $\dot{W}(t)$ can be expressed as

$$W(t) = (\varepsilon_0^2/4)\tilde{G}(\omega)\{2\omega t \sin\theta(\omega) + \cos\theta(\omega) - \cos[2\omega t + \theta(\omega)]\} \tag{9.1-96}$$

and

$$\dot{W}(t) = (\varepsilon_0^2/2)\omega\tilde{G}(\omega)\{\sin\theta(\omega) + \sin[2\omega t + \theta(\omega)]\}\ . \tag{9.1-97}$$

The energy absorption therefore consists of a linear increase with time upon which are superposed oscillations whose frequency is twice the frequency of the excitation. The rate at which energy is absorbed oscillates with the same double frequency.

Equations (9.1-95) and (9.1-94) may also be obtained readily from Eqs. (9.0-1) and (9.0-2). Taking the imaginary parts of Eqs. $(2.5\text{-}30)_2$ and $(2.5\text{-}31)_2$, since the excitation is sinusoidal, and letting $\sigma(\omega) = \sigma(t)$, the stress becomes

$$\sigma(t) = \varepsilon_0 G'(\omega)\sin \omega t + \varepsilon_0 G''(\omega)\cos \omega t \tag{9.1-98}$$

where the terms on the right represent, respectively, the components of the stress which are in phase and out of phase with the strain. Since the rate of strain is $\varepsilon_0\omega \cos \omega t$, Eq. (9.0-1) immediately leads to Eq. (9.1-95). Integration of $\varepsilon_0\omega\, \sigma(t)\cos \omega t$, subject to the initial condition, $W(0) = 0$, yields Eq. (9.1-94) by Eq. (9.0-2). One might infer, erroneously, that $(\varepsilon_0^2/2)\omega G'(\omega)\sin 2\omega t$ and $(\varepsilon_0^2/2)\omega G''(\omega)(1 + \cos 2\omega t)$ represent, respectively, the expressions for the rate at which energy is stored and dissipated, and that the integrals over these expressions, $(\varepsilon_0^2/4)G'(\omega)(1 - \cos 2\omega t)$ and $(\varepsilon_0^2/4)G''(\omega)(2\omega t + \sin 2\omega)$, furnish the expressions for the energy stored and dissipated. This reasoning, however, neglects the lack of phase coherence among the energy storing mechanisms on the one hand, and the energy dissipating mechanisms on the other.

In the steady state one is evidently not interested in the values of the various energy and power functions at $t = 0$ and $t = \infty$, respectively. For the stored energy, a useful parameter is the *average*, taken over a full cycle of the excitation. From Eqs. (9.1-54) and (9.1-61) we find

$$W_s(\omega)_{av} = (\omega/2\pi) \int_{2\pi(n-1)/\omega}^{2\pi n/\omega} W_s(t)dt = (\varepsilon_0^2/4)G'(\omega) = [\varepsilon_0\sigma_0(\omega)/4]\cos \theta(\omega) \tag{9.1-99}$$

which is one half of the maximum coherently storable energy. $G'(\omega)$ is thus seen to be proportional to the average energy stored per unit volume of the material over a cycle of the excitation. This justifies its name as storage modulus. The average energy dissipated over a full cycle increases with each cycle. However, in the steady-state, the total energy dissipated over a full cycle of the excitation is a constant. We have

$$W_d(\omega)_{cyc} = \int_{2\pi(n-1)/\omega}^{2\pi n/\omega} \dot{W}_d(t)dt = \pi\varepsilon_0^2 G''(\omega) = \varepsilon_0\sigma_0(\omega)\pi \sin \theta(\omega) \tag{9.1-100}$$

which turns out to be the same amount of energy that would be dissipated coherently. Thus, $G''(\omega)$ is directly proportional to the energy dissipated per unit volume of the material over a cycle of deformation and is therefore appropriately called the loss modulus. The ratio

$$\frac{W_d(\omega)_{cyc}}{2W_s(\omega)_{av}} = 2\pi \tan \theta(\omega) \tag{9.1-101}$$

is sometimes known as the *specific loss*. The ratio of the energy dissipated coherently to the energy stored coherently per quarter cycle is $\frac{\pi}{2} \tan \theta(\omega)$.

We note that the average rate at which energy is stored is zero, while the average rate at which energy is dissipated equals the rate at which energy is absorbed. Thus,

$$\dot{W}_d(\omega)_{av} = \dot{W}(\omega)_{av} = (\varepsilon_0^2/2)\omega G''(\omega) = [\varepsilon_0\sigma_0(\omega)/2]\omega \sin\theta(\omega) \ . \tag{9.1-102}$$

These rates are proportional to $G''(\omega)$.

The total energy absorbed over a cycle of the excitation equals the energy dissipated since the energy stored over a complete cycle is zero. The average energy absorbed per cycle, of course, increases with each cycle.

On closing this section, a comment must be made regarding Eqs. (9.1-21) and (9.1-102). From Eqs. (4.1-5) and (4.1-7) it follows that

$$\lim_{t\to\infty} \sigma(t) = \lim_{\omega\to 0} \sigma(\omega) = \varepsilon_0\{G_e\} \tag{9.1-103}$$

and from Eqs. (4.1-12) and (4.1-14) that

$$\lim_{t\to 0} \sigma(t) = \lim_{\omega\to\infty} \sigma(\omega) = \varepsilon_0 G_g \ . \tag{9.1-104}$$

Equations (9.1-103) and (9.1-104) imply that at either extremes of the time or frequency scale a linear viscoelastic material approaches the behavior of a purely elastic one. A purely elastic material does not dissipate energy. One would expect to find, therefore, that

$$\lim_{t\to\infty} \dot{W}_d(t) = \lim_{\omega\to 0} \dot{W}_d(\omega)_{av} = 0 \tag{9.1-105}$$

and this is, indeed, borne out by Eqs. (9.1-21) and (9.1-102). One would also expect $\dot{W}_d(t)$ and $\dot{W}_d(\omega)$ to vanish as $t \to 0$ or $\omega \to \infty$. However, the same equations, using Eq. (4.6-22), now give

$$\lim_{t\to 0} \dot{W}_d(t) = \varepsilon_0^2 \langle \tau_M^{-1} \rangle [G_g - \{G_e\}] \tag{9.1-106}$$

and

$$\lim_{\omega\to\infty} \dot{W}_d(\omega)_{av} = (\varepsilon_0^2/2) \langle \tau_M^{-1} \rangle [G_g - \{G_e\}] \ , \tag{9.1-107}$$

i.e. the rate of energy dissipation is predicted to be finite in response to a step excitation as $t \to 0$ or in response to a harmonic excitation as $\omega \to \infty$. It may be noted that $\dot{W}_d(0) = 0$ in slope response as expected. Thus, while the stress-strain relations of linear viscoelastic theory reduce to those of linear elastic theory for these limits, the energy relations do not necessarily do so.

9.2 Stress Excitation (Creep Behavior)

In creep behavior one observes the strain as a response to a stress excitation. The series-parallel model appropriate to the study of energy storage and dissipation is, therefore, the Kelvin model which consists essentially of Voigt units in series (cf. Figs. 3.5-3 and 3.5-4). Because the elements of a Voigt unit are coupled in parallel,

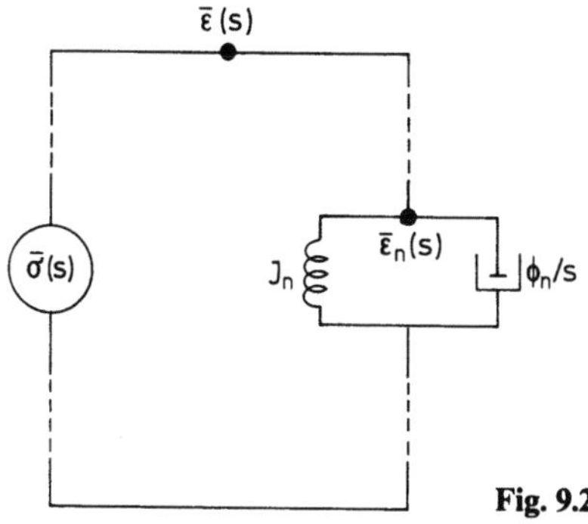

Fig. 9.2-1. The nth Voigt unit of a Kelvin model

the strain across both is the same although it is different across the different units. Analogously, the stresses acting through the two elements are not the same but the same stress acts through all Voigt units, and through the isolated elements as well. Hence it is sufficient to consider a single Voigt unit to determine the general relation which expresses how much of the total stress imposed on the model acts through the nth spring, and n^{th} dashpot, respectively. We proceed analogously to the case of a strain stimulus discussed in Sect. 9.1, and shall consider the step, slope, and harmonic stress excitation. Creep recovery forms the subject of Problem 9.2-3.

The nth Voigt unit of a Kelvin model is shown in Fig. 9.2-1. We have

$$\bar{\varepsilon}_n(s) = \bar{U}_n(s)\bar{\sigma}(s) \tag{9.2-1}$$

where $\bar{\varepsilon}_n(s)$ is the transform of the strain across the nth unit and

$$\bar{U}_n(s) = \frac{J_n}{1 + \tau_n s} \tag{9.2-2}$$

is its retardance. In Eq. (9.2-2), $\tau_n = J_n/\phi_n$ is the nth retardation time. Both elements of the unit experience the same strain. Hence, the transform of the stress through the spring is

$$\bar{\sigma}_{sn}(s) = \bar{\varepsilon}_n(s)/J_n \tag{9.2-3}$$

while that through the dashpot is

$$\bar{\sigma}_{dn}(s) = s\bar{\varepsilon}_n(s)/\phi_n \ . \tag{9.2-4}$$

Expressing $\bar{\varepsilon}_n(s)$ in terms of $\bar{\sigma}(s)$ from Eqs. (9.2-1) and (9.2-2) gives

$$\bar{\sigma}_{sn}(s) = \frac{1}{1 + \tau_n s}\bar{\sigma}(s) \tag{9.2-5}$$

and

$$\bar{\sigma}_{dn}(s) = \frac{\tau_n s}{1 + \tau_n s}\bar{\sigma}(s) \ . \tag{9.2-6}$$

The two stress transforms (and hence, the two stresses) are additive, i.e.

$$\sigma_{sn}(t) + \sigma_{dn}(t) = \sigma(t) \ . \tag{9.2-7}$$

Equations (9.2-5) and (9.2-6) may be used to find the energy stored and the energy dissipated in a material with a given distribution of retardation times, τ_n. For the isolated spring $\bar{\sigma}_{sg}(s) = \bar{\sigma}(s)$, and for the isolated dashpot, if present, $\bar{\sigma}_{df}(s) = \bar{\sigma}(s)$. These results follow from Eq. (9.2-5) by letting $\tau_n = \tau_g \to 0$ and from Eq. (9.2-6) by letting $\tau_n = \tau_f \to \infty$. The subscripts g and f have been chosen to designate the stresses in the isolated spring and dashpot, respectively, because these are associated with J_g, and with ϕ_f.

From Eq. $(9.0\text{-}12)_2$ the energy stored in the response to a stress excitation up to time t is

$$W_s(t) = \tfrac{1}{2}J_g\sigma^2(t) + \tfrac{1}{2}\sum_n J_n\sigma_{sn}^2(t) \tag{9.2-8}$$

where

$$\sigma_{sn}(t) = \mathscr{L}^{-1}\frac{\bar{\sigma}(s)}{1+\tau_n s} \tag{9.2-9}$$

τ_n being the retardation time of the nth Voigt unit (or n^{th} retardation mechanism). For a continuous distribution of retardation times, (or, more precisely, for a continuous distribution of compliances)

$$W_s(t) = \tfrac{1}{2}J_g\sigma^2(t) + \tfrac{1}{2}\int_{-\infty}^{\infty} L(\tau)\sigma_s^2(t;\tau)d\ln\tau \tag{9.2-10}$$

where

$$\sigma_s(t;\tau) = \mathscr{L}^{-1}\frac{\bar{\sigma}(s)}{1+\tau s} \ . \tag{9.2-11}$$

The rate at which energy is dissipated is, by Eq. $(9.0\text{-}8)_2$,

$$\dot{W}_d(t) = \sum_n \phi_n\sigma_{dn}^2(t) + \{\phi_f\}\sigma^2(t) \tag{9.2-12}$$

where

$$\sigma_{dn}(t) = \mathscr{L}^{-1}\frac{\tau_n s\bar{\sigma}(s)}{1+\tau_n s} \tag{9.2-13}$$

for a discrete distribution, and

$$\dot{W}_d(t) = \int_{-\infty}^{\infty} \tau^{-1}L(\tau)\sigma_d^2(t;\tau)d\ln\tau + \{\phi_f\}\sigma^2(t) \tag{9.2-14}$$

where

$$\sigma_d(t;\tau) = \mathscr{L}^{-1}\frac{\tau s\bar{\sigma}(s)}{1+\tau s} \tag{9.2-15}$$

for a continuous distribution of retardation times. Alternative formulations may be obtained (cf. Problem 9.2-10) by carrying out the transform inversions indicated in Eqs. (9.2-9), (9.2-11), (9.2-13), and (9.2-15).

We now apply these relations to the step, slope and harmonic stress excitation.

9.2.1 Step Response

The stress as a step function of time is $\sigma(t) = \sigma_0$ and, therefore, $\bar{\sigma}(s) = \sigma_0/s$. We have

$$\sigma_s(t;\tau) = \mathscr{L}^{-1}\frac{\sigma_0}{s(1+\tau s)} = \sigma_0[1 - \exp(-t/\tau)] \tag{9.2-16}$$

and

$$\sigma_d(t;\tau) = \mathscr{L}^{-1}\frac{\sigma_0\tau}{1+\tau s} = \sigma_0\exp(-t/\tau) \ . \tag{9.2-17}$$

Consequently, the energy stored up to time t is

$$W_s(t) = (\sigma_0^2/2)\left[J_g + \int_{-\infty}^{\infty} L(\tau)[1 - \exp(-t/\tau)]^2 d\ln\tau\right] \tag{9.2-18}$$

and the rate at which it is stored at time t is obtained as

$$\dot{W}_s(t) = (\sigma_0^2/2)\left[J_g\delta(t) + 2\int_{-\infty}^{\infty} \tau^{-1}L(\tau)[1 - \exp(-t/\tau)]\exp(-t/\tau)d\ln\tau\right] \tag{9.2-19}$$

by differentiating Eq. (9.2-18) using Eq. (9.1-18). The rate at which the energy is dissipated at time t becomes

$$\dot{W}_d(t) = \sigma_0^2\left[\int_{-\infty}^{\infty} \tau^{-1}L(\tau)\exp(-2t/\tau)d\ln + \{\phi_f\}\right] \tag{9.2-20}$$

while integration of Eq. (9.2-20) yields the amount dissipated up to time t as

$$W_d(t) = (\sigma_0^2/2)\left[\int_{-\infty}^{\infty} L(\tau)[1 - \exp(-2t/\tau)]d\ln\tau + 2t\{\phi_f\}\right] \ . \tag{9.2-21}$$

We immediately derive

$$W(t) = (\sigma_0^2/2)\left[J_g + 2\int_{-\infty}^{\infty} L(\tau)[1 - \exp(-t/\tau)]d\ln\tau + 2t\{\phi_f\}\right] \tag{9.2-22}$$

for the energy absorbed up to time t by adding Eqs. (9.2-18) and (9.2-21). Differentiation of this, using the appropriate analog of Eq. (9.1-18), or addition of Eqs. (9.2-19) and (9.2-20) yields

$$\dot{W}(t) = \sigma_0^2\left[J_g\delta(2t) + \int_{-\infty}^{\infty} \tau^{-1}L(\tau)\exp(-t/\tau)d\ln\tau + \{\phi_f\}\right] \tag{9.2-23}$$

for the rate at which energy is absorbed at time t. Equations (9.2-22) and (9.2-23) can also be obtained from Eqs. (9.0-2) and (9.0-1) along the lines of derivation of Eqs. (9.1-26) and (9.1-25) as outlined in Problem 9.1-4.

In terms of the creep compliance the energies become

$$W_s(t) = (\sigma_0^2/2)[2J(t) - J(2t)] \tag{9.2-24}$$

$$W_d(t) = (\sigma_0^2/2)[J(2t) - J_g] \tag{9.2-25}$$

and

$$W(t) = (\sigma_0^2/2)[2J(t) - J_g] \tag{9.2-26}$$

while the rates at which energy is stored, dissipated, or absorbed may be expressed as

$$\dot{W}_s(t) = \sigma_0^2[J_g\delta(2t) + \dot{J}(t) - \dot{J}(2t)] = \sigma_0^2[U(t) - U(2t)] \tag{9.2-27}$$

$$\dot{W}_d(t) = \sigma_0^2\dot{J}(2t) = \sigma_0^2[U(2t) - J_g\delta(2t)] \tag{9.2-28}$$

and

$$\dot{W}(t) = \sigma_0^2[J_g\delta(2t) + \dot{J}(t)] = \sigma_0^2[U(t) - J_g\delta(2t)] \ . \tag{9.2-29}$$

Equation (9.2-26) for the energy absorbed by the material up to time t when the stimulus is a step of *stress* may be compared with Eq. (9.1-25) which is valid when a step of *strain* is imposed. In the latter case the deformation is instantaneous. The energy required to do this is also absorbed instantaneously at $t = 0$ and, hence, does not change as a function of time. In the former case, the material absorbs energy equivalent to $(\sigma_0^2/2)J_g$ at $t = 0$, but then proceeds to absorb further energy as it continues to be deformed. For the energy stored at $t = 0$ and $t = \infty$, we have

$$W_s(0) = (\sigma_0^2/2)J_g \qquad \text{and} \qquad W_s(\infty) = (\sigma_0^2/2)J_e^{\{o\}} \ , \tag{9.2-30}$$

i.e. the stored energy increases from its initial value [equivalent to W(0)] to its final value representing, in the language of rheological models, the sum of the energies stored in all the springs at their maximum deformation. The initial value of the energy dissipated is zero, i.e. $W_d(0) = 0$. In a rheodictic material the dissipated as well as the absorbed energies continue to grow indefinitely. Hence $W_d^\circ(\infty) = W^\circ(\infty) = \infty$. For an arrheodictic material (denoted by the superscript $\times$) we have, however,

$$W_d^\times(\infty) = (\sigma_0^2/2)(J_e - J_g) \qquad \text{and} \qquad W^\times(\infty) = (\sigma_0^2/2)(2J_e - J_g) \ . \tag{9.2-31}$$

Thus, at long times, an arrheodictic material will have absorbed both the energy stored in the springs and the energy dissipated in the dashpots.

9.2.2 Slope Response

In slope response the stress is $\sigma(t) = \dot{\sigma}_0 t$ where $\dot{\sigma}_0$ is the (constant) rate of stress. Transformation yields $\bar{\sigma}(s) = \dot{\sigma}_0/s^2$. Substitution into Eqs. (9.2-11) and (9.2-15) leads to

$$\sigma_s(t;\tau) = \mathscr{L}^{-1}\frac{\dot{\sigma}_0}{s^2(1 + \tau s)} = \dot{\sigma}_0\{t - \tau[1 - \exp(-t/\tau)]\} \tag{9.2-32}$$

and

$$\sigma_d(t;\tau) = \mathscr{L}^{-1}\frac{\dot{\sigma}_0}{s(1 + \tau s)} = \dot{\sigma}_0\tau[1 - \exp(-t/\tau)] \ . \tag{9.2-33}$$

Hence, the energy stored up to time t in response to a slope stress is

$$W_s(t) = (\dot{\sigma}_0^2/2)\left[J_g t^2 + \int_{-\infty}^{\infty} L(\tau)\{t - \tau[1 - \exp(-t/\tau)]\}^2 d\ln\tau\right] \tag{9.2-34}$$

while the rate at which it is dissipated at time t is

$$\dot{W}_d(t) = \dot{\sigma}_0^2\left[\int_{-\infty}^{\infty} \tau L(\tau)[1 - \exp(-t/\tau)]^2 d\ln\tau + \{\phi_f\}t^2\right]. \tag{9.2-35}$$

Integration between the limits 0 and t yields the energy dissipated up to time t as

$$W_d(t) = (\dot{\sigma}_0^2/2)\left[\int_{-\infty}^{\infty} \tau L(\tau)\{2t - \tau[1 - \exp(-t/\tau)][3 - \exp(-t/\tau)]\}d\ln\tau + \tfrac{2}{3}\{\phi_f\}t^3\right], \tag{9.2-36}$$

while differentiation of Eq. (9.2-34) gives the rate at which energy is stored at time t as

$$\dot{W}_s(t) = \dot{\sigma}_0^2\left[J_g t + \int_{-\infty}^{\infty} L(\tau)\{t - \tau[1 - \exp(-t/\tau)]\}[1 - \exp(-t/\tau)]d\ln\tau\right]. \tag{9.2-37}$$

Finally, Eqs. (9.0-3) and (9.0-4) give the energy absorbed up to time t as

$$W(t) = (\dot{\sigma}_0^2/2)\left[J_e^{\{o\}}t^2 - 2\int_{-\infty}^{\infty} \tau^2 L(\tau)[1 - \exp(-t/\tau) - (t/\tau)\exp(-t/\tau)]d\ln\tau + \tfrac{2}{3}\{\phi_f\}t^3\right] \tag{9.2-38}$$

and the rate at which it is absorbed at time t as

$$\dot{W}(t) = \dot{\sigma}_0^2\left[J_g t + t\int_{-\infty}^{\infty} L(\tau)[1 - \exp(-t/\tau)]d\ln\tau + \{\phi_f\}t^2\right]. \tag{9.2-39}$$

It is clear that at $t = 0$ all energies and all rates are zero. An arrheodictic material cannot be deformed indefinitely under a constant stress. For a rheodictic material all rates and all energies approach ∞ as $t \to \infty$. Now, by Eq. (2.4-10), $\dot{\varepsilon}(t) = \dot{\sigma}_0 J(t)$, and Eq. (9.0-1) furnishes

$$\dot{W}(t) = \dot{\sigma}_0^2 t\, J(t) \tag{9.2-40}$$

which is, of course, immediately obtainable from Eq. (9.2-39) also. Integration by parts then yields

$$W(t) = \dot{\sigma}_0^2 \int_0^t uJ(u)du = \dot{\sigma}_0^2[t\chi(t) - \rho(t)] \tag{9.2-41}$$

where $\chi(t)$ is defined by Eq. $(2.4\text{-}11)_1$ and

$$\rho(t) = \int_0^t \chi(u)du\,. \tag{9.2-42}$$

The reader may show (see Problem 9.2-4) that Eqs. (9.2-41) and (9.2-38) are identical. The function $\rho(t)$ is an experimental response function just as $J(t)$ and $\chi(t)$, or $G(t)$, $\eta(t)$, and $\psi(t)$. It can be obtained as the area under the $\chi(t)$ curve, or, equivalently (see Problem 9.2-5) as the area under the strain-stress curve. In terms of these response functions the energies may be expressed as

$$W_s(t) = (\dot{\sigma}_0^2/2)[2\,t\chi(t) + 2\,\rho(t) - \rho(2t)] \tag{9.2-43}$$

and

$$W_d(t) = (\dot{\sigma}_0^2/2)[\rho(2t) - 4\,\rho(t)] \tag{9.2-44}$$

while the rates will be given by

$$\dot{W}_s(t) = \dot{\sigma}_0^2[tJ(t) + 2\,\chi(t) - \chi(2t)] \tag{9.2-45}$$

and

$$\dot{W}_d(t) = \dot{\sigma}_0^2[\chi(2t) - 2\,\chi(t)]\ . \tag{9.2-46}$$

9.2.3 Harmonic Response

Let the harmonic stress excitation be given by $\sigma(t) = \sigma_0 \sin \omega t$. Then $\bar{\sigma}(s) = \sigma_0\omega/(s^2 + \omega^2)$ by LTP (14) and Eq. (9.2-11) gives

$$\bar{\sigma}_s(s;\tau) = \frac{\sigma_0\omega}{(s^2+\omega^2)(1+\tau s)} \tag{9.2-47}$$

which consists of a transient as well as a steady-state part. As in Sect. 9.1.3, we obtain the steady-state stress by decomposing into partial fractions and retaining only the term whose poles arise from the excitation function. This leads to

$$\bar{\sigma}_s^{ss}(s;\tau) = \frac{\sigma_0\omega(1-\tau s)}{(s^2+\omega^2)(1+\omega^2\tau^2)}\ . \tag{9.2-48}$$

Retransformation to the time domain then yields

$$\sigma_s^{ss}(t;\tau) = \frac{\sigma_0}{1+\omega^2\tau^2}(\sin \omega t - \omega\tau \cos \omega t)\ . \tag{9.2-49}$$

Introduction of this equation into Eq. (9.2-10) furnishes the energy stored up to time t in a harmonic stress excitation as

$$\begin{aligned} W_s(t) = \frac{\sigma_0^2}{4}\Bigg\{ & J_g + \int_{-\infty}^{\infty} L(\tau)\frac{1}{1+\omega^2\tau^2}\, d\ln\tau \\ & - \left[J_g + \int_{-\infty}^{\infty} L(\tau)\frac{1-\omega^2\tau^2}{(1+\omega^2\tau^2)^2}\, d\ln\tau\right]\cos 2\omega t \\ & - \int_{-\infty}^{\infty} L(\tau)\frac{2\omega\tau}{(1+\omega^2\tau^2)^2}\, d\ln\tau \,\sin 2\omega t\Bigg\}\ . \end{aligned} \tag{9.2-50}$$

But

$$J'(\omega) + \frac{d\,J'(\omega)}{d\ln\omega} = J_g + \int_{-\infty}^{\infty} \frac{1 - \omega^2\tau^2}{(1 + \omega^2\tau^2)^2} d\ln\tau \tag{9.2-51}$$

and

$$J''(\omega) + \frac{d\,J''(\omega)}{d\ln\omega} = \int_{-\infty}^{\infty} L(\tau)\frac{2\omega\tau}{(1 + \omega^2\tau^2)^2} d\ln\tau\ . \tag{9.2-52}$$

With the use of these relations we obtain

$$W_s(t) = \frac{\sigma_0^2}{4} J'(\omega) - \left[\left(J'(\omega) + \frac{d\,J'(\omega)}{d\ln\omega}\right)\cos 2\omega t - \left(J''(\omega) + \frac{d\,J''(\omega)}{d\ln\omega}\right)\sin 2\omega t\right] \tag{9.2-53}$$

for the energy stored up to time t. For $t = 0$ we find

$$W_s(0) = -(\sigma_0^2/4)\frac{d\,J'(\omega)}{d\ln\omega}\ . \tag{9.2-54}$$

Differentiation of Eq. (9.2-53) yields the rate at which the energy is dissipated at time t as

$$\dot{W}_s(t) = (\sigma_0^2/2)\omega\left[\left(J'(\omega) + \frac{d\,J'(\omega)}{d\ln\omega}\right)\sin 2\omega t - \left(J''(\omega) + \frac{d\,J''(\omega)}{d\ln\omega}\right)\cos 2\omega t\right]\ . \tag{9.2-55}$$

As in Sect. 9.3, we transform these equations by setting

$$J'(\omega) + \frac{d\,J'(\omega)}{d\ln\omega} = S_s'(\omega) = \tilde{S}_s(\omega)\cos 2\,v_s(\omega) \tag{9.2-56}$$

and

$$J''(\omega) + \frac{d\,J''(\omega)}{d\ln\omega} = S_s''(\omega) = \tilde{S}_s(\omega)\sin 2\,v_s(\omega) \tag{9.2-57}$$

where

$$\tilde{S}_s(\omega) = \sqrt{[S_s'(\omega)]^2 + [S_s''(\omega)]^2} \tag{9.2-58}$$

and

$$\tan 2\,v_s(\omega) = S_s''(\omega)/S_s'(\omega)\ . \tag{9.2-59}$$

This allows us to write

$$W_s(t) = (\sigma_0^2/4)\{J'(\omega) - \tilde{S}_s(\omega)\cos 2[\omega t - v_s(\omega)]\} \tag{9.2-60}$$

and

$$\dot{W}_s(t) = (\sigma_0^2/2)\omega\,\tilde{S}_s(\omega)\sin 2[\omega t - v_s(\omega)]\ . \tag{9.2-61}$$

Thus the stored energy oscillates with twice the frequency of the excitation and leads the latter by the angle $v_s(\omega)$ which arises from the combined effect of the lack of phase coherence in the energy storing mechanisms. Storage in the nth mechanism is out of phase with the excitation by the angle $v_n(\omega) = \arctan(\omega\tau_n)$ where τ_n is the retardation time associated with the nth mechanism, or n[th] Voigt unit. The maximum *coherently* storable energy is

$$W_s(\omega)_{max} = (\sigma_0^2/2)J'(\omega) = [\sigma_0\varepsilon_0(\omega)/2]\cos\theta(\omega) . \tag{9.2-62}$$

At low and high frequencies we have, with good approximation,

$$\lim_{\omega\to 0} W_s(\omega)_{max} = (\varepsilon_0^2/2)J_e^{\{o\}} \quad \text{and} \quad \lim_{\omega\to\infty} W_s(\omega)_{max} = (\varepsilon_0^2/2)J_g . \tag{9.2-63}$$

Relations analogous to Eqs. (9.1-68), (9.1-69), (9.1-70), (9.1-74), and (9.1-75) are easily written down. Thus, e.g., the energy cycled between the material and the driving system is

$$W_s(\omega)_{cyc} = (\sigma_0^2/2)\tilde{S}_s(\omega) . \tag{9.2-64}$$

Again, $W_s(\omega)_{cyc} \le W_s(\omega)_{max}$. The ratio $W_s(\omega)_{cyc}/W_s(\omega)_{max}$, which equals $\tilde{S}_d(\omega)/J'(\omega)$, is plotted in Fig. 9.2-2 together with the loss angle, $\theta(\omega)$, and the angles $2\,v_s(\omega)$ and $2\,v_d(\omega)$ for the same hypothetical arrheodictic material whose behavior in response to a strain excitation was shown in Fig. 9.1-2 with which the present figure should be compared. The behavior was modelled by Eqs. (6.1-94) and (6.1-95), from which the equations employed in constructing Fig. 9.1-2 were derived. The parameters $\omega_0 = 1\,s^{-1}$ and $c = 0.5$ were used, together with the values $J_g = 1/G_g$ and $J_e = 1/G_e$.

Turning now to energy dissipation in a harmonic stress excitation,

$$\bar{\sigma}_d(s;\tau) = \frac{\sigma_0\omega\tau s}{(s^2+\omega^2)(1+\tau s)} \tag{9.2-65}$$

by Eq. (9.2-15). Partial fraction decomposition and rejecting the transient term yields

$$\bar{\sigma}_d^{ss}(t;\tau) = \frac{\sigma_0\omega(\tau s + \omega^2\tau^2)}{(s^2+\omega^2)(1+\omega^2\tau^2)} . \tag{9.2-66}$$

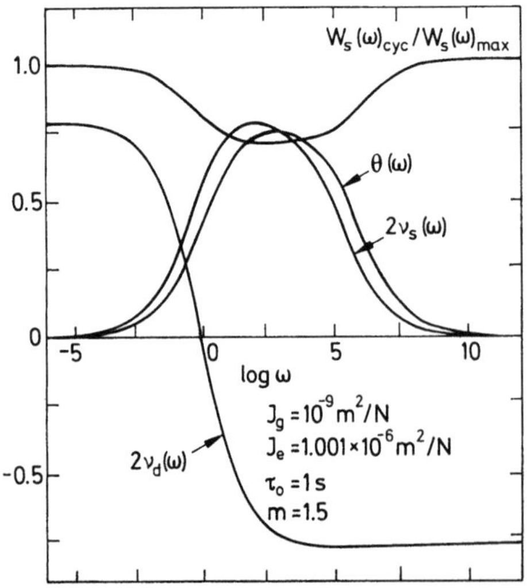

Fig. 9.2-2. $W_s(\omega)_{act}/W_s(\omega)_{max}$, $\theta(\omega)$, $2v_s(\omega)$, and $2v_d(\omega)$ as functions of log ω for typical arrheodictic behavior

Hence, in the steady-state

$$\sigma_d^{ss}(t;\tau) = \frac{\sigma_0 \omega\tau}{1+\omega^2\tau^2}(\cos \omega t + \omega\tau \sin \omega t) \tag{9.2-67}$$

and substitution into Eq. (9.2-14) leads to

$$\dot{W}_d(t) = (\sigma_0^2/2)\omega\left[\int_{-\infty}^{\infty} L(\tau)\frac{\omega\tau}{1+\omega^2\tau^2} d\ln\tau + \int_{-\infty}^{\infty} L(\tau)\frac{2\omega^2\tau^2}{(1+\omega^2\tau^2)^2} d\ln\tau \cdot \sin 2\omega t \right.$$
$$\left. + \int_{-\infty}^{\infty} L(\tau)\frac{\omega\tau(1-\omega^2\tau^2)}{(1+\omega^2\tau^2)^2} d\ln\tau \cdot \cos 2\omega t + \{\phi_f/\omega\}(1-\cos 2\omega t)\right]. \tag{9.2-68}$$

But

$$\frac{d\,J'(\omega)}{d\ln\omega} = -\int_{-\infty}^{\infty} L(\tau)\frac{2\omega^2\tau^2}{(1+\omega^2\tau^2)^2} d\ln\tau \tag{9.2-69}$$

$$\frac{d\,J''(\omega)}{d\ln\omega} = \int_{-\infty}^{\infty} L(\tau)\frac{\omega\tau(1-\omega^2\tau^2)}{(1+\omega^2\tau^2)^2} d\ln\tau - \{\phi_f\}/\omega \tag{9.2-70}$$

and, consequently,

$$\dot{W}_d(t) = (\sigma_0^2/2)\omega\left[J''(\omega) + \frac{d\,J''(\omega)}{d\ln\omega}\cos 2\omega t - \frac{d\,J'(\omega)}{d\ln\omega}\sin 2\omega t\right] \tag{9.2-71}$$

is the rate at which energy is dissipated at time t in response to a harmonic stress excitation.

To obtain the amount of energy dissipated, we integrate Eq. (9.2-71) subject to the initial condition $W_d(0) = W(0) - W_s(0)$. Allowing $W(0)$ to vanish gives

$$W_d(0) = (\sigma_0^2/4)\frac{d\,J'(\omega)}{d\ln\omega} \tag{9.2-72}$$

by Eq. (9.2-54). Hence,

$$W_d(t) = (\sigma_0^2/4)\left[2\omega\, J''(\omega) + \frac{d\,J''(\omega)}{d\ln\omega}\sin 2\omega t + \frac{d\,J'(\omega)}{d\ln\omega}\cos 2\omega t\right]. \tag{9.2-73}$$

We now let

$$-\frac{d\,J'(\omega)}{d\ln\omega} = S_d'(\omega) = \tilde{S}_d(\omega)\cos 2\,\nu_d(\omega) \tag{9.2-74}$$

and

$$\frac{d\,J''(\omega)}{d\ln\omega} = S_d''(\omega) = \tilde{S}_d(\omega)\sin 2\,\nu_d(\omega) \tag{9.2-75}$$

where

$$\tilde{S}_d(\omega) = \sqrt{[S_d'(\omega)]^2 + [S_d''(\omega)]^2} \tag{9.2-76}$$

and

$$\tan 2\nu_d(\omega) = S''_d(\omega)/S'_d(\omega) \ . \tag{9.2-77}$$

The angle $2\nu_d$ is plotted in Fig. 9.2-2 as a function of log ω for a typical arrheodictic material. We now rewrite Eqs. (9.2-73) and (9.2-71) as

$$W_d(t) = (\sigma_0^2/4)\{2\omega t\, J''(\omega) - \tilde{S}_d(\omega)\cos 2[\omega t + \nu_d(\omega)]\} \tag{9.2-78}$$

and

$$\dot{W}_d(t) = (\sigma_0^2/2)\omega\{J''(\omega) + \tilde{S}_d(\omega)\sin 2[\omega t + \nu_d(\omega)]\} \ . \tag{9.2-79}$$

These equations have the same form as Eqs. (9.1-89) and (9.1-90). The rate of energy dissipation oscillates between the limits $(\sigma_0^2/2)\omega[J''(\omega) \pm \tilde{S}_d(\omega)]$ with twice the frequency of the excitation. The maximum rate of coherent energy dissipation results as

$$\dot{W}_d(\omega)_{max} = \sigma_0^2\omega J''(\omega) = \varepsilon_0(\omega)\sigma_0 \sin\theta(\omega) \ . \tag{9.2-80}$$

Analogous remarks as those made in Sect. 9.1.3 with reference to the energy dissipated in response to a harmonic strain excitation can also be made here.

It remains to consider the total energy absorbed up to time t, W(t), and the rate at which it is dissipated at time t. These can be obtained readily as described in Sect. 9.1.3 and we shall therefore merely present the results here. We have

$$W(t) = (\sigma_0^2/4)[J''(\omega)(2\omega t - \sin 2\omega t) + J'(\omega)(1 - \cos 2\omega t)] \tag{9.2-81}$$

and

$$\dot{W}(t) = (\sigma_0^2/2)\omega[J''(\omega)(1 - \cos 2\omega t) + J'(\omega)\sin 2\omega t] \ , \tag{9.2-82}$$

or, alternatively,

$$W(t) = (\sigma_0^2/2)\tilde{J}(\omega)\{\omega t \sin\theta(\omega) + \sin\omega t \sin[\omega t - \theta(\omega)]\} \tag{9.2-83}$$

and

$$\dot{W}(t) = \sigma_0^2\omega\tilde{J}(\omega)\sin\omega t \cos[\omega t - \theta(\omega)] \ . \tag{9.2-84}$$

We note that the energy absorbed vanishes at t = 0. This is the condition we have (arbitrarily) chosen as the initial condition.

Taking the imaginary parts of Eqs. $(2.5\text{-}36)_2$ and $(2.5\text{-}37)_2$ [cf. Eq. (9.1-98)] it is seen that the steady-state strain is

$$\varepsilon(t) = \sigma_0 J'(\omega)\sin\omega t - \sigma_0 J''(\omega)\cos\omega t \tag{9.2-85}$$

where the terms on the right represent, respectively, the components of the strain which are in phase and out of phase with the stress. Again, the reader must be cautioned against identifying these as the terms associated, respectively, with energy storage and dissipation. Such identification would lead to expressions for energy stored or dissipated *coherently*, and would not take into account the fact that the various energy storing mechanisms are not in phase with each other, and neither are the dissipating mechanisms.

In concluding this section we point out that over a full cycle of deformation the energy stored averages to

$$W_s(\omega)_{av} = (\sigma_0^2/4)J'(\omega) = [\sigma_0\varepsilon_0(\omega)/4]\cos\theta(\omega) \tag{9.2-86}$$

and the total energy dissipated over a full cycle is

$$W_d(\omega)_{cyc} = \sigma_0^2\pi\, J''(\omega) = \sigma_0\varepsilon_0(\omega)\pi \sin\theta(\omega)\ . \tag{9.2-87}$$

The specific loss is still given by Eq. (9.1-101). The average rate at which energy is stored is again zero, and the average rate at which energy is dissipated or absorbed becomes

$$\dot{W}_d(\omega)_{av} = \dot{W}(\omega)_{av} = (\sigma_0^2/2)\omega J''(\omega) = [\sigma_0\varepsilon_0(\omega)/2]\omega \sin\theta(\omega)\ . \tag{9.2-88}$$

These relations are all analogous to those given in Sect. 9.1.

9.3 Hysteresis

The energy absorbed by a purely elastic material during deformation is stored elastic energy which is wholly released in the recovery from the deformation. In a viscoelastic material the storage of elastic energy during the deformation is accompanied by energy dissipation due to viscous losses. In the recovery, stored elastic energy is released but this process is again accompanied by the dissipation of energy through viscous losses. The net loss of energy over a complete cycle of deformation and recovery is referred to as the *hysteretic loss* and the phenomenon of energy loss over such a cycle as *hysteresis*. The word is derived from the Greek word *hysterein*, to be behind or to lag, and expresses the fact that the energy losses in a viscoelastic material are associated with the lag in the response to a stimulus caused by viscous dissipation. In an elastic material there is no hysteresis. Hysteretic effects (heat generation) are important in cyclic deformations.

In a *steady-state cyclic deformation* the response as well as the stimulus return at the end of the cycle to the values they had assumed at the beginning of it. In this case the energy stored during the deformation is completely released during the recovery and the hysteretic loss equals the net energy absorbed during the process of taking unit volume of the material through the closed loop. Denoting the hysteretic loss by $W_d(\text{loop})$ and the net energy absorbed over the loop by $W(\text{loop})$, we have

$$W_d(\text{loop}) = W(\text{loop})\ . \tag{9.3-1}$$

The net energy absorbed over a complete cycle is given by the contour integral around the loop formed by plotting the stress, $\sigma(t)$, as a function of the strain, $\varepsilon(t)$. By Eq. (9.0-2) we have

$$W(\text{loop}) = \oint \sigma(t)\, d\varepsilon(t) \tag{9.3-2}$$

as the fundamental equation of hysteresis. It follows from Eqs. (9.3-1) and (9.3-2) that, in a steady-state cyclic deformation the hysteretic loss can be obtained from the area of the loop.

In a *non-steady-state cyclic deformation* the response does not return to its initial value at the completion of the cycle because the energy stored in the deformation is not completely released in the recovery, and there will thus be a net energy storage over the cycle. Denoting the energy stored at the completion of the cycle by $W_s(\text{loop})$, we then have

$$W_d(\text{loop}) = W(\text{loop}) - W_s(\text{loop}) \tag{9.3-3}$$

and the hysteretic loss can *not* be determined from the area.

Creep and creep recovery, and relaxation of stress after cessation of steady-state flow discussed in Sects. 7.1.1 and 7.1.2 are deformations of this type and can be treated by the methods described in this chapter (see Problems 9.3-1 and 9.3-2). Triangular excitations (see Sects. 7.2.3 and 7.2.4) are commonly used to study hysteresis. These excitations, both in the non-steady-state and in the steady-state, will be discussed in Sect. 9.3.2. First, however, we deal with the commonest type of steady-state cyclic deformation, the response to harmonic excitations.

9.3.1 Response to Harmonic Excitations

We begin by considering a harmonic *strain* excitation. Let the strain be given by

$$\varepsilon(t) = \varepsilon_0 \sin \omega t \ . \tag{9.3-4}$$

Then the steady-state stress is

$$\sigma(t) = \sigma_0(\omega) \sin[\omega t + \theta(\omega)] \ , \tag{9.3-5}$$

by Eq. (2.5-28) and the contour integral becomes

$$\oint \sigma(t) d\varepsilon(t) = \omega \varepsilon_0 \sigma_0(\omega) \int_0^{2\pi/\omega} \cos \omega t \sin[\omega t + \theta(\omega)] dt \ . \tag{9.3-6}$$

Thus, by Eq. (9.3-2), the energy absorbed over a full cycle is

$$W(\text{loop}) = W(\omega)_{\text{cyc}} = \pi \varepsilon_0 \sigma_0(\omega) \sin \theta(\omega) = \pi \varepsilon_0^2 G''(\omega) \ . \tag{9.3-7}$$

But, by Eq. (9.1-100), this is identical with the energy dissipated over the cycle, $W_d(\omega)_{\text{cyc}}$. Hence, in a harmonic deformation the net energy dissipated is also the total net energy absorbed by the material over a cycle of the deformation and there is no net stored energy. This is in accordance with the steady-state nature of the deformation.

Equations (9.3-4) and (9.3-5) are the parametric equations of an ellipse. Elimination of ωt between them (see Problem 9.3-1) leads to

$$\frac{\sigma^2(t)}{\sigma_0^2(\omega)\sin^2\theta(\omega)} - \frac{2\,\sigma(t)\varepsilon(t)\cos\theta(\omega)}{\varepsilon_0\sigma_0(\omega)\sin^2\theta(\omega)} + \frac{\varepsilon^2(t)}{\varepsilon_0^2\sin^2\theta(\omega)} = 1 \tag{9.3-8}$$

as the equation of any of the Lissajous ellipses displayed on the right hand side of Fig. 2.5-4. The ellipse represents the hysteresis loop in a harmonic excitation. The left hand side of Fig. 9.3-1 shows one half of an ellipse drawn for $\theta(\omega) = \pi/6$. On the right are shown the stress and strain as functions of time. As indicated by the

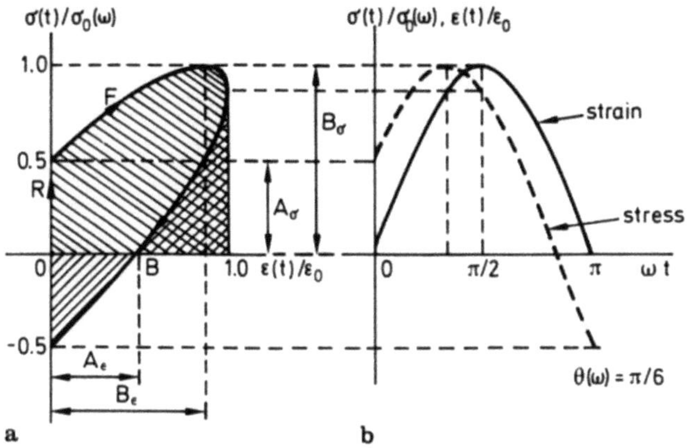

Fig. 9.3-1. Stress as a function of strain (**a**) and of time (**b**) in a harmonic excitation.

arrows, the ellipse is traversed clockwise as time increases. We also note that

$$A_\sigma = \frac{\sigma(t)|_{\varepsilon(t)=0}}{\sigma_0(\omega)} = \sin\theta(\omega) \tag{9.3-9}$$

$$B_\sigma = \frac{\sigma(t)|_{\varepsilon(t)=\varepsilon_0}}{\sigma_0(\omega)} = \cos\theta(\omega) \tag{9.3-10}$$

$$A_\sigma/B_\sigma = \frac{\sigma(t)|_{\varepsilon(t)=0}}{\sigma(t)|_{\varepsilon(t)=\varepsilon_0}} = \tan\theta(\omega) \tag{9.3-11}$$

$$A_\varepsilon = \frac{\varepsilon(t)|_{\sigma(t)=0}}{\varepsilon_0} = \sin\theta(\omega) \tag{9.3-12}$$

$$B_\varepsilon = \frac{\varepsilon(t)|_{\sigma(t)=\sigma_0(\omega)}}{\varepsilon_0} = \cos\theta(\omega) \tag{9.3-13}$$

and

$$A_\varepsilon/B_\varepsilon = \frac{\varepsilon(t)|_{\sigma(t)=0}}{\varepsilon(t)|_{\sigma(t)=\sigma_0(\omega)}} = \tan\theta(\omega)\ . \tag{9.3-14}$$

These relations may be used to determine the loss angle $\theta(\omega)$ from an ellipse obtained by displaying a harmonic stress as function of the strain.

We now consider the energy relations in more detail. Making use of the symmetry of the ellipse, we examine one half of it only. The half-loop is formed by the three sections denoted F, B, and R in Fig. 9.3-1. The section denoted F is the *forward section* during which the strain increases from zero to its maximum amplitude, ε_0. The section denoted B is the *backward section* during which the strain again decreases to zero. Finally, R marks the (fictive) path along which the stress would be returned from $-\sigma_0 \sin\theta(\omega) = -\varepsilon_0 G''(\omega)$ to $\sigma_0 \sin\theta(\omega) = \varepsilon_0 G''(\omega)$ at zero strain. Since the strain is zero, no energy would be absorbed along this section.

We obtain the energy stored and the energy dissipated along the forward section from

$$\int_F \dot{W}_s(t)dt = \int_0^{\pi/2\omega} \dot{W}_s(t)dt = W_s(\omega)_F \tag{9.3-15}$$

and

$$\int_F \dot{W}_d(t)dt = \int_0^{\pi/2\omega} \dot{W}_d(t)dt = W_d(\omega)_F \tag{9.3-16}$$

by introducing Eqs. (9.1-56) and (9.1-82), respectively. This leads to

$$W_s(\omega)_F = (\varepsilon_0^2/2)\left(G'(\omega) - \frac{d\,G'(\omega)}{d\ln\omega}\right) \tag{9.3-17}$$

and

$$W_d(\omega)_F = (\varepsilon_0^2/2)\left(\tfrac{\pi}{2}\,G''(\omega) + \frac{d\,G'(\omega)}{d\ln\omega}\right) . \tag{9.3-18}$$

The sum of the two integrals is the energy absorbed along this path. It is given by

$$W(\omega)_F = (\varepsilon_0^2/2)[\tfrac{\pi}{2}\,G''(\omega) + G'(\omega)] \tag{9.3-19}$$

and is represented by the area shaded downwards from left to right in Fig. 9.3-1, including the cross-hatched area. Equation (9.3-19) can, of course, also be obtained directly from

$$W(\omega)_F = \omega\varepsilon_0\sigma_0(\omega) \int_0^{\pi/2\omega} \cos\omega t\,\sin[\omega t + \theta(\omega)]dt . \tag{9.3-20}$$

It is important to realize that the first and second terms of Eq. (9.3-19) do not represent, respectively, the energy dissipated and stored along the forward section. It may be noted also that the net energy stored along this quarter cycle is smaller than the maximum energy cycled between the material and the driving system because that energy is not absorbed between the limits $t_1 = 0$ and $t_2 = \pi/2\omega$ but between the limits $t_1 = -\mu_s(\omega)/\omega$ and $t_2 = \pi/2\omega - \mu_s(\omega)/\omega$ [cf. Eq. (9.1-69)].

The energies stored, dissipated, and absorbed along the backward section are obtained similarly, integrating between the limits $t_1 = \pi/2\omega$ and $t_2 = \pi/\omega$. We have

$$\int_B \dot{W}_s(t)dt = \int_{\pi/2\omega}^{\pi/\omega} \dot{W}_s(t)dt = W_s(\omega)_B \tag{9.3-21}$$

and

$$\int_B \dot{W}_d(t)dt = \int_{\pi/2\omega}^{\pi/\omega} \dot{W}_d(t)dt = W_d(\omega)_B \tag{9.3-22}$$

and find

$$W_s(\omega)_B = -(\varepsilon_0^2/2)\left(G'(\omega) - \frac{d\,G'(\omega)}{d\ln\omega}\right) \tag{9.3-23}$$

$$W_d(\omega)_B = (\varepsilon_0^2/2)\left(\tfrac{\pi}{2}G''(\omega) - \frac{d\,G'(\omega)}{d\ln\omega}\right) \tag{9.3-24}$$

and

$$W(\omega)_B = (\varepsilon_0^2/2)[\tfrac{\pi}{2}G''(\omega) - G'(\omega)]\ . \tag{9.3-25}$$

The last equation gives the area under the curve marked B which consists of the area shaded upwards from left to right, including the cross-hatched area in Fig. 9.3-1. We see that the energy stored during the first quarter of the deformation is completely released during the second quarter. However, because $G'(\omega)$ is monotone non-decreasing [i.e. $G'(\omega) \geq 0$, $dG'(\omega)/d\ln \geq 0$], less energy is dissipated (or absorbed) along the backward section than along the forward one. For both sections together

$$\int_{F+B} \dot{W}(t)dt = \int_{F+B} \dot{W}_d(t)dt = \pi\varepsilon_0^2 G''(\omega)/2 \tag{9.3-26}$$

and this represents exactly half of the area of the complete ellipse as it should [cf. Eq. (9.3-7)]. We note that, by Eq. $(9.1\text{-}99)_2$,

$$W_s(\omega)_{av} = 4[W(\omega)_F - W(\omega)_B]\ , \tag{9.3-27}$$

i.e. the average energy stored per cycle can be obtained from the areas measured on the half-ellipse.

So far, we have discussed hysteresis in response to a harmonic *strain* stimulus. Analogous results may be obtained readily for a harmonic *stress* excitation. In this case the parametric equations

$$\sigma(t) = \sigma_0 \sin\omega t \tag{9.3-28}$$

and

$$\varepsilon(t) = \varepsilon_0(\omega)\sin[\omega t - \theta(\omega)] \tag{9.3-29}$$

[cf. Eqs. (2.5-34)] again lead to Eq. (9.3-8) but with σ_0 and $\varepsilon_0(\omega)$ in place of $\sigma_0(\omega)$ and ε_0. Equation (9.3-7) becomes

$$W(\omega)_{loop} = \pi\varepsilon_0(\omega)\sigma_0 \sin\theta(\omega) = \pi\sigma_0^2 J''(\omega)\ . \tag{9.3-30}$$

A plot of $\varepsilon(t)/\varepsilon_0(\omega)$ against $\sigma(t)/\sigma_0$ for the same loss angle is identical in shape to that shown in Fig. 9.3-1 but with σ substituted for ε and vice versa. Thus now

$$\sin\theta(\omega) = \frac{\varepsilon(t)|_{\sigma(t)=\sigma_0}}{\varepsilon_0(\omega)} \tag{9.3-31}$$

and the other relations follow analogously.

9.3.2 Response to Triangular Excitations

A common non-steady-state cyclic deformation results from the triangular *strain* excitations discussed in Sects. 7.1.3 and 7.3. We shall treat here only the *regular* triangular excitation in which the rate of strain is the same in the deformation and in the recovery half-cycles. Triangular excitations under constant rates of stress can be dealt with analogously. However, such excitations are less commonly used.

We first examine energy storage and dissipation in a single cycle. Extension to an arbitrary number of cycles using the relations developed in Sect. 7.3.1 is straightforward and will not be undertaken here. Instead, we shall proceed directly to the steady-state relations to contrast them with those appropriate for a single cycle.

9.3.2.1 Single Cycle Response

For a regular triangular strain pulse we have

$$\varepsilon(t) = \begin{cases} \dot{\varepsilon}_0 t, & 0 \le t \le t' \\ -\dot{\varepsilon}_0(t - 2t'), & t' \le t \le 2t' \\ 0, & t > 2t' \end{cases} \tag{9.3-32}$$

[cf. Eq. (7.1-46)] and the stress in response to it becomes

$$\sigma(t) = \begin{cases} \dot{\varepsilon}_0 \eta(t), & 0 \le t \le t' \\ \dot{\varepsilon}_0[\eta(t) - 2\eta(t - t')], & t' \le t \le 2t' \\ \dot{\varepsilon}_0[\eta(t) - 2\eta(t - t') + \eta(t - 2t')], & t > 2t' \end{cases} \tag{9.3-33}$$

[cf. Eq. (7.1-50)]. These relations are plotted in Fig. 9.3-2 below using Eq. (6.1-60) with the term $G_e t$ added to account for arrheodictic behavior. The values were $G_e = 10^6/1.001$ N/m^2, $\eta_f = 5 \times 10^6$ Ns/m^2, $t_0 = 1$ s, $c = 0.5$, and $t' = 5$ s.

Figure 9.3-2a shows the (normalized) stress, $\sigma(t)/\sigma(t')$, as a function of the (normalized) strain, $\varepsilon(t)/\varepsilon(t')$. Figure 9.3-2b shows both as functions of the time.

The energy absorbed over a complete cycle will depend on the half-period t'. To obtain $W(t')_{loop}$ we split the complete cycle into a deformation and a recovery

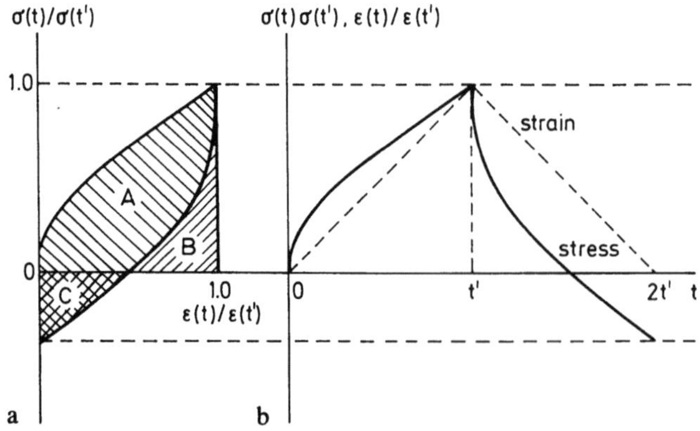

Fig. 9.3-2. Stress as a function of strain (**a**) and of time (**b**) in a (regular) triangular pulse excitation

half-cycle and obtain the contour integral, Eq. (9.3-2), from

$$W(t')_{loop} = W(t')_{def} + W(t')_{rec} \tag{9.3-34}$$

where $W(t')_{def}$ is obtained by taking the contour from $\varepsilon(0)$ to $\varepsilon(t')$, and $W(t')_{rec}$ by taking it from $\varepsilon(t')$ back to $\varepsilon(2t') = \varepsilon(0)$. No energy is absorbed along the ordinate and this portion of the contour contributes nothing to $W(t')_{loop}$.

We thus have

$$W(t')_{def} = \dot{\varepsilon}_0^2 \int_0^{t'} \eta(t)dt \tag{9.3-35}$$

and

$$W(t')_{rec} = -\dot{\varepsilon}_0^2 \int_{t'}^{2t'} [\eta(t) - 2\eta(t - t')]dt . \tag{9.3-36}$$

Letting [cf. Eq. (4.1-16)]

$$\eta(t) = \{G_e\}t + \eta_{\{f\}} - \int_{-\infty}^{\infty} \tau H(\tau)\exp(-t/\tau)d \ln \tau \tag{9.3-37}$$

we obtain

$$W(t')_{def} = (\dot{\varepsilon}_0^2/2)\left[\{G_e\}t'^2 + 2\eta_{\{f\}}t' - 2 \int_{-\infty}^{\infty} \tau^2 H(\tau)[1 - \exp(-t'/\tau)]d \ln \tau\right] \tag{9.3-38}$$

and

$$W(t')_{rec} = (\dot{\varepsilon}_0^2/2)\Bigg[2\eta_{\{f\}}t' - \{G_e\}t'^2 - 2 \int_{-\infty}^{\infty} \tau^2 H(\tau)[1 - \exp(-t'/\tau)][2 - \exp(-t'/\tau)]d \ln \tau\Bigg] . \tag{9.3-39}$$

By Eq. (9.3-34), then,

$$W(t')_{loop} = (\dot{\varepsilon}_0^2/2)\Bigg[4\eta_{\{f\}}t' - 2 \int_{-\infty}^{\infty} \tau^2 H(\tau)[1 - \exp(-t'/\tau)][3 - \exp(-t'/\tau)]d \ln \tau\Bigg] . \tag{9.3-40}$$

We note that Eq. (9.3-38) follows directly from Eq. (9.1-38) by substituting t' for t.

The next step consists in examining separately the energy storage and dissipation during the deformation and recovery half-cycles. By Eq. $(7.1\text{-}47)_2$ the strain transform is given by

$$\bar{\varepsilon}(s) = \dot{\varepsilon}_0[1 - \exp(-t's)]^2/s^2 . \tag{9.3-41}$$

Substitution into Eq. (9.1-11) yields

$$\bar{\varepsilon}_s(s;\tau) = \frac{\dot{\varepsilon}_0\tau}{s(1 + \tau s)}[1 - \exp(-t's)]^2 . \tag{9.3-42}$$

Retransformation of Eq. (9.3-41) leads to

$$\varepsilon(t) = \dot{\varepsilon}_0 t - 2\dot{\varepsilon}_0(t - t')h(t - t') + \dot{\varepsilon}_0(t - 2t')h(t - 2t') \tag{9.3-43}$$

and that of Eq. (9.3-42) to

$$\begin{aligned}\varepsilon_s(t;\tau) = {} & \dot{\varepsilon}_0\tau[1 - \exp(-t/\tau)]h(t) - 2\dot{\varepsilon}_0\tau\{1 - \exp[-(t - t')]/\tau\}h(t - t') \\ & + \dot{\varepsilon}_0\tau\{1 - \exp[-(t - 2t')]/\tau\}h(t - 2t') \ .\end{aligned} \tag{9.3-44}$$

These equations must be substituted into Eq. (9.1-10) to yield the energy stored in the two half-cycles.

During the deformation, i.e. for $0 \le t \le t'$, only the first terms on the right of Eqs. (9.3-43) and (9.3-44) survive. Substitution into Eq. (9.1-10) gives

$$W_s(0 \le t \le t') = (\dot{\varepsilon}_0^2/2)\left[\{G_e\}t^2 + \int_{-\infty}^{\infty} \tau^2 H(\tau)[1 - \exp(-t/\tau)]^2 d\ln\tau\right] . \tag{9.3-45}$$

Thus, the energy stored up to $t = t'$ becomes

$$W_s(t')_{def} = (\dot{\varepsilon}_0^2/2)\left[\{G_e\}t'^2 + \int_{-\infty}^{\infty} \tau^2 H(\tau)[1 - \exp(-t'/\tau)]^2 d\ln\tau\right] . \tag{9.3-46}$$

This result also follows directly from Eq. (9.1-34) by setting $t = t'$. During the recovery, i.e. for $t' \le t \le 2t'$, the last terms on the right of Eqs. (9.3-43) and (9.3-44) vanish, and we have

$$\begin{aligned}W_s(t' \le t \le 2t') = (\dot{\varepsilon}_0^2/2)\Bigg[& \{G_e\}(2t' - t)^2 \\ & + \int_{-\infty}^{\infty} \tau^2 H(\tau)\{1 - 2\exp[-(t - t')]/\tau + \exp(-t/\tau)\}^2 d\ln\tau\Bigg] .\end{aligned} \tag{9.3-47}$$

Accordingly, the energy stored up to $t = 2t'$, i.e. over the complete cycle, becomes

$$W_s(t')_{loop} = (\dot{\varepsilon}_0^2/2)\int_{-\infty}^{\infty} \tau^2 H(\tau)[1 - \exp(-t'/\tau)]^4 d\ln\tau \ . \tag{9.3-48}$$

The last equation represents the energy stored in the material at the instant when the deformation returns to zero. Thus, in a triangular excitation the energy stored during the deformation is not completely released during the recovery. The energy which is, in fact, released, is obtained from

$$W_s(t')_{rec} = W_s(t')_{loop} - W_s(t')_{def} \tag{9.3-49}$$

which yields

$$\begin{aligned}W_s(t')_{rec} = -(\dot{\varepsilon}_0^2/2)\Bigg[& \{G_e\}t'^2 + \int_{-\infty}^{\infty} \tau^2 H(\tau)[1 - \exp(-t'/\tau)]^2 \\ & \times [2 - \exp(-t'/\tau)]\exp(-t'/\tau) d\ln\tau\Bigg] .\end{aligned} \tag{9.3-50}$$

The negative sign is in accordance with the fact that $W_s(t')_{rec}$ represents energy released.

The energy which remains stored upon completion of the cycle diminishes from the value at $t = 2t'$ given by Eq. (9.3-47) to zero as $t \to \infty$. That this is so, can be seen by finding $W_s(t)$ for $t > 2t'$. We obtain

$$W_s(t \geq 2t') = (\dot{\varepsilon}_0^2/2) \int_{-\infty}^{\infty} \tau^2 H(\tau)[1 - \exp(t'/\tau)]^4 \exp(-2t/\tau) d \ln \tau \tag{9.3-51}$$

and it follows that $W_s(t) \to 0$ as $t \to \infty$.

The rate at which energy is stored at time t during the deformation is

$$\dot{W}_s(0 \leq t \leq t') = \dot{\varepsilon}_0^2 \left[\{G_e\} t + \int_{-\infty}^{\infty} \tau H(\tau)[1 - \exp(-t/\tau)] \exp(-t/\tau) d \ln \tau \right] \tag{9.3-52}$$

and the rate at which it is released at time t during the recovery is

$$\dot{W}_s(t' \leq t \leq 2t') = -\dot{\varepsilon}_0^2 \left[\{G_e\}(2t' - t) + \int_{-\infty}^{\infty} \tau H(\tau) \{1 - 2\exp[-(t - t')/\tau] + \exp(-t/\tau)\} \{\exp(-t/\tau) - 2\exp[-(t - t')/\tau]\} d \ln \tau \right] . \tag{9.3-53}$$

Finally, the rate at which it diminishes upon completion of the cycle is

$$\dot{W}_s(t \geq 2t') = -\dot{\varepsilon}_0^2 \int_{-\infty}^{\infty} \tau H(\tau)[1 - \exp(t'/\tau)]^4 \exp(-2t/\tau) d \ln \tau . \tag{9.3-54}$$

Turning now to energy dissipation we substitute Eq. (9.3-41) into Eq. (9.1-15) to obtain

$$\bar{\dot{\varepsilon}}_d(s; \tau) = \frac{\dot{\varepsilon}_0}{s(1 + \tau s)} [1 - \exp(-t's)]^2 \tag{9.3-55}$$

and the rate of strain follows as

$$\dot{\varepsilon}_d(t; \tau) = \dot{\varepsilon}_0[1 - \exp(-t/\tau)]h(t) - 2\dot{\varepsilon}_0\{1 - \exp[-(t - t')/\tau]\}h(t - t') + \dot{\varepsilon}_0\{1 - \exp[-(t - 2t')/\tau]\}h(t - 2t') . \tag{9.3-56}$$

This must be substituted into Eq. (9.1-14) to furnish the rate at which energy is dissipated at time t during the stages of the cycle.

For the dissipation rate during the deformation, i.e. when $0 \leq t \leq t'$, we find

$$\dot{W}_d(0 \leq t \leq t') = \dot{\varepsilon}_0^2 \int_{-\infty}^{\infty} \tau H(\tau)[1 - \exp(-t/\tau)]^2 d \ln \tau . \tag{9.3-57}$$

Integrating between the limits 0 and t and rearranging produces the energy dissipated up to time t during the deformation as

$$W_d(0 \leq t \leq t') = (\dot{\varepsilon}_0^2/2) \left[2\eta_{\{f\}} t - \int_{-\infty}^{\infty} \tau^2 H(\tau)[1 - \exp(-t/\tau)][3 - \exp(-t/\tau)] d \ln \tau \right] \tag{9.3-58}$$

and the energy dissipated during the deformation half-cycle follows as

$$W_d(t')_{def} = (\dot{\varepsilon}_0^2/2)\left[2\eta_{\{f\}}t' - \int_{-\infty}^{\infty} \tau^2 H(\tau)[1 - \exp(-t'/\tau)] \times [3 - \exp(-t'/\tau)]d \ln \tau\right] . \tag{9.3-59}$$

Equation (9.3-59) is identical with that one obtains by putting $t = t'$ in Eq. (9.1-37).

During the recovery, i.e. when $t' \le t \le 2t'$, the rate at which energy is dissipated at time t is

$$\dot{W}_d(t' \le t \le 2t') = \dot{\varepsilon}_0^2 \int_{-\infty}^{\infty} \tau H(\tau)\{1 - 2\exp[-(t - t')/\tau] + \exp(-t/\tau)\}^2 d \ln \tau . \tag{9.3-60}$$

Integrating between the limits t′ and t yields the energy dissipated in the recovery up to time t. We have

$$W_d(t' \le t \le 2t') = (\dot{\varepsilon}_0^2/2)\left[2\eta_{\{f\}}(t - t') - \int_{-\infty}^{\infty} \tau^2 H(\tau)\{2 - 2\exp[-(t - t')/\tau] + \exp(-t/\tau)\}^2 d \ln \tau + \int_{-\infty}^{\infty} \tau^2 H(\tau)\exp(-2t'/\tau)d \ln \tau\right] \tag{9.3-61}$$

from which the energy dissipated over the course of the recovery can be obtained at once by setting $t = 2t'$. We find

$$W_d(t')_{rec} = (\dot{\varepsilon}_0^2/2)\left[2\eta_{\{f\}}t' - \int_{-\infty}^{\infty} \tau^2 H(\tau)\{[1 - \exp(-t'/\tau)]^4 + [1 - \exp(-t'/\tau)][3 - \exp(-t'/\tau)]\}d \ln \tau\right] . \tag{9.3-62}$$

The energy dissipated over the complete cycle is the sum of the energies dissipated in the deformation and the recovery, i.e.

$$W_d(t')_{loop} = (\dot{\varepsilon}_0^2/2)\left[4\eta_{\{f\}}t' - \int_{-\infty}^{\infty} \tau^2 H(\tau)\{[1 - \exp(-t'/\tau)]^4 + 2[1 - \exp(-t'/\tau)][3 - \exp(-t'/\tau)]\}d \ln \tau\right] . \tag{9.3-63}$$

Finally, the rate at which energy is dissipated after completion of the recovery (i.e. when $t \ge 2t'$) is found to be

$$\dot{W}_d(t \ge 2t') = \dot{\varepsilon}_0^2 \int_{-\infty}^{\infty} \tau H(\tau)[1 - \exp(t'/\tau)]^4 \exp(-2t/\tau)d \ln \tau . \tag{9.3-64}$$

Integration yields

$$W_d(t \ge 2t') = -(\dot{\varepsilon}_0^2/2) \int_{-\infty}^{\infty} \tau^2 H(\tau)[1 - \exp(t'/\tau)]^4 \exp(-2t/\tau)d \ln \tau . \tag{9.3-65}$$

Comparison of Eqs. (9.3-64) and (9.3-65) with Eqs. (9.3-54) and (9.3-51) shows that the energy which remains stored upon completion of the recovery is dissipated as $t \to \infty$. No energy is absorbed when $t > 2t'$ and, in fact, $W_s(t \geq 2t')$ and $W_d(t \geq 2t')$ exactly balance one another.

The energies absorbed up to time t during the deformation and recovery are obtained from Eq. (9.0-3) as

$$W(0 \leq t \leq t') = (\dot{\varepsilon}_0^2/2)\left[\{G_e\}t^2 + 2\eta_{\{f\}}t - 2\int_{-\infty}^{\infty} \tau^2 H(\tau)[1 - \exp(-t/\tau)]\,d\ln\tau\right] \tag{9.3-66}$$

and

$$W(t' \leq t \leq 2t')$$
$$= (\dot{\varepsilon}_0^2/2)\left[\{G_e\}(2t' - t)^2 + 2\eta_{\{f\}}(t - t') - \int_{-\infty}^{\infty} \tau^2 H(\tau)\{3 + 2\exp(-t/\tau) - 4\exp[-(t - t')/\tau] - \exp(-2t'/\tau)\}\,d\ln\tau\right]. \tag{9.3-67}$$

The rate at which energy is absorbed at time t is

$$\dot{W}(0 \leq t \leq t') = \dot{\varepsilon}_0^2\left[\{G_e\}t + \int_{-\infty}^{\infty} \tau H(\tau)[1 - \exp(-t/\tau)]\,d\ln\tau\right] \tag{9.3-68}$$

during the deformation, and

$$\dot{W}(t' \leq t \leq 2t')$$
$$= \dot{\varepsilon}_0^2\left[\{G_e\}(t - 2t') + \int_{-\infty}^{\infty} \tau H(\tau)\{1 - 2\exp[-(t - t')/\tau] + \exp(-t/\tau)\}\,d\ln\tau\right] \tag{9.3-69}$$

during the recovery. At $t > 2t'$ there is, of course, no energy absorption. Equations (9.3-46) and (9.3-59) add to give Eq. (9.3-38). Similarly, Eqs. (9.3-50) and (9.3-62) produce Eq. (9.3-39) and Eqs. (9.3-48) and (9.3-63) yield Eq. (9.3-40) which had been derived earlier.

Comparison of Eqs. (9.3-59) and (9.3-40) shows that

$$W(t')_{\text{loop}} = 2\,W_d(t')_{\text{def}}\ , \tag{9.3-70}$$

i.e. the area of the loop represents twice the energy dissipated in the deformation. The energy dissipated in the recovery is less than the energy dissipated in the deformation by the amount remaining stored at the instant when the deformation has returned to zero.

The response to a triangular excitation allows the deformational energies to be determined graphically. Since the area under the stress-strain curve represents the total energy absorbed in the deformation (cf. Problem 9.1-6), a glance at Fig. 9.3-2a shows that

$$W(t')_{\text{def}} = A + B \tag{9.3-71}$$

$$W_d(t')_{\text{def}} = (A + C)/2 \tag{9.3-72}$$

and

$$W_s(t')_{def} = B + (A - C)/2 \tag{9.3-73}$$

where A is the area shaded downward from left to right, B is the area shaded upward from right to left, and C is the crosshatched area.

It must be noted that the exact relations derived here apply only when the material has been taken through a full cycle, i.e. from zero initial to zero final strain. If the recovery is stopped when the *stress* has reached zero, neither $W_d(t')_{def}$ nor $W_s(t')_{def}$ can be calculated because the area C is not known.

9.3.2.2 *Steady-State Response*

Let us now examine the energy absorption in response to a (regular) triangular pulse strain excitation in the *steady state.* The rate of strain will be $\dot{\varepsilon}_0$ in each deformation half-cycle and $-\dot{\varepsilon}_0$ in each recovery half-cycle. The stress in the two half-cycles follows from Eqs. (7.3-31) and (7.3-32), respectively, as

$$\sigma_{def}(\theta) = \dot{\varepsilon}_0\left[\{G_e\}\theta + \eta_{\{f\}} - 2\int_{-\infty}^{\infty} \tau H(\tau)\frac{\exp(-\theta/\tau)}{1 + \exp(-t'/\tau)}\, d\ln\tau\right] \tag{9.3-74}$$

and

$$\sigma_{rec}(\theta) = -\dot{\varepsilon}_0\left[\{G_e\}(\theta - 2t') + \eta_{\{f\}} - 2\int_{-\infty}^{\infty} \tau H(\tau)\frac{\exp[-(\theta - t')/\tau]}{1 + \exp(-t'/\tau)}\, d\ln\tau\right] \tag{9.3-75}$$

where θ is the steady-state time [cf. Eq. (7.3-30)]. The steady-state response is modeled in Fig. 9.3-3 with the same parameters used to obtain Fig. 9.3-2. With these parameters the steady state is reached sensibly after about 30 cycles. We notice that, in the steady-state, the stress, upon completion of the loop, returns to the value it

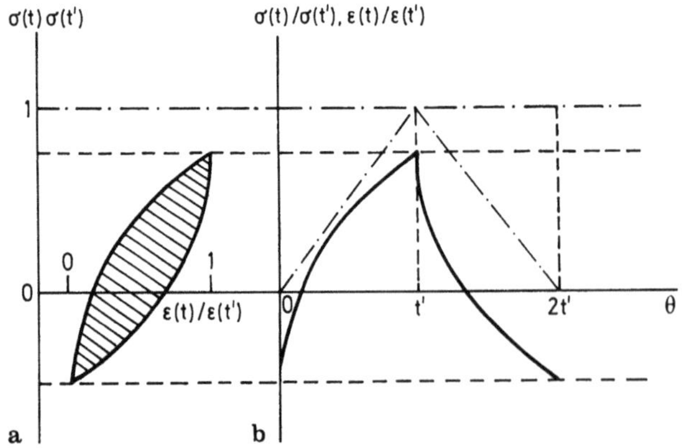

Fig. 9.3-3. Stress as a function of strain (**a**) and of time (**b**) in a steady-state (regular) triangular excitation

had at the beginning of the cycle (cf. Fig. 9.3-2). The stress is a compressive stress at the beginning of the deformation half-cycle and at the end of the recovery half-cycle.

We are now ready to derive the energy relations using Eqs. (9.3-74) and (9.3-75). Substituting the former into

$$W(\theta) = \int_{\theta_1}^{\theta_2} \sigma(\theta)\dot{\varepsilon}(\theta)d\theta \tag{9.3-76}$$

and integrating from $\theta_1 = 0$ to $\theta_2 = t'$ gives

$$W(t')_{def} = (\dot{\varepsilon}_0^2/2)\left[\{G_e\}t'^2 + 2\eta_{\{f\}}t' - 4\int_{-\infty}^{\infty} \tau^2 H(\tau)\frac{1 - \exp(-t'/\tau)}{1 + \exp(-t'/\tau)} d\ln\tau\right] \tag{9.3-77}$$

as the energy absorbed during the deformation half-cycle in the steady-state. Substituting Eq. (9.3-75) into Eq. (9.3-76) and integrating from $\theta_1 = t'$ to $\theta_2 = 2t'$ yields

$$W(t')_{rec} = (\dot{\varepsilon}_0^2/2)\left[2\eta_{\{f\}}t' - \{G_e\}t'^2 - 4\int_{-\infty}^{\infty} \tau^2 H(\tau)\frac{1 - \exp(-t'/\tau)}{1 + \exp(-t'/\tau)} d\ln\tau\right] \tag{9.3-78}$$

as the energy absorbed during the recovery half-cycle. The total energy absorbed over a complete cycle then follows as

$$W(t')_{loop} = (\dot{\varepsilon}_0^2/2)\left[4\eta_{\{f\}}t' - 8\int_{-\infty}^{\infty} \tau^2 H(\tau)\frac{1 - \exp(-t'/\tau)}{1 + \exp(-t'/\tau)} d\ln\tau\right]. \tag{9.3-79}$$

Having found the relation expressing the energy absorbed over the hysteresis loop in the steady-state triangular excitation, we must now examine whether this is equal to or different from the energy dissipated over the loop. The direct way to do this is to derive an expression for the energy dissipated. An equivalent way consists in finding the net energy stored over the loop. If this is zero, the area of the loop is a measure of the energy dissipated.

Let us examine the energy dissipation in the steady-state. Substituting Eq. $(7.3\text{-}9)_2$ into (9.1-15), we obtain

$$\dot{\varepsilon}_d(t;\tau) = \dot{\varepsilon}_0\left([1 - \exp(-t/\tau)] + 2\sum_{n=1}^{N-1} (-1)^n\{1 - \exp[-(t - nt')/\tau]\}\right) \tag{9.3-80}$$

if we stipulate that the sum be zero when $N = 1$. Equation (9.3-80) may be rearranged to

$$\dot{\varepsilon}_d(t;\tau) = \dot{\varepsilon}_0\left[1 + 2\sum_{n=1}^{N-1} (-1)^n - \left(1 + 2\sum_{n=1}^{N-1} (-1)^n\exp(nt'/\tau)\right)\exp(-t/\tau)\right] \tag{9.3-81}$$

and, since N is odd, this can be reduced to

$$\dot{\varepsilon}_d(t;\tau) = \dot{\varepsilon}_0\left(1 - \exp(-t/\tau)\frac{2\exp(Nt'/\tau) - \exp(t'/\tau) + 1}{1 + \exp(t'/\tau)}\right) \tag{9.3-82}$$

by the use of Eqs. (7.3-14) and (7.3-16). As N becomes large, $1 - (\exp t'/\tau)$ can be neglected against $2\exp(Nt'/\tau)$. We then have

$$\dot{\varepsilon}_d(\theta;\tau) = \dot{\varepsilon}_0\left(1 - 2\frac{\exp(-\theta/\tau)}{1+\exp(-t'/\tau)}\right) \tag{9.3-83}$$

where θ is given by Eq. (7.3-30). Inserting Eq. (9.3-83) into Eq. (9.1-14) expresses the rate at which energy is dissipated during the deformation half-cycle in the steady-state as

$$\dot{W}_d(\theta)_{def} = \dot{\varepsilon}_0^2 \int_{-\infty}^{\infty} \tau H(\tau)\left(1 - 2\frac{\exp(-\theta/\tau)}{1+\exp(-t'/\tau)}\right)^2 d\ln\tau \tag{9.3-84}$$

and integration from $\theta_1 = 0$ to $\theta_2 = t'$ then furnishes

$$W_d(t')_{def} = \dot{\varepsilon}_0^2\left[\eta_{\{f\}}t' - 2\int_{-\infty}^{\infty} \tau^2 H(\tau)\frac{1-\exp(-t'/\tau)}{1+\exp(-t'/\tau)} d\ln\tau\right] \tag{9.3-85}$$

for the energy dissipated over the deformation half-cycle.

Turning now to the recovery half-cycle, the upper limit of the sums in Eq. (9.3-80) becomes N, and we now find that

$$\dot{\varepsilon}_d(\theta;\tau) = -\dot{\varepsilon}_0\left(1 - 2\frac{\exp[-(\theta - t')/\tau]}{1+\exp(-t'/\tau)}\right) \tag{9.3-86}$$

in the steady state. Substituting Eq. (9.3-86) into Eq. (9.1-14) and integrating from $\theta_1 = t'$ to $\theta_2 = 2t'$ show that the energy dissipated over the recovery half-cycle is the same as that dissipated over the deformation half-cycle, i.e. $W_d(t')_{rec} = W_d(t')_{def}$. This immediately shows that there is no net energy storage over the complete cycle, so that

$$W_d(t')_{loop} = W(t')_{loop} \tag{9.3-87}$$

in accordance with Eq. (9.3-1). Thus, in the regular triangular *steady-state* excitation (and only in the steady-state) the area of the hysteresis loop represents the total energy dissipated in the deformation and recovery, just as it does in the harmonic excitation.

9.4 Problems

Problem 9.1-1. To obtain the Laplace transform of $[f(t)]^2$ where the poles of the function f(t) are simple poles only, we first combine Eqs. (A3-31) and (A3-32) to give

$$f(t) = \sum_k \lim_{z\to z_k} (z - z_k)\bar{f}(z_k)\exp(z_k t) \tag{1}$$

where we have replaced the usual transform variable s by z because we will need s in the next step. We then multiply both sides of Eq. (1) by f(t) and take the Laplace transform. This yields

$$\mathscr{L}[f(t)]^2 = \sum_k \lim_{z \to z_k} (z - z_k)\bar{f}(z_k)\mathscr{L}[f(t)\exp(z_k t)] \ . \tag{2}$$

But then

$$\mathscr{L}[f(t)]^2 = \sum_k \lim_{z \to z_k} (z - z_k)\bar{f}(z_k)\bar{f}(s - z_k) \tag{3}$$

by Eq. (A3-7), and this can be restated as

$$\mathscr{L}[f(t)]^2 = \sum_k \mathrm{Res}_k[\bar{f}(z_k)\bar{f}(s - z_k)] \tag{4}$$

where

$$\mathrm{Res}_k = \lim_{z \to z_k} (z - z_k)\bar{f}(z)\bar{f}(s - z_k) \ . \tag{5}$$

(A) Using Eq. (4), transform Eq. (9.1-8) to obtain the general equation for the Laplace transform of the energy stored in a rheodictic material in response to a strain excitation by introducing Eq. (9.1-5) into the transform relation.

(B) Then, solve the general equation for the particular case of a slope strain excitation and show that retransformation to the time domain leads to Eq. (9.1-34) as required.

Problem 9.1-2. Consider the response of the elements of a Wiechert model, Fig. 3.5-2, to a step strain. What is the history of the strain in the nth spring? In the nth dashpot? In the isolated spring, if present?

Problem 9.1-3. (A) Obtain Eq. (9.1-19) by retransforming

$$\bar{\dot{W}}(s) = s\bar{W}(s) \tag{1}$$

into the time domain. Then (B) differentiate Eq. (9.1-17) and reintegrate.

Problem 9.1-4. (A) Derive Eq. (9.1-25) from Eq. $(9.0\text{-}2)_2$. Then (B) show that here $h(0)$ must be interpreted as having the value 1/2. Finally (C) derive Eq. (9.1-24) from Eq. (9.0-1) using this result.

[*Hint*: the strain is given by $\varepsilon_0 h(t)$. The fact that the unit step function is not defined for $t = 0$ will cause trouble. Hence, use the relation

$$f(x)f'(x) = \frac{d[f(x)]^2}{2dx} \ , \tag{1}$$

replace the unit step function by the ramp function and take the limit as $t' \to 0$.]

Problem 9.1-5. Consider the response of the elements in the units of a Wiechert model, Fig. 3.5-2, to a slope strain. What is the history of the strain in the nth spring? In the nth dashpot? In the isolated spring, if present?

Problem 9.1-6. Show that the energy absorbed per unit volume of material in a constant rate of strain experiment is represented by the area under the stress-strain curve.

Problem 9.1-7. Starting from the basic relation, Eq. (9.1-10), show that the maximum energy which can be stored coherently (i.e. with all energy storing mechanisms in phase) in a harmonic deformation is given by

$$W_s(\omega)_{max} = (\varepsilon_0^2/2)G'(\omega) \ . \tag{1}$$

Problem 9.1-8. Starting from the basic relation, Eq. (9.1-14), show that the maximum rate at which energy would be dissipated coherently, is

$$\dot{W}_d(\omega)_{max} = \varepsilon_0^2\omega\, G''(\omega) = \varepsilon_0\sigma_0(\omega)\omega \sin\theta(\omega) \ . \tag{1}$$

Problem 9.1-9. Consider energy storage, dissipation, and absorption in response to a ramp strain at times greater than t′, i.e. as a function of the shifted time, $\theta = t - t'$. Compare the results with those obtained for the response to a step strain.

Problem 9.1-10. Consider the harmonic strain excitation $\varepsilon(t) = \varepsilon_0 \sin \omega t$. Let the strain produced in response to this excitation in the spring and in the dashpot of the nth Maxwell unit of a Wiechert model be $\varepsilon_{sn}(t)$ and $\varepsilon_{dn}(t)$, respectively. Now,

(A) Express these strains in terms of the loss angle, $\mu_n(\omega)$, of the Maxwell unit. Then,

(B) Indicate the relations between $\varepsilon_{sn}(t)$, $\varepsilon_{dn}(t)$, and $\varepsilon(t)$ in a *phasor* diagram.

Note: A sinusoidal oscillating quantity may be represented conveniently as a phasor. Phasors have long been used in electric circuit theory. The amplitude of a phasor is expressed by the length of an arrow while its phase is indicated by the angle formed with the reference (horizontal) axis. Phasors behave (e.g. add) generally like vectors but differ from them in definition. By convention, they rotate counter-clockwise with time.

Problem 9.1-11. (A) Show that the maximum energy which can be stored in the nth Maxwell unit of a Wiechert model in response to the strain excitation $\varepsilon(t) = \varepsilon_0 \sin \omega t$ is given by

$$w_{sn}(\omega)_{max} = (\varepsilon_0^2/2)G'_n(\omega) \tag{1}$$

where $G'_n(\omega)$ is the storage modulus of the nth Maxwell unit.

(B) Show that $w_{sn}(\omega)_{max}$ represents the energy stored during the first or third quarter of each cycle of the excitation, or, equivalently, the energy released during each second or fourth quarter cycle.

(C) Finally, show that summing of the maximum energies which may be stored in all the springs of a Wiechert model leads to Eq. $(9.1\text{-}63)_2$. Interpret this result.

[*Hint*: you will require the results of Problem 9.1-10.]

Problem 9.1-12. (A) Show that the energy dissipated over any quarter cycle of the excitation in the dashpot of the nth Maxwell unit is

$$w_{dn}(\omega)_{qtr} = (\varepsilon_0^2/4)\pi\, G''_n(\omega) \ . \tag{1}$$

where $G''_n(\omega)$ is the loss modulus of the nth Maxwell unit.

(B) What is the result of summing Eq. (1) over all dashpots of a Wiechert model?

[*Hint*: use the results of Problem 9.1-10.]

Problem 9.1-13. Calculate the heat generated per unit volume of a viscoelastic material in unit time when the material is subjected to a harmonic strain excitation under adiabatic conditions at a radian frequency ω and a strain amplitude ε_0.

Problem 9.1-14. A specimen is stretched to an elongation ε_f at a constant rate of strain $\dot{\varepsilon}_0$.

(A) How much heat is generated in the specimen per unit volume if the deformation is considered to be adiabatic and energy storage is considered to be negligible.

(B) By how much would the temperature of this specimen have risen under these conditions?

Problem 9.1-15. Discuss energy storage and dissipation in ladder models. Illustrate the general procedure by considering the response of the Gross-Marvin model (Sect. 5.1) to a step strain.

Problem 9.1-16. Show that (A)

$$W_s(t) = \tfrac{1}{2}\int_0^t\int_0^t G(2t - u - v)\dot{\varepsilon}(u)\dot{\varepsilon}(v)du\,dv \tag{1}$$

and (B)

$$\dot{W}_d(t) = -\int_0^t\int_0^t \dot{G}(2t - u - v)\dot{\varepsilon}(u)\dot{\varepsilon}(v)du\,dv \tag{2}$$

where $\dot{G}(2t - u - v)$ is obtained by differentiating $G(t)$ with respect to t and replacing the argument by $2t - u - v$.

[*Hint*: carry out the transform inversion indicated in Eqs. (9.1-11) and (9.1-15), substitute the results into Eqs. (9.1-10) and (9.1-14), and rearrange.]

Problem 9.2-1. Consider the response of the elements of a Kelvin model, Fig. 3.5-3, to a step stress. What is the history of the stress in the nth spring? In the nth dashpot? In the isolated spring and, if present, in the isolated dashpot?

Problem 9.2-2. Consider the response of the elements of a Kelvin model, Fig. 3.5-3, to a slope stress. What is the history of the stress in the nth spring? The nth dashpot? The isolated spring and, if present, the isolated dashpot?

Problem 9.2-3. Examine energy storage and dissipation in creep recovery as a function of the shifted time, $\theta = t - t'$.

Problem 9.2-4. Prove that Eqs. (9.2-38) and $(9.2\text{-}41)_2$ are identical expressions for the energy absorbed in response to a slope stress excitation.

Problem 9.2-5. Show that $\rho(t)$ may be obtained from the area under the strain-stress curve in an experiment at constant rate of stress.

Problem 9.2-6. Consider the harmonic stress excitation $\sigma(t) = \sigma_0 \sin \omega t$. Let the stress produced in response to this excitation in the spring and in the dashpot of the nth Voigt unit of a Kelvin model be $\sigma_{sn}(t)$ and $\sigma_{dn}(t)$, respectively. Now,

(A) Express these stresses in terms of the loss angle, $\nu_n(\omega)$, of the Voigt unit. Then,

(B) Indicate the relations between $\sigma_{sn}(t)$, $\sigma_{dn}(t)$, and $\sigma(t)$ in a phasor diagram.

[*Hint*: see the *Note* at the end of Problem 9.1-10.]

Problem 9.2-7. (A) Determine the ratio of the energy dissipated over the first quarter of any cycle in response to a harmonic stress excitation to that stored over the same interval.

(B) Do the same for a harmonic strain excitation.

(C) Ascertain what happens as ω approaches either 0 or ∞.

Problem 9.2-8. Prove that

$$\mu_s(\omega) + \nu_s(\omega) = \theta(\omega) = \mu_d(\omega) - \nu_d(\omega) \ . \tag{1}$$

Problem 9.2-9. Show that the ratio of the energy dissipated per second to the energy stored during steady-state flow of a rheodictic material is $2/J_e^\circ \eta_f$.

Problem 9.2-10. Show that (A)

$$W_s(t) = \sigma(t)\varepsilon(t) - \tfrac{1}{2}\int_0^t\int_0^t J(2t - u - v)\dot{\sigma}(u)\dot{\sigma}(v)du\,dv \tag{1}$$

and (B)

$$\dot{W}_d(t) = \int_0^t\int_0^t \dot{J}(2t - u - v)\dot{\sigma}(u)\dot{\sigma}(v)du\,dv \tag{2}$$

where $\dot{J}(2t - u - v)$ is obtained by differentiating $J(t)$ with respect to t and replacing the argument by $2t - u - v$.

[*Hint*: carry out the transform inversion indicated in Eqs. (9.2-11) and 9.2-15), insert the results into Eqs. (9.2-10) and (9.2-14), and rearrange.]

Problem 9.3-1. Prove Eq. (9.3-8) by eliminating ωt between

$$\varepsilon(t) = \varepsilon_0 \sin \omega t \qquad \text{and} \qquad \sigma(t) = \sigma_0(\omega)\sin[\omega t + \theta(\omega)] \ . \tag{1}$$

References (Chap. 9)

1. J.D. Ferry, *Viscoelastic Properties of Polymers*, 1st and 2nd eds. *only*, Wiley, New York, 1961, p. 433, p. 606.
2. N.G. MacCrum, B.E. Read, and W.G. Williams, *Anelastic and Dielectric Effects in Polymeric Solids*, Wiley, New York, 1967, p. 10.

10. The Modelling of Multimodal Distributions of Respondance Times

Il y a un demi siècle on le confessait franchement et on proclamait que la nature aime la simplicité; elle nous a donné depuis trop de démentis.

Henri Poincaré

Raffiniert ist der Herrgott, aber boshaft ist er nicht.

Albert Einstein

10.0 Introduction

We discussed the representation of linear viscoelastic behavior by series-parallel models in Chap. 3, by spectral response functions (canonical representations) in Chap. 4, by ladder models in Chap. 5, and by mathematical equations in Chap. 6. With the exception of the standard solid and liquid models, these representations all assumed the existence of a *distribution* of respondance times. We did not specify the precise nature of any of these distributions* but tacitly assumed that they all had their origin in a *single* viscoelastic mechanism. We took this to mean — again tacitly — that a discrete distribution of respondance times characterizing this mechanism possessed a single maximum, and that all spectral lines obeyed the relation

$$\cdots F_{m-2}\delta(\tau_{m-2}) \leq F_{m-1}\delta(\tau_{m-1}) \leq F_m\delta(\tau_m) \geq F_{m+1}\delta(\tau_{m+1}) \geq F_{m+2}\delta(\tau_{m+2}) \geq \cdots \; . \qquad (10.0\text{-}1)$$

The spectrum describing such a distribution is a line spectrum (cf. Sect. 4.5). On a plot of the spectrum, $F(\tau)$, as a function of the respondance time, τ, the spectral lines are indicated by arrows of appropriate length (cf. Appendix A2.1). The subscript m denotes the *maximal* line, i.e. the line whose strength exceeds all others. An envelope formed by connecting the end points of the arrows displays a single peak (see Problem 10.0-1). In the case of a continuous distribution of respondance times we may require that the spectral response functions possess a single peak only, i.e. that a plot of $\log H(\tau)$ or $\log L(\tau)$ vs. $\log \tau$ be bell-shaped.

A distribution of respondance times with the characteristics just described may be called a *unimodal distribution.* For some viscoelastic materials such unimodal distributions yield satisfactory representations of their viscoelastic properties. There are others, however, whose properties can be described adequately only by two or more distinct distributions of respondance times, reflecting the presence of more

* Section 6.5, in which we discussed models of the spectral response functions, forms an exception.

than one viscoelastic mechanism. Such materials are said to exhibit *multimodal distributions* of respondance times. Multiphase polymers, such as polyblends, block and graft copolymers, are examples of such materials. For the sake of simplicity, we shall consider only *bimodal distributions*, i.e. distributions with two distinct groups of relaxation or retardation times. Extension to more than two groups is generally straightforward. It is the purpose of Sect. 10.1 to discuss the modelling of arrheodictic and rheodictic bimodal distributions.

A special case is presented by a class of materials which we shall call *pseudo-arrheodictic*. These materials are, in fact, rheodictic but appear to display arrheodictic behavior over part of the time scale. Such behavior can often be represented in an acceptable manner by the models to be considered in Sect. 10.1. An alternative approach consists in describing the behavior with a single distribution in which, however, the first few respondance times are prolongated through multiplication by a constant. The result is qualitatively similar to that exhibited by a bimodal distribution. We shall refer to single distributions with such lengthened respondance times as *prolongated unimodal distributions*. Not only pseudo-arrheodictic behavior but certain types of arrheodictic behavior as well can be modelled by prolongated unimodal distributions. Finally, such distributions can be added to other unimodal and even bimodal distributions for modelling more complex behavior. Prolongated unimodal distributions form the subject of Sect. 10.2.

10.1 Bimodal Distributions

We will discuss the modelling of arrheodictic and rheodictic behavior by bimodal distributions of respondance times under the headings:

(1) series-parallel models,
(2) models for the spectral response functions,
(3) ladder models, and
(4) mathematical models.

We will consider only *standard* models, i.e. models having a finite, non-zero instantaneous modulus, G_g, and a finite, non-zero instantaneous compliance, J_g. The relations [cf. Eqs. (3.5-29) and (3.5-30), and Eqs. (4.6-20) and (4.6-21)].

$$\sum_k G_k = \int H(\tau)d\ln\tau = G_g - \{G_e\} \qquad (10.1\text{-}1)$$

or

$$\sum_k J_k = \int L(\tau)d\ln\tau = J_e^{\{o\}} - J_g \qquad (10.1\text{-}2)$$

apply to all such models. In addition, there are certain other general conditions to which all models for bimodal distributions are subject. Thus, since the relaxation modulus is a monotonic non-increasing function of time, we have the inequalities

$$\sum_m G_m > \sum_n G_n > \{G_e\} \qquad (10.1\text{-}3)$$

where the subscripts m and n designate the distributions associated respectively with the shorter and with the longer relaxation times. In addition, Eq. (10.1-1) above applies to the sum of both groups of moduli.

Similarly considerations apply to the compliances. Since the creep compliance is a monotonic non-decreasing function of time, we must have

$$J_g < \sum_m J_m < \sum_n J_n \tag{10.1-4}$$

in analogy to Inequalities (10.1-3), and Eq. (10.1-2) must apply to the sum of both groups of compliances.

We note that the sums of moduli or compliances may, of course, be replaced everywhere by the integrals over the respective spectra when modelling continuous bimodal distributions.

10.1.1 Series-Parallel Models

Let us first consider series-parallel models with a finite number of elements. The respondances of bimodal distributions can be modelled simply by the sum of two single distributions to which the appropriate viscoelastic constants are added. For the relaxance of a bimodal distribution we have [cf. Eq. (3.5-3)]

$$\bar{Q}(s) = \{G_e\} + \sum_{m=0;1}^{m=M-1} \frac{G_m \tau_m s}{1+\tau_m s} + \sum_{n=M}^{n=R} \frac{G_n \tau_n s}{1+\tau_n s} \tag{10.1-5}$$

or

$$\bar{Q}_\bullet(s) = G_g - \sum_{m=0;1}^{m=M-1} \frac{G_m}{1+\tau_m s} - \sum_{n=M}^{n=R} \frac{G_n}{1+\tau_n s} \tag{10.1-6}$$

where τ_R represents the longest relaxation time, and $m = 0$ applies when modelling rheodictic behavior, while $m = 1$ is used when the behavior is arrheodictic (see Sect. 3.5.1.1).

For the response to a stress excitation we use*

$$\bar{U}(s) = J_g + \sum_{m=1}^{m=M-1} \frac{J_m}{1+\tau_m s} + \sum_{n=M}^{n=R} \frac{J_n}{1+\tau_n s} + \{\phi_f/s\} \tag{10.1-7}$$

or

$$\bar{U}_\bullet(s) = J_e^{\{o\}} - \sum_{m=1}^{m=M-1} \frac{J_m \tau_m s}{1+\tau_m s} - \sum_{n=M}^{n=R} \frac{J_n \tau_n s}{1+\tau_n s} + \{\phi_f/s\} \tag{10.1-8}$$

where τ_R is now the longest retardation time.

The experimental response functions follow from the respondances in the usual way. As an example, for the relaxation modulus we have

$$G(t) = \{G_e\} + \sum_{m=0;1}^{m=M-1} G_m \exp(-t/\tau_m) + \sum_{n=M}^{n=R} G_n \exp(-t/\tau_n) \tag{10.1-9}$$

---

* The sets of relaxation times are, of course, quite independent of the sets of retardation times even though we have used the same symbols to denote them.

or

$$G_\bullet(t) = G_g - \sum_{m=0;\,1}^{m=M-1} G_m[1 - \exp(-t/\tau_m)] - \sum_{n=M}^{n=R} G_n[1 - \exp(-t/\tau_n)] \ . \tag{10.1-10}$$

Equations (10.1-9) and (10.1-10) represent extensions of Eq. (3.5-7) and are, therefore, bimodal generalized Maxwell models. It should be clear from the discussion in Sect. 3.5.1.2 that distinct steps in a plot of log G(t) vs. log t will appear only if the two groups of relaxation times are well separated. Figure 10.1-1 illustrates schematically the typical behavior of a rheodictic bimodal distribution.

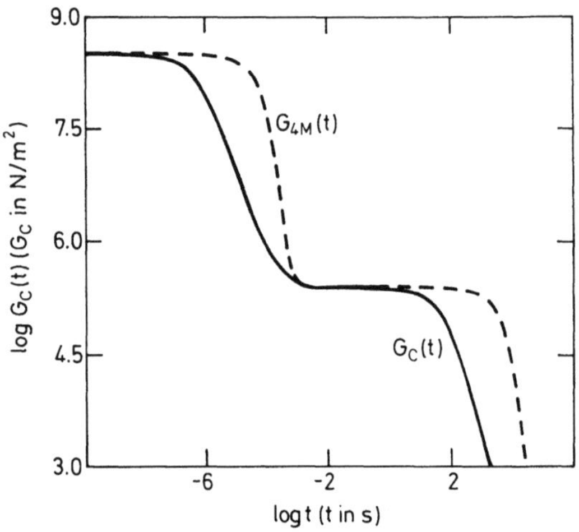

Fig. 10.1-1. Relaxation modulus calculated from a rheodictic bimodal Wiechert model compared with the modulus of the standard liquid model

The full line shows a plot of Eq. (10.1-9) based on the set of parameters, G_k and τ_k, of Problem 3.6-2 (see also Problem 10.0-1). G_e was omitted to obtain rheodictic behavior. The first distribution in Eq. (10.1-9) was calculated with $G_m = 10 \times G_k$ and $\tau_m = \tau_k$, and the second with $G_n = G_k$ and $\tau_n = \tau_k \times 10^8$. The latter factor insures a rather wide separation of the two distributions for the purpose of demonstration.

For comparison, the dashed line shows the relaxation modulus of the 4-parameter Maxwell model

$$G(t) = G_1 \exp(-t/\tau_1) + G_2 \exp(-t/\tau_2) \tag{10.1-11}$$

with $G_1 = 10 \times G_g$, $\tau_1 = 1.443 \times 10^{-6}$ seconds, $G_2 = G_g$, and $\tau_2 = \tau_1 \times 10^8$. G_g is the sums of all the moduli of the discrete rheodictic distribution described above. The full line clearly shows the "broadening" effect of *distribution*, i.e. the presence in the model of two *distributions* of relaxation times instead of merely two *discrete* relaxation times.

We chose a model in which the second distribution is derived from the first merely as a matter of convenience. We could have used an entirely independent set of moduli and/or relaxation times for the second distribution.

The appropriate equations for $\eta(t)$ can be developed from Eq. (3.5-11) (cf. Problem 10.1-1). We note especially that Eq. (3.5-12) remains valid, i.e. the steady-state viscosity of a bimodal distribution is given by

$$\eta_f = \sum_m G_m\tau_m + \sum_n G_n\tau_n \; . \tag{10.1-12}$$

Turning now to the creep compliances, we have

$$J_\bullet(t) = J_e^{\{o\}} - \sum_{n=1}^{n=M-1} J_n\exp(-t/\tau_n) - \sum_{m=M}^{m=R} J_m\exp(-t/\tau_m) + \{\phi_f t\} \tag{10.1-13}$$

or

$$J(t) = J_g + \sum_{m=1}^{m=M-1} J_m[1 - \exp(-t/\tau_m)] + \sum_{n=M}^{n=R} J_n[1 - \exp(-t/\tau_n)] + \{\phi_f t\} \tag{10.1-14}$$

which are the equations of a bimodal generalized Voigt model [cf. Eqs. (3.5-16)].

The harmonic response functions follow analogously.

The finite generalized models in all these equations can be extended to canonical models with infinite numbers of elements if care is taken that the distributions of moduli or compliances on respondance times converge suitably.

The distributions derived from the Gross-Marvin ladder models can be used to good advantage for modelling bimodal behavior. Note, however, that Eqs. (10.1-5) to (10.1-10), and (10.1-13) and (10.1-14) must be modified slightly because of the way the distributions are defined (see Problems 10.1-2 and 10.1-3).

10.1.2 Models for the Spectral Response Functions

Bimodal arrheodictic and rheodictic behavior can, of course, be modelled also by choosing suitable spectral functions. The discrete bimodal relaxation and retardation spectra become

$$H(\tau) = \sum_{m=0;\,1}^{m=M-1} G_m\tau_m\delta(\tau - \tau_m) + \sum_{n=M}^{n=R} G_n\tau_n\delta(\tau - \tau_n) \tag{10.1-15}$$

and

$$L(\tau) = \sum_{m=1}^{m=M-1} J_m\tau_m\delta(\tau - \tau_m) + \sum_{n=M}^{n=R} J_n\tau_n\delta(\tau - \tau_n) \; , \tag{10.1-16}$$

respectively. These equations represent, of course, the spectra associated with Eqs. (10.1-5) to (10.1-10) of the preceding section. Each group of respondance times is subject to Eq. (10.0-1).

Bimodal canonical representations can be modelled through a combination of any two of the mathematical models* for the spectral response functions discussed

---

* Although this subject might appear to belong more properly in Sect. 10.1.4, it seems convenient to discuss it here and to confine that section to a discussion of the modelling of the *experimental* response functions.

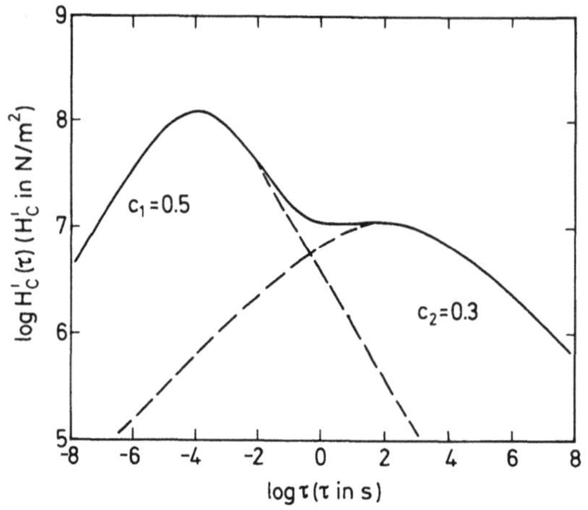

Fig. 10.1-2. Bimodal relaxation spectrum calculated from the Cole-Cole function

in Sect. 6.3. As an example, let us consider Eq. (6.3-129). The spectral function model which it embodies, is based on the Cole-Cole function, Eq. (6.1-23). For a bimodal spectrum describing a rheodictic material it takes the form

$$H_C(\tau) = \frac{(2/\pi)(G_g - G_P)\sin \pi c_1/2}{(\tau_1/\tau)^{c_1} + 2\cos \pi c_1/2 + (\tau/\tau_1)^{c_1}} + \frac{(2/\pi)G_P \sin \pi c_2/2}{(\tau_2/\tau)^{c_2} + 2\cos \pi c_2/2 + (\tau/\tau_2)^{c_2}} \tag{10.1-17}$$

where G_P is, as it were, the instantaneous modulus of the second distribution. A plot of this relation is shown in Fig. 10.1-2. The parameters were chosen as follows: $G_g = 10^9$ N/m², $\tau_1 = 10^{-4}$ s, $c_1 = 0.5$, $G_P = 1.5708 \times 10^8$ N/m², $\tau_2 = 100$ s, $c_2 = 0.3$. The broken lines represent the "invisible" parts of the two spectra. We have used here the same mathematical model for both spectra but made the second broader by using a $c_2 < c_1$.

10.1.3 Ladder Models

Bimodal distributions may, of course, also be modelled by ladder models. The regular obverse ladder model of the Gross-Marvin type which displays the characteristics of a bimodal distribution is shown in Fig. 10.1-3. This model consists of two sections: the first is simply the Gross-Marvin ladder model shown in Fig. 5.2-1. The second, mechanically in parallel with the first, has the same characteristics, but the number of elements and the value of the compliances and viscosities differ from those in the first section. This is indicated by distinguishing the material parameters of the second section from those of the first one by an apostrophe as J′, and η′. Although both sections are, in themselves, regular ladder models, the relative simplicity of a one-section regular ladder model is lost by the addition of a second section.

To derive the equations governing the behavior of such a composite ladder model we return to Chap. 5. By Eq. (5.6-34) the (mechanical) input relaxance of the

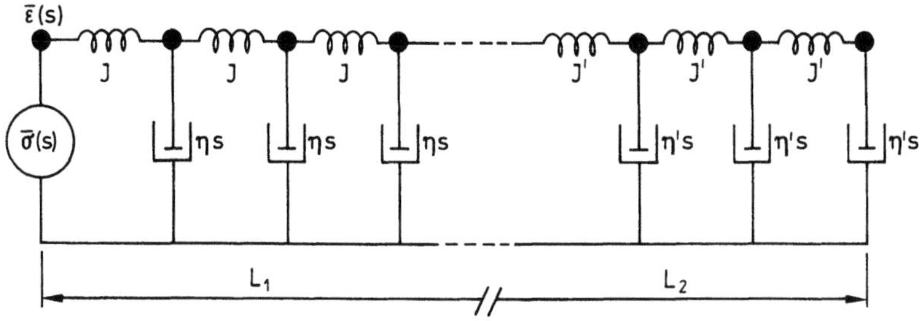

Fig. 10.1-3. Bimodal Gross-Marvin ladder model

first section is

$$Q_{m1}(s) = Q_{mC_1}(s)\frac{Q_{mL_1}(s) + Q_{mC_1}(s)\tanh[L_1\Gamma_1(s)]}{Q_{mC_1}(s) + Q_{mL_1}(s)\tanh[L_1\Gamma_1(s)]} \tag{10.1-18}$$

where we have written $\Gamma_1(s)$ for $\Gamma_{1_1}(s)$ to simplify the notation. In Eq. (10.1-18) $Q_{mC_1}(s)$, $Q_{mL_1}(s)$, and $\Gamma_1(s)$ are the characteristic relaxance, terminal relaxance, and transmittance of the first section of the model, and L_1 is its length. But the terminal relaxance of the first section is the input relaxance of the second section, and this is given by Eq. (5.6-37). Making appropriate changes in the notation and introducing the length, L_2, of the second section as $L - L_1$, we find the input relaxance to this section to be

$$Q_{mL_1}(s) = Q_{mC_2}(s)\tanh[(L - L_1)\Gamma_2(s)] \tag{10.1-19}$$

since $Q_{mL_2}(s) = 0$. Substituting Eq. (10.1-19) into Eq. (10.1-18) and suppressing the subscript m as no longer necessary, we obtain the relaxance of the composite Gross-Marvin ladder model as

$$\bar{Q}_{GM}(s) = Q_{C_1}(s)\frac{Q_{C_1}(s)\tanh[L_1\Gamma_1(s)] + Q_{C_2}(s)\tanh[(L - L_1)\Gamma_2(s)]}{Q_{C_1}(s) + Q_{C_2}(s)\tanh[L_1\Gamma_1(s)]\tanh[(L - L_1)\Gamma_2(s)]} \tag{10.1-20}$$

where $Q_{C_1}(s)$ and $Q_{C_2}(s)$ are the characteristic relaxances, and $\Gamma_1(s)$ and $\Gamma_2(s)$ are the transmittances of the two sections, and $L = L_1 + L_2$ is the total length of the composite ladder. The retardance becomes

$$\bar{U}_{GM}(s) = U_{C_1}(s)\frac{U_{C_1}(s) + U_{C_2}(s)\coth[L_1\Gamma_1(s)]\coth[(L - L_1)\Gamma_2(s)]}{U_{C_1}(s)\coth[L_1\Gamma_1(s)] + U_{C_2}(s)\coth[(L - L_1)\Gamma_2(s)]} \tag{10.1-21}$$

as the reciprocal of the relaxance. $U_{C_1}(s)$ and $U_{C_2}(s)$ are, of course, the characteristic retardances of the sections.

In principle, we could now proceed to establish the properties of the ladder along the lines of the analysis undertaken in Chap. 5. Unfortunately, this becomes rather complicated for a two-section model.* However, we may conjecture that the series-parallel model equivalent to the two-section ladder displayed in Fig. 10.1-3 will

* A somewhat simplified example is partially treated in Sect. 10.2.3.

consist of the sum of two distributions similar to those we derived in Sect. 5.1. Thus, we have essentially the respondances given by Eqs. (10.1-5) to (10.1-8), except that the distributions are known discrete distributions of respondance times instead of the unspecified sets of moduli and relaxation times, or compliances and retardation times, of the latter equations (see Problems 10.1-2 and 10.1-3).

10.1.4 Mathematical Models

It was pointed out in Sect. 6.1.3.1 that the mathematical models introduced there satisfactorily reproduce the qualitative features of the response to strain excitations of the standard linear solid but do not reproduce the characteristics of the standard linear liquid model in response to such excitations. The latter model possesses two relaxation times. Therefore, a mathematical model capable of displaying the same qualitative features must model a bimodal distribution of relaxation times. The models to be introduced here remedy the deficiency of the unimodal models considered in Chap. 6. As an example of this, in Problem 10.1-6 the complete set of experimental data assembled in Table 6.1-1 is fitted to one of the models introduced in this section. In Sect. 6.1.3.1 only the first portion of the complete set was fitted, treating that portion there as a set of arrheodictic data.

To model bimodal distributions of respondance times by mathematical equations we again make use of the *matching functions*, $\hat{Z}(x)$, $\hat{s}(x)$, and $\hat{\Lambda}(x)$ introduced in Chap. 6. As before, x stands for s, t, ω, or τ, as needed. Two terms, each with its own matching function, are needed for the representation of a bimodal distribution. The two matching functions need not be the same, but both terms should feature *either* primary, *or* associated functions. When using matching functions of the Z- and S-type, we again use upper and lower subscripts to distinguish the models based on the primary matching functions, $\hat{Z}(x)$, from those base on the associated functions, $\hat{s}(x)$. We shall use the subscripts M and m for the matching functions of the first term, and N and n for those of the second term. The two may or may not contain different steepness parameters (k, c, b, etc.). The determination of the location parameters, x_1 and x_2, and the steepness parameters*, μ_1 and μ_2, may involve trial and error. Some of the techniques described in Sect. 6.1 may be helpful.

Modelling by matching functions is straightforward. The functions are algebraically additive, both for the response to strain and to stress excitations. Z-type functions are added to $\{G_e\}$ or J_g, while S-type functions are subtracted from G_g or $J_e^{\{o\}}$. Rheodictic bimodal models for J(t) and J″(ω) carry the flow term in addition.

In response to a strain excitation we have

$$G_M(t) = \{G_e\} + G_1\hat{Z}_M(t) + G_2\hat{Z}_N(t) \tag{10.1-22}$$

or

$$G_m(t) = G_g - G_1\hat{s}_m(t) - G_2\hat{s}_n(t) \tag{10.1-23}$$

* To avoid confusion, we shall here denote the generic steepness parameter, for which we have used m in earlier chapters, by μ.

for the step responses. The harmonic responses become

$$G'_M(\omega) = G_g - G_1 \hat{Z}_M(\omega) - G_2 \hat{Z}_N(\omega) \tag{10.1-24}$$

or

$$G'_m(\omega) = \{G_e\} + G_1 \hat{s}_m(\omega) + G_2 \hat{s}_n(\omega) \tag{10.1-25}$$

for the storage functions, while the loss functions are given by

$$G''(\omega) = G''_{max\,1} \hat{\Lambda}_M(\omega) + G''_{max\,2} \hat{\Lambda}_N(\omega) \ . \tag{10.1-26}$$

The moduli must, of course, satisfy the condition $G_1 + G_2 = G_g - \{G_e\}$.

Again (cf. Sect. 6.1.3.3), models derived in this way for $G'(\omega)$ and for $G''(\omega)$ are not self-consistent. Self-consistent models can be obtained from models of the relaxances

$$\bar{Q}_M(s) = G_g - G_1 \hat{Z}_M(s) - G_2 \hat{Z}_N(s) \tag{10.1-27}$$

or

$$\bar{Q}_m(s) = \{G_e\} + G_1 \hat{s}_m(s) + G_2 \hat{s}_n(s) \tag{10.1-28}$$

by replacement of s by $j\omega$ and separation of the real and imaginary parts.

As an example, let us examine the relaxation modulus, $G_{C,K}(t)$, in which the first distribution is modelled by the Cole-Cole function, Eq. (6.1-22), while the second distribution uses the Kohlrausch function, Eq. (6.1-20). The equation is

$$G_{C,K}(t) = \{G_e\} + (G_g - G_P)\frac{1}{1 + (t/t_1)^c} + G_P \exp[-(t/t_2)^k] \tag{10.1-29}$$

and the plot is shown in Fig. 10.1-4. It was obtained with the following parameters: $G_g = 10^9$, $G_P = 10^{7.5}$, $G_e = 10^6$ N/m^2, $t_1 = 10^{-4}$, $t_2 = 10^5$ s, $c = 0.6$, and $k = 0.35$. Both arrheodictic and rheodictic behavior (broken line) is shown.

The equations applicable when the excitation is a stress follow analogously. The step responses are given by

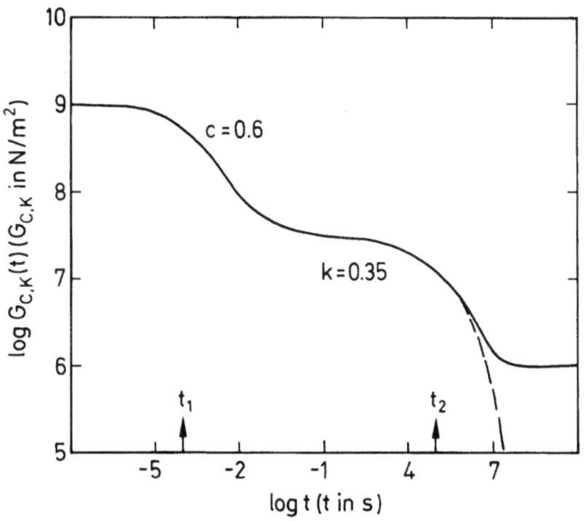

Fig. 10.1-4. Bimodal relaxation modulus calculated using the Cole-Cole and the Kohlrausch functions

$$J_M(t) = J_e^{\{o\}} - J_1\hat{Z}_M(t) - J_2\hat{Z}_N(t) + \{\phi_f t\} \tag{10.1-30}$$

or

$$J_m(t) = J_g + J_1\hat{s}_m(t) + J_2\hat{s}_n(t) + \{\phi_f t\} \tag{10.1-31}$$

while the storage functions are represented by

$$J'_M(\omega) = J_g + J_1\hat{Z}_M(\omega) + J_2\hat{Z}_N(\omega) \tag{10.1-32}$$

or

$$J'_m(\omega) = J_e^{\{o\}} - J_1\hat{s}_m(\omega) - J_2\hat{s}_n(\omega) \tag{10.1-33}$$

and the loss functions by

$$J''(\omega) = J''_{\max 1}\hat{\Lambda}_M(\omega) + J''_{\max 2}\hat{\Lambda}_N(\omega) + \{\phi_f/\omega\} \ . \tag{10.1-34}$$

The retardances become

$$\bar{U}_M(s) = J_g + J_1\hat{Z}_M(s) + J_2\hat{Z}_N(s) + \{\phi_f/s\} \tag{10.1-35}$$

or

$$\bar{U}_m(s) = J_e^{\{o\}} - J_1\hat{s}_m(s) - J_2\hat{s}_n(s) + \{\phi_f/s\} \ . \tag{10.1-36}$$

Self-consistent harmonic models can again be derived through replacement of s by $j\omega$ and separation of the real and imaginary parts. Here, the restriction $J_1 + J_2 = J_e - J_g$ applies to all models.

Figure 10.1-5 shows a plot of the self-consistent storage and loss compliances, $J'_C(\omega)$ and $J''_C(\omega)$, given by the bimodal forms of Eqs. (6.1-94) and (6.1-95). These are

$$J'_C(\omega) = J_g + \frac{J_P[(\omega/\omega_1)^{c_1} + \cos \pi c_1/2]}{(\omega/\omega_1)^{c_1} + 2\cos \pi c_1/2 + (\omega_1/\omega)^{c_1}} + \frac{(J_e - J_P)[(\omega/\omega_2)^{c_2} + \cos \pi c_2/2]}{(\omega/\omega_2)^{c_2} + 2\cos \pi c_2/2 + (\omega_2/\omega)^{c_2}} \tag{10.1-37}$$

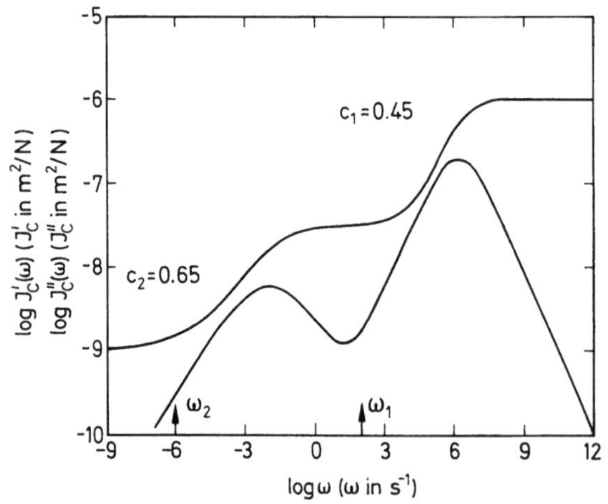

Fig. 10.1-5. Bimodal storage and loss compliances calculated using the Cole-Cole function

and

$$J''_C(\omega) = \frac{J_P \sin \pi c_1/2}{(\omega/\omega_1)^{c_1} + 2\cos \pi c_1/2 + (\omega_1/\omega)^{c_1}} + \frac{(J_e - J_P)\sin \pi c_2/2}{(\omega/\omega_2)^{c_2} + 2\cos \pi c_2/2 + (\omega_2/\omega)^{c_2}} . \tag{10.1-38}$$

The following parameters were used: $J_g = 10^{-9}$, $J_P = 10^{-7.5}$, $J_e = 10^{-6}$ m^2/N, $\omega_1 = 10^2$, $\omega_2 = 10^{-6}$ s^{-1}, $c_1 = 0.45$, and $c_2 = 0.65$. Only arrheodictic behavior is shown. Rheodictic behavior merely requires addition of the flow term, ϕ_f/ω, to $J''_C(\omega)$.

The equations discussed here may be suitably modified to suit specific applications. One such is the represention of the behavior of *entangled polymers*. These exhibit what we will call *pseudo-arrheodictic* behavior. A fuller discussion of the concepts of entanglements in polymers, and of pseudo-arrheodictic behavior in general, will follow in Sect. 10.2. Here, we merely list the equations for modelling the relaxation modulus and the creep compliance of entangled polymers. These equations are simple modifications of Eqs. (10.1-22) and (10.1-23) in which G_p is replaced by $G_N^{\{o\}}$, and of Eqs. (10.1-30) and (10.1-31) in which J_p is replaced by $J_N^{\{o\}}$. The equations for any of the other response functions can easily be constructed in analogous ways.

For the relaxation modulus of an entangled polymer we write

$$G_M(t) = \{G_e\} + (G_g - G_N^{\{o\}})\hat{Z}_M(t) + G_N^{\{o\}}\hat{Z}_N(t) \tag{10.1-39}$$

or

$$G_m(t) = G_g - (G_g - G_N^{\{o\}})\hat{s}_m(t) - (G_N^{\{o\}} - \{G_e\})\hat{s}_n(t) \tag{10.1-40}$$

and for the creep compliance we have

$$J_M(t) = J_e^{\{o\}} - (J_e^{\{o\}} - J_N^{\{o\}})\hat{Z}_M(t) - (J_N^{\{o\}} - J_g)\hat{Z}_N(t) + \{\phi_f t\} \tag{10.1-42}$$

or

$$J_m(t) = J_g + (J_N^{\{o\}} - J_g)\hat{s}_m(t) + (J_e^{\{o\}} - J_N^{\{o\}})\hat{s}_n(t) + \{\phi_f t\} \tag{10.1-41}$$

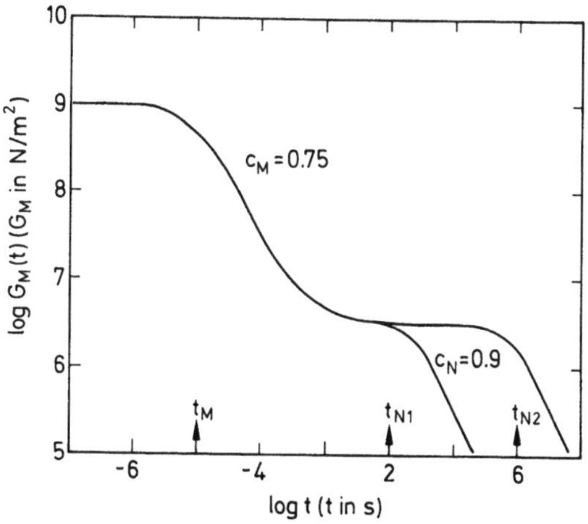

Fig. 10.1-6. Pseudo-arrheodictic relaxation moduli plotted using the Cole-Cole function

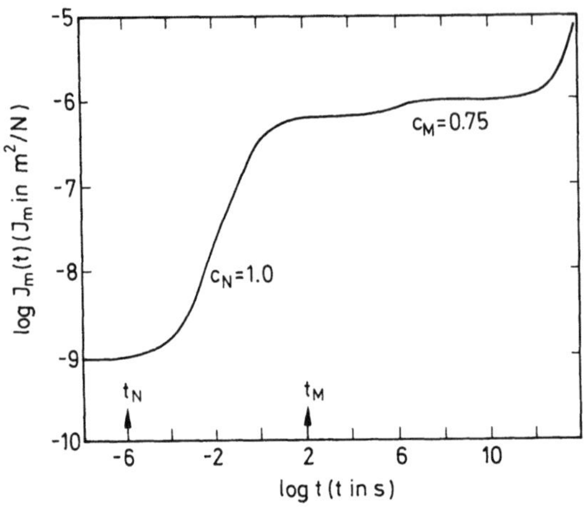

Fig. 10.1-7. Pseudo-arrheodictic creep compliance plotted using the Cole-Cole function

where $G_N^{\{o\}}$ and $J_N^{\{o\}}$ are the *plateau*, *pseudo-equilibrium*, or *entanglement modulus*, and *compliance*, respectively.

The above two figures exhibit relaxation modulus and creep compliance curves for entangled rheodictic polymers. Figure 10.1-6 shows Eq. (10.1-40) using the Cole-Cole function, Eq. (6.1-23), for $\hat{s}(t)$. It was plotted with the parameters $G_g = 10^9$, $G_N = 10^{6.5}$ N/m², $t_m = 10^{-4}$, $t_{n1} = 10^2$, $t_{n2} = 10^6$ s, $c_m = 0.75$, and $c_n = 0.9$ for both curves. It is clear that the length of the so-called entanglement plateau depends on the ratio t_m/t_n. It is the longer the smaller the ratio, and may become a mere shoulder as the ratio approaches unity.

Figure 10.1-7 shows Eq. (10.1-42), using the Cole-Cole function, Eq. (10.1-22), for $\hat{Z}(t)$. The parameters were $J_g = 10^{-9}$, $J_N = 10^{-6.2}$, $J_e^o = 10^{-6}$ m²/N, $t_M = 10^2$, $t_N = 10^6$ s, $c_M = 0.75$, $c_N = 1.0$, and $\phi_f = 10^{-18}$ m²/Ns. The plateau and steady-state compliances were chosen close together. They may, of course, also coincide completely.

Other examples of the representation of bimodal distributions by mathematical models may be found in Problems 10.1-4, 10.1-5, and 10.1-6.

10.2 Prolongated Unimodal Distributions

Figures 10.1-6 and 10.1-7 at the end of the preceding section illustrated the behavior of a *pseudo-arrheodictic* material. In a material exhibiting this behavior the relaxation modulus descends, with an increase in time, from the glassy modulus, G_g, through the transition region, to a *plateau*, characterized by the *plateau*, or *pseudo-equilibrium modulus*, G_N^o, simulating the true equilibrium modulus of arrheodictic materials. With a further increase in time, however, the modulus enters the steady-flow region typical of rheodictic materials. The behavior is qualitatively similar to that shown by $G_{4M}(t)$ in Fig. 3.4-23. In that figure, however, it arises because of the presence of two discrete relaxation times rather than two distributions of such times. By contrast, Fig. 10.1-6 shows the characteristic behavior of a bimodal distribution.

As shown by Fig. 10.1-7, the creep compliance displays, *mutatis mutandis*, analogous behavior.

Typical pseudo-arrheodictic comportment is shown by polymeric materials which may be modelled *as if* they contained temporary crosslinks (commonly referred to [1] as *entanglements*). Polymer molecules forming a three-dimensional network through covalent crosslinks are arrheodictic because the crosslinks prevent them from exhibiting steady-state flow. The temporary crosslinks pictured as entanglements between the long-chain polymer molecules mimic, as it were, arrheodictic behavior in a certain region of the time scale. At longer times, however, the entanglements disentangle and the material then exhibits its true rheodictic nature. An entangled arrheodictic polymer may exhibit a plateau as well as an equilibrium modulus. Any of the procedures introduced in the preceding section which are suitable for the modelling of bimodal viscoelastic behavior can, in principle, be used to model the behavior of entangled polymers, both rheodictic, and arrheodictic. Examples are provided by Figs. 10.1-4 to 10.1-7.

This section deals with a different approach. In this approach the material is not considered to possess two distinct distributions of respondance times. Rather, its behavior is modelled by a single unimodal distribution in which, however, the longest respondance times above a critical value, τ_c, are further lengthened, or *prolongated*, through multiplication by a factor, q. Because of the lengthening, or prolongating, of the relaxation or retardation behavior produced in this manner, we will call such a modified distribution a *prolongated unimodal distribution*. Such a distribution is not, of course, a true bimodal distribution although it has the qualitative aspects of one. It seemed appropriate, therefore, to treat the topic in the present chapter. We note that any bimodal or multimodal distribution, rheodictic or arrheodictic, can be modified by the method of *prolongation* discussed here. This can be useful in modelling more complex behavior if needed.

We proceed again by considering series-parallel models first.

10.2.1 Series-Parallel Models

The relaxance of a series-parallel model modified by prolongation becomes

$$\bar{Q}(s) = \{G_e\} + \sum_{n=0,1}^{n=c-1} \frac{G_n\tau_n s}{1+\tau_n s} + \sum_{n=c}^{n=R} \frac{G_n q\tau_n s}{1+q\tau_n s} \tag{10.2-1}$$

and the retardance follows as

$$\bar{U}(s) = J_g + \sum_{n=1}^{n=c-1} \frac{J_n}{1+\tau_n s} + \sum_{n=c}^{n=R} \frac{J_n}{1+q\tau_n s} + \{\phi_f/s\} \; . \tag{10.2-2}$$

where the respondance times, τ_n, are deemed to increase with an increase in the running index, n.

The experimental response functions can be obtained from these retardances in the usual way. The behavior of some typical "prolongated" relaxation moduli is shown in Fig. 10.2-1. The curves were generated from the series-parallel equivalents of the Gross-Marvin ladder models as discussed in Sect. 5.2.2.3. Since the relaxation

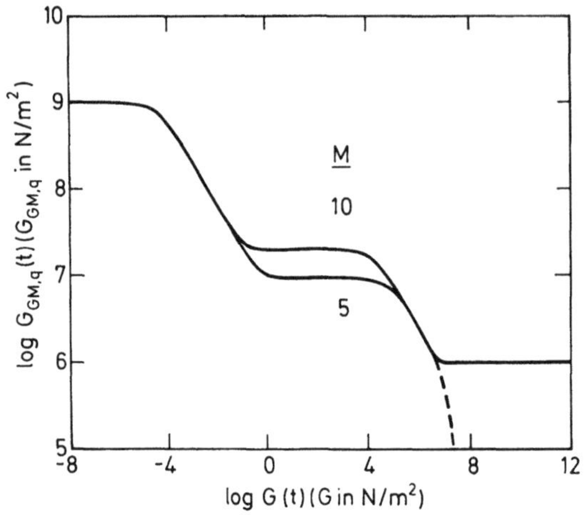

Fig. 10.2-1. Prolongated relaxation modulus calculated using the Gross-Marvin series-parallel distributions

times of these models decrease as the running index increases, our relaxation modulus becomes

$$G_{GM,q}(t) = \{G_e\} + \sum_{n=0,1}^{n=M-1} G_n \exp(-t/q\tau_n) + \sum_{n=M}^{n=N-1} G_n \exp(-t/\tau_n) \; . \tag{10.2-3}$$

The subscript q indicates that we are dealing with a relaxation modulus whose largest relaxation times were lengthened by the factor q, here 10^5. This factor determines the length of the "plateau region". Its height depends on M. Generally, only a few of the largest relaxation times need to be prolongated. The two curves in Fig. 10.2-1 were produced with $M = 5$, and $M = 10$, respectively. The moduli and relaxation times given by Eqs. (5.2-92) and (5.2-93) were used for rheodictic behavior, while those given by Eqs. (5.2-87) and (5.2-88) served for arrheodictic behavior. The number of elements in the models, N, was taken as 1000 in both cases. The other model parameters were $J = 10^{-9}$ m^2/N, and $\theta = 10^{-4}$ s. For rheodictic behavior G_e is, of course, absent and the summation begins with $n = 0$. It is this model then, which would be used to model pseudo-arrheodictic comportment.

The creep compliances and the various harmonic responses follow analogously (see Problem 10.2-1).

10.2.2 Models for the Spectral Response Functions

If a specific discrete distribution of respondance times is known, then the spectral functions can be prolongated in the manner indicated in Sect. 10.2.1. Thus, the so-called Rouse distribution, Eq. (6.3-2), would become

$$H_{R,p}(\tau) = G \sum_{n=1}^{n=M-1} q\tau_n \delta(\tau - q\tau_n) + G \sum_{n=M}^{n=N} \tau_n \delta(\tau - \tau_n) \tag{10.2-4}$$

where $\tau_n/\tau_1 = 1/n^2$.

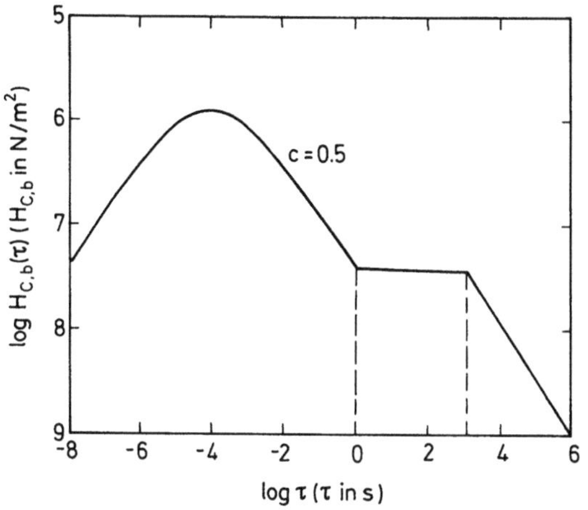

Fig. 10.2-2. Prolongated Cole-Cole relaxation spectrum

The spectra may be modelled also by any of the mathematical models of Sect. 6.3. The mathematical models do not provide a distribution of respondance times. They only simulate one, generally by a continuous function of τ. Hence, there are no distinct respondance times to prolong. Instead, the effect of prolongation is achieved by the use of the box distribution. Thus, a combination of the wedge and the box distribution has been proposed for modelling pseudo-arrheodictic behavior [2]. In the general case we write

$$F_{M,b}(\tau) = \begin{cases} F_M(\tau), & 0 < \tau < \tau_b \\ F_M(\tau_b), & \tau_b < \tau < q\tau_b \\ F_M(\tau/q), & q\tau_b < \tau < \infty \end{cases} \tag{10.2-5}$$

or, in terms of step functions, $h(\tau)$,

$$F_{M,b}(\tau) = F_M(\tau)[h(\tau) - h(\tau - \tau_b)] + F_M(\tau_b)[h(\tau - \tau_b) - h(\tau - q\tau_b)] + F_M(\tau/q)h(\tau - q\tau_b) \; . \tag{10.2-6}$$

In the equations above the subscript M symbolizes the spectral distribution to be prolonged, and b indicates the box distribution, Eq. (6.3-4), $F_M(\tau_b)$ being its height, and τ_b and $q\tau_b$ its lower and upper cut-off times.

Figure 10.2-2 displays the result of combining the continuous Cole-Cole spectrum, $H'_C(\tau)$, given by Eq. (6.3-129), with the box distribution. The curve was calculated with the parameters $G_g = 10^9$, $G_e = 10^6$ N/m², $\tau_0 = 10^{-4}$, $\tau_b = 1$ s, $q = 1000$, and $c = 0.5$. The broken lines serve to outline the box.

10.2.3 Ladder Models

Prolongation can be incorporated also into a ladder model. We demonstrate the procedure on hand of the extended Marvin-Oser model [3] (cf. Sect. 5.5.2). The model now becomes a composite ladder consisting of two sections. We let the

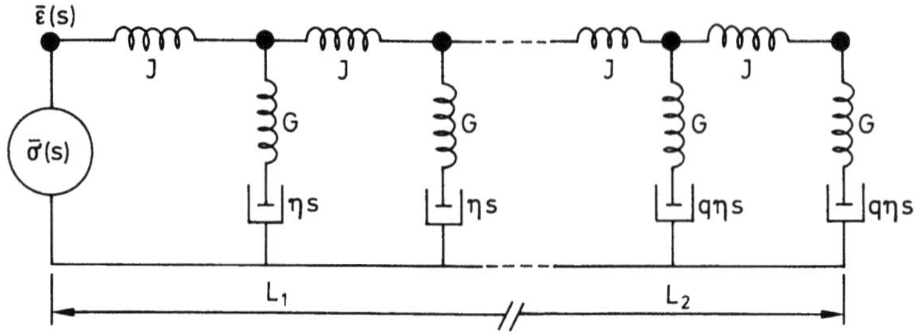

Fig. 10.2-3. Composite extended Marvin-Oser ladder model

characteristic time of the second section be that of the first, but prolongated by the factor q. This is equivalent to letting the rung vicosity of the second section be multiplied by q. The model is displayed in Fig. 10.2-3.

The relaxance of the composite model is given by Eq. (10.1-20) which we now write as

$$\bar{Q}_{MO,q}(s) = Q_{C_1}(s)\frac{Q_{C_1}(s)\tanh[L_1\Gamma_1(s)] + Q_{C_2}(s)\tanh[(L-L_1)\Gamma_2(s)]}{Q_{C_1}(s) + Q_{C_2}(s)\tanh[L_1\Gamma_1(s)]\tanh[(L-L_1)\Gamma_2(s)]} \,. \tag{10.2-7}$$

Let us look at the characteristic parameters. By Eq. (5.5-51) the characteristic relaxances become

$$Q_{C_1}(s) = \frac{1}{U_{C_1}(s)} = \frac{1}{J_T^\circ}\sqrt{\frac{\theta_\infty^\circ s}{1+\delta^\circ s}} = \frac{\Gamma_1(s)}{J_T^\circ} \tag{10.2-8}$$

and

$$Q_{C_2}(s) = \frac{1}{U_{C_2}(s)} = \frac{1}{J_T^\circ}\sqrt{\frac{q\theta_\infty^\circ s}{1+q\delta^\circ s}} = \frac{\Gamma_2(s)}{J_T^\circ} \,. \tag{10.2-9}$$

Equations (10.2-8) and (10.2-9) also define the characteristic retardances, $U_{C_1}(s)$ and $U_{C_2}(s)$, and the transmittances, $\Gamma_1(s)$ and $\Gamma_2(s)$, in terms of the ladder parameters. In these equations $\theta_\infty^\circ = J_T^\circ\eta_T^\circ$, and $\delta^\circ = \eta_T^\circ/G_T^\circ$, as in Sect. 5.5.2.

Marvin and Oser [3] had let the length of the first section of their composite ladder be unity, and designated the total length of the ladder by l. If we do the same and then let the total length be unity, Eq. (10.2-7) correctly reduces to Eq. (5.5-83).

Alternatively, we may let $L = 1$, and substitute the fraction, x, of the total length for L_1.

We can now rewrite Eq. (10.2-7) as

$$\bar{Q}_{MO,q}(s) = Q_{C_1}(s)\frac{\tanh[L_1\Gamma_1(s)] + \rho(s)\tanh[(L-L_1)\Gamma_2(s)]}{1 + \rho(s)\tanh[L_1\Gamma_1(s)]\tanh[(L-L_1)\Gamma_2(s)]} \,. \tag{10.2-10}$$

where

$$\rho(s) = \frac{Q_{C_2}(s)}{Q_{C_1}(s)} = \frac{\Gamma_2(s)}{\Gamma_1(s)} = \sqrt{\frac{q+q\delta^\circ s}{1+q\delta^\circ s}} \,. \tag{10.2-11}$$

The retardance, $\bar{U}_{MO,q}(s)$, is the reciprocal of the relaxance, $\bar{Q}_{MO,q}(s)$. Hence

$$\bar{U}_{MO,q}(s) = U_{C_1}(s)\frac{U_{C_2}(s) + U_{C_1}(s)\tanh[L_1\Gamma_1(s)]\tanh[(L-L_1)\Gamma_2(s)]}{U_{C_2}(s)\tanh[L_1\Gamma_1(s)] + U_{C_1}(s)\tanh[(L-L_1)\Gamma_2(s)]} \tag{10.2-12}$$

and

$$\bar{U}_{MO,q}(s) = U_{C_1}(s)\frac{1 + \rho(s)\tanh[L_1\Gamma_1(s)]\tanh[(L-L_1)\Gamma_2(s)]}{\tanh[L_1\Gamma_1(s)] + \rho(s)\tanh[(L-L_1)\Gamma_2(s)]} \tag{10.2-13}$$

or, alternatively,

$$\bar{U}_{MO,q}(s) = U_{C_1}(s)\frac{\rho(s) + \coth[L_1\Gamma_1(s)]\coth[(L-L_1)\Gamma_2(s)]}{\coth[(L-L_1)\Gamma_2(s)] + \rho(s)\coth[L_1\Gamma_1(s)]} \ . \tag{10.2-14}$$

We intend to use the prolonged Marvin-Oser ladder to model pseudo-arrheodictic behavior. Since this is essentially rheodictic behavior, the relaxance does not require knowledge of any viscoelastic constant. The retardance, however, does. Let us, therefore, now find the viscoelastic constants of $\bar{U}_{MO,q}(s)$. First, we have

$$J_g = \lim_{s\to\infty} \bar{U}_{MO,q}(s) \ . \tag{10.2-15}$$

But

$$\rho(\infty) = \Gamma_1(\infty) = \Gamma_1(\infty) = \Gamma_2(\infty) = \sqrt{J_T^\circ G_T^\circ} \tag{10.2-16}$$

and

$$U_{C_1}(\infty) = \sqrt{J_T^\circ/G_T^\circ} \ . \tag{10.2-17}$$

Hence,

$$J_g = U_{C_1}(s)\frac{\rho(\infty) + \coth[L_1\rho(\infty)]\coth[(L-L_1)\rho(\infty)]}{\coth[(L-L_1)\rho(\infty)] + \rho(\infty)\coth[L_1\rho(\infty)]} \ . \tag{10.2-18}$$

If we let $L_1 \to L$, we obtain

$$J_g = \sqrt{J_T^\circ/G_T^\circ}\coth L\sqrt{J_T^\circ G_T^\circ} \tag{10.2-19}$$

which reduces to Eq. (5.5-78) for $L = 1$.

The steady-flow fluidity is obtained from

$$\phi_f = \lim_{s\to 0} s\bar{U}_{MO,q}(s) \ . \tag{10.2-20}$$

After some algebraic manipulation using L'Hospitals' theorem in addition, we find (see Problem 10.2-2)

$$\phi_f = \frac{1}{L_1\eta_T^\circ + (L-L_1)q\eta_T^\circ} \ . \tag{10.2-21}$$

Again, if $q = 1$ and $L = 1$ also, $\phi_f = 1/\eta_T^\circ$ as in Sect. 5.5.2.

To obtain J_e°, we use

$$J_e^\circ = \lim_{s\to 0} [\bar{U}_{MO,q}(s) - \phi_f/s] \ . \tag{10.2-22}$$

This again requires some rather tedious algebra (see Problem 10.2-3). The result is

$$J_e^\circ = J_T^\circ\left(\frac{L_1 + (L - L_1)q}{3} + \frac{(L - L_1)^3(1 - q)q^2}{3[L_1 + (L - L_1)q]^2} + \frac{L_1 + (L - L_1)q^2}{G_T^\circ[L_1 + (L - L_1)q]^2}\right) . \tag{10.2-23}$$

As $L_1 \to L$, we have

$$J_e^\circ = \frac{L\,J_T^\circ}{3} + \frac{1}{L\,G_T^\circ} \tag{10.2-24}$$

which correctly reduces to Eq. (5.5-92) for $L = 1$.

The poles and residues can, in principle, be obtained in the usual way. To find those associated with $\bar{U}_{MO,q}(s)$, we recast Eq. (10.2-14) as

$$\bar{U}_{MO,q}(s) = \frac{J_T^\circ(1 + \delta^\circ s)^{1/2}}{(\theta_\infty^\circ s)^{1/2}}\,\frac{A(s) + B(s)}{C(s) + D(s)} \tag{10.2-25}$$

where

$$A(s) = [\Gamma_1(s) + \Gamma_2(s)]\cosh[L_1\Gamma_1(s) + (L - L_1)\Gamma_2(s)] \tag{10.2-25$_1$}$$

$$B(s) = [\Gamma_1(s) - \Gamma_2(s)]\cosh[L_1\Gamma_1(s) - (L - L_1)\Gamma_2(s)] \tag{10.2-25$_2$}$$

$$C(s) = [\Gamma_1(s) + \Gamma_2(s)]\sinh[L_1\Gamma_1(s) + (L - L_1)\Gamma_2(s)] \tag{10.2-25$_3$}$$

and

$$D(s) = [\Gamma_1(s) - \Gamma_2(s)]\sinh[L_1\Gamma_1(s) - (L - L_1)\Gamma_2(s)] . \tag{10.2-25$_4$}$$

Now, it can be shown that $(\theta_\infty^\circ s)^{1/2}$ constitutes a removable singularity. Hence, the poles, s_n, would be obtained as the roots of the equation

$$C(s_n) + D(s_n) = 0 \tag{10.2-26}$$

and the residues would follow from

$$\mathrm{Res}_n(\bar{U}_{MO,q}) = \left.\frac{A(s) + B(s)}{[C'(s) + D'(s)]}\right|_{s=s_n} \tag{10.2-27}$$

where $'$ denotes differentiation with respect to s [cf. Eq. (5.2-45)]. In principle, once this is done, the experimental response functions based on $\bar{U}_{MO,q}(s)$ could be found. The process could then be repeated for $\bar{Q}_{MO,q}(s)$. In practice, however, this proves to be a rather intractable problem. Thus, the experimental response functions would best be obtained by judicious approximations based on knowledge of the magnitudes of certain of the parameters [3].

10.2.4 Mathematical Models

Prolongation of any of the mathematical models for the experimental response functions can be accomplished with the use of the box distribution in the same way in which this was done in Sect. 10.2 with mathematical models for the spectral functions. An example is given in Problem 10.2-4.

10.3 Problems

Problem 10.0-1. Construct the line spectrum corresponding to the series-parallel model discussed in Problems 3.6-2 and 8.1-1. Use the values listed in Problem 3.6-2 as obtained by computer calculation and plot them against log τ_k.

Connect the end points of the arrows representing the location and the strength of the spectral lines by straight line segments.

Problem 10.1-1. Develop the bimodal form of $\eta(t)$, using Eq. (3.5-11).

Problem 10.1-2. You wish to use the series-parallel models furnished by the Gross-Marvin ladders to emulate bimodal behavior. Therefore, state the bimodal form of the retardance, $\bar{U}_{GM}(s)$, for arrheodictic behavior. What values of M and N will make J_e equal to 10^{-6}, and the plateau compliance, J_N, equal to about $10^{-7.4}$ m^2/N?

Problem 10.1-3. State the equations for $J'(\omega)$ and $J''(\omega)$ based on the bimodal retardance, $\bar{U}_{GM}(s)$, obtained in Problem 10.1-2.

Problem 10.1-4. Develop the bimodal form of $\eta(t)$, using Eq. (6.1-60).

Problem 10.1-5. Give the bimodal form of $G_R''(\omega)$ using Eq. (6.1-67).

Problem 10.1-6. You are asked to fit the complete set of experimental data assembled in Table 6.1-1 to a mathematical model. Which model would you choose? How would you determine its parameters?

Problem 10.2-1. You wish to simulate the behavior of a pseudo-arrheodictic material by prolongating the Gross-Marvin models. What equations would you use for the storage and and the loss compliances?

Problem 10.2-2. Derive the steady-state fluidity, ϕ_f, for the prolongated Marvin-Oser ladder model.

Problem 10.2-3. Derive the steady-state compliance, J_e°, for the prolongated Marvin-Oser ladder model.

Problem 10.2-4. Using the Kobeko equation, Eq. (6.1-24), state the equation for the prolongated relaxation modulus.

References (Chap. 10)

1. J.D. Ferry, *Viscoelastic Properties of Polymers*, 3rd ed., Wiley, New York, 1980, p. 241.
2. A.V. Tobolsky, *Properties and Structure of Polymers*, Wiley, New York, 1960.
3. R.S. Marvin and H. Oser, J. Res. Nat. Bureau Standards, *66B*:171 (1962).

11. Linear Viscoelastic Behavior in Different Modes of Deformation

11.0 Introduction

Chapters 2 to 10 discussed linear viscoelastic behavior in terms of behavior in shear (cf. Sect. 2.1). We now enlarge our discussion to cover linear viscoelastic behavior in modes of deformation other than shear. To do this, we need to establish the viscoelastic analogs of the stress-strain relations of a general *anisotropic* purely elastic linear material (see Sect. 1.4). This is the subject of the first section of this chapter. In the next section we specialize these relations for a linear viscoelastic *isotropic* material. In Sect. 11.3, then, we discuss the behavior of such a material in different modes of deformation. Finally, in Sect. 11.4 we deal with the problem of interconverting the various response functions which characterize the behavior of an isotropic linear viscoelastic material in the modes of deformation we considered in Sect. 11.3.

11.1 The General (Anisotropic) Viscoelastic Stress-Strain Relations

To establish the viscoelastic analogs of Eqs. (1.4-1) and (1.4-84) we make use of the *correspondence* (or *equivalence*) *principle*, which we had briefly introduced in Sect. 2.1.1. According to this principle, if an elastic solution to a stress analysis problem is known, substitution of the appropriate complex-plane transforms for the quantities employed in the elastic analysis immediately furnishes the viscoelastic solution in the transform plane. The time-dependent viscoelastic solution is then obtained by inverting the transform.

11.1.1 The Correspondence Principle

To obtain the viscoelastic counterparts of Eqs. (1.4-1) and (1.4-84) we substitute the Laplace transforms of the stress and strain tensors for the tensors themselves, and

Table 11.1-1. Elastic quantities and their viscoelastic analogs

Elastic quantity	Symbol	Impulse response	Step response	Ramp response	Harmonic response
Stress tensor	σ_{ij}	$\bar{\sigma}_{ij}(s)$	$\bar{\sigma}_{ij}(s)$	$\bar{\sigma}_{ij}(s)$	$\sigma_{ij}(\omega)$
Trace of stress tensor	Σ	$\bar{\Sigma}(s)$	$\bar{\Sigma}(s)$	$\bar{\Sigma}(s)$	$\Sigma(\omega)$
General modulus tensor	C_{ijkl}	$\bar{\Gamma}_{ijkl}(s)$	$s\bar{C}_{ijkl}(s)$	—	$C^*_{ijkl}(\omega)$
Strain tensor	γ_{kl}	$\bar{\gamma}_{ij}(s)$	$\bar{\gamma}_{ij}(s)$	$\bar{\gamma}_{ij}(s)$	$\gamma_{ij}(\omega)$
Trace of strain tensor	Δ	$\bar{\Delta}(s)$	$\bar{\Delta}(s)$	$\bar{\Delta}(s)$	$\Delta(\omega)$
General compliance tensor	S_{ijkl}	$\bar{\Psi}_{ijkl}(s)$	$s\bar{S}_{ijkl}(s)$	—	$S^*_{ijkl}(\omega)$
Shear modulus	G	$\bar{Q}(s)$	$s\bar{G}(s)$	$s^2\bar{\eta}(s)$	$G^*(\omega)$
Bulk modulus	K	$\bar{P}(s)$	$s\bar{K}(s)$	$s^2\bar{\xi}(s)$	$K^*(\omega)$
Stretch modulus	E	$\bar{Y}(s)$	$s\bar{E}(s)$	$s^2\bar{\zeta}(s)$	$E^*(\omega)$
Wave modulus	M	$\bar{V}(s)$	$s\bar{M}(s)$	—	$M^*(\omega)$
Shear compliance	J	$\bar{U}(s)$	$s\bar{J}(s)$	$s^2\bar{\chi}(s)$	$J^*(\omega)$
Bulk compliance	B	$\bar{A}(s)$	$s\bar{B}(s)$	$s^2\bar{\beta}(s)$	$B^*(\omega)$
Stretch compliance	D	$\bar{L}(s)$	$s\bar{D}(s)$	$s^2\bar{\psi}(s)$	$D^*(\omega)$
Wave compliance	T	$\bar{O}(s)$	$s\bar{T}(s)$	—	$T^*(\omega)$
Poisson's ratio	μ	$\bar{v}(s)$	$s\bar{\mu}(s)$	—	$\mu^*(\omega)$

the Laplace transforms of the corresponding impulse response functions for the elastic constants linking stress and strain in a purely elastic material. Thus, σ_{ij} becomes $\bar{\sigma}_{ij}(s)$, and γ_{ij} becomes $\bar{\gamma}_{ij}(s)$. Given an isotropic material, the shear modulus, G, becomes the (shear) relaxance, $\bar{Q}(s)$, and the shear compliance, J, becomes the (shear) retardance, $\bar{U}(s)$. These had been introduced earlier in Chap. 2. New symbols must be found for the impulse response functions corresponding to the other elastic constants. Table 11.1-1 assembles all these quantities for a general as well as for an isotropic material. The first and second column of the table contain the elastic quantities and the symbols we have used in Chap. 1 to represent them. The next column contains the symbols (several of them new), which we use for the Laplace transforms of the corresponding viscoelastic quantities. These are appropriate when the excitation is an impulse function of time. Since viscoelastic properties depend (in contrast to elastic properties) on the time regime employed to elicit them (see Sect. 4.0), we need different symbols when describing the response to other types of excitations. The next two columns contain those we use to represent the response to a stress or strain as a step function, and a ramp function of time, respectively. The last column lists the harmonic counterparts of the step response functions. Both these latter are in close correspondence with the symbols for the elastic constants because they have the same physical dimensions.

For example, $\bar{P}(s)$ is a new symbol for the *(operational) bulk relaxance*, i.e. the Laplace transform of the response to an isotropic stress impulse. The step response, however, is characterized by $s\bar{K}(s)$, i.e. by the Carson transform (the s-multiplied Laplace transform) of the bulk modulus, K. The corresponding function in the real time domain (not shown in the table) is simply the bulk relaxation modulus. K(t). We need no new symbol to denote it since it suffices to draw attention to its dependence on time. Similarly, the complex bulk modulus, $K^*(\omega)$, is the $j\omega$-multiplied Fourier transform of K(t). Thus, the complex functions do not require new symbols either. They simply assume the symbol for the elastic constant with the addition of an asterisk and the indication of their dependence on the radian frequency, ω. These notational details parallel, of course, those with which the reader is already familiar from our earlier discussions of the behavior in shear.

In Table 11.1-1 we have introduced the symbol T for the wave compliance of which, however, we make no use in this book. Again for the sake of completeness, we have also introduced the (operational) bulk, stretch, and wave relaxances, $\bar{P}(s)$, $\bar{Y}(s)$, and $\bar{V}(s)$, and the (operational) bulk, stretch, and wave retardances, $\bar{A}(s)$, $\bar{L}(s)$, and $\bar{O}(s)$, as well as the symbol $\bar{v}(s)$ for the (operational) lateral contraction in simple tension, or Poisson's ratio in the transform plane. All these are but rarely used, and there is no general agreement on the symbols to denote them.

The column containing the ramp response functions lacks several entries. The corresponding functions do, of course, exist. However, they again are not commonly used and no symbols for them are in general use. The symbols η, ξ, and ζ for the shear, bulk, and stretch (or tensile) viscosity we have introduced in Sect. 1.5 in our discussion of a purely viscous linear material. The symbols $\bar{\eta}(s)$, $\bar{\xi}(s)$, and $\bar{\zeta}(s)$ are the transforms of the responses to unit rates of strain in the three modes of deformation just listed. The symbol $\bar{\chi}(s)$ for the transform of the response to a unit rate of stress in shear had also been introduced earlier, in Sect. 2.4. To complete the picture, we propose here the symbols $\bar{\beta}(s)$ and $\bar{\psi}(s)$ for the transforms of the same responses in bulk and in stretch deformations, respectively.

With Eq. (1.5-3) we had defined the shear fluidity, $\phi = 1/\eta$, as the reciprocal of the shear viscosity, η. We offer no symbols here for the bulk and the stretch fluidities and will simply use $1/\xi$, and $1/\zeta$, respectively, if the need arises.

With help of Table 11.1-1 the relations between the elastic constants assembled in Tables 1.4-1 and 1.4-2 are easily rewritten for the viscoelastic case. Because of the similarities in the symbols, we obtain the relations between the harmonic quantities most easily. As an example,

$$E = 2(1 + \mu)G \tag{11.1-1}$$

becomes

$$E^*(\omega) = 2[1 + \mu^*(\omega)]G^*(\omega) \tag{11.1-2}$$

in the harmonic case. Separation of the real and imaginary parts furnishes the corresponding storage and loss moduli

$$E'(\omega) = 2[1 + \mu'(\omega)]G'(\omega) + 2\,\mu''(\omega)G''(\omega) \tag{11.1-3}$$

and

$$E''(\omega) = 2[1 + \mu'(\omega)]G''(\omega) - 2\,\mu''(\omega)G'(\omega)\ . \tag{11.1-4}$$

Again taking Eq. (11.1-1) as the departure point, Table 11.1-1 gives the relation between the impulse response functions in the transform plane as

$$\bar{Y}(s) = 2[1 + \bar{v}(s)]\bar{Q}(s)\ . \tag{11.1-5}$$

For the relations between the step response functions we substitute the Carson transforms from Table 11.1-1 and retransform the resulting relation onto the real time domain. Equation (11.1-1) thus first becomes

$$s\bar{E}(s) = 2[1 + s\bar{\mu}(s)]\, s\bar{G}(s) \tag{11.1-6}$$

in the transform plane. Cancellation of the transform variable, s, in front of $\bar{E}(s)$ and $\bar{G}(s)$ (but not $\bar{\mu}(s)$!) and retransforming then yields

$$E(t) = 2(1 + \mu_g)G(t) - 2\int_0^t G(u)\frac{d\,\mu(t-u)}{du}du \ . \tag{11.1-7}$$

We note that there are four equivalent forms of this equation (see Problem 11.1-1).

As another example, the Carson transform relation for the stretch relaxation modulus in terms of the shear and bulk relaxation moduli becomes

$$s\bar{E}(s) = \frac{9s\bar{K}(s)\,s\bar{G}(s)}{3s\bar{K}(s) + s\bar{G}(s)} \tag{11.1-8'}$$

which reduces to

$$\bar{E}(s) = \frac{9\,\bar{K}(s)\bar{G}(s)}{3\,\bar{K}(s) + \bar{G}(s)} \ . \tag{11.1-8}$$

Because of the cancellation of the transform variable, it might be thought that we obtained this relation by direct substitution of the Laplace transforms for the elastic quantities. As shown, however, by Eq. (11.1-6) where s does not cancel everywhere, unthinking replacement of the elastic constants by the corresponding Laplace transforms instead of the Carson transforms can lead to quite erroneous relations.

To retransform to the time domain, Eq. (11.1-8) must first be rewritten linearly as

$$\bar{E}(s)[3\,\bar{K}(s) + \bar{G}(s)] = 9\,\bar{K}(s)\bar{G}(s) \ . \tag{11.1-9}$$

With the help of Eq. (A3-16) we then find

$$\int_0^t E(t-u)[3\,K(u) + G(u)]du = 9\int_0^t K(t-u)G(u)du \ . \tag{11.1-10}$$

It is, unfortunately, not possible to express E(t) explicitly in terms of the two fundamental moduli, G(t) and K(t).

Equations (11.1-2) to (11.1-10) are examples of the interconversion of viscoelastic response functions obtained in different modes of deformation. This subject will be taken up more fully in Sect. 11.4.

We remark that the reciprocity relations between relaxances and retardances, as exemplified by Eqs. (2.1-19) and (2.3-13) for shear, remain valid for all other modes of deformation, together with all the consequences that flow from this validity. Thus, as an example, we have

$$\bar{A}(s)\bar{P}(s) = 1 \qquad \text{or} \qquad \bar{K}(s)\bar{B}(s) = 1/s^2 \tag{11.1-11}$$

and

$$\int_0^t K(t-u)B(u)du = \int_0^t K(u)B(t-u)du = t \ , \tag{11.1-12}$$

the last one in analogy to Eq. (2.3-14).

11.1.2 The General Stress-Strain Relations

We now formulate the stress-strain relations for a general linear viscoelastic material. Applying the correspondence principle to Eqs. (1.4-1) and (1.4-84) with the help of Table 11.1-1 furnishes

$$\bar{\sigma}_{ij}(s) = \bar{\Gamma}_{ijkl}(s)\bar{\gamma}_{kl}(s) \tag{11.1-13}$$

and

$$\bar{\gamma}_{ij}(s) = \bar{\Psi}_{ijkl}(s)\bar{\sigma}_{kl}(s) \tag{11.1-14}$$

as the basic transform relations when the excitation is a unit impulse. Here, the fourth-order tensors $\bar{\Gamma}_{ijkl}(s)$ and $\bar{\Psi}_{ijkl}(s)$ represent the (operational) relaxance and (operational) retardance of a general anisotropic linear viscoelastic material.

To derive the transform relations for the step responses we substitute $s\bar{C}_{ijkl}(s)$ for $\bar{\Gamma}_{ijkl}(s)$, and $s\bar{S}_{ijkl}(s)$ for $\bar{\Psi}_{ijkl}(s)$ and obtain

$$\bar{\sigma}_{ij}(s) = s\bar{C}_{ijkl}(s)\bar{\gamma}_{kl}(s) \tag{11.1-15}$$

and

$$\bar{\gamma}_{ij}(s) = s\bar{S}_{ijkl}(s)\bar{\sigma}_{kl}(s) \ . \tag{11.1-16}$$

On the real time axis these equations become

$$\sigma_{ij}(t) = \int_0^t C_{ijkl}(t-u)\frac{d\,\gamma_{kl}(u)}{du}du \tag{11.1-17}$$

and

$$\gamma_{ij}(t) = \int_0^t S_{ijkl}(t-u)\frac{d\,\sigma_{kl}(u)}{du}du \ . \tag{11.1-18}$$

There are, of course, again four equivalent forms of both of these equations.

In harmonic excitation the general stress-strain relations become

$$\sigma_{ij}(\omega) = C^*_{ijkl}(\omega)\gamma_{kl}(\omega) \tag{11.1-19}$$

and

$$\gamma_{ij}(\omega) = S^*_{ijkl}(\omega)\sigma_{kl}(\omega) \tag{11.1-20}$$

where

$$C^*_{ijkl}(\omega) = \bar{\Gamma}_{ijkl}(s)|_{s=j\omega} \tag{11.1-21}$$

and

$$S^*_{ijkl}(\omega) = \bar{\Psi}_{ijkl}(s)|_{s=j\omega} \ . \tag{11.1-22}$$

Making further use of the correspondence principle, we may now derive from the relations discussed in Sect. 1.4 the response of linearly viscoelastic plane-symmetric, orthotropic, axisymmetric, and isotropic materials. We shall, however, not deal in detail with the lower symmetries here but, in the next two sections, concentrate on the response of isotropic linearly viscoelastic materials in various modes of deformation. Problems 11.1-8 and 11.3-8 discuss illustrative examples for the case of axisymmetry.

11.2 The Isotropic Viscoelastic Stress-Strain Relations

In Sects. 1.4.4.1 and 1.4.4.2 we derived two relations, Eqs. (1.4-69) and (1.4-92), which we called the generalized Hooke's laws for an isotropic linear purely elastic material. These represented the general stress-strain relations for such a material. Similarly, in Sect. 1.5.1 we derived several forms of what we called the generalized Newton's law for an isotropic purely viscous material. These were Eqs. (1.5-5), (1.5-11), and (1.5-13). The corresponding equations for an isotropic linear viscoelastic material are most easily obtained by applying the correspondence principle to the equations just listed. We take the elastic relations first, and follow these with the viscous ones.

11.2.1 The Viscoelastic Generalized Hooke's Laws

Applying the correspondence principle to Eqs. (1.4-69) and (1.4-92) we obtain the impulse response forms of the generalized Hooke's laws in the transform plane as

$$\bar{\sigma}_{ij}(s) = [\bar{P}(s) - \tfrac{2}{3}\,\bar{Q}(s)]\bar{\Delta}(s)\delta_{ij} + 2\,\bar{Q}(s)\bar{\gamma}_{ij}(s) \tag{11.2-1}$$

[cf. Eq. (1.4-69)], and

$$\bar{\gamma}_{ij}(s) = [\tfrac{1}{9}\,\bar{A}(s) - \tfrac{1}{6}\,\bar{U}(s)]\bar{\Sigma}(s)\delta_{ij} + \tfrac{1}{2}\,\bar{U}(s)\bar{\sigma}_{ij}(s) \tag{11.2-2}$$

[cf. Eq. (1.4-92)], where

$$\bar{\Delta}(s) = \bar{\gamma}_{11}(s) + \bar{\gamma}_{22}(s) + \bar{\gamma}_{33}(s) \tag{11.2-3}$$

is the (operational) dilatation (or contraction), while

$$\bar{\Sigma}(s) = \bar{\sigma}_{11}(s) + \bar{\sigma}_{22}(s) + \bar{\sigma}_{33}(s) \tag{11.2-4}$$

when divided by 3, is the (operational) mean normal stress.

Retransformation of Eqs. (11.2-1) and (11.2-2) onto the real time axis leads to

$$\sigma_{ij}(t) = \int_0^t [P(t-u) - \tfrac{2}{3}\,Q(t-u)]\Delta(u)\delta_{ij}du + 2\int_0^t Q(t-u)\gamma_{ij}(u)du \tag{11.2-5}$$

and

$$\gamma_{ij}(t) = \int_0^t [\tfrac{1}{9}\,A(t-u) - \tfrac{1}{6}\,U(t-u)]\Sigma(u)\delta_{ij}du + \tfrac{1}{2}\int_0^t U(t-u)\sigma_{ij}(u)du\;. \tag{11.2-6}$$

Replacing in Eqs. (11.2-1) and (11.2-2) the Laplace by the Carson transforms (see Table 11.1-1) provides the relations

$$\bar{\sigma}_{ij}(s) = [s\bar{K}(s) - \tfrac{2}{3}\, s\bar{G}(s)]\bar{\Delta}(s)\delta_{ij} + 2\, s\bar{G}(s)\bar{\gamma}_{ij}(s) \tag{11.2-7}$$

and

$$\bar{\gamma}_{ij}(s) = [\tfrac{1}{9}\, s\bar{B}(s) - \tfrac{1}{6}\, s\bar{J}(s)]\bar{\Sigma}(s)\delta_{ij} + \tfrac{1}{2}\, s\bar{J}(s)\bar{\sigma}_{ij}(s)\ . \tag{11.2-8}$$

Retransformation of Eqs. (11.2-7) and (11.2-8) into the real time domain then yields

$$\sigma_{ij}(t) = \int_0^t K(t-u)\frac{d\,\Delta(u)}{du}\delta_{ij}du - \frac{2}{3}\int_0^t G(t-u)\frac{d\,\Delta(u)}{du}\delta_{ij}du + 2\int_0^t G(t-u)\frac{d\,\gamma_{ij}(u)}{du}du \tag{11.2-9}$$

and

$$\gamma_{ij}(t) = \frac{1}{9}\int_0^t B(t-u)\frac{d\,\Sigma(u)}{du}\delta_{ij} - \frac{1}{6}\int_0^t J(t-u)\frac{d\,\Sigma(u)}{du}\delta_{ij}du + \frac{1}{2}\int_0^t J(t-u)\frac{d\,\sigma_{ij}(u)}{du}du\ . \tag{11.2-10}$$

There are, of course, again four equivalent forms of Eqs. (11.2-9) and (11.2-10) (see Appendix A3.1.6).

Exchanging in Eqs. (11.2-1) and (11.2-2) the transforms for the harmonic response forms leads to

$$\sigma_{ij}(\omega) = [K^*(\omega) - \tfrac{2}{3}\, G^*(\omega)]\Delta(\omega)\delta_{ij} + 2\, G^*(\omega)\gamma_{ij}(\omega) \tag{11.2-11}$$

with

$$\Delta(\omega) = \gamma_{11}(\omega) + \gamma_{22}(\omega) + \gamma_{33}(\omega)\ , \tag{11.2-12}$$

and to

$$\gamma_{ij}(\omega) = [\tfrac{1}{9}\, B^*(\omega) - \tfrac{1}{6}\, J^*(\omega)]\Sigma(\omega)\delta_{ij} + \tfrac{1}{2}\, J^*(\omega)\sigma_{ij}(\omega) \tag{11.2-13}$$

with

$$\Sigma(\omega) = \sigma_{11}(\omega) + \sigma_{22}(\omega) + \sigma_{33}(\omega)\ . \tag{11.2-14}$$

In the theory of elasticity the generalized Hooke's law has only *one* form, Eq. (1.4-69), for stress excitation and *one* form, Eq. (1.4-92), for strain excitation. In the theory of linear viscoelasticity it has many forms for either excitation, depending on the time regime chosen for the excitation function.

11.2.2 The Viscoelastic Generalized Newton's Laws

The viscoelastic generalized Newton's laws represent the stress/rate-of-strain relations for isotropic linear viscoelastic liquids. Applying the correspondence principle to Eq. (1.5-7) gives

$$\bar{\sigma}_{ij}(s) = [\bar{\xi}(s) - \tfrac{2}{3}\,\bar{\eta}(s)]\bar{\Delta}(s)\delta_{ij} + 2\,\bar{\eta}(s)\bar{\dot{\gamma}}(s) \tag{11.2-15}$$

where $\bar{\dot{\gamma}}(s)$ is the rate of strain transform and

$$\bar{\dot{\Delta}}(s) = \bar{\dot{\gamma}}_{11}(s) + \bar{\dot{\gamma}}_{22} + \bar{\dot{\gamma}}_{33}(s) \tag{11.2-16}$$

is the (operational) rate of dilatation.

If the liquid is deemed to respond elastically to an isotropic pressure, we turn to Eq. (1.5-11) and obtain

$$\bar{\sigma}_{ij}(s) = [\bar{K}(s)\Delta(s) + \bar{\xi}(s)\bar{\dot{\Delta}}(s) - \tfrac{2}{3}\,\bar{\eta}(s)\bar{\dot{\Delta}}(s)]\delta_{ij} + 2\,\bar{\eta}(s)\bar{\dot{\gamma}}(s) \tag{11.2-17}$$

If the liquid is considered to be incompressible, then Eq. (11.2-17) reduces to

$$\bar{\sigma}_{ij}(s) = -p\,\delta_{ij} + 2\,\bar{\eta}(s)\bar{\dot{\gamma}}_{ij}(s) \tag{11.2-18}$$

where p is the isotropic pressure.

11.3 Linear Viscoelastic Behavior of Isotropic Materials in Different Modes of Deformation

We are now ready to consider the behavior of an isotropic linear viscoelastic material in the most frequently employed modes of deformation. We begin with the behavior in shear (Sect. 11.3.1), follow this with the behavior in isotropic compression (Sect. 11.3.2), then with the behavior in constrained uniaxial compression (Sect. 11.3.3), and finally with that in uniaxial (i.e. simple) tension and compression (Sect. 11.3.4). The latter also requires a consideration of the time-dependent lateral contraction or extension (the time-dependent Poisson's ratio). Sect. 11.3.5 treats this. In Sect. 11.3.6 we add a short discussion of the propagation of waves other then shear waves in a linear viscoelastic isotropic material.

Generally, in each subsection we begin by establishing the stress-strain relations for the mode of deformation under scrutiny. We do this by first formulating the Laplace transforms of the stress and strain tensors appropriate for each deformation, and then substitute these into the generalized Hooke's laws for an isotropic linear viscoelastic material. In the ensuing transform relations the stress and the strain transforms are linked by means of the transforms of viscoelastic material functions which characterize the behavior in the chosen deformation. Where we consider deformations other than shear, these functions are new. In some cases they are linear, in other cases they are non-linear combinations of the transforms of the functions which characterize the behavior of a linear viscoelastic material in the two *fundamental modes of deformation*, viz. shear and bulk deformation. We then discuss the continuous canonical representation of these material functions. Excepting Sect. 11.3.5 where it appears *not* to be expedient to do so, we refrain from spelling out the discrete forms since these are easily obtained from the continuous forms.

We will be dealing only with the modes of deformation which are most commonly employed in establishing linear viscoelastic properties. If needed, the behavior in some other mode of deformation (e.g. in equibiaxial tension) can be derived in analogous ways (see Problem 11.3-1).

In Chap. 1 we had followed essentially the same course, but on the basis of infinitesimal elasticity theory. We showed there that the stress and strain tensors could be split into isotropic and deviatoric tensors [cf. Eqs. (1.4-58) and (1.4-61)]. In the two fundamental deformations, bulk and shear, one pair of these tensors became "null tensors", i.e. tensors whose every component is zero. In the case of bulk deformation these were the deviatoric tensors. The deformation, therefore, was one of *size* only and was characterized by the *bulk modulus*, K, or the *bulk compliance*, B. In the case of the shear deformation the isotropic tensors became the null tensors. Thus, the deformation was one of *shape* only and was characterized by the *shear modulus*, G, or the *shear compliance*, J. In any other deformation, none of the two pairs of tensors vanished. The resulting behavior consequently required combinations of K and G, or B and J for its description, and new material constants were defined in terms of these combinations.

The essential difference between the approach used in Chap. 1 and that introduced here lies in the nature of the material descriptors. Linear viscoelastic theory being an extension of infinitesimal elastic theory to time-dependent behavior, the material constants discussed in Chap. 1 now become material functions of time (or, equivalently, of frequency).

The material functions also contain *viscoelastic constants*. Those appropriate to deformation in shear were treated in Sect. 4.6. In deformations other than shear new viscoelastic constants are needed (see Table 11.1-1) but, *mutatis mutandis*, Sect. 4.6 remains valid.

11.3.1 Behavior in Shear

We begin by considering the first of the two fundamental modes of deformation, that in shear. It is this mode which we have considered exclusively in all preceding chapters in which we examined linear viscoelastic behavior. This section contains little new information. Its purpose is, rather, to place shear deformation into the proper context among other deformations of a linear viscoelastic material, and to do this on hand of an already well understood mode.

In shear the *shape* of the body is changed *isochorically*, i.e. without a change in size. Thus, normal stresses are absent and only shear stresses act on the material. The other fundamental deformation, a pure change in size without any change in shape, will be discussed in the next section. In simple shear in the 1,2-plane, say, the isotropic tensors become null tensors. The total tensors thus become equal to the deviatoric tensors, each of which then has only two identical non-vanishing components. The matrices of their transforms become

$$[\bar{\sigma}_{ij}(s)] = \begin{bmatrix} 0 & \bar{\sigma}_{12}(s) & 0 \\ \bar{\sigma}_{12}(s) & 0 & 0 \\ 0 & 0 & 0 \end{bmatrix} \qquad (11.3\text{-}1)$$

and

$$[\bar{\gamma}_{ij}(s)] = \begin{bmatrix} 0 & \bar{\gamma}_{12}(s) & 0 \\ \bar{\gamma}_{12}(s) & 0 & 0 \\ 0 & 0 & 0 \end{bmatrix} . \tag{11.3-2}$$

Substituting the tensors into the generalized Hooke's law for isotropic bodies, Eq. (11.2-1), yields two identical equations for the stress-strain relation in simple shear. The relation is

$$\bar{\sigma}_{12}(s) = 2\,\bar{Q}(s)\bar{\gamma}_{12}(s) . \tag{11.3-3}$$

where $\bar{\sigma}_{12}(s)$ and $\bar{\gamma}_{12}(s)$ are the transforms of the shear stress, and the shear strain, respectively. By Eq. (11.2-2) we also have

$$\bar{\gamma}_{12}(s) = \tfrac{1}{2}\,\bar{U}(s)\bar{\sigma}_{12}(s) . \tag{11.3-4}$$

Exchanging the shear strain component, $\bar{\gamma}_{12}(s)$, against the *amount of shear* [$\bar{\varepsilon}(s) = 2\,\bar{\gamma}_{12}(s)$, see Eq. (1.3-19), and the introduction to Sect. 2.1.], Eqs. (11.3-3) and (11.3-4) reduce to

$$\bar{\sigma}(s) = \bar{Q}(s)\bar{\varepsilon}(s) \qquad \text{and} \qquad \bar{\varepsilon}(s) = \bar{U}(s)\bar{\sigma}(s) . \tag{11.3-5}$$

It is these equations [cf. Eqs. (2.1-15)] which formed the basis of all our discussions from Chap. 2 to Chap. 10.

The time-dependent and frequency-dependent stress-strain relations follow at once. Substituting $s\bar{G}(s)$ and $s\bar{J}(s)$ for $\bar{Q}(s)$ and $\bar{U}(s)$, respectively, in Eqs. (11.3-3) and (11.3-4) in accordance with Table 11.1-1, and retransforming to the real time axis we find

$$\sigma_{12}(t) = 2\int_0^t G(t-u)\frac{d\,\gamma_{12}(u)}{du}\,du \tag{11.3-6}$$

and

$$\gamma_{12}(t) = \frac{1}{2}\int_0^t J(t-u)\frac{d\,\sigma_{12}(u)}{du}\,du \tag{11.3-7}$$

which reduce to Eqs. $(2.3\text{-}16)_3$ and $(2.3\text{-}17)_3$ upon the introduction of the amount of shear. The frequency-dependent stress-strain relations are obtained by substitution of the appropriate quantities taken from Table 11.1-1 into Eqs. (11.3-5) and become

$$\sigma(\omega) = G^*(\omega)\varepsilon(\omega) \qquad \text{and} \qquad \varepsilon(\omega) = J^*(\omega)\sigma(\omega) , \tag{11.3-8}$$

i.e. Eqs. $(2.5\text{-}12)_2$ and $(2.5\text{-}16)_2$.

We do not need to discuss here the various forms which the shear moduli and shear compliances can take because we have already done so in quite some detail in Sect. 3.5 for the discrete, and in Sect. 4.1 for the continuous forms.

11.3.2 Behavior in Isotropic Compression

Isotropic compression is the next mode of deformation to consider. It is governed by the second fundamental material function, the (time-dependent) bulk modulus.

In isotropic compression the *size* of the material is changed by bringing to bear on it a set of three identical normal stresses (the pressure). If the resulting deformation is a *decrease* in size, it is called *isotropic compression*. An *increase* in size, called *isotropic dilatation* is, of course, also possible in principle but an isotropic tension is difficult to generate experimentally. We shall, therefore, consider here isotropic compression only. Isotropic dilatation can be accounted for by appropriate sign changes.

Isotropic compression and isotropic pressure are sometimes referred to as hydrostatic compression and hydrostatic pressure. Strictly speaking, however, the term *hydrostatic pressure* refers to the pressure generated by a liquid, more particularly by water.

In isotropic compression it is the deviatoric tensors which vanish, and the total tensors become equal to the isotropic tensors. The components of the stress tensor transform become $\bar{\sigma}_{11}(s) = \bar{\sigma}_{22}(s) = \bar{\sigma}_{33}(s)$, and those of the strain tensor transform are $\bar{\gamma}_{11}(s) = \bar{\gamma}_{22}(s) = \bar{\gamma}_{33}(s)$. The tensor matrices then result as

$$[\bar{\sigma}_{ij}(s)] = \begin{bmatrix} \bar{\sigma}_{11}(s) & 0 & 0 \\ 0 & \bar{\sigma}_{11}(s) & 0 \\ 0 & 0 & \bar{\sigma}_{11}(s) \end{bmatrix} \tag{11.3-9}$$

and

$$[\bar{\gamma}_{ij}(s)] = \begin{bmatrix} \bar{\gamma}_{11}(s) & 0 & 0 \\ 0 & \bar{\gamma}_{11}(s) & 0 \\ 0 & 0 & \bar{\gamma}_{11}(s) \end{bmatrix}, \tag{11.3-10}$$

respectively. Introducing the tensors into the generalized Hooke's laws for isotropic materials, Eqs. (11.2-1) and (11.2-2), yields (for each of the laws) three identical stress-strain transform relations which have the form

$$\bar{\sigma}_{11}(s) = 3[\bar{P}(s) - \tfrac{2}{3}\,\bar{Q}(s)]\bar{\gamma}_{11}(s) + 2\,\bar{Q}(s)\bar{\gamma}_{11}(s) \tag{11.3-11}$$

where we used $\bar{\Delta}(s) = 3\,\bar{\gamma}_{11}(s)$, and

$$\bar{\gamma}_{11}(s) = 3[\tfrac{1}{9}\,\bar{A}(s) - \tfrac{1}{6}\,\bar{U}(s)]\bar{\sigma}_{11}(s) + \tfrac{1}{2}\,\bar{U}(s)\bar{\sigma}_{11}(s) \tag{11.3-12}$$

where we used $\bar{\Sigma}(s) = 3\,\bar{\sigma}_{11}(s)$. Equations (11.3-11) and (11.3-12) reduce to

$$\frac{\bar{\sigma}_{11}(s)}{\bar{\gamma}_{11}(s)} = 3\,\bar{P}(s) = 3\,s\bar{K}(s) \tag{11.3-13}$$

and

$$\frac{\bar{\gamma}_{11}(s)}{\bar{\sigma}_{11}(s)} = \tfrac{1}{3}\,\bar{A}(s) = \tfrac{1}{3}\,s\bar{B}(s) \tag{11.3-14}$$

where $\bar{P}(s)$ and $\bar{A}(s)$ are the *bulk relaxance* and *bulk retardance*, while $\bar{K}(s)$ and $\bar{B}(s)$ are the (operational) *bulk modulus* and *bulk compliance*, respectively.

Upon transform inversion Eqs. $(11.3\text{-}13)_2$ and $(11.3\text{-}14)_2$ yield

$$\sigma_{11}(t) = 3\int_0^t K(t-u)\frac{d\,\gamma_{11}(u)}{du}\,du \tag{11.3-15}$$

and

$$\gamma_{11}(t) = \frac{1}{3}\int_0^t B(t-u)\frac{d\,\sigma_{11}(u)}{du}du \tag{11.3-16}$$

where K(t) and B(t) are the (time-dependent) *bulk relaxation modulus*, and *bulk creep compliance*, respectively.

The stress-strain relations may be restated as pressure/volume-change relations. The isotropic (or hydrostatic) pressure* is simply the negative inverse of the stress component, i.e.

$$p(t) = -\sigma_{11}(t) = \tfrac{1}{3}\Sigma(t)\ . \tag{11.3-17}$$

The (relative) volume change is the (time-dependent) cubical contraction given by

$$\Delta(t) = 3\,\gamma_{11}(t) = \frac{V_0 - V(t)}{V_0} = 1 - \frac{V(t)}{V_0} \tag{11.3-18}$$

where V(t) is the deformed, and V_0 is the undeformed volume. Thus,

$$p(t) = -\int_0^t K(t-u)\frac{d\,\Delta(u)}{du}du \tag{11.3-19}$$

and

$$\Delta(t) = -\int_0^t B(t-u)\frac{d\,p(u)}{du}du\ . \tag{11.3-20}$$

In the transform plane Eqs. (11.3-19) and (11.3-20) become

$$\bar{p}(s) = -s\bar{K}(s)\bar{\Delta}(s) \tag{11.3-21}$$

and

$$\bar{\Delta}(s) = -s\bar{B}(s)\bar{p}(s)\ . \tag{11.3-22}$$

Having established the stress-strain relations in isotropic compression, let us now turn to the material functions which characterize it. The bulk relaxance takes the forms

$$\bar{P}(s) = s\bar{K}(s) = K_e + \int_{-\infty}^{\infty} H_K(\tau)\frac{\tau s}{1+\tau s}d\ln\tau \tag{11.3-23}$$

or

$$\bar{P}_\bullet(s) = s\bar{K}_\bullet(s) = K_g - \int_{-\infty}^{\infty} H_K(\tau)\frac{1}{1+\tau s}d\ln\tau \tag{11.3-24}$$

where $H_K(\tau)$ is the *bulk relaxation spectrum*, and K_e and K_g are the equilibrium and glassy bulk moduli.

---

* The time-dependent pressure, p(t), must not be confused with the slope function, also denoted by p(t) (see Sect. A2.4 of the Appendix).

The bulk retardance becomes

$$\bar{A}(s) = s\bar{B}(s) = B_g + \int_{\infty}^{\infty} L_B(\tau)\frac{1}{1+\tau s}\, d\ln\tau \tag{11.3-25}$$

or

$$\bar{A}_\bullet(s) = s\bar{B}_\bullet(s) = B_e - \int_{-\infty}^{\infty} L_B(\tau)\frac{\tau s}{1+\tau s}\, d\ln\tau \tag{11.3-26}$$

where $L_B(\tau)$ is the *bulk retardation spectrum* and B_e and B_g are the equilibrium and glassy bulk compliances.

Finally, for the transform of the time-dependent *bulk viscosity* we have

$$\bar{\xi}(s) = \frac{\bar{K}(s)}{s} = \frac{K_e}{s^2} + \int_{-\infty}^{\infty} H_K(\tau)\frac{\tau}{s(1+\tau s)}\, d\ln\tau\ . \tag{11.3-27}$$

As indicated by the absence of the flow term in Eqs. (11.3-25) and (11.3-26), and by the absence of braces around K_e in Eqs. (11.3-23) and (11.3-27), bulk behavior is always necessarily arrheodictic. It is clearly not possible for the volume to decrease indefinitely under an applied isotropic pressure of any magnitude.

Relaxation and retardation in bulk is generally *not* identical with relaxation and retardation in shear. Phenomenological theory cannot say anything about the appearance of either the shear or the bulk relaxation spectra, except that both have the nature of a distribution function (see Sect. 4.6) and are, therefore, bell-shaped as discussed in Sects. 4.2.1 and 6.3. While a good deal of experimental evidence has been obtained concerning shear relaxation in various materials, much less is currently known concerning bulk relaxation. In many synthetic polymers the bulk relaxation modulus changes from the glassy to the equilibrium state by only a factor of about 2 to 3 while the shear modulus changes by 3 to 4 logarithmic decades. This behavior is contrasted in Fig. 11.3-1.

The curves were calculated from models for the standard linear solid [cf. Eq. (3.4-127)]. The parameters for $G(t)$ were $G_g = 10^9\ N/m^2$, $G_e = 10^6/1.001\ N/m^2$, and $\tau = 10^{-4}/1.001$ s. These are the same values with which $G_{3M}(t)$ was plotted in Fig. 3.4-20. The moduli used to calculate $K(t)$ were $K_g = 2.67\ G_g$ and $K_e = 3\ G_e$, respectively. The former corresponds to a glassy Poisson's ratio (see Sect. 11.3.5), μ_g, of 1/3, the latter to an equilibrium ratio, μ_e, of 1/2. These are typical values for a synthetic polymer. The relaxation time was taken to be the same for both calculations since we have no way of knowing what the ratio of one to the other would be in any real material whose shear behavior might conceivably be modelled by the standard linear solid model. Portions of the *first order approximations* to the two relaxation spectra, $H(\tau)$ and $H_K(\tau)$, are included in Fig. 11.3-2. $H_1(\tau)$ is given by Eq. (4.2-11).

We conclude that the shear and the bulk spectra will generally differ in magnitude as well as in shape. As a consequence of the differences between the spectra which characterize the behavior in the two fundamental modes of deformation, the spectra characterizing all other modes of deformation are also all different. Rather than introducing a new symbol for each spectrum, we continue to use H and L but

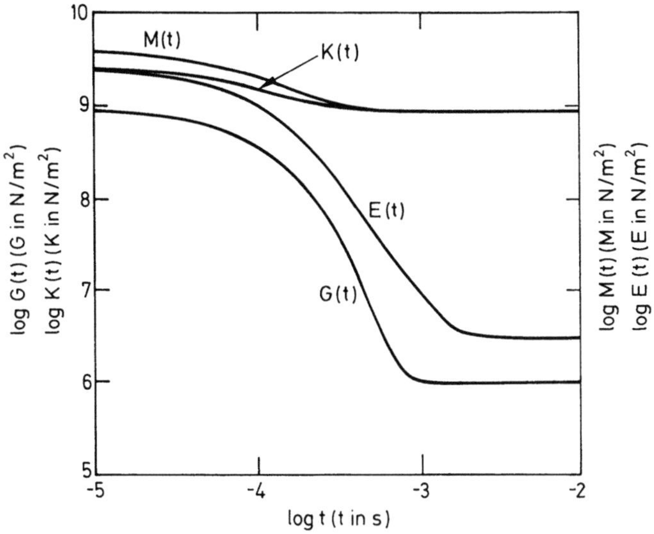

Fig. 11.3-1. Comparison of log G(t), log K(t), log M(t), and log E(t), calculated from standard linear solid models for G(t) and K(t) with $\tau_K = \tau_G$

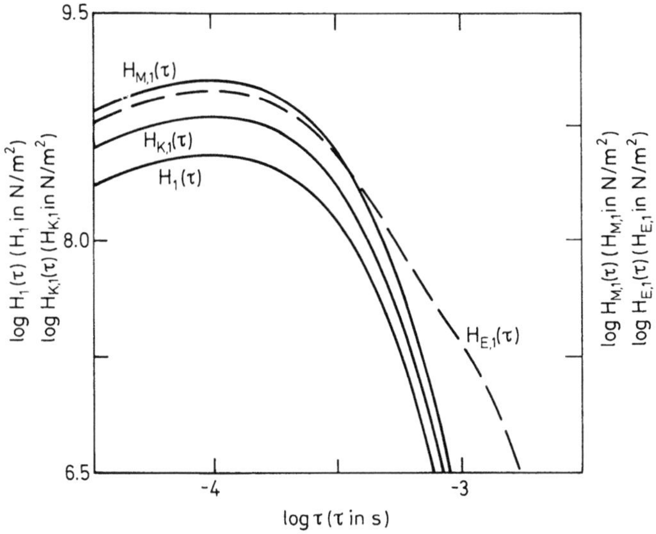

Fig. 11.3-2. Comparison of $H_1(\tau)$, $H_{K,1}(\tau)$, $H_{M,1}(\tau)$, and $H_{E,1}(\tau)$

decorate them with subscripts representing the corresponding moduli or compliances. For the spectra representating bulk behavior we use $H_K(\tau)$ and $L_B(\tau)$, respectively.* The difference between the shear and bulk spectra is then expressed

* For consistency in notation the shear spectra, $H(\tau)$ and $L(\tau)$, should have been denoted by $H_G(\tau)$ and $L_J(\tau)$, respectively. We dispensed with this purely for the sake of convenience.

by the inequalities

$$H_K(\tau) \neq H(\tau) \qquad \text{and} \qquad L_B(\tau) \neq L(\tau) \; . \tag{11.3-28}$$

We complete this section by listing the (continuous) canonical representations of the step response and the storage and loss functions in bulk deformation in the same way we have done this for behavior in shear in Sect. 4.1. For the moduli we obtain

$$K(t) = K_e + \int_{-\infty}^{\infty} H_K(\tau)\exp(-t/\tau)\mathrm{d}\ln\tau \tag{11.3-29}$$

or

$$K_\bullet(t) = K_g - \int_{-\infty}^{\infty} H_K(\tau)[1 - \exp(-t/\tau)]\mathrm{d}\ln\tau \tag{11.3-30}$$

and

$$K'(\omega) = K_e + \int_{-\infty}^{\infty} H_K(\tau)\frac{\omega^2\tau^2}{1+\omega^2\tau^2}\mathrm{d}\ln\tau \tag{11.3-31}$$

or

$$K'_\bullet(\omega) = K_g - \int_{-\infty}^{\infty} H_K(\tau)\frac{1}{1+\omega^2\tau^2}\mathrm{d}\ln\tau \tag{11.3-32}$$

and

$$K''(\omega) = \int_{-\infty}^{\infty} H_K(\tau)\frac{\omega\tau}{1+\omega^2\tau^2}\ln\tau \; , \tag{11.3-33}$$

while the time-dependent bulk viscosity becomes

$$\xi(t) = K_e t + \int_{-\infty}^{\infty} \tau H_K(\tau)[1 - \exp(-t/\tau)]\mathrm{d}\ln\tau \; . \tag{11.3-34}$$

For the compliances we find

$$B_\bullet(t) = B_g + \int_{-\infty}^{\infty} L_B(\tau)\exp(-t/\tau)\mathrm{d}\ln\tau \tag{11.3-35}$$

or

$$B(t) = B_e - \int_{-\infty}^{\infty} L_B(\tau)[1 - \exp(-t/\tau)]\mathrm{d}\ln\tau \tag{11.3-36}$$

and

$$B'(\omega) = B_g + \int_{-\infty}^{\infty} L_B(\tau)\frac{1}{1+\omega^2\tau^2}\mathrm{d}\ln\tau \tag{11.3-37}$$

or

$$B'_\bullet(\omega) = B_e - \int_{-\infty}^{\infty} L_B(\tau)\frac{\omega^2\tau^2}{1+\omega^2\tau^2}\mathrm{d}\ln\tau \tag{11.3-38}$$

and

$$B''(\omega) = \int_{-\infty}^{\infty} L_B(\tau)\frac{\omega\tau}{1+\omega^2\tau^2}\ln\tau \ . \tag{11.3-39}$$

The glassy and equilibrium constants are, of course, linked by the relations

$$\int_{-\infty}^{\infty} H_K(\tau)d\ln\tau = K_g - K_e \tag{11.3-40}$$

and

$$\int_{-\infty}^{\infty} L_B(\tau)d\ln\tau = B_e - B_g \ . \tag{11.3-41}$$

We omit the expressions for the discrete canonical representations because these are easily written down in analogy to the continuous ones, replacing the integrals by sums and the spectra by K_n and B_n, respectively (cf. Sect. 3.5).

11.3.3 Behavior in Constrained Uniaxial Compression

We now look at the first deformation that entails a change in shape as well as in size. This is uniaxial constrained compression. In this deformation a stress (pressure) is applied in one spatial direction while deformation in the two transverse directions is prevented. The constraint sets up (equal) stresses in the two transverse directions. The transforms of the stress and the strain tensor matrices, therefore, become

$$[\bar{\sigma}_{ij}(s)] = \begin{bmatrix} \bar{\sigma}_{11}(s) & 0 & 0 \\ 0 & \bar{\sigma}_{22}(s) & 0 \\ 0 & 0 & \bar{\sigma}_{22}(s) \end{bmatrix} \tag{11.3-42}$$

and

$$[\bar{\gamma}_{ij}(s)] = \begin{bmatrix} \bar{\gamma}_{11}(s) & 0 & 0 \\ 0 & 0 & 0 \\ 0 & 0 & 0 \end{bmatrix} . \tag{11.3-43}$$

Substitution into the generalized Hooke's law, Eq. (11.2-1), then yields

$$\bar{\sigma}_{11}(s) = [\bar{P}(s) + \tfrac{4}{3}\bar{Q}(s)]\bar{\gamma}_{11}(s) = [s\bar{K}(s) + \tfrac{4}{3}s\bar{G}(s)]\bar{\gamma}_{11}(s) \tag{11.3-44}$$

and

$$\bar{\sigma}_{22}(s) = [\bar{P}(s) - \tfrac{2}{3}\bar{Q}(s)]\bar{\gamma}_{11}(s) = [s\bar{K}(s) - \tfrac{2}{3}s\bar{G}(s)]\bar{\gamma}_{11}(s) \tag{11.3-45}$$

where we have used $\bar{\Delta}(s) = \bar{\gamma}_{11}(s)$.

The ratio of the two stress transforms is

$$\frac{\bar{\sigma}_{22}(s)}{\bar{\sigma}_{11}(s)} = \frac{3\,\bar{K}(s) - 2\,\bar{G}(s)}{3\,\bar{K}(s) + 4\,\bar{G}(s)} \tag{11.3-46}$$

[cf. Eq. (1.4-82)]. Adding the second equation twice to the first yields

$$\bar{\sigma}_{11}(s) + 2\,\bar{\sigma}_{22}(s) = 3\,s\bar{K}(s)\bar{\gamma}_{11}(s) \tag{11.3-47}$$

while subtraction of the second from the first leads to

$$\bar{\sigma}_{11}(s) - \bar{\sigma}_{22}(s) = 2\,s\bar{G}(s)\bar{\gamma}_{11}(s)\ . \tag{11.3-48}$$

Thus, the two (operational) fundamental moduli can only be recovered from this type of measurement if both stresses are known in addition to the strain. Since $\bar{\sigma}(s)$ is relatively difficult to determine, one normally simply uses the ratio of the stress to the strain transform

$$\frac{\bar{\sigma}_{11}(s)}{\bar{\gamma}_{11}(s)} = s\bar{K}(s) + \tfrac{4}{3}\,s\bar{G}(s) = s\bar{M}(s) \tag{11.3-49}$$

where $\bar{M}(s)$ is the (operational) *longitudinal bulk*, or *wave relaxation modulus* [cf. Eq. (1.4-81)], so called because it governs the propagation of longitudinal waves in extended linear viscoelastic media (see Sect. 11.3.6.1). Its reciprocal, the (operational) *longitudinal bulk*, or *wave creep compliance* will not be considered here.

The canonical representations of the wave modulus are derived from the *wave relaxance*

$$\bar{V}(s) = s\bar{M}(s) = M_e + \int_{-\infty}^{\infty} H_M(\tau)\frac{\tau s}{1+\tau s}\,d\ln\tau \tag{11.3-50}$$

or

$$\bar{V}_\bullet(s) = s\bar{M}_\bullet(s) = M_g - \int_{-\infty}^{\infty} H_M(\tau)\frac{1}{1+\tau s}\,d\ln\tau \tag{11.3-51}$$

where $H_M(\tau)$ is the *wave relaxation spectrum* or *longitudinal bulk relaxation spectrum*, and M_e and M_g are the equilibrium and glassy wave or longitudinal bulk moduli. The behavior in uniaxial constrained compression is always arrheodictic because the equilibrium wave modulus is given [see Eq. (4.6-4)] by

$$\lim_{s\to 0} \bar{V}(s) = \bar{P}(0) + \tfrac{3}{4}\,\bar{Q}(0) = K_e + \tfrac{4}{3}\{G_e\} = M_e \tag{1.3-52}$$

[cf. Eq. (1.4-81]. Thus, when G_e vanishes, M_e is still a constant, viz., in this case, K_e.

From Eq. $(11.3\text{-}49)_2$ we have, for the response to a strain as a step function of time,

$$M(t) = K(t) + \tfrac{4}{3}\,G(t)\ . \tag{11.3-53}$$

M(t) for the standard linear solid model is displayed together with G(t) and K(t) in Fig. 11.3-1. The parameters used in the calculation were those described in the preceding section. As expected, log M(t) differs from log K(t) only where K(t) and G(t) are roughly within an order of magnitude of one another.

Naturally, the restriction

$$\int_{-\infty}^{\infty} H_M(\tau)\,d\ln\tau = M_g - M_e \tag{11.3-54}$$

applies, and we also have the inequalities

$$H_M(\tau) \neq H(\tau) \neq H_K(\tau) \tag{11.3-55}$$

as demonstrated in Fig. 11.3-2.

Other desired representations can easily be obtained in full analogy to those developed in Sect. 11.3.2 for behavior in bulk deformation.

11.3.4 Behavior in Uniaxial Tension and Compression

Because of the relative ease with which the behavior in *uniaxial tension* and/or *uniaxial compression*, also called *simple tension* and *simple compression*, can be evaluated, these modes of deformation are probably the most widely used in the determination of material properties. In uniaxial tension or compression a stress or a strain is applied in a single direction of space. The deformed body contracts (in tension) or expands (in compression) in the two transverse directions. The stress-strain relations are the same in tension and in compression except for the negative sign of the compressive stress. We consider here simple tension only. The matrices of the transforms of the stress and the strain tensors become

$$[\bar{\sigma}_{ij}(s)] = \begin{bmatrix} \bar{\sigma}_{11}(s) & 0 & 0 \\ 0 & 0 & 0 \\ 0 & 0 & 0 \end{bmatrix} \tag{11.3-56}$$

and

$$[\bar{\gamma}_{ij}(s)] = \begin{bmatrix} \bar{\gamma}_{11}(s) & 0 & 0 \\ 0 & \bar{\gamma}_{22}(s) & 0 \\ 0 & 0 & \bar{\gamma}_{22}(s) \end{bmatrix} . \tag{11.3-57}$$

Substitution of Eqs. (11.3-56) and (11.3-57) into the generalized Hooke's law expressions, Eq. (11.2-1), yields the equations

$$\bar{\sigma}_{11}(s) = [\bar{P}(s) - \tfrac{2}{3}\,\bar{Q}(s)]\bar{\Delta}(s) + 2\,\bar{Q}(s)\bar{\gamma}_{11}(s) \tag{11.3-58}$$

and

$$0 = [\bar{P}(s) - \tfrac{2}{3}\,\bar{Q}(s)]\bar{\Delta}(s) + 2\,\bar{Q}(s)\bar{\gamma}_{22}(s) \ . \tag{11.3-59}$$

where $\bar{\Delta}(s) = \bar{\gamma}_{11}(s) + 2\,\bar{\gamma}_{22}(s)$ by Eq. (11.2-3). Elimination of $\bar{\Delta}(s)$ between Eqs. (11.3-58) and (11.3-59) furnishes

$$\bar{\sigma}_{11}(s) = 2\,\bar{Q}(s)[\bar{\gamma}_{11}(s) - \bar{\gamma}_{22}(s)] \tag{11.3-60}$$

or

$$\bar{\sigma}_{11}(s) = 2\,s\bar{G}(s)[\bar{\gamma}_{11}(s) - \bar{\gamma}_{22}(s)] \ . \tag{11.3-61}$$

Thus, if the strain in the transverse direction is known in addition to the imposed strain, the shear modulus can be recovered from measurements in simple tension. Because such measurements are experimentally difficult, it is customary to consider simply the much more easily determined ratio of the stress to the strain in the stretch direction. In the transform plane we have

$$\frac{\bar{\sigma}_{11}(s)}{\bar{\gamma}_{11}(s)} = \bar{Y}(s) \tag{11.3-62}$$

where $\bar{Y}(s)$ is the *stretch relaxance*, and

$$\frac{\bar{\sigma}_{11}(s)}{\bar{\gamma}_{11}(s)} = s\bar{E}(s) \tag{11.3-63}$$

where $\bar{E}(s)$ is the (operational) *stretch, tensile*, or *elongational relaxation modulus.* It is defined in terms of the two fundamental moduli by Eq. (11.1-8), and in terms of the shear modulus and Poisson's ratio by Eq. (11.1-6). As in the case of constrained uniaxial compression, behavior in uniaxial tension or compression is characterized by two material functions. In contrast to $\bar{M}(s)$, however, $\bar{E}(s)$ is not a linear combination of $\bar{G}(s)$ and $\bar{K}(s)$.

To obtain the canonical representations we define

$$\bar{Y}(s) = s\bar{E}(s) = \{E_e\} + \int_{-\infty}^{\infty} H_E(\tau)\frac{\tau s}{1 + \tau s} d \ln \tau \tag{11.3-64}$$

and

$$\bar{Y}_{\bullet}(s) = s\bar{E}_{\bullet}(s) = E_g - \int_{-\infty}^{\infty} H_E(\tau)\frac{1}{1 + \tau s} d \ln \tau \tag{11.3-65}$$

in which $H_E(\tau)$ is the *stretch, tensile*, or *elongational relaxation modulus*, and E_e and E_g are the equilibrium and the glassy stretch moduli.

$H_E(\tau)$ is generally close to the shear relaxation spectrum but the two are by no means identical. As shown by Eq. (11.1-10), the relation between E(t), G(t), and K(t) can only be formulated in terms of convolution integrals. As discussed in Sect. 11.3.2, the latter two functions cannot, in general, be expected to relax identically. That E(t) is not a simple function of K(t) and G(t) is strikingly demonstrated in Problem 11.3-2 in which it is calculated from the same standard linear solid models for K(t) and G(t) from which M(t) was obtained in Sect. 11.3.3. While the models for G(t) and K(t) exhibit Maxwellian relaxation (they have *singlet spectra* characterized by a single relaxation time), the stretch modulus shows a double Maxwellian relaxation, i.e. it has a *doublet spectrum*, characterized by two relaxation times. E(t) is compared with the other three in Fig. 11.3-1. For the values chosen to model G(t) and K(t) in Fig. 11.3-1, the equations derived in Problem 11.3-2 give $E_1 = 2.52 \times 10^9$ N/m^2, $E_2 = 1.39 \times 10^8$ N/m^2, $\tau_1 = 0.999 \times 10^{-4}$ s, and $\tau_2 = 0.337 \times 10^{-3}$ s. The first relaxation time, τ_1, is virtually the same as $\tau_G = \tau_K$ but there is the additional relaxation time, τ_2, which is almost 3.5 times longer than τ_1.

Clearly, one would not expect $H_E(\tau)$ to have the same shape as $H(\tau)$. This is borne out by $H_{E,1}(\tau)$ in Fig. 11.3-2. The curve for $H_{E,1}(\tau)$ distinctly shows up the double Maxwellian character of E(t). This is not so clearly seen in Fig. 11.3-1 because the two relaxation times are not separated widely enough.

Not too much should be read into the demonstration that E(t) shows a doublet spectrum when calculated from singlet spectrum models for G(t) and K(t). G(t) or K(t) would likewise acquire a doublet spectrum when derived from singlet spectrum models. In models characterized by distributions of relaxation times the effect would

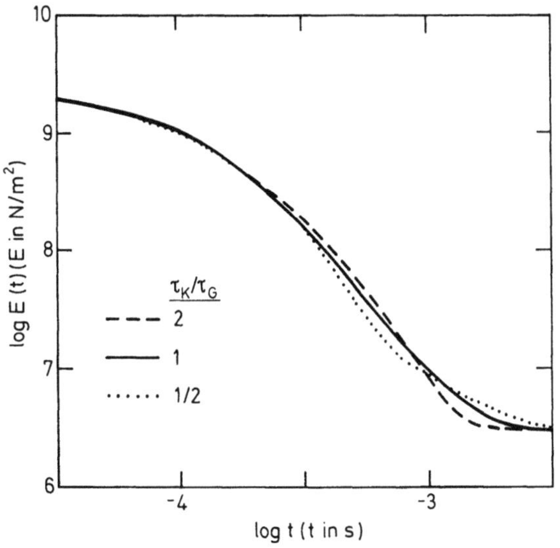

Fig. 11.3-3. Comparison of E(t) calculated with different ratios of τ_K to τ_G

surely be much less pronounced. Nevertheless, the exercise does draw attention to possible consequences of the separate nature of the two fundamental types of relaxation behavior.

All curves in Figs. 11.3-1 and 11.3-2 were obtained with $\tau_K = \tau_G$. The effect of a change in the ratio of τ_K to τ_G in these calculations is relatively small. This is demonstrated in Fig. 11.3-3 which compares E(t) calculated with three different ratios, viz., $\tau_K/\tau_G = 2$, 1, and 1/2, on a somewhat enlarged time scale.

Let us now turn to the time-dependent *stretch, tensile*, or *elongational viscosity*. The transform of its canonical form (see Table 11.1-1) is

$$\bar{\zeta}(s) = \frac{\bar{E}(s)}{s} = \frac{\{E_e\}}{s^2} + \int_{-\infty}^{\infty} H_E(\tau)\frac{\tau}{s(1+\tau s)}\, d\ln\tau \qquad (11.3\text{-}66)$$

in complete analogy to $\bar{\eta}(s)$ [cf. Eq. (4.1-15)].

As indicated by the braces around E_e in Eqs. (11.3-64) and (11.3-66), the behavior in simple tension may be rheodictic or arrheodictic. At long times, i.e. in the limit as $s \to 0$, we have, by analogy with Eq. (4.6-4),

$$\lim_{s\to 0} \bar{Y}(s) = \frac{9\,\bar{P}(0)\bar{Q}(0)}{3\,\bar{P}(0) + \bar{Q}(0)} = \frac{9\,K_e\{G_e\}}{3\,K_e + \{G_e\}} = \{E_e\}\ . \qquad (11.3\text{-}67)$$

Thus, E_e vanishes when G_e does, even though K_e does not.

We now turn to the compliances. By definition, $\bar{L}(s) = 1/\bar{Y}(s)$, and $s\bar{D}(s) = 1/\bar{E}(s)$ where $\bar{L}(s)$ is the *stretch retardance*, and $\bar{D}(s)$ is the (operational) *stretch* or *tensile compliance*, $\bar{D}(s)$, given by

$$\frac{\bar{\gamma}_{11}(s)}{\bar{\sigma}_{11}(s)} = \bar{L}(s) = s\bar{D}(s)\ . \qquad (11.3\text{-}68)$$

We have

$$\bar{L}(s) = s\bar{D}(s) = D_g + \int_{-\infty}^{\infty} L_D(\tau)\frac{1}{1+\tau s}d\ln\tau + \{1/\zeta_f s\} \tag{11.3-69}$$

or

$$\bar{L}_\bullet(s) = s\bar{D}_\bullet(s) = D_e^{\{o\}} - \int_{-\infty}^{\infty} L_D(\tau)\frac{\tau s}{1+\tau s}d\ln\tau + \{1/\zeta_f s\} \tag{11.3-70}$$

where $L_D(\tau)$ is the *stretch, tensile*, or *elongational retardation spectrum*, D_g and $D_e^{\{o\}}$ are the glassy and the {pseudo-} equilibrium compliances, and ζ_f is the *steady-state stretch, tensile*, or *elongational viscosity*.

Again, we have the inequalities

$$H_E(\tau) \neq H(\tau) \neq H_K(\tau) \qquad \text{and} \qquad L_D(\tau) \neq L(\tau) \neq L_K(\tau) \tag{11.3-71}$$

and the constants are connected by the relations

$$\int_{-\infty}^{\infty} H_E(\tau)d\ln\tau = E_g - \{E_e\} \tag{11.3-72}$$

and

$$\int_{-\infty}^{\infty} L_D(\tau)d\ln\tau = D_e^{\{o\}} - D_g\ . \tag{11.3-73}$$

The response functions can again be developed in analogy to those presented in Sect. 11.3.2, taking care to include rheodictic as well as arrheodictic behavior.

11.3.5 Lateral Contraction (Poisson's Ratio)

We now consider the viscoelastic equivalent of the elastic constant known as *Poisson's ratio* [cf. Eq. (1.4-78)]. This is the ratio of the strain in the transverse direction which develops in response to an imposed uniaxial strain, to the imposed strain. In uniaxial tension the ratio represents the relative *lateral contraction*. In uniaxial compression it becomes the relative *lateral dilatation*. In Chap. 1 we considered a purely elastic material. There, Poisson's ratio was a material constant. In a viscoelastic material Poisson's ratio becomes a time-dependent material function just as all the moduli and compliances do. Whether the material is purely elastic or viscoelastic, if it is isotropic in the undeformed state, the lateral contractions in the two transverse directions are identical.

Solving Eq. (11.3-59) of the preceding section for $\bar{\gamma}_{22}(s)$ gives

$$-\bar{\gamma}_{22}(s) = \frac{3\,\bar{P}(s) - 2\,\bar{Q}(s)}{6\,\bar{P}(s) + 2\,\bar{Q}(s)}\bar{\gamma}_{11}(s) = \frac{3\,\bar{K}(s) - 2\,\bar{G}(s)}{6\,\bar{K}(s) + 2\,\bar{G}(s)}\bar{\gamma}_{11}(s)\ . \tag{11.3-74}$$

Comparing Eq. (11.3-74) with Eq. (1.4-78) and consulting Table 11.1-1 we recognize that the first fraction represents the impulse response form of Poisson's ratio, $\bar{v}(s)$. The second fraction similarly represents the step response form. Hence, Eq. (11.3-74) may be recast as

$$-\bar{\gamma}_{22}(s) = \bar{v}(s)\bar{\gamma}_{11}(s) = s\bar{\mu}(s)\bar{\gamma}_{11}(s) \tag{11.3-75}$$

where

$$\bar{\nu}(s) = \frac{3\,\bar{P}(s) - 2\,\bar{Q}(s)}{6\,\bar{P}(s) + 2\,\bar{Q}(s)} \quad \text{and} \quad \bar{\mu}(s) = \frac{3\,\bar{K}(s) - 2\,\bar{G}(s)}{s[6\,\bar{K}(s) + 2\,\bar{G}(s)]} \,. \tag{11.3-76}$$

Poisson's ratio is the same whether the excitation is a strain or a stress (see Problem 11.3-12).

Considering now a strain in the 1-direction as a step function of time, $\bar{\gamma}_{11}(s) = \gamma_{11,0}/s$, the Laplace transform of Poisson's ratio becomes

$$\bar{\mu}(s) = \frac{-\bar{\gamma}_{22}(s)}{\gamma_{11,0}} \tag{11.3-77}$$

and retransformation yields

$$\mu(t) = -\gamma_{22}(t)/\gamma_{11,0} \,. \tag{11.3-78}$$

In this equation, $-\gamma_{22}(t)$ represents the time-dependent lateral contraction which the material experiences when stretched in the 1-direction by the application of a step strain of height $\gamma_{11,0}$, while no other strains or stresses are imposed in the other two directions. Thus, in such a stretch relaxation experiment $\mu(t)$ can be obtained directly from the measured lateral contraction. In a stretch creep experiment this is not so simple (see Problem 11.3-11).

Although the stretched piece of material experiences stress relaxation in the stretched 1-direction, the transverse contraction is delayed, i.e. it attains its ultimate state only after an infinite time has elapsed. The lateral contraction thus has the character of a retardation phenomenon (the strain in the transverse direction is retarded) and is therefore formally analogous to a compliance. We have

$$\bar{\nu}(s) = \mu_g + \sum_i \mu_i \frac{1}{1 + \tau_i s} = \mu_e - \sum_i \mu_i \frac{\tau_i s}{1 + \tau_i s} \tag{11.3-79}$$

in analogy to $\bar{U}(s)$, or

$$\bar{\mu}(s) = \frac{\bar{\nu}(s)}{s} = \frac{\mu_g}{s} + \sum_i \mu_i \frac{1}{s(1 + \tau_i s)} = \frac{\mu_e}{s} - \sum_i \mu_i \frac{\tau_i}{1 + \tau_i s} \tag{11.3-80}$$

in analogy to $\bar{J}(s)$. We call τ the *delay time*, to emphasize that it is not identical with the retardation time in simple tension*. Each μ_i may then be regarded to be the Poisson's ratio associated with the ith delay time.

The behavior of Poisson's ratio is necessarily arrheodictic just as bulk behavior is (see Sect. 11.3.2). By Eqs. (4.6-4) and (11.3-76), at long times, i.e. as $s \to 0$, we have

$$\lim_{s \to 0} \bar{\nu}(s) = \frac{3\,\bar{P}(0) - 2\,\bar{Q}(0)}{6\,\bar{P}(0) + 2\,\bar{Q}(0)} = \frac{3K_e - 2\{G_e\}}{6K_e + 2\{G_e\}} = \mu_e \tag{11.3-81}$$

while, at short times, i.e. as $s \to \infty$, we have

* No simple relation exists between the delay time and the retardation time (or the relaxation time, for that matter)

$$\lim_{s\to\infty} \bar{v}(s) = \frac{3\,\bar{P}(\infty) - 2\,\bar{Q}(\infty)}{6\,\bar{P}(\infty) + 2\,\bar{Q}(\infty)} = \frac{3K_g - 2G_g}{6K_g + 2G_g} = \mu_g \tag{11.3-82}$$

where μ_g and μ_e are the glassy, and the equilibrium Poisson's ratio, respectively. The relation

$$\sum_i \mu_i = \mu_e - \mu_g \tag{11.3-83}$$

applies to the sum of the μ_i.

In the glassy state the bulk and shear moduli are generally of roughly comparable magnitude. In equilibrium, however, the bulk modulus may well exceed the shear modulus by orders of magnitude. Thus, the largest value that Poisson's ratio can theoretically attain, is 1/2. This is the value it would assume in a (hypothetical) *incompressible* material in mechanical equilibrium as $K_e \to \infty$. The glassy Poisson's ratio is always less than 1/2 and is often around 1/3. In the latter case $K_g = 2.67G_g$.

From Eq. (11.3-80) we obtain the *time-dependent Poisson's ratio* in response to a strain as a step function of time by retransforming to the time domain. This leads to

$$\mu(t) = \mu_g + \sum_i \mu_i[1 - \exp(-t/\tau_i)] \tag{11.3-84}$$

or

$$\mu_\bullet(t) = \mu_e - \sum_i \mu_i \exp(-t/\tau_i) \;. \tag{11.3-85}$$

The *frequency-dependent*, or *complex Poisson's ratio* is the (generalized) Fourier transform of $\bar{v}(s)$. It characterizes the behavior of a linear viscoelastic material when a sinusoidally oscillating strain is applied in one direction and the lateral contraction or dilatation is measured in any of the two transverse directions. We have

$$\mu^*(\omega) = \bar{v}(s)|_{s=j\omega} = \mu'(\omega) - j\,\mu''(\omega) \;. \tag{11.3-86}$$

The real and imaginary parts of $\mu^*(\omega)$ become

$$\mu'(\omega) = \mu_g + \sum_i \mu_i \frac{1}{1+\omega^2\tau_i^2} \tag{11.3-87}$$

or

$$\mu'_\bullet(\omega) = \mu_e - \sum_i \mu_i \frac{\omega^2\tau_i^2}{1+\omega^2\tau_i^2} \tag{11.3-88}$$

and

$$\mu''(\omega) = \sum_i \mu_i \frac{\omega\tau_i}{1+\omega^2\tau_i^2} \tag{11.3-89}$$

for a discrete distribution of delay times.

Figure 11.3-4 displays $\mu(t)$, $\mu'(\omega)$, and $\mu''(\omega)$ for the standard linear solid (see Problem 11.3-9) with $\mu_e = 0.5$, $\mu_g = 0.333$, and $\mu = \mu_e - \mu_g$, plotted against log t, and $-\log\omega$, respectively. Note that $\mu''(\omega)$ is shifted upward by 0.333 for a more convenient plot. The general characteristics of these plots are analogous to those of the compliances of the 3-parameter Voigt model in Fig. 3.4-22.

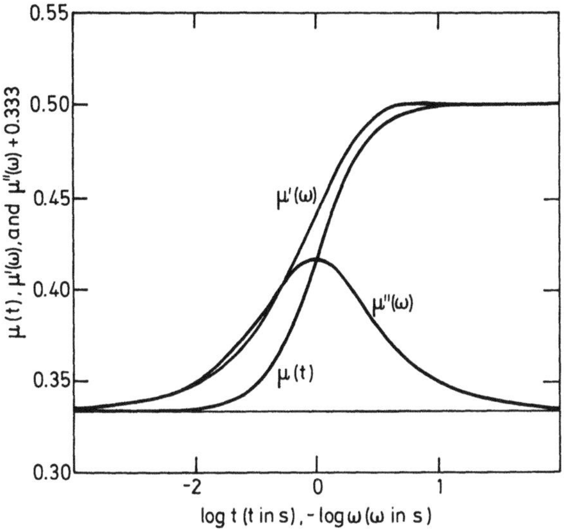

Fig. 11.3-4. The Poisson's ratios μ(t), μ'(ω), and μ''(ω) for the standard linear solid model

To introduce the canonical representations we first define a *distribution*, or *spectrum*, of *delay times*, m(τ). This then allows us to write

$$\bar{\nu}(s) = s\bar{\mu}(s) = \mu_g + \int_{-\infty}^{\infty} m(\tau)\frac{1}{1+\tau s} d \ln \tau \tag{11.3-90}$$

or

$$\bar{\nu}_\bullet(s) = s\bar{\mu}_\bullet(s) = \mu_e - \int_{-\infty}^{\infty} m(\tau)\frac{\tau s}{1+\tau s} d \ln \tau \tag{11.3-91}$$

where now the relation

$$\int_{-\infty}^{\infty} m(\tau) d \ln \tau = \mu_e - \mu_g \tag{11.3-92}$$

replaces Eq. (11.3-83).

In the real time domain Eqs. (11.3-89) and (11.3-90) take the form

$$\mu(t) = \mu_g + \int_{-\infty}^{\infty} m(\tau)[1 - \exp(-t/\tau)] d \ln \tau \tag{11.3-93}$$

or

$$\mu_\bullet(t) = \mu_e - \int_{-\infty}^{\infty} m(\tau) \exp(-t/\tau) d \ln \tau \tag{11.3-94}$$

and in the frequency domain they become

$$\mu'(\omega) = \mu_g + \int_{-\infty}^{\infty} m(\tau)\frac{1}{1+\omega^2\tau^2} d \ln \tau \tag{11.3-95}$$

or

$$\mu'_{\bullet}(\omega) = \mu_e - \int_{-\infty}^{\infty} m(\tau)\frac{\omega^2\tau^2}{1+\omega^2\tau^2} d \ln \tau \tag{11.3-96}$$

and

$$\mu''(\omega) = \int_{-\infty}^{\infty} m(\tau)\frac{\omega\tau}{1+\omega^2\tau^2} d \ln \tau \tag{11.3-97}$$

for a continuous distribution of delay times.

The real part of $\mu^*(\omega)$ is the proportionality constant between the applied harmonic strain and the in-phase component of the lateral contraction or dilatation. The imaginary part is, similarly, the proportionality constant between the applied strain and the out-of-phase component of the lateral contraction or dilatation.

The sign of the imaginary parts in these equations has been the subject of some controvery in the literature.* The negative sign is, however, a direct consequence of the compliance nature of Poisson's ratio, and is open to experimental verification. It implies that the lateral contraction lags the imposed sinusoidally oscillating strain (Problem 11.3-10). It also implies that the damping (i.e. the loss tangent) in shear is larger than the damping in extension which, in turn, is larger than the damping in bulk (Problems 11.4-6 and 11.4-7). Furthermore, the latter conclusion can be derived without any consideration of Poisson's ratio (Problem 11.4-5). A positive sign for $\mu''(\omega)$ would thus create an inconsistency in the theory.

11.3.6 Wave Propagation

In Sect. 5.6 we introduced the concept of the *material transmission line* to help us understand the way in which a mechanical disturbance propagates along a piece of linear viscoelastic material. We discussed the propagation of a shear disturbance along a thin rod of uniform circular cross-section. We showed [cf. Eqs. (5.6-48) and (5.6-60)] that a harmonic displacement is transmitted along an *inertial line* in the form of a travelling wave. We also showed (see the discussion towards the end of Sect. 5.6.2.2, and in Sect. 7.3.2.2) that the transmission of an arbitrary excitation (and the response to it) can be understood in terms of a Fourier decomposition of the disturbance. Our earlier deliberations were restricted to the propagation of a shear deformation. We now extend these earlier deliberations to a consideration of the propagation of deformations other than shear.

As shown in Sect. 5.6, the propagation of a disturbance along an inertial material transmission line is governed by two characteristic material line functions. In operational form these are: the *propagation function per unit length*, $\gamma(s)$, and the *characteristic mechanical relaxance per unit area*, $\mathcal{Q}_C(s)$, or its reciprocal, the *characteristic mechanical retardance per unit area*, $\mathcal{U}_C(s)$. We distinguished the *lossless* and the *lossy* line as two subcases of the inertial material transmission line. In the

* See, e.g., Waterman [5]

first case the two functions were

$$\gamma(s) = s\sqrt{\rho J} \qquad \text{and} \qquad \mathcal{Q}_C(s) = s\sqrt{\rho G} \tag{11.3-98}$$

[cf. Eqs. (5.6-40) and (5.6-41)]. In the second case we had

$$\gamma(s) = s\sqrt{\rho\,\bar{U}(s)} \qquad \text{and} \qquad \mathcal{Q}_C(s) = s\sqrt{\rho\,\bar{Q}(s)} \tag{11.3-99}$$

[cf. Eqs. (5.6-70) and (5.6-71)]. The presence of the density, ρ, in these characteristic functions immediately reminds us that we are dealing with an inertial line in either case. The lossless line describes the propagation of a mechanical disturbance through a purely elastic material, while the lossy line does so for a linear viscoelastic material. The material constants, J and G, respectively the material functions, $\bar{U}(s)$ and $\bar{Q}(s)$, express this difference for a shear disturbance in accordance with Table 11.1-1.

Now, if the propagation of disturbances other than a shear deformation are to be accounted for, the general theory discussed in Sect. 5.6 remains valid. We merely need to substitute for the shear quantities from Table 11.1-1 the quantities appropriate for the type of disturbance to be considered. It is necessary, however, to consider just what type of disturbances *can* propagate through a material. We distinguish two cases: that of an *extended* (or *infinite*, or *unbounded*) medium, and that of a *bounded* (or *finite*) medium. The basic theory has been discussed, e.g., by Kolsky [1], and by others [2, 3].

11.3.6.1 Wave Propagation in an Extended Medium

A disturbance propagating along an inertial transmission line of finite length is reflected at the end of the line (see Sect. 5.6.2.2). In an extended medium we do not need to concern ourselves with the complications arising from reflection at the boundaries of the medium. In an elastic material such a medium must be pictured to extend indefinitely in the three dimensions of space. In a viscoelastic material it may be sufficient to consider the medium to have dimensions such that any wave is effectively attenuated to zero before it reaches a boundary. The attenuation function depends on the properties of the medium itself [cf. Eq. (5.6-76)] and must be carefully appraised by the experimenter.

To consider wave propagation in an extended medium the equations we developed earlier for treating the transmission of a disturbance along a *line* must be suitably modified. We return to one of the two fundamental equations of the continuous transmission line, Eq. (5.6-9). For the lossless inertial line (cf. Sect. 5.6.2.2) this equation takes the form

$$-\frac{\partial\,\bar{f}(s,l)}{\partial l} = I_l s^2 \bar{x}(s,l) \tag{11.3-100}$$

where $I_l s^2$ is the *inertance per unit length*. Replacing the force by the stress transform gives

$$-\frac{\partial\,\bar{\sigma}(s,l)}{\partial l} = \rho s^2 \bar{x}(s,l) \tag{11.3-101}$$

where ρ is the density. To enable us to apply this equation to an extended medium we change the transform of the stress imposed at the input end of the line, $-\bar{\sigma}(s,l)$, to the transform of the general stress tensor in the interior of an extended medium, $\bar{\sigma}_{ij}(s)$. In accordance with this, we replace $\bar{x}(s,l)$ by the transform of the displacement vector, $\bar{u}_i(s)$, and the length, l, by the coordinate, x_j. This leads to

$$\frac{\partial\,\bar{\sigma}_{ij}(s)}{\partial\,x_j} = \rho s^2 \bar{u}_i(s)\ , \tag{11.3-102}$$

which is the transform of the equation of motion for a homogeneous, isotropic, deformable body in the absence of body forces (Newton's Second Law).

We now make use of the generalized Hooke's law to replace the components of the stress tensor by the displacements. To treat the case of an elastic (lossless) material, we would use Eq. (1.4-69). Proceeding directly to the case of a viscoelastic (lossy) material, we substitute Eq. (11.2-1). This yields

$$\rho s^2 \bar{u}_i(s) = [\bar{P}(s) - \tfrac{2}{3}\,\bar{Q}(s)]\frac{\bar{\Delta}(s)}{\partial\,x_i} + 2\,\bar{Q}(s)\frac{\partial\,\bar{\gamma}_{ij}(s)}{\partial\,x_j}\ . \tag{11.3-103}$$

But the components of the strain tensor transform are given by

$$\bar{\gamma}_{ij}(s) = \frac{1}{2}\left(\frac{\partial\,\bar{u}_i(s)}{\partial\,x_j} + \frac{\partial\,\bar{u}_j(s)}{\partial\,x_i}\right) \tag{11.3-104}$$

[cf. Eq. (1.3-15)]. Substitution into Eq. (11.3-103) furnishes

$$\rho s^2 \bar{u}_i(s) = [\bar{P}(s) + \tfrac{1}{3}\,\bar{Q}(s)]\frac{\bar{\Delta}(s)}{\partial\,x_i} + \bar{Q}(s)\frac{\partial^2 \bar{u}_i(s)}{\partial\,x_j \partial\,x_j} \tag{11.3-105}$$

because

$$\frac{\partial\,\bar{u}_j(s)}{\partial\,x_j} = \bar{\Delta}(s)\ . \tag{11.3-106}$$

Let us now differentiate both sides of Eq. (11.3-105) with respect to x_i. This enables us to introduce the dilatation as our measure of deformation. We obtain

$$\rho s^2 \frac{\partial\,\bar{u}_i(s)}{\partial\,x_i} = [\bar{P}(s) + \tfrac{1}{3}\,\bar{Q}(s)]\frac{\partial^2 \bar{\Delta}(s)}{\partial\,x_i \partial\,x_i} + \bar{Q}(s)\frac{\partial^2}{\partial\,x_j \partial\,x_j}\left(\frac{\partial\,\bar{u}_i(s)}{\partial\,x_i}\right)\ , \tag{11.3-107}$$

which, by virtue of Eq. (11.3-106) reduces to

$$\rho s^2 \bar{\Delta}(s) = [\bar{P}(s) + \tfrac{3}{4}\,\bar{Q}(s)]\frac{\partial^2 \bar{\Delta}(s)}{\partial\,x_j \partial\,x_j} = \bar{V}(s)\frac{\partial^2 \bar{\Delta}(s)}{\partial\,x_j \partial\,x_j}\ . \tag{11.3-108}$$

where $\bar{V}(s)$ is the *wave relaxance*.

Equation (11.3-108) is a partial differential equation known as a *wave equation*. It may be rewritten as

$$\frac{\partial^2 \bar{\Delta}(s)}{\partial\,x_j \partial\,x_j} = s^2 \rho\,\bar{O}(s)\bar{\Delta}(s) \tag{11.3-109}$$

where $\bar{O}(s)$ is the *wave retardance*. The equation is formally analogous to Eq.

(5.6-12). The (operational) propagation function is seen to be

$$\gamma(s) = s\sqrt{\rho\,\bar{O}(s)} \tag{11.3-110}$$

[cf. Eq. (5.6-70)]. Equation (11.3-109) describes the propagation of a dilatation (or compression) through an extended medium. If the disturbance is the result of a harmonic displacement, the (operational) propagation function is replaced by its complex form

$$\gamma^*(\omega) = \gamma(s)|_{s=j\omega} = j\omega\sqrt{\rho\,O^*(\omega)} = j\omega\sqrt{\rho/M^*(\omega)} \tag{11.3-111}$$

where $M^*(\omega)$ is the complex wave modulus, or complex longitudinal bulk modulus. The disturbance then propagates as a *dilatational* or a *compressional wave.* A dilatational wave is also called a *longitudinal bulk wave.* Yet another name is *irrotational.* This derives from the fact that the propagation of a dilatational wave involves no rotation of an elemental volume of the material.

In the propagation of a plane dilatational wave the motion of a material particle is *parallel* to the direction of propagation. Experimentally, a disturbance approximating a plane dilatational wave can be set up by agitating a plate in contact with the medium in the longitudinal direction.

A dilatational wave, however, is not the only wave that can propagate through an extended medium. If the dilatation vanishes, Eq. (11.3-105) reduces to

$$\frac{\partial^2 \bar{u}_i(s)}{\partial x_j \partial x_j} = s^2 \rho\,\bar{U}(s)\bar{u}_i(s)\ . \tag{11.3-112}$$

This, too, is a *wave equation.* The (operational) propagation function now is

$$\gamma(s) = s\sqrt{\rho\,\bar{U}(s)}\ , \tag{11.3-113}$$

or, in complex form,

$$\gamma^*(\omega) = \gamma(s)|_{s=j\omega} = j\omega\sqrt{\rho\,U^*(\omega)} = j\omega\sqrt{\rho/G^*(\omega)}\ . \tag{11.3-114}$$

We are therefore now dealing with the propagation of a shear deformation through the extended medium. In the propagation of a plane shear wave the motion of a material particle is *perpendicular* to the direction of propagation. Experimentally, the propagation of a shear disturbance can be approximated by vibrating a plate in contact with the medium in a direction transverse to that in which a longitudinal disturbance would be produced. A shear wave is also called a *distortional wave,* or an *equivoluminal wave* because an (infinitesimal) shear deformation does not involve a change in volume.

Shear waves and dilatational waves are the only types of wave that can propagate through an extended medium [1]. In an experiment one generally tries to generate one to the exclusion of the other. The concept of the inertial transmission line can be applied to the propagation of a plane wave simply by considering the length of the line to be the direction of propagation of the wave.

The two types of waves which can propagate through an extended medium could have been expected to be shear and bulk waves rather than shear and *longitudinal* bulk waves. It is clear, however, that a bulk wave, i.e. a wave whose propagation function would contain the complex bulk modulus, $K^*(\omega)$, rather than the complex longitudinal bulk modulus, $M^*(\omega) = K^*(\omega) + \frac{3}{4} G^*(\omega)$, cannot be realized physically because in the propagation of a compression through an extended medium an elemental volume of the material undergoes not only a change in size, but a change in shape also, precisely as in uniaxial constrained compression and dilatation. If it were not for the constraint imposed by the extended medium itself, the change in the size and shape of the elemental volume would be the same as in uniaxial extension and contraction. A deformation with these propagational properties can, however, only be produced in a bounded medium.

11.3.6.2 Wave Propagation in a Bounded Medium

A bounded medium is typically represented by most of the specimens of finite dimensions which are normally used in the determination of material properties. The geometry of such a specimen is generally kept simple to minimize the "end-effects" presented by the boundaries. In all but the plainest cases the stress analysis required to extract the material properties from the measurements of the various possible propagation characteristics discussed in Sect. 5.6 is a formidable task. The simplest geometry is unquestionably that of a thin rod (cylinder) of uniform circular cross-sectional area. This geometry is normally also a good approximation to that of a fiber which may clearly be regarded as a very thin rod. The disturbance is effectively applied at one end of the rod or fiber and propagates along its axis. Ideally, the dimensions of the transverse directions (the radius) are such that any disturbance would be "damped out" before it reaches the boundaries of the specimen (the mantle of the cylinder). Stated in a different way this means that the length of any wave propagated along the axis of the cylindrical specimen must be large compared with its diameter. Reflection at the other end can usually be accounted for (and, indeed, is often used in a positive way) as detailed in Sect. 5.6.

The excitation employed to produce a suitable disturbance can be applied essentially in three ways. If we apply either a torque or an angular displacement at one end so that each section of the cylinder remains in its own plane and merely rotates around the axis, the disturbance will constitute a *torsional* wave. If the excitation is a uniaxial force or displacement applied in the direction of the length axis so that elemental volumes of the cylinder contract and expand, the disturbance will constitute a *longitudinal* wave. In either case the axis of the cylinder suffers no lateral displacement. A *flexural* wave is generated if elements of the axis are displaced laterally.

Torque and force on the one hand, and angular and linear displacement on the other are converted into stress and strain via the appropriate shape factor, specimen coefficient, or cell constant (see Sect. 3.1.2) produced by stress analysis, the details of which are outside the scope of this book. Torsional waves are shear waves characterized operationally by the (shear) relaxance, $\bar{Q}(s)$, and in the harmonic case by the complex shear modulus, $G^*(\omega)$. Longitudinal and flexural waves are both characterized operationally by the stretch relaxance, $\bar{Y}(s)$, and in the harmonic case by the complex stretch or Young's modulus, $E^*(\omega)$. As an example, the propagation of harmonic longitudinal *or* flexural waves is characterized by the complex propagation constant

$$\gamma^*(\omega) = j\omega\sqrt{\rho\, D^*(\omega)} \tag{11.3-115}$$

and the complex characteristic mechanical relaxance per unit area

$$\mathcal{Q}_C^*(\omega) = j\omega\sqrt{\rho\, E^*(\omega)}\ . \tag{11.3-116}$$

We note in closing that shear waves may propagate either in an extended or in a bounded medium. By contrast, the nature of waves other than shear waves depends on the medium in which they propagate.

11.4 Interconversion of the Isotropic Response Functions in Different Modes of Deformation

We now turn to the problem of interconverting the responses of a linear viscoelastic material elicited in various modes of deformation. We begin by recapitulating the fundamental features of the theory which the book attempted to describe in its preceding sections and chapters. We summarize these features as follows.

The infinitesimal theory of elasticity describes the deformational properties of materials with the aid of two fundamental material constants, the *bulk modulus*, K, and the *shear modulus*, G. The first governs changes in size, the second, changes in shape. The reciprocals of these moduli are the *bulk compliance*, B, and the *shear compliance*, J. Other moduli (such as the *stretch modulus*, E, and the *wave modulus*, M), compliances (e.g. the *stretch compliance*, D), or combinations of these (such as *Poisson's ratio*, μ), may be defined when needed or simply when convenient. Any of these elastic material constants can be expressed in terms of any *two* of the others. In particular, every one of the constants defined "as needed" may be expressed in terms of the fundamental moduli, K and G, or the fundamental compliances, B and J.

The theory of linear viscoelasticity is an extension of the infinitesimal theory of elasticity to time-dependent behavior. The various moduli, compliances, and other elastic constants of infinitesimal elasticity theory become functions of time (or, equivalently, of frequency). In analogy to G and K of the theory of elasticity, the two fundamental linear viscoelastic response functions are the (operational) shear relaxance, $\bar{Q}(s)$, the (operational) bulk relaxance, $\bar{P}(s)$, and their reciprocals, the (operational) shear retardance, $\bar{J}(s)$, and the (operational) bulk retardance, $\bar{A}(s)$. The relaxances and retardances constitute, as it were, the "parent" members of the families of functions describing linear viscoelastic changes in shape [$\bar{Q}(s)$, $\bar{G}(s)$, $G(t)$, $G^*(\omega)$, etc., and $\bar{U}(s)$, $\bar{J}(s)$, $J(t)$, $J^*(\omega)$, etc.] and linear viscoelastic changes in size [$\bar{P}(s)$, $\bar{K}(s)$, $K(t)$, $K^*(\omega)$, etc., and $\bar{A}(s)$, $\bar{B}(s)$, $B(t)$, $B^*(\omega)$, etc.] depending on the time regime chosen to elicit them.

The interconversion of linear viscoelastic responses obtained in different modes of deformation is complicated by the fact that we must account not only for differences in the nature of the material functions (i.e. whether, e.g., the response in uniaxial tension is to be converted to that in shear) but also for differences in the time-dependence of the properties described by these functions. The phenomenon of linear time-dependence manifests itself as *stress relaxation* if it arises in response to a strain excitation, and as *strain retardation* if it results from a stress excitation. The two fundamental modes of deformation, bulk deformation, and shear deformation, must be considered to be independent of one another. Linear viscoelastic theory is a phenomenological theory. As such, it cannot answer the question whether bulk relaxation (or retardation) and shear relaxation (or retardation) are in any way related to, or derivable from, each other. In the absence of any experimental evidence or theoretical guidance to the contrary, we must assume that they are not, and must accept that there exist two basically different types of relaxation and/or retardation behavior, viz., that in *shear* (relating to changes in *shape*) and that in *bulk* (relating to changes in *size*).

Those material functions that do not belong to the families of shear or bulk deformation, such as, e.g., $E(t)$, $M^*(\omega)$, etc., are functions that intermingle shape and size changes and, therefore, also different relaxation or retardation behavior. Just as the material constants of the theory of elasticity, any of the material functions of the linear theory of viscoelasticity can likewise always be expressed in terms of two others, notably, the fundamental ones. The form of these expressions depends, however, on the time regime which characterizes the functions under consideration. Basically, the interrelations between the functions of linear viscoelastic theory are relations between the "parent" transforms, i.e. the Laplace transforms of the responses to impulse excitations. Commonly, however, they are expressed as relations between the Laplace transforms of the responses to step excitations, i.e. as Carson transforms of the responses to impulse excitations. In the frequency domain they become interrelations between the complex harmonic responses, obtained by replacing the transform variable s in the "parent" transform relations by the complex variable $j\omega$ as exemplified by Eqs. (2.5-13) and (2.5-17). The use of Table 11.1-1 in conjunction with Tables 1.4-1 or 1.4-2 facilitates the procedure.

Examples of what we have just outlined have already been given in Sect. 11.1. We attempt here a more systematic treatment of the problem of interconversion between linear viscoelastic response functions obtained in different modes of deformation. The need for such interconversions obviously arises because we may wish to know the characteristic viscoelastic properties of a material in a deformation other than those in which such information is available. There are considerable differences in the ease with which viscoelastic properties can be measured. In Chap. 8 we discussed the interconversion of responses elicited in different time regimes of excitation, undertaken perhaps in order to extend the time scale of the response. For experimental reasons, the measurements in different time regimes may have had to be carried out also in different modes of deformation. Interconversion is then needed to combine the information. The second main reason for interconverting responses obtained in different modes of deformation is the frequent desire to reconstruct the behavior in a given mode of deformation from measurements in modes that are more easily measured experimentally. This is true particularly of behavior in bulk which is experimentally difficult to access directly.

We cannot deal here with all possible interconversions. The conversion to E(t) when G(t) and either μ(t) or K(t) are known has already been considered in Sect. 11.1. In addition, we deal here with the following:

(1) behavior in shear from measurements in uniaxial tension including lateral contraction (Sect. 11.4.1);
(2) behavior in bulk from measurements in uniaxial tension and in shear (Sect. 11.4.2);
(3) behavior in bulk from measurements in uniaxial constrained compression and in shear (Sect. 11.4.3); and
(4) behavior in bulk from measurements in uniaxial tension including lateral contraction (Sect. 11.4.4).

Sections 11.4.1 to 11.4.4 serve as models for the procedure to be followed when other types of interconversion are needed. (See Problems 11.4-1 through 11.4-4 and 11.4-8).

In each section we shall first present the basic interrelation equation formulated between the Carson transforms. We obtain this through the use of Table 11.1-1 from the corresponding relation between the elastic material constants, using Tables 1.4-1 or 1.4-2. We shall then discuss the interrelation equation derived in a similar manner between the frequency-dependent functions, and finally, the interrelation between the time-dependent functions. The latter generally involve convolution integrals. All these interrelations will be *exact interrelations* requiring two source functions for each target function. Most interconversions (particularly those trying to establish bulk behavior) require highly precise measurements to be successful. If at all feasible, the two source functions should be measured simultaneously to increase the precision of the measurements.

Because of experimental and/or computational difficulties the exact interrelations are often difficult to use in practice. Where possible, we will therefore also consider *approximate* interconversion relations.

11.4.1 Behavior in Shear from Measurements in Uniaxial Tension Including Lateral Contraction

We begin with a discussion of the conversion to shear of data obtained in uniaxial tension. We assume that both the stress in the direction of the imposed strain and the contingent lateral contraction in one of the transverse directions have been measured. We shall deal here only with relaxation behavior. Retardation behavior is easily accommodated (see Problem 11.4-2).

11.4.1.1 Exact Interconversion

By Tables 1.4-1 and 11.1-1 the relation between the Carson transforms becomes

$$s\bar{G}(s) = \frac{s\bar{E}(s)}{2[1 + s\bar{\mu}(s)]} \quad . \tag{11.4-1}$$

where the transform variable s before $\bar{G}(s)$ and $\bar{E}(s)$ [but not before $\bar{\mu}(s)$!] may be cancelled.

In complex form Eq. (11.4-1) becomes

$$G^*(\omega) = \frac{E^*(\omega)}{2[1 + \mu^*(\omega)]} , \tag{11.4-2}$$

and separation of the real and imaginary parts yields

$$G'(\omega) = \frac{1}{2} \frac{E'(\omega)[1 + \mu'(\omega)] - E''(\omega)\mu''(\omega)}{[1 + \mu'(\omega)]^2 + [\mu''(\omega)]^2} \tag{11.4-3}$$

and

$$G''(\omega) = \frac{1}{2} \frac{E''(\omega)[1 + \mu'(\omega)] + E'(\omega)\mu''(\omega)}{[1 + \mu'(\omega)]^2 + [\mu''(\omega)]^2} . \tag{11.4-4}$$

These are relatively simple algebraic expressions. However, they generally require data of rather high precision for successful interconversion.

Let us now turn to the corresponding interconversion between the time-dependent functions. Equation (11.4-1) cannot be retransformed into the real time domain in its present form. We must first rewrite it as

$$2[1 + s\bar{\mu}(s)]\bar{G}(s) = \bar{E}(s) . \tag{11.4-5}$$

Retransformation with the help of Eq. $(A3\text{-}17)_4$ now leads to

$$2(1 + \mu_g)G(t) - 2 \int_0^t G(u) \frac{d\,\mu(t-u)}{du} du = E(t) \tag{11.4-6}$$

[cf. Eq. (11.1-5)]. Unfortunately, this equation cannot be made explicit in G(t).

We can introduce the measured strains by recognizing that

$$\bar{E}(s) = [\bar{\sigma}_{11}(s)]/\gamma_{11,0} \qquad \text{and} \qquad \bar{\mu}(s) = -[\bar{\gamma}_{22}(s)]/\gamma_{11,0} . \tag{11.4-7}$$

Substituting into Eq. (11.4-5) and retransforming again, yields

$$\sigma_{11}(t) = 2[\gamma_{11,0} - \gamma_{22}(0)]G(t) + 2 \int_0^t G(u) \frac{d\,\gamma_{22}(t-u)}{du} du . \tag{11.4-8}$$

Thus, G(t) can be determined, at least in principle, from measurements of the tensile stress in response to a strain applied as a step function of time, if one simultaneously observes the time-dependent lateral contraction. A method for evaluating the convolution integral in Eq. (11.4-6) is presented in Sect. 11.4.1.2.

In Sect. 11.4.4 we shall show that K(t) can be determined from the same measurements in an analogous way. Simultaneous measurements of the lateral contraction and the stress in uniaxial tension imposed as a step function of time is thus a powerful experiment which yields — at least in principle — complete information on the relaxation behavior of a linear viscoelastic material. Analogous information on the retardation behavior can be obtained from simultaneous measurements of the lateral contraction and the strain in a uniaxial creep experiment (see Problem 11.4-3).

11.4.1.2 Numerical Interconversion

A numerical method for evaluating the convolution integral in Eq. (11.4-6) is provided by any of the various adaptations of the method of Hopkins and Hamming which we discussed in Sect. 8.1.3.2. Here, and elsewhere in this chapter, we shall use a straightforward version of the method and leave it to the reader to modify this in any way he sees fit. We begin by rewriting Eq. (11.4-6) for a discrete set of n values of t and, simultaneously, subdivide the interval of integration into n subintervals. This gives

$$E(t_n) = 2(1 + \mu_g)G(t_n) - 2 \sum_{i=1}^{i=n} \int_{t_{i-1}}^{t_i} G(u) \frac{d\,\mu(t_n - u)}{du} du \tag{11.4-9}$$

where $t_0 = 0$. We now assume that we have made each subinterval small enough so that we may take out G(t) from under the integral sign with its midvalue in it. The integration can then be carried out immediately and we obtain

$$E(t_n) = 2(1 + \mu_g)G(t_n) - \sum_{i=1}^{i=n} [G(t_i) + G(t_{i-1})][\mu(t_n - t_i) - \mu(t_n - t_{i-1})] \ . \tag{11.4-10}$$

But

$$\sum_{i=1}^{i=n} G(t_{i-1})[\mu(t_n - t_i) - \mu(t_n - t_{i-1})] = \sum_{i=0}^{i=n-1} G(t_i)[\mu(t_n - t_{i+1}) - \mu(t_n - t_i)] \tag{11.4-11}$$

since we merely changed the summation index. Using this relation, and rearranging, leads to the recurrence formula

$$G(t_n) = \frac{E(t_n) - G_g[\mu(t_n) - \mu(t_n - t_1)] - \sum_{i=1}^{i=n-1} G(t_i)[\mu(t_n - t_{i-1}) - \mu(t_n - t_{i+1})]}{2 + \mu_g + \mu(t_n - t_{n-1})} \tag{11.4-12}$$

with

$$G(t_1) = \frac{E(t_1) - G_g[\mu(t_1) - \mu_g]}{2 + \mu_g + \mu(t_1)} \tag{11.4-13}$$

and

$$G_g = \frac{E_g}{2(1 + \mu_g)} \ . \tag{11.4-14}$$

The method can equally well be used on the measured data themselves. To find the form of the recurrence formula in that case, we merely need to substitute $\sigma_{11}(t)/\gamma_{11,0}$ for E(t), and $-\gamma_{22}(t)/\gamma_{11,0}$ for $\mu(t)$, recognizing that μ_g becomes $-\gamma_{22}(0)/\gamma_{11,0}$, where $-\gamma_{22}(0)$ is the instantaneous lateral contraction at $t = 0$.

This derivation assumes that the data are equally spaced on the (arithmetic) time scale. If, as is usually the case, they are equally spaced on the logarithmic time scale, values of $\mu(t_n - t_k)$, where $k = i - 1, i, i + 1, n - 1$, must be obtained by linear interpolation between the two closest values. Care must be taken when the intervals on the logarithmic time scale become too large on the arithmetic scale.

11.4.1.3 Approximate Interconversion

Because it is difficult to determine either $\mu(t)$ [or $K(t)$, see Sect. 11.4.4], or their frequency-dependent counterparts experimentally, it is often desirable to be able to convert data in tension to data in shear (or *vice versa*) in an approximate manner. A common approach is to assume that Poisson's ratio is a time-invariant constant, say, μ_c. We then have $\bar{\mu}(s) = \mu_c/s$, and Eq. (11.4-1) at once yields

$$E(t) = 2(1 + \mu_c)G(t) \ . \qquad (11.4\text{-}15)$$

As pointed out in Sect. 11.3.5, if the material is considered to be incompressible, its Poisson's ratio becomes 0.5. Introducing this into Eq. (11.4-15) immediately leads to

$$G(t) = \tfrac{1}{3}\, E(t) \qquad (11.4\text{-}16)$$

and it also follows that

$$G'(\omega) = \tfrac{1}{3}\, E'(\omega) \quad \text{and} \quad G''(\omega) = \tfrac{1}{3}\, E''(\omega) \ . \qquad (11.4\text{-}17)$$

Even if the ratio were, in fact, as low as $\mu_c = 1/5$, on the usual logarithmic plot this would corresponds to a difference of 0.1 logarithmic units and would frequently be within the experimental error.

11.4.2 Behavior in Bulk from Measurements in Uniaxial Tension and in Shear

Determination of the properties of a linear viscoelastic material in bulk (i.e. pure size) deformation is difficult experimentally. It is, therefore, of interest to investigate whether the wanted information cannot be recovered from measurements in experimentally more easily accessible deformations. One of these, evidently, has to be a deformation which does contain information on bulk behavior. The three main options for doing this are the options numbered (11.4-2) to (11.4-4) in the introduction to this section. We take number (11.4-2) first.

The relations between the stress response functions are particularly simple. By Tables 1.4-2 and 11.1-1 we have

$$\bar{B}(s) = 9\,\bar{D}(s) - 3\,\bar{J}(s) \ , \qquad (11.4\text{-}18)$$

a simple additive relation. The frequency-dependent and the time-dependent relations are so easily derived from Eq. (11.4-18) that is unnecessary to state them explicitly here.

The real and imaginary components of $B^*(\omega)$ are readily converted to the corresponding components of $K^*(\omega)$ (see Sect. 8.1.2). The conversion of $B(t)$ into $K(t)$ is, however, a much more difficult task (see Sect. 8.1.3). Hence, there may be some interest in determining $K(t)$ from measurements of $E(t)$ and $G(t)$. The equation for $\bar{K}(s)$ in terms of $\bar{E}(s)$ and $\bar{G}(s)$, however, is not a simple additive expression. The derivation of the relation between the relaxation moduli is left as an exercise (cf. Problems 11.4-1 and 11.4-4).

11.4.3 Behavior in Bulk from Measurements in Uniaxial Constrained Compression and Shear

Option (11.4-3) requires the Carson transform relation

$$\bar{K}(s) = \bar{M}(s) + \tfrac{4}{3}\bar{G}(s) \tag{11.4-19}$$

which had been derived already in Sect. 11.3.3. We notice immediately that $\bar{K}(s) \simeq \bar{M}(s)$ if $\bar{K}(s) \gg \bar{G}(s)$. Consequently, $K_e \simeq M_e$. Generally, however, separate measurements must be made in shear and subtracted form the measurements in uniaxial constrained compression. Again, rather precise measurements are needed. Because the terms in Eq. (11.4-19) are additive, the frequency-dependent as well as the time-dependent relations are easily derived and will not be explicitly stated here.

11.4.4 Behavior in Bulk from Measurements in Uniaxial Tension or Compression Including Lateral Contraction or Dilatation

In Sect. 11.4.1 we discussed the combination of measurements in uniaxial tension/compression with measurements of the contingent lateral contraction/dilatation to predict the behavior in shear. The behavior in bulk can be derived from the *same* measurements. This makes option (11.4-4) a rather powerful experiment — at least in principle.

11.4.4.1 Exact Interconversion

Using Tables 1.4-1 and 11.1-1 again, we find

$$s\bar{K}(s) = \frac{s\bar{E}(s)}{3[1 - 2\,s\bar{\mu}(s)]} \tag{11.4-20}$$

as the basic Carson transform relation from which we derive

$$K^*(\omega) = \frac{E^*(\omega)}{3[1 - 2\,\mu^*(\omega)]} \tag{11.4-21}$$

as the relation between the complex response functions. Separation of the real and imaginary parts yields

$$K'(\omega) = \frac{1}{3}\,\frac{[1 - 2\,\mu'(\omega)]E'(\omega) + 2\,\mu''(\omega)E''(\omega)}{[1 - 2\,\mu'(\omega)]^2 + [2\,\mu''(\omega)]^2} \tag{11.4-22}$$

and

$$K''(\omega) = \frac{1}{3}\frac{E''(\omega)[1-2\,\mu'(\omega)] - E'(\omega)[2\,\mu''(\omega)]}{[1-2\,\mu'(\omega)]^2 + [2\,\mu''(\omega)]^2} \quad . \tag{11.4-23}$$

To obtain the relation between the step response functions we multiply both sides of Eq. (11.4-20) by the denominator, and retransform to the real time domain. We find

$$3(1-2\mu_g)K(t) + 6\int_0^t K(u)\frac{d\,\mu(t-u)}{du}du = E(t) \tag{11.4-24}$$

which may be compared with Eq. (11.4-6).

11.4.4.2 Numerical Interconversion

The convolution integral in Eq. (11.4-24) can be solved numerically by the same procedure that was used in Sect. 11.4.1.2. The resulting recurrence formula is

$$K(t_n) = \frac{E(t_n)/3 + K_g[\mu(t_n) - \mu(t_n - t_1)] + \sum_{i=1}^{i=n-1} K(t_i)[\mu(t_n - t_{i-1}) - \mu(t_n - t_{i+1})]}{1 - \mu_g - \mu(t_n - t_{n-1})} \tag{11.4-25}$$

with

$$K(t_1) = \frac{E(t_1)/3 + K_g[\mu(t_1) - \mu_g]}{1 - \mu_g - \mu(t_1)} \tag{11.4-26}$$

and

$$K_g = \frac{E_g}{3(1-2\mu_g)} \quad . \tag{11.4-27}$$

The remarks we made at the end of Sect. 11.4.1.3 with respect to the spacing of the data apply here also.

11.5 Problems

Problem 11.1-1. Retransform Eq. (11.1-6)

$$s\bar{E}(s) = 2[1 + s\bar{\mu}(s)]s\bar{G}(s) \tag{1}$$

into the real time domain.

Problem 11.3-1. In Problem 1.4-1 we defined an elastic modulus, H, as the ratio of the stress in one of the directions of stretch to the strain in the same direction in an equibiaxial (uniform biaxial) deformation.

(A) Express the corresponding Carson transform, $s\bar{H}(s)$, in terms of the fundamental moduli.

(B) State the relation which links $\bar{E}(s)$ to $\bar{H}(s)$.

(C) What is the relation between G(t) and H(t), and between E(t) and H(t) for an incompressible material?

Note: H(t) here has nothing to do, of course, with the shear relaxation spectrum, $H(\tau)$.

Problem 11.3-2. Obtain E(t) from standard linear solid models for G(t) and K(t).

Problem 11.3-3. At a constant rate of strain in uniaxial extension it is often useful to define [4] a *constant strain rate modulus* (cf. Problem 4.1-8)

$$F(t) = \frac{\sigma_{11}(t)}{\gamma_{11}(t)} = \frac{\sigma_{11}(t)}{\dot{\gamma}_{11,0}t} = \frac{\sigma(t)}{\dot{\gamma}_0 t} \,. \tag{1}$$

F(t) is also referred to as the *secant modulus*, because on a plot of $\sigma(t)$ against $\gamma(t)$ it is the secant of the angle between the abscissa, $\gamma(t)$, and a straight line drawn from the origin to $\sigma(t)$. In a linear material F(t) is independent of the magnitude of the rate of strain.

(A) Express F(t) in terms of E(t), and E(t) in terms of F(t); then

(B) plot the course of both F(t) and E(t) in doubly logaritmic coordinates, modelling the behavior using the standard linear solid model with $E_g = 2.67G_g$, $E_e = 3G_e$, and $\tau_E = \tau_G$, using the values $G_g = 10^9$ N/m^2, $G_e = 10^6/1.001$ N/m^2, and $\tau_G = 10^{-4}/1.001$ s [cf. Eqs. (3.4-93)].

Problem 11.3-4. Develop the canonical representation of F(t), defined in Problem 11.3-3 and of $\bar{F}(s)$ for an arrheodictic material.

Problem 11.3-5. Show that

$$E^*(\omega) = \omega^2 \left.\frac{d\,\bar{F}(s)}{ds}\right|_{s=j\omega} \tag{1}$$

where $\bar{F}(s)$ is the (operational) *constant strain rate modulus* defined in Problem 11.3-3.

Problem 11.3-6. Write out the complete relaxance matrix, $[\bar{\Gamma}_{kl}(s)]$, for an isotropic linear viscoelastic material. Then relate the shear relaxance, $\bar{Q}(s)$, and the bulk relaxance, $\bar{P}(s)$, to the two independent components of the matrix.

{*Hint*: $[\bar{\Gamma}_{kl}(s)]$ is the 6 × 6 form of $[\bar{\Gamma}_{ijkl}(s)]$, cf. Eq. (1.4-5).}

Problem 11.3-7. Express $\bar{V}(s)$, $\bar{Y}(s)$, and $\bar{\nu}(s)$ in terms of $\bar{\Gamma}_{11}(s)$ and $\bar{\Gamma}_{12}(s)$.

Problem 11.3-8. As discussed in Sect. 1.4.3, an axisymmetric (or transversely isotropic) purely elastic material requires five material constants for a complete characterization of its properties. If the material is viscoelastic, the five material constants become material functions of time. In a transversely isotropic body, changes in size and changes in shape cannot be neatly separated as in an isotropic body. Upon imposition of an isotropic pressure on an axisymmetric body the compression in the axial direction will be different from the compression in the two transverse directions. Thus, both the size and the shape of the specimen will change. We may, therefore, choose the five fundamental material functions purely on the grounds of experimental convenience. This might depend, e.g., on the shape of the specimen: whether fiber or film, for example. The following represents one possible choice.

Let the axis of symmetry be the third axis of the coordinate system as in Problems 1.4-10 to 1.4-12. Let the experimentally determined functions be three relaxation moduli and two Poisson's ratios. Let the first relaxation modulus, the *axial modulus*, $E_a(t)$, be measured in uniaxial tension in the axial direction, so that

$$E_a(t) = \sigma_{33}(t)/\gamma_{33,0} \ . \qquad (1)$$

Let the second be measured in the 1-direction. This *transverse modulus* thus becomes

$$E_t(t) = \sigma_{11}(t)/\gamma_{11,0} \ . \qquad (2)$$

The third will be a shear modulus measured in the plane perpendicular to the 1-direction, and will be defined by

$$G_t(t) = \sigma_{23}(t)/2\gamma_{23,0} \ . \qquad (3)$$

Finally, let the two Poisson's ratios be associated with $E_a(t)$ and $E_t(t)$, respectively. We have

$$\mu_a(t) = \mu_{13}(t) = \frac{-\gamma_{11}(t)}{\gamma_{33,0}} \qquad (4)$$

and

$$\mu_t(t) = \mu_{21}(t) = \frac{-\gamma_{22}(t)}{\gamma_{11,0}} \ . \qquad (5)$$

As indicated by the subscripts, the first is the ratio of the time-dependent lateral contraction in the 1-direction when the specimen is stretched in the 3-direction. The second similarly is the lateral contraction in the 2-direction when the specimen is deformed in the 1-direction.

Now, relate these material functions to the relaxances, $\bar{\Gamma}_{ijkl}(s)$, and verify that they reduce correctly to the functions E(t), G(t), and $\mu(t)$ of an isotropic body.

Problem 11.3-9. State the equations needed to calculate $\mu(t)$, $\mu'(\omega)$, $\mu''(\omega)$, and $\theta_\mu(\omega)$ using the standard linear solid as a model. Use $\mu_e = 1/2$, $\mu_g = 1/3$.

Problem 11.3-10. Show that it follows from the compliance nature of Poisson's ratio that the lateral contraction, $-\gamma_2(\omega)$, lags an imposed sinusoidal strain

$$\gamma_1(\omega) = \gamma_{1,0}\exp(j\omega t) \ . \qquad (1)$$

Problem 11.3-11. Suppose that the strain, $\gamma_1(t)$, has been determined in a uniaxial creep experiment together with the lateral contraction, $-\gamma_2(t)$. Derive the expression for Poisson's ratio in terms of these experimental quantities.

Problem 11.3-12. Show that the time-dependent Poisson's ratio is the same whether the lateral contraction results from a strain, or a stress excitation.

Problem 11.3-13. A strip specimen is stretched at a constant rate of strain, $\dot{\gamma}_{1,0}$, in uniaxial extension in such a manner that the strain, $\gamma_1(t)$, remains below the limit above which linear theory would no longer be applicable. Concurrently, the lateral contraction, $\gamma_2(t)$, is also determined. Establish the relation between $\gamma_2(t)$ and $\mu(t)$.

Problem 11.4-1. (A) Develop equations for calculating the response in shear from data obtained in bulk and in uniaxial tension.

(B) Suggest approximations based on these equations.

Problem 11.4-2. (A) Develop the relations for $J(t)$ in terms of $D(t)$ and $\mu(t)$, and for the real and imaginary parts of $J^*(\omega)$ in terms of the real and imaginary parts of $D^*(\omega)$ and of $\mu^*(\omega)$.

(B) How would you calculate the shear creep compliance numerically?

(C) State the equations for the shear storage and loss compliances.

(D) Derive approximate relations between the shear and the tensile compliances.

Problem 11.4-3. Section 11.4.4 discussed the theory underlying the determination of both the shear and the bulk relaxation properties of a linear viscoelastic material from the lateral contraction measured simultaneously with the *stress* in uniaxial tension/compression experiments. The retardation properties can be obtained similarly from simultaneous measurements of the lateral contraction and the *strain*. Develop the required equations.

Problem 11.4-4. Obtain $G(t)$ numerically from $E(t)$, assuming that the bulk modulus can be represented by a constant, K.

Problem 11.4-5. Show that it follows directly from the "compliance nature" of Poisson's ratio that the damping in shear is larger than the damping in bulk, i.e. that

$$\tan\theta(\omega) > \tan\theta_K(\omega) \tag{1}$$

where

$$\tan\theta(\omega) = \frac{G''(\omega)}{G'(\omega)} \quad \text{and} \quad \tan\theta_K(\omega) = \frac{K''(\omega)}{K'(\omega)} \;. \tag{2}$$

Problem 11.4-6. Show (A) that

$$\tan\theta(\omega) \geq \tan\theta_E(\omega) \geq \tan\theta_K(\omega) \tag{1}$$

and (B) that

$$\tan\theta(\omega) \geq \tan\theta_M(\omega) \geq \tan\theta_K(\omega) \;. \tag{2}$$

Problem 11.4-7. (A) Prove inequalities (1) of Problem 11.4-6 starting from the compliance relation

$$3\,J^*(\omega) = 9\,D^*(\omega) - B^*(\omega) \tag{1}$$

rather than from the Poisson's ratio. Then

(B) derive an approximate relation between $\tan\theta(\omega)$ and $\tan\theta_E(\omega)$ based on the assumption that $K'(\omega) \simeq K$, and $K''(\omega) = 0$.

Problem 11.4-8. If the measurements can be made with sufficient precision, a combination of shear and tensile measurements is an experimentally attractive way for determining Poisson's ratio. Develop the necessary relations.

References (Chap. 11)

1. H. Kolsky, *Stress Waves in Solids*, Clarendon Press, Oxford, (1953); Dover Publications, New York (1963).
2. R.W. Whorlow, *Rheological Techniques*, Halstead Press, New York (1980).
3. H.J. Skimmin, *Ultrasonic Methods for Measuring the Mechanical Properties of Liquids and Solids*, in: W.P. Mason, ed., *Physical Acoustic, Principles and Methods*, Vol. 1, Pt. A, Academic Press, New York (1964).
4. T.L. Smith, Trans. Soc. Rheol., *6*, 61 (1962).
5. H.A. Waterman, Rheol. Acta, *16*, 31 (1977).

Appendix: Transformation Calculus

... come with your appendix.

Shakespeare: The Taming of the Shrew

Extensive use is being made in this book of integral transformation calculus. Although the Laplace transformation is used primarily, several closely related transforms also occur. This Appendix attempts a short review of the Laplace and related transformations. It should by no means be considered an exposition of the subject on which many excellent textbooks are available (see, e.g., Ref. [1], [2], or [3]). It is meant to serve simply as an introduction to the notation used in the main body of the book and as a handy reference. Theorems and formulae are generally stated without proof, in the form in which they are applicable in the context of the book. The reader who has no familiarity with the subject is certainly urged to become acquainted with it.

A1 Introduction

We call F(p) an *(integral) transform* of f(t) if

$$F(p) = \int_0^\infty f(t)K(t,p)dt \tag{A1-1}$$

and F(p) exists. The operation of obtaining the transform entails integration over the product of the given, so-called *source function*, f(t), with the *kernel function*, K(t, p), to obtain the *image function*, F(p). Thus, a function of the variable t (which becomes a dummy variable in the definite integral) is replaced by a function of the variable p in the transformation. The practical reason for undertaking such a transformation is, of course, that some simplicity is expected in operating with F(p) instead of f(t). Thus, for instance, the Laplace transformation (see Sect. A3) in which K(t, p) takes the form exp(−pt), transforms the linear differential equations with constant coefficients which describe linear viscoelastic behavior, into algebraic equations (cf. Sect. 2.1.1). These image functions are more readily manipulated than the source functions themselves. Once the desired result is obtained, inversion of the transform yields the same result one would have obtained by manipulating the original source functions. The rules for transforming f(t) and retransforming F(p) on the basis of Eq. (A1-1) and its inverse, constitute the *(integral) transformation calculus*. The particular type of calculus obviously depends on the nature of the kernel function.

Equation (A1-1) may be expressed alternatively by

$$F(p) = \mathscr{P}_K[f(t)] \tag{A1-2}$$

where

$$\mathscr{P}_K = \int_0^\infty [\ ]K(t,p)dt \tag{A1-3}$$

is an integral *operator*. Accordingly, transformation calculus is sometimes referred to as *operational calculus*.

We define the *inverse transformation* implicitly by

$$f(t) = \mathscr{P}_K^{-1}[F(p)] \tag{A1-4}$$

where $\mathscr{P}_K^{-1}$ is the *inverse transform operator* which must be given the meaning appropriate to the definition of the *direct transformation*, Eq. (A1-1).

A familiar example of a (non-integral) transformation calculus is the use of logarithms. Instead of *multiplying* x and y, their logarithms, log x and log y, are *added* to obtain log xy which is then retransformed to xy. Here, the simplification obtained through the application of the logarithmic transformation arises from the fact that the order of algebraic manipulation of numbers is lowered when their logarithmic transforms are manipulated. Thus, addition replaces multiplication, multiplication replaces exponentiation, subtraction replaced division, and division replaces root taking. The logarithmic transformation is carried out with the help of logarithm tables. In the integral transformation calculus use is made similarly of tables of pairs of corresponding source and image functions.

Discussion of the Laplace transformation in Sect. A3 is preceded by a discussion of certain special functions. Section A3 is followed by brief discussions of the Fourier, Stieltjes, and Hilbert transformations.

A2 Special Functions

In the context of this book the need for the special functions discussed in this section arises mainly because the excitations assumed to elicit linear viscoelastic response are generally restricted to certain regions of the time scale. In particular, the origin of the time scale is selected as the point in time prior to which the material is deemed to have experienced no deformation history. Thus, the lower limit of integration in Eq. (A1-1) becomes zero and the variable t is restricted to positive values only. Any function f(t) of time can be so restricted through multiplication with the *unit step function* (see Sect. A2.2) whose role is thus central to the application of transformation calculus to linear viscoelastic theory. In fact, all of the special functions discussed in this section are related to, and may be expressed in terms of, the unit step function. For reasons of convenience we begin, however, with a discussion of the unit impulse or delta function which may be regarded formally as the derivative of the unit step function.

A2.1 The Delta Function

Consider the function

$$\delta(t;\varepsilon) = \frac{1}{\pi}\int_0^\infty \exp(-\varepsilon\omega)\cos\omega t \cdot d\omega = \frac{\varepsilon}{\pi(\varepsilon^2 + t^2)} \tag{A2-1}$$

where ε is a parameter which may take arbitrary positive values. The function is plotted in Fig. A2-1 for various values of ε. The curves are bell-shaped and become progressively higher and narrower as ε decreases. However, the area under the curve, given by the integral

$$\int_{-\infty}^{\infty} \delta(t;\varepsilon)dt = \frac{\varepsilon}{\pi}\int_{-\infty}^{\infty}\frac{dt}{\varepsilon^2 + t^2} = \frac{1}{\pi}\tan^{-1}\frac{t}{\varepsilon}\bigg|_{-\infty}^{\infty} = 1 \ , \tag{A2-2}$$

is unity for every $\varepsilon > 0$. As $\varepsilon \to 0$, the width of the bell-shaped curve vanishes and its height increases indefinitely. Thus, in the limit as $\varepsilon \to 0$, we write

$$\delta(t) = \frac{1}{\pi}\lim_{\varepsilon\to 0}\frac{\varepsilon}{\varepsilon^2 + t^2} \tag{A2-3}$$

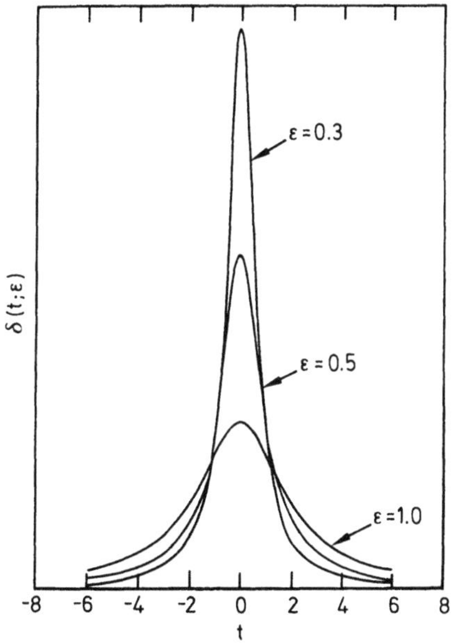

Fig. A2-1. The function $\delta(t;\varepsilon)$ for several values of ε

where δ(t) has the properties that

$$\delta(t) = \begin{cases} \infty, & t = 0 \\ 0, & t \neq 0 \end{cases} \tag{A2-4}$$

and

$$\int_{-\infty}^{\infty} \delta(t)dt = 1 \ . \tag{A2-5}$$

We call δ(t) the *delta function* or *unit impulse function.* Mathematically, δ(t) is *not* a function since the limit as $\varepsilon \to 0$ does not exist when $t = 0$. Because, with some caution, δ(t) may, however, be manipulated as if it were an ordinary function, it is sometimes called a *symbolic function.*

Alternative formulations are possible. Thus, we may write

$$\delta(t) = \lim_{\varepsilon \to 0} \frac{1}{\pi} \int_0^{\infty} \exp(-\varepsilon\omega)\cos \omega t \cdot d\omega \ . \tag{A2-6}$$

The functions $\delta(t;\varepsilon)$ represented by Eqs. (A2-1) are continuously differentiable approximations to the delta function. A discontinuous function which approximates δ(t) is the function

$$\delta(t;\varepsilon) = \begin{cases} 1/\varepsilon, & 0 \le t \le \varepsilon \\ 0, & t > \varepsilon \end{cases} \tag{A2-7}$$

whose graph appears in Fig. A2-2a. As $\varepsilon \to 0$ the height of the shaded rectangular area increases indefinitely while its width shrinks to zero. At the same time, however, the area, $\varepsilon \cdot 1/\varepsilon$, is always equal to unity. In Appendix A2.3 we shall recognize Fig. A2-2a as representing what we shall there call the *gate function.* In Appendix A2.2 we shall also see that the delta function may be considered formally as the derivative of the *unit step function.*

When using δ(t) it should always be borne in mind that formulations in which this symbol occurs have meaning in the usual mathematical sense only when the formulations are integrated over an interval which includes the argument of the delta function. Quite generally,

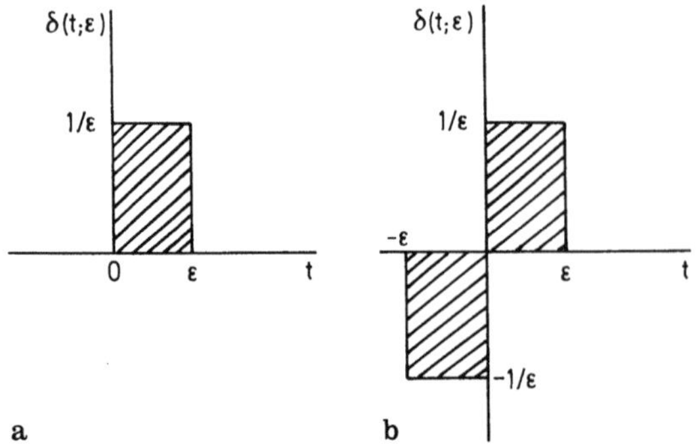

Fig. A2-2. Discontinuous functions approximating (**a**) the delta function, (**b**) its first derivative

$$\int_{-\infty}^{\infty} f(t)\delta(t)dt = f(0) \qquad \text{(A2-8)}$$

for any continuous function f(t). This important property of the delta function is known as the *sifting property*. It is called that because the operation of integrating over f(t)δ(t) sifts out the values of f(t) when its argument is zero, from among all other possible values.

The symbolic function δ(t) may be regarded as an *even function*, i.e.

$$\delta(-t) = \delta(t) \ . \qquad \text{(A2-9)}$$

The definition of the delta function may be generalized to

$$\delta(t - t') = \frac{1}{\pi} \lim_{\varepsilon \to 0} \frac{\varepsilon}{\varepsilon^2 + (t - t')^2} \qquad \text{(A2-10)}$$

so that

$$\delta(t - t') = \begin{cases} \infty, & t = t' \\ 0, & t \neq t' \end{cases} \qquad \text{(A2-11)}$$

where

$$\int_{-\infty}^{\infty} \delta(t - t')dt = 1 \ . \qquad \text{(A2-12)}$$

We call $\delta(t - t') = \delta(t' - t)$ the *shifted delta function*. It has the sifting property

$$\int_{-\infty}^{\infty} f(t)\delta(t' - t)dt = \int_{-\infty}^{\infty} f(t' - t)\delta(t)dt = f(t') \ . \qquad \text{(A2-13)}$$

We note that Eqs. (A2-8) and (A2-13) remain valid when the interval of integration is changed from $-\infty \leq t \leq \infty$ to $0 \leq u \leq t$. Thus, we have

$$\int_0^t f(u)\delta(u)du = f(0) \qquad \text{(A2-14)}$$

and

$$\int_0^t f(u)\delta(t - u)du = \int_0^t f(t - u)\delta(u)du = f(t) \ . \qquad \text{(A2-15)}$$

The two forms of Eqs. (A2-13) and (A2-15) are obtained through a change of variable. Sometimes the lower limit of integration is written as 0^- to emphasize that $u = 0$ is considered to fall within the limits of integration.

Through a change of variable in Eq. (A2-5) it is easily shown that

$$\delta(t/a) = a\,\delta(t) \ . \qquad \text{(A2-16)}$$

We call $a \cdot \delta(t)$ an impulse of strength 'a'. Formally, the dimension of the delta function is the reciprocal of the dimension of its argument. If this is the time, t, an impulse of strength 'a' has the dimensions of a/t. Graphically, it may be represented by an arrow of height 'a' or one of height a/t, as convenient, located along the time axis at the point where the argument of the (shifted) delta function is zero (cf. Figs. 2.2-1, 4.2-2, 4.3-3, and 10.0-1).

Derivatives of the delta function, $\delta'(t)$, $\delta''(t)$, etc., can be defined. The symbolic function $\delta'(t)$, or unit doublet impulse, can be interpreted as the limit as $\varepsilon \to 0$ of the function $\delta'(t;\varepsilon)$ sketched in Fig. A2-2b. The derivatives of the delta function of order n, $\delta^{(n)}(t)$, are discussed in various text (see, e.g., Goldman [2]), but are needed in this book at most in a purely formal sense.

A2.2 The Step Function

Consider now the function $h(t;\varepsilon)$ obtained by integrating the function $\delta(t;\varepsilon)$ given by Eq. $(A2\text{-}1)_2$, subject to the condition $h(\infty;\varepsilon) = 1$. We find

$$h(t;\varepsilon) = \frac{1}{2} + \frac{1}{\pi}\arctan\frac{t}{\varepsilon} \; . \qquad \text{(A2-17)}$$

The function is plotted in Fig. A2-3 for various values of ε.

As ε decreases, $h(t;\varepsilon)$ approaches a straight line at $h(t;\varepsilon) = 1$ extending from $t = 0$ to $t = \infty$. We call the function resulting from going to the limit as $\varepsilon \to 0$ the *unit step function* and denote it by h(t). The symbol h is taken from the name of Heaviside* who used the function extensively in his application of operational calculus to electrical problems. It is sometimes referred to as the Heaviside unit step function or the Heaviside function. Other designations of the function use the letter u (for unit step function) in various forms (i.e., u, U, $\mathcal{U}$, $\mathcal{U}_1$, etc.). We have

$$h(t) = \frac{1}{2} + \frac{1}{\pi}\lim_{\varepsilon\to 0}\arctan\frac{t}{\varepsilon} = \frac{1}{2} + \frac{1}{\pi}\lim_{\varepsilon\to 0}\int_0^\infty \exp(-\varepsilon\omega)\frac{\sin\omega t}{\omega}d\omega \; . \qquad \text{(A2-18)}$$

Equation $(A2\text{-}18)_2$ is obtained by integrating Eq. (A2-6) with respect to t and using the same integration constant as above.

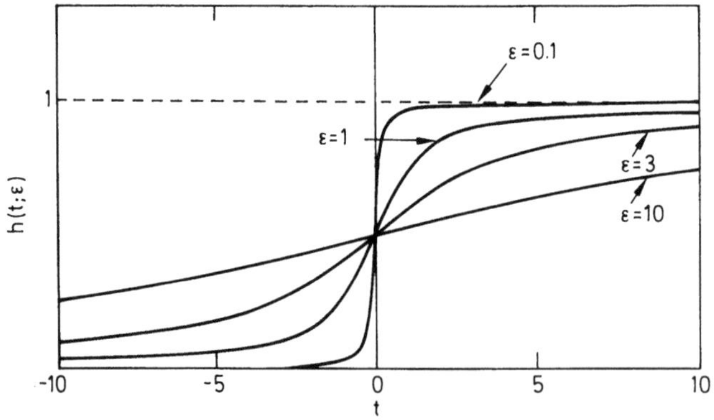

Fig. A2-3. The function $h(t;\varepsilon)$ for several values of ε

* Oliver Heaviside, 1850–1925, English physicist, who suggested the presence of the ionosphere (Heaviside layer) in the upper layers of the atmosphere, and worked extensively on the theoretical aspects of electrical problems in telegraphy and telephony.

Although from these two definitions it would seem to follow that h(0) = 1/2, the assignment of a value to h(0) is arbitrary and does not affect the Laplace transform of h(t). In the applications considered in this book a value for h(0) can be specified if and when needed. Consequently we define the unit step function as

$$h(t) = \begin{cases} 0, & t < 0 \\ 1, & t > 0 \; . \end{cases} \tag{A2-19}$$

We see from Eqs. (A2-18) that

$$\delta(t) = \frac{d\,h(t)}{dt} \; , \tag{A2-20}$$

i.e. the delta function may formally be regarded as the derivative of the step function. We also recognize that the step function is an *odd* function of its argument, i.e.

$$h(t) = -h(-t) \; . \tag{A2-21}$$

Equation (A2-20) may be generalized to the definition of the *shifted unit step function*,

$$h(t - t') = \begin{cases} 0, & t < t' \\ 1, & t > t' \; . \end{cases} \tag{A2-22}$$

Again, $h(t - t')$ is *not* defined for $t = t'$. The two functions are shown in Figs. A2-4a and 4b. Both are examples of a *sectionally*, or *piecewise*, *continuous function.*

The function h(t) is useful as a device for restricting a given function to its values for $t > 0$ as shown in Fig. A2-4c where

$$f(t) = \phi(t)h(t) \tag{A2-23}$$

is plotted as a function of t. Multiplication of $\phi(t)$ by h(t) thus results in a new function, f(t), in which that portion of $\phi(t)$ for which $t < 0$ (this appears as the

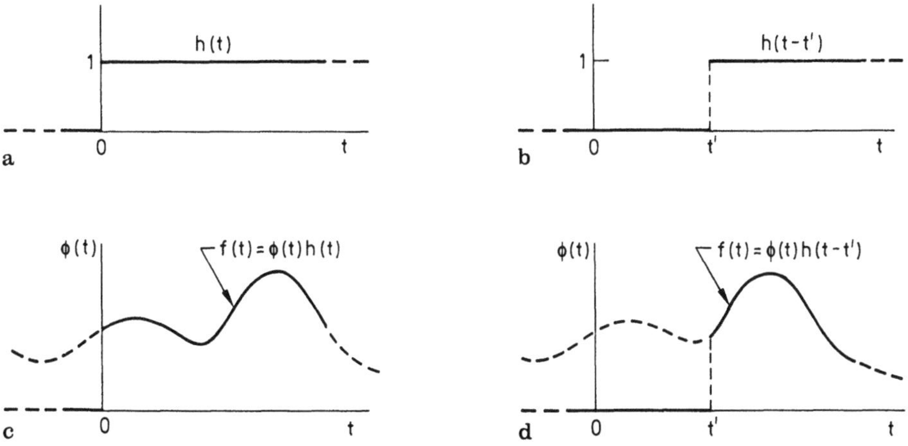

Fig. A2-4a–d. The unit step function and its use as a cut-off function. **a** Step function; **b** shifted step function; **c** $\phi(t)$ cut off by step function; **d** $\phi(t)$ cut off by shifted step function

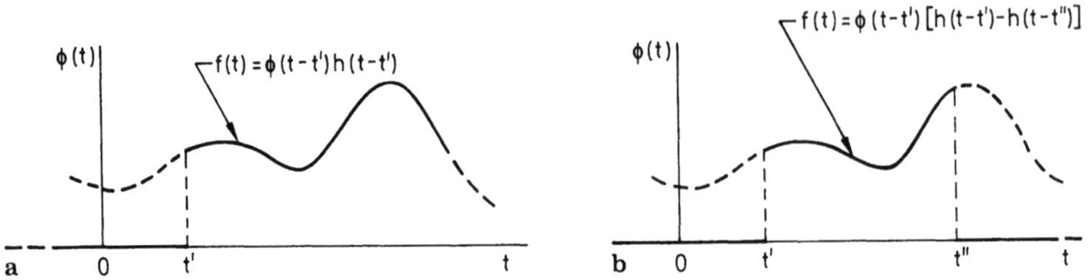

Fig. A2-5. Cut-off and translation with the aid of (**a**) the unit step function and (**b**) the unit gate function

curve consisting of short dashes) is cut off. Since the viscoelastic response functions introduced in Chaps. 2 and 7 of this book are all zero for $t < 0$, they should all be written as $f(t)h(t)$. However, we shall usually consider the symbol $h(t)$ as implied and write it explicitly only when it needs special emphasis.

The function $h(t - t')$ serves in a similar way to cut off another function for values of $t < t'$ as illustrated in Fig. A2-4d, where

$$f(t) = \phi(t)h(t - t') \ . \tag{A2-24}$$

Cut-off and translation along the t-axis is achieved by letting

$$f(t) = \phi(t - t')h(t - t') \ . \tag{A2-25}$$

This function appears in Fig. A2-5.

In Eqs. (A2-23), (A2-24), and (A2-25) $\phi(t)$ is a continuous function defined on the interval $-\infty < t < \infty$. The function $f(t)$, by contrast, is non-zero only in the interval $0 < t < \infty$ in Eq. (A2-23), and in the interval $t' < t < \infty$ in Eqs. (A2-24) and (A2-25). These functions are, therefore, sectionally continuous.

We note that the unit step function is dimensionless. When multiplied by a constant, say a, it represents a step of height a and the product evidently has the dimensions of a.

A2.3 The Gate Function

In applying the *unit gate function*

$$h_{t'}(t) = h(t) - h(t - t') \tag{A2-26}$$

a unit step function is imposed at $t = 0$ and this is followed by the imposition of another unit step function of opposite sign at $t = t'$. The second term on the right hand side of Eq. (A2-26) annihilates the effect of the first. The function is shown in Fig. A2-6a.

Multiplication of a given function by $h_{t'}(t)$ has the effect of restricting, or confining, the function to its value in the interval $0 < t < t'$, as illustrated in Fig. A2-6c where

$$f(t) = \phi(t)[h(t) - h(t - t')] \ . \tag{A2-27}$$

This property is useful in discussing periodic functions (see Sects. A3.1.8 and A3.1.9).

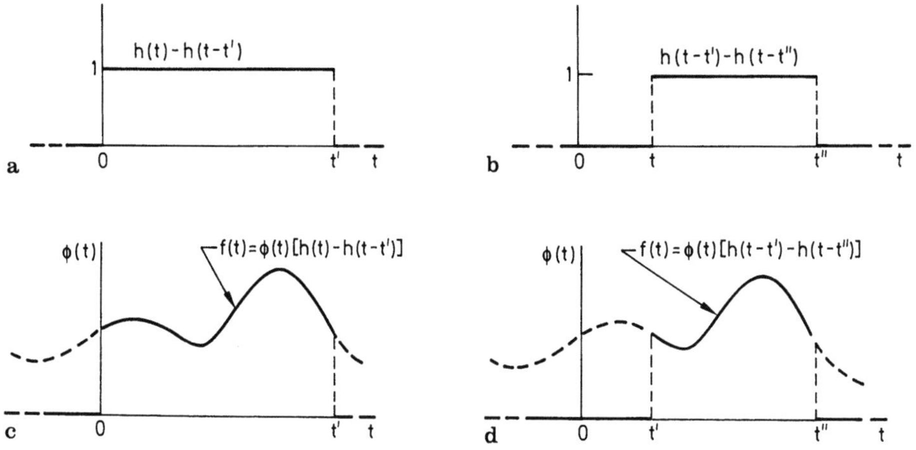

Fig. A2-6a–d. The unit gate function and its use as a cut-off function. **a** Gate function; **b** shifted gate function; **c** $\phi(t)$ cut off by gate function; **d** $\phi(t)$ cut off by shifted gate function

The unit gate function is expressed explicitly as

$$h_{t'}(t) = \begin{cases} 0, & t < 0 \\ 1, & 0 < t < t' \\ 0 & t > t' \end{cases} \tag{A2-28}$$

and is, therefore, *not* defined at $t = 0$ and at $t = t'$. It is, thus, another example of a sectionally continuous function. Its name derives from its application in the theory of electric circuits where it is used to model the *gating* of a signal by the opening and closing of a switch. For this reason it is sometimes also referred to as the *unit pulse function.* It is used in Chap. 7 for the imposition and removal of an excitation.

In analogy to the shifted unit step function we define a *shifted unit gate function* as

$$h_{t''}(t - t') = h(t - t') - h(t - t'') \tag{A2-29}$$

or, explicitly, as

$$h_{t''}(t - t') = \begin{cases} 0, & t < t' \\ 1, & t' < t < t' + t'' \\ 0, & t > t' + t'' . \end{cases} \tag{A2-30}$$

The function appears in Fig. A2-6b. The function $f(t) = \phi(t)h_{t''}(t - t'')$ is shown in Fig. A2-6d. Translation *and* gating with its help is illustrated in Fig. A2-5b where

$$f(t) = \phi(t - t')[h(t - t') - h(t - t' - t'')] \tag{A2-31}$$

is compared with $f(t) = \phi(t - t')h(t - t')$ in Fig. A2-6a.

The unit gate function may also be used to introduce the delta function. Let

$$\delta(t;t') = (1/t')h_{t'}(t) = \frac{h(t) - h(t-t')}{t'} . \tag{A2-32}$$

Then,

$$\delta(t) = \lim_{t' \to 0} \frac{h(t) - h(t-t')}{t'} . \tag{A2-33}$$

This is also another way of showing that $\delta(t)$ is the (formal) derivative of h(t).

With respect to dimensions the same considerations apply as to the step function. The product $a \cdot h_{t'}(t)$ is a gate function of height "a" having the dimensions of "a".

A2.4 The Slope Function

We distinguish the *slope function*

$$p(t) = t\,h(t) \tag{A2-34}$$

from simple proportionality to t. The unit step function restricts p(t) to positive values of t. The slope function appears in Fig. A2-7a. We denote it by the letter p taken from the French word *pente* for slope because we use the letter s for the Laplace transform variable. The slope function is defined explicitly by

$$p(t) = \begin{cases} 0, & t \leq 0 \\ t, & t > 0 \end{cases} \tag{A2-35}$$

and is thus a continuous function. The slope function has the dimensions of its argument. The function $a \cdot p(t)$, where "a" is a constant, is a slope function of slope $a = \tan \alpha$. The slope function is the integral of the step function. We have

$$p(t) = \int_0^t h(u)du = t\,h(t) . \tag{A2-36}$$

The *shifted slope function*

$$p(t - t') = (t - t')h(t - t') \tag{A2-37}$$

is illustrated in Fig. A2-7b.

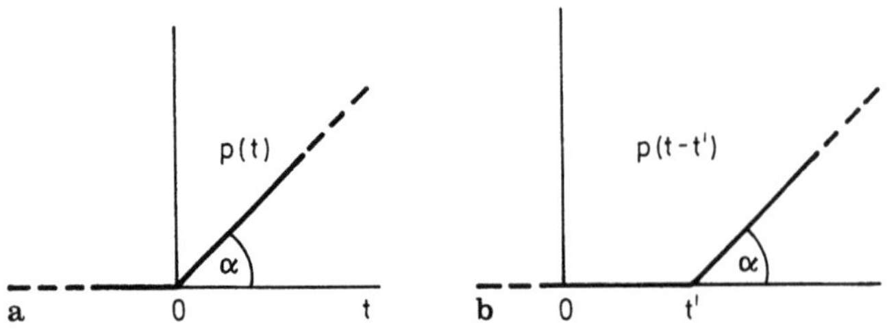

Fig. A2-7. The slope function (**a**) and the shifted slope function (**b**)

A2.5 The Ramp Function

Consider the function illustrated in Fig. A2-8a. The solid line represents the slope function, p(t), from which, at times $t \geq t'$, the function $p(t - t')$, i.e. the same slope function shifted to the right by t', is subtracted. The function $p(t) - p(t - t')$ may be represented by writing

$$r_{t'}(t) = p(t) - p(t - t') = \begin{cases} 0, & t \leq 0 \\ t, & 0 \leq t \leq t' \\ t', & t \geq t' \, . \end{cases} \tag{A2-38}$$

This function will be called a *ramp function* of height t'. The *unit ramp function*

$$r_{t'}(t) = (1/t')[p(t) - p(t - t')] \tag{A2-39}$$

is illustrated in Fig. A2-8b. Substitution of Eqs. (A2-34) and (A2-37) yields

$$r_{t'}(t) = (1/t')[t\, h(t) - (t - t')h(t - t')] \tag{A2-40}$$

which expresses the unit ramp function in terms of the unit step function. Equation (A2-40) may be rewritten as

$$r_{t'}(t) = (t/t')[h(t) - h(t - t')] + h(t - t') \, . \tag{A2-41}$$

The first term on the right represents the slope function normalized by t'. The second term annihilates the effect of the first for times $t > t'$. The two terms in brackets represent the gate function, $h_{t'}(t)$. Thus, the slope function is restricted to the interval $0 \leq t \leq t'$. The third term then imposes a unit step function for $t > t'$. The interpretation of the unit ramp function in terms of step functions is, therefore, different from, but equivalent to, its interpretation in terms of slope functions. Explicitly, the unit ramp function is defined by

$$r_{t'}(t) = \begin{cases} 0, & t \leq 0 \\ t/t', & 0 \leq t \leq t' \\ 1, & t \geq t' \, . \end{cases} \tag{A2-42}$$

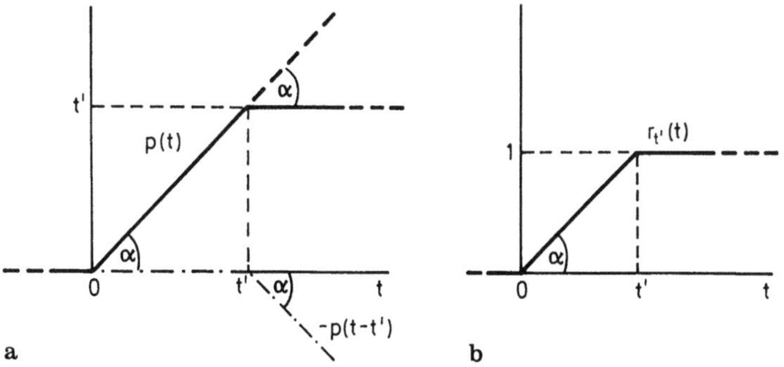

Fig. A2-8. The ramp function (**a**) and the unit ramp function (**b**)

It is, therefore, a continuous, dimensionless function. The function $a \cdot r_{t'}(t)$ is a ramp function of height 'a', having the dimension of 'a'. We note that

$$a\, r_{t'}(t) = \mathring{a}\, p_{t'}(t) \tag{A2-43}$$

where $\mathring{a} = a/t'$. Finally, we remark (see Problem 7.3-6) that

$$\lim_{t' \to 0} r_{t'}(t) = h(t) \ , \tag{A2-44}$$

i.e. for a sufficiently short *rise time*, t', the unit ramp function is an approximation to the unit step function to which it reduces for an infinitely short, i.e. zero, rise time.

A3 The Laplace Transform

We call

$$\bar{f}(s) = \int_0^\infty f(t)\exp(-st)dt = \mathscr{L}[f(t); s] \tag{A3-1}$$

the *Laplace* transform* of f(t) when the integral $\int_0^\infty |f(t)| \exp(-st)dt$ exists. When it is not necessary to indicate the transform variable explicitly, we shall write $\mathscr{L}$ f(t) for $\mathscr{L}[f(t); s]$. The Laplace transform variable, s, is in general complex, i.e.

$$s = -\alpha \pm j\beta \tag{A3-2}$$

although s may be regarded as real ($\beta = 0$) when the only singularities of the transform lie on the real axis of the complex plane.

We write the *inverse transform* as

$$\mathscr{L}^{-1}\bar{f}(s; t)(=)f(t) \tag{A3-3}$$

but use $\mathscr{L}^{-1}\bar{f}(s)$ for short whenever possible. The symbol (=) means "equals almost everywhere" (cf. Sect. A3.1.2).

We shall generally refer to the *source function*, f(t), as being in the *time domain*, and to the *image function*, $\bar{f}(s)$, as being in the *complex domain*. Alternatively we may refer to the former as being on the real time axis, and the latter as being in the transform or complex plane.

A3.1 Properties

A brief review follows of those properties of the Laplace transformation which are used in this book.

---

* The transform thus defined is the *one-side* Laplace transform. A two-sided transform

$$\bar{f}_2(s) = \int_{-\infty}^{\infty} f(t)\exp(-st)dt$$

can be defined but is not used in this book (cf. Sect. 2.1.1).

A3.1.1 Linearity

The Laplace transformation is a linear operation. The *transform operator* $\mathscr{L}$ is thus a *linear operator*. Specifically, if $f_1(t)$ and $f_2(t)$ are functions with Laplace transforms $\bar{f}_1(s)$ and $\bar{f}_2(s)$, respectively, and c_1 and c_2 are constants, then, by the *linearity property*

$$\mathscr{L}[c_1 f_1(t) + c_2 f_2(t)] = c_1 \mathscr{L} f_1(t) + c_2 \mathscr{L} f_2(t) = c_1 \bar{f}_1(s) + c_2 \bar{f}(s). \quad \text{(A3-4)}$$

We note that *scalar multiplication* of a function in the time domain corresponds to scalar multiplication of its transform in the transform plane.

A3.1.2 Uniqueness

The direct Laplace transformation is unique. The inverse transformations may not be unique. Two functions $f_1(t)$ and $f_2(t)$ may have identical continuous portions but may be defined differently at points of discontinuity. Both functions will then have the same transform but care must be taken with the definition of the inverse transform at the points of discontinuity. The unit step function is a case in point.

A3.1.3 Translation

Because of the way in which the Laplace transformation is defined, restriction of a function f(t) to positive values of t in the time domain through multiplication by the unit step function h(t) does not change its transform, i.e.

$$\mathscr{L} f(t)h(t) = \mathscr{L} f(t) = \bar{f}(s) \; . \quad \text{(A3-5)}$$

Translation of the function along the real time axis by the amount t′, however, corresponds to multiplication of its transform by $\exp(-t's)$. Thus,

$$\mathscr{L} f(t - t')h(t - t') = \bar{f}(s)\exp(-t's) \; . \quad \text{(A3-6)}$$

Equation (A3-6) is the mathematical expression of the so-called *real translation theorem*. The corresponding *complex translation theorem*

$$\bar{f}(s + a) = \mathscr{L} f(t)\exp(-at) \quad \text{(A3-7)}$$

states that translation in the transform plane by the amount a is tantamount to multiplication of the function f(t) by the factor $\exp(-at)$ in the time domain.

A3.1.4 Real Differentiation

Differentiation of a function in the time domain (*real differentiation*) corresponds in the complex plane to multiplication of its transform by the transform variable. We need not be concerned here with *complex differentiation*, i.e. with differentiation with respect to s in the transform plane. For real differentiation

$$\mathscr{L}\frac{d\,f(t)}{dt} = \mathscr{L}\,f'(t) = s\bar{f}(s) - f(0) \tag{A3-8}$$

where f(0) is the *initial condition* needed to integrate f′(t). If f(t) is discontinuous at t = 0 but the limit as t → 0 exists, then

$$f(0) = \lim_{t\to 0} f(t) \;. \tag{A3-9}$$

The inverse transform of $s\bar{f}(s)$ is given by

$$\mathscr{L}^{-1} s\bar{f}(s) = \frac{d\,f(t)}{dt} + f(0)\delta(t) \;. \tag{A3-10}$$

Equation (A3-8) and (A3-10) can be generalized to the transforms of higher derivatives and their inverses. Thus, for the second derivative

$$\mathscr{L}\frac{d^2f(t)}{dt^2} = s^2\bar{f}(s) - s\,f(0) - f'(0) \tag{A3-11}$$

and

$$\mathscr{L}^{-1} s^2\bar{f}(s) = \frac{d^2f(t)}{dt^2} + f(0)\delta'(t) + f'(0)\delta(t) \;. \tag{A3-12}$$

If f(t) is discontinuous at t = 0, but the limits exist, f(0) and f′(0) must again be interpreted as the limits of f(t) and f′(t), respectively, as t → 0. If f(t) and/or its derivatives are discontinuous elsewhere in the time domain, suitable modifications can be introduced.

Generalization to still higher derivatives is straight-forward. In this book they only occur for *quiescent* (*zero*) *initial conditions* for which

$$\mathscr{L}\frac{d^nf(t)}{dt^n} = s^n\bar{f}(s) \;. \tag{A3-13}$$

A3.1.5 Real Integration

Integration of a function between the limits 0 and t corresponds to division of its transform by the transform variable. Thus*

$$\mathscr{L}\int_0^t f(u)du = \frac{\bar{f}(s)}{s} \;. \tag{A3-14}$$

Repeated integration entails repeated division, e.g.,

---

* If desired, the transform of an indefinite integral can be written as

$$\mathscr{L}\int f(u)du = \frac{\bar{f}(s)}{s} + \frac{f^{-1}(0)}{s}$$

where $f^{-1}(0)$ is shorthand for the value of the indefinite integral as t → 0. This value represents the *initial condition.*

$$\mathscr{L}\int_0^t\int_0^v f(u)du\,dv = \frac{\bar{f}(s)}{s^2}\ . \tag{A3-15}$$

Integration in the transform plane (*complex integration*) does not need to concern us here. (But see Problem 4.1-8).

A3.1.6 Real Convolution

Multiplication of two transforms in the transform plane corresponds to *convolution* in the time domain. We have

$$\bar{f}_1(s)\bar{f}_2(s) = \begin{cases} \mathscr{L}\int_0^t f_1(t-u)f_2(u)du \\ \\ \mathscr{L}\int_0^t f_1(u)f_2(t-u)du\ . \end{cases} \tag{A3-16}$$

The two forms of the convolution integral exist because the operation of convolution is commutative. Their equivalence is easily proved by a change of variable.

The inverse transform of $s\bar{f}_1(s)\bar{f}_2(s)$ has altogether four forms because each of the two commutative forms of $\mathscr{L}^{-1}\bar{f}_1(s)\bar{f}_2(s)$ have again two equivalents according to whether multiplication by s (which is distributive in the complex plane) is deemed to imply differentiation of $f_1(t)$ or of $f_2(t)$ in the time domain. Thus

$$s\bar{f}_1(s)\bar{f}_2(s) = \begin{cases} \mathscr{L}\left[f_1(0)f_2(t) + \int_0^t \frac{d\,f_1(u)}{du}f_2(t-u)du\right] \\ \mathscr{L}\left[f_1(0)f_2(t) + \int_0^t \frac{d\,f_1(t-u)}{d(t-u)}f_2(u)du\right] \\ \mathscr{L}\left[f_1(t)f_2(0) + \int_0^t f_1(t-u)\frac{d\,f_2(u)}{du}du\right] \\ \mathscr{L}\left[f_1(t)f_2(0) + \int_0^t f_1(u)\frac{d\,f_2(t-u)}{d(t-u)}du\right]\ . \end{cases} \tag{A3-17}$$

We note that the double integral, Eq. (A3-15), can be expressed as a convolution integral in two equivalent forms. We have

$$\int_0^t\int_0^v f(u)du\,dv = \int_0^t u\,f(t-u)du = \int_0^t (t-u)f(u)du \tag{A3-18}$$

so that the inverse transform $\mathscr{L}^{-1}\bar{f}(s)/s^2$, has these three equivalent forms.

Complex convolution will not be required (but see the end of Sect. 9.0).

A3.1.7 Limit Values

A property of the Laplace transformation which is important to applications in this book is embodied in the *limit value theorems*. The *initial* value theorem states that, if the limit exists,

$$\lim_{t \to 0} f(t) = \lim_{s \to \infty} s\bar{f}(s) = f(0) \ . \tag{A3-19}$$

The *final value* theorem states analogously that, if the limit exists,

$$\lim_{t \to \infty} f(t) = \lim_{s \to 0} s\bar{f}(s) = f(\infty) \ . \tag{A3-20}$$

A3.1.8 Cut-off. The Finite Laplace Transform

Let a function f(t) be restricted to values of its argument in the interval $0 \le t \le t'$. We then have

$$f_{t'}(t) = f(t)h_{t'}(t) = f(t)[h(t) - h(t - t')] \tag{A3-21}$$

where $h_{t'}(t)$ is the unit gate function introduced in Sect. A2.3. The Laplace transform of the restricted function becomes

$$\mathscr{L} f(t)h_{t'}(t) = \int_0^\infty f(t)\exp(-ts)dt - \int_{t'}^\infty f(t)\exp(-ts)dt \tag{A3-22}$$

or

$$\bar{f}_{t'}(s) = \int_0^{t'} f(t)\exp(-ts)dt \ . \tag{A3-23}$$

We shall call the transform of the *cut-off function*, $f_{t'}(t)$, the *finite Laplace transform.* It is useful in the treatment of *periodic functions* (see Sect. A3.1.9).

A3.1.9 Repetition

Consider the function

$$\begin{aligned} f(t) &= f_r(t)h(t) + f_r(t - t')h(t - t') + \cdots + f_r(t - nt')h(t - nt') \\ &= \sum_{n=0}^{n=\infty} f_r(t - nt')h(t - nt') \end{aligned} \tag{A3-24}$$

where $f_r(t)$ is called the *repeat function.* The function f(t) consists of the superposition, at regular intervals of duration t', of the functions $f_r(t - nt')$. If the Laplace transform of $f_r(t)$ exists, the transform of f(t) is

$$\bar{f}(s) = \bar{f}_r(s) \sum_{n=0}^{n=\infty} \exp(-nt's) = \frac{\bar{f}_r(s)}{1 - \exp(-t's)} \ . \tag{A3-25}$$

Equations (A3-25) represent the *repetition theorem* which states that the transform of a function consisting of the superposition of functions $f_r(t - nt')$is obtained by dividing the transform of $f_r(t)$, if it exists, by $1 - \exp(-t's)$. Thus, it is only necessary to know $\bar{f}_r(s)$ to obtain $\bar{f}(s)$.

If the repeat function is restricted to the interval $0 \leq t \leq T$, then

$$f(t) = f(t + nT) \; , \tag{A3-26}$$

and f(t) is said to be *periodic* with *period* T. In this case the repetition theorem takes the form

$$\bar{f}(s) = \bar{f}_T(s) \sum_{n=0}^{n=\infty} \exp(-nTs) = \frac{\bar{f}_T(s)}{1 - \exp(-Ts)} \tag{A3-27}$$

and is called the *periodicity theorem*. The transform of the repeat function $f_T(t)$ is then the *finite Laplace* transform introduced in Sect. A3.1.8.

A3.2 Inversion

The inverse Laplace transformation is linear just as the direct transformation is. It takes the form

$$f(t) = \frac{1}{2\pi j} \int_{c-j\infty}^{c+j\infty} \bar{f}(s)\exp(st)ds = \mathscr{L}^{-1}[\bar{f}(s); t] \tag{A3-28}$$

where c denotes an abscissa located to the right of any singularity that $\bar{f}(s)$ may possess. If $\bar{f}$ does not contain branch points or essential singularities and f(t) satisfies certain growth conditions which are always fulfilled in the context of this book, then the integration along the ordinate at c from $-j\infty$ to $j\infty$ can be replaced by any contour enclosing the poles of $\bar{f}(s)$. We have

$$f(t) = \frac{1}{2\pi j} \oint \bar{f}(s)\exp(st)ds = \mathscr{L}^{-1}[\bar{f}(s); t] \tag{A3-29}$$

and it turns out that f(t) is then the sum of the residues of $\bar{f}(s)\exp(st)$.

A3.2.1 The Residue Theorem

If the only singularities which $\bar{f}(s)$ possesses are a finite number of poles, it can be expressed as a rational algebraic fraction. Making use of the calculus of residues, the inversion can then be accomplished conveniently with the aid of *Cauchy's residue theorem* [1, 4] (see also Problem 4.1-1). Let

$$\bar{f}(s) = \frac{a(s)}{b(s)} = \frac{a_p s^p + a_{p-1} s^{p-1} + \cdots + a_1 s + a_0}{s^q + b_{q-1} s^{q-1} + \cdots + b_1 s + b_0} \tag{A3-30}$$

in which the a's and b's are real constants and p and q are nonnegative integers. Thus, a(s) and b(s) are polynomials in s. Without loss of generality we may assume that $p < q$. If it is not, a(s) can always be divided by b(s) until the remainder has this property. Any constant and/or terms in powers of s can be inverted separately, invoking the linearity property. We further assume that the zeros of b(s) are distinct. This means that the poles are simple poles. A double pole at the origin is the only multiple pole likely to arise in the applications discussed in this book. The residue theorem* states that f(t) is the sum of the residues of $\bar{f}(s)\exp(st)$ which are associated with the poles of $\bar{f}(s)$. The latter are given by the zeros of b(s), i.e. by the roots of the equation $b(s) = 0$. Thus, the residue theorem may be expressed here in the form

$$f(t) = \sum_k \mathrm{Res}_k[\bar{f}(s)\exp(st)] \tag{A3-31}$$

where $\mathrm{Res}_k[\ ]$ may be considered to be an operator requiring the determination of the residue associated with the pole at $s = s_k$ of the function displayed between the brackets. The residue is obtained [1, 4] from

$$\mathrm{Res}_k = \lim_{s \to s_k} (s - s_k)\bar{f}(s)\exp(st) \tag{A3-32}$$

if the pole at $s = s_k$ is a pole of order one, and from

$$\mathrm{Res}_l = \frac{1}{(n-1)!} \lim_{s \to s_k} \frac{d^n}{ds^{n-1}} (s - s_k)^n \bar{f}(s)\exp(st) \tag{A3-33}$$

if it is a pole of order n.

Equations (A3-31), (A3-32), and (A3-33) allow us to study the behavior associated in the time domain with poles located at different points in the complex transform plane. Several such poles are indicated by crosses in Fig. A3-1. The

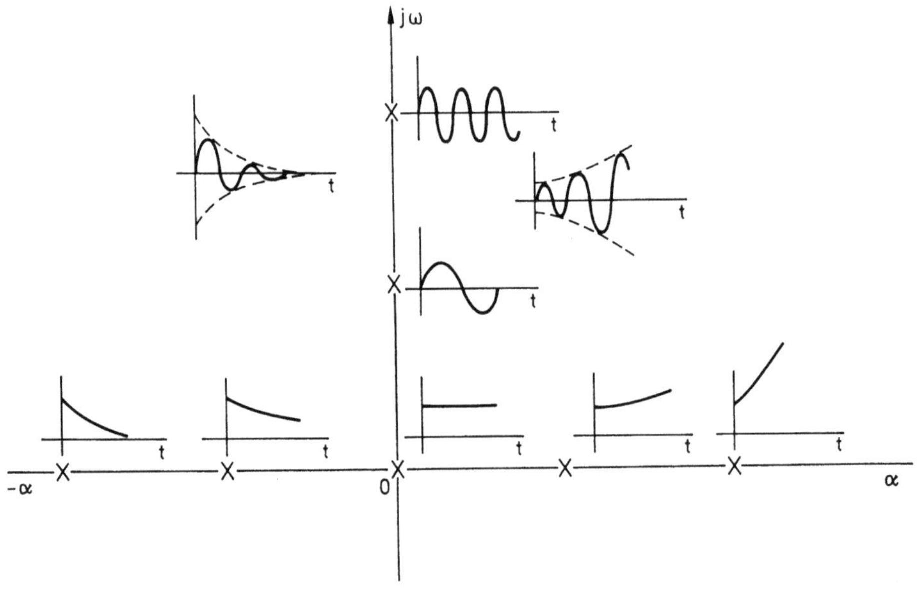

Fig. A3-1. Behavior in the time domain at different locations in the complex domain

* The *(Heaviside) expansion theorem* [1, 2, 3] is a special form of the residue theorem. We find it more convenient to use the residue theorem directly.

corresponding behavior in the time domain is illustrated by small inserts placed as close as feasible to the poles.

Let us first consider simple poles lying on the real axis of the complex plane excluding the origin. For such poles $b(s) = s \pm \alpha$, where $\alpha > 0$. Then $s_k = \mp\alpha$. For simplicity, and without loss of generality, we may let $a(s) = 1$. By Eq. (A3-32), then,

$$\mathrm{Res}_k = \lim_{s \to s_k} (s - s_k)\frac{\exp(st)}{s \pm \alpha} = \exp(s_k t) \tag{A3-34}$$

and, by Eq. (A3-31)

$$f(t) = \exp(\mp\alpha t) \ , \qquad t \geq 0 \ , \tag{A3-35}$$

i.e. the behavior in the time domain corresponding to a simple pole along the real axis of the complex plane is represented by an exponential. If the pole lies on the negative real axis, so that $s_k = -\alpha$, the exponential is a decreasing one, and it decreases the faster the farther the pole is located from the origin. This is indicated schematically in Fig. A3-1. If, on the other hand, the pole lies on the positive real axis, $s_k = \alpha$, and the exponential is an increasing one. For a single pole at the origin, $b(s) = s$, so $s_k = 0$, and this gives

$$f(t) = h(t) \tag{A3-36}$$

since t is restricted to positive values (cf. Sect. A2.2).

To examine the behavior of simple poles not on the real axis, we note that such poles must occur in conjugate pairs. Hence we have $b(s) = (s \pm \alpha)^2 + \beta^2$, $s_1 = \mp\alpha + j\beta$, and $s_2 = \mp\alpha - j\beta$. We find

$$\mathrm{Res}_1 = \frac{\exp(s_1 t)}{s_1 - s_2} \qquad \text{and} \qquad \mathrm{Res}_2 = \frac{\exp(s_2 t)}{s_2 - s_1} \tag{A3-37}$$

so that Eq. (A3-31) gives

$$f(t) = \beta^{-1} \exp(\mp\alpha t)\sin \beta t \ , \qquad t > 0 \tag{A3-38}$$

where $\beta > 0$. For poles lying away from the imaginary axis ($\alpha \neq 0$) the behavior in the negative half-plane ($s = -\alpha$), and by exponentially increasing oscillations if it lies in the positive half-plane ($s = \alpha$). For a pole located on the imaginary axis lies in the positive half-plane ($s = a$). For a pole located on the imaginary axis ($\alpha = 0$), but not at the origin, the behavior in the time domain consists of undamped oscillations. The behavior at the conjugate poles in the lower half of the complex plane is not shown in Fig. A3-1 because negative frequencies do not occur experimentally.

To examine the behavior of a double pole we consider $b(s) = (s \pm \alpha)^2$. We could apply Eqs. (A3-30) and (A3-32) but the result is obtained more simply from Eq. (A3-38) by letting $\beta \to 0$. This gives

$$f(t) = t \exp(\mp\alpha t) \ , \qquad t \geq 0 \tag{A3-39}$$

which shows that the behavior in the time domain is the product of a linear factor and an exponential. Setting $\alpha = 0$ in Eq. (A3-39) gives the behavior associated with a double pole at the origin as

$$f(t) = p(t) = t\,h(t)\ . \tag{A3-40}$$

It is clear that poles lying in the right half of the transform plane lead to functions in the time domain which increase in magnitude. The poles of transforms describing passive systems thus cannot be located in the right half-plane.

A3.2.2 Inversion on the Real Axis

When s is real, the transform can be inverted on the real axis of the complex plane by use of the Post-Widder formula [5]. Letting $s = x$, we then have

$$f(t) = \lim_{k\to\infty} \frac{(-1)^k}{k!} x^{k+1} \frac{d^k}{dk^x} \bar{f}(x)|_{x=k/t}\ . \tag{A3-41}$$

We shall find it convenient (cf. Sect. 4.2) to bring Eq. (A3-41) into another form. It is easily shown that

$$x^k \frac{d^k}{dx^k} = D_x(k) \tag{A3-42}$$

where

$$D_x^{(k)} = D_x(D_x - 1)(D_x - 2)\ldots(D_x - k + 1) \tag{A3-43}$$

is the *factorial function* used extensively in finite difference calculus [6] in which, here,

$$D_x^k = \frac{d^k}{d\ \ln^k x}\ . \tag{A3-44}$$

Substitution of Eq. (A3-42) into Eq. (A3-41) gives

$$f(t) = \lim_{k\to\infty} \frac{(-1)^k x}{k!} D_x^{(k)} \bar{f}(x)|_{x=k/t} \tag{A3-45}$$

which yields f(t) in terms of the logarithmic derivatives of $\bar{f}(x)$. Equation (A3-45) is put to use in Sect. 4.2.1.

A3.3 Laplace Transform Pairs

This section contains the Laplace Transform pairs occuring in this book where they are identified as LTP followed by the number under which they are listed below. Extensive tabulations can be found in a number of publications (see, e.g., Refs. [1], [3], [5], or [7]). Table A3-1 contains the pairs involving the special functions discussed in Sect. A.2. Table A3-2 lists other pairs. It is clear that both members of a pair may be multiplied by the same arbitrary constant without affecting the transformation.

Table A3-1. Laplace transform pairs involving the special functions

Function	f(t)	$\bar{\bar{f}}(s)$	No.
Unit impulse	$\delta(t)$	1	(1)
Multiplet unit impulse	$\delta^{(n)}(t)$	s^n	(2)
Shifted unit impulse	$\delta(t-t')$	$\exp(-t's)$	(3)
Unit step	$h(t)$	s^{-1}	(4)
Shifted unit step	$h(t-t')$	$s^{-1}\exp(-t's)$	(5)
Slope	$p(t)$	s^{-2}	(6)
Shifted slope	$p(t-t')$	$s^{-2}\exp(-t's)$	(7)
Unit gate	$h_{t'}(t)$	$[1-\exp(-t's)]/s$	(8)
Unit ramp	$r_{t'}(t)$	$[1-\exp(-t's)]/t's^2$	(9)

Table A3-2. Other Laplace transform pairs

f(t)	$\bar{f}(s)$	No.
$\exp(-t/\tau)$	$\frac{\tau}{1+\tau s}$	(10)
$1-\exp(-t/\tau)$	$\frac{1}{s(1+\tau s)}$	(11)
t^{n-1}	$\frac{\Gamma(n)}{s^n}, \quad n>0$	(12)
$\exp(j\omega t)$	$\frac{1}{s-j\omega}$	(13)
$\sin \omega t$	$\frac{\omega}{s^2+\omega^2}$	(14)
$\cos \omega t$	$\frac{s}{s^2+\omega^2}$	(15)
$[1-\exp(-t/\tau)]/t$	$\ln(1+1/\tau s)$	(16)
$\delta(t)-\tau^{-1}\exp(-t/\tau)$	$\frac{\tau s}{1+\tau s}$	(17)
$t-\tau[1-\exp(-t/\tau)]$	$\frac{1}{s^2(1+\tau s)}$	(18)
$[1-\exp(-t/\tau)]^2$	$\frac{2}{s(1+\tau s)(2+\tau s)}$	(19)
$\frac{\exp(-t/\tau_1)-\exp(-t/\tau_2)}{\tau_1-\tau_2}$	$\frac{1}{(1+\tau_1 s)(1+\tau_2 s)}$	(20)
$\frac{\omega^2\tau^2}{1+\omega^2\tau^2}\sin\omega t+\frac{\omega\tau}{1+\omega^2\tau^2}[\cos\omega t-\exp(-t/\tau)]$	$\frac{\omega\tau s}{(1+\tau s)(s^2+\omega^2)}$	(21)
$\frac{\omega^2\tau^2}{1+\omega^2\tau^2}\cos\omega t-\frac{\omega\tau}{1+\omega^2\tau^2}[\sin\omega t-\exp(-t/\tau)]$	$\frac{\tau s^2}{(1+\tau s)(s^2+\omega^2)}$	(22)
$t\exp(-t/\tau)$	$\frac{\tau^2}{(1+\tau s)^2}$	(23)
$(t+a)^{-m}$	$s^{m-1}\Gamma(1-m;as)\exp(as)$	(24)

(Continued)

Table A3-2 (*continued*)

f(t)	$\bar{f}(s)$	No.
$\mathrm{Erfc}(a/2\sqrt{t})$	$s^{-1}\exp(-a\sqrt{s})$	(25)
$1 - \exp(a^2t)\mathrm{Erfc}(a\sqrt{t})$	$(a/s)/(\sqrt{s} + a)$	(26)
$(1/\sqrt{a})\mathrm{Erf}(\sqrt{at})$	$1/s\sqrt{s + a}$	(27)
$\mathrm{Ei}(t/a)$	$s^{-1}\ln(1 + as)$	(28)
$\gamma(m; t/a)t^{-m}$	$s^{m-1}B(m, 1 - m; 1/as)$	(29)
$\Gamma^{-1}(m)t^{m-1}\exp(-at)$	$(s + a)^{-m}$	(30)
$(1/\sqrt{4\pi a t^3})\exp(-1/4at)$	$\exp(-\sqrt{s/a})$	(31)
$(\pi t)^{-1/2} - a\exp(a^2t)\mathrm{Erfc}(a\sqrt{t})$	$1/(a + \sqrt{s})$	(32)

Erf(x) and Erfc(x) are the normal and the complementary normal error integrals, see Eqs. (6.1-106) and (6.1-108); Ei(x) is the exponential integral, see Eq. (6.3-12); $\Gamma(m)$ is the gamma function, $\gamma(m; x)$ is the incomplete, and $\Gamma(m; x)$ is the complementary incomplete gamma function, see Eqs. (6.3-34) and (6.3-99). $B(p, q; x)$ is the incomplete beta function, see Eq. (6.3-32).

The transform of the *shifted unit gate function*, $h_{t''}(t - t')$ and the *shifted unit ramp function*, $r_{t''}(t - t')$, are easily derived through the translation theorem, Eq. (A3-6).

A3.4 The s-Multiplied Laplace Transform (Carson Transform)

We call

$$\check{f}(s) = s\int_0^\infty f(t)\exp(-ts)dt \tag{A3-46}$$

the *s-multiplied Laplace transform* of f(t). Although it is occasionally referred to in the main body of this book, its properties need not be detailed here since they can always be obtained through the corresponding Laplace transform.

The early development of a transformation calculus based on Eq. (A3-46) is due to Cauchy. Heaviside, who may have discovered it independently, used it in his work on electrical systems and it is therefore sometimes referred to as the *Heaviside calculus*. The transform is occasionally called the *Carson transform* although historically the name *Cauchy-Heaviside transform* would appear to be more appropriate.

A4 The Fourier Transform

We call

$$f^*(\omega) = \int_0^\infty f(t)\exp(-j\omega t)dt = \mathscr{F}[f(t); j\omega] \tag{A4-1}$$

the *Fourier transform* of f(t). When it is not necessary to designate the transform variable explicitly, we shall simply write $\mathscr{F} f(t)$ for $\mathscr{F}[f(t); j\omega]$. The integral converges and, hence, the transform exists, if the integral $\int_0^\infty |f(t)|dt$ exists.

Separating the real and imaginary parts of $\exp(-j\omega t)$ we may further define

$$\mathscr{F}[f(t); j\omega] = \mathscr{F}_c[f(t); \omega] - j\,\mathscr{F}_s[f(t); \omega] \tag{A4-2}$$

where

$$f'(\omega) = \int_0^\infty f(t)\cos \omega t \cdot dt = \mathscr{F}_c[f(t); \omega] \tag{A4-3}$$

is the *Fourier cosine transform*, and

$$f''(\omega) = \int_0^\infty f(t)\sin \omega t \cdot dt = \mathscr{F}_s[f(t); \omega] \tag{A4-4}$$

is the *Fourier sine transform* of f(t). These transforms directly yield the real and imaginary parts of the image function, respectively. The image function obtained by the exponential, cosine, or sine Fourier transformation is said to be in the *frequency domain.*

A4.1 Properties

The existence conditions for the Fourier transform are more stringent than those for the Laplace transform. A Laplace transform has a corresponding Fourier transform only if it has no singularity anywhere on the imaginary axis (including the origin) or to the right of it because otherwise the integral $\int_0^\infty |f(t)|dt$ does not exist. In fact, the Laplace transform may be considered to be a Fourier transform of a function which has been multiplied by the *integrating factor* $\exp(-\alpha t)$ (where $\alpha > 0$) which insures convergence of the integral.

A4.2 Inversion

Since, conversely, the Fourier transform may be regarded as a Laplace transform in which the transform variable is purely imaginary, we may expect that the formulation in which $j\omega$ has the positive sign is also a Fourier transform. This is indeed true. For the two-sided transform (see the footnote at the beginning of Sect. A3), letting

$$f_2^*(\omega) = \frac{1}{\sqrt{2\pi}} \int_{-\infty}^{\infty} f(t)\exp(-j\omega t)dt \tag{A4-5}$$

we also have

$$f_2(t) = \frac{1}{\sqrt{2\pi}} \int_{-\infty}^{\infty} f_2^*(\omega)\exp(j\omega t)d\omega \ . \tag{A4-6}$$

In this so-called *symmetric form* of the Fourier transformation $f_2(t)$ and $f_2^*(\omega)$ are each other's transforms and it is a matter of indifference which is regarded the transform and which its inverse. The normalization constants in front of the integrals can take any form as long as their product is $1/2\pi$.

Because of our use of the one-sided Laplace transform, we prefer to use the Fourier transform in the closely related form given by Eq. (A4-1). This is really the same as the two-sided transform since, in our usage (see Sect. 2.1.1), $f(t) = 0$ when $t < 0$. Thus, our *inverse Fourier transform* remains two-sided and takes the form

$$f(t)(=)\frac{1}{2\pi}\int_{-\infty}^{\infty} f^*(\omega)\exp(j\omega t)d\omega(=)\mathscr{F}^{-1}[f^*(\omega);t] \tag{A4-7}$$

This should be compared with the Laplace inversion formula, Eq. (A3-28). The symbol $(=)$ means "equals except at points of discontinuity" (cf. Sect. A3.1.2).

The inverse Fourier transform exists only when the integral $\int_{-\infty}^{\infty}|f^*(\omega)|d\omega$ exists. We then have

$$\mathscr{F}^{-1}[f^*(\omega);t] = \mathscr{F}_c^{-1}[f^*(\omega);t] + j\,\mathscr{F}_s^{-1}[f^*(\omega);t] \tag{A4-8}$$

where

$$\mathscr{F}_c^{-1}[f^*(\omega);t] = \frac{1}{2\pi}\int_{-\infty}^{\infty} f^*(\omega)\cos\omega t\cdot d\omega \tag{A4-9}$$

is the *inverse Fourier cosine transform*, and

$$\mathscr{F}_s^{-1}[f^*(\omega);t] = \frac{1}{2\pi}\int_{-\infty}^{\infty} f^*(\omega)\sin\omega t\cdot d\omega \tag{A4-10}$$

is the *inverse Fourier sine transform*.

A special situation arises in the applications considered in this book. If the corresponding Laplace transforms, $\bar{f}(s)$, is a rational algebraic fraction (cf. Sect. 2.1.1), then, in

$$f^*(\omega) = f'(\omega) - j\,f''(\omega)\ , \tag{A4-11}$$

$f'(\omega)$, the real part of $f^*(\omega)$, is an *even*, and $f''(\omega)$, the imaginary part, is an *odd function* of its argument. We then have

$$\mathscr{F}^{-1}f^*(\omega) = \frac{1}{\pi}\int_0^{\infty} f'(\omega)\cos\omega t\cdot d\omega + \frac{1}{\pi}\int_0^{\infty} f''(\omega)\sin\omega t\cdot d\omega \tag{A4-12}$$

because only the even products survive the integration over symmetric limits, and the value of these integrals is twice the value of the integrals from 0 to ∞. But $\mathscr{F}^{-1}f^*(\omega) = f(t)$ and if $f(t)$ is zero whenever $t < 0$ we have

$$0 = \frac{1}{\pi}\int_0^{\infty} f'(\omega)\cos\omega t\cdot d\omega + \frac{1}{\pi}\int_0^{\infty} f''(\omega)\sin\omega t\cdot d\omega\ . \tag{A4-13}$$

Now replace t by $-t$. Then for $t > 0$ we have

$$\int_0^{\infty} f'(\omega)\cos\omega t\,d\omega = \int_0^{\infty} f''(\omega)\sin\omega t\,d\omega \tag{A4-14}$$

and it follows that

$$\mathscr{F}^{-1}f^*(\omega) = \mathscr{F}_c^{-1}f'(\omega) = \mathscr{F}_s^{-1}f''(\omega) \tag{A4-15}$$

where

$$\mathscr{F}_c^{-1}f'(\omega) = \frac{2}{\pi}\int_0^\infty f'(\omega)\cos\omega t\cdot d\omega = f(t) \tag{A4-16}$$

and

$$\mathscr{F}_s^{-1}f''(\omega) = \frac{2}{\pi}\int_0^\infty f''(\omega)\sin\omega t\cdot d\omega = f(t)\ . \tag{A4-17}$$

Thus, given a complex function whose real part is an even, while its imaginary part is an odd function of its argument, the cosine transform of the real part equals the sine transform of the imaginary part, and either is equal to the Fourier transform itself. Consequently, the source function, $f(t)$, may be recovered either from the complex image function, $f^*(\omega)$, or from its real part, $f'(\omega)$, or from its imaginary part, $f''(\omega)$.

A4.3 Generalized Fourier Transforms

The Fourier transform of many functions, including the standard excitation functions $\delta(t)$, $h(t)$, t[or $p(t)$], $\exp j\omega t$ (or $\sin\omega t$; $\cos\omega t$) does not exist in the sense of Eq. (A4-1). However, we may define a *generalized Fourier transform*

$$f^*(\omega) = \lim_{\alpha\to 0}\int_0^\infty f(t)\exp[-(\alpha + j\omega)t]dt \tag{A4-18}$$

by taking the limit as the real part of the complex transform variable, $s = \alpha - j\omega$, vanishes. Since the integral on the right of Eq. (A4-18) is the Laplace transform, $\bar{f}(s)$, the equation may be written in the form

$$f^*(\omega) = \bar{f}(s)|_{s=j\omega}\ . \tag{A4-19}$$

Thus, the (generalized) Fourier transform is obtained by replacing the transform variable, s, of the Laplace transform, by its imaginary part, $j\omega$. Much use is being made in this book of this relation [see, in particular, Eqs. (2.5-13) and (2.5-17)].

The *generalized cosine* and *sine transforms* similarly take the form

$$f'(\omega) = \lim_{\alpha\to 0}\int_0^\infty f(t)\exp(-\alpha t)\cos\omega t\cdot dt \tag{A4-20}$$

and

$$f''(\omega) = \lim_{\alpha\to 0}\int_0^\infty f(t)\exp(-\alpha t)\sin\omega t\cdot dt\ . \tag{A4-21}$$

Because of the limit process involved in obtaining the generalized Fourier transforms, the triple identity, Eq. (A4-15), may not always hold in its entirety. It frequently reduces to either

Table A4-1. Pairs of generalized Fourier transforms

f(t)	f*(ω)	f′(ω)	f″(ω)	No.
$\delta(t)$	1	1	0	(1)
$h(t)$	$1/j\omega$	0	$1/\omega$	(2)
t	$-1/\omega^2$	$-1/\omega^2$	0	(3)
t^2	$-2/j\omega^3$	0	$-2/\omega^3$	(4)
$\exp(-t/\tau)$	$\dfrac{\tau}{1+j\omega\tau}$	$\dfrac{\tau}{1+\omega^2\tau^2}$	$\dfrac{\omega\tau^2}{1+x^2\tau^2}$	(5)
$\sin\beta t$	$\dfrac{\beta}{\beta^2-\omega^2}$	$\dfrac{\beta}{\beta^2-\omega^2}$	0	(6)
$\cos\beta t$	$\dfrac{j\omega}{\beta^2-\omega^2}$	0	$\dfrac{\omega}{\omega^2-\beta^2}$	(7)

$$\mathscr{F}^{-1}f^*(\omega) = \mathscr{F}_c^{-1}f'(\omega) \tag{A4-22}$$

or

$$\mathscr{F}^{-1}f^*(\omega) = \mathscr{F}_s^{-1}f''(\omega)\ . \tag{A4-23}$$

Thus, in almost all of the relations assembled in Table A4-1 either the sine or the cosine transform vanishes.

A4.4 Generalized Fourier Transform Pairs

Only a few generalized Fourier transform pairs are required and these occur in Chapter 8. They are readily obtained from the corresponding Laplace transforms using Eq. (A4-19) and then separating the real and imaginary parts. Thus, in Table A4-1,

$$f^*(\omega) = \bar{f}(j\omega) \tag{A4-24}$$

$$f'(\omega) = \mathscr{R}e\, f^*(\omega) \tag{A4-25}$$

and

$$f''(\omega) = -\mathscr{I}m\, f^*(\omega)\ . \tag{A4-26}$$

We note that $f'(\omega)$ represents the cosine, and $f''(\omega)$ the sine transform. Where they exist, the pairs are each other's transforms. Thus, if, e.g., $\omega\tau^2/(1+\omega^2\tau^2)$ is the (generalized) sine transform of $\exp(-t/\tau)$, then $\exp(-t/\tau)$ is the (generalized) sine transform of $\omega\tau^2/(1+\omega^2\tau^2)$. The transform pairs are referred to in the text by FTP followed by the number of the pair as with the Laplace transform pairs.

A5 The Stieltjes Transform

We call

$$\tilde{f}(s) = \int_0^\infty \frac{f(\sigma)}{s+\sigma}\,d\sigma = \mathscr{S}[f(\sigma); s] \tag{A5-1}$$

the *Stieltjes transform* of $f(\sigma)$. When it is not necessary to designate the transform variable explicity, we shall write $\mathscr{S}\, f(\sigma)$ for $\mathscr{S}[f(\sigma); s]$. The transform variable, s, is generally complex, i.e.,

$$s = -\alpha \pm j\beta \ , \tag{A5-2}$$

although, just as in the case of the Laplace transform, it may also be real. The integral converges, and, hence, the transform exists, if $\int_0^\infty |f(\sigma)/(s+\sigma)|\, d\sigma$ exists. The case when s is purely imaginary is discussed in Sect. A5.3.

A5.1 Properties

The Stieltjes transform is an iterated Laplace transform. Let

$$\bar{f}(t) = \int_0^\infty f(\sigma)\exp(-\sigma t)d\sigma \tag{A5-3}$$

be the Laplace transform of $f(\sigma)$. Now carry out a second Laplace transformation to obtain

$$\tilde{f}(s) = \int_0^\infty \bar{f}(t)\exp(-ts)dt \ . \tag{A5-4}$$

Substituting Eq. (A5-3) into Eq. (A5-4) and interchanging the order of integration gives

$$\tilde{f}(s) = \int_0^\infty f(\sigma) \int_0^\infty \exp[-(s+\sigma)t]dt\, d\sigma \ . \tag{A5-5}$$

Equation (A5-1) follows immediately upon carrying out the integration with respect to t.

A5.2 Inversion

Just as the Laplace transform, the Stieltjes transform can be inverted in the complex plane, or, when s is real, on the real axis. The inversion of the Stieltjes transform in the complex plane is simpler than the inversion of the Laplace transform.

A5.2.1 The Jump

The only singularities of $\tilde{f}(s)$ are branch points at $s = 0$ and $s = \infty$ equivalent to a cut along the entire real axis of the complex plane including the origin (cf. Fig. 4.1-1). The value of $\tilde{f}(s)$ on the negative real axis will therefore depend on whether the cut is approached from above or below, i.e. whether we take the limit

$$\lim_{\beta\to 0} \tilde{f}(-\alpha + j\beta) = \lim_{\beta\to 0,\, \beta>0} \tilde{f}(-\alpha + j\beta) \tag{A5-6}$$

or the limit

$$\lim_{\beta \to 0} \tilde{f}(-\alpha - j\beta) = \lim_{\beta \to 0, \beta < 0} \tilde{f}(-\alpha + j\beta) \ . \tag{A5-7}$$

The difference between the two limits (i.e. the *jump* between the two Riemann surfaces joined at the cut) can be used to invert the transform. Substituting Eq. (A5-2) into Eq. (A5-1) we have

$$\tilde{f}(-\alpha \pm j\beta) = \int_0^\infty \frac{f(\sigma)}{\sigma - \alpha \pm j\beta} d\sigma \ . \tag{A5-8}$$

Separating the real and imaginary parts gives

$$\tilde{f}(-\alpha \pm j\beta) = \int_0^\infty \frac{(\sigma - \alpha) f(\sigma)}{(\sigma - \alpha)^2 + \beta^2} d\sigma \mp j \int_0^\infty \frac{f(\sigma)\beta}{(\sigma - \alpha)^2 + \beta^2} d\sigma \ . \tag{A5-9}$$

By Eq. (A2-10),

$$\lim_{\beta \to 0} \frac{\beta}{(\sigma - \alpha)^2 + \beta^2} = \pi\, \delta(\sigma - \alpha) \tag{A5-10}$$

where $\delta(x)$ is the delta function. Hence, taking the limit gives

$$\lim_{\beta \to 0, \beta > 0} \tilde{f}(-\alpha + j\beta) = \fint_0^\infty \frac{f(\sigma)}{\sigma - \alpha} d\sigma - j\pi\, f(\alpha) \tag{A5-11}$$

and

$$\lim_{\beta \to 0, \beta < 0} \tilde{f}(-\alpha - j\beta) = \fint_0^\infty \frac{f(\sigma)}{\sigma - \alpha} d\sigma + j\pi\, f(\alpha) \ . \tag{A5-12}$$

Subtracting Eq. (A5-11) from Eq. (A5-12) yields

$$f(\alpha) = \frac{1}{2\pi j} \lim_{\beta \to 0, \beta < 0} \tilde{f}(-\alpha - j\beta) - \lim_{\beta \to 0, \beta > 0} \tilde{f}(-\alpha + j\beta) \ . \tag{A5-13}$$

But this is valid for all values of $\alpha > 0$. Hence, we may write

$$f(\sigma) = \frac{1}{2\pi j} \left(\lim_{\beta \to 0, \beta < 0} \tilde{f}(-\alpha - j\beta) - \lim_{\beta \to 0, \beta > 0} \tilde{f}(-\alpha + j\beta) \right)_{\alpha = \sigma} = \mathscr{S}^{-1}[\tilde{f}(s); \sigma] \tag{A5-14}$$

where $\sigma > 0$ also. Eq. (A5-14) can be written in a slightly simpler form by condensing Eqs. (A5-11) and (A5-12) into

$$f(\sigma) = \mp \frac{1}{\pi} \lim_{\beta \to 0} \mathscr{I}m\, \tilde{f}(-\alpha \pm j\beta)|_{\alpha = \sigma} = \mathscr{S}^{-1}[\tilde{f}(s); \sigma] \ . \tag{A5-15}$$

Equations (A5-14) and (A5-15) are the general inversion formulae of the Stieljes transform. Again, we use $\mathscr{S}^{-1}\tilde{f}(s)$ for short when no confusion is likely to arise.

We now further observe that

$$-\alpha \pm j\beta = \sqrt{\alpha^2 + \beta^2} \exp[\pm j(\pi - \arctan \beta/\alpha)] \tag{A5-16}$$

and, therefore,

$$\lim_{\beta \to 0} (-\alpha \pm j\beta) = \alpha \exp(\pm j\pi) \ . \tag{A5-17}$$

When $\tilde{f}(s)$ is *not* a rational function of s, it is admissible to "anticipate" the limiting operation. We can then write

$$f(\sigma) = \frac{1}{2\pi j} \{\tilde{f}[\alpha \exp(-j\pi)] - \tilde{f}[\alpha \exp(j\pi)]\}|_{\alpha=\sigma} \tag{A5-18}$$

or, alternatively,

$$f(\sigma) = \mp\frac{1}{\pi} \mathscr{I}m\, \tilde{f}[\sigma \exp(\pm j\pi)] \tag{A5-19}$$

where we have replaced α by σ in the argument. This can, of course, be done also in Eq. (A5-18). Thus, we simply replace the transform variable by $\sigma \exp \pm j\pi$ and compute the jump (cf. Sect. 6.3.2.2). Equations (A5-18) or (A5-19) fail when $\tilde{f}(s)$ is rational because then the limit cannot be anticipated. In fact, Eq. (A5-18) then yields zero.

We also note that the inversion equations remain valid even if the transform variable is confined to the real or the imaginary axis of the complex plane.

A5.2.2 Inversion on the Real Axis

When the transform is given on the real axis but is not available in analytic form, the inversion can be accomplished, in principle, from

$$f(\sigma) = \lim_{k \to \infty} \frac{(-1)^k}{(k-1)!\,k!} \frac{d^k}{dx^k}\left[x^{2k} \frac{d^k}{dx^k} \tilde{f}(x)\right]_{x=\sigma} \tag{A5-20}$$

or

$$f(\sigma) = \lim_{k \to \infty} \frac{(-1)^k}{(k-1)!\,(k+1)!} \frac{d^{k+1}}{dx^{k+1}}\left[x^{2k+1} \frac{d^k}{dx^k} \tilde{f}(x)\right]_{x=\sigma} \ . \tag{A5-21}$$

where we have let s = x. Equation (A5-20) is obtained from the formula of Hirschmann and Widder [8]

$$f(\sigma) = \lim_{k \to \infty} \frac{(-1)^k}{2\pi} \left(\frac{e}{k}\right)^{2k} \frac{d^k}{dx^k}\left[x^{2k} \frac{d^k}{dx^k} \tilde{f}(x)\right]_{x=\sigma} \tag{A5-22*}$$

making use of the fact that

$$\lim_{k \to \infty} \frac{(-1)^k}{2\pi} \left(\frac{e}{k}\right)^{2k} = \lim_{k \to \infty} \frac{(-1)^k}{(k-1)!\,k!} \tag{A5-23*}$$

by Stirling's approximation.* Equation (A5-21) is given by Widder [5] together with an equivalent form

* In Eqs. (A5-22) and (A5-23) e is the basis of the natural logarithm.

$$f(\sigma) = \lim_{k\to\infty} \frac{(-x)^k}{(k-1)!\,(k+1)!} \frac{d^{2k+1}}{dx^{2k+1}} [x^{k+1}\tilde{f}(x)]_{x=\sigma} \;. \tag{A5-24}$$

The parameter k in Eqs. (A5-20) through (A5-24) takes on positive integral values only.

Equations (A5-20) and (A5-21) will be used in Chap. 4 for obtaining approximations to the spectral distribution functions from the harmonic response functions. They may be brought into a more convenient form. We take Eq. (A5-20) first. Let

$$x^k \frac{d^k}{dx^k} f(x) = g(x,k) \;. \tag{A5-25}$$

We may then write

$$\frac{d^k}{dx^k}\left[x^{2k}\frac{d^k}{dx^k}\right] f(x) = \frac{d^k}{dx^k}[x^k g(x,k)] \;. \tag{A5-26}$$

Now, by Leibnitz's rule for the higher derivatives of products of functions

$$\frac{d^k}{dx^k}[x^k g(x,k)] = \sum_{m=0}^{m=k} \binom{k}{m} \left[\frac{d^{k-m}}{dx^{k-m}} x^k\right]\left[\frac{d^m}{dx^m} g(x,k)\right] \tag{A5-27}$$

where the zeroth derivative is the function itself. But

$$\frac{d^{k-m} x^k}{dx^{k-m}} = \frac{k!\, x^m}{m!} \;. \tag{A5-28}$$

That this is so is easily seen by letting $k - m = n$. Then

$$\frac{d^n x^{m+n}}{dx^n} = (m+n)(m+n-1)\cdots(m+1)x^m = \frac{(m+n)!\, x^m}{m!} \;. \tag{A5-29}$$

Separating out the first term of the sum and using Eq. (A5-28) gives

$$\frac{d^k}{dx^k}[x^k g(x,k)] = k!\left[1 + \sum_{m=1}^{m=k} \binom{k}{m} \frac{1}{m!} x^m \frac{d^m}{dx^m}\right] g(x,k) \;. \tag{A5-30}$$

But

$$x^m \frac{d^m}{dx^m} = D_x(D_x - 1)\cdots(D_x - m + 1) = \prod_{i=0}^{i=m-1} (D_x - i) = D_x^{(m)} \tag{A5-31}$$

where $D_x^{(m)}$ is the factorial function of D_x [cf. Eq. (A3-43)], and

$$D_x^m = \frac{d^m}{d \ln^m x} \;. \tag{A5-32}$$

We may now write

$$\begin{aligned}\frac{d^k}{dx^k}[x^k g(x,k)] &= \left[k! + \sum_{m=1}^{m=k} \binom{k}{m} \frac{k!}{m!} \prod_{i=0}^{i=m-1} (D_x - i)\right] g(x,k) \\ &= \left[\prod_{i=1}^{i=k} (D_x + i)\right] g(x,k) \;.\end{aligned} \tag{A5-33}$$

But, by Eq. (A5-31),

$$g(x,k) = x^k \frac{d^k}{dx^k} f(x) = \left[\prod_{j=0}^{j=k-1} (D_x - j)\right] f(x) = \left[\prod_{i=-k+1}^{i=0} (D_x + i)\right] f(x) \tag{A5-34}$$

where the second of Eqs. (A5-34) is obtained by introducing a new running index $i = j - k + 1$. Writing $(D_x + i)$ for $(D_x - i - k + 1)$ merely changes the order in which the terms appear in the product. Combining Eq. $(A5\text{-}34)_2$ with Eq. (A5-26) and using Eq. $(A5\text{-}33)_2$ we obtain

$$\frac{d^k}{dx^k}\left[x^{2k}\frac{d^k}{dx^k} f(x)\right] = \prod_{i=-k+1}^{i=k} [(D_x + i)] f(x) \tag{A5-35}$$

and Eq. (A5-20) becomes

$$f(\sigma) = \lim_{k\to\infty} C_1(k)\left[\prod_{i=-k+1}^{i=k} (D_x + i)\right]\tilde{f}(x)|_{x=\sigma} \tag{A5-36}$$

where we have written

$$C_1(k) = \frac{(-1)^k}{(k-1)!\,k!} \tag{A5-37}$$

for simplicity.

Developing Eq. (A5-21) in a similar manner, Eq. (A5-25) yields

$$\frac{d^{k+1}}{dx^{k+1}}\left[x^{2k+1}\frac{d^k}{dx^k}\right] f(x) = \frac{d^{k+1}}{dx^{k+1}}[x^{k+1} g(x,k)] \ . \tag{A5-38}$$

By Leibnitz's rule,

$$\frac{d^{k+1}}{dx^{k+1}}[x^{k+1} g(x,k)] = \sum_{m=0}^{m=k+1} \binom{k+1}{m}\left[\frac{d^{k+1-m}}{dx^{k+1-m}} x^{k+1}\right]\left[\frac{d^m}{dx^m} g(x,k)\right] . \tag{A5-39}$$

But

$$\frac{d^{k+1-m}}{dx^{k+1-m}} x^{k+1} = \frac{(k+1)!\,x^m}{m!} \tag{A5-40}$$

and therefore,

$$\frac{d^{k+1}}{dx^{k+1}}[x^{k+1} g(x,k)] = (k+1)!\left[1 + \sum_{m=1}^{m=k+1}\binom{k+1}{m}\frac{1}{m!} x^m \frac{d^m}{dx^m}\right] g(x,k) \ . \tag{A5-41}$$

With Eq. (A5-31) we obtain

$$\begin{aligned}\frac{d^{k+1}}{dx^{k+1}}[x^{k+1} g(x,k)] &= \left[(k+1)! + \sum_{m=1}^{m=k+1}\binom{k+1}{m}\frac{(k+1)!}{m!}\prod_{i=0}^{i=m-1}(D_x - i)\right] g(x,k)\\ &= \left[\prod_{i=1}^{i=k+1}(D_x + i)\right] g(x,k) \ .\end{aligned} \tag{A5-42}$$

Using Eq. (A5-26) again, we finally have

$$\frac{d^{k+1}}{dx^{k+1}}\left[x^{2k+1}\frac{d^k}{dx^k}f(x)\right]=\left[\prod_{i=-k+1}^{i=k+1}(D_x+i)\right]f(x)\ . \tag{A5-43}$$

Thus, Eq. (A5-21) becomes

$$f(\sigma)=\lim_{k\to\infty}C_2(k)\left[\prod_{i=-k+1}^{i=k+1}(D_x+i)\right]\tilde{f}(x)|_{x=\sigma}\ . \tag{A5-44}$$

where

$$C_2(k)=C_1(k)/(k+1)\ . \tag{A5-45}$$

Changing the running index from i to j − k, the new index j will run from 1 to 2k in Eq. (A5-36), and from 1 to 2k + 1 in Eq. (A5-44). Thus we obtain

$$f(\sigma)=\lim_{k\to\infty}C_1(k)\left[\prod_{i=1}^{i=2k}(D_x+i-k)\right]\tilde{f}(x)|_{x=\sigma} \tag{A5-46}$$

from Eq. (A5-36) and

$$f(\sigma)=\lim_{k\to\infty}C_2(k)\left[\prod_{i=1}^{i=2k+1}(D_x+i-k)\right]\tilde{f}(x)|_{x=\sigma} \tag{A5-47}$$

from Eq. (A5-44). In the limit as $k\to\infty$ the two equations become identical. If the equations are used in an approximate inversion, i.e. without going to the limit (cf. Sect. 4.3), Eq. (A5-46) will yield approximations of even order, and Eq. (A5-47) will yield approximations of odd order. Both equations may conveniently be rewritten with the help of the factorial function as

$$f(\sigma)=\lim_{k\to\infty}C_1(k)(D_x+k)^{(2k)}\tilde{f}(x)|_{x=\sigma} \tag{A5-48}$$

and

$$f(\sigma)=\lim_{k\to\infty}C_2(k)(D_x+k+1)^{(2k+1)}\tilde{f}(x)|_{x=\sigma}\ . \tag{A5-49}$$

It is in the form of these last two equations that we used the real-axis Stieltjes transform inversion in Sect. 4.3.3.

A5.3 The Fourier-Laplace Transform

It was shown at the beginning of this section that the Stieltjes transform arises from a repeated application of the Laplace transformation. The Laplace transform variable is, in general, complex. In fact, the Laplace transform with purely imaginary variable is the generalized Fourier transform. Thus, we may ask what results from following a Laplace transformation not by another one but by a Fourier transformation. The answer is, of course, that this leads again to the Stieltjes transform but with a purely imaginary transform variable. We shall find it convenient to distinguish this subclass of Stieltjes transforms as the *Fourier-Laplace transform.* We denote it by $\mathscr{I}$ and define it as

$$\tilde{f}^*(\omega) = \int_0^\infty \frac{f(\sigma)}{\sigma + j\omega} d\sigma = \mathscr{J} f[(\sigma); j\omega] = \mathscr{F}\{\mathscr{L}[f(\sigma); \sigma']; j\omega\} \ . \tag{A5-50}$$

In the notation $\tilde{f}^*(\omega)$, the tilde refers to the fact that this transform is essentially a Stieltjes transform. The asterisk distinguishes it as having the harmonic operator, $j\omega$, as the transform variable.

We note that the Fourier-Laplace transform is complex, and write

$$\tilde{f}^*(\omega) = \tilde{f}'(\omega) - j\,\tilde{f}''(\omega) \ . \tag{A5-51}$$

The Fourier-Laplace transform is obtained from the Stieltjes transform by replacing the Stieltjes transform variable by $j\omega$. We thus have

$$\mathscr{S}[f(\sigma); s]|_{s=j\omega} = \mathscr{J}[f(\sigma); j\omega] \tag{A5-52}$$

or

$$\tilde{f}(\sigma)|_{s=j\omega} = \tilde{f}^*(\omega) \ . \tag{A5-53}$$

The relation between the Fourier-Laplace and the Stieltjes transforms is thus analogous to the relation between the Fourier and the Laplace transforms [cf. Eq. (A4-19)].

The inversion formulae for the Fourier-Laplace transform are the same as those for the Stieltjes transform.

A5.4 Stieltjes Transform Pairs

Only a very few Stieltjes transform pairs are needed in this book. They are identified by STP followed by the number.

Table A.5-1. Stieltjes transform pairs

$f(t)$	$\tilde{f}(s)$	No.
$t^{m-1}\exp(-at)$	$s^{m-1}\Gamma(m)\Gamma(1-m; as)\exp(as)$	(1)
$t^{-m-1}\exp(-1/at)$	$s^{-m-1}\Gamma(1+m)\Gamma(-m; 1/as)\exp(1/as)$	(2)
$tf(t)$	$\int_0^\infty f(t)dt - s\tilde{f}(s)$	(3)

$\Gamma(m; x)$ is the complementary incomplete gamma function, see Eqs. (6.3-99).

A6 The Hilbert Transform

We call

$$\hat{f}(\nu) = \frac{1}{\pi} \fint_{-\infty}^{\infty} \frac{f(\mu)}{\mu - \nu} d\mu = \mathscr{H}[f(\mu); \nu] \tag{A6-1}$$

the *Hilbert transform* of $f(\mu)$. When it is not necessary to designate the transform variable explicitly, we may write $\mathscr{H}\, f(\mu)$ for $\mathscr{H}[f(\mu); \nu]$. Both ν and μ are real variables and

$$\fint_{-\infty}^{\infty} = \lim_{\varepsilon \to 0} \int_{-\infty}^{\mu-\varepsilon} + \int_{\mu+\varepsilon}^{\infty} \tag{A6-2}$$

is the Cauchy principal value of $\int_{-\infty}^{\infty}$. The Hilbert transform is clearly related to the Stieltjes transform and, hence, to the Laplace transform [9].

A6.1 Inversion

The inversion formula for the Hilbert transform is

$$f(\mu) = \frac{1}{\pi} \fint_{-\infty}^{\infty} \frac{\hat{f}(\nu)}{\nu - \mu} d\nu = \mathscr{H}[\hat{f}(\nu); \mu] \ . \tag{A6-3}$$

References (Appendix)

1. M.R. Spiegel, *Theory and Problems of the Laplace Transform*, Schaum's Outline Series, McGraw-Hill, New York, 1965.
2. S. Goldman, *Transformation Calculus and Electrical Transients*, Constable and Co., Ltd., London, 1949.
3. M.F. Gardner and J.L. Barnes, *Transients in Linear Systems*, Vol. 1, Wiley, New York, 1963.
4. T.M. Apostol, *Mathematical Analysis*, 2nd ed., Addison-Wesley, Reading, Mass., 1974.
5. D.V. Widder, *The Laplace Transform*, Princeton University Press, Princeton, New Jersey, 1946.
6. M.R. Spiegel, *Finite Differences and Difference Equations*, Schaum's Outline Series, McGraw-Hill, New York, 1971.
7. M. Abramowitz and I.A. Stegun, *Handbook of Mathematical Functions*, Applied Mathematics Series 55, National Bureau of Standards, U.S. Government Printing Office, Washington, D.C., 1964.
8. I.I. Hirschman and D.V. Widder, *The Convolution Transform*, Princeton University Press, Princeton, New Jersey, 1955.
9. Bateman Manuscript Project, *Tables of Integral Transforms*, Vol. II, Chap. IV, MacGraw-Hill, New York, 1954.

Solutions to Problems

Οἰδίπους. ὡς πάντ' ἄγαν αἰνικτὰ κἀσαφῆ λέγεις.
Τειρεσίας. οὔκουν σὺ ταῦτ' ἄριστος εὑρίσκειν ἔφυς;

Οἰδίπους Τύραννος Σοφοκλέους

Chapter 1

Problem 1.1-1

By Eq. (1.1-1), the elongation is given by

$$l = l_0(1 + f/A_0E) \qquad (1)$$

where l and l_0 are the extended and unextended length, respectively, f is the force in newtons represented by the weight, A_0 is the cross-sectional area in m^2, and E is the modulus in Nm^{-2}. We have $f = 9.81$ M $kgms^{-2}$, M being the mass of the weight in kg. Thus $f = 0.4905$ N. We find

l_0(m)	$A_0(10^3 \times m^2)$	l(m)
0.20	0.250	0.239
0.15	0.0250	0.179
0.15	0.0125	0.209

Problem 1.1-2

By Eq. (1.1-2) $E = E\varepsilon$. Hence $\varepsilon = 1$, i.e. the elongation is 100%.

Problem 1.2-1

The equation for the unit vector is

$$n_i = ae_1 + ae_2 + ae_3 \qquad (1)$$

where the e_i are the unit base vectors in the directions of the coordinates. The number a can be obtained from the fact that n_i is a *unit* vector. Hence, $n_in_i = 1$, and we find $3a^2 = 1$, i.e. $a = 1/\sqrt{3}$. Therefore,

$$n_i = e_1/\sqrt{3} + e_2/\sqrt{3} + e_3/\sqrt{3} \ . \qquad (2)$$

Problem 1.2-2

The equation of the unit vector on the octahedral plane is given by Eq. (2) of Problem 1.2-1. For the stress vector to be zero we must have $\sigma_{ij}n_j = 0$. In matrix form

$$\begin{bmatrix} \sigma & a\sigma & b\sigma \\ a\sigma & \sigma & c\sigma \\ b\sigma & c\sigma & \sigma \end{bmatrix} \begin{bmatrix} 1/\sqrt{3} \\ 1/\sqrt{3} \\ 1/\sqrt{3} \end{bmatrix} = \begin{bmatrix} 0 \\ 0 \\ 0 \end{bmatrix} . \tag{2}$$

Hence

$$\begin{aligned} a + b &= -1 \\ a + c &= -1 \\ b + c &= -1 \end{aligned} \tag{3}$$

and solution of these equations yields $a = b = c = -1/2$. The matrix of the stress tensor therefore is

$$[\sigma_{ij}] = \begin{bmatrix} \sigma & -\sigma/2 & -\sigma/2 \\ -\sigma/2 & \sigma & -\sigma/2 \\ -\sigma/2 & -\sigma/2 & \sigma \end{bmatrix} . \tag{4}$$

Problem 1.3-1

The matrix of the displacement gradient tensor is

$$[\partial u_i/\partial x_j] = \begin{bmatrix} 2(x_1 - x_3) & 0 & -2(x_1 - x_3) \\ 0 & 2(x_2 + x_3) & 2(x_2 + x_3) \\ -x_2 & -x_1 & 0 \end{bmatrix} . \tag{2}$$

Decomposing this into the symmetric and antisymmetric parts [cf. Eq. (1.3-13)] gives

$$[\gamma_{ij}] = \begin{bmatrix} 2(x_1 - x_3) & 0 & -(x_1 + x_2/2 - x_3) \\ 0 & 2(x_2 + x_3) & -(x_1/2 - x_2 - x_3) \\ -(x_1 + x_2/2 - x_3) & -(x_1/2 - x_2 - x_3) & 0 \end{bmatrix} \tag{3}$$

and

$$[\omega_{ij}] = \begin{bmatrix} 0 & 0 & -x_1 + x_2/2 + x_3 \\ 0 & 0 & x_1/2 + x_2 + x_3 \\ x_1 - x_2/2 - x_3 & -x_1/2 - x_2 - x_3 & 0 \end{bmatrix} . \tag{4}$$

The matrix of the spherical strain tensor becomes

$$[\gamma'_{ij}] = \begin{bmatrix} 2(x_1 + x_2)/3 & 0 & 0 \\ 0 & 2(x_1 + x_2)/3 & 0 \\ 0 & 0 & 2(x_1 + x_2)/3 \end{bmatrix} \tag{5}$$

and that of the deviator is

$$[\gamma''_{ij}] = \begin{bmatrix} 2(2x_1 - x_2 - 3x_3)/3 & 0 & -(x_1 + x_2/2 - x_3) \\ 0 & -2(x_1 - 2x_2 - 3x_3)/3 & -(x_1/2 - x_2 - x_3) \\ -(x_1 + x_2/2 - x_3) & -(x_1/2 - x_2 - x_3) & -2(x_1 + x_2)/3 \end{bmatrix} . \tag{6}$$

Problem 1.4-1

In a uniform biaxial deformation perpendicular to the 3-dimension, the stress and strain matrices are

$$[\sigma_{ij}] = \begin{bmatrix} \sigma_{11} & 0 & 0 \\ \cdot & \sigma_{11} & 0 \\ \cdot & \cdot & 0 \end{bmatrix} \tag{1}$$

and

$$[\gamma_{ij}] = \begin{bmatrix} \gamma_{11} & 0 & 0 \\ \cdot & \gamma_{11} & 0 \\ \cdot & \cdot & \gamma_{33} \end{bmatrix}, \tag{2}$$

respectively. We define H as

$$H = \sigma_{11}/\gamma_{11} \ . \tag{3}$$

Substituting Eqs. (1) and (2) into Eq. (1.4-69) gives

$$\sigma_{11} = (K - 2G/3)(2\gamma_{11} + \gamma_{33}) + 2G\gamma_{11} \tag{4}$$

and

$$0 = (K - 2G/3)(2\gamma_{11} + \gamma_{33}) + 2G\gamma_{33} \ . \tag{5}$$

From Eq. (5)

$$\mu_H = \frac{-\gamma_{33}}{\gamma_{11}} = \frac{6K - 4G}{3K + 4G} \ . \tag{6}$$

Elimination of γ_{33} from Eqs. (4) and (5) gives

$$H = \frac{18KG}{3K + 4G} \ . \tag{7}$$

As $K \to \infty$, $H \to 6G$, and $\mu_H \to 2$.

Problem 1.4-2

We have

$$[\sigma_{ij}] = \begin{bmatrix} \sigma_{11} & 0 & 0 \\ \cdot & \sigma_{22} & 0 \\ \cdot & \cdot & 0 \end{bmatrix} \tag{1}$$

and

$$[\gamma_{ij}] = \begin{bmatrix} \gamma_{11} & 0 & 0 \\ \cdot & 0 & 0 \\ \cdot & \cdot & \gamma_{33} \end{bmatrix} . \tag{2}$$

We define the modulus U as

$$U = \sigma_{11}/\gamma_{11} \; . \tag{3}$$

Substituting Eqs. (1) and (2) into Eq. (1.4-69) gives

$$\sigma_{11} = (K - 2G/3)(\gamma_{11} + \gamma_{33}) + 2G\gamma_{11} \tag{4}$$

and

$$\sigma_{22} = (K - 2G/3)(\gamma_{11} + \gamma_{33}) \tag{5}$$

$$0 = (K - 2G/3)(\gamma_{11} + \gamma_{33}) + 2G\gamma_{33} \; . \tag{6}$$

From Eq. (6)

$$\mu_U = \frac{-\gamma_{33}}{\gamma_{11}} = \frac{3K - 2G}{3K + 4G} \; . \tag{7}$$

Equations (4) and (6) then yield

$$U = \frac{4G(3K + G)}{3K + 4G} \; . \tag{8}$$

Finally, using Eqs. (5) and (7)

$$\sigma_{22}/\sigma_{11} = \frac{3K - 2G}{6K + 2G} \; . \tag{9}$$

As $K \to \infty$, $\mu_U \to 1$, $U \to 4G$, and $\sigma_{22}/\sigma_{11} \to 1/2$.

Problem 1.4-3

By Eqs. $(1.4\text{-}58)_2$, (1.4-90), and (1.4-91)

$$\gamma_{ij} = (B/3)\sigma'_{ij} + (J/2)\sigma''_{ij} \; . \tag{2}$$

But by Eq. $(1.4\text{-}61)_1$

$$\sigma'_{ij} = \tfrac{1}{3} \textstyle\sum \delta_{ij} \; , \tag{3}$$

and, thus,

$$\sigma''_{ij} = \sigma_{ij} - \tfrac{1}{3} \textstyle\sum \delta_{ij} \; . \tag{4}$$

Equation (1.4-92) then follows by substituting Eqs. (3) and (4) into Eq. (2).

Problem 1.4-4

Young's modulus is given by

$$E = \sigma_1/\gamma_1 \; . \tag{1}$$

We have, by Eq. (1.4-85),

$$\gamma_1 = S_{11}\sigma_1 \tag{2}$$

since $\sigma_2 = \sigma_3 = 0$. Hence,

$$E = 1/S_{11} \; . \tag{3}$$

Poisson's ratio is given by

$$\mu = -\gamma_2/\gamma_1 \; . \tag{4}$$

Equation (1.4-85) also gives

$$\gamma_2 = S_{12}\sigma_1 \; . \tag{5}$$

Substitution of Eqs. (2) and (5) into (4) yields

$$\mu = -S_{12}/S_{11} \; . \tag{6}$$

Problem 1.4-5

By Eqs. (1.4-5) and (1.4-85) the matrices $[C_{kl}]$ and $[S_{kl}]$ are reciprocal. We invert $[C_{kl}]$ to obtain $[S_{kl}]$. For an isotropic body

$$[C_{kl}] = \left[\begin{array}{ccc|ccc} C_{11} & C_{12} & C_{12} & & & \\ C_{12} & C_{11} & C_{12} & & \bigcirc & \\ C_{12} & C_{12} & C_{11} & & & \\ \hline & & & C_{44} & 0 & 0 \\ & \bigcirc & & 0 & C_{44} & 0 \\ & & & 0 & 0 & C_{44} \end{array}\right] . \tag{1}$$

The two diagonal submatrices are uncoupled. Hence, they may be inverted separately. The inversion of a symmetric matrix yields a symmetric matrix. Thus

$$S_{44} = 1/C_{44} \; . \tag{2}$$

The determinant of the first submatrix is

$$\Delta = (C_{11} + 2C_{12})(C_{11} - C_{12})^2 \; . \tag{3}$$

The cofactor* of the C_{11}-term is $C_{11}^2 - C_{12}^2$. Hence

$$S_{11} = \frac{C_{11} + C_{12}}{(C_{11} + 2C_{12})(C_{11} - C_{12})} \; . \tag{4}$$

The cofactor of the C_{12}-term is $C_{12}(C_{12} - C_{11})$ and, therefore,

$$S_{12} = \frac{C_{12}}{(C_{11} + 2C_{12})(C_{12} - C_{11})} \; . \tag{5}$$

This completes the inversion.

Because of the symmetry of the problem,

$$C_{11} = \frac{S_{11} + S_{12}}{(S_{11} + 2S_{12})(S_{11} - S_{12})} \tag{6}$$

* The cofactor is the signed minor of the component. For a symmetric matrix, any component of its inverse is the cofactor divided by the determinant.

$$C_{12} = \frac{S_{12}}{(S_{11} + 2S_{12})(S_{12} - S_{11})} \tag{7}$$

and

$$C_{44} = 1/S_{44} \ . \tag{8}$$

Problem 1.4-6

We have

$$[C_{kl}] = \left[\begin{array}{ccc|ccc} \lambda + 2G & \lambda & \lambda & & & \\ \lambda & \lambda + 2G & \lambda & & \bigcirc & \\ \lambda & \lambda & \lambda + 2G & & & \\ \hline & & & 2G & 0 & 0 \\ & \bigcirc & & 0 & 2G & 0 \\ & & & 0 & 0 & 2G \end{array}\right]$$

by Eq. (1.4-54). Since the submatrices are uncoupled, they can be inverted separately to yield the corresponding submatrices of $[S_{kl}]$. Thus, since $G = 1/J$,

$$S_{44} = J/2 \tag{1}$$

follows immediately. The determinant of the first submatrix is $12KG^2$. The cofactor of the $(\lambda + 2G)$-term is $4G(\lambda + G)$. Hence

$$S_{11} = \frac{\lambda + G}{3KG} = \frac{3K + G}{9KG} = \frac{B}{9} + \frac{J}{3} \tag{2}$$

because $K = 1/B$. The cofactor of the λ-term is $-2\lambda G$. Hence

$$S_{12} = -\frac{\lambda}{6KG} = \frac{2G - 3K}{18KG} = \frac{B}{9} - \frac{J}{6} \ . \tag{3}$$

Note that, by Eq. (1.4-77),

$$S_{11} = \frac{1}{E} = D \ . \tag{4}$$

Problem 1.4-7

These relations follow immediately from Eqs. (1.4-87) and (1.4-88), using Eqs. (2) and (3) of Problem 1.4-6.

Problem 1.4-8

From Eq. (1.4-92) we obtain

$$\gamma_{11} = (\tfrac{1}{9}B + \tfrac{1}{3}J)\sigma_{11} + (\tfrac{1}{9}B - \tfrac{1}{6}J)(\sigma_{22} + \sigma_{33}) \ . \tag{2}$$

But, from Table 1.4-2,

$$\tfrac{1}{9}B + \tfrac{1}{3}J = D = 1/E \tag{3}$$

and

$$\tfrac{1}{6}J - \tfrac{1}{9}B = D\mu \ . \tag{4}$$

Substitution into Eq. (2) yields Eq. (1).

Problem 1.4-9

In terms of the fundamental moduli, K and G, we have

$$M = K + 4G/3 \ . \tag{1}$$

From Table 1.4-1 we have

$$G = \frac{3K(1 - 2\mu)}{2(1 + \mu)} \ . \tag{2}$$

Substituting this expression in the equation for M we find

$$M = 3K\frac{1 - \mu}{1 + \mu} \ . \tag{3}$$

Hence, $M = K$ if $\mu = 1/2$ (i.e. the material is incompressible) but $M = 1.5K$ if $\mu = 1/3$.

Problem 1.4-10

The required relations are most easily obtained from Eq. (1.4-85). By analogy with Eqs. (1.4-31) and (1.4-35) we must have, for an axisymmetric material,

$$\begin{bmatrix} \gamma_1 \\ \gamma_2 \\ \gamma_3 \\ \gamma_4 \\ \gamma_5 \\ \gamma_6 \end{bmatrix} = \begin{bmatrix} S_{11} & S_{12} & S_{13} & 0 & 0 & 0 \\ S_{12} & S_{11} & S_{13} & 0 & 0 & 0 \\ S_{13} & S_{13} & S_{33} & 0 & 0 & 0 \\ 0 & 0 & 0 & S_{11} - S_{12} & 0 & 0 \\ 0 & 0 & 0 & 0 & S_{55} & 0 \\ 0 & 0 & 0 & 0 & 0 & S_{55} \end{bmatrix} \begin{bmatrix} \sigma_1 \\ \sigma_2 \\ \sigma_3 \\ \sigma_4 \\ \sigma_5 \\ \sigma_6 \end{bmatrix} \tag{1}$$

where the 3-direction is the axis of symmetry.

To obtain the axial compliance, D_3, let a stress, σ_3, be applied in the direction of this axis. Then $\sigma_1 = \sigma_2 = \sigma_4 = \sigma_5 = \sigma_6 \equiv 0$, and we obtain at once

$$D_3 = \gamma_3/\sigma_3 = S_{33} \ . \tag{2}$$

With the convention that the first index denotes the axis of contraction, the axial Poisson's ratio becomes

$$\mu_{13} = -\frac{\gamma_1}{\gamma_3} = -\frac{S_{13}}{S_{33}} = -\frac{\gamma_2}{\gamma_3} = \mu_{23} \ . \tag{3}$$

To obtain the transverse compliance, D_1, we apply a stress in one of the transverse directions. Let this be the 1-direction. Then all stresses except σ_1 equal zero, and we find at once

$$D_1 = \gamma_1/\sigma_1 = S_{11} \ . \tag{4}$$

The transverse Poisson's ratio becomes

$$\mu_{21} = -\frac{\gamma_2}{\gamma_1} = -\frac{S_{12}}{S_{11}} \ . \tag{5}$$

If the stress is applied in the other transverse direction, we find $D_2 = \gamma_2/\sigma_2 = S_{11} = D_1$, and $\mu_{12} = -\gamma_1/\gamma_2 = S_{12}/S_{11} = \mu_{21}$ because of the axial symmetry. D_1, D_2, μ_{13}, and μ_{21} determine S_{11}, S_{33}, S_{12}, and S_{13}. To obtain the remaining elastic coefficient, S_{55}, the sample must be twisted around one of the transverse directions. The torsional compliance is thus obtained as

$$J = 2\gamma_5/\sigma_5 = 2S_{55} \ . \tag{6}$$

[cf. Eq. (1) of Problem 1.4-6].

Problem 1.4-11

By Problem 1.4-10, $\mu_{13} = -S_{13}/S_{33}$, and $\mu_{21} = -S_{12}/S_{11}$. A third ratio may be defined and is given by $\mu_{31} = -S_{13}/S_{11}$. We therefore first need the expressions for S_{11}, S_{12}, S_{13}, and S_{33} in terms of the components of $[C_{kl}]$. We then form the required ratios and substitute Eqs. (1.4-36) through (1.4-40). Finally, we find the limit as $\lambda \to \infty$.

Since the matrix is symmetric, each component S_{mn} of $[S_{kl}]$ is equal to the cofactor of C_{mn} in $[C_{kl}]$, divided by the determinant, $\Delta = \det[C_{kl}]$. Hence, from Eq. (1.4-31), using Eq. (1.4-35),

$$S_{11} = C_{55}^2(C_{11} - C_{12})(C_{11}C_{33} - C_{13}^2)/\Delta \tag{1}$$

$$S_{12} = -C_{55}^2(C_{11} - C_{12})(C_{12}C_{33} - C_{13}^2)/\Delta \tag{2}$$

$$S_{13} = -C_{55}^2(C_{11} - C_{12})^2 C_{13}/\Delta \tag{3}$$

and

$$S_{33} = C_{55}^2(C_{11} - C_{12})^2(C_{11} + C_{12})/\Delta \ . \tag{4}$$

We obtain immediately

$$\mu_{13} = \frac{C_{13}}{C_{11} + C_{12}} \tag{5}$$

$$\mu_{21} = \frac{C_{12}C_{33} - C_{13}^2}{C_{11}C_{33} - C_{13}^2} \tag{6}$$

and

$$\mu_{31} = \frac{(C_{11} - C_{12})C_{13}}{C_{11}C_{33} - C_{13}^2} \ . \tag{7}$$

Upon substitution of Eqs. (1.4-36) through (1.4-40) we obtain

$$\mu_{13} = \frac{\lambda + \delta}{2(\lambda + G + \rho)} \tag{8}$$

$$\mu_{21} = \frac{(\lambda + 2\rho)(\lambda + \beta + 2G) - (\lambda + \delta)^2}{(\lambda + 2G)(\lambda + \beta + 2G) - (\lambda + \delta)^2} \tag{9}$$

and

$$\mu_{31} = \frac{2(\lambda + \delta)(G - \rho)}{(\lambda + 2G)(\lambda + \beta + 2G) - (\lambda + \delta)^2} . \tag{10}$$

As $\beta, \rho, \delta \to 0$, all three equations reduce to

$$\mu = \frac{\lambda}{2(\lambda + G)} = \frac{3K - 2G}{6K + 2G} \tag{11}$$

as required. Finally,

$$\lim_{\lambda \to \infty} \mu_{13} = \tfrac{1}{2} \tag{12}$$

$$\lim_{\lambda \to \infty} \mu_{21} = \frac{2G + \beta - 2\delta + 2\rho}{4G + \beta - 2\delta} \tag{13}$$

and

$$\lim_{\lambda \to \infty} \mu_{31} = \frac{2(G - \rho)}{4G + \beta - 2\delta} . \tag{14}$$

Problem 1.4-12

By Problem 1.4-10, $E_1 = 1/D_1 = 1/S_{11}$ and $E_3 = 1/D_3 = 1/S_{33}$. In Problem 1.4-11 we found S_{11} and S_{33} in terms of the C_{kl} but the expressions contained the determinant Δ which was not required explicitly there. We now find

$$\Delta = \det[C_{kl}] = C_{55}^2(C_{11} - C_{12})[C_{33}(C_{11} + C_{12}) - 2C_{13}^2] \tag{1}$$

and, from Eqs. (1) and (4) of Problem 1.4-11 we obtain

$$E_1 = (C_{11} - C_{12})\frac{(C_{11} + C_{12})C_{33} - 2C_{13}^2}{C_{11}C_{33} - C_{13}^2} \tag{2}$$

and

$$E_3 = C_{33} - \frac{2C_{13}^2}{C_{11} + C_{12}} . \tag{3}$$

Substituting Eqs. (1.4-36) through (1.4-40) gives

$$E_1 = 4(G - \rho)\frac{(\lambda + G + \rho)(\lambda + \beta + 2G) - (\lambda + \delta)^2}{(\lambda + 2G)(\lambda + \beta + 2G) - (\lambda + \delta)^2} \tag{4}$$

and

$$E_3 = \lambda + \beta + 2G - \frac{(\lambda + \delta)^2}{\lambda + G + \rho} . \tag{5}$$

As $\beta, \rho, \delta \to 0$, both expressions reduce to

$$E = G\frac{3\lambda + 2G}{\lambda + G} = \frac{9KG}{3K + G} \tag{6}$$

as required. Finally,

$$\lim_{\lambda\to\infty} E_1 = 4(G - \rho)\frac{3G + \beta - 2\delta + \rho}{4G + \beta - 2\delta} \tag{7}$$

and

$$\lim_{\lambda\to\infty} E_3 = 3G + \beta - 2\delta + \rho \ . \tag{8}$$

Problem 1.5-1

For a shear in the 1, 2-directions Eq. (1.5-5) becomes

$$\sigma_{12} = 2\eta\dot{\gamma}_{12} \ . \tag{1}$$

But $\sigma_{12} = f/a$, where f is the shearing force and a is the area over which it acts. Now, by Eq. $(1.3\text{-}20)_2$,

$$\dot{\gamma}_{ij} = \frac{1}{2}\left(\frac{\partial v_i}{\partial x_j} + \frac{\partial v}{\partial x_i}\right) \ . \tag{2}$$

Hence, we have

$$f = a\eta\left(\frac{\partial v_1}{\partial x_2} + \frac{\partial v_2}{\partial x_1}\right) \ . \tag{3}$$

But in one-dimensional laminar shear $\partial v_2/\partial x_1 = 0$. Thus, Eq. (1.1-4) follows.

Problem 1.5-2

For an elongational flow in the 1-direction we have

$$[\sigma_{ij}] = \begin{bmatrix} \sigma_{11} & 0 & 0 \\ \cdot & 0 & 0 \\ \cdot & \cdot & 0 \end{bmatrix} \tag{1}$$

and

$$[\dot{\gamma}_{ij}] = \begin{bmatrix} \dot{\gamma}_{11} & 0 & 0 \\ \cdot & \dot{\gamma}_{22} & 0 \\ \cdot & \cdot & \dot{\gamma}_{22} \end{bmatrix} \ . \tag{2}$$

Inserting these into Eq. (1.5-11) gives

$$\sigma_{11} - K\Delta = (\xi + \tfrac{4}{3}\eta)\dot{\gamma}_{11} + 2(\xi - \tfrac{2}{3}\eta)\dot{\gamma}_{22} \tag{3}$$

and

$$-K\Delta = (\xi - \tfrac{2}{3}\eta)\dot{\gamma}_{11} + 2(\xi + \tfrac{1}{3}\eta)\dot{\gamma}_{22} \ . \tag{4}$$

Solving for $\dot{\gamma}_{11}$ yields

$$\dot{\gamma}_{11} = \frac{(3\xi + \eta)\sigma_{11} - 3K\Delta\eta}{9\xi\eta} \tag{5}$$

from which Eq. (1.5-12) follows by using Eq. (1.5-8).

Chapter 2

Problem 2.3-1

By Eq. (2.3-11)

$$Q(t) = \frac{d\,G(t)}{dt} + G(0)\delta(t) \ . \qquad (2)$$

Hence

$$\int_0^t Q(u)du = \int_0^t \frac{d\,G(u)}{du} du + G(0) \int_0^t \delta(u)du$$

$$= G(t) - G(0) + G(0) = G(t) \ . \qquad (3)$$

Omission of the delta function term would have given the wrong result, $G(t) - G_g$, for the integral.

Problem 2.3-2

The equations are all equivalent. We take $(2.3\text{-}16)_4$

$$\sigma(t) = -\int_0^t G(u) \frac{d\,\varepsilon(t-u)}{du} du \qquad (2)$$

and insert the step excitation $\varepsilon(t) = \varepsilon_0 h(t)$. This gives

$$\sigma(t) = -\varepsilon_0 \int_0^t G(u) \frac{d\,h(t-u)}{du} du \ . \qquad (3)$$

But $d\,h(t-u)/du$ is $-\delta(t-u)$ by applying the chain rule of differentiation to Eq. (A2-21). Therefore

$$\sigma(t) = \varepsilon_0 \int_0^t G(u)\delta(t-u)du = \varepsilon_0 G(t) \qquad (4)$$

as required.

Problem 2.3-3

A series of discrete steps of strain can be represented with the help of the unit step function as

$$\varepsilon(t) = \sum_k \varepsilon_k h(t - t_k) \ , \qquad k = 1, 2, \ldots \qquad (1)$$

where the ε_k are the heights of the steps imposed at the times t_k. Inserting Eq. (1) into Eq. $(2.3\text{-}16)_3$ gives

$$\sigma(t) = \sum_k \varepsilon_k \int_0^t G(t-u)\delta(u-t_k)du \tag{2}$$

since the order of summation and integration can be interchanged. By the *sifting property* of the delta function [cf. Eq. (A2-13)] we have then

$$\sigma(t) = \sum_k \varepsilon_k G(t-t_k) , \qquad k = 1, 2, \dots . \tag{3}$$

According to the procedure we generally follow in this book we would first seek the Laplace transform of the excitation. By LPT(5), this is

$$\bar{\varepsilon}(s) = \sum_k (\varepsilon_k/s)\exp(-t_k s) , \qquad k = 1, 2, \dots . \tag{4}$$

Introducing this into Eq. $(2.3\text{-}15)_1$ produces

$$\bar{\sigma}(s) = \sum_k \varepsilon_k \bar{G}(s)\exp(-t_k s) \tag{5}$$

and retransformation gives Eq. (3).

Problem 2.3-4

The stress transform in response to a strain transform is

$$\bar{\sigma}(s) = s\bar{G}(s)\bar{\varepsilon}(s) . \tag{1}$$

The transform of a step strain imposed at $t = 0$ is ε_0/s where ε_0 is the height of the step. Hence,

$$\bar{\sigma}_1(s) = \bar{\varepsilon}_0\bar{G}(s) . \tag{2}$$

If the excitation is shifted from 0 to t_1, the strain transform becomes $[\varepsilon_0\exp(-t_1 s)]/s$ by the translation theorem (cf. Appendix A3.13). Hence,

$$\bar{\sigma}_2(s) = \varepsilon_0\bar{G}(s)\exp(-t_1 s) . \tag{3}$$

Retransformation yields

$$\sigma_1(t) = \varepsilon_0 G(t)h(t) \tag{4}$$

and

$$\sigma_2(t) = \varepsilon_0 G(t-t_1)h(t-t_1) . \tag{5}$$

This is an example of time-shift invariance. The response to the two stimuli is the same except for a shift along the time axis.

Problem 2.4-1

Using Eq. $(2.4\text{-}8)_1$ we have (cf. Problem 2.3-1)

$$\int_0^v Q(u)du = \int_0^v \frac{d\,G(u)}{du}du + G(0)\int_0^v \delta(u)du = G(v) \tag{2}$$

A second integration yields

$$\int_0^t \int_0^v Q(u)du\, dv = \int_0^t G(v)dv = \eta(t) \tag{3}$$

by Eq. (2.4-6).

Problem 2.4-2

Since $\eta(t)$ is the response to a slope strain, the excitation is

$$\varepsilon(t) = \dot{\varepsilon}_0 p(t) = \dot{\varepsilon}_0 t \ , \qquad t \geq 0 \ . \tag{3}$$

Inserting this into

$$\sigma(t) = \int_0^t Q(u)\varepsilon(t-u)du \tag{4}$$

gives

$$\eta(t) = \int_0^t (t-u)Q(u)du = t\int_0^t Q(u)du - \int_0^t uQ(u)du \ . \tag{5}$$

Integrate the second integral by parts. Let $x = u$, and $dy = Q(u)du$. Then $dx = du$, and

$$y = \int_0^u Q(w)dw \ . \tag{6}$$

Therefore

$$\int_0^t uQ(u)du = u\int_0^u Q(w)dw\Big|_0^t - \int_0^t \int_0^u Q(w)dw\, du$$

$$= t\int_0^t Q(w)dw - \int_0^t \int_0^v Q(u)du\, dv \ . \tag{7}$$

Hence,

$$\int_0^t (t-u)Q(u)du = \int_0^t \int_0^v Q(u)du\, dv \tag{8}$$

from which Eq. (2) follows by Eq. $(5)_1$.

Problem 2.5-1

From Eqs. (2.5-39) and (2.5-38)

$$\sigma(\omega) = \eta^*(\omega)\dot{\varepsilon}(\omega) \ . \tag{1}$$

Now,

$$\dot{\varepsilon}(\omega) = -\varepsilon_0\omega(\sin \omega t - j\cos \omega t) \tag{2}$$

[cf. Eq. (2.5-33)]. Hence, by Eq. $(2.5\text{-}40)_1$, Eq. (2.5-50) follows.

Problem 2.5-2

By the Residue Theorem, Eq. (A3-31),

$$\sigma_{ss}(t) = \sum_k \mathrm{Res}_k[\bar{\sigma}(s)\exp(st)] \tag{1}$$

where the residues are those associated with the pole(s) of the excitation transform since we wish to obtain the *steady-state* response. Now, by Eq. (2.5-4)

$$\bar{\sigma}(s) = \frac{\varepsilon_0 \bar{Q}(s)}{s - j\omega} \tag{2}$$

where $\varepsilon_0/(s - j\omega)$ is the excitation transform. It has a single, simple pole at $s = j\omega$. The corresponding residue is obtained from Eq. (A3-32) as

$$\mathrm{Res} = \lim_{s \to j\omega} \varepsilon_0 Q(s)\exp(st) = \varepsilon_0 \bar{Q}(j\omega)\exp(j\omega t) \ . \tag{3}$$

Hence, Eq. (2.5-11) follows.

Problem 2.5-3

Let the stress be given by $\sigma(t) = \sigma_0 \sin \omega t$. Since (see Fig. 2.5-3) the strain *lags* the stress by the *loss angle*, $\theta(\omega)$, we have $\varepsilon(t) = \varepsilon_0(\omega)\sin[\omega t - \theta(\omega)]$. Differentiation with respect to time gives the rate of strain as

$$\dot{\varepsilon}(t) = \varepsilon_0(\omega)\omega \cos[\omega t - \theta(\omega)] = \varepsilon_0(\omega)\omega \sin[\omega t - \theta(\omega) + \tfrac{\pi}{2}]. \tag{1}$$

But, by Eq. (2.5-45), $\theta(\omega) = \frac{\pi}{2} - \Theta(\omega)$, and, therefore,

$$\dot{\varepsilon}(t) = \varepsilon_0(\omega)\omega \sin[\omega t + \Theta(\omega)] \ . \tag{2}$$

Thus, the rate of strain *leads* the stress by the *storage angle*, $\Theta(\omega)$.

Problem 2.6-1

Introducing the generalized harmonic rate of strain

$$\dot{\varepsilon}(t) = j\omega\varepsilon_0 \exp(j\omega t) \tag{1}$$

into

$$\sigma(t) = \int_0^t G(u)\dot{\varepsilon}(t - u)du \tag{2}$$

[cf. Eq. $(2.3\text{-}16)_4$], the steady-state response becomes

$$\sigma(\omega) = j\omega\varepsilon(\omega) \lim_{t \to \infty} \int_0^t G(u)\exp(-j\omega u)du \ . \tag{3}$$

But $j\omega\,\varepsilon(\omega) = \dot{\varepsilon}(\omega)$ and, changing the dummy variable of integration,

$$\sigma(\omega)/\dot{\varepsilon}(\omega) = \eta^*(\omega) = \int_0^\infty G(t)\exp(-j\omega t)dt \ . \tag{4}$$

The right hand side of Eq. (4) is the Fourier transform of G(t) [cf. Eq. (A4-1)].

Problem 2.6-2

Equations (2.3-16) are all equivalent. Inserting $\varepsilon(t) = \dot{\varepsilon}_0 t$ into Eq. $(2.3\text{-}16)_3$, we obtain

$$\sigma(t) = \dot{\varepsilon}_0 \int_0^t G(t-u)du = \dot{\varepsilon}_0 \int_0^t G(u)du \tag{1}$$

whence Eq. $(1)_2$ follows by a change of variable.

Problem 2.6-3

(A) First we rewrite Eq. $(2.3\text{-}16)_3$ extending it into the indefinite past. This gives

$$\sigma(t) = \int_{-\infty}^{t} G(t-u)\frac{d\,\varepsilon(u)}{du}du\ , \qquad -\infty \le u \le t\ . \tag{2}$$

We then introduce the *elapsed time*, $s = t - u$. This change of variable leads to

$$\sigma(t) = -\int_0^\infty G(s)\frac{d\,\varepsilon(t-s)}{ds}ds\ , \qquad 0 \le s \le \infty\ . \tag{3}$$

Finally, we substitute Eq. (1) and obtain

$$\sigma(t) = -\int_0^\infty G(s)\frac{d\,\varepsilon_t(s)}{ds}ds\ , \qquad 0 \le s \le \infty\ . \tag{4}$$

(B) For a step strain Eq. (1) furnishes

$$\varepsilon_t(s) = \varepsilon_0[h(t-s) - h(t)]\ . \tag{5}$$

Now

$$\frac{d\,\varepsilon_t(s)}{ds} = -\varepsilon_0\delta(t-s) \tag{6}$$

and, therefore

$$\sigma(t) = \varepsilon_0 \int_0^\infty G(s)\delta(t-s)ds = \varepsilon_0 G(t) \tag{7}$$

[cf. Eq. (A2-13)].

(C) For a slope strain we have

$$\varepsilon_t(s) = \dot{\varepsilon}_0[(t-s)h(t-s) - t\,h(t)] \tag{8}$$

and

$$\frac{d\,\varepsilon_t(s)}{ds} = -\dot{\varepsilon}_0 h(t-s) - \dot{\varepsilon}_0(t-s)\delta(t-s)\ . \tag{9}$$

Hence,

$$\sigma(t) = \dot{\varepsilon}_0 \int_0^\infty G(s)h(t-s)ds + \dot{\varepsilon}_0 \int_0^\infty (t-s)\delta(t-s)ds\ . \tag{10}$$

The second term is zero. The first term is non-zero only when $s < t$. Thus

$$\sigma(t) = \dot{\varepsilon}_0 \int_0^t G(s)ds = \dot{\varepsilon}_0\ \eta(t)\ . \tag{11}$$

Chapter 3

Problem 3.1-1

(A) The mechanical model representing the damped vibration absorber is shown in Fig. P3.1-1b. The same diagram is also shown in Fig. 3.2-2 but there the nodes represent velocities, not displacements, and the elements are shown as mechanical impedances. In Fig. P3.1-1b the elements are shown as mechanical relaxances.

(B) The transform equation of motion at the first node is

$$\bar{f}(s) = (Is^2 + 2E + F_1 s + 2E_1)\bar{x}(s) - (F_1 s + 2E_1)\bar{x}_1(s) \tag{1}$$

and that at the second node is

$$0 = -(F_1 s + 2E_1)\bar{x}(s) + (I_1 s^2 + F_1 s + 2E_1)\bar{x}_1(s) \ . \tag{2}$$

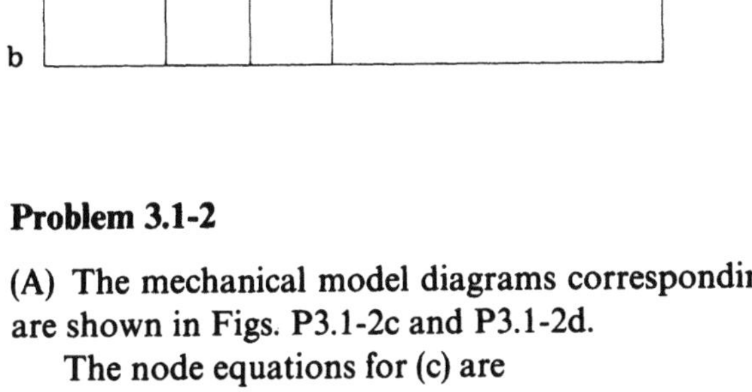

Fig. P3.1-1b. Mechanical model diagram representing damped vibration absorber

Problem 3.1-2

(A) The mechanical model diagrams corresponding to Figs. P3.1-2a and P3.1-2b are shown in Figs. P3.1-2c and P3.1-2d.

The node equations for (c) are

$$\bar{f}(s) = \left[\frac{E_a F_a s}{E_a + F_a s} + E\right]\bar{x}(s) - E\,\bar{x}_1(s) \tag{1}$$

and

$$0 = -E\,\bar{x}(s) + (Is^2 + Fs + E)\bar{x}_1(s) \ . \tag{2}$$

Elimination of $\bar{x}_1(s)$ leads to

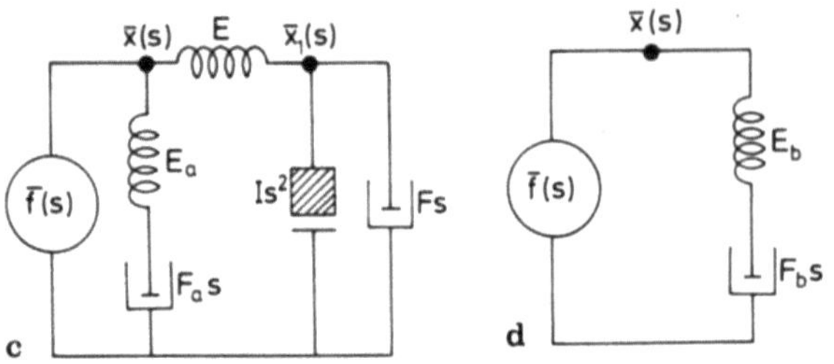

Fig. P3.1-2c,d. Mechanical model diagrams corresponding to Figs. P3.1-2a,b

$$\bar{Q}_{ma}(s) = \frac{E_a F_a s}{E_a + F_a s} + \frac{(Is^2 + Fs)E}{Is^2 + Fs + E} . \tag{3}$$

For (d) we obtain directly

$$\bar{Q}_{mb}(s) = \frac{E_b F_b s}{E_b + F_b s} . \tag{4}$$

(B) We now rewrite Eq. (3) as

$$\bar{Q}_{ma}(s) = \frac{E_a s}{s - s_a} + \frac{Es(Is + F)}{I(s - s_1)(s - s_2)} \tag{5}$$

where s_a, s_1, and s_2 are the poles of $\bar{Q}_{ma}(s)$, and

$$s_a = -E_a/F_a \tag{6}$$

while

$$s_{1,2} = -\frac{F}{2I} \pm \sqrt{\left(\frac{F}{2I}\right)^2 - \frac{E}{I}} . \tag{7}$$

Partial fraction decomposition of the second term in Eq. (5) leads to

$$\bar{Q}_{ma}(s) = \frac{E_a s}{s - s_a} + \frac{Es}{\sqrt{F^2 - 4EI}}\left(\frac{Is_1 + F}{s - s_1} - \frac{Is_2 + F}{s - s_2}\right) . \tag{8}$$

We have thus expressed $\bar{Q}_{ma}(s)$ as the sum of three terms, each associated with one of its poles. If we now set

$$\frac{E_a s}{s - s_c} = \frac{Es(Is_c + F)}{(s - s_c)\sqrt{F^2 - 4EI}} \tag{9}$$

where $s_c = s_a = s_2$, then

$$\bar{Q}_{ma}(s) = \frac{Es(Is_1 + F)}{(s - s_1)\sqrt{F^2 - 4EI}} , \tag{10}$$

i.e. $\bar{Q}_{ma}(s)$ then possesses a single pole only, and its behavior is qualitatively the same as that of $\bar{Q}_{mb}(s)$. We must, therefore, first establish the relations between the parameters of $\bar{Q}_{ma}(s)$ which ensure that Eq. (9) is satisfied. These relations are obtained by setting, in turn, $s_c = s_2$ and $s_c = s_a$ in Eq. (9). By Eq. (7)

$$s_2 = -(F + R)/2I \tag{11}$$

where $R^2 = F^2 - 4EI$. Substituting first Eq. (11), and then Eq. (6) into Eq. (9) gives

$$E_a = E(F - R)/2R \tag{12}$$

and

$$E_a = E(F - E_a I/F_a)/R \ . \tag{13}$$

Elimination of E_a between these two equations yields

$$F_a = \frac{EI(F - R)}{R(F + R)} \ . \tag{14}$$

Thus, if E_a and F_a are chosen according to Eqs. (12) and (14), then Eq. (9) is satisfied and we have

$$\bar{Q}_{ma}(s) = \frac{E(F + R)s}{2R(s - s_1)} \ . \tag{15}$$

since $Is_1 + F = (F + R)/2$. Now

$$\bar{Q}_{mb}(s) = \frac{E_b s}{s - s_b} \tag{16}$$

where $s_b = -E_b/F_b$. For the two mechanical relaxances to be the same we must have

$$E_b = \frac{E(F + R)}{2R} \tag{17}$$

and, from $s_b = s_1$, using Eq. (17),

$$F_b = \frac{EI(F + R)}{R(F - R)} \ . \tag{18}$$

(C) The foregoing analysis shows that the mechanical behavior of two different combinations of mechanical elements may be identical.

Problem 3.1-3

The mechanical model diagram representing the dynamic shear rheometer sketched in Fig. P3.1-3a is shown in Fig. P3.1-3b.

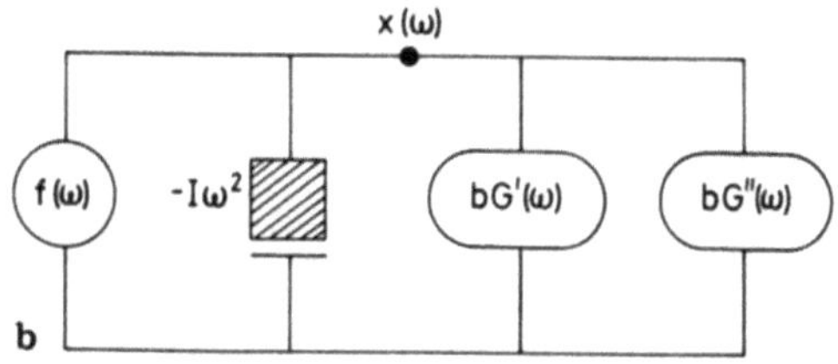

Fig. P3.1-3b. Mechanical model diagram for dynamic shear rheometer

The force, as measured by the load cell, is applied to the rigid shearing member having the harmonic inertance $-I\omega^2$, and through it to the specimens having the elastance $bG'(\omega)$ and frictance $j\omega b\eta'(\omega) = jbG''(\omega)$. The displacement will oscillate with the same frequency but will be shifted in phase. Hence we write

$$x(\omega) = x_0(\omega)\exp\{j[\omega t + \delta(\omega)]\} \ . \tag{3}$$

There is only a single node, and the node equation is

$$f(\omega) = [-I\omega^2 + bG'(\omega) + jbG''(\omega)]x(\omega) \ . \tag{4}$$

Introducing Eqs. (1) and (3) into Eq. (4) and cancelling $\exp(j\omega t)$ leads to

$$A_r(\omega)[\cos\delta(\omega) + j\sin\delta(\omega)] = [-I\omega^2 + bG'(\omega) + jbG''(\omega)] \tag{5}$$

where

$$A_r(\omega) = f_0/x_0(\omega) \ . \tag{6}$$

Separation of the real and imaginary parts gives

$$G'(\omega) = HA_r(\omega)\cos\delta(\omega) + HI\omega^2 \tag{7}$$

and

$$G''(\omega) = HA_r(\omega)\sin\delta(\omega) \tag{8}$$

where $H = 1/b$ is the shape factor.

The loss tangent becomes

$$\tan\theta(\omega) = \frac{A_r(\omega)\sin\delta(\omega)}{A_r(\omega)\cos\delta(\omega) + I\omega^2} \ . \tag{9}$$

The loss angle, $\theta(\omega)$, must not be confused with the *apparatus phase angle*, $\delta(\omega)$.

Problem 3.1-4

The mechanical model diagram is shown in Fig. P3.1-4 and the node equations are

$$f_1(\omega) = bE^*(\omega)x_1(\omega) - bE^*(\omega)x_2(\omega) \tag{1}$$

and

$$0 = -bE^*(\omega)x_1(\omega) + [bE^*(\omega) + E_T]x_2(\omega) \ . \tag{2}$$

Since $x_2(\omega) = f_2(\omega)/E_T$, and $x_1(\omega)$ and $f_2(\omega)$ are measured, only the second of these

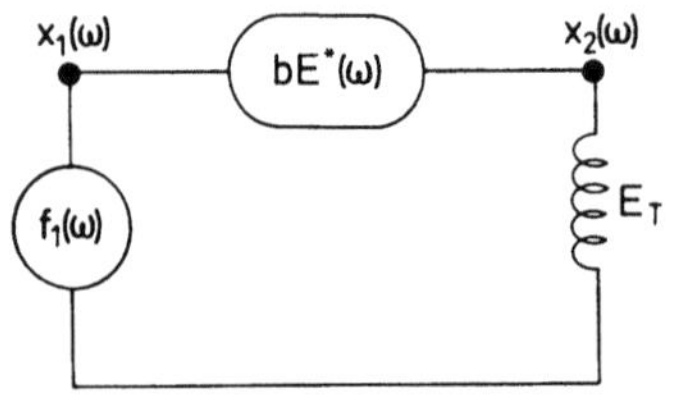

Fig. P3.1-4. Mechanical model diagram for dynamic tensile rheometer

equations is needed. We have

$$E^*(\omega) = \frac{HE_T f_2(\omega)}{E_T x_1(\omega) - f_2(\omega)} \tag{3}$$

where H is the shape factor. For a specimen with uniform cross-sectional area $H = L_0/A_0$ where L_0 and A_0 are the undeformed effective length and undeformed cross-sectional area, respectively. Now $x_1(\omega) = x_0 \exp(j\omega t)$, and $f_2(\omega) = f_0(\omega)\exp\{j[\omega t + \delta(\omega)]\}$, and, therefore,

$$f_2(\omega)/x_1(\omega) = A_r(\omega)[\cos \delta(\omega) + j \sin \delta(\omega)] \tag{4}$$

where $A_r(\omega) = f_0(\omega)/x_0$. After some complex algebra, we find

$$E'(\omega) = \frac{HE_T A_r(\omega)[E_T \cos \delta(\omega) - A_r(\omega)]}{A_r^2(\omega) - 2A_r(\omega)E_T \cos \delta(\omega) + E_T^2} \tag{5}$$

and

$$E''(\omega) = \frac{HE_T^2 A_r(\omega) \sin \delta(\omega)}{A_r^2(\omega) - 2A_r(\omega)E_T \cos \delta(\omega) + E_T^2} \; . \tag{6}$$

Thus, determination of the tensile storage and loss moduli from the amplitude ratio and phase angle between the displacement and force signals also requires knowledge of the internal elastance of the load cell.

Normally, E_T will be high enough so that it can be considered to be virtually infinite. In that case, Eqs. (5) and (6) reduce to

$$E'(\omega) = HA_r(\omega)\cos \delta(\omega) \tag{7}$$

and

$$E''(\omega) = HA_r(\omega)\sin \delta(\omega) \; . \tag{8}$$

Problem 3.1-5

(A) By Eq. (2) of Problem 3.1-4

$$x_2(\omega) = \frac{bE^*(\omega)}{E_T + bE^*(\omega)} x_1(\omega) \; . \tag{3}$$

This equation relates the displacement $x_2(\omega)$ to the displacement $x_1(\omega)$ in terms of the elastances E_T and $bE^*(\omega)$. Since $E^*(\omega)$ is complex, we may write

$$\frac{bE^*(\omega)}{E_T + bE^*(\omega)} = R(\omega)\exp[j\, \nu(\omega)] \tag{4}$$

where $R(\omega) = x_{20}(\omega)/x_{10}$ is the ratio of the peak amplitudes of the two displacements, and $\nu(\omega)$ is the phase angle between them. We find that

$$R(\omega) = \frac{b\tilde{E}(\omega)}{E_T\sqrt{1 + 2bE'(\omega)/E_T + [b\tilde{E}(\omega)/E_T]^2}} \tag{5}$$

and

$$\tan \nu(\omega) = \frac{E''(\omega)}{E'(\omega) + b\tilde{E}^2(\omega)/E_T} \tag{6}$$

where $\tilde{E}(\omega)$ is the absolute tensile modulus of the material. If E_T is large compared with $b\tilde{E}(\omega)$, then

$$\frac{x_{20}(\omega)}{x_{10}} \simeq \frac{b\tilde{E}(\omega)}{E_T}\,, \tag{7}$$

i.e. $x_2(\omega)$ can be made negligibly small with respect to $x_1(\omega)$ if E_T is made large enough with respect to $b\tilde{E}(\omega)$.

(B) We have

$$\frac{x_{20}(\omega)}{x_{10}} = \frac{A_r(\omega)}{E_T}\,. \tag{8}$$

Therefore, $A_r(\omega)/E_T \to 0$ as $x_{20}(\omega) \to 0$. Thus, Eqs. (1) and (2) follow from Eqs. (5) and (6) of Problem 3.1-4 if the displacement $x_2(\omega)$ is small compared with the displacement $x_1(\omega)$.

Problem 3.1-6

The mechanical model diagram shown in Fig. P3.1-4 must be modified as shown in Fig. P3.1-6.

Here, I_1 and I_2 represent the inertance (mass) associated with the motions at the driver and the pick-up, respectively. The node equation for the motion at the pick-up becomes

$$(1 - \omega^2 I_2/E_T)f_2(\omega) = bE^*(\omega)[x_1(\omega) - f_2(\omega)/E_T] \tag{1}$$

Hence, the equations for $E'(\omega)$ and $E''(\omega)$ derived in Problems 3.1-4 and 3.1-5 are modified simply by multiplying the right hand sides with the factor $1 - \omega^2 I_2/E_T$. Evidently, this correction in negligible if $E_T \gg \omega^2 I_2$. Only the mass of the moving parts between the specimen and the pick-up, but not that between the driver and the specimen, need to be taken into account. This mass will be largely that of the specimen holder (clamp) and the connecting rod, and may, therefore, be determined approximately by weighing. (Note that weight is a force, not a mass.)

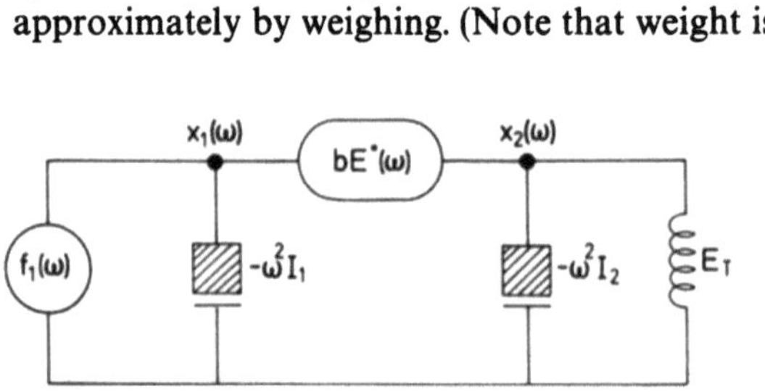

Fig. P3.1-6. Modification of mechanical model diagram shown in Fig. P3.1-4

Problem 3.1-7

We first draw the mechanical model diagram. The apparatus is *displacement driven*, i.e. the excitation is an angular displacement, not a torque. The driving transform

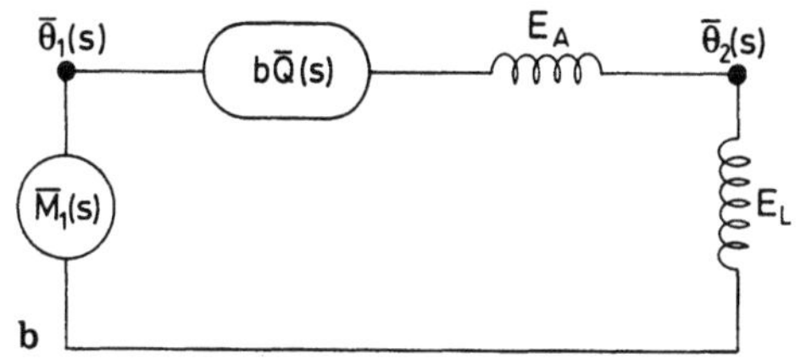

Fig. P3.1-7b. Mechanical model diagram for torsional rheometer

is $\bar{\theta}_1(s)$. The torque, $\bar{M}_1(s)$, required to produce the displacement, is transmitted to the load cell through the series combination of the specimen and the added elastance. The mechanical model diagram is shown in Fig. P3.1-7b.

Since the motion at the junction between the specimen and the added elastance is not of interest, the corresponding node has been omitted from the diagram. The load cell is represented by its internal elastance, E_L, and we have

$$\bar{\theta}_2(s) = \bar{M}_2(s)/E_L \tag{2}$$

where $\bar{M}_2(s)$ is the transform of the torque measured by the load cell, and $\bar{\theta}_2(s)$ is the transform of the motion (angular displacement) at the cell.

Except for the added elastance, the diagram is the same as that shown in Fig. P3.1-4. The fact that now torque and angular displacement replace force and displacement, that the respondances are rotational ones, and that the dimension of the specimen coefficient differs from that in Problem 3.1-4 is of no consequence to the analysis.

Again, only the equation for the second node is required. It is

$$0 = \left(E_L + \frac{E_A b\bar{Q}(s)}{E_A + b\bar{Q}(s)}\right)\bar{\theta}_2(s) - \frac{E_A b\bar{Q}(s)}{E_A + b\bar{Q}(s)}\bar{\theta}_1(s) \ . \tag{3}$$

Using Eq. (2) to eliminate $\bar{\theta}_2(s)$, Eq. (3) becomes

$$\bar{M}_2(s) + (1/E_L + 1/E_A)b\bar{Q}(s)\bar{M}_2(s) = b\bar{Q}(s)\bar{\theta}_1(s) \ . \tag{4}$$

Assume that the internal elastance of this load cell is very much larger than that of the added elastance. Then Eq. (4) reduces to

$$[HE_A + \bar{Q}(s)]\bar{M}_2(s) = \bar{Q}(s)E_A\bar{\theta}_1(s) \ . \tag{5}$$

This equation can now be specialized to the two kinds of excitation proposed in the problem.

(A) For a twist imposed as a step function of time, $\bar{\theta}_1(s) = \theta_0/s$. Inserting this, as well as $\bar{Q}(s) = s\bar{G}(s)$, and retransforming to the time domain leads to

$$G(t) - (E_A\theta_0)^{-1} \int_0^t G(t-u)d\,M_2(u) = (H/\theta_0)M_2(t) \tag{6}$$

which requires the numerical solution of a convolution integral. If the added elastance is replaced by a rigid connection, we have simply

$$G(t) = (H/\theta_0)M_2(t) \ . \tag{7}$$

(B) The harmonic solution is obtained immediately from Eq. (5). It is

$$[HE_A + G^*(\omega)]M_2(\omega) = G^*(\omega)E_A\theta_1(\omega) \,. \tag{8}$$

Letting $\theta_1(\omega) = \theta_1 \exp(j\omega t)$, we have $M_2(\omega) = M_2 \exp\{j[\omega t + \delta(\omega)]\}$. Hence

$$G^*(\omega) = \frac{HE_A M_2 \exp[j\,\delta(\omega)]}{E_A\theta_1 - M_2 \exp[j\,\delta(\omega)]} \,. \tag{9}$$

Separation of the real and imaginary parts yields the storage and loss modulus.

Problem 3.2-1

To obtain the equivalent mechanical model by the electrostatic analogy, a dot is placed inside each mesh of the electric circuit, one is placed outside, and the dots are connected with each other as illustrated in Fig. P3.2-1b. Finally, the mechanical model is redrawn as shown in Fig. P3.2-1c.

The mass with inertia I_1 which corresponds to the inductance, L_1, in Fig. P3.2-1c is in a physically unrealisable position. Its motion would have to be observed with respect to either node 1 or node 2 instead of ground which is physically impossible. Thus, while mechanical models can generally be represented by equivalent electrical circuits, the converse is not necessarily true.

(B) When the electromechanical analogy is used, the interconnection of the elements may be left as it is in the electric circuit. The node diagram is shown in Fig. P3.2-1d.

The mass with inertia I_2 and the spring with reciprocal elastance $1/E_2$ could be interchanged without disturbing the formal behavior of the model, referring the

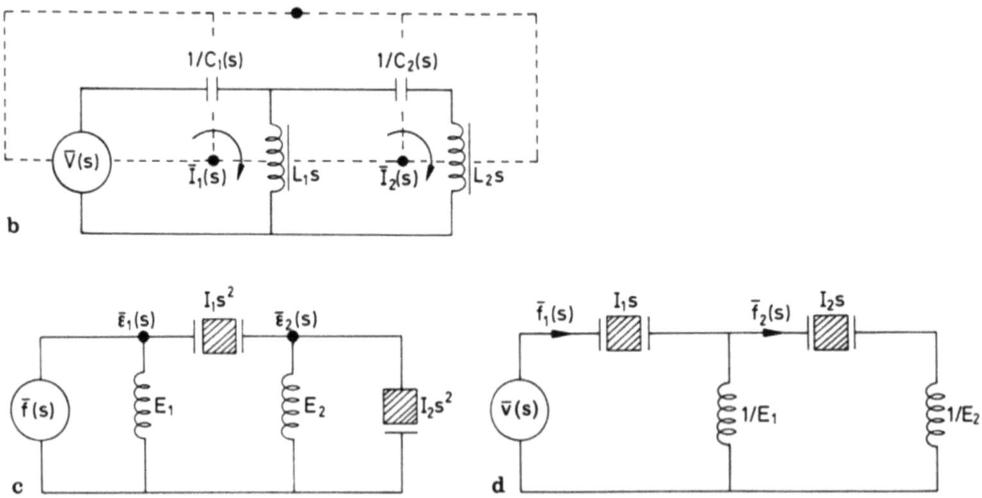

Fig. P3.2-1. b Two-mesh electrical circuit and equivalent mechanical model diagram. **c** Equivalent mechanical model diagram (electrostatic analogy). **d** Equivalent mechanical model diagram (electromagnetic analogy)

force acting through the mass to ground. However, the mass I_1 is still in a physically unrealizable position.

(C) The defective analogies cannot be used because either the capacitances (defective electrostatic analogy) or the inductances (defective electromagnetic analogy) cannot be represented in the mechanical model.

It should be noted that the model diagrams of Figs. P3.2-1c and P3.2-1d would lead to differential equations which are of the same form as that corresponding to the electrical circuit shown in Fig. P3.2-1a. It would not be possible, however, to construct a mechanical model conforming to the equation.

Problem 3.3-1

The relaxance of the Maxwell unit is

$$\bar{Q}(s) = \frac{G_M \tau s}{1 + \tau_M s} \tag{1}$$

while the retardance of the Voigt unit is given by

$$\bar{U}(s) = \frac{J}{1 + \tau_V s} \ . \tag{2}$$

Substituting these expressions into $\bar{Q}(s)\bar{U}(s) = 1$ leads to

$$(1 + \tau_V s)(1 + \tau_M s) = GJ\tau_M s \ . \tag{3}$$

This equation cannot be satisfied for either $s = -1/\tau_V$ or for $s = -1/\tau_M$ because either substitution would lead to $GJ = 0$ which violates the initial assumption that the models exist.

Problem 3.4-1

(A) By Eq. (3.4-127)

$$G_{3M}(t) = G_e + G \exp(-t/\tau) \ . \tag{1}$$

Hence, using Eq. (2.3-11),

$$Q_{3M}(t) = G_g \delta(t) - (G/\tau)\exp(-t/\tau) \ . \tag{2}$$

From Eq. (2.4-6) (see footnote) $\eta_{3M}(t)$ is obtained as

$$\eta_{3M}(t) = \int_0^t (t-u) Q_{3M}(u) du \ . \tag{3}$$

Substitution of Eq. (2) leads to

$$\eta_{3M}(t) = G_g \int_0^t (t-u)\delta(u) du - (G/\tau) \int_0^t (t-u)\exp(-u/\tau) du \ . \tag{4}$$

But the first integral is simply $G_g t$ and

$$\int_0^t (t-u)\exp(-u/\tau) du = \tau\{t - \tau[1 - \exp(-t/\tau)]\}. \tag{5}$$

Hence

$$\eta_{3M}(t) = G_e t + G\tau[1 - \exp(-t/\tau)] \quad . \tag{6}$$

(B) Now,

$$G_{3M}(t) = \int_0^t Q_{3M}(u)du \tag{7}$$

by Eq. (2.3-7). Hence

$$G_{3M}(t) = G_g \int_0^t \delta(u)du - (G/\tau) \int_0^t \exp(-u/\tau)du \tag{8}$$

which clearly leads to Eq. (1).

By Eq. (2.4-9)

$$G_{3M}(t) = \frac{d\,\eta_{3M}(t)}{dt} + \eta(0)\delta(t) \quad . \tag{9}$$

But $\eta(0) = 0$ by Eq. (6) and, therefore, Eq. (1) follows again.

(C) Finally, by Eq. $(2.4\text{-}8)_2$, with $\eta(0) = 0$, and $\eta'(0) = G(0) = G_g$, we have

$$Q_{3M}(t) = \frac{d^2\eta(t)}{dt^2} + G_g\delta(t) \tag{10}$$

which directly leads to Eq. (2).

Problem 3.4-2

We have $\varepsilon(t) = \varepsilon_0 h(t)$ for the excitation. Hence,

$$\frac{d\,\varepsilon(u)}{du} = \varepsilon_0\delta(u) \quad . \tag{2}$$

Substitution into Eq. (1) gives

$$\sigma(t) = \varepsilon_0 \int_0^t G(t-u)\delta(u)du = \varepsilon_0 G(t) \tag{3}$$

by Eq. $(\text{A2-15})_2$. Hence,

$$\sigma(t) = \varepsilon_0[G_e + G\exp(-t/\tau)] \tag{5}$$

as required [cf. Eq. $(3.4\text{-}13)_1$].

Problem 3.4-3

(A) By Eq. (3.4-153)

$$J_{4V}(t) = J_g + J[1 - \exp(-t/\tau)] + \phi_f t \quad . \tag{1}$$

Hence, using Eq. (2.3-12),

$$U_{4V}(t) = J_g\delta(t) + (J/\tau)\exp(-t/\tau) + \phi_f \quad . \tag{2}$$

(B) By the use of

$$J_{4v}(t) = \int_0^t U_{4v}(u)du \tag{3}$$

[cf. Eq. (2.3-10)] we readily recover Eq. (1).

(C) Similarly, we have

$$\chi_{4v}(t) = \int_0^t J_{4v}(u)du \ . \tag{4}$$

The integration gives

$$\chi_{4v}(t) = J_e^{\{o\}}t - J\tau[1 - \exp(-t/\tau)] + \tfrac{1}{2}\phi_f t^2 \tag{5}$$

since $J_g + J = J_e^{\{o\}}$.

(D) In the complex plane

$$\bar{\chi}_{4v}(s) = \bar{J}_{4v}(s)/s = \frac{Jg}{s} + \frac{J}{s^2(1 + \tau s)} + \frac{\phi_f}{s^3} \ . \tag{6}$$

The use of LTPs (12) and (18) immediately leads to Eq. (5).

Problem 3.4-4

Since we want to find the stress as a function of the strain, we use the 3-parameter Maxwell model as the appropriate standard arrheodictic model.

(A) By Eq. (3.7-5) the operator equation of the model is

$$\sigma(t) + \tau\frac{d\,\sigma(t)}{dt} = G_e\varepsilon(t) + G_g\tau\frac{d\,\varepsilon(t)}{dt} \ . \tag{1}$$

Division by τ and multiplication by the integrating factor $\exp(t/\tau)$ gives

$$(1/\tau)\exp(t/\tau)\sigma(t) + \exp(t/\tau)\frac{d\,\sigma(t)}{dt} = (G_e/\tau)\exp(t/\tau)\varepsilon(t) + G_g\exp(t/\tau)\frac{d\,\varepsilon(t)}{dt} \ . \tag{2}$$

The left hand side is a total differential. Hence

$$\exp(t/\tau)\sigma(t) = (G_e/\tau)\int_0^t \exp(u/\tau)\varepsilon(u)du + G_g\int_0^t \exp(u/\tau)\frac{d\,\varepsilon(u)}{du}du \ . \tag{3}$$

Integrating the first integral on the right hand side by parts gives

$$\int_0^t \exp(u/\tau)\varepsilon(u)du = \tau\,\varepsilon(t)\exp(t/\tau) - \int_0^t \exp(u/\tau)\frac{d\,\varepsilon(u)}{du}du \ . \tag{4}$$

Inserting this into Eq. (3) and multiplying both sides by $\exp(-t/\tau)$ leads to

$$\sigma(t) = G_e\varepsilon(t) + (G_g - G_e)\int_0^t \exp[-(t-u)/\tau]\frac{d\,\varepsilon(u)}{du}du \ . \tag{5}$$

(B) The step response of the 3-parameter Maxwell model is

$$G(t) = G_e + (G_g - G_e)\exp(-t/\tau) \tag{6}$$

by Eq. (3.4-127). Introducing this into Eq. $(2.3\text{-}16)_1$ immediately gives Eq. (5). The two approaches are thus equivalent.

Problem 3.4-5

The Laplace transform of the excitation is $\varepsilon_0\omega/(s^2+\omega^2)$. Substituting this into $\bar{\sigma}(s) = \bar{Q}_{3M}(s)\bar{\varepsilon}(s)$ gives

$$\sigma(s) = \varepsilon_0\left(G_e + \frac{G\tau s}{1+\tau s}\right)\frac{\omega}{s^2+\omega^2} \; .$$

From Eq. (1) the total stress results as

$$\sigma(t) = \varepsilon_0\left[\left(G_e + \frac{G\omega^2\tau^2}{1+\omega^2\tau^2}\right)\sin\omega t + \frac{G\omega\tau}{1+\omega^2\tau^2}\cos\omega t - \frac{G\omega\tau}{1+\omega^2\tau^2}\exp(-t/\tau)\right] \tag{2}$$

by the use of LTP (21). The last term on the right is the transient term which will have decayed to negligible proportions in the steady state. We therefore find the in-phase component of the steady-state response to be

$$\sigma'(\omega) = \varepsilon_0\left(G_e + \frac{G\omega^2\tau^2}{1+\omega^2\tau^2}\right)\sin\omega t = \varepsilon_0 G'(\omega)\sin\omega t \tag{3}$$

and the out-of-phase component to be

$$\sigma''(\omega) = \varepsilon_0\left(\frac{G\omega\tau}{1+\omega^2\tau^2}\right)\cos\omega t = \varepsilon_0 G''(\omega)\cos\omega t \; . \tag{4}$$

The total response in the steady state, therefore, is given by

$$\sigma(\omega) = \varepsilon_0[G'(\omega)\sin\omega t + G''(\omega)\cos\omega t] \; . \tag{5}$$

Problem 3.4-6

From Eq. (3.4-23),

$$\varepsilon_{as}(t) = \varepsilon_0 + \lim_{t\to\infty}\varepsilon[1-\exp(-t/\tau)] + \dot{\varepsilon}_0 t = \varepsilon_e^{\{o\}} + \dot{\varepsilon}_0 t \tag{1}$$

where we used Eq. $(3.4\text{-}24)_2$. Eq. (3.4-39) gives

$$\varepsilon_{as}(t) = \lim_{t\to\infty}\varepsilon[1-\exp(-t/\tau)] + \dot{\varepsilon}_0 t = \varepsilon_e^{\{o\}} + \dot{\varepsilon}_0 t \; . \tag{2}$$

The difference is simply that the equilibrium strain, $\varepsilon_e^{\{o\}}$, consists of the instantaneous strain, ε_0, plus the delayed strain, ε, in the first case, and of the delayed strain alone in the second case.

Problem 3.4-7

(A) The relaxance of the model is

$$\bar{Q}_{4a}(s) = \frac{1}{1/G' + 1/\bar{Q}'(s)} \tag{1}$$

where

$$\bar{Q}'(s) = \frac{G\tau s}{1 + \tau s} + \eta' s \tag{2}$$

and where $\tau = \eta/G$. Introducing Eq. (2) into Eq. (1) gives the relaxance, after some rearrangement, as

$$\bar{Q}_{4a}(s) = \frac{[(\eta + \eta')GG' + \eta\eta' G' s]s}{(G + \eta s)(G' + \eta' s) + G\eta s} . \tag{3}$$

(B) By Eq. (3.4-121)

$$\eta_f = \lim_{s \to 0} \bar{Q}_{4a}(s)/s = \eta + \eta' . \tag{4}$$

Hence, the model represents rheodictic behavior.

(C) By Eq. (3.4-112)

$$G_g = \lim_{s \to \infty} \bar{Q}_{4a}(s) = G' . \tag{5}$$

Thus, its instantaneous modulus is finite.

(D) The same conclusion can be reached simply by inspection. Since the model cannot be traversed by rigid connections and springs only, it represents *rheodictic* behavior. Since it cannot be traversed by rigid connections and dashpots only, it is a *standard* model, i.e. its instantaneous modulus is finite.

Problem 3.4-8

The relaxance of the model was found in Problem 3.4-7. Division by GG′ gives

$$\bar{Q}_{4a}(s) = \frac{(\eta\eta'/G)s^2 + (\eta + \eta')s}{(\eta\eta'/GG')s^2 + (\eta/G + \eta/G' + \eta'/G')s + 1} . \tag{1}$$

This must be brought into the form of the relaxance of the standard linear liquid,

$$\bar{Q}_{4M}(s) = \frac{G_0\tau_0 s}{1 + \tau_0 s} + \frac{G_1\tau_1 s}{1 + \tau_1 s} . \tag{2}$$

$\bar{Q}_{4M}(s)$ has poles at $s_0 = -1/\tau_0$ and $s_1 = -1/\tau_1$. $\bar{Q}_{4a}$ also has two poles and these must be identical to those of $\bar{Q}_{4M}(s)$ if the two relaxances are to be identical. The poles of $\bar{Q}_{4a}(s)$ are found by solving the quadratic equation obtained by equating the denominator to zero. We find

$$s_0 = -(b/2a)(1 - \sqrt{1 - 4a/b^2}) \tag{3}$$

and

$$s_1 = -(b/2a)(1 + \sqrt{1 - 4a/b^2}) \tag{4}$$

where

$$a = \eta\eta'/GG' \tag{5}$$

and

$$b = \eta/G + \eta/G' + \eta'/G' \ . \tag{6}$$

Thus, the relaxation times become

$$\tau_0 = \frac{2a}{b - \sqrt{b^2 - 4a}} \tag{7}$$

and

$$\tau_1 = \frac{2a}{b + \sqrt{b^2 - 4a}} \ . \tag{8}$$

It is easily seen that $b^2 > 4a$. We must have

$$(\eta/G + \eta/G' + \eta'/G')^2 > 4\eta\eta'/GG' \ . \tag{9}$$

This will certainly be true if

$$(\eta/G + \eta'/G')^2 > 4\eta\eta'/GG' \tag{10}$$

because $\eta\eta'/GG'$ is necessarily greater than zero. Squaring and simplifying gives

$$(\eta/G)^2 + (\eta'/G')^2 > 2\eta\eta'/G'G' \ . \tag{11}$$

But we know this is true because

$$(\eta/G - \eta'/G') \geq 0 \ , \tag{12}$$

the equality applying only when $\tau_0 = \tau_1$. Thus, the square root is real and $\tau_0 > \tau_1$.

The moduli are found by a partial fraction decomposition of Eq. (1) expressed in the form

$$\bar{Q}_{4a}(s) = \frac{(\eta\eta'/G)s^2 + (\eta + \eta')s}{(1 + \tau_0 s)(1 + \tau_1 s)} \ . \tag{13}$$

This yields

$$G_0 = \frac{(\eta + \eta')\tau_0 - \eta'\tau}{\tau_0 - \tau_1}$$

and

$$G_1 = \frac{\eta'\tau - (\eta + \eta')\tau_1}{\tau_0 - \tau_1} \ . \tag{15}$$

Finally, the viscosities of the dashpots are obtained from $\eta_0 = G_0\tau_0$, and $\eta_1 = G_1\eta_1$, respectively.

Problem 3.4-9

The relaxance of the model is

$$\bar{Q}_{4c}(s) = \frac{G_2(G_1 + \eta_1 s)}{G_1 + G_2 + \eta_1 s} + \eta_2 s \tag{1}$$

and we find

$$G_g = \lim_{s \to \infty} \bar{Q}_{4c}(s) = \infty \tag{2}$$

and

$$\eta_f = \lim_{s \to 0} \bar{Q}_{4c}(s)/s = \infty \ . \tag{3}$$

Hence, the instantaneous modulus is infinite and the model represents arrheodictic behavior. $\bar{Q}_{4c}(s)$ has a single pole. The equivalent series-parallel model, therefore, is that given by Fig. 3.4-9. Writing Eq. (3.4-59) in the form

$$\bar{Q}_{4M}(s) = \frac{G' + (G' + G)\tau s}{1 + \tau s} + \eta' s \tag{4}$$

and comparing it with

$$\bar{Q}_{4c}(s) = \frac{\dfrac{G_1 G_2}{G_1 + G_2} + \dfrac{G_2 \eta_1}{G_1 + G_2} s}{1 + \dfrac{\eta_1}{G_1 + G_2} s} + \eta_2 s \tag{5}$$

we immediately have

$$G' = \frac{G_1 G_2}{G_1 + G_2} \tag{6}$$

and

$$\eta' = \eta_2 \ . \tag{7}$$

Further,

$$\frac{(G' + G)\eta}{G} = \frac{G_2 \eta_1}{G_1 + G_2} \tag{8}$$

and

$$\frac{\eta}{G} = \frac{\eta_1}{G_1 + G_2} \ . \tag{9}$$

Solving for G and η gives

$$G = \frac{G_2^2}{G_1 + G_2} \ . \tag{10}$$

and

$$\eta = \left(\frac{G_2}{G_1 + G_2}\right)^2 \eta_1 \ . \tag{11}$$

Problem 3.4-10

Laplace transformation of Eq. (1) yields

$$\bar{G}(s) = \frac{G_e}{s} + \frac{G\tau_M}{1 + \tau_M s} \ . \tag{3}$$

Now, by $s\bar{J}(s) = 1/s\bar{G}(s)$ [cf. Eq. (2.3-13)]

$$\bar{J}(s) = \frac{1 + \tau_M s}{s(G_e + G_g \tau_M s)} \tag{4}$$

where $G_g = G_e + G$. Division of the numerator by the denominator yields

$$\bar{J}(s) = \frac{1}{G_g s} + \frac{1/G_e - 1/G_g}{s(1 + G_g \tau_M s/G_e)} = \frac{J_g}{s} + \frac{J}{s(1 + \tau_V s)} . \tag{5}$$

Retransformation then leads to Eq. (2).

Problem 3.5-1

Since it can be traversed along springs only, this model describes arrheodictic behavior. Because it has only one dashpot, one of the springs is redundant. Hence, it must be equivalent to the 3-parameter Voigt model shown in Fig. 3.4-1, whose retardance is

$$\bar{U}_{3V}(s) = \frac{J_e + J_g \tau_V s}{1 + \tau_V s} . \tag{1}$$

The retardance of the redundant model is

$$\bar{U}_4(s) = J_1 + [1/J_2 + 1/(J' + \phi'/s)]^{-1} \tag{2}$$

which may be expressed as

$$\bar{U}_4(s) = \frac{J_1 + J_2 + [J_1(J_2 + J') + J_2 J']s/\phi'}{1 + (J_2 + J')s/\phi} . \tag{3}$$

Comparing the two retardances, we have

$$J_e = J_1 + J_2 \tag{4}$$

$$\tau_V = (J_2 + J')/\phi' \tag{5}$$

and

$$J_g = J_1 + \frac{J_2 J'}{J_2 + J'} . \tag{6}$$

But $\tau_V = J/\phi = (J_e - J_g)/\phi$ and, therefore,

$$\phi = \left(\frac{J_2}{J_2 + J'}\right)^2 \phi' . \tag{7}$$

Problem 3.5-2

The relaxance of the redundant model is

$$\bar{Q}_4(s) = G_1 + \frac{G_2(G' + \eta' s)}{G_2 + G' + \eta' s} \tag{4}$$

Now, if the two models are to represent identical behavior, the glassy and equilibrium moduli must be the same. Hence, by the initial and final value theorems,

$$\lim_{s \to 0} \bar{Q}_{3M}(s) = \lim_{s \to 0} \bar{Q}_4(s) \tag{5}$$

and

$$\lim_{s \to \infty} \bar{Q}_{3M}(s) = \lim_{s \to \infty} \bar{Q}_4(s) \ . \tag{6}$$

The expressions for G_e and G follow immediately from Eqs. (5) and (6), respectively. To find the expression for η, consider that $\bar{Q}_4(s)$ has a single pole which lies at

$$s = s_\infty^4 = -(G_2 + G')/\eta' \ . \tag{7}$$

The standard solid model also possesses a single pole, this one at

$$s = s_\infty^{3M} = -G/\eta \ . \tag{8}$$

Because each model has a single pole only, they will describe the same behavior only if the two poles are identical. We have, therefore,

$$s_\infty^{3M} = s_\infty^4 \ . \tag{9}$$

Equations (7), (8) and (2) then yield Eq. (3).

Problem 3.5-3

(A) The relaxance of the model is

$$\bar{Q}_6(s) = \frac{1}{J_1 + J_2/(1 + \tau_2 s) + \phi_2/s} + G_3 + \eta_3 s \ . \tag{1}$$

The values of the equilibrium and instantaneous moduli are obtained from the limit value theorems. By Eqs. (3.4-115) and (3.4-112)

$$G_e = \lim_{s \to 0} \bar{Q}_6(s) = G_3 \tag{2}$$

and

$$G_g = \lim_{s \to \infty} \bar{Q}_6(s) = \infty \ . \tag{3}$$

(B) Adding a spring $G = 1/J$ in series with the Voigt unit on the right changes the relaxance to

$$\bar{Q}_7(s) = \frac{1}{J_1 + J_2/(1 + \tau_2 s) + \phi_2/s} + \frac{1}{J + J_3/(1 + \tau_3 s)} \ . \tag{4}$$

where $\tau_3 = \eta_3/G_3$. Now

$$G_e = \lim_{s \to 0} \bar{Q}_7(s) = GG_3/(G + G_3) \tag{5}$$

where $G_3 = 1/J_3$, and

$$G_g = \lim_{s\to\infty} \bar{Q}_7(s) = G_1 + G \tag{6}$$

where $G_1 = 1/J_1$. Thus, the addition of the spring has resulted in a finite instantaneous modulus for the model.

Problem 3.6-1

For a rheodictic material, rearrangement of Eq. (3.5-11) gives

$$\log[\eta_f - \eta(t)] = \log\left[\sum_n \eta_n \exp(-t/\tau_n) + \eta_0 \exp(-t/\tau_0)\right] . \tag{1}$$

Hence, a plot of $\log[\eta_f - \eta(t)]$ vs. t has the asymptote

$$\log[\eta_f - \eta(t)] = \log \eta_0 - t/2.303\ \tau_0 \tag{2}$$

and τ_0 can be estimated from its slope.

Problem 3.6-2

The G_k are obtained from the matrix equation $A_j = B_{jk}G_k$, A_j and B_{jk} being given by Eqs. (3.6-8) and (3.6-15), and by Eqs. (3.6-13) and (3.6-16), respectively. Dividing A_j by 10^6 for simplicity, the matrix equation becomes

$$\begin{bmatrix} 194.9 \\ 180.9 \\ 127.7 \\ 49.0 \\ 9.86 \\ 1.09 \\ 0.19 \end{bmatrix} = \begin{bmatrix} 1.000 & 1.000 & 1.000 & 1.000 & 1.000 & 1.000 & 1.000 \\ 0.001 & 0.500 & 0.933 & 0.993 & 0.999 & 1.000 & 1.000 \\ 0 & 0.001 & 0.500 & 0.933 & 0.993 & 0.999 & 1.000 \\ 0 & 0 & 0.001 & 0.500 & 0.933 & 0.993 & 0.999 \\ 0 & 0 & 0 & 0.001 & 0.500 & 0.933 & 0.993 \\ 0 & 0 & 0 & 0 & 0.001 & 0.500 & 0.933 \\ 0 & 0 & 0 & 0 & 0 & 0.001 & 0.500 \end{bmatrix} \begin{bmatrix} G_0 \\ G_1 \\ G_2 \\ G_3 \\ G_4 \\ G_5 \\ G_6 \end{bmatrix} \tag{1}$$

where the 7 × 7 matrix is a straightforward extension of the 4 × 4 matrix in Eq. (3.6-17).

Approximating the matrix as specified in the problem statement, we obtain the set of equations

$$0.5\ G_6 = 0.19 \qquad (2)_1$$

$$0.5\ G_5 + 0.933\ G_6 = 1.09 \qquad (2)_2$$

$$0.5\ G_4 + 0.933\ G_5 + 0.993\ G_6 = 9.86 \qquad (2)_3$$

$$0.5\ G_3 + 0.933\ G_4 + 0.993\ G_5 + G_6 = 49.0 \qquad (2)_4$$

$$0.5\ G_2 + 0.933\ G_3 + 0.993\ G_4 + G_5 + G_6 = 127.7 \qquad (2)_5$$

$$0.5\ G_1 + 0.933\ G_2 + 0.993\ G_3 + G_4 + G_5 + G_6 = 180.9 \qquad (2)_6$$

$$G_0 + G_1 + G_2 + G_3 + G_4 + G_5 + G_6 = 194.9 \qquad (2)_7$$

which can be solved for the G_k's one after another. The results are: The values of

k	$G_k \times 10^{-6}$(N/m^2)	τ_k (second)
0	0.90	1.443×10^{-8}
1	11.91	1.443×10^{-7}
2	99.97	1.443×10^{-6}
3	64.05	1.443×10^{-5}
4	16.22	1.443×10^{-4}
5	1.47	1.443×10^{-3}
6	0.38	1.443×10^{-2}
$G_g - G_e$	194.90	

the τ_k needed in reconstructing G(t) are included. They are obtained from Eq. (3.6-12) with $m = -8$ which follows from Eq. (3.6-11).

The values calculated by approximating B_{jk} by a triangular matrix are quite close to the exact values obtained by computer inversion of B_{jk}. In this case, therefore, the approximation is acceptable.

The plot of log G(t) vs. log t with the experimental values shown as full circles is displayed in Fig. P3.6-2. The fit is quite satisfactory.

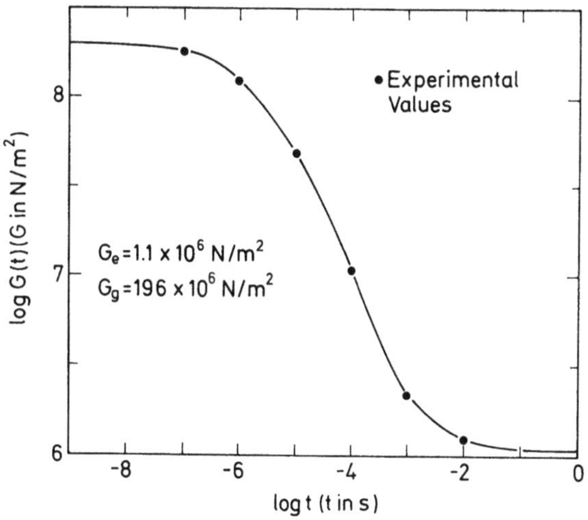

Fig. P3.6-2. Plot of experimental and reconstructed data of log G(t) vs. log t

Problem 3.7-1

Introducing Eq. (3.4-21) into Eq. $(2.1\text{-}15)_2$ and clearing of fractions yields

$$s\bar{\varepsilon}(s) + \tau s^2\bar{\varepsilon}(s) = \phi_f\bar{\sigma}(s) + (J_e + \phi_f\tau)s\bar{\sigma}(s) + J_g\tau s^2\bar{\sigma}(s) \; . \tag{1}$$

Retransformation results in the operator equation

$$\frac{d\,\varepsilon(t)}{dt} + \tau\frac{d^2\varepsilon(t)}{dt^2} = \phi_f\sigma(t) + (J_e + \phi_f\tau)\frac{d\,\sigma(t)}{dt} + J_g\tau\frac{d^2\sigma(t)}{dt^2} \; . \tag{2}$$

Chapter 4

Problem 4.1-1

Our task is to determine the contribution which the pole at s = jω alone makes to $\bar{Q}(s)/(s - j\omega)$, excluding the branch cut responsible for the transient behavior.

According to Cauchy's Integral Formula, if f(s) is a function of the complex variable s, and f(s) is analytic everywhere inside a closed contour of C in the complex plane, then the value of f(s) at the point a within the contour, i.e., f(a), is given by

$$f(a) = \frac{1}{2\pi j} \oint_C \frac{f(s)}{s - a} ds \ . \tag{3}$$

The integral is taken counterclockwise around the contour C.

By Cauchy's Residue Theorem the residue Res(g) associated with a simple pole of the function g(s) is given by

$$\mathrm{Res}(g) = \frac{1}{2\pi j} \oint_{C'} g(s) ds \ , \tag{4}$$

where C′ is a contour which encloses a region of the complex plane within which that pole is the only singularity.

Consider the contour traced out by the lines marked with arrows in Fig. P4.1-1. $\bar{Q}(s)$ is analytic everywhere inside this contour because the branch cut along the negative real axis is by-passed by detouring around the branch point at the origin. The contour includes the pole at s = jω. Hence, by Cauchy's Integral Formula, Eq. (3), the value of $\bar{Q}(s)$ at the point s = jω is given by Eq. (2) taking the integral along the indicated contour.

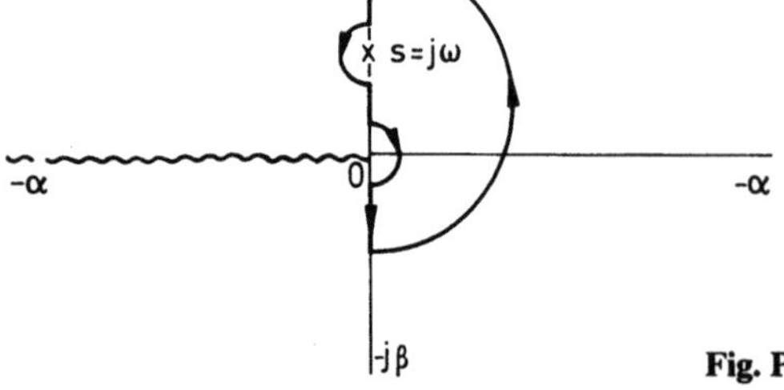

Fig. P4.1-1. Contour of integration for $\bar{Q}(s)/(s - j\omega)$

By Cauchy's Residue Theorem, Eq. (4), the integral taken around the same ($C = C'$) or any topologically equivalent contour which excludes all other singularities of $\bar{Q}(s)/(s - j\omega)$, represents the contribution arising solely from the pole at $s = j\omega$. Thus $\bar{Q}(j\omega)$ is seen to be the material function representing the response of the material to a harmonic strain excitation in the steady-state. Equation (1) follows through the equivalence of the notations $G^*(\omega)$ and $\bar{Q}(j\omega)$.

The derivation remains valid if the branch cut is replaced by a finite or infinite set of poles along the negative real axis, i.e. when one considers a discrete distribution of relaxation times or line spectrum (see Sect. 4.7). The derivation given here is therefore a more general one including that presented in Sect. 2.5.

An actual integration along the contour above will be undertaken in Sect. 8.3.1.1.

Problem 4.1-2

When the retardance is that of an arrheodictic material, $\bar{U}^{\times}(s)$, its singularities are the same as those of $\bar{Q}(s)$, shown in Fig. 4.1-1. When the retardance is that of a rheodictic material, $\bar{U}^{\circ}(s)$, there is a pole at the origin in addition to the branch point there. These singularities are shown in Fig. 4.1-2.

In either case, the contour of integration is the same. Hence, the derivation of Eq. (1) is entirely analogous to that of Eq. (2) of Problem 4.1-1. Again, it remains valid also for a discrete distribution of retardation times.

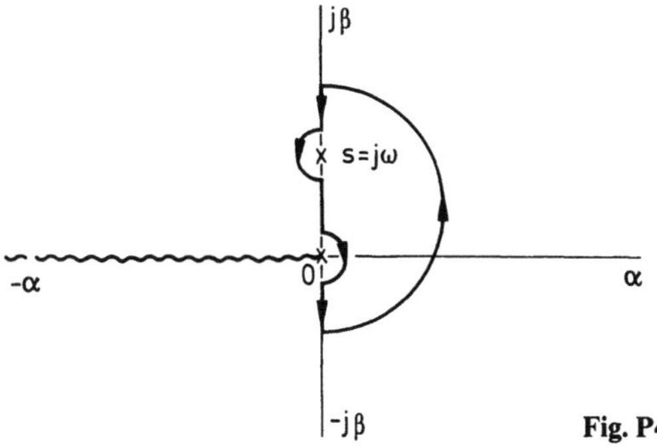

Fig. P4.1-2. Contour of integration for $\bar{U}^{\circ}(s)/(s - j\omega)$

Problem 4.1-3

We have

$$\bar{\dot{G}}(s) = \mathscr{L}\,\dot{G}(t) = \mathscr{L}\frac{d\,G(t)}{dt} \ . \tag{2}$$

But, by Eq. (4.1-5), the canonical representation of G(t) is

$$G(t) = \{G_e\} + \int_{-\infty}^{\infty} H(\tau)\exp(-t/\tau)d\ln\tau \ . \tag{3}$$

Hence,

$$\dot{G}(t) = -\int_{-\infty}^{\infty} \tau^{-1}H(\tau)\exp(-t/\tau)d\ln\tau \tag{4}$$

and

$$\bar{\dot{G}}(s) = -\int_{-\infty}^{\infty} H(\tau)\frac{1}{1+\tau s}d\ln\tau \ . \tag{5}$$

Now, by Eq. (4.1-10), one of the two canonical representations of $\bar{Q}(s)$ is

$$\bar{Q}(s) = G_g - \int_{-\infty}^{\infty} H(\tau)\frac{1}{1+\tau s}d\ln\tau \ , \tag{6}$$

and, therefore, Eq. (1) follows.

Problem 4.1-4

By Eqs. (2.5-39), (2.5-40) and (2.5-41),

$$\eta^*(\omega) = \eta'(\omega) - j\,\eta''(\omega) = G^*(\omega)/j\omega = G''(\omega)/\omega - j\,G'(\omega)/\omega \ . \tag{1}$$

Hence, Using Eq. (4.1-6)

$$\eta^*(\omega) = \{G_e\}/j\omega + \int_{-\infty}^{\infty} \tau H(\tau)\frac{1}{1+j\omega\tau}d\ln\tau \ . \tag{2}$$

Further, by using Eq. (4.1-8)

$$\eta'(\omega) = \int_{0}^{\infty} \tau H(\tau)\frac{1}{1+\omega^2\tau^2}d\ln\tau \tag{3}$$

and, using Eq. (4.1-7)

$$\eta''(\omega) = \{G_e\}/\omega + \int_{0}^{\infty} \tau H(\tau)\frac{\omega\tau}{1+\omega^2\tau^2}d\ln\tau \ . \tag{4}$$

Problem 4.1-5

By Eq. (4.1-2)

$$\tau H(\tau) = \tau^2\hat{Q}(\tau) = \tau\hat{G}(\tau) = \hat{\eta}(\tau) \ . \tag{1}$$

Since $\hat{\eta}(\tau)$ has the dimensions of a viscosity, so does $\tau H(\tau)$. This proves the first part of the problem. Now

$$\tau L(\tau) = \tau^2\hat{U}(\tau) = \tau^2\hat{\phi}(\tau) \tag{2}$$

and it is seen that $\tau L(\tau)$ does not have the dimensions of a fluidity. Hence, $\tau L(\tau)$ cannot be considered a fluidity density on retardation time.

Problem 4.1-6

The Laplace transform of the response is

$$\bar{\chi}(s) = \bar{U}(s)/s^2 \ . \tag{1}$$

by Eq. $(2.4\text{-}10)_2$. The canonical representation analogous to Eq. (4.1-16) will be obtained by inversion of $\bar{\chi}(s) = \bar{U}_\bullet(s)/s^2$, where $\bar{U}_\bullet(s)$ is given by Eq. (4.1-26). Division by s^2 yields

$$\bar{\chi}(s) = \frac{J_e^{\{o\}}}{s^2} - \int_{-\infty}^{\infty} \tau L(\tau) \frac{d \ln \tau}{s(1 + \tau s)} + \frac{\{\phi_f\}}{s^3} \tag{2}$$

and transformation into the time domain using LTPs (11) and (12) leads to

$$\chi(t) = J_e t - \int_{-\infty}^{\infty} \tau L(\tau)[1 - \exp(-t/\tau)] d \ln \tau + \{\phi_f\} t^2/2 \tag{3}$$

which may be compared with Eq. (4.1-16).

Problem 4.1-7

Division of Eq. (1) by s and transformation into the time domain gives

$$G(t) = G_g - \int_0^{\infty} \zeta^{-1} M(\zeta)[1 - \exp(-\zeta t)] d\zeta \ . \tag{4}$$

Using Eq. (2) then leads directly to Eq. (3).

Problem 4.1-8

(A) By Eqs. (2.4-6)

$$R(t) = \eta(t)/t = t^{-1} \int_0^t G(u) du \tag{2}$$

Multiplication of both sides by t and differentiating yields

$$G(t) = R(t)\left[1 + \frac{d \log R(t)}{d \log t}\right] . \tag{3}$$

(B) We take the Laplace transform on both sides of Eq. $(2)_1$. But division by t in the time domain is equivalent to integration in the complex plane from s to ∞ (complex integration). Thus,

$$\bar{R}(s) = \int_s^{\infty} \bar{\eta}(x) dx = \int_s^{\infty} \frac{\bar{G}(x)}{x} dx = \int_s^{\infty} \frac{\bar{Q}(x)}{x^2} dx \ . \tag{4}$$

(C) Using Eq. (4.1-15), Eq. $(4)_1$ yields

$$\bar{R}(s) = \frac{\{G_e\}}{s} + \int_s^{\infty} \int_{-\infty}^{\infty} \tau H(\tau) \frac{d \ln \tau}{x(1 + \tau x)} dx \tag{5}$$

which gives

$$\bar{R}(s) = \{G_e\}/s + \int_{-\infty}^{\infty} \tau H(\tau) \ln(1 + 1/\tau s) d \ln \tau \ . \tag{6}$$

Retransformation into the time domain using LTP (16) leads to

$$R(t) = \{G_e\} + t^{-1} \int_{-\infty}^{\infty} \tau H(\tau)[1 - \exp(-t/\tau)] d \ln \tau \ . \tag{7}$$

(D) Equation (7) can be obtained directly from Eq. (4.1-16) through dividing by t.

Problem 4.2-1

We have

$$G(t) = \{G_e\} + \int_{-\infty}^{\infty} H(\tau) \exp(-t/\tau) d \ln \tau \tag{2}$$

A glance at Fig. 4.1-2 shows that

$$\exp(-t/\tau) \simeq \begin{cases} 0 & \text{when } \log t/\tau > 0, \text{ i.e. } \tau < t \\ 1 & \text{when } \log t/\tau < 0, \text{ i.e. } \tau > t \ . \end{cases} \tag{3}$$

Thus $\exp(-t/\tau)$ may be approximated by the step function, $h(\tau - t)$. It follows that an approximation, $H_1(\tau)$, to $H(\tau)$ can be obtained by writing

$$G(t) = \{G_e\} + \int_{-\infty}^{\infty} H_1(\tau) h(\tau - t) d \ln \tau \tag{4}$$

which may be recast as

$$G(t) = \{G_e\} + \int_{t}^{\infty} \tau^{-1} H_1(\tau) d\tau \ . \tag{5}$$

Differentiation yields

$$\frac{d\,G(t)}{dt} = -t^{-1} H_1(t) \tag{6}$$

and this is identical with Eq. (1) as asserted.

Problem 4.2-2

The plot is shown in Fig. P4.2-2.

The true spectrum is the *doublet line spectrum* indicated by the arrows. Because of the rather wide spacing of the relaxation times (they are five logarithmic decades apart) even the first approximation is capable of resolving the two peaks which represent the location of the lines. As the order of the approximation increases, the trough between the two peaks deepens. Each peak, as is obvious from Eq. (3), essentially reflects the shape of the intensity functions. These, as shown in Fig. 4.2-1, are asymmetric. Figure P4.2-2 should be compared also with Fig. 3.4-23 which displays $\log G''_{4M}(\omega)$ as a function of $-\log \omega$. Qualitatively, $G''(\omega)$ and the reflection of $H(\tau)$ in the $\log(\tau)$-axis, i.e. $H(1/\tau)$, exhibit the same features. For another useful comparison, see Problem 4.3-3.

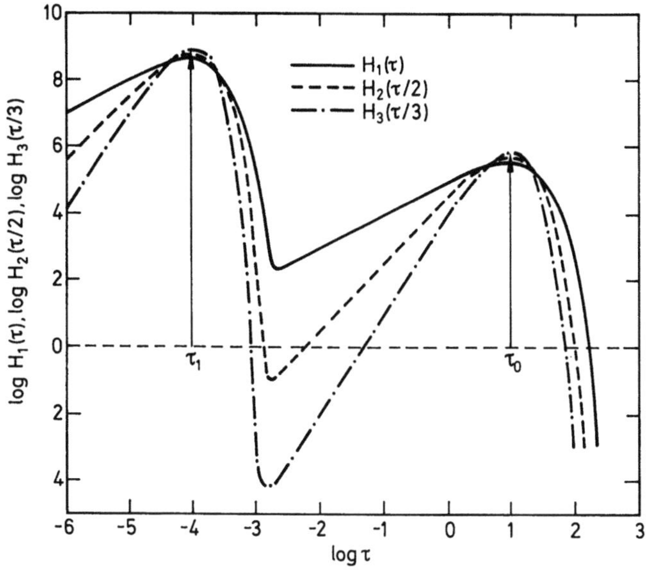

Fig. P4.2-2. log $H_1(\tau)$, log $H_2(\tau/2)$, and log $H_3(\tau/3)$ as functions of log τ for the four-parameter Maxwell model

Problem 4.2-3

(A) If the intensity function is to consist of at least two terms, no more than k-2 of the coefficients b_n can vanish. Thus, at most k-1 relations exist between these coefficients in the case of the multi-term intensity functions, the additional relation being provided by the normalization requirement. Since there are altogether k coefficients b_n (and, hence, coefficients a_n), the system of coefficients is necessarily underdetermined when the intensity function consists of more than one term. In Sect. 4.2.1 we have imposed additional requirements by demanding that $(t/\tau)^m$ = constant when $1 < m < k$. These requirements are sufficient but not necessary to generate intensity functions of the required form.

(B) By the normalization requirement, Eq. (4.2-41), we have

$$\sum_{n=1}^{n=k} b_n \int_0^\infty x^{n-1} \exp(-x)dx = 1 \ . \tag{2}$$

But

$$\int_0^\infty x^{n-1} \exp(-x)dx = (n-1)! \tag{3}$$

when n is a positive integral or zero. Hence, the requirement that the intensity functions be normalized furnishes the relation

$$\sum_{n=1}^{n=k} (n-1)!\, b_n = 1 \tag{4}$$

where at most $k-2$ of the b_n may vanish, between the coefficients b_n.

The general two-term equation of order $k = 2$ is

$$I_2(x) = (b_1x + b_2x^2)\exp(-x) \tag{5}$$

and the coefficients b_1 and b_2 can be freely chosen as long as $0 < b_1 < 1$ and $b_1 + b_2 = 1$.

Problem 4.2-4

Writing $t/\tau = x$ for simplicity, Eq. (4.2-36) gives

$$I_2(x) = [-(a_1 + a_2)x + a_2x^2]\exp(-x) \tag{1}$$

for $k = 2$. By the condition that all but the highest term in x vanish, $a_2 = -a_1 = 1$. This leads to Eq. (4.2-38).

For $k = 3$ we find

$$I_3(x) = [-(a_1 + a_2 + a_3)x + (a_2 + 3a_3)x^2 - a_3x^3]\exp(-x) \ . \tag{2}$$

Our condition now yields $a_2 + a_3 = 1$ and $a_2 + 3a_3 = 0$. Thus, $a_3 = -1/2$ and we obtain Eq. (4.2-39) as required.

Problem 4.2-5

Equation (4.2-49) gives

$$I_2(x) = [(x + x^2)I_2^{(1)}(0) + 0.5x^2I_2^{(2)}(0)]\exp(-x) \tag{1}$$

which, using Eq. $(4.2\text{-}48)_2$ and rearranging, becomes

$$I_2(x) = [-(a_1 + a_2)x + a_2x^2]\exp(-x) \ . \tag{2}$$

We note, parenthetically, that this, as it should, is the same equation as Eq. (1) of Problem 4.2-4 which was derived from Eq. (4.2-36).

Letting $I_2^{(1)}(0) = 0$ leads to Eq. (4.2-38). On the other hand, letting $I_2^{(2)}(0) = 0$ gives $a_1 + a_2 = 0$ and, hence, $a_2 = 1$ since $a_1 = -1$. We obtain the second order intensity function

$$I_2(x) = 0.5(x + x^2)\exp(-x) \ . \tag{3}$$

Now, the expansion of $I_3(x)$ is already given by Eq. (4.2-52). We obtain two new intensity functions by equating either $I_3^{(1)}(0)$ and $I_3^{(3)}(0)$, or $I_3^{(2)}(0)$ and $I_3^{(3)}(0)$ to zero. In the first case, from Eqs. (4.2-56) and (4.2-54), we find $a_1 = -1$, $a_2 = 4/3$, and $a_3 = -1/3$. Consequently,

$$I_3(x) = (1/3)(x^2 + x^3)\exp(-x) \ . \tag{4}$$

In the second case, Eqs. (4.2-55) and (4.2-54) furnish $a_1 = -1$, $a_2 = 5/11$, and $a_3 = -1/11$. Hence,

$$I_3(x) = (1/11)(7x + 2x^2 + x^3)\exp(-x) \ . \tag{5}$$

We note that Eq. (3) differs from both Eq. (4.2-38), and Eq. (4.2-47). However, Eqs. (3), (4), and (5) are all linear combinations of Eqs. (4.2-37), (4.2-38), and (4.2-39).

Problem 4.2-6

Comparing Eqs. (4.2-81) and (4.2-84) we recognize that

$$\sum_{m=0}^{m=k} b_m = \sum_{n=1}^{n=k} \frac{a_n}{\ln^n h} \sum_{m=0}^{m=n} (-1)^{n-m} \binom{n}{m} . \tag{2}$$

But, by the binomial theorem,

$$\sum_{m=0}^{m=n} (-1)^{n-m} \binom{n}{m} = 0 \tag{3}$$

because Eq. (3) forms a symmetric series of coefficients with alternating signs. Since Eq. (3) is true for any m, Eq. (1) follows.

Problem 4.2-7

$H_{2h}(\tau/\beta_{2h})$ is given by Eq. (4.2-118) as

$$H_{2h}(\tau/\beta_{2h}) = \left. \frac{h\,G(h^{-1}t) - (h+1)G(t) + G(ht)}{(h-1)\ln h} \right|_{t=\tau} . \tag{2}$$

Taking the limit as $h \to 1$ results in 0/0. Hence, we apply L'Hospital's theorem. Differentiation of the numerator and of the denominator with respect to h gives

$$G(h^{-1}t) - h^{-1}t\frac{d\,G(h^{-1}t)}{dh^{-1}t} - G(t) + t\frac{d\,G(ht)}{dht} \tag{3}$$

and

$$\ln h + 1 - h^{-1} . \tag{4}$$

The limit of the ratio of these quantities is still indeterminate. A second differentiation yields

$$h^{-3}t^2\frac{d^2G(h^{-1}t)}{d(h^{-1}t)^2} + t^2\frac{d^2G(ht)}{d(ht)^2} \tag{5}$$

and

$$h^{-1} + h^{-2} . \tag{6}$$

We now have

$$\lim_{h \to 1} H_{2h}(\tau/\beta_{2h}) = H_2(\tau/\beta_2) = t^2\frac{d^2G(t)}{dt^2} . \tag{7}$$

But $\beta_2 = 2$ (see the remark at the end of Sect. 4.2.1) and

$$t^2\frac{d^2}{dt^2} = \frac{d^2}{d\ln^2 t} - \frac{d}{d\ln t} \tag{8}$$

by Eq. (4.2-33). We thus recover Eq. (4.2-12) and this completes the proof.

Problem 4.3-1

We have

$$G'(\omega) = \{G_e\} + \int_{-\infty}^{\infty} H(\tau)\frac{\omega^2\tau^2}{1+\omega^2\tau^2} d \ln \tau \tag{2}$$

Figure 4.1-2 reveals that

$$\frac{\omega^2\tau^2}{1+\omega^2\tau^2} \simeq \begin{cases} 0 & \text{when } \log \omega\tau < 0, \text{ i.e. } \tau < 1/\omega \\ 1 & \text{when } \log \omega\tau > 0, \text{ i.e. } \tau > 1/\omega \ . \end{cases} \tag{3}$$

Thus, an approximation, $H'_1(\tau)$, to $H(\tau)$ may be obtained by replacing the kernel of Eq. (2) by $h(\tau - 1/\omega)$. We thus have

$$G'(\omega) = \{G_e\} + \int_{-\infty}^{\infty} H'_1(\tau)h(\tau - 1/\omega) d \ln \tau \tag{4}$$

which may be written in the form

$$G'(\omega) = \{G_e\} + \int_{1/\omega}^{\infty} \tau^{-1}H'_1(\tau)d\tau \ . \tag{5}$$

Differentiation with respect to $1/\omega$ yields

$$\frac{d\, G'(\omega)}{d(1/\omega)} = -(1/\omega)^{-1}H'_1(1/\omega) \tag{6}$$

which is readily shown to be identical with Eq. (1).

Problem 4.3-2

The operator $\mathscr{P}'_k$ is defined by Eq. (4.3-4). To prove Eq. (1) it is enough to show that

$$\frac{d^n}{d \ln^n\omega}\left(\frac{\omega^2\tau^2}{1+\omega^2\tau^2}\right) = \frac{d^n}{d \ln^n\omega}\left(-\frac{1}{1+\omega^2\tau^2}\right) \tag{2}$$

But $d^n/d \ln^n\omega$ requires the repeated application of $d/d \ln \omega$. Thus Eq. (1) is proven if

$$\frac{d}{d \ln \omega}\left(\frac{\omega^2\tau^2}{1+\omega^2\tau^2}\right) = \frac{d}{d \ln \omega}\left(-\frac{1}{1+\omega^2\tau^2}\right) \tag{3}$$

This is easily shown to be true.

Problem 4.3-3

The approximations are obtained from

$$H'_k(\tau/\beta'_k) = G_0 I'_k(\omega\tau_0) + G_1 I'_k(\omega\tau_1)|_{\omega=1/\tau} \tag{1}$$

and

$$H''_k(\tau/\beta''_k) = G_0 I''_k(\omega\tau_0) + G_1 I''_k(\omega\tau_1)|_{\omega=1/\tau} \ . \tag{2}$$

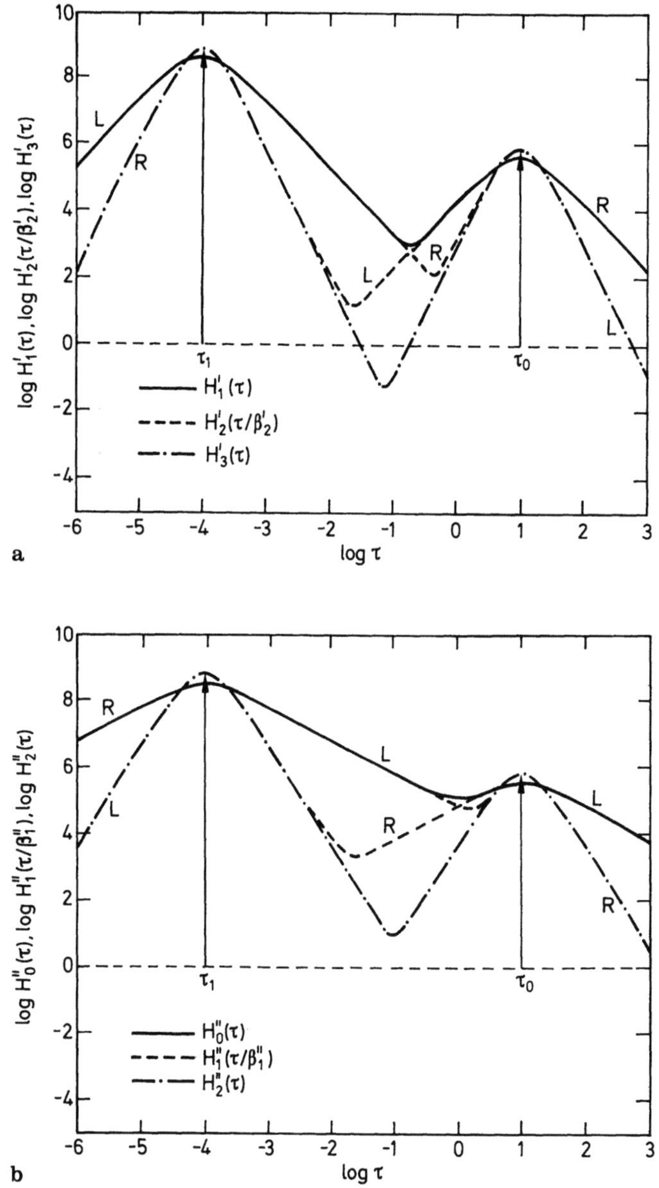

Fig. P4.3-3. a log $H'_1(\tau)$, log $H'_{2L}(\tau/\sqrt{2})$, log $H'_{2R}(\sqrt{2}\tau)$, and log $H'_3(\tau)$ as functions of log τ for the 4-parameter Maxwell model. **b.** log $H''_0(\tau)$, log $H''_{1L}(\tau/\sqrt{3})$, log $H''_{1R}(\sqrt{3}\tau)$, and log $H''_2(\tau)$ as functions of log τ for the 4-parameter Maxwell model

using Eq. (4.3-11) for $I'_1(\omega\tau)$, Eqs. (4.3-15) and (4.3-16) for $I'_2(\omega\tau)$, Eq. (4.3-22) for $I'_3(\omega\tau)$, Eq. (4.3-41) for $I''_0(\omega\tau)$, Eqs. (4.3-45) and (4.3-46) for $I''_1(\omega\tau)$, and Eq. (4.3-53) for $I''_2(\omega\tau)$. The approximations and the true spectra are plotted in Figs. P4.3-3a and P4.3-3b.

The true line spectrum is again represented by the arrows (cf. Problem 4.2-2). The approximation of lowest order, $I''_0(\omega\tau)$ barely succeeds in resolving the two peaks which represent the location of the lines of the true spectrum. The asymetric approximations virtually coincide with portions of the symmetric ones over major sections of the plots. When this happens, they are indicated by the letters L and R, respectively. For example, $\log H'_{2L}(\tau)$ lies very close to $\log H'_1(\tau)$ from $\log \tau = -6$ to -4, then lies just above $\log H'_3(\tau)$ to about $\log \tau = -1.75$, follows its own course up to about $\log \tau = -0.5$, then lies close to $\log H'_1(\tau)$ again up to about $\log \tau = 1$, and finally follows $\log H'_3(\tau)$ again.

Thus, the asymetric approximations are nearly as good as the approximations of next higher order in some parts of the calculated spectrum and not noticeably better than the approximation of the next lower order in other parts.

Comparison with the approximations obtained from the relaxation modulus shows that, in all cases, the peaks of the calculated spectra do lie where the lines of the true spectra are located. The height of the peaks is, of course, affected by the order of the approximation, and an estimation of the modulus associated with the nth relaxation time from the height of the peak will be only approximate. It should be noted that the resolution of the true line spectrum by the approximations is as good as it is only because the lines are quite widely spaced from each other. In this respect, see also the discussion in Sect. 4.4.

Problem 4.3-4

We have

$$\sum_{m=0}^{m=k} \frac{b'_m h^{2m}\omega^2\tau^2}{1 + h^{2m}\omega^2\tau^2} = \sum_{m=0}^{m=k} b'_m \left(1 - \frac{1}{1 + h^{2m}\omega^2\tau^2}\right) . \tag{2}$$

Equation (1) follows because

$$\sum_{m=0}^{m=k} b'_m = 0 \tag{3}$$

in analogy to Eq. (4.2-89).

Problem 4.3-5

By comparing Eqs. (4.3-145) and (4.3-146) we find that

$$\sum_{m=0}^{m=k} b''_m = \sum_{n=0}^{n=k} \frac{a''_n}{\ln^n h} \sum_{m=0}^{m=n} (-1)^{n-m} \binom{n}{m} . \tag{2}$$

This is identical with Eq. (2) of Problem 4.2-6, except for the lower limit of summation of the first sum on the right. Hence, by Eq. (3) of Problem 4.2-6,

$$\sum_{m=0}^{m=k} b''_m = a''_0 . \tag{3}$$

But $a''_0 = 2/\pi$ as shown by Eq. (4.3-153). Thus, Eq. (1) follows.

Problem 4.5-1

(A) We have

$$\bar{Q}_{\bullet}(s) = \frac{G_g - G_1 + G_g\tau_1 s}{1 + \tau_1 s} \; . \tag{2}$$

The only pole of $\bar{Q}_{\bullet}(s)$ is the pole at $s = -1/\tau_1$. By Eq. (4.5-17), then

$$\mathrm{Res}_1(\bar{Q}_{\bullet}) = \frac{G_g - G_1 + G_g\tau_1(-1/\tau_1)}{\tau_1} = -G_1/\tau_1 \tag{3}$$

and, by Eq. (4.5-21),

$$H(\tau) = G_1\tau_1\delta(\tau - \tau_1) \; . \tag{4}$$

Substitution into

$$\bar{Q}_{\bullet}(s) = G_g - \int_{-\infty}^{\infty} \frac{H(\tau)}{1 + \tau s} d \ln \tau \tag{5}$$

gives

$$\bar{Q}_{\bullet}(s) = G_g - \int_0^{\infty} \frac{G_1\tau_1}{(1 + \tau s)\tau} \delta(\tau - \tau_1) d\tau \tag{6}$$

which becomes Eq. (1) as required.

(B) Now, by Eq. (4.1-36),

$$\tilde{M}(s) = \frac{G_1}{1 + \tau_1 s} = \frac{G_1\zeta_1}{s + \zeta_1} \; . \tag{7}$$

We see that the only pole of $\tilde{M}(s)$ is the pole at $s = -\zeta_1$. Hence, by Eq. (4.5-10),

$$\mathrm{Res}_1(\tilde{M}) = G_1\zeta_1 \; . \tag{8}$$

The frequency spectrum follows as

$$M(\zeta) = G_1\zeta_1\delta(\zeta - \zeta_1) \tag{9}$$

from Eq. (4.5-14). Substituting this into

$$\bar{Q}_{\bullet}(s) = G_g - \int_0^{\infty} \frac{M(\zeta)}{s + \zeta} d\zeta \tag{10}$$

[cf. Eq. (4.1-35)], gives

$$\bar{Q}_{\bullet}(s) = G_g - \int_0^{\infty} \frac{G_1\zeta_1}{s + \zeta} \delta(\zeta - \zeta_1) d\zeta = G_g - \frac{G_1\zeta_1}{s + \zeta_1} \tag{11}$$

which, with $\zeta_1 = 1/\tau_1$, leads to Eq. (1).

Problem 4.5-2

(A) We have

$$\bar{Q}(s) = \frac{G_e + (G_e + G_1)\tau_1 s}{1 + \tau_1 s} \; . \tag{2}$$

The only pole of $\bar{Q}(s)$ is the pole at $s = -1/\tau_1$. By Eq. (4.5-17), then

$$\mathrm{Res}_1(\bar{Q}) = \frac{G_e + (G_e + G_1)\tau_1(-1/\tau_1)}{\tau_1} = -G_1/\tau_1 \ , \tag{3}$$

and, by Eq. (4.5-21).

$$H(\tau) = G_1\tau_1\delta(\tau - \tau_1) \ . \tag{4}$$

Substitution into

$$\bar{Q}(s) = \{G_e\} + \int_{-\infty}^{\infty} H(\tau)\frac{\tau s}{1 + \tau s} d\ln\tau \tag{5}$$

gives

$$\bar{Q}(s) = \{G_e\} + \int_0^{\infty} \frac{G_1\tau_1 s}{1 + \tau s}\delta(\tau - \tau_1)d\tau \tag{6}$$

which becomes Eq. (1) as required.

(B) Now, by Eq. (4.1-42),

$$\bar{\bar{\Xi}}(s) = \frac{G_1\tau_1}{1 + \tau_1 s} = \frac{G_1}{s + \zeta_1} \ . \tag{7}$$

The only pole of $\bar{\bar{\Xi}}(s)$ is the pole at $s = -\zeta_1$. But

$$\mathrm{Res}_n(\bar{\bar{\Xi}}) = \lim_{s \to -\zeta_n} (s + \zeta_n)\bar{\bar{\Xi}}(s) \ , \tag{8}$$

and, therefore,

$$\mathrm{Res}_1(\bar{\bar{\Xi}}) = G_1 \ . \tag{9}$$

The frequency spectrum then becomes

$$\Xi(\zeta) = G_1\delta(\zeta - \zeta_1) \tag{10}$$

by Eq. $(4.5\text{-}34)_1$.

Substitution into

$$\bar{Q}(s) = G_e + s\int_0^{\infty} \frac{\Xi(\zeta)}{\zeta + s} d\zeta \tag{11}$$

[cf. Eq. (4.1-41)] leads to

$$\bar{Q}(s) = G_e + \frac{G_1 s}{\zeta_1 + s} \tag{12}$$

which, with $\tau_1 = 1/\zeta$, becomes Eq. (1).

Problem 4.5-3

Since $\bar{Q}(s)\bar{U}(s) = 1$, we have

$$\bar{U}_{4v}(s) = \frac{(1 + \tau_0 s)(1 + \tau_1 s)}{(\eta_0 + \eta_1)(1 + \tau' s)s} \tag{1}$$

where

$$\tau' = \frac{1/G_0 + 1/G_1}{1/\eta_0 + 1/\eta_1} \tag{2}$$

[cf. Eqs. (3.4-33) and (3.4-34)]. $\bar{U}_{4v}(s)$ has poles at $s_0 = 0$ and $s_1 = -1/\tau'$. The residues are obtained from

$$\mathrm{Res}_n(\bar{U}) = \frac{\bar{u}(s_n)}{\bar{q}'(s_n)} = \frac{(1 + \tau_0 s_n)(1 + \tau_1 s_n)}{(\eta_0 + \eta_1)(1 + 2\tau' s_n)} \; . \tag{3}$$

where n = 0, 1. We find

$$\mathrm{Res}_0(\bar{U}) = 1/(\eta_0 + \eta_1) = \phi_f \tag{4}$$

[cf. Eqs. (4.5-29) and (3.7-23)], and

$$\mathrm{Res}_1(\bar{U}) = \frac{(G_0\eta_1 - G_1\eta_0)^2}{(G_0 + G_1)^2\eta_0\eta_1(\eta_0 + \eta_1)} = \phi \tag{5}$$

[cf. Eqs. (4.5-28 and (3.7-29)]. But $J = \phi\tau'$ and, therefore,

$$J = \frac{(G_0\eta_1 - G_1\eta_0)^2}{G_0G_1(G_0 + G_1)(\eta_0 + \eta_1)^2} \tag{6}$$

[cf. Eq. (3.7-28)].

The glassy compliance, J_g, must be obtained from the final value theorem. By Eq. (2), using L'Hospital's theorem, we have

$$\lim_{s\to 0} \bar{U}(s) = \lim_{s\to 0} \frac{\tau_0 + \tau_1 + 2\tau_0\tau_1 s}{(\eta_0 + \eta_1)(1 + 2\tau' s)} \tag{7}$$

which yields

$$J_g = 1/(G_0 + G_1) \tag{8}$$

[cf. Eq. (3.7–22)].

Problem 4.5-4

In Fig. 4.5-4 symbols have been assigned to the springs and dashpots of Fig. 3.4-17a. In terms of these, the relaxance of this non-series-parallel model becomes

$$\bar{Q}(s) = \frac{GG'(\eta + \eta')s + G'\eta\eta's^2}{GG' + (G\eta + G\eta' + G'\eta)s + \eta\eta's^2} \; . \tag{1}$$

The pole equation may be written as

$$s^2 + 2As + B^2 = 0 \tag{2}$$

where

$$2A = G/\eta + G/\eta' + G'/\eta' \tag{3}$$

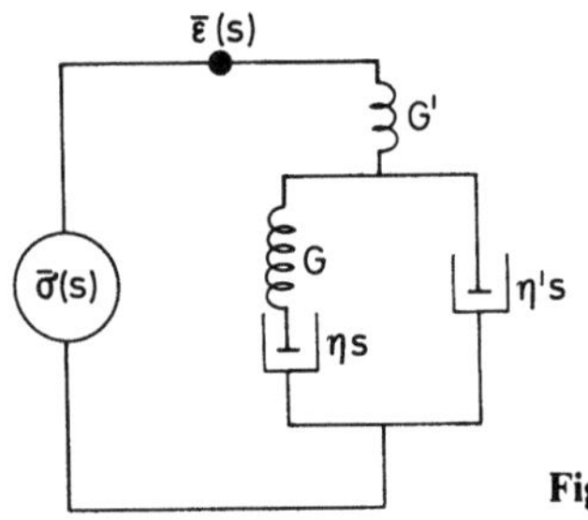

Fig. P4.5-4. 4-parameter non-series-parallel model

and

$$B^2 = GG'/\eta\eta' \ . \tag{4}$$

Hence, $\bar{Q}(s)$ has two poles which we denote as s_0 and s_1, respectively. They are

$$s_{0,1} = -A \pm \sqrt{A^2 - B^2} \tag{5}$$

and the relaxation times follow as

$$\tau_{0,1} = 1/(A \mp \sqrt{A^2 - B^2}) \ . \tag{6}$$

The moduli are obtained from

$$G_{0,1} = \mathrm{Res}_{0,1}(\bar{Q})/s_{0,1} = \frac{GG'(\eta + \eta') + G'\eta\eta' s_{0,1}}{G\eta + G\eta' + G'\eta + 2\eta\eta' s_{0,1}} \tag{7}$$

as

$$G_{0,1} = \frac{G'}{2} \pm \frac{B^2(\eta + \eta') - AG'}{2\sqrt{A^2 - B^2}} \ . \tag{8}$$

through division by $\eta\eta'$ and the use of Eq. (3) after substitution of Eq. (5).

Problem 4.5-5

The relaxance of the model is

$$\bar{Q}(s) = \frac{GG'\eta s + (G + G')\eta\eta' s^2}{GG' + (G\eta' + G'\eta + G'\eta')s + \eta\eta' s^2} \ . \tag{1}$$

The pole equation is again given by Eq. (2) of Problem 4.5-4 but now

$$2A = G/\eta + G'/\eta + G'/\eta' \ . \tag{2}$$

The poles and relaxation times follow from Eqs. (5) and (6) of the preceding problem but the moduli are obtained from

$$G_{0,1} = \mathrm{Res}_{0,1}(\bar{Q})/s_{0,1} = \frac{GG'\eta + (G + G')\eta\eta' s_{0,1}}{G\eta' + G'\eta + G'\eta' + 2\eta\eta' s_{0,1}} \tag{3}$$

and division by $\eta\eta'$ and using Eq. (2) yields

$$G_{0,1} = \frac{G + G'}{2} \pm \frac{B^2\eta - (G + G')A}{2\sqrt{A^2 - B^2}} \quad . \tag{4}$$

Problem 4.6-1

By the definition of the Stieltjes transform, Eq. (A5-1),

$$\tilde{M}(s) + s\tilde{\Xi}(s) = \int_0^\infty \frac{M(\zeta)}{s + \zeta} d\zeta + s \int_{-\infty}^\infty \frac{\Xi(\zeta)}{s + \zeta} d\zeta \tag{3}$$

But $\Xi(\zeta) = \zeta^{-1}M(\zeta)$ and we have

$$\int_0^\infty M(\zeta)\frac{1 + s\zeta^{-1}}{s + \zeta} d\zeta = \int_{-\infty}^\infty M(\zeta) d \ln \zeta \ . \tag{4}$$

Problem 4.6-2

By STP (3)

$$s\tilde{\Xi}(s) = -\mathscr{S}\, \zeta\Xi(\zeta) + \int_0^\infty \Xi(\zeta) d\zeta \ . \tag{3}$$

But $\zeta\Xi(\zeta) = M(\zeta)$, and $\mathscr{S}\, \zeta\Xi(\zeta) = \tilde{M}(s)$. Hence, Eq. (1) follows.
The proof of Eq. (2) is analogous.

Problem 4.6-3

By Eqs. (4.6-18)

$$\int_{-\infty}^\infty \tau^{1+\alpha} H(\tau) d \ln \tau = \int_{-\infty}^\infty \zeta^{-\alpha-1} M(\zeta) d \ln \zeta = \int_{-\infty}^\infty \zeta^{-\alpha}\Xi(\zeta) d \ln \zeta \tag{3}$$

Letting $\alpha = -1$ and comparing the result with Eq. (4.6-20), Eq. (1) follows at once.
Equation (2) is obtained in an analogous manner from Eqs. (4.6-19) and (4.6-21).

Problem 4.6-4

Equations (1) to (4) follow directly from Eqs. (4.1-75) to (4.1-78).

Problem 4.6-5

We have [cf. Eq. (4.1-5)]

$$G(t) - \{G_e\} = \int_0^\infty \tau^{-1} H(\tau) \exp(-t/\tau) d\tau \ . \tag{2}$$

Multiplying both sides by $t^{\alpha-1}$ and integrating gives, after interchanging the order of integration on the right,

$$\int_0^\infty t^{\alpha-1}[G(t) - \{G_e\}] dt = \int_0^\infty \tau^{-1} H(\tau) \int_0^\infty t^{\alpha-1} \exp(-t/\tau) dt \, d\tau \ . \tag{3}$$

We can integrate the second integral on the right through the change of variable $t = k\tau$ [cf. Eq. (4.2-8)]. This gives

$$\int_0^\infty t^{\alpha-1}\exp(-t/\tau)dt = \tau^\alpha \int_0^\infty k^{\alpha-1}\exp(-k)dk = \tau^\alpha\Gamma(\alpha) \tag{4}$$

and Eq. (1) follows.

Problem 4.6-6

By Eq. (1)

$$J'(\omega) = \frac{G'(\omega)}{[G'(\omega)]^2 + [G''(\omega)]^2} \ . \tag{3}$$

For a rheodictic material

$$G'(\omega) = \int_{-\infty}^{\infty} H(\tau)\frac{\omega^2\tau^2}{1+\omega^2\tau^2}d\ln\tau \tag{4}$$

and

$$G''(\omega) = \int_{-\infty}^{\infty} H(\tau)\frac{\omega\tau}{1+\omega^2\tau^2}d\ln\tau \ . \tag{5}$$

Substituting into Eq. (3), cancelling ω^2, and then seeking the limit as $\omega \to 0$ gives

$$J_e^\circ = \frac{\int_0^\infty \tau H(\tau)\, d\tau}{\left(\int_0^\infty H(\tau)\, d\tau\right)^2} \ . \tag{6}$$

But the numerator is $G_g\langle\tau_M\rangle$ by Eq. $(4.6\text{-}22)_2$, and the denominator is η_f^2 by Eq. $(4.6\text{-}24)_2$. Hence, Eq. (4.6-39) follows.

Problem 4.6-7

The steady-state compliance, J_e°, is given by Eq. (6) of Problem 4.6-6 in terms of the relaxation spectrum, $H(\tau)$. This expression is easily converted into one containing the frequency spectrum, $M(\zeta)$, using Eqs. (4.6-18). We find

$$J_e^\circ = \frac{\int_{-\infty}^{\infty} \zeta^{-2}M(\zeta)d\ln\zeta}{\left(\int_{-\infty}^{\infty} \zeta^{-1}M(\zeta)d\ln\zeta\right)^2} \ , \tag{1}$$

and, since $M(\zeta) = \zeta\Xi(\zeta)$, we also have

$$J_e^\circ = \frac{\int_{-\infty}^{\infty} \zeta^{-1}\Xi(\zeta)d\ln\zeta}{\left(\int_{-\infty}^{\infty} \Xi(\zeta)d\ln\zeta\right)^2} \ . \tag{2}$$

Problem 4.6-8

By Eq. (4.6-40)$_3$

$$\eta_f = (2/\pi) \int_0^\infty \omega^{-2} G'(\omega) d\omega \ . \tag{1}$$

But, for the standard linear liquid,

$$G'(\omega) = \frac{G_0 \omega^2 \tau_0^2}{1 + \omega^2 \tau_0^2} + \frac{G_1 \omega^2 \tau_1^2}{1 + \omega^2 \tau_1^2} \tag{2}$$

[cf. Eq. (3.4-162)]. Substituting and integrating yields

$$\eta_f = G_0 \tau_0 + G_1 \tau_1 = \eta_0 + \eta_1 \ . \tag{3}$$

Problem 4.6-9

By Eq. (4.6-41)$_3$

$$J_e^\circ \eta_f^2 = (2/\pi) \int_0^\infty \omega^{-1} [G''(\omega) - d\, G''(\omega)/d \ln \omega] d\omega \ . \tag{1}$$

Now, for the standard linear liquid,

$$G''(\omega) - \frac{d\, G''(\omega)}{d \ln \omega} = \frac{2 G_0 \omega^3 \tau_0^3}{(1 + \omega^2 \tau_0^2)^2} + \frac{2 G_1 \omega^3 \tau_1^3}{(1 + \omega^2 \tau_1^2)^2} \tag{2}$$

Substituting and integrating yields

$$J_e^\circ \eta_f^2 = G_0 \tau_0^2 + G_1 \tau_1^2 \tag{3}$$

Hence,

$$J_e^\circ = \frac{\eta_0 \tau_0 + \eta_1 \tau_1}{(\eta_0 + \eta_1)^2} \ . \tag{4}$$

Problem 4.6-10

The square of the standard deviation, the variance, is given by Eq. (4.6-26) as

$$\sigma_\tau^2 = \langle \tau_M^2 \rangle - \langle \tau_M \rangle^2 \tag{2}$$

which, using Eqs. (4.6-38) and (4.6-39), becomes

$$\sigma_\tau^2 = \eta_f^2 (J_e^\circ - J_g) J_g \tag{3}$$

from which Eq. (1) follows.

Chapter 5

Problem 5.1-1

The node equations are:

$$\bar{\sigma}_0(s) = \quad (1/J_0)\bar{\varepsilon}_0(s) - (1/J_0)\bar{\varepsilon}_1(s) \tag{1}$$

$$0 \quad = -(1/J_0)\bar{\varepsilon}_0(s) + (1/J_0 + 1/J_1 + \eta_1 s)\bar{\varepsilon}_1(s) - (1/J_1)\bar{\varepsilon}_2(s) \tag{2}$$

$$0 \quad = -(1/J_1)\bar{\varepsilon}_1(s) + (1/J_1 + 1/J_2 + \eta_2 s)\bar{\varepsilon}_2(s) - (1/J_2)\bar{\varepsilon}_3(s) \tag{3}$$

and

$$0 = -(1/J_2)\bar{\varepsilon}_2(s) + (1/J_2 + \eta_3 s)\bar{\varepsilon}_3(s) \ . \tag{4}$$

We must now express $\bar{\varepsilon}_n(s)$ in terms of $\bar{\varepsilon}_{n-1}(s)$ until we arrive at a relation in $\bar{\varepsilon}_1(s)$ only. To lighten the algebra we introduce characteristic times $\theta_k = J_{k-1}\eta_k$. From Eq. (4) we have

$$\bar{\varepsilon}_3(s) = \frac{1}{1 + \theta_3 s}\bar{\varepsilon}_2(s) \ . \tag{5}$$

Substituting this into Eq. (3) and solving for $\bar{\varepsilon}_2(s)$ yields

$$\bar{\varepsilon}_2(s) = \frac{1 + \theta_3 s}{1 + (\theta_2 + a\theta_3)s + \theta_2\theta_3 s^2}\bar{\varepsilon}_1(s) \tag{6}$$

where

$$a = 1 + J_1/J_2 \ . \tag{7}$$

Repeating the process for $\bar{\varepsilon}_1(s)$ leads to

$$\bar{\varepsilon}_1(s) = \frac{1 + (\theta_2 + a\theta_3)s + \theta_2\theta_3 s^2}{1 + (\theta_1 + b\theta_2 + c\theta_3)s + (\theta_1\theta_2 + a\theta_1\theta_3 + b\theta_2\theta_3)s^2 + \theta_1\theta_2\theta_3 s^3}\,\varepsilon_0 s \tag{8}$$

where

$$b = 1 + J_0/J_1 \tag{9}$$

and

$$c = 1 + J_0/J_2 + J_1/J_2 \ . \tag{10}$$

Substitution of Eq. (8) into Eq. (1) finally yields

$$\bar{Q}(s) = \frac{(\eta_1 + \eta_2 + \eta_3)s + [\eta_1\theta_2 + (J_1 + J_2)\eta_1\eta_3 + \eta_2\theta_3]s^2 + \eta_1\theta_2\theta_3 s^3}{1 + (\theta_1 + b\theta_2 + c\theta_3)s + (\theta_1\theta_2 + a\theta_1\theta_3 + b\theta_2\theta_3)s^2 + \theta_1\theta_2\theta_3 s^3} \ . \tag{11}$$

Both the numerator and the denominator are polynominals in s of the third degree because N = 3. Clearly, the process becomes quite unwieldy when N is large.

Problem 5.1-2

For convenience, we redraw the model in the alternative form of Fig. P5.1-2. Using the combination rules that relaxances add in parallel while retardances add in series, we find

$$\bar{Q}(s) = \cfrac{1}{J_0 + \cfrac{1}{\eta_1 s + \cfrac{1}{J_1 + \cfrac{1}{\eta_2 s + \cfrac{1}{J_2 + \cfrac{1}{\eta_3 s}}}}}} . \tag{1}$$

The relaxance of a ladder model is thus seen to be expressible as a *Stieltjes continued fraction.* The continued fraction can be "rolled up" from the bottom. For convenience we again introduce characteristic times, $\theta_k = J_{k-1}\eta_k$. Starting at the bottom with θ_3, we have

$$\bar{Q}(s) = \cfrac{1}{J_0 + \cfrac{1}{\eta_1 s + \cfrac{1}{J_1 + \cfrac{1}{\eta_2 s + \cfrac{\eta_3 s}{1 + \theta_3 s}}}}} . \tag{2}$$

Continuation of the process eventually leads to Eq. (11) of Problem 5.1-1.

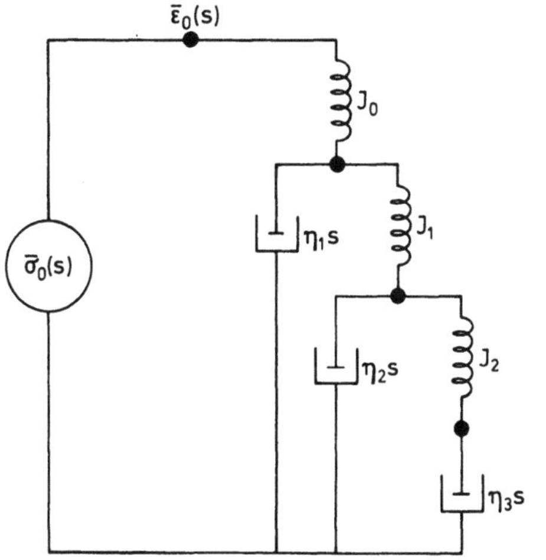

Fig. P5.1-2. General obverse ladder model with 6 elements

Problem 5.2-1

(A) The Gross model is particularly easy to analyze because it consists of N identical T-sections only. Thus, it has no initial section and it terminates into a rigid connection, i.e. into an infinite final relaxance, $Q_F(s)$. We have

$$\begin{bmatrix} \bar{\sigma}_0(s) \\ \bar{\varepsilon}_0(s) \end{bmatrix} = \begin{bmatrix} \cosh N\, a(s) & Q_T(s)\sinh N\, a(s) \\ U_T(s)\sinh N\, a(s) & \cosh N\, a(s) \end{bmatrix} \begin{bmatrix} Q_F(s) \\ 1 \end{bmatrix} \bar{\varepsilon}_{N+1}(s) \ . \tag{1}$$

Since $Q_F(s) = \infty$, matrix multiplication gives the retardance as

$$\bar{U}_G^{\times}(s) = U_T(s)\tanh N\, a(s) = \frac{J\sqrt{4+\theta s}\,\sinh N\, a(s)}{2\sqrt{\theta s}\,\cosh N\, a(s)} \ . \tag{2}$$

The retardance of the Gross-Marvin model which has the same number of nodes is

$$\bar{U}_{GM}^{\times}(s) = U_T(s)\tanh[(N+\tfrac{1}{2})a(s)] + J/2 \ . \tag{3}$$

The second term on the right, J/2, represents the compliance of the spring which must be placed in series with the first spring of the Gross model to convert it to the Gross-Marvin model. Similarly, the appearance of $N + \frac{1}{2}$ in the first term results from adding another spring of compliance J/2 in series at the end of the model.

By the method discussed in Sect. 5.2 the poles and residues of the retardance become

$$s_n = -\frac{4}{\theta}\sin^2\frac{\pi(2n-1)}{4N} \ , \qquad n = 1, 2, \ldots, N \tag{4}$$

and

$$\mathrm{Res}_n(\bar{U}_G^{\times}) = \frac{2J}{N\theta}\cos^2\frac{\pi(2n-1)}{4N} \ , \qquad n = 1, 2, \ldots, N \tag{5}$$

respectively. Thus, the retardation times and compliances of the equivalent Kelvin model result as

$$\tau_n = \frac{\theta}{4}\csc^2\frac{\pi(2n-1)}{4N} \ , \qquad n = 1, 2, \ldots, N \tag{6}$$

and

$$J_n = \frac{J}{2N}\cot^2\frac{\pi(2n-1)}{4N} \ , \qquad n = 1, 2, \ldots, N \ . \tag{7}$$

As $s \to \infty$, $a(s) \to \infty$, $\tanh a(s) \to 1$. Hence, $J_g = J/2$. We have*

$$\bar{U}_G^{\times}(s) = \frac{J}{2} + \sum_{n=1}^{N} \frac{J_n}{1+\tau_n s} \tag{8}$$

and it follows that

---

* Note that the summation now extends to N.

$$J_e = \frac{J}{2} + \frac{J}{2N}\sum_{n=1}^{N} \cot^2\frac{\pi(2n-1)}{4n} = JN \tag{9}$$

because the sum equals N(2N − 1). Hence, $J_e/J_g = 2N$.

The relaxance of the arrheodictic Gross model,

$$\bar{Q}_G^{\times}(s) = Q_T(s)\coth N\, a(s)\ , \tag{10}$$

has the poles

$$s_n = -\frac{4}{\theta}\sin^2\frac{\pi n}{2N}\ , \qquad n = 1, 2, \ldots, N \tag{11}$$

and residues

$$\begin{aligned} \text{Res}_n(\bar{Q}_G^{\times}) &= -\frac{8}{JN\theta}\sin^2\frac{\pi n}{2N}\ , \qquad n = 1, 2, \ldots, N-1 \\ \text{Res}_N(\bar{Q}_G^{\times}) &= -\frac{4}{JN\theta}\ . \end{aligned} \tag{12}$$

Consequently, the relaxation times and moduli of the equivalent Wiechert model become

$$\tau_n = \frac{\theta}{4}\csc^2\frac{\pi n}{2N}\ , \qquad n = 1, 2, \ldots, N \tag{13}$$

and

$$\begin{aligned} G_n &= \frac{2}{JN}\ , \qquad n = 1, 2, \ldots, N-1 \\ G_N &= \frac{1}{JN}\ . \end{aligned} \tag{14}$$

Thus, in the equivalent model, N − 1 of the springs have identical moduli. The modulus associated with the shortest relaxation time is different. Its value is one-half of that of the others.

(B) The rheodictic form is obtained simply by letting $Q_F(s) = 0$. The last spring of the last T-section then becomes ineffective and the model terminates effectively in a dashpot. We find

$$\bar{Q}_G^{\circ}(s) = Q_T(s)\tanh N\, a(s) \tag{15}$$

and

$$\bar{U}_G^{\circ}(s) = U_T(s)\coth N\, a(s) \tag{16}$$

for the respondances.

Problem 5.2-2

The representation of the models by Π-section is shown in Fig. P5.2-2. The matrix equation for the initial section becomes

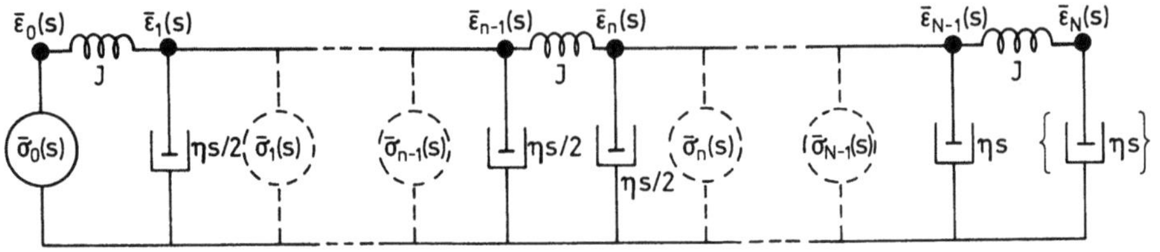

Fig. P5.2-2. Representation of Gross-Marvin model by Π-sections

$$\begin{bmatrix} \bar{\sigma}_0(s) \\ \bar{\varepsilon}_0(s) \end{bmatrix} = \begin{bmatrix} 1 & \eta s/2 \\ J & 1 + \theta s/2 \end{bmatrix} \begin{bmatrix} \bar{\sigma}_1(s) \\ \bar{\varepsilon}_1(s) \end{bmatrix}, \tag{1}$$

that of the Π-section is

$$\begin{bmatrix} \bar{\sigma}_{n-1}(s) \\ \bar{\varepsilon}_{n-1}(s) \end{bmatrix} = \begin{bmatrix} 1 + \theta s/2 & \eta s(1 + \theta s/4) \\ J & 1 + \theta s/2 \end{bmatrix} \begin{bmatrix} \bar{\sigma}_n(s) \\ \bar{\varepsilon}_n(s) \end{bmatrix}, \tag{2}$$

and the relaxance of the final section results as

$$Q_F(s) = \frac{\eta s}{2} + \frac{1}{J + \{1/\eta s\}} . \tag{3}$$

The propagation function, a(s), is still as given by Eq. (5.2-13). Equations (5.2-14), (5.2-15) and (5.2-23) also remain unchanged. The characteristic relaxance of the Π-section is

$$Q_\Pi(s) = \sqrt{\theta s(4 + \theta s)}/2J \tag{4}$$

and we have

$$J = U_\Pi(s) \sinh a(s) \tag{5}$$

$$\eta s = 2Q_\Pi(s) \tanh a(s)/2 \tag{6}$$

and

$$\eta s(1 + \theta s/4) = Q_\Pi(s) \sinh a(s) . \tag{7}$$

The last four equations should be compared with Eqs. (5.2-16), (5.2-17), (5.2-25), and (5.2-18). We note that

$$Q_\Pi(s)/Q_T(s) = U_T(s)/U_\Pi(s) = \cosh^2 a(s)/2 \tag{8}$$

and that

$$Q_\Pi(s)/\cosh a(s)/2 = \eta s/\sqrt{\theta s} = \sqrt{\theta s}/J \tag{9}$$

and

$$U_\Pi(s) \cosh a(s)/2 = \sqrt{\theta s}/\eta s = J/\sqrt{\theta s} . \tag{10}$$

Since there are now N-2 Π-sections, we obtain

$$\begin{bmatrix} \bar{\sigma}_0(s) \\ \bar{\varepsilon}_0(s) \end{bmatrix} = \begin{bmatrix} 1 & Q_\Pi(s)\tanh a(s)/2 \\ U_\Pi(s)\sinh a(s) & \cosh a(s) \end{bmatrix}$$

$$\times \begin{bmatrix} \cosh[(N-2)a(s)] & Q_\Pi(s)\sinh[N-2)a(s)] \\ U_\Pi(s)\sinh[(N-2)a(s)] & \cosh[(N-2)a(s)] \end{bmatrix} \begin{bmatrix} Q_F(s) \\ 1 \end{bmatrix} \bar{\varepsilon}_{N-1}(s) \ . \tag{11}$$

We have

$$Q_F^\circ(s) = Q_\Pi(s)\tanh 3a(s)/2 \tag{12}$$

and

$$Q_F^\times(s) = Q_\Pi(s)\coth a(s) \ . \tag{13}$$

Substitution of these equations into Eq. (11), matrix multiplication and the use of Eq. (9) eventually lead to Eqs. (5.2-27) and (5.2-32) for the rheodictic and the arrheodictic models, respectively.

We remark that increasing the number of Π-sections by one and letting the relaxance of the final section be $\eta s/2 = Q_\Pi(s)\tanh a(s)/2$ would have produced the same result for $\bar{Q}^\circ_{GM}(s)$.

Problem 5.2-3

Equation (1) is easily proved by mathematical induction. We have

$$\begin{bmatrix} \cosh xa & U_T\sinh xa \\ Q_T\sinh xa & \cosh xa \end{bmatrix} \begin{bmatrix} \cosh ya & U_T\sinh ya \\ Q_T\sinh ya & \cosh ya \end{bmatrix}$$

$$= \begin{bmatrix} \cosh(x+y)a & U_T\sinh(x+y)a \\ Q_T\sinh(x+y)a & \cosh(x+y)a \end{bmatrix} . \tag{2}$$

Now let y = x. Then the argument of the product matrix becomes 2xa. Multiply this matrix by the ya-matrix again, and again let y = x. The new argument of the product matrix is 3xa. This process can be continued any desired number of times.

Problem 5.2-4

Using the formula $\cosh(x-y) = \cosh x \cosh y - \sinh x \sinh y$ and realizing that $\coth(\infty) = 1$, it is readily shown that

$$\lim_{s\to\infty} \frac{\sqrt{\theta s}\cosh[(N-\frac{1}{2})a(s)]}{J \sinh N\, a(s)} = \frac{1}{J} \lim_{s\to\infty} \frac{\sqrt{\theta s}}{\exp[a(s)/2]} = 1 \ . \tag{3}$$

By contrast,

$$\lim_{s\to\infty} \frac{\sqrt{\theta s}}{J} \coth N\, a(s) = \frac{1}{J} \lim_{s\to\infty} \sqrt{\theta s} = \infty \ . \tag{4}$$

Thus, letting $(N-\frac{1}{2}) \to N$ changes the model from a standard to a non-standard one.

Problem 5.2-5

Because $\bar{G}(s) = \bar{Q}(s)/s$, we have

$$\bar{G}^{\circ}_{GM}(s) = \frac{\eta \sinh N\,a(s)}{\sqrt{\theta s}\cosh[(N+\frac{1}{2})a(s)]} \; . \tag{2}$$

The inversion is accomplished by Eq. (A3-31) which becomes

$$G^{\circ}_{GM}(t) = \sum_n \mathrm{Res}_n[\bar{G}^{\circ}_{GM}\exp(st)] \; . \tag{3}$$

Now, $\bar{G}^{\circ}_{GM}(s)$ is regular at the origin and the poles of $\bar{G}^{\circ}_{GM}(s)\exp(st)$ therefore are the same as those of $\bar{Q}^{\circ}_{GM}(s)$ and are given by Eq. (5.2-67) as

$$s_n = -1/\tau_n = -(4/\theta)\sin^2\frac{\pi(2n+1)}{2(2N+1)} \, , \qquad n = 0, 1, \ldots, N-1 \; . \tag{4}$$

The residues are obtained [cf. Eq. $(5.2\text{-}45)_2$] from

$$\mathrm{Res}_n[\bar{G}^{\circ}_{GM}\exp(st)] = \frac{\bar{g}(s_n)\exp(s_n t)}{\bar{j}'(s_n)} \tag{5}$$

where $\bar{g}(s)$ and $\bar{j}(s)$ are the numerator and denominator, respectively, of $\bar{G}^{\circ}_{GM}(s)$. Calculating the residues as in Sect. 5.2 we find

$$\mathrm{Res}_n[\bar{G}^{\circ}_{GM}\exp(st)] = \frac{4+\theta s_n}{J(2N+1)}\exp(s_n t) \tag{6}$$

which we could also have found directly from Eq. (5.2-70) by realizing that

$$\mathrm{Res}_n(\bar{G}) = \mathrm{Res}_n(\bar{Q})/s_n \; . \tag{7}$$

Substitution of Eq. (4) into Eq. (6) leads to

$$\mathrm{Res}_n[\bar{G}^{\circ}_{GM}\exp(st)] = \frac{4\exp(-t/\tau_n)}{J(2N+1)}\cos^2\frac{\pi(2n+1)}{2(2N+1)} \tag{8}$$

which, with Eq. (5.2-93), i.e. with

$$G_n = \frac{4}{J(2N+1)}\cos^2\frac{\pi(2n+1)}{2(2N+1)} \tag{9}$$

finally gives

$$G^{\circ}_{GM}(t) = \sum_n G_n\exp(-t/\tau_n) \; . \tag{10}$$

This result is identical with that which we would have obtained by inverting Eq. (5.2-94) after division by s.

Problem 5.2-6

Application of L'Hospital's theorem shows that the sought-for limit is the reciprocal of the limit

$$L = \lim_{x \to 0} \frac{d \tanh[M \sinh^{-1}(x/2)]}{dx} \tag{2}$$

where M = 2N. We have

$$L = \lim_{x \to 0} M \operatorname{sech}^2[M \sinh^{-1}(x/2)] \frac{d \sinh^{-1}(x/2)}{dx} . \tag{3}$$

Now, $\sinh^{-1}(0) = 0$ and $\operatorname{sech}(0) = 1/\cosh(0) = 1$. Hence, letting $x/2 = y$, we have

$$L = \frac{M}{2} \lim_{y \to 0} \frac{d \sinh^{-1} y}{dy} = \frac{M}{2} \lim_{y \to 0} \frac{1}{\sqrt{y^2 + 1}} = \frac{M}{2} \tag{4}$$

and Eq. (1) follows.

Problem 5.2-7

We begin by examining the points $s = 0$ and $s = -4/\theta$. Let, as shown in Fig. P5.2-7, a contour be drawn around each point. For the contour around the point $s = 0$ we write

$$s_0 = r_0 \exp(j\phi_0) \tag{3}$$

and for that around $s = -4/\theta$ we have

$$s_N = -4/\theta + r_N \exp(j\phi_N) . \tag{4}$$

We first examine the point $s = 0$. If this is not a branch point, then, as we move around it counterclockwise along the contour in a full circle, i.e. from $r_0\exp(j\phi_0)$ to $r_0\exp[j(\phi_0 + 2\pi)]$, the value of $f(s_0)$ must not change. Substitution of Eq. (3) into Eqs. (1) and (2) gives

$$f(r_0, \phi_0) = [r_0\theta \exp(j\phi_0)]^{\mp 1/2} \sqrt{4 + r_0\theta \exp(j\phi_0) \frac{R_0^{2N} + 1}{R_0^{2N} - 1}} \tag{5}$$

where R_0 is an abbreviation for $\exp[a(s_0)]$ and the upper and lower signs apply to Eqs. (1) and (2) respectively. By Eq. (5.2-13),

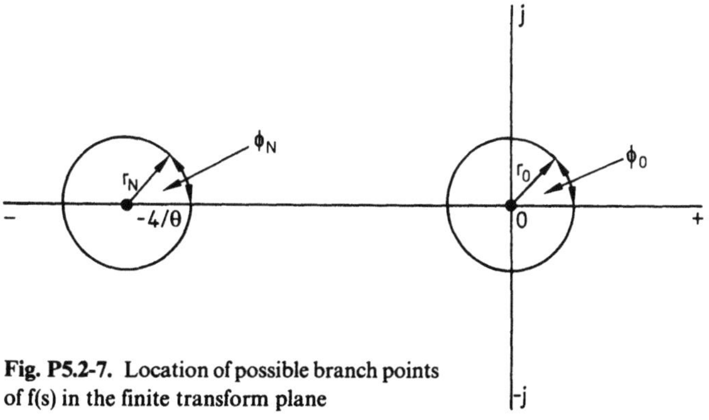

Fig. P5.2-7. Location of possible branch points of f(s) in the finite transform plane

$$R_0 = 1 + \frac{r_0\theta}{2}\exp(j\phi_0) + \tfrac{1}{2}\exp(j\phi_0/2)\sqrt{r_0\theta[4 + r_0\theta\exp(j\phi_0)]}\ . \tag{6}$$

Similarly,

$$f(r_0, \phi_0 + 2\pi) = -[r_0\theta\exp(j\theta_0)]^{\mp 1/2}\sqrt{4 + r_0\theta\exp(j\phi_0)}\,\frac{R_{2\pi}^{2N} + 1}{R_{2\pi}^{2N} - 1} \tag{7}$$

where

$$R_{2\pi} = 1 + \frac{r_0\theta}{2}\exp(j\phi_0) - \tfrac{1}{2}\exp(j\phi_0/2)\sqrt{r_0\theta[4 + r_0\theta\exp(j\phi_0)]}\ . \tag{8}$$

Multiplying both the numerator and the denominator in Eq. (7) by R_0^{2N}, and realizing that $R_{2\pi}R_0 = 1$ leads to

$$f(r_0, \phi_0 + 2\pi) = f(r_0, \phi_0)\ . \tag{9}$$

Hence, $s = 0$ is not a branch point. An entirely similar argument shows that

$$f(r_N, \phi_N + 2\pi) = f(r_N, \phi_N) \tag{10}$$

and, therefore, $s = -4/\theta$ is not a branch point either.

It is easy to show that f(s) does not have a branch point at $s = \infty$. We have

$$\lim_{s\to\infty} f(s) = \lim_{s\to\infty}\frac{R^{2N} + 1}{R^{2N} - 1} = 1\ . \tag{10}$$

Since the limit is neither 0 nor ∞, $s = \infty$ cannot be a branch point.

Problem 5.2-8

By Eqs. (5.2-9) and (5.2-19)

$$\begin{bmatrix}\bar{\sigma}_{n-\frac{1}{2}}(s)\\ \bar{\varepsilon}_{n-\frac{1}{2}}(s)\end{bmatrix}\begin{bmatrix}\cosh a(s) & Q_T(s)\sinh a(s)\\ U_T(s)\sinh a(s) & \cosh a(s)\end{bmatrix}\begin{bmatrix}\bar{\sigma}_{n+\frac{1}{2}}(s)\\ \bar{\varepsilon}_{n+\frac{1}{2}}(s)\end{bmatrix} \tag{1}$$

and the input relaxance becomes

$$\bar{Q}_{n-\frac{1}{2}}(s) = Q_T(s)\frac{\bar{Q}_{n+\frac{1}{2}}(s)\cosh a(s) + Q_T(s)\sinh a(s)}{Q_T(s)\cosh a(s) + \bar{Q}_{n+\frac{1}{2}}(s)\sinh a(s)} \tag{2}$$

where $\bar{Q}_{n\pm\frac{1}{2}}(s) = \bar{\sigma}_{n\pm\frac{1}{2}}(s)/\bar{\varepsilon}_{n\pm\frac{1}{2}}(s)$. Now let $\bar{Q}_{n+\frac{1}{2}}(s) = Q_T(s)$, the characteristic relaxance. Then $\bar{Q}_{n-\frac{1}{2}}(s) = Q_T(s)$ as asserted.

Problem 5.2-9

By Eq. (5.2-46), $a(-4/\theta) = j\pi$. Hence, by Eq. (5.2-34),

$$\bar{U}^{\times}_{GM}(s_N) = \frac{J}{2}\lim_{x\to\pi}\frac{\sin Nx}{\cos[(N - \frac{1}{2})x)]}\ . \tag{1}$$

Since N is an integer, the limit is indeterminate. Applying L'Hospital's theorem gives

$$\bar{U}^{\times}_{GM}(s_N) = -\frac{JN}{2N-1}\lim_{x\to\pi}\frac{\cos Nx}{\sin[(N-\frac{1}{2})x]} \tag{2}$$

and the limit becomes -1. Therefore,

$$\bar{U}^{\times}_{GM}(s_N) = \frac{JN}{2N-1} \ . \tag{3}$$

Problem 5.2-10

We have

$$G_g = \lim_{s\to\infty} \bar{Q}_{GM}(s) \ . \tag{1}$$

Equations (5.2-28) and (5.2-33) may be recast in the forms

$$\bar{Q}^{\circ}_{GM}(s) = (1/2J)\{\sqrt{\theta s(4+\theta s)}\tanh[(N+\tfrac{1}{2})a(s)] - \theta s\} \tag{2}$$

and

$$\bar{Q}^{\times}_{GM}(s) = (1/2J)[\sqrt{\theta s(4+\theta s)}\coth N\,a(s) - \theta s] \ , \tag{3}$$

respectively. But as $s \to \infty$, $a(s) \to \infty$, and $\tanh(\infty) = \coth(\infty) = 1$. Hence, both equations reduce to

$$G_g = (1/2J)\lim_{s\to\infty}[\sqrt{\theta s(4+\theta s)} - \theta s] = 1/J \tag{4}$$

because

$$\lim_{x\to\infty}[\sqrt{x(4+x)} - x] = \lim_{x\to\infty}[\sqrt{(x+2)^2} - x] = 2 \ . \tag{5}$$

Problem 5.2-11

By Eq. (3.4-121)

$$\eta_f = \lim_{s\to 0} \bar{Q}^{\circ}_{GM}(s)/s \ . \tag{1}$$

From Eq. (5.2-28) we have

$$\bar{Q}^{\circ}_{GM}(s)/s = (\eta/2)\{\sqrt{1+4/\theta s}\,\tanh[(N+\tfrac{1}{2})a(s)] - 1\} \ . \tag{2}$$

Now, by Eq. $(5.2\text{-}23)_2$, $a(s) = 2\sinh^{-1}\sqrt{\theta s}/2$, and, letting $\sqrt{\theta s} = x$, we obtain

$$\eta_f = \eta\left(\lim_{x\to 0}\frac{\tanh[(2N+1)\sinh^{-1}(x/2)]}{x} - \frac{1}{2}\right) = \eta N \tag{3}$$

since the limit (see Problem 5.2-6) is $N + \frac{1}{2}$.

Problem 5.2-12

The eigenvalues are the roots of the *secular equation*

$$\begin{vmatrix} 1+\theta s/2-\lambda & \eta s \\ J(1+\theta s/4) & 1+\theta s/2-\lambda \end{vmatrix} = 0 \ . \tag{1}$$

Since $J\eta = \theta$, we have

$$\lambda^2 - 2\lambda(1 + \theta s/2) + 1 = 0 \ . \tag{2}$$

Letting $1 + \theta s/2 = \cosh a(s)$, we find the roots to be given by

$$\lambda_{1,2} = \cosh a(s) \pm \sqrt{\cosh^2 a(s) - 1} = \cosh a(s) \pm \sinh a(s) \ . \tag{3}$$

Hence, the eigenvalues are

$$\lambda_1 = \exp[a(s)] = \cosh a(s) + \sinh a(s) \tag{4}$$

and

$$\lambda_2 = \exp[-a(s)] = \cosh a(s) - \sinh a(s) \tag{5}$$

where $\sinh a(s) = \sqrt{\theta s(4 + \theta s)}/2$ [cf. Eqs. (5.2-14) and (5.2-15)].

Problem 5.2-13

The diagonal terms of the matrix, M', are dimensionless. The off-diagonal terms are not. We therefore first rewrite Eq. (5.2-9) in the equivalent form

$$\begin{bmatrix} \bar{\sigma}_{n-\frac{1}{2}}(s) \\ \bar{\varepsilon}_{n-\frac{1}{2}}(s)/J \end{bmatrix} = \begin{bmatrix} 1 + \theta s/2 & \theta s \\ 1 + \theta s/4 & 1 + \theta s/2 \end{bmatrix} \begin{bmatrix} \bar{\sigma}_{n+\frac{1}{2}}(s) \\ \bar{\varepsilon}_{n+\frac{1}{2}}(s)/J \end{bmatrix} \tag{2}$$

in which all elements of the matrix, M, are dimensionless. To raise it to the Zth power we use the equation $M^Z = N\Lambda^Z N^{-1}$ in which Λ is the diagonal matrix formed from the eigenvalues of M, and N is the matrix of the eigenvectors. The secular equation now is

$$\begin{vmatrix} 1 + \theta s/2 - \lambda & \theta s \\ 1 + \theta s/4 & 1 + \theta s/2 - \lambda \end{vmatrix} = 0 \ . \tag{3}$$

The eigenvalues are $\exp[a(s)]$ and $\exp[-a(s)]$ (see Problem 5.2-12) and are, therefore, distinct. We note that $\cosh a(s) = 1 + \theta s/2$ and $\sinh a(s) = \sqrt{\theta s(4 + \theta s)}/2$. The eigenvector corresponding to the first eigenvalue is obtained from

$$(-\sqrt{\theta s(4 + \theta s)}/2)N_1^{(1)} + \theta s\, N_2^{(1)} = 0 \tag{4}$$

which follows from the first of Eqs. (2), and from

$$[N_1^{(1)}]^2 + [N_2^{(1)}]^2 = 1 \ . \tag{5}$$

These relations give

$$N^{(1)} = \begin{bmatrix} \pm 2\sqrt{\theta s} \\ \pm\sqrt{4 + \theta s} \end{bmatrix} \frac{1}{\sqrt{4 + 5\theta s}} \ . \tag{6}$$

The second eigenvector is obtained similarly from

$$(\sqrt{\theta s(4 + \theta s)}/2)N_1^{(2)} + \theta s\, N_2^{(2)} = 0 \tag{7}$$

and

$$[N_1^{(2)}]^2 + [N_2^{(2)}]^2 = 1 \tag{8}$$

and becomes

$$N^{(2)} = \begin{bmatrix} \pm 2\sqrt{\theta s} \\ \mp\sqrt{4+\theta s} \end{bmatrix} \frac{1}{\sqrt{4+5\theta s}} \ . \tag{9}$$

Thus, the eigenvalue matrix takes the form

$$N = \frac{1}{C}\begin{bmatrix} \pm A & \pm A \\ \pm B & \mp B \end{bmatrix} \tag{10}$$

where $A = 2\sqrt{\theta s}$, $B = \sqrt{4+\theta s}$, and $C = \sqrt{4+5\theta s}$. The matrix is not symmetric. To invert it, we form the transpose of its cofactor matrix and divide it by the determinant, det[N]. Then

$$N^{-1} = \frac{C}{2}\begin{bmatrix} \pm 1/A & \pm 1/B \\ \pm 1/A & \mp 1/B \end{bmatrix} \tag{11}$$

and we have

$$M^Z = \frac{1}{2}\begin{bmatrix} \pm A & \pm A \\ \pm B & \mp B \end{bmatrix}\begin{bmatrix} \exp[Z\,a(s)] & 0 \\ 0 & \exp[-Z\,a(s)] \end{bmatrix}\begin{bmatrix} \pm 1/A & \pm 1/B \\ \pm 1/A & \mp 1/B \end{bmatrix} . \tag{12}$$

Matrix multiplication leads to

$$M^Z = \begin{bmatrix} \cosh Z\,a(s) & (A/B)\sinh Z\,a(s) \\ (B/A)\sinh Z\,a(s) & \cosh Z\,a(s) \end{bmatrix} . \tag{13}$$

But

$$A/B = JQ_T(s) \qquad \text{and} \qquad B/A = U_T(s)/J \tag{14}$$

where $Q_T(s)$ and $U_T(s)$ are the characteristic respondances introduced by Eq. (5.2-16). The last two equations establish the relation with the original matrix, M′.

Problem 5.3-1

By Eq. (5.3-39) the residue associated with the pole at the origin is ϕ and this represents the steady-state fluidity, ϕ_f. The other poles are given by Eq. (5.3-35) and give rise to the retardation times

$$\tau_n = -1/s_n = 4\hat{\theta}\cos^2\frac{\pi n}{2N} , \qquad n = 1, 2, \ldots, N-1 \ . \tag{1}$$

The instantaneous compliance, J_g, is 1/GN and the other compliances follow from Eqs. (1) and (5.3-40) as

$$J_n = \tau_n \mathrm{Res}_n(\bar{U}^{\circ}_{RC}) = \frac{2}{GN}\sin^2\frac{\pi n}{2N} , \qquad n = 1, 2, \ldots \ . \tag{2}$$

We thus find

$$\bar{U}_{RC}(s) = \frac{1}{GN} + \sum_{n=1}^{N-1}\frac{J_n}{1+\tau_n s} + \frac{\phi}{s} \tag{3}$$

and recognize $\bar{U}^{\circ}_{RC}(s)$ to be equivalent to the retardance of a Kelvin model with 2N elements in which the springs have the compliances given by Eq. (2) and the dashpots have fluidities given by

$$\phi_n = J_n/\tau_n = \frac{\phi}{2N} \tan^2 \frac{\pi n}{2N} , \qquad n = 1, 2, \ldots, N - 1 \tag{4}$$

in addition to $\phi_f = \phi$.

Problem 5.4-1

We have

$$G(t) = G_e + \sum_{n=1}^{N-1} G_n \exp(-t/\tau_n) \tag{1}$$

where

$$G_n = 2G_e \cos^2(\pi n/2N) \tag{2}$$

by Eq. (5.2-88) and

$$\tau_n = (\theta/4) \csc^2(\pi n/2N) \tag{3}$$

by Eq. (5.2-87) for $G_{GM}(t)$ and

$$G_n = [G_e/(2N - 1)]\tan^2[\pi n/(2N - 1)] \tag{4}$$

by Eq. (5.3-34) and

$$\tau_n = 4\hat{\theta} \cos^2[\pi n/(2N - 1)] \tag{5}$$

by Eq. (5.3-33) for $G_{RC}(t)$. Hence,

$$G_{GM}(t) = G_e\left[1 + 2 \sum_{n=1}^{N-1} \cos^2 \frac{\pi n}{2N} \exp\left(-\frac{4t}{\theta} \sin^2 \frac{\pi n}{2N}\right)\right] \tag{6}$$

and

$$G_{RC}(t) = G_e\left[1 + \frac{1}{2N - 1} \sum_{n=1}^{N-1} \tan^2 \frac{\pi n}{2N - 1} \exp\left(-\frac{t}{4\hat{\theta}} \sec^2 \frac{\pi n}{2N - 1}\right)\right] . \tag{7}$$

Problem 5.5-1

By Eq. (2) of Problem 5.2-1 the retardance of the Gross model is

$$\bar{U}^{\times}_{G}(s) = \sqrt{J/\eta s + (J/2)^2} \tanh N\, a(s) . \tag{1}$$

We must replace J and η by J_T/N and η_T/N, and must find the total compliance and total viscosity first. The total compliance in the Gross model is $J(N - 1) + 2(J/2) = JN$. By Eq. (9) of Problem 5.2-1, $JN = J_e$. Thus, $J = J_e/N$. The number of dashpots is N, and by Eqs. (13) and (14) of the problem,

$$\sum_n \eta_n = \sum_n G_n\tau_n = \frac{\eta}{2N} \sum_{n=1}^{N-1} \csc^2 \frac{\pi n}{2N} + \frac{\eta}{4N} . \tag{2}$$

But the sum is $2(N^2 - 1)/3$. Hence

$$\sum_n \eta_n = \eta(4N^2 - 1)/12N \simeq \eta N/3 \ . \tag{3}$$

Using $\eta_\times$ for the sum of viscosities, we have $\eta = 3\eta_\times/N$. Thus,

$$\bar{U}_G^\times(s) = \sqrt{\frac{J_e}{3\eta_\times s} + \left(\frac{J_e}{2N}\right)^2} \ \tanh N\, a(s) \ .$$

As $N \to \infty$, the limit of $N\,a(s)$ is again given by Eq. (5.5-6), and we find that $\bar{U}_{G\infty}^\times(s) = \bar{U}_{GM\infty}^\times(s)$.

Problem 5.5-2

We replace J by J_T/N and η by η_T/N in both relaxances. Equation (5.2-16) becomes

$$Q_T(s) = 1/\sqrt{J_T/\eta_T s + (J_T/2N)^2} \tag{1}$$

while Eq. (4) of Problem 5.2-2 gives

$$Q_\Pi(s) = \sqrt{\eta_T s/J_T + (\eta_T s/2N)^2} \ . \tag{2}$$

Clearly, then,

$$\lim_{N\to\infty} Q_T(s) = \lim_{N\to\infty} Q_\Pi(s) = \sqrt{\eta_T s/J_T} \ . \tag{3}$$

Problem 5.5-3

The desired functions are obtained through division of the relaxances and the retardances by s, followed by inversion of the transform. Thus, for the rheodictic model we find

$$G_{GM\infty}(t) = \frac{2}{3J_e^\circ} \sum_{n=0}^{\infty} \exp(-t/\tau_n)$$

from Eq. (5.5-43), and inversion with the help of LTP (10). The use of Eq. (5.5-41) then gives

$$G_{GM\infty}(t) = (2G_e^\circ/3) \sum_{n=1}^{\infty} \exp[-(2n+1)^2\pi^2 t/4\theta_\infty^\circ] \tag{2}$$

where $G_e^\circ = 1/J_e^\circ$ is the steady-state modulus. Similarly, from Eq. (5.5-31) we have

$$G_{GM\infty}(t) = \frac{1}{J_e} + \frac{2}{J_e} \sum_{n=0}^{\infty} \exp(-t/\tau_n) \tag{3}$$

for the arrheodictic model. The use of Eq. (5.5-28) gives

$$G_{GM\infty}(t) = G_e + 2G_e \sum_{n=1}^{\infty} \exp(-n^2\pi^2 t/\theta_\infty^\times) \ . \tag{4}$$

To obtain $\bar{J}_{GM\infty}(s)$ for the rheodictic form we divide Eq. (5.5-37) by s. Transform inversion yields

$$J_{GM\infty}(t) = \sum_{n=1}^{\infty} J_n[1 - \exp(-t/\tau_n)] + t/\eta_f \ . \tag{5}$$

Equations (5.5-35), (5.5-36), and (5.5-39) then lead to

$$J_{GM\infty}(t) = J_e^\circ[1 - (6/\pi^2) \sum_{n=1}^{\infty} n^{-2}\exp(-n^2\pi^2 t/\theta_\infty^\circ)] + t/\eta_f \ . \tag{6}$$

Finally, for the arrheodictic form we obtain

$$J_{GM\infty}(t) = \sum_{n=1}^{\infty} J_n[1 - \exp(-t/\tau_n)] \tag{7}$$

from Eq. (5.5-25). Substitution using Eqs. (5.5-21), (5.5-24) and (5.5-45) furnishes

$$J_{GM\infty}(t) = J_e\left\{1 - (8/\pi^2) \sum_{n=1}^{\infty} (2n-1)^{-2}\exp[-(2n-1)^2\pi^2 t/4\theta_\infty^\times]\right\} \ . \tag{8}$$

Problem 5.5-4

(A) With the substitution $\theta_\infty^\times s = j\omega\theta$, Eq. $(5.5\text{-}14)_1$ becomes

$$\bar{Q}_{GM\infty}^\times(j\omega) = (\sqrt{j\omega\theta}/J_e)\coth\sqrt{j\omega\theta} = G_{GM\infty}^*(\omega) \ . \tag{1}$$

Let $\sqrt{j\omega\theta} = \alpha(1 + j)$ where $\alpha = \sqrt{\omega\theta/2}$. Then

$$\coth\sqrt{j\omega\theta} = \left.\frac{\sinh 2\alpha - j\sin 2\alpha}{\cosh 2\alpha - \cos 2\alpha}\right|_{\alpha=\sqrt{\omega\theta/2}} \tag{2}$$

and it follows that

$$G'_{GM\infty}(\omega) = \left(\frac{\sqrt{2\omega\theta}}{2J_e}\right)\left(\frac{\sinh\sqrt{2\omega\theta} + \sin\sqrt{2\omega\theta}}{\cosh\sqrt{2\omega\theta} - \cos\sqrt{2\omega\theta}}\right) \tag{3}$$

and

$$G''_{GM\infty}(\omega) = \left(\frac{\sqrt{2\omega\theta}}{2J_e}\right)\left(\frac{\sinh\sqrt{2\omega\theta} - \sin\sqrt{2\omega\theta}}{\cosh\sqrt{2\omega\theta} - \cos\sqrt{2\omega\theta}}\right) \ . \tag{4}$$

(B) By Eq. $(4.6\text{-}4)_3$

$$G_e = \lim_{\omega\to 0} G'(\omega) = \frac{1}{J_e}\lim_{\alpha\to 0}\frac{\alpha(\sinh 2\alpha + \sin 2\alpha)}{\cosh 2\alpha - \cos 2\alpha} \ . \tag{5}$$

Repeated application of L'Hospital's theorem gives $G_e = 1/J_e$.

Problem 5.5-5

(A) We have $\theta_\infty^\circ = 3J_e^\circ\eta_f$. Hence, using Eq. (4.6-39) for J_e° and Eq. (4.6-38) for η_f, Eq. (1) follows.

(B) Using Eqs. $(4.6\text{-}41)_2$ and $(4.6\text{-}40)_2$ gives

$$\theta_\infty^\circ = \frac{3\int_0^\infty t\,G(t)dt}{\int_0^\infty G(t)dt}\,. \tag{2}$$

Using Eqs. $(4.6\text{-}41)_3$ and $(4.6\text{-}40)_3$ yields

$$\theta_\infty^\circ = \frac{3\int_0^\infty \omega^{-3}[G''(\omega) - d\,G''(\omega)/d\ln\omega]d\omega}{\int_0^\infty \omega^{-2}G'(\omega)d\omega}\,. \tag{3}$$

Problem 5.5-6

(A) To obtain the mechanical model we represent each bead by a dashpot whose one end is connected to the ground node. The dashpots are connected by the springs. The dashpots all have the same viscosity and the springs have the same compliance. The diagram turns out to be the ladder model shown in Fig. P5.5-6b. Thus, the attempt to model the viscoelastic behavior of long chain molecules quite naturally leads to its representation by a regular ladder model. We note that the model does not have to contain an infinite number of nodes (i.e. beads). We treat it here in this fashion merely for simplicity.

The model displayed above is clearly the extended Gross-Marvin model. Hence, the relaxance is given by Eq. (5.5-13) and we have

$$\bar{Q}_{GM\infty}(s) = \frac{\sqrt{\theta_\infty^\circ s}}{3J_e^\circ}\tanh\sqrt{\theta_\infty^\circ s} \tag{2}$$

for the relaxance in closed form.

(B) We now rewrite Eq. (1) in the form

$$\bar{Q}_R(s) = \frac{2\theta_R s}{5J_e^\circ}\sum_{n=1}^{\infty}\frac{1}{n^2\pi^2 + \theta_R s}\,. \tag{3}$$

Then, by Eq. (5.5-29), we obtain

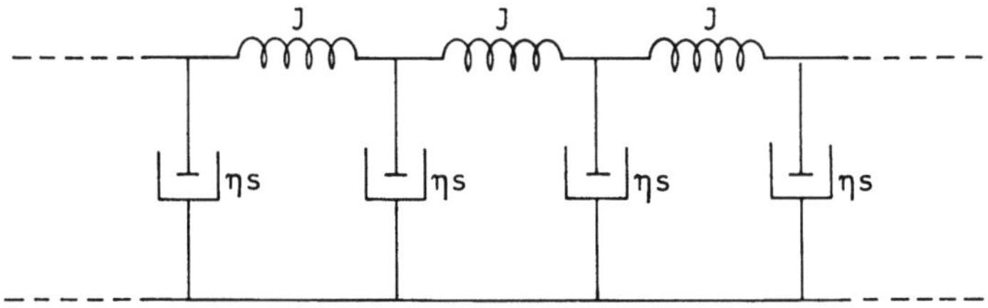

b

Fig. P5.5-6b. Ladder model corresponding to bead-and-spring assembly

$$\bar{Q}_R(s) = \frac{\sqrt{\theta_R s}}{5J_e^\circ}\left(\coth\sqrt{\theta_R s} - 1/\sqrt{\theta_R s}\right) . \tag{4}$$

again in closed forms.

To compare Eqs. (2) and (4) we use the identity $\tanh\alpha = \coth 2\alpha - \operatorname{csch} 2\alpha$ to write Eq. (2) in the form

$$\bar{Q}_{GM\infty}(s) = \frac{\sqrt{\theta_\infty^\circ s}}{3J_e^\circ}\left(\coth(2\sqrt{\theta_\infty^\circ s}) - \frac{1}{\sinh(2\sqrt{\theta_\infty^\circ s})}\right) . \tag{5}$$

But $\theta_R = 5\theta_\infty^\circ$. Hence,

$$\bar{Q}_R(s) = \frac{1.342\sqrt{\theta_\infty^\circ s}}{3J_e^\circ}\left(\coth(2.236\sqrt{\theta_\infty^\circ s}) - \frac{1}{2.236\sqrt{\theta_\infty^\circ s}}\right) . \tag{6}$$

The two models thus show qualitatively the same behavior.

Problem 5.5-7

(A) We have first

$$\tanh(\alpha + j\beta) = \frac{\tanh\alpha \sec^2\beta + j\tan\beta\,\operatorname{sech}^2\alpha}{1 + \tanh^2\alpha\tan^2\beta} \tag{1}$$

and the real and imaginary parts of $T^* = (\alpha + j\beta)\tanh(\alpha + j\beta)$ become

$$T' = \frac{\alpha\sinh 2\alpha - \beta\sin 2\beta}{\cosh 2\alpha + \cos 2\beta} \tag{2}$$

and

$$T'' = \frac{\beta\sinh 2\alpha + \alpha\sin 2\beta}{\cosh 2\alpha + \cos 2\beta} . \tag{3}$$

Next we have

$$\coth(\alpha + j\beta) = \frac{\coth\alpha\csc^2\beta - j\cot\beta\,\operatorname{csch}^2\alpha}{\coth^2\alpha + \cot^2\alpha} \tag{4}$$

and obtain the real and imaginary parts of $C^* = (\alpha + j\beta)\coth(\alpha + j\beta)$ as

$$C' = \frac{\alpha\sinh 2\alpha + \beta\sin 2\beta}{\cosh 2\alpha - \cos 2\beta} \tag{5}$$

and

$$C'' = \frac{\beta\sinh 2\alpha - \alpha\sin 2\beta}{\cosh 2\alpha - \cos 2\beta} . \tag{6}$$

(B) For the extended Marvin-Oser models

$$\Gamma(j\omega) = \sqrt{\frac{j\omega\theta}{1 + j\omega\delta}} = \sqrt{\omega\theta\sin\phi}\,\exp(j\phi/2) \tag{7}$$

where $\cot\phi = \omega\delta$. Hence,

$$\alpha = \sqrt{\omega\theta \sin\phi}\,\cos\phi/2 \qquad \text{and} \qquad \beta = \sqrt{\omega\theta \sin\phi}\,\sin\phi/2\ . \tag{8}$$

The storage modulus then becomes

$$G'_{MO\infty}(\omega) = \mathscr{R}e\,\bar{Q}^{\times}_{MO\infty}(j\omega) = G_e \mathscr{R}e\,\Gamma(j\omega)\coth\Gamma(j\omega)\ . \tag{9}$$

(C) For the extended Gross-Marvin models $\delta = 0$. Thus $\phi = \frac{\pi}{2}$ and $\alpha = \beta = \sqrt{\omega\theta/2}$ in accordance with the value of α used in Problem 5.5-4.

Problem 5.5-8

(A) For convenience, Let us set

$$x^2 = \frac{\theta^\circ_\infty s}{1 + \delta^\circ s} = \frac{J^\circ_T \eta^\circ_T s}{1 + \eta^\circ_T s/G^\circ_T}\ . \tag{2}$$

But $\eta^\circ_T = \eta_f$. Therefore,

$$\frac{1}{\eta_f s} = \frac{J^\circ_T}{x^2} - \frac{1}{G^\circ_T} \tag{3}$$

and using Eq. (5.5-75), Eq. (2) becomes

$$J^\circ_e = 1/G^\circ_T + J^\circ_T \lim_{x\to 0}\left(\frac{\coth x}{x} - \frac{1}{x^2}\right)\ . \tag{4}$$

Now

$$\coth x = \frac{1}{x} + \frac{x}{3} - \frac{x^3}{45} \pm \text{terms in higher powers of } x\ . \tag{5}$$

Clearly, then, the limit equals 1/3 and we recover Eq. (1)

(B) By Eq. (5.5-81)

$$J^\circ_e = J_g + 2J^\circ_T \sum_{n=1}^{\infty} \frac{1}{n^2\pi^2} - 2J^\circ_T \sum_{n=1}^{\infty} \frac{1}{J^\circ_T G^\circ_T + n^2\pi^2}\ . \tag{6}$$

Using Eqs. $(5.5\text{-}78)_1$ and (5.5-39) we are again led to Eq. (1).

(C) By Eqs. (5.5-89) and (5.5-90), using $\eta^\circ_T = \eta_f$ again, we find

$$J^\circ_e = (\eta^\circ_T)^{-2} \sum_{n=0}^{\infty} \left[\delta^\circ + \frac{4\theta^\circ_\infty}{(2n+1)^2\pi^2}\right] \frac{8\eta^\circ_T}{(2n+1)^2\pi^2}\ . \tag{7}$$

But $\delta^\circ = \eta^\circ_T/G^\circ_T$ and $\theta^\circ_\infty = J^\circ_T\eta^\circ_T$. Hence, we have

$$J^\circ_e = (8/G^\circ_T) \sum_{n=0}^{\infty} \frac{1}{(2n+1)^2\pi^2} + 32 J^\circ_T \sum_{n=0}^{\infty} \frac{1}{(2n+1)^4\pi^4}\ . \tag{8}$$

The first sum equals 1/8 by Eq. (5.5-45). The second [4c] is 1/96. Thus we again recover Eq. (1).

Problem 5.5-9

The matrix equation becomes

$$\begin{bmatrix} \bar{\sigma}_0(s) \\ \bar{\varepsilon}_0(s) \end{bmatrix} = \begin{bmatrix} \cosh\Gamma(s) & Q_C(s)\sinh\Gamma(s) \\ U_C(s)\sinh\Gamma(s) & \cosh\Gamma(s) \end{bmatrix} \begin{bmatrix} \bar{\sigma}_\infty(s) \\ \bar{\varepsilon}_\infty(s) \end{bmatrix}\ . \tag{1}$$

Equation (5.5-47) follows at once by letting $\bar{\sigma}_0(s)/\bar{\varepsilon}_0(s) = Q_0(s)$ and $\bar{\sigma}_\infty(s)/\bar{\varepsilon}_\infty(s) = Q_F(s)$.

Problem 5.5-10

The retardance, $\bar{U}_\infty(s)$, is simply the reciprocal of the relaxance, $\bar{Q}_\infty(s)$, given by Eq. (5.5-47). We obtain

$$\bar{U}_\infty(s) = U_C(s)\frac{U_F(s)\cosh\Gamma(s) + U_C(s)\sinh\Gamma(s)}{U_C(s)\cosh\Gamma(s) + U_F(s)\sinh\Gamma(s)} \tag{1}$$

where the transmittance, $\Gamma(s)$, is still given by Eq. (5.5-48), $U_F(s) = 1/Q_F$, and the characteristic retardance

$$U_C(s) = 1/Q_C(s) = \sqrt{U_s(s)/Q_r(s)} = \Gamma(s)/Q_r(s) \tag{2}$$

follows from Eq. (5.5-49). The rheodictic form is obtained by letting $U_F(s) = \infty$, and the arrheodictic form results from selecting $U_F(s) = 0$.

Problem 5.6-1

The redrawn diagram is shown in Fig. P5.6-1a. Next, broken-line rectangles have been drawn around each node in such a way that they pass through all those

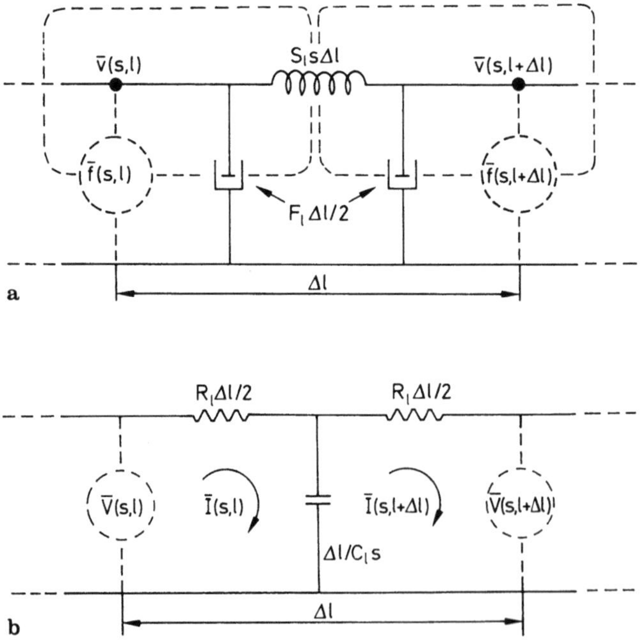

Fig. P5.6-1. a Mechanical model diagram of an elemental section of the inertia-less viscoelastic transmission line. **b** Electric circuit diagram of an elemental section of the non-inductive, leakage-free electrical transmission line

mechanical elements, active or passive, that are connected to it. Finally, the broken-line rectangles were redrawn substituting the appropriate electrical elements for the mechanical ones according to the electrostatic analogy, using Table 3.2-1. The result is the elemental section of the non-inductive leakage-free electrical transmission line shown in Fig. P5.6-1b.

Problem 5.6-2

The electromagnetic analogy preserves the interconnection of the system elements. Hence, it is merely necessary to replace the mechanical elements in Fig. P5.6-1a by the corresponding electric ones according to the electromagnetic analogy using Table 3.2-1. The result is the line section shown in Fig. P5.6-2 which was used by Blizard [5].

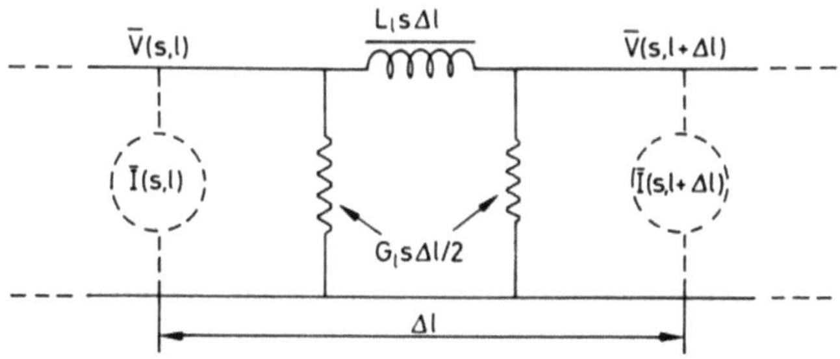

Fig. P5.6-2. Electric circuit diagram of an elemental section of Blizard's transmission line

Problem 5.6-3

(A) The initial and final sections are shown in Fig. P5.6-3. The propagation section is identical with that shown in Fig. 5.6-6, when this is rewritten for the harmonic steady-state case.

(B) The pertinent canonical equation is the harmonic analog of Eq. (5.6-23), using $\gamma^*(\omega)$ instead of $\Gamma_l^*(\omega)$. This gives

$$f(\omega, l) = f(\omega, 0)\cosh l\,\gamma^*(\omega) - Q^*_{mC}(\omega)x(\omega, 0)\sinh l\,\gamma^*(\omega) \ . \qquad (1)$$

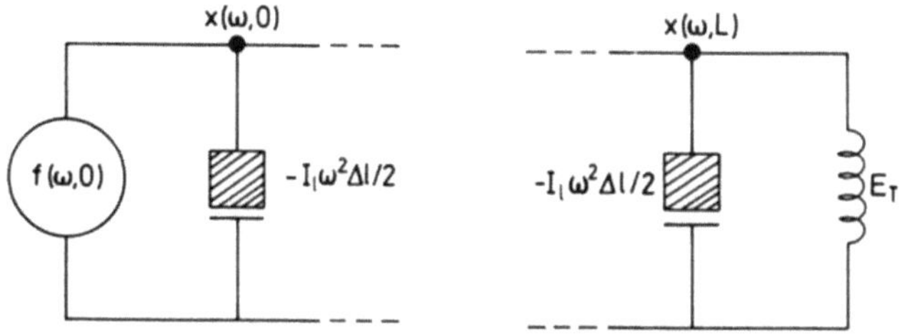

Fig. P5.6-3. Initial and final sections for Problem 3.1-4 with sample inertia

Substituting $f(\omega, 0) = Q^*_{mo}(\omega)x(\omega, 0)$ [cf. Eq. (5.6-29)] and letting $l = L$ leads to

$$f(\omega, L) = x(\omega, 0)[Q^*_{mo}(\omega)\cosh L\,\gamma^*(\omega) - Q^*_{mC}(\omega)\sinh L\,\gamma^*(\omega)] \tag{2}$$

which is the required equation. The brackets contain the material properties. Elimination of $Q^*_{mo}(\omega)$ by the harmonic analog of Eq. (5.6-31) and using $Q^*_{mL}(\omega) = E_T$ furnishes

$$f(\omega, L) = \frac{E_T Q^*_{mC}(\omega)x(\omega, 0)}{Q^*_{mC}(\omega)\cosh L\,\gamma^*(\omega) + E_T \sinh L\,\gamma^*(\omega)}\,. \tag{3}$$

But $Q^*_{mC}(\omega) = j\omega A\sqrt{\rho E^*(\omega)}$ by Eq. (5.6-73) and $\gamma^*(\omega) = j\omega\sqrt{\rho/E^*(\omega)}$ by Eq. (5.6-72) where A is the cross-sectional area of the sample, and $E^*(\omega)$ is the complex tensile modulus as required by the problem. With these substitutions we find

$$f(\omega, L) = \frac{E_T\omega\sqrt{\rho\, E^*(\omega)}x(\omega, 0)}{\omega\sqrt{\rho\, E^*(\omega)}\cos[\omega L\sqrt{\rho/E^*(\omega)}] + (E_T/A)\sin[\omega L\sqrt{\rho/E^*(\omega)}]}\,. \tag{4}$$

There are ways of solving Eq. (4) but it cannot be made explicit in $E^*(\omega)$ or its components.

(C) When the sample inertia is negligible the inertances per unit length in the initial, propagation, and final sections of the model diagram vanish. The strut retardances add, being now in series. Their sum, $LU^*_{ml}(\omega)$ equals $U^*_m(\omega)$, the harmonic mechanical retardance of the sample. The model diagram thus reduces to that shown in Fig. P3.1-4 where $bE^*(\omega) = 1/U^*_m(\omega)$, $f_2(\omega) = f(\omega, L)$, and $x_1(\omega) = x(\omega, 0)$.

Equation (4) is reduced by retaining only the first terms in expanding the cosine and sine terms. The first becomes unity and the second is replaced by its argument. This leads to

$$f(\omega, L) = \frac{E_T E^*(\omega)x(\omega, 0)}{E^*(\omega) + HE_T} \tag{5}$$

where $H = L/A$ is the shape factor of the sample, L being its length. Equation (5) is identical with Eq. (3) of Problem 3.1-4.

Problem 5.6-4

Our point of departure is the relation between $\mathcal{Q}^*_0(\omega)$ and $G^*(\omega)$. By

$$\mathcal{Q}^*_0(\omega) = \frac{\sigma(\omega, 0)}{x(\omega, 0)} = -\frac{G^*(\omega)\varepsilon(\omega, 0)}{x(\omega, 0)} \tag{1}$$

where the negative sign must be used because of the way in which the stress is defined. Retaining the first two terms of the sine functions in Eq. (5.6-59) the steady-state displacement along the line is

$$x(\omega, l) = x_0\frac{L - l}{L}\left[1 + \frac{\beta^2(\omega)(2L - l)l/6}{1 - \beta^2(\omega)L^2/6}\right]\exp(j\omega t) \tag{2'}$$

which we further simplify to

$$x(\omega, l) = x_0 \frac{L - l}{L}[1 + \beta^2(\omega)(2L - l)l/6] \exp(j\omega t) \ . \quad (2)$$

We have, then,

$$\varepsilon(\omega, 0) = \left.\frac{\partial\, x(\omega, l)}{\partial l}\right|_{l=0} = -x_0 L^{-1}[1 - \beta^2(\omega)L^2/3]\exp(j\omega t) \ . \quad (3)$$

Letting $l = 0$ in Eq. (2) and recognizing that

$$\mathscr{Q}_0^*(\omega) = A^{-1}Q_{m0}^*(\omega) = bA^{-1}Q_0^*(\omega) = L^{-1}Q_0^*(\omega) \quad (4)$$

we obtain

$$Q_0^*(\omega) = G^*(\omega)[1 - \beta^2(\omega)L^2/3] \ . \quad (5)$$

But $\beta^2(\omega) = \omega^2/v^2 = \omega^2\rho/G^*(\omega)$. Hence,

$$G^*(\omega) = Q_0^*(\omega) + \omega^2\rho L^2/3 \ . \quad (6)$$

Thus, to obtain the complex modulus when inertial effects are not too severe, one may measure it in the usual way but apply the mass correction just derived. We call this a *lumped mass* approximation because the distributed mass of the sample is replaced by a lumped mass in parallel with the relaxance. Note that $\rho L^2 = Hm$ is the *inertivity* discussed in Sect. 3.3. Clearly, for $m = 0$ Eq. (6) reduces to $G^*(\omega) = Q_0^*(\omega)$.

Problem 5.6-5

From Fig. P5.6-5 we have

$$\bar{x}_1(s, 0) = \frac{\bar{f}_D(s)}{Q_{m0}(s) + Q_{mD}(s)} \ . \quad (1)$$

But $Q_{m0}(s) = A\mathscr{Q}_C(s)$ because the input relaxance of a quiescent line is equal to its characteristic impedance for a wave initiated at its entrance. Introducing the abbreviation

$$R(s) = \frac{\bar{f}_D(s)}{Q_{mD}(s) + A\mathscr{Q}_0(s)} \quad (2)$$

we have

$$\bar{x}_1(s, l) = R(s)\exp[-l\,\gamma(s)] \quad (3)$$

at the point l. On reaching the end of the line there will be a reflection so that the returning wave will have the transform

$$\bar{x}_2(s, l) = R(s)\exp[-L\,\gamma(s)]r_{xL}(s)\exp[-(L - l)\gamma(s)] \quad (4)$$

where $r_{xL}(s)$ is the transform of the reflection coefficient at the far end of the line given by Eq. (5.6-56). At the input end this wave will again be reflected, this time with reflection coefficient

$$r_{x0}(s) = \frac{Q_{mD} - AQ_C(s)}{Q_{mD} + AQ_C(s)} \ , \tag{5}$$

and, when it returns, we have

$$\bar{x}_3(s,l) = R(s)\exp[-L\,\gamma(s)]r_{xL}(s)\exp[-L\,\gamma(s)]r_{x0}(s)\exp[-l\,\gamma(s)] \ . \tag{6}$$

After an indefinite time a steady state is reached in which $\bar{x}(s,l)$ is the sum (superposition) of an infinite number of such terms, i.e.

$$\begin{aligned}\bar{x}(s,l) = &\{R(s)\exp[-l\,\gamma(s)] + r_{xL}(s)\exp[-2\,L\,\gamma(s)]\exp[l\,\gamma(s)]\} \\ &\times \{1 + r_{xL}(s)r_{x0}(s)\exp[-2\,L\,\gamma(s)] \\ &+ [r_{xL}(s)r_{x0}(s)\exp(-2\,L\,\gamma(s))]^2 + \cdots\} \ .\end{aligned} \tag{7}$$

Equation (7) may be written in closed form as

$$\bar{x}(s,l) = R(s)\frac{\exp[-l\,\gamma(s)] + r_{xL}(s)\exp[-2\,L\,\gamma(s)]\exp[l\,\gamma(s)]}{1 - r_{xL}(s)r_{x0}(s)\exp[-2\,L\,\gamma(s)]} \tag{8}$$

and this is identical with Eq. (5.6-57), showing that

$$x_1(s) = \frac{R(s)}{1 - r_{xL}(s)r_{x0}(s)\exp[-2L\,\gamma(s)]} \tag{9}$$

when it is interpreted in terms of multiple reflections.

Problem 5.6-6

For an open line $\mathcal{Q}_L(s) = 0$ and Eq. (5.6-52) gives

$$x_1(s) = \frac{\bar{x}(s,0)\exp[L\,\gamma(s)]}{2\cosh L\,\gamma(s)} \tag{1}$$

while Eq. (5.6-56) furnishes $r_{xL}(s) = 1$. By Eq. (5.6-57) then

$$\bar{x}(s,l) = \bar{x}(s,0)\frac{\cosh[(L-l)\gamma(s)]}{\cosh L\,\gamma(s)} \ . \tag{2}$$

If the excitation is harmonic, this becomes

$$x(t,l) = x_0\exp(j\omega t)\frac{\cos\beta(L-l)}{\cos\beta L} \tag{3}$$

and, taking the real part of $\exp(j\omega t)$, we have

$$x(t,l) = x_0'\{\cos[\omega t + \beta(L-l)] + \cos[\omega t - \beta(L-l)]\} \tag{4}$$

where $x_0' = x_0/2\cos\beta L$.

The forward and backward waves now have equal amplitude and equal signs. At the far end of the line

$$x(t,L) = 2x_0'\cos\omega t \ . \tag{5}$$

Using $d = L - l$, Eq. (4) becomes

$$x(t',d) = 2x_0' \cos \omega t' \cos \beta d \tag{6}$$

for any time t'. The node condition, $\cos \beta d = 0$ is satisfied for $d_n = \pi(n + \frac{1}{2})/\beta$ ($n = 0, 1, 2, \ldots$). For the antinodes we have $|\cos \beta d| = 1$ and $d_n = \pi n/\beta$. Thus, the far end of the open line is necessarily an antinode.

Problem 5.6-7

Equation (1) follows directly from Eq. (5.6-66) using Eqs. (5.6-52), (5.6-53), and (5.6-54).

Equation (2) may be obtained from Eq. (5.6-51) using Eq. $(5.6\text{-}44)_1$ and then substituting Eq. (5.6-43).

Problem 5.6-8

If the liquid is purely viscous, $G'(\omega) = 0$ and $G''(\omega) = \omega\eta$. Equation (5.6-72) then becomes

$$\gamma^*(\omega) = \sqrt{j\omega\rho/\eta} \tag{2}$$

and we have

$$\alpha^2(\omega) + 2\,j\alpha(\omega)\beta(\omega) - \beta^2(\omega) = j\omega\rho/\eta\ . \tag{3}$$

from Eq. (5.6-75). Hence, $\alpha(\omega) = \beta(\omega)$, and Eq. (1) follows. It can also, of course, be obtained directly from Eqs. (5.6-81) and (5.6-82).

Problem 5.6-9

By Eq. (5.6-42) the relation between $\gamma^*(\omega) = \alpha(\omega) + j\,\beta(\omega)$ and $\mathscr{Z}_C^*(\omega) = \mathscr{R}_C(\omega) + j\,\mathscr{X}_C(\omega)$ is

$$\gamma^*(\omega)\mathscr{Z}_C^*(\omega) = j\omega\rho\ . \tag{1}$$

Substituting the real and imaginary parts, multiplying, and separating again leads to

$$\alpha(\omega)\mathscr{R}_C(\omega) - \beta(\omega)\mathscr{X}_C(\omega) = 0 \qquad \text{and} \qquad \beta(\omega)\mathscr{R}_C(\omega) + \alpha(\omega)\mathscr{X}_C(\omega) = \omega\rho \tag{2}$$

and, hence, to

$$\alpha(\omega) = \frac{\omega\rho\,\mathscr{X}_C(\omega)}{\mathscr{R}_C^2(\omega) + \mathscr{X}_C^2(\omega)} \qquad \text{and} \qquad \beta(\omega) = \frac{\omega\rho\,\mathscr{R}_C(\omega)}{\mathscr{R}_C^2(\omega) + \mathscr{X}_C^2(\omega)} \tag{3}$$

on the one hand, and to

$$\mathscr{R}_C(\omega) = \frac{\omega\rho\,\beta(\omega)}{\alpha^2(\omega) + \beta^2(\omega)} \qquad \text{and} \qquad \mathscr{X}_C(\omega) = \frac{\omega\rho\,\alpha(\omega)}{\alpha^2(\omega) + \beta^2(\omega)} \tag{4}$$

on the other.

Problem 5.6-10

The general form of the Fourier series expansion for a periodic function with repeat function f(t) and period T is

$$f_F(t) = \frac{a_0}{2} + \sum_{k=1}^{\infty}\left(a_k \cos\frac{2\pi kt}{T} + b_k \sin\frac{2\pi kt}{T}\right) \quad (1)$$

where

$$a_k = \frac{2}{T}\int_c^{c+T} f(t)\cos\frac{2\pi k\theta}{T}\,d\theta \quad (2)$$

and

$$b_k = \frac{2}{T}\int_c^{c+T} f(t)\sin\frac{2\pi k\theta}{t}\,d\theta\ , \quad (3)$$

where c is an arbitrary real constant. We have $c = 0$ and $f(t) = h(t) - h(t - t')$. Hence $f(t) = 1$ for $0 \leq t \leq t'$ and zero elsewhere and we have

$$a_k = \frac{2}{T}\int_0^{t'} \cos\frac{2\pi k\theta}{T}\,d\theta = \frac{1}{\pi k}\sin\frac{2\pi kt'}{T} \quad (4)$$

and

$$b_k = \frac{2}{T}\int_0^{t'} \sin\frac{2\pi k\theta}{T}\,d\theta = \frac{1}{\pi k}\left(1 - \cos\frac{2\pi kt'}{T}\right) \quad (5)$$

with $a_0 = 2t'/T$. Then

$$f_F(t) = \frac{t'}{T} + \sum_{k=1}^{\infty}(\pi k)^{-1}\left[\sin\frac{2\pi kt'}{T}\cos\frac{2\pi kt}{T} + \left(1 - \cos\frac{2\pi kt'}{T}\right)\sin\frac{2\pi kt}{T}\right] \quad (6)$$

and rearranging yields

$$f_F(t) = t'/T + \sum_{k=1}^{\infty}(\pi k)^{-1}[\sin\omega_k t - \sin\omega_k(t - t')] \quad (7)$$

where $\omega_k = 2\pi k/T$.

Chapter 6

Problem 6.1-1

By Eq. (6.1-46) we have

$$\eta(t) = \{G_e t\} + [G_g - \{G_e\}] \int_0^t \exp[-(u/t_0)^k] du \; . \tag{1}$$

With the change of variable, $(u/t_0)^k = z$, we find

$$\eta(t) = \{G_e t\} + \eta_{\{f\}} \frac{\gamma[1/k; (t/t_0)^k]}{\Gamma(1/k)} \tag{2}$$

where $\gamma[1/k; (t/t_0)^k]$ is the incomplete gamma function defined by Eq. (6.3-34), and where $[G_g - \{G_e\}](t_0/k)$ has been replaced by $\eta_{\{f\}}/\Gamma(1/k)$. Since $\gamma(m; \infty) = \Gamma(m)$, Eq. (6.1-55) is satisfied because

$$\lim_{t\to\infty} [\eta(t) - \{G_e t\}] = \eta_{\{f\}} < \infty \; . \tag{3}$$

Now, with $k = 1/2$, Eq. (6.3-35) becomes

$$\gamma(2; x) = \sum_{n=0}^{\infty} \frac{(-1)^n x^{2+n}}{(2+n)n!} = 1 - (1 + x)\exp(-x) \tag{4}$$

and, when $G_e = 0$, we obtain

$$\eta(t) = \eta_f[1 - (1 + \sqrt{t/t_0})\exp(-\sqrt{t/t_0})] \tag{5}$$

since $\Gamma(2) = 1$. It cannot be expected that η_f will generally be equal to $G_g t_0 \Gamma(1/k)/k$ although it may be reasonably close to it.

Problem 6.1-2

With $c = 1/2$, Eq. (6.1-42) becomes

$$G_C(t) = \{G_e\} + \frac{G_g - \{G_e\}}{1 + \sqrt{t/t_0}} \tag{1}$$

[cf. Eq. (6.1-50)]. Let $t/t_0 = u$. Then

$$t_0 \int_0^{t/t_0} \frac{du}{1 + \sqrt{u}} = 2t_0[\sqrt{t/t_0} - \ln(1 + \sqrt{t/t_0})] \; . \tag{2}$$

Hence, setting $2[G_g - \{G_e\}]t_0 = \eta_{\{f\}}$, we have

$$\eta(t) = \eta_{\{f\}}[\sqrt{t/t_0} - \ln(1 + \sqrt{t/t_0})] \ . \tag{3}$$

But

$$\lim_{x\to\infty} [\sqrt{x} - \ln(1 + \sqrt{x})] = \lim_{x\to\infty} \ln\frac{\exp(\sqrt{x})}{1 + \sqrt{x}} = \lim_{x\to\infty} \sqrt{x} = \infty \tag{4}$$

and, therefore, $\eta_{\{f\}} = \infty$, which violates Eq. (6.1-55).

Problem 6.1-3

From Eq. (6.1-58) we have

$$\eta_K(t) = \{G_e t\} + \eta_{\{f\}}[1 - \exp(-(t/t_0)^k)] \ . \tag{1}$$

Differentiation yields

$$G_K(t) = \{G_e\} + [G_g - \{G_e\}](t/t_0)^{k-1}\exp[-(t/t_0)^k] \tag{2}$$

where we have replaced $\eta_{\{f\}}k/t_0$ by $G_g - \{G_e\}$. Now, if k is greater than 0 but smaller than 1,

$$\lim_{t\to 0} G_K(t) = \infty \ , \tag{3}$$

i.e. Eq. (2) is a non-standard model. If k is larger than 1,

$$\lim_{t\to 0} G_K(t) = 0 \tag{4}$$

and this would not furnish a model for G(t).

Again, by Eq. (6.1-58),

$$\eta_C(t) = \{G_e t\} + \eta_{\{f\}}[1 + (t_0/t)^c]^{-1} \tag{5}$$

and differentiation leads to

$$G_C(t) = \{G_e\} + \frac{[G_g - \{G_e\}]t_0/t}{(t_0/t)^c + 2 + (t/t_0)^c} \tag{6}$$

where $\eta_{\{f\}}\, c/t_0$ has now been replaced by $G_g - \{G_e\}$. The behavior of Eq. (6) as $t \to 0$ is the same as that of Eq. (2).

Problem 6.1-4

We begin by showing that

$$Z_F(s) = s\mathscr{L}\, S_F(t) = 1 - s\mathscr{L}\, Z_F(t) \ . \tag{5}$$

Equations $(5)_1$ follow because Eqs. (6.1-72) and (6.1-75) are the Carson transforms of Eqs. (6.1-4) and (6.1-5), respectively. Equations $(5)_2$ then result through the application of the relation $S_F(t) = 1 - Z_F(t)$. It can be shown analogously that

$$S_F(s) = s\mathscr{L}\, Z_F(t) = 1 - s\mathscr{L}\, S_F(t) \ . \tag{6}$$

But, if Eqs. (5) and (6) are true, then the matching functions and the associated matching functions must be related in the same way. Substitution of

$$\hat{Z}_M(s) = 1 - s\mathscr{L}\,\hat{Z}_M(t) \tag{7}$$

into Eqs. (6.1-76) and (6.1-77) then produces Eqs. (1) and (2), and substitution of

$$\hat{s}_M(s) = 1 - s\mathscr{L}\,\hat{s}_M(t) \tag{8}$$

into Eqs. (6.1-78) and (6.1-79) yields Eqs. (3) and (4).

Problem 6.1-5

Since $\tau/s(1 + \tau s) = \tau/s - \tau^2/(1 + \tau s)$, Eq. (4.1-15) may be written as

$$\bar{\eta}(s) = \eta_f/s - \int_{-\infty}^{\infty} \tau^2 H(\tau)\frac{1}{1 + \tau s}\,d\ln\tau\;, \tag{2}$$

and we have [cf. Eq. (6.1-56)]

$$\left\langle \frac{1}{1 + \tau s} \right\rangle_{\tau^2 H} = \frac{\int_{-\infty}^{\infty} \tau^2 H(\tau)(1 + \tau s)^{-1} d\ln\tau}{\int_{-\infty}^{\infty} \tau^2 H(\tau)d\ln\tau}\;. \tag{3}$$

Hence, using Eq. (2) in the numerator, and Eq. $(4.6\text{-}41)_1$ in the denominator, we obtain

$$Z_H(s) = \left\langle \frac{1}{1 + \tau s} \right\rangle_{\tau^2 H} = \frac{\eta_f/s - \bar{\eta}(s)}{J_e^\circ \eta_f^2}\;. \tag{4}$$

Equation (1) follows from matching $Z_H(s)$ with $\hat{Z}_M(s)$.

Problem 6.1-6

By Eq. (2) of Problem 4.1-4

$$\eta^*(\omega) = \{G_e/j\omega\} + \int_{-\infty}^{\infty} \tau H(\tau)\frac{1}{1 + j\omega\tau}\,d\ln\tau\;. \tag{2}$$

Using Eq. (6.1-1) we obtain the dimensionless function

$$Z_{\tau H}(\omega) = \langle 1/(1 + j\omega\tau)\rangle_{\tau H} = \frac{\int_{-\infty}^{\infty} \tau H(\tau)(1 + j\omega\tau)^{-1}\,d\ln\tau}{\int_{-\infty}^{\infty} \tau H(\tau)d\ln\tau} = \frac{\eta^*(\omega) - \{G_e/j\omega\}}{\eta_{\{f\}}} \tag{3}$$

and the matching $Z_{\tau H}(\omega) = \hat{Z}_M^*(\omega)$ then yields

$$\eta_M^*(\omega) = \{G_e/j\omega\} + \eta_{\{f\}}\hat{Z}_M^*(\omega)\;. \tag{4}$$

Now, for a rheodictic material

$$\eta_M^*(\omega) = \eta_f \hat{Z}_M^*(\omega)\;, \tag{5}$$

and Eq. (1) follows at once upon substitution of $\hat{Z}_C^*(\omega)$.

Problem 6.1-7

$G''_C(\omega)$ has a maximum when $\omega = \omega_0$. Letting this frequency be ω_{max}, we find $G''_C(\omega_{max})$ to be given by

$$G''_{max} = [G_g - \{G_e\}]\frac{\sin \pi c/2}{2(1 + \cos \pi c/2)} = 0.5[G_g - \{G_e\}]\tan \pi c/4 \ . \tag{1}$$

Introducing this into Eq. (6.1-91) yields

$$G''_C(\omega) = \frac{2G''_{max}(1 + \cos \pi c/2)}{(\omega_{max}/\omega)^c + 2 \cos \pi c/2 + (\omega/\omega_{max})^c} \ . \tag{2}$$

Problem 6.1-8

To obtain $J'_B(\omega)$ and $J''_B(\omega)$ we combine Eqs. (6.1-82) and (6.1-85) and separate the real and imaginary parts. This yields

$$J^*_B(\omega) = J_g + [J_e^{\{o\}} - J_g]\hat{Z}'_B(\omega) - j[J_e^{\{o\}} - J_g]\hat{Z}''_B(\omega) - j\{\phi_f/\omega\} \ . \tag{1}$$

But $\hat{Z}'_B(\omega)$ and $\hat{Z}''_B(\omega)$ are the real and imaginary parts of

$$\hat{Z}^*_B(\omega) = \frac{1}{[1 + (\omega/\omega_0)\exp(j\pi/2)]^b} \ . \tag{2}$$

The real part of the bracketted expression is 1 and the imaginary part is ω/ω_0. Hence, the expression becomes $\sqrt{1 + (\omega/\omega_0)^2}\exp[j\,\phi(\omega)]$ where $\tan \phi(\omega) = \omega/\omega_0$, and we have

$$\hat{Z}^*_B(\omega) = \frac{\exp[-jb\,\phi(\omega)]}{(1 + \omega^2/\omega_0^2)^{b/2}} \ . \tag{3}$$

Substitution of the real and imaginary parts into Eq. (1) then yields

$$J'_B(\omega) = J_g + [J_e^{\{o\}} - J_g]\cos^b\phi(\omega)\cos b\phi(\omega) \tag{4}$$

and

$$J''_B(\omega) = [J_e^{\{o\}} - J_g]\cos^b\phi(\omega)\sin b\phi(\omega) + \{\phi_f/\omega\} \tag{5}$$

But $\eta''_C(\omega)$ reaches its maximum when $\omega = \omega_0 = \omega_{max}$. Hence

Problem 6.1-9

When $\eta_f = \infty$, Eq. (6.1-96) becomes

$$G'_{C'}(\omega) = \frac{A(\omega)}{D(\omega)} = \frac{G_e(\omega/\omega_0)^c + (G_g + G_e)\cos \pi c/2 + G_g(\omega_0/\omega)^c}{(G_e/G_g)(\omega/\omega_0)^c + 2 \cos \pi c/2 + (G_g/G_e)(\omega_0/\omega)^c} \ . \tag{1}$$

Now, Eq. (6.1-90) can be recast in the form

$$G'_C(\omega) = \frac{G_g(\omega/\omega_0)^c + (G_g + G_e)\cos \pi c/2 + G_e(\omega_0/\omega)^c}{(\omega_0/\omega)^c + 2 \cos \pi c/2 + (\omega_0/\omega)^c} \tag{2}$$

Upon multiplication of ω_0 in Eq. (1) by $(G_g/G_e)^c$, we find $G'_{C'}(\omega) = G'_C(\omega)$ as asserted.

Similarly, Eq. (6.1-97) becomes

$$G''_{C'}(\omega) = \frac{C(\omega)}{D(\omega)} = \frac{(G_g - G_e)\sin \pi c/2}{(G_e/G_g)(\omega/\omega_0)^c + 2\cos \pi c/2 + (G_g/G_e)(\omega_0/\omega)^c} \tag{3}$$

and multiplication of ω_0 leads to $G''_{C'}(\omega) = G''_C(\omega)$.

Problem 6.1-10

Equation (2) is obtained immediately upon substitution of $\omega/\omega_0 = \exp(z)$ into Eq. (6.1-91) and realizing that $\exp(z) + \exp(-z) = 2\cosh z$. The same substitution into Eq. (6.1-90) first yields

$$G'_C(\omega) = G_g - \frac{G_g - \{G_e\}}{2} \frac{\exp(-cz) + \cos \pi c/2}{\cosh cz + \cos \pi c/2} \tag{5}$$

which we can rewrite as

$$G'_C(\omega) = \frac{G_g + \{G_e\}}{2} + \frac{G_g - \{G_e\}}{2}\left(1 - \frac{\exp(-cz) + \cos \pi c/2}{\cosh cz + \cos \pi c/2}\right) \tag{6}$$

and then reduce to Eq. (1) using $\exp(z) - \exp(-z) = 2\sinh z$.

Equations (3) and (4) are derived analogously.

Problem 6.1-11

Separating the real and imaginary parts in

$$\eta^*_C(\omega) = \frac{\eta_f}{1 + (j\omega/\omega_0)^c} \tag{3}$$

yields

$$\eta'_C(\omega) = \eta_f \frac{(\omega_0/\omega)^c + \cos \pi c/2}{(\omega_0/\omega)^c + 2\cos \pi c/2 + (\omega/\omega_0)^c} \tag{4}$$

and

$$\eta''_C(\omega) = \eta_f \frac{\sin \pi c/2}{(\omega_0/\omega)^c + 2\cos \pi c/2 + (\omega/\omega_0)^c}\ . \tag{5}$$

But $\eta''_C(\omega)$ reaches its maximum when $\omega = \omega_0 = \omega_{max}$. Hence

$$\eta''_{C,max} = \frac{\eta_f \sin \pi c/2}{2(1 + \cos \pi c/2)} = \tfrac{1}{2}\eta_f \tan \pi c/4 \tag{6}$$

and Eq. (1) follows. Letting $\omega = \omega_{max}$ in Eq. (4) and using Eq. $(6)_2$ then leads to Eq. (2) since $\omega_0/\omega_{max} = 1$.

Problem 6.2-1

Retransformation of Eqs. (6.2-1) and (6.2-2) into the time domain using Eq. (6.2-3), gives

$$\sigma(t) = G\tau^m \frac{d^m\varepsilon(t)}{dt^m} \quad , \quad 0 < m < 1 \ . \tag{1}$$

We may now use Eq. (1) to derive Eq. $(2\text{-}7)_1$ for $G_D(t)$. Letting $\varepsilon(t) = \varepsilon_0 h(t)$ and using Eq. (6.2-17) gives

$$\sigma(t) = \frac{G\tau^m\varepsilon_0}{\Gamma(1-m)} \int_0^t u^{-m}\delta(t-u)du = \frac{G\varepsilon_0(t/\tau)^{-m}}{\Gamma(1-m)} \tag{2}$$

which reduces to Eq. (6.2-7) if we let τ equal t_0 and write $G_D(t)$ for $\sigma(t)/\varepsilon_0$.

Problem 6.3-1

To derive Eq. (1), we apply the normalization condition, Eq. $(6.3\text{-}5)_1$, to Eq. (6.3-3) with $F(\tau) = H(\tau)$. This leads to

$$\int_{-\infty}^{\infty} H(\tau)d \ln \tau = \frac{r+r'}{r'} H_{max} \int_0^{\infty} \frac{x^{\rho r - 1}dx}{1 + (r/r')x^{\rho(r+r')}} = G_g - \{G_e\} \tag{3}$$

where we have used the abbreviation $x = \tau/\tau_{max}$. Using Eq. (2) we find

$$\frac{\alpha^{-\rho r}}{\rho r'} H_{max} \int_0^{\infty} \frac{u^{-r'/(r+r')}}{1+u} du = G_g - \{G_e\} \tag{4}$$

by the change of variable $(\alpha x)^{\rho(r+r')} = u$. But the integral is the (complete) beta function [cf. Eq. (6.3-55) in which $p = r/(r+r')$ and $q = r'/(r+r')$]. Thus, $q = 1 - p$, and, since $B(p, 1-p) = \pi/\sin \pi p$, we obtain

$$\int_0^{\infty} \frac{u^{-r'}/(r+r')}{1+u} du = \frac{\pi}{\sin[\pi r/(r+r')]} = \frac{\pi}{\sin[\pi r'/(r+r')]} \ . \tag{5}$$

Hence,

$$H_{max} = \alpha^{\rho r}(\rho r'/\pi)[G_g - \{G_e\}]\sin[\pi r/(r+r')] \ . \tag{6}$$

Introducing this into Eq. (6.3-3) leads to Eq. (1)

We recognize α as a parameter which adjusts the time scale so that the spectrum always has a maximum at $\tau = \tau_{max}$, regardless of the magnitude of r and r′.

When $r = r'$, the spectrum takes the simple form

$$H(\tau) = \frac{(2m/\pi)[G_g - \{G_e\}]}{(\tau_{max}/\tau)^m + (\tau/\tau_{max})^m} \tag{7}$$

where $m = \rho r$.

An analogous derivation yields

$$L(\tau) = \frac{(\rho/\pi)(r+r')[J_e^{\{o\}} - J_g]\sin[\pi r/(r+r')]}{(\tau_{max}/\alpha\tau)^{\rho r} + (\alpha\tau/\tau_{max})^{\rho r'}} \ . \tag{8}$$

Problem 6.3-2

To separate the real and imaginary parts of $G_D^*(\omega)$, Eq. (6.3-15), we recognize that

$$\frac{1 + j\omega\tau_{max}}{1 + j\omega\tau_{min}} = \frac{1 + \omega^2\tau_{max}\tau_{min}}{1 + \omega^2\tau_{min}^2} + \frac{j\omega(\tau_{max} - \tau_{min})}{1 + \omega^2\tau_{min}^2} \ . \tag{1}$$

But

$$\ln(a + jb) = \tfrac{1}{2}\ln(a^2 + b^2) + j\tan^{-1}(b/a) \ . \tag{2}$$

Hence

$$\ln\frac{1 + j\omega\tau_{max}}{1 + j\omega\tau_{min}} = \frac{1}{2}\ln\frac{1 + \omega^2\tau_{max}^2}{1 + \omega^2\tau_{min}^2} + j\tan^{-1}\frac{\omega(\tau_{max} - \tau_{min})}{1 + \omega^2\tau_{max}\tau_{min}} \ . \tag{3}$$

Problem 6.3-3

To derive Eq. (6.3-38) we note that, by Eq. (6.3-32),

$$j^m B(m, 1 - m; 1/j\omega\tau) = j^m \int_0^{1/j\omega\tau} \frac{u^{m-1}}{1 + u} du \ . \tag{1}$$

Let $ju = \sqrt{x}$. Then

$$j^m \int_0^{1/j\omega\tau} \frac{u^{m-1}}{1 + u} du = \frac{1}{2}\int_0^{1/\omega^2\tau^2} \frac{x^{m/2-1}}{1 + x}(1 + j\sqrt{x})dx \tag{2}$$

and, hence,

$$j^m \int_0^{1/j\omega\tau} \frac{u^{m-1}}{1 + u} du = \frac{1}{2}\int_0^{1/\omega^2\tau^2} \frac{x^{m/2-1}}{1 + x} dx + \frac{j}{2}\int_0^{1/\omega^2\tau^2} \frac{x^{m/2-1/2}}{1 + x} dx \tag{3}$$

which is Eq. (6.3-38).

Problem 6.3-4

To show that Eq. (6.3-36) is the equation of a standard model, we must show that

$$\lim_{t\to 0} G_w(t) = G_g \tag{1}$$

where G_g is finite and non-zero. As $t \to 0$, so does $\gamma(m; x)$. Consequently, we must apply L'Hospital's theorem. Now,

$$\frac{d}{dt}\int_0^{t/\tau_m} u^{m-1}\exp(-u)du = \tau_m^{-m} t^{m-1}\exp(-t/\tau_m) \tag{2}$$

where τ_m stands for either τ_{min} or τ_{max}. Hence

$$\lim_{t\to 0} \frac{\gamma(m; t/\tau_{min}) - \gamma(m; t/\tau_{max})}{(t/\tau_{min})^m} = \frac{1 - (\tau_{min}/\tau_{max})^m}{m} \tag{3}$$

and Eq. (6.3-36) gives $G_w(0) = G_g$ as a finite, non-zero quantity.

Problem 6.3-5

Attempting to derive $J_w(t)$ from

$$J_w(t) = \mathscr{L}^{-1}\bar{U}_w(s)/s \tag{1}$$

by using Eq. (4.1-18) for $\bar{U}(s)$, one finds that the inverse Laplace transform is not tabulated. However, using Eq. (4.1-26), i.e. $\bar{U}_{\bullet}(s)$, the inversion may be accomplished with the help of LTP (29).

Substituting Eq. (6.3-66) into $\bar{U}_{\bullet}(s)$ yields

$$\bar{U}_w(s) = J_e^{\{o\}} - \frac{[J_e^{\{o\}} - J_g]^m}{1 - (\tau_{min}/\tau_{max})^m}(\tau_{max}s)^{-m}$$

$$\times [B(-m, 1 + m; 1/\tau_{min}s) - B(-m, 1 + m; 1/\tau_{max}s)] + \{\phi_f/s\} \tag{2}$$

with the use of the change of variable $\tau s = 1/x$. Division by s and using LTP (29) in the form

$$\mathscr{L}^{-1}[s^{-m-1}B(-m, 1 + m; 1/\tau s)] = \gamma(-m; t/\tau)t^m \tag{3}$$

then leads to Eq. (6.3-76) for $J_w(t)$.

Problem 6.3-6

To demonstrate that Eq. (3-76) yields $J_e^{\{o\}}$ in the limit as $t \to \infty$, we show that

$$\lim_{t\to\infty} [\theta(m; t/\tau_{max}) - (\tau_{min}/\tau_{max})^m\theta(m; t/\tau_{min})] = [1 - (\tau_{min}/\tau_{max})^m]/m \ . \tag{1}$$

Using Eq. $(6.3\text{-}75)_1$ the bracketted expression becomes

$$[1 - (\tau_{min}/\tau_{max})^m]/m + [\gamma(-m; t/\tau_{max}) - \gamma(-m; t/\tau_{min})](t/\tau_{max})^m \ .$$

But, as $t \to \infty$, the incomplete gamma-functions become $\Gamma(-m)$. Hence, the second term vanishes and we obtain Eq. (1).

Problem 6.3-7

Letting $\tau_0 = b\tau_{max}$, Eq. (6.3-97) yields

$$H_B(\tau) = \Gamma^{-1}(b)[G_g - \{G_e\}](b\tau_{max}/\tau)^b \exp(-b\tau_{max}/\tau) \ . \tag{1}$$

With the change of variable, $\tau = b\tau_{max}/x$, we find

$$\int_{-\infty}^{\infty} (b\tau_{max}/\tau)^b \exp(-b\tau_{max}/\tau) d \ln \tau = \Gamma(b) \ . \tag{2}$$

Hence Eq. (6.3-27) is satisfied.

Problem 6.3-8

We use Eqs. (4.1-11). The differentiation [cf. Eq. (6.1-30)] yields

$$H_{C,1}(\tau) = -\left.\frac{d\,G_C(t)}{d \ln t}\right|_{t=\tau} = \frac{(G_g - G_e)^c}{(\tau_0/\tau)^c + 2 + (\tau/\tau_0)^c} \tag{1}$$

where t_0 has become τ_0.

Problem 6.3-9

To obtain $H'_C(\tau)$ and $H''_C(\tau)$ we must substitute $\hat{Z}'_C[\tau^{-1}\exp(\pm j\pi/2)]$ into Eq. (6.3-115) (in which we change $\hat{Z}_M$ to $\hat{Z}'_M$), and $\hat{Z}''_C[\tau^{-1}\exp(\pm j\pi/2)]$ into Eq. (6.3-133). This requires that we first replace ω by $\tau^{-1}\exp(\pm j\pi/2)$, and ω_0 by τ_0^{-1} in Eqs. (6.1-88) and (6.1-89). Writing r for τ/τ_0 to simplify the notation we have

$$\hat{Z}'_C[r^{-1}\exp(\pm j\pi/2)] = \frac{(r^c + 1)\cos \pi c/2 \mp jr^c \sin \pi c/2}{(r^c + 2 + r^{-c})\cos \pi c/2 \mp j(r^c - r^{-c})\sin \pi c/2} . \tag{1}$$

But then

$$\mathscr{Im}\, \hat{Z}'_C[\tau^{-1}\exp(\pm j\pi/2)] = \mp \frac{\frac{1}{2}\sin \pi c}{(\tau/\tau_0)^c + 2\cos \pi c + (\tau_0/\tau)^c} . \tag{2}$$

Introduction of this into Eq. (6.3-115) and comparing the resulting expression with Eq. (6.3-109) shows at once that $H^*_C(\tau) = H'_C(\tau)$ as it should.

For the imaginary part we find

$$\hat{Z}''_C[r^{-1}\exp(\pm j\pi/2)] = \frac{\sin \pi c/2}{(r^c + 2 + r^{-c})\cos \pi c/2 \mp j(r^c - r^{-c})\sin \pi c/2} \tag{3}$$

and

$$\mathscr{Re}\, \hat{Z}''_C[\tau^{-1}\exp(\pm j\pi/2)] = \frac{\frac{1}{2}\sin \pi c}{(\tau/\tau_0)^c + 2\cos \pi c + (\tau_0/\tau)^c} . \tag{4}$$

Substitution into Eq. (6.3-133) then shows that $H^*_C(\tau) = H''_C(\tau)$ as well.

Problem 6.3-10

To obtain the relaxation spectrum in terms of the complex viscosity, $\eta^*(\omega)$, and its components, $\eta'(\omega)$, and $\eta''(\omega)$, we substitute $G^*(\omega) = j\omega\eta^*(\omega)$ into Eq. (6.3-102), $G'(\omega) = \omega\eta''(\omega)$ into Eq. (6.3-114), and $G''(\omega) = \omega\eta'(\omega)$ into Eq. (6.3-120). Replacing $j\omega$ by $\tau^{-1}\exp(\pm j\pi)$, or ω by $\tau^{-1}\exp(\pm j\pi/2)$, and using the relations $\mathscr{Im}\, j(\alpha + j\beta) = \mathscr{Re}(\alpha + j\beta)$, and $\mathscr{Re}\, j(\alpha + j\beta) = -\mathscr{Im}(\alpha + j\beta)$, we obtain

$$H^*(\tau) = \mp(1/\pi\tau)\mathscr{Im}\, \eta^*[\tau^{-1}\exp(\pm j\pi)] \tag{1}$$

$$H'(\tau) = \mp(2/\pi\tau)\mathscr{Im}\, \eta'[\tau^{-1}\exp(\pm j\pi/2)] \tag{2}$$

and

$$H''(\tau) = (2/\pi\tau)\mathscr{Re}\, \eta''[\tau^{-1}\exp(\pm j\pi/2)] . \tag{3}$$

$H''(\tau)$ has been changed to $H'(\tau)$ in Eq. (2), and $H'(\tau)$ to $H''(\tau)$ in Eq. (3), to indicate that the spectrum is obtained from $\eta'(\omega)$ in the first, and from $\eta''(\omega)$ in the second case.

To introduce the matching functions $\hat{Z}^*_M(\omega) = \hat{Z}'_M(\omega) - j\,\hat{Z}''_M(\omega)$ [cf. Eq. (6.1-85)] we use Eq. (4) of Problem 6.1-6, and its real and imaginary parts. This yields

$$H^*_M(\tau) = \mp(\eta_{\{f\}}/\pi\tau)\mathscr{Im}\, \hat{Z}^*_M[\tau^{-1}\exp(\pm j\pi)] \tag{4}$$

$$H'_M(\tau) = \mp(2\eta_{\{f\}}/\pi\tau)\mathscr{Im}\, \hat{Z}'_M[\tau^{-1}\exp(\pm j\pi/2)] \tag{5}$$

and

$$H''_M(\tau) = (2\eta_{\{f\}}/\pi\tau)\mathscr{R}e\,\hat{Z}''_M[\tau^{-1}\exp(\pm j\pi/2)] \ . \tag{6}$$

Note that, when the imaginary part is taken, the $\{Ge/j\omega\}$-term vanishes in Eq. (4) upon substitution of $\tau^{-1}\exp(j\pi)$ for $j\omega$, and in Eq. (5) upon substitution of $\tau^{-1}\exp(j\pi/2)$ for ω.

Problem 6.3-11

To find the relaxation spectrum if the complex viscosity is modelled by Eq. (1) of Problem 6.1-8, we introduce it into Eq. (1) of Problem 6.3-10. This gives

$$H_C^*(\tau) = \mp(\eta_f/\pi\tau)\mathscr{I}m\frac{1}{1 + (\tau_0/\tau)^c\exp(\pm j\pi c)} \tag{1}$$

which yields

$$H_C^*(\tau) = (\eta_f/\pi\tau)\frac{\sin \pi c}{(\tau/\tau_0)^c + 2\cos \pi c + (\tau_0/\tau)^c} \ . \tag{2}$$

This should be compared with Eq. (6.3-109). The two spectra become the same if we equate η_f with $G_g\tau$.

Problem 6.3-12

The retardation spectrum associated with the complex Kobeko equation is easily derived from Eq. (6.3-106). We need the imaginary part of $\hat{Z}_B^*[\tau^{-1}\exp(\pm j\pi)]$. Substituting $\tau^{-1}\exp(\pm j\pi)$ for $j\omega$, and τ_0^{-1} for ω_0 in Eq. (6.1-92) gives

$$\hat{Z}_B^*[\tau^{-1}\exp(\pm j\pi)] = \left(\frac{\exp(\mp j\pi)}{\tau_0/\tau + \exp(\mp j\pi)}\right)^b = \left(\frac{\tau}{\tau_0 - \tau}\right)^b \exp(\mp j\pi b) \ . \tag{1}$$

Consequently,

$$L_B^*(\tau) = \frac{[J_e^{\{o\}} - J_g]\sin \pi b}{\pi(\tau_0/\tau - 1)^b} \ , \quad \tau < \tau_0 \ . \tag{2}$$

The restriction is necessary because the spectrum diverges when $\tau = \tau_0$ and it assumes physically unrealistic values when $\tau > \tau_0$.

Problem 6.3-13

It will be shown in Chap. 8 that $J^*(\omega)$ is essentially the Fourier transform of $J(t)$. Unfortunately, the inverse Fourier transform of Eq. (1) does not possess a convenient closed form. $J(t)$ may, however, be obtained by finding the retardation spectrum corresponding to Eq. (1) and using it in the canonical representation for $J(t)$. We have

$$J(t) = J_g + \int_{-\infty}^{\infty} L_C^*(\tau)[1 - \exp(-t/\tau)]d\ln\tau + \{\phi_f t\} \tag{2}$$

where

$$L_C^*(\tau) = \pi^{-1}(J_e^{\{o\}} - J_g)\frac{\sin \pi c}{(\tau_0/\tau)^c + 2\cos \pi c + (\tau/\tau_0)^c} \tag{3}$$

by Eq. (6.3-106).

Chapter 7

Problem 7.1-1

From Eq. (2.3-16)$_3$ the stress in response to a ramp strain becomes

$$\sigma_{t'}(t) = \int_0^t G(t-u)\frac{d\,\varepsilon(u)}{du}du \ . \tag{1}$$

We can now proceed in two different ways, according to how we handle $\varepsilon(u)$.

(A) Let this be given by Eq. (7.1-18). Differentiation yields

$$\frac{d\,\varepsilon(u)}{du} = \dot{\varepsilon}_0[h(u) + u\,\delta(u) - h(u-t') + (u-t')\delta(u-t')] \tag{2}$$

and substitution into Eq. (1) leads to

$$\sigma_{t'}(t) = \dot{\varepsilon}_0 \int_0^t G(t-u)[h(u) - h(u-t')]du \tag{3}$$

since the delta function terms do not contribute to the value of the integral. Equation (3) is equivalent to

$$\sigma_{t'}(t) = \dot{\varepsilon}_0 \int_0^t G(t-u)du - \dot{\varepsilon}_0 \int_{t'}^t G(t-u)du \ , \tag{4}$$

i.e. to

$$\sigma_{t'}(t) = \dot{\varepsilon}_0 \int_0^{t'} G(t-u)du = \dot{\varepsilon}_0 \int_{t-t'}^t G(w)dw \tag{5}$$

where the last integral follows through a change of variable. If t' is small enough, we can write, with good approximation,

$$\sigma_{t'}(t) \simeq \dot{\varepsilon}_0 t' G(t - t'/2) = \varepsilon_0 G(t - t'/2) \ , \quad t \geqq t' \tag{6}$$

by taking $G(w)$ out from under the integral with its mid-value, i.e. the mean of the values it assumes at the ends of the range of integration. Changing to the shifted time, $\theta = t - t'$, we recover Eq. (7.1-33).

(B) Equation (5) can also be reached from Eq. (1) directly by rewriting Eq. (7.1-18) as

$$\varepsilon(t) = \begin{cases} \dot{\varepsilon}_0 t, & 0 \leq t \leq t' \\ \dot{\varepsilon}_0 t', & t \geqq t' \ . \end{cases} \tag{7}$$

Problem 7.1-2

The strain may be considered to consist of a ramp strain of height ε_0 imposed at $t = 0$, from which another ramp strain of the same height is subtracted at $t = t'$. Thus

$$\varepsilon(t) = \varepsilon_0[r_{t'}(t) - r_{t''}(t - t')] \ . \tag{1}$$

But $\tan\alpha = \dot{\varepsilon}_0' = \varepsilon_0/t'$ and $\tan\beta = \dot{\varepsilon}_0'' = \varepsilon_0/t''$. Hence, using LTP (9) for $r_t'(t)$, and the translation theorem, Eq. (A3-6), in addition for $r_{t''}(t - t'')$, we find

$$\bar{\varepsilon}(s) = \dot{\varepsilon}_0'/s^2)[1 - \exp(-t's)] - (\dot{\varepsilon}_0''/s^2)[1 - \exp(-t''s)]\exp(-t's) \ . \tag{2}$$

When $t'' = t'$, then $\dot{\varepsilon}_0' = \dot{\varepsilon}_0''$, and Eq. (2) reduces to Eq. (7.1-47) as required.

Problem 7.1-3

(A) In terms of unit ramp functions, the trapezoidal strain excitation becomes

$$\varepsilon(t) = \varepsilon_0[r_{t'}(t) - r_{t'''}(t - t' - t'')] \ . \tag{2}$$

By LTP (9)

$$\mathscr{L}\, r_{t'}(t) = [1 - \exp(-t's)]/t's^2 \ . \tag{3}$$

Using the translation theorem to find the transform of the shifted unit ramp function, we obtain

$$\mathscr{L}\, r_{t'''}(t - t' - t'') = (1/t'''s^2)[1 - \exp(-t'''s)]\exp[-(t' + t'')s] \ . \tag{4}$$

Thus, letting $\varepsilon_0 = \dot{\varepsilon}_0't' = \dot{\varepsilon}_0'''(t''' - t'')$, we have

$$\bar{\varepsilon}(s) = (\dot{\varepsilon}_0'/s^2)[1 - \exp(-t's)] - (\dot{\varepsilon}_0''/s^2)[1 - \exp(-t'''s)]\exp[-(t' + t'')s] \tag{5}$$

which is the solution to the first part of the problem.

(B) When $t''' = t'' = t'$, then $\dot{\varepsilon}_0' = \dot{\varepsilon}_0''$, and therefore

$$\bar{\varepsilon}(s) = (\dot{\varepsilon}_0'/s^2)[1 - \exp(-t's)][1 - \exp(-2t's)] \tag{6}$$

which completes the problem.

Problem 7.1-4

By Eq. $(4.6\text{-}41)_1$

$$J_e^\circ \eta_f^2 = \int_0^\infty \tau H(\tau)d\tau \tag{1}$$

and by Eq. (7.1-26)

$$\eta_{t'}(\theta) = \int_0^\infty H(\tau)\exp(-\theta/\tau)d\tau \ . \tag{2}$$

The area under the curve is

$$\int_0^\infty \eta_{t'}(\theta)d\theta = \int_0^\infty \tau H(\tau)d\tau \tag{4}$$

which equals $J_e^\circ \eta_f^2$ by Eq. (1).

Problem 7.1-5

Insertion of

$$G(t) = \int_{-\infty}^{\infty} H(\tau)\exp(-t/\tau)d\ln\tau \tag{2}$$

into Eq. (1) leads to

$$\begin{aligned}\sigma(\theta) &= \dot{\varepsilon}_0 \int_{-\infty}^{\infty} H(\tau)d\ln\tau \int_{\theta}^{\infty} \exp(-t/\tau)d\,t \\ &= \dot{\varepsilon}_0 \int_{-\infty}^{\infty} \tau H(\tau)\exp(-\theta/\tau)d\ln\tau \ .\end{aligned} \tag{3}$$

Equation $(3)_2$ is identical with Eq. (7.1-26) since $\sigma(\theta)/\dot{\varepsilon}_0 = \eta(\theta)$.

Problem 7.1-6

We must show that

$$\lim_{t' \to 0} r_{t'}(t) = h(t) \ . \tag{1}$$

Substituting for $r_{t'}(t)$ from Eq. (A2-40) and using L'Hospital's theorem we obtain

$$\begin{aligned}\lim_{t' \to 0} \frac{th(t) - (t - t')h(t - t')}{t'} &= \lim_{t' \to 0} \frac{d(t - t')h(t - t')}{d(t - t')} \\ &= \lim_{t' \to 0} [h(t - t') + (t - t')\delta(t - t')] \\ &= h(t) + t\delta(t) \ .\end{aligned} \tag{2}$$

But $t\delta(t)$ is zero whenever an integration is performed, because $\delta(t)$ is zero unless t is zero but then $t\delta(t)$ is zero again. Hence, Eq. (1) follows.

Problem 7.1-7

(A) The relaxance of the non-standard 3-parameter Maxwell model is given [cf. Eq. (3.4-40)] by

$$\bar{Q}(s) = \eta' s + \frac{G\tau s}{1 + \tau s} \ . \tag{1}$$

The Laplace transform of the strain as a ramp function of time is

$$\bar{\varepsilon}(s) = \varepsilon_0 \bar{r}_{t'}(s) = \dot{\varepsilon}_0[1 - \exp(-t's)]/s^2 \tag{2}$$

by LTP (9). Substitution of these expressions into $\bar{\sigma}(s) = \bar{Q}(s)\bar{\varepsilon}(s)$ yields

$$\bar{\sigma}(s) = \dot{\varepsilon}_0\left(\frac{\eta'}{s} + \frac{G\tau}{s(1 + \tau s)}\right)[1 - \exp(-t's)] \tag{3}$$

and transformation to the time domain gives the total response as

$$\sigma(t) = \dot{\varepsilon}_0(\eta'[h(t) - h(t - t')] + G\tau[1 - \exp(-t/\tau)]h(t)$$
$$- G\tau\{1 - \exp[-(t - t')]/\tau\}h(t - t')) \ . \tag{4}$$

Introducing the *shifted time*, $\theta = t - t'$, we have

$$\sigma(\theta) = \dot{\varepsilon}_0 G\tau[1 - \exp(-t'/\tau)]\exp(-\theta/\tau) \tag{5}$$

for the response at times $t > t'$. This response is seen to be indistinguishable from that of a Maxwell unit.

(B) The response to a step strain can be obtained from Eq. (3) by writing

$$\bar{\sigma}(s) = \varepsilon_0\left(\frac{\eta'}{s} + \frac{G\tau}{s(1 + \tau s)}\right)\lim_{t' \to 0}\frac{1 - \exp(-t's)}{t'} \ . \tag{6}$$

Taking the limit yields

$$\bar{\sigma}(s) = \varepsilon_0\left(\eta' + \frac{G\tau}{1 + \tau s}\right) \tag{7}$$

and retransformation gives

$$\sigma(t) = \varepsilon_0[\eta'\delta(t) + G\exp(-t/\tau)] \tag{8}$$

which [cf. Eq. $(3.4\text{-}42)_1$] is the response of the model to a step strain. The same result (albeit requiring a little more algebra) can, of course, be obtained (cf. Problem 7.1-6) by applying the limiting operation to Eq. (4).

Problem 7.1-8

(A) The behavior is illustrated in Fig. P7.1-8. Thus the stress history is

$$\sigma(t) = \varepsilon_0 G(t)h_{t'}(t) \tag{1}$$

where $h_{t'}(t)$ is the unit gate function. By Eqs. (A3-21) and (A3-23) the transform becomes

$$\bar{\sigma}(s) = \varepsilon_0\bar{G}_{t'}(s) \tag{2}$$

where

$$\bar{G}_{t'}(s) = \int_0^{t'} G(t)\exp(-st)dt \ . \tag{3}$$

The stress transform must be substituted into the general relation for the strain transform, $\bar{\varepsilon}(s) = s\bar{J}(s)\bar{\sigma}(s)$. However, the recovery begins at $t = t'$. Hence the strain must be shifted along the time-axis by this amount. We have

$$\bar{\varepsilon}(s)\exp(-t's) = \varepsilon_0 s\bar{J}(s)\bar{G}_{t'}(s) \tag{4}$$

as the sought-for transform relation.

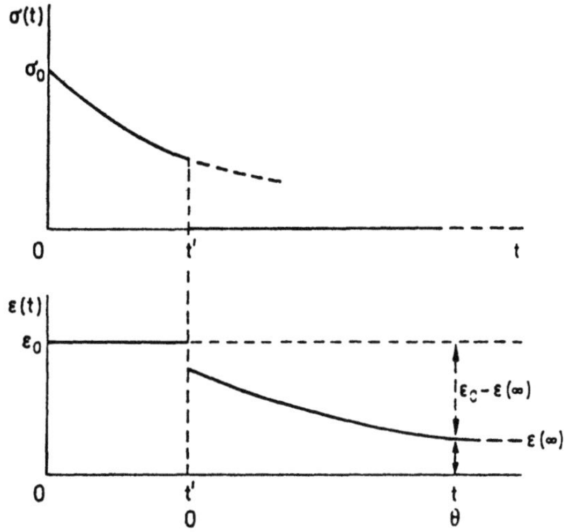

Fig. P7.1-8. Strain recovery after partial stress relaxation

Retransformation is simple. We obtain

$$\varepsilon(t - t') = \varepsilon_0 J_g G(t) h_{t'}(t) - \varepsilon_0 \int_0^t G(u) h_{t'}(u) \frac{d\, J(t - u)}{du} du \tag{5}$$

which is equivalent to

$$\varepsilon(t - t') = \varepsilon_0 J_g G(t) h_{t'}(t) - \varepsilon_0 \int_0^{t'} G(u) \frac{d\, J(t - u)}{du} du \ . \tag{6}$$

Introducing the shifted time, $\theta = t - t'$, we find

$$\varepsilon(\theta) = -\varepsilon_0 \int_0^{t'} G(u) \frac{d\, J(\theta + t' - u)}{du} du \ . \tag{7}$$

(B) The residual strain or permanent set is obtained as

$$\varepsilon(\infty) = \lim_{\theta \to \infty} \varepsilon(\theta) = -\varepsilon_0 \int_0^{t'} G(u) \lim_{\theta \to \infty} \frac{d\, J(\theta + t' - u)}{du} du \ . \tag{8}$$

But

$$\lim_{\theta \to \infty} \frac{d\, J(\theta + t' - u)}{du} = -\{\phi_f\} \tag{9}$$

and therefore,

$$\varepsilon(\infty) = \varepsilon_0 \{\phi_f\} \int_0^{t'} G(u) du = \varepsilon_0 \{\phi_f\} \eta(t') \ . \tag{10}$$

For an arrheodictic material $\phi_f = 0$ and there is no permanent set. Hence the strain is completely recovered. For a rheodictic material, $\varepsilon(\infty)$ is seen to represent the deformation resulting from the flow of the material in the interval from $t = 0$ to $t = t'$. Thus the strain is not recovered completely. The recovered strain is

$$\varepsilon_0 - \varepsilon(\infty) = \varepsilon_0 \phi_f \left(\eta_f - \int_0^{t'} G(u)du \right) . \tag{11}$$

But

$$\eta_f = \lim_{t\to\infty} \eta(t) = \int_0^{\infty} G(u)du \tag{12}$$

and, therefore,

$$\varepsilon_0 - \varepsilon(\infty) = \varepsilon_0 \phi_f \int_{t'}^{\infty} G(u)du . \tag{13}$$

(C) Using u for the time variable, the relaxation modulus [cf. Eq. (3.4-127)] and the creep compliance [cf. Eq. (3.4-140)] of the standard linear models become

$$G(u) = G_e + G \exp(-u/\tau_M) \tag{14}$$

and

$$J(u) = J_e - J \exp(-u/\tau_V) \tag{15}$$

so that

$$\frac{d\,J(\theta + t' - u)}{du} = -\frac{J}{\tau_V}\exp[-(\theta + t' - u)/\tau_V] . \tag{16}$$

Inserting Eqs. (14) and (16) into Eq. (7) leads to

$$\varepsilon(\theta) = \frac{J\varepsilon_0}{\tau_V}\int_0^{t'} [G_e + G\exp(-u/\tau_M)]\exp[-(\theta + t' - u)/\tau_V]du . \tag{17}$$

Evaluation of the integral then gives

$$\varepsilon(\theta) = JG_e\varepsilon_0\exp[-(\theta + t')/\tau_V]$$
$$\times \left\{\exp(t'/\tau_V - 1) + \frac{\tau_M}{\tau_M - \tau_V}\exp\left[-\left(\frac{1}{\tau_M} - \frac{1}{\tau_V}\right)\right]t' - \frac{\tau_M}{\tau_M - \tau_V}\right\} . \tag{18}$$

But (see Sect. 3.7) $\tau_V = G_g\tau_M/G_e$, $G = G_g - G_e$, and $J = J_e - J_g$. Hence

$$\varepsilon(\theta) = \varepsilon_0(1 - G_e/G_g)[1 - \exp(-t'/\tau_M)]\exp(-G_e\theta/G_g\tau_M) . \tag{19}$$

Problem 7.2-1

By Eq. (7.2-1)

$$\bar{\varepsilon}(s) = \frac{\varepsilon_0 \bar{r}_{t'}(s)}{1 - \exp(-mt's)} \tag{1}$$

where $r_{t'}(t)$ is the unit ramp function and the duration of the interval is mt'. The transform $\bar{r}_{t'}(s)$ is given by LTP (9). Substituting this into Eq. (1) and using $\varepsilon_0 = \dot{\varepsilon}_0 t'$ gives

$$\bar{\varepsilon}(s) = \frac{\dot{\varepsilon}_0[1 - \exp(-t's)]}{s^2[1 - \exp(-mt's)]} = (\dot{\varepsilon}_0/s^2)[1 - \exp(-t's)] \sum_{n=\infty}^{\infty} \exp(-mnt's) . \tag{2}$$

Problem 7.2-2

The strain excitation is given by Eq. (7.2-5). This is to be substituted into the Boltzmann superposition integral

$$\sigma(t) = \int_0^t G(t-u)\dot{\varepsilon}(u)du \; . \qquad (1)$$

Differentiation of Eq. (7.2-5) yields

$$\dot{\varepsilon}(u) = \varepsilon_0 \sum_{n=0}^{N} \delta(t-nt') \qquad (2)$$

and the substitution leads to

$$\sigma(t) = \varepsilon_0 \sum_{n=0}^{N} G(t-nt') \qquad (3)$$

by the sifting property of the delta function (cf. Appendix A2.1). Since G(t) is not defined for $t < 0$, Eq. (3) is identical with Eq. (7.2-8).

Problem 7.3-1

Upon developing the sum, the left hand side becomes

$$1 - 2E + 2E^2 - 2E^3 + \cdots + 2E^{N-1} \; . \qquad (2)$$

Now,

$$2E^{N-1} : (E^{-1} + 1) = 2E^N - 2E^{N+1} + 2E^{N+2} + \cdots \qquad (3)$$

and

$$(E^{-1} - 1) : (E^{-1} + 1) = 1 - 2E + \cdots + 2E^{N-1} - 2E^N + 2E^{N+1} + \cdots \; . \qquad (4)$$

Addition of these two infinite series leads to Eq. (2). Hence, Eq. (1) follows.

Problem 7.3-2

(A) The Laplace transform of the repeat function is the finite transform (cf. Appendix A3.1.8) of the sine function. Thus

$$\bar{\varepsilon}_{t'}(s) = \varepsilon_0 \int_0^{t'} \sin \omega t \exp(-ts)dt \qquad (1)$$

where $\omega = \pi/t'$ because t' is the half-period of the sine wave. Integration yields

$$\bar{\varepsilon}_{t'}(s) = \frac{\varepsilon_0}{s^2 + \omega^2}[\omega - (s \sin \omega t' + \omega \cos \omega t')\exp(-st')] \; . \qquad (2)$$

(B) The Laplace transform of the complete periodic excitation is simply

$$\bar{\varepsilon}(s) = \frac{\bar{\varepsilon}_{t'}(s)}{1 - \exp(-Ts)} \qquad (3)$$

through replacing $\bar{\varepsilon}_T(s)$ by $\bar{\varepsilon}_{t'}(s)$ in Eq. (7.3-2).

(C) With $T = 2t'$ and $\omega = \pi/t'$ we have

$$\bar{\varepsilon}(s) = \frac{\pi\varepsilon_0 t'}{(s^2t'^2 + \pi^2)[1 - \exp(-t's)]} \quad . \tag{4}$$

(D) With $T = t'$ and $\omega = \pi/t'$ we find

$$\bar{\varepsilon}(s) = \frac{\pi\varepsilon_0 t'\cosh(t's/2)}{(s^2t'^2 + \pi^2)\sinh(t's/2)} \quad . \tag{5}$$

Problem 7.3-3

(A) The repeat function is given by

$$\varepsilon_{t'}(t) = \varepsilon_0 h_{t'}(t) = \varepsilon_0[h(t) - h(t - t')] \tag{1}$$

the Laplace transform of which is

$$\bar{\varepsilon}_{t'}(s) = \varepsilon_0[1 - \exp(-t's)]/s \quad . \tag{2}$$

The Laplace transform of the strain then becomes

$$\bar{\varepsilon}(s) = \frac{\bar{\varepsilon}_{t'}(s)}{1 - \exp(-2t's)} = \frac{\varepsilon_0}{s[1 + \exp(-t's)]} = \frac{\varepsilon_0 \exp(t's/2)}{2s\cosh(t's/2)} \quad . \tag{3}$$

(B) The steady-state response is obtained from Eq. (7.3-38). The poles of the excitation transform are the same as in the case of the regular triangular wave. The residue associated with the pole at the origin is

$$\mathrm{Res}_0 = \varepsilon_0 \lim_{s\to 0} \frac{\bar{Q}(s)\exp(t's/2)\exp(ts)}{2\cosh(t's/2)} = \tfrac{1}{2}\varepsilon_0\{G_e\} \quad . \tag{4}$$

The residues associated with the poles at $s = s_k$ ($k = \pm 1, \pm 3, \ldots$) are given by

$$\begin{aligned}\mathrm{Res}_k &= \varepsilon_0 \lim_{s\to s_k} \frac{(s - s_k)\bar{Q}(s)\exp(t's/2)\exp(ts)}{2s\cosh(t's/2)} \\ &= \frac{\varepsilon_0 \bar{Q}(s_k)\exp(ts_k)\exp(t's_k/2)}{2s_k} \lim_{s\to s_k} \frac{s - s_k}{\cosh(t's/2)} \quad .\end{aligned} \tag{5}$$

But

$$\lim_{s\to s_k} \frac{s - s_k}{\cosh(t's/2)} = \frac{2}{t'\sinh(t's_k/2)} \tag{6}$$

by L'Hospital's theorem. But $s_k = j\omega k = j\pi k/t'$ where $k = \pm 1, \pm 3, \ldots$ [cf. Eq. (7.3-43)]. But then $\exp(j\pi k/2) = j\sin(\pi k/2)$, and thus

$$\mathrm{Res}_k = \varepsilon_0 \frac{\bar{Q}(j\omega k)\exp(j\omega kt)}{j\pi k} \quad . \tag{7}$$

By Eq. (7.3-37) then

$$\sigma_{ss}(t) = \tfrac{1}{2}\varepsilon_0\{G_e\} - j(\varepsilon_0/\pi)\sum_k k^{-1}[\bar{Q}(j\omega k)\exp(j\omega kt) - \bar{Q}(-j\omega k)\exp(-j\omega kt)] \tag{8}$$

where $k = 1, 3, \ldots$. On taking the products of the complex quantities, the real terms cancel. Hence

$$\sigma_{ss}(t) = \tfrac{1}{2}\varepsilon_0\{G_e\} + (2\varepsilon_0/\pi)\sum_k k^{-1}[G''(\omega k)\cos\omega kt + G'(\omega k)\sin\omega kt]_{\omega=\pi/t'} \;. \tag{9}$$

Thus, the coefficients in the Fourier expansion become

$$\begin{aligned} a_0 &= \varepsilon_0\{G_e\} \\ a_k &= (2\varepsilon_0/\pi k)G''(\omega k)\cos\omega kt \\ b_k &= (2\varepsilon_0/\pi k)G'(\omega k)\sin\omega kt \;. \end{aligned} \tag{10}$$

Problem 7.3-4

The repeat function is represented by the regular triangular pulse shown in Fig. P7.3-4b. We have

$$\sigma(t) = \begin{cases} 0, & t < 0 \\ \dot{\sigma}_0(t - t'/2), & 0 \le t \le t' \\ -\dot{\sigma}_0(t - 3t'/2), & t' \le t \le 2t' \\ 0, & t \ge 2t' \end{cases} \tag{1}$$

where

$$\dot{\sigma}_0 = \tan\alpha = 2\sigma_0/t' \;. \tag{2}$$

In terms of step functions the pulse can be expressed as

$$\sigma(t) = \dot{\sigma}_0[(t - t'/2)h(t) - 2(t - t')h(t - t') + (t - 3t'/2)h(t - 2t')] \;. \tag{3}$$

Laplace transformation yields the transform of the repeat function as

$$\bar{\sigma}_{2t'}(s) = (\dot{\sigma}_0/s^2)[1 - \exp(-t's)]^2 - (\sigma_0/s)[1 - \exp(-2t's)] \;. \tag{4}$$

Consequently, the transform for the pulse train becomes

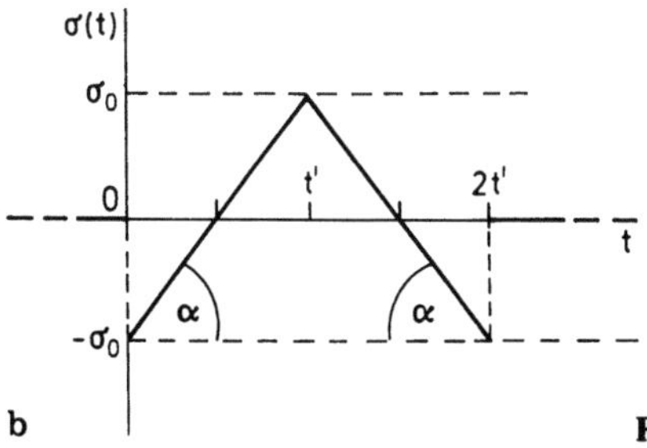

b

Fig. P7.3-4b. Regular triangular stress pulse

$$\bar{\sigma}(s) = \frac{\bar{\sigma}_{2t'}(s)}{1 - \exp(-Ts)} = \frac{\dot{\sigma}_0[1 - \exp(-t's)]^2}{s^2[1 - \exp(-2t's)]} - \frac{\sigma_0}{s} \tag{5}$$

or

$$\bar{\sigma}(s) = (\dot{\sigma}_0/s^2)\left[1 + 2\sum_{n=1}^{\infty}(-1)^n \exp(-nt's) - \sigma_0/s\right] \tag{6}$$

which may be compared with Eq. (7.3-22). Substitution into

$$\bar{\varepsilon}(s) = s^2\bar{\chi}(s)\bar{\sigma}(s) = s\bar{J}(s)\bar{\sigma}(s) \tag{7}$$

followed by retransformation yields

$$\varepsilon(t) = \dot{\sigma}_0\left[\chi(t)h(t) + 2\sum_{n=1}^{\infty}(-1)^n\chi(t - nt')h(t - nt')\right] - \sigma_0 J(t)h(t) \tag{8}$$

from which the strain during the Nth half cycle results as

$$\varepsilon_N(t) = \dot{\sigma}_0\left[\chi(t) + 2\sum_{n=1}^{N-1}(-1)^n\chi(t - nt') - \sigma_0 J(t)\right] . \tag{9}$$

This is identical with Eq. (7.3-23) except for the $\sigma_0 J(t)$-term. This term is given by

$$\sigma_0 J(t) = (\dot{\sigma}_0 t'/2)\left[J_e^{\{o\}} - \int_{-\infty}^{\infty} L(\tau)\exp(-t/\tau)d\ln\tau + \{\phi_f\}t\right] . \tag{10}$$

In the steady state, i.e. as $t \to \infty$, the integral vanishes. Hence, the $J_e^{\{o\}}$ terms in Eqs. (7.3-33) and (7.3-34) become $J_e^{\{o\}}(\theta - t'/2)$ and $-J_e^{\{o\}}(\theta - 3t'/2)$, respectively, and the term $\{\phi_f\}tt'/2$ must be subtracted from both equations. The results is

$$\chi_{def}(\theta) = J_e^{\{o\}}(\theta - t'/2) - \chi + 2\int_{-\infty}^{\infty} \tau L(\tau)\frac{\exp(-\theta/\tau)}{1 + \exp(-t'/\tau)}d\ln\tau + \{\phi_f\}\theta(\theta - t')/2 \tag{11}$$

and

$$\chi_{rec}(\theta) = -J_e^{\{o\}}(\theta - 3t'/2) + \chi - 2\int_{-\infty}^{\infty} \tau L(\tau)\frac{\exp[-(\theta - t')/\tau]}{1 + \exp(-t'/\tau)}d\ln\tau + \{\phi_f\}(\theta - t')(\theta - 2t')/2 . \tag{12}$$

The two equations are now free of any term in t for a rheodictic material. Consequently, a steady-state can be reached.

Problem 7.3-5

The excitation transform, $\bar{\sigma}(s)$, for the triangular stress wave has already been obtained in Problem 7.3-4 as Eq. (5). Using Eq. $(2)_2$ of the same Problem, we may rewrite Eq. (5) as

$$\bar{\sigma}(s) = \frac{\dot{\sigma}_0 \sinh(t's/2)}{s^2\cosh(t's/2)} - \frac{\dot{\sigma}_0 t'}{2s} . \tag{1}$$

The residue associated with the simple pole at the origin becomes

$$\mathrm{Res}_0 = \dot{\sigma}_0 \lim_{s\to 0} \bar{U}(s)\left(\frac{\sinh(t's/2)}{s} - \frac{t'}{2}\right) \tag{2}$$

or

$$\mathrm{Res}_0 = (\dot{\sigma}_0 t'/2) \lim_{s\to 0} \bar{U}(s)[\cosh(t's/2) - 1] \tag{3}$$

through application of L'Hospital's theorem. For an arrheodictic material $\bar{U}(0) = J_e$. Consequently, $\mathrm{Res}_0 = 0$. For a rheodictic material, however, $U(0) = \infty$ because of the flow term, ϕ/s. Hence we use $\bar{U}(s) = 1/\bar{Q}(s)$ and apply L'Hospital's theorem a second time. This gives

$$\mathrm{Res}_0 = (\sigma_0 t'/2) \lim_{s\to 0} \frac{\sinh(t's/2)}{\bar{Q}'(s)} . \tag{4}$$

But differentiation of Eq. (4.1-3) and using Eq. $(4.6\text{-}40)_1$ shows that $\bar{Q}'(0) = \eta_f$. Thus, $\mathrm{Res}_0 = 0$ in this case also.

The residues associated with the poles at $s = s_k$ are given by

$$\mathrm{Res}_k = \lim_{s\to s_k} \frac{(s - s_k)\dot{\sigma}_0 \bar{U}(s)\sinh(t's/2)\exp(st)}{s^2\cosh(t's/2)} = \frac{2\dot{\sigma}_0 \bar{U}(s_k)\exp(s_k t)}{t's_k^2} \tag{5}$$

since the contribution of the second term in Eq. (1) vanishes. Using $\dot{\sigma}_0 = 2\sigma_0/t'$ again and, in addition, Eq. (7.3-43), we obtain

$$\sum_k \mathrm{Res}_k = -\sum_k \frac{4\sigma_0 \bar{U}(j\omega k)\exp(j\omega kt)}{\pi^2 k^2}, \qquad k = \pm 1, \pm 3, \ldots \tag{6}$$

in analogy to Eq. (7.3-46). Proceeding further in the same manner as in Sect. 7.3.2.2, and remembering that $\bar{U}(\pm j\omega k) = J'(\omega k) \mp jJ''(\omega k)$, the steady-state strain now takes the form

$$\varepsilon_{ss}(t) = -(8\sigma_0/\pi^2)\sum_k k^{-2}[J'(\omega k)\cos \omega kt + J''(\omega k)\sin \omega kt]_{\omega=\pi/t'}, \qquad k = 1, 3, \ldots \tag{7}$$

[cf. Eq. (7.3-52)], and the terms in the Fourier expansion become

$$a_0 = 0$$

$$a_k = -(8\sigma_0/\pi^2 k^2)J'(\omega k) \tag{8}$$

and

$$b_k = -(8\sigma_0/\pi^2 k^2)J''(\omega k)$$

[cf. Eqs. (7.3-54)]. Since $\mathrm{Res}_0 = 0$, $a_0 = 0$ also. Thus, there is no "D.C. component" superposed on the oscillations, and a steady state can, therefore, develop even if the material subjected to this excitation is rheodictic.

Chapter 8

Problem 8.0-1

The time-dependent response functions are included in Table 8.0-1 as their Laplace transforms while the frequency-dependent ones are found in the complex form. The Laplace transform of G(t) is $\bar{G}(s)$. $J'(\omega)$ is the real part of $J^*(\omega)$. The appropriate entry in the table reads $1/j\omega\bar{G}(\omega)$. Thus,

$$J^*(\omega) = 1/j\omega\bar{G}(\omega) = 1/j\omega[\bar{G}(s)]|_{s=j\omega} \tag{2}$$

making use of Eq. $(8.0\text{-}4)_1$. It then follows that

$$J'(\omega) = \mathscr{Re}\{1/j\omega[\bar{G}(s)]|_{s=j\omega}\} \; . \tag{3}$$

But

$$\bar{G}(s) = \frac{G_e}{s} + \frac{G\tau}{1+\tau s} = \frac{G_e + G_g\tau s}{s(1+\tau s)} \; . \tag{4}$$

Hence, after substitution of $j\omega$ for s, separation of the real and imaginary parts, and using $J_g = 1/G_g$, $J_e = 1/G_e$, $J = J_e - J_g$, we obtain

$$J'(\omega) = J_g + \frac{J}{1+\omega^2\tau'^2} \tag{5}$$

where $\tau' = G_g\tau/G_e$.

Problem 8.0-2

By Table 8.0-1,

$$\bar{\eta}(s) = \frac{G^*(s)}{s^2} \; . \tag{2}$$

Hence,

$$\eta(t) = \mathscr{L}^{-1}\frac{[G^*(\omega)]|_{j\omega=s}}{s^2} \; . \tag{3}$$

Now, for the standard linear solid model,

$$G^*(\omega) = G_e + \frac{Gj\omega\tau}{1+j\omega\tau} \; , \tag{4}$$

and, therefore,

$$\eta(t) = \mathscr{L}^{-1}\left[\frac{G_e}{s^2} + \frac{G\tau}{s(1+\tau s)}\right] = G_e t + G\tau[1 - \exp(-t/\tau)] \ . \tag{5}$$

Problem 8.1-1

The G_k''s and τ_k''s are tabulated below.

k	$G_k \times 10^{-6}$ (N/m²)	τ_k (second)
0	0.92	1.443×10^{-8}
1	11.80	1.443×10^{-7}
2	100.16	1.443×10^{-6}
3	64.05	1.443×10^{-5}
4	16.15	1.443×10^{-4}
5	1.44	1.443×10^{-3}
6	0.38	1.443×10^{-2}
$G_g - G_e$	194.90	

The instantaneous and equilibrium moduli are $G_g = 196 \times 10^6$, and $G_e = 1.1 \times 10^6$ N/m², respectively.

For the collocation we require the column vector A_j which, by Eqs. (3.6-59) and (3.6-60) is

$$A_0 = J_e - J_g = 1/G_e - 1/G_g \tag{3}$$

$$A_j = \bar{U}(s_j) - J_g = 1/\bar{Q}(s_j) - 1/G_g \ , \quad j = 1, 2, \ldots, 6 \tag{4}$$

and the matrix B_{jk}, which is simply an extension of the 4×4 matrix in Eq. (3.6-63) to a 7×7 matrix.

A_j is to be obtained from $\bar{Q}(s_j)$ where $s_j = 10^{m+j}$ by Eq. $(3.6\text{-}51)_1$. The value of m can be chosen by plotting $\log \bar{Q}(s)$ as a function of log s. In the present case, however, the G_k''s and τ_k''s were obtained from the collocation to $G(t_j)$ described in Problem 3.6-2. There, $t_j = 10^{m+j}$ [cf. Eq. (3.6-11)], with $m = -8$ and $j = 1, 2, \ldots, 6$. Since $s_j = 1/t_j$, we obtain $\log s_j = 2, 3, 4, 5, 6, 7$. Thus

$\log s_j$ (s^{-1})	$\log \bar{Q}(s_j)$ (N/m²)	$A_j \times 10^{-9}$ (N/m²)
2	6.265	537.58
3	6.732	180.10
4	7.340	40.60
5	7.837	9.47
6	8.145	2.07
7	8.263	0.35

The values of $\log \bar{Q}(s_j)$ are shown with a 3-digit mantissa. However, the corresponding values of A_j were calculated with a 5-digit mantissa.

The matrix equation now follows as

$$\begin{bmatrix} 903.99 \\ 537.58 \\ 180.10 \\ 40.60 \\ 9.47 \\ 2.07 \\ 0.35 \end{bmatrix} = \begin{bmatrix} 1.000 & 1.000 & 1.000 & 1.000 & 1.000 & 1.000 & 1.000 \\ 0.091 & 0.500 & 0.909 & 0.990 & 1.000 & 1.000 & 1.000 \\ 0.010 & 0.091 & 0.500 & 0.909 & 0.990 & 1.000 & 1.000 \\ 0 & 0.010 & 0.091 & 0.500 & 0.909 & 0.990 & 1.000 \\ 0 & 0 & 0.010 & 0.091 & 0.500 & 0.909 & 0.990 \\ 0 & 0 & 0 & 0.010 & 0.091 & 0.500 & 0.909 \\ 0 & 0 & 0 & 0 & 0.010 & 0.091 & 0.500 \end{bmatrix} \begin{bmatrix} J_0 \\ J_1 \\ J_2 \\ J_3 \\ J_4 \\ J_5 \\ J_6 \end{bmatrix} . \tag{5}$$

Inversion of the matrix on a computer yields the J_k's shown below.

k	$J_k \times 10^4$ (m^2/N)	τ_k (second)
0	12.19	10^{-1}
1	675.66	10^{-2}
2	193.05	10^{-3}
3	11.96	10^{-4}
4	9.02	10^{-5}
5	1.95	10^{-6}
6	0.16	10^{-7}
$J_e - J_g$	903.99	

The τ_k''s follow from Eq. (3.6-51) according to which $m = 1$ and $\tau_k = 10^{-k-1}$. The plot of $\log \bar{U}(s)$ vs. $\log s$ is compared with the collocation points $\bar{U}(s_j) = 1/\bar{Q}(s_j)$ in Fig. 8.1-1.

Problem 8.1-2

Rewrite Eq. $(8.1\text{-}13)_2$ in the form

$$G_g J(t) + \int_0^t \dot{G}(u)\{J(t) - [J(t) - J(t-u)]\}du = 1 \ . \tag{1}$$

Separating out the first term of the integral gives

$$J(t)\left(G_g + \int_0^t \dot{G}(u)du\right) - \int_0^t \dot{G}(u)\{[J(t) - J(t-u)]\}du = 1 \tag{2}$$

which at once reduces to

$$J(t)G(t) - \int_0^t \dot{G}(u)\{[J(t) - J(t-u)]\}du = 1 \tag{3}$$

since $G(0) = G_g$. But $J(t) - J(t-u) \geq 0$ because $0 \leq u \leq t$. Further, by differentiating Eq. (4.1-5), we find

$$\dot{G}(u) = -\int_{-\infty}^{\infty} \tau^{-1} H(\tau) \exp(-u/\tau) d\ln\tau \leq 0 \tag{5}$$

because both $H(\tau)$ and τ are ≥ 0. Thus, the second term in Eq. (3) is positive despite the minus sign, and the proposition follows.

Problem 8.1-3

The Laplace transforms, $\bar{\eta}(s)$ and $\bar{J}(s)$, are interrelated, according to Table 8.0-1, by

$$\bar{\eta}(s) = 1/s^2\bar{J}(s) \; . \qquad (1)$$

Retransformation yields

$$\int_0^t \eta(t-u)J(u)du = \tfrac{1}{2}t^2 \qquad (2)$$

as the desired relation.

Problem 8.1-4

We evaluate the integral in Eq. (8.1-57), using Eq. (6.3-91) and the change of variable $u = \tau x$. The integral becomes

$$\int_{-\infty}^{\infty} \frac{\tau L(u)}{\tau - u} d\ln u = L(\tau) \int_0^{\infty} \frac{x^{m-1}}{1-x} dx = L(\tau)\pi \cot \pi m \; . \qquad (2)$$

Upon substitution into Eq. (8.1-57), with Jg and $\{\phi\tau\}$ omitted, we obtain Eq. (1), which, for $m = 1/2$, reduces to $H(\tau)L(\tau) = 1/\pi^2$.

Problem 8.2-1

The basic relations are Eqs. (8.2-4) and (8.2-5), and Eqs. (8.2-7) and (8.2-8). To insure that the relations to be derived will be quite generally applicable, we use Eqs. (8.2-16) to specialize the basic relations. We find

$$G'(\omega) = \{G_e\} + \omega \int_0^{\infty} [G(t) - \{G_e\}] \sin \omega t \, dt \qquad (1)$$

and

$$G''(\omega) = \omega \int_0^{\infty} [G(t) - \{G_e\}] \cos \omega t \, dt \qquad (2)$$

for the frequency-dependent functions, and

$$G(t) = \begin{cases} \{G_e\} + \dfrac{2}{\pi} \displaystyle\int_0^{\infty} \frac{G'(\omega) - \{G_e\}}{\omega} \sin \omega t \, dt \\[2ex] \{G_e\} + \dfrac{2}{\pi} \displaystyle\int_0^{\infty} \frac{G''(\omega)}{\omega} \cos \omega t \, dt \end{cases} \qquad (3)$$

for the time-dependent ones.

Problem 8.2-2

For the standard linear solid model we have

$$G(t) = G_e + G\exp(-t/\tau) \qquad \text{and} \qquad G'(\omega) = G_e + \frac{G\omega^2\tau^2}{1+\omega^2\tau^2} \, . \tag{3}$$

Using Eq. (3) in Eqs. (1) and (2) we find

$$G'(\omega) = \omega \int_{-\infty}^{\infty} [G_e + G\exp(-t/\tau)]\sin\omega t \, dt \tag{4}$$

and

$$G'(\omega) = \omega \int_{-\infty}^{\infty} G\exp(-t/\tau)\sin\omega t \, dt \tag{5}$$

Applying FTPs (2) and/or (3) we recover Eq. $(3)_2$ in either case.

Problem 8.2-3

The basic relations are again Eqs. (8.2-4) and (8.2-5), and Eqs. (8.2-7) and (8.2-8). As in Problem 8.2-1, we use Eqs. (8.2-16) to specialize the basic relations. We obtain

$$J'_\bullet(\omega) = J_e^{\{o\}} - \omega \int_0^{\infty} [J_e^{\{o\}} - J_\bullet(t) + \{\phi_f t\}]\sin\omega t \, dt \tag{1}$$

and

$$J''(\omega) = \omega \int_0^{\infty} [J_e^{\{o\}} - J_\bullet(t) + \{\phi_f t\}]\cos\omega t \, dt + \{\phi_f t\} \tag{2}$$

for the frequency-dependent functions. For the time-dependent functions we have

$$J_\bullet(t) = \begin{cases} J_e^{\{o\}} - \dfrac{2}{\pi}\displaystyle\int_0^{\infty} \dfrac{J_e^{\{o\}} - J'_\bullet(\omega) + \{\phi_f/\omega\}}{\omega}\sin\omega t \, d\omega + \{\phi_f t\} \\[2ex] J_e^{\{o\}} - \dfrac{2}{\pi}\displaystyle\int_0^{\infty} \dfrac{J''(\omega) - \{\phi_f/\omega\}}{\omega}\cos\omega t \, d\omega + \{\phi_f t\} \, . \end{cases} \tag{3}$$

Problem 8.2-4

(A) Equation (8.2-17a) is obtained from Eq. $(3)_1$ of Problem 8.2-1 by introducing Eq. (4.6-20). Using Eq. $(4.6\text{-}20)_2$ we obtain

$$G_\bullet(t) = G_g - \sum_k G_k + \frac{2}{\pi}\int_0^{\infty} \frac{G'_\bullet(\omega) - G_g + \sum_k G_k}{\omega}\sin\omega t \, d\omega \, . \tag{1}$$

Since the sum under the integral sign cancels the one before it, we recover Eq. (8.2-17a). The inverse relationship,

$$G'_{\bullet}(\omega) = G_g - \omega \int_0^\infty [G_g - G_{\bullet}(t)] \sin \omega t \, dt \tag{2}$$

is obtained in much the same way from Eq. (1) of Problem 8.2-1.

An entirely analogous derivation starting from Eq. $(3)_1$ of Problem 8.2-3 and introducing Eq. $(4.6\text{-}21)_2$, leads to Eq. (8.2-17b). Its inverse,

$$J'(\omega) = J_g + \omega \int_0^\infty [J(t) - J_g - \{\phi_f t\}] \sin \omega t \, dt \tag{3}$$

is obtained from Eq. (1) of Problem 8.2-3 analogously.

(B) Now, for the standard solid models we have

$$G'_{\bullet}(\omega) = G_g - \frac{G}{1 + \omega^2\tau^2} \quad \text{and} \quad J'(\omega) = J_g + \frac{J}{1 + \omega^2\tau^2} \; . \tag{4}$$

Substituting Eqs. (4) into, respectively, Eqs. (8.2-17a) and (8.2-17b), we find, after partial fraction decomposition,

$$G_{\bullet}(t) = G_g - \frac{2}{\pi} \int_0^\infty G\left(\frac{1}{\omega} - \frac{\omega\tau^2}{1 + \omega^2\tau^2}\right) \sin \omega t \, d\omega \tag{5}$$

and

$$J(t) = J_g + \frac{2}{\pi} \int_0^\infty J\left(\frac{1}{\omega} - \frac{\omega\tau^2}{1 + \omega^2\tau^2}\right) \sin \omega t \, d\omega \; . \tag{6}$$

Either direct integration, or application of FTP's (2) and (5), respectively, yields

$$G_{\bullet}(t) = G_g - G[1 - \exp(-t/\tau)] \quad \text{and} \quad J(t) = J_g + J[1 - \exp(-t/\tau)] \; . \tag{7}$$

Problem 8.2-5

(A) To derive Eq. (8.2-17c) we substitute

$$\{G_e\} = G_g - \frac{2}{\pi} \int_0^\infty \frac{G''(\omega)}{\omega} \, d\omega \tag{1}$$

[cf. Eq. (4.6-31)] into Eq. $(3)_2$ of Problem 8.2-1.

Analogously, Eq. (8.2-17d) is obtained by substitution of

$$J_e^{\{o\}} = J_g + \frac{2}{\pi} \int_0^\infty \frac{J''(\omega) - \{\phi_f/\omega\}}{\omega} \, d\omega \tag{2}$$

[cf. Eq. (4.6-35)] into Eq. $(3)_2$ of Problem 8.2-3.

(B) Introducing the expressions for $G''(\omega)$ and $J''(\omega)$ for the standard linear solid models,

$$G''(\omega) = \frac{G\omega\tau}{1 + \omega^2\tau^2} \quad \text{and} \quad J''(\omega) = \frac{J\omega\tau}{1 + \omega^2\tau^2} + \{\phi_f/\omega\} \; , \tag{3}$$

into Eqs. (8.2-17c) and integrating using the definite integrals

$$\int_0^\infty \frac{1}{1+\omega^2\tau^2}\, d\,\omega\tau = \tfrac{\pi}{2} \quad \text{and} \quad \int_0^\infty \frac{\cos \omega\tau(t/\tau)}{1+\omega^2\tau^2}\, d\,\omega\tau = \tfrac{\pi}{2}\exp(-t/\tau) \tag{4}$$

we recover Eqs. (7) of Problem 8.2-4.

Problem 8.2-6

The basic relations to be used to derive Eqs. (1) are Eqs. (8.2-6) to (8.2-8), together with Eqs. (8.2-16). By Table 8.01-1, $G^*(\omega) = j\omega\,\eta^*(\omega)$. Separation of the real and imaginary parts gives $G'(\omega) = \omega\eta''(\omega)$, and $G''(\omega) = \omega\eta'(\omega)$. Using these substitutions in Eqs. (8.2-16), Eqs. (1) immediately follow from Eqs. (8.2-6) to (8.2-8).

Problem 8.2-7

The creep compliance of the standard linear liquid model is given by

$$J(t) = J_g + J[1 - \exp(-t/\tau)] + \phi_f t \tag{3}$$

and, therefore,

$$\dot{J}(t) = (J/\tau)\exp(-t/\tau) + \phi_f\ . \tag{4}$$

For the complex compliance we have

$$J^*(\omega) = J_g + \frac{J}{1+j\omega\tau} + \phi_f/j\omega\ , \tag{5}$$

so that, by Eq. $(2)_1$,

$$\dot{J}(t) = \mathscr{F}^{-1}\left(\frac{J}{1+j\omega\tau}\right) + \phi_f\ . \tag{6}$$

Equation (4) then follows from FTP (5).

Problem 8.2-8

We let $\omega = 1$, $t = 1/1.44$, rewrite the equation in the form

$$G(t) = G'(\omega) - [G'(\omega) - G'(0.694\omega)] - 0.40[G'(1.104\omega) - G'(0.134\omega)]|_{\omega=1/t} \tag{2}$$

and find the difference kernel approximation to be given by

$$K'_{\Delta 2}(\omega\tau) = \frac{(\omega\tau)^2}{1+(\omega\tau)^2} - \frac{(0.694\omega\tau)^2}{1+(0.694\omega\tau)^2} + 0.40\left(\frac{(1.104\omega\tau)^2}{1+(1.104\omega\tau)^2} - \frac{(0.134\omega\tau)^2}{1+(0.134\omega\tau)^2}\right)\ . \tag{3}$$

The approximation is therefore of the second order and the subscript is justified.

Problem 8.3-1

We first derive Eq. (8.3-21) from Eq. (8.3-17). By Eq. (4.1-14)

$$G_g - G'_\bullet(\beta) = \int_{-\infty}^{\infty} H(\tau)\frac{1}{1+\beta^2\tau^2} d \ln \tau \tag{1}$$

and, by Eq. (4.1-7),

$$G'(\beta) - \{G_e\} = \int_{-\infty}^{\infty} H(\tau)\left(1 - \frac{1}{1+\beta^2\tau^2}\right) d \ln \tau \ . \tag{2}$$

Adding the two equations and using Eq. $(4.6\text{-}20)_2$ gives

$$G_g - G'_\bullet = -[G'(\omega) - \{G_e\}] + G_g - \{G_e\} \ . \tag{3}$$

Substituting this into Eq. (8.3-17) and using Eq. (8.3-22) we find Eq. (8.3-21).
The derivation of Eq. (8.3-20) from (8.3-11) is analogous.

Problem 8.3-2

By Eq. (8.2-4) we have (see Problem 8.2-1)

$$G'(\omega) = \{G_e\} + \omega \int_0^{\infty} [G(t) - \{G_e\}] \sin \omega t \, dt \ , \tag{1}$$

and, by Eq. (8.2-8),

$$G(t) = \{G_e\} + \frac{2}{\pi}\int_0^{\infty} \frac{G''(\beta)}{\beta} \cos \beta t \, d\beta \tag{2}$$

where we have changed the dummy variable of integration from ω to β to keep the two integrals separate. Substituting Eq. (2) into Eq. (1) and changing the order of integration yields

$$G'(\omega) = \{G_e\} + \frac{2\omega}{\pi}\int_0^{\infty} \frac{G'(\beta)}{\beta} \int_0^{\infty} \cos \beta t \sin \omega t \, dt \, d\beta \ . \tag{3}$$

But the integral over t is the generalized Fourier sine transform of $\cos \beta t$, which FTP (7) gives as $\omega/(\omega^2 - \beta^2)$. Substitution of this immediately leads to Eq. (8.3-19).
If we start from Eq. (2) of Problem 8.2-4, i.e. from

$$G'_\bullet(\omega) = G_g - \omega \int_0^{\infty} [G_g - G_\bullet(t)] \sin \omega t \, dt \tag{4}$$

and use Eq. (8.2-17), i.e.

$$G_\bullet(t) = G_g - \frac{2}{\pi}\int_0^{\infty} \frac{G''(\beta)}{\beta}(1 - \cos \beta t) d\beta \ , \tag{5}$$

we obtain

$$G'_\bullet(\omega) = G_g - \frac{2\omega}{\pi}\int_0^{\infty} \frac{G''(\beta)}{\beta} \int_0^{\infty} (1 - \cos \beta t) \sin \omega t \, dt \, d\beta \ . \tag{6}$$

But

$$\frac{2}{\pi}\int_0^\infty (1-\cos\beta t)\sin\omega t\, dt = \frac{\beta^2}{\omega(\beta^2-\omega^2)} \tag{7}$$

by FTPs (2) and (7). Thus Eq. (8.3-20) follows.

A similar derivation furnishes Eqs. (8.3-13) and (8.3-24). However, these derivations shed no light on the lack of analogy between the Kronig-Kramers relations for $G'(\omega)$ and $J'(\omega)$. The contour integration employed in Sect. 8.3.1.1 shows that this arises from the difference in the kernels of $\bar{Q}(s)$ and $\bar{U}(s)$. The latter, $1(1+\tau s)$, vanishes as $s \to \infty$. Under the same circumstances, however, the former, $\tau s(1+\tau s)$, approaches unity.

Problem 8.3-3

Substituting Eq. (1) into Eq. (8.3-17) gives

$$G''(\omega) = \frac{2G\omega}{\pi} \fint_{-\infty}^{\infty} \frac{d\beta}{(\omega^2-\beta^2)(1+\beta^2\tau^2)} \; . \tag{2}$$

Decomposition into partial fractions yields

$$G''(\omega) = \frac{2G\omega}{\pi(1+\omega^2\tau^2)} \fint_{-\infty}^{\infty} \left(\frac{1}{\beta^2-\omega^2} + \frac{1}{\beta^2+\tau^{-2}}\right) d\beta \; . \tag{3}$$

The first integral is zero. The second is $\pi\tau/2$. Hence,

$$G''(\omega) = \frac{G\omega\tau}{1+\omega^2\tau^2} \tag{4}$$

as required.

Problem 8.3-4

The storage and loss compliances of the standard linear solid are

$$J'(\omega) = \frac{J_e + J_g\omega^2\tau^2}{1+\omega^2\tau^2} = J_g\frac{b^2+y^2}{1+y^2} \tag{2}$$

and

$$J''(\omega) = \frac{J\omega\tau}{1+\omega^2\tau^2} = J_g\frac{ay^2}{1+y^2} \tag{3}$$

where $J = J_e - J_g$, and we have introduced the abbreviations

$$y = \omega\tau \; , \qquad J_e/J_g = b^2 \; , \qquad \text{and} \qquad J/J_g = a \; . \tag{4}$$

Using the same abbreviations, the absolute compliance and the loss tangent become

$$\tilde{J}(\omega) = J_g\sqrt{\frac{(b^2+y^2)^2 a^2 y^2}{(1+y^2)^2}} \tag{5}$$

and

$$\theta(\omega) = \arctan\frac{ay}{b^2 + y^2} \; . \tag{6}$$

Inserting Eq. (6) into the integral in Eq. (8.3-77) yields

$$\ln\frac{\tilde{J}(\omega)}{J_g} = \frac{2}{\pi} \fint_{-\infty}^{\infty} \frac{x \arctan\dfrac{ax}{b^2 + x^2} - y \arctan\dfrac{ay}{b^2 + y^2}}{x^2 - y^2} dx \; . \tag{7}$$

where $x = \beta\tau$. The integral can be split into two. The second is then seen to vanish by Eq. (8.3-22) and the first represents the integral in Eq. (1). Hence, in view of Eq. (6), Eq. (8.3-77) is indeed correct.

Chapter 9

Problem 9.1-1

(A) For a rheodictic material $G_e = 0$. Hence, Eq. (9.1-8) becomes

$$\overline{W}_s(s) = \tfrac{1}{2}\sum_n G_n \mathscr{L}\, \varepsilon_{sn}^2(t) \tag{6}$$

Applying Eq. (4) gives

$$\overline{W}_s(s) = \tfrac{1}{2}\sum_n G_n \mathrm{Res}_n[\bar{\varepsilon}_{sn}(z)\bar{\varepsilon}_{sn}(s-z)] \ . \tag{7}$$

Using $\tau_n = 1/\zeta_n$ for convenience, we write Eq. (9.1-5) as

$$\bar{\varepsilon}_{sn}(z) = \frac{z\bar{\varepsilon}(z)}{z+\zeta_n} \ . \tag{8}$$

The general equation then becomes

$$\overline{W}_s(s) = \tfrac{1}{2}\sum_n G_n \mathrm{Res}_n\left[\frac{z\bar{\varepsilon}(z)(s-z)\bar{\varepsilon}(s-z)}{(z+\zeta_n)(s-z+\zeta_n)}\right] \tag{9}$$

where the residues are calculated from the pole(s) at $z = -\zeta_n$ and, in addition, the pole(s) of $\bar{\varepsilon}(z)$. By Eq. (5), the formula for generating the residues in the general equation for the stored energy in a rheodictic material becomes

$$\mathrm{Res}_n = \lim_{z\to z_n}(z-z_n)\frac{z\bar{\varepsilon}(z)}{(z+\zeta_n)}\frac{(s-z)\bar{\varepsilon}(s-z)}{(s-z+\zeta_n)} \ . \tag{10}$$

(B) For a slope strain excitation, $\bar{\varepsilon}(\eta) = \dot{\varepsilon}_0/z^2$. Hence,

$$\bar{\varepsilon}_{sn}(z) = \frac{\dot{\varepsilon}_0}{z(z+\zeta_n)} \tag{11}$$

and there is a pole at $z = 0$ in addition to those at $z = -\zeta_n$. Substituting $\dot{\varepsilon}_0/z^2$ for $\bar{\varepsilon}(z)$, and $\dot{\varepsilon}_0/(s-z)^2$ for $\bar{\varepsilon}(s-z)$, into Eq. (10) leads to

$$\mathrm{Res}_0 = \dot{\varepsilon}_0^2 \lim_{z\to 0}\frac{1}{s(z+\zeta_n)(s+\zeta_n)} = \frac{\dot{\varepsilon}_0^2}{s(s+\zeta_n)\zeta_n} \tag{12}$$

and

$$\mathrm{Res}_n = \dot{\varepsilon}_0^2 \lim_{z \to -\zeta_n} \frac{1}{z(s+\zeta_n)(s+2\zeta_n)} = -\frac{\dot{\varepsilon}_0^2}{(s+\zeta_n)(s+2\zeta_n)\zeta_n}\,, \tag{13}$$

and, therefore,

$$\sum_n \mathrm{Res}_n = \frac{\dot{\varepsilon}_0^2}{(s+\zeta_n)\zeta_n}\left[\frac{1}{s} - \frac{1}{s+2\zeta_n}\right] = \dot{\varepsilon}_0^2 \frac{2\tau_n^2}{s(1+\tau_n s)(2+\tau_n s)} \tag{14}$$

Finally,

$$\overline{W}_s(s) = \frac{\dot{\varepsilon}_0^2}{2}\sum_n \eta_n \tau_n \frac{2}{s(1+\tau_n s)(2+\tau_n s)}\,. \tag{15}$$

Retransformation using LTP (19) leads to

$$W_s(t) = (\dot{\varepsilon}_0^2/2)\sum_n \eta_n \tau_n [1 - \exp(-t/\tau)]^2 \tag{16}$$

which, by Eq. (4.6-24), is the discrete distribution form of Eq. (9.1-34) for a rheodictic material.

Problem 9.1-2

For a step strain, $\varepsilon(t) = \varepsilon_0$, and hence, $\bar{\varepsilon}(s) = \varepsilon_0/s$ [cf. Eq. 2.3-2].

Inserting into Eqs. (9.1-9) and (9.1-13), multiplying the latter by s, and retransforming yields

$$\varepsilon_{sn}(t) = \varepsilon_0 \exp(-t/\tau_n) \tag{1}$$

and

$$\varepsilon_{dn}(t) = \varepsilon_0[1 - (\exp - t/\tau_n)]\,. \tag{2}$$

Thus, the strain in the spring assumes its initial value instantaneously, i.e. $\varepsilon_{sn}(0) = \varepsilon_0$. In the dashpot the strain is initially zero, i.e. $\varepsilon_{dn}(0) = 0$, but increases to its final value, $\varepsilon_{dn}(\infty) = \varepsilon_0$, as strain is being transferred from the spring to the dashpot. Hence the strain in the dashpot is retarded while the strain in the spring relaxes from its initial value to zero, i.e. $\varepsilon_{sn}(\infty) = 0$. The transfer of strain proceeds differently in each unit, the rate at which the transfer takes place being governed by the relaxation time, τ_n, of the unit. No strain is transferred from an isolated spring. In such a spring, therefore, the strain is given by $\varepsilon_{se}(t) = \varepsilon_0$ at all times after the imposition of the step. This result can be obtained formally from Eq. (1) by letting $\tau_n = \tau_e \to \infty$.

Problem 9.1-3

(A) Equation (9.1-17) transforms to

$$\overline{W}(s) = \frac{\varepsilon_0^2}{2}\left(\{G_e\}/s + \int_{-\infty}^{\infty} H(\tau)\frac{\tau}{2+\tau s}\, d\ln\tau\right)\,. \tag{2}$$

Consequently,

$$\overline{\dot{W}}(s) = \frac{\varepsilon_0^2}{2}\left[\{G_e\} + \int_{-\infty}^{\infty} H(\tau)\frac{\tau s}{2+\tau s} d\ln\tau\right] \tag{3}$$

which, using Eq. (4.6-20), can be rearranged to

$$\overline{\dot{W}}(s) = \frac{\varepsilon_0^2}{2}\left[G_g + 2\int_{-\infty}^{\infty} \tau^{-1}H(\tau)\frac{\tau}{2+\tau s} d\ln\tau\right] . \tag{4}$$

Retransformation then yields Eq. (9.1-19) as required.

(B) Differentiation of Eq. (9.1-17) yields

$$\dot{W}_s(t) = -\varepsilon_0^2 \int_{-\infty}^{\infty} \tau^{-1}H(\tau)\exp(-2t/\tau) d\ln\tau . \tag{5}$$

Reintegration then gives

$$W_s(t) = (\varepsilon_0^2/2)\int_{-\infty}^{\infty} H(\tau)\exp(-2t/\tau) d\ln\tau + C . \tag{6}$$

With the initial condition, $W_s(0) = (\varepsilon_0^2/2)G_g$, using Eq. (4.6-20), we find that $C = \{G_e\}$, and thus recover Eq. (9.1-17).

Problem 9.1-4

(A) We wish to use Eq. (9.0-2) to obtain the energy absorbed by the material, W(t), when a strain is applied as a step function of time at $t = 0$. We write

$$\sigma(u) = \varepsilon_0 G(u)h(u) \tag{2}$$

and we also have

$$\dot{\varepsilon}(u) = \varepsilon_0 \delta(u) \tag{3}$$

since $\varepsilon(u) = \varepsilon_0 h(u)$. Substitution into Eq. $(9.0\text{-}2)_2$ gives

$$W(t) = \varepsilon_0^2 \int_{0^-}^{t} G(u)h(u)\delta(u)du \tag{4}$$

where we have changed the lower limit of integration to 0^-, the instant just preceding the imposition of the strain. This does not, of course, change the value of the integral. Ordinarily, $h(t) = 1$ but we cannot use this relation here because the strain is applied at $t = 0$ and $h(t)$ is not defined at that point. However, (cf. Appendix A2.1), the delta function is the symbolic derivative of the step function and, therefore,

$$h(u)\delta(u) = h(u)h'(u) . \tag{5}$$

By Eq. (1) we then obtain

$$W(t) = \frac{\varepsilon_0^2}{2}\int_{0^-}^{t} G(u)d[h(u)]^2 . \tag{6}$$

Now, h(u) is still not defined for $u = 0$. We therefore approximate the unit step function by the ramp function (see Appendix A2.5) which, in the limit that $t' \to 0$, approaches the unit step function [cf. Eq. (A2-44)]. This yields

$$W(t) = \frac{\varepsilon_0^2}{2} \lim_{t' \to 0} \int_0^{t'} G(t) d[r_{t'}(t)]^2 \tag{7}$$

where we have changed the variable and the limits of integration because the value of the integral is not affected by this change. Now, as $t' \to 0$, the interval of integration narrows. We take G(t) out from under the integral sign with its midvalue (i.e. the mean over the range of integration) and write

$$W(t) = (\varepsilon_0^2/2) \lim_{t' \to 0} \left(\frac{G(0) + G(t')}{2} \right) \int_0^{t'} d[r_{t'}(t)]^2 \ . \tag{8}$$

But

$$\int_0^{t'} d[r_{t'}(t)]^2 = [r_{t'}(t')]^2 - [r_{t'}(0)]^2 = 1 \tag{9}$$

and this is true whatever the value of t' is. Thus, proceding to the limit, Eq. (8) yields $W(t) = (\varepsilon_0^2/2)G_g$ as required.

(B) The same result would follow from Eq. (4) if we assigned the value of 1/2 to h(0). This would be consistent with Eqs. (A2-18) and (A2-19). The value arises here, however, from taking G(u) out from under the integral with its *mid-value* and does not depend on any interpretation of the value of h(t) at t = 0.

(C) To obtain $\dot{W}(t)$ from Eq. (9.0-1) we again substitute Eqs. (2) and (3). This gives

$$\dot{W}(t) = \varepsilon_0^2 G(0) h(0) \delta(t) \tag{10}$$

where we have written G(0) h(0) for G(t) h(t) because only these values survive integration over the product with $\delta(t)$. Using the value of 1/2 for h(0) as established above, immediately leads to Eq. $(9.1\text{-}24)_2$.

Problem 9.1-5

For a slope strain, $\varepsilon(t) = \dot{\varepsilon}_0 t$, and hence, $\bar{\varepsilon}(s) = \dot{\varepsilon}_0/s^2$ [cf. Eqs. 2.4-2]. Inserting into Eq. (9.1-9) and retransforming gives

$$\varepsilon_{sn}(t) = \dot{\varepsilon}_0 \tau_n [1 - \exp(-t/\tau_n)] \ . \tag{1}$$

Substitution into Eq. (9.1-13), dividing by s (this is equivalent to integration in the time domain), and using LTP (18) yields

$$\varepsilon_{dn}(t) = \dot{\varepsilon}_0 t - \dot{\varepsilon}_0 \tau_n [1 - (\exp - t/\tau_n)] \ . \tag{2}$$

Thus, upon imposition of a slope excitation, the strain in a Maxwell unit is initially zero in both the spring and the dashpot. At long times (i.e. as $t \to \infty$), the strain eventually reaches the finite value, $\dot{\varepsilon}_0 \tau_n$, in the spring, but increases without limit in the dashpot. In the isolated spring the strain is, of course, simply $\varepsilon(t)$. This result can be obtained formally by letting $\tau_n = \tau_e \to \infty$ in Eq. (1). We have

$$\varepsilon_{se}(t) = \dot{\varepsilon}_0 \lim_{\tau_e \to \infty} \tau_e [1 - \exp(-t/\tau_e)] = \dot{\varepsilon} \lim_{x \to 0} \frac{1 - \exp(-tx)}{x} = \dot{\varepsilon}_0 t \tag{3}$$

and, thus, $\varepsilon_{se}(t) = \varepsilon(t)$.

Problem 9.1-6

(A) We start from Eq. (9.0-2)$_2$, i.e. from

$$W(t) = \int_0^t \sigma(u)\dot{\varepsilon}(u)du \ . \tag{1}$$

In a constant rate of strain experiment $\varepsilon(u) = \dot{\varepsilon}_0 u$. Hence, $\dot{\varepsilon}(u) = \dot{\varepsilon}_0$ and $\dot{\varepsilon}_0 du = d\varepsilon(u)$. On making these substitutions in Eq. (1), the limits must be changed from $0 \leftarrow u \rightarrow t$ to $0 \leftarrow \varepsilon(u) \rightarrow \varepsilon(t)$. This leads to

$$W(t) = \int_0^{\varepsilon(t)} \sigma(u)d\,\varepsilon(u) \ . \tag{2}$$

But the strain is directly proportional to the time. We may thus consider the strain rather than the time as the variable of interest. Letting ε denote the strain at the given constant rate, and ε_f its final value, we have

$$W(\varepsilon_f) = \int_0^{\varepsilon_f} \sigma(\varepsilon)d\varepsilon \tag{3}$$

which is the desired result.

(B) Alternatively we may start from Eq. (9.1-43)$_1$ rewriting it as

$$W(t) = \int_0^{\varepsilon(t)} \dot{\varepsilon}_0\eta(u)d\,\dot{\varepsilon}_0 u \ . \tag{4}$$

Since $\dot{\varepsilon}_0\eta(u) = \sigma(u)$, Eq. (2) and, hence, Eq. (3) follows.

Problem 9.1-7

Let the excitation be given by $\varepsilon(t) = \varepsilon_0 \sin \varepsilon t$. The maximum value that $\varepsilon(t)$ can take up is clearly ε_0. To find the maximum value of $\varepsilon_s(t;\tau)$ we differentiate Eq. (9.1-50) with respect to t and set the result equal to 0. This leads to $\tan \omega t = \omega\tau$ as the relation for those values of t at which $\varepsilon_s(t;\tau)$ becomes a maximum for a given ω and τ. By the relations between the sine, cosine, and tangent of an angle, $\tan \omega t = \omega\tau$ implies $\sin \omega t = \omega\tau/(1 + \omega^2\tau^2)^{1/2}$ and $\cos \omega t = 1/(1 + \omega^2\tau^2)^{1/2}$. Substituting these values into Eq. (9.1-50) and squaring yields

$$\varepsilon_s^2(t_{max};\tau) = \frac{\varepsilon_0^2\omega^2\tau^2}{1 + \omega^2\tau^2} \ . \tag{2}$$

Equation (2) represents the maximum value which $\varepsilon_s^2(t;\tau)$ can take up for a given ω and τ. Introducing $\varepsilon(t) = \varepsilon_0$ and Eq. (2) into Eq. (9.1-10) is equivalent to summing the contributions of all energy storing mechanisms without regard to their phase relations. The result is

$$W_s(\omega)_{max} = \frac{\varepsilon_0^2}{2}\left(\{G_e\} + \int_{-\infty}^{\infty} H(\tau)\frac{\omega^2\tau^2}{1 + \omega^2\tau^2} d\ln\tau\right) \tag{3}$$

which is identical with Eq. (1) by the definition of $G'(\omega)$.

Problem 9.1-8

We need to find the maximum value that $\dot{\varepsilon}_d(t;\tau)$ can take up for a given ω and τ. For a harmonic strain excitation $\dot{\varepsilon}_d(t;\tau)$ is given by Eq. (9.1-78). Differentation with respect to t and setting the result equal to zero furnishes $\tan \omega t = \omega\tau$ and hence, $\cos \omega t = 1/(1 + \omega^2\tau^2)^{1/2}$ and $\sin \omega t = \omega t/(1 + \omega^2\tau^2)^{1/2}$. Inserting these expressions into Eq. (9.1-78) and squaring leads to

$$\dot{\varepsilon}_d^2(t_{max};\tau) = \frac{\varepsilon_0^2\omega^2}{1 + \omega^2\tau^2} \tag{2}$$

as the contribution to the maximum rate of energy dissipation of the mechanism associated with the relaxation time τ at the frequency ω. Substitution of Eq. (2) into Eq. (9.1-14) sums all contributions without regard to the phase relations between them. We have

$$\dot{W}_d(\omega)_{max} = \varepsilon_0^2\omega \int_{-\infty}^{\infty} H(\tau)\frac{\omega\tau}{1 + \omega^2\tau^2} d \ln \tau = \varepsilon_0^2\omega\, G''(\omega) \ . \tag{3}$$

The second part of Eq. (1) follows from the recognition that $G''(\omega) = \tilde{G}(\omega) \sin \theta(\omega)$ and $\tilde{G}(\omega) = \sigma_0(\omega)/\varepsilon_0$.

Problem 9.1-9

The ramp strain

$$\varepsilon(t) = \dot{\varepsilon}_0[th(t) - (t - t')h(t - t')] \tag{1}$$

[cf. Eq. (7.1-18)] consists of a slope strain imposed at $t = 0$ which is removed at $t = t'$. The energy relations for $0 \leq t \leq t'$, i.e. those arising from the slope strain, have been discussed in Sect. 9.2.2. To obtain those for $t \geq t'$, we substitute the transform of Eq. (1),

$$\varepsilon(s) = \dot{\varepsilon}_0[1 - \exp(-t's)]/s^2 \tag{2}$$

into Eqs. (9.1-11) and (9.1-15) to obtain

$$\varepsilon_s(t;\tau) = \dot{\varepsilon}_0\big(\tau[1 - \exp(-t/\tau)]h(t) - \tau\{1 - \exp[-(t - t')]/\tau\}h(t - t')\big) \tag{3}$$

and

$$\dot{\varepsilon}_d(t;\tau) = \dot{\varepsilon}_0\big([1 - \exp(-t/\tau)]h(t) - \{1 - \exp[-(t - t')/\tau]\}h(t - t')\big) \ . \tag{4}$$

Introducing the shifted time yields

$$\varepsilon_s(\theta;\tau) = \dot{\varepsilon}_0\tau[1 - \exp(-t'/\tau)]\exp(-\theta/\tau) \tag{5}$$

and

$$\dot{\varepsilon}_d(\theta;\tau) = \dot{\varepsilon}_0[1 - \exp(-t'/\tau)]\exp(-\theta/\tau) \ . \tag{6}$$

In addition, since the strain is constant for $t \geq t'$, we also have

$$\varepsilon(\theta) = \dot{\varepsilon}_0 t' \ . \tag{7}$$

Substituting Eqs. (7) and (5) into Eq. (9.1-10) gives

$$W_s(\theta) = \frac{\dot{\varepsilon}_0^2}{2}\left[\{G_e\}t'^2 + \int_{-\infty}^{\infty} \tau^2 H(\tau)[1 - \exp(-t'/\tau)]^2 \exp(-2\theta/\tau) d \ln \tau\right] \tag{8}$$

and the rate results as

$$\dot{W}_s(\theta) = -\dot{\varepsilon}_0^2 \int_{-\infty}^{\infty} \tau H(\tau)[1 - \exp(-t'/\tau)]^2 \exp(-2\theta/\tau) d \ln \tau \ . \tag{9}$$

Substitution of Eq. (6) into Eq. (9.1-14) furnishes

$$\dot{W}_d(\theta) = \dot{\varepsilon}_0^2 \int_{-\infty}^{\infty} \tau H(\tau)[1 - \exp(-t'/\tau)]^2 \exp(-2\theta/\tau) d \ln \tau \ . \tag{10}$$

Now, when $\theta = 0$, $t = t'$. Equation (9.1-37), rearranged for convenience, gives

$$W_d(t') = \frac{\dot{\varepsilon}_0^2}{2} \int_{-\infty}^{\infty} \tau H(\tau)\{2t' - 2\tau[1 - \exp(-t'/\tau)] - \tau[1 - \exp(-t'/\tau)]^2\} d \ln \tau \ . \tag{11}$$

and this serves as the initial condition for the integration of $\dot{W}_d(\theta)$ to obtain $W_d(\theta)$. We find

$$W_d(\theta) = \frac{\dot{\varepsilon}_0^2}{2} \int_{-\infty}^{\infty} \tau H(\tau)\{2t' - 2\tau[1 - \exp(-t'/\tau)] - \tau[1 - \exp(-t'/\tau)]^2 \exp(-2\theta/\tau)\} d \ln \tau \ . \tag{12}$$

Further, by Eqs. (9.0-3) and (9.0-4), we find that

$$W(\theta) = \frac{\dot{\varepsilon}_0^2}{2}\left[\{G_e\}t'^2 + 2 \int_{-\infty}^{\infty} \tau H(\tau)\{t' - \tau[1 - \exp(-t'/\tau)]\} d \ln \tau\right] = \dot{\varepsilon}_0^2 \int_0^{t'} \eta(u) du \ . \tag{13}$$

and

$$\dot{W}(\theta) = 0 \ . \tag{14}$$

Problem 9.1-10

(A) Rewriting Eq. (9.1-50) for the nth relaxation times leads to

$$\varepsilon_{sn}(t) = \varepsilon_0 \left(\frac{\omega\tau_n}{1 + \omega^2\tau_n^2} \cos \omega t + \frac{\omega^2\tau_n^2}{1 + \omega^2\tau_n^2} \sin \omega t\right) \ . \tag{1}$$

Substitution of $\bar{\varepsilon}(s) = \varepsilon_0/(s^2 + \omega^2)$ into Eq. (9.1-6), rejection of the transient part after partial fraction decomposition, and retransformation yields

$$\varepsilon_{dn}(t) = -\varepsilon_0 \left(\frac{\omega\tau_n}{1 + \omega^2\tau_n^2} \cos \omega t - \frac{1}{1 + \omega^2\tau_n^2} \sin \omega t\right) \ . \tag{2}$$

Now, the phase angle $\mu_n(\omega)$ is given by

$$\tan \mu_n(\omega) = 1/\omega\tau_n = G''_n(\omega)/G'_n(\omega) \tag{3}$$

where

$$G''_n(\omega) = \frac{G_n\omega\tau_n}{1+\omega^2\tau_n^2} \quad \text{and} \quad G'_n(\omega) = \frac{G_n\omega^2\tau_n^2}{1+\omega^2\tau_n^2} \tag{4}$$

are the loss and storage moduli of the nth Maxwell unit. Hence,

$$\frac{\omega\tau_n}{\sqrt{1+\omega^2\tau_n^2}} = \cos\mu_n(\omega) \quad \text{and} \quad \frac{1}{\sqrt{1+\omega^2\tau_n^2}} = \sin\mu_n(\omega)\ . \tag{5}$$

Substitution of these relations into Eqs. (1) and (2) leads to

$$\varepsilon_{sn}(t) = \varepsilon_0\cos\mu_n(\omega)\sin[\omega t + \mu_n(\omega)] = \varepsilon_{sn}\sin[\omega t + \mu_n(\omega)] \tag{6}$$

and

$$\varepsilon_{dn}(t) = -\varepsilon_0\sin\mu_n(\omega)\cos[\omega t + \mu_n(\omega)] = -\varepsilon_{dn}\cos[\omega t + \mu_n(\omega)] \tag{7}$$

where ε_{sn} and ε_{dn} are the respective peak amplitudes.

(B) The phasor diagram is shown in Fig. P9.1-10. In the steady-state the strain in the nth spring, $\varepsilon_{sn}(t)$, leads the overall strain imposed on the unit, $\varepsilon(t)$, by the phase angle $\mu_n(\omega)$. Since the dashpot is a purely viscous element, the strain in it, $\varepsilon_{dn}(t)$, will be ninety degrees out of phase with the strain in the purely elastic spring. Hence $\varepsilon_{dn}(t)$ will lag behind $\varepsilon(t)$ by the angle $\pi/2 - \mu_n(\omega)$. We note that this is the same as leading by $3\pi/2 + \mu_n(\omega)$. When $\mu_n(\omega) = 0$, $\varepsilon_{sn}(t) = \varepsilon(t)$, and $\varepsilon_{dn}(t) = 0$. The behavior of the unit is then purely elastic. Formally, one may picture the dashpot to be so viscous that it amounts to a rigid rod and the unit therefore degenerates to a simple spring. When $\mu_n(\omega) = \pi/2$, $\varepsilon_{sn}(t) = 0$, and $\varepsilon_{dn}(t) = \varepsilon(t)$. In this case the unit shows purely viscous behavior and one may now consider the spring to behave as a rigid rod. The unit then degenerates to a single dashpot. Letting $m = 0, 1, 2, \ldots$, the imposed strain reaches its peak amplitudes, $\pm\varepsilon_0$, when $\omega t = (2m+1)\pi/2$, the strain in the spring reaches $\pm\varepsilon_{sn} = \pm\varepsilon_0\cos\mu_n(\omega)$ when $\omega t = (2m+1)\pi/2 - \mu_n(\omega)$, and the strain in the dashpot attains $\pm\varepsilon_{dn} = \pm\varepsilon_0\sin\mu_n(\omega)$ when $\omega t = m\pi - \mu_n(\omega)$.

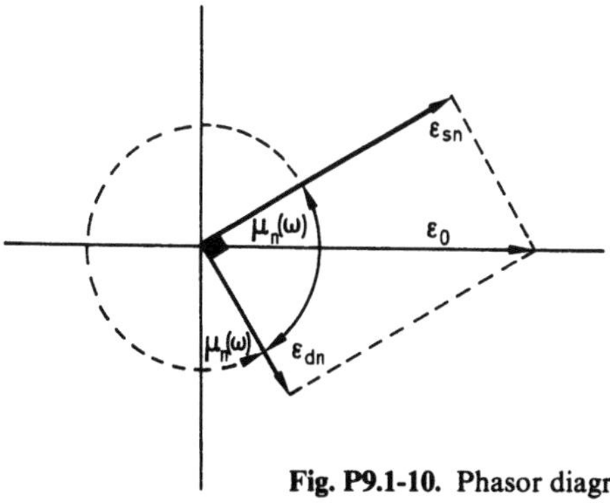

Fig. P9.1-10. Phasor diagram for harmonic strain excitation

Thus, although the strain reaches its peak values in all Maxwell units simultaneously, the strains in the individual springs and dashpots do not. Only in an isolated spring, if present, is the strain in phase with the excitation, i.e. $\varepsilon_{se}(t) = \varepsilon_0 \sin \omega t$. Formally, this result can be obtained from Eq. (1) by letting $\tau_n = \tau_e \to \infty$.

Problem 9.1-11

(A) We need the expression for the energy stored in the spring of the Maxwell unit as a function of time. By Eq. $(9.0\text{-}11)_1$ this becomes

$$w_{sn}(t) = \tfrac{1}{2} G_n \varepsilon_{sn}^2(t) \tag{2}$$

where [see Eq. (6) of Problem 9.1-10]

$$\varepsilon_{sn}(t) = \varepsilon_0 \cos \mu_n(\omega) \sin[\omega t + \mu_n(\omega)] \ . \tag{3}$$

Substitution of Eq. (3) into Eq. (2) and using the relation $G_n'(\omega) = G_n \cos^2 \mu_n(\omega)$ leads to

$$w_{sn}(t) = (\varepsilon_0^2/4) G_n'(\omega) \{1 - \cos 2[\omega t + \mu_n(\omega)]\} \tag{4}$$

as the desired relation. Clearly, $w_{sn}(t)$ has maxima when the cosine term equals -1, i.e. when $\omega t = (2m + 1)\pi/2 - \mu_n(\omega)$ where $m = 0, 1, 2, \ldots$. Thus, Eq. (1) follows.

(B) The rate at which energy is stored in the nth Maxwell unit follows from Eq. (4) as

$$\dot{w}_{sn}(t) = (\varepsilon_0^2/2) \omega G_n'(\omega) \sin[\omega t + \mu_n(\omega)] \ . \tag{5}$$

Integration between the limits $m\pi/\omega - \mu_n(\omega)/\omega$ and $(2m + 1)\pi/2\omega - \mu_n(\omega)/\omega$ yields the energy stored during each first quarter of any cycle of the excitation. Increasing the limits by π/ω furnishes the energy stored during the third quarter cycle. Both are identical with Eq. (1). Increasing the limits by $\pi/2\omega$ or $3\pi/2\omega$ yields the energy dissipated during the second and fourth quarter of each cycle. This differs only in sign from $(\varepsilon_0^2/2)\, G_n'(\omega)$.

(C) Summing the stored energies over all Maxwell elements of a Wiechert model and adding the energy stored in the isolated spring gives

$$W_s(\omega)_{max} = \frac{\varepsilon_0^2}{2}\left[\{G_e\} + \sum_n G_n'(\omega)\right] = \frac{\varepsilon_0^2}{2} G'(\omega) \tag{6}$$

in accordance with Eq. $(9.1\text{-}63)_2$. $W_s(\omega)_{max}$ represents the maximum *coherently* storable energy because the energies stored in the various springs are added without regard to the phase differences arising from the differences in the relaxation times associated with each Maxwell unit.

Problem 9.1-12

By Eq. (9.0-7), the rate at which energy is dissipated in the dashpot of the nth Maxwell unit is

$$\dot{w}_{dn}(t) = \eta_n \dot{\varepsilon}_{dn}^2(t) = G_n \tau_n \dot{\varepsilon}_{dn}^2(t) \ . \tag{2}$$

But [see Eq. (7) of Problem 9.1-10]

$$\varepsilon_{dn}(t) = -\varepsilon_0 \sin \mu_n(\omega) \cos[\omega t + \mu_n(\omega)] \ . \tag{3}$$

Differentiating and substituting into Eq. (2) yields

$$\dot{w}_{dn}(t) = (\varepsilon_0^2/2)\omega G''_n(\omega)\{1 - \cos 2[\omega t + \mu_n(\omega)]\} \ . \tag{4}$$

Integration of $\dot{w}_{dn}(t)$ over any quarter cycle of the excitation then yields Eq. (1).

(B) Summing over all dashpots of a Wiechert model gives

$$W_d(\omega)_{qtr} = (\varepsilon_0^2/4)\pi \, G''(\omega) \ . \tag{5}$$

This represents energy dissipated *coherently*, i.e. without regard to the phase differences between the dissipating mechanisms.

Problem 9.1-13

The heat generated under adiabatic conditions is equivalent to the energy dissipated. By Eq. $(9.1\text{-}100)_2$, the energy dissipated per cycle per unit volume of material, is given by

$$W_d(\omega)_{cyc} = \pi\varepsilon_0^2 G''(\omega) \tag{1}$$

where $G''(\omega)$ is the loss modulus. The number of cycles per second is $f = \omega/2\pi$. The heat generated per unit time per unit volume therefore becomes

$$Q = W_d(\omega)_{cyc} f = (\varepsilon_0^2/2)\omega G''(\omega) \ . \tag{2}$$

Problem 9.1-14

(A) Under the stated conditions the heat generated per unit volume is equal to the total energy, $W(\varepsilon_f)$, absorbed per unit volume in the course of the deformation. By Problem 9.1-6 this energy can be obtained from the area under the stress-strain curve.

(B) Let ρ be the density of the material and C_P its heat capacity at constant pressure. The temperature rise, ΔT, is then

$$\Delta T = W(\varepsilon_f)/\rho C_P \ . \tag{1}$$

Problem 9.1-15

Energy storage and dissipation in a ladder model is conveniently treated in terms of the equivalent series-parallel model. The *primary* energy functions for such a model are given by Eqs. (9.1-8) and (9.1-12).

For a step strain excitation

$$\varepsilon_{sn}(t) = \varepsilon_0 \exp(-t/\tau_n) \tag{1}$$

and

$$\dot{\varepsilon}_{dn}(t) = \varepsilon_0 \tau_n^{-1} \exp(-t/\tau_n) \tag{2}$$

[cf. Eqs. (9.1-16) and (9.1-20)]. Thus, for the Gross-Marvin model,

$$W_s(t) = \frac{\varepsilon_0^2}{2}\left[\{G_e\} + \sum_{n=1}^{n=N-1} G_n \exp(-2t/\tau_n)\right] \tag{3}$$

and

$$\dot{W}_d(t) = \varepsilon_0^2 \sum_{n=1}^{n=N-1} (G_n/\tau_n)\exp(-2t/\tau_n) \tag{4}$$

where the G_n and τ_n are given by Eqs. (5.2-88) and (5.2-87) if the model is arrheodictic, and by Eqs. (5.2-93) and (5.2-92) if it is rheodictic.

Problem 9.1-16

(A) By Eq. (9.1-11)

$$\varepsilon_s(t;\tau) = \mathscr{L}^{-1}\frac{\tau}{1+\tau s}\bar{\varepsilon}(s) = \int_0^t \exp[-(t-u)/\tau]\dot{\varepsilon}(u)du \tag{3}$$

and, therefore,

$$\varepsilon_s^2(t;\tau) = \int_0^t \exp[-(t-u)/\tau]\dot{\varepsilon}(u)du \int_0^t \exp[-(t-v)/\tau]\dot{\varepsilon}(v)dv$$

$$= \int_0^t\int_0^t \exp[-(2t-u-v)/\tau]\dot{\varepsilon}(u)\dot{\varepsilon}(v)du\,dv \tag{4}$$

because the two integrations are independent of each other. Substitution into Eq. (9.1-10) yields

$$W_s(t) = \tfrac{1}{2}\int_0^t\int_0^t\left[\{G_e\} + \int_{-\infty}^{\infty} H(\tau)\exp[-(2t-u-v)/\tau]d\ln\tau\right]\dot{\varepsilon}(u)\dot{\varepsilon}(v)du\,dv \tag{5}$$

and Eq. (1) follows by Eq. (4.1-5).

(B) By Eq. (9.1-15)

$$\dot{\varepsilon}_d(t;\tau) = \mathscr{L}^{-1}\frac{\bar{\varepsilon}(s)}{1+\tau s} = \int_0^t \tau^{-1}\exp[-(t-u)/\tau]\dot{\varepsilon}(u)du \tag{6}$$

and we have

$$\dot{\varepsilon}_d^2(t;\tau) = \int_0^t\int_0^t \tau^{-2}\exp[-(2t-u-v)/\tau]\dot{\varepsilon}(u)\dot{\varepsilon}(v)du\,dv\ . \tag{7}$$

Introducing Eq. (7) into Eq. (9.1-14) then leads to

$$\dot{W}_d(t) = \int_0^t\int_0^t\left[\int_{-\infty}^{\infty}\tau^{-1}H(\tau)\exp[-(2t-u-v)/\tau]d\ln\tau\right]\dot{\varepsilon}(u)\dot{\varepsilon}(v)du\,dv \tag{8}$$

from which Eq. (2) follows with

$$\dot{G}(t) = -\int_{-\infty}^{\infty}\tau^{-1}H(\tau)\exp(-t/\tau)d\ln\tau\ . \tag{9}$$

The last equation is obtained by differentiating Eq. (4.1-5).

Problem 9.2-1

For a step strain, $\sigma(t) = \sigma_0$, and hence, $\bar{\sigma}(s) = \sigma_0/s$. Equations (9.2-9) and (9.2-13) therefore yield

$$\sigma_{sn}(t) = \sigma_0[1 - \exp(-t/\tau_n)] \tag{1}$$

and

$$\sigma_{dn}(t) = \sigma_0 \exp(-t/\tau_n) \ . \tag{2}$$

The stress in each Voigt unit is σ_0. However, in the springs of the Voigt units the stress is initially zero, i.e. $\sigma_{sn}(0) = 0$ at $t = 0$. Each spring carries the full value of the stress, σ_0, only after it is fully extended, i.e. at $t = \infty$. Hence, for $0 < t < \infty$. the stress in the spring is retarded. By contrast, the stress in the dashpot relaxes away from its initial value, σ_0, as stress is being transferred to the spring. The rate at which this reapportioning of stress takes place is governed by the retardation time, τ_n. In the isolated spring the stress assumes the final value, σ_0, instantaneously. In the isolated dashpot, if present, the stress never loses the initial value, σ_0. These results can be obtained formally from Eq. (9.2-5) by letting $\tau_n = \tau_g \to 0$ and from Eq. (9.2-6) by letting $\tau_n = \tau_f \to \infty$.

Problem 9.2-2

For a slope stress, $\sigma(t) = \dot{\sigma}_0 t$, and, hence, $\bar{\sigma}(s) = \dot{\sigma}_0/s^2$. Equations (9.2-9) and (9.2-13) therefore lead to

$$\sigma_{sn}(t) = \dot{\sigma}_0 t - \dot{\sigma}_0 \tau_n[1 - \exp(-t/\tau_n)] \tag{1}$$

and

$$\sigma_{dn}(t) = \dot{\sigma}_0 \tau_n[1 - \exp(-t/\tau_n)] \tag{2}$$

for the spring and the dashpot, respectively, of the nth Voigt unit. The stress through the unit as a whole is $\dot{\sigma}_0 t$. The same stress acts, of course, also through the isolated spring and the isolated dashpot, if the latter is present. By letting $\tau_n = \tau_g \to 0$ in Eq. (1) and $\tau_n = \tau_f \to \infty$ in Eq. (2), we find $\sigma_{sg}(t) = \sigma_{df}(t) = \dot{\sigma}_0 t$. In each Voigt unit the stress is zero in both elements at $t = 0$. At $t = \infty$, we would have $\sigma_{sn}(\infty) = \infty$, $\sigma_{dn}(\infty) = \dot{\sigma}_0 \tau_n$ if the stress through a Voigt unit could increase indefinitely. This, of course, is not possible because at some time t_{bn} the stress which the nth spring can sustain would be exceeded, i.e. the spring would break. Long before this happens, however, the linear limit would be exceeded.

Problem 9.2-3

The stress history [cf. Eq. (7.1-1)] is given by $\sigma(t) = \sigma_0 h_{t'}(t)$. By LTP (8) the stress transform results as $\sigma_0[1 - \exp(-t's)]/s$, and this must be substituted into Eqs. (9.2-11) and (9.2-15). We obtain

$$\bar{\sigma}_s(s;\tau) = \frac{\sigma_0[1 - \exp(-t's)]}{s(1 + \tau s)} \tag{1}$$

and

$$\bar{\sigma}_d(s;\tau) = \frac{\sigma_0\tau[1 - \exp(-t's)]}{1 + \tau s} \ . \tag{2}$$

Retransformation gives

$$\sigma_s(t;\tau) = \sigma_0([1 - \exp(-t/\tau)]h(t) - \{1 - \exp[-(t - t')/\tau]\}h(t - t')) \tag{3}$$

and

$$\sigma_d(t;\tau) = \sigma_0\{\exp(-t/\tau)h(t) - \exp[-(t - t')/\tau]h(t - t')\} \ . \tag{4}$$

The energy relations in creep, i.e. for $0 \leq t \leq t'$, have been discussed in Sect. 9.2.1 and need not be repeated here. In the recovery part of the response $h(t) = h(t - t') = 1$. Introducing the shifted time, $\theta = t - t'$, Eqs. (3) and (4) become

$$\sigma_s(\theta;\tau) = \sigma_0[1 - \exp(-t'/\tau)]\exp(-\theta/\tau) \tag{5}$$

and

$$\sigma_d(\theta;\tau) = -\sigma_0[1 - \exp(-t'/\tau)]\exp(-\theta/\tau) \ . \tag{6}$$

In addition, we have

$$\sigma(\theta) = \sigma_0[h(\theta + t') - h(\theta)] = 0 \tag{7}$$

for the stress through the isolated spring, and through the isolated dashpot if the material is rheodictic. Substitution of these equations into Eqs. (9.2-10) and (9.2-14) yields

$$W_s(\theta) = \frac{\sigma_0^2}{2} \int_{-\infty}^{\infty} L(\tau)[1 - \exp(-t'/\tau)]^2 \exp(-2\theta/\tau) d \ln \tau \tag{8}$$

and

$$\dot{W}_d(\theta) = \sigma_0^2 \int_{-\infty}^{\infty} \tau^{-1} L(\tau)[1 - \exp(-t'/\tau)]^2 \exp(-2\theta/\tau) d \ln \tau \tag{9}$$

for the energy stored up *to* time θ, and for the rate at which energy is dissipated *at* time θ, respectively.

Comparing $W_s(t = t')$ from Eq. (9.2-18) with $W_s(\theta = 0)$ from Eq. (8), we see that the difference, $W_s(t = t') - W_s(\theta = 0) = (\sigma_0^2/2)J_g$, is stored energy which is instantaneously released when the stress is removed at $t = t'$, i.e. at $\theta = 0$. The remaining stored energy then decreases to zero.

To obtain the rate at which energy is stored, we differentiate Eq. (8) according to Eq. (9.1-18). This gives

$$\dot{W}_s(\theta) = \frac{\sigma_0^2}{2} \int_{-\infty}^{\infty} L(\tau)[1 - \exp(-t'/\tau)] d \ln \tau \ \delta(\theta)$$

$$- \sigma_0^2 \int_{-\infty}^{\infty} \tau^{-1} L(\tau)[1 - \exp(-t'/\tau)]^2 \exp(-2\theta/\tau) d \ln \tau \ . \tag{10}$$

To obtain the amount of energy dissipated up to time θ, Eq. (9) must be integrated subject to the initial condition $W_d(\theta = 0) = W_d(t = t')$ [cf. Eq. (9.2-21)]. To keep our equations consistent with the rules of transformation calculus, we incorporate this condition in the equation for $\dot{W}_d(\theta)$, i.e. we write

$$\dot{W}_d(\theta) = \frac{\sigma_0^2}{2}\left[\int_{-\infty}^{\infty} L(\tau)[1 - \exp(-2t'/\tau)]d\ln\tau + 2\{\phi_f\}t'\right]\delta(\theta)$$

$$+ \sigma_0^2 \int_{-\infty}^{\infty} \tau^{-1}L(\tau)[1 - \exp(-t'/\tau)]^2\exp(-2\theta/\tau)d\ln\tau \tag{11}$$

where the first term on the right is $W_d(t')\delta(t)$. Integration of Eq. (11) between the limits 0 and θ then yields

$$W_d(\theta) = \frac{\sigma_0^2}{2}\left[\int_{-\infty}^{\infty} L(\tau)[1 - \exp(-t'/\tau)]\right.$$

$$\left.\times\{2 - [1 - \exp(-t'/\tau)]\exp(-2\theta/\tau)\}d\ln\tau + 2\{\phi_f\}t'\right] . \tag{12}$$

Thus the energy dissipated decreases from

$$W_d(\theta = 0) = \frac{\sigma_0^2}{2}\left[\int_{-\infty}^{\infty} L(\tau)[1 - \exp(-2t'/\tau)]d\ln\tau + 2\{\phi_f\}t'\right] \tag{13}$$

at $\theta = 0$ ($t = t'$) to

$$W_d(\theta = \infty) = \sigma_0^2\left[\int_{-\infty}^{\infty} L(\tau)[1 - \exp(-t'/\tau)]d\ln\tau + \{\phi_f\}t'\right] \tag{14}$$

at $\theta = t = \infty$.

Adding Eqs. (8) and (12) yields

$$W(\theta) = \sigma_0^2\left[\int_{-\infty}^{\infty} L(\tau)[1 - \exp(-t'/\tau)]d\ln\tau + \{\phi_f\}t'\right] . \tag{15}$$

This is the amount of energy absorbed at $\theta = 0$, and is constant with time, no further energy being absorbed during creep recovery. $W(\theta)$ differs by the amount $(\sigma_0^2/2)J_g$ from $W(t')$. Comparing Eqs. (14) and (15) shows that all of the energy absorbed at $\theta = 0$ is eventually dissipated. The rate at which energy is absorbed at time θ results as

$$\dot{W}(\theta) = \sigma_0^2\left[\int_{-\infty}^{\infty} L(\tau)[1 - \exp(-t'/\tau)]d\ln\tau + \{\phi_f\}t'\right]\delta(\theta) . \tag{16}$$

Problem 9.2-4

Equation (9.2-38) expresses $W(t)$ in terms of the retardation spectrum, $L(\tau)$. Equation $(9.2\text{-}41)_2$ does it in terms of the viscoelastic functions

$$\chi(t) = \mathscr{L}^{-1}\bar{J}(s)/s \qquad \text{and} \qquad \rho(t) = \mathscr{L}^{-1}\bar{J}(s)/s^2 . \tag{1}$$

Using the equation [cf. Eq. (4.1-27)]

$$\bar{J}(s) = J_e^{\{o\}}/s - \int_{-\infty}^{\infty} L(\tau)\frac{\tau}{1 + \tau s}d\ln\tau + \{\phi_f\}/s^2 \tag{2}$$

for the transform of the creep compliance, we have

$$\chi(t) = J_e^{\{o\}}t - \int_{-\infty}^{\infty} \tau L(\tau)[1 - \exp(-t/\tau)]d \ln \tau + \{\phi_f\}t^2/2 \tag{3}$$

and

$$\rho(t) = J_e^{\{o\}}t^2/2 - \int_{-\infty}^{\infty} \tau L(\tau)\{t - \tau[1 - \exp(-t/\tau)]\}d \ln \tau + \{\phi_f\}t^3/6 \tag{4}$$

by LTPs (12), (11), and (18). Combining Eqs. (3) and (4) according to Eq. $(9.2\text{-}41)_2$ leads to Eq. (9.2-38).

Problem 9.2-5

As shown by Eq. (9.2-42), $\rho(t)$ is the integral over $\chi(t)$, and $\chi(t) = \varepsilon(t)/\dot{\sigma}_0$ by Eq. $(2.4\text{-}11)_1$. In a constant rate of stress experiment $\dot{\sigma}(t) = \dot{\sigma}_0$. Substitution into Eq. (9.2-42) therefore yields

$$\rho(t) = \dot{\sigma}_0^{-1} \int_0^t \varepsilon(u)du = \dot{\sigma}_0^{-2} \int_0^t \dot{\sigma}_0\varepsilon(u)du = \dot{\sigma}_0^{-2} \int_0^{\sigma(t)} \varepsilon(u)d\,\sigma(u) \ . \tag{1}$$

Note that

$$\int_0^{\sigma(t)} \varepsilon(u)d\,\sigma(u) = \int_0^t \varepsilon(u)\dot{\sigma}(u)du = W_c(t) \ . \tag{2}$$

$W_c(t)$ is known as the *complementary energy* per unit volume [cf. Eq. $(9.0\text{-}2)_2$].

Problem 9.2-6

Rewriting Eqs. (9.2-49) and (9.2-67) for the nth relaxation time gives

$$\sigma_{sn}(t) = \sigma_0\left(\frac{1}{1+\omega^2\tau_n^2}\sin \omega t - \frac{\omega\tau_n}{1+\omega^2\tau_n^2}\cos \omega t\right) \tag{1}$$

and

$$\sigma_{dn}(t) = \sigma_0\left(\frac{\omega\tau_n}{1+\omega^2\tau_n^2}\cos \omega t + \frac{\omega^2\tau_n^2}{1+\omega^2\tau_n^2}\sin \omega t\right) \ . \tag{2}$$

The phase angle $\nu_n(\omega)$ of the nth Voigt unit is given by

$$\tan \nu_n(\omega) = \omega\tau_n = J_n''(\omega)/J_n'(\omega) \tag{3}$$

where

$$J_n''(\omega) = \frac{J_n\omega\tau_n}{1+\omega^2\tau_n^2} \tag{4}$$

is the loss compliance, and

$$J_n'(\omega) = \frac{J_n}{1+\omega^2\tau_n^2} \tag{5}$$

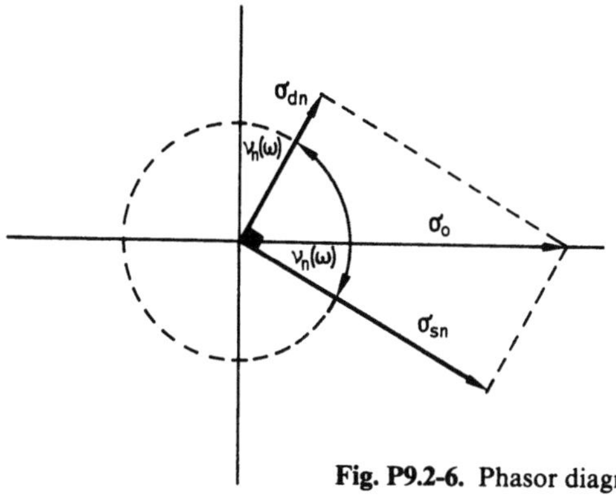

Fig. P9.2-6. Phasor diagram for harmonic stress excitation

is the storage compliance of the nth Voigt unit. Hence,

$$\cos \nu_n(\omega) = \frac{1}{\sqrt{1 + \omega^2\tau_n^2}} \qquad \text{and} \qquad \sin \nu_n(\omega) = \frac{\omega\tau_n}{\sqrt{1 + \omega^2\tau_n^2}} \tag{6}$$

and substitution of these relations into Eqs. (1) and (2) gives

$$\sigma_{sn}(t) = \sigma_0 \cos \nu_n(\omega) \sin[\omega t - \nu_n(\omega)] = \sigma_{sn} \sin[\omega t - \nu_n(\omega)] \tag{7}$$

and

$$\sigma_{dn}(t) = \sigma_0 \sin \nu_n(\omega) \cos[\omega t - \nu_n(\omega)] = \sigma_{dn} \cos[\omega t - \nu_n(\omega)] \tag{8}$$

where σ_{sn} and σ_{dn} are the respective peak amplitudes.

(B) The phasor diagram is shown in Fig. P9.2-6. The stress in the nth spring, $\sigma_{sn}(t)$, lags the overall stress imposed on the unit, $\sigma(t)$, by the phase angle $\nu_n(\omega)$. The stress in the nth dashpot, $\sigma_{dn}(t)$, being ninety degrees out of phase with the stress in the spring, leads $\sigma(t)$ by the angle $\frac{\pi}{2} - \nu_n(\omega)$. While the overall stress attains its peak amplitudes, $\pm\sigma_0$, when $\omega t = (2m + 1)\pi/2$, where $m = 0, 1, \ldots$, the stress in the nth spring reaches $\pm\sigma_{sn} = \pm\sigma_0 \cos \nu_n(\omega)$ when $\omega t = (2m + 1)\pi/2 + \nu_n(\omega)$, and the stress in the dashpot attains $\pm\sigma_{dn} = \pm\sigma_0 \sin \nu_n(\omega)$ when $\omega t = m\pi + \nu_n(\omega)$. The stress in the isolated spring is in phase with $\sigma(t)$, i.e. $\sigma_{sg}(t) = \sigma_0 \sin \omega t$. The stress in the isolated dashpot, if the latter is present, is also in phase with $\sigma(t)$, i.e. $\sigma_{df}(t) = \sigma_0 \sin \omega t$. Formally, these results follow from setting $\tau_n = \tau_g \to 0$ and $\tau_n = \tau_f \to \infty$ in Eqs. (1) and (2), respectively.

Problem 9.2-7

(A) The ratio is given by

$$r(\omega) = \frac{\int_0^{\pi/2\omega} \dot{W}_d(t)dt}{\int_0^{\pi/2\omega} \dot{W}_s(t)dt} \,. \tag{1}$$

Integration of either Eqs. (9.2-71) and (9.2-55), or Eqs. (9.2-79) and (9.2-61), leads to

$$r_J(\omega) = \frac{\frac{\pi}{2}J''(\omega) - d\,J'(\omega)/d\ln\omega}{J'(\omega) + d\,J'(\omega)/d\ln\omega}\ . \tag{2}$$

(B) Integration of either Eqs. (9.1-82) and (9.1-56), or Eqs. (9.1-90) and (9.1-62) results in

$$r_G(\omega) = \frac{\frac{\pi}{2}G''(\omega) + d\,G'(\omega)/d\ln\omega}{G'(\omega) - d\,G'(\omega)/d\ln\omega}\ . \tag{3}$$

(C) The two ratios become identical in the limit that ω approaches either 0 or ∞. In that case the logarithmic derivatives vanish and the ratios approach $\frac{\pi}{2}\tan\theta(\omega)$.

Problem 9.2-8

By Eqs. (9.1-85) and (9.1-86), and Eqs. (9.2-74) and (9.2-75)

$$\tan 2\mu_d = DG''/DG' \qquad \text{and} \qquad \tan 2\nu_d = -DJ''/DJ' \tag{2}$$

where $D = d/d\ln\omega$ and the dependence on ω has been suppressed everywhere for simplicity's sake. Hence,

$$\tan 2(\mu_d - \nu_d) = \frac{\tan 2\mu_d - \tan 2\nu_d}{1 + \tan 2\mu_d \tan 2\nu_d} = \frac{DJ'DG'' + DJ''DG'}{DJ'DG' - DJ''DG''} \tag{3}$$

and, making use of $J^*(\omega) = 1/G^*(\omega)$ (cf. Sect. 8.7.2), we also have

$$DJ' = [(G''^2 - G'^2)DG' - 2G'G''DG'']/\tilde{G}^4 \tag{4}$$

and

$$DJ'' = [(G'^2 - G''^2)DG'' - 2G'G''DG']/\tilde{G}^4. \tag{5}$$

Substitution yields

$$\tan 2(\mu_d - \nu_d) = \frac{2G''/G'}{1 - (G''/G')^2} = \frac{2\tan\theta}{1 - \tan^2\theta} = \tan 2\theta \tag{6}$$

from which Eq. $(1)_2$ follows. Next,

$$\tan 2\mu_s = \frac{G'' - DG''}{G' - DG'} \qquad \text{and} \qquad \tan 2\nu_s = \frac{J'' + DJ''}{J' + DJ'}\ . \tag{7}$$

Thus,

$$\tan 2(\mu_s + \nu_s) = \frac{\tan 2\mu_s + \tan 2\nu_s}{1 - \tan 2\mu_s \tan 2\nu_s}\ . \tag{8}$$

Substitution of Eqs. (7) into Eq. (8) and using Eqs. (4) and (5) eventually leads to

$$\tan 2(\mu_s + \nu_s) = \frac{2G''/G'}{1 - (G''/G')^2} = \tan 2\theta \tag{9}$$

from which Eq. $(1)_1$ follows.

Problem 9.2-9

The energy dissipated per second in steady-state flow is the rate at which energy is dissipated in response to a slope strain as $t \to \infty$. By Eq. (9.1-35), then

$$\lim_{t\to\infty} \dot{W}_d(t) = \dot{\varepsilon}_0^2 \int_{-\infty}^{\infty} \tau H(\tau) d\ln\tau = \dot{\varepsilon}_0^2 \eta_f \ . \tag{1}$$

The energy stored follows from Eq. (9.1-34) as

$$\lim_{t\to\infty} W_s(t) = \frac{\dot{\varepsilon}_0^2}{2} \int_{-\infty}^{\infty} \tau^2 H(\tau) d\ln\tau = \dot{\varepsilon}_0^2 J_e \eta_f^2/2 \tag{2}$$

and the ratio of the two quantities gives $2/J_e\eta_f$ as required.

Problem 9.2-10

(A) By Eq. (9.2-11)

$$\sigma_s(t;\tau) = \mathscr{L}^{-1}\frac{\bar{\sigma}(s)}{1+\tau s} = \mathscr{L}^{-1}\frac{\bar{\dot{\sigma}}(s)}{s(1+\tau s)} = \int_0^t \{1 - \exp[-(t-u)/\tau]\}\dot{\sigma}(u)du$$

$$= \sigma(t) - \int_0^t \exp[-(t-u)/\tau]\dot{\sigma}(u)du \tag{3}$$

and, therefore,

$$\sigma_s^2(t;\tau) = \sigma^2(t) - 2\sigma(t)\int_0^t \exp[-(t-u)/\tau]\dot{\sigma}(u)du$$

$$+ \int_0^t\int_0^t \exp[-(2t-u-v)/\tau]\dot{\sigma}(u)\dot{\sigma}(v)du\, dv \ . \tag{4}$$

The third term on the right results because

$$\left(\int_0^t \exp[-(t-u)/\tau]\dot{\sigma}(u)du\right)^2$$

$$= \int_0^t \exp[-(t-u)/\tau]\dot{\sigma}(u)du \int_0^t \exp[-(t-v)/\tau]\dot{\sigma}(v)dv \ . \tag{5}$$

[cf. Eq. (4) of Problem 9.1-16]. Substitution of Eq. (4) into Eq. (9.2-10) yields

$$W_s(t) = \frac{1}{2}\Bigg[J_g\sigma^2(t) + \sigma^2(t)\int_{-\infty}^{\infty} L(\tau)d\ln\tau$$

$$- 2\sigma(t)\int_{-\infty}^{\infty} L(\tau)\int_0^t \exp[-(t-u)/\tau]\dot{\sigma}(u)du\, d\ln\tau$$

$$+ \int_{-\infty}^{\infty} L(\tau)\int_0^t\int_0^t \exp[-(2t-u-v)/\tau]\dot{\sigma}(u)\dot{\sigma}(v)du\, dv\, d\ln\tau\Bigg] \tag{6}$$

But

$$\tfrac{1}{2}\left[J_g\sigma^2(t) + \sigma^2(t)\int_{-\infty}^{\infty} L(\tau)d\ln\tau\right] = \tfrac{1}{2}J_e^{\{o\}}\sigma^2(t)$$

$$= \sigma(t)J_e^{\{o\}}\int_0^t \dot{\sigma}(u)du - \tfrac{1}{2}J_e^{\{o\}}\int_0^t\int_0^t \dot{\sigma}(u)\dot{\sigma}(v)du\,dv\ . \tag{7}$$

Hence

$$W_s(t) = \sigma(t)\int_0^t\left[J_e^{\{o\}} - \int_0^t L(\tau)\exp[-(t-u)/\tau]d\ln\tau\right]\dot{\sigma}(u)du$$
$$-\tfrac{1}{2}\int_0^t\int_0^t\left[J_e^{\{o\}} - \int_{-\infty}^{\infty} L(\tau)\exp[-(2t-u-v)/\tau]d\ln\tau\right]\dot{\sigma}(u)\dot{\sigma}(v)du\,dv\ . \tag{8}$$

Using Eq. (4.1-28), and considering that the flow term does not contribute to the stored energy, we find

$$W_s(t) = \sigma(t)\int_0^t J(t-u)\dot{\sigma}(u)du - \tfrac{1}{2}\int_0^t\int_0^t J(2t-u-v)\dot{\sigma}(u)\dot{\sigma}(v)du\,dv \tag{9}$$

which, in view of Eq. $(2.3\text{-}17)_3$, reduces to Eq. (1).

(B) By Eq. (9.2-15)

$$\sigma_d(t;\tau) = \mathscr{L}^{-1}\frac{\tau}{1+\tau s}\bar{\dot{\sigma}}(s) = \int_0^t \exp[-(t-u)/\tau]\dot{\sigma}(u)du \tag{10}$$

and, therefore,

$$\sigma_d^2(t;\tau) = \int_0^t\int_0^t \exp[-(2t-u-v)/\tau]\dot{\sigma}(u)\dot{\sigma}(v)du\,dv \tag{11}$$

[cf. Eq. (5)]. Substitution of Eq. (11) into Eq. (9.2-14) then yields

$$\dot{W}_d(t) = \int_0^t\int_0^t\left[\int_{-\infty}^{\infty}\tau^{-1}L(\tau)\exp[-(2t-u-v)/\tau]d\ln\tau + \{\phi_f\}\right]\dot{\sigma}(u)\dot{\sigma}(v)du\,dv \tag{12}$$

and Eq. (2) results using

$$\dot{J}(t) = \int_{-\infty}^{\infty}\tau^{-1}L(\tau)\exp(-t/\tau)d\ln\tau + \{\phi_f\} \tag{13}$$

which follows from Eq. (4.1-28).

Problem 9.3-1

From Eqs. (1) we have

$$\theta(\omega) = \arcsin[\sigma(t)/\sigma_0(\omega)] - \arcsin[\varepsilon(t)/\varepsilon_0]\ . \tag{2}$$

But

$$\arcsin a \pm \arcsin b = \arcsin\left(a\sqrt{1-b^2} \pm b\sqrt{1-a^2}\right)\ . \tag{3}$$

Hence

$$\sin^2\theta(\omega) = \frac{\sigma^2(t)}{\sigma_0^2(\omega)} + \frac{\varepsilon^2(t)}{\varepsilon_0^2} - \frac{2\,\sigma(t)\varepsilon(t)}{\sigma_0(\omega)\varepsilon_0}X \tag{4}$$

where

$$X = \frac{\sigma(t)\varepsilon(t)}{\sigma_0(\omega)\varepsilon_0} + \left(1 - \frac{\varepsilon^2(t)}{\varepsilon_0^2}\right)^{1/2}\left(1 - \frac{\sigma^2(t)}{\sigma_0^2(\omega)}\right)^{1/2}. \tag{5}$$

Substituting from Eqs. (1) we find

$$X = \sin\omega t \sin[\omega t + \theta(\omega)] + \cos\omega t \cos[\omega t + \theta(\omega)] = \cos\theta(\omega) \tag{6}$$

and, therefore, Eq. (9.3-8) follows.

Chapter 10

Problem 10.0-1

The spectrum is represented by

$$H(\tau) = \sum_k G_k \tau_k \delta(\tau - \tau_k) , \qquad k = 0, 1, \ldots, 6 \tag{1}$$

[cf. Eq. (4.5-21)]. The G_k are listed in the solution section of Problem 8.1-1. The τ_k are given by $\tau_k{}^{-(8-k)}$. The arrows representing the location, $\log \tau_k$, and strength, G_k, of the spectral lines are displayed in Fig. P10.0-1.

The straight line segments connecting the end points of the arrows form an envelope which is a crude approximation to a bell-shaped curve.

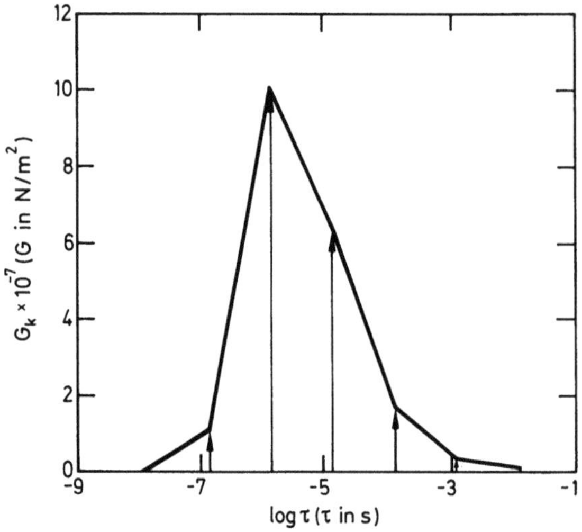

Fig. P10.0-1. Line spectrum and envelope

Problem 10.1-1

We split the sum in Eq. (3.5-11) into two parts. This gives

$$\eta(t) = \{G_e\}t + \eta_{\{f\}} - \sum_{m=1}^{M-1} \eta_m \exp(-t/\tau_m) - \sum_{n=M}^{R} \eta_n \exp(-t/\tau_n) \tag{1}$$

where the viscosities must obey the relation

$$\eta_{\{f\}} = \sum_{m=1}^{M-1} \eta_m + \sum_{n=M}^{R} \eta_n \tag{2}$$

in accordance with Eq. (10.1-12).

Problem 10.1-2

The necessary relations may be found in Sect. 5.2.2.3. We add a second sum in Eq. (5.2-75) and obtain

$$\bar{U}_{GM}(s) = J + \sum_{m=1}^{m=M-1} \frac{J_m}{1 + \tau_m s} + \sum_{n=1}^{n=N-1} \frac{J_n}{1 + \tau_n s} \tag{1}$$

where

$$J_m = \frac{J}{2M-1} \cot^2 \frac{\pi(2m-1)}{2(2M-1)}, \qquad m = 1, 2, \ldots, M-1 \tag{2}$$

and

$$\tau_m = \frac{\theta_m}{4} \csc^2 \frac{\pi(2m-1)}{2(2M-1)}, \qquad m = 1, 2, \ldots, M-1 \tag{3}$$

for the first sum, and

$$J_n = \frac{J}{2N-1} \cot^2 \frac{\pi(2n-1)}{2(2N-1)}, \qquad n = 1, 2, \ldots, N-1 \tag{2}$$

with

$$\tau_n = \frac{\theta_n}{4} \csc^2 \frac{\pi(2n-1)}{2(2N-1)}, \qquad n = 1, 2, \ldots, N-1 \tag{3}$$

for the second sum. Note that the second summation must begin with $n = 1$, not $n = M$. M and N are obtained from the relations $J_N = J_g M$ and $J_e = J_g(M + N - 1)$. Here, selecting the nearest integer, M becomes 25, and N then is 976.

Problem 10.1-3

Substituting $j\omega$ for s in Eq. (1) of Problem 10.1-2 and separating the real and imaginary parts gives

$$J'(\omega) = J_g + \sum_{m=1}^{m=M-1} \frac{J_m}{1 + \omega^2\tau_m^2} + \sum_{n=1}^{n=N-1} \frac{J_n}{1 + \omega^2\tau_n^2} \tag{1}$$

$$J''(\omega) = \sum_{m=1}^{m=M-1} \frac{J_m\omega\tau_m}{1 + \omega^2\tau_m^2} + \sum_{n=1}^{n=N-1} \frac{J_n\omega\tau_n}{1 + \omega^2\tau_n^2} \tag{2}$$

where the J_m, J_n, τ_m, and τ_n are identical with those obtained in Problem 10.1-2.

Problem 10.1-4

Equation (6.1-60) becomes

$$\eta_c(t) = \frac{\eta_{f1}}{1 + (t_1/t)^{c_1}} + \frac{\eta_{f2}}{1 + (t_2/t)^{c_2}} \tag{3}$$

where $\eta_f = \eta_{f1} + \eta_{f2}$.

Problem 10.1-5

The bimodal form of Eq. (6.1-67) is

$$G_R''(\omega) = \frac{2G''_{max,1}}{(\omega_{max,1}/\omega)^{m_1} + (\omega/\omega_{max,1})^{m_1}} + \frac{2G''_{max,2}}{(\omega_{max,2}/\omega)^{m_2} + (\omega/\omega_{max,2})^{m_2}} . \tag{1}$$

The height and the broadness of each peak can be varied independently through G''_{max} and m. The location of the second peak is shifted to the right along the abscissa by an increase in the $\omega_{max,2}/\omega_{max,1}$ ratio.

Problem 10.1-6

The data in Table 6.1-1 were obtained on an entangled, uncrosslinked, polymer and thus represent pseudo-arrheodictic behavior. We therefore use Eq. (10.1-39) as the modelling equation. For the first, pseudo-arrheodictic, part of the data the Cole-Cole function had been used as the matching function [cf. (Eq. (6.1-50)]. We now select the same function, Eq. (6.1-22), as the matching function for the second, rheodictic, part of the data also. Thus we have

$$G_{CC} = \frac{G_g - G_N^\circ}{1 + (t/t_1)^{c_1}} + \frac{G_N^\circ}{1 + (t/t_2)^{c_2}} . \tag{1}$$

Since the behavior is pseudo-arrheodictic, $G_e = 0$, and the entanglement plateau modulus, G_N°, takes the place of G_e in Eq. (6.1-50). Equation (1) reduces to the latter (with $G_N^\circ = G_e$) when, in the second part of Eq. (1), t vanishes. For the parameters of Eq. (1) we already have $\log G_g = 9.00$, $\log G_N^\circ = 5.40$ (in N/m^2), $\log t_1 = -8.413$ (in seconds), and $c_1 = 0.680$. It remains to determine $\log t_2$ and c_2.

To obtain these, we plot the second portion of the data in Table 6.1-1 also according to Eq. (6.1-51) which, however, now takes the form

$$\log r(t) = \log[G_N^\circ/G(t) - 1] . \tag{2}$$

Using only the data in the region $0.16 \leq \log t \leq 4.96$, the plot (not shown) yields $\log t_2 = 2.606$, and $c_2 = 0.424$. The fit (left as an exercise) is again quite satisfactory, although another matching function might have furnished an equally good or perhaps an even better fit.

Problem 10.2-1

The necessary relations are found in Sect. 5.2.2.3. From Eq. (5.2-75) we obtain

$$J'_{GM,q}(\omega) = J_g + \sum_{n=1}^{n=M-1} \frac{J_n}{1 + \omega^2 q^2 \tau^2} + \sum_{n=M}^{n=N-1} \frac{J_n}{1 + \omega^2 \tau^2} \tag{1}$$

and

$$J''_{GM,q}(\omega) = \sum_{n=1}^{n=M-1} \frac{J_n\omega\tau}{1+\omega^2q^2\tau^2} + \sum_{n=M}^{n=N-1} \frac{J_n\omega\tau}{1+\omega^2\tau^2} \tag{2}$$

where the τ_n and J_n are given by Eqs. (5.2-73) and (5.2-74) for arrheodictic behavior.

For rheodictic behavior, Eq. (1) remains the same, but the flow term must now be added to Eq. (2). We have

$$J''_{GM,q}(\omega) = \sum_{n=1}^{n=M-1} \frac{J_n\omega\tau}{1+\omega^2q^2\tau^2} + \sum_{n=M}^{n=N-1} \frac{J_n\omega\tau}{1+\omega^2\tau^2} + \phi_f/\omega \ . \tag{3}$$

In addition, the retardation times and compliances must now be obtained from Eqs. (5.2-79) and (5.2-80). The discussion surrounding Eq. (10.2-3) remains relevant.

Problem 10.2-2

We apply Eq. (10.2-20) to Eq. (10.2-13). Since $\rho(0) = q^{1/2}$, $\tanh(0) = 0$, and

$$\lim_{s\to 0} U_{C_1}(s) = (J_T^\circ/\eta_T^\circ)^{1/2} \lim_{s\to 0} 1/s^{1/2} \tag{1}$$

we immediately obtain

$$\phi_f = (J_T^\circ/\eta_T^\circ)^{1/2} \lim_{s\to 0} \frac{s^{1/2}}{\tanh[L_1(\Gamma_1(s)] + q^{1/2}\tanh[(L-L_1)\Gamma_1(s)]} \ . \tag{2}$$

We now apply L'Hospital's theorem. But

$$ds^{1/2}/ds = 1/2s^{1/2} \ , \tag{3}$$

$$\lim_{s\to 0} \frac{d\tanh[L_1\Gamma_1(s)]}{ds} = \lim_{s\to 0} \frac{L_1(J_T^\circ\eta_T^\circ)^{1/2}}{2s^{1/2}} \tag{4}$$

and

$$\lim_{s\to 0} \frac{d\tanh[(L-L_1)\Gamma_2(s)]}{ds} = \lim_{s\to 0} \frac{(L-L_1)(J_T^\circ q\eta_T^\circ)^{1/2}}{2s^{1/2}} \ . \tag{5}$$

Substitution into Eq. (2) then yields Eq. (10.2-21).

Problem 10.2-3

The steady-state compliance for the composite Marvin-Oser model is obtained [see Eq. (4.6-6)] from

$$J_e^\circ = \lim_{s\to 0} [\bar{U}(s) - \phi_f/s] \tag{1}$$

in which $\bar{U}(s)$ is given by Eq. (10.2-12), and ϕ_f by Eq. (10.2-21). To simplify the notation in what follows, we introduce the abbreviations $\delta = \delta^\circ, \theta = \theta_\infty^\circ, J = J_T^\circ$, and

$$\psi(s) = Q_{c_1}(s) = \frac{1}{J}\sqrt{\frac{\theta s}{1+\delta s}} \ . \tag{2}$$

With these, Eq. (10.2-12) becomes

$$\bar{U}(s) = \frac{1}{\psi(s)} \frac{\dfrac{1}{\psi(qs)} + \dfrac{1}{\psi(s)} \tanh[L_1 J\, \psi(s)] \tanh[(L - L_1) J\, \psi(qs)]}{\dfrac{1}{\psi(qs)} \tanh[L_1 J\, \psi(s)] + \dfrac{1}{\psi(s)} \tanh[(L - L_1) J\, \psi(qs)]} \tag{3}$$

or

$$\bar{U}(s) = \frac{1 + \dfrac{\psi(qs)}{\psi(s)} \tanh[L_1 J\, \psi(s)] \tanh[(L - L_1) J\, \psi(qs)]}{\psi(s) \tanh[L_1 J\, \psi(s)] + \psi(qs) \tanh[(L - L_1) J\, \psi(qs)]} = \frac{A(s)}{B(s)} . \tag{4}$$

We now develop A(s) and B(s) in power series in s in the neighborhood of $s = 0$. We obtain

$$\bar{U}(s) = \frac{1 + a_1 s + a_2 s^2 + \dots}{b_1 s + b_2 s^2 + \dots} = \frac{1}{b_1 s} \frac{1 + a_1 s + a_2 s^2 + \dots}{1 + (b_2/b_1)s + \dots} . \tag{5}$$

The pole at the origin, arising from $1/b_1 s$, will eventually cancel against the flow term, $1/\phi_f s$.

Let us now recast Eq. (5) in the form

$$\begin{aligned} \bar{U}(s) = \frac{1}{b_1 s}&(1 + a_1 s + a_2 s^2 + \dots)[1 - (b_2 s/b_1 + b_3 s^2/b_1 + \dots) \\ &+ (b_2 s/b_1 + b_3 s^2/b - \dots) \dots] \end{aligned} \tag{6}$$

or

$$\begin{aligned} \bar{U}(s) &= \frac{1}{b_1 s}(1 + a_1 s + a_2 s^2 + \dots)(1 - b_2 s/b_1 + \dots) \\ &= \frac{1}{b_1 s}[1 + (a_1 - b_2/b_1)s + \dots] \end{aligned} \tag{7}$$

or

$$\bar{U}(s) = \frac{1}{b_1 s} + \left(a_1 - \frac{b_2}{b_1}\right)\frac{1}{b_1} + c_1 s + c_2 s^2 + \dots \tag{8}$$

where the c's are coefficients in powers of s which will not be needed.

It then follows that

$$\lim_{s \to 0} s\bar{U}(s) = \frac{1}{b_1} \tag{9}$$

and

$$\lim_{s \to 0}\left(\bar{U}(s) - \frac{1}{b_1 s}\right) = \left(a_1 - \frac{b_2}{b_1}\right)\frac{1}{b_1} = \frac{a_1 b_1 - b_2}{b_1^2} . \tag{10}$$

We proceed to assign meaning to the coefficients in the power expansions of A(s) and B(s). To this end we make use of the power expansion of tanh x at $s = 0$. We need the first two terms only. Thus,

$$\tanh x = x - x^3/3 + \ldots \quad (11)$$

For A(s), Eq. (4) gives

$$A(s) = 1 + \frac{\psi(qs)}{\psi(s)}\left[\left(L_1 J\, \psi(s) - \frac{L_1^3 J^3 \psi^3(s)}{3} + \ldots\right)\right.$$
$$\left.\times\left((L - L_1)J\, \psi(qs) - \frac{(L - L_1)^3 J^3 \psi^3(qs)}{3} + \ldots\right)\right] . \quad (12)$$

After some further manipulation we obtain

$$A(s) = 1 + L_1(L - L_1)J^2\psi^2(qs) - \frac{L_1(L - L_1)^3 J^4 \psi^4(qs)}{3}$$
$$- \frac{L_1^3(L - L_1)J^4\psi^2(s)\psi^2(qs)}{3} + \ldots .$$

The functions $\psi(s)$ and $\psi(qs)$ occur in even powers only. By Eq. (2)

$$\psi^2(s) = \frac{1}{J^2}\frac{\theta s}{1 + \delta s} = \frac{\theta}{J^2}(s - \delta s^2 + \delta^2 s^3 - \ldots) \quad (14)$$

and

$$\psi^2(qs) = \frac{\theta}{J^2} q(s - \delta q s^2 + \delta^2 q s^3 - \ldots) . \quad (15)$$

No more than the three listed terms will be needed in each series. We now have

$$A(s) = 1 + L_1(L - L_1)\theta q s - \ldots \quad (16)$$

and find

$$a_1 = L_1(L - L_1)\theta q . \quad (17)$$

Proceeding analogously with B(s) yields, after some algebra,

$$B(s) = \frac{\theta}{J}[L_1 + (L - L_1)q]s$$
$$- \left[\frac{\theta\delta}{J}[L_1 + (L - L_1)q^2] + \frac{\theta^2}{3J}[L_1^3 + (L - L_1)^3 q^2]\right]s^2 + \ldots \quad (18)$$

so that

$$b_1 = \frac{\theta}{J}[L_1 + (L - L_1)q] \quad (19)$$

and

$$b_2 = \frac{\theta\delta}{J}[L_1 + (L - L_1)q^2] + \frac{\theta^2}{3J}[L_1^3 + (L - L_1)^3 q^2] . \quad (20)$$

Inserting these values into Eq. (9) gives Eq. (10.2-21) while inserting them into Eq. (10) eventually leads to Eq. (10.2-23).

Problem 10.2-4

Using Eqs. (6.1-24) and (6.1-42) gives

$$G_B(t) = \{G_e\} + \frac{G_g - \{G_e\}}{(1 + t/t_0)^b} \tag{1}$$

for the unimodal modulus. By analogy to Eq. (10.2-6) we then find

$$G_{B,b}(t) = G_B(t)[h((t) - h(t - t_q)] + G_B(t_q)[h(t - t_q) - h(t - qt_q)]$$
$$+ G_B(t/q)h(t - qt_q) \tag{2}$$

where t_q is the time at which the prolongation begins, and q is the prolongation factor.

Chapter 11

Problem 11.1-1

We recast Eq. (1) as

$$\bar{E}(s) = 2\,\bar{G}(s) + 2\,s\bar{\mu}(s)\bar{G}(s)\ . \tag{2}$$

In the real time domain the second term on the right hand side of Eq. (2) becomes a convolution integral which takes on four equivalent forms [cf. Eq. (A3-17)]. Letting $\mu(t)$ and $G(t)$ be $f_1(t)$ and $f_2(t)$, respectively, the four forms become

$$E(t) = \begin{cases} 2(1+\mu_g)G(t) + 2\int_0^t \frac{d\,\mu(u)}{du} G(t-u)du \\ 2(1+\mu_g)G(t) - 2\int_0^t \frac{d\,\mu(t-u)}{du} G(u)du \\ 2\,G(t) + 2\,G_g\mu(t) + 2\int_0^t \mu(t-u)\frac{d\,G(u)}{du}du \\ 2\,G(t) + 2\,G_g\mu(t) - 2\int_0^t \mu(u)\frac{d\,G(t-u)}{du}du\ . \end{cases} \tag{3}$$

Problem 11.3-1

(A) Applying the correspondence principle to Eq. (7) of Problem 1.4-1 yields

$$\bar{H}(s) = \frac{18\,\bar{K}(s)\bar{G}(s)}{3\,\bar{K}(s) + 4\,\bar{G}(s)} \tag{1}$$

after cancelling s.

(B) Dividing Eq. (11.1-8) by Eq. (1) above leads to

$$\bar{E}(s) = \frac{3\,\bar{K}(s) + 4\,\bar{G}(s)}{6\,\bar{K}(s) + 2\,\bar{G}(s)}\bar{H}(s)\ . \tag{2}$$

(C) In an incompressible material $K(t) \gg G(t)$. If this approximation is acceptable, then it follows immediately that

$$H(t) = 2\,E(t) = 6\,G(t) \tag{3}$$

the last relation following by Eq. (11.4-16).

Problem 11.3-2

We start from Eq. (11.1-8), i.e. from

$$\bar{E}(s) = \frac{9\,\bar{K}(s)\bar{G}(s)}{3\,\bar{K}(s) + \bar{G}(s)} \tag{1}$$

and substitute

$$\bar{G}(s) = \frac{G_e + G_g\tau_G}{s(1 + \tau_G s)} \tag{2}$$

and

$$\bar{K}(s) = \frac{K_e + K_g\tau_K}{s(1 + \tau_K s)} \; . \tag{3}$$

After some algebra, this leads to

$$\bar{E}(s) = E_g\frac{s^2 + m_1 s + m_0^2}{s(s^2 + \mu_1 s + \mu_0^2)} = E_g\frac{s^2 + m_1 s + m_0^2}{s[(s+\alpha)^2 - \beta^2]} \tag{4}$$

where

$$E_g = \frac{9K_gG_g}{3K_g + G_g} \tag{5}$$

$$m_1 = \frac{G_e}{G_g\tau_G} + \frac{K_e}{K_g\tau_K} \tag{6}$$

$$m_0^2 = \frac{G_e}{G_g\tau_G}\frac{K_e}{K_g\tau_K} \tag{7}$$

$$\mu_1 = \frac{3K_g + G_e}{(3K_g + G_g)\tau_G} + \frac{3K_e + G_g}{(3K_g + G_g)\tau_K} \tag{8}$$

$$\mu_0^2 = \frac{3K_e + G_e}{(3K_g + G_g)\tau_G\tau_K} \tag{9}$$

$$\alpha = \mu_1/2 \tag{10}$$

and

$$\beta^2 = \alpha^2 - \mu_0^2 = \mu_1^2/4 - \mu_0^2 \; . \tag{11}$$

Now,

$$\frac{s^2 + m_1 s + m_0^2}{s[(s+\alpha)^2 - \beta^2]} = \frac{s^2 + m_1 s + m_0^2}{s(s+\alpha+\beta)(s+\alpha-\beta)} \; . \tag{12}$$

After partial fraction decomposition and a little rearrangement this yields

$$\frac{s^2 + m_1 s + m_0^2}{s(s+\alpha+\beta)(s+\alpha-\beta)}$$

$$= \left(\frac{R}{s} + \frac{\beta + \alpha - m_1 - (\beta - \alpha)R}{2\beta(s+\alpha+\beta)} + \frac{\beta - \alpha + m_1 - (\beta + \alpha)R}{2\beta(s+\alpha-\beta)}\right) \tag{13}$$

where $R = m_0^2/\mu_0^2$. Since $E_g R = E_e$, we have

$$\bar{E}(s) = \frac{E_e}{s} + \frac{E_1}{s+\alpha+\beta} + \frac{E_2}{s+\alpha-\beta} \tag{14}$$

where

$$E_1 = E_g \frac{\beta + \alpha - m_1 - (\beta - \alpha)R}{2\beta} \tag{15}$$

and

$$E_2 = E_g \frac{\beta - \alpha + m_1 - (\beta + \alpha)R}{2\beta} . \tag{16}$$

In the real time domain then

$$E(t) = E_e + E_1 \exp(-t/\tau_1) + E_2 \exp(-t/\tau_2) \tag{17}$$

where

$$\tau_1 = \frac{1}{\alpha+\beta} \quad \text{and} \quad \tau_2 = \frac{1}{\alpha-\beta} . \tag{18}$$

Problem 11.3-3

(A) Substituting $\gamma(t) = \dot{\gamma}_0 t$ into the Boltzmann superposition integral

$$\sigma(t) = \int_0^t E(u) \frac{d\,\gamma(t-u)}{du} du \tag{2}$$

and using Eq. $(1)_3$ gives

$$F(t) = \frac{1}{t} \int_0^t E(u) du . \tag{3}$$

Multiplying by t and differentiating yields

$$E(t) = \frac{d\,tF(t)}{dt} = F(t) + t\frac{d\,F(t)}{dt} = F(t)\left[1 + \frac{d \log F(t)}{d \log t}\right] . \tag{4}$$

(B) We have

$$E(t) = \{E_e\} + (E_g - \{E_e\})\exp(-t/\tau) \tag{5}$$

and obtain

$$F(t) = \{E_e\} + \frac{E_g - \{E_e\}}{t/\tau}[1 - \exp(-t/\tau)] \tag{6}$$

using Eq. (3). F(t) and E(t) calculated using the values given in the problem statement are plotted in Fig. P11.3-3.

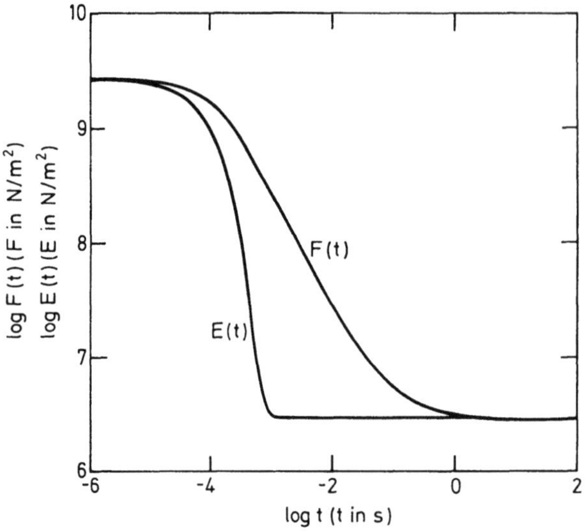

Fig. P11.3-3. Comparison of F(t) and E(t) calculated from the standard linear solid model

Problem 11.3-4

To obtain the canonical representation of F(t), we substitute

$$E(t) = \{E_e\} + \int_{-\infty}^{\infty} H_E(\tau)\exp(-t/\tau)d\ln\tau \tag{1}$$

into Eq. (3) of Problem 11.3-3. This gives

$$F(t) = \{E_e\} + \int_{-\infty}^{\infty} H_E(\tau)\int_0^t \exp(-u/t)du\ln\tau \tag{2}$$

which becomes

$$F(t) = \{E_e\} + \frac{1}{t}\int_{-\infty}^{\infty} \tau H_E(\tau)[1 - \exp(-t/\tau)]d\ln\tau\ . \tag{3}$$

Using LTP (16) then leads to

$$\bar{F}(s) = \{E_e/s\} + \int_{-\infty}^{\infty} \tau H_E(\tau)\ln(1 + 1/\tau s)d\ln\tau\ . \tag{4}$$

Problem 11.3-5

Differentiation of Eq. (4) of Problem 11.3-4 yields

$$\frac{d\bar{F}(s)}{ds} = -\frac{\{E_e\}}{s^2} - \int_{-\infty}^{\infty} H_E(\tau)\frac{\tau}{s(1 + \tau s)}d\ln\tau\ . \tag{2}$$

Inserting into Eq. (1) gives

$$E^*(\omega) = \{E_e\} + \int_{-\infty}^{\infty} H_E(\tau)\frac{j\omega\tau}{1 + j\omega\tau} d \ln \tau \tag{3}$$

which is the correct result.

Problem 11.3-6

The relaxance matrix is defined by the equation

$$[\bar{\sigma}_l(s)] = [\bar{\Gamma}_{kl}(s)][\bar{\gamma}_l(s)] \tag{1}$$

where $[\bar{\sigma}_k(s)]$ and $[\bar{\gamma}_l(s)]$ are 6-component column vectors, containing the components of the transforms of the stress and the strain, respectively, and the 6×6 relaxance matrix is

$$[\bar{\Gamma}_{kl}(s)] = \begin{bmatrix} \bar{\Gamma}_{11}(\tau) & \bar{\Gamma}_{12}(s) & \bar{\Gamma}_{12}(s) & 0 & 0 & 0 \\ \bar{\Gamma}_{12}(s) & \bar{\Gamma}_{11}(s) & \bar{\Gamma}_{12}(s) & 0 & 0 & 0 \\ \bar{\Gamma}_{12}(s) & \bar{\Gamma}_{12}(s) & \bar{\Gamma}_{11}(s) & 0 & 0 & 0 \\ 0 & 0 & 0 & \bar{\Gamma}_{11}(s) - \bar{\Gamma}_{12}(s) & 0 & 0 \\ 0 & 0 & 0 & 0 & \bar{\Gamma}_{11}(s) - \bar{\Gamma}_{12}(s) & 0 \\ 0 & 0 & 0 & 0 & 0 & \bar{\Gamma}_{11}(s) - \bar{\Gamma}_{12}(s) \end{bmatrix} . \tag{2}$$

The matrix is obtained simply by applying the correspondence principle to Eq. (1.4-50) making use of Eq. $(1.4\text{-}49)_2$. Its components contain only two independent quantities: $\bar{\Gamma}_{11}(s)$ and $\bar{\Gamma}_{12}(s)$.

Now, by Eq. (11.3-3),

$$\bar{\sigma}_{12}(s) = 2\,\bar{Q}(s)\bar{\gamma}_{12}(s) . \tag{3}$$

But $\bar{\sigma}_{12}(s) = \bar{\sigma}_4(s)$, $\bar{\gamma}_{12}(s) = \bar{\gamma}_4(s)$ [cf. Eq. (1.4-50)], and all other components of the stress and strain column vectors vanish in simple shear or torsion. Hence,

$$\bar{\sigma}_4(s) = [\bar{\Gamma}_{11}(s) - \bar{\Gamma}(s)]\bar{\gamma}_4(s) \tag{4}$$

and it follows that

$$\bar{Q}(s) = \tfrac{1}{2}[\bar{\Gamma}_{11}(s) - \bar{\Gamma}_{12}(s)] . \tag{5}$$

By Eq. (11.3-13),

$$\bar{\sigma}_{11}(s) = 3\,\bar{P}(s)\bar{\gamma}_{11}(s) . \tag{6}$$

But now $\bar{\sigma}_{11}(s) = \bar{\sigma}_{22}(s) = \bar{\sigma}_{33}(s) = \bar{\sigma}_1(s)$, $\bar{\gamma}_{11}(s) = \bar{\gamma}_{22}(s) = \bar{\gamma}_{33}(s) = \bar{\gamma}_1(s)$, and all other components of the column vectors vanish in isotropic compression. Hence,

$$\bar{\sigma}_1(s) = [\bar{\Gamma}_{11}(s) + 2\,\bar{\Gamma}_{12}(s)]\bar{\gamma}_1(s) \tag{7}$$

and we have

$$\bar{P}(s) = \tfrac{1}{3}[\bar{\Gamma}_{11}(s) + 2\,\bar{\Gamma}_{12}(s)] . \tag{8}$$

Problem 11.3-7

We merely need to substitute the expressions for $\bar{P}(s)$ and $\bar{Q}(s)$ established in Problem 11.3-6 into the expressions for $\bar{V}(s)$, $\bar{Y}(s)$, and $\bar{v}(s)$ which we obtain from the equations to be cited below, changing, if necessary, the Carson to the Laplace transforms. For $\bar{V}(s)$ we have, by Eq. $(11\text{-}3\text{-}49)_2$,

$$\bar{V}(s) = \bar{P}(s) + \tfrac{4}{3}\bar{Q}(s) = \bar{\Gamma}_{11}(s) \ . \tag{1}$$

For $\bar{Y}(s)$, Eq. (11.1-8) leads to

$$\bar{Y}(s) = \frac{9\ \bar{P}(s)\bar{Q}(s)}{3\ \bar{P}(s) + \bar{Q}(s)} = \bar{\Gamma}_{11}(s) - \frac{2\ \bar{\Gamma}_{12}^2(s)}{\bar{\Gamma}_{11}(s) + \bar{\Gamma}_{12}(s)} \ . \tag{2}$$

Finally, for $\bar{\nu}(s)$, Eq. (11.3-74) gives

$$\bar{\nu}(s) = \frac{3\ \bar{P}(s) - 2\ \bar{Q}(s)}{6\ \bar{P}(s) + 2\ \bar{Q}(s)} = \frac{\bar{\Gamma}_{12}(s)}{\bar{\Gamma}_{11}(s) + \bar{\Gamma}_{12}(s)} \ . \tag{3}$$

Problem 11.3-8

For an axisymmetric linear viscoelastic body the relaxance matrix, $[\bar{\Gamma}_{kl}(s)]$, (see Problem 11.3-6) takes the form

$$[\bar{\Gamma}_{kl}(s)] = \begin{bmatrix} \bar{\Gamma}_{11}(s) & \bar{\Gamma}_{12}(s) & \bar{\Gamma}_{13}(s) & 0 & 0 & 0 \\ \bar{\Gamma}_{12}(s) & \bar{\Gamma}_{11}(s) & \bar{\Gamma}_{13}(s) & 0 & 0 & 0 \\ \bar{\Gamma}_{13}(s) & \bar{\Gamma}_{13}(s) & \bar{\Gamma}_{33}(s) & 0 & 0 & 0 \\ 0 & 0 & 0 & \bar{\Gamma}_{11}(s) - \bar{\Gamma}_{12}(s) & 0 & 0 \\ 0 & 0 & 0 & 0 & \bar{\Gamma}_{55}(s) & 0 \\ 0 & 0 & 0 & 0 & 0 & \bar{\Gamma}_{55}(s) \end{bmatrix} \tag{6}$$

when the symmetry axis is the 3-direction [cf. Eqs. (1.4-31) and (1.4-35)].

We should now proceed to find $\bar{E}_a(s)$, $\bar{E}_t(s)$, $\bar{G}_t(s)$, $\bar{\mu}_a(s)$, and $\bar{\mu}_t(s)$. Fortunately, most of the work has already been done in Problems 1.4-10 to 1.4-12. E_3 and E_1 of Problem 1.4-12 correspond to $s\bar{E}_a(s)$ and $s\bar{E}_t(s)$, and μ_{13} and μ_{21} of Problem 1.4-11 correspond to $s\bar{\mu}_a(s)$ and $s\bar{\mu}_t(s)$, respectively. Applying the correspondence principle to the appropriate equations and cancelling the transform variable, s, wherever possible, leads to

$$\bar{E}_a(s) = \bar{C}_{33}(s) - \frac{2\ \bar{C}_{12}^2(s)}{\bar{C}_{11}(s) + \bar{C}_{12}(s)} \tag{7}$$

$$\bar{E}_t(s) = [\bar{C}_{11}(s) - \bar{C}_{12}(s)]\frac{[\bar{C}_{11}(s) + \bar{C}_{12}(s)]\bar{C}_{33}(s) - 2\ \bar{C}_{13}^2(s)}{\bar{C}_{11}(s)\bar{C}_{33}(s) - \bar{C}_{13}^2(s)} \tag{8}$$

$$s\bar{\mu}_a(s) = \frac{\bar{C}_{13}(s)}{\bar{C}_{11}(s) + \bar{C}_{12}(s)} \tag{9}$$

and

$$s\bar{\mu}_t(s) = \frac{\bar{C}_{12}(s)\bar{C}_{33}(s) - \bar{C}_{13}^2(s)}{\bar{C}_{11}(s)\bar{C}_{33}(s) - \bar{C}_{13}^2(s)} \ . \tag{10}$$

Finally, since $\bar{\sigma}_{23}(s) = \bar{\sigma}_5(s)$ and $\bar{\gamma}_{23}(s) = \bar{\gamma}_5(s)$ [cf. Eq. (1.4-6)], we have

$$\bar{G}_t(s) = \tfrac{1}{2}\bar{C}_{55}(s) \ . \tag{11}$$

It remains to show that Eqs. (7) to (11) reduce properly to the equations describing an isotropic body. Comparing Eq. (7) with Eq. (2) of Problem 11.3-6 reveals that, to obtain the relaxance matrix for an isotropic body, we must replace $\bar{\Gamma}_{13}(s)$ by $\bar{\Gamma}_{12}(s)$, $\bar{\Gamma}_{33}(s)$ by $\bar{\Gamma}_{11}(s)$, and $\bar{\Gamma}_{55}(s)$ by $\bar{\Gamma}_{44}(s) = \bar{\Gamma}_{11}(s) - \bar{\Gamma}_{12}(s)$. Carrying out these substitutions using $\bar{C}_{kl}(s) = \bar{\Gamma}_{kl}(s)/s$, Eqs. (7) and (8) reduce to Eqs. (2), and Eqs. (9) and (10) reduce to Eq. (3) of Problem 11.3-6, as required. Equation (11), of course, also reduces correctly. In Problem 11.3-6 the equations represent the transforms of the responses to impulse excitations. In Problem 11.3-7, on the other hand, they are couched in terms of the transforms of the responses to excitations that are step functions of the time.

Problem 11.3-9

By Eq. $(11.3\text{-}80)_3$

$$\bar{\mu}(s) = \frac{\mu_e}{s} - \frac{\mu\tau}{1 + \tau s} \ . \tag{1}$$

By Eq. (11.3-82) $\mu = \mu_e - \mu_g = 1/6$. Hence, using Eq. (11.3-85)

$$\mu(t) = \tfrac{1}{2} - \tfrac{1}{6}\exp(-t/\tau) \ . \tag{2}$$

By Eqs. (11.3-87) and (11.3-89)

$$\mu'(\omega) = \frac{3 + 2\omega^2\tau^2}{6(1 + \omega^2\tau^2)} \tag{3}$$

$$\mu''(\omega) = \frac{\omega\tau}{6(1 + \omega^2\tau^2)} \tag{4}$$

and

$$\theta_\mu(\omega) = \frac{\omega\tau}{3 + 2\omega^2\tau^2} \tag{5}$$

follows.

Problem 11.3-10

The demonstration follows closely on the proof that the stress *leads* the strain in the same experiment (see Sect. 2.5). From the definition of Poisson's ratio it ensues that

$$\mu^*(\omega) = -\gamma_2(\omega)/\gamma_1(\omega) \ . \tag{2}$$

Rewriting Eq. (11.3-86) in polar coordinates gives

$$\mu^*(\omega) = \tilde{\mu}(\omega)\exp[-j\,\theta_\mu(\omega)] \tag{3}$$

where

$$\tilde{\mu}(\omega) = \{[\mu'(\omega)]^2 + [\mu''(\omega)]^2\}^{1/2} \tag{4}$$

is the *absolute* Poisson's ratio, and

$$\tan \theta_\mu(\omega) = \mu''(\omega)/\mu'(\omega) \tag{5}$$

is the phase angle between the lateral contraction and the strain excitation. Inserting Eq. (3) into Eq. (2) yields

$$-\gamma_2(\omega) = \gamma_1(\omega)\tilde{\mu}(\omega)\exp\{j[\omega t - \theta_\mu(\omega)]\} \ . \tag{6}$$

Introducing the peak amplitude of the lateral contraction from

$$\tilde{\mu}(\omega) = -\gamma_{2,0}(\omega)/\gamma_{1,0} \tag{7}$$

we obtain

$$-\gamma_2(\omega) = -\gamma_{2,0}(\omega)\exp\{j[\omega t - \theta_\mu(\omega)]\} \ . \tag{8}$$

Equation (8) shows that the lateral contraction reaches its peak amplitude later than the imposed strain reaches *its* peak. Hence, $-\gamma_2(\omega)$ *lags* $\gamma_1(\omega)$. If Poisson's ratio had been defined as a material property with the nature of a modulus rather than a compliance, the complex ratio would have taken the form

$$\mu^*(\omega) = \mu'(\omega) + j\,\mu''(\omega) \tag{9}$$

and $-\gamma_2(\omega)$ would then have *led* $\gamma_1(\omega)$ just as the stress leads it.

Problem 11.3-11

We start from Eq. $(11.3\text{-}75)_2$ and obtain

$$\gamma_1(0)\mu(t) - \int_0^\infty \mu(u)\frac{d\,\gamma_1(t-u)}{du}\,du = -\gamma_2(t) \tag{1}$$

through retransformation to the real time axis using Eq. $(A3\text{-}17)_4$. The time-dependent Poisson's ratio can then be recovered by an adaptation of the technique described in Sect. 11.4.1.2. This results in the recurrence formula

$$\mu(t_n) = \frac{-2\,\gamma_2(t_n) + \mu(t_{n-1})[\gamma_1(0) - \gamma_1(t_n - t_{n-1})] + \sum_{i=1}^{i=n-1}[\mu(t_i) + \mu(t_{i-1})][\gamma_1(t_n - t_i) - \gamma_1(t_n - t_{i-1})]}{\gamma_1(0) + \gamma_1(t_n - t_{n-1})} \tag{2}$$

with

$$\mu(t_1) = -\frac{2\,\gamma_2(t_1) + \mu_g[\gamma_1(t_1) - \gamma_1(0)]}{\gamma_1(t_1) + \gamma_1(0)} \ . \tag{3}$$

Problem 11.3-12

The transforms of the stress and the strain tensors in uniaxial extension are the same whether the excitation is a strain or a stress. We derived Eq. $(11.3\text{-}74)_2$, i.e.

$$-\bar{\gamma}_{22}(s) = \frac{3\,\bar{K}(s) - 2\,\bar{G}(s)}{6\,\bar{K}(s) + 2\,\bar{G}(s)}\bar{\gamma}_{11}(s) \tag{1}$$

by substituting Eqs. (11.3-56) and (11.3-57) into the strain excitation (or modulus) form of the generalized Hooke's law, Eq. (11.2-7). If we introduce the same equations into the stress excitation (or compliance) form, Eq. (11.2-8), we obtain

$$\bar{\gamma}_{11}(s) = [\tfrac{1}{9}\,\bar{B}(s) + \tfrac{1}{3}\,\bar{J}(s)]s\bar{\sigma}_{11}(s) \tag{2}$$

and

$$\bar{\gamma}_{22}(s) = [\tfrac{1}{9}\,\bar{B}(s) - \tfrac{1}{6}\,\bar{J}(s)]s\bar{\sigma}_{11}(s) \ . \tag{3}$$

Elimination of $s\bar{\sigma}_{11}(s)$ between these equations yields

$$-\bar{\gamma}_{22}(s) = \frac{3\,\bar{J}(s) - 2\,\bar{B}(s)}{6\,\bar{J}(s) + 2\,\bar{B}(s)}\bar{\gamma}_{11}(s) \ . \tag{4}$$

But $s\bar{J}(s) = 1/s\bar{G}(s)$ and $s\bar{B}(s) = 1/s\bar{K}(s)$. Substitution then leads directly to Eq. (1), showing that the two derivations produce the same result.

Problem 11.3-13

We use Eq. (11.3-75) in the form

$$\bar{\gamma}_2(s) = -s\bar{\mu}(s)\bar{\gamma}_1(s) \ . \tag{1}$$

But for constant rate of strain

$$\bar{\gamma}_1(s) = \dot{\gamma}_{1,0}/s^2 \tag{2}$$

and, therefore,

$$\bar{\mu}(s) = -s\bar{\gamma}_2(s)/\dot{\gamma}_{1,0} \ . \tag{3}$$

Retransformation gives

$$\mu(t) = -\dot{\gamma}_{1,0}\, d\,\gamma_2(t)/dt \ . \tag{4}$$

Problem 11.4-1

(A) From Tables 1.4-1 and 11.1-1 we obtain the basic transform relation as

$$\bar{G}(s) = \frac{3\,\bar{K}(s)\bar{E}(s)}{9\,\bar{K}(s) - \bar{E}(s)} \tag{1}$$

and the relation between the complex functions as

$$G^*(\omega) = \frac{3\,K^*(\omega)E^*(\omega)}{9\,K^*(\omega) - E^*(\omega)} \ . \tag{2}$$

The real and imaginary parts follow as

$$G'(\omega) = 3\frac{9\,E'(\omega)[\tilde{K}(\omega)]^2 - K'(\omega)[\tilde{E}(\omega)]^2}{[9\,K'(\omega) - E'(\omega)]^2 + [9\,K''(\omega) - E''(\omega)]^2} \tag{3}$$

and

$$G''(\omega) = 3\frac{9\,E''(\omega)[\tilde{K}(\omega)]^2 - K''(\omega)[\tilde{E}(\omega)]^2}{[9\,K'(\omega) - E'(\omega)]^2 + [9\,K''(\omega) - E''(\omega)]^2} \tag{4}$$

where

$$[\tilde{K}(\omega)]^2 = [K'(\omega)]^2 + K''(\omega)]^2 \tag{5}$$

$$[\tilde{E}(\omega)]^2 = [E'(\omega)]^2 + [E''(\omega)]^2 \tag{6}$$

and $\tilde{K}(\omega)$ and $\tilde{E}(\omega)$ are the *absolute* bulk and tensile moduli, respectively.

To obtain the relations between the step responses, we first rewrite Eq. (1) as

$$9\,\bar{G}(s)\bar{K}(s) - \bar{G}(s)\bar{E}(s) = 3\,\bar{K}(s)\bar{E}(s)\ . \tag{7}$$

Retransformation to the real time axis then gives

$$9\int_0^t G(u)K(t-u)du - \int_0^t G(u)E(t-u)du = 3\int_0^t K(u)E(t-u)du\ . \tag{8}$$

(B) Approximation to these exact relations can be obtained by assuming that the bulk modulus is very much larger (effectively infinite) compared with the shear modulus. This leads to the same results which we have already derived in Sect. 11.4.1.3.

Another approach consists in assuming that the time-dependence of the modulus may be neglected. Substituting the constant K/s for $s\bar{K}(s)$ in Eq. (1) gives

$$\bar{G}(s) = \frac{3K\,\bar{E}(s)}{9K - s\bar{E}(s)} \tag{9}$$

and we have

$$G^*(\omega) = \frac{3K\,E^*(s)}{9K - E^*(\omega)} \tag{10}$$

with

$$G'(\omega) = \frac{3K\{E'(\omega)[9K - E'(\omega)] - [E''(\omega)]^2\}}{[9K - E'(\omega)]^2 + [E''(\omega)]^2} \tag{11}$$

and

$$G''(\omega) = \frac{3K\{E''(\omega)[9K - E'(\omega)] - E'(\omega)E''(\omega)\}}{[9K - E'(\omega)]^2 + [E''(\omega)]^2} \tag{12}$$

for the harmonic responses and

$$9K\,G(t) - E_g G(t) + \int_0^t G(u)\frac{d\,E(t-u)}{du}du = 3K\,E(t) \tag{13}$$

for the step responses. A numerical solution of Eq. (13) is the subject of Problem 11.4-4.

Problem 11.4-2

(A) Using Table 1.4-2 and applying the correspondence principle we obtain the basic relation we need as

$$\bar{J}(s) = 2\,\bar{D}(s)[1 + s\bar{\mu}(s)] \ . \tag{1}$$

From this we obtain

$$J(t) = 2(1 + \mu_g)D(t) - \int_0^t D(u)\frac{d\,\mu(t-u)}{du}\,du \tag{2}$$

for the shear creep compliance and

$$J^*(\omega) = 2\,D^*(\omega)[1 + \mu^*(\omega)] \ . \tag{3}$$

for the complex shear compliance.

(B) Equation (2) is identical with Eq. (11.4-6) if we interchange E(t) with D(t), and G(t) with J(t). Hence, J(t) may be calculated from D(t) and $\mu(t)$ using Eqs. (11.4-12) to (11.4-14) with the same interchanges.

(C) $J'(\omega)$ and $J''(\omega)$ are obtained simply by separating the real and imaginary parts of $J^*(\omega)$. This yields

$$J'(\omega) = 2\,D'(\omega)[1 + \mu'(\omega)] - 2\,D''(\omega)\mu''(\omega) \tag{4}$$

and

$$J''(\omega) = 2\,D''(\omega)[1 + \mu'(\omega)] + 2\,D'(\omega)\mu''(\omega) \tag{5}$$

for the shear storage and loss compliances.

(D) If we let Poisson's ratio assume the value 0.5, Eq. (2) reduces to

$$J(t) = 3\,D(t) \tag{6}$$

and Eqs. (4) and (5) become

$$J'(\omega) = 3\,D'(\omega) \qquad \text{and} \qquad J''(\omega) = 3\,D''(\omega) \tag{7}$$

respectively. These equations are the analogs of Eqs. (11.4-16) and (11.4-17) of Sect. 11.4.1.3.

Problem 11.4-3

Using Table 1.4-2 and the correspondence principle we obtain the basic equation as

$$\bar{D}(s) = \frac{\bar{B}(s)}{3[1 - 2\,s\bar{\mu}(s)]} \ . \tag{1}$$

This equation is identical with Eq. (11.4-20) if we replace the moduli by the corresponding compliances. Thus, all the equations derived in Sect. 11.4.4 are applicable immediately to the problem at hand simply by effecting the same interchanges everywhere.

Problem 11.4-4

We simply need to evaluate Eq. (13) of Problem 11.4-1 numerically. We rewrite it as

$$E(t) = (3 - E_g/3K)G(t) + (1/3K)\int_0^t G(u)\frac{d\,E(t-u)}{du}du \tag{1}$$

and proceed as in Sect. 11.4.1.2. The recurrence formula then becomes

$$G(t_n) = \frac{(6K + K_g)E(t_n) - G_g E(t_n - t_1) - \sum_{i=1}^{i=n-1} G(t_i)[E(t_n - t_{i+1}) - E(t_n - t_{i-1})]}{18\,K - E_g - E(t_n - t_{n-1})} \tag{2}$$

with

$$G(t_1) = \frac{(6K + G_g)E(t_1)}{18K - E_g - E(t_1)} \tag{3}$$

and

$$G_g = \frac{3KE_g}{9K - E_g}\ . \tag{3}$$

Problem 11.4-5

Because of the nature of Poisson's ratio as a viscoelastic function with the character of a compliance, the complex ratio becomes [see Eq. (11.3-86)]

$$\mu^*(\omega) = \mu'(\omega) - j\,\mu''(\omega)\ , \tag{2}$$

i.e. the imaginary part is negative. For this to be true it is necessary that

$$\mu''(\omega) \geq 0\ . \tag{3}$$

To show this, we separate

$$\mu^*(\omega) = \frac{3\,K^*(\omega) - 2\,G^*(\omega)}{6\,K^*(\omega) + 2\,G^*(\omega)} \tag{4}$$

into its real and imaginary parts. The latter becomes

$$\mu''(\omega) = \frac{4.5[K'(\omega)G''(\omega) - K''(\omega)G'(\omega)]}{[3\,K'(\omega) + G'(\omega)]^2 + [3\,K''(\omega) + G''(\omega)]^2}\ . \tag{5}$$

For this to be positive we must have

$$K'(\omega)G''(\omega) \geq K''(\omega)G'(\omega) \tag{6}$$

that is

$$\frac{G''(\omega)}{G'(\omega)} \geq \frac{K''(\omega)}{K'(\omega)}\ . \tag{7}$$

This proves Inequality (1).

Problem 11.4-6

(A) We have already shown, in Problem 11.4-5, than $\tan\theta(\omega) \geq \tan\theta_K(\omega)$. We must now show that $\tan\theta(\omega) \geq \tan\theta_E(\omega)$, and that $\tan\theta_E(\omega) \geq \tan\theta_K(\omega)$. To prove the first inequality we use

$$\mu^*(\omega) = \frac{E^*(\omega)}{2\,G^*(\omega)} - 1 \ . \tag{3}$$

To prove the second, we have

$$\mu^*(\omega) = \frac{1}{2} - \frac{E^*(\omega)}{6\,K^*(\omega)} \ . \tag{4}$$

Just as in Problem 11.4-5, separation of the real and imaginary parts and requiring that $\mu''(\omega)$ be ≥ 0 leads to the desired result in either of the two cases.

(B) Here we need to show that $\tan\theta(\omega) \geq \tan\theta_M(\omega)$ and that $\tan\theta_M(\omega) \geq \tan\theta_K(\omega)$. This time we use

$$\mu^*(\omega) = \frac{M^*(\omega) - 2\,G^*(\omega)}{2\,M^*(\omega) - 2\,G^*(\omega)} \tag{5}$$

for the first, and

$$\mu^*(\omega) = \frac{3\,K^*(\omega) - M^*(\omega)}{3\,K^*(\omega) + M^*(\omega)} \tag{6}$$

for the second inequality. The two relations are easily established with the aid of Table 1.4-1.

Problem 11.4-7

(A) Dividing the imaginary parts on both sides of Eq. (1) by the real ones gives

$$\frac{J''(\omega)}{J'(\omega)} = \frac{9\,D''(\omega) - B''(\omega)}{9\,D'(\omega) - B'(\omega)} \ . \tag{2}$$

Now, since $J''(\omega)/(J'(\omega) = G''(\omega)/G'(\omega)$, $D''(\omega)/D'(\omega) = E''(\omega)/E'(\omega)$, and $B''(\omega)/B'(\omega) = K''(\omega)/K'(\omega)$, we may rearrange Eq. (2) to read

$$\frac{G''(\omega)}{G'(\omega)} = \frac{9\,D'(\omega)}{9\,D'(\omega) - B'(\omega)}\,\frac{E''(\omega)}{E'(\omega)} - \frac{9\,D'(\omega)}{9\,D'(\omega) - B'(\omega)}\,\frac{K''(\omega)}{K'(\omega)} + \frac{K''(\omega)}{K'(\omega)} \ . \tag{3}$$

But

$$\frac{9\,D'(\omega)}{9\,D'(\omega) - B'(\omega)} = 1 + \frac{B'(\omega)}{3\,J'(\omega)} \tag{4}$$

and, hence

$$\left[\frac{G''(\omega)}{G'(\omega)} - \frac{E''(\omega)}{E'(\omega)}\right] = \frac{B'(\omega)}{3\,J'(\omega)}\left[\frac{E''(\omega)}{E'(\omega)} - \frac{K''(\omega)}{K'(\omega)}\right] \ . \tag{5}$$

Both $B'(\omega)$ and $J'(\omega)$ are positive quantities. Thus, since $\tan\theta(\omega) \geq \tan_K(\omega)$ by Inequality (1) of Problem 11.4-5, Eq. (5) can only be true if both bracketed expressions are positive. This proves Inequalities (1) of Problem 1.4-6.

(B) Inserting the assumptions $K''(\omega) = 0$, and $K'(\omega) \simeq K \simeq 1/B('\omega)$ into Eq. (5) immediately leads to

$$\frac{G''(\omega)}{G'(\omega)} \simeq \left[1 + \frac{1}{3K\,J'(\omega)}\right]\frac{E''(\omega)}{E'(\omega)} \tag{6}$$

or

$$\tan\theta(\omega) \simeq \left[1 + \frac{1}{3K\,J'(\omega)}\right]\tan\theta_E(\omega) \tag{7}$$

as an occasionally useful approximate relation between the loss tangents in shear and in extension.

Problem 11.4-8

Tables 1.4-1 and 1.4-2 offer the relations

$$\mu = \tfrac{1}{2}EJ - 1 = \tfrac{1}{2}GD - 1\ . \tag{1}$$

for a purely elastic material. By the correspondence principle, these become

$$\bar{\mu}(s) = \tfrac{1}{2}s\bar{E}(s)\bar{J}(s) - 1/s = \tfrac{1}{2}s\bar{G}(s)\bar{D}(s) - 1/s\ . \tag{2}$$

for a linear viscoelastic material. Inversion yields four different but equivalent convolution integral relations for each of the two equations. One of these is

$$\mu(t) = \frac{1}{2}\left[E_g J(t) + \int_0^t \frac{d\,E(u)}{du} J(t-u)du\right] - 1 \tag{3}$$

which we present here merely by way of illustration.

Epilogue

ТРУД

Миг вожделенный настал: окончен мой труд многолетний.
Что ж непонятная грусть тайно тревожнт меня?
Илн, свой подвиг свершив, я стою, как поденщнк ненужный,
Плату приявший свою, чуждый работе другой?
Или жаль мне труда, молчалнвого спутннка ночн,
Друга Авроры златой, друга пенатов святых?

А.С. Пушкин

The Work

Come is the moment I craved: my work of long years is completed.
    Why then this strange sense of woe secretly harrowing me?
Having my high task performed, do I stand as a useless day laborer
    Stands, with his wages received, foreign to all other toil?
Or am I sorry to part with my work, night's silent companion,
    Golden Aurora's friend, friend of the household gods?

A.S. Pushkin

Notes on Quotations

The prefatory quotation is from the *Stundenbuch* (the Book of Hours) of the German poet Rainer Maria Rilke (1875–1926). The words:

> 'Do not do wonders for me,
> See that your laws prevail
> which, with each generation,
> unfold more clearly.'

are the words of a Russian monk praying in his cell.

The Prologue, attributed as an excerpt from *Viajes de varones prudentes* (Travels of Praiseworthy Men), libro cuatro, cap. XIV, Lérida, 1658, to one 'Suarez Miranda', was written by the Argentinian writer Jorge Luis Borges with Adolfo Bioy Casares and is reproduced here as it appears in the *Obras Completas* published by Emecé Editores, S.A., Buenos Aires, 1954, in the volume entitled *Historia universal de la infamia*. The translation (and these notes) are due to Norman Thomas di Giovanni. They are taken from the 1975 Penguin Books edition of the *Universal History of Infamy*.

The motto to the Preface is from J. Willard Gibbs's letter of acceptance on the occasion of the award of the Rumford Medal to him. It is quoted in L.P. Wheeler's *Josiah Willard Gibbs, the History of a Great Mind*, published by Yale University Press in 1962.

The motto to the Introduction: 'I have given you the concept; now listen how it has been developed' comes from the Prologue of the opera *I Pagliacci* by Ruggiero Leoncavallo (1858–1919) who wrote both the libretto and composed the music.

Chapter 1 carries as motto the tongue-in-cheek quotation from Stanza VII, Canto 1, of Byron's *Don Juan*.

The first of the mottos to Chap. 2 is taken from the Foreword to Herbert Leaderman's book (see Ref. [5] of that chapter) by Harold deWitt Smith, then a member of the Advisory Committee for Scientific Research of the Textile Foundation which published the book. The second motto is a grafitto which I found on the wall of a washroom in Caltech's Millikan Library. The third: 'What is time? If no-one asks me, I know; if I wish to explain, I do not know' is from Chapter IV, Book XI, of the *Confessions* of St. Augustine (354–430).

Chapter 6 is headed by a motto taken from Chapter V of *A Voyage to Laputa* of Jonathan Swift (1667–1745). The popular saying which follows it is said to be from advertisements of prize fights stating that a celebrated boxer named McCoy would participate, not an inferior boxer of the same name (Webster's Dictionary, 2nd ed.). The real McCoy is thus the real person or thing.

The motto to Chap. 8 'The more it changes, the more it is the same', is the best-known of many epigrams from the pen of the French novelist and journalist Alphonse Karr (1808–1890).

Chapter 10 is prefixed by the observation: 'Half a century ago it was freely admitted and proclaimed that Nature loves simplicity; since then she has given us too many disavowals' from Jules Henri Poincaré's (1854–1912) *La science et l'hypothèse* (1901 edition). Albert Einstein's (1879–1955) aperçu: 'The Lord is sophisticated, but He is not malicious' is carved over the fireplace in Fine Hall at the Institute for Advanced Studies in Princeton. He made it during a discussion following a lecture he had given in Princeton (see Ronald Clark, *Einstein, The Life and Times*, The World Publishing Company, 1971, p. 390).

The motto to Chap. 11 is the beginning of the twentieth verse of the *Dàodè Jīng* of Lǎozi (Tao Te Ching of Lao Tse in the most common English spelling before the introduction of the now official *pinyin* spelling). Lǎozi, an older contemporary of Kǒngfuzi (Confucius), was keeper of the imperial archives in Luòyáng (Loyang) in the province of Hénan (Honan) in the sixth century B.C. Lǎozi's style is terse and laconic. The four characters of the quotation read 'jué xué wú yōu' and mean 'end learning not-have troubles'. In the 1972 translation of Gia-Fu Feng and Jane English (Vintage Books, Random House) they become: 'Give up learning, and put an end to your troubles'. In the brief comments on Lǎozi's philosophy on the back of the book one reads: '... we will see that work proceeds more quickly and easily if we stop "trying", if we stop putting in so much extra effort ...' Since I wrote Chap. 11 last, I thought the quotation particularly apt.

The motto heading the Solution to Problems is from Sophokles's *Oedipus the King*. Oedipus says to the blind prophet Teiresias: 'You speak in riddles and obscurities'. The seer (evidently referring to Oedipus's famous solving of the 'Riddle of the Sphinx') answers: 'If so, are you not the best person to solve them?'

The motto in front of the Appendix is from Scene iv of Act IV of the *Taming of the Shrew*. Biondello is telling Lucentio that he is going to alert the priest to prepare for Lucentio's wedding. He says: 'My master hath appointed me to go to Saint Luke's, to bid the priest be readie to come against you come with your appendix.'

The poem which I added as an Epilogue was written by Aleksandr Sergeevich Pushkin in 1830 at the time when he finished writing *Evgeniy Onegin*. The translation is by Vladimir Nabokov and can be found at the end of his commentary to *Onegin* which were published by Princeton University Press in 1964.

List of Symbols

Symbols that occur only in isolated instances and are not referred to elsewhere, have not been included in this list. When the physical property denoted by a symbol may occur under different names, only that preferred in this book is listed. A *number* following an entry refers to the page where the symbol itself or a closely related variant is defined. When part of a meaning is given in *parentheses*, this part may or may not appear in the definition or in the usage of the term.

To avoid unnecessary duplications, the following scheme was adopted: a viscoelastic material function is generally listed as the time-dependent response, F(t). The corresponding harmonic responses, $F^*(\omega)$, $F'(\omega)$, and $F''(\omega)$, which have the same dimensions and are therefore represented by the same letter symbol, are not listed separately. The corresponding respondances, which do not have the same dimensions and are therefore denoted by different letter symbols, are entered in operational form, i.e. as Laplace transforms, say, $\bar{R}(s)$.

Duplication of some symbols could not be entirely avoided because of the limited number of Roman and Greek letters.

Roman Letters

$a(s)$; $\hat{a}(s)$	transform of the propagation function of a ladder model 252, 264
$\bar{A}(s)$	bulk retardance 518
b	specimen constant 77
B	bulk compliance 29
B_e	equilibrium bulk compliance 520
B_g	glassy bulk compliance 520
$B(p, q)$	beta function 351
$B(p, q; m)$	incomplete beta function 348
$B(t)$	bulk creep compliance 522
C	electrical capacitance 70
C_{kl}, C_{ijkl}	general modulus tensor 14
D	stretch compliance 29
D_e	equilibrium stretch compliance 528
D_e°	steady-state stretch compliance 528
D_g	glassy stretch compliance 528
$D(t)$	stretch creep compliance 527
E	elastance 73, 76
E	stretch modulus 26
E_e	equilibrium stretch modulus 526
E_e°	steady-state stretch modulus 528
E_g	glassy stretch modulus 526
$E(t)$	stretch relaxation modulus 511

$Ei(x)$	exponential integral function	344
$Erf(x)$	(Gaussian) error function	338
$Erfc(x)$	complementary error function	338
f	force	73
F	frictance	73
$F(t)$	constant strain rate modulus	545
G	conductance	78
G	shear modulus	25
G_e	equilibrium (shear) modulus	91
G_e°	steady-state (shear) modulus	94
G_g	glassy (shear) modulus	48
G_N	(shear) plateau modulus	499
$G(t)$	(shear) relaxation modulus	48
$h(t)$	unit step function	554
H	shape factor	77
$H(\tau)$	(shear) relaxation spectrum	158
$H_E(\tau)$	stretch relaxation spectrum	526
$H_K(\tau)$	bulk relaxation spectrum	519
$H_M(\tau)$	wave relaxation spectrum	524
I	current	70
I	inertance	73
$I_k(x)$	intensity function	172
J	shear compliance	29
J_e	equilibrium (shear) compliance	90
J_e°	steady-state (shear) compliance	94
J_g	glassy (shear) compliance	48
J_N	(shear) plateau compliance	499
J_T	total (shear) compliance	271
$J(t)$	(shear) creep compliance	48
K	bulk modulus	25
K_e	equilibrium bulk modulus	519
K_g	glassy bulk modulus	519
$K(t)$	bulk relaxation modulus	521
$K(x,\tau)$	kernel function	315
$K_\Delta(x)$	difference kernel	418
l_{ij}	direction cosine	16
L	electrical inductance	70
L	length	2
$L(\tau)$	(shear) retardation spectrum	161
$L_B(\tau)$	bulk retardation spectrum	519
$L_D(\tau)$	stretch retardation spectrum	528
$L_M(\tau)$	wave retardation spectrum	524
$\bar{L}(s)$	stretch retardance	510
m	momentum	78
$m(\tau)$	delay time spectrum	531
M	torque	78
M	wave modulus	26
M_e	equilibrium wave modulus	26
M_g	glassy wave modulus	26
$M(t)$	wave relaxation modulus	524
$M(\zeta)$	(shear) relaxation frequency spectrum	163
$N(\zeta)$	(shear) retardation frequency spectrum	164
N_D	Deborah number	35
$\bar{O}(s)$	wave retardance	510

p	isotropic pressure	28
p(t)	time-dependent isotropic pressure	519
p(t)	slope function	52, 558
$\bar{P}(s)$	bulk relaxance	518
q	electrical charge	78
q	prolongation factor	501
$\bar{Q}(s)$	(shear) relaxance	40
$Q_C(s)$	characteristic (shear) relaxance of a ladder model	272
$Q_F(s)$	(shear) relaxance of the final section of an obverse ladder model	250
$\bar{Q}_M(s)$	mechanical relaxance	286
$Q_T(s)$	$Q_C(s)$ of the T-section of an obverse ladder model	252
$Q_\Pi(s)$	$Q_C(s)$ of the Π-section of an obverse ladder model	252
$\hat{Q}_\Pi(s)$	$Q_C(s)$ of the Π-section of a converse ladder model	264
$\hat{Q}(\tau)$	modulus density on relaxation time	158
$\mathcal{Q}_C(s)$	characteristic mechanical relaxance per unit cross-sectional area	296
$\mathcal{Q}_O(s)$	input mechanical relaxance per unit cross-sectional area	296
$\mathcal{Q}_L(s)$	output mechanical relaxance per unit cross-sectional area	296
$\bar{Q}_\infty(s)$	relaxance of an (extended) ladder model with an infinite number of nodes	276
r(t)	ramp function	559
R	electrical resistance	70
s	Laplace transform variable	560
s	Stieltjes transform variable	575
S	mechanical compliance	76
S_x^y	Sterling number of the second kind	175
S_{kl}, S_{ijkl}	general compliance tensor	28
S(x)	monotone non-decreasing dimensionless experimental response function	315
$\hat{s}(x)$	associated matching function of the S-type	318
$\hat{S}(x)$	matching function of the S-type	317
t	present time	36
T	period	381
T(t)	wave creep compliance	510
u	past time	36
$\bar{U}(s)$	(shear) retardance	40
$U_C(s)$	characteristic (shear) retardance of a ladder model	272
$\bar{U}_M(s)$	mechanical retardance	286
$U_T(s)$	$U_C(s)$ of the T-section of an obverse ladder model	252
$\hat{U}_\Pi(s)$	$U_C(s)$ of the Π-section of a converse ladder model	264
$\hat{U}(\tau)$	compliance density on retardation time	161
$\mathcal{U}_C(s)$	characteristic mechanical retardance per unit cross-sectional area	532

$\bar{U}_\infty(s)$	retardance of an (extended) ladder model with an infinite number of nodes 656
v	velocity 3, 78
V	voltage 70
V	volume 24
$\bar{V}(s)$	wave relaxance 524
W	free energy of deformation 14
W(t)	deformational energy per unit volume absorbed *up to* time t 443
$W_d(t)$	deformational energy per unit volume dissipated *up to* time t 443
$W_s(t)$	deformational energy per unit volume stored *up to* time t 443
$\dot{W}(t)$	rate of energy absorption per unit volume *at* time t 444
$\dot{W}_d(t)$	rate of energy dissipation per unit volume *at* time t 444
$\dot{W}_s(t)$	rate of energy storage per unit volume *at* time t 444
x	displacement 73
Y	electrical admittance 78
$\bar{Y}(s)$	stretch relaxance 526
$\bar{Y}(s)$	(operational) electrical admittance 79
$\bar{Y}_M(s)$	(operational) mechanical admittance 292
Z	electrical impedance 71
$\bar{Z}(s)$	(operational) electrical impedance 72
$Z(\omega)$	harmonic electrical impedance 71
$\bar{Z}_M(s)$	(operational) mechanical impedance 292
Z(x)	monotone non-increasing dimensionless experimental response function 315
$\hat{z}(x)$	associated matching function of the Z-type 318
$\hat{Z}(x)$	matching function of the Z-type 317

Greek Letters

$\alpha(\omega)$	attentuation function 301
β	bulk fluidity 30
β	phase constant 297
$\beta(\omega)$	phase function 302
$\beta(t)$	strain response to unit rate of stress in a bulk deformation 510
γ	$= \gamma_{12}$, (shear) strain 13
γ_{ij}	strain tensor 12
$\gamma_{ij}(t)$	time-dependent strain tensor 514
$\dot{\gamma}$	rate of (shear) strain 13
$\dot{\gamma}_{ij}$	rate of strain tensor 13
$\gamma(s)$	propagation function of an inertial transmission line 296, 301
$\gamma(m; x)$	incomplete gamma function 348
$\gamma(t)$	time-dependent (shear) strain
$\Gamma(m)$	gamma function
$\Gamma^{-1}(m)$	reciprocal gamma function 233
$\Gamma(s)$	transmittance of a ladder model 271

$\Gamma_l(s)$	propagation function per unit length of a continuous (distributed) ladder 288	$\theta, \hat{\theta}$	characteristic time of a ladder model 251, 263
$\bar{\Gamma}_{ijkl}(s)$	relaxance of a general anisotropic linear viscoelastic material 512	$\theta_\infty, \hat{\theta}_\infty$	characteristic time of an extended ladder model 271, 282
δ	η_T/G_T 276	$\theta(\omega)$	loss angle 58
$\delta(t)$	delta function 550, 551	$\Theta(\omega)$	storage angle 62
δ_{ij}	Kronecker's delta (unit tensor) 23	$\kappa(t)$	response of rate of stress to unit step of (shear) strain 54
Δ	trace of the strain tensor matrix; dilatation 23; 24	λ	(first) Lamé constant 23
$\Delta(t)$	time-dependent cubical contraction or dilatation 519	λ	wave length 297
ε	$= \varepsilon_{12}$, (amount of shear) strain 13	$\Lambda(\zeta)$	$N(\zeta)/\zeta$ 164
$\varepsilon_e^{\{o\}}$	{pseudo-}equilibrium strain 94	$\bar{\Lambda}(\omega)$	dimensionless experimental response function possessing a single maximum 316
$\varepsilon(t)$	time-dependent (amount of shear) strain 37	$\bar{\Lambda}(\omega)$	normalized dimensionless experimental response function possessing a single maximum 317
$\dot{\varepsilon}$	time derivative (rate) of the amount of shear 13	$\hat{\Lambda}(x)$	matching function of the Λ-type 322
ζ	stretch viscosity 30	μ	angular momentum 78
$\zeta(t)$	time-dependent stretch viscosity 527	μ	ω^{-2} 169
ζ_f	steady-state stretch viscosity 528	μ	Poisson's ratio 27
η	shear viscosity 3, 30	μ_e	equilibrium Poisson's ratio 529
η_f	steady-state shear viscosity 94	μ_g	glassy Poisson's ratio 530
$\eta_{\{f\}}$	η_f *or* $\eta_\times$ 232	$\mu(t)$	time-dependent lateral contraction (Poisson's ratio) 529
η_T	total viscosity 271	ν	ω^2 167
$\eta_\times$	sum of viscosities of an arrheodictic model 271	$\bar{\nu}(s)$	(operational) lateral contraction 510, 529
$\eta(t)$	time-dependent (shear) viscosity 52	ξ	bulk viscosity 3
θ	angular displacement 76	$\xi(t)$	time-dependent bulk viscosity 520
θ	shifted time 368	$\Xi(\zeta)$	$M(\zeta)/\zeta$ 164

ρ	density 296
σ	(shear) stress 3
$\sigma(t)$	time-dependent (shear) stress 37
σ_{ij}	stress tensor 9
$\sigma_{ij}(t)$	time-dependent stress tensor 514
Σ	trace of the stress tensor matrix 24
τ	relaxation *or* retardation time 86
τ	delay time 529
τ_M	relaxation time 84
τ_V	retardation time 83
ϕ	shear fluidity 30
ϕ_f	steady-state shear fluidity 94
$\phi(t)$	response of rate of strain to unit step of (shear) stress, or time-dependent (shear) fluidity 54
Φ	glidance (reciprocal frictance) 76
$\Phi(x)$	general symbol for the experimental response functions 169, 414, 430
φ	flux 78
χ	$\Sigma_m J_n \tau_n$ 233
$\chi(t)$	strain response to (constant) rate of stress in shear 53
$\psi(t)$	$\int \eta(u)du$ 451
$\psi(t)$	strain response to unit rate of stress in a stretch deformation 510
$\bar{\Psi}_{ijkl}(s)$	retardance of a general anisotropic linear viscoelastic material 512
ω	radian frequency 55
ω_{ij}	rotation tensor 12
$\dot{\omega}_{ij}$	vorticity tensor 13

Subscripts

•	denotes 'dot-form' 121
o	input
C	characteristic
D	function derived by fractional derivation 339
e	equilibrium modulus or compliance
f	(steady-state) flow viscosity or fluidity
F	final (also output)
g	glassy (instantaneous) modulus or compliance
l	(lower case L), signifies 'per unit length'
L	terminal (end of the line), output
m	mechanical (respondance or immittance)
T	T-section
Π	Π-section
ss	steady state 56

Superscripts

ss	steady state 452
o	indicates rheodictic behavior 124, 253
×	indicates arrheodictic behavior 124, 254

$'$	the real part of a complex function, or a function derived from it		
$''$	the imaginary part of a complex function, or a function derived from it		
*	a complex function, or a function derived from it		

Operators

(=)	equals except at points of discontinuity 572	$L^{\{n\}}f(x)$	{repeated} logarithmic shift operator 182
[ ]	around indexed quantities denotes a matrix	$L_m f(x)$	multiplication operator 182
$\langle\ \rangle$	denotes averaging 233, 315	$\mathscr{L}$	Laplace transform operatosr 560
{ }	denotes the absence or presence of the expression in braces according to whether the behavior is rheodictic or arrheodictic 121–123	$\mathscr{L}^{-1}$	inverse Laplace transform operator 560
$⨍$	Cauchy principal value of integral 582	$\mathscr{P}_k[\ ]$	(logarithmic) differential operator of order k 172
$\oint$	(counterclockwise) circular or contour integral 237	$\mathscr{P}_{kh}[\ ]$	(logarithmic) finite difference operator of order k 183
det	determinant of a matrix	$\mathscr{P}_{\Delta k}[\ ]$	difference kernel operator of order k 418
$\mathscr{F}$	Fourier transform operator 570	$\mathscr{P}_K[\ ]$	integral (transform) operator 550
$\mathscr{F}^{-1}$	inverse Fourier transform operator 570	$\mathscr{P}_K^{-1}[\ ]$	inverse integral (transform) operator 550
$\mathscr{H}$	Hilbert transform operator 581	$\mathrm{Res}_n(F)$	residue associated with the nth pole of the complex function F 566
$\mathscr{I}m\ F^*(\omega)$	imaginary part of the complex function $F^*(\omega)$	$\mathscr{R}e\ F^*(\omega)$	real part of the complex function $F^*(\omega)$
$\mathscr{J}$	Fourier-Laplace transform 580	$\mathscr{S}$	Stieltjes transform operator 574
$\mathscr{J}^{-1}$	inverse Fourier-Laplace transform 580	$\mathscr{S}^{-1}$	inverse Stieltjes transform operator 576

Author Index

Numerals refer to the page where the author's work is cited. **Bold numerals** refer to the page where the full reference is given.

Subject Index

If a secondary entry begins with a capital letter, it precedes the primary one when the entries are read together. Thus, 'Absorber, Damped vibration' should be read as 'Damped vibration absorber'. Similar considerations apply, *mutatis mutandis*, when a tertiary or quaternary entry begins with a capital letter. The general rule is: work backwards from the 'farthest down' or 'deepest in' capitalized entry. With longer concatenated entries this rule may not always work but the context will generally make it clear how the combined entries should be read.

Expressions in parentheses are optional, or serve to clarify the meaning of an entry. Thus, 'Model(s)' stands for either 'Model' or 'Models'. Under '(shear) relaxation modulus', one will probably find only 'relaxation modulus' but this entry then properly refers to the 'shear relaxation modulus'.

Capitalized personal names refer to the pages on which footnotes containing biographical notes and/or help with pronounciations may be found.

If a section or chapter is devoted primarily to a given subject, the Section or Chapter number is printed in **bold** numerals, and is followed immediately by the number of the page on which that Section or Chapter begins. Similarly, references to Tables are followed by the number of the page on which they appear.